Diamond and Diamond-like Films and Coatings

NATO ASI Series
Advanced Science Institutes Series

A series presenting the results of activities sponsored by the NATO Science Committee, which aims at the dissemination of advanced scientific and technological knowledge, with a view to strengthening links between scientific communities.

The series is published by an international board of publishers in conjunction with the NATO Scientific Affairs Division

A	**Life Sciences**	Plenum Publishing Corporation
B	**Physics**	New York and London
C	**Mathematical and Physical Sciences**	Kluwer Academic Publishers
D	**Behavioral and Social Sciences**	Dordrecht, Boston, and London
E	**Applied Sciences**	
F	**Computer and Systems Sciences**	Springer-Verlag
G	**Ecological Sciences**	Berlin, Heidelberg, New York, London,
H	**Cell Biology**	Paris, Tokyo, Hong Kong, and Barcelona
I	**Global Environmental Change**	

Recent Volumes in this Series

Volume 261—Z° Physics: *Cargèse 1990*
edited by Maurice Lévy, Jean-Louis Basdevant, Maurice Jacob, David Speiser, Jacques Weyers, and Raymond Gastmans

Volume 262—Random Surfaces and Quantum Gravity
edited by Orlando Alvarez, Enzo Marinari, and Paul Windey

Volume 263—Biologically Inspired Physics
edited by L. Peliti

Volume 264—Microscopic Aspects of Nonlinearity in Condensed Matter
edited by A. R. Bishop, V. L. Pokrovsky, and V. Tognetti

Volume 265—Fundamental Aspects of Heterogeneous Catalysis Studied by Particle Beams
edited by H. H. Brongersma and R. A. van Santen

Volume 266—Diamond and Diamond-like Films and Coatings
edited by Robert E. Clausing, Linda L. Horton, John C. Angus, and Peter Koidl

Volume 267—Phase Transitions in Surface Films 2
edited by H. Taub, G. Torzo, H. J. Lauter, and S. C. Fain, Jr.

Series B: Physics

Diamond and Diamond-like Films and Coatings

Edited by

Robert E. Clausing
Oak Ridge National Laboratory
Oak Ridge, Tennessee

Linda L. Horton
Oak Ridge National Laboratory
Oak Ridge, Tennessee

John C. Angus
Case Western Reserve University
Cleveland, Ohio

and

Peter Koidl
Fraunhofer Institut für
Angewandte Festkörperphysik
Freiburg, Germany

Plenum Press
New York and London
Published in cooperation with NATO Scientific Affairs Division

Proceedings of a NATO Advanced Study Institute on Diamond and Diamond-like Films and Coatings, held July 22-August 3, 1990, in Castelvecchio, Pascoli, Italy

Library of Congress Cataloging-in-Publication Data

NATO Advanced Study Institute on Diamond and Diamond-like Films and Coatings (1990 : Castelvecchio Pascoli, Italy)
Diamond and diamond-like films and coatings / edited by Robert E. Clausing ... [et al.].
p. cm. -- (NATO ASI series. Series B, Physics ; v. 266)
"Proceedings of a NATO Advanced Study Institute on Diamond and Diamond-Like Films and Coatings, held July 22-August 3, 1990, in Castelvecchio Pascoli, Italy"--T.p. verso.
"Published in cooperation with NATO Scientific Affairs Division."
Includes bibliographical references and index.
ISBN-13: 978-1-4684-5969-2 e-ISBN-13: 978-1-4684-5967-8
DOI: 10.1007/978-1-4684-5967-8
1. Diamonds, Artificial--Congresses. 2. Diamond thin films--Congresses. I. Clausing, Robert E. II. North Atlantic Treaty Organization. Scientific Affairs Division. III. Title. IV. Series.
TP873.5.D5N38 1990
666'.88--dc20 91-20196
CIP

ISBN-13: 978-1-4684-5969-2

Softcover reprint of the hardcover 1st edition 1991

A Division of Plenum Publishing Corporation
233 Spring Street, New York, N.Y. 10013

The major support for the Institute was provided by the Scientific Affairs Division of NATO. Significant additional support was provided by:

Metals and Ceramics Division, Oak Ridge National Laboratory, USA
Fraunhofer Institut für Angewandte Festkörperphysik, FRG
National Science Foundation, USA
Consorzlo Instituto Nazionale Fisica della Materia, Italy
Politecinico di Torino, Italy
Office of Naval Research, USA
Nano Instruments, Inc., USA
European Research Office, United States Army
Applied Science and Technology, Inc., USA
Genasystems, USA
Crystallume, Inc., USA

The home institutions of the directors provided much direct and indirect support for the planning and organization of the Institute. Their assistance, cooperation, and leniency are gratefully acknowledged.

The editors

Dr. R. E. Clausing
Metals and Ceramics Division,
Oak Ridge National Laboratory,
Oak Ridge TN, USA

Dr. L. L. Horton
Metals and Ceramics Division,
Oak Ridge National Laboratory,
Oak Ridge TN, USA

Prof. J. C. Angus
Chemical Engineering Dept.,
Case Western Reserve University,
Cleveland OH, USA

Prof. Peter. Koidl
Fraunhofer Institut fur
Angewandte Festkorperphysik
Freiburg, FRG

PREFACE

Diamond films grown by activated chemical vapor deposition have superlative thermal, mechanical, optical, and electronic properties combined with a very high degree of chemical inertness to most environments. These properties, together with the ability to fabricate films and shapes of considerable size, promise an exciting new material with many applications. Some applications are on the verge of commercialization but many await a few more technological developments. Diamond-like films are already employed in both commercial and military applications. The popular press, as well as the scientific and technological and industrial communities, are increasingly interested in the potential for future development of these materials.

Although there are many technical papers and review articles published, there is no single comprehensive introduction to these technologies. The Scientific Affairs Division of NATO recognized the need and the future importance of these technologies and authorized an Advanced Study Institute on diamond and diamond-like films. NATO Advanced Study Institutes are high level teaching activities at which a carefully defined subject is presented in a systematic and coherently structured program. The subject is treated in considerable depth by lecturers eminent in their fields and of international standing. The presentations are made to students who are scientists in the field or who possess an advanced general scientific background. The primary objectives of an Advanced Study Institute are to contribute to dissemination of advanced knowledge not yet available in standard university courses and to the formation of contacts among scientists from different countries. Participation in the ASI is the primary instrument, however, publication of the work presented in the basic lectures and status reports also contributes to these objectives.

This book is the record of the proceedings of "The NATO Advanced Study Institute on Diamond and Diamond-like Films and Coatings," held at Il Ciocco International Tourist and Conference Center, Castelvecchio, Pascoli, (Tuscany), Italy from July 22 to August 3, 1990. The Institute brought together scientists and engineers from various disciplines to review, discuss, and clarify our current understanding of (1) the growth of diamond and diamond-like films, (2) the characterization of these materials, (3) the optical, physical, and mechanical properties of these films, (4) the relationship of these properties to structure and processing, and (5) the current and potential technological applications of these new diamond and diamond-like materials.

The book has been divided into parts according to the topics discussed. In each part the tutorial lectures are presented first, followed by the contributed papers. The tutorial lectures by world renowned experts review fundamental concepts and the state of the art. At the Institute, they provided the background for seminar discussions and the presentation and examination of current research findings. Most of the review articles contain extensive bibliographies. Many of the contributed papers present significant new results in this rapidly changing field, others are simply progress reports reflecting the current state of the art in the author's laboratory. It is hoped that this book, with its mixture of papers will serve as an excellent introduction to the field. It combines reviews of the basic science and technology needed to understand the creation and application of diamond and diamond-like materials with the current status of the technology.

We thank all of the chairmen, lecturers, and participants who contributed to the discussions at the meeting and who also provided manuscripts for this book.

Much of the success of this meeting is owed to the experience and organizational expertize of Dr. C. J. McHargue. He first suggested the need for this ASI and helped the directors develop the program and conference format. He also took care of the innumerable details of arrangements and communications with participants and the hotel.

The International Advisory Committee provided guidance and assistance in many ways, including program and financial matters. We thank them for all their help.

The hotel, its facilities and staff contributed much to the success of the Institute in spite of the unusually hot weather. The hotel is a first class resort but is sufficiently isolated to remove outside distractions. It provides excellent facilities for an institute such as ours and easy access to the cultural and historical attractions of Tuscany. The entire staff, especially Mr. Bruno Giannasi, deserves our special thanks.

We also express our thanks to others whose efforts made the Institute work so well: Lee Heatherly who helped with the audio-visual equipment, and Ms. Lou Pyatt who provided expert secretarial help. Ms. Julia Clausing and Ms. Linda Heatherly were always willing, and helped at registration, poster sessions and in other ways e.g., by providing a program of activities for accompanying persons.

R. E. Clausing

Synopsis of the Institute

The Institute was divided into three parts: (I) fundamental concepts, (II) diamond-like films and coatings and (III) CVD diamond films. The first part, on fundamentals, provided a background for understanding of: (1) the structure and basic properties of diamond and diamond-related materials, (2) the basic physics and chemistry of plasmas and the plasma-surface interactions involved in film synthesis, and (3) the characterization methods appropriate for the films and the manufacturing processes. The second part treated the deposition, properties and use of diamond-like films and coatings. The third part included; (1) the practical and theoretical aspects of nucleation and growth of CVD diamond films; (2) specific preparation techniques; (3) characterization, properties and applications of these films; and (4) some of the directions of current research and development including those specifically for electronic devices and high rate deposition processes. Although the theme of institute was the science and technology of diamond materials, there was also discussion of boron nitride, silicon carbide, and other phases in the composition tetrahedron C-B-N-Si.

The technology and applications for diamond and diamond-like materials continue to develop rapidly. Understanding of the synthesis, structure, and properties of these new materials, however, lags behind the technology. There was a strong sense of excitement and anticipation concerning crystalline CVD diamond materials. All of the aspects of growth, characterization and use were eagerly discussed, but the critically important nucleation and growth processes and the roles atomic hydrogen and oxygen in these processes received most attention.

"State of the art," high rate, CVD diamond deposition processes were reviewed and discussed to ascertain the common characteristics which lead to deposition rates of 100 to 1000 microns per hour (compared with the usual one micron per hour deposition rates for films made by the more common low pressure, activated CVD methods.) Film quality and uniformity were also discussed.

A panel discussion at the end of the meeting addressed the needs and opportunities in the synthesis of CVD diamond. The key scientific needs are those related to modeling the nucleation and growth processes, to elucidation of the critical roles of atomic hydrogen, and to the mechanisms of carbon addition to the growing surfaces. The development and more extensive use of in situ diagnostics for both surface and gas phases are important to solving these issues. The more immediate practical questions concern the identification of the growth rate limiting steps, the relation of growth parameters to the resulting film structure, and the dependence of properties on structure.

The meeting was successful from all points of view. The response to the initial meeting announcements was very good; the 105 selected participants formed a well-balanced group from industry, government laboratories, university faculty and graduate students. Seventeen countries were represented. The lecturers, who were chosen from the pioneers and current leading experts in the world, prepared excellent presentations and were very well received. The extended discussion periods were fully utilized with the exchange of information at all levels on an informal basis.

R. E. Clausing

CONTENTS

Properties

Applications

PART III - DIAMOND FILMS

Nucleation and Growth Mechanisms

Nucleation and Growth Technology

Diagnostics

Characterization

Applications

Electronic Devices and Applicaitons

Special Topics

DIAMOND: POTENTIAL AND STATUS

Max N. Yoder

Electronics Division
Office of Naval Research
Arlington, Virginia 22217

DIAMOND: THE WONDER MATERIAL

Consider virtually any material property; the value of that property in diamond is virtually certain to represent an extremest value. Although the superlative properties of diamond were known for centuries, knowledge of its wholly carbon constitution came much later. It was the French chemist Antoine L. Lavoisier who found in 1772 that the products of combustion of diamond were singular: CO_2. This knowledge of the carbon constituency did not, however, precipitate a rush toward artifact diamond. The first written claim to artifact diamond was made sixty years later by C. Cagniard de la Tour, but the products of his endeavors are no longer to be found. In 1880, J. B. Hanney allegedly synthesized diamond from lithium, bone meal, and mineral oil heated together in a wrought iron tube. His products are on display in the British Museum. Thirteen years after that Henry Moisson developed a procedure that heated sugar charcoal and iron to a molten mixture in an electric furnace and followed by a water quench. The Nobel award winning high pressure work of Percy Bridgman of Harvard University in the 1940s pointed the way toward combining the electric furnace with high pressure to create quality artifact diamond. This approach first became publicly known in the 1950s by a team at General Electric but similar documented but unreported work may have occurred earlier in Sweden. From this time forward the wonder material improved in quality and size until today it is possible to produce artifact diamond films and to coat various materials with polycrystalline diamond.

The rate of increase of knowledge concerning the properties of diamond is at an all time high and continues to increase. In Japan alone, over 180 different organizations have engaged in the study or synthesis of diamond during the past decade. Significant advances continue to be made in 1990: thermal conductivity of artifact diamond has significantly surpassed that of natural diamond[1] and the first negative electron affinity (NEA) material not requiring photostimulation or cesiating of the surface was demonstrated using diamond[2]. Although much is known about diamond, a detailed understanding of diamond nucleation and growth leaves much to be desired. This is particularly true in understanding the nucleation of single crystalline diamond on non-diamond substrates. The most promising of the new growth procedures cyclically subject the growth surface to different reagent gases. These procedures not only appear to increase growth rate, but improve uniformity and material quality as well.

MECHANICAL/PHYSICAL PROPERTIES AND POTENTIAL OF DIAMOND

Gemstones are the most commonly known application of diamond. Of even greater commercial value than the gemstones are those applications exploiting the unmatched hardness of diamond -- primarily grinding, cutting, abrading, and indenting. Second to gemstones -- and most probably the fastest growing of all applications of diamond -- are those exploiting its unmatched thermal conductivity. With a coefficient of friction similar to that of Teflon, the advent of diamond thin films will witness many new tribological applications. The velocity of sound in diamond exceeds that of all other materials. The wavelength of any given frequency in diamond is therefore the longest of any material. As such, the largest area (and most efficient) transducers may be made with diamond -- an application already being exploited. Applications exploiting the inherent strength and toughness of diamond are currently in their embryonic stages.

Structure

Gemstones and all other applications of diamond use the most common structure of diamond -- the face-centered cubic crystal shown in figure 1. Other common materials having this structure are silicon, germanium, gold, and silver. In the crystal structure of figure 1 it is seen that in addition to the 8 corner atoms and the 6 face-centered atoms, that there are four other atoms (2 of which are connected by the dotted lines).

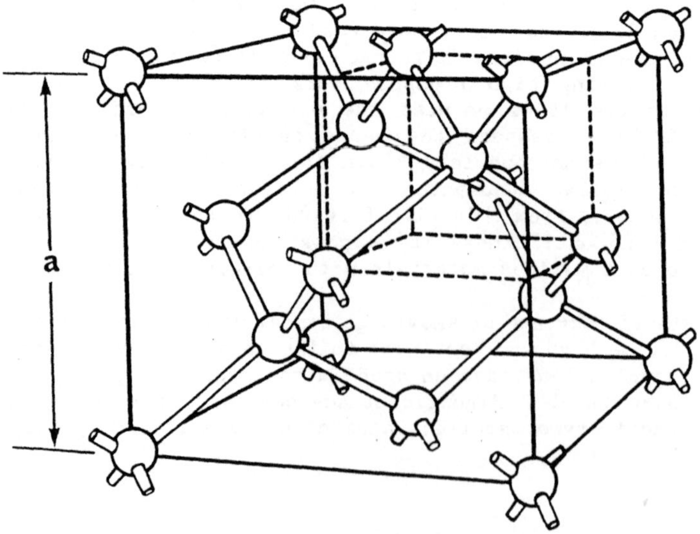

Figure 1. Cubic Structure of Diamond

These four atoms are spaced a quarter of a cube diagonal from the former and represent adjacent interpenetrating lattices. Another form of diamond is the hexagonal structure (not shown) know as lonsdaleite and was first found in meteorites. Still other forms or "polytypes" of diamond are thought to exist -- particularly in very small particles.[3]

The potential for exploiting the crystalline structure (lattice constant and cubic form) of diamond (apart from its other properties) is limited. The most likely such exploitation would be as a substrate for the heteroepitaxial growth of cubic boron nitride or other possible materials whose lattice constants (or multiples thereof) match that of diamond.

Hardness

Diamond, with a Knoop hardness of up to 10,400 kg/mm^2, has long been known to be the hardest of all materials. Only boron nitride at 4,500 kg/mm^2 comes close. This dominance may change, however, as the C-N bond is known to be shorter and stronger than the C-C bond of diamond.[4] While it may not be possible to create C_3N_4, it may be possible to introduce limited amounts (e.g., a few per cent) of nitrogen into diamond to increase its hardness. Many immediate applications exist to exploit the inherent hardness of diamond films being grown today.

A head crash on a magnetic disk used for data storage can be disastrous. These crashes are most prone to occur in operating environments wherein power failures are frequent or wherein the storage units may be subjected to sudden acceleration forces several times that of gravity. For diamond films to be effective in protecting the surface from these deleterious effects, the underlying storage media must be very hard itself and the film (1) must not exceed 25 nm in thickness (to avoid poor signal to noise ratio), (2) should be conformally applied, and (3) must possess a low coefficient of friction.

Natural and artifact diamond chips along with sintered diamond powders currently proliferate in the cutting and grinding market. The annual cutting tool market exceeds 2×10^5 and is growing at a rate of 7%. Over 10 million such chips are used each year as abrasives. These applications are not without problems. As an example, sintered diamond cutters used in high speed circular saws must be adhered by a costly electroplating process to improve their adhesion and the safety of the tool. The efficacy of the diamond tool is not only its hardness, but the adaptability to be honed to a very fine edge. Micromachining operations find the latter attribute particularly attractive because the profile of the tool edge is transferred to the finished product. Finishes as smooth as 1 nm can be obtained with a diamond cutting tool without resorting to post-cutting polishes. The advent of conformal diamond coatings is expected to significantly lower the production cost of cutting tools. While adhesion of diamond coatings to stock such as Si_3N_4 is comparatively simple, adhesion to such common tool stocks as tungsten carbide requires technology improvement. Even though diamond is exceptionally hard and therefore exhibits a much longer operating life than do competitive tools, it is not useful for cutting ferrous materials and other materials that easily form carbides. The reason for this is that under the very high pressures exerted on the tools and the heat generated in the cutting process, the diamond edge will convert to a carbide.

High performance wire dies and water jet nozzles are currently fabricated of single crystalline diamond and of sintered diamond. Both exhibit problems that are expected to be overcome with the advent of artifact polycrystalline diamond. The crystalline diamond dies wear asymmetrically as the hardness of diamond varies as a function of the crystalline plane. To overcome this, sintered diamond powders exhibit efficacy in symmetrical wear, but are softer and wear faster. Artifact polycrystalline diamond not only satisfies the hardness and asymmetry problems, but can be manufactured on a sacrificial mandrel for lowered production costs.

The potential for exploiting the hardness of diamond is virtually unlimited. While early applications of thin film diamond will mimic those already in use for diamond chips (e.g., abrasives and cutting tools), many new conformal coating applications will be devised. This type of application will significantly expand as halocarbon deposition and other growth techniques are further developed to permit deposition at low temperatures.

Thermal Conductivity

At 20 watts/cm-°C natural diamond exhibits thermal conductivity greater than that of any other material. Again, only cubic BN at 13 watts/cm-°C approaches this figure and this value is theoretical as cBN has not been grown in sizes sufficiently large so as to experimentally ascertain the value.

The thermal conductivity of diamond is greater than 4 times that of silicon carbide, 13 times that of silicon, and 43 times that of gallium arsenide. While it would appear that diamond is unexcelled in thermal conductivity, this conductivity derives from phonon (elastic wave) propagation.[1] Any scattering of these elastic waves will reduce the thermal conductivity. Since carbon is found with a natural abundance of 1% C^{13} and 99% of C^{12}, diamond synthesized from this material (also natural diamond) will exhibit a scattering from the random and substantial (per cent wise) mass differential of its constituent atoms. If diamond were to be synthesized from isotopically pure carbon, the thermal conductivity would be substantially increased.[5] Indeed, the experiment has been performed and the initial experimental results substantially exceed theoretical expectations.[6]

Since diamond has an exceptionally high thermal conductivity, it is exceptionally efficient in use as a heat spreader. Molar heat capacity, however, is proportional to the atomic mass of the elements comprising the substance and being comprised solely of carbon, diamond is NOT a good heat sink. While being 50% better than water in this respect, it is scarcely 25% that of silver, gold, or copper. The judicious application of diamond as a thermal material must, therefore, be as a heat exchanger or heat spreader and there are numerous applications requiring these properties. Many of these applications are in electronics. GaAs power transistors in particular can benefit from an underlying diamond heat spreader/exchanger. Currently the individual "fingers" of a GaAs power transistor (either bipolar or field effect) must be spaced about 5 times further apart than would otherwise be necessary so as to prevent hot spots and thermal run away. This additional spacing creates significant additional parasitic capacitance and resistance in the device thus lowering its upper operating frequency. High performance computers are currently at the stage of development where their speed is limited not by the inherent switching speed of the transistors within the integrated circuits (IC), but by the length of the interconnecting signal lines between ICs. The lengths of these interconnecting lines is largely dictated by the need for adequate (on board) thermal dissipation means. If the heat could be much more efficiently transferred to the *edge* of the circuit board, then the ICs could be more densely populated and the boards

---

[1]. See subsequent section on acoustic properties.

could be larger to reduce the need to transport signals off board. It can be shown that artifact diamond circuit boards would be ideal in this respect. Not only do they represent unmatched thermal properties, but their low dielectric constant and loss tangent also provide significant advantages.[2] Although diamond has a thermal expansion coefficient of but 30% of that of silicon and 13% of that of GaAs, this may be largely irrelevant. Recently a technique has been devised to lift off epitaxial layers of integrated circuits (from their native substrates) and attach them to a foreign, but smooth, surface with Van der Waals bonds.[7] In this manner, the semiconductor die is relatively free to move in the horizontal plane to accommodate thermally induced motion, but will not lift off or otherwise separate. As artifact diamond films become more economical to produce, this application is certain to become very pervasive of the electronics industry. Diamond is already in widespread use for mounting of IMPATT diodes. The efficacy of diamond in this application not only accrues not only from its excellent thermal conductivity, but from its low dielectric constant and loss tangent as well. Since IMPATT diodes are typically used as individual units, the advent of diamond films and coatings is not expected to impact this application. Another primarily thermal application for diamond in electronics is that of the mounting of the helices of traveling wave tubes. While this traditional application may not be directly impacted by thin film diamond, it may be possible to design traveling wave structures in diamond films thereby enabling new approaches to these devices.

The exceptional hardness of diamond renders it very resistant to erosion by moving fluids. This property when combined with its exceptional thermal conductivity renders it an ideal material for conventional heat exchangers. Since it is much stronger and lighter in weight than copper or other competing materials, not only could the heat exchanger walls be made thinner than in competing materials, but the overall greater heat transfer efficiency would permit a much smaller overall area between the two fluids. The resulting savings in both weight and volume make it exceptionally attractive for use in heat exchangers in spacecraft and aircraft.

The potential for exploitation of the thermal conductivity of diamond is enormous. Isometrically pure carbon will significantly enhance the already superlative position of diamond. A primary application will be in dissipating the heat from and increasing the operating speed of high performance electronics. Fluidic heat exchangers will also see applications of diamond -- particularly where decreased volume and weight are required.

Friction

The friction between two surfaces is defined as

$$f_s = \frac{\text{(magnitude of maximum frictional force)}}{\text{(magnitude of the normal thrust)}}$$

and is dependant on many variables. The nature of the material, its crystallographic orientation, its temperature, its surface finish, the relative velocity between the surfaces, environmental atmosphere, temperature, and the degree of cleanliness are among the most pronounced of these variables. Nevertheless, under similar conditions, diamond exhibits a coefficient similar to that of Teflon: 0.05. This compares with a coefficient of 0.10 for lubricated metals and graphite. This low coefficient is believed to result from a hydrogen termination of the

---

[2]. See subsequent section on optical properties.

surface. At temperatures above which hydrogen desorbs, the coefficient approaches unity.[8] The low coefficient of friction is particularly relevant in extending the tool life of diamond cutting tools by reducing heat generation. Gas turbines and certain other applications require operation of bearings at elevated temperatures. At operating temperatures below 500 Celsius unlubricated diamond film bearings should see a significant market. The ability to deposit diamond films at temperatures below 300 Celsius will permit the coating of softer materials -- even certain plastics -- with diamond. Non-ferrous bearing surfaces will also benefit from such a capability. Ferrous bearing journals will require a noncarbide forming interface barrier.

Potential applications for exploiting the low friction properties of diamond are very extensive. This application will most assuredly experience an exponential growth with the advent of conformal diamond coatings. Aside from such a simple but massive consumer market in diamond coated cookware with the first coating that is both hard and slippery, applications abound for oilless bearings requiring no maintenance. Other applications wherein operating temperatures are marginal for lubricant compatibility will also witness increased use of diamond films. System benefits accrue not only from elimination of the lubricant system, but its associated cooling as well. The pervasiveness of diamond bearing surface applications will depend in large part on the development of viable interface materials between diamond and underlying ferrous metals.

Acoustic

Elastic waves in solids are of two types -- longitudinally polarized and orthogonally polarized (shear) waves. The elastic wave velocities in diamond exceed those of any other material. Since the wavelength of an elastic wave at any given frequency is proportional to its propagation velocity, the wavelength (at a given frequency) in diamond is longer than in any other material. While this makes diamond the poorest of all candidates for an acoustic delay line, it renders it exceptionally good as an acoustic transducer in that the transducer area may be made much larger before wavefront cancellation effects occur. To augment the credentials of diamond in this respect, it is also the strongest of all materials (excepting graphite fibers) so that a transducer may be made very thin and yet remain rigid. Diamond is also comparatively light weight thereby lowering the inertia of the transducer.

While Sony has already exploited this attribute of diamond for acoustic headphones and hi-fidelity speakers, the market potential for significant expansion in this and related applications is guarded. Certain ultrasonic transducers, could, however, profit from diamond films.

Young's Modulus

The elastic region is that region over which a material subjected to a deforming force reverts to its original condition after the deforming force is removed. The ratio of stress to strain is approximately a constant over this region as expressed by Hooke's law. Young's modulus is the ratio of tensile stress to linear strain in the region over which Hooke's law applies. Young's modulus for diamond is 105×1^{11} dynes/cm^2. This value is three times that of runner up tungsten and 5.25 times that of the finest of steel. A potential exploitation of this property is in certain high power optical devices where distortion from energy loading is unwanted.

Fracture Toughness

Fracture toughness is calculated as

$$K = (0.16)\ E\ P\ /\ H\ c^{1.5}$$

where E is Young's modulus, P is the indentation load, H is hardness, and c is crack length. Definitive measurements have yet to be made and since diamond exhibits both the highest E and the highest H of all materials, predictions are uncertain. Polycrystalline diamond (whose grain size is now easily controllable) will certainly exhibit crack lengths shorter than those of natural or single crystalline diamond. High power optics may be a potential user of this property should it prove meritorious.

CHEMICAL PROPERTIES AND POTENTIAL OF DIAMOND

At temperatures below 427 Celsius single crystalline diamond is virtually immune to chemical attack by all substances unless they are energetic (e.g., ion beams). At 427 Celsius molten sodium nitrate is known to attack the diamond surface. At temperatures as low as 625 Celsius the diamond surface will become blackened in the presence of atomic oxygen. Below 1000 Celsius, oxidizing substances appear to be the only effective means of attacking the diamond surface (in the absence of energetic particles). At higher temperatures and pressures there are two other groups of materials known to attack diamond. The first of these are the carbide formers such as tungsten, tantalum, zirconium, titanium, and silicon. The second group are the solvents such as molten nickel, iron, cobalt, manganese, and platinum. Diamond cutting tools are inappropriate for use with any of these materials. Although boiling KOH and similarly corrosive materials will not attack diamond, they can permeate diamond coatings and attack the underlying material if the grain size of the diamond is too large. Grain sizes of less than 30 nanometers are generally adequate for such protection. Many potential applications for corrosion proof coatings of diamond abound. Most prosthetic implants are candidates for diamond coatings. Not only is the constituent carbon content of diamond compatible with most tissue, but artificial joints could also benefit from the its coefficient of friction. Artificial heart valves may also benefit from diamond coatings as blood is thought not to clot on a diamond surface as it does on vitreous carbon. Underwater objects and indeed all objects exposed to water and oxygen at temperatures below 427 Celsius could be protected from their environment by diamond coatings. The potential for diamond as a corrosive prophylactic is extremely large, but remains a virtually unexploited resource.

Controlled etching of diamond is best accomplished by the use of nitrogen dioxide and 50 eV noble gas ions.[9] This technique can be further developed for the precise atomic layer by atomic layer removal of carbon atoms from the diamond surface.[10] Unlike most etchant residues, the residue of diamond etched with oxygen is a gas. In this manner deep trenches may be etched to very precise depths and no residue is left. Using these well controlled chemical etching processes, diamond may be patterned very controllably into intricate patterns for various optical and x-ray applications.

The chemical inertness of diamond is thought to have a significant potential in prosthetic implant devices and in corrosive environments such as a marine environment. Should economical deposition processes become available and should a viable interface between diamond and iron become available, the coating of ships would not only significantly reduce friction, but eliminate barnacle attachment and otherwise be enormously cost effective in reducing dry docking requirements.

Diamond was originally classified primarily by its optical properties. These properties derived from impurity induced optical color centers as will be noted below. Type I has large amounts of nitrogen included. When the nitrogen is present in concentrations sufficient to aggregate, it is further classified as type Ia. When nitrogen is disbursed (as in most artifact diamond) the designation Ib applies. Diamond virtually free of nitrogen is classified as type II. If electrically insulating, it is further classified as type IIa. When semiconducting (p-type, and with boron concentration exceeding the background nitrogen concentration), it is classified as type IIb.

Optical Centers

Diamond has many optical centers. Virtually all of them accrue from nitrogen, boron, vacancies, or complexes thereof. A comprehensive listing has been published by Davies.[11] Although diamond is not a direct bandgap semiconductor, diamond lasers have been demonstrated.[12] These lasers have been electron beam pumped with optical generation from impurity centers. When pumped with extremely large amounts of power, stress is built up within the crystal and the resultant strain may adversely affect the mirror-to-mirror spacing. The unexcelled Young's modulus of diamond minimizes this unwanted effect. The inherent strength of 272 kg/mm^2 for diamond renders it the strongest of all optical materials and thus most resistant to shattering from large internal stress. The manner in which a laser host crystal interacts with the impurity dopants within that crystal is a critical factor in all pumped lasers. The manner in which the diamond crystal fields lift the degeneracy of the impurity atom electrons embedded within is largely unknown. Until this becomes known, prospects for optimizing diamond lasers are dim. Even more questionable is whether impurity ions may be introduced into the small lattice of diamond without seriously disrupting the lattice structure. Smaller, transition metal impurities may be considered, but their efficacy as luminescence centers in diamond is unknown.

Refractive Index

Diamond is characterized by an index of refraction of 2.41 - 2.44 in the spectrum between 656 and 486 nm. This is considerably less than that of most of the common semiconductors, but nearly twice that of most common insulators. Any material considered for use in matching the refractive indices between any two materials must have a refractive index that is the geometric mean of the two:

$$n_2 = (n_1 * n_3)^{-.5}$$

It can be shown that a diamond window may be optically matched to air or vacuum by a coating having a refractive index of 1.55. In such an arrangement, the reflectivity of a normal incident wave would be nil. Quartz is such a material, but its use would severely reduce the efficacy of diamond to combat microparticle abrasion, corrosion, and other properties for which the diamond was originally chosen. There are solutions to this problem. A primitive method would be to render the surface of the diamond film porous for a depth of 1/4 wavelength. The porous film would exhibit a lower dielectric constant and refractive index; unfortunately, it would also be softer. A procedure developed by Allen Cox for use in silicon[13] and adapted for use in diamond by this author is shown in figure 2. Small depressions are etched by processes described above. The diamond surface is structured much like that of the eye of a moth. In designing the structure, the following should be observed:

- PERIOD $P = \frac{\lambda}{n_2}$

  (ONLY ZERO ORDER PROPAGATES)
- FILL FACTOR = 50% BY AREA

  $\left(\frac{W}{P}\right)^2 = 0.5$

  (POLARIZATION INDEPENDENCE)
- DEPTH $d = \frac{\lambda}{4\sqrt{n_1 n_2}}$

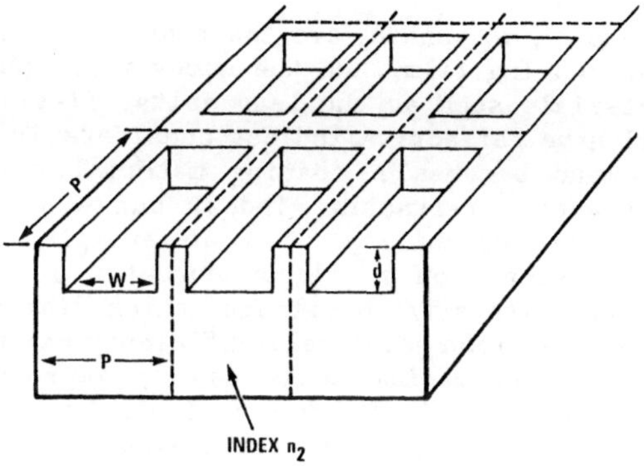

Figure 2. Moth's Eye Lens

$$P < \text{wavelength} / n_2$$

$$w = (0.5 * P)^{.5}$$

$$d = 0.25 * \text{wavelength} / (n_1 * n_2)^{.5}$$

The effective spectrum over which the procedure is applicable may be increased by rendering the walls of the depressions curved rather than straight. The procedure may be used as either a single or a double layer coating. In the latter case the diamond next to the material to be matched is continuous and superimposed on it is the moth's eye structure.

Windows, Lenses, and Mirrors

The hardness of diamond makes it very attractive as a protective window for softer optical materials. The magnitude of reflection from an optical coating of low refractive index on that of on underlying material of higher refractive index is given by

$$R = [\ (n_1^2 - n_0 n_2) / (n_1^2 - n_o n_2)\ ]^2$$

where n_0, n_1, and n_2 are the respective refractive indices of air (vacuum), the coating film, and the underlying substrate. Conventional IR detector materials such as the lead salts, II-VI materials, germanium, and silicon all have refractive indices considerably higher than that of diamond. Thus diamond becomes a coating material of choice not only because of its attractive refractive index, but also because of its hardness and its chemical durability. As an example, at a wavelength of 0.6 nm, the reflection from a silicon detector or solar cell is about 30%, but when coated with a 1/4 wavelength thick diamond film reflectivity is reduced to 6.6%. Silicon solar cell efficiency can typically be increased by 40% while that of germanium cells can be improved by as much as 88% by diamond coatings. Reliability (particularly in an out space environment) is also improved by virtue of the diamond resistance to microparticle abrasion. Even greater improvements are engendered to longer wavelength (higher refractive index) detectors when given diamond window coatings.

Germanium windows are widely used for IR detectors as they are isotropic and exhibit virtually no distortion. Being soft, however, they are vulnerable to microparticle abrasion and rain droplet erosion. This vulnerability is significantly exacerbated when such windows are moving at high velocities with respect to their environment. Diamond, by virtue of its hardness and low optical loss is useful in these applications -- either as a coating or as a stand alone window. The great strength of diamond may be used to advantage in this application to create a larger diameter window without significantly increasing its thickness.

When heat is a prime factor, there may be no substitute for a diamond window. The window used for the IR emission sensor in the Venus explorer is a prime example. High power lasers can also benefit from the use of diamond windows, lenses, and mirrors. The very low optical absorption of diamond can be combined with its unexcelled thermal conductivity.[14] In this manner, that heat that is absorbed by the diamond window can be quickly and efficiently distributed to the window edge where it may be dissipated by water cooling or other appropriate heat sinking techniques. Optical mirrors are also vulnerable to high power lasers and may be improved by diamond coatings or backings. Diamond mirrors may substantially reduce the operating length of free electron lasers by reducing the need to diverge the beam and reduce its power density.

Convex and concave lenses of diamond must currently be made on and removed from a double mandrel. Each half must then be bonded together. This is a cumbersome procedure at best, but alternatives are few since diamond is not readily polished and shaped as are softer materials. One alternative is the Fresnel lens. In such a lens, the circular grooves are etched with oxygen by one of the methods noted above. The positioning and sizing of the rings are such that from a point source of light located on axis and at a focal distance away from the surface, alternate half wave zones are in phase with each other. To obtain optimum phasing, the radii of theses zones are made proportional to the square root of their integer number. The depth of the grooves must be such as to introduce a half wavefront advance. To reduce reflection, the moth's eye superstructure may be superimposed upon the structure of the Fresnel lens.

Optical windows subjected to marine environments must meet rather demanding requirements. Periscope windows not only require strength to resist very high pressure differentials, but they must also resist corrosion and the attachment of barnacles and other marine life. Diamond not only meets these requirements, but also has the attribute that as the window emerges from its underwater environment, its tribological properties ensure that water droplets do not adhere to its surface and obscure vision.

The potential optical market for diamond is expected to be primarily military in nature. Low optical losses, high temperature tolerance, and microparticle abrasion advantages will be the features most sought.

ELECTRICAL PROPERTIES AND POTENTIAL OF DIAMOND

Excepting for electron velocity, the electrical properties of diamond exceed those of virtually all other semiconductors. Table I below provides the details.

Table I. Electrical Comparisons

Property	diamond	GaAs	silicon
Bandgap (meV)	5450	1430	1100
Hole mobility#	1800	400	600
Electron mobility#	2000	8500	1500
Resistivity (ohm-cm)	10^{16}	10^{9}	10^{3}
Breakdown (V/cm x 10^{6})	10	6	5
Dielectric constant	5.7	12.5	11
Work function (Ev)	*	4.7	4.8
High field electron velocity (cm/sec x 10^{7}))	2.7	1	1

cm^2/volt-sec
* 4.8 eV (p-type). Diamond is thought to exhibit a negative electron affinity on its {111} surface.

In natural diamond, the Hall coefficient extends from 10^{12} at 140 K to 10^{5} at room temperature.[15] Relaxation time disparities among the three hole bands lead to widely differing values depending on measurement technique used. The mobility of holes is even higher than stated in Table I if magnetoresistance data are accepted. Although germanium and indium antimonide exhibit higher hole mobility, they exhibit poor saturated hole velocity and low dielectric strength. Hole mobility in artifact diamond homoepitaxial films has been measured as high as 850 cm^2/V-sec.[16] The electrical resistivity of natural diamond is 7 orders of magnitude greater than that of GaAs and that of artifact homoepitaxial diamond films grown by RF immersed plasma and remote plasma using carbon monoxide as well as those grown by thermal halogen CVD are equally high.[16] In natural diamond, the impurity distribution is not homogeneous and therefore the measured charge carrier transport values are open to question.

There is no known shallow substitutional donor impurity for diamond. Both lithium and sodium are thought to be shallow when confined to interstitial sites.[17] While lithium will diffuse at moderate temperatures (thereby canceling the high temperature attraction of diamond), sodium is thought to be stable. Degenerately implanted carbon is useful in creating shallow n+ conductivity regions in diamond.

Diodes

Natural diamond has been used as a **radiation detector** for over 40 years.[18] The resistivity of both natural and artifact single crystal diamond has been measured at over 10^{14} ohm-cm. This resistivity derives from its bandgap and absence of impurities and defects. Not all diamond qualifies. A very high electric field may be applied across a diamond film without producing significant leakage current. In contrast, most common semiconductors require reverse biased junctions to avoid leakage current. Even then, leakage current fluctuations typically dominate the signal in reversed biased p-n junction devices. This is unacceptable for the detection of very small signals. Diamond has other advantages as a radiation detector as well. In diamond, the hole mobility is virtually as high as the electron mobility; both exceed the electron mobility of silicon. Consider this in the context of a 200 micrometer thick semiconductor detector. The maximum practical voltage that can be placed across a reverse biased silicon diode is 1000 thereby yielding a maximum electric field of less than 10^5 V/cm. In the silicon detector the hole collection time is 60 ns and the electron collection time is 20 ns. In the diamond metal-insulator-metal (MIM) device, a voltage of 2×10^5 may be applied yielding an electric field of 10^7 V/cm and a collection time of 1 ns for both electrons and holes.

The intrinsic radiation hardness of a semiconductor relates primarily to (1) its chemical bonding energy and (2) the manner in which radiation-induced lattice dislocations cause fluctuations in the diode current. The chemical bonding energy of diamond is the strongest of all semiconductors and therefore the most immune to radiation damage. Not only are the number and severity of radiation induced lattice dislocations smaller in diamond than in silicon, but many self-anneal at room temperature. Radiation-induced lattice dislocations do, however, produce energy states within the forbidden band of any semiconductor. These states can act as generation centers and will emit holes into the valence band and electrons into the conduction band. Once in the valence or conduction band, the charges are swept out of the depletion region by the externally applied electric field. In a detector device requiring a p-n junction (e.g., most common semiconductors), the carrier concentration is not in equilibrium, that is:

$$pn < n_i^2$$

where p, n, and n_i are the hole, electron, and intrinsic carrier concentrations. Since it is known that the generation current is proportional to the intrinsic carrier concentration n_i [19], diamond diodes (having no p or n regions) having many orders of magnitude lower carrier concentrations, accrue a significant advantage. Stated differently, in the diamond diodes the carrier concentrations are in equilibrium ($pn = n_i^2$), the generation rate is equal to the recombination rate, and there is no net generation/recombination current. Since $n_i \propto e^{-Eg/2kT}$, is nearly 30 orders of magnitude smaller in diamond than in silicon, diamond accrues another significant advantage as a radiation detector. To test the validity of this assumption, Schottky diodes of diamond and of silicon were fabricated at the M.I.T. Lincoln Laboratories and exposed to 1.5 MeV electrons at doses up to 10^{16} $/cm_2$ (about 1 GigaRad).[20] The silicon devices were characterized by pre-radiation leakage currents 100 times those of the diamond devices and increased by another factor of 40 after irradiation. The diamond devices were unchanged by the radiation exposure.

The radiation capture cross section of diamond for neutrons is 25 times less than that of silicon thereby rendering diamond much less sensitive to neutron damage. Such damage may cause trapping centers that lead to a decrease in carrier lifetime. Collection efficiency (and therefore pulse height) may be low if carrier lifetime is small compared to collection time. Experimental evidence[21] shows that natural diamond detectors survive several times 10^{14} neutrons/cm^2 or about 100 times equivalent silicon devices. This is rather remarkable in that natural diamond is usually characterized by an abundance (e.g., $10^{20}/cm^3$) of nitrogen impurities that generate trapped charges and significantly reduce carrier lifetime.

<u>Transistors</u>

As an RF power amplifier, a leading figure-of-merit is the power frequency squared impedance product.[22] For diamond this is 8200 times that of silicon, over 500 times that of GaAs, and nearly 8 times that of silicon carbide. Practical considerations, however, must also be taken into account. Paramount among them is that of a high quality ohmic contact. Another consideration is that of midgap nitrogen traps. Until these two issues are resolved, high performance diamond transistors will lag those of silicon carbide. The lack of a shallow donor impurity in diamond does not necessarily mean that one can not design an efficient n-channel diamond transistor. Two approaches are appropriate. The first is to implant n+ source and drain regions in insulating, nitrogen-free, diamond. By virtue of the argument given above for diamond diodes, the channel need not be impurity doped with acceptors. As such, the ionized impurity scattering will be much less than that experienced in conventional semiconductors field effect transistors (FETs). A second possibility is that of a heterojunction device similar to that of a High Electron Mobility Transistor (HEMT). Since boron nitride is virtually lattice-matched to diamond, can be doped both p and n type, and has a higher bandgap than diamond, it is a candidate for such a device.

In a conventional single gate FET, the electrons enter the channel at near zero velocity, are accelerated by a comparatively low electric field until they emerge on the drain side of the gate and then accelerate under a high electric field until collected at the drain. In GaAs and all materials with a negative electron conductivity region, the electrons in the high field region are transferred to the subsidiary conduction band valley, become heavy, and slow. In silicon and the III-V materials, the average (i.e., effective) velocity is much less than the peak velocity. The introduction of a second control gate (between the first and the drain) in the channel is ineffective in III-V materials because of the lower electron velocity in the high field region and is ineffective in silicon because the saturated velocity is only 10^7 cm/sec. In diamond, however, the presence of a second gate is efficacious. Unlike the III-V FET case, electrons entering the region of the second gate are entering at maximum velocity (e.g., $> 2 \times 10^7$ cm/sec). Therefore, their velocity under this second control gate is MUCH larger than is the velocity of electrons under the first gate in III-V materials. The corollary is that the effective gate length (of the second gate) in diamond is about four times shorter than is that of the first gate in a III-V FET. The average (effective) overall velocity of charge carriers in a dual gate diamond FET is significantly larger than that in other materials.

Integrated Circuits

In densely packed integrated circuits, charge carrier transit time is not the only factor dominating performance. Parasitic capacitance (i.e. dielectric constant) and thermal conductivity are also parameters to be considered. The figure of merit expressing this[23] places diamond 32 times greater than silicon and 64 times greater than GaAs. To be able to capitalize on these superlative properties, however, a means must first be found to economically nucleate and grow single crystalline diamond on large area substrates, to create ohmic contacts, and develop a processing technology for diamond devices.

Cathodes

The intriguing prospect exists that the [111] diamond surface may be a negative electron affinity (NEA) surface. If this is the case, vacuum tube devices can not only be simplified, but made far more reliable. In current high power vacuum tubes, the product of amperes times operating life of the cathode is approximately a constant. The typical 1100 Celsius operating temperature of these cathodes also means that they can not quickly be brought to full operating power and, in many operating conditions, must therefore be left on for prolonged periods on time when the tube is otherwise inactive. Microminiature vacuum tubes currently require field effect emission cathodes. The availability of NEA material would significantly reduce fabrication complexity.

There is an immediate potential for the use of diamond as a radiation detector. In the superconducting super collider (SSC) alone, very large quantities are envisioned. In x-ray tomography, the availability of diamond detector arrays would alleviate problems associated with dark current and signal to noise (contrast) ratio. A potential mass consumer market also exists in the vehicle of Radon detectors. At such time as good ohmic contacts and heteroepitaxial nucleation become available, discrete diamond power transistors will enjoy a market in VHF/UHF transmitters, air search radar phased array antennas, and, perhaps, the replacement of certain vacuum tubes. The market potential for diamond integrated circuits with diamond being the active element is dependant on several technological advances and is guarded -- at least until the turn of the century.

STATUS

Opaque polycrystalline diamond can currently be grown at rates exceeding 100 micrometers/hr and over comparatively large areas (e.g., 75 millimeters diameter). Visually transparent polycrystalline diamond films can be routinely grown over the same areas, but at rates less than 1 micrometer/hr. Single crystalline diamond films can be homoepitaxially grown by several methods and with properties equal to or exceeding those of the underlying substrate and at temperatures as low as 500 Celsius. Large area heteroepitaxial growth of single crystalline diamond is yet to be demonstrated. Quality polycrystalline diamond films have been grown at temperatures as low as 200 Celsius.

REFERENCES

1. W. Pryor and T. Anthony, private communications

2. M. Geis, private communications

3. M. Frenklach, P. Kematick, D. Huang, W. Howard, K. E. Spear, A. W. Phelps, and R. Koba, "Homogenous Nucleation of Diamond Powder in the Gas Phase" J. App'l. Phys. 66, 395, (1989)

4. Status and Applications of Diamond and Diamond-like Materials: An Emerging Technology, p 94, National Academy Press, Washington, D.C. 1990

5. R. Seitz, The End of the Age of Sand, Presented at the Gorham Advanced Materials Institute, Marco island, FA 15-17 October (1989)

6. R. Pryor and T. Anthony, private communications

7. E. Yablonovitch, E. Kapon, T. J. Gmitter, C. P. Yun, and R. Bhat, "Double Heterostructure GaAs/AlGaAs Thin Film Diode Lasers on Glass Substrates", IEEE Photonics Tech. Lett 1,41 (1989)

8. M. N. Gardos, Evidence of Reconstruction of (111)/(100)-Textured, Polycrystalline Diamond Films by SEM Tribology, presented at Workshop of the Science and Technology of Diamond Thin Films, Quail Hollow, concord, Ohio, may 1990

9. M. W. Geis, D. D. Rathman, D. J. Ehrlich, R. A. Murphy, and W. T. Lindley, presented at the SDIO/ONR Diamond Technology Initiative Workshop, Lexington, MA (1987)

10. M. Yoder, Atomic Layer Etching, U.S. Patent No. 4,756,794

11. G. Davies, Chem. Phys. Carbon 13, 1-143 (1977)

12. S. C. Rand and L. G. DeShazer, Visible color-centered laser in diamond, Optics Ltrs. Vol 10 No. 10 (1986) p 481-3

13. Allen Cox, Honeywell, Inc. (used by permission)

14. J. R. Seitz, Laser Systems With Diamond Optical Elements, U. S. Patent No. 3,895,313

15. A. W. S. Williams, E. C. Lightowlers, and A. T. Collins, J. Phys. C: 3, 1727 (1970)

16. R. J. Markunas, private communications

17. J. Bernholc, A. Antonelli, and S. Kajihara, presented at the Fourth S.D.I.O./IST - ONR Diamond Technology Initiative Symposium, Crystal City, VA 12 Jul 1989

18. D. Wooldridge, A. Ahearn, J. Burton, "Conductivity Pulses Induced in Diamond by Alpha-Particles", Phys. Rev. 71, 913 (1948)

19. S. M. Sze, Physics of Semiconductor Devices, 2nd ed. J.Wiley & Sons, New York (1981)

20. M. Geis, MIT Lincoln Labs, private communications

21. S. F. Kozlov et al., Preparation and Characteristics of Natural Diamond Nuclear Radiation detectors, IEEE Trans. Nuc. Sci. **NS-22**, 160 (1975)

22. A. Johnson, RCA Rev. **26**, 163 (1965)

23. R. W. Keyes, Proc. IEEE **60**, 225 (1972)

NATURAL DIAMOND: THE STANDARD

J. E. FIELD
PCS, Cavendish Laboratory
Cambridge, CB3 OHE, UK

INTRODUCTION

The extreme rigidity of the diamond lattice is the key to many of its exceptional properties. This review first discusses the main physical properties of diamond including its structure, stability, the impurities diamonds contain and their effect on its thermal, optical and electrical behaviour. With this as background, recent results and data are then presented on the elastic, cleavage, strength, fracture, hardness, frictional and wear properties of diamond.

Diamond is an exciting material with many outstanding properties. It is pre-eminent as a gemstone, an industrial tool and as a material for solid state research. Since natural diamonds grew deep below the Earth's surface before their ejection to mineable levels, they also contain valuable information for geologists. The key to many of diamond's properties is the rigidity of its structure which explains, for example, its exceptional hardness and its high thermal conductivity. Since 1953 it has been possible to grow synthetic diamond. Techniques are now available to grow gem quality synthetic diamonds greater than 1 carat (0.2gm). However, the costs are high, and the largest commercially available industrial diamonds are about 0.01 carat in weight or about 1mm in linear dimension. The bulk of synthetic diamonds used industrially are 600μm or less. Over 75% of diamond used for industrial purposes today is synthetic material. In recent years there have been two significant developments. The first is the production of composites based on diamond; these materials have a significantly greater toughness than diamond while still maintaining very high hardness and reasonable thermal conductivity. The second is the production of thin film diamond coatings which are the main subject of the present conference. These developments have added further to the versatility of diamond.

Diamond Stability

Diamond and graphite are both forms of carbon. In 1955, Berman and Simon[1] considered the thermodynamics of the phase change and obtained the diamond/graphite equilibrium diagram. Subsequent recalculation, using improved data, for specific heats, molar volumes and compressibilities has modified the line slightly above pressures of 40kbar (0.4GPa)[2]. It is important to appreciate that although the line marks the boundary between regions of thermodynamic stability it says nothing quantitatively about the rate of change from one phase to another. Diamond it will be noted is **not** the stable form under environmental conditions; experience suggests that the transformation to graphite must be a slow process. Similarly when synthesizing diamond it is not sufficient to simply take graphite to a P/T point above the Berman/Simon line; if this were so many of the early attempts at diamond synthesis would have been successful.

Diamond and Diamond-Like Films and Coatings
Edited by R.E. Clausing *et al.*, Plenum Press, New York, 1991

Origin of Diamond in Nature

The high temperatures and pressures required to crystallise diamond occur at depths of greater than about 150km and in theory formation is still possible down to about 1500km. Diamond is usually, though not always, associated with the rock Kimberlite in which it is thought to crystallise when the Kimberlite is in magma form and to be transported when the Kimberlite is driven upwards by excess pressure exerted by CO_2. After their ejection into a region near the Earth's surface, diamonds frequently remain in the Kimberlite eruption but it is also possible for natural erosion processes to distribute them into alluvial deposits near rivers and the sea, often at large distances from their points of emergence. All diamonds are different, though there are interesting similarities between diamonds found in particular mines[3].

Table 1

Structural Properties of Cubic and Hexagonal Diamond and Graphite

	Cubic	**Hexagonal**	**Graphite**
Point group of atoms	Td	Td	
Space group	Oh^7;Fd3m	D^4-$_{6h}$;$P6_3/mmc$	D_{6H}^4
Unit cell constants	Cubic	a=0.252nm	a=0.246nm
	0.357nm	c=0.412nm	c=0.671nm
C-C nearest neighbour distance	0.154nm	0.154nm	0.142nm
Bonding	sp^3	sp^3	sp^2
Atoms located at	$(000)(\frac{1}{2}\frac{1}{2}0)\ (0\frac{1}{2}\frac{1}{2})$	$(000)\ (00\frac{3}{8})$	$(\frac{1}{3}\frac{2}{3}0)\ (\frac{2}{3}\frac{1}{3}0)$
	$(\frac{1}{2}0\frac{1}{2})\ (\frac{1}{4}\frac{1}{4}\frac{1}{4})\ (\frac{3}{4}\frac{3}{4}\frac{1}{4})$	$(\frac{1}{3}\frac{2}{3}\frac{1}{2})\ (\frac{1}{3}\frac{2}{3}\frac{7}{8})$	$(\frac{1}{3}\frac{2}{3}\frac{1}{2})\ (00\frac{1}{2})$
	$(\frac{1}{4}\frac{3}{4}\frac{3}{4})\ (\frac{3}{4}\frac{1}{4}\frac{3}{4})$		
General crystal	Zinc Blende type	2H-wurtzite	
Density	$3.52 \times 10^3 kgm^{-3}$	$3.52 \times 10^3 kgm^{-3}$	$2.3 \times 10^3 kgm^{-3}$

Synthesis of Diamond

The problem of diamond synthesis was first solved in the ASEA Laboratory in Sweden in 1953. However, this early success was not published at the time. In 1954 the General Electric (GE) Company successfully synthesized and proved that they had produced diamond Very soon thereafter diamonds were made in South Africa at the De Beers Adamant Research Laboratory. All of these groups used a carbon solvent/catalyst to overcome the large kinetic barrier and act as a transport agent for carbon. The production of thin film diamond by metastable growth is a major theme of the present Conferences.

Crystallographic Structures

Diamond has tetrahedral, covalent bonds between each atom and its four nearest neighbours. In diamond, the commonest way of linking the tetrahedral units into a three-dimensional structure is the cubic form. An alternative structure is the hexagonal wurtzite lattice which was predicted as a possible polymorph of diamond by Ergun and Alexander[4]. Bundy[5] synthesized this form in the laboratory and it has now also been produced in various shock loading experiments. The cubic form of diamond occurs most commonly in nature as octahedra with {111} planes or in dodecahedra with {110} planes as a dissolution form. The simple cube with {100} planes is relatively rare with natural diamonds, but is common synthetically if particular conditions of T and P have been used in the growth capsule. Diamond invariably contains an individual and complex internal structure which reflects the growth conditions, impurity content and any deformation while at high P and T. Table 1 summarises the structural data.

Impurities: Inclusions

The observed properties of diamonds show unambiguously that elements other than carbon must be present in the diamond lattice. Much of recent diamond research has had the objectives of identifying these elements, their chemical relationship, their effect on optical, electrical, thermal and mechanical properties and their geochemical significance. The major impurity in diamond is nitrogen, but recent research has shown that there are significant amounts of oxygen and hydrogen both in the bulk and at the surface of diamond. Table 4.1 in Sellschop[6] details, 60 elements have now been identified as occurring in diamond. Some of these have been found using mass spectrometry or optical spectroscopy, but most (> 50) have been detected by nuclear probe techniques.

Some diamonds contain visibly quite large inclusions of up to millimeter dimensions. The inclusions may be gaseous, liquid or solid. The diarist John Evelyn records having seen a diamond with a solid inclusion in 1645, while in the period 1820 to 1862 Sir David Brewster made an extensive study of fluid inclusions. In 1904 Bauer and Spencer inferred the presence of liquid and gaseous inclusions of H_2O and CO_2 in diamond. Solid inclusions occur either as totally enclosed single crystals or as crystalline material lying within fracture systems in the diamond. Two broad mineral groupings are recognized by the geologist. The first is the syngenetic minerals, which crystallize under the same physical conditions as diamond and are then enclosed and protected by the surrounding diamond, and the epigenetic minerals which formed subsequent to diamond growth, either while the diamond was in the kimberlite or later during transport to an alluvial deposit. It is the first category which has had the more detailed study (for a review, see Ref. 3). The considerable interest in the mineralogy and chemistry of the syngenetic inclusions is due to their value in evaluating ideas on the origin of diamonds, xenoliths and kimberlites and their inter-relation.

Platelets: Voidites

The nature of the platelike features found in type 1a diamonds is one of the most intriguing mysteries of diamond science. The main contenders are nitrogen impurity or carbon itself. In the last few years the consensus of opinion about the amount of nitrogen in the platelets has fluctuated wildly. There is only space here to comment in the briefest terms. Direct observation of platelets by electron microscopy was made in 1962 by Evans and Phaal[7]. They showed that they formed on {100} planes in type I diamonds but were absent in type II. They found that the platelet size and density varied markedly from specimen to specimen and also in different regions of each specimen. Platelet densities were in the range (1 to 3.5) x 10^{15} cm^{-3} and diameters ranged up to ~ 0.1 μm. More recent observations[8] have found "giant" platelets of up to 5 μm diameter. Platelets contain extra material which forces apart the diamond matrix (ie are intrinsic defects). They also have an effect on strength, for example, they pin dislocations and so affect the resistance of type I diamond to plastic deformation.

So-called "voidites" were first observed by Stephenson and Evans[9,10]. They are confined to sheets parallel to {100} and frequently occur in partial platelets where they replace part of the original platelet area. Their formation appears to be associated with the decomposition of platelets[11].

Diamond Types: The Role of Nitrogen

Measurements of the absorption of diamond in the infrared, visible and ultraviolet has led to a classification into types I and II[12-14]. Most diamonds are type I and have their absorption edge at ca. 330nm, while a much smaller group, type II, have this edge at ca. 220nm. The adsorption edge at 220nm (5.48eV) corresponds to the fundamental edge from the valence to the conduction band[14], and hence the edge at 330nm has been attributed to defects. As would be expected from the size of the band gap, most diamonds are good insulators. However, Custers[15,16] found that a small proportion of type II diamonds are semiconductors, and proposed classifying the non-conducting type II as IIa and the semiconductors as IIb.

The major factor in explaining the absorption spectra of diamond is the presence of nitrogen in amounts up to ca. 0.3%[17]. In these experiments, Kaiser and Bond graphitized diamonds and measured the gases evolved. The amount of nitrogen was several orders of magnitude higher than that of any other known impurity of diamond. Another valuable observation was that there is a direct correlation between the 7.8μm peak and the nitrogen content, thus providing a non-destructive technique for measuring the amount of nitrogen in a particular diamond. This is forbidden in a perfect diamond lattice because of its symmetry but the presence of the nitrogen defect lowers the local symmetry and provides coupling between the electromagnetic radiation and lattice vibration.

It would be wrong to give the impression that there is a rigid distinction between the various types of diamond. Some diamonds are 'mixed' in the sense that they contain regions of both type I and II. This is almost certainly associated with the growth history. Tables 2 and 3 summarise some of the main points of classification. As shown later, it is also possible to make correlations with various mechanical properties and the diamond types.

Thermal Properties

The very high thermal conductivity of diamond is a major factor in explaining its performance in many technological applications. As is well known, diamond is frequently used in processes such as machining, grinding, cutting and polishing which generate high temperatures at the work interface. It is paradoxical that a material which graphitizes in air at ca. 900K should survive such conditions. The answer is the thermal conductivity which has a

Table 2
Diamond Types and Nitrogen Content

Type	Abundance	Nitrogen
Ia	ca. 98% of natural	Nitrogen up to 0.3% (i.e. 10^{19} to 5×10^{20} atoms cm^{-3}) A and B aggregates; low levels of single N atoms ($<10^{16}$ atoms cm^{-3}); platelets.
Ib	ca. 0.1% of natural; main synthetic type	N in single substitutional sites; paramagnetic (10^{17} to 10^{20} atoms cm^{-3}).
IIa	ca. 2%	Very low Nitrogen level
IIb		Extremely low Nitrogen level; semiconducting due to Boron acceptors.

Table 3
Diamond Types and Absorption

Type	U-V Absorption edge	Infra-red	Comments
Ia	ca. 330 nm	Absorption 2.5µm to 10µm	A and B spectral features related to A and B Nitrogen aggregates. Intensity of 7.8 µm peak relates to total Nitrogen content. Intensity of 7.3 µm peak relates to X-ray spike intensity and to platelet size and concentration.
II	220 nm; at fundamental (5.48 eV)	Absorption 2.5µm to 6µm; transparent > 6µm	-------

value at room temperature about 5 times greater than that of copper. The heat is conducted by phonons so that the rigidity of the lattice which gives diamond its great hardness also favours a high conductivity.

The nearest approach to ideal behaviour is with type IIa diamonds. As would be expected, Umklapp and boundary processes are factors and in addition there is an isotope effect (natural diamond has ca. 1% of C13) but there are only small effects from impurities and dislocations. Good type IIs have a peak in their conductivity at about 80K and a value of ca. 1.5 x 104 W m^{-1} K^{-1}. Other types of diamond (see Table 4) have lower conductivities because of dislocations, impurities, nitrogen aggregates and platelets. Good quality synthetic diamonds have a similar conductivity to natural material. A diamond composite, such as the De Beers product "Syndite" can have a conductivity of up to ca. 400 W m^{-1} K^{-1} but lower than type IIa diamond (ca. 2100 W m^{-1} K^{-1}). A review of thermal properties has been given by Berman [2].

A major breakthrough was reported in July 1990 by General Electric [66] with the preparation of high-quality diamond crystals with low C^{13} isotopic impurity. A value of 3320 W m^{-1} K^{-1} was reported for the purest samples (see table 4): this is about 50% higher than for pure IIa natural diamond. It is thought that the reduction is caused primarily by the effect on Umklapp processes and not simply by a reduction in scattering from the C^{13} isotopic impurity.

Table 4

Thermal conductivity of Diamond in W m^{-1} K^{-1}

Type	At 293K	At 80 K
Ia	600-1000	2000 - 4000
IIa	2000-2200	Up to 15,000
Synthetic (0.07% ^{13}C)	3320	---

Electrical Properties

Pure diamond, with an energy gap of 5.48eV is a good insulator. However, a very small number of diamonds (type IIb) are p-type semiconductors, and it has been shown that the acceptor centre is substitutional boron. Nitrogen is a common impurity in diamond and it has been suggested that substitutional nitrogen is the donor centre in semi-conducting diamond. For an account of the electrical properties of diamond, the reader is referred to Ref. 19.

Optical Properties

Type II diamonds are transparent in the ranges 220nm to 2.5μm and > 6μm, and type I from ca. 330nm to 2.5μm and > 10μm (see also Table 3). Colourless (white) diamonds are amongst the rarest and most valued as gems. The most common colours are yellow and brown, though other colours found include orange, pink, mauve, green, blue, red and black. Colour is due to absorption bands in the visible region of the spectrum or in, for example, the case of the Premier Over Blue (type IIb) to the tail of the I.R. absorption spectra associated with the boron acceptor centre. The yellow colour of type Ib diamonds results from the tail of the U-V absorption associated with the substitutional nitrogen impurity. The yellow colour deepens as the nitrogen content increases and above ca. 100ppm a new absorption band is present at 670nm. The combination of absorption in the blue and at 670nm causes a greenish tinge. At nitrogen contents of 300-400 ppm, the diamond is green [20].

Diamond as a Gem-Stone

The pre-eminence of diamond as a gem-stone is due to a combination of optical and mechanical properties. The important optical properties are its high refractive index, its zero birefringence (if unstressed) and its dispersion. These give diamond its so-called "brilliance", "sparkliness" and "fire". Brilliance is the ability of a gem-stone to reflect light and depends on its cut and refractive index; in quantitative terms for diamond it is the percentage of light that, after entering the stone, is scattered out of the crown facets. Sparkliness is a measure of the variation in intensity of this scattered light and fire is a popular term to describe the dispersion of white light into the colours of the spectrum. A high refractive index means that the solid has a high critical angle. By faceting the back surface of a diamond (i.e. by polishing a regular pattern of small flat surfaces of chosen design) it is possible to have a high brilliance and to give the reflected light "sparkle". In assessing the quality of a diamond, the gemologist will consider its brilliance, sparkliness and fire and other qualities such as size, colour, presence of flaws etc. Many minerals exhibit double-refraction with refracted light split into the ordinary and extraordinary rays; birefringence is a measure of the difference between the two refractive indices. Double-refraction is a disadvantage with a gem-stone since it gives a fuzzy appearance due to the eye seeing two images of the back surface. Since diamond is cubic, and so optically isotropic, it does not suffer from this disadvantage. Of the various gem-stones, diamond does **not** have the highest refractive index or dispersion. It is the combination of good performances from these properties, combined with its hardness, strength, chemical stability and zero birefringence which make it pre-eminent.

Chemical Reactivity

Diamond is inert to most forms of chemical attack and is not affected by acids or other chemicals except those which at high temperatures act as oxidising agents. In oxygen itself diamond reacts at temperatures above ca. 900K. A common etchant for diamond is molten potassium nitrate at a temperature in the range 800-1100K. This technique has been used extensively for surface studies of diamond (see reviews by Tolansky[21]; Moore[22]). At high temperatures two distinct groups of metals react with diamond. The first react chemically to form their respective carbides and include W, Ta, Ti and Zr. The second includes Fe, Co, Mn, Ni, Cr and the platinum groups of metals. In the molten state these metals are true solvents of carbon and some of these are used in the solvent/catalyst synthesis of diamond.

The diamond surface physically and chemically adsorbs various gases[65]. Apart from its considerable scientific interest, there are implications to technology through the effects on adhesion and friction processes.

The Diamond Surface

Studies of the topography of the diamond surface are very much associated with the name of Tolansky. In his work he either developed, or significantly improved, a whole range of optical techniques, including two-beam, multiple-beam and white-light interferometry. His work on diamond is summarised in the reviews of Ref. 21, 22.

MECHANICAL PROPERTIES

Elastic Constants

The extreme rigidity of the diamond lattice is the key to many of its mechanical properties. Tables 5 and 6 summarize recent determinations of its elastic constants and values for the moduli and Poisson ratio. Poisson's ratio varies markedly with orientation, with values ranging from 0.01 to greater than 0.2. The ν_{21} value equals 0.104. the isotropic aggregate value is given in Table 6.

Table 5

Elastic constants, moduli and pressure and temperature coefficients

Source	C_{11}	C_{12}	C_{44}	$E_{11}/10^{11}Nm^{-2}$	$K/10^{11}Nm^{-2}$
23	10.76	1.25	5.76	10.50	4.42
24	10.79	1.24	5.78	10.53	4.42
25	10.764	1.252	5.774	10.50	4.42
26	-	-	-	-	5.6
dC/dP	5.98±0.7	3.06±0.7	2.98±0.3	-	-
dC/CdT	-1.37±0.2	-5.70±1.5	1.25±0.1	-	-

Table 6

Isotropic aggregate values computed from single crystal data[27]

Material	$E/10^{11}Nm^{-2}$	$G/10^{11}Nm^{-2}$	$K/10^{11}Nm^{-2}$	υ
Ge	1.316	0.545	0.76	0.207
Si	1.629	0.666	0.979	0.223
Diamond	11.41	5.53	4.42	0.07

Cleavage Plane

The dominant cleavage plane is the (111) but others have been observed. Sutton[28] notes that (110) cleavage and conchoidal fractures are possible. Ramachandran[29] found (211) cleavage, while Ramaseshan[30], in a study of 15 crystal fragments from Raman's collection, observed (111) (211) (110) (322) (331) (221) and (322). The (111) was most common while the (221) and (110) were not uncommon. These results are consistent with the order given in Table 7 based on Ramaseshan's calculations. However, it is not clear from these figures why the (111) should dominate, particularly if the largest tensile stresses are normal to other planes. The answer is more likely to be due to defects on the (111) plane giving preferential weakening. Since the (111) plane is the growth plane it seems possible that it will contain layers of faulted growth. Evidence for this has come from the studies of Pandya and Tolansky[31], Ramage[32] and Wilks[33]. Ansell[34] has pointed out that the calculations of cleavage energy have only taken into account the numbers of bonds broken and have not included the relief of strain due to non-bonded interactions. His paper shows that once a cleavage starts in a particular plane it will be encouraged to continue on that plane since it produces a maximum relief of non-bonded interactions. Tolansky[21] has emphasized the irregular nature of most diamond cleavage surfaces. The traditionally perfect cleavage is usually on a micro-scale. Type II diamonds in general produce smoother cleavage surfaces than type I.

Table 7

Theoretical cleavage energy for diamond

Plane	Angle between plane and (111) plane	Cleavage energy/ J m^{-2}
111	0° and 70° 32'	10.6
332	10° 0'	11.7
221	15° 48'	12.2
331	22° 0'	12.6
110	35^o16' and 90'	13.0
322	11° 24'	13.4
321	22° 12'	14.3
211	19° 28'	15.0
320	36° 48'	15.3
210	39° 14'	16.4
311	29° 30'	16.6
100	54° 44'	18.4

Note: to obtain a fracture surface energy, γ, divide by 2.

Velocity of Cleavage

Not surprisingly there are few experimental observations on the velocity of cleavage. From her study of ten cleavage surfaces Ramage[32] estimated that typical cleavage velocities could reach about one fifth the distortional wave speed i.e. about 2000m s^{-1}.

Field[35] has used high-speed photography at sub-microsecond framing intervals to observe crack growth in diamonds. Velocities up to ca. 7200m s^{-1} were recorded. Since relatively few specimens were fractured, this does not rule out the possibility of even higher crack velocities. The value of 7200m s^{-1} is about 0.7 of the Rayleigh wave velocity (the theoretical limit for brittle fracture). Since diamond transmits stress waves faster than any other material it is no surprise that it also has the highest measured fracture velocity. However, under conditions similar to that used in diamond cleaving, velocities of 1000 to 3000m s^{-1} are more typical. A fuller discussion of fast crack growth in brittle solids and data comparing diamond with other brittle materials can be found in Field[35].

Cleavage Energy

Several authors have approached this problem by calculating the number of bonds which cross unit area of a chosen plane and multiplying this by the C-C bond strength. Table 7 is similar to one given by Ramaseshan[30], but with a value of 83 Kcal mol^{-1} (i.e. 5.8×10^{-19} J for the C-C bond[36]). Since the cleavage process in advancing by unit area produces two unit areas of surface, the surface energy would be half the cleavage energies listed in Table 7. The theoretical fracture surface energy for cleavage on the {111} plane, assuming there is no plastic deformation at the crack tip, is therefore 5.3Jm^{-3}.

By monitoring the extent of crack growth beneath a loaded indenter Field and Freeman[37] have been able to measure the fracture energy γ, of diamond as 5.50 ± 0.15 J m^{-2} for (111) cleavage. Experimental values of γ for most materials are usually greater than, often very much greater than, the theoretical value. The fact that the measured value is only slightly greater than the theoretical value for diamond suggests that little, if any, dislocation movement occurs in the stress field at the crack value. The factor most likely to cause an increase in γ for diamond is secondary cleavage or conchoidal fracture. The limited amount of data obtained by Field and Freeman for 'cones' produced for (110) surfaces supports this view since γ equals 7.1± 1.0 Jm^{-2}. No significantly different values were obtained between the different diamond types.

The critical stress intensity factor K_{IC} is related to γ and G by $K_{IC}^2 = 2E\gamma = EG$. For diamond K_{IC} equals 3.4 $MNm^{-3/2}$ and this compares with 0.75 $MNm^{-3/2}$ for soda-lime glasses and sapphire, silicon carbides and silicon nitrides which all fall in the range 3 to 5 $MNm^{-3/2}$. Tungsten carbide can have a K_{IC} value three to five times the diamond figure. A value for diamond composites measured at the Diamond Research Laboratory, Johannesburg, was 7.3 and 8.6 $MNm^{-3/2}$ for specimens made with particles of average diameter of 10μm and 25μm respectively.

Fracture Strength

Rather different methods have been used for measuring the strength of macro-diamonds (those of millimeter dimensions) and micro-diamonds (sub-millimeter particles, grits, powders, etc...)

(a) **Macro-diamonds:** Fracture data has invariably been obtained using blunt indenters (see the review by Field[35], for detailed references) One of the most comprehensive studies was by Howes and Tolansky. Their study began in 1955 with the application of sensitive interferometric techniques to the observation of ring cracks in octahedral, cube and dodecahedral diamond surfaces. A later paper showed that ring cracks could be induced in diamond with indenters softer than diamond, for example, tungsten carbide and sapphire. In 1959 Howes reported theoretically predicted and experimentally measured values of strengths for a variety of surfaces. This was attributed to a size effect, but interfacial frictional stresses are also likely to be important[35]. A summary of strength data is given in Table 8. p_c is the average stress obtained by dividing the load by the crack area, and p_c' is the average stress obtained from the load and observed, or calculated, area. Most workers agree that the observed area, and that calculated from Hertz theory, are in close agreement. It should be emphasized that the data of Table 8 are not necessarily values given in the original papers. To obtain consistency, the data was all recalculated using $E = 10.5 \times 10^{11}$ Nm^{-2} and $\nu = 0.1$. For example, Bowden and Tabor initially obtained a value for p_c' of 19.62GPa. The change to 31.73GPa is due to using better values for E and ν and less rounding-off of figures during calculation. Bell et al. used diamond indenters to compare different diamonds. They have found that diamonds with strong birefringence are usually weaker. Their results are given in terms of the critical loads applied, but if we take the mean load value for their best diamond (SC2B) and calculated an area of contact using Hertz theory the value of p_c quoted in Table 8 is found. The work of Chrenko and Strong is of interest since they investigated both synthetic and natural diamonds. Highest strengths were obtained with high quality, low nitrogen content, colourless, synthetic gem-stones. The presence of atomically dispersed substitutional nitrogen in synthetic diamond they claimed reduced the strength. As it stands this statement sounds suspect, and it is more likely that a high nitrogen content in their synthetics also meant higher concentrations of metal inclusions, dislocations and other defects which would be more likely to affect strength. The work of Ruoff and Wanagel[38], and Ikawa et al.[39,40] shows a strength/size effect with much higher strengths with small indenters. The Japanese work also shows that slightly higher strengths are achieved with fine-polished surfaces, though the effect is only significant for the 50 and 200μm indenters.

Table 8

Strength data for diamond. Data recalculated with $E = 10.5 \times 10^{-11} N m^{-2}$ and $\nu = 0.1$

Author		Indenter radius /mm	p_c/GPa	p_c'/GPa
(a)	Octahedral		10.30	
	Cube	0.5, tungsten carbide	13.25	
	Dodecahedral		11.77	
(a)	Octahedral		10.79	13.46
	Cube	0.39, diamond	13.73	31.73
	Dodecahedral		17.66	
(b)	Cube	0.28, diamond	13.73	31.73
(c)	Cube	0.245, diamond	29.32	
(d)	Cube	0.25, diamond		16.06 - 31.73
(d)	Synthetic octahedral and cube	0.25, diamond		21.15 - 31.73
(e)	Not given	0.02, diamond		>64<71
		0.008, diamond		>71<86
		0.0065, diamond		>86<94
		0.0025, diamond		>86<100
(f)	Octahedral	0.50, diamond		16.66
	Octahedral	0.20, diamond		19.60
	Octahedral	0.05, diamond		31.18
	Octahedral	0.014, diamond		47.58
	Octahedral	0.005, diamond		63.96

(a) Howes (1965); (b) Bowden and Tabor (1965); (c) Bell et al. (1977); (d) Chrenko and Strong (1975); (e) Ruoff and Wanagel (1977); (f) Ikawa and Shimada (1983).

The main conclusions from Table 8 are that the strength of diamond depends on the crystallographic face indented, that there is a strength/size effect, and that for good quality diamonds, indented by similar sized indenters, the results from different studies are encouragingly consistent.

From the above data, it is possible to calculate the tensile strength of diamond . The value is 3.75 GPa for $\nu = 0.1$. This compares with a theoretical strength[41] of 121 GPa which suggests that good quality diamonds have sharp ended defects of about 0.5μm. Other results, to be discussed below, show that there is a strength-size effect with diamond so higher strength values would be expected if smaller areas are stressed.

(b) Micro-diamonds Various methods have been devised for assessing the strength of diamond grits and references are given Ref. 35,37. An example of an approach used in industry is the Friatest machine developed by De Beers which involves putting a sample of two carats in a cylindrical steel capsule with a steel ball which is vibrated to and fro along the axis for a set number of 2400 cycles. The amount of particle fracture which occurs is measured by grading (screening) the grit. In the latest version of the tests, the impact strength is expressed in terms of the number of cycles to produce the 'half-life' strength value.

Single grit compression is more laborious but allows a systematic study of particle strength and factors affecting it. It involves placing grits between anvils and loading in compression until failure occurs. Particles which fracture at low stress usually break into two fragments, but those which withstand a high load shatter into a fine micron-sized powder. In this latter situation there are often several small 'clicks' during loading as small high regions fracture off. [42,37]. The work on diamond grits shows that smaller sizes, though they withstand lower loads, have higher strengths.

Synthetic material frequently has a pair of opposing flat crystal faces. This allows the contact area to be measured and the compressive strength calculated, from the recorded load in a crushing experiment. In experiments on synthetic material, Field et al.[35] showed that particles with a high magnetic impurity had only about 35% of the strength of similar material with a low magnetic impurity. Other important factors were particle shape and quality of the surface.

Recently, Feng and Field[43] have shown that the dynamic strength of diamond grits at a strain rate $\dot{\varepsilon}$ of 10^4 s^{-1} is similar to that found in low strain rate tests ($\dot{\varepsilon} \approx 10^{-2}$ s^{-1}). This suggests that diamond, unlike many glasses and ceramics, is not susceptible to atmospheric fatigue (i.e. stress corrosion) effects.

Plastic Flow

Diamond has the greatest resistance to plastic flow of any material. Its exceptional yield strength arises essentially for two reasons; the first is that the theoretical strength is high because of the high density of strong atomic bonds, and secondly because of the strongly directional nature of the bonding only allows the presence of narrow dislocations which are difficult to move. There has been considerable interest in determining the conditions under which flow is possible.

Natural diamonds give clear evidence from their internal structure of dislocation and shear deformation while the material was at high T and P in the earth. In the high temperature experiments on bulk plastic deformation by Evans and Co-workers (See Ref.44 for a review) diamond plates were stressed in 3-point bending. The plates were of approximate dimensions 5 x 3 x 0.5mm. The results showed that bulk deformation, in this loading situation, occurred at temperatures above 1900 K. Further, type II diamonds deformed at significantly lower stresses than type I. This was explained by the different initial defect distributions. Type I diamonds have a low dislocation density and a high density of platelets. In contrast, type II diamonds have a high dislocation density but with no platelets present.. Electron micrographs of deformed type I diamonds showed dislocation pinning at the platelets. The platelets clearly play an important role in hindering dislocation movement in type I diamonds and this explains the higher stresses needed to obtain permanent deformation. The question of plastic deformation at room temperature has been more controversial. It now appears that it can take place, if there is sufficient constraint to eliminate tensile fracture. Research to generate high pressures by pressing a sample between diamond anvils has led to a deeper understanding of the ultimate compressive and shear strength of diamond.

Ruoff[27] has assessed the yield strength of Ge, Si and diamond by considering (i) dynamic yield strengths obtained from Huoniot elastic limits, (ii) yield strengths obtained from hardness data, and (iii) yield strengths obtained by measuring the first sign of permanent deformation to an indenter. Table 9 summarises the results obtained from an analysis of (i). Note that the G and E used for diamond are the isotropic aggregate values given in Table 6. For a ductile solid, the yield strength σ is ~ 1/3 the hardness value, H. According to Marsh[45] this fraction increases slowly as the ratio σ_o/E increases. The hardness of diamond depends on the crystal orientation but if we take H = 90 GPa then σ_o equals ~ 30 GPa if $H = 3\sigma_o$ and ~ 36 GPa if $H = 2.5\ \sigma_o$.

Table 9

Dynamic yield stress (After Ruoff[27])

Material	σ_o(GPa)	σ_o/G	σ_o/E
Ge	3.61	0.0662	0.0274
Si	4.33	0.0650	0.0266
Diamond	35*		0.0308+

* Computed from average value of σ_o/G for Ge and Si and E for diamond.

+ Computed from average value of σ_o/E for Ge and Si and E for diamond.

The maximum load diamond anvils can withstand is clearly of practical interest, since diamond cells are used in a variety of high pressure optical and X-ray work. the first objective of any design is to keep the tensile stresses below the tensile failure strength of diamond. Assuming this can be done the question then arises as to the ultimate pressures which can be achieved without extensive plastic yielding of the anvils. The relation between the maximum pressure p_o in the applied pressure distribution and σ_o the yield strength depends on the details of the pressure distribution. For uniform pressure $p_o = 1.28\ \sigma_o$ and for a spherical indenter (Hertz situation) $p_o = 1.36\sigma_o$. For these pressure distributions plastic deformation would be expected to begin at ~ 45 and ~ 48 GPa respectively. Ruoff[27] estimates that plastic dents would become visible at ~ 70 GPa and that limiting pressures of up to ~ 85 GPa could be achieved. Mao et al.[46,47] have indeed observed plastic dents in diamond anvils and claim to have achieved pressures of 172 GPa.

Hardness of Diamond

Early measurements on the indentation hardness of diamond have been reviewed by Brookes[48]. Particularly noteworthy is the apparent lack of reproducibility in the quoted values from this early work. It is now known that the hardness of diamonds depends on a great many factors such as orientation, diamond type, impurity content, load and shape of indenter. Added to the fact that the indents themselves are small, thus giving relatively large experimental errors, the early scatter is understandable. Recent measurements, with greater control of the variables listed above, show good agreement. Table 10 from Brookes[48] gives data on diamonds, indented at room temperature with a 1 kg load, on mechanical polished surfaces.

Note that the type II values are consistently higher than those for type I and that there is a marked anisotropy by a factor of about 1.5. Few crystals in fact show such a large variation with orientation as diamond. Brookes et al.[48] and Pospeich and Gryziechi[49] have developed models, based on the resolved shear stresses developed on the primary slip systems in the bulk of the crystal beneath the indenter which successfully explain and predict the anisotropy of hardness in most crystals. The fact that diamond follows the predictions of these modes, assuming a {111} <$\bar{1}$10> slip system has been argued by Brookes that the indents are evidence of plastic flow at room temperature. Chrenko and Strong[20] have published important data on General Electric laboratory grade diamonds, and a selection of the results are given in Table 11.

Note that as the nitrogen content decreases the tensile strength (cleavage pressure), abrasion resistance and Vickers hardness all increase. Interestingly GE results on natural diamonds show high hardness but rather modest abrasion resistances. The different types of indenter and lower load explain the higher hardness found for natural diamond in the GE work compared with that of Brookes.

Table 10

Knoop hardness (H_K) of diamond - GPa (after Brookes[48])

Plane	Direction	Type I	Type II
(001)	[110]	81.0	89.0
(001)	[100]	96.0	101.0
(110)	[110]	86.0	92.0
(110)	[001]	106.0	113.0
(111)	[110]	55.0	75.0
(111)	[112]	62.0	108.0

Data taken from Brookes[48] Bakul et al.[50] show that there is a clear evidence of higher hardness values with decreasing load. At first sight this is surprising since geometric similarity implies that, for pyramidal indenters, the hardness of a crystal should be independent of normal load. However, this trend is not confined to diamond but has been found for a whole range of crystalline solids[48]. An important additional observation is that the effect becomes more pronounced as the crystal's hardness increases. The true explanation of the hardness increase at low loads is most likely associated with the early stages of plastic deformation when small volumes are stressed. However, whether it is due to the small number of dislocation sources involved, or due to the initial stages of slip being on a fine scale with increased dislocation interaction, and thus greater hardening (a mechanism favoured by Brookes and his co-workers) needs further study.

Table 11

Strength, hardness, abrasion resistance of laboratory grade diamond (after Chrenko and Strong[20])

Diamond	Colour	Nitrogen/ ppm atomic	Cleavage/ pressure[a]/GPa	Hardness[b]/ GPa	Abrasion resistance[c]
70	Yellow	70	20.3	95	0.84
12	Yellow	20	---	106	1.31
82	Pale green	~2	24.5	---	1.72
15	Colourless	~0.3	---	110	3.05
87	Pale green	~0.1	~30.0	131	3.11

(a) Indentation on a (001) face with a spherical indenter
(b) Vickers indenter on a (001) face, with the diagonals of the indenter parallel to <100> direction.
(c) Grinding wheel test with 70 Al_2O_3 grit and with grinding ratio expressed as volume of wheel wear in m^3 divided by weight of diamond lost in gm.

There is no space here to review all the high temperature hardness work on diamond. However, mutual indentation studies with octahedra by Evans and Sykes[51] showed that type I hardnesses above 1650°C were higher than type II. This result is consistent with Evans' other work on **bulk** plastic deformation (discussed above). It conflicts, however, with the room temperature indentation data given in Table 10. The Evans and co-workers' measurements on yield stresses and creep behaviour are consistent with the fact that type II has little nitrogen, but many dislocations, while type I has a relatively high level of nitrogen including platelets which hinder dislocation movement. The surprise is really the reversal of behaviour when diamonds are indented at room temperature, which suggests that when small volumes are stressed, and localised flow commences, that type I diamonds are softer than type II. This result taken with the dramatic increase in hardness at low load suggests that further research on indentation plasticity in hard crystals is needed.

The usual behaviour when a pure crystalline solid is irradiated by fast neutrons is to introduce point defects and dislocation loops which increase the density. With diamond Brookes and Vance[52] have shown that both the crystal density and the Knoop indentation hardness decrease. These results are consistent with the view that any change from the perfect diamond lattice at room temperature, be it by addition of nitrogen or point defects, decreases its hardness.

Wear

There is a considerable practical interest in the abrasion and wear properties of diamond (for reviews, see Wilks and Wilks[53-55]. Firstly there are the situations where diamond is ground and polished to make jewelry or industrial tools. Secondly, there are the many industrial applications where diamond is used for polishing, grinding, cutting and machining.

The work of Wilks and Wilks has demonstrated three main results about the abrasion resistance of diamond: (i) its great dependence on the orientation of the facet and the direction of abrasion; (ii) the very sharp changes in magnitude which occur near certain planes and directions. A change of a few Å can more than double the material removal rate; (iii) the impossibility of representing the variation of abrasion resistance by a simple vector quantity. In other words, if the direction of traversal is reversed a different wear rate can result.

The technique developed by Wilks and Wilks involves rotating a cast iron or bonded wheel of 1 cm radius and allowing the U-shaped edge to be loaded against the diamond under test. The cuts produced are then measured using multiple-beam interferometry at spacings of $\lambda/2$, where λ is the wavelength of a monochromatic light source (546.1nm). An important result is that each revolution of the wheel removes approximately the same amount of material over a wide speed range. This suggests that thermal effects at 'hot spots' are not important. An example from their extensive data is given in Table 12. Note that there is an order of 10 difference in the depth of cuts which implies a difference of over 100 in the rate of removal of material!

The first attempt at explaining the mechanism of material removal was that of Tolkowsky[56]. His model, subsequently refined and developed by Wentorf[57] and Wilks and Wilks[53,54] is based on micro-cleavage with cleavage on the {111} planes dominating. The model has proved amazingly successful in explaining the main features of the results. It is unfortunate that some authors have introduced the phrase 'abrasion hardness'. The 'hard' and 'soft' planes and directions for abrasion are not the same as with indentation hardness, which is not surprising since abrasion of diamond is primarily a brittle process. For this reason the phrase 'abrasion resistance' is to be preferred.

Table 12

Depths of cuts made by bonded wheels, normalised to a depth of 10 ($\lambda/2$) in the softest direction (Wilks and Wilks[54])

	Plane and Direction					
Plane	Dodecahedron (011)		Cube (001)		Octahedron (111)	
Direction	[100]	[011]	[100]	[110]	$[11\bar{2}]$	$[\overline{11}2]$
Depth of cut ($\lambda/2$)	10.0	1.8	8.5	1.1	3.9	2.1

Friction

Tabor[58] has reviewed research on the friction of diamond. The main results, in air, can be summarised as follows:
(a) The friction depends on crystal face and orientation. On the octahedral face the friction is low and there is practically no anisotropy. On the cube face, polished in the <100> direction there is a four-fold symmetry in the friction. The lowest friction is along the <011> or cube diagonal direction: the highest along the <100> or cube edge direction.

(b) If the cube face is polished in the <110> direction the four-fold symmetry is replaced by two-fold symmetry.
(c) In some situations repeated traversal over the same track on the cube face of diamond reverses the order of frictional anisotropy. After a few hundred traversals the friction is higher in the <011> direction and low in the <100> direction.
(d) The frictional anisotropy on the cube face becomes less marked if a conical stylus of large apical angle is used.
(e) The coefficient of friction may or may not depend on load depending on the geometry of the sliding system.
(f) The friction is scarcely changed by the addition of a lubricant though recent work by Feng and Field (unpublished) has shown that water can reduce the friction coefficient.
(g) Over a wide range of experimental conditions the coefficient of friction lies between μ = 0.05 and 0.15; it rarely exceeds μ=0.2, except under vacuum conditions.
(h) Finally we note that, although the wear rate on the (100) and (110) faces of diamond may depend on the sense of sliding, the friction in such situations does not depend on the sense of sliding.

For diamond sliding on diamond under vacuum conditions, the friction can reach a high value (μ ~1). For example if the samples are cleaned by heating or bombardment or alternatively if there are repeated traversals over the same track. The high friction is accompanied by a very high wear rate[59]. If a mono-molecular oxygen film is allowed to reform the friction decreases dramatically. Buckley and co-workers at NASA have recently made a detailed study of the adhesion and friction of diamond in contact with various metals. The results showed that μ was related to the relative chemical activity of the metals in high vacuum. All the metals studied transferred to the surface of diamond during sliding[60,61]under a vacuum of 10^{-8}N m $^{-2}$. Several authors have considered the dissipative processes when diamond slides on diamond. There is still no entirely convincing answer to explain the coefficients of friction which are observed. The loads at the asperities are high but will this cause fracture (which does **not** expend much energy) or plastic deformation (which might) ? Samuels and Wilks[62] argued that the vibration of asperities can cause energy losses but argue that an adhesive term is unlikely. Other workers argue that there must be some adhesion and that some plastic deformation at high loads is possible.(Feng and Field, unpublished).

High Modulus Coatings

There is a growing interest in using high modulus coatings to protect brittle materials from contact and impact damage. Recently we have completed finite-element analyses on the effect of thin hard coatings on the Hertzian stress field[63,64]. Of particular interest is the reduction of the maximum (radial) tensile stresses. The amount of reduction in the substrate solid increases with the Young's modulus and thickness of the coating. The reduction is very large with diamond-like coatings since the coating modulus, E_c, can be typically ten times the substrate modulus E_s. Further, films of a few micrometers thickness can have a beneficial effect[64].

REFERENCES

1. Berman, R. and Simon, F. 1955, Z Electrochem. 59 333-338
2. Berman, R. 1979, The Properties of Diamond. Ed. J.E. Field. Academic Press, London, pp 3-22
3. Harris, J.W. and Gurney, J.J. 1979, the Properties of Diamond. Ed. J.E. Field. Academic Press, London, pp 555-594
4. Ergun, S. and Alexander, L. 1962, Nature Lond. 195 765-767
5. Bundy, F.P. 1962, J. Chem. Phys. 38 631-643
6. Sellschop, J.P.F. 1979, Properties of Diamond. Ed. J.E. Field. Academic Press, London, pp 107-163.

7. Evans, T. and Phaal, C. 1962, Proc. Fifth Biennial Conf. on Carbon, Penn. State Univ. 147-153
8. Woods, G.S. 1976, Phil. Mag. 34 993-1012
9. Stephenson R. F. 1977, PhD Thesis, Univ of Reading
10. Evans T. 1978, In "Diamond Resarch 1978" pp17-22 Ascot: De Beers Industrial Diamond Division (Pty)
11. Barry et al. 1987 Phil Trans Roy Soc Lond A321 361-401
12. Robertson, R., Fox, J.J. and Martin, A.E. 1934, Phil. Trans. R. Soc. A232 463-535 Seal, M. 1958b, Proc. Roy. Soc. Lond. A248 379-393
13. Sutherland, G.B.B.M., Blackwell, D.E. and Simeral, W.G. 1954, Nature, Lond. 174 901-904
14. Clark, C.D., Ditchburn, R.W. and Dyer, H.B. 1956, Proc. Roy. Soc. Lond. A234 363-381
15. Custers, J.F.H. 1952, Physica 18 489-496
16. Custers, J.F.H. 1954, Physica 20 183-184
17. Kaiser, W. and Bond. W. 1959, Phys. Rev. 115, L857-863
18. Messmer, R.P. and Watkins, G.D. 1970, Phys. Rev. Lett. 25 656-659
19. Collins, A.T. and Lightowlers, E.C. 1979, The Properties of Diamond. Ed. J.E. Field. Academic Press, London, pp 79-106
20. Chrenko, R.M. and Strong, H.M. 1975, General Electric Report 75CRDO89 1-46
21. Tolansky, S. 1965, Physical Properties of Diamond. Ed. R. Berman. Clarendon Press, Oxford, pp 135-173
22. Moore M. 1979, The Properties of Diamond Ed. J. E. Field, Academic Press, London pp 245-277
23. McSkimin, H.J. and Bond, W.L. 1957, Phys. Rev. 105 116-987
24. McSkimin, H.J. and Andretch, P. 1972, J. Appl. Phys. 43 985, 2944-2948
25. Grimsditch, M.H. and Ramdas, A.K. 1975, Phys. Rev. B11 3139-3148
26. Drickamer, H.G., Lynch, R.W., Clenenden, R.L. and Perez-Albuerne, E.A. 1966, Solid State Physics. Eds. F. Seitz and D. Turnbull. Vol 19 pp 135-228
27. Ruoff, A.L. 1979, High Pressure Science and Technology. Eds. K.D. Timmerhaus and M.S. Barber. Vol. 2, Plenum, pp 525-548
28. Sutton, J.R. 1928, Diamond. Murby and Co. London
29. Ramachandran, G.N. 1946, Proc. Ind. Acad. Sci. A221 163-198
30. Ramaseshan, S. 1946, Proc. Ind. Acad. Sci. A24 114-121
31. Pandya, N.S. and Tolansky, S. 1954, Proc. Roy. Soc. Lond. A225 40-48
32. Ramage, J.B. 1956, PhD thesis, University of London
33. Wilks, E.M. 1958, Phil Mag. 3 1074-1080
34. Ansell M. F. 1964, Chemistry in Britain, 20, 1017-1021
35. Field, J.E. ed. 1979, The Properties of Diamond. Academic Press, London, pp 281-324
36. Pauling, L. 1960, The Nature of the Chemical Bond. Cornell University Press, p 85
37. Field, J.E. and Freeman, C.J. 1981, Phil. Mag. 43 595-618
38. Ruoff A. L. and Wanagel J. 1977 Science 198, 1037-1038
39. Ikawa N. and Shimada S. 1983, Tech. Rep. of Osaka Univ, 33 343-348
40. Ikawa N, Shimada S. and Tsuwa H. 1985, Annals of the CIRP; 34 117-120
41. Lawn, B.R. and Wilshaw, T.R. 1975, Fracture of Brittle Solids, C.U.P.
42. Tsypin, N.V. and Gargin, V.G. 1973, Sint. Almazy 3 32-35
43. Feng Z. and Field J. E. 1979
44. Evans, T. 1967, Science and Technology of Diamonds. Ed. J. Burls. Ind. Dia. Info. Bureau, London, pp 105-110
45. Marsh, D.H. 1964, Proc. Roy. Soc. Lond. A279 420-435
46. Mao, H.K., Bell, P.M., Shaner, J. and Steinberg, D.J. 1978, J. Appl. Phys. 49 3276-3283
47. Mao, H.K. and Bell, P M 1978 Science, 2000 1145-1147
48. Brookes, C.A. 1979, The Properties of Diamond. Ed. J.E. Field. Academic Press, London, pp 383-402
49. Brookes, C.A. 1971, Indus. Diam. Rev. 21-24
50. Bakul, V.N., Loshak, M.G. and Mal 'nev, V.I. 1973, Sint. Almazy 6 16-19

51. Evans, T. and Sykes, J. 1974, Phil. Mag. 29 135-147
52. Brookes, C.A. and Vance, E.R. 1972, Diamond Conference, Bristol (unpublished) reported in Brookes, 1979
53. Wilks, E.M. and Wilks, J. 1965, Physical Properties of Diamond. Ed. R. Berman, Clarendon Press, Oxford, pp 221-256
54. Wilks, E.M. and Wilks, J. 1972, J. Phys. D: Appl. Phys. 6 1772-1781
55. Wilks, J. and Wilks, E.M. 1979, The Properties of Diamond. Academic Press, London, pp 351-382
56. Tolkowsky, M. 1920, PhD thesis, University of London
57. Wentorf, R.H. 1959, J. Appl. Phys. 30 1765-1768
58. Tabor, D. 1979, The Properties of Diamond. Ed. J.E. Field. Academic Press, London, pp 325-350
59. Bowden, F.P. and Hanwell, A.E. 1966, Proc. Roy. Soc. Lond. A295 233-243
60. Miyoshi, K. and Buckley, D.H. 1980, Appl. Surf. Sci. 6 161-172
61. Pepper, S.V. 1981, NASA Tech. Memo. 82741, 1-13
62. Samuels B. and Wilks J. 1988, J of Mater Sci, 23, 2846-2864.
63. Zwagg, van de S. and Field, J.E. 1982, Phil. Mag. 46 133-150
64. Zwagg, van de S. and Field, J.E. 1983, Phil. Mag. 48 767-777
65. Pate B. B. 1986 Surface Science 165, 83-142
66. Anthony, T. R., Banholzer W. F., Fleischer J.F., Wei L., Kuo P.K., Thomas, R.L. and Pryor, R.W. 1990 Phys. Rev B 42, 1104-111

ELECTRONIC STRUCTURE OF DIAMOND, ITS DEFECTS AND SURFACES

J. Robertson

National Power Labs
Leatherhead, Surrey, KT22 7SE, UK

INTRODUCTION

Diamond has the potential to become an important high temperature electronic material because it has the highest thermal conductivity, electron velocity and breakdown voltage of any semiconductor[1]. Additionally, the vapour phase deposition techniques offer the promise of thin film devices and controlled impurity contents[2]. However, a number of problems must first be resolved, such as an improved n-type doping facility and a further development of the vapour deposition process. To this end, the electronic structure of diamond, its vacancies, interstitials, substitutional impurities and surfaces is reviewed. It is noted that vacancies mediate atomic diffusion, that P is a shallow but low solubility donor and that hydrogen abstraction from the hydrogenated diamond surface is often the rate-limiting step in diamond deposition.

BAND STRUCTURE

The band structure of diamond is well established[3-6] (Fig. 1). The bands around the gap are like those of Si. The gap of 5.5 eV is indirect, from a triply degenerate pσ-like Γ_{25}' state at the top of the valence band to a state about 76% along the ΓX direction. The effective mass of the holes is about $0.7m_o$, where m_o is the free electron mass[7,8]. The electron states form six degenerate valleys with surfaces in the form of prolate spheroids. The electron mass tensor has $m_{//} = 1.4m_o$ and $m_\perp = 0.36m_o$ giving $m_e = 0.48m_o$[9]. The band structure also has a similar ordering of bands to Si away from the gap, with the sσ*-like Γ_2' state lying well above the pσ*-like Γ_{15}' state in the conduction band.

Carbon's electronic configuration differs from that of Si in two ways. Firstly, the absence of 1p core states means that there is no 2p pseudopotential so the 2p states, unlike the 2s states, feel the full nuclear potential. This allows the 2p states to be as compact as the 2s states, unlike the 3s and 3p states of Si[10]. A second more important difference is the absence of a 2d shell in carbon comparable to the 3d shell in Si. This allows the 8 lowest bands of diamond and graphite to be well represented by simple tight-binding or molecular orbital models with only s and p states. Such models provide an easily understood picture of bonding of bulk, defect and surface states. The absence of 2d states also restricts carbon's ability to form coordinations above four. This allows life to exist because it stabilises tetravalent carbon against hydroysis, which occurs via a 6-coordinated intermediate stage. It means that higher coordinated, metallic phases of carbon are less stable, so diamond itself is the stable phase of carbon to very high pressures[11].

Diamond and Diamond-Like Films and Coatings
Edited by R.E. Clausing *et al.*, Plenum Press, New York, 1991

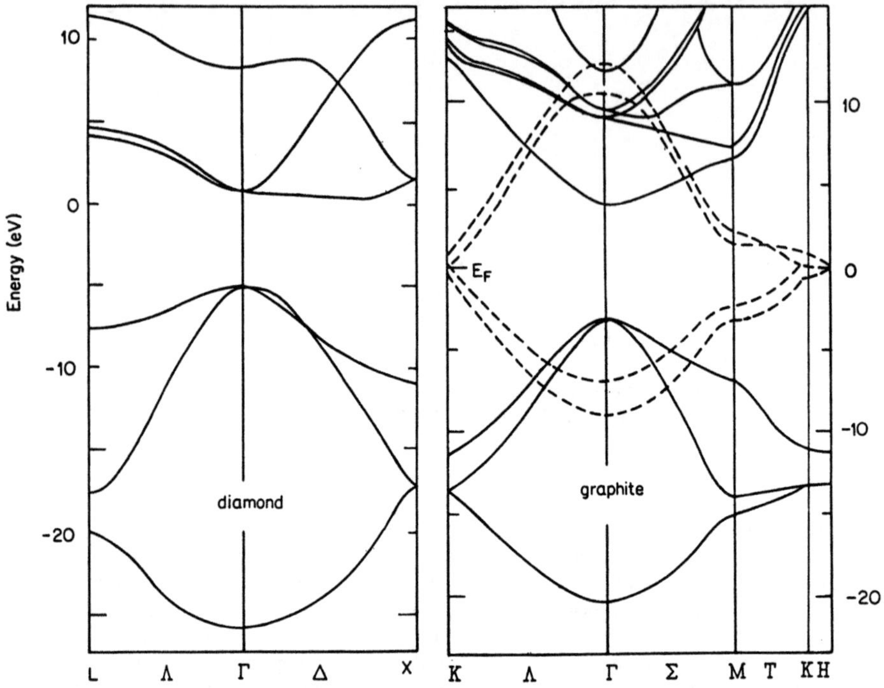

Fig. 1. Band structure of diamond[3] and graphite[13]

It also means that unlike other group IV elements, liquid carbon has an unusual low density, 2- and 3-coordinated structure, whereas liquid Si has a higher density 6-coordinated structure[12].

INTRINSIC DEFECTS

The key properties of point defects of interest are the need to identify possible shallow donor species and to identify the dominant intrinsic defect mediating atomic diffusion. We consider first the intrinsic defects; the vacancy and the various interstitials (Fig. 2). The vacancy is formed by removing one atom from a bulk site, which creates four 'dangling bonds', one each on the adjacent atoms, which tend to form states at the position of the original sp^3 hybrid, ie near midgap. The interstitials are created by adding an atom to the lattice, either in one of the lattice interstices of tetrahedral (T) or hexagonal (H) symmetry, or at a bond centre (B) or by forming a split interstital (S). The T and H interstitials can be accommodated with little relaxation of the adjacent lattice, but the bond centre interstitial requires the outward relaxation of the two nearest atoms, while at the split interstitial one bulk atom and the extra atom each become trivalent. The most stable site for the interstitial changes with its charge state.

The activation energy of self-diffusion Q is given by the sum of the defect formation energy E_D and the defect migration energy E_M

$$Q = E_D + E_M \tag{1}$$

Generally, E_D is the dominant term in diamond and Si. Table 1 shows the formation energy for the neutral state of the vacancy and the most stable

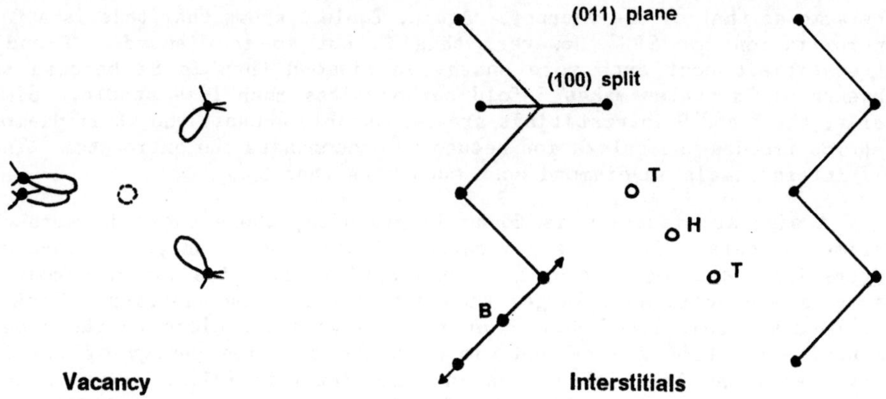

Fig. 2. Vacancy and interstitial configurations in diamond.

Table 1. Comparison of the formation energies of neutral defects and cohesive energies for Si, diamond and graphite (eV).

	Vacancy	Interstitial	Z Interstitial	E_{coh}	ref.
Si	4.2	5.7		4.64	(16)
Diamond	7.2	15.8		7.41	(14)
Graphite	7.0	$\simeq$18	7.0	7.43	(17,18)

interstitial[14,15]. In Si the vacancy and the interstitial have similar formation energies, so that atomic diffusion is mediated by a combination of defects depending on factors such as temperature[16]. Diamond is very different, the vacancy is found to cost less than half the energy of any interstitial, so the vacancy is dominates atomic diffusion[14]. Migration energies are 1.5-2.0 eV for the diamond vacancy.

It is possible to illustrate the energetics of intrinsic defects very simply. A vacancy is created by taking one atom from the bulk and placing it on the crystal surface, the reservoir of bulk sites. This breaks four bonds at the vacancy but creates two new bonds on average at the surface. The cohesive energy E_{coh} is the heat of volatilization and requires breaking two bonds per atom. Thus, neglecting relaxation, the formation energy is

$$E_D \simeq 2E_b = E_{coh} \tag{2}$$

where E_b is the bond energy. Table 1 shows that this is a fair approximation for both diamond and Si.

The T interstitial is formed by taking one atom from the surface, so it involves forming four new bonds at the interstitial and loosing two bonds at the surface. The interstitial atom changes the coordination of its four neighbours to 5-fold. Roughly speaking, the cohesive energy of a Si atom does not change much as its coordination changes from 4 to 5; that is the new bonds at a 5-fold site will be only 4/5 as strong as the bulk bonds. Thus the formation energy of the interstitial is mainly due to the loss of the two surface bonds, giving

$$E_D \simeq 2E_b \tag{3}$$

the same as that of the vacancy. Again, Table 1 shows that this is a fair approximation for Si. However, this is not so in diamond. T and H interstitials cost much more energy in diamond than in Si because the absence of 2d states makes 5-fold carbon sites much less stable. Similarly, the B and S interstitials are less stable because the stiff diamond lattice impedes the relaxation needed to accommodate the extra atom. Thus all interstitials in diamond cost much more that E_{coh}.

A similar situation is found in graphite, where there is more experimental data. There are now three defects, the vacancy, the in-plane interstitial and the inter-plane interstitial (Z). The defect formation energies are again much larger than defect migration energies. Kaxiras and Pandey[17] calculated $E_D \simeq 7.5$ eV for the vacancy, close to the experimental estimate[18] 7.0 eV and close to the cohesive energy of 7.4 eV. They also found $E_D \simeq 18$ eV for an in-plane interstitial, over twice that of the vacancy. In graphite, though, the inter-plane interstitial has a similar E_D to that of the vacancy and is very mobile[18].

The dangling bonds of a vacancy create states deep in the band gap. Their position can be understood using the defect molecule consisting of just the four sp^3 hybrids pointing into the vacancy and their associated electrons. The vacancy has T_d symmetry so the four hybrids combine to give a fully symmetric state of A_1 symmetry and a triply degenerate state of T_2 symmetry. Calculations find that the A_1 state lies at -2 eV below the valence band edge, E_v, and T_2 lies at 1.5-2.0 eV in the gap[5,14]. Hence the A_1 state of the neutral vacancy V^0 will be filled with 2 electrons and 2 electrons will lie in T_2. The vacancy can lower its energy if its atoms can move from their ideal positions. The relaxation at the diamond vacancy is actually small because of the stiffness of bonds in diamond[14]. However, the two electrons only partially fill the T_2 level, leaving it orbitally degenerate and unstable to the Jahn-Teller effect. V^0 undergoes a tetragonal distortion which has the effect of partially rebonding dangling bonds in pairs. One must also consider multiplet splittings. These various effects imply that the vacancy creates states across the lower half of the gap. There are numerous cases of deep donor (0/+) levels at $\simeq 1.5$ eV[19,20] which may be due to the T_2 vacancy states.

IMPURITIES

Boron is an effective p-type dopant of diamond, forming a shallow acceptor level 0.37 eV above E_v. However, there is a problem with donors. Shallow donor levels are expected at 0.19 eV below the conduction band edge E_c from effective mass theory with the dielectric constant $\varepsilon = 5.8$ and $m_e = 0.48 m_0$. Nitrogen in column V of the same Period as carbon might be expected to be a shallow donor, as occurs in other semiconductors, but it forms a deep level at 1.7 eV. It is therefore necessary to consider other elements such as P, Li and Na. Li and Na are possible interstitial donors that could be introduced by diffusion or ion-implantation[20]. Presently, it is possible to create n-type layers by C^+ ion bombardment and careful annealling out of the radiation damage, and indeed Prins was able to make a diamond transistor[20,21]. P is a possible donor to be introduced from PH_3 in vapour deposited diamond. Setaka[22] has found that P is only able to reduce the conductivity activation energy to 0.84 eV, which is still too high. This could be due to P also forming a deep level or to a poor incorporation efficiency. Spitsyn[23] found quite efficient P doping, but further details of sample characterisation are desireable here. Clearly theory could give useful guidance here. Indeed, Bernholc et al[15] find that low solubility is the problem in all cases, because of the large sizes of the atoms.

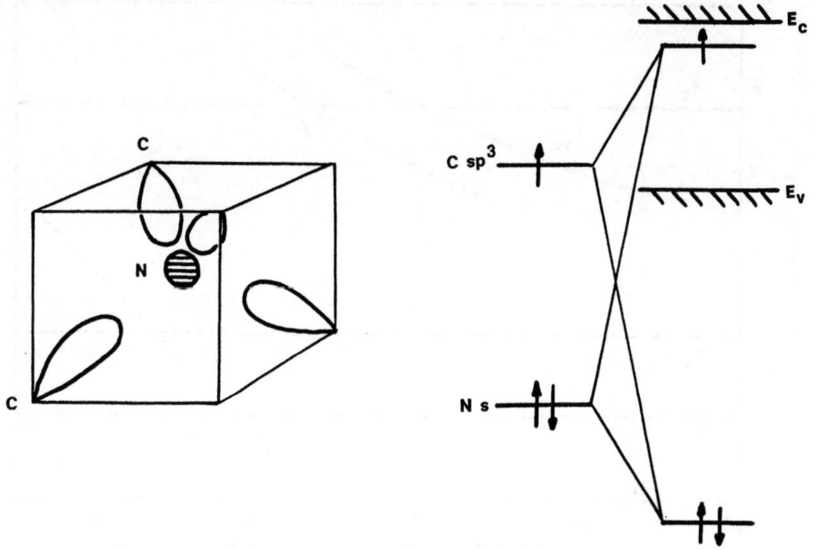

Fig. 3. The impurity molecule and its molecular orbitals.

Consider first subsititutional impurities. The long range part (eg Coulombic) of the defect potential will always bind shallow states, but the short range part of the potential must exceed some threshold before it can bind a deep state in the gap. Deep states can be treated by the defect molecule model as used earlier for the vacancy. This now consists of the four sp^3 hybrids of the vacancy and the s and p orbitals of the impurity, Fig. 3(a). The impurity s orbital has A_1 symmetry and interacts with the A_1 combination of hybrids, ie the A_1 level of the vacancy, to form bonding and antibonding levels at

$$E = 1/2(E_I + E_V) \pm (1/4(E_I - E_V)^2 + V_2{}^2)^{1/2} \tag{4}$$

as in Fig 3(b), where $E_V = E(A_1)$, E_I is the impurity s orbital energy E(s), and V_2 is the appropriate impurity-host two-centre interaction. Similarly, the impurity p orbitals are T_2 and interact with the T_2 vacancy levels. The impurity levels form hyperbolic curves when plotted as a function of E(s) or E(p). Donors correspond to the upper, antibonding branch of the A_1 curve and acceptors to the lower bonding branch of the T_2 curve. Concentrate for the moment on the A_1 donor levels, Fig. 4. The level lies outside the gap for $E(s) \simeq E_V$ so these impurities produce only shallow states. As the donor becomes more electronegative and E(s) becomes more negative, at some critical value the antibonding level enters the gap, giving a deep gap state.

Nitrogen forms a deep, trigonally-distorted substitutional centre in diamond. Bernholc et al[15] found that the ideal substitutional N site gives a deep state 1.2 eV below E_c. Fig. 4 suggests that this occurs because N is too electronegative compared to C to be shallow, despite being adjacent to C in the Periodic table. Thus, the deepness of N arises primarily from its electronegativity and not its distorted geometry.

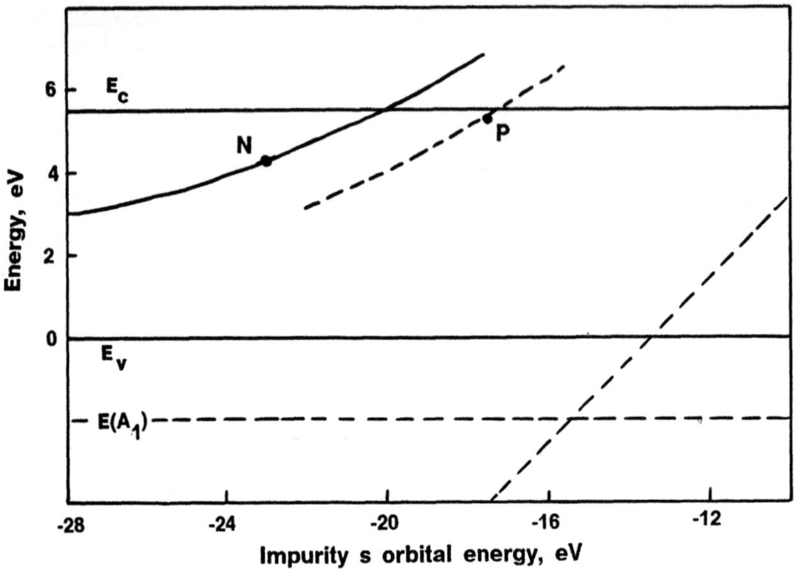

Fig. 4. Dependence of substitutional donor levels on the donor's orbital energy.

Substitutional P was found to give a shallow donor level about $E_C - 0.2$ eV[15]. This position depends on two factors in our model. First, P is much less electronegative than N, so Fig. 4 suggests that an unrelaxed P site would give a level well above E_C. Second, P is much larger than C with a tetrahedral radius of 1.13Å compared to carbon's 0.774 Å. This would normally cause a large outward relaxation to give a host-P bond length near the sum of the covalent radii. As the two-centre interaction V_2 in eqn (4) varies[24] with bond length d as d^{-2}, this reduces V_2 and lowers the donor level in Fig. 4. A full relaxation would make P in diamond borderline deep. In fact, the bond stiffness of diamond is found to allow only 0.15 Å relaxation[15], which keeps P shallow. The corollary of this small relaxation, however, is that the free energy of P in diamond is very high[15], $\simeq$10.7 eV, which causes the solubility of P to be very low, even at high temperatures.

Li and Na are possible interstitial donors. They are calculated to favour the T interstitial site and to produce shallow states. The ionic radii of Li^+ and Na^+, 0.68Å and 0.96 Å respectively, are similar to the size of the T interstice, 0.77 Å. Nevertheless, Li and Na each cause a small outward relaxation. This again gives high free energies (10 and 16 eV) and low solubilities[15]. These results led Bernholc et al[15] to conclude that P, Li and Na are shallow donors but must be introduced into diamond in a strongly athermal manner in order to achieve reasonable incorporation efficiencies.

A further 'impurity' of importance in diamond is hydrogen[25], present in concentrations of up to 1%. Hydrogen is important because of its ability to passivate dangling bonds and impurity states in semiconductors. It could be present as either H or H_2. H is expected to be a rapid diffusing species while H_2 is much slower. However, the spin signal of neutral H, H^0, is not observed. Similar effects occur in Si, and for

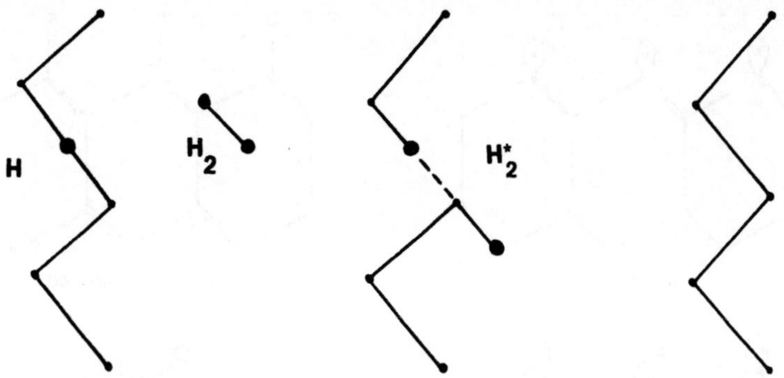

Fig. 5. Hydrogen configurations in diamond; (a) H at bond-centre, (b) H_2 at T interstitial and (c) H_2* in the (110) plane

muonium, the nuclear analogue of H. It is now known that the stable location of H in Si and diamond depends on its charge, Fig. 5; H^+ is stable at the bond-centre site, as is H^0, while H^- has similar energies at many interstice sites[26]. These degeneracies give H a low migration energy of 0.1-0.3 eV in Si. Secondly, molecular H_2 can exist at the T site. This is more stable than 2H but its migration energy is much higher because of its size (0.8 eV in Si). A second configuration of H_2 is a partly dissociated pair[27,28], denoted H_2*, of one H at a bond centre and a second H at that bond's antibonding site. The lower C relaxes away from the bond centre to create two almost ≡CH configurations. This is more stable than 2H and more stable than H_2 at T in diamond, but less stable in Si. The stability of H_2* arises because the two hydrogens tend to have a partial positive and negative charge respectively. H_2 * has a lower migration energy, because each H can, to a degree, jump independently. In this way, diamond and Si can accommodate considerable hydrogen in diamagnetic form with potentially low migration energies.

SURFACES

Understanding the behaviour of diamond surfaces is crucial to vapour deposition because the growth rate is generally believed to be controlled by a surface reaction and because surface properties may also control the selectivity of the deposition between diamond and graphitic phases. A key point is the interaction of the surface with atomic hydrogen. H is known to be necessary to promote the selective depostion of diamond, but the precise mechanism of this is debated.

Diamond surfaces have many similarities to those of Si but with two important differences; the greater stability of C-H bonds causes the surfaces to be hydrogenated below ≃950°C while above ≃1250°C the surfaces become graphitic. The ideal (111) surface shown in Fig. 6(a) consists of an array of trivalent atoms with one dangling bond on each site directed normal to the surface. Each dangling bond contains one electron so the ideal (111) surface introduces a surface band in the bulk band gap which is half-filled and metallic[4]. This situation is unstable with respect to a reconstruction which can partially rebond the surface atoms and open up a gap at the Fermi level, E_F. An early model of the (2x1) reconstruction of Si(111) was the buckling model in which alternate surface atoms move into or out of the surface, inducing a charge transfer, Fig. 6(c). This reconstruction was found to be unstable. The accepted

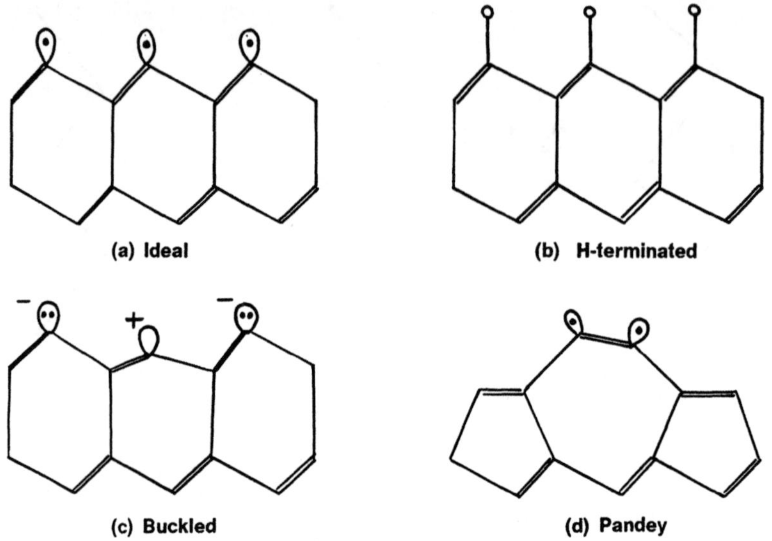

Fig. 6. Surface structures of diamond (111)

model of Si(111)-(2x1) is the π-bonded chain model of Pandey[29], Fig. 6(d). This is a topological reconstruction of the ideal surface with 5- and 7-membered rings. The dangling bonds are moved to adjacent atoms which then form π-bonded chains, opening a gap at E_F in the surface bands.

The diamond (111) surface is more complicated than Si(111). It shows a (1x1) cell at low temperatures and an apparent (2x2) cell above 950 °C[30]. The (2x2) structure is a clean surface and actually consists of domains of three differently oriented (2x1) reconstructions. Its structure consists of π bonded chains[30–34] and is basically similar to that of Si(111)-(2x1). While the π-bonded chains can be considered to be precursors of graphitic surfaces, angle resolved photoemission spectra[32] show that the (2x1) diamond surface is distinct from graphite in that the dispersion of the surface band is much less than the π band of graphite.

The (1x1) structure is the hydrogenated surface in which H allows each surface C atom to remain sp^3, Fig 6(b). This surface has no surface states near the gap as the C-H bonds tie up all of the dangling bonds. The presence of hydrogen is detected by photon- and electron-stimulated desorption[30,34]. The surface can be rehydrogenated by atomic hydrogen but not by molecular hydrogen[30]. This is because the C-H bond is slightly weaker than the H-H bond. The hydrogenated surface has a negative electron affinity but the clean surface does not[32].

The surface energy γ of diamond is the energy lost in breaking the surface bonds and is 5.3 Jm^{-2} for the (111) face for E_b=3.7 eV. Cleavage creates two new surfaces. The experimental cleavage energy of 11.0 Jm^{-2} shows that diamond cleaves with little plastic deformation[35]. Hydrogenation lowers the effective surface energy to near zero because of the energy regained in forming the C-H bonds. This greatly lowers the minimum critical nucleation radius for diamond growth[36]. However, hydrogenation does not lower the cleavage energy because cleavage is the faster process.

The structure of the partially hydrogenated surface is interesting. Hamza et al[34] showed that dehydrogenation occurs at at slightly lower temperature than the (2x1) structure forms. This may be because the topological character of the reconstruction allows some hysteresis. Northrup and Cohen[37] found that the ideal (111)Si surface could convert into the chain structure with only a small energy barrier per atom. However, the barrier may be higher for diamond, because the absence of 2d states may raise the energy of the overcoordinated intermediate stage. More importantly, the topological character of the reconstruction requires the concerted movement of many surface atoms. Thus, single atoms and possibly small groups of atoms on the diamond surface should be able to remain in a sp^3 configuration after dehydrogenation, rather than immediately reverting to a π-bonded or ultimately a graphitic structure.

The selective deposition of diamond from methane etc. requires the presence of atomic hydrogen[2,36]. Its role has been attributed to

1. the preferential etching of any deposited graphitic phase[38],
2. the maintainence of a sp^3-like rather than graphitic surface,
3. the reduction in the minimum critical nucleation size[36]
4. its use in hydrogen abstraction from hydrogenated surface atoms

$$\equiv C - H + H = \equiv C. + H_2 \quad (5)$$

Spitsyn et al[38] argued that graphitic and diamond crystallites were deposited simultaneously and that the selective deposition of diamond depended primarily on the selective etching of graphitic carbon by atomic hydrogen. Frenklach and Spear[39] argued that the etching rates were too low and proposed that the selective deposition of diamond arose from the selective suppression of poly-aromatic species, the precursors of graphitic carbon, by H_2 in the gas phase, and that the role of atomic hydrogen was to maintain a hydrogenated sp^3 surface. Frenklach and Spear[39,40] also proposed that the hydrogen abstraction reaction (5) was the rate-limiting reaction at higher deposition temperatures. This is consistent with their acetylene intermediate mechanism in which it was found that hydrogen abstraction was the only endothermic step in the reaction of acetylene and a hydrogenated surface[41].

Taking hydrogen abstraction as the rate limiting step provides a definite model of diamond growth with interesting consequences. Quite low substrate temperatures (500-800°C) are now favoured for good quality diamond[42]. Only isolated surface C dangling bonds are likely to arise at these temperatures, and they would then react quickly with a growth species. The supression of graphitic surface character should not then be a problem. Secondly, the growth rate of diamond would be proportional to the arrival rate of atomic H in the depositing flux. It should be possible to test this proposition by observing growth rates in different plasma conditions, some of which are quite high rates[43], in which the concentration of H and that of possible growth species, C_2H_2, C etc, vary opppositely, assuming that their gas phase lifetime is sufficient for them to cross the boundary layer at the surface. Hydrogen abstraction is not expected to be rate limiting for very high temperature plasmas if C is found to be the growth species, as suggested by Bachmann[43].

CONCLUSIONS

Theoretical calculations suggest that vacancies mediate atomic diffusion in diamond and that P, Li and Na are shallow but low solubility donors. The diamond surface is terminated by hydrogen below 1000°C. Atomic hydrogen is necessary to maintain this condition but a key role of atomic H may be initiating the growth reaction by abstracting hydrogen from the C-H surface bonds.

References

1. M. W. Geis, J Vac Sci Technol A 6 1953 (1988); this volume
2. J. C. Angus and C. C. Hayman, Science 241 913 (1988)
3. G. S. Painter, D. E. Ellis and A. R. Lubinsky, Phys Rev B 4 3610 (1971)
4. J. Ihm, S. G. Louie and M. L. Cohen, Phys Rev B 17 769 (1978)
5. G. B. Bachelet, G. A. Baraff and M. Schluter, Phys Rev B 24 4736 (1981)
6. F. J. Himpsel, J. F. van der Veen and D. E. Eastman, Phys Rev B 22 1967 (1980)
7. L. Reggiani, S. Bosi, C. Canali, and S. F. Kovlov, Solid State Commun 30 333 (1979)
8. A. T. Collins and E. C. Lightowlers, in 'Properties of Diamond', ed. J. E. Field, Academic Press, London (1979)
9. F. Nava, C. Canali, C. Jacoboni, L. Reggiani and S. F. Kozlov, Solid State Commun 33 475 (1980)
10. M. T. Yin and M. L. Cohen, Phys Rev B 24 6121 (1981); 29 6996 (1984)
11. M. T. Yin and M. L. Cohen, Phys Rev Lett 50 2006 (1983)
12. G. Galli, R. M. Martin, R. Car, M. Parrinello, Phys Rev Let 63 988 (1989)
13. N. A. W. Holzwarth, S. G. Louie and S. Rabii, Phys Rev B 26 5382 (1982)
14. J. Bernholc, A. Antonelli, T. M. DelSole, Y. Bar-Yam and S. T. Pantelides, Phys Rev Lett 61 2689 (1988)
15. S. A. Kajihara, A. Antonelli and J. Bernholc, Mat Res Soc Symp Proc 162 xxx (1990)
16. R. Car, P. J. Kelly, A. Oshiyama and S. T. Pantelides, Phys Rev Lett 52 1814 (1984); 54 360 (1985)
17. E. Kaxiras and K. C. Pandey, Phys Rev Lett 61 2693 (1988)
18. P. A. Thrower and R. M. Mayer, Phys Stat Solidi A 47 11 (1978)
19. M. I. Landstrass and K. V. Ravi, App Phys Lett 55 1391 (1989)
20. J. Prins, Phys Rev B 38 5576 (1988)
21. J. Prins, App Phys Lett 41 950 (1982)
22. N. Setaka, Mat Res Soc Symp Proc 152 3 (1989)
23. A. E. Alexenko and B. V. Spitsyn; this volume.
24. W. A. Harrison, 'Electronic Structure' (Freeman, San Francisco, 1979)
25. J. P. F. Sellschop, C. C. P. Madiba and H. J. Annegarn, Nucl Inst Methods 168 529 (1980)
26. C. G. Van de Walle, P. Denteneer, Y. Bar-Yam and S. T. Pantelides, Phys Rev B 39 10791 (1989)
27. K. J. Chang and D. J. Chadi, Phys Rev Lett 62 937 (1989); Phys Rev B 40 11644 (1989)
28. P. Briddon, R. Jones and G. M. S. Lister, J Phys C 21 L1027 (1988)
29. K. C. Pandey, Phys Rev Lett 47 1913 (1981); 49 223 (1982); Phys Rev B 25 4338 (1982)
30. B. B. Pate, Surf Sci 165 83 (1986)
31. D. Vanderbilt and S. G. Louie, Phys Rev B 30 6118 (1984)
32. F. J. Himpsel, D. E. Eastman, P. Heimann and J. F. van der Veen, Phys Rev B 24 7270 (1981)
33. G. D. Kubiak and K. W. Kolasinski, Phys Rev B 39 1381 (1989)
34. A. V. Hamza, G. D. Kubiak and R. H. Stulen, Surf Sci 206 L833 (1988)
35. J. E. Field and C. J. Freeman, Phil Mag A 43 595 (1981); this volume
36. T. R. Anthony, Mat Res Soc Symp Proc 162 xxx (1990); this volume
37. J. E. Northrup and M. L. Cohen, Phys Rev Lett 49 1349 (1982)
38. B. I. Spitsyn, L. L. Bouilov and B. V. Deryagin, J Crystal Growth 52 219 (1981)
39. M. Frenklach and K. E. Spear, J Mater Res 3 133 (1988)
40. M. Frenklach, this volume; J App Phys 65 5142 (1989)
41. D. Huang, M. Frenklach, M. Marconcelli, J Am Chem Soc 92 6379 (1988)
42. Y. Liou, A. Inspektor and R. Messier, App Phys Let 55 631 (1990)
43. P. Bachmann, this volume

PHYSICS OF GLOW DISCHARGE PLASMAS AND PLASMA/SURFACE INTERACTIONS DURING THIN FILM GROWTH

Jan-Eric Sundgren

Thin Film Laboratory, Dept. of Physics, Linköping University
S-581 83 Linköping, Sweden

1. Introduction

Low pressure glow discharge plasmas are today frequently used in materials processing. For example, sputter deposition and other ion assisted growth techniques are essential for production of semiconductor devices, hard wear resistant coatings and for various optical thin films. Also plasma assisted etching techniques and plasma polymerization are extensively used in many industrial processes. A glow discharge plasma can be defined as a partially ionized low pressure gas in a quasi-neutral state sustained by the presence of energetic electrons. The character of such plasmas is basically a consequence of the mass difference between the electrons and the ions. Due to this mass difference, energy is transfered more rapidly to the electrons than to the ions. These energetic electrons then accumulate sufficient energy to have a high probability of causing ionization and excitation events when colliding with heavier particles. The generation of these particles and their interactions with surfaces and growing films are the most important reasons why glow discharge plasmas have become of such importance in materials science.

The purpose of this paper is to present a brief review of some of the physical phenomena occurring in glow discharge plasmas. Since there have been several review papers[1] and books[2] written about fundamental characteristics of glow discharge plasmas the emphasis in this chapter is placed on the importance of plasma/surface interactions during growth of thin films from the vapor phase. However, a brief description of particle motion and collective phenomena of plasmas is first given. Among the different existing types of plasma discharges only the magnetron discharge is treated in this paper due to the increasing importance of such discharges especially for ion assisted depositions. There are naturally several important aspects of plasmas and usage of plasmas that not are treated in this chapter. However, some of these aspects are discussed in other papers in these proceedings.

Diamond and Diamond-Like Films and Coatings
Edited by R.E. Clausing *et al.*, Plenum Press, New York, 1991

2. Plasma Science

2.1 Particle motion and gas phase collision processes

Since the generation, existence and characteristics of glow discharge plasmas relies upon energy transfer during collisions in the gas phase, the knowledge about probabilities for various type of collisions is of fundamental importance. These probabilities are usually described in terms of cross sections and mean free paths.

The mean free path λ between collisions of gas molecules or atoms is in the most simple way calculated by assuming that the involved particles can be treated as hard spheres in which the collision cross section σ is simply related to the diameter of the spheres. For example, in the case of particles in a gas with a density N;

$$\lambda = (\sigma N)^{-1} \qquad (1)$$

However, in many cases the situation is considerably more complicated. For example, in sputter deposition we have a situation where one particle type (the sputtered atoms) is travelling through a gas of another type. For such a case we can, from kinetic gas theory[3] and by assuming diffusing particles of the hard sphere type, write the mean free path of a particle with mass M_s travelling through a gas of particles with mass M_g as;

$$\lambda^{-1} = 2^{1/2} N_s \phi_s^2 + 1/4\, \pi N_g(\phi_s + \phi_g)^2(1 + M_s/M_g)^{1/2} \qquad (2)$$

where ϕ_s and ϕ_g are the diameters of the particles. Thus from this simple formula one can see that λ decreases with the size of the particles, as expected, and also with the mass ratio between the two types of particles. Consequently, if two types of particles with different masses are travelling through a gas consisting of a third type of particles the heavier particle will undergo more collisions in a given distance than the lighter particle. This effect can in alloy sputtering lead to a depletion of the number of heavier species reaching the growing film surface. The λ-values given by eq. 2 are valid for diffusing gases and it can be expected that λ will be higher for atoms with a high velocity, as for example sputtered atoms which typically have energies of some eV. From classical kinetic gas theory λ will be a maximum of $\approx$ 1.4 higher for fast travelling atoms[4].

At each collision the energy and direction of the particles will change. Westwood[4] has calculated, using kinetic gas theory and assuming hard spheres, the average energy retained by an atom of mass M_s in collisions with an atom of mass M_g as well as the average angle of deviation due to the collision. The results from these calculations for different M_g/M_s ratios are shown in Fig. 1. It can be seen that the largest energy loss occurs for cases where the mass ratio is close to 1 whereas the scattering angle increases with M_g/M_s. For example, in the case of Ta atoms travelling through an Ar gas would lose 30 % of their energy and be deflected 10° in each collision and for Al atoms in Ar, the values would be 60 % and 65°.

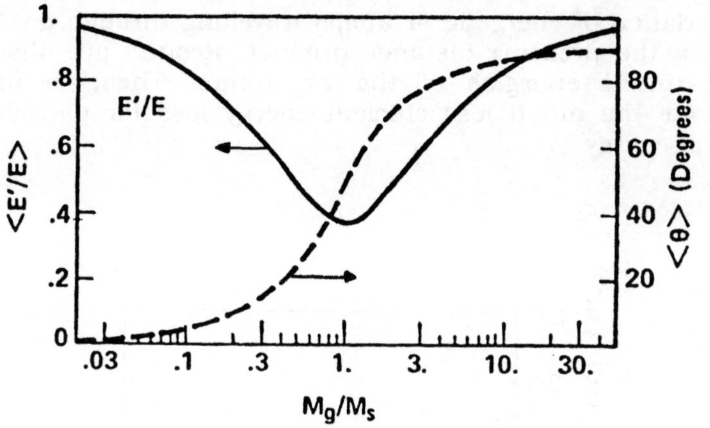

Fig. 1 The average fraction of energy retained by atoms of mass M_s after collisions with atoms of mass M_g and the average scattering angle due to the collision. (From Westwood[4]).

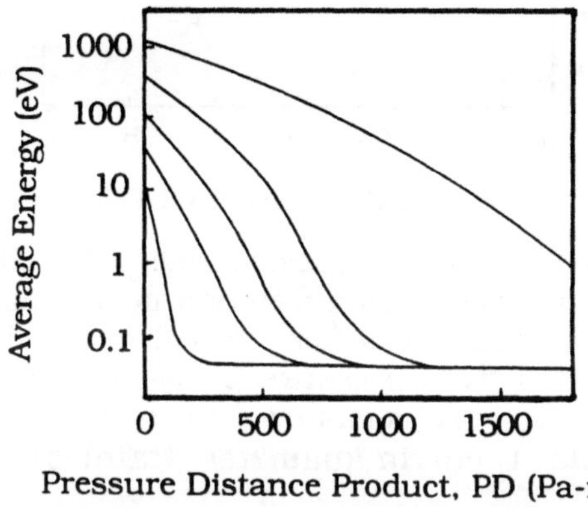

Fig. 2 Energy variation of energetic Ar atoms travelling through a thermal Ar gas as a function of the pressure-distance product. Results for 5 different initial energies of the energetic Ar atoms are shown (from Somekh[7]).

However, since the hard sphere assumption is not necessarily correct, particularly not for high energies, the amount of scattering estimated using the above simple models are only good as a first approximation. For example, Robinson[5] has shown theoretically how atom scattering cross sections varies with energy. For energies in the range 10 to 100 eV a considerable reduction is found compared to hard sphere models. Somekh[6,7] has used these energy dependent cross sections and a Monte Carlo method to calculate the energy loss of energetic particles travelling through a thermal gas. Fig. 2 shows, from the work of Somekh[7], the energy variation of energetic Ar atoms travelling through an Ar gas as a function of the pressure-distance product. Results are shown for five different initial energies of the Ar atoms. These results clearly demonstrate the much less efficient energy loss for particles with an high initial energy.

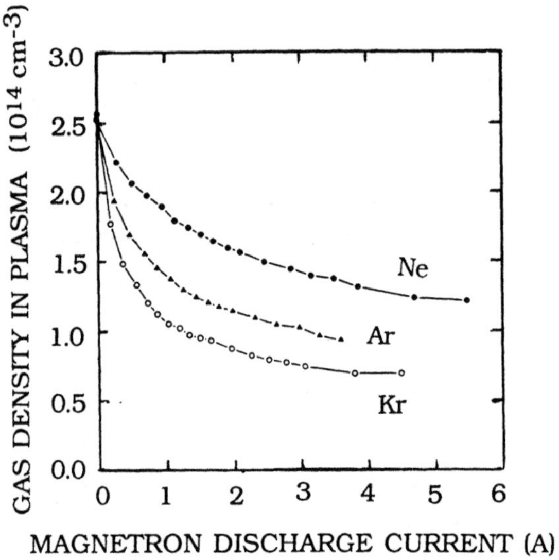

Fig. 3 Gas density in the hot plasma region for the cases of Ne, Ar, and Kr gases and a Cu cathode as a function of the magnetron discharge current. The system pressure was kept at 7.6 mTorr (1 Pa) and the gas densities were calculated from pressure measurements 5.3 cm from the cathode. The pressures were measured by a capacitance manometer fitted with a movable 1 cm in diameter stainless steel tube. (from Rossnagel, ref. 9).

If the number of energetic atoms travelling through a thermal gas is high enough, the energy transfered from the energetic atoms to the gas atoms will result in an increase of the average energy of the thermal gas atoms and thus effectively to a reduction in the gas pressure. This has for example been demonstrated by Hoffman[8] and by Rossnagel[9] to be the case in magnetron sputtering at high rates. The energy of sputtered particles are usually in the order of 10 eV and the thermallization procedure of these sputtered atoms can cause a considerably reduction in the gas pressure in front of the sputtering source. A typical example of the reduction in the gas density in front of a Cu sputtering cathode versus the magnetron discharge current is shown in Fig. 3. Result for three different sputtering gases are shown. As the discharge current is increased, the number of sputtered Cu atoms increases and thus the number of collisions with gas atoms. Lighter sputtering gas atoms have a less pronounced pressure reducing effect due to both a lower sputtering yield and a higher thermal conductivity.

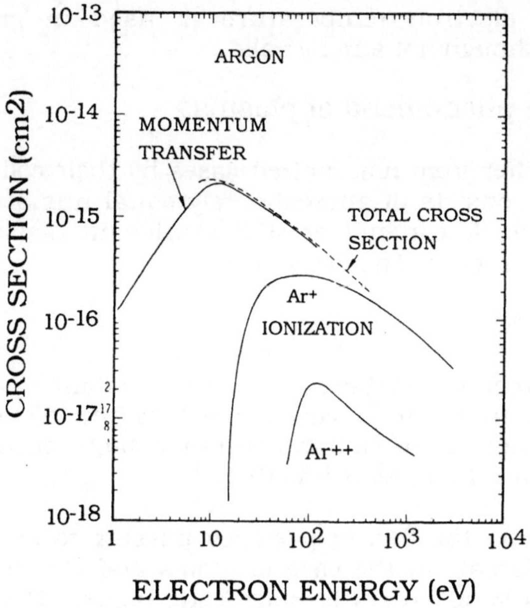

Fig. 4 Collision cross section for electrons in an Ar gas as a function of the electron energy (from ref. 10).

For electrons passing through a gas the total collision cross section can be written as;

$$\sigma_t = \sigma_{el} + \sigma_{ex} + \sigma_{ion} + \sigma_a + \sigma_{oth} \tag{3}$$

where the subscripts el, ex, ion, a and oth denotes the particular type of collision, namely elastic (momentum exchange), excitation, ionization, attachment and other processes, respectively. A typical example for electrons in Ar gas[10] is shown in Fig. 4. As can be seen from this figure σ is a strong function of the electron energy. At low electron energies the primary collision process is elastic collisions ($\sigma_t \approx \sigma_{el}$), while at energies larger than the ionization potential (15.75 eV for Ar) ionization collisions dominate ($\sigma_t \approx \sigma_{ion}$).

For most purposes, a glow discharge plasma can be characterized by the density of heavy particles, N, the electron density n_e and the electron energy distribution function $F_e(E)$. Due to the large mass difference between electrons and atoms or molecules, the energy gained by a plasma electron moving in an electric field will be much larger than the energy lost in collisions with the heavy neutral particles in the plasma. Typically the electrons have energies of several eV while the heavy particles only have energies of a few percent of an eV. Under conditions of local thermodynamic equilibrium the electrons will have a Maxwellian velocity distribution and their state can be defined by an electron temperature T_e. However, in most cases there are deviations from such Maxwellian distributions due to the presence of electric fields, electron-electron collisions and inelastic collision processes. The latter type of processes will generally deplete the high energy tail of the distribution. However, since energy distribution functions not easily are measured, the electron temperature is used in most engineering calculations[1], although not strictly valid.

2.2 Collective phenomena of plasmas

Plasmas differ from non-ionized gases by their collective behaviour. Since a plasma consists of approximately equal number of positive and negative particles, it can not exist if significant departures from this charge neutrality occurs. The Debye length,

$$\Lambda_D = K\,(kT_e/n_e)^{1/2} \qquad (4)$$

where k is Boltzmanns constant and K is a constant equal to 743 if T_e is given in eV and n_e in cm^{-3}, corresponds to the distance over which significant deviations from charge neutrality might occur. A plasma can not exist in a space if its size is less than Λ_D.

For a plasma the flux of particles passing to an adjacent wall is considerably different in the case of atoms and electrons. The electron flux is usually several orders of magnitude higher. This higher electron flux will result in the build up of a space charge region adjacent to the wall in which one of the species is largely excluded. The potential variation between the surface and the plasma will predominantly be confined to this region, called the sheath. Sheaths are usually several Debye lengths thick. The sheath is thus a barrier that adjusts itself so that the more mobile species (which except in the case of the presence of strong magnetic fields is the electrons) are electrostatically reflected. When the surface is electrically isolated, the electron flux is reduced to a point where it equals the ion flux. The potential developed on such a surface with respect to the plasma is the floating potential V_f. The value of V_f is typically only a few volts for dc and magnetron sputtering[11] but may be as high as a few hundred volts in the rf case depending on the gas pressure and system geometry[12]. The plasma potential V_p is the potential between the plasma and ground. The potential variation between the cathode and the anode in a diode glow discharge is illustrated schematically in Fig. 5.

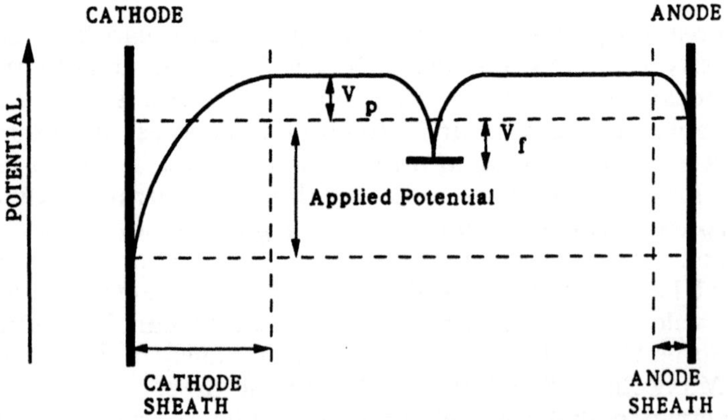

Fig. 5 Schematic illustration of a typical potential variation in a diode glow discharge where the anode is large compared to the cathode.

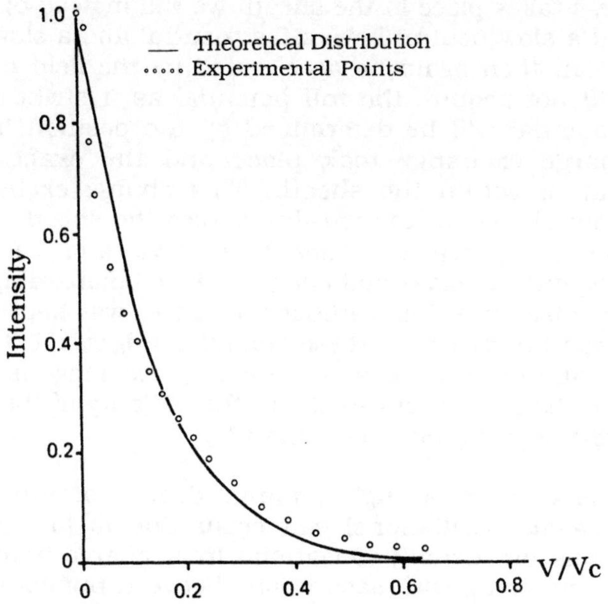

Fig. 6 Energy distribution for Ar^+ ions impinging on the cathode in an Ar glow discharge. The cathode voltage was -600 V and the Ar pressure 60 mTorr. Under these conditions a sheath thickness d_s of 1.3 cm was measured. Both the experimental data points (o) and a theoretical distribution (----) are shown. The theoretical distribution was obtained by assuming a cross section for charge exchange collisions of 5.3×10^{-15} cm^2. This gave a ratio between d_s and the mean free path for charge exchange collisions of 15. (From the work of Davis and Vanderslice, ref. 13).

At the cathode a large potential difference V_s between the plasma and the cathode surface occurs. V_s equals the sum of the applied potential V_a and V_p. The sheath thickness in this region d_s is related to V_s and the ion current density to the cathode J_i through the Child-Langmuir law which can be written as[1];

$$J_i = 0.273\ (40/m_i)^{1/2}(V_s^{3/2}/d_s^2) \quad \text{mA cm}^{-2} \qquad (5)$$

where V_s is in kV, d_s in cm and m_i is the ion molecular weight in g. At low pressures the ions will bombard the cathode with energies equal to eV_s while at higher pressures collisions might take place in the sheath region and the flux of bombarding energetic species will consist of both ions and neutrals with energies considerably less than eV_s.

The collision process that will dominate in the sheath is the resonant charge exchange process, e.g.,

$$Ar^+ + Ar \rightarrow Ar + Ar^+ \qquad (6)$$

If such a process takes place in the sheath we will instead of having a fast moving ion and a slow neutral have a fast neutral and a slow moving ion. This slow ion can then again be accelerated by the field in the sheath. However, it will not acquire the full potential as it strikes the cathode surface. Its potential will be determined by the position in the sheath where the charge exchange took place and the exact form of the potential variation within the sheath. This charge exchange process results in no net change in the ion flux across the sheath. Fig. 6 shows the result from a classical experiment by Davis and Vanderslice[13] in which they analysed the mass and energy of ions bombarding the cathode in a Ar dc glow discharge. The cathode voltage V_c was kept at 600 V and the pressure was 60 mTorr. At these conditions d_s was 1.3 cm and they estimated the ratio between d_s and the charge exchange mean free path to be 15. As can be seen from this figure the majority of the ions striking the target have energies much lower than V_c.

For plasmas with a high enough degree of ionization, wave phenomena (plasma oscillations) can occur due to long range forces. These oscillations are usually deviations from charge neutrality in the form of charge bunching with separations between bunches in the order of the Debye length. However, for most glow discharge plasmas, with the exception of some highly ionized magnetron discharges, plasma oscillations are not of any significant importance.

2.3 Magnetrons

As charged particles move in the presence of an uniform magnetic field B, their motion will be strongly affected. The charged particles will orbit the field lines with a cyclotron frequency that is given by[1];

$$\omega_c = eB/m \qquad (8)$$

The cyclotron radius or Lamour radius is then given by

$$r_c = m/3(v_\perp/B) \qquad (9)$$

where $v_\perp$ is the velocity perpendicular to the magnetic field. Any manipulation and confinement of a plasma requires that r_c is small compared to the apparatus size. For electrons with an average energy of 10 eV in a magnetic field with B = 100 G, r_g will be ≈ 0.1 cm. Magnetic fields between 100 and 1000 G are typically used in magnetron discharges.

When an electric field E is present and directed parallel to the magnetic field line the electrons will be accelerated along the field lines. However, if the field has a component that is perpendicular to B there will be a force acting on the particle making it drift in a direction that is perpendicular to both E and B (ExB drift). The drift of electrons along a magnetic field line can also be influenced by gradients in the magnetic field as exemplified in Fig. 7. An electron moving in a magnetic field will conserve its magnetic moment μ_M defined by[14];

$$\mu_M = W_\perp / B \qquad (10)$$

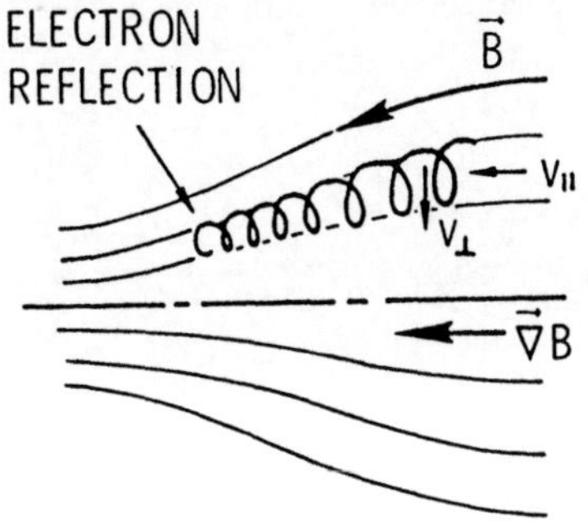

Fig. 7 Electron reflection in a magnetic field gradient (from ref. 1).

where $W_\perp$ is the average energy of the electron perpendicular to the field. Thus when the electron moves in a direction of increasing field strength, $W_\perp$ must increase. Since the conservation of energy implies that $W_{//} + W_\perp$ must be constant, $W_{//}$ must decrease and the electron will be reflected at some point. Inhomogeneous magnetic fields and electron reflections are common in magnetron devices.

Thus we can conclude that the net result of an magnetic field in a glow discharge is to change the motions of the charged particles. This changed motion results in a higher degree of ionization in the plasma since the electrons transverse a larger volume of the plasma before escaping to the anode. The possibility of making ionization collisions thus increases.

The plasmas can either be confined close to the cathode surface or extended in a wide "plasma beam" by altering the magnetic field configuration. Window and Savides[15] have studied the operation of planar magnetrons with different magnetic field configurations. They have classified the different magnetron configurations in three classes depending on the magnetic field configuration, type I, type II and an intermediate class as is shown schematically in Fig. 8. In type I magnetrons the central magnetic pole is much stronger than the annular magnet and the discharge is confined in the magnetic tunnel between the cathode surface and the surface defined by the last magnetic field line which cuts the anode (in their case the shield used to mask the edges of the cathode surface). Since the anode ring usually is located in a high magnetic field some of the electrons approaching the anode will be reflected back towards the central pole. In this configuration the fluxes of both electrons and ions reaching the substrates will be low.

In type II magnetrons the annular magnet is the strong magnet and all the field lines originate on this magnet. The nature of this discharge is considerably more complex than the type I discharge. There will be a high current to the substrates. For example Window and Savides[15] have, for this type of magetron design, reported electron currents to the

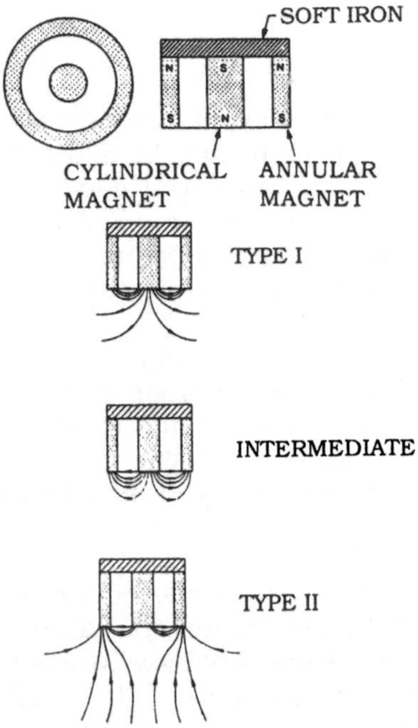

Fig. 8 Classification according to Window and Savvides[15] of different magnetic field configurations in planar magnetron devices

substrates as high as the discharge currents. The electron flux to the substrates will also increase the ion flux at the substrate surface because of the ambipolar nature of diffusion[2]. In fact the difference in ion current densities to substrates, placed at a given distance from the cathode, between type I and type II discharges might be several orders of magnitude.

Kadlec et al.[16] have used a magnetron where the magnetic fields were provided by means of electromagnets. By this configuration they were able to vary the field strength continously. Fig. 9 shows the variation in the substrate ion current I_s as a function of the current I_2 in the electromagnet producing the field corresponding to that from the annular magnet in Fig. 9. For I_2 equal to zero the magnetron works in a type I mode. Results for three different substrate bias voltages are shown. As can be seen from this figure the substrate ion currents can be increased by a factor of more than 50 depending on I_2 and the substrate bias voltage.

3. Plasma-Surface Interactions

Surfaces in contact with plasmas are bombarded with electrons, ions and energetic neutrals and photons all of which can have important effects during etching and growth. While a large body of experimental results exist on the effects of ion and electron irradiation less is known about the effects of photon irradiation. In this part of the paper we give some examples of how ion and electron irradiation from the plasma affect the growth of thin films.

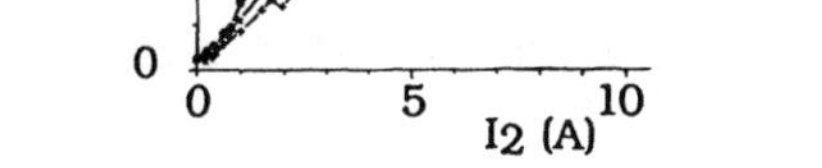

Fig. 9 The substrate ion current versus the current through the outer magnetic coil I_2, in an unbalanced magnetron based on electromagnets. An increase in I_2 implies an increased field strength to the annular magnetic pole shown in Fig. 8. For the case when $I_2 = 0$ the magetron behaves like a type I magnetron. Results are shown for three different substrate bias voltages. (from Kadlec et al. ref 16)

3.1 Ion bombardment

Bombardment of solid surfaces by ions or energetic neutrals gives rise to a variety of both elastic and inelastic effects as is illustrated in Fig. 10. The effects which dominate in a particular case depends upon several parameters such as ion energy E_i, mass of the ions, ion flux j_i, as well as the substrate and film materials, orientation, film growth rate R, and growth temperature T_s. In the case of ion bombardment during film growth low energies typically below 500 eV are most commonly used, and it has been demonstrated that low-energy ion irradiation has an important role in controlling nucleation and growth kinetics of films grown from the vapor phase. For example, many workers have shown that in order to obtain dense polycrystalline films at low growth temperatures T_s, ion irradiation is necessary (low T_s in this context is

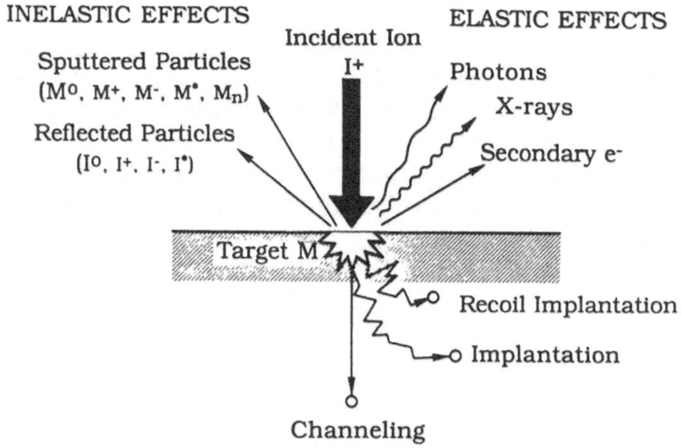

Fig. 10 Schematic diagram of ion-surface interactions.

usually $\leq 0.3\ T_m$ where T_m is the melting temperature of the deposited material). A typical example has been given by Håkansson et al.[17] for $Ti_{0.5}Al_{0.5}N$ films grown by reactive magnetron sputtering. By using transmission electron microscopy (TEM), they have shown that films grown on substrates kept at ground potential were columnar in nature with a high amount of pores and voids. However, as a negative substrate bias V_s was applied the number of voids decreased and for $V_s > 100$ V dense films were obtained. (The ion-to-vapor flux ratio was ≈ 0.9 in these experiments.) As V_s was increased the grain size also decreased and for $V_s \geq 150$ V the columnar growth was interrupted. The results are in good agreement with predictions by Müller[18-22] using Monte Carlo and molecular dynamic simulations.

In the following part of the paper we summarize some other recent results that demonstrate the importance of low-energy ($\leq$ 500 eV) ion bombardment for nucleation kinetics, defect structure within as-deposited films, and elemental incorporation probabilities. A detailed review of this topic can be found in ref. 23.

The nucleation kinetics were studied utilizing both thermal In and partially ionized In^+ beams (ion-to-neutral ratio = 0.35) to deposit In on amorphous Si_3N_4 substrates kept at T_s = 70 °C[24]. The depositions were carried out in an ultra-high vacuum MBE system and the partially ionized beams were provided by a metal-ion source described in detail in ref. (4). Films with different nominal thicknesses t from 1.5 to 10 nm were grown at acceleration energies E_{In} of 0 (thermal beams), 150, 200, 300 and 400 eV. TEM was then used to study average island sizes <d> and island size distributions. Fig. 11 shows typical TEM micrographs at a nominal thicknesses of 10 nm and at different E_{In} values. For a given film thickness the ion bombardment clearly results in larger average island sizes. For example, at t = 10 nm, <d> increased from 6.5 to 50 nm as E_{In} increased from 0 to 300 eV. Another striking feature is the decrease in secondary nucleation at higher energies until at 300 eV a uniform island size distribution is obtained and essentially no islands with a diameter less than ≈10 nm are observed.

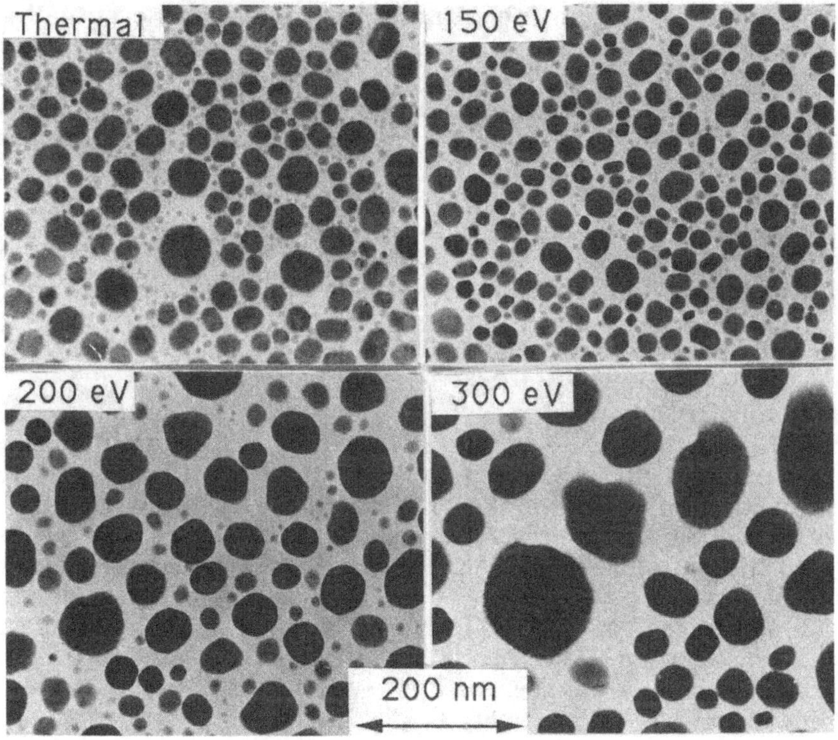

Fig. 11. TEM micrographs of In islands on Si_3N_4 substrates. The films were grown to a nominal thickness of 10 nm. The incident In beams were either thermal or partially ionized and accelerated to E_{In^+} = 150, 200 or 300 eV. (from ref. 24).

These results can be explained based on enhanced adatom diffusivity and/or depletion of small clusters by sputtering. Ion-irradiation enhanced surface diffusion can occur through: **i)** Direct coupling of energy to the adatoms via the ion-induced phonon cascade; **ii)** Forward sputtering of atoms from edges of condensed clusters; and **iii)** Ion-induced dissociation of small clusters. The latter mechanism will lead to an apparent enhancement of the adatom diffusivity since the number of adatoms which can feed large clusters will increase. Also as small clusters are reduced in size they become more mobile on the surface and can also diffuse to feed larger clusters. Moreover, clusters reduced to subcritical size will spontaneously dissociate to form single adatoms. A schematic illustration of these three mechanisms for ion enhanced surface diffusion is given in Fig. 12. Based on experiments carried out using both thermal and partially-ionized beams simultaneously, mechanisms ii) and iii) were found to be dominating in the case of In^+ growth on Si_3N_4 substrates.

Another example of ion irradiation enhanced adatom surface mobility has been shown in the case of epitaxial TiN(100) films grown on MgO substrates by reactive magnetron sputtering in pure N_2 discharges[25]. In this case TEM analyses were used to determine the dislocation density n_d in thick (a few μm) epitaxial films grown at different negative substrate bias voltages V_s. For a given T_s, n_d decreased rapidly with increasing V_s until a minimum was reached at a specific bias voltage V_s^*. For $V_s > V_s^*$, n_d then increased rapidly and eventually the films became polycrystalline. V_s^* was found to increase with increasing T_s and increased from ≈ 150 to ≈ 400 V as T_s increased from 550 to 850 °C. A typical example of the variation in n_d with V_s is shown in Fig. 13 for films grown at T_s = 650 °C. Thus, for low voltages ($V_s < V_s^*$) the effect of the ion bombardment was defect annihilation due to enhanced adatom diffusivities during growth. For bias voltages larger than V_s^* the increased projected range of the impinging ions resulted in a larger fraction of the irradiation-induced defects being trapped in the growing film. The residual defect density in the films was thus a function of both V_s and T_s.

In the case of the epitaxial TiN films an increased N/Ti ratio, from 1.0 to 1.2 was also observed[25-27] for bias voltages higher than ≈ V_s^*. Implantation of the impinging N_2^+ ions, to depths large enough that N was not desorbed before becoming buried by subsequent deposited material, resulted in a trapping of the impinging ions and to a supersaturation of N in the TiN lattice. However, even though the mobility of the trapped excess nitrogen in the TiN lattice was not large enough to allow it to desorb before being buried and trapped it was sufficiently large for $550 \leq T_s \leq 800$ °C to allow precipitation and formation of nitrogen gas bubbles in films. The most stable shape of these bubbles were observed to be a cube bounded by (100) planes with some (110) facets and to a smaller extent (111) facets. This indicates an anisotropy in the TiN surface energies γ with $\gamma^{100} < \gamma^{110} < \gamma^{111}$.

Also Moire´ fringes were observed in some bubbles and their spacings indicated that solid nitrogen was present in the face-centered-cubic phase in the bubles.

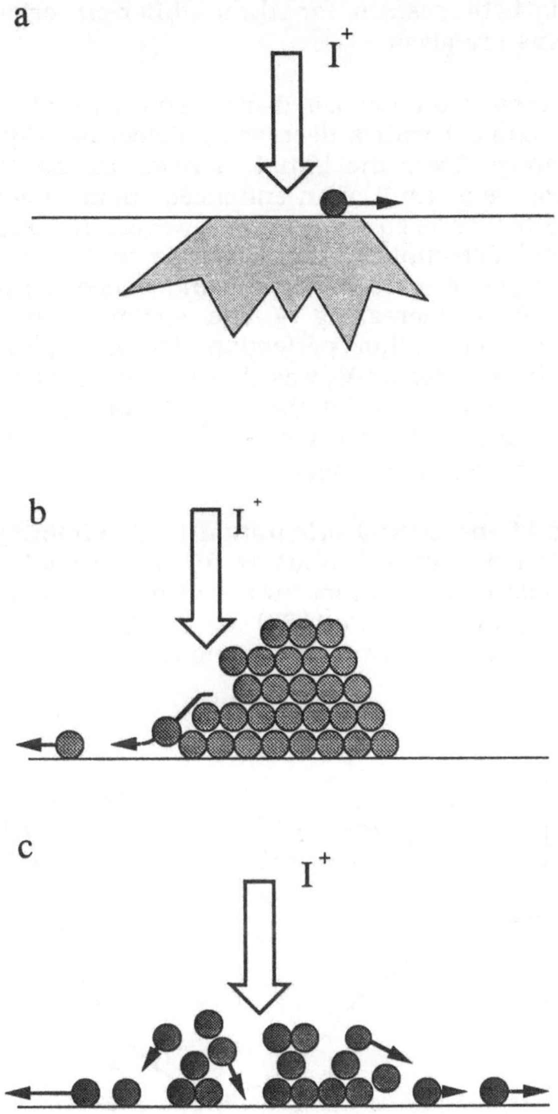

Fig. 12. Schematic illustration of three possible mechanisms for ion enhanced surface diffusion. a). Direct coupling of energy to single adatoms through the ion generated phonon cascades. b). Forward sputtering from edges of three dimensional islands. c). Ion induced dissociation of small clusters.

In order to further investigate trapping of the impinging ions, experiments were also carried out in mixed Ar/N_2 discharges[27]. The Ar concentration in the films, measured using energy-dispersive x-ray analyses, is shown in Fig. 14a and 14b both as a function of T_s at a constant V_s - value and as a function of V_s at a constant T_s - value. In the latter case (Fig.14b) results for three different orientations of the epitaxial TiN films are given.

TEM analyses of the deposited films showed that films grown at $T_s \geq 550$ °C were epitaxial with a decreasing defect density with increasing growth temperature. Thus the initial increase in the Ar concentration, shown in Fig. 14a, is a result of an enhanced channeling of the Ar-ions in increasingly better ordered crystals. However, for higher T_s- values thermal activated detrapping processes due to Ar surface segregation and desorption became more important and hence the Ar concentration decreased with further increasing T_s. The maximum observed in Fig. 14b was also related to crystalline perfection. The initial increase in the Ar concentration with increasing V_s was due to additional trapping resulting from longer projected ranges of the ions. However, at $V_s \geq 400$ V the production of ion-induced defects decreased the crystalline order in the films and thus the projected ranges.

The importance of the crystal orientation in determining the entrapped gas concentration in epitaxial films is further exemplified in Fig. 14b. The increased size of the surface unit cell of the B1-NaCl structure of TiN as we go from (111) to (100) to (110) oriented films is also manifested as an increased Ar concentration.

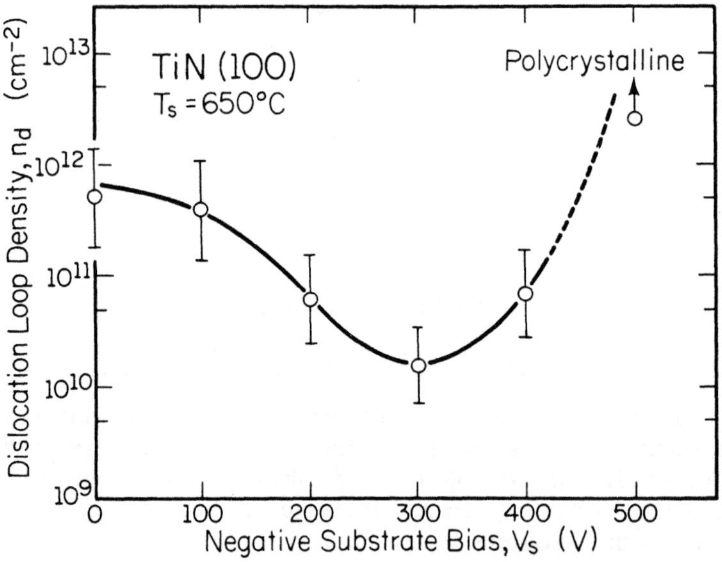

Figure 13. Dislocation loop density n_d in TiN (100) films as a function of negative substrate bias V_s during reactive sputter deposition on MgO (100) substrates held at T_s = 650 °C. The films grown at V_s = 0 - 400 V were epitaxial while those grown at 500 V were polycrystalline. (from Hultman et al. ref. 25)

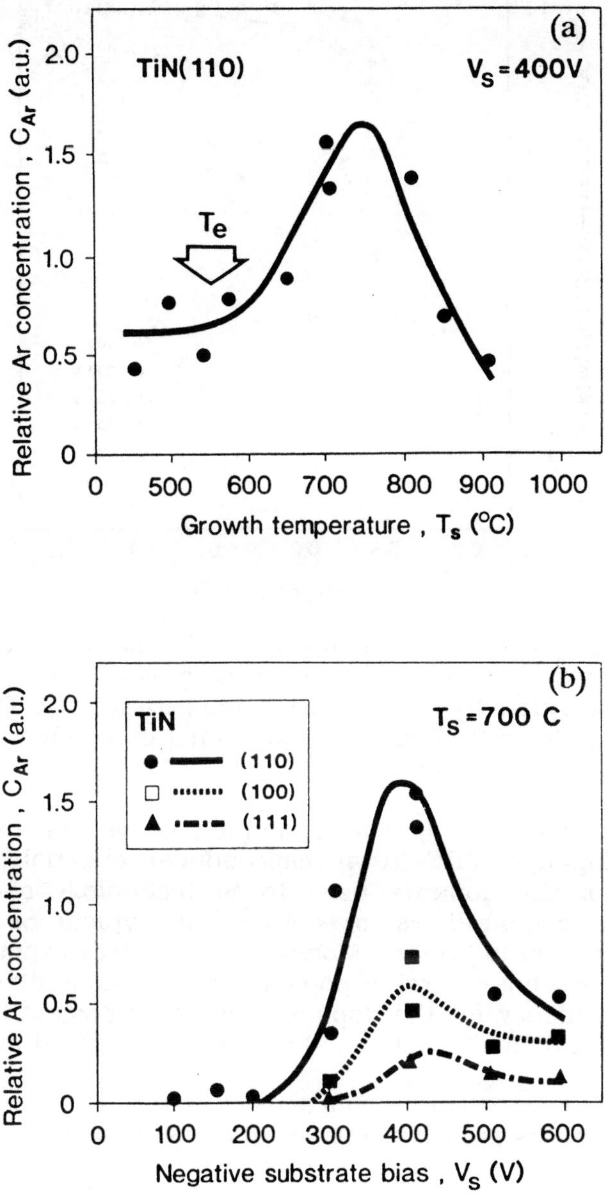

Fig. 14. a) Trapped Ar Concentration C_{Ar} versus deposition temperature T_s for TiN films deposited on (110) oriented MgO by reactive sputtering in mixed Ar/N_2 atmospheres with an applied negative substrate bias V_s = 400 V. b) C_{Ar} versus V_s for TiN films deposited on (110), (100), and (111) oriented MgO held at T_s = 700 °C.(from ref. 27)

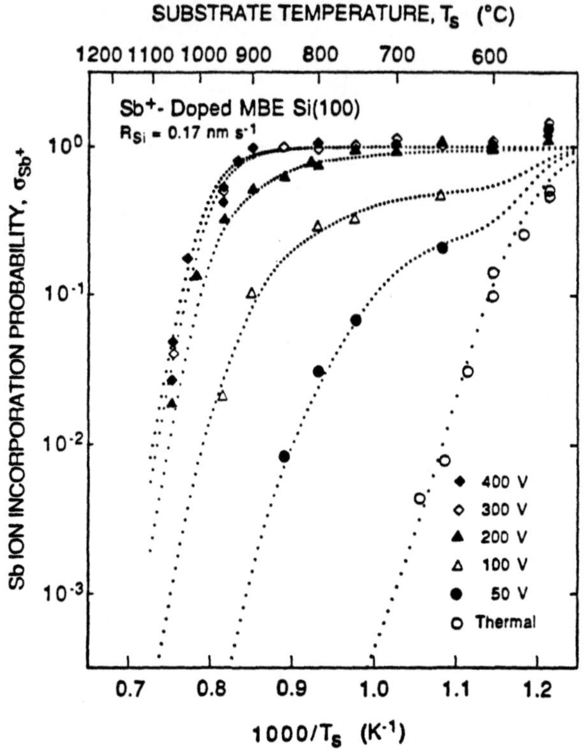

Fig. 15. Incorporation probability σ for thermal and accelerated Sb-ion beam doping during Si(100) MBE growth as a function of acceleration potential V_{Sb^+} and growth temperature T_s. The films were grown at a rate R_{Si} of 0.17 nm s^{-1}. The dotted curve show fitted results using a model described in ref. 29.

An increased entrappment of impinging ions is also frequently used during doping of MBE-grown semiconductor materials, in particular Si[28]. Many of the dopants used in Si technology show very low incorporation probabilities σ ($< 10^{-3}$) at typical Si-MBE growth temperatures of 600 to 800 °C. Moreover, σ decreases exponentially with increasing T_s making control of dopant concentrations difficult. Also a pronounced tendency for the dopant atoms to segregate to the growth surface makes it difficult to obtain sharp and well defined profiles.

Fig. 15 shows σ-values, measured using secondary ion mass spectroscopy, versus growth temperature for Sb doping during growth of Si (100) films by MBE[29]. Results are shown for films grown using both thermal Sb and accelerated Sb^+ ion beams. For thermal Sb beams σ decreases exponentially with increasing T_s and is only $5x10^{-3}$ at T_s = 700 °C. In the case of accelerated beams σ is much higher compared to the thermal beam doping case and also shows a much slower decrease with T_s for $T_s < 900$ °C and acceleration potentials $V_{Sb^+} > 100$ V . (Since the ion source emits Sb_x clusters, x = 4, 2, and 1, with an average size of 2.2, the energy of the impinging ions can be approximated with $V_{Sb^+}/2.2$.) In fact, for $V_{Sb^+} \geq 200$ V σ is constant at 1 up to a temperature of ≈ 950 °C. At this temperature Sb bulk diffusion in Si is high enough to allow the shallow implanted Sb_x^+ ions to reach the growth surface and desorb before being buried and trapped in the film.

The dotted lines in Fig.15 were calculated from a model taking into account surface segregation and desorption, bulk diffusion, implantation and the dynamic situation of growth. The model is based on exchange processes for dopant atoms moving between potential wells corresponding to different sites in the near surface region. At least five different binding energy sites, including surface, bulk and three intermediate sites were needed in order to coherently describe both thermal and accelerated beam doping.

It should also be stressed that due to the possibility of introducing ion-induced radiation damage in the films, acceleration energies as low as possible should be used. In the case of Sb^+- ion beam doping, electrical characterization of as-deposited films grown with V_{Sb^+} = 150 V (Even for V_{Sb^+} = 150 V, σ is close to unity over a large T_s) did not show any signs of damage within the films[30].

Low-energy ion bombardment during film growth has also been used to form a variety of thermodynamically metastable phases including diamond-like carbon films[31], epitaxial $(III\text{-}V)_{1-x}(IV_2)_x$ semiconductors[32-36] and ternary nitrides such as $Ti_{0.5}Al_{0.5}N$.[17] Depending upon the material and growth technique, the primary role of ion bombardment in stabilizing the growth of these compounds ranges from ion-assisted chemistry to collisional mixing.

A variety of new single-crystal thermodynamically-metastable semiconductors have been synthesized, for example $(GaAs)_{1-x}(Ge_2)_x$[33], $(GaSb)_{1-x}(Ge_2)_x$[34] and $(GaSb)_{1-x}(Sn_2)_x$[35]. A key feature in the growth of most of these alloys is the use of self-ion or inert-ion bombardment of the growing film. The ion bombardment provides a continual collisional mixing of the upper one or two atomic layers during deposition. Higher ion energies, and thus larger ion penetration depths, give rise to creation of defects in the bulk of the growing film and to effects such as enhanced bulk diffusion in turn leading to phase separation.

A typical example of the importance of ion bombardment is given in Fig. 16 which shows x-ray diffraction (XRD) spectra of $(GaSb)_{0.92}(Sn_2)_{0.08}$ films grown by sputtering at T_s = 300 °C and at two different negative substrate bias voltages 25 and 75 respectively.[35] Whereas the film grown with V_s = 25 V shows peaks from both the zinc-blend GaSb phase and the tetragonal β-Sn phase the film grown with V_s = 75 V shows only a (220) peak located at slightly lower 2θ values than the (220) peak from pure GaSb. These results show that while the films grown with V_s = 25 V are two-phase, those grown with V_s = 75 V are highly oriented single phase films of a metastable $(GaSb)_{0.92}(Sn_2)_{0.08}$ alloy. Thermodynamically, GaSb and Sn exhibit essentially no mutual solubility. Crystalline films grown with V_s = 175 V were both depleted in Sn and showed two-phase XRD spectra. The phase separation caused by the ion enhanced bulk diffusion at these energies also resulted in a preferential sputtering of Sn from the incipient Sn-rich precipitates.

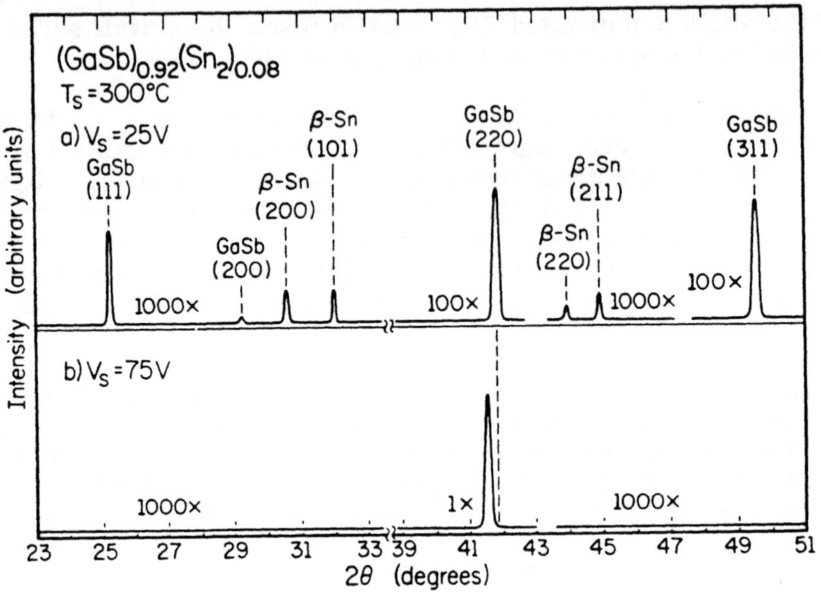

Fig. 16. X-ray diffraction patterns from $(GaSb)_{0.92}(Sn_2)_{0.08}$ films grown on glass substrates with negative substrate biases of a) V_s = 25 V and b) V_s = 75 V. The films in (a) are two-phase while those in (b) are single-phase with a strong (220) prefered orientation (from ref. 35).

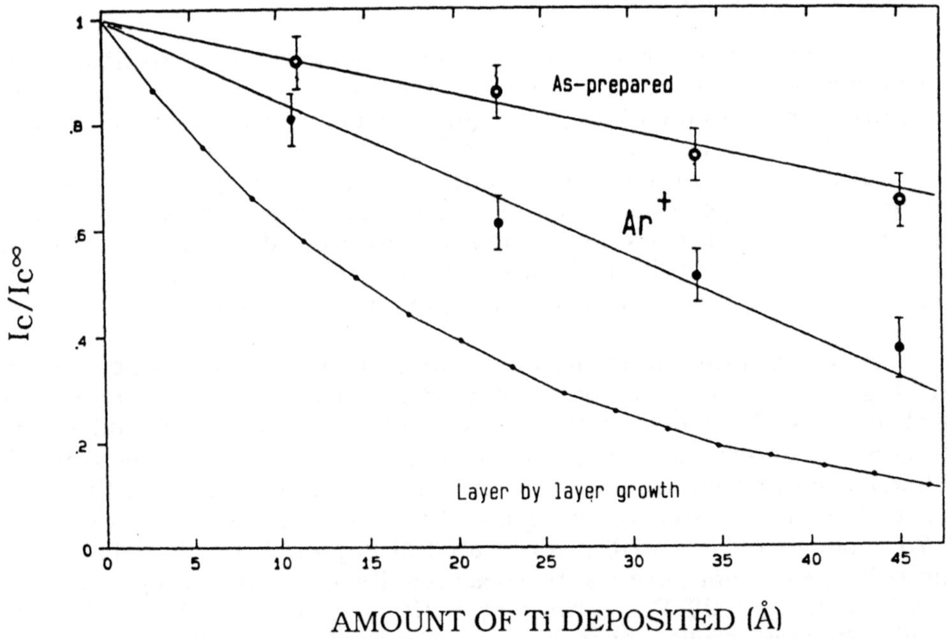

Fig. 17. Relative C1s peak intensity as a function of the amount of Ti deposited on as-prepared and Ar^+ ion bombarded surfaces of polydimethylsiloxane. Also a calculated curve for a layer-by-layer growth is shown. This curve was calculated using an electron escape depth determined from the relation given by Seah and Dench[53]. (from Bodö and Sundgren, ref 46).

3.2 Electron bombardment

Electron irradiation is the primary source of substrate heating during film deposition in which the substrates are exposed to a large flux of energetic electrons as is the case in dc and rf diode sputtering[37, 38]. Energetic electron irradiation of ionically bonded substrates has also been shown to affect nucleation kinetics through creation of charged surface vacancies which act as preferential nucleation sites[39,40]. Also reduced epitaxial temperatures have been observed for several film/substrate combinations[39, 41] as well as enhanced dopant incorporation during MBE growth[41]. However, no detailed explanation for these effects has yet been given. Electron irradiation can also give rise to several chemical effects such as electron stimulated desorption, bond rearrangement and dissociation. In the case of polymers large structural changes might also occur through cross-linking. Bodö and Sundgren[42] have observed strong structural and morphological changes in the case of electron irradiation of polyethylene substrates.

3.3 Glow Discharge Surface Cleaning and Activation

Cleaning and activation of substrates immersed in glow discharge plasmas is extensively used in many applications, particularly in the case of insulating substrates that can not be sputter etched using dc discharges. Glow discharge treatments of polymer substrates prior to metal deposition have been found to be highly effective in increasing the adhesion strength of metal deposits. X-ray photoelectron spectroscopy (XPS) studies have shown that O_2 plasma treatments of several different polymers [44-48] leads to the formation of C-O bonds. This in turn results in a higher adhesion through the formation of oxygen bridge bonds between the metal and C atoms. Also Ar^+ ion bombardment has been shown by Bodö and Sundgren[43,46,48] to improve the adhesion of various metals to both polyethylene[43], polydimethylsiloxane[46] and polyimide[48] substrates. The Ar^+ ion bombardment was found to remove low molecular weight impurities, promote cross-linking of the outermost polymer surface layer and to allow an increased number of metal-carbon bonds to form.

Also the growth mode of the metal films on the polymer substrates was changed by the ion bombardment substrate pretreatment. Fig. 17 shows results of XPS studies of the attenuation of the C1s XPS signal as a function of the amount of Ti deposited onto polydimethylsiloxane substrates[46]. Results for both as-prepared substrates and ion-bombardment pretreated substrates are given. In all cases the behaviour is typical for a pronounced three dimensional island nucleation type of growth with only a weak decay of the substrate signal with an increasing amount of metal deposited. As a comparison a calculated curve for a layer-by-layer growth is also shown in the figure. However, as the substrates are ion bombarded prior to Ti deposition, the decay of the C1s signal is more rapid compared to as-prepared substrates. This behaviour indicates that the initial growth is changed to give an island distribution consisting of smaller more densely packed islands, indicative of an increased number of nucleation sites due to an increased interaction between the Ti atoms and the substrate.

Glow discharge and ion bombardment surface cleaning can also result in damage in the outermost surface layer[49]. However, recent work on damage production and sputter-cleaning of substrate surfaces prior to epitaxial growth[50-52] suggest that the ion-irradiation-induced damage can be continously annealed out if the substrates are held at an elevated temperature during the cleaning procedure. Yu[51] has used low energy electron diffraction to show that the temperature required to maintain a Si(111)7x7 surface reconstruction during Ne ion irradiation decreased from ≈ 450 °C to 150 °C as the ion energy was decreased from 500 to 80 eV. Garverick *et al*[52] used a low-power rf Ar plasma with an accelerating potential of 100 V to clean Si(100) substrates held at 750 °C prior to chemical vapor deposition of Si. By a combination of Rutherford backscattering and TEM analyses they showed that both the films and substrates were defect free.

4. Acknowledgements

The author gratefully acknowledges the financial support, over the course of several years, from the Swedish Natural Science Research Council (NFR). The stimulating and fruitful cooperation with Prof. J. E. Greene at the Materials Science Division at University of Illinois, Urbana, USA and his co-workers has played a major role in this work. The author would also like to express his sincere thanks to his collegues Dr M.A. Hasan, Mr G. Håkansson, Mr W.-X. Ni, Dr L. Hultman, Dr P. Bodö and Dr U. Helmersson at Linköping University.

References

1. J.A. Thornton and J.E. Greene, *"Plasmas used in deposition processes"* in *Deposition Technologies for Films and Coatings* ed. R.F. Bunshah, Noyes Publ. New York, (1990) ch 2.
2. B. Chapman, *Glow Discharge Processes* , John Wiley & Sons, New York, (1980).
3. see for example, E. W. McDaniel, *Collision Phenomena in Ionized Gases*, Wiley, New York, (1964).
4. W.D. Westwood, *"Calculations of deposition rates in diode sputtering systems"*,J. Vac. Sci. Technol. **15**, 1, (1978).
5. R. S. Robinson, *"Energetic binary collisions in rare gas plasmas"*, J. Vac. Sci. Technol. **16**, 185, (1979).
6. R. E. Somekh, *"The thermalization of energetic atoms during the sputtering process"*, J. Vac. Sci. Technol. **A 2**, 1285, (1984).
7. R. E. Somekh, *"Calculations of thermalization during the sputter deposition process"*, Vacuum, **34**, 987, (1984).
8. D.W. Hoffman, *"A sputtering wind"*, J. Vac. Sci. Technol. **A 3**, 561, (1985).
9. S.M. Rossnagel, *"Gas density reduction effects in magnetrons"*,J. Vac. Sci. Technol. **A 6**, 19, (1988).

10. J.A. Thornton and A.S. Penfold, *"Cylindrical magnetron sputtering"* in *Thin Film Processes*, eds. J.L. Vossen and W. Kern, Academic Press, New York (1978) p 75.
11. J.L. Vossen and J.J. Cuomo, *"Glow discharge sputter deposition"* in *Thin Film Processes*, eds. J.L. Vossen and W. Kern, Academic Press, New York (1978) p 11.
12. J.W. Coburn and E. Kay, *"Positive ion bombardment of substrates in rf-diode glow-discharge sputtering"*, J. Appl. Phys. **43**, 4965, (1972).
13. W.D. Davis and T. A., Vanderslice, *"Ion energies at the cathode of a glow discharge"*, Phys Rev. **131**, 219 (1963).
14. F.F. Chen, *Introduction to plasma physics*, Plenum Press, New York (1974).
15. B. Window and N. Savides, *"Charged particle fluxes from planar magnetron sources"*, J. Vac. Sci. Technol. **A 4**, 196, (1986).
16. S. Kadlec, J. Musil, W.D. Münz, G. Håkansson and J.-E. Sundgren, Surf. Coat. Technol. *"Reactive deposition of TiN films using an unbalanced magnetron"*, **39/40**, 487, 1989.
17. G. Håkansson, J.-E. Sundgren, D. McIntyre, J.E. Greene and W.D. Munz, *"Microstructure and physical properties of polycrystalline metastable $Ti_{0.5}Al_{0.5}N$ alloys grown by magnetron sputter deposition"*, Thin Solid Films, **153**, 55 (1987).
18. K.-H. Müller, *"Ion-beam-induced epitaxial vapor phase growth: A molwcular-dynamics study"*, Phys. Rev. B **35**, 7906, (1987).
19. K.-H. Müller, *"Role of incident kinetic energy of adatoms in thin film growth"*. Surf. Sci. Lett, **184**, L357, (1987).
20. K.-H. Müller, *"Model for ion assisted thin-film densification"*, J. Appl. Phys. **59**, 2803, (1986).
21. K.-H. Müller, *"Modelling of ion-assisted deposition of CeO_2 films"*, Appl. Phys. **A40**, 209, (1986).
22. K.-H. Müller, *"Dependence of thin-film microstructure on deposition rate by means of a computer simulation"*, J. Appl. Phys. 58, 2573, (1986).
23. see for example, J.E. Greene, S.A. Barnett, J.-E. Sundgren and A. Rockett, *"Low-energy ion/surface interactions during film growth from the vapor phase"*, in *Ion-Beam Assisted Film Growth*, Ed. T.Itoh, Elseviere, New York, (1988) Ch 5.
24. M.A. Hasan, S.A.Barnett, J.-E. Sundgren and J.E. Greene, *"Nucleation and initial growth of In on Si_3N_4 using low-energy (<300 eV) accelerated beams in ultra high vacuum"*, J. Vac. Sci. Technol. **A5**, 1883 (1987).
25. L. Hultman, S.A. Barnett, J.-E. Sundgren and J.E. Greene, *"Growth of TiN(100) films by reactive magnetron sputtering: Effects of low-energy ion bombardment"*, J. Crystal Growth **92**, 639 (1988).
26. L. Hultman, B.O. Johansson, J.-E. Sundgren, L.C. Markert and J.E. Greene, *"Ar incorporation in epitaxial TiN films deposited by reactive magnetron sputtering"*, Appl. Phys. Lett. **53**, 1175 (1988).
27. L. Hultman, J.-E. Sundgren and J.E. Greene, *"Formation of N_2 polyhedral gas bubbles in reactively sputtered TiN"*, J. Appl. Phys. **66**, 536, (1989).
28. see for example, J.-E. Sundgren, J. Knall, W.-X. Ni, M.A. Hasan, L.C. Markert and J.E. Greene, *"Dopant kinetics and abrupt transitions during Si-MBE"*, Thin Solid Films, **183**, 281, (1989).

29. W.-X. Ni, J. Knall, M.A. Hasan, G.V. Hansson, J.-E. Sundgren, S.A. Barnett, L.C. Markert and J.E. Greene, *"Low-energy ion beam doping of Sb during growth of Si MBE"*, Phys. Rev. B **40**, 10449, (1989).
30. P. Fons, N. Hirashita, L.C. Markert, Y.-W. Kim, J.E. Greene, W.-X. Ni, J. Knall, G.V. Hansson and J.-E. Sundgren, *"Electrical properties of Si(100) films doped with low-energy ions during growth by MBE"*, Appl. Phys. Lett. **53**, 1732, (1988).
31. H. Tsai and D. B. Bogy, *"Characterization of diamond-like carbon films and their application as over-coats in thin film media for magnetic recording"*, J. Vac. Sci. Technol. **A 5**, 3287, (1987).
32. J.E. Greene, *"Crystal growth, atomic ordering and physical properties of epitaxial metastable semiconductors"*, J. Vac. Sci. Technol. **A 5**, 1947, (1987).
33. L.T. Romano, J.-E. Sundgren, S.A. Barnett and J.E. Greene, *"Domain structure in epitaxial metastable zincblend $(GaAs)_{1-x}(ge_2)_x$ (100) alloys grown by sputter deposition"*,Phys Rev. **B36**, 7523, (1987).
34. K.C. Cadien, A.H. Eltoukhy and J.E. Greene, Appl Phys. Lett **38,** 773, (1981).
35. L.T. Romano, J.-E. Sundgren, S.A. Barnett and J.E. Greene, Superlattices and Microstructures, *"Metastable $GaSb_{1-x}(Sn_2)_x$ alloys; Crystal growth and phase stability of single crystal and polycrystal layers"* **2**, 233, (1986).
36. S.I. Shah, B.Kramer, S.A. Barnett and J.E. Greene, *"Direct evidence for an order/disorder phase transition at $x \approx 0.3$ in single crystal metastable $(GaSb)_{1-x}(Ge_2)_x$ alloys"*, J. Appl. Phys. **59**, 1482, (1986).
37. D.J. Ball, *"Plasma diagnostics and energy transport of a dc discharge used for sputtering"*, J. Appl. Phys. **43**, 3047, (1972).
38. S.S. Lau, R.H. Mills and D.G. Muth, *"Temperature rise during film deposition by rf and dc sputtering"*, J. Vac. Sci. Technol. **9**, 1196, (1972).
39. M.R. Jordan and D.J. Stirland, *"Changes in epitaxy produced by electron bombardment"*, Thin Solid Films, **8,** 221, (1971).
40. D.G. Lord and M. Prutton, *"Electrons and epitaxial-growth of metals on alkali-hallides"*, Thin Solid Films, **21**, 341, (1974).
41. G. Shimaoka, *"Modifications of epitaxy by electric charge effects"*, Cryst. Growth, **31**, 92, (1975).
42. F.A. d'Avitaya, S. Delage, and F. Rosencher, *"Si-MBE; Recent developments"*, Surf. Sci. **168**, 483, (1986).
43. P. Bodö and J.-E. Sundgren, *"Adhesion of evaporated Ti films to ion bombarded polyethylene"*, J Appl. Phys **60**, 1161, (1986).
44. P.S. Ho, P.S. Hahn, J.W. Bartha, G.W. Rubloff, F.K. LeGoues and B. Silverman, *"Chemical bonding and reactions at metal/polymer interfaces"*, J. Vac. Sci. Technol. **A 3**, 739, (1985).
45. J.M. Burkstrand, *"Formation of metal-oxygen-polymer complexes of polystyrene with nickel and chromium"*, J. Vac. Sci. Technol. **16**, 1072, (1979).
46. P. Bodö and J.-E. Sundgren, *"Titanium deposition onto plasma-treated polydimethylsiloxane: Surface modification and adhesion"*, Thin Solid Films, **136**, 147, (1986).

47. N.J. Chou and C.H. Tang, *"Interfacial reaction during metallization of cured polyimide: An XPS study"*, J.Vac. Sci. Technol. **A 4**, 751, (1984).
48. P. Bodö and J.-E. Sundgren, *"Ion bombardment and Titanium film growth on polyimide"*, J. Vac. Sci. Technol. **A 6**, 2396, (1988).
49. J.C. Bean, G.E. Becker, P.M. Petroff and T.E. Seidel, *"Residual damage on temperature during sputter cleaning of Si"*, J. Appl. Phys. **48**, 907, (1977).
50. R. Roussille, R. Boch, G.L. Destefanis and J.L. Tissot, *"32x32 planar IR photovoltaic mosaic arrays on sputtered $Cd_xHg_{1-x}Te$ epilayers"*, Appl. Phys. Lett. **44**, 679, (1984).
51. M.L. Yu, *"Thermal regrowth of Si(111) surface during ion bombardment"*, Appl. Phys. Lett. **40**, 986, (1982).
52. L.M. Gaverick, J.H. Comfort, T.R. Uyeh, R. Reif, F.A. Baiocchi, and H.S. Luftman, *"Silicon surface cleaning by low dose argon-ion bombardment for low temperature (750 °C) epitaxial deposition; II Epitaxial quality"*, J. Appl. Phys. **62**, 3398, (1987).
53. M.P. Seah and W. A. Dench, *"Quantitative photoelectron spectroscopy of surfaces"*, Surf. Interface Anal. **1**, 2, (1979).

REACTIVE GAS GLOW DISCHARGES

J.W. Coburn

IBM Almaden Research Center
650 Harry Road
San Jose, CA 95120-6099, USA

INTRODUCTION

In the last twenty years or so there has been an enormous growth in the use of reactive gas glow discharges for the processing of surfaces. Surface processing in this context can be divided into three major categories: deposition, etching and surface modification. Deposition and etching are terms which are well understood by most technically trained people but surface modification is a less familiar expression. The term surface modification is used to describe those processes which alter the physical and/or chemical characteristics of a pre-existing solid surface and in which this alteration is limited to a region relatively near the surface of the solid. Examples of surface modification processes are plasma oxidation, plasma nitriding and surface texturing. Some areas of technology in which reactive gas glow discharges play a major role are microelectronics, magnetic and optical recording technologies, photovoltaics, architectural glass, and machine tool fabrication.

What are the reasons for this rapidly growing interest in plasma processing of surfaces in the face of the relatively large capital costs associated with plasma processing equipment? Clearly there are surface processing tasks which can be performed readily in plasmas but which can not be easily accomplished in other environments. In recent years plasma processing of surfaces has been used in some instances to avoid the chemical disposal problems associated with wet processing. The effluent from plasma reactors which use reactive gases is often toxic and/or corrosive and must be carefully processed but the much smaller quantities involved relative to wet processing make this problem much less severe in plasma processing.

What are the characteristics of plasma processing which make it so attractive? There are two characteristics of plasmas which dominate in the this application area. One is the ability of certain kinds of plasmas to deliver a uniform flux of energetic positive ions to a surface with a controllable energy and flux. The second characteristic of some plasmas, which is equally valuable, is the ability to deliver spatially uniform, relatively large fluxes of atoms and or molecular radicals to surfaces. In both cases the surfaces can be near room temperature (external temperature control can be provided if required) and electrically insulating surfaces can be processed with no difficulty. In some applications, only the energetic ion bombardment is required, in others only the radical fluxes are important; but in more and more cases it is the combination of energetic ions and reactive radicals or atoms which are responsible for the unique surface processing capabilities of certain reactive gas glow discharges. When fluxes of both reactive radicals and energetic ions are required, it is difficult to find an approach which is superior to a reactive gas glow discharge.

The method is not without its problems. The large capital equipment costs mentioned previously are a consequence of this being a vacuum (or more accurately, a reduced pressure) process. In addition there is a general lack of understanding concerning the details of many of the critical physical and/or chemical phenomena which determine the outcome of the plasma-surface interactions. This lack of detailed knowledge when coupled with the large, easily accessible parameter space associated with most plasma processing equipment, has led to great difficulties in the development and optimization of plasma-based manufacturing processes. Plasma process development procedures are largely empirical, are usually very time-consuming and costly and are rarely optimized. Hopefully this situation will improve as a result of much research activity in the area of plasma-surface interactions, plasma diagnostics and plasma modelling.

In the following discussion, the emphasis will be on describing some of the more important plasma diagnostic methods and reviewing some of the interesting phenomena occurring at surfaces in reactive gas glow discharges. No attempt will be made to survey the large parameter space associated with plasma apparatus nor will any discussion of plasma modelling be included.

PLASMA DIAGNOSTICS

1. Optical Emission Spectroscopy (OES)

Optical emission spectroscopy, the spectral analysis of the light emanating from a glow discharge, is probably the most widely used method for monitoring and diagnosing plasma processes, including plasmas of interest in diamond film synthesis.[1,2] The major reason for the popularity of OES is its non-intrusive nature and its relative ease of implementation. All that is needed is a monochromator and optical access to the glow discharge, either via a window in the vacuum wall of the reactor or by means of an optical fibre. In some specific applications a narrow bandpass filter can be used in place of the monochromator. In addition to the ease of implementation and the non-intrusive nature of OES, the method has good sensitivity, information can be extracted with spatial and temporal resolution from various regions of the glow discharge and both neutral and ionic species contribute to the observed signals.

There are several factors which complicate the use of OES in plasma diagnostics not the least of which is the extreme complexity of the spectra. Although optical emission from glow discharges spans a very wide range of frequencies from the infrared out to the soft X-ray range, by far the most widely used part of the spectra in OES is the wavelength range between about 200 and 900 nm (6.2 and 1.4 eV respectively). This range of the spectrum contains many of the electronic transitions of greatest interest in plasma processing. The transmission characteristics of the optical viewport in the vacuum system, or of any other optical components used, must be considered in that glass cuts off radiation with wavelengths shorter than about 300 nm (4.1 eV) whereas fused quartz and sapphire transmit down to at least 200 nm (6.2 eV). Shorter wavelengths require fragile window materials such as lithium fluoride which can transmit down to about 110 nm (11 eV). However absorption of the radiation by the air complicates the study of spectral emission with wavelengths shorter than about 180 nm (7 eV). OES studies of the infrared region of the spectra (wavelengths longer than 900 nm) are troubled by typically weak emissions, interference with black body radiation and low sensitivity detectors. Consequently relatively little work has been carried out in this area.

Emission from atomic and diatomic species provides the most useful information in plasma processing applications primarily because of the simpler, narrower spectra. Atomic emission lines, free from the complexity of vibrational and rotational transitions, are very sharp, often quite intense and usually easily identifiable. Also, atomic species are often of greatest interest in plasma processing. Typically the larger polyatomic species either do not radiate significantly in the spectral range used for OES or display very broad spectra which are difficult to identify conclusively and are not very useful for plasma diagnostics or process control. This great species selectivity is a disadvantage of OES if one wishes to fully characterize a plasma environment, but for the vast majority of applications in plasma processing this is not a serious limitation.

The most common way in which OES is used in the diagnostics of processing plasmas is to monitor the emission from atom X, ion X^+ or molecule XY as a function either of time or as a function of an externally adjustable plasma parameter (eg. pressure, excitation power, etc.). This emission data is then used to determine the changes in the concentration of the species being observed. In order that this can be done, it is essential that the emission be a result of the electron impact excitation of the the ground state of the species:

$$\text{e(electron)} + \text{X(ground state)} \longrightarrow \text{X*(excited state)} + \text{e}$$

$$X^* \longrightarrow X^{**} + h\nu$$

where X^{**} is either the ground state or a second excited state with lower energy than X^*. If the same excited state X^* is created by a dissociative process:

$$e + XYZ \longrightarrow X^* + YZ + e \longrightarrow X^{**} + h\nu + YZ + e$$

then the interpretation of the emission becomes very difficult and a direct correlation between the emission intensity and the density of species X cannot be readily made. This ambiguity concerning the processes responsible for creating the electronically excited states responsible for the optical emission is a major weakness of OES in plasma diagnostics. It is often difficult to determine the extent of this problem. Furthermore, even if it is known that more than one source of a specific emission exists, it is equally difficult to interpret the resulting emission signal with confidence.

A second important limitation of OES is the dependence of the optical emission intensity on the ability of the plasma to excite the ground state species X into the electronically excited emitting state X*. If it is assumed that the optical emission is caused entirely by the electronic excitation of the ground state species X, then the intensity of the optical emission will depend on both the density of species X and the excitation efficiency of the plasma. If no significant changes are being made to the plasma parameters (ie. as encountered during a study the emission signal as a function of time only), then the plasma excitation efficiency probably can be assumed to be constant. However if any changes in power, pressure, gas composition, etc. are made, this assumption is no longer valid and the intensity of the optical emission signal cannot be assumed to be proportional to the density of the ground state species X.

Fortunately, unlike the situation with the undefined emission discussed previously, there is a relatively simple procedure for removing the influence of the plasma excitation efficiency from the OES data. The approach is known as actinometry[3,4] and involves injecting a small quantity (1-2%) of an inert gas (eg. argon) with the reactive plasma feedgas. The concentration of this actinometer gas is kept constant and therefore the optical emission from the actinometer is a measure of the excitation efficiency of the plasma. It is important to have the energy of the emitting species X* and the excited state of the actinometer Ar* roughly comparable so that the same group of electrons in the plasma are involved in the excitation of both X and Ar. Then the OES data can be interpreted as:

$$I(X^*)/I(Ar^*) = K\ N(X)$$

where I(X*) and I(Ar*) are the intensities of the OES signal from X* and Ar* respectively, K is an arbitrary constant and N(X) is the density of ground state species X. This actinometric method is an approximate correction only, but direct comparisons with independently measured species concentrations indicate that the approximation can be a good one.[5] It should be emphasized that actinometry does not address the problem of emission originating from a dissociative process as discussed earlier. Other possible applications of actinometry include its use as a process scaling parameter (ie. the quantity I(X*)/I(Ar*) is easily measured and could be an important scaling parameter), and the quantity I(Ar*), spatially resolved, could be used to assess the uniformity of the plasma density.

2. Laser-Induced Fluorescence (LIF)

Laser-induced fluorescence involves the use of a pulsed

laser to provide the excitation of a specific transition of a particular ground state atom, molecule or ion. A monochromator is used to monitor the resulting optical emission (fluorescence) usually at right angles to the exciting laser beam. High intensity, very short laser pulses combined with gated detection ensures that the optical emission resulting from the laser excitation is much larger than the emission from species which were excited by plasma electrons. Thus LIF overcomes the two major problems associated with OES; namely the undefined source of the emitting species and the dependence of the emission on the plasma density. In addition, whereas OES has only two-dimensional spatial resolution, LIF provides spatial resolution in three dimensions. Furthermore, LIF can be used with the discharge off, unlike OES. LIF is capable of providing quantitative measurements of the density of a large variety of species with high spatial and temporal resolution in a wide range of environments. These capabilities have led to the widespread use of LIF in clarifying the basic processes in plasmas of all types.[6,7]

The major disadvantage of LIF in routine plasma processing applications is the complexity of the apparatus required. Unlike OES which is relatively easily implemented by those unfamiliar with lasers, spectroscopy and optics in general, the proper implementation of LIF requires significant skills in these areas. Cost, space and appropriate optical access to the plasma reactor are also factors to be considered. The fact that many of the species of interest in plasma processing have only high energy excited states poses a problem for LIF in that lasers with enough photon energy to excited these species are not available. Multiphoton excitation has been used to observe O and Cl atoms[8-10] but the detection of F atoms is still a challenge.

3. Other Optical and Laser-Based Diagnostic Techniques

Whereas OES and LIF have been used most often in studying reactive gas glow discharges, there are many other laser-based and optical methods which have been used less frequently. This brief review allows only a definition of these methods and the reader is referred to more comprehensive review articles for more detailed information.[11-14]

Optical absorption spectroscopy is a very well known technique for the analysis of gases, liquids and solids but its use in the study of reactive gas plasmas has not been extensive. Problems include the need for vacuum ultraviolet light sources for many of the elements of interest, low sensitivity in the infrared region of the spectrum, spectral overlaps in multicomponent mixtures and the broad structureless character of many molecular absorption bands in the visible and ultraviolet regions of the spectrum. However useful results have been obtained using absorption spectroscopy.[15,16]

The critical importance of small particles in integrated circuit manufacturing has resulted in the use of laser light scattering in processing plasmas to measure gas phase particulates.[17,18] The observation that the processing plasmas may be the source of particles previously attributed to external wafer handling will certainly stimulate activity in this area.

Although in most situations the laser probing of a plasma does not perturb the plasma appreciably, perturbations of the plasma induced by laser radiation can be observed. The observable is a change in the electrical impedance of the plasma and the method which uses this approach is known as laser optogalvanic spectroscopy. The interpretation of the data obtained with this technique is not trivial and its application to processing plasmas has not been extensive. The method is well suited to the detection of negative ions.[19]

Laser Raman spectroscopy has also been used to probe reactive gas plasmas usually using ultraviolet laser radiation to obtain higher sensitivities.[20] This method is quantifiable and can detect most species but suffers from a low sensitivity, interferences from the exciting laser radiation and the high laser powers needed can perturb the plasma.

Whereas the above methods have involved probing the gas phase species, lasers have also been used for in-situ characterization of the surfaces being processed. This, of course, is an extremely important capability in that the short mean free path and the highly reactive environment prevailing in reactive gas glow discharges preclude in-situ surface studies with electron- or ion-based surface analytical methods. Laser interferometry has been extensively used to measure the thickness of transparent overlayer films as a function of time.[21,22] Laser-induced photoemission[23,24] and laser-induced photoluminescence[25] are two approaches which have been recently applied to probing surfaces during plasma processing. Ellipsometry[26] and second harmonic generation[27] are other approaches to characterizing surfaces which are compatible with a plasma processing environment.

4. Mass Spectrometry

Mass spectrometric sampling of plasmas has been used by many people with a wide range of objectives. Mass spectrometers can be used in several ways to study plasmas: (1) Direct sampling of positive ions formed in the plasma, (2) Line-of-sight sampling of the neutral species from the plasma using modulated molecular beam methods, and (3) Analysis of the plasma gas without a line-of-sight, usually through a leak valve. All of these approaches require an auxiliary vacuum pumping system in the chamber containing the mass spectrometer in order to provide the low pressure needed for collisionless transit of the extracted species through the mass spectrometer (10^{-3} Pa is adequate). In addition, the first two approaches require an intimate connection to the plasma (ie. line-of-sight) which is usually incompatible with vendor-supplied plasma reactors. Thus, the first two methods have been limited primarily to research studies whereas non-line-of-sight gas sampling has been used frequently in development and manufacturing environments.

Mass spectrometric ion sampling from plasmas has been carried out extensively for many kinds of plasmas including those relevant to diamond film synthesis.[28-31] Much of the interest in mass spectrometric ion sampling is in the area of the analysis of solids, and glow discharge mass spectrometry[32] is now a commercially available analytical method.[33] A second major motivation is the acquisition of mechanistic insight about plasma processes; the reason for such studies in diamond film synthesis plasmas. It is certainly the case that large amounts of complicated mass spectra can be acquired quickly by mass spectrometric ion sampling of reactive gas glow discharges. It is also the case that there are many artifacts and complications which must be recognized before meaningful interpretations of these measurements can be made.[34,35]

One aspect of the mass spectrometric sampling of ions which must be understood concerns the energy of the ions. In most studies of this type, quadrupole mass spectrometers are used. These instruments operate optimally with ions in the energy range of about 5 to 25 eV. If the ion sampling is carried out through a grounded wall and the plasma potential is less than 25 eV, then the quadrupole mass spectrometer can be operated at ground potential. Even in this simplest case, there may be ion-neutral collisions or ion formation processes in the sheath between the bulk plasma and the sampling aperture which will result in ions with a range of energies arriving at the mass spectrometer. These collision or ion formation processes can distort the relative abundances of the ions observed relative to the true abundance present in the bulk plasma since cross-sections for various ion-molecule reactions vary wildly depending on the species involved. The point can be made that, in mechanistic studies of processing plasmas, the ions formed in the sheath should be included in that they do arrive at the surface being processed. If the ions are being sampled through a negatively biased electrode, or if the plasma potential is large, then it becomes almost essential to incorporate energy filtering of the ions prior to the mass spectrometer. This author believes the approach which minimizes the problems of data interpretation involves using a deflection energy filter and a mass spectrometer with a variable dc potential (including all orifices and lenses seen by the extracted ion beam).[34] It should be noted that if ions are sampled through a negatively biased electrode, the potential of the mass spectrometer must at or slightly below the potential of the sampling orifice in order that ions formed near the orifice can be observed. The points discussed here are obvious consequences of the electrostatics of the ion sampling environment.

There are several factors which determine the size of the sampling orifice besides the need to provide a differential pumping conductance compatible with the upstream pressure and the downstream pump speed. The diameter of the orifice should be less than the Debye length of the plasma[36] so that the presence of the orifice does not perturb the plasma, the orifice should be approximately "zero length" to eliminate the effect of collisions with the orifice wall and it should be small enough in diameter to eliminate collisions between the extracted ions and the plume of gas passing through the orifice with the ions.[37]

The above discussion has applied primarily to positive ion extraction issues. Negative ion extraction is much more difficult to accomplish in capacitively coupled rf plasmas because of the fact that the plasma potential usually is some tens (sometimes hundreds) of volts positive with respect to the grounded walls of the plasma reactor. Thus negative ions created in the bulk plasma are not able to reach a grounded sampling orifice. One cannot apply a positive voltage to the sampling orifice in rf plasmas as this will grossly perturb the plasma in most instances- large electron currents are drawn to the orifice probably generating negative ions which are not present in the unperturbed plasma[38] and the plasma potential will usually increase in response to the loss of electrons to the positively biased orifice.[39] Negative ions should be more easily observed at the anode of a dc plasma because in this situation the plasma potential is very close to the anode potential (ie. no retarding field for electrons is required at the anode of a dc plasma). One approach to observing negative ions in capacitively coupled rf plasmas is to use a pulsed discharge, sampling the negative ions during the off cycle of the of the applied voltage.[40] One might question the value of negative ion sampling in mechanistic studies of rf processing plasmas since negative ions cannot reach the surface being processed in most situations because of the same electrostatic considerations which complicate the negative ion sampling process.

Once the experimental issues discussed above have been resolved, the measured positive ion fluxes should be very representative of the ion fluxes arriving at a surface exposed to the plasma. However, these measured ion fluxes will often bear no resemblance to the fluxes of neutral species arriving at surfaces in the plasmas. This is immediately obvious in many instances in which the ions are formed by ion-molecule reactions. For example, CH_5^+ is a very major ionic species in a methane plasma; ArH^+ is a very major ion in an argon plasma when trace amounts of water, hydrogen, etc. are present. The opposite is is also true; that is one cannot deduce relative ion concentrations from considerations of the concentration of various species in the feed gas. For example, the positive ions in a 5% CCl_4/95% He mixture will consist almost entirely of ions derived from CCl_4 and very few He^+ ions will be present. The preponderance of low energy electrons in plasmas strongly favors ionization of species with low ionization potentials.

In the sampling of neutral species from plasmas, the problems associated with electrostatics no longer arise and particle energies are not as large an issue, although certain collisional processes in the sheath region can produce energetic neutral particles. Instead, one now must cope with the interpretational complexity introduced by dissociative ionization. This, of course, is a pervasive problem in all forms of analytical mass spectrometry. The issue is simply that most molecules undergo dissociation upon ionization and often the ion representing the parent molecule does not exist. For example, the ion CH_3^+ observed upon ionization of the neutral species emanating from a methane plasma could be indicative of the presence of methyl radicals, methane, ethane, propane, etc. Various approaches to resolving this issue have been used:
1. Introduce a time-of-flight measurement capability by chopping the neutral beam downstream of the sampling orifice and measuring the flight time from the chopper to the ionizer in the mass spectrometer. Thus if all the paricles have the same energy, the low mass ones will arrive at the ionizer before the higher mass particles and one can therefore identify the parent

molecules. This approach is very complicated and assumes a knowledge of the energies of the neutral species.
2. One can take advantage of the fact that in certain situations the ionization potential of the species of interest might be significantly lower than the ionization potential of the interfering species. One then uses ionizing electrons with energy above the ionization potential of the species of interest but below the ionization potentials of the interfering species. This approach has been used recently to identify methylene and methyl radicals extracted from a methane plasma.[41] The concerns with this approach are inherently low sensivity as a result of the very low energy electrons used and the possibility that vibrational excitation of the effusing interfering molecules may lower their ionization potentials.[42]
3. Finally one can separately measure the fragmentation spectra of all the species in the unknown gas and deconvolute the observed spectra using computer codes written for this purpose. This approach is not reliable if there are large differences in the concentrations of the various species and one cannot readily measure the fragmentation spectra of radicals which are often of greatest interest in plasma mass spectrometry.

The advantages of using a line-of-sight sampling of the neutral species instead of the much easier leak valve approach are: (a) Very reactive and/or condensible species can be detected with the line-of-sight approach whereas these species condense on or react with the walls enroute to or following their passage through the leak valve into the mass spectrometer, and (b) The partial pressure of species measured with the leak valve approach will depend on the pumping speed for the species in the mass spectrometer region whereas with a line-of-sight method, modulated beam approaches can be used to detect only those species coming directly from the sampling orifice.

It should be emphasized that having a mass spectrometer associated with a plasma reactor allows one to accomplish more mundane but important tasks such as leak detection, feed gas purity monitoring and feed gas mixture control. Often mass spectrometry is the easiest way to ascertain the degree to which the feed gas is dissociated by the plasma; an important quantity to be aware of in many plasma processing applications.

PLASMA-SURFACE INTERACTIONS

As was mentioned in the introduction, reactive gas glow discharges are used for etching, surface modification and deposition. In an etching application, the feed gas is chosen so that atoms or radicals formed by the dissociation of the plasma gas react with the surface to be processed to form VOLATILE products. The products then spontaneously evaporate from the surface and are pumped away. In a surface modification application such as oxidation or nitriding, the feed gas is chosen so that the atoms or radicals formed by dissociation of the plasma gas react with the surface to form an INVOLATILE product which remains on the surface and changes the surface characteristics of the processed material. In a deposition application such as plasma-enhanced chemical vapor deposition (PECVD) or plasma polymerization, the feed gas contains the constituents of the film to be deposited and the dissociation of the plasma gas creates radicals which bond to the surface and form a thin film.

In all these applications, energetic ion bombardment plays an important role although in many instances this role has not been fully explored and is not well understood. The effects of ion bombardment on thin film growth during sputter deposition will be discussed elsewhere in this proceedings and will not be included here. The discussion will be divided according to the three application areas described above as the physical and chemical phemomena have significantly different character depending on the products of the plasma-surface interaction.

1. Plasma Etching and Reactive Ion Etching (RIE)

The importance of plasma etching and, in particular, reactive ion etching in the manufacturing of VLSI circuits has led to many studies of the surface science associated with this technology. The emphasis has been placed on the silicon-fluorine and silicon-chlorine chemical systems. As a result, our understanding of the surface science aspects of the etching of Si with fluorine or chlorine is more complete than it is for the the etching of other materials and plasma processing in general.

Most of the materials of importance in VLSI circuit manufacturing are compatible with plasma etching and RIE; that is they react with species easily formed in a plasma (usually halogen atoms) to form volatile products. Such materials include Si and its compounds, Al, W, Mo, Ti, photoresist, polyimide, GaAs, etc. Materials which are incompatible with this dry etching technology include the elements of Groups I and II in the periodic table, most of the rare earths, the noble metals Cu and Ag, and many of the technologically important transition metals including Fe, Ni, Co and Mn.

It is important to note that an etching process consists of three sequential steps:[43] (1) Adsorption or reaction of the gas phase species with the solid; (2) Formation of a product molecule, and; (3) Desorption of the product molecule from the surface. In many gas-solid systems, these three sequential steps proceed with a reasonable rate at room temperature. Examples are F atoms or H atoms reacting with Si and O atoms reacting with various organic solids. Halogen atoms and O atoms are the most widely used etchants. Hydrogen atoms are also occasionally used for etching. The methyl radical has long been known to be capable of etching some metals presumably forming metal alkyls. This process is thought to be active in the etching of various materials in CH_4/H_2 and related plasmas.[44-46] In some instances, molecules are known to be effective etchants- Cl_2 on Al is one example.[47] This author is not aware of any low temperature etching process involving the formation of metal carbonyls.

The halogenation of materials in an etching process has been compared to the oxidation of similar materials in that it is believed that the halogen atom, upon adsorption on many solids, abstracts negative charge from the solid and becomes similar to a negative ion. If a surface atom is to be fully halogenated to form a product molecule such as SiF_4, some penetration of the adsorbed halogen into the lattice must occur.

This motion is thought to be field-assisted[48] as in the Mott-Cabrera model for oxidation. The physical size of the etchant specie also determines the extent to which lattice penetration and subsequent etching occurs. For example, both H and F atoms etch Si quite effectively, Cl atoms etch Si rather slowly and it is suspected that Br atoms will not etch Si at all at room temperature unless energetic particle bombardment is simultaneously present.

Energetic (tens to hundreds of eV) positive ion bombardment of surfaces is known to play a very important role in RIE. In a typical RIE process the surface is simultaneously exposed to both energetic positive ions and thermal energy neutral radicals or atoms. The ion bombardment often accelerates the overall etching process dramatically; in some instances the increase can be as large as two orders of magnitude. The increase is much larger than that expected from the sputtering of the surface by the ions.[49] This so-called ion-assisted gas-surface chemistry or chemical sputtering has been extensively studied for both the Si-F system[50-52] and the Si-Cl system.[53-55] Ion-assisted chemical etch yields can be very large. Up to 25 or 30 Si atoms leaving a fully fluorinated Si surface (predominantly as SiF_4 molecules) per 1 keV incident ion have been observed.[56] Si surfaces which have been subjected to various doses of fluorine have been studied with synchrotron x-ray photoemission and have been found to consist of several layers of partially fluorinated Si species; SiF_3 being the dominant specie.[57,58] The energetic ions are able to accelerate the rate limiting product formation step by forming SiF_4 from this partially reacted layer. The microscopic details of this ion-assisted chemical process are not well understood but ion-induced nuclear motion is believed to be involved in that electrons at the same energy are about two orders of magnitude less efficient in this process. The mechanistic picture for the Si-Cl system is somewhat similar except that ion-induced mixing of the Cl into the Si lattice is thought to be more important.[59] Cl, because of its large size, is less mobile than F in Si.

The rate at which reactive species react with solid surfaces is accelerated by energetic ion bombardment in most situations. Exceptions are Si-H,[60] Al-Cl[51] and Cu-Cl.[51] It is very fortunate that there are relatively few exceptions because the directionality of etching which can be accomplished with RIE methods is due to this ion-assisted gas-surface chemistry. Directional or anisotropic etching is the ability to etch downward into the substrate without etching laterally under the masking layer and this capability is required to maintain dimensional control in VLSI circuit fabrication. This asymetry of etching is a result of the fact that positive ions are normally incident onto surfaces in plasma environments; arriving at the bottom surface of an etched feature but not the sidewalls. Thus the etch rate of the bottom surface is enhanced over the sidewall etch rate and anisotropic etching is observed.

Whereas the ion-assisted gas-surface chemistry is the most important and interesting of the surface phenomena observed in the RIE environment, there are many other surface processes taking place. In the deposition of diamond films, hydrogen atom or oxygen atom etching processes are believed to be important in inhibiting the growth of graphitic carbon.[2,61,62] Conversely, carbon deposition during etching influences both the selectivity and the directionality of specific etching processes. The way in which the etching directionality is affected by the arrival of carbon-containing species follows from the above discussion of ion-assisted gas-surface chemistry. The etch rate of carbon-containing species is strongly increased by energetic ion bombardment and as the stoichiometry of the feed gas is made halogen deficient, a condition can be reached in which carbon-containing species, arriving from the gas phase, form a protecting thin film on the sidewalls of the etched feature stopping the lateral etching completely. The ion-assisted etching on the bottom surface of the feature keeps the carbonaceous film from forming and the downward etching of the surface proceeds essentially unaffected. This process is referred to as directionality by "sidewall blocking"[63,64] and is responsible for the etching directionality which is obtained in a wide variety of RIE processes. Etching selectivity between oxides and non-oxides is also accomplished because of carbon-containing species arriving at the etched surface.[65,66] As the feed gas is made more halogen deficient, the etch rate of non-oxides (eg. Si) is decreased because of the competition between the carbon and the Si for the halogen atoms. On oxide surfaces (eg. SiO_2), the oxygen atoms in the lattice of the solid react with the carbon forming volatile CO or CO_2. Thus the competition between the carbon and the Si for halogen is alleviated and the etching of the oxide continues while the etching of the non-oxide is blocked. Usually energetic ion bombardment is needed for this selective etch process to be fully effective.

Two very obvious phenomena which are often important in etching processes are "backscattering" and "redeposition."[67] Backscattering is the reflection of material, previously evolved from a surface, back to the surface following numerous collisions with molecules of the plasma gas. This process has a high probability once the pressure exceeds about 1 Pa. Backscattering is not an issue if the species are volatile molecules, as encountered in RIE, because the molecules will not condense on the surface. However if the species are condensible entities as evolved in physical sputtering, they will deposit on the surface. If the surface on which they condense is undergoing a chemical etching process (ie. RIE), the condensed species tend to aggregate and form a "micromask." Thus a small region of the surface is protected from the chemical etching process, leading to severe topography, often with conical shapes and sometimes called "grass." Redeposition is the term used to describe material evolved from the bottom surface of a high aspect ratio feature which collides with and condenses upon the sidewall of the feature. Typically all features in microstructure fabrication processes are very small (ie. micrometers) with respect to the collisional mean free path in most processing plasmas (ie. tens of micrometers to millimeters). Consequently species evolved from the bottom surface of high aspect ratio features have a very high probability of colliding with a sidewall. Again if these species are volatile molecules they will not condense and there is no problem. However if they are condensible species they will probably not escape from the feature. This redeposition process severely limits the pattern transfer capabilities of physical sputtering.[68]

The etching of carbon films (including diamond films) has been studied in considerable detail. Carbon is somewhat unique in that it forms volatile compounds with hydrogen, oxygen, fluorine, chlorine, etc. Much of the motivation for studies of the etching of graphite with hydrogen arises from the use of graphite as a first wall material in controlled thermonuclear fusion reactors.[69] In this application, etching of the wall material by hydrogen atoms or ions can introduce unacceptable levels of methane contamination into the very high density fusion plasma. Molecular hydrogen ions have be shown to etch graphite somewhat more effectively than diamond albeit with rather low absolute yields.[70] The etching of polymeric materials is routinely carried out in microelectronic processing and in most instances oxygen is used as the etch gas. Oxygen has been shown to be an effective etchant for diamond films as well, both in ion beam assisted etching[71] and with RIE.[72] Whereas the etching of polymeric materials can be accomplished with fluorine-based chemistries, it is well established that mixtures of fluorine and oxygen are more much more effective than either etchant separately.[73] The reasons for this synergism are: (a) The addition of CF_4 to an oxygen plasma tends to increase the electron density and/or the electron temperature resulting in significantly more dissociation of the oxygen[9], and (b) F atoms and O atoms tend to work together in reacting with organic solids in that F atoms abstract H from the solid leaving the C more susceptible to oxidation.[74] The author is not aware of any studies of the etching of diamond films using oxygen-fluorine mixtures but one would expect that only the first of these two enhancement mechanisms would be important because of the absence of hydrogen in the diamond lattice. Furthermore, the increase in the electron density and/or temperature caused by the CF_4 addition to the oxygen plasma might be expected for a wide range of additive gases. The etching of hydrocarbon solids with fluorine alone is not very effective probably because of the fact that a stable, rather unreactive, teflon-like layer tends to form before the volatile products CF_4 and C_2F_6 are formed.[73,75] Generally, the etching of carbon-containing solids is strongly enhanced by energetic ion bombardment. Although the studies are not complete, it is believed that this statement applies to all forms of carbon in all etchants leading to volatile products. Therefore the reactive ion etching of carbon-containing solids should be anisotropic. Thus it is expected that the etching of diamond films in an oxygen plasma should be reasonably straight-forward.

2. Surface Modification with Plasmas

There are many examples of surfaces being exposed to reactive gas glow discharges so that some property or properties of the surface can be modified. As is the case often in plasma processing, the development of these surface modification processes is largely empirical with limited understanding of the fundamental surface science involved. The use of oxygen glow discharges to grow oxide films on existing metal or semiconductor surfaces[76] is one of the oldest plasma-based surface processing techniques. As is the case in most of these plasma processes, a wide variety of experimental configurations have been used ranging from dc plasmas in which a positive potential is applied to the surface being oxidized (plasma anodization) to rf sputter etching kinds of discharges where the surface is subjected to extensive energetic ion bombardment. In the latter configurations, the process is identical to RIE with the only difference being the surface being oxidized does not form volatile products in the plasma oxidation application.

Mechanistically this process is somewhat less complex than PECVD or RIE but still there are aspects of plasma oxidation which are not well understood. Intuitively one would expect atomic oxygen to be one of the most important species particularly in situations where there is no appreciable energetic ion bombardment. Whether or not gas phase negative atomic oxygen ions play any role is not clear. Whereas oxide films are readily grown when the surface is positively biased with respect to ground, it is not certain that under these conditions the surface is actually positive with respect to the glow discharge because of the tendency of glow discharges to follow the potential of positively biased surfaces. The growth of the oxide film under anodic growth conditions is, however, believed to be controlled by the drift of negative O ions through the film. It is possible that these negative ions arrive at the surface as neutral atoms where, upon adsorption, they acquire a negative charge. Under these growth conditions the oxidation rate decreases with time approaching a final thickness which is determined by the bias voltage. When the surface is placed on the negatively biased electrode, the polarity of the electric field through the growing oxide prohibits negative oxygen ion drift to the oxide-metal interface. Thus the oxide thickness is determined by diffusive processes which can be greatly enhanced by the accompanying energetic ion bombardment of the oxide surface. Additionally, these atomic and molecular oxygen ions provide another source of oxygen for the oxidation process.

Whereas thermal oxidation has been dominant in silicon technology up to now, the constant need for lower processing temperatures has led to plasma oxidation being considered in microelectronic processing.[77] Plasma oxidation has been proven to be very useful in the growth of very thin oxide films on Pb, Nb or related alloys for Josephson junctions. These films are typically grown under energetic ion bombardment conditions and relatively quickly reach a steady state thickness after which the oxide growth rate and the removal rate by sputter etching are equal. Very precise control of film thickness can be realized by adjusting the key plasma parameters (power, pressure, gas composition- ie percentage of oxygen in argon) in that each set of parameters results in a different steady state oxide thickness. Very precise, reproducible thin oxide films can be produced in this way.[78]

Plasma nitriding is very similar to plasma oxidation as might be expected. The major difference is, whereas oxygen tends to become a negative ion upon arrival at many metal or semiconductor surfaces, nitrogen does not. Therefore film growth by the field assisted motion of negative ions to the film-substrate interface is not an important process in plasma nitriding. Plasma nitriding is usually best accomplished as a cathodic process with relatively energetic positive ion bombardment and can be expected to behave similarly to cathodic oxidation. Thus atomic nitrogen atoms as well as energetic atomic and molecular nitrogen ions play a role in the nitride film formation process. At low temperatures where thermal diffusion rates are very low, the nitride film thickness will be determined by the enhanced diffusion often associated with energetic ion bombardment of surfaces.

One of the most important commercial applications of plasma nitriding is the case-hardening of machine tools and other mechanical components.[79] This process is often carried out at high temperatures (~ 500° C) where diffusive processes determine the thickness of the nitride films which is in the range of 0.5 mm. In the microelectronics industry there is interest in the plasma nitriding (under much more gentle conditions) of semiconductor surfaces such as Si and GaAs but the applications are sparse relative to those associated with plasma oxidation. Recently,[80] high voltage (up to 40 kV) pulses have been used to accomplish plasma nitriding of surfaces with encouraging results. This approach is referred to as plasma source ion implantation and offers the possibility of replacing conventional ion implantation in certain specialized applications.

Another surface modification process is plasma texturing. It has long been recognized that physical sputter etching of surfaces can lead to the evolution of microscopically very rough surfaces. Most of the studies of this phenomenon have been carried out using ion beams at very high fluences.[81] The dependence of the sputtering yield on the angle of incidence and the presence of foreign species with anomalously low sputtering yields are believed to be two of the major contributing factors to this topography evolution. The crystalline structure of the surface also is very relevant. This ion-induced microtopography is an important effect in plasma processing of surfaces, producing both beneficial and deleterious effects. In those applications in which the objective of a surface process is to introduce texture or microroughness on a surface, the "micromasking" phenomenon described previously is very effective. Recall the conditions leading to "micromasking" were the presence of an impurity on a surface with the chemistry such that the surface is forming volatile products and is etching but the impurity does not form volatile products. That is, small regions on the surface will be protected from the plasma etching process giving rise to severe microtopography.[82] The backscattering phenomenon, discussed earlier in this paper, provides an excellent source of the foreign species which are so important in both ion-induced microtopography and micromask-induced roughness. This poses a problem in RIE applications and it is well-known that energetic ion bombardment of surfaces which form involatile products with the etching gas should be avoided. This in turn limits the allowable plasma potential in etching systems. However for those instances in which the texturing of a surface is the primary objective of a surface process, this approach to texturing surfaces is very effective. In many cases the surface is an organic solid on which adhesion is generally poor. The improvement in adhesion can be a result of the increased interface area resulting from the surface roughness, from surface cleaning or from chemical modification of the surface. Organic solids are particularly easy to texture because they are readily etched in an oxygen discharge and there are many ways of introducing small fluxes of foreign species which have involatile oxides.

Hydrogen plasmas are often used to process surfaces for a wide variety of applications and probably with less understanding than is usual even for plasma processing. "Plasma annealing" is sometimes used to describe this process.[83] Some physical or chemical characteristics of surfaces or films which are influenced by various plasma treatments include photoconductivity,[83] dielectric breakdown strength,[84] nucleation characteristics,[85] wettability of fabrics,[86] chemical reactivity of metals,[87] liquid crystal alignment,[88] photoresist flow[89] and optical reflectivity.[90]

3. Plasma Enhanced Chemical Vapor Deposition (PECVD) and Plasma Polymerization

In a PECVD process, the film forming elements are injected into the reactor as gases, just as in thermal CVD. The plasma dissociates the stable molecules into reactive radicals which condense on surfaces and form good quality films at temperatures which are significantly lower than those used in thermal CVD. This ability to deposit useful thin films at lower temperatures is one of the major advantages of PECVD in microelectronic processing. Some of the more important PECVD processes, besides the formation of diamond films, are the deposition of hydrogenated amorphous Si from silane, the deposition of silicon nitride from silane and ammonia and/or nitrogen, the deposition of silicon oxide from silane and oxygen or nitrous oxide and the deposition of tungsten from WF_6/H_2 mixtures.

PECVD is probably the most complex of all the plasma-based surface processing technologies and there exists relatively little detailed understanding of the key physical and chemical mechanisms determining the characteristics of the deposited films. Very little is known about the nature and relative abundances of the species arriving at the growing film and the parameter space in PECVD is awesome. The equipment variety itself is very formidable. Downstream (ie out of the plasma) deposition is currently of interest including some arrangements where only one of the injected gases is passed through the discharge; the other(s) being added downstream. Another approach is to pass only the inert carrier gas through the discharge relying on the metastable atoms to dissociate the film-forming gases which are injected downstream. Needless to say, process development is, by necessity, largely empirical. In the implementation of PECVD, there are often problems with uniformity, both in deposition rate and, more importantly, in film composition. Whereas in thermal CVD, thermodynamic considerations are largely responsible for determining the surface reactions, in PECVD the plasma is the key factor. Most radicals react with surfaces and consequently the film composition is largely determined by the relative fluxes of all the various film-forming species; none of which are known with any certainty. This can be an advantage and a disadvantage- one can fabricate completely unique material quite easily but making the same material a second time is often a real challenge.

Energetic ion bombardment of the growing film has a large influence on films deposited by PECVD but again the mechanistic understanding is missing. It has been shown that low temperature epitaxial deposition can be accomplished by low energy ion bombardment during PECVD.[91] Recently it has been shown that very directional deposition can be obtained with PECVD under energetic ion bombardment conditions.[92] However, currently most PECVD processes do not rely heavily on energetic ion bombardment and, in the opinion of this author, there is much opportunity for invention and creativity in this area. The reader is referred to review articles for further discussion of PECVD.[93,94]

Plasma polymerization is very similar to PECVD, the primary difference being that in PECVD one is concerned with depositing inorganic films (diamond and hard carbon films are considered to be inorganic), whereas in plasma polymerization the resulting film is an organic polymer. An example of plasma polymerization is the formation of a teflon-like film (often highly cross-liked) by injecting an unsaturated fluorocarbon gas (eg tetrafluoroethylene, hexafluoropropene, etc) into a low power density glow discharge. High quality films are formed which cannot be readily deposited by other means. The mechanistic and parametric complexities are similar to those encountered in PECVD. One significant difference in the parameter situation is the much reduced importance of ion bombardment in plasma polymerization. Energetic ion bombardment of polymeric materials is usually not a beneficial treatment as the most common result is graphitization of the polymer. There are various applications of plasma polymerization but it is probably the least technologically important of the plasma-based deposition methods at the present time. Application areas include passivation and isolation, fabrication of semi-permeable membranes, deposition of video-disk media, deposition of dry developable resists for lithography, etc. More discussion of plasma polymerization can be obtained elsewhere.[95,96]

Possibly one of the most significant applications of plasma polymerization arises in the RIE process discussed previously. It is quite easy to promote polymer formation when saturated halocarbon gases are injected into plasma systems provided there is a process that will consume the excess halogen in the plasma. Etching is a large consumer of excess halogens and polymerization is commonly observed in RIE. More significantly, this tendency for polymerization to take place is responsible for some critically important characteristics of the reactive ion etching process; namely directionality in some cases and silicon dioxide-to-silicon etch rate selectivity as discussed previously. One factor which determines whether etching or deposition dominates in halogen deficient reactive ion etching systems is the extent to which a surface is subjected to ion bombardment and often one observes etching on biased surfaces (ie. cathode or target) and polymer deposition on grounded surfaces. It is relatively easy to arrange things so that the material which is etched from the ion bombarded target becomes incorporated in the polymeric film being deposited on the substrates (or grounded surfaces). The thin films formed in this way can be electrically insulating or conducting, depending on their metal content, and in general are uniform good quality films.[97]

REFERENCES

1. T. Kokubo, F. Tochikubo and T. Makabe, Diagnostics of low-frequency CH_4 discharge by optical emission spectroscopy, J. Phys.D 22:1281 (1989).
2. J.A. Mucha, D.L. Flamm and D.E. Ibbotson, On the role of oxygen and hydrogen in diamond-forming discharges, J. Appl. Phys. 65:3448 (1989).
3. J.W. Coburn and M. Chen, Optical emission spectroscopy of reactive plasmas: A method for correlating emission intensities to reactive particle density, J. Appl. Phys. 51:3134 (1980).
4. R. d'Agostino, F. Cramarossa, S. De Benedictus and G. Ferraro, Spectroscopic diagnostics of CF_4-O_2 plasmas during Si and SiO_2 etching processes, J. Appl. Phys. 52:1259 (1981).
5. V.M. Donnelly, D.L. Flamm, W.C. Dautremont-Smith and D.J. Werder, Anisotropic etching of SiO_2 in low frequency CF_4/O_2 and NF_3/Ar plasmas, J. Appl. Phys. 55:242 (1984).
6. R.M. Roth, K.G. Spears and G. Wong, Spatial concentrations of silicon atoms by laser-induced fluorescence in a silane glow discharge, Appl. Phys. Lett. 45:28 (1984).
7. R.A. Gottscho, R.H. Burton, D.L. Flamm, V.M. Donnelly and G.P. Davis, Ion dynamics of rf plasmas and plasma sheaths: A time-resolved spectroscopic study, J. Appl. Phys. 55:2707 (1984).
8. G.S. Selwyn, Spatially resolved detection of O atoms in etching plasmas by two-photon laser-induced fluorescence, J. Appl. Phys. 60:2771 (1986).
9. R.E. Walkup, K.L. Saenger and G.S. Selwyn, Studies of atomic oxygen in O_2 + CF_4 rf discharges by two-photon laser-induced fluorescence and optical emission spectroscopy, J. Chem. Phys. 84:2668 (1986).
10. G.S. Selwyn, L.D. Baston and H.H. Sawin, Detection of Cl and chlorine-containing negative ions in rf plasmas by two-photon laser-induced fluorescence, Appl. Phys. Lett. 51:898 (1987).
11. R.A. Gottscho and T.A. Miller, Optical techniques in plasma diagnostics, Pure and Appl. Chem. 56:189 (1984).
12. W.R. Harshbarger, Plasma diagnostics and end-point detection, in: "VLSI Electronics- Microstructure Science, Vol. 8- Plasma Processing for VLSI," N.G. Einspruch and D.M. Brown, eds., Academic Press, Orlando (1984), pp. 411-446.
13. R.W. Dreyfus, J.M. Jasinski, R.E. Walkup and G.S. Selwyn, Optical diagnostics of low pressure plasmas, Pure and Appl. Chem. 57:1265 (1985).
14. V.M. Donnelly, Optical diagnostic techniques for low pressure plasmas and plasma processing, in: "Plasma Diagnostics, Vol. 1- Discharge Parameters and Chemistry," O. Auciello and D.L. Flamm, eds., Academic Press, San Diego (1989), pp. 1-46.
15. L.A. Farrow, Infrared laser spectroscopy of BCl_3 in an rf discharge, J. Chem. Phys. 82:3625 (1985).
16. J.A. O'Neill, J. Singh and G.G. Gifford, In situ infrared diagnostics of particle forming etch plasmas, J. Vac. Sci. Technol.A 8:1716 (1990).

17. K.G. Spears, T.J. Robinson and R.M. Roth, Particle distributions and laser-particle interactions in an rf discharge of silane, IEEE Trans. Plasma Sci. PS-14:179 (1986).

18. G.S. Selwyn, J.S. McKillop, K.L. Haller and J.J. Wu, In situ plasma contamination measurements by HeNe laser light scattering: A case study, J. Vac. Sci. Technol.A 8:1726 (1990).

19. R.A. Gottscho and C.E. Gaebe, Negative ion kinetics in rf glow discharges, IEEE Trans. Plasma Sci. PS-14:92 (1986).

20. P.J. Hargis, Trace detection of N_2 by KrF-laser-excited spontaneous Raman spectroscopy, Appl. Optics 20:149 (1981).

21. P.J. Marcoux and P.D. Foo, Methods for endpoint detection for plasma etching, Solid State Technol. 24-4:115 (1981).

22. P.A. Heimann, Optical etch-rate monitoring using active device areas: Lateral interference effects, J. Electrochem. Soc. 132:2003 (1985).

23. G.S. Selwyn, B.D. Ai and J. Singh, Real-time measurements of plasma/surface interaction by plasma-amplified photoelectron detection, Appl. Phys. Lett. 52:1953 (1988).

24. S.W. Downey, A. Mitchell and R.A. Gottscho, Photoemission optogalvanic spectroscopy: An in-situ method for plasma electrode surface characterization, J. Appl. Phys. 63:5280 (1988).

25. A. Mitchell and R.A. Gottscho, Plasma power dissipation at wafer surfaces measured using pulsed photoluminescence spectroscopy, J. Vac. Sci. Technol.A 8:1712 (1990).

26. D.E. Aspnes and R.P.H. Chang, Spectroscopic ellipsometry in plasma processing, in: "Plasma Diagnostics, Vol. 2- Surface Analysis and Interactions," O. Auciello and D.L. Flamm, eds., Academic Press, San Diego (1989), pp. 67-108.

27. T.F. Heinz, M.M.T. Loy and W.A. Thompson, Study of symmetry and disordering of Si(111)-7x7 surfaces by optical second harmonic generation, J. Vac. Sci. Technol.B 3:1467 (1985).

28. H.E. Evans and P.P. Jennings, Mass spectrometric study of the ionic species in a radiofrequency discharge in methane, J. Phys. Chem. 70:1265 (1966).

29. G. Smolinsky and M.J. Vasile, Ionic and neutral products of an rf discharge in methane, Intern. J. Mass Spectrom. Ion Phys. 16:137 (1975).

30. M.J. Vasile and G. Smolinsky, Mass spectrometric sampling of the ionic and neutral species present in different regions of an rf discharge in methane, Intern. J. Mass Spectrom. Ion Phys. 18:179 (1975).

31. R.B. Lockwood, R.E. Miers, L.W. Anderson, J.E. Lawler and C.C. Lin, Effect of water vapor on a CH_4-H_2 discharge plasma, Appl. Phys. Lett. 55:1385 (1989).

32. J.W. Coburn, E. Taglauer and E. Kay, Glow-discharge mass spectrometry- Technique for determining elemental composition profiles in solids, J. Appl. Phys. 45:1779 (1974).

33. VG9000- Glow Discharge Mass Spectrometer, VG Isotopes Ltd.

34. J.W. Coburn, Mass spectrometric studies of positive ions in rf glow discharges, Thin Solid Films 171:65 (1989).

35. M.J. Vasile and H.F. Dylla, Mass spectrometry of plasmas, in: "Plasma Diagnostics, Vol. 1- Discharge Parameters and

Chemistry," O. Auciello and D.L. Flamm, eds., Academic Press, San Diego (1989), pp. 185-238.

36. B. Chapman, "Glow Discharge Processes," John Wiley & Sons, New York (1980), p. 59.

37. J.W. Coburn and E. Kay, Pressure considerations associated with ion sampling from glow discharges, J. Vac. Sci. Technol. 8:738 (1971).

38. H. Lergon, M. Venugopalan and K.G. Muller, Mass spectrometer-wall probe diagnostic of Ar discharges containing SF_6 and/or O_2: Reactive ions in etching plasmas, Plasma Chem. Plasma Process. 4:107 (1984).

39. J.W. Coburn and E. Kay, Positive-ion bombardment of substrates in rf diode glow discharge sputtering, J. Appl. Phys. 43:4965 (1972).

40. L.J. Overzet, J.H. Beberman and J.T. Verdeyen, Enhancement of the negative ion flux to surfaces from radio-frequency processing discharges, J.Appl. Phys. 66:1622 (1989).

41. H. Kojima, H. Toyoda and H. Sugai, Observation of CH_2 radical and comparison with CH_3 radical in a rf methane discharge, Appl. Phys. Lett. 55:1292 (1989).

42. T.A. Milne, J.E. Beachey and F.T. Greene, Study of relaxation in free jets using temperature dependence of n-butane mass spectra, J. Chem. Phys. 56:3007 (1972).

43. H.F. Winters, The role of chemisorption in plasma etching, J. Appl.Phys. 49:5165 (1978).

44. J.E. Spencer, J.H. Dinan, P.R. Boyd, H. Wilson and S.E. Buttrill, Stoichiometric dry etching of mercury cadmium telluride using a secondary afterglow reactor, J. Vac. Sci. Technol.A 7:676 (1989).

45. T.R. Hayes, M.A. Dreisbach, P.M. Thomas, W.C. Dautremont-Smith and L.A. Heimbrook, Reactive ion etching of InP using CH_4/H_2 mixtures: Mechanisms of etching and anisotropy, J. Vac. Sci. Technol.B 7:1130 (1989).

46. S.J. Pearton, W.S. Hobson and K.S. Jones, Etch rates and surface chemistry of GaAs and AlGaAs reactive ion etched in C_2H_6/H_2, J. Appl. Phys. 66:5009 (1989).

47. D.L. Smith and R.H. Bruce, Si and Al etching and product detection in a plasma beam under ultrahigh vacuum, J. Electrochem. Soc. 129:2045 (1982).

48. H.F. Winters, J.W. Coburn and T.J. Chuang, Surface processes in plasma-assisted etching environments, J. Vac. Sci. Technol.B 1:469 (1983).

49. J.W. Coburn and H.F. Winters, Ion- and electron-assisted gas-surface chemistry- An important effect in plasma etching, J. Appl. Phys. 50:3189 (1979).

50. Y-Y. Tu, T.J. Chuang and H.F. Winters, Chemical sputtering of fluorinated silicon, Phys. Rev.B 23:823 (1981).

51. H.F. Winters and J.W. Coburn, Plasma-assisted etching mechanisms: The implications of reaction probability and halogen coverage, J. Vac. Sci. Technol.B 3:1376 (1985).

52. F.A. Houle, A reinvestigation of the etch products of Si and XeF_2; Doping and pressure effects, J. Appl. Phys. 60:3018 (1986).

53. R.A. Barker, T.M. Mayer and W.C. Pearson, Surface studies of and a mass balance model for Ar^+ ion-assisted Cl_2 etching of Si, J. Vac. Sci. Technol.B 1:37 (1983).

54. A.W. Kolfschoten, R.A. Haring, A. Haring and A.E. de Vries, Argon-ion assisted etching of silicon by molecular chlorine, J. Appl. Phys. 55:3813 (1984).

55. R.A. Rossen and H.H. Sawin, Time-of-flight and surface residence time measurements for ion-enhanced Si-Cl_2 reaction products, J. Vac. Sci. Technol.A 5:1595 (1987).

56. U. Gerlach-Meyer, J.W. Coburn and E. Kay, Ion-enhanced gas-surface chemistry: The influence of the mass of the incident ion, Surf. Sci. 103:177 (1981).

57. F.R. McFeely, J.F. Morar and F.J. Himpsel, Soft x-ray photoemission study of the silicon-fluorine etching reaction, Surf. Sci. 165:277 (1986).

58. J.A. Yarmoff and F.R. McFeely, Mechanism of ion-assisted etching of silicon by fluorine atoms, Surf. Sci. 184:389 (1987).

59. T. Mizutani, C.J. Dale, W.K. Chu and T.M. Mayer, Surface modification in plasma-assisted etching of silicon, Nucl. Instrum. Meth.B 7:825 (1985).

60. J. Abrefah and D.R. Olander, Reaction of atomic hydrogen with crystalline silicon, Surf. Sci. 209:291 (1989).

61. J.C. Angus, H.A. Will and W.S. Stanko, Growth of diamond seed crystals by vapor deposition, J. Appl. Phys. 39:2915 (1968)

62. Y. Hirose and Y. Terasawa, Synthesis of diamond thin films by thermal CVD using organic compounds, Jpn. J. Appl. Phys. 25:L519 (1986).

63. D.L. Smith and P.G. Saviano, Plasma beam studies of Si and Al etching mechanisms, J. Vac. Sci. Technol. 21:768 (1982).

64. R.H. Bruce and G.P. Malafsky, High rate anisotropic aluminum etching, J. Electrochem. Soc. 130:1369 (1983).

65. R.A.H. Heinecke, Control of relative etch rates of SiO_2 and Si in plasma etching, Solid State Electron. 18:1146 (1975).

66. J.W. Coburn and H.F. Winters, Plasma etching- A discussion of mechanisms, J. Vac. Sci. Technol. 16:391 (1979).

67. J.W. Coburn, Pattern transfer, Superlattices and Microstructures 2:17 (1986).

68. H.W. Lehmann, L. Krausbauer and R. Widmer, Redeposition- A serious problem in rf sputter etching of structures with micrometer dimensions, J. Vac. Sci. Technol. 14:281 (1977).

69. O. Auciello, Recent progress in understanding ion bombardment-induced synergism in the erosion of carbon due to multispecies impact, Nucl. Instrum. Meth.B 13:561 (1986).

70. R. Yamada, Chemical sputtering of sintered diamond compacts and diamond film, J. Vac. Sci. Technol.A 5:2222 (1987).

71. N.N. Efremow, M.W. Geis, D.C. Flanders, G.A. Lincoln and E.P. Economou, Ion-beam-assisted etching of diamond, J. Vac. Sci. Technol.B 3:416 (1985).

72. G.S. Sandu and W.K. Chu, Reactive ion etching of diamond, Appl. Phys. Lett 55:437 (1989).

73. F.D. Egitto, F. Emmi, R.S. Horwath and V. Vukanovic, Plasma etching of organic materials. I. Polyimide in O_2-CF_4, J. Vac. Sci. Technol.B 3:893 (1985).

74. D.L. Flamm and V.M. Donnelly, The design of plasma etchants, Plasma Chem. Plasma Process. 1:317 (1981).

75. H. Kasai, M. Kogoma, T. Moriwaki and S. Okazaki, Surface structure estimation by plasma fluorination of amorphous carbon, diamond, graphite and plastic film surfaces, J. Phys.D 19:L225 (1986).

76. P. Friedel and S. Gourrier, Review of oxidation processes in plasmas, J. Phys. Chem. Solids 44:353 (1983).

77. V.Q. Ho and T. Sugano, Plasma anodization of silicon and its application to the fabrication of devices and integrated circuits, Thin Solid Films 95:315 (1982).

78. J.H. Greiner, Oxidation of lead films by rf sputter etching in an oxygen plasma, J. Appl. Phys. 45:32 (1974).

79. P.C. Jindal, Ion nitriding of steels, J. Vac. Sci. Technol. 15:313 (1978).

80. J.R. Conrad, J.L. Radke, R.A. Dodd, F.J. Worzala and N.C. Tran, Plasma-source ion implantation technique for surface modification of materials, J. Appl. Phys. 62:4591 (1987).

81. I.H. Wilson, The topography of ion bombarded surfaces, Surface Topography 2:289 (1989).

82. J.L. Vossen, Inhibition of chemical sputtering of organics and C by trace amounts of Cu surface contamination, J. Appl. Phys. 47:544 (1976).

83. T. Makino, H. Nakamura and T. Nakashita, Increase in photoconductivity of polysilicon by plasma annealing, J. Appl. Phys. 51:5868 (1980).

84. S. Iwamatsu, Effects of plasma cleaning on the dielectric breakdown in SiO_2 film on Si, J. Electrochem. Soc. 129:224 (1982).

85. C.J. Robinson, The effects of a glow discharge on the nucleation characteristics of Au on polymer substrates, Thin Solid Films 57:285 (1979).

86. T.L. Ward, H.Z. Jung, O. Hinojosa and R.R. Benerito, Effect of cold plasmas on polysaccharides, Surf. Sci. 76:257 (1978).

87. D.F. Klemperer and D.J. Williams, Changes in the chemical reactivity of metals exposed to an inert gas glow discharge, Vacuum 33:301 (1983).

88. G.J. Sprokel and R.M. Gibson, Liquid crystal alignment produced by rf plasma deposited films, J. Electrochem. Soc. 124:557 (1977).

89. J.M. Moran and G.M. Taylor, Plasma pretreatment to improve resist flow properties by reduction of resist flow during postbake, J. Vac. Sci. Technol. 19:1127 (1981)

90. J-S. Maa, D. Meyerhofer, J.J. O'Neill, L. White and P.J. Zanzucchi, Reflectivity reduction by oxygen plasma treatment of capped metallization layer, J Vac. Sci. Technol.B 7:145 (1989).

91. T.J. Donahue and R. Reif, Silicon epitaxy at 650-800 C using low pressure chemical vapor deposition with and without plasma enhancement, J. Appl. Phys. 57:2757 (1985).

92. J.J. Hsieh, D.E. Ibbotson, J.A. Mucha and D.L. Flamm, Directional deposition of silicon oxide by a plasma enhanced TEOS process, MRS Symposium Proceedings 165:XXX (1989).

93. D.W. Hess, Plasma-surface interactions in plasma-enhanced chemical vapor deposition, Ann. Rev. Mater. Sci. 16:163 (1986).

94. S. Veprek, Applications of low pressure plasmas in materials science: Especially chemical vapor deposition, in: "Current Topics in Materials Science," Vol.4, E. Kaldis, ed., North-Holland, Amsterdam (1980), pp. 151-236.

95. H. Yasuda, "Plasma Polymerization," Academic Press, New York (1985).

96. J.F. Evans and G.W. Prohaska, Preparation of thin polymer films of predictable chemical functionality using plasma chemistry, Thin Solid Films 118:171 (1984).

97. E. Kay and A. Dilks, Metal-containing polymer films produced by simultaneous plasma etching and polymerization: The series of perfluoroalkanes C_nF_{2n+2} (n=1,2,3,4), Thin Solid Films 78:309 (1981).

MODELLING AND SIMULATION OF PARTICLE-SURFACE INTERACTIONS

Roger Smith

Department of Mathematical Sciences
Loughborough University
Leicestershire, LE11 3TU

1. INTRODUCTION

This paper will describe different approaches to modelling the interaction of energetic particles with surfaces. In the first part of the paper the primary focus will be towards developing models that can explain how surface topography changes on the **macroscopic** scale both as a result of sputtering (erosion) and deposition or redeposition (growth). The development of the theory will be general and not specifically related to diamond. Two approaches will be described. The simplest model will be to consider the substrate as consisting of a regular cellular structure[1-5]. Deposition occurs as a result of adding cells to the surface and erosion by their removal. The rules for adding or removing cells from the surface will be specified in terms of the local environment of each cell. This approach leads to the possibility of parallel computer algorithms for large scale systems although all the calculations presented here have been carried out on serial machines.

A second method of calculating macroscopic scale changes will be to use differential equation models to track the motion of the surface as it grows or erodes[6-13]. The simplest continuum model is to regard the surface as a wavefront whose speed at any point is known as a function of the bombarding particle and substrate species, surface orientation, position and time. For example, in the case of a uniform beam sputtering a homogeneous substrate, the shape of the eroding surface can be calculated if the sputtering yield Y is known as a function of the incidence angle θ.

The second part of the paper will consider growth and erosion processes on the **atomic** scale. Molecular dynamics[14-22] simulations will be used to describe erosion

Diamond and Diamond-Like Films and Coatings
Edited by R.E. Clausing *et al.*, Plenum Press, New York, 1991

and deposition of individual atomic impacts. These simulations involve the solution of Newton's equations of motion for thousands of particles and are increasingly realistic as a result of the development of many-body interaction potentials for a number of different covalent materials including carbon. The primary focus of this work will be towards materials which have the diamond lattice structure, in particular natural diamond but also Si and SiC. The mechanisms by which atoms and molecules are ejected as a result of energetic particle bombardment will be described and comparisons made between the different crystalline faces of natural diamond and graphite. In addition some recent results[23] using molecular dynamics to model the initial stages of epitaxial growth on diamond lattice structures will be described.

2. CELLULAR MODELS OF DEPOSITION AND EROSION

Cellular automata are discrete (dynamical) systems whose behaviour is completely specified in terms of a local relation. They are the computer scientist's counterpart to the physicist's concept of 'field'[24]. Over recent years, specialist computers called Cellular Automata Machines have been built to compute efficiently models built using the discrete particle approach. Such machines consist of large numbers of parallel processors which exploit the simplicity and the local nature of the model. For large field problems the speed gained by this parallelism is essential. For simple erosion problems parallelism may be desirable but is not necessary for small scale problems and the simulations described in this section have all been carried out on a MacIntosh II with 2mb of RAM using MPW Pascal.

The cellular method assumes that the erosion which occurs is dependent on the local geometry of the surface at the point of impact. The erosion region is described in terms of a large number of particles (cells). A surface cell is one which has a missing neighbour and only surface cells are assumed to be eroded. For the simplest model we assume a square two-dimensional lattice. Figure 1 gives a schematic representation of a two dimensional lattice. A cell lies in the surface if any one of the squares labelled 1-4 in (a) is missing.

Consider first a simple model of isotropic erosion. The erosion of a cell from the surface is assumed to be dependent only on the number of neighbouring cells which are missing. Assuming a simple power law relationship for erosion[2], the probability for removal of a cell is given by

$$p = \left(\frac{b}{4}\right)^n \qquad n = 0, 1, 2......... \quad (1)$$

Here b is the number of missing neighbouring cells and n is the assumed power law dependence. If we regard the cells as atoms then the number of missing neighbours corresponds to the number of broken bonds. The local shape of the surface depends on n and also on the form of lattice chosen. If $n = 0$ all exposed particles have the same probability of removal. As n increases, the surface becomes locally smoother since particles with more missing neighbours are preferentially eroded, but whatever the value of n the macroscopic surface morphology which develops corresponds to isotropic erosion.

Some preliminary tests of erosion of an initially flat surface using a Monte-Carlo model, corresponding to $n = 0$ and $n = 1$ were carried out. A smoother surface develops with increasing n although the erosion rate decreases. The surface roughness which develops, stabilizes after the removal to a depth of about 30 cells.

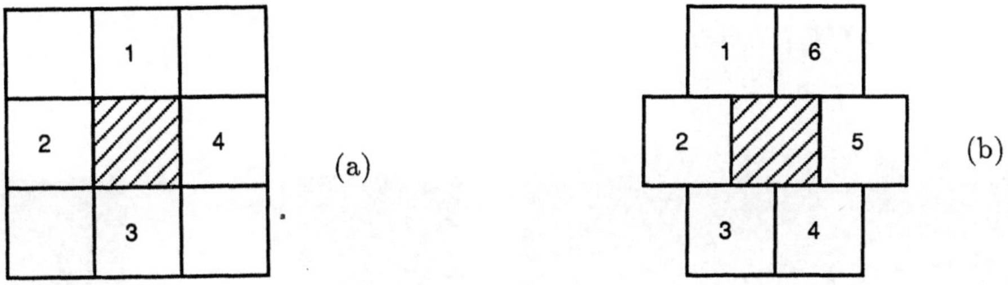

Figure 1. Cellular structures used in the computer simulations. (a) A regular square lattice. (b) An offset or hexagonal lattice.

The simplicity of the computational model means that it is very easy to incorporate different physical effects. For example, the effect of gravity can be included by preferentially removing cells which have a missing neighbour in direction 3. The dependence of the erosion rate on surface orientation can be achieved by weighting the removal probability according to the local incidence angle but this angle needs to be defined by averaging locally over the surface.

In order to demonstrate the method, two examples are chosen. The first shown in figure 2, is the isotropic erosion of a rectangular structure which might represent the cross-section of a wire in microelectronic device manufacture. The second shown in figure 3 is the isotropic erosion of an initially flat surface with an intersecting strata of material of a different erosion rate. This model was chosen in order to explore the problem of media changes. The problems were programmed with the four neighbour model of figure 1a although the model in 1b produces smoother surfaces. A pseudo-random number generator was used to choose cells from the surface string. Cells were added to this string as they became exposed. Another random number generator selected a number between 0 and 1 and the cell was eroded if this number was less than the removal probability according to equation (1). The simulations were carried out with a value of $n = 2$, and the the angle ϕ that the strata in figure 3 makes with the horizontal is given by $\tan \phi = 0.62$. The ratio of the relative erosion speeds for the two media is given by $c_1/c_2 = 0.5$. Here the cell size was chosen as the pixel size on the computer screen in order to obtain good visual images. Simulations corresponding to different erosion times are presented. As the erosion continues an equilibrium shape develops which consists of a curved front edge, a flat top of "soft" material and a trailing slope of a well defined angle. This particular simulation could be used as a model of geomorphological processes, where a strata of hard rock might intersect the earth's surface.

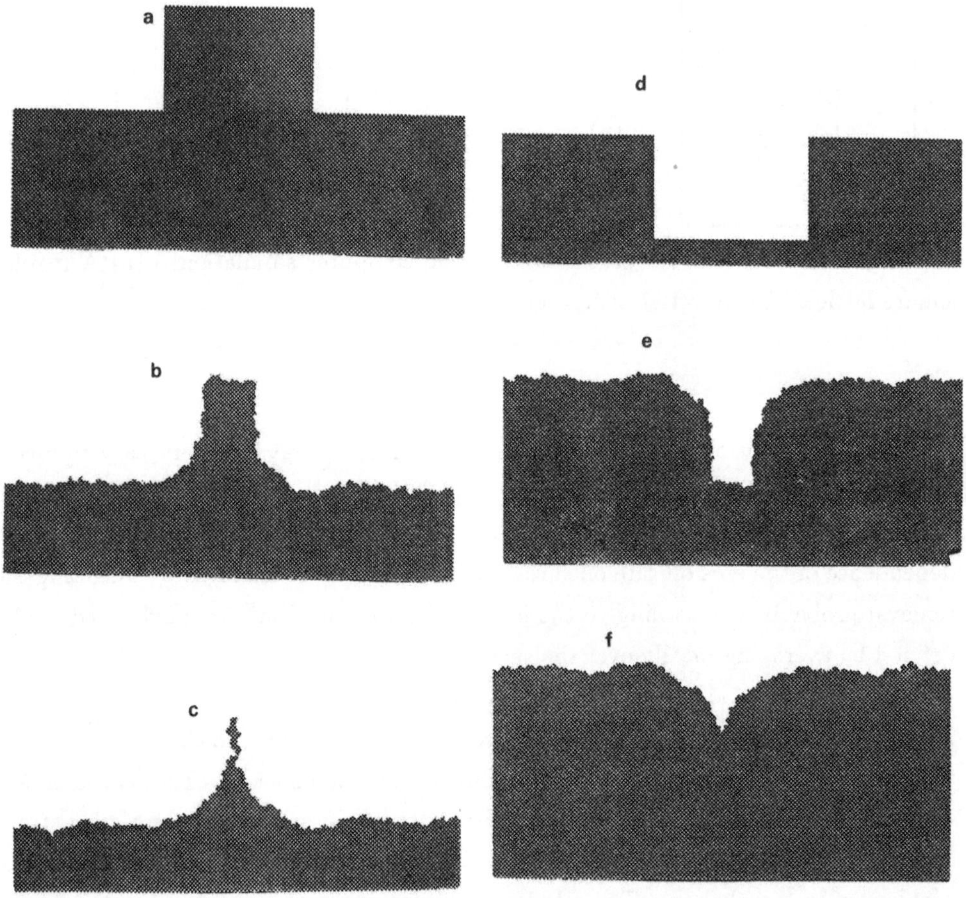

Figure 2. Isotropic erosion and deposition on rectangular structures using the cellular model with $n = 2$. (a)-(c) Erosion of a rectangular profile. (d)-(f) Deposition in a rectangular groove.

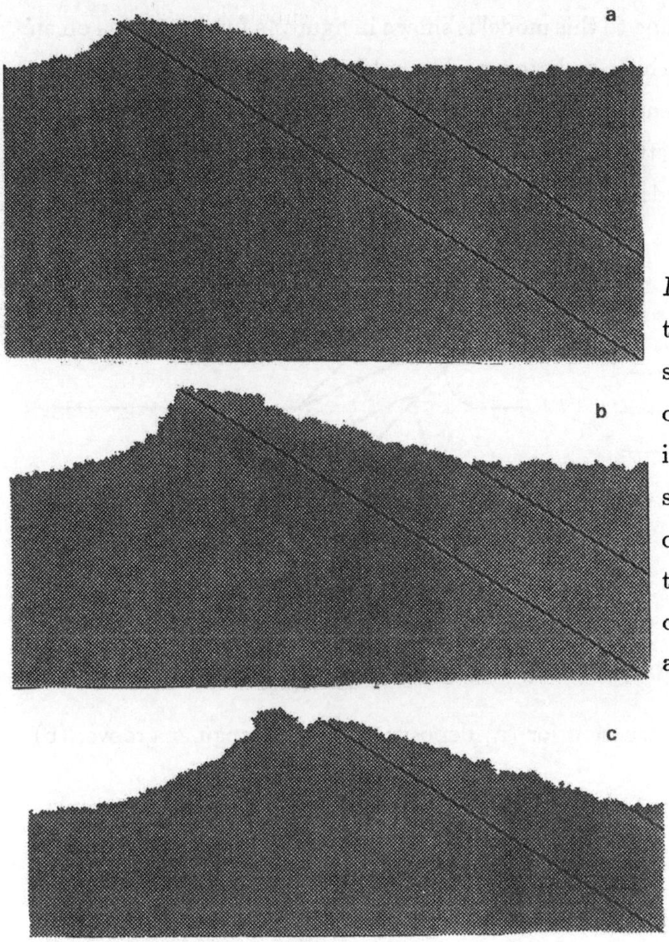

Figure 3. Various stages in the erosion of an initially flat surface composed of material of different erosion rates. The intersecting strata has an erosion rate which is a half that of the surrounding media. Eventually a structure develops which consists of a "soft" top with a "hard" trailing slope.

Figure 2 illustrates the erosion of a rectangular profile, 150 cells in width and 100 cells in height. If we arbitrarily assume that the cell is equivalent to an atom then the structure is of the order of magnitude of 600$\mathring{A}$ in width and statistical effects could be an important contribution to the overall eroded shape on that scale. The continuum kinematic wave models of surface shape changes equivalent to the cellular method are shown in figure 4 for the same initial surface structures and are described in detail in the next section. Of note here is that the overall shapes are the same as calculated using the cellular approach. Cellular deposition models based on similar criteria can also easily be programmed. However isotropic deposition is more likely to take place at sites which have more missing neighbours so that the probability of deposition at a site could be given by given by

$$p = (1 - \frac{b}{4})^n \qquad (2).$$

The wave solution corresponding to this model is shown in figure 4a for deposition on an initially rectangular profile. This simulation could model the problem of planarisation in microelectronic device manufacture whereby multilayer electronic structures are produced by depositing dielectric material to form a smooth surface upon which the next electronic circuit is laid down.

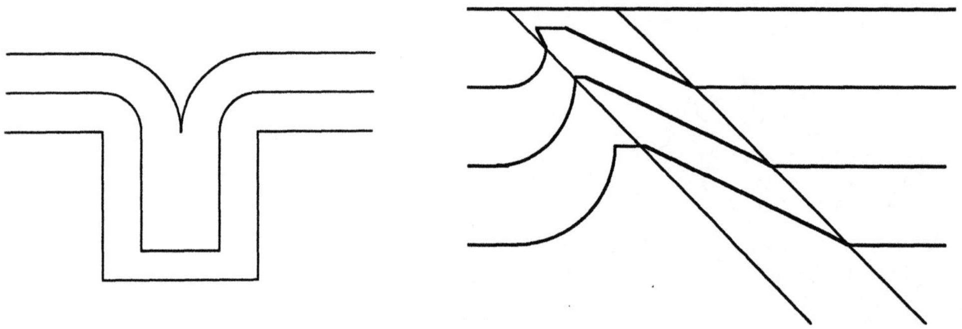

(a) (b)

Figure 4. The Huyghens construction for (a) deposition in a rectangular groove, (b) an eroding strata.

The models described above refer to isotropic processes but it may well be that erosion or deposition is directional in nature. The case of erosion or deposition by low energy particles arriving in a well-defined direction can also be simulated by altering the local rules for attachment. In the case of deposition, these are usually known as ballistic aggregation models[5]. Figure 5a illustrates a cellular simulation carried out for normal incidence deposition. Particles are assumed to arrive randomly at the surface in the normal direction and attach themselves to the first adjacent surface cell they meet. As columns begin to grow, arriving particles can either attach themselves to the tops or the sides of the structures. This can then shadow lower portions of the surface from the growth and a deposition develops which consists of particles and holes. The effect is more marked at non-normal incidence. Figure 5b illustrates the growth of columnar morphology at an angle which is different from the angle at which particles arrive at the surface. The relationship between the angle of incidence and the angle at which the columns grow has been investigated in detail by Meakin, Ramanlal, Sander and Ball[3]. Careful experimental measurements of these angles have also been carried out on vapour deposited aluminium films and on vapour-deposited and sputter-deposited rare-earth-metal-transition-metal thin films[25,26].

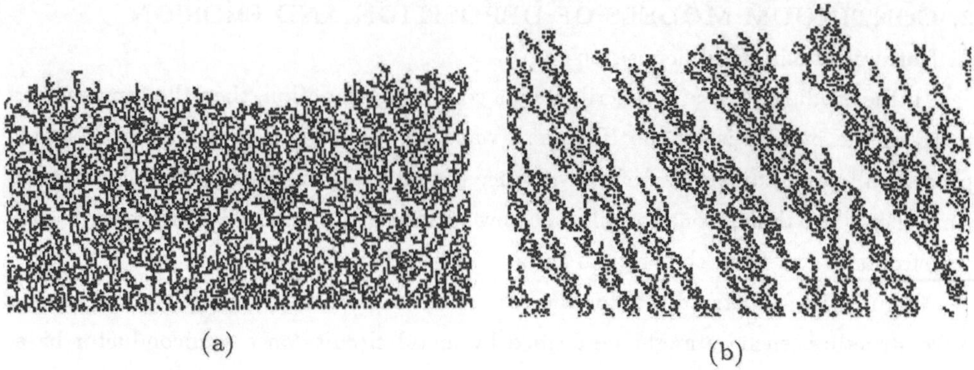

Figure 5.(a) Deposition at normal incidence using a square lattice as shown in figure 1a. (b) Deposition at an angle 62° to the normal. The structures grow at an angle of about 35° to the normal.

Directional erosion models have also been considered[4]. Figure 6 illustrates a simulation carried out by randomly removing surface cells which have a missing neighbour in direction 1 of figure 1. This is effectively the sequential layer sputtering model due to Benninghoven[27]. Benninghoven developed a statistical analysis based on this model (the so-called sequential layer sputtering model) to predict that the depth resolution after the removal of depth d during depth-profiling in surface analysis was proportional to $d^{\frac{1}{2}}$. However the simulation shown in figure 6 shows that the model is unphysical because it predicts peaks and troughs in the surface which are one cell in width. Diffusion or other physical effects would prevent this occuring in practice. This is a clear example of a computer simulation demonstrating the inapplicability of a mathematical model.

Cellular models described above have also been used to model other similar phenomena. The case where the erosion or growth is dependent on the surface gradient or curvature can also be computed using the cellular approach. The case of curvature dependent growth has been considered in the context of diffusion limited aggregation problems such as viscous fingering and cluster growth. Realistic simulations of these phenomena have been carried out. See reference [5] for a readable review. Monte-Carlo models of surface erosion which include the effect of particle reflection and redeposition in addition to primary beam effects have been developed by Rossnagel and Robinson[28]

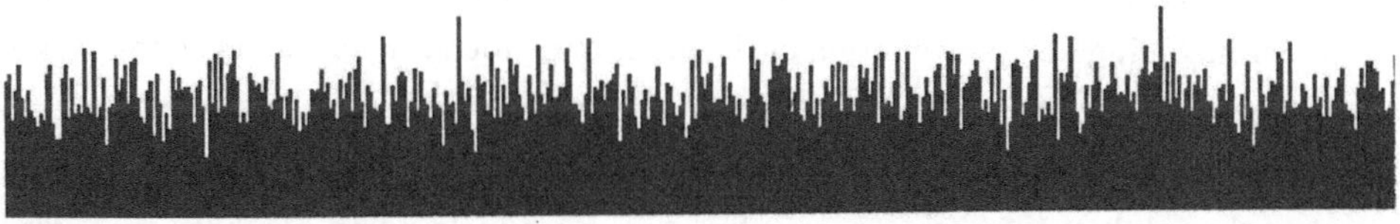

Figure 6. Random removal of cells with a missing neighbour in direction 1 of figure 1a.

3. CONTINUUM MODELS OF DEPOSITION AND EROSION

A. Isotropic etching and deposition

If the eroding surface is described by a continuous function, then the simplest way to track the motion is by the Huyghens wavefront construction[12]. For an isotropic process, all points on the wavefront advance (or retard) a fixed distance, *d*, in a given time. For a two dimensional profile, the new surface is the envelope of all the spherical wavefronts drawn from the original surface with radius *d*.

Figures 2a-f illustrate this for the case of the erosion and deposition on a square wave structure, such as might be formed by metal circuits on a semiconductor base. Note that the convex corners, in the case of deposition, figures 2d-f, expand into circular arcs, whereas the concave corners progress as concave corners. The situation is reversed, figures 2a-c, for the erosion process. For the case of deposition, on an initially angular structure, the surface rapidly planarises. Isotropic deposition is therefore an efficient procedure for planarising semiconductor devices. For a rectangular wave, a simple geometrical calculation shows that if the variation in the level of the surface is Δd after a thickness d has been deposited on the upper flats of the profile then

$$\frac{\Delta d}{d} = 1 - \left[1 - \left[\frac{w}{2d}\right]^2\right]^{\frac{1}{2}}, \qquad (d > w/2)$$

where w is the width of the groove bottom. This equation illustrates the increasing relative planarisation of the surface with increasing deposited thickness and could be used for specified planarity requirements to predict the total deposited thickness needed.

The Huyghens wavefront construction for the erosion of a surface consisting of media of different erosion rates is carried out in a similar way to that described above. The difference in this case is that the surface must be continuous at the boundary between two media. This leads to a law equivalent to Snell's law in optics along the boundary. The combination of Huyghens wavefront construction and Snell's law enable the rapid determination of surface shape subject to an isotropic erosion or deposition process although this can be complicated to evaluate in the case of multiple composition.

B. Directional erosion

The surface of an amorphous material subjected to inert gas ion bombardment does not necessarily erode isotropically. The sputtering yield Y, the number of particles ejected per incoming ion, is a function of the angle of incidence θ, of the incoming beam. For amorphous or polycrystalline materials, Y increases with θ up to a maximum at about 65° and then drops to zero at 90° as more ions are reflected from the surface than penetrate the solid and eject atoms, see figure 7. This yield curve has been measured experimentally for a number of ion-target combinations[29] and its determination by modelling can be carried out using techniques described in the next section.

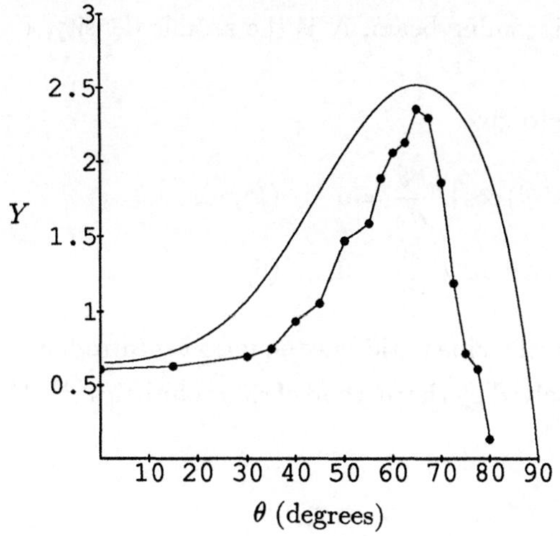

Figure 7. The yield curve, as a function of incidence angle for 1keV Ar bombardment of Si. The smooth curve represents a fit to experimental data and has been taken from reference [6]. The points joined by the broken curve are data calculated for normal incidence bombardment of the dimer reconstructed $Si\{100\}(2\times 1)$ face[21].

It is also possible for the erosion rate to be dependent on the surface curvature, particularly where diffusion is important, in which case in two dimensions

$$\frac{\partial y}{\partial t} = f\left[\frac{\partial y}{\partial x}, \frac{\partial^2 y}{\partial x^2}\right] ,$$

where y is the beam direction, x is perpendicular to this and t is time.

Differentiating with respect to x gives

$$\frac{\partial^2 y}{\partial x \partial t} = f_p \frac{\partial^2 y}{\partial x^2} + f_q \frac{\partial^3 y}{\partial x^3}$$

where $\tan\theta = p = \partial y/\partial x$ and $q = \partial^2 y/\partial x^2$.
Thus

$$\frac{\partial p}{\partial t} = f_p \frac{\partial p}{\partial x} + f_q \frac{\partial^2 p}{\partial x^2} \qquad (3).$$

A similar equation using t and y as the independent variables gives

$$\frac{\partial p}{\partial t} = g_p \frac{\partial p}{\partial y} + g_q \frac{\partial^2 p}{\partial y^2} \qquad (4).$$

where $g = f/p$ and subscripts denote partial differentiation.
However, the dependence of the erosion rate on second derivative terms will not be considered further in this paper, see the papers by Sethian[30,31] for further details. For erosion which depends only on the surface gradient and not its curvature, f takes the form

$$f = -\frac{J}{N} Y(\theta)$$

where J is the current flux density of the incoming beam, N is the atomic density, Y the sputtering yield and t is time.
Equations (3) and (4) simplify in this case to give

$$\frac{\partial\theta}{\partial t} + \frac{J}{N}Y'(\theta)\cos^2\theta\frac{\partial\theta}{\partial x} = 0 \qquad (5),$$

$$\frac{\partial\theta}{\partial t} + \frac{J}{N}(-Y(\theta) + Y'(\theta)\sin\theta\cos\theta)\frac{\partial\theta}{\partial y} = 0 \qquad (6).$$

Each of equations (3) and (4) are in the form of a kinematic wave equation as introduced by Lighthill and Whitham[32]. They can be solved by the method of characteristics[7,11–13] whose equations are

$$\frac{dx}{dt} = \frac{J}{N}\cos^2\theta \qquad (7),$$

$$\frac{dy}{dt} = \frac{J}{N}[Y'(\theta)\sin\theta\cos\theta - Y(\theta)] \qquad (8),$$

$$\theta = constant \qquad (9).$$

The characteristics are lines of constant θ and from a given initial contour defined by the set of n points (x_i, y_i) $i = 1, ...n.$, the contour at time t, (x'_i, y'_i) can then be constructed by a direct mapping according to

$$x'_i = x_i + \frac{J}{N}\cos^2 Y(\theta)t \qquad (10)$$

$$y'_i = y_i + \frac{J}{N}[Y'(\theta)\sin\theta\cos\theta - Y(\theta)]t \qquad (11).$$

Equations (5) and (6) are non-linear and after a time t the new surface may contain cusps and folds due to characteristic intersections which occur from sections of the initial surface from which the characteristics converge. When these intersections occur, the surface develops an edge due to the discontinuous change in θ. On the other hand from sections of the surface where the characteristics diverge a facet forms. These effects are illustrated in detail in figure 9. If the initial surface itself contains edges, then the edge may continue to progress as an edge or a characteristic expansion fan can radiate from these edges. Thus the relatively simple mapping given by equations (10) and (11) can lead to a complicated surface due to characteristic intersections and expansions. At a point of initial surface discontinuity where the characteristics diverge, the shape of the eroded surface is found from the envelope of the locus of the characteristics of all the included angles. The locus curve for characteristics corresponding to all surface orientations after unit time is plotted in figure 8 for a typical $Y - \theta$ curve corresponding to $1keV$ argon ion bombardment of amorphous silicon shown in figure 7. Figure 8 is analogous to the spherical wavefront in the case of isotropic erosion. The eroded surface can be drawn from the envelope of these curves in a similar way as for isotropic erosion[33] but is more complicated to do since the characteristics at different surface points are now of unequal lengths.

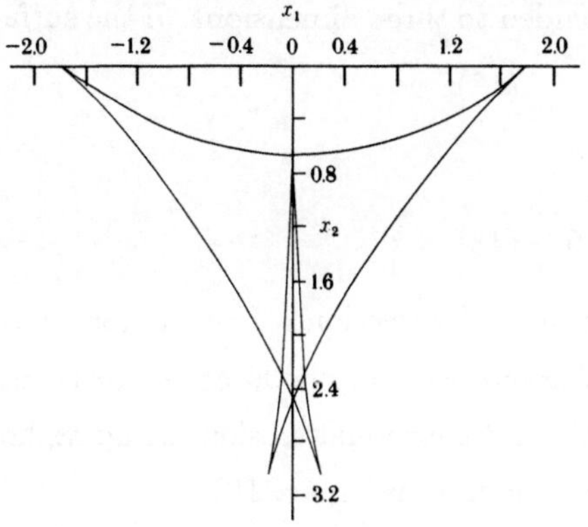

Figure 8. The characteristic locus curve, for 1keV *Ar* bombardment of *Si*. This curve is traced out by characteristics emanating from the origin of all surface orientations after a time corresponding to $\frac{Jt}{N} = 1$.

To illustrate the effect of characteristic intersections and expansions, consider the erosion of two planes at right angles bombarded along the negative y direction. If the planes are joined by a circular section, then a shock forms due to characteristic intersections. Eventually a smooth section of surface develops, joined by angular discontinuities to the two perpendicular planes. If, on the other hand, the planes meet with an angular discontinuity, then an expansion fan emanates from the corner, figure 9. Ultimately the end result is the same for both initial conditions. The corner fans out into a smooth portion of surface as before. The angles of orientation of this surface lie between those corresponding to the maximum sputtering yield at the point Q_R, to that whose erosion rate in the y direction is the same as for the flat plane, at the point Q_L.

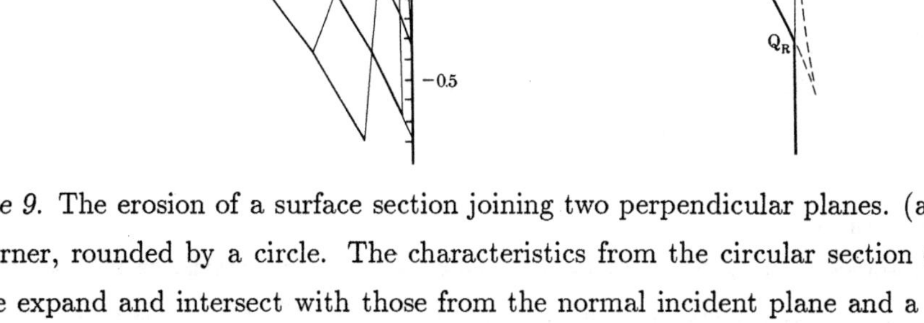

Figure 9. The erosion of a surface section joining two perpendicular planes. (a) The 90° corner, rounded by a circle. The characteristics from the circular section of the surface expand and intersect with those from the normal incident plane and a shock develops (discontinuity of θ). A surface gradient discontinuity also forms where the

(continued)

The characteristic method can also be extended to three dimensions. If the surface at time t has equation

$$S(\mathbf{r},t) = 0 \qquad (12)$$

then the equation of motion of the surface is

$$\frac{\partial S}{\partial t} + (\mathbf{c}.\nabla)S = 0 \qquad (13)$$

where **c** is the velocity of the surface in the normal direction. The motion of the surface can be determined as before using characteristic methods once the normal erosion rate is known. A detailed analysis of three dimensional erosion including how edge formation and expansions can take place is given in reference [13].

The computational method used to construct the surfaces in [13] was to use the characteristics to map the original surface into the eroded surface and to use the computer graphics to produce the visualisations. Other computational methods have been used in both two and three dimensions, usually based on line or surface segment ideas. Recently Katardjiev[34] has developed a segment algorithm based on triangulating the surface. This and similar methods use a numerical integration with small timesteps and the angle of incidence θ has to be recalculated for each segment after each erosion step and are therefore more subject to rounding errors. However they have the advantage of automatically deleting surface folds which occur when a direct mapping using characteristics is employed.

This section has shown how erosion can be simulated once the surface speed is known as a function of the spatial co-ordinates, surface orientation and time. Crystal growth can be similarly modelled by reversing the sign of the surface speed in the equations. The kinematic theory has been useful to explain a wide range of phenomena including cone and pyramid formation on ion-bombarded surfaces in addition to the shapes observed using ion beams to etch patterns in microelectronic device manufacture. The key to good modelling lies in knowing the erosion or growth rate as a function of incidence angle. This can be measured experimentally or calculated using the techniques described in the next section. The effects of redeposition can also be determined using continuum ideas. This can be a problem near to steep sided surface structures such as might exist in microelectronic device fabrication. The redeposited flux can be calculated by integrating the flux of particles ejected from the surface. In general an analytic solution does not exist and this can only be done numerically. Specific examples illustrating the effects of redeposition on groove walls in microelectronic device fabrication have been published[35].

Figure 9. (cont.)
circle intersects the grazing incidence plane, due to characteristics intersecting this plane. (b) The expansion wave at a convex 90° corner. The beam is incident in both (a) and (b) in the direction $P_R Q_R$.

4. MOLECULAR DYNAMICS MODELS OF ATOMIC SCALE PROCESSES IN DIAMOND LATTICES AND CRYSTALLINE CARBON

The previous section has highlighted how macroscopic changes in surface shapes can occur. Molecular dynamics simulations can be used to simulate atomic processes. Molecular dynamics computer simulations are not new. They were pioneered in the 1960s by Gibson et al[36] and applied by Harrison and co-workers[14–22] particularly for the case of particle-surface interactions. The motion of the particles is modelled using classical dynamics. A single energetic particle is aimed at the substrate. If the energy of the bombarding particle is sufficiently high a collision cascade is set up. The motion of the cascade is calculated by integrating numerically Newton's equations of motion. The simulations usually assume that the crystal is perfect before the start of each trajectory and statistics from a large number of trajectories corresponding to different impact points on the crystal are generated. A particle is counted as having ejected when it lies above the original surface and the sum of its binding energy with the surface and the upward component of kinetic energy is positive. The cascade is terminated when the maximum energy of any particle left in the target is smaller than a prespecified threshhold. The final velocities of the ejecting species can be used to calculate the energy and angular distributions. In the work involving keV particle bombardment described here, the calculation uses an integration algorithm which evaluates the forces only once per timestep[37] and also uses neighbour lists and a moving atom approximation[14] whereby only the trajectories of those particles which have been sufficiently hard hit by a moving particle are integrated. The moving atom approximation can improve the speed of the calculation by a factor four compared to a full molecular dynamics simulation. A fuller description of the calculation procedure and the moving atom approximation and the assumptions and limitations involved in the model is given elsewhere[14].

The molecular dynamics simulations are material specific. Different materials have different crystal structures, cohesive energies and interatomic force fields. These forces can be determined by carrying out complex quantum mechanical calculations but the forces calculated in this way are not suitable for use in a simulation code and so semi-empirical interatomic potentials are used. For carbon, these must be sufficiently accurate to model stable diamond lattice and graphite structures in addition to the correct cohesive energies. There have been a number of semi-empirical many-body interatomic potentials for carbon reported in the literature[38,39] derived using the Tersoff/Abell[40] approach. In references [38,39], the potential is fit not only to the bulk materials properties but also to over- and underco-ordinated configurations. The potential of [38] is used in the simulations described here. The carbon potential is also short ranged so is ideal for use in a molecular dynamics simulation. The potential gives a convenient and relatively accurate description of the structural properties and energetics of carbon, including elastic properties, phonons, polytypes and defects and migration barriers in diamond and graphite and has also been applied to study amorphous carbon. A further advantage of this potential is that it has minima for both the graphite and diamond lattice structures although in the graphite structure there is no bonding between adjacent hexagonal layers. It is important in simulations such as these described here that the potential models the initial crystal structure as a stable lattice. This is not achieved using radial pairwise additive potentials.

In the Tersoff[38] formulation, the total potential energy E has the form

$$E = \frac{1}{2}\sum_{i \neq j} f_C(r_{ij})(V_R(r_{ij}) + b_{ij}V_A(r_{ij})) \ ,$$

where r_{ij} is the distance between atoms i and j, V_R is a repulsive term, V_A is an attractive term, b_{ij} is a many-body term that depends on the positions of atoms i and j and the neighbours of atom i and $f_C(r_{ij})$ is a function which switches off the potential at a short range. b_{ij} is not invariant in the interchange of the labels i and j and incorporates the many-body features of the potential by modifying its attractive part depending on the arrangement of the near neighbours of atom i. The potential is insufficiently repulsive close in and the repulsive part is splined to a Molière potential[41] with a screening length of 0.83 times the Firsov value. The splining to a Molière potentialis carried out using a similar but improved procedure to that described in Smith, Harrison and Garrison[20,21]. For $r_a < r < r_b$ a function of the form $V_{fit} = A + exp(ar + b)$ with A, a and b constants is fit to the forces of the Molière potential at r_a and V_R and its derivative at r_b. The values of r_a and r_b are then adjusted to match the Molière potential at r_a.

The parameters used for both the Tersoff carbon potential and the Molière potential are as follows.

$$\begin{aligned}
V_R(r) &= A\exp(-\lambda_1 r) \\
V_A(r) &= -B\exp(-\lambda_2 r) \\
f_C(r) &= \begin{cases} 1\ , & \text{r} < R - D \\ \frac{1}{2} - \frac{1}{2}\sin[\pi(r-R)/(2D)]\ , & \text{R-D} \leq r \leq R + D \\ 0\ , & \text{r} > R + D\ , \end{cases} \\
b_{ij} &= (1 + \beta^n \xi_{ij}^n)^{-1/2n}\ , \\
\xi_{ij} &= \sum_{k \neq i,j} f_C(r_{ik}) g(\theta_{ijk}), \\
g(\theta) &= 1 + c^2/d^2 - c^2/[d^2 + (h - \cos\theta)^2].
\end{aligned}$$

Here i, j and k label the atoms of the system, r_{ij} is the length of the ij bond and θ_{ijk} is the bond angle between bond ij and ik. The values of the constants for carbon are: A = 1393.6 eV, B = 346.74 eV, λ_1 = 3.4879 Å, λ_2 = 2.2119 Å, β = 1.5724 × 10^{-7}, n = 0.72751, c = 38049.0, d = 4.3484, h = -0.57058, R = 1.95 Å, D = 0.15Å.

R and D were chosen to ensure the short-range nature of the potential and were not systematically optimised in the fitting process. The Molière potential[41] to which the repulsive part of the Tersoff potential is joined takes the form

$$V(r_{ij}) = [Z_1 Z_2 e^2 / r_{ij}] \sum_{k=1}^{3} a_{Mk} \exp(-b_{Mk} r_{ij}/a)$$

where $a = 0.83 \times 0.885 a_o [Z_1^{2/3} + Z_2^{2/3}]^{-1/2}$ and a_o is the Bohr radius (0.529Å), Z_1 and Z_2 are the atomic numbers of the two atoms, e is the electronic charge, a_{M1} =0.35, a_{M2} =0.55, a_{M3} =0.1, b_{M1} =0.3, b_{M2} =1.2 and b_{M3} =6.0.

Potentials for other materials with the diamond lattice structure such as Si, SiC, and $GaAs$ have also been published[42,43]. The SiC potential was obtained directly from the Si and C potentials by an averaged combination of the parameters which describe Si and C. The resulting material properties of SiC are accurately modelled.

Because the only assumptions made in the molecular dynamics are those of Newtonian mechanics, simulations of particle-surface interactions can be used with much wider application than many competing techniques. Models of particle-surface interactions based on the solution of the linear Boltzmann transport equations[44,45] are not valid at low energies where the number of particles involved in the collision cascade are too few to randomize the initial momentum of the incoming particle. Similarly multiple interactions and crystal effects dominate at low energies making the binary collision approximation[46–48] invalid. Molecular dynamics is valid for all energies and simulations of particle bombardment in the keV energy range have shown for example, how the energy and angular distributions of ejected material can be accurately predicted for both metals and covalent materials[16–20], how single particle impacts can form craters[49], and how molecular ejection can take place[50,51]. The next section describes results of the simulations of keV carbon bombardment of diamond and graphite using the potentials described above.

A. Normal incidence keV carbon bombardment of diamond and graphite

Figure 10 shows the crystal faces together with a description of the azimuthal angle used in the text. Carbon has a high binding energy and this is reflected in the calculated yields which are low. These are shown in table 1 for a variety of bombardment energies. For the graphite lattice no ejection occurred until the energy of the bombarding ion reached around 600eV. Ejection occured at lower energies for the diamond structure but the ejection times for both structures were found to be quite short. Very little material was ejected after 50 fs (1fs=10^{-15} s). These yields are calculated assuming a perfect crystal before each impact and thus are likely to be less than those measured after large doses of irradiation where implantation of carbon within the crystal will cause amorphisation by the breaking of bonds and the creation of defects. This will decrease the cohesive forces and increase the ejection probability. It should be noted that there is not a monotonic relationship between surface binding energy and the calculated yields. Although the high surface binding energy of graphite gives a smaller yield than the diamond structures there is less ejected from the diamond {100} face than the {111} face despite the higher surface binding energy for the {111} face.

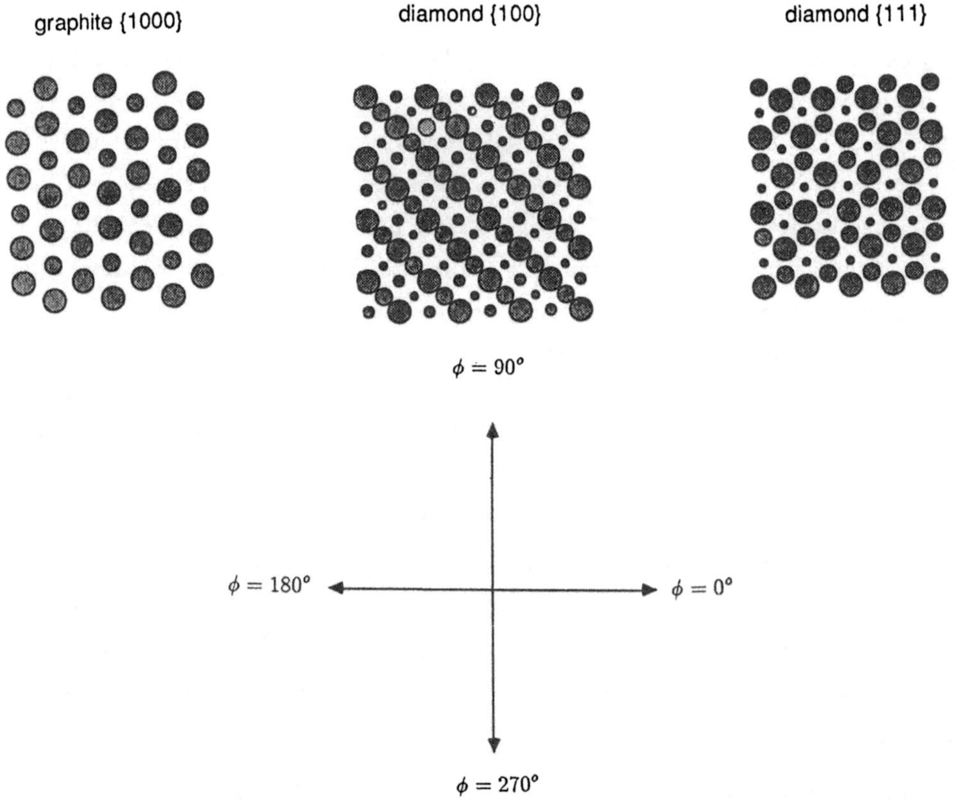

Figure 10. Carbon surface arrangements, together with a definition of the azimuthal angle ϕ used in the text. The largest circles represent the top layer atoms and the sizes of the other atoms are drawn decreasing with increasing depth. Viewed from normal incidence only two layers are visible for the graphite {1000} surface, four for the diamond bulk terminated {100} surface and three for the diamond {111} surface. For the diamond {111} surface, the first second and fourth layer atoms are visible. The third layer atoms lie directly under the second layer.

TABLE 1

Calculated number of particles that eject as a function of bombardment energy.

Force	Energy (eV)	Number of Impacts	No of C ejected	Yield
graphite{1000}	600	300	2	0.01
	700	300	8	0.03
	800	300	16	0.05
	1000	2700	198	0.07
diamond{111}	100	300	0	0.00
	200	300	9	0.03
	600	300	48	0.16
	700	300	51	0.17
	800	300	55	0.18
	1000	5400	1024	0.19
diamond{100}	1000	2700	323	0.12

TABLE 2

The percentage contribution of the different layers of the overall yield for normal incidence bombardment at 1 keV.

	diamond{100}	diamond{111}	graphite	Cu {100}
Layer 1	64%	59%	94%	90%
Layer 2	29%	40%	4.5%	9%
Layer 3	7%	1%	1.5%	1%

Table 2 shows the percentage contribution from each layer to the yields for 1keV bombardment. As with previous work involving diamond lattice structures, these results show a relatively large contribution from the subsurface layers compared to results for fcc metals[15]. The interplanar spacing for the graphite structure is relatively high (3.4Å) and the contribution to the yield from the second layer for the graphite face is less than for the diamond faces.

TABLE 3

Percentage of atoms ejected per single impact.

number emitted per impact	diamond{100}	diamond{111}	graphite{1000}
0	91.5%	86.2%	93.3%
1	5.8%	9.7%	6.0%
2	2.2%	3.1%	0.6%
3	0.3%	0.8%	0.03%
4	0.1%	0.2%	0%
>4	0.1%	0.05%	0%
percentage emitted as dimers	22%	23%	5%
percentage emitted as trimers	1%	1%	0%

Table 3 shows the distribution of atoms ejected per single impact. The percentage of trajectories that emit no atoms is high for all structures but there are fewer trajectories that emit multiple atoms for the graphite structure. Also of note is the high percentage of atoms ejected as dimers ($> 20\%$, expressed as a percentage of the total number of atoms ejected) due to the high binding energy of the carbon dimer. For diamond bombardment over 60% of trajectories which eject two atoms do so as dimers. Examination of individual trajectories shows that for normal incidence bombardment, these molecules are ejected as neighbouring pairs within the crystal and are not formed by recombination in the vacuum after ejection. For the diamond {100} face these are mostly neighbouring surface atoms. For the diamond {111} face some neighbouring surface atoms are ejected as dimers but a majority originate as a first and second layer pair. Some of the processes by which molecular ejection occurs from the diamond {111} face are described in detail in figure 11 by the examination of an individual trajectory. In figure 11 six atoms are ejected as a result of a single impact of which four are in the form of two dimers. A sequence of snapshots viewed at normal incidence from above the surface layer are shown at 2,8,15,33 and 40fs. After 2fs the incoming particle collides with a surface atom (A) pushing it downwards underneath a neighbouring surface atom (B). Atom B ejects after 15fs along an azimuth close to the direction of its original bond with atom A ($\phi = 120^o$). After 15fs we can also see that an energetic particle formed by the cascade is passing below the two second layer atoms which will ultimately form ejected dimers. After 33fs one of the dimers

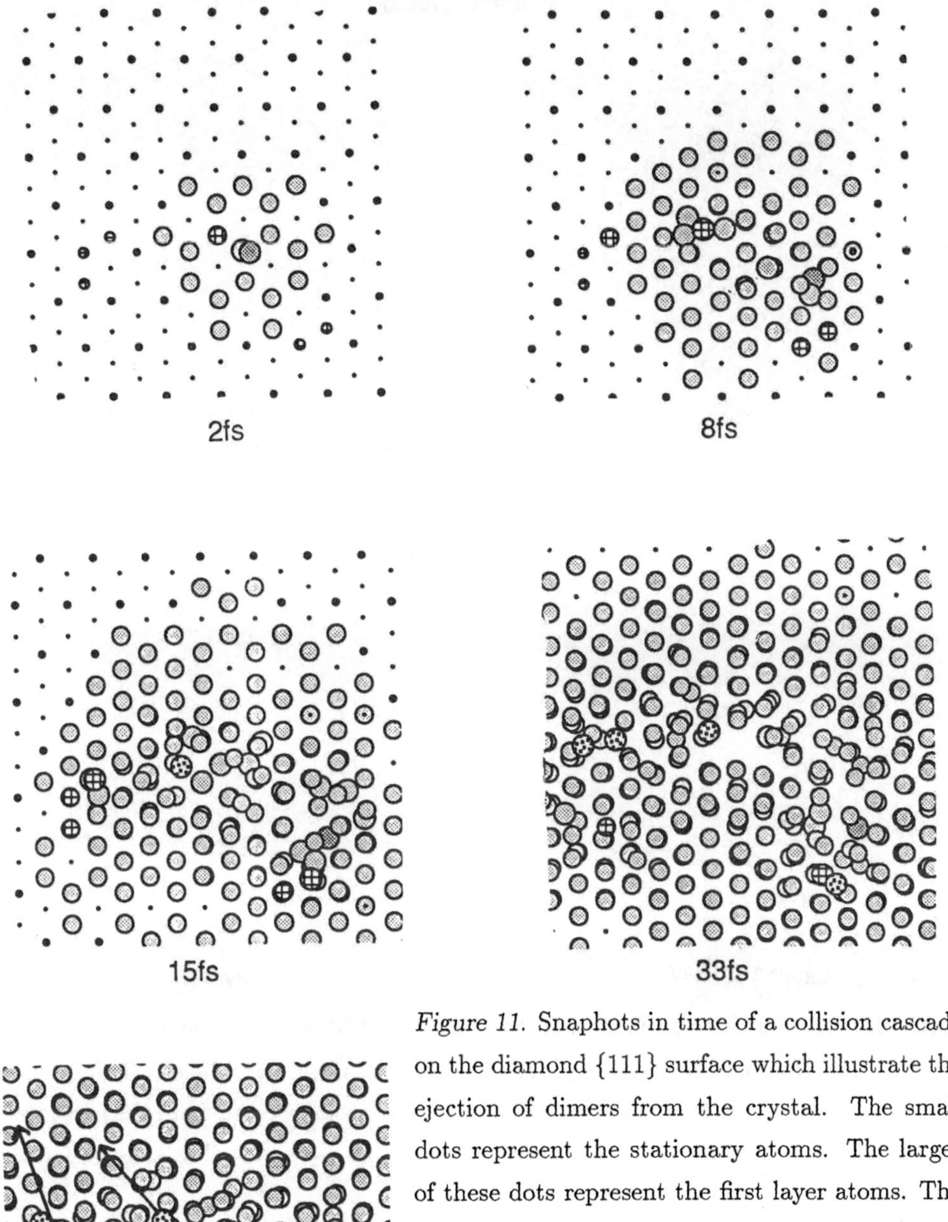

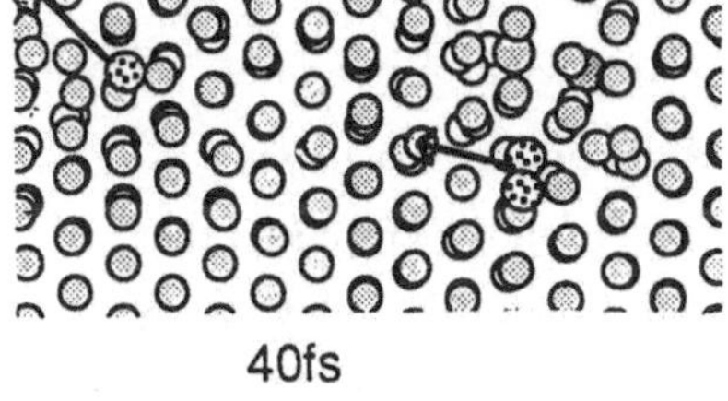

Figure 11. Snaphots in time of a collision cascade on the diamond {111} surface which illustrate the ejection of dimers from the crystal. The small dots represent the stationary atoms. The larger of these dots represent the first layer atoms. The atom sizes of the moving atoms are shown larger and increase with increasing kinetic energy for three discrete energy ranges of <10eV, <100eV and >100eV. The atoms that ultimately ejected are flagged with hatching before ejection and with a stippled pattern after ejection. Frames are shown after 2, 8 15, 33 and 40 fs by which time all ejection has taken place.

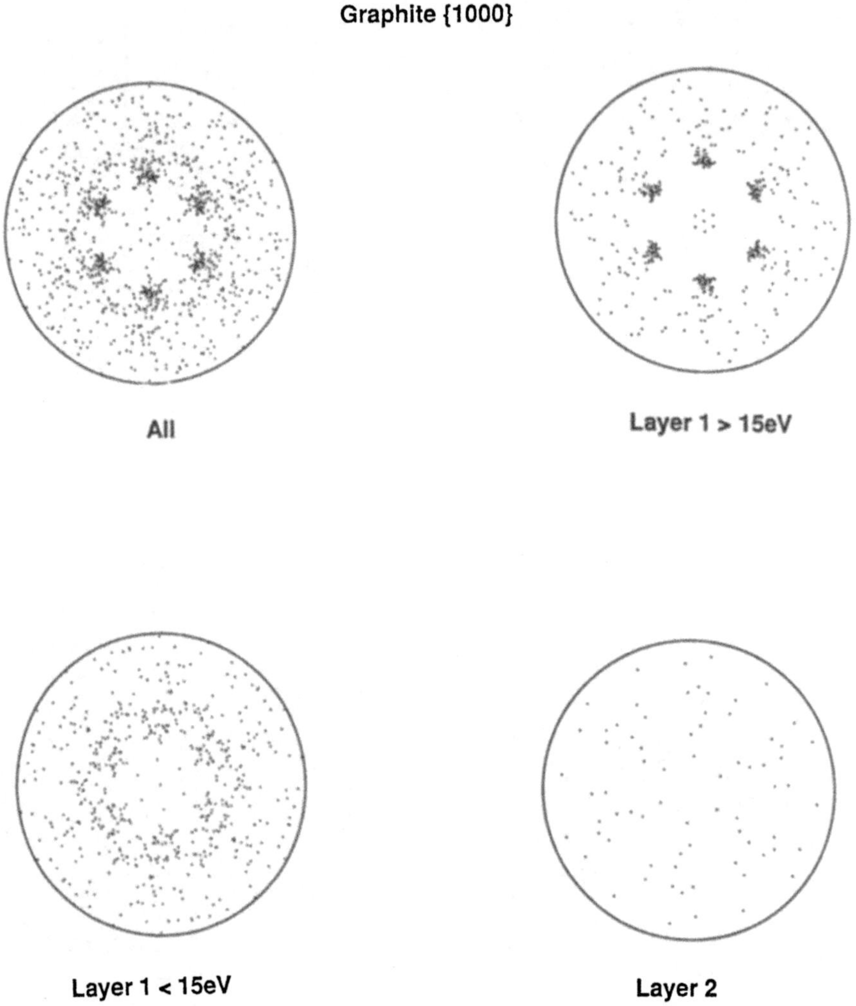

Figure 12. Angular distributions from the graphite {1000} surface orientated as shown in figure 10. Bombardment takes place at normal incidence at 1000eV. The intense features are located at $\theta = 23^o, \phi = 30^o, 90^o, ...$ The atoms ejected from the first layer are separated according to energy whereas the second layer distributions are for all energies.

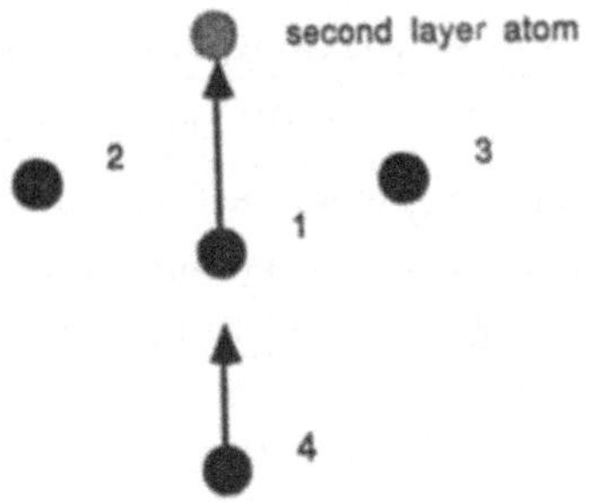

Figure 13. A dominant mechanism for ejection of atoms from the graphite {1000} surface. Atom 1 is pushed downward into the adjacent second layer atom and ejects after the collision.

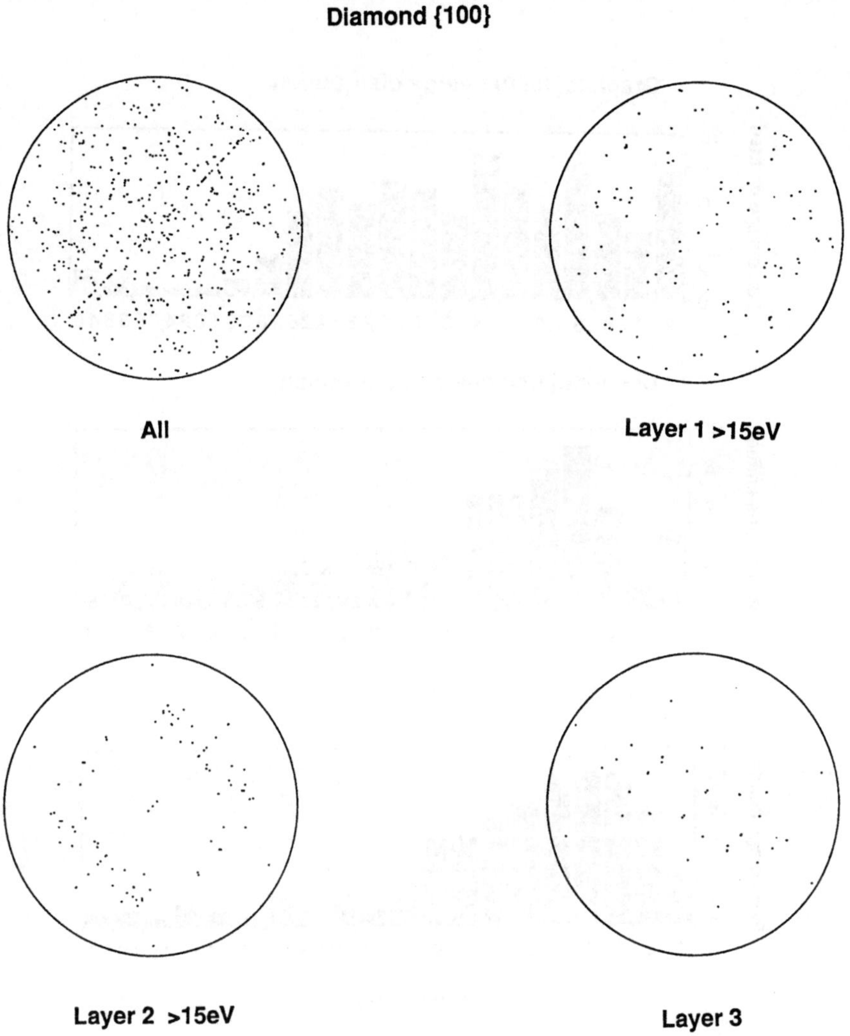

Figure 14. Angular distributions the diamond {100} surface. Note the preferred azimuth near to $\phi = 45^o$ for the first and second layer atoms. The higher energy atoms ejected from the second layer preferentially eject at a polar angle of $\theta \approx 40^o$. The third layer atoms preferentially eject close to $\phi = 135^o$.

has already been ejected and after 40fs all the ejection has taken place. Note that a neighbouring surface atom of one of the ejected dimers is also ejected close to the $\phi = 270^o$ azimuth. Both of the ejected dimers formed as a result of a second layer atom being hit from below and picking up the surface layer atom at the point of ejection. The arrows indicate the azimuthal ejection directions.

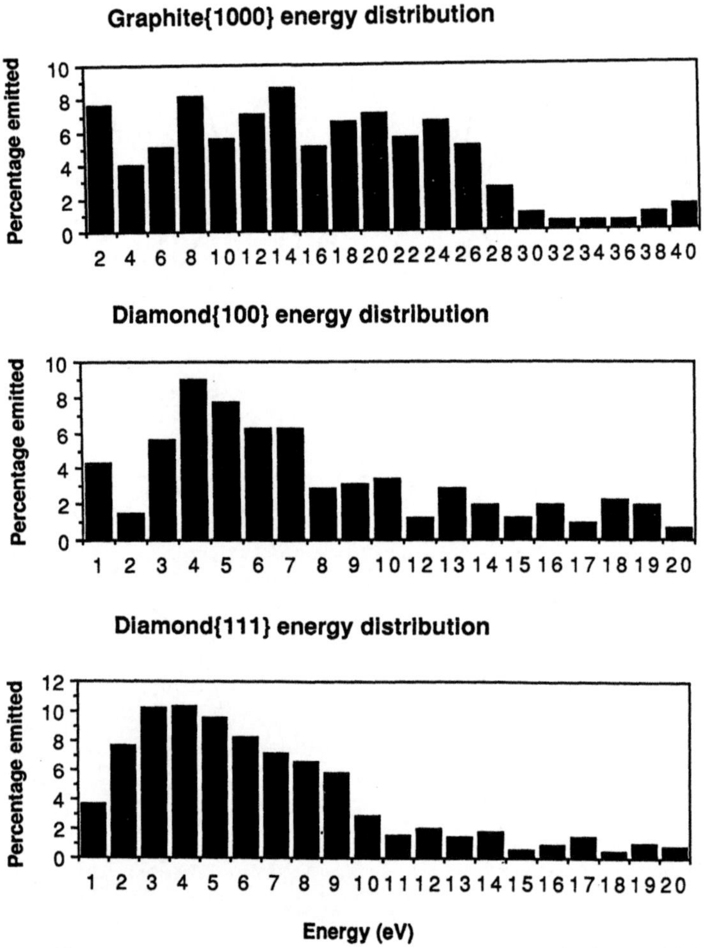

Figure 15. Energy distributions of ejected atoms. The graphite distribution is flat to statistical accuracy up to 25eV. Both diamond surfaces peak at 4eV.

There are many mechanisms by which ejection can take place from the diamond structure and these lead to complex angular distributions of ejected material. However, for graphite, the angular distributions show distinct features. Figure 12 shows the ejection patterns for the graphite {1000} face. These patterns are drawn by calculating the points at which the trajectories of the ejecting atoms cross a sphere and then projecting these points back onto the crystal surface. Thus the centre of the pattern corresponds to a polar angle of $\theta = 0^o$, the edges correspond to $\theta = 90^o$ and the radial distance from the centre of the circles is proportional to $\sin\theta$. The pattern shows that very little material is ejected for small values of θ and intense features at $\theta = 23^o$ and $\phi = 0^o, 60^o, 120^o$... .This contrasts with results for both fcc metals and for Si where a variety of faces generally show ejection angles at higher values of θ, typically[20], $\theta = 40^o$. The intense features in the ejection pattern can be seen to be predominantly due to first layer atoms ejected with energies > 15eV. For energies between 0 and 15eV the azimuthal dependence is less marked but the polar dependence peaks between $\theta = 16^o$ and $\theta = 32^o$. The results are in general agreement with simulations carried out for fcc metals which indicate that it is the angular distributions of the higher energy ejected atoms[16] which are most indicative of the underlying crystal structure. Figure 12 show some evidence to indicate the preferred azimuthal ejection direction for the second layer atoms is the same as for the first layer but better ejection statistics are required to confirm this fully and to ascertain any preferred polar angle dependence for the second layer atoms.

There is a distinct mechanism associated with the six intense features in the angular distributions for graphite. This is described in figure 13. Atom 4 is pushed towards atom 1 breaking its bonds. Atoms 2 and 3 then channel atom 1 in the direction of the original 4-1 bond. Atom 1 is pushed downwards into the second layer atom and reflects back out of the crystal.

The angular distributions for the diamond {100} face shown in figure 14 contain a complex variety of features. There is clearly a preferred ejection direction along the $\phi = 45^o$ azimuth. This direction is favoured by the higher energy ejected atoms from the first two layers. Second layer atoms which eject along this direction are channelled between adjacent surface layer atoms but when $\phi = 135^o$ surface layer atoms act to block ejection of second layer atoms. Atoms from the third layer which eject do so close to the $\phi = 135^o$ azimuth, a direction which is unblocked by first and second layer atoms. The preferred ejection directions for the surface layer atoms are not well defined. If the explanations of channelling and blocking which have been used to describe the ejection directions for fcc metals were to be correct then we would expect to see preferred ejection along the $\phi = 0^o$ and 90° azimuths. Ejection would be inhibited along $\phi = 45^o$ and 135°, the directions along which the surface atoms are most closely packed. However the higher energy atoms from the top layer eject preferentially close to the $\phi = 45^o$ azimuth. Although some of the lower energy atoms also eject along this azimuth there are a variety of mechanisms to account for the ejection process and some evidence also of ejection along the $\phi = 0^o$ and 90° azimuths. These results contrast with those for the {100} face of Si subjected to 1keV Ar bombardment[20] which showed preferential ejection only along the $\phi = 45^o$ and 135° azimuths.

The energy distributions of the ejected atoms are shown in figure 15. Of note is that for the diamond structures the peak in the ejection energy is around 4eV whereas for the graphite structure the distribution is fairly flat until much higher energies. Transport theories of sputtering in amorphous materials would predict a peak in the energy distribution at about half the cohesive energy; for carbon the cohesive energy is 7.37 eV. The agreement between energy predictions of transport theory and the diamond results suggests that trajectories of ejected atoms are randomised much more by the the diamond lattice structure than by graphite. This is also confirmed by the angular distributions of the ejected material which show less dramatic peaks for the diamond structure compared to graphite where there is a dominant ejection mechanism.

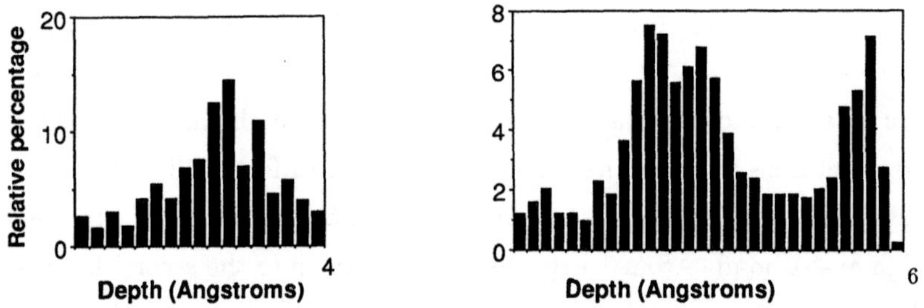

Figure 16. Normal incidence implant distributions at 100eV. (a) diamond {100} , (b) graphite {1000}.

It is also of interest to examine briefly the location of implanted material within the crystal as this information is also readily obtainable from the code and can help give some clues as to how radiation damage can destroy the crystal structure. In order to make sure that the cascades were totally contained within the material only simulations at a bombardment energy of 100eV have been carried out. Figure 16 shows that for graphite {1000} the distribution of the implanted particles peaks between the top two surface layers (the interlayer spacing is $3.3\AA$). All incident particles implant and no implanted particles reach as far as the second layer. The atoms are more closely packed in the diamond lattice and for bombardment of the {100} face the peak in the implantation distribution is between the first and second layers (interlayer spacing $1.4\AA$). Atoms implanted through the diamond {100} face penetrate only just over half the distance than through the graphite structure although the dominant peak in each distribution is approximately at the same depth.

B. Simulations of epitaxial growth using molecular dynamics

Simulations of epitaxial growth are more computationally intensive than those involving keV particle bombardment. Although we have to run a large number of cascades corresponding to different impact points on the crystal to generate good ejection statistics, the cascades can usually be terminated when the total energy of the most energetic particle in the crystal drops below about 1 eV. A new crystal is then assumed for thc next impact. For a target of 1280 atoms arranged in 10 layers, each trajectory for the dimer reconstructed $Si\{100\}(2 \times 1)$ surface under normal incidence Ar^+ bombardment at 1 keV took approximately 117 seconds of computer time on an IBM 3090[20]. For epitaxial growth on Si or C particles arrive at the surface at sub eV energies and kinetic energy is imparted to the crystal as the arriving particles bind with the lattice. Realistic simulations have been carried out for deposition on the dimer reconstructed $Si\{100\}(2 \times 1)$ surface using a lattice consisting of 10 layers of atoms with 32 particles in each layer and with periodic boundary conditions. Deposition of 4-5 monolayers required 800 hours of computer time on an IBM 3090[23] which is pushing current computing power to its limits. Nonetheless these simulations have revealed a *novel* mechanism of cooperative motion where dimers simulataneously open in adjacent rows. This mechanism predicts that there should be single rows of dimers rather than isolated dimers in the first epitaxial layer. Concurrently with these calculations the same isotropic growth of $Si\{100\}$ during MBE has been observed in the scanning tunnelling microscope[52]. The reaction dynamics indicate that epitaxial growth grows preferentially in a direction perpendicular to the original dimer rows on the surface. As the growth continues the simulations have shown that the top layer constantly reconstructs and the atoms in the next to the top layer constantly return to the bulk sites. Figures 17 and 18 illustrate these effects. The success with Si illustrates the power of molecular dynamics simulations as a possible tool for the prediction of the optimum conditions for diamond growth and in particular the conditions under which diamond structures would grow preferentially over graphite structures.

Figure 17. STM pictures of epitaxial growth on $Si\{100\}(2 \times 1)^{52}$.

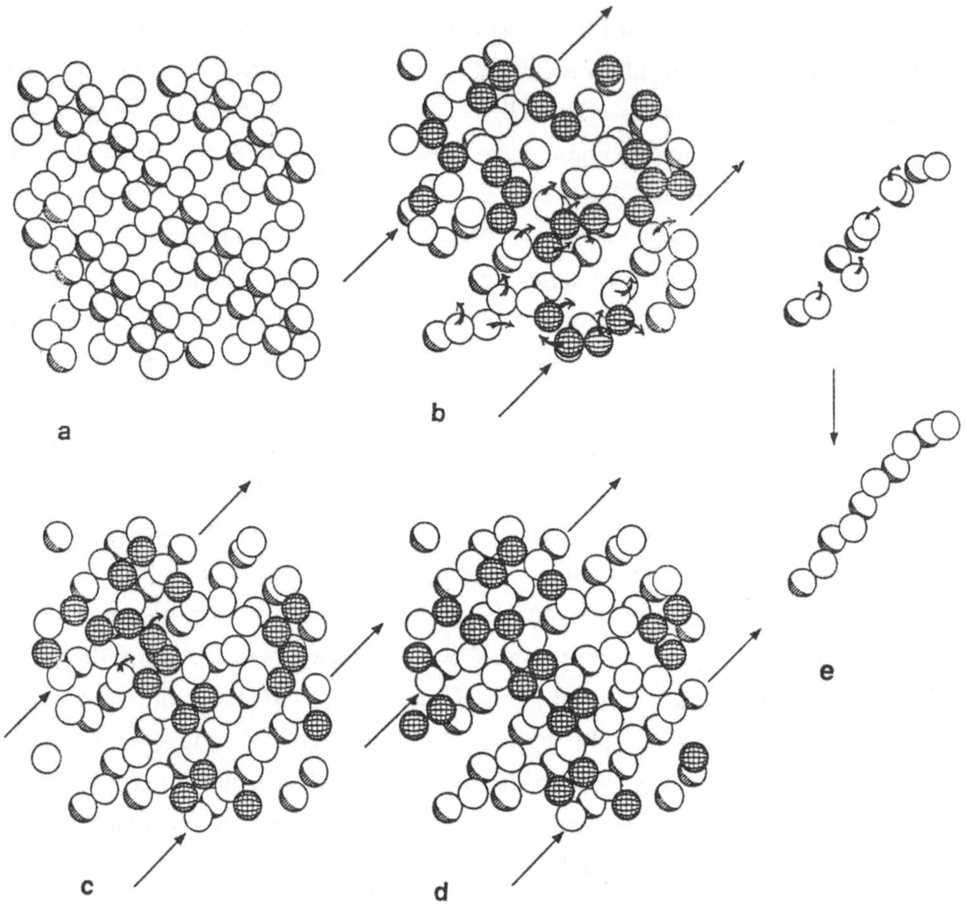

Figure 18. Molecular dynamics simulations of epitaxial growth on $Si\{100\}(2 \times 1)$. (a) Original surface at 800K. The shaded circles represent the surface dimer atoms. The open circles represent the next two subsurface layers. Periodic boundary conditions connect the top and bottom of the surface and the left and the right. The dimer rows proceed from the upper left to the bottom right. (b) The surface after deposition of 1.5ML. The original surface dimer atoms are shown as shaded cicles. The subsurface atoms shown in (a) are no longer displayed. The hatched circles are the topmost deposited atoms and the open circles are the remaining deposited atoms. The curly arrows indicate the directions that the atoms move in the next 512 ps. The straight arrows on the sides of the crystal are a guide to the periodic boundary conditions and the correlated motion mechanism. (c) The surface after 512 ps of equilibration at 800K. The atom representations are the same as in (b). (d) The surface after another 476 ps of equilibration. (e) The dimer opening mechanism. The original dimer atoms are the shaded circles and grow into chains as the deposition continues.

5. CONCLUSIONS

Computer simulation, a scientific newcomer having both theoretical and experimental aspects, provides an alternative approach to scientific investigation. The ever increasing capabilities of modern computers means that realistic modelling of particle surface interactions can be achieved with increasing sophistication. Such modelling is now an essential tool which complements theoretical and experimental investigation. Both chemical and physical effects are amenable to analysis using molecular dynamics. All that is required is an accurate description of the interatomic forcefields and a powerful computer.

The calculation of ejection statistics requires that data from a large number of individual cascades are collected. For such simulations there is no advantage to be gained in using parallelism. The continual addition or removal of particles from surfaces can be modelled more quickly using parallel algorithms either with the cellular approach or molecular dynamics. So far these have not been attempted but when implemented will mean that large scale realistic problems will be solvable.

The models described above fall into three basic categories, categorised in order of increasing computational intensity : (i) the continuum models (ii) the Monte-Carlo models and (iii) the molecular dynamics models. The continuum models can give an accurate description of topographical changes on the scale of a few hundred $\AA$ up to geomorphological scales. They can predict many features in observed in microelectronic device fabrication using beams, in particular the formation of edges and facets from initially smooth surfaces. The cellular models make simplifying assumptions about the erosion or deposition process based on the local environment at the point at which the interacting particle reaches the substrate and have had success in modelling some of the features of film growth under ballistic deposition conditions. However, the most complete picture of morphological changes is provided by the molecular dynamics simulations. These simulations involve the interaction of single particles with the substrate and calculate the subsequent motion of all the atoms in the collision cascade and provide an accurate prediction of atomic scale processes. It is the author's belief that it is with these simulations that the future lies.

5. References

[1] P. Ramanlal and L. M. Sander, *Phys. Rev. Lett.* **54(16)**, 1828, (1985).

[2] G. E. Blonder, *Phys. Rev. B.* **33(9)**, 6157, (1986).

[3] P. Meakin, P. Ramanlal, L. M. Sander and R. C. Ball, *Phys. Rev. A.* **34(6)**, 5091, (1986).

[4] R. Smith, A. H. Osbaldestin, G. Carter and M. J. Nobes, in *Materials modification by high-fluence beams*, R. Kelly and M. Fernanda Da Silva eds NATO ASI Series E no 155, Kluwer, Dordrecht, Netherlands 205,(1989).

[5] F. Family and T. Vicsek, *Computers in Physics*, **4(1)**, 44, (1990).

[6] J. P. Ducommun, M. Cantagrel and M. Moulin, *J. Mater. Sci.* **16**, 52, (1975).

[7] R. Smith and J. M. Walls, *Philos. Mag. A* **42**, 235, (1980).

[8] C. H. Ting and A. R. Neureuther, *Solid State Technol.*, **25**, 115, (1982).

[9] G. Carter, In *Erosion and growth of solids stimulated by atom and ion beams* (Proceedings of NATO Advanced Study Institute, Heraklion, Crete, Greece, 16–27 September 1985), p70, Holland: Martinus Nijhoff, (1986).

[10] M. J. Nobes, *Ibid*, 103.

[11] R. Smith and M. A. Tagg, *Ibid*, 121.

[12] R. Smith, S. J. Wilde, G. Carter, I. V. Katardjiev and M. J. Nobes, *J. Vac. Sci. Technol. B*, **5** (2), 579, (1987).

[13] R. Smith, G. Carter and M. J. Nobes, *Proc. Roy. Soc. London, Ser A*, **407**, 405, (1986).

[14] D. E. Harrison Jr., *Critical Reviews in Solid State and Materials Sciences* **14**, S1, (1988).

[15] R. Smith and D.E. Harrison Jr., *Phys. Rev. B.*, **40**, 2090, (1989).

[16] N. Winograd, B. J. Garrison and D. E. Harrison Jr., *Phys. Rev. Lett.* **41**, 1120, (1978).

[17] N. Winograd, P. H. Kobrin, G. A. Shick, J. Singh, J. P. Baxter and B. J. Garrison, *Surf. Sci. Lett.*, **176**, L817, (1986).

[18] B. J. Garrison, C. T. Reimann, N. Winograd and D. E. Harrison Jr., *Phys. Rev B* **36**, 3516, (1986).

[19] C. T. Reimann, M. El-Maazawi, K. Walzl, B. J. Garrison, N. Winograd and D. M. Deaven, *J. Chem. Phys.* **90**, 2027, (1989).

[20] R. Smith, D. E. Harrison Jr. and B. J. Garrison, *Phys. Rev. B.* **40**, 93, (1989).

[21] R. Smith, D. E. Harrison Jr. and B. J. Garrison, *Nucl. Instrum. Meth. B* **46**, 1, (1990).

[22] B. J. Garrison and N. Winograd, Science **216**, 805, (1982).

[23] D. Srivastava, B. J. Garrison and D. W. Brenner, *Phys. Rev. Lett.* **63**, 302, (1989).

[24] T. Toffoli and N. Margolus, *Cellular Automata Machines* MIT Press p5, (1987).

[25] H. J. Leamy, G. H. Gilmer and A. G. Dirks, in *Current Topics in Materials Science*, E. Katdis (ed), North Holland, Amsterdam 309, (1980).

[26] J. M. Nieuwenheuzen and H. B. Haanstra, *Philips Tech. Rev.*, **27**, 87, (1976).

[27] A. Benninghoven, Z Phys, **230**, 403, (1971).

[28] S. M. Rossnagel and R. S. Robinson, *J. Vac. Sci. Technol. A*, **1**, 426, (1983).

[29] H. H. Andersen and H. L. Bay, in *Sputtering by Particle Bombardment 1* 144, R Behrisch (Ed), Springer-Verlag, Berlin, (1981).

[30] J. A. Sethian, *Commun. Math. Phys.*, **101**, 487, (1985).

[31] J. A. Sethian, *J. Diff. Geom.*, in press, (1990).

[32] M. J. Lighthill and G. B. Whitham, *Proc Roy Soc London*, Ser A, **229**, 218, (1955).

[33] I. V. Katardiev, *J. Vac. Sci. Technol. A*, **7(6)**, 3222, (1989).

[34] I. V. Katardiev, private communication, (1989).

[35] R. Smith, S. S. Makh and J. M. Walls, *Phil. Mag.*, **A47**, 453, (1983).

[36] J. B. Gibson, A. N. Goland, M. Milgrim and G. H. Vineyard, *Phys. Rev.* **120**, 1229, (1960).

[37] R. Smith and D.E. Harrison Jr., *Computers in Physics*, **3(5)**, 68, (1989).

[38] J. Tersoff, *Phys. Rev. Lett.* **61**, 2879, (1988).

[39] D. W. Brenner, *Mater. Res. Soc. Symp. Proc.* **141**, 59, (1989).

[40] G. C. Abell, *Phys. Rev. B.*, **35**, 6184, (1985).

[41] G. Molière, *Z. Naturforsch.*, **2a**, 133, (1947).

[42] J. Tersoff, *Phys. Rev. B.* **39**, 5566, (1989).

[43] T. Ito, K. E. Kohr and S. Das Sarma, *Phys. Rev. B.* **41**, 3893, (1990).

[44] M. W. Thompson, *Phil. Mag.* **18**, 377, (1968).

[45] P. Sigmund, *Phys. Rev.* **184**, 383, (1969).

[46] M. T. Robinson, in *Sputtering by Particle Bombardment*, ed by R. Behrisch, Chapter 3, pp.73-144, Springer-Verlag, Berlin, (1981).

[47] J. P. Biersak and W. Eckstein, *Appl. Phys.* **A34**, 73, (1984).

[48] M. T. Robinson and I. M. Torrens, *Phys. Rev. B.* **9**, 5008, (1974).

[49] R. P. Webb, and D. E. Harrison Jr., *Phys. Rev. Lett.* **50**, 1478, (1983).

[50] R. Smith, *Proc. Roy. Soc. A* in press, (1990).

[51] R. Smith and R. P. Webb, *Nucl. Instrum. Meth. B* in press, (1990).

[52] R. J. Hamers, U. K. Köhler and J. E. Demuth, *J. Vac. Sci Technol. A*, **8(1)**, 195, (1990).

CHARACTERISATION OF DIAMOND AND DIAMOND-LIKE FILMS

P.R. Chalker

Materials & Manufacturing Technology Division
AEA Industrial Technology, Harwell Laboratory
Didcot, Oxfordshire, OX11 0RA, United Kingdom

1. INTRODUCTION

The development of both CVD and PVD technologies for the deposition of diamond and diamond-like carbon (DLC) films has provided a wide range of materials for potential applications in optical, electronic and tribological fields. The range of materials produced by these deposition technologies arise partly from the allotropy of carbon and also from the possibility of forming hydrogenated materials which are widely classified as α-C:H ceramics. Analytical techniques for characterising these materials are important in as much as they may define the product of a set of particular process parameters and that they relate macroscopic properties of the films such as hardness, electrical and thermal conductivity, or optical transparency to microscopic properties such as phase composition, defect structures, impurities and microstructure.By understanding the relationships between macroscopic and microscopic properties, films with optimal behaviour can be produced to meet the requirements of a specific application.

An extensive range of analytical methods now exist for the characterisation of diamond and DLC films many of which are beyond the scope of this review. The purpose of this paper is to introduce the type of information that can be obtained about carbon films using a range of analytical techniques that have been applied to these materials.

2. OPTICAL CHARACTERISATION

2.1 Absorption and Luminescence

The optical characterisation of diamond and DLC films focuses on the reflection/transmission or absorption of photons from the infrared region (0.05 -1.5 eV) through the visible region (1.5 - 3.0 eV) to the ultraviolet (3.0 - 7.3eV). Interactions in the IR region involve the excitation of phonon modes whereas the visible-UV radiation is energetic enough to excite electronic transitions. Measurement of optical absorption in these ranges therefore provides both structural and chemical bonding information about the film. The optical properties of carbon films can be described generally in terms of the complex dielectric function $\varepsilon = \varepsilon_1 + i\varepsilon_2$

Diamond and Diamond-Like Films and Coatings
Edited by R.E. Clausing *et al.*, Plenum Press, New York, 1991

and the complex refractive index N = n+ik (n is the refractive index and k the optical extinction coefficient) of the carbon films. The relationships between these constants at a given frequency are;

$$\varepsilon_1 = n^2 - k^2 \qquad (1)$$

$$\varepsilon_2 = 2nk \qquad (2)$$

$$\alpha = 4\pi k/\lambda \qquad (3)$$

where α is the absorption coefficient and λ the photon wavelength. The absorptive part of the dielectric function ε_2 dictates photon induced electronic transitions and is proportional to the sum of optical transitions between filled valence band and unoccupied conduction band states providing information about the densities of states. The optical bandgap is determined by the electronic densities of states and can be measured from optical absorption studies or by ellipsometric methods[1,2]. Natural diamond has an indirect bandgap of 5.45eV in contrast to sp^2 bonded graphite which has a no bandgap. Diamond and diamond-like carbon films have varying sp^2-sp^3 bonding configurations and consequently have bandgaps ranging from 0.5 to 5.3 eV[3].

When donor, acceptor or defect states exist within the band gap optical absorption spectra can be used to probe electronic excitations between these states. At 300K optical absorption spectra generally consist of rather broad peaks of several hundred meV halfwidth. For absorption spectra of bulk diamond and diamond film samples recorded at temperatures below 77K these broad bands may be resolved into fine structure due to coupling between the excited states and the phonon modes of the lattice[4].The absorption spectra of DLC films arise from the excitation of π and σ-type electrons to the conduction band via $\pi - \pi^*$ and $\sigma - \sigma^*$ transitions.

Complimentary emission or luminescence can occur on relaxation of the excited state by electron transfer back to the valence band or an acceptor level. Luminescence can be studied using various means of excitation,optical stimulation leads to photoluminescence, excitation by an electron beam is referred to as cathodoluminescence and radiative recombination at a Schottkey barrier or pn-junction is called injection luminescence. The luminescent decay is therefore also characteristic and has been used to investigate the nature of defects and dopants within carbon films having a bandgap.

Cathodoluminescence has the advantages that the exciting electron beam can supply a high current density producing intense luminescence in comparison with other excitation sources and that spatially resolved information can be obtained. Disadvantages include specimen charging and considerable sample heating which may be deleterious for sensitive materials. Cathodoluminescence is a near surface sensitive technique depending on the energy of the exciting electron beam which may produce luminescence from several tens of microns within a sample although surface states may have a significant contribution to the emission spectrum. Cathodoluminescence of diamond films is frequently characterised by a broad band (Band A)[5] emission assigned to a transition between closely separated donor-acceptor states or quasi-one-dimensional donor - acceptor pairs localised along a dislocation core [6]. This is illustrated in figure 1 which shows spectra of annealed (350°C, 3 hours) and boron doped diamond films. The band A emission is centred at approximately 2.8eV which is similar to natural diamond [7]. The presence of appreciable boron doping results in a shifted broad band emission centred at 2.4eV and the narrow colour centre related luminescence bands associated with the other CVD

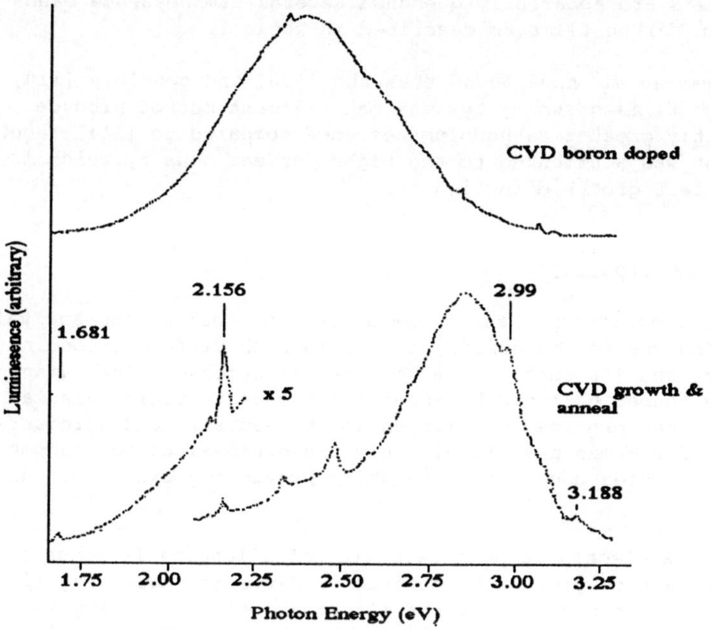

Figure 1. Cathodoluminsence of diamond films after different treatments.

Table 1. Luminescence bands observed in diamond

Band position (eV)	Assignment	Reference
1.673	GR1:isolated neutral vacancy (rad)	6,9,12
1.681/1.682	Si related	9,12
2.0	Nitrogen (N)-vacany complex	6
2.156	N-vacany complex (rad)	9,12
2.2		9
2.322 & 2.335	Related to N	9
2.326/2.368	Second N-vacancy complex	6,12
2.462/2.466	H3 vacancy - N2 complex (rad)	9,11
2.476		13
2.479		9
2.567		9
2.599		9
2.642/2.638	TR12 (rad)	9,11
2.670		9
2.756		9
2.786		9
2.807	N-C(interstitial)(rad)	6,9
2.82	dislocation related	6
2.985	N3 (IIa natural diamond)	13
2.988		9
3.092		9
3.186/3.188	Interstitial N or N-C complex (rad)	9,10,11
3.272		9
3.853		13
4.582	5RL C interstitial zero phonon & 4 lines seperated by 235meV	9
(rad) - induced during irradiation, N - nitrogen		

diamond films are apparently quenched.Several luminescence bands have been observed in carbon films as described in Table 1.

Feldman et al[8] have found that the [100] and possibly [110] facets of CVD diamond films grown by the thermal filament method produce significantly greater cathodoluminescence compared to [111] facets. This observation was attributed to the higher defect density which is expected along the fast growth direction.

2.2 Raman Spectroscopy

Raman spectroscopy has become a powerful tool in the analysis of carbon films due to its ability to distinguish different bonding types, domain size and its sensitivity to internal stresses. The Raman peak shift, the shape, half-width, intensity and polarisation are each parameters that provide information about chemistry and structure. The intensity of a Raman peak is directly proportional to the concentration of scattering species under analysis which provides a basis for quantitative analysis

When the electronic polarizibility of a lattice is constant or static during the lattice vibration, elastic or Rayleigh scattering of an incident photon can occur. If the electronic polarizibility is modulated by vibrational or rotational motions the change in polarisation may lead to Raman scattering which is an inelastic process. The inelastically scattered Raman photon has a frequency $\nu \pm \Delta\nu$ where $\Delta\nu$ is the frequency associated with a change in either the vibrational, rotational (or electronic) state of the chemical system. Photons emerging with frequency $(\nu - \Delta\nu)$ are said to have undergone Stokes Raman scattering and photons of frequency $(\nu + \Delta\nu)$ anti-Stokes Raman scattering. In practice the thermal population of anti-Stokes initial states is relatively small and consequently the anti-Stokes Raman spectrum is very weak. The Raman scattering process has a low cross section and is orders of magnitude (typically 10^{6}-10^{10}) less probable than elastic Rayleigh scattering. The frequency of scattered radiation is independent of the energy difference between the initial and final vibrational states and consequently sources for Raman instrumentation can be coherent or incoherent with frequencies in the ultraviolet to near infrared spectrum.

The face centred cubic crystal symmetry of diamond results in a triply degenerate first order phonon and as the wave vector of the incident radiation is considerably smaller than the Brillouin zone extension, single photon or first order Raman scattering produces information about phonons near the zone centre (k=0). The first order Raman peak for diamond appears at $1332cm^{-1}$ and is diagnostic of the presence of diamond in carbon films. The structure of graphite leads to two lattice vibrations that meet the requirements for Raman scattering and these modes occur within the sp^2-bonded plane giving rise to peaks at 42 and $1581cm^{-1}$(the so called G - line, E_{2g}). The low wavenumber peak is broadened by disorder within the lattice and becomes very weak in intensity . As the level of disorder increases a new peak at $1360cm^{-1}$ (the D-line) appears which is associated with microcrystalline domains. Decreasing domain sizes lead to increases in the intensity and width of the $1360cm^{-1}$ peak with an increase in the wavenumber of the G-line due to a breakdown in the selection rules and sampling of other optical phonon modes . In addition to the influence of microstructure the graphitic G-line increases in intensity and shifts to higher wavenumber when the exciting photon energy is increased[14]. For sp^2- bonded clusters in

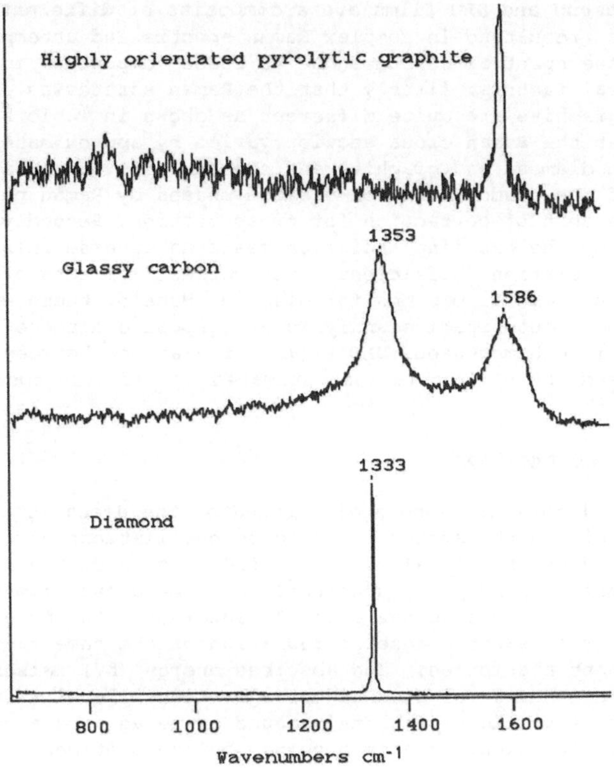

Figure 2. Raman spectra of highly orientated pyrolytic graphite, glassy carbon and natural diamond.

amorphous carbon resonantly enhanced emission occurs at photon energies close to the π -π* electronic transition (3.5 - 6.5eV)[15] Similarly Yoshikawa[16] has observed this effect in α-C films.

Two photon or second order Raman scattering may occur involving a pair of photons with equal and opposite wavevectors. The second order Raman spectrum of diamond reflects the two photon combined density of states and this spectrum is very weak below 1800cm^{-1} with a cuts-off at 2680cm^{-1}. In CVD diamond films the second order spectrum is characterised by a sharp peak at 2450cm^{-1}.[17] The second order Raman spectra of films showing the D-line exhibits a peak at 2710cm^{-1}.

Table 2. Raman scattering efficiencies for diamond and graphite modes.

Sample	Raman scattering efficiency (cm^{-1}sr^{-1})	Reference
diamond F_{2g}	6.06 x 10^{-7}	18
	9 x 10^{-7}	19
graphite E_{2g}	306 x 10^{-7}	20
	500 x 10^{-7}	19

Many diamond and DLC films are a composite of differently bonded carbon phases resulting in complex Raman spectra and attempts to deconvolute the spectral features due to these components must take into account several factors; firstly that the Raman scattering efficiency for diamond and graphite are quite different as shown in Table 2:which indicates that the Raman cross section varies by approximately a factor of fifty between diamond and graphite and consequently quantitative assessment of the diamond and non-diamond phases by Raman peak area requires some sort of correction for cross section. Secondly the volume of film sampled by the exciting radiation bears an inverse relationship to the optical absorption coefficient which varies from tens of nanometres for graphite to several microns for diamond. Both of these effects currently hinder quantitative analysis of composite diamond films. However, for more homogeneous DLC films correlations between Raman peak intensities and the sp^3/sp^2 ratios estimated by infrared spectroscopy show more promise.[16]

2.3 Infrared Spectroscopy

Lattice vibrations cause a distortion of the electric charge distribution within the system leading to oscillations of the electric charge governed by the normal vibration frequencies of the lattice. An oscillating molecular dipole, quadrupole or some other component of the oscillating electric charge can directly interact with the oscillating electric vector of electromagnetic radiation of the same frequency, giving rise to resonant absorption. The absorbed energy ($h\nu$) raises the system from one vibrational level to a higher vibrational level which will occur statistically from the vibrational ground state as a consequence of the population of vibrational states governed by the Boltzmann distribution.

The interaction of chemical bonds with radiations of infrared frequencies is of predominantly dipole character (quadrupole and higher terms are only weakly interacting) and only transitions involving vibrational states coupled by a dipole operator are observed in general. Not all vibrational modes lead to a changing electric dipole and will not absorb or be "infrared active". Measurement of the spectrum of absorbed radiation is the basis of infrared spectroscopy and the infrared spectrum is a measure of the radiation either transmitted or reflected by the specimen as a function of frequency. Infrared spectra can be measured either by dispersive or interferometric methods: the dispersive method involves dispersing radiation from an incoherent source via a prism or grating and is comparatively inefficient of source intensity requiring monochromator entrance and exit slits. In contrast, Fourier Transform IR (FTIR) samples all the frequencies of radiation from the source simultaneously using an interferometric method resulting in improved signal-to-noise ratios for the same acquisition time as dispersive IR.

The infrared spectra of bulk diamond and thick CVD diamond films[21] are comprised of broad bands due to the phonon absorption modes. Although the single phonon mode is forbidden on symmetry grounds, the presence of impurities such as nitrogen and boron or other defects cause the mode to become infrared active. The presence of the single phonon absorption has been used to measure the concentration of nitrogen present in bulk diamonds and even to infer its chemical bonding[22].

Other absorptions in diamond films have been related to the presence of intentional dopants such as boron [23] and to hydrogen incorporated during growth. C-H stretching and deformation modes tend to be weak absorptions in IR spectra and are complicated by the presence of interference bands due to the films finite thickness and scattering due to grain boundaries and surface roughness.These problems make quantitative absorption measurements

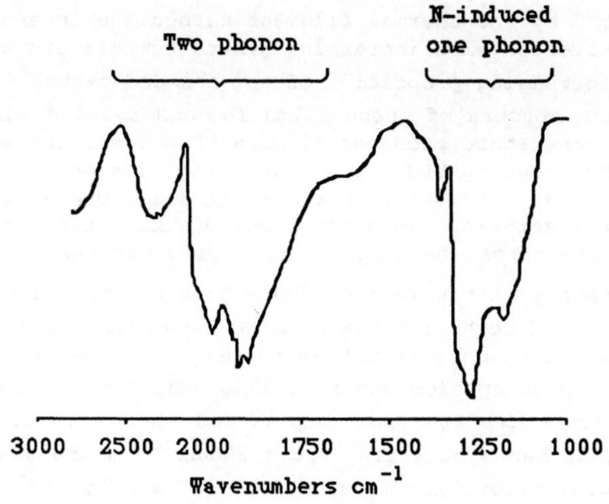

Figure 3. Infrared absorption spectra of 140μm diamond powder

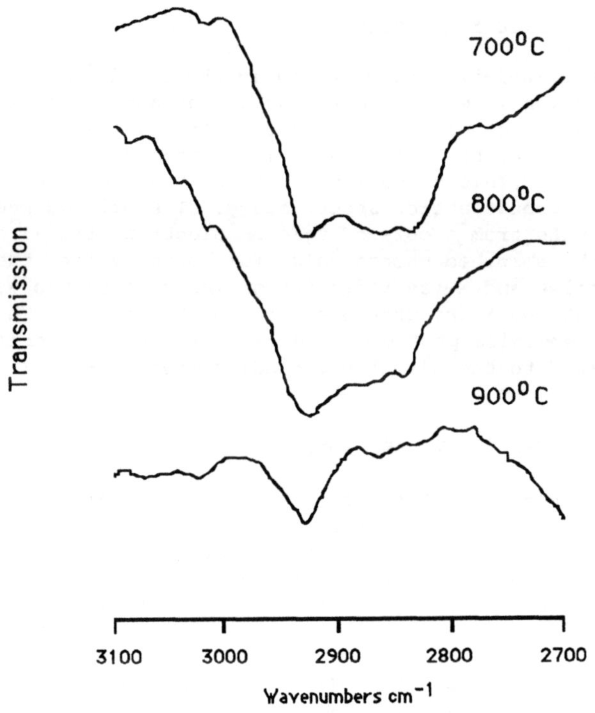

Figure 4 Infrared absorption spectra of C-H streching bands of CVD diamond films grown at different temperatures.

quite difficult in crystalline diamond films and require integrating optics to detect all of the scattered radiation. Figure 4 shows a portion of the absorption band due to C-H stretching modes from CVD diamond films grown between 700 and 900°C by the thermal filament method.The intensity of the absorption band decreases with increasing growth temperature which is accompanied by an increasing proportion of sp^3- bonded carbon indicated at 1332cm^{-1} in the Raman spectra of these films.The observation indicates that the higher growth temperature produces films with a lower concentration of chemically bonded hydrogen.In DLC films containing considerably higher hydrogen concentrations (to 50%) IR absorptions are comparatively more intense and complex absorption bands can occur due to a range of aromatic and aliphatic $C\text{-}H_x$ stretches, bending or other deformations. The stretching modes fall approximately in the region 2700 - 3310cm^{-1} of the infrared spectrum and Coudrec and Catherine[24] have analysed broad bands in this region recorded from films deposited from 50 kHz discharges of methane. C-H stretches from sp^1-bonded species occur at 3300 - 3310cm^{-1}. However, the sp^2 and sp^3 C-H stretching modes are more complex and lie in the range 3300 - 2700 cm^{-1}. In general, bands occurring above 2960cm^{-1} can be assigned to sp^2 (aromatic or olefinic) stretching modes and those below to sp^3 or aliphatic modes. Consequently, by summing the areas of the absorption bands in these two regions, the proportion of sp^2 and sp^3 bonded carbons can be deduced. Carbon - carbon stretch and C-H deformation modes tend to be less intense than the C-H stretch in DLC films and fall in the region 1650 - 800 cm^{-1}.

3.ELECTRON SPECTROSCOPIES

3.1 Photoelectron Spectroscopies (PES)

Electromagnetic radiation with energy in the 1 - 10^4 eV range may undergo photoabsorption in which the energy of the photon is transferred to an atom resulting in electronic excitation. If the quantum of energy absorbed is greater than the first ionisation potential of the atom an electron may be emitted.This mechanism is called the photoelectric effect and is the basis of photoelectron spectroscopy. The emitted photoelectron may originate from a core or valence electron orbital depending on the energy of the absorbed photon which fall broadly into the two categories ultraviolet and X-ray radiation giving rise to the acronyms UPS and XPS (ultraviolet and X-ray photoelectron spectroscopy). The kinetic energy (E_k) of the emerging photoelectron is an important parameter because it is related to the electronic binding energy (E_b)

$$E_k = h\nu - E_b - e\phi \qquad (4)$$

where h is Planck's constant, ν the frequency of the incident radiation and ϕ is the work function of the material. The term $e\phi$ is the minimum energy required to move an electron from the material into free space and is typically of the order of a few electron volts. The vacant electron hole left after photoemission may undergo decay via radiative fluorescence or an Auger process.

Ultraviolet radiation (e.g. the He(I) = 21.22 eV and He(II)= 40.81 eV)) has only sufficient energy to excite photoemission from the valence bands or shallow core levels of many elements providing a detailed insight into the electronic bonding configurations or valence densities of states of atoms in the surface region of a specimen. Measurement of the higher

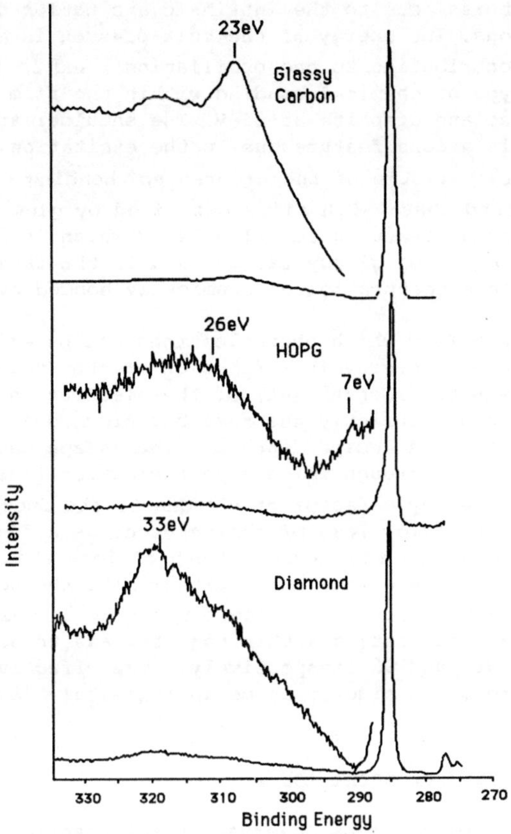

Figure 5 Plasmon loss features in XPS C(1s) spectra of highly orientated pyrolytic graphite, glassy carbon and natural diamond.

binding energy core levels requires photons with energies of hundreds of electron volts and to obtain this information X-rays must be used (e.g. $AlK_{\alpha 1,2}$ (1486.6 eV) and $MgK_{\alpha 1,2}$ (1253.6 eV)) which affects the measured kinetic energy of a photoelectron as a consequence of equation 4. The width of the X-ray line also contributes to the ultimate resolution of photoelectron kinetic energies which may overlap for core levels of different elements or obscure chemical shift information.

The binding energy of a photoelectron emitted from an atom in a solid may be modified by the surrounding chemical environment. The change in binding energy of a particular electronic orbital due to different chemical environments is called the "chemical shift".The chemical shift of the C(1s) photoelectron peak is relatively invariant in different forms of carbon shifting by only about 1eV[25] which is further complicated by surface charging effects found in insulating films. The photoelectron spectrum reflects aspects of both initial and final electronic states of a specimen and subtle spectral features due to electronic configuration interactions, multiplets, plasmon excitation and other processes can occur. Plasmon losses in the XPS C(1s) spectrum can be used to distinguish various carbon phases. For example, figure 5 shows the C(1s) spectra of diamond, glassy carbon and highly orientated pyrolytic graphite with their associated plasmon loss structures due to the intrinsic excitation of collective electron oscillations. The energy of the bulk plasmon is a function of the electron density contributing to the oscillation[26] and is therefore sensitive to the type of chemical bonding within the film. Diamond has a plasmon at 31eV loss and graphite at 26eV. The shoulder at 7eV in the graphite spectrum is a loss feature due to the excitation of a electronic $\pi - \pi^*$ transition characteristic of the ordered sp^2 bonding of graphite. Catherine [27] has found that α-C:H films deposited by glow discharge techniques show plasmon features at 21 - 24 eV which is in a similar position to the spectrum of glassy carbon, and in the case of the films was attributed to a high concentration of chemically bonded hydrogen.

The sampling depth from which photoelectrons can be emitted from the specimen without loss of energy is a function of the Inelastic Mean Free Path (IMFP) or Attenuation Length (AL) of the electron in the solid. These parameters are not strictly the same but provide a physical basis for modelling an effective escape depth λ. The escape depth is defined as the thickness of sample through which a photoelectron flux will pass before it is attenuated by a factor of e^{-1} due to the energy loss processes. Different energy loss processes occur as a function of kinetic energy and result in a minimum escape depth of less than 1nm at 50-100 eV. The incident X-rays penetrate much further into the solid (typically 10^5nm) generating photoelectrons and the proportion of measured photoelectron signal originating within one, two and three escape depths correspond to 63%, 86% and 95% respectively. The effective sampling depth in PES may therefore be considered to be approximately 3λ at normal emission[28].

3.2 Auger Electron Spectroscopy (AES)

Auger electron peaks can be observed in PES spectra; Auger spectra may also be generated using ions or most commonly energetic electrons. The Auger process arises from the non-radiative decay of a vacancy in a core-level and can be thought of as a three electron process. The kinetic energy of the Auger electron is independent of the energy of the incident probe unlike photoelectrons. The complimentary nature of the Auger and X-ray emission processes has led to the use of X-ray notation in describing Auger transitions. A transition involving the transfer of an electron from the carbon C(2p) band to a C(1s) vacancy with the subsequent emission of an Auger electron from the C(2p) band is denoted $KL_{2,3}L_{2,3}$ and the Auger electron kinetic energy E_{aug}

$$E_{aug} = E'_K - E'_{L23} - E'_{L23} - e\phi \qquad (5)$$

where the prime E' indicates a binding energy.The typical shape of an Auger spectrum generated by an electron probe is dominated by an intense peak of elastically scattered electrons of the primary beam energy, an intense background of inelastically scattered electrons and a broad band of true secondary electrons below ~50eV. The Auger peaks are imposed on this intense background and are readily detectable if the first derivative is plotted as dE.dN(E)/d(E) versus E (sometimes referred to as the N'(E)/E mode). When one or two of the Auger final state holes reside in the valence band, the shape of the transition reflects the VB DOS. The C $KL_{23}L_{23}$ transition reflects a convolution of the occupied valence band states and can be used to distinguish diamond, graphite and amorphous carbons[29].Figure 6 shows the C $KL_{23}L_{23}$ in the N'(E)/E mode of highly orientated pyrolytic graphite,a CVD diamond film and a natural diamond reference.

Electron induced AES also has the advantage that the primary beam

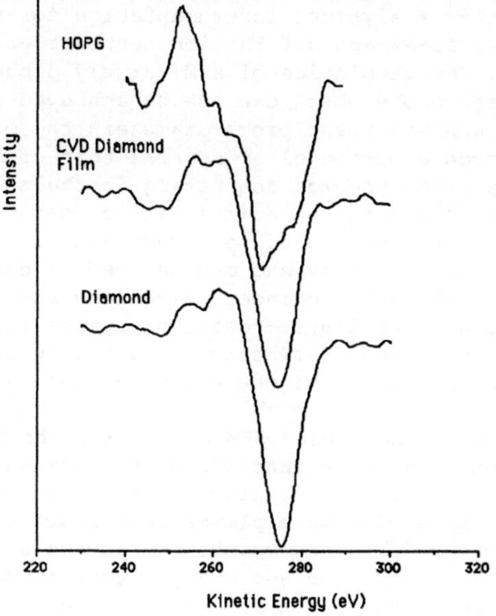

Figure 6 'N(E)/(E) Auger spectra of graphite and natural diamond compared with a CVD diamond film.

can be collimated into a probe of diameters approaching 10 nm. To achieve these fine probes high brightness field emission sources and high beam voltages must be employed to overcome spacecharge effects within the probe. Resolution can be improved at lower beam currents with a corresponding decrease in Auger signal and sensitivity. The primary beam voltage also has an affect on the Auger cross-section and may lead to specimen degradation or electron impact induced desorption. The focused electron probe can be used to map elemental distributions across the surface of a specimen by rastering the probe, counting at pixel elements until sufficient signal has been acquired. Because the background function consists of both backscattered and inelastic electrons, it is common practise to record signals not only of electrons from an Auger peak but also from one or two points in the nearby background. Although this slows the image acquisition process, digital maps of the ratio of peak P to background B or P/(P-B) may be reconstructed which serve to eliminate some of the topographically induced effects on signal intensity.

Like PES, Auger spectroscopy is quantitative with a variation of less than two orders of magnitude in the Auger cross-section throughout the Periodic table. An important factor that must be considered in electron induced AES is the backscattering correction coefficient[30]. Energetic primary electrons cause secondary electron emission below the surface of the specimen, some of which have sufficient energy to ionise atoms in the surface region. Bishop[31] has shown that the backscattering coefficient is a function of specimen atomic number which may have a significant effect on quantification and the signal contrast observed in Auger microscopy.

4. ANALYTICAL ELECTRON MICROSCOPIES

One of the most widely available analytical tools is Scanning Electron Microscopy (SEM) which principally involves imaging the secondary electrons which are stimulated by the primary beam. The conventional SEM may often include other analytical instrumentation for the examination of X-ray emission, backscatter and diffraction contrast effects or possibly cathodoluminescence. The resolution of SEM largely depends upon the formation of a focused probe which can now be achieved with a diameter of less than 5 nm. To achieve these probe diameters the primary electrons are accelerated through a potential of several tens of keV and the incident primary electrons undergo scattering in the solid losing energy by various processes including the stimulation of secondary electron emission. The energy distribution of secondary electrons is sharply peaked at low energies of below a few eV and can be used to examine the topography of a film. The slow secondary electrons are extracted by a high potential applied to a scintillator- detector system and secondary electron images are generated by synchronising the optical output of the detector system with the raster of the electron probe across the sample.

Williams and Glass[32] have used SEM to examine the variation in morphology of diamond films as a function of gas composition and position within the deposition system. Films grown using 1% methane (balance hydrogen) source gas in a microwave plasma were shown to exhibit marked variation in phase composition and morphology across a 2cm x 1cm substrate. Figure 7 (a) shows the secondary electron image from the centre of the sample which consists of fourfold {100} faceted crystallites. In the region 0.25 cm from the specimen edge smaller facets appear to be superimposed on a "cauliflower" morphology. Within 0.05cm of the edge of the specimen only the cauliflower morphology is observed which is attributed to an increase in the sp2-bonded carbon within the film. The variation in the local growth mode is principally due to temperature variations across the substrate during growth.

As in Auger microscopy, some of the primary electrons are backscattered, and the efficiency of this process is a function of the target atom atomic number Z. The backscattering yield varies significantly more than the secondary electron yield across the periodic table and consequently the backscatter image will have stronger contrast for samples with grossly varying volumes of different atomic number[33]. As the backscattered electrons have the energy of the incident beam they are not as surface sensitive as the low energy secondary electrons and their escape depth may be of the order of 100 times greater. The extra scattering that the backscattered electrons undergo within the bulk of the specimen results in poorer lateral resolution compared with the secondary electron micrograph.

Another analytical tool which is employed in electron microscopy is analysis of the X-rays which are generated by the radiative relaxation of a core-level hole excited by the primary beam. The X-rays emerge from 0.5 µm or greater depths below the specimen surface and are characteristic of the target atom's electronic structure. X-ray analysis can be achieved by energy dispersive or wavelength dispersive spectroscopies (EDS and WDS, respectively). The EDS system typically uses a lithium doped silicon detector in conjunction with a Be window for analysis of elements of $Z>9$. This system is faster than WDS but has a poorer energy resolution. The range of EDS can be extended by using a windowless detector allowing analysis of elements down to $Z=5$(B). The WDS method uses a range of crystals of materials with different lattice parameters. The crystals are rotated to focus X-rays of different wavelengths onto the X-ray detector. X-rays from elements of atomic number >6 can be analysed. Both EDS and WDS provide quantitative methods of evaluating the elemental composition of thin films although care must be taken to account for the volume of specimen from which the X-rays are excited. The sampling depth of these methods can be varied by changing the primary beam energy so that signal

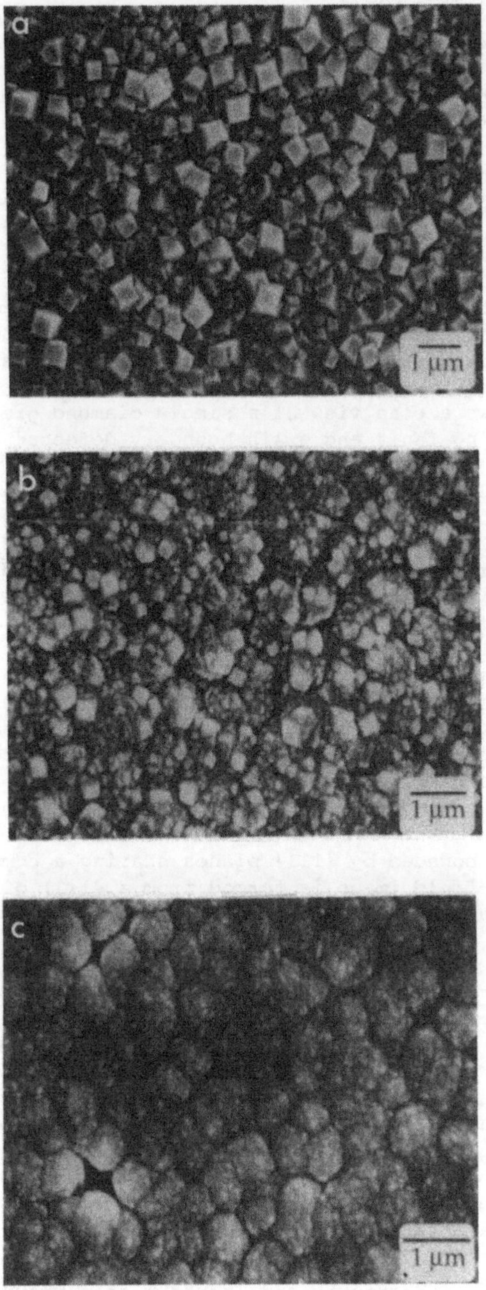

Figure 7 (a) SEM micrograph of the center region of a 2cm x 1cm diamond sample grown with 1% CH_4 in H_2. (b) SEM micrograph of the diamond film 2.5mm from the edge of the sample. (c) SEM micrograph of the diamond film within 0.5mm of the edge of the sample.
(With permission of B.E. Williams and J.T. Glass, Copyright of the Materials Research Society)

from the substrate can be minimised if necessary. Like the secondary electron micrograph, the X-ray signal can be recorded in synchronisation with the beam raster and elemental maps can be generated with a resolution governed by the excitation volume rather than the electron probe diameter.

In Transmission Electron Microscopy (TEM) the electrons transmitted through a specimen are examined for diffraction effects. This has two consequences: firstly that the primary beam voltage used in TEM has to be of the order of a few hundred kV and secondly that the specimen must be thinned to dimensions of the order of a micron or less to make the sample electron transparent. The thinning process may be carried out by a combination of mechanical polishing, chemical etching and ion beam milling and samples can be thinned either through their thickness for plan view TEM or across their width providing a sample for cross- sectional or XTEM.

The electron beam transmitted through a thinned sample does not undergo the same scattering as in a bulk material and the emerging electrons can be imaged to show lattice features with resolutions of 0.2 nm. TEM can therefore be used to image defects which is illustrated in Figure 8 which shows a plan view of a single diamond grain showing twin, dislocation, stacking fault and grain boundary defects[32].

Williams et al[34] have used high resolution electron microscopy (HREM) to image the lattice within the core of a multiply twinned zone within a CVD diamond film. The five-fold multiply twinned geometry in figure 9 must accommodate approximately 7.5^{o} of misfit which is accomplished by the formation of a series of dislocations on twin boundaries.

The directly transmitted image is usually referred to as a bright-field image, however an image can also be formed from the diffracted electron beam which gives rise to the so-called dark-field image. In the dark-field image bright regions correspond to features satisfying the Bragg diffraction criteria and this method can be used to study microstructural phase relationships. Figure 10 shows a pair of complimentary bright and dark-field images from a multiply twinned diamond crystallite[34]. The fine growth in the individual sectors consist of twinned tetrahedra bounded by {111} planes sharing a common [110] axis. The weak beam dark-field image is formed from a (111) diffraction spot and shows the twinned regions as bright streaks.

The TEM can be coupled with analytical tools such as microdiffraction, core-level electron energy loss spectrosocpy (CEELS)[35] and X-ray analysis. Detailed microdiffraction can be obtained by either focussing the beam onto the area of specimen investigated (convergent beam diffraction) or by placing a selective area aperture after the objective lens which is often referred to as selective area diffraction or SAD. The CEELS analysis requires energy analysis of the high energy transmitted electrons. At such high energies a magnetic sector electron spectrometer is required to disperse the electrons with sufficient resolution. The transmitted electrons lose energy by core level ionisation processes which are again characteristic of the electronic configuration of the target atoms. Fine loss structures arise due to excitation of plasmons which may provide a fingerprint loss spectrum reflecting the chemical environment of the target atom.

5 X-RAY DIFFRACTION ANALYSIS

The diffraction of electromagnetic radiation by periodic atomic arrays provides the basis of a range of techniques for structural analysis. X-rays are predominantly scattered by the electrons within

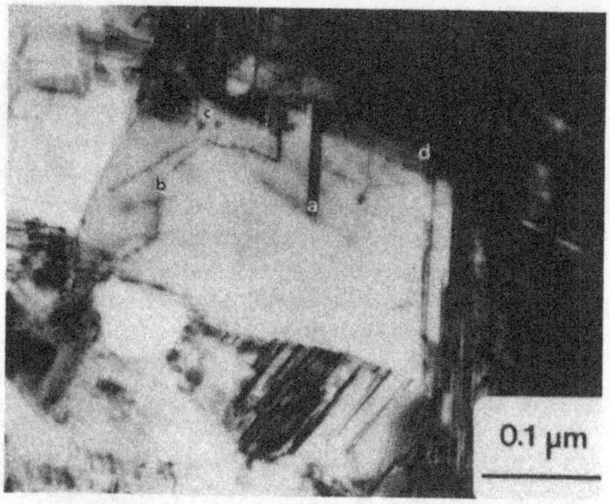

Figure 8. Electron micrograph of single diamond grain exhibiting (a) twins, (b) dislocations, (c) stacking fault tetrahdra, and (d) a grain boundary.(With permission of B.E. Williams, and J.T.Glass, Copyright of the Materials Research Society)

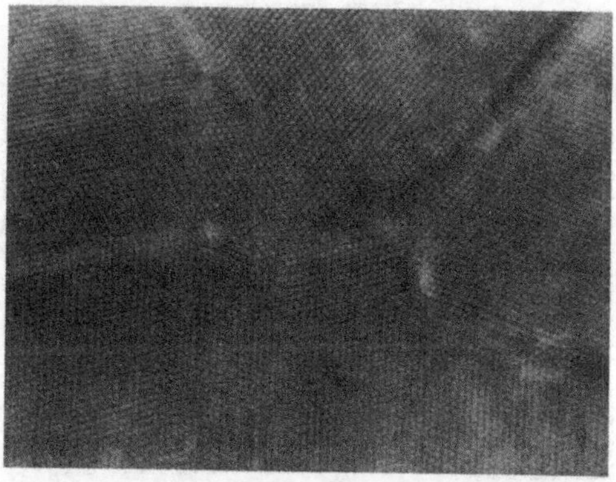

(a)

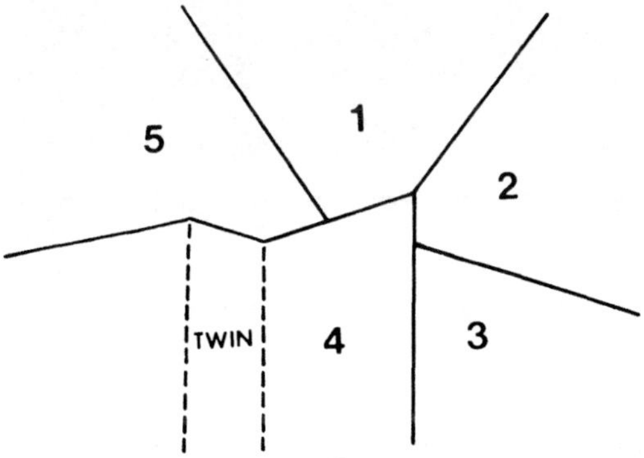

(b)

Figure 9. (a) A high resolution electron micrograph of the core of a fivefold multiply twinned particle.(MTP) (b) A schematic of the growth sectors of the MTP in (a).(With permission of B.E. Williams, H.S.Hong and J.T.Glass, Copyright of the Materials Research Society)

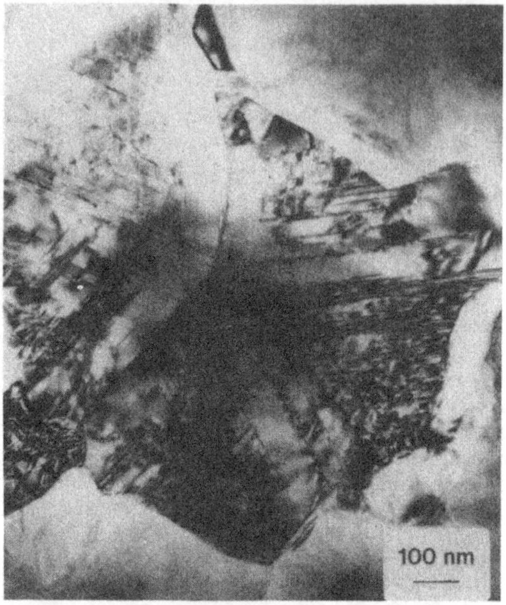

(a)

(b)

Figure 10 (a) A Bright-field micrograph of a multiply twinned particle comprised of tetrahedra bounded by {111} planes sharing a common [110] axis and twinned relative to each other. (b) A centred weak beam dark-field micrograph imaged by using a (111) diffracted spot.
(With permission of B.E. Williams, H.S.Hong and J.T.Glass, Copyright of the Materials Research Society)

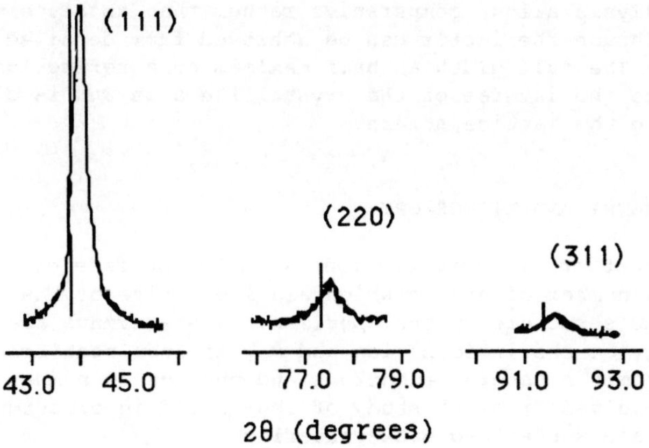

Figure 11. X-ray diffraction pattern of a CVD diamond film compared with the stick pattern of a randomly oriented diamond sample.

atoms. The scattering factor of X-rays is smaller than that of electrons; therefore, X-ray diffraction provides structural information from a larger volume of material compared with complimentary electron diffraction techniques. Constructive scattering from an atomic array of interplanar spacing d gives rise to the diffraction condition when the Bragg equation

$$n\lambda = 2\ d\ \sin\ \theta \qquad (6)$$

is satisfied where the parameter n is an integer, and λ the wavelength of the radiation. Diffraction leads to maxima in the intensity of the scattered X-rays and the position of the maxima provides information about the size and the shape of the unit cell whilst the width of the maxima can be used to evaluate the size, orientation and strain of grains in polycrystalline materials.

The preferred orientation or "texture" of a polycrystalline material can also be investigated by X-ray diffraction[14] as certain sets of crystalline planes will be aligned with respect to the sample surface. Figure 11 shows diffraction peaks for a polycrystalline CVD diamond film grown by the thermal filament method. The "stick" pattern shows the diffraction angles (2θ) and the relative intensities from a randomly orientated standard which closely resembles the reflection positions and intensities of the diamond film. The texture of films may be influenced by growth conditions and Kobashi et al[36] have made a detailed study of the morphology of diamond films deposited onto silicon and microcrystalline diamond. Films grown directly onto polished silicon using microwave assisted CVD (30 Torr and 800°C) were shown to exhibit a morphological dependence on the partial pressure (C%) of methane used. For C% of 1.6 the films consisted of microcrystalline diamond with some amorphous-graphitic carbon; however, for C% of 0.2 to 1.2 sub-μm faceted films were grown. At C% > 0.4 the films showed a preferred (100) texture in SEM. At C% , 0.4 a (111) texture was adopted. If the film is grown on the plasma etched microcrystalline layer, predominantly (110) facetted material was deposited with C% < 1.0 which was illustrated by the predominant (220) reflection in the XRD spectrum.

The width of Bragg reflection peaks can be used to determine crystallite size and lattice strain. In its simplest form, line broadening analysis allows comparative rather than absolute measurements to be made although the latter can be achieved from detailed diffraction measurements. The full width at half maximum of a reflection peak is proportional to the inverse of the crystallite size and is directly proportional to the lattice strain.

6. ION SPECTROMETRY AND MICROSCOPY

The interaction of energetic ions with the surface of a solid may give rise to a number of events which are indicative of the chemical composition and structure of the specimen. These events are dependent upon the energy of the incident ion and may include scattering or the emission of atomic clusters, electrons and photons. A number of analytical methods have evolved from the study of ion- solid interactions including Secondary Ion and Rutherford Backscattering .

6.1 Secondary Ion Mass Spectrometry (SIMS)

The energy of incident primary ions used in SIMS analysis typically falls in the range 0.5-50 keV. Structural and chemical information is derived from the mass and energy analysis of atomic or molecular ions emitted from the surface as a consequence of the energy transfer that occurs between the primary ion and the solid. The emission of charged and neutral clusters from the surface is called sputtering and almost all of the emerging or secondary ions originate from the outermost atomic layers of the bombarded specimen, giving rise to the surface sensitivity of SIMS.

As a heavy primary ion enters into the surface region of a sample it transfers energy to the surrounding target atoms. The interaction gives rise to elastic collisions, transfer of translational energy to target atoms and also electronic excitation of target atoms. The increased translational energy of the target atoms along the length of the primary ions' trajectory result in collisions with neighbouring target atoms leading to a collisional cascade. The lifetime of the cascade is of the order of picoseconds and extends through a volume of the target with a diameter of the order of 10 nm. As the cascade propagates to the surface atoms receiving enough translational energy may be sputtered and ejected from the material. The collisional cascade has other consequences in that it induces damage and disorder in the selvedge region as target atoms recoil into the substrate. This mixing process is a limiting factor in the depth resolution which can be achieved with ion analysis.

The fluence of primary ions and the efficiency with which sputtering occurs (sputter yield) are two important parameters in determining how fast the bombarded material is eroded. This erosion process offers the opportunity of limiting the analysis to the outer surface, termed Static SIMS, or to dynamically sputter the specimen to examine the secondary ionyield as a function of depth. Primary ion probes of the order of a micron in dimensions can be generated from duoplasmatron or Penning sources using condenser lenses and collimating apertures. Liquid metal ion sources can produce probes of 20 nm diameter by field emission from needle sources wetted with low melting point metals such as gallium and gold. The emitted secondary ions are extracted in an electrostatic field and passed to the input aperture of the spectrometer to be analysed. Spatially resolved information can be obtained by monitoring the secondary ion signal in synchronisation with a focused primary beam or by using secondary ion microscope optics which are similar in principle to the light microscope.

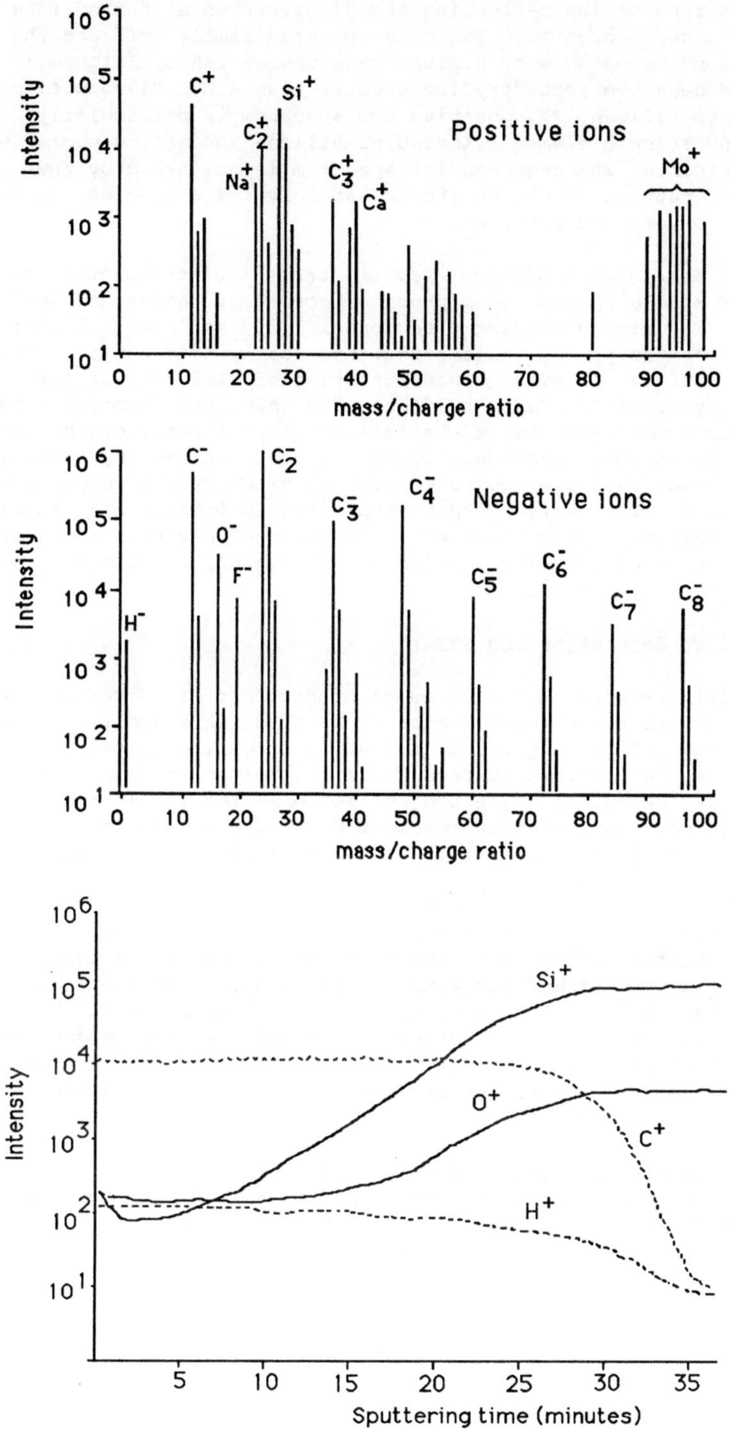

Figure 12. SIMS spectra and depth profile of a CVD diamond film.

The data from SIMS may be in the form of a mass spectrum, an image of a particular type of ion reflecting its distribution at the surface of the specimen or a depth profile. The mass spectrum simply reflects the numbers of ions detected with a given mass/charge ratio. Figure 12 shows positive and negative secondary ion spectra from a CVD diamond film deposited onto silicon. The positive ion spectrum is particularly sensitive to cationic elements including silicon and molybdenum as well as other contaminants. The negative ion spectrum is dominated by the fragmentation pattern of carbon atomic and polyatomic species, in addition to hydrogen, oxygen and fluorine.

SIMS is generally a highly sensitive analytical technique; however, detection of some elements is a problem because of their poor ionisation efficiency. The widely ranging electropositivity and negativity of elements across the period table affect how easily they can be ionised and so their sensitivity in SIMS. Quantification of SIMS data is further complicated by so-called "matrix effects" by which the chemical nature of the host matrix may also change the ionisation efficiency of the sputtered particle. The yield of secondary metal ions may increase by several orders of magnitude when they are sputtered from an oxide matrix compared with elemental metal. This effect can be exploited to improve upon sensitivity by using an oxygen primary ion beam. The yield of negative secondary ions can also be enhanced by using an electropositive primary ion such as Cs^+ to stimulate secondary ion emission.

6.2 Rutherford Backscattering (RBS)

Ions with energies in the range of hundreds of eV to MeV may be backscattered within the first micron or so of a solid specimen. The energy spectrum and the angular distribution of backscattered ions which re-emerge from the specimen surface contain information about its composition and structure. The projectile ions used in RBS are typically helium or hydrogen ions with energies of MeV. The nuclear stopping cross-sections of these particles is relatively small with the consequence sputtering yields are low and the technique is non- destructive in comparison with SIMS. The backscattering process occurs after a nuclear collision process with a cross-section that can be calculated from the Rutherford potential after which the technique is named. Although the inelastic collisions of MeV projectile ions with valence electrons results in approximately 10^4 times less energy loss than interactions with atomic nuclei the cross-section for electronic scattering is some eight orders of magnitude higher than nuclear scattering. Consequently the predominant mechanism of energy loss for the projectile ion is through electronic stopping.

The backscattered projectile ions undergo a deflection through an angle θ from the incidence angle. For ions detected at a deflection angle of 90° the relationship between the projectile`s final energy E_f, its mass and the mass of the target atom (M_t) is simply given by the kinematic factor K

$$K = \frac{E_f}{E_i} = \frac{(M_t - M_i)}{(M_t + M_i)} \qquad 7$$

otherwise

$$K = \left[\frac{M_i.\cos\theta + \sqrt{M_t^2 - M_i^2.\sin^2\theta}}{M_i + M_t}\right]^2 \qquad 8$$

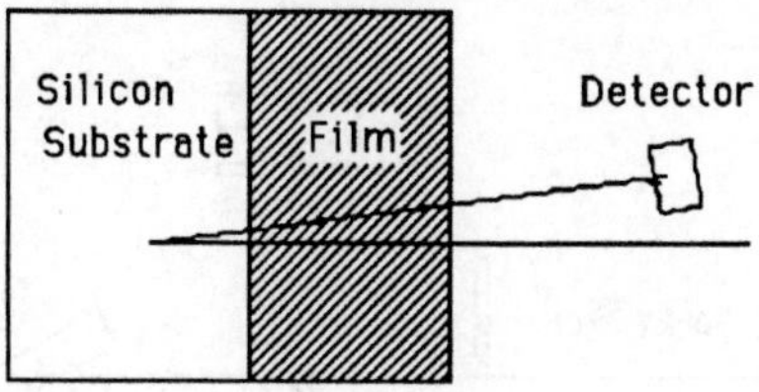

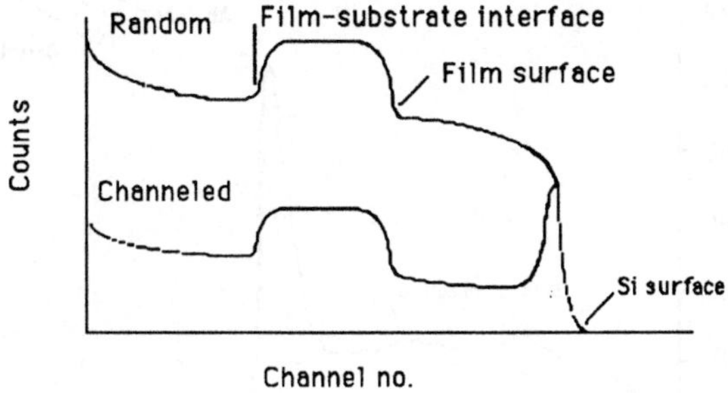

Figure 13. Schematic RBS spectrum of a thin film (M_1) on a crystalline substrate (M_2) under random and channeling conditions ($M_1<M_2$)

The energy of the deflected projectile can either be measured using an electrostatic analyser or a solid state detector with a mylar window to reduce the flux of secondary electrons and low energy ions. From this energy the mass of the target atom is readily calculated. The spectrum of deflected ion signal versus E_0 of a thin film specimen deposited on a crystalline substrate is schematically shown in Figure 13 and shows a peak at KE_0 representing the inelastically scattered projectile from the surface of the specimen. In addition to the nuclear, collision the projectile ion also looses energy through electron scattering. More scattering occurs for ions deeper in the substrate and so the tail of projectile ion energy distribution is a function of electron scattering losses and represents the target ion distribution as a function of depth. The electronic loss per unit length dE/dx is characteristic for a given target and the backscatter spectrum can be converted to a depth scale. The MeV projectile ions can be steered by the screening Coulomb forces of rows or planes of atoms along channels in a crystalline lattice and by orientation of the sample with the incident projectile beam the channelling effect can be optimised to minimise the contribution of the crystalline substrate and enhance the signal from the thin film layer. RBS is relatively insensitive to light elements (a few percent) in heavy matrices but has a higher sensitivity to heavy elements in light matrices.

Prawer et al[37] have used RBS to monitor the loss of hydrogen from an a-C:H film deposited onto highly orientated pyrolytic graphite during irradiation with 50keV C^+ ions at 100°C. Using a 0.32 MeV proton beam and backscatter angle of 165°, the measured hydrogen concentration of the as-deposited film was 36 at% (± 5 at%). But after a threshold dose of 10^{15} C^+ cm^{-2} the evolution of hydrogen was detected which continued up to a dose of 10^{17} C^+ cm^{-2}, resulting in complete removal of the incorporated hydrogen.

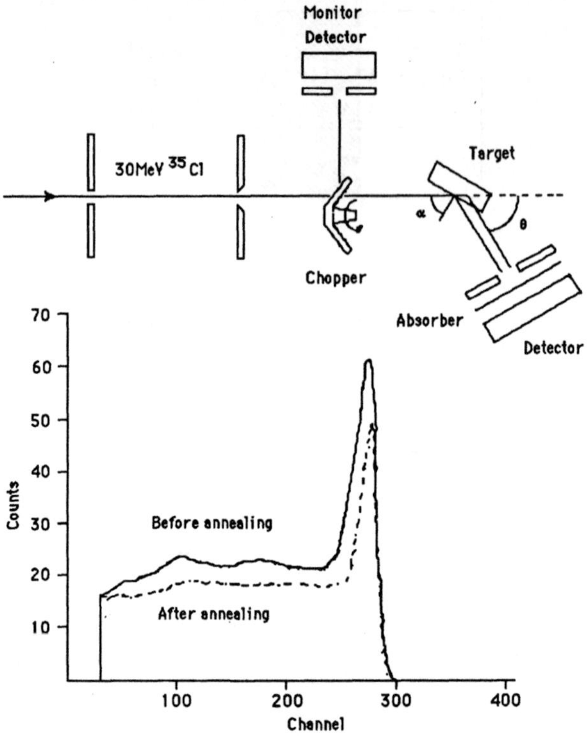

Figure 14. ERDA spectra of hydrogen incorporated in a CVD diamond film before and after annealing at 350°C.

Better sensitivity to light element analysis can be achieved by adjusting the geometry of the ion probe and detector to glancing angles so that the incident ion causes recoil of the light element from the target. This technique is termed Elastic Recoil Detection and uses 35MeV Cl ions as the incident probe. Figure 13 shows the hydrogen depth profile of a CVD diamond film deposited onto silicon before and after annealing at 350°C. The levels of incorporated hydrogen in good quality diamond films is typically of the order of a few percent after growth. However the modest annealing conditions of 350°C were found to reduce the hydrogen concentration by 0.17 at% resulting in a three to five order of magnitude increase in the electrical resistivity of the film.

Nuclear by-products of the heavy ion interactions with the film such as gamma rays or alpha particles are also characteristic and can be exploited quantitatively. Nuclear Reaction Analysis has been used to measure hydrogen and deuterium concentrations using the following nuclear reactions: $^{1}H(^{15}N,\alpha\gamma)^{12}C$, $^{1}H(^{19}F,\alpha\gamma)^{16}O$, $^{1}H(^{18}O,\alpha)^{15}N$ and $^{2}D(^{3}He,p)\alpha$.

7.CONCLUSIONS

A diverse range of analytical techniques exist which can be exploited to probe the chemistry and microstructure of carbon films. Other techniques which have been used to investigate diamond and DLC such as Nuclear Magnetic Resonance (NMR)[38], Electron Spin Resonance (ESR)[39], Extended X-ray Absorption Fine Structure (EXAFS) and Neutron Scattering can also contribute further information about these materials. The complementary nature of many of these analytical techniques combine to create a model of the microstructure within the films, the defect and impurity distributions, as well as the character of the surface and interfaces.

8. ACKNOWLEDGEMENTS

I would like to thank the Diamond Research Team at AEA Technology, especially Ian Buckley-Golder, Colin Johnston. Ruth Harper and Dave Totterdell, Koji Kobashi (Kobe Steel), Jeff Glass (North Carolina State University) and Sophia Chalker (University of Reading) for some of the samples and their helpful discussion with the analysis. I would like to acknowledge the Corporate Research Programme of AEA Technology and the MITI-NEDO (Japan) for their provision of funding for this work.

9. REFERENCES

(1) N. Savvides J. Appl. Phys. 59:4133 (1986).
(2) A.Rengan, A.R. Srivatsa, J. Krishnaswamy, J. Narayan, K.Vedam, K.V. Ravi and M.A. Caolo. 89-12:456 in "Proc. of the first Int. Symp. on Diamond and Diamond-like Films,". J.P.Dismukes, A.J.Purdes, K.E. Spear, B.S. Meyerson, K.V. Ravi, T.D. Moustakas and M.Yoder eds., The Electrochemical Society, Pennington (1989).
(3) N. Savvides Materials Science Forum 52/53:407 (1989).
(4) C.D. Clarke, E.W.J. Mitchell and B.J. Parsons Chapter Two. Colour Centres and Optical Properties, in: "The Properties of Diamond," J.E. Field ed., Academic Press New York (1979).
(5) P.J.Dean Phys.Rev. 139A: 588 (1965).
(6) N. Yamamoto, J.C.H. Spence and D. Fathy Phil. Mag B49:609 (1984).
(7) J. Walker Rep. Prog. Phys. 42:1605 (1979).
(8) A. Feldman, L.H.Robins, P.R.Frederikse, E.N. Farabaugh and E. Etz.. 89-12:162 in "Proc. of the first Int. Symp. on Diamond and Diamond-like Films,". J.P.Dismukes, A.J.Purdes, K.E. Spear, B.S. Meyerson, K.V. Ravi, T.D. Moustakas and M.Yoder eds.,The Electrochemical Society, Pennington (1989).
(9) A.R. Collins, M.Kamo and Y Sato J Phys D: Appl Phys 22:1402 (1989).
(10) A.T. Collins and S.C. Lawson J.Phys. Condens. Matter 1:6929 (1989).
(11) G.Davies, C.Foy and K.O'Donnell J.Phys.C 14:4153 (1981).
(12) V.S. Vavilov, A.A. Gippius, B.V. Zaitsev, B.V. Deryaguin, B.V. Spitsyn and A.E.Aleksenko Sov. Phys. Semicond 14:1078 (1980).
(13) A.T. Collins, M.Kamo and Y.Sato J.Phys. Condens. Matter 1:4029 (1989).
(14) Ch. Wild, N.Herres, J.Wagner, P.Koidl and T.R.Athony 89-12:283 in "Proc. of the first Int. Symp. on Diamond and Diamond-like Films,". J.P.Dismukes, A.J.Purdes, K.E. Spear, B.S. Meyerson, K.V. Ravi, T.D. Moustakas and M.Yoder eds.,The Electrochemical Society, Pennington (1989).
(15) M.Ramsteiner and J. Wagner Appl. Phys. Lett 51:1355 (1987).
(16) M.Yoshikawa Materials Science Forum Vols 52/53:365 (1990).
(17) S.A. Solin and A.K. Ramdas Phys Rev B1:1687 (1970).
(18) M.H. Grimdirch and A.K. Ramdas Phys Rev B11:3139 (1975).
(19) N.Wada and S.A. Solin Physica 105 B:353 (1981).
(20) N. Wada, P.J. Gaczi and S.A. Solin J. Non-Cryst. Solids 35/36:543 (1980).
(21) Xiang Xi Bi, P.C. Eklund, J.G. Zhang, A.M. Rao, T.A. Perry and C.P. Beetz, Jr. SPIE Vol. 1146 Diamond Optics II:192 (1989).
(22) G.B.B.M. Sutherland, D.E. Blackwell and W.G. Simeral Nature Lond. 174 :901 (1954).
(23) K.A. Snail, L.M. Hannsen, A.A. Morrish and W.A. Carrington SPIE Vol. 1146 Diamond Optics II:144 (1989).
(24) P.Coudrec and Y. Cathrine Thin Solid Films 146:93 (1987).

(25) F.R. McFeely, S.P. Kowalczyk, L, Ley, R.G. Cavell, R.A. Pollak and D.A. Shirley Phys. Rev B9 (12):5268 (1974).
(26) J. Fink, Th. Muller-Heinzerling, J. Pfluger, A. Bubenzer, P. Koidl and G. Grecelius. Solid. St. Commun. 47:687 (1983).
(27) Y. Catherine Materials Science Forum Vols 52/53:175 (1989).
(28) D. Briggs, Chapter 4, in "Handbook of X-ray and Ultraviolet Photoelectron Spectroscopy", D. Briggs ed., Heyden & Son, (1978).
(29) P.G. Laurie and J.M. Wilson Surf. Sci. 65:476 (1977).
(30) H.E. Bishop and J.C. Rivière, J. Appl. Phys., 40:1740 (1969).
(31) H.E. Bishop, Br. J. Appl. Phys., 18:703 (1967).
(32) B.E. Williams and J.T.Glass J. Mater. Res. 4(2):373 (1989).
(33) T. Koshikawa and R. Shimizu, J. Phys. D: Appl. Phys., 7:1303 (1974).
(34) B.E. Williams, H.S.Hong and J.T. Glass J. Mater. Res. 5(4):801 (1990).
(35) R.F. Egerton and M.J. Whelan Philos. Mag. 30:739 (1974).
(36) K. Kobashi, K. Nishimura, K. Miyata, K. Kumagai and Y. Kawate 89-12:296 in "Proc. of the first Int. Symp. on Diamond and Diamond-like Films,". J.P.Dismukes, A.J.Purdes, K.E. Spear, B.S. Meyerson, K.V. Ravi, T.D. Moustakas and M.Yoder eds.,The Electrochemical Society, Pennington (1989).
(37) S. Prawer, R. Kalish, M.E.Adel and V. Richter J. Appl. Phys. 61:4492 (1987).
(38) M.A. Petrich Materials Science Forum Vols 52/53:387 (1990).
(39) R.J. Gambino and J.A. Thompson Solid State Commun. 34:15 (1970).
(40) J. Fink, Th. Muller-Heinzerling, J. Pfluger, A. Bubenzer, P. Koidl and G. Grecelius. Solid. St. Commun. 47:887 (1983).

A REVIEW OF CUBIC BN AND RELATED MATERIALS

R. C. DeVries

17 Van Vorst Drive
Burnt Hills, N.Y.

Introduction

Scope

The composition tetrahedron B-N-Si-C is a convenient way to define the scope of this review (Fig. 1). Within this system are the hardest materials known, diamond and cubic BN (Borazon), and if hydrogen were added to each of these, the scheme would then include the amorphous hydrogenated diamondlike phases which are also a subject of this seminar. Less well characterized are the ternary compositions in the system B-C-N, and the interesting phase, C_3N_4, which is known in the graphitic form but is still a theoretical concept as far as a 3-D harder-than-diamond material is concerned. SiC and Si_3N_4 are well-known materials in the hard materials and ceramic industries. The primary emphasis here will be on the synthesis of cubic BN and other phases in the B-N-C composition triangle from both high temperature-high pressure (HPHT) and low pressure processes. Synthesis of films and crystals of SiC will be considered briefly.

Historical

Although neither the hexagonal, rhombohedral, wurtzite (wBN) or sphalerite (cBN) forms of BN have yet to be found in nature, the original notion that a hard form of BN might exist occurred to several people based on the graphite-diamond analogy. R. H. Wentorf, Jr., showed this could be done under HPHT conditions in 1956 (1) in an unequivocal manner devoid of the controversial history of diamond synthesis. He also doped cBN to be both p- and n-type and demonstrated rectification in combinations of the two types (2). He tried to make p-n junctions by overgrowth, but the crystals were too small for measurements.

Properties

Although it is the unique combination of properties of diamond, diamond-like materials, cBN, and SiC that is at the heart of this meeting, for the purposes of this review of the phases in the B-N-C-Si system the emphasis will be on where I think we are with respect to making these materials to achieve their potential. Synthesis of adequate form and

Diamond and Diamond-Like Films and Coatings
Edited by R.E. Clausing *et al.*, Plenum Press, New York, 1991

quality is still the rate -limiting step for progress especially in electronic applications. For more information on properties see references 3 to 10. An abbreviated listing of properties is shown in Table I.

Synthesis of Cubic BN (Borazon)

HPHT
Direct conversion

Hexagonal BN can be converted directly to both cubic BN (cBN, sphalerite structure) and wBN (wurtzite form) by both static and dynamic techniques (11,12). The pressure and temperature conditions are illustrated in the P-T diagrams of Fig.2, where wBN is shown as a metastable phase. The P-

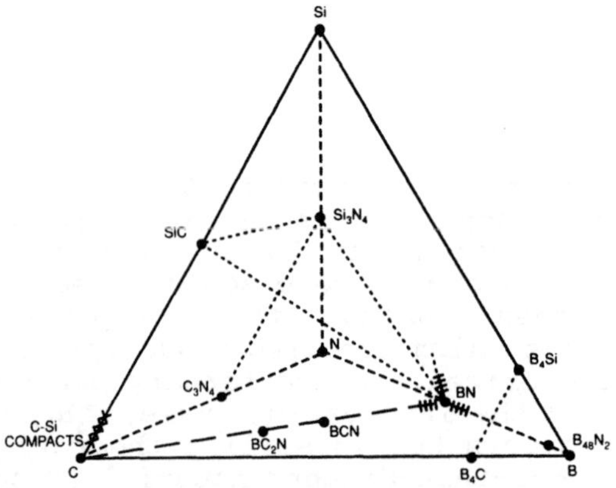

Fig. 1 The composition tetrahedron C-B-N-Si. (Hash marks indicate unknown amount of solid solution).

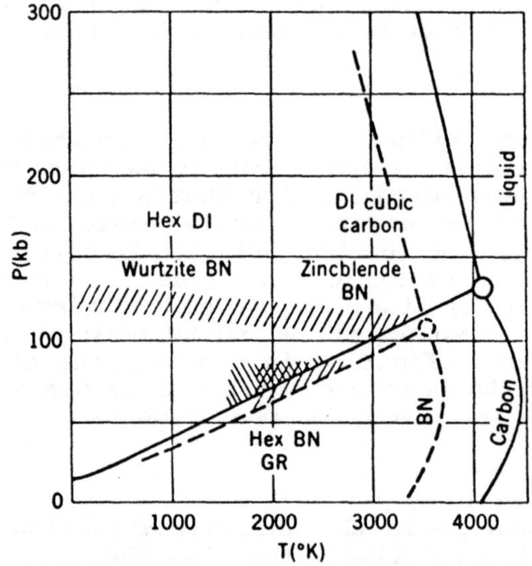

Fig. 2 P - T diagram for the system BN compared to diamond (By analogy to new data on the diamond system, the melting curve for cBN probably also has a positive slope instead of negative as shown).

TABLE I

SELECTED PROPERTIES OF DIAMOND, BORON NITRIDE, SILICON CARBIDE

	cBN	Diamond	SiC	
STRUCTURAL				
Structure	cubic F43m sphalerite	cubic Fd3m diamond	cubic F43m beta	hex P6mc alpha 6H
Unit Cell (A)	3.65	3.567	4.3596	a=3.08065 c=15.1174
A-X (A)	1.57	1.54		1.89
Density (gm/cm^3)	3.48	3.52	3.21	3.21
Cleavage	(011)	(111)		
CHEMICAL				
Mol. wt.	24.8177	12.011	40.1	40.1
Lattice dopants	B, S, Si, CN-, Al, P	B. N. (Be)	N, P, Al, B	
MECHANICAL				
Hardness (kg/mm^2)	4500	9000+	2670	2130 2755
Bulk Mod. GPa	367	435		212
Compress (cm^2/dyne) x 10^{-12}	0.34	0.23		
THERMAL				
Conductivity W/cm/°C (25°C)	13 calc 8 (polyxln)	20 (IIa)	4.9	
Expansion cm/cm/°C (400°C)	4.8	3.5	4.3	
Stability oxidation	1200°C protective	600°C	1000°C protective	
Graphitize	>1550°C hBN	1400-1800°C	dissociates	
OPTICAL				
RI 5893 Å	2.117	2.417	2.48	2.69 2.647
Band Gap eV direct	6.1 - 6.6	5.47	2.2	2.86
ELECTRICAL				
Resistivity ohm-cm	10^2 10^{10}	10 10^{16}	10^{-2} 10^6	10^{-2} 10^{10}
Dielec. K	4.5	5.58		6.7

T conditions are somewhat lower than for the direct conversion of diamond, but commercially the shock mode of direct synthesis predominates over the static method because of the strain on equipment of the latter process. The wurtzite form is available from shock synthesis as fine powders which can be used for polishing or for sintering into polycrystalline tool materials. In static HP experiments all attempts to increase the size of individual wBN crystals beyond a few hundred angstroms fail because increasing temperature to promote diffusion only leads to formation of the stable phase, cBN. During a sintering process some or all of the wBN may be converted to cubic without extensive grain growth. The phase boundary between liquid and cubic BN is now considered by analogy with new work on the carbon phase diagram (13) to be of positive slope instead of negative as shown in Figure 2.

Indirect conversion

Wentorf converted hexagonal BN to cubic BN by using alkali and alkaline earth metals and their nitrides (1,2), and these are still the principal solvents used today by most manufacturers either singly or in combination. A considerable body of literature exists on these systems and representative T-X and P-T diagrams are shown in Figures 3a,3b,4a,4b. Cubic BN can also be precipitated from metal alloys which include Al-Si (14), lanthanides and actinides (15), and some of the same ones used for diamond synthesis like Fe, Ni, Co, Mn (16,17). The latter seem to require a pinch of Al, however. This interesting phenomenon has been demonstrated, both in the synthesis and sintering of cubic BN, to be closely related to a catalytic or nucleation agent. Hirano has demonstrated the latter mechanism very clearly (18), and the effect is related to the structural similarity of AlN and BN and also may be related to the relative free energies of AlN and BN as a function of temperature. AlN is more stable at lower temperatures, BN at higher temperatures.

For most of the T-X diagrams for BN versus alkali or alkaline earth nitrides, it is common to have a eutectic between BN and a compound of BN and the alkali or alkaline earth nitride being used (Fig.3). When the temperature is raised with the system at a pressure where cubic BN is stable, it precipitates from solution above the eutectic temperature. When starting with the graphitic form of BN, the solution is immediately supersaturated with respect to the cubic form so nucleation is rapid. Nucleation is very sensitive to pressure both in the diamond and cubic BN systems, and kinetic effects are obvious in the P-T diagrams as broad U-shaped extensions of the stability region of the cubic form toward lower temperatures. At the high pressure/lower temperature part of the diagram, there will be many small crystals; at higher temperatures and lower pressures nucleation and growth are somewhat more controllable, and therefore the region of synthesis of most abrasive grain.

Cubic BN has the sphalerite structure and therefore has no center of symmetry like diamond, so the growth forms tend to be a mixture of positive and negative tetrahedra with one form usually dominating (Fig.5). If equally developed,

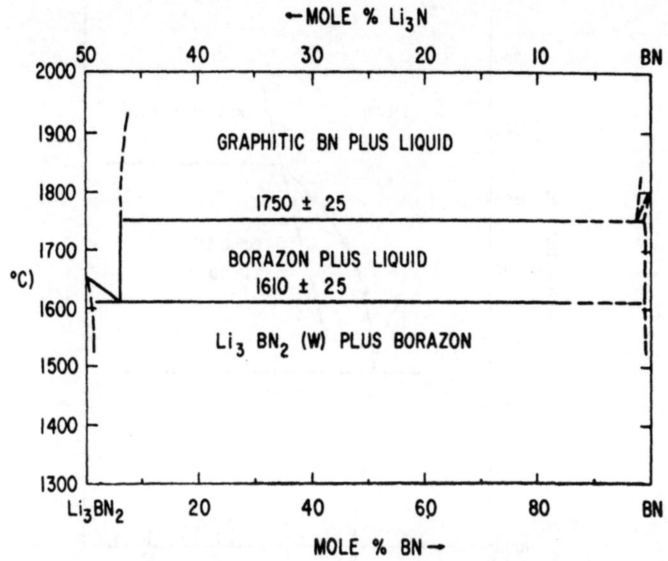

Fig. 3a T - X diagram for the system Li_3BN_2 - BN.

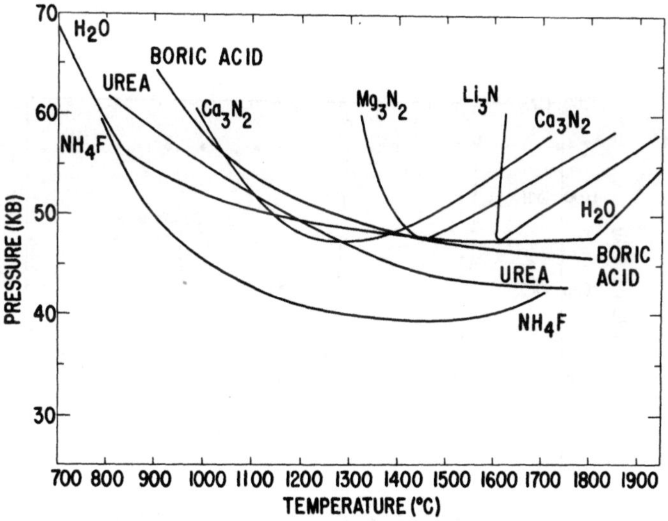

Fig. 3b P - T diagram for several solvent/catalysts for BN (cBN stable above each curve, hBN below).

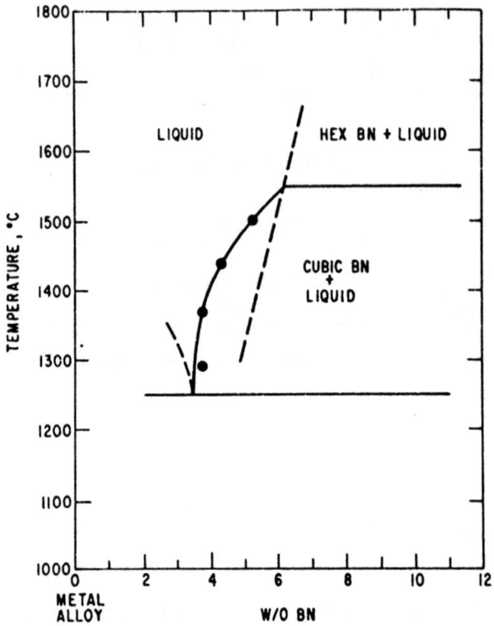

Fig. 4a T - X diagram for the system metal alloys A and B with BN.

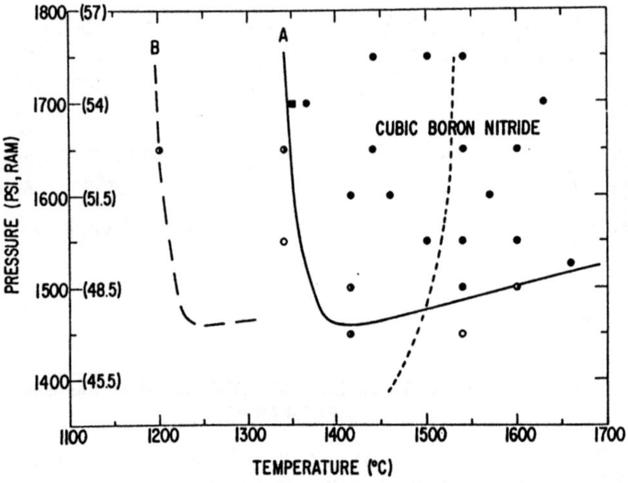

Fig. 4b P - T diagram for the system metal alloy - BN.

octahedra would result. It is less easy to control the growth forms of cubic BN than for diamond where the combination of cube and octahedra is predominantly temperature dependent for most of the alloy systems. Cubic BN often grows as flat plates, twinning is prevalent, and both (100) and (110) forms are seen as for diamond. Moreover, the internal structure of most cBN grains is complex and zoned with respect to chemistry and mechanical behavior such as wear (Fig.6). In general it has also proved more difficult to grow larger crystals of cubic BN than for diamond.

Growth of large crystals of cubic BN

The delta T technique developed by Wentorf and Strong and Tuft at GE (19,20) for growth of large gem quality crystals of diamond can be used to grow larger crystals of cubic BN, but the success has been noticeably less spectacular than for diamond. The delta T technique is shown for an actual crystal in its growth environment (a metal alloy) is seen in Fig.7. In this mode with a bit of Al added, some clear AlN crystals grow simultaneously with the cBN. Both Sumitomo and DeBeers have scaled up the GE process for growing large crystals of colorless (IIa), blue (IIb), and yellow (Ib) diamond up to maximum dimensions of 17 mm (21-23). For cubic BN a large crystal is about 3 mm from either alkali/alkaline earth nitrides or from metal alloy solutions (24-28), but rumors persist of larger, as yet unreported, crystals. The two atom problem, charge transfer, plus a strong tendency to twin probably all contribute to the difficulties in a manner similar to the relatively easy growth of Si compared to GaAs. In accord with the Wentorf prediction (2), a p-n junction has been made by epitaxial growth of one type on top of the other at HPHT conditions (26). The difficulties and limitations of this process are clearly a strong driving force for developing a successful low pressure CVD process for cBN.

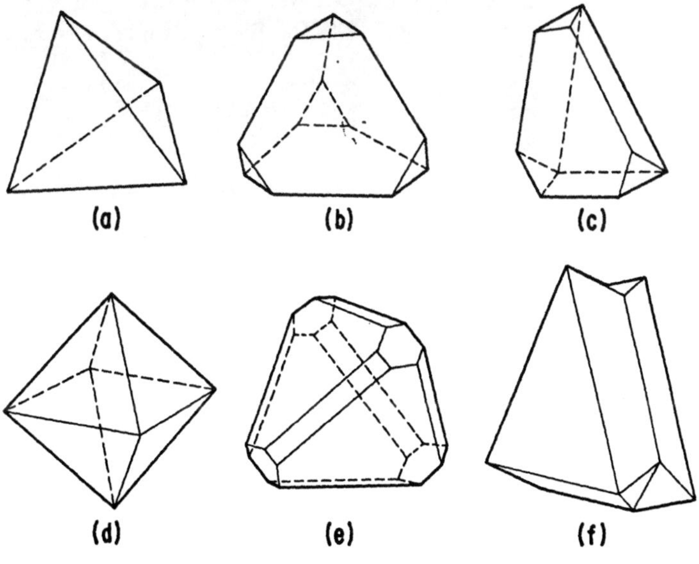

Fig. 5 Morphology of cBN crystals.

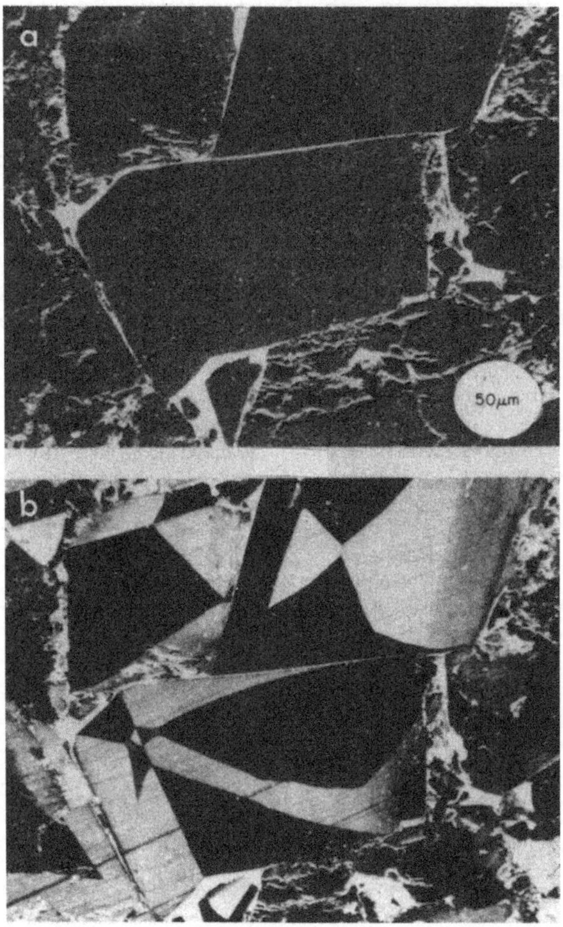

Fig. 6 Sector growth in cBN crystals.
a. Backscattered electron image.
b. Secondary electron image.

Polycrystalline aggregates of cBN

These can be made as a single phase compact by direct conversion of the other forms of BN by static or dynamic pressure or as a polyphase (predominantly cBN) composite by sintering cBN grains on cemented tungsten carbide with sintering aids such as alloys of Ni, Co, Fe with Al (29). cBN-cBN bonding between grains is generated by this liquid phase sintering process. The sintered cBN/carbide composite has a higher thermal stability than the diamond equivalent.

The direct conversion process can produce translucent to transparent polycrystalline aggregates of near theoretical density and rather high thermal conductivity (9 W/cm/C) (30). Corrigan has shown that preferred orientation can be obtained in the cBN aggregate by using oriented hBN (12). This suggests there may be possibilities for this mode to produce larger single crystals from the solid state under HPHT conditions.

LPHT

The effort to grow cubic BN by low pressure CVD or by ion beam techniques also has not yet achieved the success of CVD diamond in the sense that there are not published accounts of visible facetted crystals on a variety of substrates -- the criterion that triggered the current interest in diamond. The intensity of effort has picked up noticeably since about 1983 (see Table II), and it is reasonable to expect considerable success soon; in fact this may have happened already. An oversimplified summation of much of the work reported to date is that perhaps some very small regions (100 - 200 Å) of cBN have been made in a matrix of amorphous BN (a-BN) and /or hexagonal BN and/or amorphous hydrogenated BN (a-BN:H). For a recent review, see reference 31. This skeptical evaluation will now be qualified by choosing a few studies which appear to be, or have the potential of being, exceptions to the above evaluation and which point the new direction.

The first is a curious one by Vickery (32) where it was claimed that a layer of BP (isostructural with cubic BN) can be converted to cBN by reaction with ammonia in nitrogen. This simple approach probably cannot produce very thick films because of limitations imposed by kinetics of diffusion through the layer, but a recent patent (32a) claims layers to 3 microns. BP is not the most convenient intermediary, but the concept of chemical replacement is interesting and possibly could lead to some BP/BN composites.

Gashtol'd et al., (33) used high frequency sputtering of boron in a nitrogen atmosphere to deposit BN layers on SiO_2-Si base for sandwich transistors. Films were also deposited on NaCl. From the properties it is not clear that cBN was obtained, but modifications of this approach suggest this is worth more effort. Satou and Fujimoto (34) evaporated boron simultaneously with nitrogen ion beam bombardment (30 keV N^{2+}) onto substrates at room temperature and claimed a mixture of cBN in a-BN. Weissmantel et al., published extensively on use of high energy ion beams to make hard

Fig. 7 Crystal growth of cBN by the thermal gradient method. Top section (cracked): nutrient cBN; hotter portion of cell middle section (light): metal solvent bottom center: cBN crystal which was nucleated on seed; cooler end.

TABLE II

NUMBERS OF PAPERS (*) AND PATENTS ($) ON cBN FROM CVD

1959-78	****$	
1979	***	
1980	*$	
1981	***	
1982	*$$	
1983	******$$$$	
1984	**$$$$$$	
1985	******$$$$$$$$$$$$	
1986	****$$$$$$$$$$$$$$$$$$$$$$$	
1987	****************$$$$$$$$$$$$	28 total
1988	*********$$	
1989	*****	
1990	**	

coatings of both diamond and cBN (i-C and i-BN) (35). Sokolowski and colleagues in a series of papers on reactive pulse plasma crystallization (36-38) also pursued these processes involving high energy levels of particles to deposit the metastable phases of wBN and cBN in a coaxial plasma generator with a boron electrode in N_2/H_2. The presence of these high pressure phases, including a high pressure form of boron and the absence of hBN is essentially explained by transient HPHT conditions similar to an explosion. So strictly speaking, if this is indeed the mechanism, these techniques are HPHT without the press and without the recovery problems of shock synthesis -- sort of a multiplicity of small shocks on a controlled surface. The Sokolowska paper (38) actually presents the stability regions of cBN and E-BN (a phase known from explosion experiments) in terms of a P, T and E diagram where E is the electron field intensity on a W tip as a third parameter of state. Perhaps it is really a part of the P and T parameters.

Perhaps related mechanistically to the high energy systems described above are the very recent developments using lasers for sputtering and evaporation or for rapid heating and cooling. Fedoseev et al., have demonstrated that small particles of carbon black, hBN, and quartz can be converted to high pressure polymorphic forms when rapidly heated in a laser beam and cooled (39). This has been confirmed for SiO_2 by Adam et al. (40). Fedoseev et al., have done similar syntheses of hard metastable phases of carbon with hydrocarbons falling through a laser beam (41). The yield is low and the particle size very small, but there is no question about the phases formed. Using a Q-switched Nd glass laser directed at two different forms of BN for targets (wBN and hBN) in a nitrogen atmosphere, Kessler et al., (42) demonstrated that the coating produced was the same as the target. Would a cBN target have yielded a cBN deposit?

The results reported by Doll et al., (43) are potentially the most exciting to date -- pending reproducibility and confirmation of the results. They used a pulsed excimer laser to ablate hBN in a nitrogen atmosphere and produced films which seem to be cBN by several different characterization methods. The film was not stoichiometric, which is probably related to the tendency of BN to lose nitrogen when exposed to such high energy (perhaps a cBN substrate would help). The Kessler and Doll results are very encouraging and are certainly even now receiving considerable attention because the potential of growth and doping of single crystal films of cBN on a common, easily obtained substrate like Si.

The only claim for the metastable formation of actual crystalline particles of cBN is the work of Matsui (44). He irradiated some specially treated hBN with the electron beam in an electron microscope and found the new crystals had grown about 20-30 microns away from the collapsed hBN fragment. The crystals had the morphology of cBN and showed multiple twinning. The special treatment of the hBN consisted of pressing in an ethanol/water mixture for 15 minutes at 130 kb in a diamond anvil cell, and it was stated that the material was still hBN. The reproducibility of this experiment is not known. It would be interesting to know if the pressed sample had been converted to the rhombohedral form of BN and thereby an easier step to cBN.

Saitoh et al., (45,46) have combined the RF plasma CVD with the tungsten filament for thermally activated electrons in a process more like "diamond type CVD" to make cBN deposits. They have demonstrated that the filament temperature is critical: if below about 1600°C, hBN forms and above this temperature, cBN deposits. This temperature is related to the work function of W, i.e., whether there is thermoelectric emission, and is important for the production of atomic hydrogen just as in the CVD diamond synthesis. The starting materials are NH_3 + B_2H_6 + H_2 or solid-gas combinations BH_3NH_3-H_2, H_3BO_3-NH_3, and $NaBH_4$-NH_3. This system appears to be very controllable and has the potential of producing the same results achieved for CVD diamond films and therefore is of considerable interest.

The LPHT techniques are obviously very attractive for production of large area single crystal n- and p-type semiconductors based on cBN. The doping can be achieved simply by adding the dopant to the target material or via the gas phase. The HPHT process wouldn`t be able to compete if the claims for LPHT synthesis are confirmed. Success here coupled with the difficulty of achieving n-type diamond speaks for itself with respect to the direction to pursue for electronic applications at least. The recent success with SiC might provide some competition.

Phases in the System B-C-N

The interest in the B-N-C system (Fig. 8) has the same motivation once given to the B-Si-C system: the possibilities of new hard phases or new semiconductors, of modifying mechanical properties of existing phases by solid solution or precipitation. Although the studies of the B-Si-C system (47) have not produced phases harder than diamond, B_4C, or SiC, nor even any improvement in the deficiencies of B_4C itself, there is support for such studies in B-N-C. The new sophistication with metastable synthesis clearly increases the probability of making some phases which elude the more nearly equilibrium approaches. In addition there is the well-known case of toughening of diamond (albeit small in magnitude) by precipitation as done by nature with the Ib to Ia transformation. Most of this discussion can take place in the context of the BN-C system where we consider the extent of solid solution in each end member, the possibility of a 1:1:1 BCN compound, and the possibility of modification of properties of cBN by solution or precipitation.

BN vs C

In the diamond system the variability with respect to chemistry and properties is more restricted than for the BN system. Only B and N and possibly Be are clearly established in diamond as substitutional atoms and these to only a few ppm on an equilibrium basis. Thus B-doped diamond (IIb) is well known as a p-type semiconductor; N-doping makes the crystal yellow if substitutional (Ib) and colorless if precipitated as platelets as in most natural diamonds (Ia) or in annealed synthesized diamonds (48). n-type doping of diamonds is somewhat controversial, and the reader is referred to Watkins and Messmer (49) for one analysis as to why it is unlikely. No data are available on the simultaneous addition of B and N to diamond.

Although the doping levels in cBN are not well documented, it was shown by Wentorf (2) that Be additions made a blue p-type crystal and that S, Si, CN groups produced n-type semiconducting behavior. In addition the crystals as grown without deliberate doping are amber-colored and are probably non-stoichiometric -- generally thought to be boron-rich because of loss of nitrogen -- and are n-type (Fig. 9). These crystals are thermochromic and turn completely opaque at about 450°C and return to yellow on cooling (5). The conductivity increases markedly at the same time. No quantitative information on the deviation from stoichiometry could be found; this is a rather difficult analysis.

BN-C

Carbon additions have been made in the synthesis of cBN at HPHT (50-52) for alteration of mechanical properties; in a patent (52), 0.02-2.0 % C is reported while others report as much as 5 atom % C in cBN (51). Butylenko reported a single phase to about 6.5 % with a maximum in hardness at 3.5 % (50). Sumya (53) reported 10-1000 ppm of C in cBN reduced cleavage and increased hardness and toughness. The concept of some solid solution of C in BN is expressed by the hash marks on the B-N-C diagram (Fig. 8) and as an exaggerated solution region in Fig. 10. Because of the solvent/catalysts used for synthesis, there are undoubtedly some alkali or alkaline earth ions incorporated in the crystals grown at high pressure, and this will complicate the understanding. The crystals are zoned (Fig.6) and there are electrical conductivity and mechanical differences between zones that are probably attributable to other atom species or combination of atoms.

Ternary compounds in the B-N-C system

The evidence for new ternary compositions is discussed in the context of three modes of synthesis: LPHT, HPHT, or the combination of the two in attempts to produce high pressure forms. For the graphitic forms made by CVD or high energy processes, it is not correct to call this synthesis under metastable conditions since it is indeed in the equilibrium mode.

BC_2N (LPHT)

There seems to be little doubt that the ternary phase, BC_2N, has been made by CVD processes (54,55) in the graphitic form, and structure calculations have been made of the possible atomic arrangements of the layers (54). Band structure calculations for this phase and related graphite-BN alloys have also been reported (56). There is no report of a high pressure modification of this phase as yet. There is one claim that this phase was used as a precursor to making a high pressure modification of BCN (55), but there's

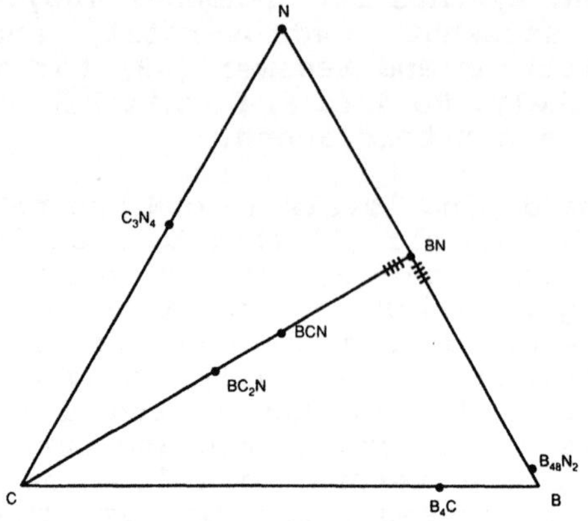

Fig. 8 The composition triangle B - C - N (Hash marks indicate unknown amount of solid solution).

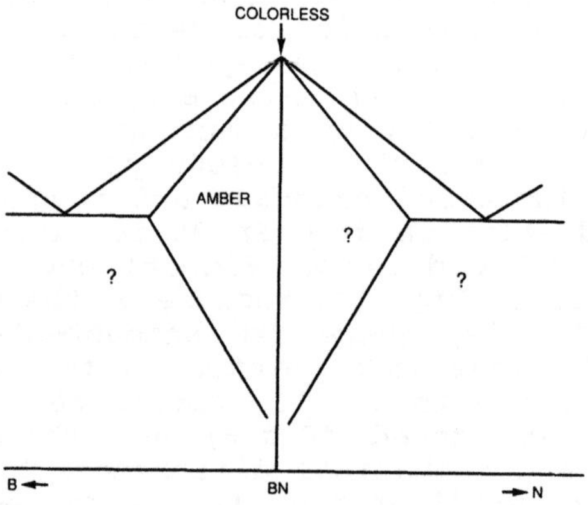

Fig. 9 Schematic representation of the binary joins N - BN and BN - B.

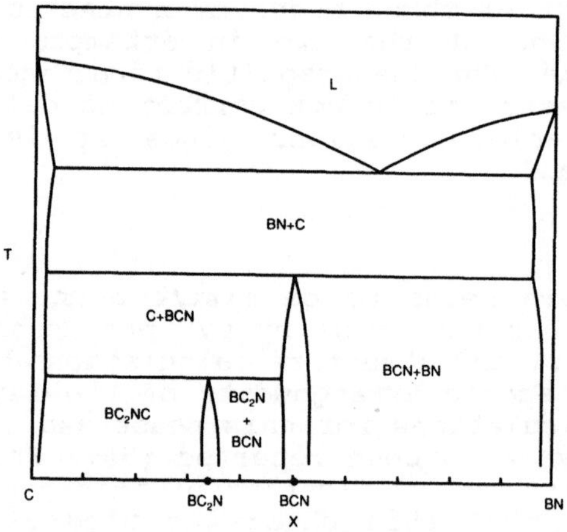

Fig. 10 Schematic representation of the system BN - C.

been no further acclaim of what would be a discovery of some interest in this field.

$B_xC_yN_z$ (LPHT)

There's also some convergence among reliable researchers for synthesis of a single phase solid solution of graphitic structure with an overall composition close to BCN (57-59) by codeposition in a gas phase containing BCl_3, a hydrocarbon, and ammonia or nitrogen/hydrogen mixtures. Badzian found the phase to be stable to about 2000°C in argon; the decomposition was to B_4C, graphite, and nitrogen (57). His material was deposited on the surface of a hot (1950°C) carbon rod. Kaner et al., (58) deposited their material (B 0.35 C 0.30 N 0.35) in the range 400 to 700°C. Their x-ray photoelectron spectra indicate the material is not a mixture of phases but a single phase with boron having both carbon and nitrogen neighbors and nitrogen having carbon atom neighbors as well as boron. Yamada (59) describes semiconductor devices having covalently bonded BNC passivation films. Moore made codeposited BN and C materials over a wide composition range (3 to 60 wt. %) and concluded they were possibly single phase solid solutions of the graphitic form (60). The term boron (or boro) carbonitride is rather common in the Russian literature, and there's one claim of a pure BNC single phase material (cubic, a=6.845) by nitridation of a mixture of amorphous boron and carbon black (61). This phase decomposed at 2200°C to BN and graphite and could not be sintered. There's considerably more activity on substitution of C in the hBN lattice for coatings (62), luminescence studies (63-66), and even for gaskets in high pressure devices (67). Besides CVD the pyrolysis or decomposition of complex polymers (68) provides another route to mixed crystals at low temperatures.

$B_xC_yN_z$ (HPHT/LPHT)

With the exception of the one claim of a cubic phase of BNC from CVD (61), nothing else could be found to indicate the formation of a hard form of this composition under LPHT conditions. The obvious step is to subject the graphitic solid solutions to high pressure. No published work could be found on subjecting the graphitic BC_2N phase to HPHT conditions even from one of the laboratories that made this phase and has the HPHT capability. There is a patent on the conversion of B carbonitride in the hexagonal or amorphous form under HPHT conditions in the presence of a metal solvent in the same process for making diamonds (69). A picture of the sphalerite lattice for this $B_xC_yN_z$ (where x = z, and does not equal y, or x = y = z) shows each site occupied by a different atom with no near neighbors being the same. There has not been a follow-up publication of this interesting result. The implication of some Japanese patents (70) is the synthesis of a cubic B carbonitride from amorphous $B_{0.35}N_{0.32}C_{0.32}$ under HPHT conditions and the subsequent incorporation of this material into a composite tool. Badzian put his graphite-BN solid solution made by CVD (57) into a HPHT apparatus and obtained a direct transformation to a cubic phase at 140 kb and 3300°K (71-72). The lattice parameter of this sphalerite structure was 3.582, intermediate between diamond and cBN with a composition

$c(BN)_{0.26}C_{0.74}$. The general formula for phases claimed in the work is $(BN)_xC_{1-x}$, where $0<x<1$. The structure seems to fit best a model of short range order with substitution of B-N pairs by C-C pairs.

Pure BNC does not appear to have been made yet at HPHT conditions, and the mixed crystals with the sphalerite structure have not yet appeared in the marketplace. While this would be a scientifically interesting achievement, there's some question whether such a phase would have any advantages over cBN. The question is simple: cBN is about half as hard as diamond; would adding a third element with the three different bonds that would be necessary be expected to increase the hardness or bulk modulus? The overall integrated bonding is no better than its weakest link. Those who calculate the structural configurations and their properties should address this question and help the experimentalist because the window for synthesis may be small and difficult to find. Perhaps the more effective result of these mixed phase and solid solution studies will come from the influence of these phenomena on the mechanical properties of the currently available phases. For example, because of its nonsymmetric crystal structure, cBN cleaves on the (110) planes of which there are six compared to the 4 easy cleavage planes (111) in diamond. On this basis cBN is intrinsically a more friable crystal than diamond, and the ability to modify that friability with precipitation or solution would be useful.

The diamond-cBN interface

The capabilities for growth of diamond and cBN by both HPHT and LPHT processes raises the possibility of cBN-diamond interfaces. CVD diamond has been deposited on crystals of cBN (73), and there are theoretical calculations on the nature of such an interface (74,75).

B-N-C-Si

Within this composition tetrahedron (Fig. 1), a decrease in hardness of beta-SiC from 2600 (normal) to a minimum of 2300 for 40 mole % BN, followed by an increase to 3700 kg/mm^2 at 80 % BN is claimed by solid solution of B and N in the lattice. The result differed somewhat depending on the details of the synthesis, which was done in the temperature range of 1300-2600°C and 10 - 7000 atmospheres pressure (76).

Growth of SiC Films and Crystals

The formation of SiC grains adequate for abrasives and ceramics goes back to Moissan. A good reference for the early committment to the problem of producing good quality single crystals for electronic devices is the Proceeding of the 1959 Conference on SiC (77) with subsequent meetings in 1968 and 1973. So the interest in this material has continued at varying degrees of intensity for decades. From the late 70's and early 80's there has been an obvious marked step in progress for techniques to produce both cubic (beta) and 6H SiC. Two techniques will be described briefly: (1)

sublimation of SiC (principally for 6H); and (2) chemical vapor deposition (principally for beta).

Sublimation of SiC

Most trace the development of modern sublimation technique back to Lely (78) for the basic system with random nucleation. Hergenrother et al., (79) reported the use of seeding and compartmentalization of the source area and growth chamber. The seed was at the bottom, and crystals up to a few cubic millimeters were grown. Tairov (80,81) put the seed at the top in a more simple chamber with a porous graphite cylinder for a growth chamber and with a controlled temperature gradient produced ingots of 6H. The seed temperature ranged from 1800-2600°C, and the crystals grew at up to 2 mm/hour. Similar 6H ingots up to 20 mm diameter by 24 mm long were grown by Ziegler et al., (82) with the seed at the bottom in a chamber which resembles that of Hergenrother. At a recent meeting, R. F. Davis (83) described growth of crystals about 40 mm in diameter by 60 - 75 mm long presumably grown by a similar sublimation technique. The general principle seems to be the transport of Si, Si_2C, SiC_2 gas species, and doping can be done by mixing the dopant with the source SiC. Slices are cut from the boules or ingots with a diamond saw. In the sublimation process the temperature range seems to favor the 6H polytype, but there are suggestions that there's some control of this by proper seed orientation.

CVD of SiC

This process also has a long history and was considered in some detail in the 1959 conference. The results reported by Nieberding (84), Nishino et al., (85), Powell (86) and more recently by Davis (83) clearly demonstrate a new degree of achievement in producing films of beta-SiC by this method. From the gas phase pyrolysis at 1400°C of flowing SiH_4 + C_3H_8 + H_2 over a (100) Si single crystal substrate, large area (5 cm^2) single crystal (100) films (to 34 microns) have been produced and devices made from them. The successful procedure described by the NASA Lewis people consists of three stages: (1) etching of the Si substrate; (2) formation of a thin buffer layer of SiC; (3) growth stage.

These developments support the conclusion that finally SiC may fulfill its potential as a high temperature semiconductor. The use of high quality crystals of SiC as substrates for heteroepitaxy of diamond and cBN also seems ripe for further study.

Summary

The conclusions of this review of cubic BN and related materials are:

(1) Synthesis of cBN grain and compacts will continue to be from HPHT conditions for some time yet, and there may be some new developments in the physical metallurgy of the BN-C system in terms of new phases and control of properties by solution and precipitation;

(2) Growth of larger crystals of cBN at HPHT will be achieved but will it be worth it in the context of the potential of LPHT processes?;

(3) There are some encouraging developments in the synthesis of single crystal cBN by LPHT techniques which could be a major breakthrough in the electronic applications of this material; perhaps diamond will become the heat sink for cBN devices;

(4) There will be an increasing need to do better characterization of cBN for electronic applications -- particulary composition -- and to better control that parameter;

(5) Because of the improvement in processes for producing both beta- and 6H SiC, this material seems to have attained its long awaited potential as a high temperature semiconductor;

(6) The techniques for synthesis under metastable conditions probably have a better chance of making new hard phases like C_3N_4 (87), possibly in solution with Si_3N_4 (88) that might otherwise dissociate to other phases under equilibrium conditions.

References

1. R. H. Wentorf, Jr., Cubic form of boron nitride, J. Chem. Phys. 26:956 (1957)
2. R. H. Wentorf, Jr., Preparation of semiconducting cubic boron nitride, J. Chem. Phys. 36:1990-91 (1962)
3. M. Neuberger, Boron Nitride, Data Sheet DS-158, Nov., 1967; Electronic Properties Information Center, Hughes Aircraft Co., Culver City, CA.; AFML Contract AF 33(615)-2460, Project 7381, Task 738103
4. V. N. Bakul, A. I. Brikhna, Composition, structure and properties of cubic boron nitride, Sin. Almazy 1:36-42 (1971)
5. R. C. DeVries, Cubic boron nitride. Handbook of properties, GE CRD Report No. 72CRD178, June 1972
6. Gmelin Handbuch der anorganishen Chemie, Band 13, Boron nitride, pp. 9-11, 41-43, 50-54, 63-65, 83 (1974)
7. A. V. Kurdyumov, A. N. Pilyankevich, Phase Transformations of Carbon and Boron Nitride, Naukova Dumka, Kiev (1979)
8. E. Rapoport, Cubic boron nitride, Ann. Chim. 10:607-38 (1985)
9. R. R. Wills, Wurtzitic boron nitride-a review, Int. J. High Technol. Ceram. 1:139-53 (1985)
10. C. A. Brookes, The mechanical properties of cubic boron nitride - a perspective view, Inst. Phys. Conf. Ser. No. 75: Chapter 3. Paper presented at 2nd Int. Conf. Sci. Hard Mater., Rhodes; Adam Hilger Ltd, 1986
11. F. P. Bundy, R. H. Wentorf, Jr., Direct transformation of hexagonal boron nitride to denser forms, J. Chem. Phys. 38:1144-49 (1963)
12. F. R. Corrigan, Epitaxial relationships for the high pressure transitions of boron nitride, GE internal report, 1975; Direct conversion process for making cubic boron

nitride from pyrolytic boron nitride, U.S. Patent 4,188,194; 2/12/1980; filed 7/1/1977.
13. F. P. Bundy, Pressure-temperature phase diagram of elemental carbon, Physica A 156:169-78 (1989)
14. T. Kabayama, T. Ikezawa, Synthesis of cBN, paper given at 15th High Pressure Meeting, Fukuoka, Japan, 10/21/73
15. R. H. Wentorf, Jr., Process for making cubic crystal boron nitride, U.S. Patent 3,150,929; 9/20/64; filed 1/17/63
16. R. C. DeVries, J. F. Fleischer, Solution-precipitation process for manufacturing cubic boron nitride abrasive tools, U.S. Patent 3,918,931; 11/11/75; filed 12/17/73
17. D. G. Flom, R. C. DeVries, W. R. Reed, CBN/Diamond Tool Science, Part 2, Characterization and testing of single point cutting tools, GE R and D Center Report 89CRD90, April, 1989
18. S. Hirano, T. Yamaguchi, S. Naka, Effects of AlN additions and atmosphere on the synthesis of cubic boron nitride, J. Am. Ceram. Soc., 64:734-36 (1981)
19. H. M. Strong, R. H. Wentorf, Jr., The growth of large diamond crystals, Naturwissenschaften 59:1-7 (1972)
20. H. M. Strong, R. E. Tuft, Method and high pressure reaction vessel for quality control of diamond growth on diamond seed, U.S. Patent 4,034,066; 7/5/77; filed 11/2/73
21. J. I. Koivula, C. W. Fryer, Identifying gem-quality synthetic diamonds: and update, Gems and Gemology XX:146-158 (1984)
22. J. E. Shigley, E. Fritsch, C. M. Stockton, J. I. Koivula, C. W. Fryer, R. E. Kane, The gemological properties of the Sumitomo gem-quality synthetic yellow diamonds, Gems and Gemology XXII: 192-208 (1986)
23. J. E. Shigley, E. Fritsch, C. M. Stockton, J. I. Koivula, C. W. Fryer, R. E. Kane, D. R. Hargett, C. W. Welch, The gemological properties of the De Beers gem-quality synthetic diamonds, Gems and Gemology XXIII:187-206 (1987)
24. T. Kobayashi, K. Susa, A method for growing cubic boron nitride single crystals, Japanese Patent No. Sho 53(1978)-106698; 9/16/78; filed 3/2/77
25. O. Mishima, S. Yamaoka, O. Fukunaga, Crystal growth of cubic boron nitride by temperature difference method at 55 Kb and 1800°C., J. Appl. Phys. 61:2822-25 (1987)
26. O. Mishima, J. Tanaka, S. Yamaoka, O. Fukunaga, High-temperature cubic boron nitride p-n junction diode made at high pressure, Science 238:181-83 (1987)
27. S. Yazu, H. Sumiya, J. Degawa, Method of synthesizing cubic system boron nitride, EP 220,462; 5/6/87; filed 9/24/85
28. M. Kagamida, H. Kanda, M. Akaishi, A. Nukui, T. Osawa, S. Yamaoka, Crystal growth of cubic boron nitride using Li_3BN_2 solvent under high temperature and pressure, J. Crystal Growth 94:261-9 (1989)
29. R. H. Wentorf, Jr., W. A. Rocco, Cubic boron nitride/sintered carbide abrasive bodies, U. S. Patent 3,767,371; 10/23/73; filed 7/1/71
30. F. R. Corrigan, Thermal conductivity of polycrystalline cubic boron nitride compacts, pp 994-999 in High Pressure Science and Technology, 6th AIRAPT Conf., Vol. 1, Eds. K. D. Timmerhaus, M. S.Barber, Plenum Press, N.Y. (1977)
31. S.P.S. Arya and A. D'Amico, Preparation, properties, and applications of boron nitride thin films, Thin Solid Films 157:267-82 (1988)

32. R. C. Vickery, Synthesis of boron phosphide and nitride, Nature 184:268 (1959); 32a, C.Hauser, Coating metallic or ceramic articles with cubic boron nitride, Swiss Patent CH 645,596; 10/15/84; filed 11/14/80
33. V. N. Gashtold et al, Preparation of films of cubic boron nitride by high frequency sputtering, Elektronnaya Tekhnika 4 (Series XII):58-68 (1970)
34. M. Satou, F. Fujimoto, Formation of cubic boron nitride films by boron evaporation and nitrogen ion beam bombardment, Jpn. Journ. Appl. Phys. 22:L171 (1983)
35. C. Weissmantel, K. Bewilogue, K. Breuer, D. Dietrich, U. Ebersbach, H.-J. Erler, G. Rau, G. Reisse, Preparation and properties of hard i-C and i-BN coatings, Thin Solid Films 96:31-44 (1982)
36. M. Sokolowski, Deposition of wurtzite type boron nitride layers by reactive pulse plasma crystallization, J. Crystal Growth 46:136-138 (1979)
37. M. Sokolowski, A. Sokolowska, A. Rusek, Z. Romanowski, B. Gokieli, M. Gojewska, Properties and growth of beta-BN (Borazon) layers from a pulsed plasma under reduced pressure, J. Crystal Growth 52:165-167 (1981)
38. A. Sokolowska, M. Wronikowski, The phase diagram (p, T, E) of boron nitride, J. Crystal Growth 76:511-13 (1986)
39. D. V. Fedoseev, I. G. Varshavskaya, A. V. Lavrent'ev, B. V. Deryaguin, V. L. Bukhovets, V. V. Matveev, V. L. Ruzinow, T. A. Karpukhina, Phase transitions of small solid particles when heated in a laser, Dokl. Akad. Nauk SSSR 270:918-22 (1983)
40. M. Alam, T. Deb Roy, R. Roy, High pressure phases of SiO2 made in air by Fedoseev=Derjaguin laser process, Appl. Phys. Lett. 53:1687-89 (1988)
41. D. V. Fedoseev, B. V. Derjaguin, I. G. Varshavskaya, A. V. Lavrent'ev, Homogeneous formation of metastable phases of carbon at high supersaturation, Carbon 21:243-46 (1983)
42. G. Kessler, H.-D. Bauer, W. Pompe, H.-J. Scheike, Laser pulse vapour deposition of polycrystalline wurtzite-type BN, Thin Solid Films 147:L45-50 (1987)
43. G. L. Doll, J. A. Sell, L. Salamanca-Riba, A. K. Ballal, Laser deposited cubic boron nitride films, submitted to Appl. Phys. Letters, 5/1/90
44. Y. Matsui, Small particles of cubic boron nitride prepared by electron irradiation of hexagonal boron nitride in a transmission electron microscope, J. Crystal Growth 66:243-47 (1984)
45. H. Saitoh, T. Ishiguro, Y. Ichinose, Synthesis of cubic boron nitride films by RF plasma CVD thermally activated with tungsten filament, ISPC-8 Tokyo, 1987, Paper P-086
46. H. Saitoh, T. Hirose, H. Matsui, Y. Hirotsu, Y. Ichinose, Synthesis of BN films by the plasma CVD with various solids: BH_3NH_3, H_3BO_3, and $NaBH_4$, Surface and Coatings Technol. 39/40:265-73 (1989)
47. P. T. B. Shaffer, The SiC phase in the system SiC-B_4C-C, Mater. Res. Bull. 213-20 (1969)
48. R. M. Chrenko, R. E. Tuft, H. M. Strong, The transformation of the state of nitrogen in diamond, Nature 27:141-44 (1977)
49. R. P. Messmer, G. D. Watkins, Linear combination of atomic orbital - molecular orbital treatment of the deep defect level in a semiconductor: nitrogen in diamond, Phys. Rev. Lett. 25:656-59 (1970)

50. A. K. Butylenko, G. V. Samsonov, I. I. Timofeeva, G. N. Makarenko, Doping of cubic boron nitride with carbon, Pis'ma Zh. Tekh. Fiz. 3:186-8 (1977)
51. N. N. Sirota, M. M. Zhuk, Solid solutions in the boron nitride-carbon system at high pressures and temperatures, Vestsi Akad. Navuk BSSR, Ser. Fiz.-Mat. Navuk 1979:122-5
52. Showa Denko K. K., Cubic boron nitride powder, JP 83,120,505; 7/18/83; filed 1/6/82
53. H. Sumya, High-hardness, cubic boron nitride sintered body and its preparation, JP 89,208,371; 8/22/89; filed 2/15/88
54. A. L. Liu, R. M. Wentzcovitch, M. L. Cohen, Atomic arrangement and electronic structure of boron carbide nitride (BC2N), Phys. Rev. B:39:1760-5 (1989)
55. N. Setaka, private communication (1988)
56. J. P. LaFemina, Electronic band structure of graphite-boron nitride alloys, J. Phys. Chem. 94:4346-51 (1990)
57. A. Badzian, T. Niemyski, S. Appenheimer, E. Olkusnik, Crystal structure in the boron-carbon-nitrogen system, Khim. Svyaz Poluprov. Polumetallakh 1972:362-6
58. R. B. Kaner, J. Kouvetakis, C. E. Warble, M. L. Sattler, N. Bartlett, Boron-carbon-nitrogen materials of graphite-like structure, Mat. Res. Bull. 22:399-404 (1987)
59. M. Yamada, Semiconductor device having covalently bonded boron carbide nitride passivation film, JP 88,129,631: 6/2/88; filed 11/2/86
60. A. W. Moore, S. L. Strong, G. L. Doll, M. S. Dresselhaus, I. L. Spain, C. W. Bowers, J. P. Issi, L. Piraux, Properties and characterization of codeposited boron nitride and carbon materials, J. Appl. Phys. 65:5109-18 (1989)
61. T. Ya. Kosolapova, G. N. Makarenko, T. I. Serebryakova, E. V. Prilutskii, O. T. Khorpyakov, O. I. Chernysheva, Boron carbonitride. I. Conditions for preparing boron carbonitride, Porosh. Met. 11:27-33 (1971)
62. T. M. Besmann, Chemical vapor deposition in the silicon-carbon and boron-carbon-nitrogen systems, Report (D3REP3) 1988, ORNL/TM-10884, 26 pp.
63. T. W. Klaus, Phototropy of boron nitride, Z. Physik. Chem. 228:1-4 (1965)
64. Anon., Multiband luminescence of layer-structured hexagonal boron nitride crystals, Muki Zaiken Nyusu 41:3 (1976)
65. E. Tiede, H. Tomaschek, The activating element in luminous boron nitride, Z. anorg. allgem. Chem. 147:111-22 (1925)
66. A. W. Moore, L. S. Singer, Electron spin resonance in carbon doped boron nitride, J. Phys. Chem. Solids 33:343:56 (1972)
67. T. V. Andreeva, T. V. Dubovik, Electrically insulating materials for linings and holders of high-pressure chambers, Dielektr. Poluprovodn. 5:7-12 (1974)
68. W. S. Rees, Jr., D. Seyferth, Preparation, characterization and pyrolysis of decaborane (14)-based polymers: boron carbide/boron nitride and boron nitride procedures, Ceram. Eng. Sci. Proc. 10:837-45 (1989)
69. R. J. Wedlake, A. L. Penny, Hard material, Ger. Offen. 2,806,070; 8/17/78; filed 2/16/77
70. Mitsubishi Metal Corp., Sintered tool material at an ultrahigh pressure JP 80 67,570 (571,574,575,576); 5/21/80; filed 11/14/78

71. A. Badzian, Cubic boron nitride-diamond mixed crystals, Mater. Res. Bull. 16:1385-93 (1981)
72. A. Badzian, High-pressure synthesis of diamond-type crystals and their atomic structure, Pr. ITME (PRITED) 1984, 12, 65 pp.
73. R. Haubner, Deposition of thermal CVD low pressure diamond on cubic BN, RM and HM, June 1990:70-76
74. W. E. Pickett, Thin superlattices and band-gap discontinuities: the (110) diamond-boron nitride interface, Phys. Rev. B 38:1316- 22 (1988)
75. W. R. L. Lambrecht, B. Segall, Electronic structure of (diamond C)/(sphalerite BN) (110) interfaces and superlattices, Phys. Rev. B 40 9909-19 (1989)
76. P. T. B. Shaffer, E. D. Whitney, Silicon carbide containing boron and nitrogen in solid solution, US Patent 3,554,717: 1/12/71; filed 1/30/68
77. Silicon Carbide, A High Temperature Semiconductor, Eds. J. R. O'Connor, J. Smiltens, Proc. of Conf. on SiC, Boston, MA., 4/2-3/59, Pergamon Press, N. Y. (1960)
78. J. A. Lely, Darstellung von Einkristallen von Siliciumcarbide und Beherrschung von Art und Menge der eingebauten Verunreinigungen, Deut. Ker. Gesellschaft. e.V., 32:229-50 (1955)
79. K. M. Hergenrother, S. E. Mayer, A. I. Mlavsky, Epitaxial and single crystal growth on to silicon carbide seeds, pp. 60-66 in Silicon Carbide, A High Temperature Semiconductor, Eds. J. R. O'Connor, J. Smiltens; Proc. of Conf. on Silicon Carbide, Boston, MA. 4/2-3/59; Pergamon Press, N.Y., 1960
80. Yu. M. Tairov, V. F. Tsvetkov, Investigation of growth process of ingots of silicon carbide single crystals, J. Crystal Growth 43:209-12 (1978)
81. Yu. M. Tairov, V. F. Tsvetkov, General principles of growing large single crystals of various silicon carbide polytypes, J. Crystal Growth 52:146-150 (1981)
82. G. Ziegler, P. Lanig, D. Theis, C. Weyrich, Single crystal growth of SiC substrate material for blue light emitting diodes, IEEE Transactions on Electron Devices, Vol. ED-30:277-81 (1983)
83. R. F. Davis, Silicon carbide and diamond semiconductor thin films: growth, defect analysis and device development, paper presented at Workshop on Science and Technology of Diamond Films, Quail Hollow Resort and Conference Center, Concord, Ohio, 5/20-24/90
84. W. C. Nieberding, Researchers develop long-sought SiC crystal growth techniques, Ind. Res. and Dev. Sept. 1983, pp. 148-50
85. S. Nishino, J. A. Powell, H. A. Will, Production of large area single crystal wafers of cubic SiC for semiconductor devices, Appl. Phys. Lett. 42:460-62 (1983)
86. J. A. Powell, Silicon carbide, a high temperature semiconductor, Tech. Memo for CECOM '83, Cleveland, Ohio, 10/4-6/83
87. A. Y. Liu, M. L. Cohen, Prediction of new low compressibility solids, Science 245:841-2(1989)
88. A. Y. Liu, M. L. Cohen, Structural properties and electronic structure of low-compressibility materials: β-Si_3N_4 and hypothetical β-C_3N_4, Phys. Rev. B. 41:10727-34 (1990)

DIAMOND - LIKE HYDROCARBON AND CARBON FILMS

John C. Angus and Yaxin Wang

Case Western Reserve University

Cleveland, OH 44106
USA

INTRODUCTION

The name "diamondlike" was apparently first coined by Sol Aisenberg[1] to describe the hard carbon films that he prepared by direct deposition from low energy carbon ion beams. It has been subsequently used rather indiscriminately to refer to many types of non–crystalline, hard carbonaceous films.

It is of some interest that Aisenberg used the term "diamondlike" because of the opprobrium attached to any claim for low pressure diamond synthesis in that era. It is quite likely that Aisenberg, in fact, produced films that had a polycrystalline diamond component. Aisenberg's results were confirmed by Spencer, Schmidt, Joy and Sansalone, who deposited films by a similar method, and clearly showed the presence of polycrystalline diamond.[2] The diamondlike materials are slowly being recognized as a new class of materials that are useful and interesting in their own right. There has been a great deal of confusion, much of it continuing, about the nature of these unusual materials. Some of this confusion has arisen because very different types of materials have been called "diamondlike." Also, the presence of hydrogen profoundly influences the type of structures that can be present. Despite this, many reports of diamondlike materials have not contained an explicit analysis for hydrogen.

The historical relationship of diamondlike films and the chemical vapor deposition of crystalline diamond has been described in several review papers and will not be repeated here.[3 4 5]

The diamondlike films have been closely associated with the formation of polycrystalline diamond by chemical vapor deposition. There are many reasons for this association. For example, both crystalline diamond and the diamondlike films can be made by similar processes. Also, a diamondlike component apparently appears as a second phase in polycrystalline diamond grown by chemical vapor deposition. The properties of diamondlike and diamond films also show some similarities. For example, the diamondlike films can be made with extreme hardness. Also, both diamond and diamondlike materials have remarkably high number densities. Diamond has the highest atom number density of any known terrestrial material and the diamondlike hydrocarbons have number densities greater than any conventional hydrocarbon polymer.[4 6] These high number densities arise from their maximally cross–linked structure and give rise to a host of extreme properties, including high hardness and elastic modulus.

Diamond and Diamond-Like Films and Coatings
Edited by R.E. Clausing *et al.*, Plenum Press, New York, 1991

Diamond is the proto–typical diamond–cubic crystal. The chemical and physical properties of diamond are most easily understood by recognizing that the lattice is a three dimensional array of six–membered carbon rings. The layers of rings are stacked in an ABC/ABC/ABC sequence along <111> directions. In diamond each six–membered ring is in the "chair" conformation and all C–C bonds are staggered. The diamond–cubic lattice is the ultimately cross–linked crystalline carbon polymer.

Diamondlike phases, on the other hand, are amorphous, not crystalline. They may be divided into two broad categories: hydrogenated and non–hydrogenated. For convenience we denote the former as "diamondlike hydrocarbons" or by the symbol a–C:H. The non–hydrogenated diamondlike materials are, in this paper, called "diamondlike carbons" and are given the symbol a–C. (It should be noted that no universally accepted, systematic nomenclature has been agreed upon to describe these materials.)

The diamondlike hydrocarbons are better understood than the diamondlike carbons, primarily because the presence of hydrogen has facilitated their study by NMR and infrared spectroscopy. Also, the hydrogenated films are the ones produced by the simplest and most easily accessible deposition techniques, i.e., rf–self bias deposition and direct low energy ion beam deposition using hydrocarbon source gases.[7]

PREPARATION AND CHEMICAL COMPOSITION OF DIAMONDLIKE FILMS

Diamondlike hydrocarbons (a–C:H) are typically prepared by impact of hydrocarbon ions in the energy range from several tens to 100 eV. Direct deposition from hydrocarbon ion beams and by rf self–bias plasma assisted chemical vapor deposition are the most commonly employed methods. These methods provide significantly higher impact energies than are achieved in inductively coupled plasmas, which are used for the production of the so–called plasma polymers.

Diamondlike carbon films (a–C) containing little hydrogen can be prepared by sputtering of carbon,[8] by direct ion beam deposition from carbon ion beams[9] or by condensation of carbon plumes produced by laser vaporization.[10] These films have received less attention than the hydrogenated films. It is clear, however, that the films are an unusual form of carbon with mass densities greater than graphite, but less than diamond, and with very high hardness.

The hydrogen content is the primary independent variable which determines the structure, and hence the properties, of the diamondlike films. The atomic percent hydrogen is very low in the non–hydrogenated diamondlike carbons, nominally less than one atomic percent. On the other hand, the hydrogen content of the hydrogenated diamondlike carbons ranges from approximately 16 to 60 atomic percent hydrogen.

PHYSICAL PROPERTIES OF DIAMONDLIKE MATERIALS

Diamondlike Hydrocarbons (a–C:H)

The diamondlike hydrocarbons typically have mass densities ranging from 1.4 to 2.0 g/cm^3, with the least dense films found at the highest hydrogen content. These densities are much higher than those of hydrocarbons of similar elemental composition.[6] See, for example, Table 1 and Figure 1, which is reproduced from reference 4.

The empirical scratch hardness of the hydrogenated films is very high, in some cases greater than SiC. However, this empirical measure of hardness depends on many factors including the friction coefficient and the adhesive strength between the film and substrate. The hardness decreases as the hydrogen content increases.

The hydrogenated diamondlike films are usually found in a state of high compressive stress. The stress level increases with increasing hydrogen content.[11] Addition of fluorine increases compressive stress, while nitrogen has been reported to reduce the compressive stress.

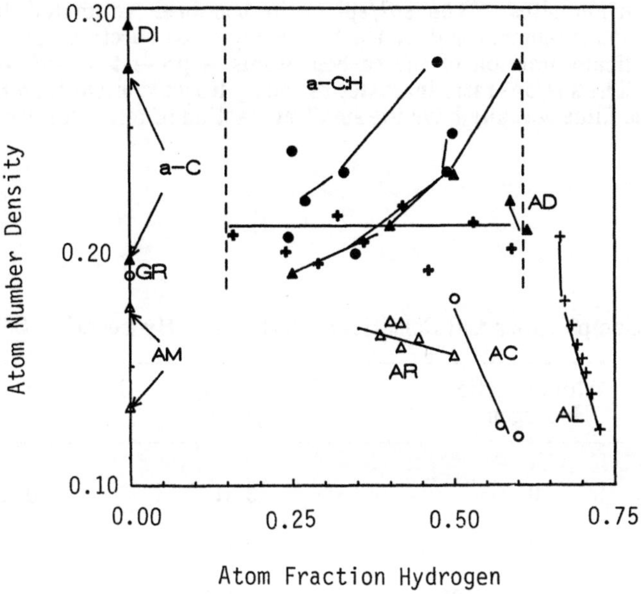

Figure 1

Atom number density versus atom fraction hydrogen. Atom number density is expressed in gram–atoms per cubic centimeter, which is the total number of atoms per cubic centimeter divided by Avogadro's number. This type of plot provides a convenient way for categorizing different classes of hydrocarbons and carbons.[4] Similar compounds, for example the n–alkanes (AL), fall in tight groupings on the diagram. Diamond, DI, is a the upper left–hand corner. The adamantanes, AD, which are 10 and 14 carbon atom molecular diamonds, are at the lower right of the diamondlike hydrocarbon (a–C:H) grouping. The diamondlike hydrocarbons (filled symbols) range from approximately 0.20 to 0.60 atom fraction hydrogen and have number densities greater than 0.19 gram–atom per cubic centimer. The vertical dashed lines show the theoretical composition range predicted by Eq. (22) for fully constrained, random hydrocarbon networks. The atom number density of a–C:H increases with hydrogen content in contrast to the behavior of conventional hydrocarbons. This increase is caused by the increase in average coordination number of carbon from 3 to 4 as the amount of hydrogen is increased. The diamondlike carbons (a–C) appear to be a different class of material with little or no hydrogen and with atom number densities from approximately 0.19 to 0.28 gram–atom per cubic centimer. These densities indicate significant sp^3, tetrahedral, bonding. The amorphous carbons (AM) are dominated by sp^2, trigonal, bonding and can be formed by laser or electron–beam evaporation. Other symbols are: AC, oligomers of acetylene; AD, adamantanes; AR, polynuclear aromatics; and GR, graphite.

The optical transmission and band–gap of the hydrogenated films both increase with increasing hydrogen content. The band–gap increases from approximately 0.5 eV to over 2.0 eV for the highly hydrogenated films. The most transparent films are the softest.

The unusual nature of the films must arise from their structure and bonding network. The films are amorphous so little has been learned from conventional diffraction experiments. However, much attention has been paid to the relative numbers of sp^3 and sp^2 carbon sites. The sp^3/sp^2 ratio has been estimated by infra–red absorption, NMR measurements and by electron energy loss spectroscopy. It has been found that a significant fraction of the carbon atoms is present as sp^3, tetrahedrally coordinated sites. The sp^3/sp^2 ratio increases as the hydrogen content increases, i.e., the hardest, most dense films actually have the smallest fraction of tetrahedrally coordinated carbon.

Table 1. Compositions and Densities of Carbon and Hydrocarbon Solids

Solid	Atom Fraction Hydrogen	Mass Density g/cm^3	Number Density $gatom/cm^3$
diamond	0	3.51	0.29
graphite	0	2.26	0.19
a–C:H	0.16–0.60	1.4–2.0	0.19–0.26
a–C	≈0	≈2.0–2.9	0.17–0.24
adamantane	0.615	1.07	0.21
di–adamantane	0.588	1.21	0.22
high density polyethylene	0.67	0.97	0.21
polyacetylene	0.5	1.16	0.18
naphthalene	0.44	1.145	0.16
anthracene	0.42	1.25	0.17

Diamondlike Carbons (a–C)

The densities of the non–hydrogenated diamondlike carbons have been reported[8] to be as high as 3.1 g/cm, which is closer to the density of diamond (3.51 g/cm^3) than that of graphite (2.26 g/cm^3). This unusually high density alone indicates an unusual structure. The density of the films and the fraction of sp^3 bonding increases as the temperature of the substrate is decreased.[8] There are well documented reports that indicate these films possess exceptional hardness, in some cases greater than that of diamond.[10]

STRUCTURAL MODELS FOR DIAMONDLIKE MATERIALS

Any model of the structure of the diamondlike hydrocarbons must be consistent with the observation that the films have high empirical hardness and can support high internal compressive stress. Furthermore, the structural model must explain the relatively low observed optical band–gap and very high electrical resistivity.

Electronic Structure of Diamondlike Hydrocarbons

The electronic properties of the films, as indicated by the band–gap and electrical resistivity, were attributed by Bredas and Street[12] and Robertson[13-14] to the presence of small graphitic–like clusters of carbon atoms within a more highly hydrogenated matrix. Using electronic–structure calculations they showed that as the overall hydrogen content decreases, these π bonded clusters increase in size, giving rise to increased optical absorption and decreased band–gap. The lack of conductivity is explained by the fact that the clusters are sufficiently spatially isolated so that there is no π electron orbital overlap.

Fully Constrained Network Model (FCN Model)

The hard, rigid mechanical nature of the films must arise from the structure of the matrix that makes up the bulk of the film. Angus and Jansen[15] modeled the diamondlike hydrocarbon films as a fully constrained, non–crystalline network using the constraint counting arguments of Phillips,[16] Thorpe[17] and Döhler.[18] The predictions of sp^3/sp^2 ratio and the average coordination number are in general agreement with the available experimental observations, particularly at the upper end of the hydrogen composition range. The model, as originally derived, predicts a range of compositions, from 0.17 to 0.62 atom fraction hydrogen, over which a fully constrained network comprised of sp^3 and sp^2 coordinated carbon atoms and hydrogen atoms can exist. The predicted stability range is in agreement with observation.

The fully constrained, non–crystalline network model (FCN model) is described in detail in the following major section.

Defective Graphite Model

Tamor has recently proposed a "defective graphite" model for the structure of the diamondlike hydrocarbons.[19] In this model an initially perfect graphite crystal is modified by removing carbon atoms and saturating the resulting dangling bonds with hydrogen. Percolation theory is used to predict the upper and lower limits of hydrogen concentration. The defective graphite model predicts an existence range of diamondlike hydrocarbons from 0.2 to 0.6 atom fraction hydrogen. It is remarkable that this model, which is based on different assumptions, predicts a similar existence range as the constrained network model.

Diamondlike Carbons

The constraint counting arguments used to describe the hydrogenated films show that a non–hydrogenated, non–crystalline network made up solely of sp^3 and sp^2 carbon sites will be highly over–constrained. Bonding constraints can be reduced by formation of clusters or microcrystallinity and this may be the source of stability for these unusual materials. In fact, recent molecular dynamics studies[11] show that several types of medium range order may be present in the non–hydrogenated amorphous carbons including five, six and seven–membered rings and local clustering of sp^3 sites.[20] Tamor has concluded on the basis of his defected graphite model that it is unlikely if more than 10% of the carbon sites in non–hydrogenated diamondlike carbons are four–fold coordinated.[19] However, recent K–edge absorption studies by Cuomo indicate up to 40 percent of the carbon atoms in a–C may be tetrahedral, sp^3 sites.[8]

FULLY CONSTRAINED NETWORK MODEL (FCN MODEL)

General Description of FCN Model

The average, nearest–neighbor bonding environment can be described in terms of the theories of random covalent networks of Phillips[16] and Thorpe.[17] According to Phillips, a random covalent network is completely constrained when the number of constraints per atom is just equal to the number of mechanical degrees of freedom per atom. The number of degrees of freedom of a single atom is just equal to the dimensionality, in this case three. The number of constraints per atom can be calculated from its coordination number, i.e., the number of directed valence bonds to nearest neighbors.

A qualitative picture of the constrained network model is instructive. Increasing the number of bonds (crosslinks) in a non–crystalline network contributes to stabilization because of the bond energy. However, increasing the number of bonds in a non–crystalline network also leads to increase of the strain energy because of bond stretching and bond angle distortion. If the network is underconstrained, accommodations in atom positions can be made to minimize the strain energy. However, once the network is fully constrained mechanically, such accommodations can not be made. Additional bonding (cross–linking) will therefore dramatically increase the strain energy. This qualitative argument indicates that the optimal coordination number that just balances these two effects should be near that of a fully constrained network.

The role of hydrogen in diamondlike hydrocarbons is easily understood in this context. Since hydrogen is univalent, it can only serve as a network terminator, thereby reducing bonding constraints. Increasing the hydrogen content therefore permits the fraction of tetrahedrally coordinated (sp^3) carbon, to increase at the expense of trigonally coordinated (sp^2) carbon, while still keeping the overall average coordination number constant.

Application of Constraint Counting to Films Without Clustering

Döhler[18] showed that for a three–dimensional random network, the number of constraints, C_i, introduced by one and two coordinated atoms (H and O, for example) is given by

$$C_m = m_i^2/2, \qquad m_i \leq 2, \tag{1}$$

where m_i is the coordination number. For elements with coordination numbers $m_i \geq 2$, the number of constraints introduced per atom is given by

$$C_m = (5m_i/2) - 3, \qquad m_i \geq 2. \tag{2}$$

Equations (1) and (2) are valid for three–dimensional random networks whose bonding is dominated by nearest–neighbor, directed valence bonds. Setting the average number of constraints per atom equal to three, Döhler obtained the condition for a completely constrained random covalent network.

$$\sum x_i C_i = 3, \tag{3}$$

where the x_i are the atom fractions of the several types of atoms making up the solid. For this discussion three types of atoms are distinguished: hydrogen with $m_1 = 1$, sp^2 bonded carbon with $m_3 = 3$, and sp^3 bonded carbon with $m_4 = 4$. The presence of divalent atoms and sp^1 bonded carbon, both with $m_2 = 2$, is ignored.

From Eqs. (1) and (2) one immediately obtains $C_1 = 0.5$, $C_3 = 4.5$, and $C_4 = 7.0$. Therefore, Eq. (3) becomes

$$0.5x_1 + 4.5x_3 + 7.0x_4 = 3. \tag{4}$$

One also has

$$x_1 + x_3 + x_4 = 1. \tag{5}$$

Combining Eqs. (4) and (5) one can solve for the ratio of tetrahedrally bonded (sp^3) carbon to trigonally (sp^2) bonded carbon in a completely constrained random hydrocarbon network.

$$\frac{x_4}{x_3} = \frac{8x_1 - 3}{8 - 13x_1} \tag{6}$$

Equation (6) indicates that there is a composition range over which completely constrained random hydrocarbon networks exist. For example, setting the denominator equal to zero one obtains the condition for a fully constrained non–crystalline network in which the carbon skeleton has only sp^3 bonding. This occurs at a hydrogen atom fraction of $x_1 = 8/13 = 0.615$, which is identical to the result obtained by Döhler for a–Si:H.[18] At hydrogen concentrations above this value, the number of bonds is insufficient to completely use up all of the degrees of freedom and the network will be "under–constrained" or "floppy."

If the numerator of Eq. (6) is set equal to zero, one obtains the theoretical lower limit of hydrogen content for a random hydrocarbon network. One obtains $x_1 = 3/8 = 0.375$. Below this hydrogen concentration the network will be "over–constrained." In fact, when the constraints per atom significantly exceed the number of dimensions, the film may reconstruct in some way, e.g., small clusters or microcrystals may appear. Another way of looking at it is that medium–range or long–range ordering processes (in addition to the short–range, nearest–neighbor, covalent bonds) can also play a role in determining the film structure.

One can also use the above results to calculate the average coordination number, m, which is defined by

$$m \equiv \sum x_i m_i. \tag{7}$$

In the present case m is

$$m = x_1 + 3x_3 + 4x_4. \tag{8}$$

Combining Eqs. (4), (5), and (8) one can solve for the optimal coordination number in terms of the hydrogen atom fraction x_1.

$$m = 2.4 - 0.4x_1. \tag{9}$$

Connectivity of Network

The connectivity of a non–crystalline network is shown schematically in Figure 2. Figure 2a shows a schematic representation of the network including terminal hydrogen atoms. The coordination number, averaged over all atoms, is given by Equation (8). If the network is fully constrained the average coordination number, m, is given by Equation (9).

Figure 2b shows the same network with the hydrogen atoms and dangling bonds removed. The carbon skeletal structure of the network remains.

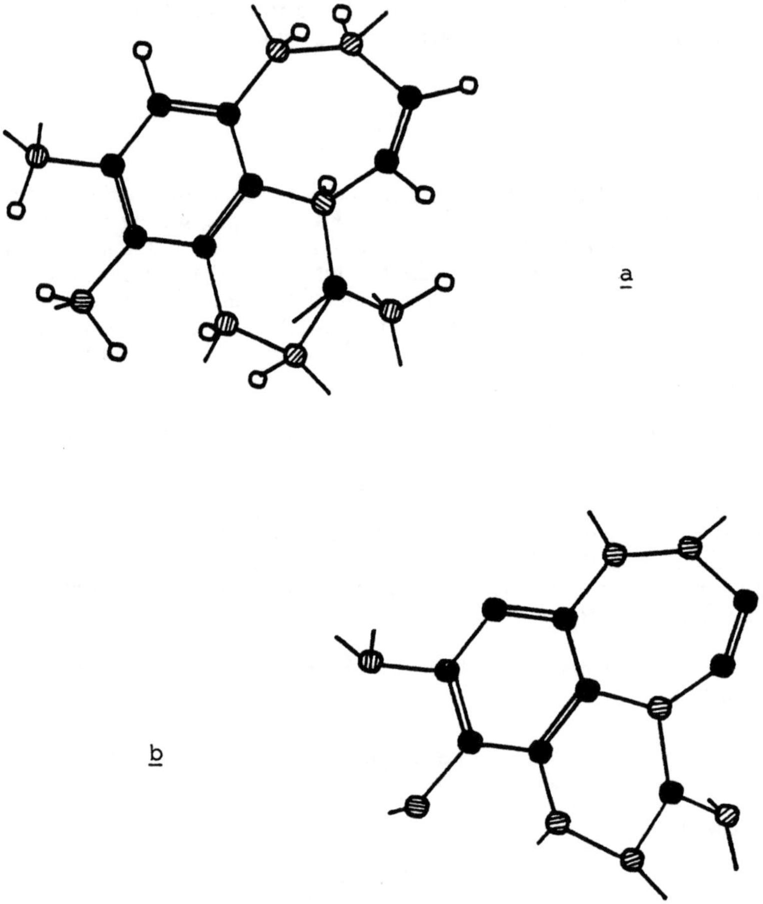

Figure 2

Schematic representation of non–crystalline network models. 2a) network comprised of hydrogen (O), sp^2 carbon atoms (●) and sp^3 carbon atoms (⊜). The sp^2 carbon atoms are shown in three environments, i.e., in a six–membered aromatic ring, in a π bonded doubled bond and as an isolated free radical site (dangling bond). 2b) skeletal network comprised only of carbon atoms after hydrogen atoms and dangling bonds are removed.

Hydrogen atoms, being monovalent, can serve only as terminating atoms along the skeletal carbon network. They cannot contribute to the connectivity of the network itself. If the contribution of the hydrogen atoms to the overall average coordination number is removed, one obtains the coordination number m_n for the carbon skeletal network alone. This can be accomplished by subtracting off the contribution of the monovalent atoms and normalizing to the total number of carbon atoms. We note that removal of each hydrogen atom removes two nearest neighbors from the counting. Therefore the average coordination number of the carbon skeletal structure alone is:

$$m_n = \frac{m - 2x_1}{1 - x_1} \tag{10}$$

Using Eqs. (8), (9), and (10) one finds

$$m_n = 2.4 \tag{11}$$

This result shows that the optimal average coordination number of the carbon skeleton alone is just equal to the value predicted by Thorpe[17] for monatomic random networks.

Constraint Counting for Films with Clustering

The above analysis is appropriate only when the triply coordinated atoms, $m_i = 3$, are randomly distributed. This might be the situation, for example, if these sites were boron atoms. Isolated sp^2 sites are free radical sites. (See Figure 2 for example.) The energy of the system will be lowered if these sites are bonded together forming π bonded carbon pairs, i.e., >C = C<. The sp^2 sites are further stabilized by condensation into aromatic ring structures. The presence of these clusters of atoms requires a modification of the above analysis.

First consider the effect only of double bonded pairs of carbon atoms. Each double–bonded carbon pair, >C = C<, makes four bonds with the rest of the network. Therefore, the pair may be considered as a single four–fold coordination site. This is shown schematically in Figure 3b. The constraint equation, corresponding to Equation (4) therefore becomes

$$0.5x_1' + 7.0\left(\frac{1}{2}x_3'\right) + 7.0\,x_4' = 3 \tag{12}$$

where x_1', $x_3'/2$ and x_4' are the atom fractions of hydrogen, sp^2 pairs and sp^3 sites respectively. They are calculated on the basis that the total number of network sites is: $N_1 + \frac{N_3}{2} + N_4$. The relationship between the fraction of network sites, x_i', and the true atomic fractions, x_i, is simply

$$x_i' = \frac{x_i}{x_1 + \frac{x_3}{2} + x_4} \tag{13}$$

or alternatively

$$x_i' = \frac{x_i}{1 - \frac{x_3}{2}} \tag{14}$$

Using Equation (14) in Equation (12) one obtains the constraint equation written in terms of the true atom fractions.

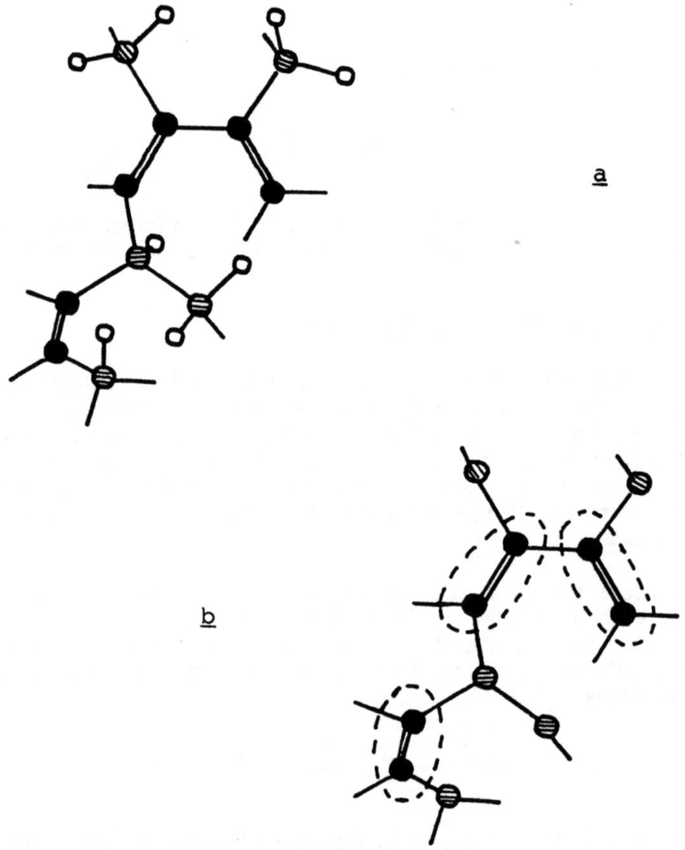

Figure 3

3a) Schematic representation of a network model with all sp² carbon atoms present as π bonded carbon atom pairs; 3b) skeletal network after removal of hydrogen atoms and showing π bonded pairs as single network sites.

$$0.5\left[\frac{x_1}{1-\frac{x_3}{2}}\right] + 7.0\left[\frac{x_3/2}{1-\frac{x_3}{2}}\right] + 7.0\left[\frac{x_4}{1-\frac{x_3}{2}}\right] = 3 \qquad (15)$$

Equation (15) differs from Equation (13) used by Angus and Jansen in their original paper.[15] In the original paper the approximation $x_i = x_i'$ was used, which is strictly valid only in the limit as $x_3 \rightarrow 0$. The authors are indebted to Dr. Michael Tamor of the Ford Motor Company for bringing this to their attention.

Combining Equations (5) and (15) one obtains a new expression for the ratio of sp^3 to sp^2 carbon sites.

$$\frac{x_4}{x_3} = \frac{9x_1 - 4}{8 - 13x_1} \qquad (16)$$

Equation (16) predicts a narrower range of hydrogen atom fraction over which fully constrained non–crystalline hydrocarbon networks can exist. The upper limit of hydrogen atom fraction, $x_1 = 8/13 = 0.615$, remains the same; however, the lower limit is now $x_1 = 4/9 = 0.444$.

The influence of double bonds distributed throughout the network may be explained as follows. When two isolated sp^2 sites form a π bonded pair, i.e., >C = C<, the number of bonds with network atoms is reduced from 6 to 4, with a corresponding reduction in network constraints form 4.5 to 3.5 per sp^2 site. However, the total number of network sites is reduced by $N_3/2$. Therefore, the change in number of constraints per network site caused by double bond formation is just

$$\frac{(3.5 - 4.5)N_3}{\frac{N_3}{2}} = -2.0$$

Since a fully constrained network has three constraints per site, double bond formation results in an increase in constraints per network site. Therefore, more hydrogen is required to reduce the average coordination number of the network. At the upper limit of hydrogen concentration, $x_1 = 8/13$, no sp^2 sites are present and the prediction is unchanged.

Further insight into the influence of clustering may be obtained by considering the effect of clusters of aromatic rings. First, consider the simplest case where the cluster is comprised of one six–membered aromatic ring.

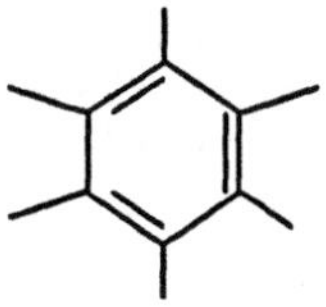

The entire cluster shown above is taken as a single network site. We assume all sp^2 carbon atoms are present only in these rings. The network is shown schematically in Figure 4a. For the ring we have m = 6 and therefore from Equation (2), $C_6 = 12$. In this case the constraint equation corresponding to Equation (15) becomes.

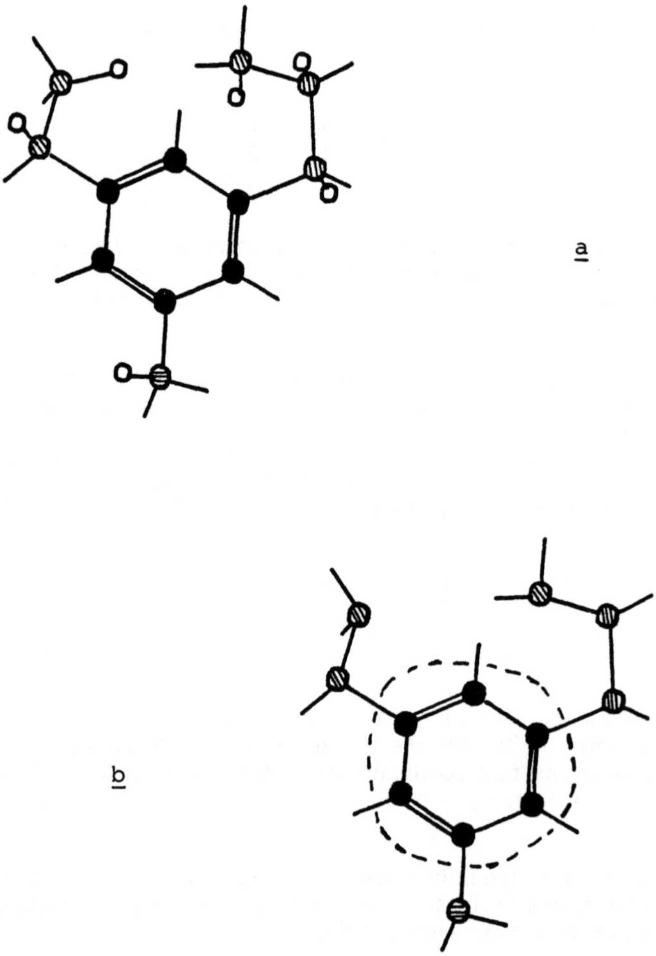

Figure 4

4a) Schematic representation of a network model with all sp^2 carbon atoms within six–membered aromatic rings. 4b) skeletal network after removal of hydrogen atoms and treatment of aromatic rings as single network sites with m=6.

$$0.5\left[\frac{x_1}{1-\frac{5x_3}{6}}\right] + 12\left[\frac{x_3/6}{1-\frac{5x_3}{6}}\right] + 7.0\left[\frac{x_4}{1-\frac{5x_3}{6}}\right] = 3 \qquad (18)$$

Equation (18) reduces immediately to

$$0.5\, x_1 + 4.5x_3 + 7x_4 = 3 \qquad (19)$$

Equation (19) is identical to Equation (14). Therefore, the predicted sp^3/sp^2 ratio is identical to that given by Equation (6).

The above result shows that condensation of isolated sp^2 sites into single six–membered aromatic rings has no net effect on the network stress. The reduction in network constraints caused by clustering into six–membered aromatic rings is exactly equal to 3.0 times the reduction in number of network sites. Therefore, the average number of constraints per network site remains the same, i.e., equal to 3.0.

In the above model the network is assumed to be comprised solely of hydrogen atoms, six–membered aromatic rings (each acting as a single network site), and sp^3 carbon atoms. The skeletal network is obtained by excluding the terminal hydrogen atoms and by treating each aromatic ring as a single network site. This is shown schematically in Figure 4b. The average coordination number of the skeletal network alone is computed by subtracting off the contribution of the terminal hydrogen atoms and normalizing to the number of network sites.

$$m_n = \frac{x_1 + 6(\frac{x_3}{6}) + 4x_4 - 2x_1}{1 - x_1 - \frac{5x_3}{6}} \qquad (20)$$

Using Equations (19) and (20) one finds after much manipulation

$$m_n = 2.4 \qquad (21)$$

In other words, the average coordination number of the fully constrained network itself, is equal to 2.4, the value obtained by Thorpe for a fully constrained monatomic network.[17]

The above model is attractive in its simplicity. However, real diamondlike films exhibit more complex structures, with a distribution of cluster sizes and types. Here we discuss briefly the influence of other types of medium range order.

Another form of medium range order in amorphous carbon networks is believed to be five–membered rings, i.e.,

Calculations analogous to those presented above for six–membered rings show that five–membered rings increase the number of constraints per network site. Consequently, they serve to increase strain energy. More hydrogen is therefore required in the network to offset the presence of five–membered rings.

The most common cluster appears to be condensed ring aromatic structures. We consider structures of the general form

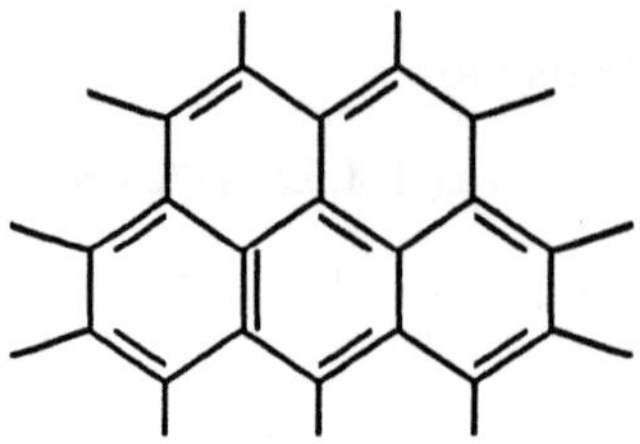

Calculations similar to those described above were performed for a series of aromatic clusters. We assume that each aromatic cluster can be treated as a single network site and that the number of constraints may be calculated using Equation (2). One finds that clusters of two or more aromatic rings serve to reduce the number of constraints per network site. Consequently, the larger aromatic clusters reduce the number of network terminators (hydrogen atoms) required to achieve a fully constrained network.

We note, however, that treating the aromatic clusters as single, rigid network sites is an over–simplified approximation, especially for the larger clusters. Real condensed ring aromatic clusters should contribute more to relief of network constraints than the present approximation indicates. We further note that much earlier Döhler invoked large clusters to explain the fact that amorphous hydrogenated silicon contains less hydrogen than predicted by simple application of a fully constrained network model.[18]

We present the results for one condensed cluster containing 76 aromatic rings. This is in the size range predicted by other means.[13 14] For this cluster one finds $m = 34$ and from Equation (2) cluster $C_{34} = 82$. Using methods analogous to those described above we find

$$\frac{x_4}{x_3} = \frac{539x_1 - 79}{92\ (8-13x_1)} \tag{22}$$

From Equation (22) the minimum and maximum hydrogen atom fractions are respectively: $x_1 = 0.15$ and $x_1 = 8/13 = 0.615$.

Comparison of FCN Model with Experiment

A plot of sp^3/sp^2 ratio versus hydrogen atom fraction is shown in Figure 5. The predicted values of sp^3/sp^2 ratio, i.e., x_4/x_3, from Equations (6) and (22), are shown along with the available experimental data.

Although the data points are limited in number, the predicted increase in sp^3/sp^2 ratio at $x_1 = 8/13 = 0.615$ is observed. Also, it is quite clear that the predicted sp^3/sp^2 ratio for large clusters, i.e., Equation (22) for clusters containing 76 aromatic rings, is closer to the measured values than the predicted values for clusters comprised of a single aromatic ring, i.e., Equation (6). A closer fit of the predicted values to the experimental data points can be obtained by adjusting the assumed size of the aromatic cluster. However, we believe that the accuracy of the experimental data and the inherent simplifications built into the model do not warrant detailed curve fitting. The FCN model does show that aromatic clusters of the size estimated by independent methods [13 14] leads to predicted sp^3/sp^2 ratios reasonably close to observed values.

We note that a number of factors prevent a precise comparison of the model with the experimental observations. First, a distribution of cluster sizes will be present in any

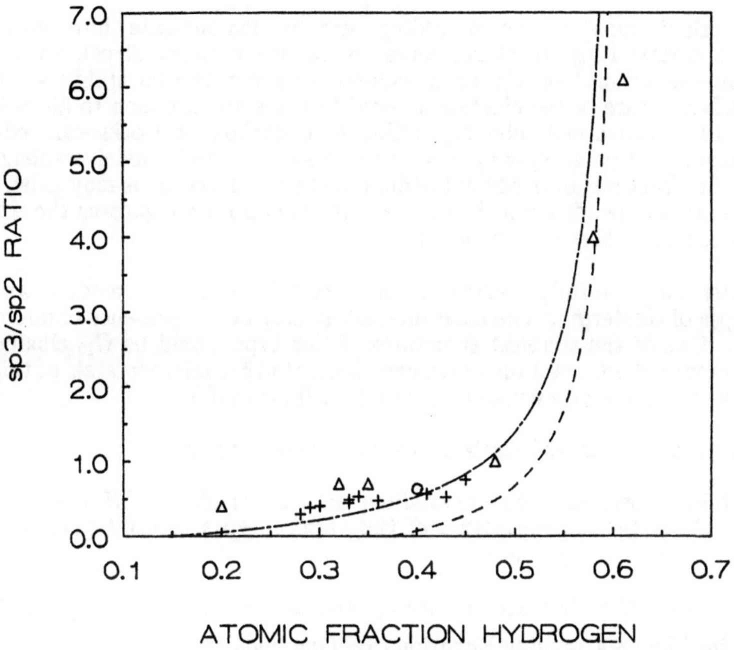

Figure 5

The ratio of sp^3 to sp^2 sites in a–C:H; The points labeled Δ are from reference (23), the point labeled o from reference (24) and the point labeled + from reference (22). The first two were obtained from NMR measurements, the last one from analysis of the electron energy loss spectrum. The lines are the theoretical expressions for fully constrained, random hydrocarbon networks, i.e., equations (6) and (22). The right hand curve was derived assuming all sp^2 carbon atoms are present in single six–membered aromatic rings. The left hand was derived assuming the aromatic clusters contained 76 rings.

single film. Furthermore, it is highly likely that the average size of the aromatic cluster will increase as the hydrogen content is decreased. This may, in fact, be the reason for the increased (negative) deviation of the predicted curve from the data points as the hydrogen concentration is decreased.

We also note that other network terminating groups that do not contribute to the network skeleton should be discounted when computing network constraints. Most prominent of these will be methyl (CH_3) and phenyl (C_6H_5) groups. There is, however, no way to easily estimate their concentrations in the films.

CONCLUDING REMARKS

Effect of Clusters

The calculations of the preceding section demonstrate how clustering into condensed aromatic ring structures serves to reduce network stress, with consequent reduction in the amount of hydrogen required to promote film stability. Despite the over-simplified nature of the clusters, several features are common to all of the models. First, no fully constrained, non-crystalline hydrocarbon networks can exist above a hydrogen atom fraction of $x_1 = 8/13 = 0.615$. Secondly, each model predicts a range of hydrogen atom fractions over which the diamondlike hydrocarbons may exist. The lower limit of this range depends upon the nature of the assumption regarding the clustering (or lack of clustering) for the sp^2 carbon sites.

Furthermore, the fully constrained non-crystalline network models can be applied to other types of clustering. The most interesting may be the presence of microcrystals of sp^3 carbon. One of the simplest structures of this type would be C_{10} clusters with the skeletal structure of adamantane. One can speculate that microcrystals of this type may help stabilize the non-hydrogenated diamondlike films (a-C).

Categorization of Carbon and Carbon/Hydrogen Solid Phases

The fully constrained non-crystalline network model (FCN model) provides a useful method for the categorization of the various types of solid phases that can be formed from carbon and hydrogen.

The fraction of carbon atoms with sp^3 coordination number, $f_{sp^3} = x_4/(x_3 + x_4)$, can be obtained by re-arranging Equation (6). One finds

$$f_{sp^3} = \frac{8x_1 - 3}{5 - 5x_1} \tag{23}$$

where f_{sp^3} is the fraction of carbon atoms with sp^3 hybridization and x_1 is the atom fraction of hydrogen. Similarly, one can obtain f_{sp^3} for the situation described by Equation (22). In this case one finds

$$f_{sp^3} = \frac{539x_1 - 79}{657 - 657x_1} \tag{24}$$

Equations (23) and (24) and the available experimental data are plotted in Figure 6. The dashed lines on Figure 6 are Equations (23) and (24) given above. The lines divide the figure into two fields. To the right are the conventional, underconstrained polymers, e.g., polyethylene, polyacetylene and the polycyclic aromatic hydrocarbons (PAH). The points show the available data for the diamondlike hydrocarbons (a – C:H). Within the experimental accuracy of the measurements, they fall near the lines for the fully constrained solids. The fully constrained network model predicts that no diamondlike

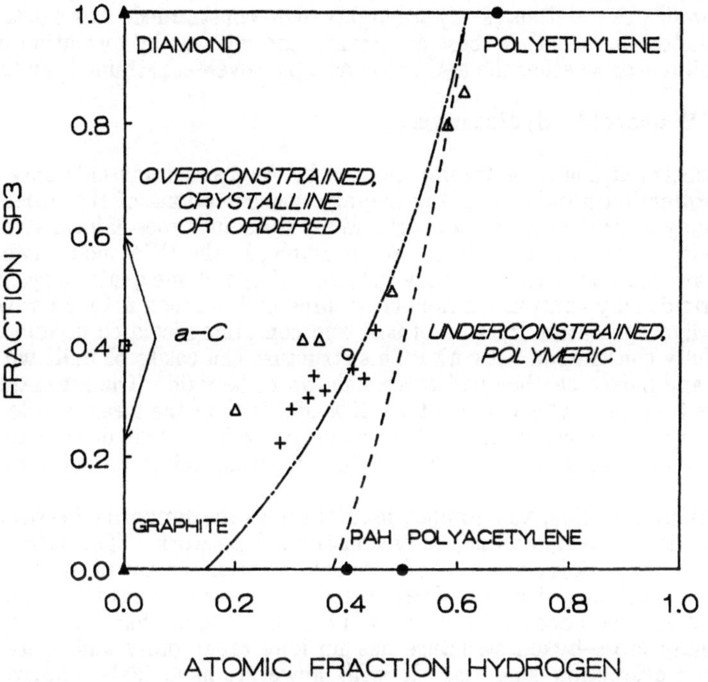

Figure 6

Fraction of carbon atoms with sp^3 hybridization versus atomic fraction hydrogen. Conventional underconstrained (floppy) polymers are to the right of the lines. Overconstrained, crystalline or ordered solids are to the left. The data points are for hydrogenated diamondlike hydrocarbons (a–C:H). The points Δ and ∘ were obtained by nuclear magnetic resonance[23 24] and the points + by electron energy loss spectroscopy[22]. a–C refers to diamondlike carbons and PAH for polycyclic aromatic hydrocarbons. The point □ is for an a–C sample reported by Cuomo.[8]

materials will be found to the right of the boundary lines shown on Figure 6. Random networks to the left of the boundary on Figure 6 will be over–constrained.

The boundary dividing Figure 6 into an "over–constrained" and "under–constrained" region is fixed ultimately by the condition on the network skeleton that $m_n = 2.4$ (Equations 11 and 21). The hard, over–constrained solids will be to the left of the line and the soft, under–constrained solids to the right. The condition that $m_n = 2.4$ was originally used to delineate the point where an amorphous solid became glassy, i.e., could flow. Solids that are significantly "over–constrained" will have increased energies because of bond length and bond angle distortions. This strain energy can be relieved by formation of medium–range or long–range order. Medium–range order would include preferential clustering of sp^3 and sp^2 carbon sites, as described earlier, or voids. Long range order would imply crystallinity. In the absence of hydrogen the counting of constraints indicates the films (a–C) are highly over–constrained. More detailed study is required to determine whether these constraints are released by formation of clusters or microcrystallinity or whether the a–C exists as a true over–constrained random network.

Structure of Diamondlike Hydrocarbons

The constraint counting arguments used by Angus and Jansen[15] only describe the average coordination number and the average concentrations of the different types of atoms that may be present in the network. Many different types of local structure can be consistent with these average values. For example, in the FCN model two ternary sp^3 sites (≡CH) are equivalent to one binary sp^3 site $>CH_2$ and one quaternary sp^3 site $>C<$. Consider a rigid, fully constrained non–crystalline hydrocarbon network with an sp^3/sp^2 ratio of 1, a hydrogen atom fraction of 0.52 and containing isolated double bonds. This will give a fully constrained network. This structure can easily be built with a ball and stick model and indeed can be qualitatively shown to be rigid. The estimated density of this model is 1.9g/cm^3, which is about 10–20% higher than the measured densities of the diamondlike hydrocarbons at this hydrogen concentration. This discrepancy may arise from the presence of voids or the neglect of the condensation into aromatic rings.

Construction of this, and similar, models shows the similarity between crystalline diamond and the non–crystalline, fully constrained network. The fully constrained, non–crystalline network is comprised of a three–dimensional array of rings, dominated by six–membered rings, but with some five and seven–membered rings present as well. Both staggered and eclipsed bonds are present, while cubic diamond has only staggered bonds. This maximally cross–linked structure has no long range order and therefore appears amorphous to diffraction analysis. It will, however, have high rigidity and should support large internal compressive stresses. The extreme hardnesses of both crystalline diamond and the non–crystalline material very likely arise from this dense, cross–linked structure.

The diamondlike hydrocarbons may be related to the "hypercross–linked" polymers recently reported by Whitesides and co–workers.[23,24]

Formation Mechanism

The chemical mechanism of formation of these unusual structures is not known. However, a likely hypothesis is simply that the energetic surface layer, exposed to the ion flux, is quenched by the underlying cold substrate.[8,15] The metastable, constrained structure is "frozen–in" by this process. A quenching mechanism is supported by the observations that the diamondlike character of the films is not obtained at high substrate temperature, e.g., 400C.[8] At the higher substrate temperatures, relaxation processes may take place which lead to formation of a more graphitic–like structure. Also, Cuomo et al. observed increased densities and sp^3 fraction of a–C as the substrate temperature was lowered and as the thermal conductivity of the substrate was increased.[8]

The evidence indicates that the presence of hydrogen during film condensation facilitates the formation of a network structure with an average coordination number close to that of a fully constrained network. This may arise because hydrogen acts as a network terminator thus enhancing the rate of re–arrangement of network structure.

ACKNOWLEDGEMENTS

The support of the National Science Foundation through a Materials Research Group grant (DMR #89–03527), the continuing interest and support of Dr. Paul H. Schmidt and the Digital Equipment Corporation and comments and discussion by Michael Tamor and John Robertson are gratefully acknowledged.

REFERENCES

1. S. Aisenberg and R. Chabot, J. Appl. Phys. 42, 2953 (1971).

2. E. G. Spencer, P. H. Schmidt, D.H. Joy and F.J. Sansalone, Appl. Phys. Lett. 29, 118 (1976).

3. R. C. DeVries, Ann. Rev. Mater. Sci. 17, 161 (1987).

4. J. C. Angus and C. C. Hayman, Science 241, 913 (1988).

5. K. E. Spear, J. Am. Ceram. Soc. 72, 171 (1989).

6. J. C. Angus, Thin Solid Films 142, 145 (1986).

7. J. C. Angus, P. Koidl and S. Domitz, in "Plasma Deposited Thin Films," edited by J. Mort and F. Jansen, CRC Press, Boca Raton, FL, 1986, Chapter 4.

8. J. J. Cuomo, "Proceedings of Workshop on the Science and Technology of Diamond Thin Films," Concord, OH, May 20–24, 1990. to be published by the American Carbon Society.

9. P. J. Martin, S. W. Filupczuk, R. P. Netterfield, J. S. Field, D. F. Whitnall and D. R. McKenzie, J. Mat. Sci. Lett. 7, 410 (1988).

10. F. Davanloo, E. M. Juengerman, D. R. Jander, T. J. Lee and C. B. Collins, J. Appl. Phys. 67, 2081 (1990).

11. J. C. Angus, C.C. Hayman and R.W. Hoffman, in "Diamond Optics," Proceedings of SPIE, 969, 2 (1988).

12. J. L. Bredas and G.B. Street, J. Phys. C. 18, L651 (1985).

13. J. Robertson, Adv. Phys. 35, 317 (1986).

14. J. Robertson and E.P. O'Reilly, Phys. Rev. B 35, 2946 (1987).

15. J. C. Angus and F. Jansen, J. Vac. Sci. Technol. A6, 1778 (May/June 1988).

16. J. C. Phillips, Phys. Rev. Lett. 42, 153 (1979).

17. M. F. Thorpe, J. Noncryst. Solids 57, 355 (1983).

18. G. H. Döhler, R. Dandaloff and H. Bilz, J. Noncryst. Solids 42, 87 (1980).

19. M. A. Tamor and C.H. Wu, J. Appl. Phys. 67, 1007 (1990).

20. G. Galli, R. M. Martin, R. Car and M. Parrinello, Phys. Rev. Lett 62, 555 (1989).

21. J. Perrin and R. Dandaloff, J. Noncryst. Solids 86, 179 (1986).

22. Y. Wang, H. Chen, R.W. Hoffman and J.C. Angus, to appear, J. Mat. Res. Nov. 1990.

23. S. Kaplan, F. Jansen and M. Machonkin, Appl. Phys. Lett. 47, 750 (1985).

24. A. Grill, B. S. Meyerson, V. V. Patel, J. A. Reimer and M. A. Petrich, J. Appl. Phys. 61, 2874 (1987).

25. T. X. Neenan, M. R. Callstrom, L. M. Scarmotzos, K. Randall Stewart and G.M. Whitesides, Macromolecules 21, 3525 (1988).

26. M. R. Callstrom, T. X. Neenan and G. M. Whitesides, Macromolecules 21, 3528 (1988).

PREPARATION TECHNIQUES FOR DIAMOND-LIKE CARBON

Yves Catherine

Laboratoire des Plasmas et des Couches Minces
- UMR 110, Institut de Physique et Chimie des Materiaux
2 rue de la Houssiniere - 44072 - Nantes Cedex 03

1. INTRODUCTION

There have recently been many publications reporting efforts to produce carbon thin films with diamond-like properties. It appears to be possible to make mechanically hard, high resistivity (>10^{12} Ωcm) semi-transparent carbon films by a variety of ion beam and plasma techniques. This kind of films can be deposited by low-energy carbon ion beam[1-3] or by dual beam[4] and ion plating techniques,[5] by RF sputtering[6,7] or RF and DC plasma decomposition of a hydrocarbon gas or other alkanes.[8-10]

The purpose of this presentation is to describe these various techniques and to show their advantages and limitations.

After a general overview of the deposition techniques, we shall first discuss the ion beam method because it is, at least conceptually, easiest to understand.

The second part will concern the various discharge techniques including plasma sputtering, DC and RF glow discharge deposition, ECR discharges and remote discharge deposition.

By far the most widespread deposition process for DLC films comprises the use of a low pressure DC or RF glow discharge in hydrocarbons or their mixture with hydrogen or inert gases. Therefore, a large part of this presentation will cover the study of the electrical behaviour of low and high frequency discharges in conjunction with the deposition process, the film structure and properties.

In conclusion a comparison of the carbon layers obtained with the different deposition process will be made and future trends in deposition technology will be outlined.

2. GENERAL REVIEW OF THE DEPOSITION PROCESSES

Generally, film deposition processes are divided into ion beam, reactive plasma, and laser techniques. This is shown in Fig. 1. However, it is often more convenient to divide these processes into those that use hydrocarbon gases as a source of carbon and those that use solid carbon itself.[11] The characteristic feature of most processes for growing dense carbon films is that energetic species, usually in the range 50 to 500 eV hit the surface of the growing films.

Diamond and Diamond-Like Films and Coatings
Edited by R.E. Clausing *et al.*, Plenum Press, New York, 1991

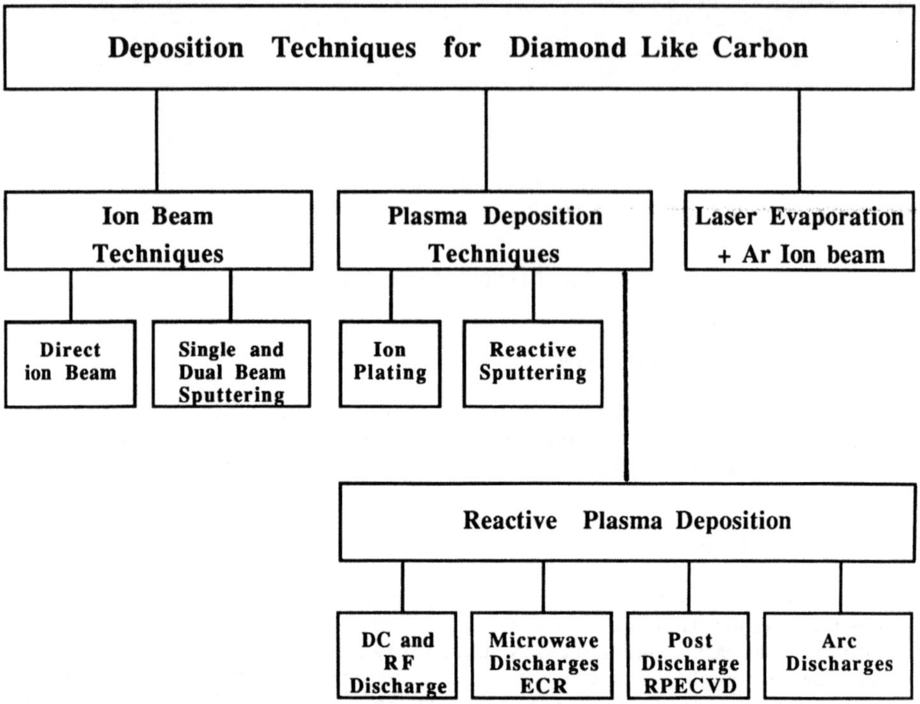

Fig. 1. General review of the diamond-like carbon deposition processes.

2.1. Processes Using a Solid Carbon Source

Figure 2 shows several carbon deposition techniques using solid carbon sources. In Fig. 2a, a primary argon ion beam impinges at about 0.5 to 1 keV onto a pyrolitic graphite target and sputters carbon onto the substrate where a second Ar ion beam impinges at about 0.1 to 0.3 keV.[12] The second ion beam is used to increase the resistivity and the optical transmittance of the films.[13] Figure 2b shows a single beam which is used both to sputter the target and to bombard the substrate.[14] In Fig. 2c and 2d an argon arc is used to erode the carbon electrodes which are the sources of the carbon atoms.[3,15] These carbon atoms are ionized in the plasma and extracted through a differentially pumped aperture along an axial magnetic field. In Fig. 2e, a CO_2 laser is used as a heat source to vaporize powdered graphite and powdered diamond.[13] Figure 2f shows a sputtering discharge using carbon cathode and an argon-hydrogen mixture.[16,17] In order to increase the sputtering rate magnetron systems have been used.[18,19] Sometimes hybrid sputtering devices using both solid carbon and a hydrocarbon gas have been employed.[20]

2.2. Processes Using a Hydrocarbon Gas Source

Both RF and DC discharge reactors and ion beams have been used to grow dense carbon films from hydrocarbon gases. The main hydrocarbon source gases have been methane, ethane, butane, propane, acetylene, ethylene, and benzene although other sources have been occasionally employed.

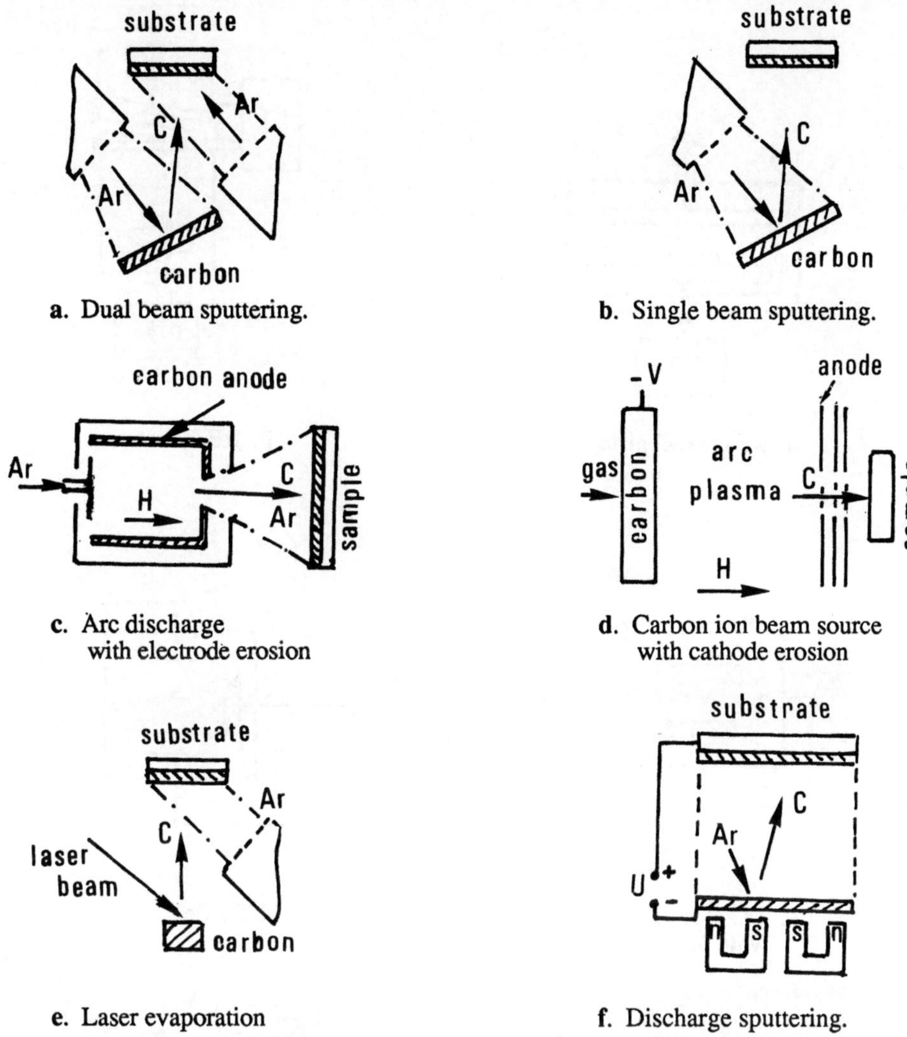

Fig. 2. Deposition processes using solid carbon.

Figure 3 shows several hydrocarbon discharge and ion beam deposition schemes. The capacitively coupled RF glow discharge (Fig. 3a) is certainly the most widespread technique for deposition of dense hydrogenated carbon films.[8-10,21-24] In this technique the substrate is placed on the powered electrode where it acquires a negative self-bias.

The inductively coupled RF glow discharge as shown in Fig. 3b has been used less extensively.[25] In this case the floating substrate bias is only a few volts below the plasma potential. Generally softer films are obtained with this coupling technique.

The DC glow discharge has also been used but generally with the bias screen (Fig. 3c) or triode (Fig. 3d) configuration.[26,27] DC discharges with a hot filament for electron emission have also been used.[28] It enables low

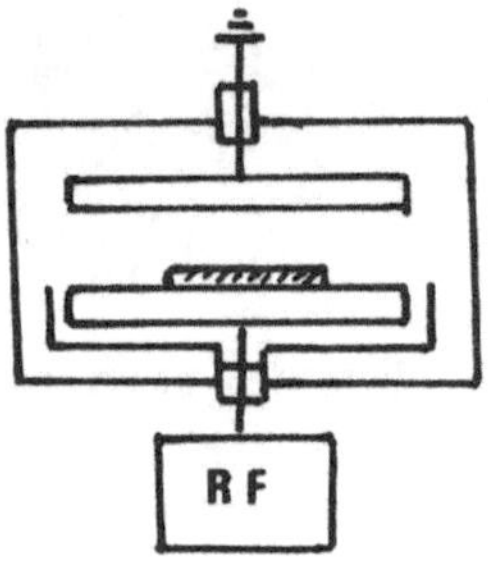

a. RF parallel plate

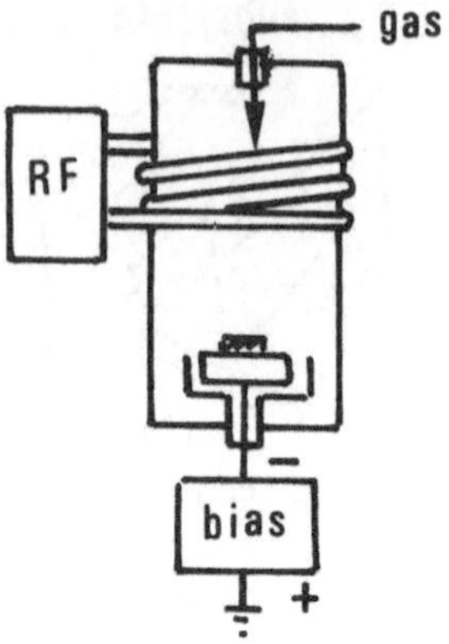

b. RF inductive

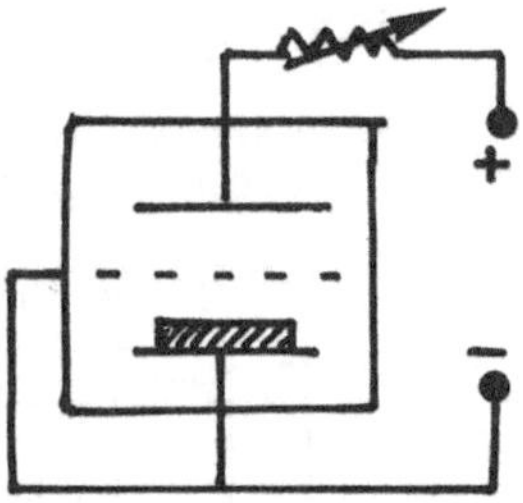

c. DC glow discharge
with bias screen

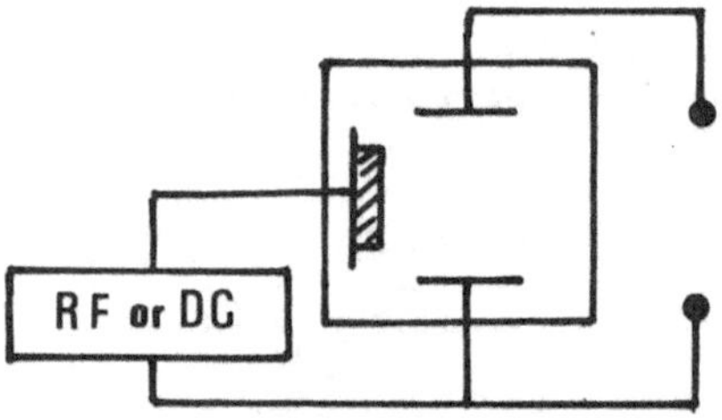

d. Triode

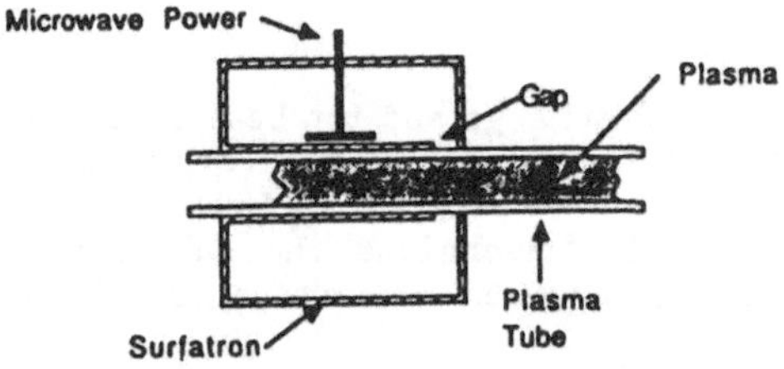

e. Microwave discharge
with surfatron applicator

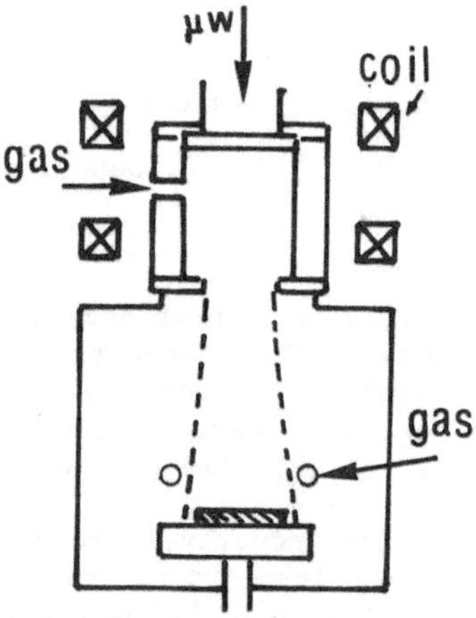

f. ECR discharge

Fig. 3. Deposition processes using a hydrocarbon gas.

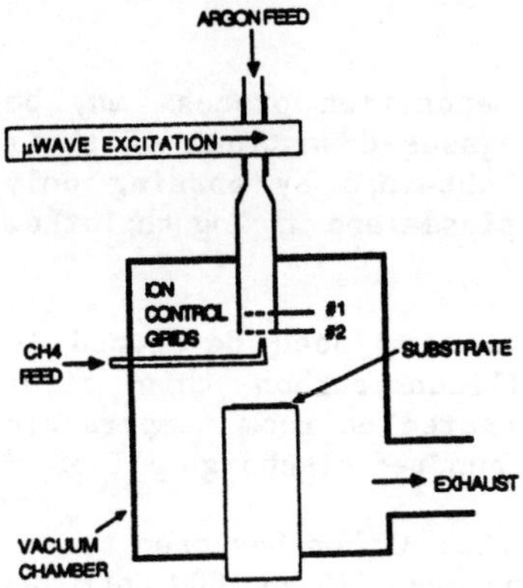

g. Downstream microwave discharge.

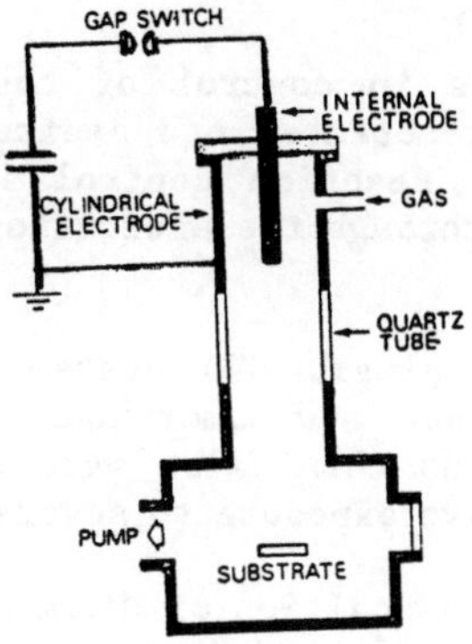

h. Pulsed discharge

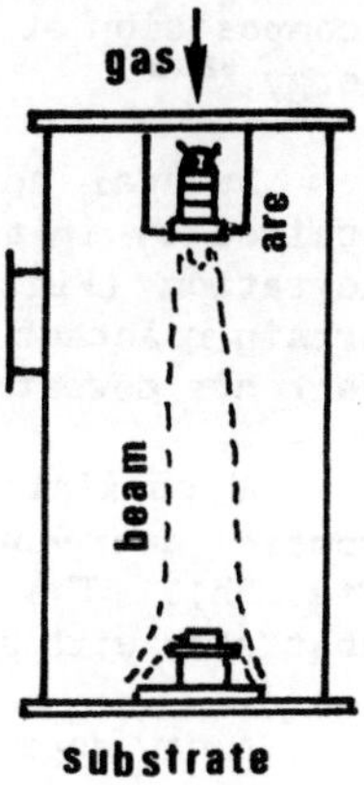

i. Cascaded arc

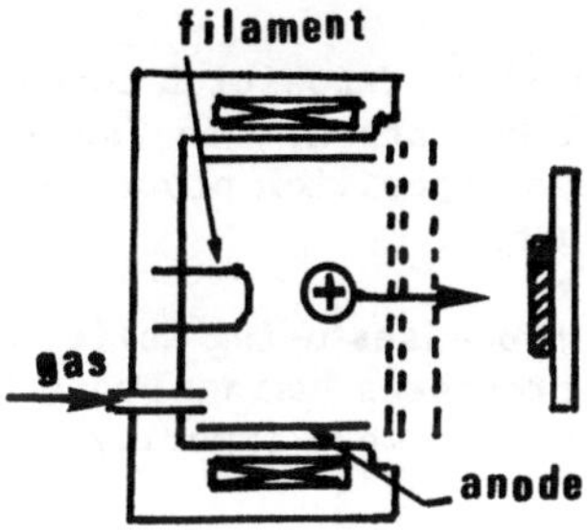

j. Direct ion beam using hydrocarbon ions

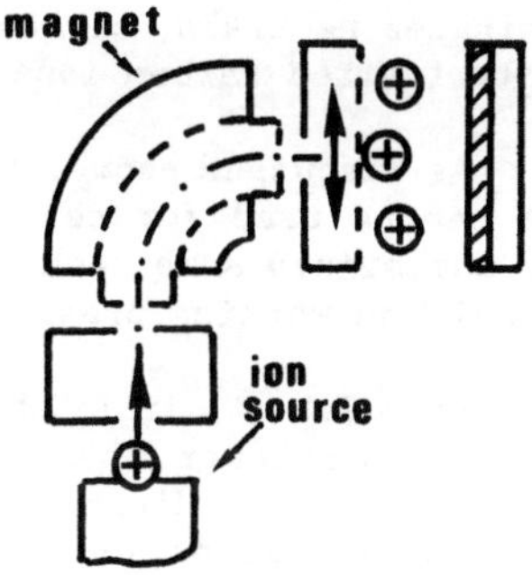

k. Ion beam source with mass selection

Fig. 3. Deposition processes using a hydrocarbon gas.

operating voltages (50 v). The substrate can be outside the discharge and negatively biased. Sometimes grids have been added in order to accelerate the ions impinging onto the substrate.[5]

Microwave discharges are also of interest for diamond-like carbon deposition. The microwave power at a frequency of 2450 MHz is transferred into a quartz tube through a rectangular waveguide,[29] a microwave cavity[30] or a surfatron applicator[31] (Fig. 3e). More recently CVD enhanced by an electron cyclotron resonance (ECR) plasma has been proposed (Fig. 3f) it is attractive due to the high degree of ionization and working gas decomposition at low pressures (10^{-4}-10^{-3} torr) and controllability of ion energy.[32,33]

Several advantages in control of the deposition process may be obtained by injecting reactants into excited gases downstream of their excitation (Fig. 3g). Reaction control is obtained by passing only certain reactant gases through the excitation plasma and mixing the other reactants downstream.[34]

A coaxial pulsed plasma CVD system has also been developed to prepare amorphous carbon and amorphous silicon carbon thin films (Fig. 3h). The amorphous thin films were prepared on room temperature substrates with successive exposure to several hundred discharge pulses.[35]

A new approach of deposition of diamond-like carbon has been to use a cascaded arc operating at an Ar/CH_4 pressure between 225 and 600 torr and expanding in a vacuum chamber.[36] Figure 3j is an outline of the cascaded arc used for deposition. The main advantage of this technique is the fast deposition rate that can be achieved (900 nm/min with CH_4 and up to 2700 nm/min with C_2H_2).

Direct ion beams using hydrocarbon gases as carbon sources have also been investigated.[4,37] Such a technique is illustrated in Fig. 3j. Mass selection can be achieved using a magnetic field (Fig. 3k). This system eliminates neutrals, clusters and impurities and results in a pure beam of singly ionized carbon ions.[38,39]

As discussed above, there are large variety of deposition techniques that can be used for deposition of diamond-like carbon. In Table 1 we have summarized some technical aspects of these deposition processes with emphasis on working pressure and impurity sources.

From Table 1 it is obvious that all the processes using solid carbon sources need lower chamber pressures than processes using hydrocarbon gases (except for direct ion beam). There are also more impurity sources in the solid carbon source processes.

Table 1. Comparison of basic deposition pressure and possible contamination sources in C:H deposition processes

	Processes from hydrocarbon gases				Processes using solid carbon		
	PECVD (DC or RF)	Hydrocarbon ion beam	ion plating	Arc discharge	DC or RF sputtering	ion beam sputtering	Laser evaporation
Pressure	0.01-0.5torr	Deposition chamber 10^{-7}-10^{-5}torr	0.1 torr	chamber 0.75 torr	$<5010^{-3}$torr	$<10^{-5}$ torr chamber	$<10^{-6}$ torr
Gases	CH_4,C_2H_2 C_6H_6,C_4H_{10}	CH_4,C_2H_2 C_4H_{10}/Ar CO,CO_2	C_6H_6	CH_4 C_2H_2	Ar,Ar+H_2 Ar + CH_4	Ar Ar + H_2	–
Solid target	–	–	–	–	Carbon (graphite)	Carbon (graphite)	Powder diamond or graphite
impurity source	gas reactor walls	ion source walls grids gas	gas reactor walls grids	gas reactor walls	gas (Ar) reactor walls target	gas (Ar) walls target	powder crucible reactor walls

As it is not possible to review all the deposition processes, we shall focus our attention on direct ion beam deposition, ion beam sputtering, RF discharge deposition, microwave and ECR discharge deposition and magnetron reactive sputtering -- which are the most commonly used deposition techniques.

3. DEPOSITION OF CARBON FILMS FROM ION BEAMS

In this section, two methods for the deposition of carbon films are discussed, i.e., the growth of films from direct ion beam using hydrocarbon and also from single and dual beam sputtering.

3.1. The Ion Source

The ion source shown in Fig. 4 is one of the simplest and most commonly used configurations.[40] A plasma discharge is sustained between the thermionic cathode and the anode. Electrons emitted from the cathode must cross the axial magnetic field to reach the anode. Thus the electrons suffer many collisions with working gas atoms, and high ionization rates can be achieved. Axial losses are minimized by maintaining the screen and chamber walls at cathode potential. Consequently, ion densities in the 10^{10} cm^{-3} range can be maintained at low (≈1 mtorr) working pressures. The potential distribution is shown in Fig. 4b. The plasma potential V_p is typically a few volts above the anode potential which is maintained above the ground potential by an amount equal to the desired ion energy, v_a ≈500 to 1000 V. The accelerator grid is maintained at a potential that is about 100 V negative relative to ground potential. The accelerator grid controls beam divergence and provides a negative potential barrier which prevents the passage of electrons from the beam back to the discharge plasma. The ion impact energy is the net voltage V_n. Typical beam current densities are 1 to 2 mA/cm^2 with 500 eV ion energies and beam sizes up to 30 cm. A source of electrons is often used to neutralize the ion beam. Neutralization reduces beam divergence and minimizes the charging of non-conducting substrates.

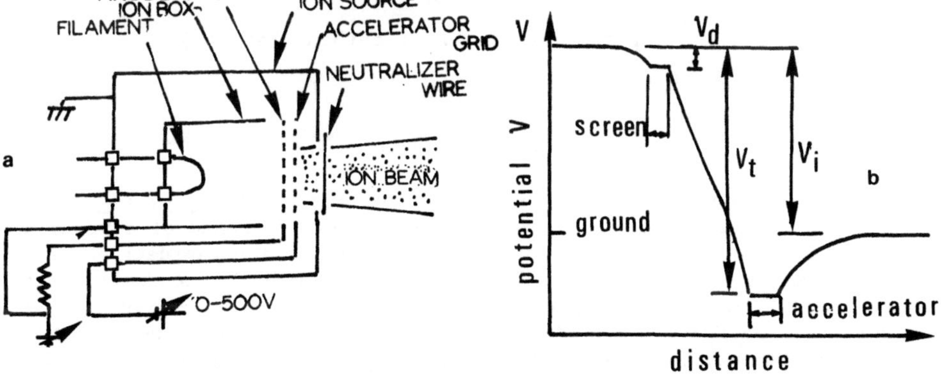

Fig. 4. Ion beam deposition system (a) and potential distribution (b).

3.2. Deposition of Hard Carbon by Direct Ion Beams Using Hydrocarbon Gas

Angus and coworkers[41] used a primary beam which was derived from a source gas containing 28 vol % CH_4 in argon. The primary beam energy and current density were 115 eV and 300 $\mu A/cm^2$.

In some experiments a second ion source was used to direct a beam of energetic argon ions (200-600 eV, 25 $\mu A/cm^2$) onto the substrate.

The species arriving at the surface can be divided into two categories: the primary ions Ar^+ and C^+ and hydrocarbon ions of the general form $C_mH_n^+$. There is also a flux of uncharged particles arising from the residual background gas. The background pressure, arising from neutral methane and argon, is 2 x 10^{-4} torr near the target during operation. At this operating pressure the ratio J_i/J_R of the ion flux over the neutral flux impinging on the substrate is ~2 x 10^{-2}. In other words the vast majority of species hitting the surface are actually neutral molecules rather than ions.

Various hydrocarbon gases have also been used to show the formation of large neutral fragments.[42] For example the cracking of 80% CH_4 in Ar gives H_2, C_2H_2, C_2H_3, and C_2H_4 and small amounts of ions such as C^+, CH_2^+ and $C_2H_4^+$, and C_2H_2 cracks into CH_2, C_2H_4, C^+ and $C_2H_4^+$.

Mass separation has been used in conjunction with a duo-pigatron ion source using CO or CO_2 gas. In this case only 100 eV C^+ ions impinge onto the grounded substrate.[38]

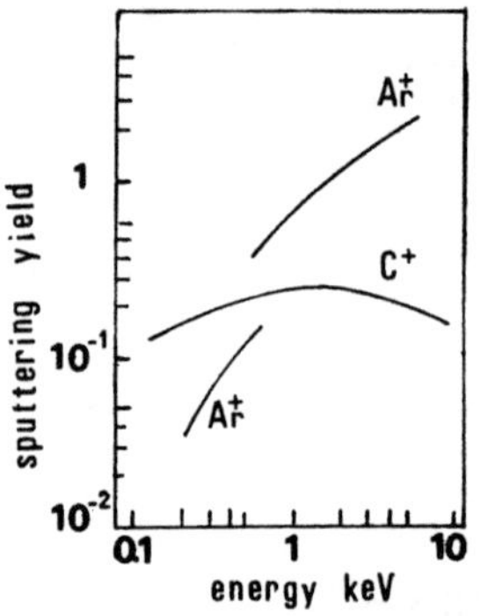

Fig. 5. Sputtering yields of carbon (ref. 44).

3.3. Deposition of Carbon from Ion Beam Sputtering

Ion beam sputter deposition systems consist generally of a graphite disk target bonded to a water-cooled holder. Ion currents of several mA are used and the Ar ion energy is typically ~1 keV. The incident angle of the ion beam is ~30°/45° off of the target normal in order to increase the sputtering yield. If hydrogen is to be incorporated in the film an Ar/H_2 mixture is used.[14,43] A second argon ion beam can irradiate the substrate to sputter loosely bound species. Figure 5 gives the sputtering yield of graphite under normal argon ion and carbon ion bombardment.[44] This sputtering yield is quite low and only low deposition rates can be achieved.

3.4. Influence of Beam Parameters on Deposition and Properties of the Carbon Films

The major parameters in beam deposition are: (1) the nature of the ion beam and the beam energy and (2) the nature of the substrate and the deposition temperature.

We shall only examine here the influence of the nature of the ion beam, of the ion energy and of the deposition temperature. In the last paragraph comparison of ion beam and discharge deposition processes will be made.

Table 2 shows the deposition rate, the hydrogen content, the impurity content, the refractive index, the density and the resistivity of carbon films deposited by different ion beam techniques.

Table 2. Variation of the deposition rate and of some carbon film properties for several ion beam deposition processes

	Direct ion beams using carbon gas source			Ion Plating	Direct ion beam using solid carbon		Ion beam sputtering			
Ion energy (eV)	100-200	300-600	–	$C_6H_5^+$ 100-600	10-1000	40-100	Ar^+ 150	–	–	–
Deposition rate(Å/min)	20-70	6-12	30-100	600	~ 80	60-300	90	5-7	17	10
Hydrogen content (at %)	Total 30-50 Bound	No	Total 40	25-50	–	No	No	H_2 added	35 H_2 added	17-30 CH_4 10^{-4}torr
Impurities (at %)	O : 6-9 N and Ar	–	–	0 < 7	–	–	–	–	–	–
Refractive index (550 nm)	1,75-2,3	2,3-2,4	–	2,8	–	2	2,9	2,4	2	–
density g/cm^3	1,7-1,8	~ 2	–	–	–	–	–	–	2,3 1,6(H_2)	–
resistivity Ω cm	8,7 10^6	1,7 10^6	10^8-10^9	> 10^{10}		10^{11}	5-10^{12}	–	–	
reference	4	38	37	5	1	3	12	14	43	45

The main parameters which influence the DLC carbon film in the ion beam techniques are: the ion composition and energy, more particularly the use of hydrocarbon and of hydrogen in the ion beam. When hydrocarbon ions or when hydrogen are used the deposited films are highly hydrogenated (up to 50%). Only few authors give the impurity content of the film, merely oxygen and nitrogen. We did not find in the literature measurements of argon incorporation in beam deposited carbon film, although argon must be trapped in the films as it is often found in sputtered layers. Films which are deposited without any hydrogen in the gas phase have larger refractive index, higher transmittance, lower optical gap (0,5-1,3 eV) and lower resistivity.[43] For H-free systems there appears to be an optimum beam energy at which amorphous carbon with about 75% tetrahedrally bonded carbon forms (E_{ion} ~ 100 eV).[46] In this condition the grain size of the tetrahedrally bonded regions is of the order of 10-100 Å.[47] If the beam energy is too high trigonal(graphitic) bonding among C atoms becomes dominant. Since transformation from four to three coordination is temperature dependent, it seems reasonable to associate high beam energy to high temperature at least locally. In order to produce polycrystalline carbon, the nucleation rate must be high and the surface mobility low.

When hydrocarbon ion sources are used or hydrogen is added to the argon ion beam a-C:H films contain as much as 50 at.% H. These alloys are not diamond but mixed phases with varying amounts of sp^3 and sp^2. Most contain hydrogen unless either substrate temperature or ion beam energies to remove it in situ are used. This is shown in Fig. 6 which shows the variations of the optical gap as a function of the hydrogen content of the ion beam and of the substrate temperature during deposition using liquid nitrogen or water cooling. The influence of some ion beam parameters on a-C:H film deposition will also be compared to the influence of glow discharge parameters in the next paragraphs.

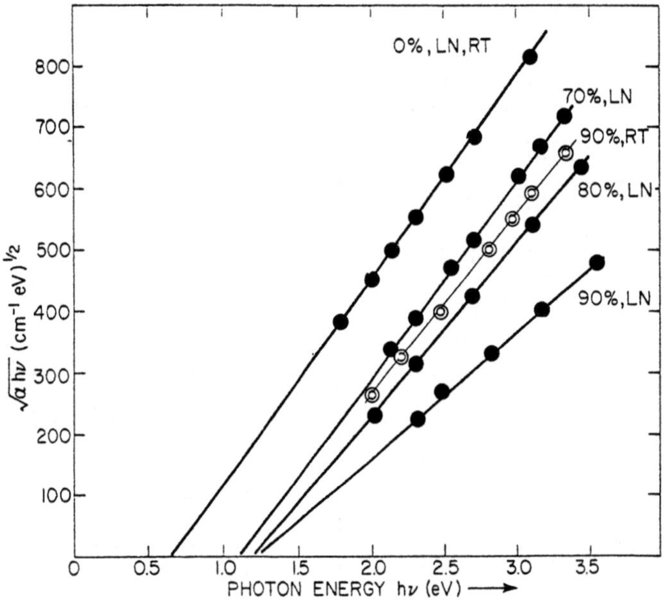

Fig. 6. Influence of hydrogen and substrate cooling on the optical gap of carbon films deposited by ion beam sputtering.[43]

4. GLOW DISCHARGE DEPOSITION OF DIAMOND-LIKE CARBON

Many different discharges have been explored for their applicability in the field of plasma deposition. We shall restrict ourselves to the most widely utilized systems: the RF discharges, the reactive sputtering discharge which combine sputtering and gas cracking with the microwave and ECR discharges.

4.1. The RF Discharge

The parallel plate reactor as illustrated in Fig. 7 is most commonly used for material processing. Many variables must be controlled in plasma deposition, such as power, total pressure, reactant partial pressures, gas flow rates, pumping speed, sample temperature, discharge frequency, electrode spacing, electrode materials and reactor geometry.

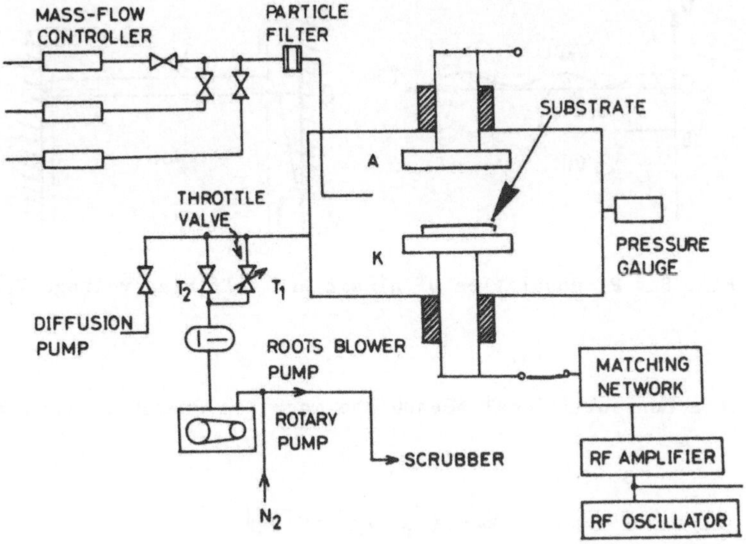

Fig. 7. A schematic of the RF discharge apparatus.

Regarding the frequency of the input power, there are two distinct regions, i.e., low frequency and high frequency. The boundary between the two regions is given by the ion plasma frequency:

$$f_{pi} = \frac{1}{2\pi}\left(\frac{n_i.e}{m_i\varepsilon_o}\right)^{1/2} \tag{1}$$

where n_i and m_i are respectively the ion density and mass.

Assuming that in hydrocarbon plasmas $n_i \simeq n_e \simeq 10^{10}$ cm^{-3} the calculated values of f_{pi} are found to be 5,4 MHz for CH_4 ions and 2.4 MHz for C_6H_6 ions.

4.1.1. The 13.6 MHz discharge and the self bias voltage

In the 13.6 MHz glow discharge, the RF field frequency is sufficiently high so that the ions require many RF cycles to cross the sheath while electrons are able to follow the RF field. Because the powered electrode is capacitively coupled, the steady-state DC current must be zero. If the powered electrode area is smaller than the grounded part of the system, it develops a negative self bias voltage V_B relative to ground. Such a situation is shown in Fig. 8.

The net result is ion bombardment of the substrate which is placed on the powered cathode during the film growth. If one assumes that the current and the anode surface are equal, thus:[48]

$$V_B/V_A = (S_A/S_B)^2 \tag{2}$$

where S_B and S_A are the respective area of cathode and anode.

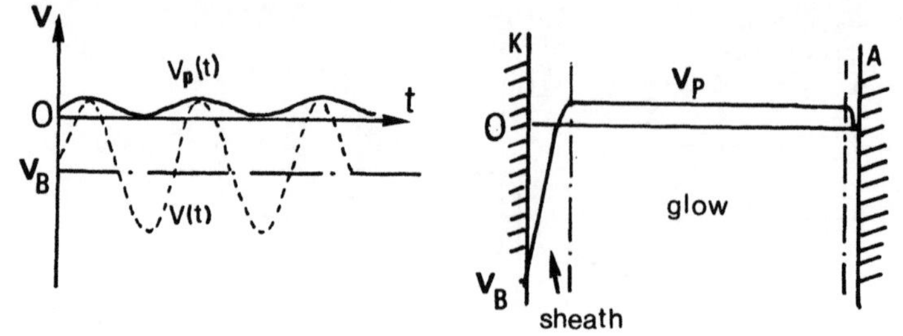

Fig. 8. RF modulation of plasma and self bias voltage V_B.

For a non-collisional sheath the mean ion energy is thus given by Eq. (3):

$$E_i - e(\bar{V}_p - V_B) \simeq e|V_B| \quad (3)$$

where $\overline{V_p}$ is the mean plasma potential.

Values of the self bias voltage V_B for CH_4, C_6H_6 and their mixture with He are reported in Fig. 9. It is found that V_B is a linear function of the square root of the ratio of the RF power over the discharge pressure

$$V_B \sim (W/P)^{1/2} \quad (4)$$

Bubenzer and coworker[9] demonstrated that in their benzene discharge the ion energy depends on both bias potential and pressure with the proportionality:

$$E_i \sim V_B P^{-1/2} \quad (5)$$

Combining Eqs. (4) and (5) leads to the relation (6)

$$E_i \sim \frac{W^{1/2}}{P} \quad (6)$$

This result is similar to the relation deduced by Zarowin[49] from an equivalent electric network (capacitive sheaths in series with a resistive plasma body). Thus the key variables which control the energy of the ions impinging on the substrate are the self bias voltage V_B (or power) and the discharge pressure. Obviously if the geometry of the discharge changes, there will be a modification of the bombarding energy of ions.

4.1.2. The low frequency discharge

In the low frequency discharge the transit time for ions in the sheath is much shorter than the period of the RF field and the ion flux

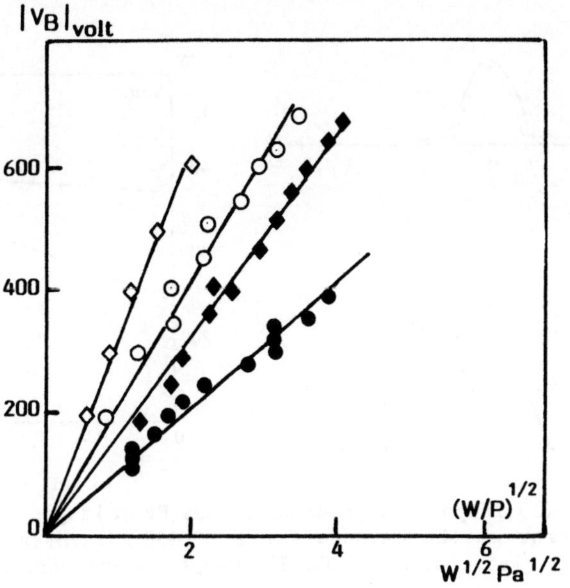

Fig. 9. Plot of the self bias potential against the parameter (power/pressure)$^{1/2}$.
• CH_4, ♦ C_6H_6, o CH_4 - He, ◇ C_6H_6 - He.

will reflect the temporal dependence $V_p(t)$ of the plasma potential.[48,50] This behaviour is illustrated in Fig. 10.

In these discharges the mean current density is an increasing function of the RF power. The $\overline{V}(\overline{j})$ characteristics of such discharges are reported in Fig. 10b. Similar variations have been observed in a 380 kHz Ar discharge.[50]

In this type of discharge the secondary electrons accelerated in the cathodic sheath are thought to take an important part in sustaining the discharge itself. It appears that the current density and the pressure are the most significant parameters of the low frequency discharge.

4.1.3. Some characteristics of the low and high frequency discharges in hydrocarbons

4.1.3.1. The ion flux at the active electrode

The knowledge of the ion flux and energy at the substrate surface is of major importance for the control of the growth of the carbon films and their properties. The charged particle flux arriving at the substrate surface can be measured by techniques such as electrostatic analysers.[51] Typical values of ion fluxes at the active electrode of low and high frequency discharges are shown in Fig. 11.

The ion fluxes are lower in the low frequency discharge than in the high frequency discharge. Typical ion fluxes are in the range 2×10^{13}-2×10^{14} $cm^{-2}s^{-1}$ for the 25 kHz discharge and in the range 2×10^{13}-5×10^{14} $cm^{-2}s^{-1}$ for the 13.6 MHz discharge and decrease as the methane pressure increases. The lower value obtained in the low frequency discharge may result from the fact it is a mean temporal value so that ions are attracted towards the electrode only during the negative half period of

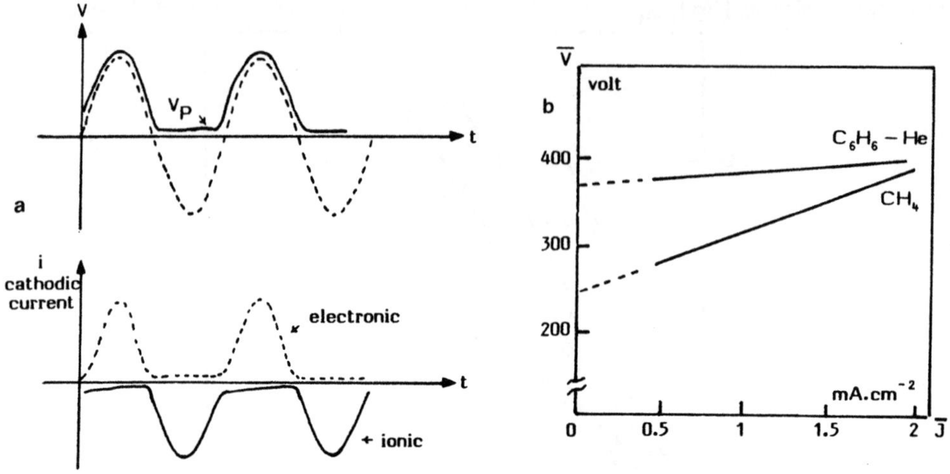

Fig. 10. (a) Temporal dependence of the RF voltage, plasma potential V_p and RF current at the cathode for a 50 kHz discharge. (b) Current voltage characteristic of a 50 kHz discharge in hydrocarbon gases.

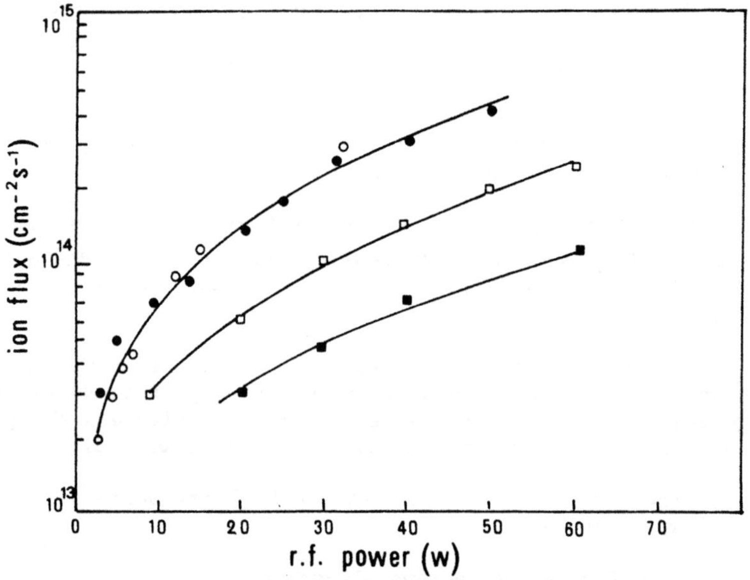

Fig. 11. Ion flux in the RF electrode as a function of the RF power: CH_4, 25 kHZ, 10 mtorr (□), 40 mtorr (■); 13,6 MHz, CH_4: 80 mtorr (○), 60% CH_4-40% H_2: 30 mtorr (•).

the low frequency field. In contrast in the 13.6 MHz discharge, due to the large negative self bias voltage which develops on the RF electrode ions are continuously accelerated towards the substrate. The ion flux (13.6 MHz) scales with V_B^n, $n \simeq 1.5 - 1.6$ which is consistent with the high field mobility version of the "Child Langmuir Law."[52]

It is interesting to know the contribution of ions to the growth of the film. Typically the ratio F_i/F_0 of the ion flux to the deposited carbon flux is in the range 0.1-0.2, i.e., the ion flux does not exceed 20% of the total deposited carbon flux.[53]

4.1.3.2. The asymmetry of the 13.6 MHz discharge

In order to obtain large bias voltage the 13.6 MHz discharge is generally generated between a large grounded area and a small cathode area. This configuration results in an heterogeneous discharge. This heterogeneity can be observed by spatially resolved optical emission spectroscopy.[54-56]

Figure 12 shows the evolution of the CH emission intensity as a function of the distance of sampling from cathode.

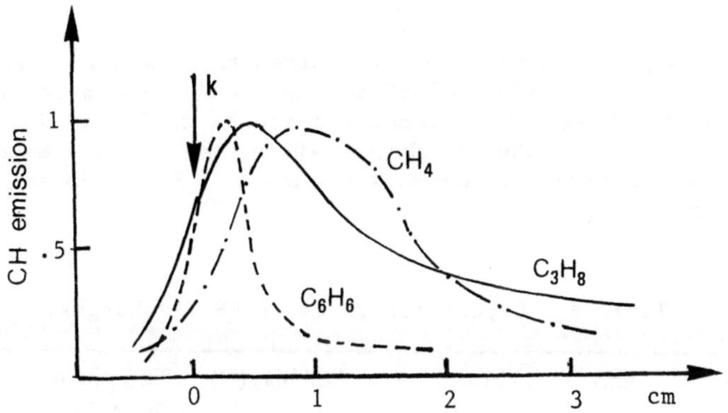

Fig. 12. Emission of the CH($A^2\Delta \rightarrow X^2\pi$) band as a function of distance from cathode in C_6H_6 (ref. 54), C_3H_8 (ref. 55), and CH_4 (ref. 56).

There are two possible explanations: (a) A large part of the CH emission in the cathode region results from excited fragments of the hydrocarbon ions impinging on the substrate. (b) Secondary electrons from ion impact on the cathode are accelerated in the cathodic sheath making excitation collision with hydrocarbon molecules at the edge of the sheath. This heteroepitaxy results in different growth rates (deposition rate on the RF electrode may be four times larger than that on the ground electrode) and in different structures and properties of the films.

4.1.4. Deposition Rates in RF Discharges

Table 3 shows the deposition rates obtained for various RF discharges and hydrocarbons.

The growth rate from hydrocarbon sources increases with the molecular weight of the source species, other parameters being held constant. The deposition rate also increases with the self bias voltage (the ion flux) but a large part of the deposited carbon comes from neutral species.

4.2. The Microwave and ECR Discharges

Microwave discharges have become an area of intense interest and have been widely utilized in recent years. Their electrodeless nature together with their ability to create high densities of excited and charged species have made low pressure microwave discharges an attractive technology for many plasma processing applications.[61] Diamond-like carbon films have been prepared by the decomposition of CH_4 in a microwave discharge (2.45 GHz).[29,30]

An important development in low pressure and low temperature microwave plasma processing is the electron cyclotron resonance (ECR) discharge. As the name ECR suggests, microwave energy is coupled to the natural resonant frequency of the electron gas in the presence of static magnetic field. This resonant frequency occurs when the electron cyclotron frequency

$$\omega_c = \frac{eB}{m}$$

(where e = charge on an electron, B = strength of static magnetic field, and m = mass of an individual electron) equals the excitation frequency ω. In an actual discharge, this condition can be satisfied in a volume or surface layer within the discharge where the static magnetic field strength is adjusted to resonance, i.e., $\omega = \omega_c$. Figure 13 shows a typical ECR discharge configuration.

Table 3. Deposition rates in RF discharges.

Frequency	Gas	Power (W)	Self bias (V)	Deposition rate (Å/min)	Reference
30 kHz	CH_4 C_4H_{10}	50 – 240	—	80 - 310 80 - 200	57
50 kHz 13.6 MHz	CH_4	25 - 75 10 - 100	 - 100 ; - 400	30 - 100 150 - 450	58
2 - 3 MHz	C6H6		- 400 ; - 2000	100 - 3000	9
13.6 MHz	CH_4 C_2H_6 C_3H_8 C_4H_{10}	160	- 900	120 220 290 400	59
13.6 MHz	C_2H_4	100 - 1000	- 10 ; - 1300	10 - 350	60
13.6 MHz	CH_4 C_2H_2	150 - 1000	- 300 ; - 1000	35 - 100	61

The main advantage of using ECR is that it allows the microwave electric field to accelerate free electrons continuously (between collisions) throughout the full wave period, which means that dense plasmas can be generated efficiently at low pressure.[62] For example, plasma densities of (6-8) x 10^{11} cm^{-3} are obtained in Ar 30 to 50 cm from the source at a microwave power of 500-600 W and pressures of 10^{-4}-10^{-3} torr.[63] Other advantages include the absence of internal electrodes and direct control over the shape (through confinement), position (through resonance) and flow of the plasma with magnetic field. Advantages for film fabrication are that gas phase nucleation can be reduced without heavy noble gas dilution and that numerous undesirable powered electrode effects (such as contamination) are eliminated. Also the high degree of control over the shape system can easily be confirmed in a variety of

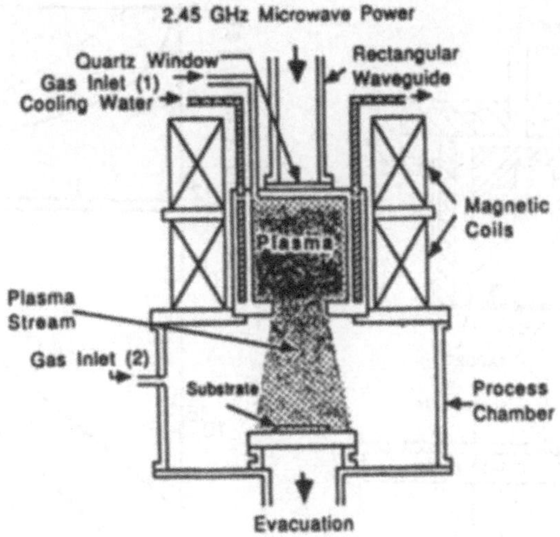

Fig. 13. Schematic of an ECR deposition set up.

ways. For example, samples can either be immersed directly in the plasma or can be located downstream. Matsumoto and coworker[64] used an ECR apparatus to prepare carbon films from methane. The deposit could be changed from graphitic-like carbon to polyethylene-like polymer as the substrate is moved away from the ECR resonance point. Due to large charged particle densities, the deposited layer may suffer from overheating. A careful cooling of the substrate holder is therefore needed.

4.3. Remote or Downstream Plasma Deposition

Several advantages in control of the deposition process may be obtained by injecting reactants into excited gases downstream of their excitation, as opposed to carrying out the deposition with the substrate immersed in the plasma.[65] By reacting the afterglow of a microwave excited Ar discharge with CH_4 it is possible to deposit glassy carbon with typical refractive index of 1.55. Interactions of metastable He atoms with CH_4 molecules have also been investigated.[66]

4.4. Reactive Magnetron Sputtering

Magnetron sputter deposition of carbon is of interest due to the ability to deposit at high rates with relatively low levels of substrate heating, facilitating deposition on organic and other heat sensitive substrates. Amorphous carbon produced by sputtering from a graphite target and using inert ion bombardment of the growing film offers a pure amorphous carbon network containing no hydrogen and may be related to sputter ion beam deposition.[67]

Hydrogenation of this material is possible by reactive sputtering of the graphite target in an argon-hydrogen or argon-hydrocarbon plasma.[67,68] This is an hybrid deposition method. A schematic diagram of the magnetron deposition system is shown in Fig. 14a. When a magnetic field is superimposed on a plasma, the electrons have a circular motion of Larmor radius r_e. The influence of the magnetic field will be particularly effective when r_e will be equal to ℓ_e the mean free path of electrons in

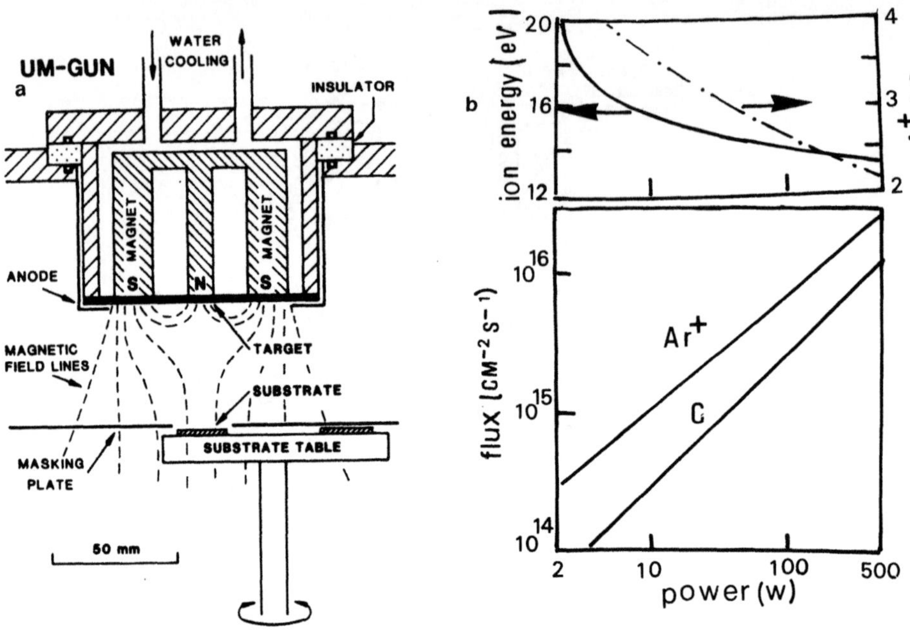

Fig. 14. (a) Schematic of the reactive magnetron sputtering system. (b) Variation with power of the particle fluxes ion energy and Ar^+/C arrival rate ratio (from ref. 67).

the gas. [For a CH_4 plasma ℓ_e (cm) $\simeq$ 3.5 $10^{-2}/P$ (torr) for 3 eV electrons]. The electronic mobility in a direction perpendicular to the magnetic field will be reduced:

$$\mu_\perp = \mu/[1 + (\omega_c/\nu_e)^2] \tag{7}$$

where ω_c is the cyclotron frequency and ν_e the electron-molecule collision frequency. Typically at P = 0.1 torr and B = 100 G a decrease of 50% of $\mu_\perp$ is observed. Thus, the electron flux to the wall is diminished, which increases the ionization, the dissociation and the sputtering process. Deposition rates of ~450 Å/min have been obtained in reactive sputter deposition using $Ar-C_2H_2$.[20]

5. INFLUENCE OF SOME DEPOSITION PARAMETERS ON THE a-C:H FILM GROWTH RATES AND PROPERTIES

5.1. The Ion Energy and the Self Bias Voltage

The impact energy and ion flux are physically the most important parameters in the deposition of dense hydrocarbon films.

The ion energy can be easily controlled in ion beam deposition but is generally not directly measurable in discharge systems and more particularly in RF systems. Most workers used the discharge power or the self bias voltage to monitor the growth rate and properties of a-C:H films. Many authors have shown that the film properties, including density, hydrogen content, refractive index, and short-wavelength optical absorption can be reproducibly varied by changing V_B and P.[9,58,60,69,70,71]

As examples, Figs. 15a and b show the evolution of the CH stretch and the refractive index of films as a function of the self bias voltage.[69]

Figure 16 gives the optical absorption data obtained from a-C:H films deposited under various bias conditions.

An advantage of the ECR deposition technique is that the ion energy can be controlled separately. Figure 17 shows the evolution of the hydrogen content of a-C:H deposited from methane as a function of the energy of ions.

Useful applications of the hard carbon films are sometimes complicated by adhesion problems and spontaneous buckling, partly due to the large internal compressive stress in the films.

Table 4 shows the variation range of the compressive stress for carbon films deposited in different systems.

Recently, Nir[76] described experiments which were designed to relate stresses to some deposition parameters. The deposition of the films was performed with ion beams because in this case the deposition parameters are almost independent, thus the dependence of the stress on any parameters can be studied separately. The deposition current was reduced to minimize substrate heating and resulting stresses of thermal origin. Figures 18a and b show, respectively, the internal stress of carbon films as a function of the ion energy and the H/C atomic ratio in the film.

The stress in the films depends on the energy of the ion beam. In the region of low energy, below the threshold for carbon recoil, the film stress is minimal and has a small dependence on the energy. In the region of higher energies, the compressive stress may be related to the ion energy, the beam current (ion flux) and the molecular weight (deposition rate) of the species. The H content cannot explain the behavior of the stresses in the DLC films. The hydrogen content in reactive sputter deposited DLC films is sometimes comparable to the content in other methods. The RF films and sputter deposited films usually have higher stresses than ion beam deposited films.

5.2. Influence of the Gas Phase

5.2.1. Nature of the hydrocarbon

Keeping the deposition parameters such as V_B and P constant and varying the process gas, no change in the film properties is observed as shown in Table 5.[77]

This is rather surprising, more particularly as far as the hydrogen content and the sp^3/sp^2 ratio are concerned. Going from benzene to methane, the number of hydrogen atom per carbon increases from 1 to 4 while the hydrogen content of the film does not change. This structural "lost of memory" effect for hard a-C:H has been observed by many authors.[58,69,78]

5.2.2. Effect of dilution H_2 and Ar and of the introduction of O_2 or H_2O

The effect of dilution of a hydrocarbon in H_2 and Ar on the deposition process of diamond-like carbon has been studied in microwave discharges,[79] and RF discharges.[80] The dilution in hydrogen results in a chemical etching of graphite phases and H adsorption on layers. In a mixture of C_2H_2 and H_2 the deposition rate increases for hydrogen content

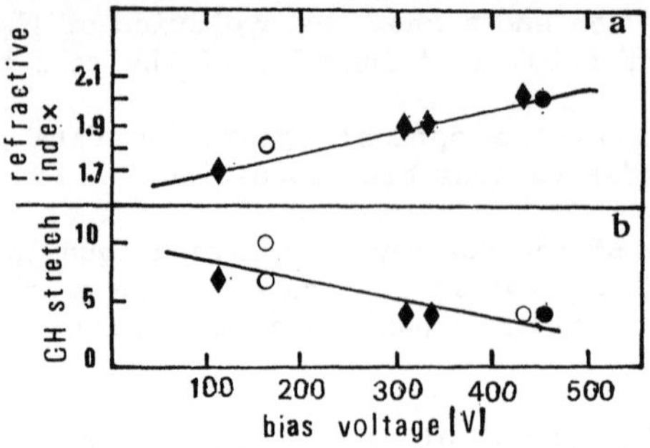

Fig. 15. (a,b) Properties of a-C:H films deposited from various hydrocarbon gases. [• CH_4 (100 mtorr), ○ C_2H_4 (150 mtorr), ◇ C_6H_6 (50-80 mtorr)] onto germanium substrates as a function of the self bias voltage V_B (from ref. 69).

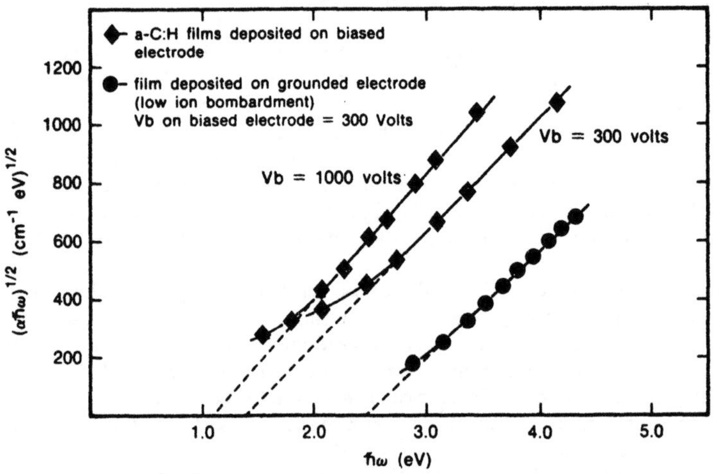

Fig. 16. Optical absorption from a-C:H films as a function of the bias voltage (from ref. 70).

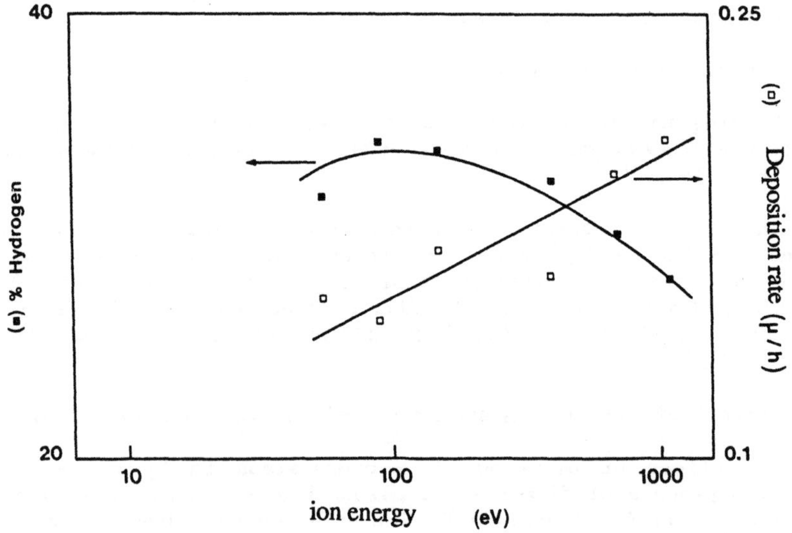

Fig. 17. Hydrogen content and deposition rate for carbon films deposited in an ECR discharge in CH_4 as a function of the ion energy (from ref. 72).

Table 4. Compressive stress in carbon films deposited from glow discharges

Deposition Technique	Reactive sputtering (C_2H_2)	Reactive sputtering C_4H_{10}	RF plasma $C_2H_4 - C_2H_2$	RF plasma $C_2H_4 - C_2H_2$ (He)	RF plasma CH_4
Type of the film	DLC	polymer	from polymer to DLC	from polymer to DLC	from polymer to DLC
compressive stress (GPa)	1.5 – 4.5	0.75	0.6 to 7	1.2 to 4.7	0.9 to 2.5
Ref.	73	74	60	58	75

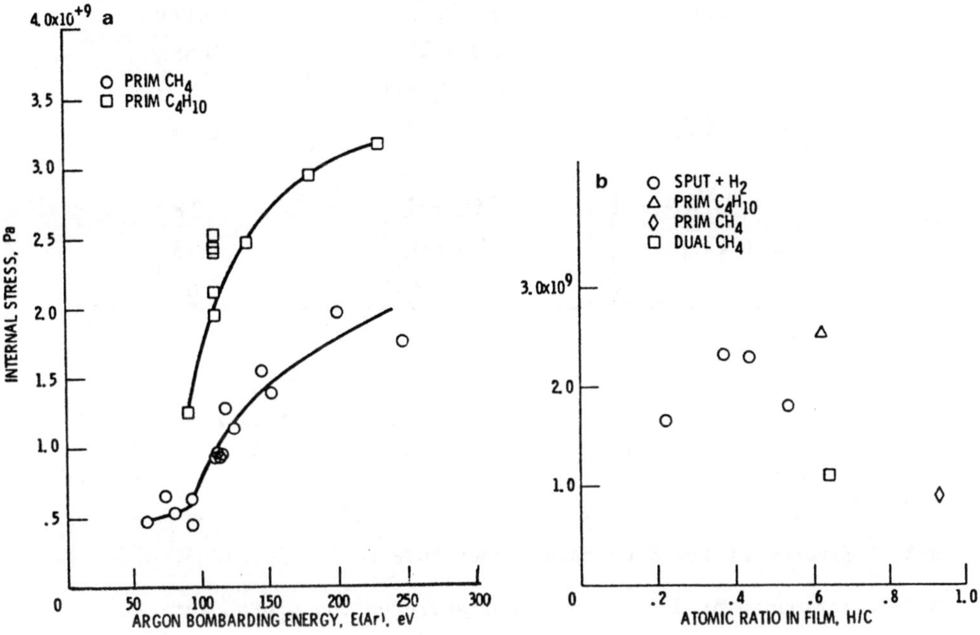

Fig. 18. (a) Internal stress as a function of Ar^+ ion energy (from ref. 76). (b) Internal stress as a function of the ratio H/C.

Table 5. Dependence of a-C:H film properties on type of hydrocarbon (from ref. 77)

Process Gas	V_B (V)	n	E_{opt} (eV)	[H] a.u.	sp^3/sp^2
Benzene	400	2	1.39	24	71/29
Cyclo-hexane	400	2.02	1.44	25.1	72/28
n Hexane	400	2.06	1.41	26.1	72/28
Methane	400	2.08	1.36	23	72/28

(from reference[77])

up to 90% and then decreases.[80] The incorporation of hydrogen reduces the internal stress of the carbon film.

Adding H_2 to CH_4 is thought to favor the creation of CH_3 radicals and thus the formation of sp^3 bonds.

When the hydrocarbon gas is diluted in Ar completely different gas phase chemistry can occur (see Table 6). The dissociation of CH_4 by metastable argon atoms may completely change the concentration of the various radicals. Recently, Amaratunga et al.[81] observed the growth of polycrystalline diamond in thin films obtained from the RF decomposition of a CH_4/Ar plasma on a Si substrate kept at 20°C.

Table 6. Dissociation of CH_4 by Ar* and electron collisions

Reaction	Products	Fraction
$Ar^* + CH_4$	$CH_2 + 2H$	0.65
	$CH + H_2 + H$	0.35
	$CH_3 + H$	
	$CH_2 + H_2$	
$e + CH_4$	$CH_3 + H$	0.8
	$CH_2 + H_2$	0.13
	$CH + H_2 + H$	0.07

Oxygen and water vapor are also known to influence the growth and structure of the carbon films. Although the precise reaction mechanisms are not known, the pressure of oxygen affects the CH and H concentration.[82,83]

5.3. Influence of the Substrate Temperature

5.3.1. Influence of deposition temperature on the growth rate

All the investigations concerning the influence of the substrate temperature on the growth rate show that it decreases as the temperature increases (Fig. 19

An activation energy of $E_a \simeq 0.09$ eV is obtained with the low frequency discharge and of $E_a \simeq 0.05$ eV with the high frequency discharge. A higher activation energy of $E_a \simeq 0.54$ eV has been obtained by vacuum arc deposition.[87] A positive temperature slope would indicate that the particles are directly incorporated in the film upon chemisorption. A negative slope indicates that the incorporation is intermediated by an adsorbed layer.

5.3.2. Effect on film properties

In order to obtain hard and nearly transparent diamond-like carbon films, it is necessary to keep the substrate temperature as low as possible because the electron and ion fluxes cause substrate heating. At

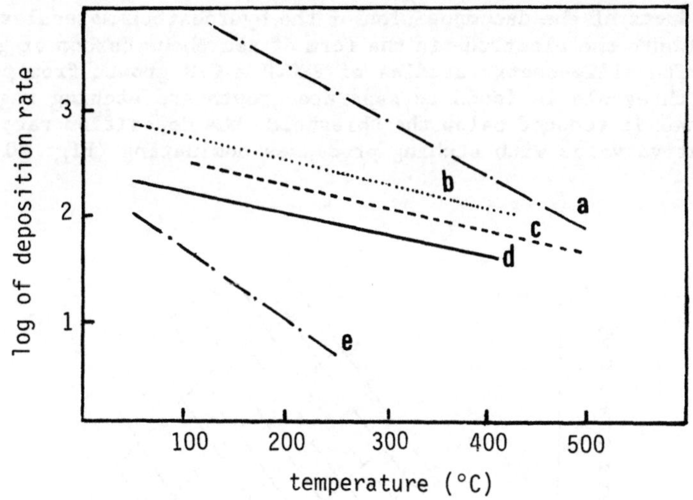

Fig. 19. Logarithm of the growth rate of carbon films as a function of the substrate temperature: (a) 30 kHz, CH_4 (ref. 84); (b) 30 kHz, C_4H_{10} (ref. 84); (c) 5 MHz, C_2H_4/He (ref. 85); (d) ECR 2.4 GHz, CH_4 (ref. 72); and (e) 13.6MHz, CH_4 (ref. 86).

high power densities this can lead to partial graphitization of the deposited film as shown in Table 7.

Table 7. Surface plasmon position and ratio of σ/π bonds as a function of deposition temperature (ref. 66)

Gas	T_S (°C)	σ_p (eV)	σ/π
CH_4/He	25	22.6	3
2/100	450	26.7	1.7
	650	27.7	1.5

5.4. Sputtering and Etching During Deposition

During deposition of carbon from a glow discharge of C_4H_{10} it has been observed that the hydrocarbon fragments impinging on the substrate initially do not deposit, but rather etch the surface of the substrate.[88] Gradually, the etching process changes over deposition after a few minutes. This experiment clearly shows that sputtering occurs during a-C:H deposition, although the sputtering yield of carbon is low. Loosely bound phases will be preferentially sputtered.

Furthermore, large amounts of hydrogen atoms and hydrogen ions are created in the hydrocarbon discharges. The reactive etching component in hydrocarbon plasmas may be atomic hydrogen. This atomic hydrogen, which is a product of the decomposition of hydrocarbon molecules in an RF electric field, can under some circumstances react very strongly with other products of the decomposition of the hydrocarbon molecules which are deposited onto the electrode in the form of amorphous carbon or graphite.[89] From in situ ellipsometry studies of PECVD a-C:H growth from pure CH_4, a RF power threshold is found to separate growth and etching regime.[90] As the RF power is reduced below the threshold, the deposition rate decreases to a negative value with etching processes dominating (Fig. 20).

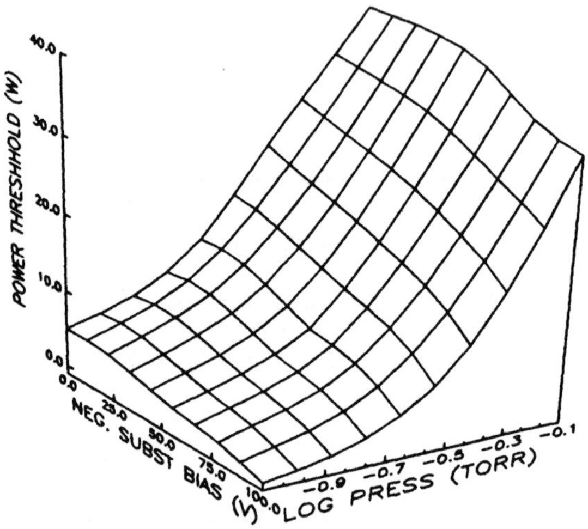

Fig. 20. RF power threshold separating deposition [above surface and etching (below surface)] regimes as a function of substrate bias and CH_4 pressure (0.08-0.8 torr) as determined by ellipsometry (ref. 90).

6. GROWTH ENVIRONMENT AND GROWTH MECHANISMS

6.1. Growth from Solid Carbon Sources (Ion Beam and Sputtering)

When no hydrogen is present, the mass spectra of ionized carbon vapor show the presence of carbon clusters.[11] For clusters of up to nine carbon atoms, the odd-number species are more abundant than the adjacent even-number species.

After C_{10}^+, there is a recurring periodic maximum with every four additional carbon atoms, i.e., at C_{14}^+, C_{18}^+, C_{22}^+, and C_{26}^+.

There seems to be general agreement that the C_9^+ and smaller species are linear chains. The maxima at C_{10}^+, C_{14}^+, etc., may be caused by the presence of monocyclic or polycyclic ring structure. Moreover, since the bond energies are on the order of several electron volts, clusters with an energy of 100 eV would almost certainly break up upon impact with the surface. Therefore, it does not appear that the mechanism of film formation from ionized carbon depends in any major way upon the structures present in the ion beam.

6.2. Growth from Hydrocarbon Gases (Glow Discharge)

Very similar considerations hold when considering hydrocarbon discharges. Neutrals and ions of the general form C_mH_n and $C_mH_n^+$ are present. At the pressures used, these species diffuse to the substrate where they contribute to the film formation.

6.2.1. Mass transfer from the gas phase

In most deposition discharges the chemical species created in the volume are transported towards the substrate by a diffusional mechanism. The local concentration nj of the specie j may be obtained from the relation:[91,92]

$$\frac{dn_j}{dt} = P - \frac{n_j}{\tau_g} + \nabla^2 n_j - v.\nabla\, n_j \tag{8}$$

where P represents the sum of all the production terms, n_j/τ_g the gas phase reaction (non linear term), $\nabla^2 n_j$ is the diffusion term and - $v.\nabla n_j$ the convection loss term. From these concentrations it is possible to calculate the flux of each neutral or ion towards the electrode. Such results are given as a function of time (position in the reactor) in Fig. 21.

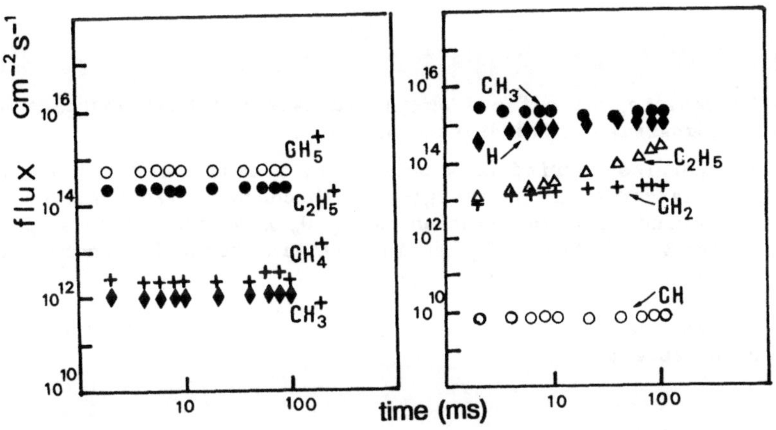

Fig. 21. Calculated neutral and ion fluxes in a CH_4 discharge. P = 0.3 torr, n_e = 10^9 cm^{-3}, T_e = 3 eV, Q = 5 sccm.

The calculated ion fluxes in the range of 10^{14}-10^{15} cm^{-2}s^{-1} are consistent with the experimental observations and contribute 10-30% to the total particle flux.

6.2.2. Surface phenomena

Let us consider a surface in contact with a plasma and at a potential V_B below the plasma potential V_p. It will be subject to a flux of high kinetic energy electrons, ions and neutral atoms, together with metastable atoms, active free radicals and UV radiation. At low gas pressures the kinetic energy of the bombarding ions will be about equal to $e(V_B - V_p)$. At higher pressures charge exchange collisions in the sheath will create a flux of both ions and neutral species with energies that can be large. Typical kinetic energies are in the 50 to 1000 eV range.

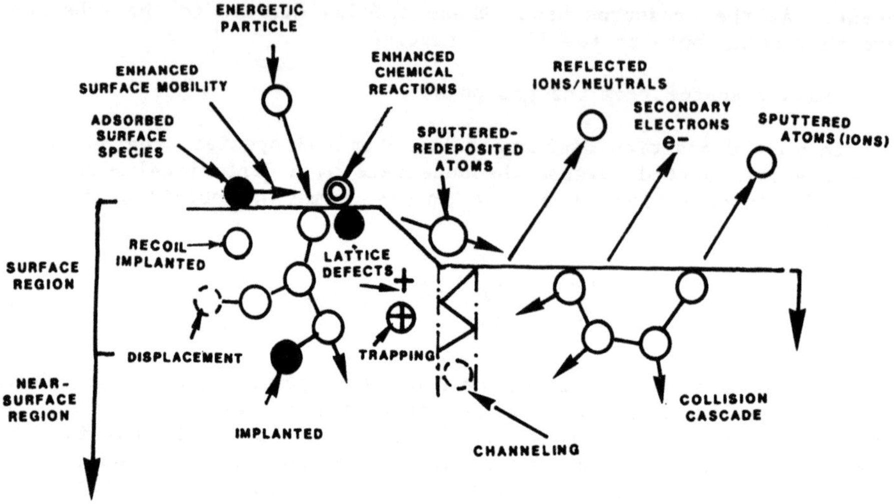

Fig. 22. Schematic representation of the energetic particle bombardment effects on surface and growing films (after ref. 93).

Figure 22 depicts the effects on the surface and the substrate surface region by bombardment by energetic species.

The energies carried to a surface by plasma generated species are equal to or larger than the energies which characterize a solid, i.e., a surface atom binding or sublimation energy U_s of about 9-10 eV a sputtering threshold energy of about 4 U_s, an interstitial formation energy of about 5 eV and a vacancy formation energy of 1 eV.

Typical energies of various processes in carbon deposition are reported in Table 8.

Table 8. Energies of various processes for carbon[11]

Item	Energy eV
Sputtering yield by C^+ ions : 0,15	500
Sputtering yield by Ar^+ ions : 0,07	500
Displacement energy of carbon atoms in diamond	80
Displacement energy of carbon atoms in graphite	25
Bond energy of diamond	7.41
Intraplanar bond energy in graphite	7.43
Interplanar bond energy in graphite	0.86
C – H bond energy	3.5

The ion flux at the surface of the substrate will enhance the mobility of atoms. Increases in diffusion coefficients can be as large as 5 orders of magnitude. As already mentioned, reactive ion etching and sputtering occurs: this is the mechanism by which hydrogen and other loosely bound graphitic phases can be removed from the material.

6.3. Proposed Growth Mechanisms

6.3.1. Deposition from an adsorbed layer

According to the mechanism of plasma deposition one can distinguish between incorporation of particles from the gas phase (flow-in-model) and deposition from an adsorbed layer (adsorbed-layer model).[87]

The flow-in-model is based on the assumption that the probability of incorporation ε^{inc} dominates. The deposition is given by:

$$R = \sum_i \frac{M_i}{\rho} J_i \varepsilon_i^{inc} \tag{9}$$

with M_i/ρ the mean volume per incorporated particle, J_i the flux of particle i. Because direct incorporation is a unidirectional chemical reaction its cross section can be written in the exponential form:

$$\varepsilon_i^{inc} = n_o \sigma_{oi}^{inc} \exp(-E_i^{inc}/kT) \tag{10}$$

with n_o the concentration of bounding sites and E_i^{inc} the activation energy for direct incorporation. When direct incorporation dominates, the deposition rate increases with increasing temperature and E_i^{inc} is about 1 eV.

Carbon film deposition cannot proceed by direct incorporation since the deposition rate decreases with increasing temperature (Fig. 19) and activation energies in the 0.05-0.5 eV range have been found.

Another type of growth is the deposition of a layer due to interaction between the surface of the substrate and a physisorbed layer. During a characteristic residence time τ, the particles either diffuse over the surface or they react with the substrate to be incorporated in the film. The deposition rate now takes on the form:

$$R = \sum_i \frac{n_o \sigma_i jx J_i \varepsilon_i^{phys}}{j_i \varepsilon_i^{phys} + n_o(1/\tau_i + \sigma_i j_x)} \tag{11}$$

where σ_i = cross section for bonding aided by particle bombardment; j_x = impingement rate of these particles. The residence time τ_i depends on the surface temperature T:

$$\tau_i = \tau_{io} \exp(E_i^{des}/kT) \tag{12}$$

where E_i^{des} = desorption energy. This model explains both the increase of deposition rate with ion bombardment and the negative slope of the temperature dependence. Also the low apparent activation energies agree with desorption energies of physisorbed particles. The growth mechanism through an adsorbed layer cannot explain by itself the low amount of hydrogen and the initial nucleation of sp^3 structure. We shall examine some mechanisms which try to explain the diamond-like carbon network formation.

6.3.2. Preferential sputtering

We have already mentioned that sputtering and etching take place during carbon deposition. The preferential sputtering was first advanced by Spencer et al.[47] Tetrahedrally bonded sp^3 structures are assumed to be more resistant to sputtering than trigonally bonded graphitic structures. The ion flux to the growing surface serves as an agent for resputtering non sp^3 structure. This hypothesis is supported by the relative energies of the various processes (Table 8). Both sputtering (etching) and deposition occurs simultaneously during film growth. Since the C-H bond energy is significantly less than the C-C bond energies, low mass hydrogen should therefore be preferentially sputtered from a carbon surface. It should be emphasized that carbon has a low sputter yield compared to other materials (Fig. 5) which permits a wide range of conditions under which preferential sputtering of hydrogen and hydrocarbon structures can be expected to occur.

6.3.3. Hot carbon atom chemistry

During plasma deposition ions are implanted with the kinetic energies ranging from 10^2 to 10^3 eV into carbon surfaces. The molecular ions CH_3^+, CH_2^+, CH_5^+, $C_2H_5^+$ will strip off their hydrogen and multicarbon species will fragment, respectively. Fast neutralization processes by charge exchange are responsible for the fact that projectiles are atoms when undergoing a chemical reaction. High energy chemistry of particles with energies exceeding at least a few eV is characterized by the fact that endothermic reactions and high activation energy channels become possible. Hot carbon atoms undergo reactions by insertion into C-H bonds and formation of unsaturated C-C bonds. The effect of primary implant are multiplied by the formation of secondary hot carbon atoms by knock-on processes of the primaries with the substrate. Simulation of collision cascades in amorphous carbon have been done.[94] Table 9 shows the number of secondary projectiles per cascade induced by primary carbon atoms.

Table 9. Number of secondary projectiles induced by a primary carbon atom (from ref. 94)

Primary energy eV	a – C	a – C : H
10^2	3.1 C	2.7 C ; 1.4 H
10^3	14.2 C	10.2 C ; 7.5 H
10^4	40.3 C	23.7 C ; 24.1 H

Taking into account the potential tertiary hot carbon atoms, a 1-keV carbon primary will produce up to 30 new hot carbon atoms in a-C and approximately 20 in a-C:H. A high number of secondary hot carbon atoms is concentrated in a very small volume and direct interaction of unsaturated structures with other cold or hot carbon atoms may occur. They may provide the pathway for ring formation which is in agreement with the most recent network models of diamond-like carbon.[95-97]

7. LIMITATION OF CURRENT TECHNIQUES

Diamond-like carbon films have been deposited using different techniques. An ideal deposition process should satisfy at least the following requirements:

- A high deposition rate on large areas with required thickness uniformity.
- A good control of the desired properties such as optical transparency, smooth surface topography, etc.
- Low stresses.

Deposition techniques currently used for the deposition of amorphous diamond-like carbon films are compared in terms of the above considerations to highlight their respective limitation in Table 10.[98]

Table 10. Comparison of various deposition techniques for a-C and a-C:H deposition

Processes	Type and quality of films produced	Advantages	Limitations
Physical vapor deposition :			
Ion beam sputtering	smooth, transparent a – C films	Extremely good operational control	very low deposition rate Usable only for small area
Dual sputtering beam	smooth transparent a – C and – C/micro-cristalline films	same as above	same as above
Laser - ion beam technique	a – C films limited data to comment on film quality	High deposition rates are possible	Difficult to scale up to large area, control/optimization of film properties and quality yet to be demonstrated
Plasma assisted chemical vapor deposition			
R.F. and D.C. plasma assisted	smooth transparent a – C : H films	high deposition rates on large area is possible	sometimes difficult to control. Large internal stress.
Microwave and ECR plasmas	Transparent a – C : H films	High deposition rates Good control of the substrate bias and position relative to plasma. No powder formation.	Over heating at the resonance point.
Remote plasma assisted deposition	Limited data available to comment on film quality	Clean process, low film damage	Difficult to control extremely low deposition rate usable only with small area
Cascaded arc plasma deposition	Smooth a – C : H films	Very high deposition rates on large area substrates Low contamination	Large internal compressive stresses

The inherent limitations of the aforementioned processes are related to the coupling of the different process parameters controlling the growth and properties of the films. For example, in plasma assisted processes the parameters governing the rate of generation of species also affect the substrate bombardment: the increase in RF power also increases the electron energy which in turn increases the electron energy which in turn increases the floating potential of the substrate. As bombardment of the films during growth by energetic ions and electrons from the plasma depends on the floating potential, an increase in RF power increases the film bombardment, thus changing the structure and properties of the films. This coupling also makes it difficult to obtain an independent control of hydrogen.

Although dual and single ion beam techniques provide advantages as regards independent control of substrate bombardment, deposition rate and hydrogen content, they suffer from low deposition rates due to the low sputtering yield of carbon and limitation for large area deposition because of the available sizes of the ion sources.

8. CONCLUSION

We have reviewed the processing technology being developed for a-C and C:H films deposition. In view of the above discussion, it is clear that processes need improvements which will overcome the above limitations. In order to realize the industrial applications, smooth surface topography, optical transmission in UV and IR, high thermal conductivity, and high electrical resistivity must be achieved. Moreover, the processes should have the following attributes: (a) high deposition rates, (b) capability to coat large as well as complex shapes, (c) flexibility and control of the properties of the film. Therefore, several issues related to the control of optical and electrical and mechanical properties, morphology, nucleation, and deposition rates of the films still need to be resolved in these processes.

9. REFERENCES

1. S. Aisenberg and R. Chabot, Ion Beam Deposition of Thin Films of Diamondlike Carbon, *J. Appl. Phys.*, ***42*** *: 2953 (1971).*

2. C. Weissmantel, K. Bewilogua, D. Dietrich, H.J. Erler, H.J. Hinnerberg, S. Klose, W. Nowick and G. Reisse, Structure and properties of quasi-amorphous fFilms prepared by ion beam techniques, *Thin Solid Films,* ***72*** *: 19 (1980).*

3. T.J. Moravec and T.W. Orent, Electron spectroscopy of ion beam and hydrocarbon plasma generated diamond like carbon films, *J. Vac. Sci. Technol.,* ***18*** *: 226 (1981).*

4. M.J. Mirtich, D.M. Swec and J.C. Angus, Dual ion beam deposition of carbon films with diamond-like properties, *Thin Solid Films,* ***131*** *: 245 (1985).*

5. K. Bewilogua, D. Dietrich, L. Pagel, C. Schürer and C. Weissmantel, Structure and properties of transparent and hard carbon films, *Surface Science,* ***86*** *: 308 (1979).*

6. N. Savvides and B. Window, Diamond-like amorphous carbon films prepared by magnetron sputtering of graphite, *J. Vac. Sci. Technol.,* ***A3*** *: 2386 (1985).*

7. S. Craig and G.L. Harding, Structure, Optical properties and decomposition kinetics of sputtered hydrogenated carbon, *Thin Solid Films,* ***97*** *: 345 (1982).*

8. L. Holland and S.M. Ohja, Deposition of hard and insulating carbonaceous films on a RF target in a butane plasma, *Thin Solid Films,* ***48*** *: L 21-23 (1978).*

9. A. Bubenzer, B. Dischler, G. Brandt and P. Koidl, R.F. Plasma deposited amorphous hydrogenated hard carbon thin Films : Preparation, Properties and Applications, *J. Appl. Phys., 54 : 4590 (1983).*

10. Y. Catherine and P. Couderc, Electrical characteristics and growth kinetics in discharges used for plasma deposition of amorphous carbon, *Thin Solid Films, **144** : 265 (1986).*

11. J.C. Angus, P. Koidl and S. Domitz, Carbon thin films, in plasma deposited thin Films, J. Mort and F. Jansen ed., CRC Press, Boca Raton (1986).

12. S.F. Pellicori, C.M..Peterson and T.P. Henson, Transparent carbon films ; Comparison of properties between ion and plasma deposition processes, *J. Vac. Sci. Technol., **A4** : 2350 (1986).*

13. S. Fujimori, T. Kasai and T. Inamura, Carbon film formation by laser evaporation and ion beam sputtering, *Thin Solid Films, **92** : 71 (1982).*

14. M. Kitabatake and K. Wasa, Growth of diamond at room temperature by an ion-beam sputter deposition under hydrogen ion bombardment, *J. Appl. Phys., **58** : 1693 (1985).*

15. S. Aisenberg, Properties and Applications of diamond like carbon films, *J. Vac. Sci. Technol. **A2** : 369 (1984).*

16. N. Wada, P.J. Gaczi and S.A. Solin, Diamond like 3 - fold coordinated amorphous carbon, *J. Non Crystal. Solids, **35-36** : 543 (1980).*

17. A. Hiraki, T. Kawano and Y. Kawakami, Tetrahedral Carbon Film by Hydrogen Gas Reactive RF - sputtering of Graphite onto Low Temperature Substrate, *Solid State Comm., **50** : 713 (1984).*

18. D.R. Mc Kenzie, R.C. Mc Phedran, N. Savvides and L.C. Botten, Properties and Structure of amorphous hydrogenated carbon films, *Phil. Mag., **B 48** : 341 (1983).*

19. C. Wyon, R. Gillet and L. Lombard, Properties of amorphous carbon films produced by magnetron sputtering, *Thin Solid Films, **122** : 203 (1984).*

20. D.R. Mc Kenzie and L.M. Briggs, Properties of hydrogenated carbon films produced by reactive magnetron sputtering, *Solar Energy Mat., **6** : 97 (1981).*

21. S. Berg and L.P. Andersson, Diamond-like carbon films produced in a butane plasma, *Thin Solid Films, 58 : 117 (1979).*

22. S.M. Ojha, M. Norström and D. Mc Culluch, The growth kinetics and properties of hard and insulating carbonaceous films grown in an RF discharge, *Thin Solid Films, **60** : 213 (1979).*

23. K. Kobayashi, N. Mutsukura and Y. Machi, Deposition of hard carbon films by the RF glow discharge method, *Thin Solid Films, **158** : 233 (1988).*

24. J.D. Warner, J.J. Pouch, S.A. Alterovitz, D.C. Liu and W.A. Lanford, Plasma deposited hydrogenated carbon on GaAs and InP, *J. Vac. Sci. Technol., **A3** : 900 (1985).*

25. Y. Catherine, P. Couderc and A. Zamouche, Dépôt de films minces de carbone amorphe hydrogéné par décharge RF dans les mélanges $CH_4 - He$ et $C_2H_4 - He$. 4eme Coll. Int. sur les Plasmas et la Pulvérisation cathodique, CIP 82, Nice (1982).

26. D. Nir, R. Kalish and G. Lewin, Diamond-like carbon films of low hydrogen contents mode with a mixture of hydrocarbon and reactive gas, *Thin Solid Films, **117** : 125 (1984).*

27. K. Fujii, S.H. Yokata and N. Shŏhata, Properties of diamondlike carbon films, *Mat. Res. Soc. Symp. Proc., 75 : 765 (1987).*

28. T. Mori and Y. Namba, Hard diamond like carbon films deposited by ionized deposition of methane gas, *J. Vac. Sci. Technol.*, ***A1*** *: 23 (1983).*

29. O. Matsumoto, H. Toshima and Y. Kanzaki, Effect of dilution gases in methane on the deposition of diamond-like carbon in a microwave discharge, *Thin Solid Films,* ***128*** *: 341 (1985).*

30. J. Kammermaier, G. Rittmayer and P. Schulte, Emission spectra relevant to plasma deposition of methane in microwave and radiofrequency induced discharges, ISPC 8, Tokyo, 1987, *Symposium Proc., Vol.* ***3***, K. Akashi and A. Kinbara ed., *1542 (1987).*

31. M. Moisan, C. Barbeau, R. Claude, J. Margot-Chaker and G. Sauvé, RF or microwave plasma reactors ? Factors determining the optimum frequency of operation, *CIPG 89, Antibes, 1989, suppt. Le Vide Les Couches Minces,* ***246*** *: 145 (1989).*

32. O. Matsumoto, T. Fujita and M. Uyama, Application of an ECR plasma apparatus to the deposition of carbonaceous films, I S P C 9, Pugnochiuso, 1989, *Symposium Proc. vol.* ***3***, *R. d'Agostino ed.*, 1804 (1989).

33. M. Morin and L. Vandenbulke, "Quelques aspects physiques et chimiques de la phase gazeuse dans un plasma multipolaire microonde pour le dépôt de films de a – C:H", *Proceed. of the 7th European conference on CVD, Perpignan, 1989,* M. Ducarroir, C. Bernard and L. Vandenbulke ed., *Coll. de Physique, suppt* ***C5, 50*** *: 929 (1989).*

34. W.D. Partlow, R.R. Mitchell, P.J. Chantry and L.E. Kline, Deposition of carbon films in a microwave excited flowing afterglow reactor, I S P C 9, Pugnochiuso, Italie, 1989, *Symp. Proc. Vol.* ***3*** ; R. d'Agostino ed., 269 (1989).

35. K. Ebihara, T. Ikegami, T. Matsumoto, S. Maeda and K. Harada, Amorphous C:H and amorphous SiC:H thin film deposition using a pulsed plasma process, I S P C 9, Pugnochiuso, Italie, 1989, *Symp. Proc. Vol.* ***3*** ; R. d'Agostino ed., 1479 (1989).

36. G.M.W. Kroesen, Plasma deposition : investigations on new approach, Physics Doctor Thesis, Technische Universiteit, Eindhoven, 1988.

37. J.M. Mackowski, M. Berton, P. Ganau and Y. Touzé, Comparaison physico-chimique des couches de carbone à large bande interdite (I–C) élaborées par différentes techniques plasma, C I P 82, Nice 1982, *Le Vide - Les Couches Minces, suppl n°212 : 99 (1982).*

38. T. Miyazawa, S. Misawa, S. Yoshida and S. Gonda, Preparation and structure of carbon films deposited by a mass-separated C^+ ion beam, *J. Appl. Phys.,* ***55*** *: 188 (1984).*

39. J. Ishikawa, K. Ogawa, Y. Takeiri and T. Takagi, Preparation of transparent carbon films by negative-carbon-ion-beam deposition, *Proc. of the 9th Symp. on ISIAT, Tokyo : 255 (1985).*

40. J.A. Thornton, Physical Vapor Deposition in semiconductor Materials and Process Technology Handbook, G. Mc Guire ed. Noyes publication, Park Ridge, 329 (1988).

41. J.C. Angus, M.J. Mirtich and E.G. Wintucky, Ion beam deposition of amorphous carbon films with diamond like properties in Metastable Materials Formation by Ion Implantation, S.T. Picraux and W.J. Choyke ed. *Elsevier Science Publishing Company, 433 (1982).*

42. T. Shingyoji, N. Kikuchi, *Proc. of the 9th Int. Vac. Cong. and 5th Int. Cong. on Solid Surf., Extended Abstracts, Madrid, TFP 13 A (1983).*

43. F. Jansen, M. Machonkin, S. Kaplan and S. Hark, The effects of hydrogenation on the properties of ion beam sputter deposited amorphous carbon, *Amer. Vac. Soc. Meet., Boston, TFTu MO8 (1984).*

44. H.H. Andersen and H.L. Bay, Sputtering yield measurements in Sputtering by particle bombardment, I.R. Behrish ed., *Topics in Appl. Phys.* ***47***, *Springer Verlag, Berlin,* ***166*** *(1981).*

45. J.C. Angus, J.E. Stultz, P.J. Shiller, J.R. Mc Donald, M.J. Mirtich and S. Domitz, Composition and properties of the so-called "Diamond-like" amorphous carbon films, *Thin Solid Films, **118** : 311 (1984).*

46. R.C. de Vries, Synthesis of diamond under metastable conditions, *Ann. Rev. Mat. Sci., **17** : 161 (1987).*

47. E.G. Spencer, P.H. Schmidt, D.C. Joy and F.J. Sansalone, Ion-beam-deposited polycristalline diamond like films, *Appl. Phys. Lett., **29** : 118 (1976).*

48. K. Köhler, D.E. Horne and J.W. Coburn, Frequency dependence of ion bombardment of grounded surfaces in rf argon glow discharges in planar systems, *J. Appl. Phys., 58 : 3950 (1985).*

49. C.B. Zarowin, Relation between the RF discharge parameters and plasma etch rates, selectivity, and anisotropy, *J. Vac. Sci. Technol. A2 : 1537 (1984).*

50. M.D. Gill, Sustaining mechanisms in r.f. plasmas, *Vacuum, **34** : 357 (1984).*

51. P. Briaud, G. Turban and B. Grolleau, Energy distribution of ions in plasma etching reactors, *Mat. Res. Symp. Proc. Vol. **68** : 109 (1986).*

52. J. Perrin, Pere Roca i Cabarrocas, B. Allain and J.M. Friedt, a – Si : H deposition from SiH_4 and Si_2H_6 rf discharges : pressure and temperature dependence of film growth in relation to the α-γ discharge transition, *Jap. J. Appl. Phys., 27 : 2041 (1988).*

53. Y. Catherine and A. Pastol, Ion flux measurements in discharges used for amorphous carbon film deposition, *E-MRS Meeting Strasbourg 1987, Vol. XVII, 145.*

54. J. Wagner, Ch. Wild, F. Pohl and P. Koidl, Optical studies of hydrogenated amorphous carbon plasma deposition, *Appl. Phys. Lett., **48** : 106 (1986).*

55. J.L. Flouttard, Synthèse par décharge luminescente RF de films composites or-carbone à propriétés modulables, *Thèse de Doctorat, Toulouse, 1989.*

56. L. Athouel, Analyse spatiale par spectrométrie d'émission de plasmas de méthane utilisés pour le dépôt de couches minces de carbone, *Rapport DEA, Nantes, Juillet 1989.*

57. J.P. Warner, J.J. Pouch, S.A. Alterovitz, D.C. Liu and W.A. Lanford, Plasma deposited hydrogenated carbon on GaAs and InP, *Amer. Vac. Soc., October 1984.*

58. P. Couderc and Y. Catherine, Structure and physical properties of plasma-grown amorphous hydrogenated carbon films, *Thin Solid Films, **146** : 93 (1987).*

59. L.P. Anderson, A review of recent trends on hard i – c films, *Thin Solid Films, 86 : 193 (1981).*

60. K. Enke, Some new results on the fabrication of and the mechanical, electrical and optical properties of i – carbon layers, *Thin Solid Films, **80** : 227 (1981).*

61. J. Asmussen, Electron cyclotron resonance microwave discharges for etching and thin film deposition, *J. Vac. Sci. Technol., A7 : 883 (1989).*

62. P.K. Shufflebotham, D.J. Thompson and H.C. Card, Behavior of downstream plasmas generated in microwave plasma chemical vapor deposition reactor, *J. Appl. Phys., **64** : 4398 (1988).*

63. O.A. Popov and H. Waldron, Electron cyclotron resonance plasma stream source for plasma enhanced chemical vapor deposition, *J. Vac. Sci. Technol., A7 : 914 (1989).*

64. O. Matsumoto, T. Fujita and H. Uyama, Application of an ECR plasma apparatus to the deposition of carbonaceous films, *9th Int. Symp. on Plasma Chemistry, Pugnochiuso, Italy (1989), Symp. Proc. Vol. **III**,* edited by R. d'Agostino : 1804 (1989).

65. W.D. Partlow, P.R. Mitchell, P.J. Chantry and L.E. Kline, Deposition of carbon films in a microwave-excited flowing afterglow reactor, 9th *Int. Symp. on Plasma Chemistry, Pugnochiuso, Italy, 1989, Symp. Proc. III,* edited by R. d'Agostino, L70 (1989).

66. D.J. Vitkavage, R.A. Rudder, G.G. Fountain and R.J. Markunas, Plasma enhanced chemical vapor deposition of polycristalline diamond and diamond like films, *J. Vac. Sci. Technol., **A6** : 1812 (1988).*

67. N. Savvides, Low energy ion assisted deposition of a – C:H films : A study of optical parameters with increasing hydrogenation, in E-MRS Symp. Proc. Amorphous hydrogenated Carbon films, edited by P. Koidl and P. Oelhafen, Les Editions de Physique - Strasbourg, France, XVII : **275** (1987).

68. D.R. McKenzie, R.C. Mc Phedran, N. Savvides and D.J.M. Cockayne, Analysis of films prepared by plasma polymerization of acetylene in a D.C. magnetron, *Thin Solid Films, **108** : 247 (1983).*

69. A. Bubenzer, B. Dischler and A. Nyaiesh, Optical properties of hydrogenated hard carbon thin films, *Thin Solid Films, **91** : 81 (1987).*

70. C.J. Robinson, R.N. Payne and A.E. Bell, Hydrogenated amorphous carbon dielectric coatings for magneto-optic data storage media, *J. Appl. Phys., **64** : 4646 (1988).*

71. P. Wood, T. Wydeven and O. Tsuji, Characterization of diamond like hydrocarbon films grown using the RF self bias technique, 9th *Int. Symp. on Plasma Chemistry, Pugno-chiuso, Italy (1989)* Symp. Proc. Vol. edited by R. d'Agostino, Vol. **III**, 1485 (1989).

72. M. Morin, Contrôle du dépôt de films de carbone amorphe hydrogéné en plasma multipolaire microonde, *Thèse de Doctorat, Orléans, 22 septembre 1989.*

73. J.W. Zou, K. Schmidt, K. Reichelt and B. Dischler, The Properties of a – C:H films deposited by plasma decomposition of C_2H_2, *J. Appl. Phys., **67** : 487 (1990).*

74. J. Zelez, Low stress diamond-like carbon films, *J. Vac. Sci. Technol., **A1** : 305 (1983).*

75. E.T. Prince and M.M. Romach, Thin films of hydrogenated amorphous carbon prepared by rf plasma decomposition of methane, *J. Vac. Sci. Technol., **A3** : 694 (1985).*

76. D. Nir, Intrinsic stress in diamond like carbon films and its dependence on deposition parameters, *Thin Solid Films, **146** : 27 (1987).*

77. C. Wild, P. Koidl and J. Wagner, Plasma deposition of a – C:H films : the role of process gas, plasma chemistry and plasma surface interaction, in E-MRS Meeting Proceed. Vol. XVII. Amorphous Hydrogenated Carbon Films, edited by P. Koidl and P. Oelhafen, Les Editions de Physique - Strasbourg, France, **137** (1987).

78. L.P. Anderson, S. Berg, H. Norström, R. Olaison and S. Towta, Properties and coating rates of diamond-like carbon films produced by RF glow discharge of hydrocarbon gases, *Thin Solid Films, **63** : 155 (1979).*

79. O. Matsumoto and T. Katagiri, Effect of dilution gases in methane on the deposition of diamond like carbon in a microwave discharge II : effect of hydrogen, *Thin Solid Films, **146** : 283 (1987).*

80. K. Enke, Influence of argon and hydrogen admixture on properties of plasma deposited a–C:H films, E-MRS Meeting Strasbourg, in Amorphous Hydrogenated carbon films, edited by P. Koidl and P. Oelhafen, Vol. XVII, **117** (1987).

81. G. Amaratunga, A. Tutnis, K. Clay and W. Milne, Crystalline diamond growth in thin films deposited from CH_4/Ar rf plasma, *Appl. Phys. Lett., **55** : 634 (1989).*

82. S. Morita, Y.Takai, S. Hattori, M. Ieda, K. Hida, S. Ikeda, S. Ishibashi and A.Ishitani, Effect of oxygen on amorphous carbon film formation, *Proc. 10th Symp. on ISIAT'86, Tokyo, 457 (1986).*

83. R.B. Lockwood, R.E. Miers, L.W. Anderson, J.E. Lawler and C.C. Lin, Effect of water vapor on a $CH_4 - H_2$ discharge plasma, *Appl. Phys. Lett., 55 : 1385 (1989).*

84. J.J. Pouch, J.A. Alterovitz and J.D. Warner, Auger Electron Spectroscopy, Secondary Ion Mass spectroscopy and optical characterization of a – C:H and BN films, NASA Technical Memorandum 87258, 1986.

85. Y. Catherine, P. Couderc and A. Zamouche, Dépôt de films minces de carbone amorphe hydrogéné par décharge RF dans les mélanges CH_4–He et C_2H_6 – He, *CIP 82, suppl. n°212, Le Vide - Les couches minces : 126 (1982).*

86. V.N. Polunin, R.P. Stanchula, A.A. Andreev and B.T. Melekh, Growth rate and properties of amorphous carbon layers produced in low-temperature plasma, *ISV Akad. Nauk SSSR Neorg. Mater. 22 : 1288 (1983).*

87. H. Kersten and G.M.W. Kroesen, On the temperature dependence of the deposition rate of amorphous hydrogenated carbon films, *J. Vac. Sci. Technol. A8 : 38 (1990).*

88. L.P.Andersson and S. Berg, Initial etching in an rf butane plasma, *Vacuum : 28, 449 (1978).*

89. S. Mitura, L. Klimek and Z. Has, Etching and deposition phenomena in an RF CH_4 plasma, *Thin Solid Films, 147 : 83 (1987).*

90. R.W. Collins, In situ ellipsometry as a diagnostic of thin-film growth : Studies of amorphous carbon, *J. Vac. Sci. Technol. A7 : 1378 (1989).*

91. K. Tachibana, M. Nishida, H. Harima and Y. Urano, Diagnostic and modelling of a methane plasma used in the chemical vapour deposition of amorphous carbon films, *J. Phys. D : Appl. Phys. 17 : 1727 (1984).*

92. L.E. Kline, W.D. Partlow and W.E. Bies, Electron and chemical kinetics in methane rf glow-discharge deposition plasmas, *J. Appl. Phys. 65 : 70 (1989).*

93. D.M. Mattox, Particle bombardement effects on thin film deposition : A review, *J. Vac. Sci. Technol. A7 : 1108 (1989).*

94. K. Roessler and G. Eich, Hot carbon atoms in the formation of amorphous hydrogenated carbon films, E-MRS Meeting - Strasbourg, Symposia Proc. Vol. XVII, Amorphous hydrogenated carbon films edited by P. Koidl and P. Oelhafen, Les Editions de Physique, **167** (1987).

95. J. Robertson, Clustering and gap states in amorphous carbon, *Phil. Mag., 57 : 143 (1988).*

96. J.L. Bredas and G.B. Street, Electronic Properties of amorphous carbon films, *J. Phys. C : Solid State Phys., 18 : L651 (1985).*

97. M.A. Tamor and C.H. Wu, Graphitic, network models of "diamond like" carbon, *J. Appl. Phys. 67 : 1007 (1990).*

98. C.V. Deshpandey and R.F. Bunshah, Diamond and diamond like films : Deposition processes and properties, *J. Vac. Sci. Technol. A7 : 2294 (1989).*

COMPUTER MODELING OF C:H FILM GROWTH

Wolfhard Möller

Max-Planck-Institut für Plasmaphysik, EURATOM-Association
D-8046 Garching (Fed. Rep. of Germany)

INTRODUCTION

Hydrogenated amorphous carbon films[1-5] exhibit specific properties which render them suitable for a wide field of applications, such as optical coatings, hard mechanical coatings, protective coatings, electrically insulating films, and low-Z inner wall coatings in fusion devices.[6] For their production, a number of different processes are available. Glow-discharge plasma-enhanced chemical vapour deposition is most widely employed. However, such films result also from hydrocarbon ion deposition[7,8] and hydrogen implantation into graphite.[9]

Despite this large field of activities, the understanding of the basic mechanisms underlying the layer growth is still at the beginning stage. Such a knowledge, however, appears indispensable if the properties of the resulting layers are to be adjusted in a predictable way by, e.g, the choice of the process parameters or the method of deposition. Some crude ideas have been forwarded to qualitatively describe the different aspects of the layer growth, such as a hypothesis of preferential sputtering[10] promoting the formation of tetragonal bonds in the growing film. Recently,[11] similar ideas have been given on a preferential displacement of carbon atoms which might influence the growth of diamond films. The formation of this sp^3 bond, which is essential for the properties also of hard hydrocarbon films, has also been argued[1] to be assisted by a negative surface charge during plasma deposition. The rearrangement of bonds can also be promoted by collisions cascades or spikes generated by energetic ion impact.[7] The chemical structure of precursor molecules may also influence the structure of the resulting film, as proposed for the growth of diamond films.[12]

In addition to the mechanisms influencing the structure of the growing film, one has to learn about the mechanisms which define the growth itself, and establish the bulk composition (i.e. the concentrations of carbon and hydrogen) in the growing film. Both the film growth and the film properties result from effects of ion bombardment when C:H films are grown by ion beam deposition or by hydrogen implantation into graphite. In the case of plasma deposition, the situation is more complicated as both hydrocarbon radicals and different ionic species impinge on the surface. Experimental studies with varying process parameters[13] indicate that the plasma-assisted film growth cannot be explained by the deposition of ions alone. In contrast, recent work[14] clearly indicates that the film growth is mainly due to neutral radicals, with the methyl radical as the most probable dominant species, as suggested by plasma modeling as well as mass-spectrometric and optical diagnostic studies.[14-19] On the other hand, ionic bombardment is essential in defining the mechanical, optical and electrical properties of the layer, as shown by many parameter studies (e.g. refs.[1,20-25]) with varying self-bias of the substrate in an RF plasma, i.e. with varying mean ion energy. Recently[14], the effect of ion bombardment on the hardness has been demonstrated conclusively by means of a pulsed bias technique. The magnitude of the ion energy also influences the strength of hydrogen bonds in the films, as indicated both by thermal effusion[26] and ion bombardment[27,28] studies.

The hydrogen content of the films may vary in a rather large range from H:C ratios below 0.2 up to larger than 1.[1,23,27-29] Typical a-C:H films contain about 0.3 hydrogen atoms per carbon atom, whereas films with an hydrogen-to-carbon ratio above about 0.7 may be classified as polymerlike. A smooth transition between these types often prevents a clear classification. The distinction between "hard" a-C:H and "soft" polymerlike films indicates a correlation between the film properties and their hydrogen content. This has been confirmed in several systematic studies.[1,23,30]

The hydrogen concentration in films generated by ion implantation into graphite has been modeled successfully on the basis of ion implantation and ion-induced release.[31] In that model, implanted hydrogen ions were allowed to be trapped chemically at carbon atoms up to a maximum trap concentration of 0.5 traps per carbon atom. Trapped hydrogen atoms may be released either thermally (only above about 500^{0}C) or by elastic knockon in collision cascades generated by postimplanted hydrogen atoms. Released atoms may either become retrapped or recombine into H_2 molecules. The latter will diffuse quickly towards the surface and be reemitted into the vacuum. (The formation of molecules in the bulk and their quick diffusion in the a-C:H network has been demonstrated both for implanted graphite[32,33] and plasma-deposited layers.[34,35])

Based on these ideas, the present paper presents a first attempt to simulate the stoichiometry of a hydrocarbon film being deposited from a glow discharge plasma. After a description of the computational model, theoretical results will be presented with different conditions for the incident fluxes of radicals and ions. Finally, the results will be compared to experimental findings in a mainly qualitative sense, in view of the simplifying assumptions of the calculations and the poor experimental information on the impinging species and fluxes.

COMPUTATIONAL MODEL

Physical Processes

The deposition of hydrogenated carbon layers by plasma-enhanced chemical vapour deposition is governed by a large number of different mechanisms being the subjects of very different fields of physics and chemistry. Plasma physics and plasma chemistry determine the properties of the plasma through a large number of individual mechanisms. For a methane plasma, some modeling studies are available which demonstrate this complexity.[15,16] Nevertheless, this area is far from being understood. For RF plasmas, which are mainly employed in experimental and practical studies, in principle time and space dependent calculations[36] in a multispecies plasma are required. A prerequisite is a very detailed knowledge of the rate coefficients involved as high-energy electron beams might occur especially at the boundary layer. A further critical area is the formation of the self-bias[37] voltage at a substrate exposed to the plasma. The energy distribution of ions arriving at the surface is determined by the bias voltage and the probability of collisions in the boundary layer. In the following, we will leave aside the complicated processes in the plasma and the boundary layer by assuming simplified but reasonable boundary conditions given by the neutral and ion fluxes from the plasma onto the substrate.

To our knowledge, the processes occuring at the surface of the growing film are even less identified. One may only speculate about the adsorption of radicals and chemical reactions under the influence of ion bombardment and a probably large surface coverage of hydrogen. These mechanisms may be rate-determining for the growth of the film. Again, we will neglect these processes by fixing as a boundary condition those fractions of the incident fluxes which finally contribute to the layer growth, i.e. after possible surface reactions including sticking, desorption or cross-linking with the growing film. In this sense, the "incident" fluxes in the model will be identical to the "sticking" fluxes in reality.

Finally, the surface-near bulk layer of the growing film is exposed to ion-bombardment effects approximately within the projected range of the impinging ionic species. This layer is the main subject of our simulation, which will treat ion implantation, the sputtering of atoms, and the ion-induced reemission of hydrogen. Ion-induced hydrogen reemission can occur via ballistic release from hydrogen-carbon bonds in the film. However, we will alternatively allow for a release due to electronic interaction of fast atoms in collision cascades with the atoms of the substance. (Indeed, some experimental results[27,28,38] indicate that electronic mechanisms may also play a role in ion-induced hydrogen release from hydrocarbon films.) For simplicity, we will not allow for retrapping of released atoms and assume that locally released atoms will be reemitted through the surface immediately. This implies that the release process is rate-controlling rather than local recombination or molecular diffusion.

Analytical Model

The problem can be sketched in a simplified picture assuming a homogeneous composition of the growing layer. In the stationary state of the layer growth, mass conservation requires the ratio of the hydrogen concentration c_H and the carbon concentration c_C to be given by the ratio of the net deposited fluxes:

$$\frac{c_H}{c_C} = \frac{j_H^0 - j_H^r}{j_C^0 - j_C^r} \tag{1}$$

In eq.(1), j^0 denote the incident fluxes of the constituents (both from neutrals and ions) and j^r their reemitted fluxes (due to sputtering and induced reemission). With the simplified relations

$$j_i^r = (\sigma_i^i j_i^+ + \sigma_k^i j_k^+) c_i, \tag{2}$$

where i, k stand for H or C, σ_k^i denote cross sections of reemission of component i by bombardment with component k ions, and j^+ the corresponding ion fluxes, eq.(1) can be solved with the normalization relation

$$c_H + c_C = 1, \tag{3}$$

in order to obtain the bulk composition c_H/c_C. For this purpose, the cross sections σ_k^i have to be known. They can, e.g., be obtained from TRIM[39]-type computer simulations.

In the above simple analytical treatment, it has implicitly been assumed that deposition and release occur at the very surface. In reality, different ranges of depth are affected by different species: Radicals are deposited at the surface, whereas ions affect different depths according to their mass and energy. This will result in one or more altered layers with probably severe deviations from a homogeneous concentration of hydrogen and carbon. Therefore, we prefer a fully dynamic computer simulation of the film growth. Its essential features will be described in the following section.

Dynamic Computer Simulation

Dynamic computer simulation codes based on the binary collision model had originally been developed in order to describe high-fluence implantation as well as atomic mixing and preferential sputtering phenomena.[40–46] Our code TRIDYN[43,45] is based on the widely spread TRIM[39] code and its sputtering[47] and multicomponent[48] extensions. A recent version (4.0) of TRIDYN allows the simultaneous incidence of atoms of different type. This has been applied successfully to the simulation of ion assisted deposition of boron nitride.[49] A modified version (4.2) of this code was also used for the present calculations.

A detailed description of TRIDYN has been given in ref.[45]. Therefore, we will only describe the main mechanisms and the modifications and parameters for the present work. In a semi-infinite medium which is subdivided into thin slabs (thickness 5Å in the present calculations), the histories of both incident projectiles and knock-on cascade atoms are traced as sequences of free-flight paths and elastic binary collisions. For the latter, the universal averaged "Kr-C" potential[50] is chosen. The inelastic energy loss between and during the collisions is treated as an equipartition of nonlocal (Lindhard/Scharff[51]) and local (Oen/Robinson[52]) interactions.

The bulk binding energy during cascade formation is set to zero[46]. Cascade atoms are regarded as being permanently displaced when their initial energy exceeds a damage threshold E_d. This energy has been chosen to be 25 eV for carbon atoms in agreement with the mean displacement energy in graphite[53]. For hydrogen, 2.5 eV has been chosen close to typical energies of chemical C-H bonds and in rough agreement with 2.7 eV employed for earlier modeling of hydrogen implantation into graphite.[31] We are aware that the assumption of a binding energy which is independent of hydrogen concentration represents again a simplification in view of experimental observations[26] during thermal effusion. A planar potential model is adopted for the binding of surface atoms. Quite arbitrarily, the surface binding energies E_s of both hydrogen and carbon atoms are set equal to 4.5 eV. They determine critically the yields of collisional sputtering, which, however, are negligible as demonstrated below. The history of any projectile or recoil atom is terminated when its energy falls below a cutoff energy, E_{co}, which is chosen as 2.5 eV for hydrogen (i.e. at the displacement threshold) and 4.5 eV for carbon (i.e. at the surface binding energy).

As the main mechanism of hydrogen reemission, the present version of TRIDYN allows to remove hydrogen atoms from the substance after ion-induced release from their binding sites. In the case of "nuclear" release, each hydrogen recoil atom with an initial energy E_{in} above the displacement threshold may be removed after coming to rest. (In this way, all possible secondary knockon atoms are treated correctly). With an efficiency factor f_n which may be predefined between 0 and 1, the probability of reemission per hydrogen recoil atom becomes

$$w_n^{reem} = \begin{cases} f_n & \text{if } E_{in} > E_d^H \\ 0 & \text{else.} \end{cases} \tag{4}$$

Alternatively, hydrogen atoms may be reemitted through electronic energy deposition with a probability per collision event

$$w_e^{reem} = f_e \frac{\Delta E_e}{E_d^H} c_H, \tag{5}$$

where ΔE_e denotes the total (local plus nonlocal) energy loss during the collision and f_e an efficiency factor.

The dynamic alteration of the substance, which causes a continuous buildup of the film in the computer model, is accomplished in TRIDYN by a local relaxation after each termination of a pseudoprojectile history, according to a density which depends on the local composition.[43] The atomic volumes are chosen to reproduce correctly the density of pure graphite and a density of 1.7 g/cm^3 for a hydrocarbon layer of composition $CH_{0.4}$.

The present simulation runs were performed with typically $5 \cdot 10^5$ incident pseudoatoms, with typical computing times of 10 min on a CRAY XMP vector computer.

Incident Fluxes

For our simulation, we restrict ourselves to the most simple case of C:H deposition from a methane plasma. For this, simplifying, we assume that the CH_3 radical is the only neutral species which contributes to the layer growth, in accordance with the knowledge described above. With a typical growth rate in the order of 10 Å/s, a typical sticking CH_3 flux is in the order of $10^{16}/(\text{cm}^2\text{s})$. A typical ion density in the plasma is around $10^9/\text{cm}^3$, at an electron temperature around 5 eV. This results, using the Bohm criterion[54] for the velocity of the ions when entering the sheath, in a total ion flux in the order of $10^{15}/(\text{cm}^2\text{s})$. Thus, we assume a 10% ion to radical ratio. Little experimental[14,18] or theoretical[16] information is available on the type or the relative fluxes of incident ions. Calculations have been performed for both hydrogen ions, represented by H_2^+, and hydrocarbon ions, represented by CH_3^+, although we are aware of the fact that other ions, e.g. CH_5^+ or $C_2H_5^+$, might be dominant under certain conditions.

As a typical sheath potential for, e.g., deposition on the driven electrode of an asymmetric RF reactor, we choose 400 V. We note, however, that the choice of monoenergetic 400 eV ions again represents a crude approximation. For incident neutrals, zero energy is assumed which causes 100% sticking through the attractive surface potential. In the case of energetic polyatomic ions, the total energy is split for the individual atoms by assuming an identical velocity. According to the desired proportions of incident atomic or ionic fluxes, their species are chosen by a random generator.

Nuclear and Electronic Energy Deposition

In order to gain some information on the role of the different effects of ion bombardment in the present energy range, we performed several static simulations. Fig.1 displays the near-surface stopping power for hydrogen and carbon atoms incident on a typical a-C:H layer as function of their energy. Whereas the electronic stopping of carbon and hydrogen is nearly

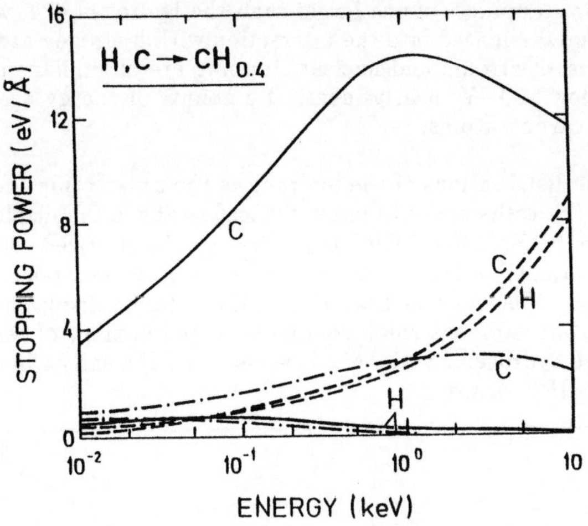

Fig.1. Stopping powers of hydrogen and carbon ions incident on an a-C:H layer. The incident ions are denoted by their chemical symbols. Dashed lines: electronic stopping; solid lines: nuclear stopping due to collisions with carbon atoms; dashed-dotted lines: nuclear stopping due to collisions with hydrogen atoms.

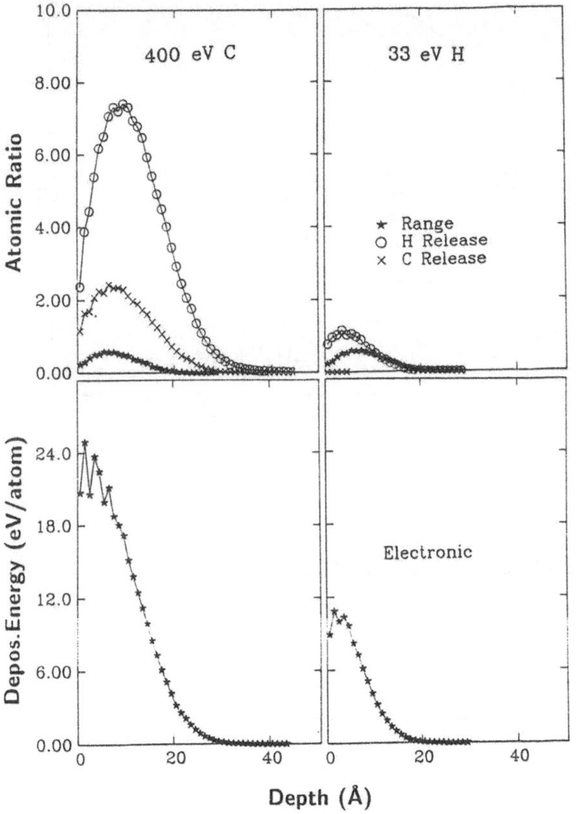

Fig.2. Range distributions, nuclear displacement distributions and electronic energy loss distributions in $CH_{0.4}$ for the atomic constituents of a 500 eV CH_3^+ ion. The nuclear displacements include secondary cascade atoms. For the displacement thresholds, 25 eV has been assumed for carbon and 2.7 eV for hydrogen.

identical, the carbon nuclear stopping is much larger than the hydrogen one, with only a relatively small fraction of energy dissipated into the interaction with hydrogen atoms of the layer. For incident hydrogen atoms, electronic and nuclear stopping are of similar magnitude over a large range of energies. Below 100 eV, nearly identical amounts of energy are dissipated into collisions with hydrogen or carbon atoms.

Fig.2 shows the depth distributions of the ion ranges, their electronic energy deposition, and nuclear release events (i.e. collisions with energy transfers above the displacement threshold) for the atomic components of a 500 eV CH_3^+ ion. Whereas the electronic energy deposition due to incident hydrogen (3 atoms!) is larger than the carbon one, nuclear release of hydrogen is mainly caused by carbon ions. We also note that, due to the different displacement thresholds, the number of displaced carbon atoms is small compared to the number of released hydrogen atoms. Thus, the release of hydrogen by the displacement of adjacent carbon atoms can be neglected for sufficiently low H:C ratios.

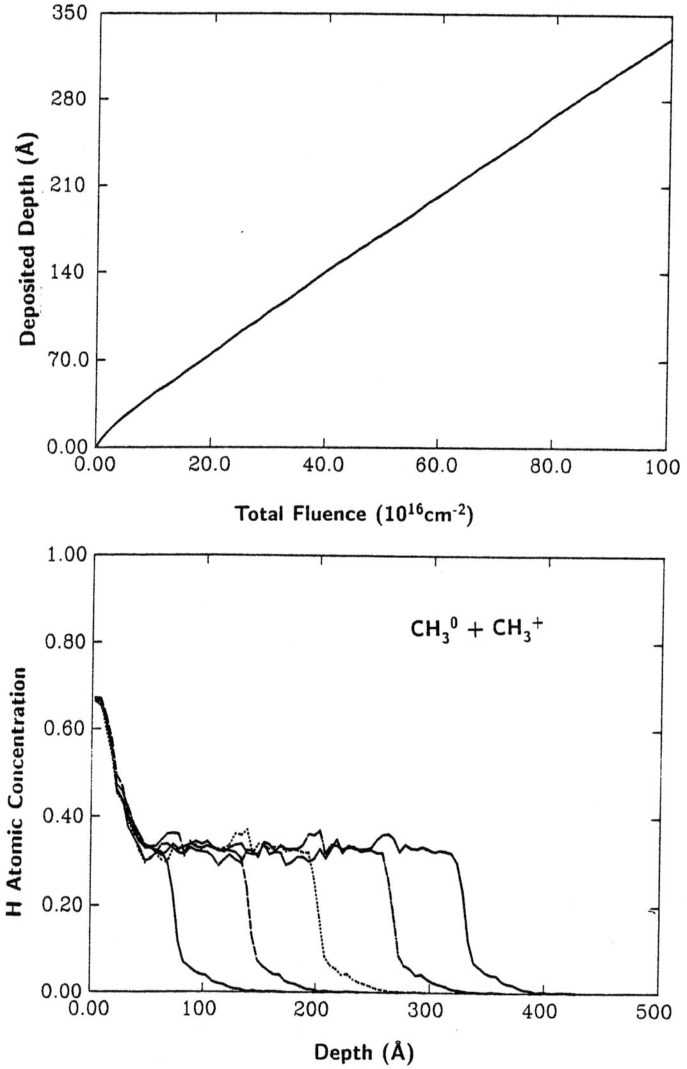

Fig.3. (Top) Thickness of a plasma-deposited hydrocarbon film on an initially plain Si surface as function of the total atomic fluence (i.e. the total number of hydrogen and carbon atoms and ions incident on the growing film). (Bottom) Hydrogen depth profiles in the growing film for fluences of 2,4,6,8, and $10 \cdot 10^{17}$ cm^{-2}. 10% ions relative to the neutral radicals are assumed. The ion energy is 400 eV.

RESULTS OF THE CALCULATIONS

The computer simulations allow very detailed statistics on the individual phenomena occuring during the layer growth. In the evaluations to be shown below, we restrict ourselves to the analysis of composition profiles during the layer growth, and to an evaluation of the reemitted fluxes. Fig.3 demonstrates the ability of the code to simulate the time-dependent growth of a hydrocarbon layer. For the incident species, CH_3^0 radicals plus 10% CH_3^+ ions have been assumed. Silicon has been chosen as the substrate material for all simulations described below. Lacking better knowledge, an equipartition between hydrogen release by nuclear knockon and by electronic energy deposition (i.e., $f_e = f_n = 0.5$) has been chosen. As expected in view of the ion ranges in the order of 10 Å, we observe a constant growth rate except for an initial transient. The hydrogen profiles of Fig.3 show an atomic concentration of about 32 at% corresponding to an H:C ratio of 0.47. Further, a considerable enrichment of hydrogen close to the surface is predicted. A tail is found extending into the Si substrate which cannot be explained by ion implantation only (see Fig.2) but rather by multiple relocation and mixing effects during the initial phase of the layer growth. The intimate mixing at the film/substrate interface is further demonstrated for the final depth profiles of all components given in Fig.4.

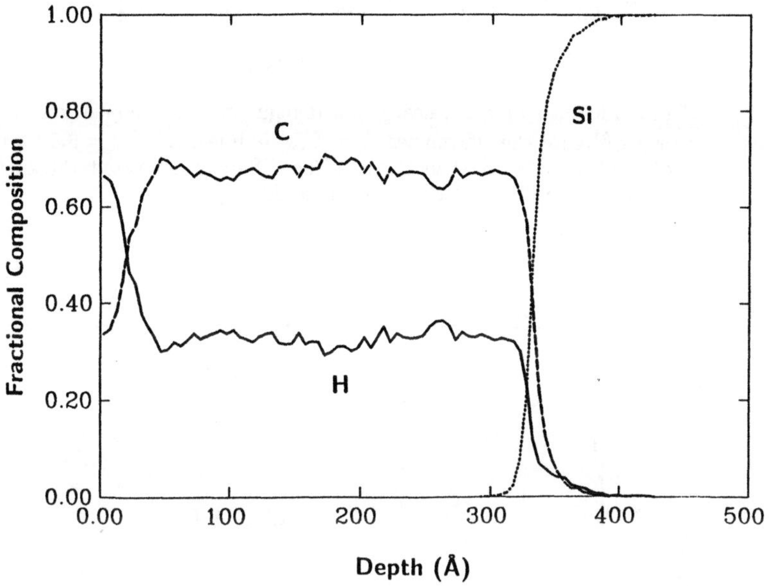

Fig.4. Depth profiles of the individual atomic constituents for the final fluence of Fig.3.

Fig.5 compares the contribution of nuclear and electronic effects to the hydrogen release. Apparently, nuclear release is much more efficient than electronic energy deposition. (This statement, however, does not hold for the bombardment with hydrogen rather than hydrocarbon ions, see below.) It should also be noted that the nuclear release depends rather critically on the displacement energy. Thus, it is always possible to adjust a desired bulk concentration of hydrogen by varying f_n, f_e, or E_d^H in reasonable ranges. However, in view of all the uncertainties mentioned above we proceed with our initial choice of parameters, i.e. $E_d^H = 2.5eV$, and $f_n = f_e = 0.5$.

Fig.6 demonstrates that the reemission of hydrogen by sputtering is in the order of the directly reflected fraction and negligible compared to the ion-induced release. Also, the flux of sputtered carbon atoms is small (less than 10^{-2} compared to the incident flux).

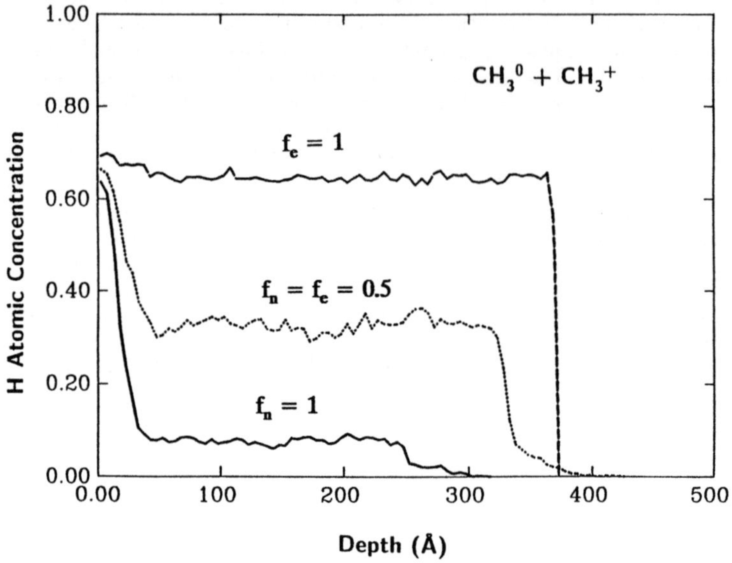

Fig.5. Influence of the release mechanism on the composition of the growing film for the efficiencies $f_n = 1$ (solid line), $f_n = f_e = 0.5$ (dotted line), and $f_e = 1$ (dashed line). (The different deposited thicknesses result from the different densities of the films.)

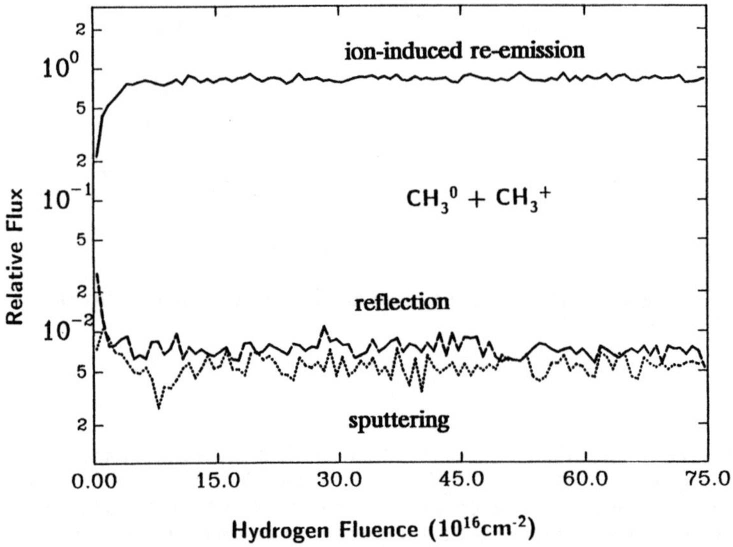

Fig.6. Outward fluxes of hydrogen relative to the incident hydrogen flux by ion-induced reemission (solid line), sputtering (dotted line), and reflection (dashed line). Here, the fluence of incident hydrogen atoms is plotted on the abscissa.

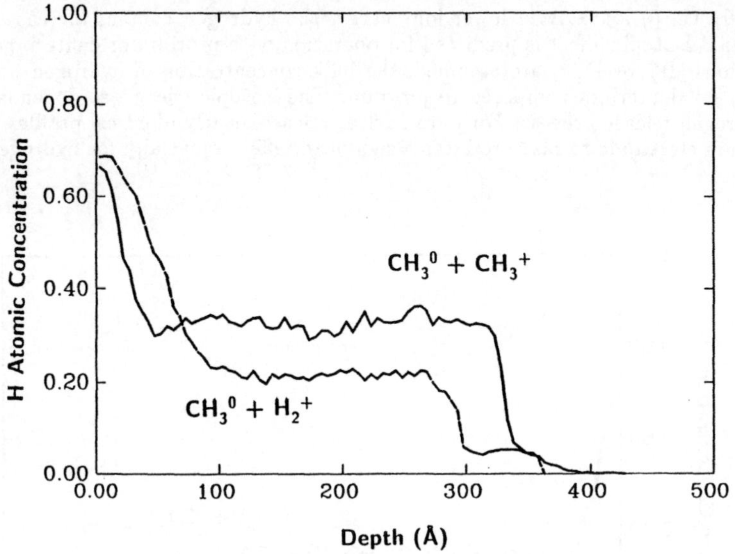

Fig.7. Hydrocarbon films deposited under ion bombardment with two different ionic species, both at 400 eV. As above, 10% ions have been assumed.

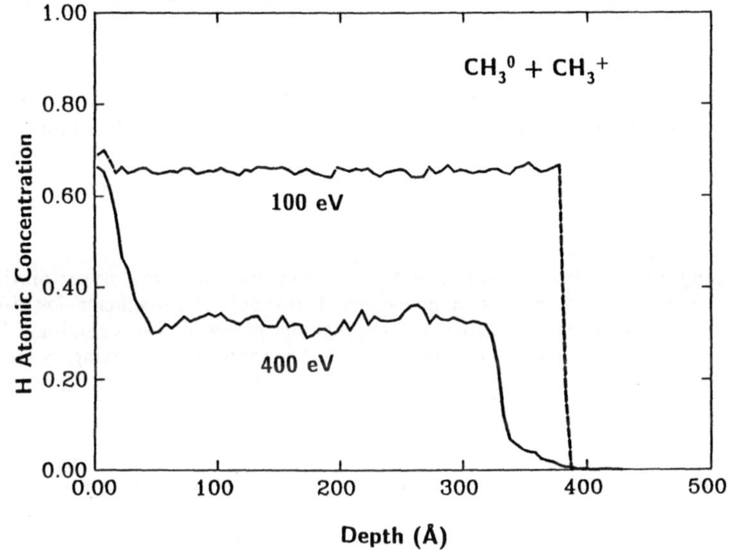

Fig.8. Hydrogen concentration resulting from plasma deposition at two different ion energies.

For Fig.7, the two different species of impinging ions have been chosen. For H_2^+ in comparison to CH_3^+ ions, the resulting hydrogen concentration is lower, but still in a reasonable range. Due to the larger ion ranges, the enriched zone at the surface becomes broader, as does the affected depth in the substrate.

Figs.8 and 9 show the influence of the ion energy and the isotopic species of hydrogen in the plasma, respectively, on the composition of the growing film, again for the case of CH_3^+

(respectively CD_3^+) ions. At a lower ion energy, the hydrogen content increases remarkably. No significant isotopic effect is predicted for operation with protium or deuterium. When only hydrogen ions (H_2^+ or D_2^+) are assumed, the bulk concentration of hydrogen becomes about 50% larger for deuterium compared to protium. The isotopic effect also depends on the type of the hydrogen release process: For pure nuclear release, nearly identical profiles are obtained, whereas pure electronic release produces a significant effect especially for hydrogen ions.

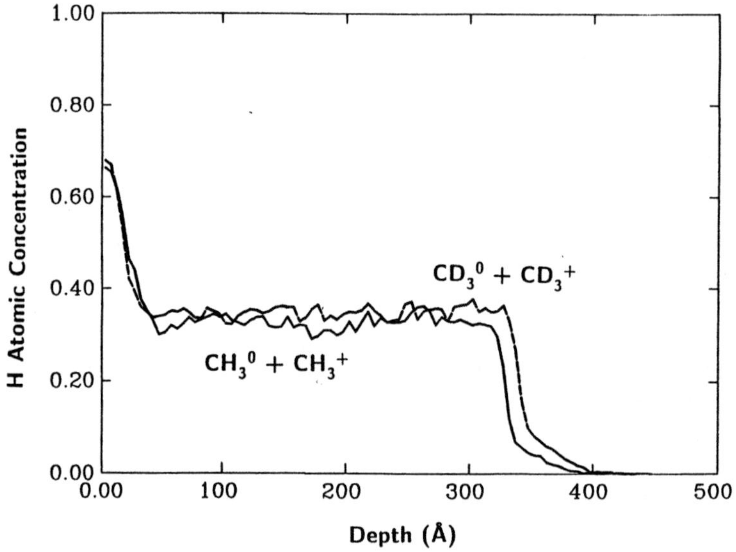

Fig.9. Hydrogen concentration resulting from plasma deposition with standard methane (solid line) or deuterated methane (dashed line).

DISCUSSION

The computer simulations described above are subject to several uncertainties connected with their input parameters, such as displacement thresholds or the efficiencies of hydrogen release, and with the boundary conditions, i.e. the incident fluxes of atoms and ions. Nevertheless, by assuming simplified but reasonable conditions rather than by pure fitting, reasonable results have been obtained with compositions of the resulting C:H films which agree quantitatively with typical compositions known from experiment.

In addition, information can be derived on the transients during initial film growth, and the atomic mixing at the film/substrate interface. For the latter we find an intermixed depth of about 50 Å, in agreement with experimental observations.[55] The collisional intermixing might be a prerequisite for chemical reactions forming carbidic phases at the interface.[55–57] In this way, it may also influence the adhesion of the films.

For lower hydrogen bulk concentrations, a strong hydrogen enrichment at the surface is predicted, which is due to the deposition of the main hydrogen-carrying species at the very surface. A similar, but less pronounced hydrogen enrichment is indeed observed in high-resolution nuclear resonance depth profiling experiments with a-C:H.[58] The width of the enriched zone increases with the ion range according to the simulations. Therefore, the enrichment predicted by the simulation would probably be reduced for a more realistic broad energy distribution of ions instead of the idealized monoenergetic beam.

The dependence of the resulting film composition on the ion energy is also in qualitative agreement with experimental experience. In our simulations, a polymer-like composition results for a sheath potential of 100 V, whereas the composition at 400 V is typical for an a-C:H layer.

Recently, a detailed experimental study has been performed[59] with isotopic mixtures using standard methane, deuterated methane, protium, and deuterium as process gases. For plasmas in pure methane, CD_4 produces films with a slightly higher hydrogen concentration than CH_4. Under the assumption that the plasma conditions and boundary fluxes are identical, this isotopic effect is in agreement with the predictions from the simulation. The simulation results show a small isotopic effect for CH_3^+ ions compared to a significant one for H_2^+ ions, and a small one for nuclear interaction compared to a significant one for electronic interaction. However, the complexity of the problem prevents from drawing any conclusions on the electronic or nuclear release efficiencies, or the species of the impinging ions. In the same set of experiments, the total hydrogen concentration of the resulting films was found to remain approximately constant over a large range of hydrogen dilution (up to 9:1) of the methane process gas at constant total pressure. This seems to be in contradiction with the above model, as increased hydrogen ion bombardment at a decreasing flux of radicals should favour the ion-induced release and result in lower hydrogen concentration in the films. (This effect has been confirmed in corresponding simulation runs.) However, the absolute fluxes of hydrogen ions from such plasmas are not known. Also, it cannot be excluded that neutral atomic hydrogen would be adsorbed on the surface of the growing film and contribute to the bulk concentration.

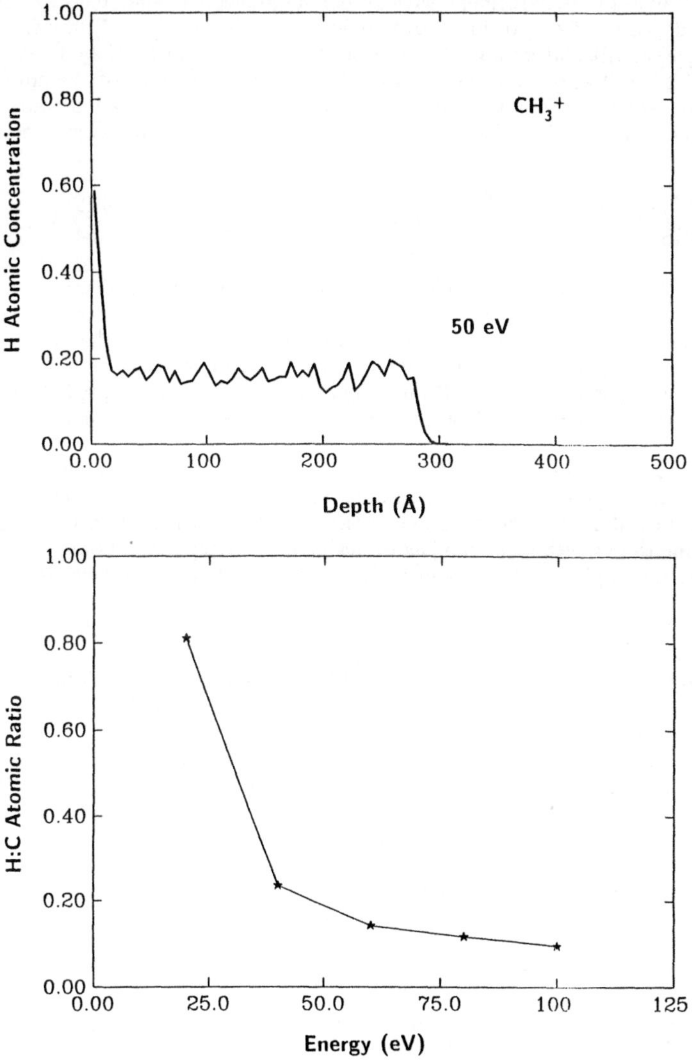

Fig.10. Hydrogen profile resulting from ion beam deposition on Si (top), and dependence of the bulk hydrogen concentration on the ion energy (bottom).

With the above computer model, one should also be able to reproduce experimental data on ion-induced release and isotopic mixing in a-C:H or a-C:D layers during keV deuterium or protium bombardment, respectively.[60] Simulation runs with the above parameters, however, predict a depletion which is much too rapid in comparison with experiment. This discrepancy might be due to the neglect of retrapping in the present model. Whereas during layer deposition a near-surface depth interval of only a few nm is affected by ion bombardment, ions of several keV cover a depth range of several 100 nm so that retrapping might become more important. This is consistent with the essential role of retrapping in the modeling of C:H layer formation by keV hydrogen implantation into graphite.[31]

Finally, in order test the consistency of the model predictions with other than plasma processes of C:H layer growth, some simulation runs were performed for hydrocarbon ion beam deposition. As a model ion, CH_3^+ was chosen at energies between 20 and 100 eV (see Fig.10). Compared to plasma deposition, films with a low hydrogen concentration, which is indicative for a-C:H rather than polymerlike material, can be produced at much lower energies. The hydrogen concentration of the resulting films varies only slowly as function of the energy above about 50 eV, but increases steeply towards a polymeric concentration at lower energies. This prediction is in qualitative agreement with recent photoelectron spectroscopy investigations[61] of C:H layers deposited from an ion source operated with methane. However, the transition from hard to polymerlike layers was observed at about 110 eV rather than 30 eV in the present calculations. This difference may be ascribed to uncertainties both of the simulation and of the experiment. For the latter, the substrate had been exposed to the ion source without any separation of neutrals, so that also neutral radicals might contribute as in the case of plasma deposition.

CONCLUSIONS

A model has been described which predicts the composition of growing hydrogenated carbon films on the basis of collisional effects including ion-induced release of hydrogen atoms. Solutions are obtained by means of dynamic computer simulations of the binary collision type with simplified assumptions on the precursor neutral and ionic fluxes as boundary conditions. For both plasma and ion beam deposition of hydrocarbon films, some predictions of the model are in qualitative and partly quantitative agreement with experimental experience.

At present, the poor experimental knowledge of the species impinging on the growing film, and their fluxes and energy distributions are prohibitive for any detailed critical tests of the model. In order to learn more about the mechanisms, including the surface processes which determine the growth rate, one has to design special experiments providing quantitative information on the impinging fluxes and surface concentrations. Such work is in preparation.

REFERENCES

1. J.C.Angus, P.Koidl, and S.Domitz in: "Plasma Deposited Thin Films", J.Mort and F.Jansen, eds., CRC Press, Boca Rayton (1985)
2. P.Couderc and Y.Catherine, Thin Sol.Films 146:93 (1987)
3. H.C.Tsai and D.B.Bogy, J.Vac.Sci.Techn. A5:3287 (1987)
4. C.Weissmantel, E.Ackermann, K.Bewilogua, G.Hecht, H.Kupfer, and B.Rau, J.Vac.Sci.Techn. A4:2892 (1986)
5. C.V.Deshpandey and R.F.Bunshah, J.Vac.Sci.Technol. A7:2294 (1989)
6. J.Winter, J.Nucl.Mater. 145-147:131 (1987)
7. C.Weissmantel, J.Vac.Sci.Technol. 18:179 (1981)
8. D.Ugolini, J.Eitle, P.Oelhafen, and M.Wittmer, Appl.Phys. A48:549 (1989)
9. W.Möller, J.Nucl.Mater. 162-164:138 (1989)
10. E.G.Spencer, P.H.Schmidt, D.C.Joy, and F.J.Sansalone, Appl.Phys.Lett. 29:118 (1976)
11. Y.Lifshits, S.R.Kasi, and J.W.Rabalais, Phys.Rev.Lett. 62:1290 (1989)
12. M.Frenklach and K.E.Spear, J.Mater.Res. 3:133 (1988)
13. Y.Catherine and P.Couderc, Thin Sol.Films 144:265 (1986)
14. G.L.Vandentop, M.Kawasaki, R.M.Nix, I.G.Brown, M.Salmeron, and G.A.Somorjai, Phys.Rev. B41:3200 (1990)

15. K.Tachibana, M.Nishida, H.Harima, and Y.Urano, J.Phys.D 17(1984)1727
16. L.E.Kline, W.D.Partlow, and W.E.Bies, J.Appl.Phys. 65(1989)70
17. G.Smolinsky and M.J.Vasile, Int.J.Mass Spectr.Ion Phys. 16(1975)137
18. M.J.Vasile and G.Smolinsky, Int.J.Mass Spectr.Ion Phys. 18(1975)179
19. H.Toyoda, H.Kojima, and H.Sugai, Appl.Phys.Lett. 84(1979)157
20. A.Bubenzer, B.Dischler, G.Brandt, and P.Koidl, J.Appl.Phys. 54:4590 (1983)
21. M.A.Tamor, J.A.Haire, C.H.Wu, and K.C.Hass, Appl.Phys. Lett. 54:123 (1989)
22. J.B.Pethica, P.Koidl, J.Gobrecht, and C.Schüler
J.Vac.Sci.Technol. A3:2391 (1985)
23. J.W.Zou, K.Reichelt, K.Schmidt, and B.Dischler, J.Appl.Phys. 65:3914 (1989)
24. X.Jiang, K.Reichelt, and B.Stritzker, J.Appl.Phys. 66:5805 (1989)
25. W.J.Varhue, K.A.Pandelisev, and B.S.Shinseki, J.Appl.Phys. 67:3835 (1990)
26. Ch.Wild and P.Koidl, Appl.Phys.Lett. 51:1506 (1987)
27. D.Boutard, W.Möller, and B.M.U.Scherzer, Phys.Rev. B38:2988 (1988)
28. D.Boutard, B.M.U.Scherzer, and W.Möller, J.Appl.Phys. 65:3833 (1989)
29. H.Baumann, T.Rupp, K.Bethge, P.Koidl, and C.Wild in:
"Amorphous Hydrogenated Carbon Films", P.Koidl and P.Oelhafen, eds.,
Les Editions de Physique, Les Ulis (1987)
30. W.Dworschak, R.Kleber, A.Fuchs, B.Scheppat, G.Keller, K.Jung, and H.Ehrhardt
Thin Sol.Films (in press)
31. W.Möller and B.M.U.Scherzer, J.Appl.Phys. 64(1988)4860
32. W.Möller, P.Børgesen, and B.M.U.Scherzer, Nucl.Instrum.Meth. B19/20:826 (1987)
33. W.Möller and B.M.U.Scherzer, Appl.Phys.Lett. 50:1870 (1987)
34. C.Wild and P.Koidl, see ref. 29
35. J.Pillath, J.Winter, and F.Waelbroeck, J.Nucl.Mater. 162-164:1046 (1989)
36. J.P.Boeuf and Ph.Belenguer, in: "Non-Equilibrium Processes in Partially
Ionized Gases", M.Capitelli and J.M.Bardsley, eds., NATO ASI Series
(in press)
37. K.U.Riemann, J.Appl.Phys. 65:999 (1989)
38. H.Bergsåker, S.Nagata, M.Rubel, and B.Emmoth, see ref. 29
39. J.P.Biersack and L.G.Haggmark, Nucl.Instrum.Meth. 174:257 (1980)
40. S.T.Kang, R.Shimizu, and T.Okutani, Jap.J.Appl.Phys. 18:1717 (1979)
41. M.Hautala, Radiat.Eff. 51:35 (1980)
42. M.L.Roush, T.D.Andreadis, and O.F.Goktepe, Radiat.Eff. 55:119 (1981)
43. W.Möller and W.Eckstein, Nucl.Instrum.Meth. B2:814 (1984)
44. A.Schönborn, N.Hecking, and E.H.teKaat, Nucl.Instrum.Meth. B43:170 (1989)
45. W.Möller, W.Eckstein, and J.P.Biersack, Comp.Phys.Comm. 51:355 (1988)
46. W.Möller in "Materials Modification by High-Fluence Ion Beams",
R.Kelly and M.F.daSilva, eds., Kluwer Academic Publ., London (1989)
47. J.P.Biersack and W.Eckstein, Appl.Phys. A34:73 (1984)
48. W.Eckstein and W.Möller, Nucl.Instrum.Meth. B7/8:727 (1985)
49. W.Möller, D.Bouchier, O.Burat, and V.Stambouli, Surf.Coat.Technol. (in press)
50. W.D.Wilson, L.G.Haggmark, and J.P.Biersack, Phys.Rev. B15:2458 (1977)
51. J.Lindhard and M.Scharff, K.Dan.Vidensk.Selsk.Mat.Fys.Medd. 27 no.15 (1953)
52. O.S.Oen and M.T.Robinson, Nucl.Instrum.Meth. 132:647 (1976)
53. B.T. Kelly: "Physics of Graphite", Applied Science Publishers (1981)
54. D.Bohm in "The Characteristics of Electrical Discharges in Magnetic Fields",
A.Guthrie and R.Wakerling, eds., McGraw-Hill, New York (1949)
55. P.Sander, M.Altebockwinkel, W.Storm, L.Wiedmann, and A.Benninghoven,
J.Vac.Sci.Technol. B7:517 (1989)
56. D.Ugolini, P.Oelhafen, and M.Wittmer, see ref.16
57. K.G.Tschersich,in: "Interfaces between Polymers, Metals, and Ceramics",
B.M.DeKoven, A.J.Gellman, R.Rosenberg, eds., Materials Research Soc.,
Pittsburgh (1989)
58. H.Baumann, priv. commun.
59. D.Boutard and W.Möller, J.Mater.Res. (in press)
60. D.Boutard, W.Möller, and B.M.U.Scherzer, J.Appl.Phys. 67:163 (1990)
61. D.Ugolini, J.Eitle, and P.Oelhafen, Vacuum (in press)

AMORPHOUS, HYDROGENATED CARBON FILMS AND RELATED MATERIALS: PLASMA DEPOSITION AND FILM PROPERTIES

P. Koidl, C. Wild, R. Locher, and R.E. Sah

Fraunhofer-Institut für Angewandte Festkörperphysik,
Tullastrasse 72, D-7800 Freiburg, Federal Republic of Germany

Plasma deposition and properties of a-C:H are reviewed and recent results on the process characterization are reported. The role of process gas, plasma chemistry and plasma surface interaction is discussed. It is shown that the deposition energy of the hydrocarbons is the most critical deposition parameter in determining the film properties. The ion energy distribution in rf-glow discharge deposition has been investigated. For comparison, a-C:H films have been deposited from monoenergetic ion beams. Hard a-C:H films with precursor-independent properties are obtained if the deposition energy is large enough to allow for efficient fragmentation of the energetic hydrocarbons at the film surface. Otherwise, soft polymerlike films are obtained at low deposition energies. As a key for the understanding of the a-C:H properties, hydrogen incorporation, carbon bonding and network structure are discussed. Chemical substitution of hydrogen and carbon in a-C:H are shown to result in related materials with an increased range of physical properties.

I. INTRODUCTION

Todays research in the metastable growth of carbon films is concentrated in two fields: low pressure deposition of diamond and the preparation of amorphous "diamondlike" films [1]. Due to the attraction and outstanding properties of diamond as a rare material, the former of these activities is certainly the more spectacular one. Nevertheless, since the publications of Aisenberg and Chabot [2] and Spencer et al. [3] amorphous carbon films with diamondlike properties have found enormous interest. These noncrystalline films are still of considerable importance since large

Diamond and Diamond-Like Films and Coatings
Edited by R.E. Clausing *et al.*, Plenum Press, New York, 1991

area deposition of homogeneous films with smooth surfaces is easily achieved on any substrate and at low temperatures. Thus many applications as a durable diamondlike coating have been realized which are not (yet) possible with polycrystalline CVD diamond films.

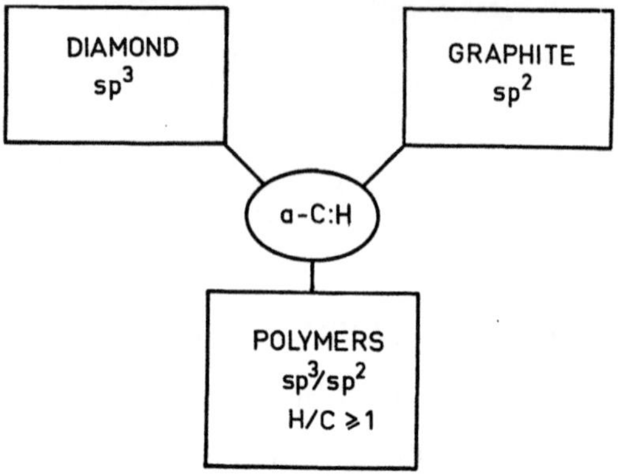

Fig. 1. Limitations of a-C:H properties by the properties of diamond, graphite and hydrocarbon polymers.

Amorphous diamondlike carbon may contain appreciable concentrations of hydrogen. The hydrogen content influences essentially the bonding and network structure of these films. We will concentrate on this class of amorphous hydrogenated carbon (a-C:H) films. In contrast to diamond with its outstanding but nevertheless fixed properties, a-C:H constitutes a class of materials with properties covering a wide range which is limited by the properties of the structurally ordered solids diamond, graphite, and hydrocarbon polymers (Fig. 1).

A number of relevant reviews on amorphous and diamondlike carbon have been published [1, 4-7, 10]. In addition overviews and recent data on the deposition, structure and properties of these materials are collected in a number of conference proceedings and monographs including the present volume [4, 8, 9].

The preparation of a-C:H films with diamondlike properties requires non-thermal-equilibrium growth conditions. Most important is growth under ion bombardment at low substrate temperatures (< 200°C). There are many deposition techniques suitable for the growth of diamondlike a-C:H films [10]. We will concentrate on the condensation of energetic hydrocarbon ions. This can be achieved by direct ion beam deposition [10] or - more easily - by the deposition from rf - excited hydrocarbon glow discharges onto biased substrates [11, 12].

Rf-plasma deposition of a-C:H will be reviewed and recent results on process characterization will be reported. It will be shown how the film properties can be tuned by changing the deposition conditions, most important being the impact energy of the film forming hydrocarbon ions.

Bonding and structure of the a-C:H films and their temperature induced changes are discussed to get information on the structure/property relationships. Finally the substitution of hydrogen in a-C:H by fluorine and the alloying of carbon with germanium will be discussed as a way to realize related materials with an increased range of physical properties.

II. a-C:H PLASMA DEPOSITION

II.1 Deposition of a-C:H in rf glow discharges

For the growth of a-C:H films, several techniques which provide energy deposition to the growing a-C:H film are appropriate. Examples are direct ion beam deposition, dual ion beam sputtering, rf and dc glow discharges [4]. Among these techniques the deposition in a capacitively coupled rf glow discharge system is the most common and the most important one. Capacitively coupled rf glow discharges are of central importance in thin film and surface technology [13]. They are commonly used for processes such as sputtering, reactive ion etching and thin film deposition. One of the key features of rf excited glow discharges is the interaction of energetic ions with solid surfaces. This interaction has a very beneficial effect on the properties of thin films [14]. Before discussing the specific properties of a-C:H plasma deposition the basic features of asymmetric, capacitively coupled rf glow discharges will briefly be described.

Fig. 2 shows a schematic of a conventional capacitively coupled, asymmetric rf glow discharge system [12]. The rf generator is connected with a small electrode via a capacitor. The counter electrode is formed by the grounded walls of the metal chamber. As a result of the asymmetry of the two electrode areas, the small electrode develops a high negative self-bias voltage (U_B) with respect to the large grounded electrode. Therefore the ion bombardment in asymmetric rf discharge systems is confined to the small electrode which is often called cathode. The bias voltage U_B can easily be measured by means of an inductively coupled dc voltmeter.

Fig. 3 shows the spatial distribution of the time-averaged potential, and the temporal variation of the cathode and of the plasma potential in an asymmetric rf discharge system. The time-averaged potential is constant in the plasma and decreases in the sheaths which are formed in the vicinity of the two electrodes. At the cathode (small electrode) the sheath potential V_{sh}, i.e. the time-averaged difference between the plasma potential and the cathode

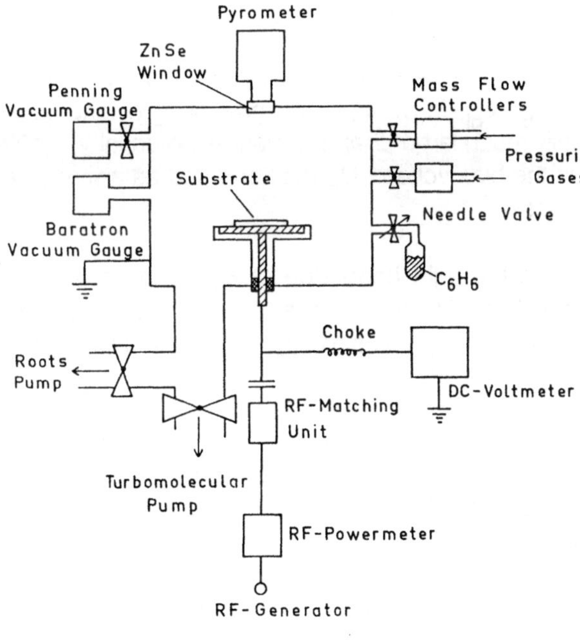

Fig. 2. Rf-plasma deposition system [12].

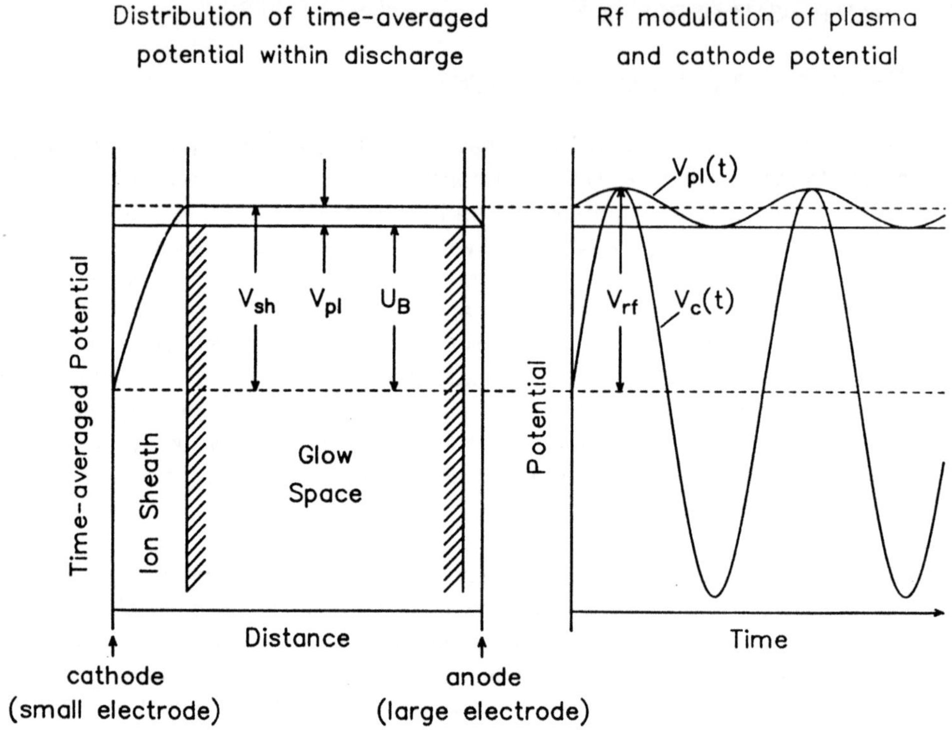

Fig. 3. Spatial and temporal potential distribution in a capacitively coupled rf-discharge system with asymmetric electrode areas [12].

potential, is given as the sum of bias voltage U_B and plasma potential V_{pl}: $V_{sh} = U_B + V_{pl}$. In strongly asymmetric rf discharge systems the bias voltage U_B is usually much larger than the plasma potential V_{pl} (100-1500 V compared to 10-30 V). Thus the bias voltage U_B is often used as a measure for the sheath-potential at the cathode.

On the right hand side of Fig. 3 the cathode potential $V_c(t)$ and the plasma potential $V_{pl}(t)$ are plotted as a function of time. The sheath-potential $V_{sh}(t) = V_{pl}(t) + V_c(t)$ is strongly modulated. During one rf cycle it oscillates between 0 and approximately $2U_B$.

The energy of the ions impinging on the cathode depends primarily on two factors: the time-averaged sheath potential V_{sh} and inelastic collisions between ions and neutrals in the sheath. Hence the ion energies can be controlled macroscopically by adjusting the bias voltage and the pressure. In Sec. III.2 it will be shown, how these and further parameters such as rf frequency, ion mass etc. influence the distribution of ion energies at the cathode.

For the preparation of a-C:H films the rf plasma deposition techniques offers a variety of advantages:

In capacitively coupled rf glow discharges the net current equals zero, i.e. the flux of positive ions on the growing a-C:H film is balanced by an equivalent flux of electrons. This is a prerequisite for the deposition of insulating films or for the deposition on insulating substrates. The latter requires high rf frequencies (> 1 MHz) because at low rf frequencies the charging of the substrate surface during one rf period may lead to a significant reduction of the ion energies.

Rf glow discharges offer the possibility to deposit a-C:H films on very large areas. For example in fusion research [15] several square meters have been coated homogeneously with amorphous carbon. The deposition rates which can be achieved are of the order 0.3 μm/min (see Sec. III.1).

For optical applications the monitoring of the film growth process plays an important role. With rf glow discharges this is possible by measuring the reflectivity of the sample during film growth [16]. As an example Fig. 4 shows the intensity of a reflected HeNe-laser beam as a function of the deposition time of an a-C:H film on germanium substrate. By analyzing the observed interference oscillations, the optical properties (refractive index and absorption coefficient) and the actual thickness of the a-C:H film can be determined on-line, i.e. during the deposition process. The right hand side of Fig. 4 depicts the intensity of the reflected HeNe-laser beam during the removal of the a-C:H film by reactive ion etching in an oxygen discharge. In this case, the intensity of the reflected laser light shows exactly the same interference oscillations. Hence this technique offers a very sensitive control of both, the deposition and etching of a-C:H films.

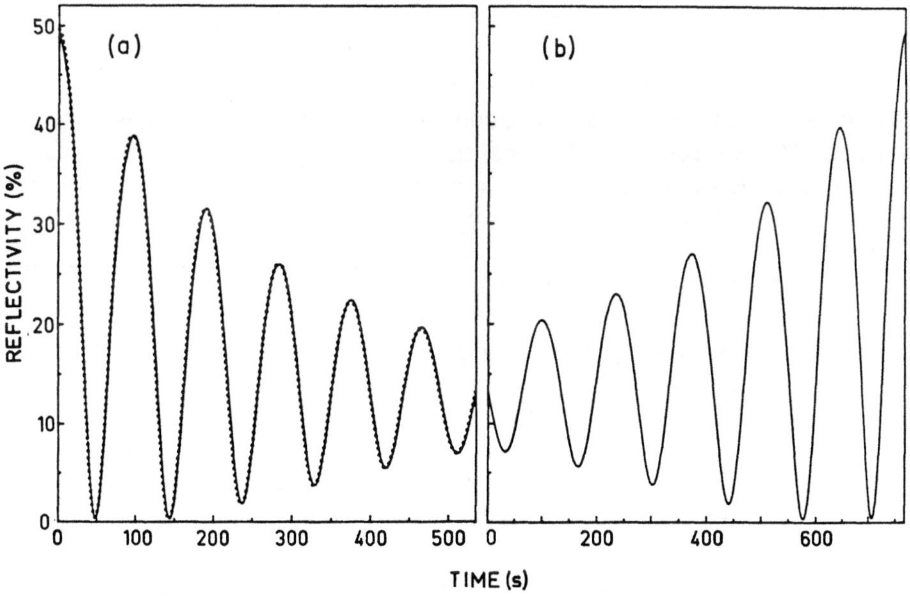

Fig. 4. In situ measured reflectivity at 633 nm of an a-C:H film on Ge (solid line) as a function of (a) deposition time and (b) etching time. Dotted curve represent theoretical fit [16].

Another feature of the rf plasma deposition is the possibility to deposit amorphous alloy films of varying composition. This can easily be done by mixing different process gases. Examples will be given in section IV.3.

II.2 Variation of a-C:H film properties

The chemical, mechanical, optical and electrical properties of a-C:H have been discussed in many recent papers [4, 8]. We will not review in detail the general properties of a-C:H films. We will rather concentrate on the dependence of the materials properties on the deposition conditions to get insight into the structure/property relationships.

By varying process parameters such as bias voltage U_B or pressure P, a-C:H films with different physical and structural properties can be deposited. Especially the bias voltage U_B, which determines the energy of the film forming hydrocarbon ions, was found to have a strong influence on the properties and structure of the growing a-C:H films [17]. In Fig. 5 various film properties such as hydrogen content, optical gap, refractive index, mass density and microhardness, are plotted as a function of the bias voltage U_B. The data were taken from a-C:H films that have been deposited in benzene discharges at 30 μbar pressure. For the determination of the hydrogen content nuclear reaction analysis has been applied [18]. Using the $^{1}H(^{15}N,\alpha\gamma)^{12}C$ reaction, which exhibits a sharp resonance in the cross section at 6.4 MeV, the hydrogen concentration can be determined quantitatively and by varying the energy of the probing ^{15}N ion beam even depth profiling is possible. The optical gap and the refractive index were determined by measuring the optical transmission and reflection spectra of a-C:H films deposited on glass substrates. The analysis of these spectra allows one to evaluate the refractive index and the absorption coefficient α as a function of the photon energy E. Thus the optical gap E_{opt} can be inferred from the Tauc relation $\sqrt{\alpha E} \sim (E-E_{opt})$. Mass density data were obtained by weighing the substrate

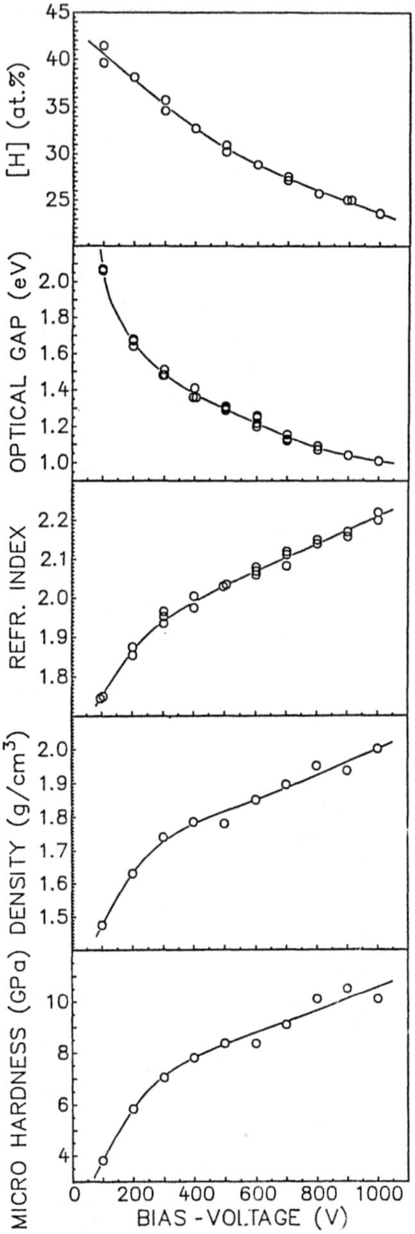

Fig. 5. Properties of a-C:H films deposited from a 13.56 MHz glow discharge of benzene at 30 μbar vs. substrate bias voltage [17].

before and after deposition and the a-C:H film hardness was determined by evaluating indentation depth vs. load curves measured with a nanoindenter .

The variation of the bias voltage U_B from 100 to 1000 V leads to a strong variation of the individual film properties. A detailed analysis of the local bonding and structure of these films, which has been performed using IR absorption spectroscopy [19] and thermal desorption spectroscopy [20, 21], revealed that there is a transition from "polymerlike" to "diamondlike" a-C:H. At low bias voltages (< 100 eV) the films contain a large concentration of hydrogen (> 40 at.%), they are transparent (E_{opt} > 2 eV) and soft. On the other hand, at high bias voltages (> 300 V) the films are dense, hard and strongly cross-linked.

In Sec. III.2 it will be shown that the transition region (100-300 V bias voltage) corresponds to average ion impact energies between 40 and 120 eV. This range of energies compares well with the energy required to dissociate a benzene molecule into small fragments such as C and CH. This finding suggests that the impact induced fragmentation of energetic hydrocarbon ions is an important feature of the a-C:H growth process. This idea will be further supported in Sec. III.1, where we will show that hard a-C:H films with properties independent of the hydrocarbon process gas can be deposited at sufficiently high bias voltages.

III. PROCESS CHARACTERIZATION OF a-C:H PLASMA DEPOSITION

III.1 The role of process gas, plasma chemistry and plasma-surface interaction

In order to elucidate the mechanisms leading to the growth of a-C:H films and having influence on their properties, the rf plasma deposition process has been investigated using mass spectroscopy and spatially resolved optical emission spectroscopy. The experimental setup has been described elsewhere [16, 22]. To get insight into the plasma-chemical processes occuring in hydrocarbon rf discharges, the mass distribution of positive ions in the plasma has been measured by means of a differentially pumped quadrupole mass-spectrometer (QMA). The QMA was opterated without electron beam ionizer thus detecting directly the positive ions in the discharge. By comparing different hydrocarbon discharges it was found that the mass spectra are specific for the type of hydrocarbon process gas and that they are strongly influenced by plasma-chemical reactions such as dissociation, ion-molecule reactions and polymerization in the gas phase [16, 23].

To clarify the importance of these plasma-chemical processes, a-C:H films have been deposited using nine different types of hydrocarbon gases. Thereby all other process parameters were kept constant (U_B=500 V, P=30 μbar). Subsequently, the thickness and the properties of these a-C:H films were analyzed using optical absorption spectrosopy in the visible and IR and by nuclear reaction analysis (NRA). The growth rate was found to depend strongly on the type of process gas. In Fig. 6 the deposition rate for different hydrocarbon gases is plotted as a function of the ionization energy. Going from methane to benzene, the deposition rate increases by more than one order of magnitude. Only part of this increase is due to the higher mass of the benzene ions. If we take into account the average mass of the ions in the plasma, there remains a significant increase of the deposition flux [23]. A reasonable explanation for this finding may be the increased ionization rate in benzene discharges due to the low ionization energy of the benzene molecules.

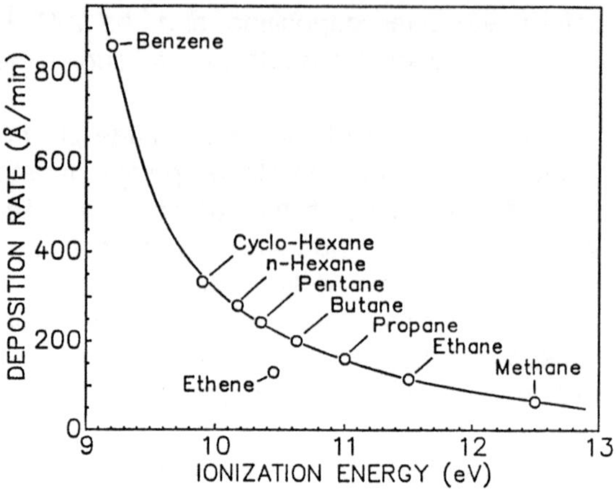

Fig. 6. Deposition rate of a-C:H vs. ionization energy of hydrocarbon precursor molecule; different gases were used at 30 μbar pressure and 400 V bias voltage.

In spite of the strong dependence of the deposition rate on the type of hydrocarbon gas, the properties of hard a-C:H films were found to be precursor-independent. In Table I several film properties, such as refractive index, optical gap, hydrogen content, and the ratio between sp^2 and sp^3 hybridized carbon atoms are given for a-C:H films that have been deposited using benzene, cyclohexane, n-hexane and methane as process gas. The variation of these film properties is negligibly small compared to the variation that can be achieved by varying the bias voltage (see Fig. 5). Thus we conclude that the properties of hard a-C:H films do not depend on the type of hydrocarbon gas used for the deposition. The structural and chemical properties of the precursor molecules (carbon hybridization, carbon to hydrogen ratio) get completely lost. Since the mass distribution of positive ions depends critically on the type of hydrocarbon, this loss of structure must occur at the plasma-substrate interface.

Table I. a-C:H film properties for a-C:H films deposited with different hydrocarbon process gases ($U_B = 400$ V, $P = 30$ μbar)

Process Gas	n	E_{opt}(eV)	[H] (at.%)	sp^3:sp^2
Benzene	2.00	1.39	31.9	71:29
Cyclo-Hexane	2.02	1.44	31.0	72:28
n-Hexane	2.04	1.41	31.7	72:28
Methane	2.08	1.36	30.3	72:28

In order to investigate the interaction between the hydrocarbon plasma and the growing a-C:H films, spatially resolved optical emission spectroscopy has been applied [22, 23]. Optical emission spectra were taken from different hydrocarbon discharges used for the deposition of a-C:H films. Thereby it was found that small fragments such as C or CH contribute strongly to the emission spectra. The emission intensity of these small fragments exhibits a pronounced maximum in the vicinity of the cathode irrespective of the type of hydrocarbon. To clarify the origin of the excited fragments, time-resolved optical emission experiments were performed [23], showing that the fragments are mainly due to the impact-induced dissociation of energetic hydrocarbon ions.

The emission intensity of the small fragments in the vicinity of the cathode was found to depend strongly on the bias voltage U_B. At low bias voltages ($U_B < 100$ V) no or only very weak emission from C or CH could be detected [17].

This finding indicates, that the fragmentation probability is very low for $U_B < 100$ V. The transition from "polymerlike" to "diamondlike" a-C:H films in Fig. 5 thus appears to be a result of the energy dependent fragmentation probability. Consequently, one would expect that the properties of "polymerlike" a-C:H films deposited at low bias voltages should depend on the type of hydrocarbon gas. This is exactly what we found. Fig. 7 shows the IR absorption bands from C-H stretch vibrations for a-C:H films that have been deposited at different bias voltages and with different hydrocarbon gases [23]. By analyzing these absorption bands, information about the hybridization of the carbon atoms and about the local hydrogen bonding can be obtained [19]. Fig. 7b depicts the C-H stretch absorption bands of two a-C:H films that have been deposited at $U_B = 400$ V using benzene and n-hexane as process gas, respectively. Apparently the absorption bands are almost identical. In contrast, at 100 V bias voltage (Fig. 7c) large differences can be observed. A detailed analysis of the absorption revealed that these low bias films do contain structural units of the precursor molecule.

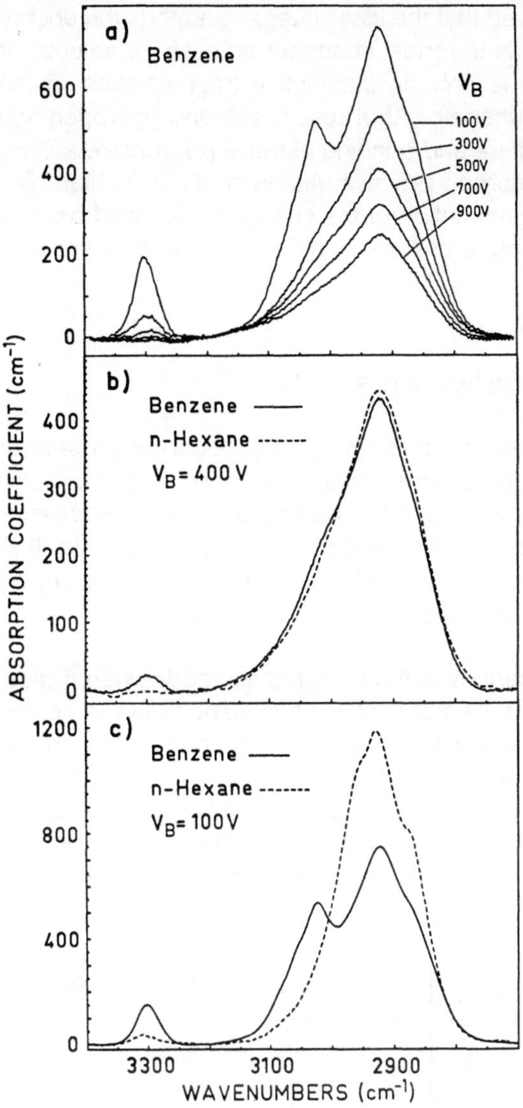

Fig. 7. CH stretch vibrational absorption spectra of various a-C:H films deposited at 30 μbar pressure from
(a) benzene at bias voltages as indicated,
(b) benzene and hexane at 400 V bias,
(c) benzene and n-hexane at 100 V bias [23].

The results described so far showed that the bias voltage - and thus the energy of the film forming hydrocarbon ions - is the most important process parameter. At low bias voltages (U_B < 100 V) there is only an incomplete fragmentation of the hydrocarbon ions impinging on the substrate. As a result, soft and hydrogen-rich a-C:H films are formed. The local structure and bonding of these polymerlike a-C:H films do depend on the type of hydrocarbon process gas. In contrast, at high bias voltages (U_B > 300 V) the efficient fragmentation of the energetic hydrocarbon ions leads to the formation of hard, strongly cross-linked a-C:H films with precursor-independent properties.

III.2 Distribution of ion energies in rf glow discharges

The relation between ion energies and bias voltage U_B depends on several parameters such as pressure, type of process gas, rf frequency and reactor geometry. For a fundamental understanding of the rf plasma deposition process, a direct correlation between ion energies and film properties appears desirable. It is therefore appropriate to investigate the energy distribution of the ions impinging on the cathode of an asymmetric, capacitively coupled rf glow discharge system.

Since the measurement of ion energies at the powered electrode of an rf glow discharge system is very difficult, we modified our system according to Fig. 8, i.e. we used an inverted electrode geometry where the rf generator is connected with the large electrode (anode) whereas the small electrode (cathode) is grounded. This experimental setup takes advantage of the fact that the distribution of the electric field within the discharge chamber does not depend on wether the rf signal is connected with the large or with the small electrode. Hence the distribution of ion energies at the cathode is the same in both cases.

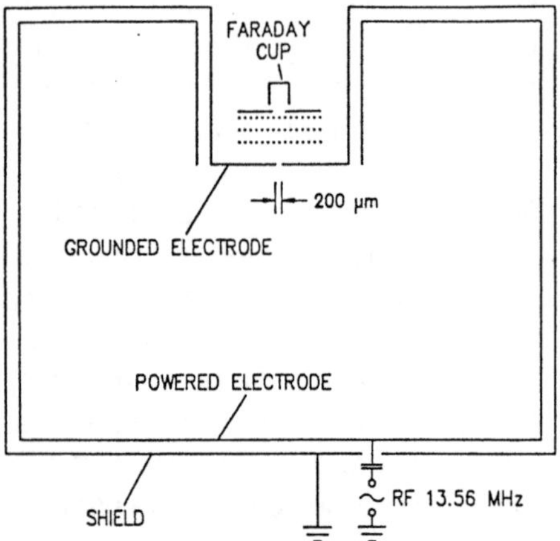

Fig. 8. Experimental setup to measure the ion energy distribution at the (grounded) cathode. Rf power is capacitively connected to the larger electrode [24].

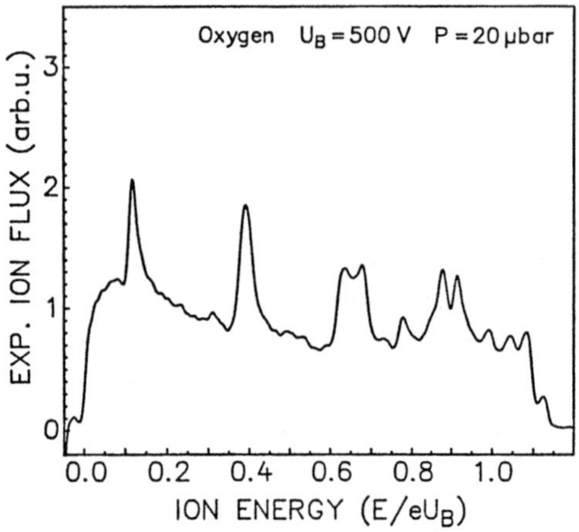

Fig. 9. Typical ion energy distribution measured in an oxygen discharge at 500 V bias and 20 μbar pressure.

Our first experiments were carried out with argon and oxygen discharges [24]. Fig. 9 shows a typical ion energy distribution (IED) measured in an oxygen discharge at 500 V bias voltage and 20 μbar pressure. The ion energy in Fig. 9 has been normalized with respect to the bias voltage U_B. It turned out that the IEDs at the cathode of capacitively coupled rf glow discharges exhibit pronounced features, namely a characteristic series of peaks and double-peaks. These peaks have been observed with various process gases (argon, oxygen, hydrogen, benzene), in a large range of bias voltages (100-800 V) and pressures (3-130 μbar). Hence it can be concluded that these peaks are an inherent property of capacitively coupled rf glow discharges. Fig. 10 shows a variety of IEDs of argon and oxygen discharges measured at different bias voltages and pressures. Apparently the position of the individual peaks depends clearly on process parameters such as bias voltage or pressure. The question arises: what is the origin of these peaks and what can we learn about the physics of rf glow discharges by analyzing these structures ?

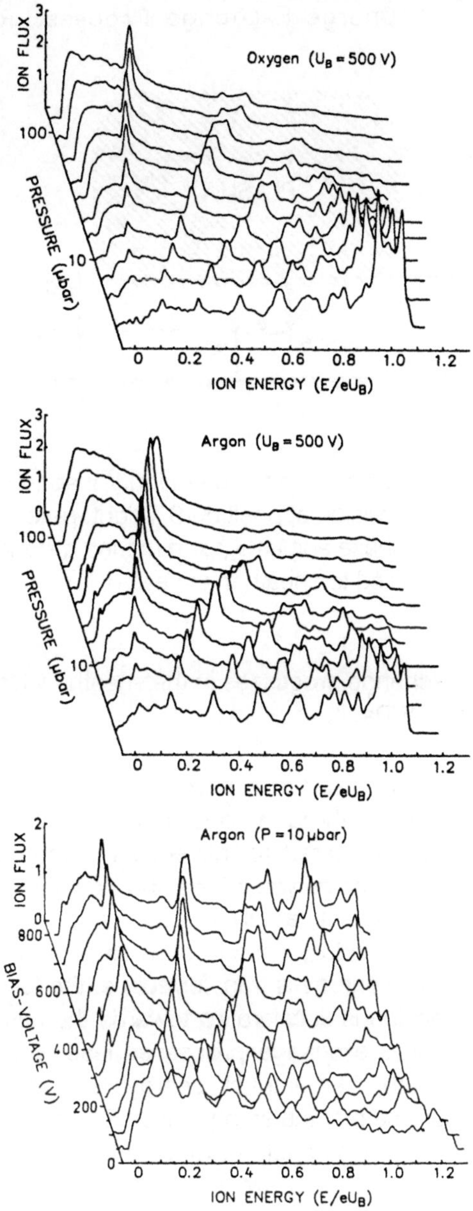

Fig. 10. Ion energy distributions measured in argon and oxygen discharges at various pressures and bias voltages.

Two main factors are known to influence the distribution of ion energies in rf glow discharges: the rf modulation of the sheath potential and the interaction between ions and neutrals in the sheath. In the case of collision-free rf sheaths the rf modulation of the sheath potential leads to saddle-shaped IEDs centered about the time-averaged sheath potential V_{sh} [25]. In the case of collisional dc sheaths, on the other hand, it is known that inelastic collisions between ions and neutrals, especially charge exchange processes, lead to broad and featureless IEDs [26]. To date, it has been general opinion that in the technologically important case of collisional rf sheaths, collisions would have the same effect, namely a featureless broadening of the IED towards lower energies. However, as we will show in the following, it is the combination of both effects - the rf modulation of the sheath potential and charge

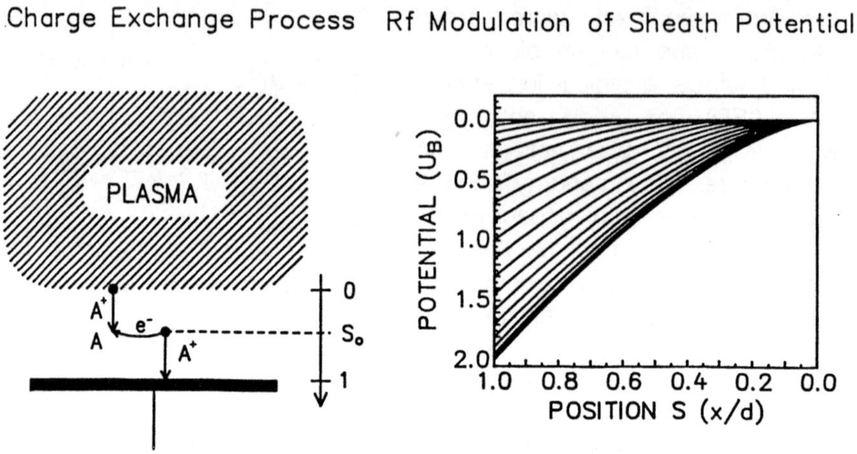

Fig. 11. Illustration of ion transport through rf sheaths; left: charge exchange collision at position S_o, right: potential distribution in the sheath for various rf phase angles φ_o; x/d means distance from plasma/sheath boundary normalized to sheath thickness d.

exchange processes in the sheath - which is responsible for the observed features in the IEDs.

In order to explain and describe the experimental IEDs, a model for the ion transport through collisional rf sheaths has been developed [27]. Two main effects are considered to influence the motion of ions in the sheath: charge exchange processes between ions and neutrals and the rf modulation of the sheath potential. These effects are illustrated in Fig. 11. During a charge exchange collision an ion is neutralized by capturing an electron from the neutral collision partner. Thereby no kinetic energy is transferred, i.e. the former ion does not change its velocity and the new ion is accelerated towards the cathode with a thermal starting velocity. Hence charge exchange processes lead to the creation of thermal ions throughout the sheath. The impact energy E_{imp} of an ion at the cathode depends on two parameters: the start position S_o and the start rf phase angle φ_o, i.e. the position and the rf phase angle at which the charge exchange process took place. A peak in the IED is expected when the first derivatives of E_{imp} with respect to S_o and φ_o vanish: $dE_{imp}/dS_o = dE_{imp}/d\varphi_o = 0$.

In order to quantify this model four main assumptions were made, which are justified by the present experimental condition:

1) The ion density $n_i(x)$ is constant in time whereas the electrons exhibit an infinite mobility.

This assumption is justified since the rf frequency ω (13.56 MHz) is high compared with the ion plasma frequency ω_{p+} but low compared with the electron plasma frequency ω_{p-}. Hence, it is reasonable to assume that the ion density is determined by time-averaged potentials in the discharge and that the electrons react instantaneously to the oscillating electric field.

2) The spatial distribution of the ion density $n_i(x)$ in the sheath is proportional to $n_i \sim x^{\nu-1}$, where x is the distance from the plasma-sheath boundary.

In the case of dc sheaths the Child-Langmuir equation can be applied to describe the spatial variation of the electric field E(x). Neglecting the initial velocity of an ion entering the sheath, the electric field is proportional to $E(x) \sim x^{\nu}$, where the exponent ν is a fixed constant equal to 1/3 in the case of space charge limited ion currents (collision-free sheaths) and 1/2 in the case of mobility limited ion currents (collisional sheaths) [13]. This implies that the ion density $n_i(x)$ is proportional to $n_i(x) \sim x^{\nu-1}$. Since the ion density is determined by time-averaged potentials (assumption 1) the same relation is assumed for rf sheaths as well.

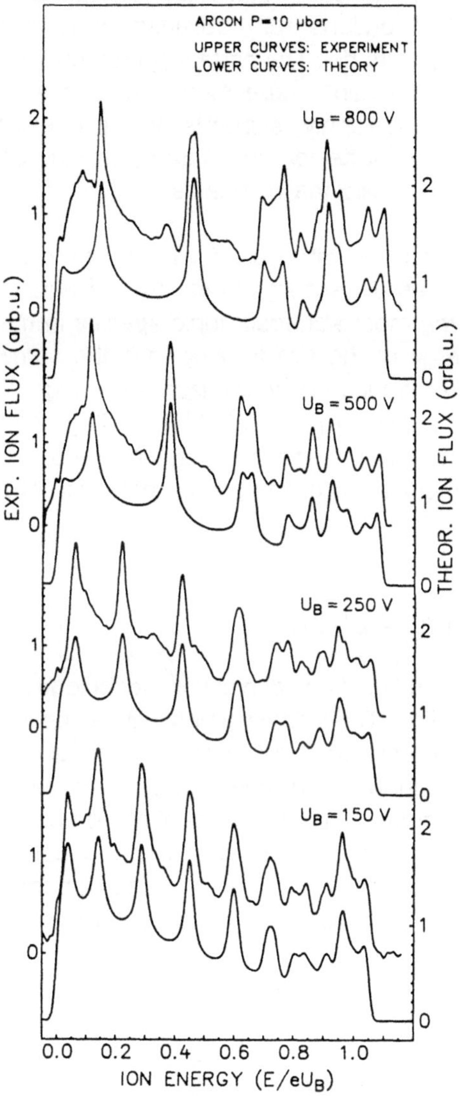

Fig. 12. Comparison between experimental and theoretical ion energy distributions for an argon discharge at different bias voltages.

The third and the fourth assumption are:

3) The sheath potential V_{sh} is described by $V_{sh}(t) = U_B(\cos(\omega t)-1)$.
4) Charge exchange is the dominant interaction process between ions and neutrals in the sheath.

Assumption 3 is a consequence of the electrode asymmetry and the capacitive coupling and assumption 4 is justified by the large cross-sections for charge exchange processes, especially when the ions drift through a background of neutrals of the same species [28, 13].

With these assumptions it is possible to determine the spatial and temporal distribution of the electric field in the sheath and the periodic variation of the sheath thickness due to the penetration of electrons into the ion sheath [27].

The model contains only two parameters that determine the theoretical positions of peaks in the IED: the exponent ν which describes the spatial variation of the ion density ($n_i(x) \sim x^{\nu-1}$), and a plasma scaling parameter η which is given by the expression $\eta = eU_B/m\omega^2 d^2$, where m is the ion mass, ω the rf frequency and d the thickness of the ion sheath. It should be noted that these two parameters are not really free parameters. The exponent ν can be varied only in a very limited range (0.33-0.5) and the

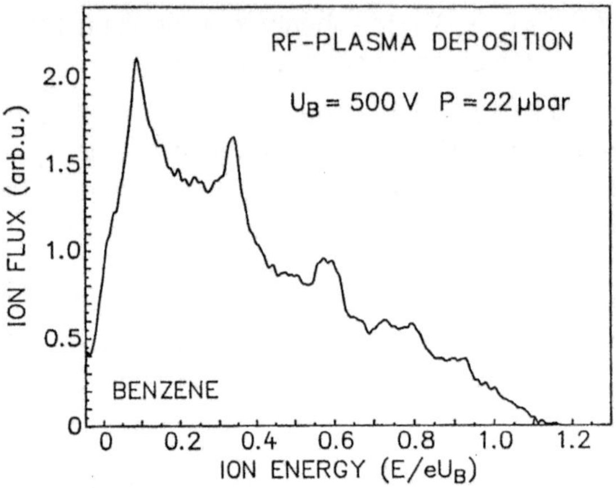

Fig. 13. Ion energy distribution at the cathode of a benzene discharge ($U_B = 500$ V, $P = 22$ μbar) used for a-C:H deposition [17].

parameter η is determined by known (U_B, m, ω) or measureable quantities (d). Hence the model does not exhibit any fitting ambiguity. Nevertheless it turned out that this model describes the experimental data very well. This is shown in Fig. 12 where experimental IED's from argon discharges at different bias voltages are compared to theoretical distributions as obtained with our model. The perfect agreement between experiment and theory strongly supports the proposed model for the ion transport through collisional rf sheaths.

For the deposition of a-C:H films, benzene is often used as process gas. Besides the high deposition rate (Fig. 6) benzene discharges offer the advantage that ionized benzene molecules are by far the most abundant ionic species in the plasma [16]. In order to determine the energy of the film forming ions the same experiment as described above were performed with benzene discharges. Fig. 13 displays a typical IED measured in a benzene discharge at 500 V bias voltage and 22 μbar pressure. The IED clearly exhibits the characteristic features, which are due to charge exchange processes and the rf modulation of the sheath potential. The average ion energy was found to be $0.4 \cdot eU_B$.

III.3 Comparison with direct ion-beam deposited a-C:H films

So far we could conclude that the rf plasma deposition technique is conceptually similar to direct ion-beam deposition. The plasma simply acts as an ion source for hydrocarbon ions, the ions are accelerated in the sheath and at the cathode there is a flux of energetic hydrocarbon ions. However, there is one basic difference between these two deposition techniques: in the case of direct ion-beam deposition a-C:H films are formed by a monoenergetic beam of hydrocarbon ions, whereas in the case of rf plasma deposition the energy of the hydrocarbon ions covers a broad energy range, namely from zero to about eU_B. The question is: does that make any difference ?

To clarify the influence of different IEDs on the properties of a-C:H films we deposited a-C:H films by direct ion-beam deposition using a broad-beam (2.5 cm diam.) Kaufmann-type ion source with a two grid extraction system [29]. The ion source was fed with methane as process gas. Emphasis was put on the dependence of the a-C:H film properties on the energy of the film forming hydrocarbon ions. In Fig. 14 the refractive index and the optical gap of direct ion-beam deposited a-C:H films are plotted (circles) as a function of the hydrocarbon ion energy. The increase of the ion energy from 100 to 1600 eV leads to a decrease of the optical gap from 1.3 to 0.3 eV and to an increase of the refractive index from 1.9 to 2.3. For comparison, data from rf plasma deposited a-C:H films have been included in Fig. 14 (crosses). In this case the refractive index and the optical gap are plotted as a function of the "average ion energy" ($0.6 \cdot U_B$) in the benzene discharge. For both types of a-C:H films the increase of the ion energy has qualitatively the same effect, namely an increase of the refractive index and a decrease of the optical gap. Under the condition, that the average ion energy in the plasma equals the ion energy in the beam it was found that the optical gap of rf plasma deposited a-C:H films is similar to the optical gap of direct ion-beam deposited a-C:H films. However, for the refractiv index this simple relationship is not valid. In this case the high energy portion of the IED appears to be much more important.

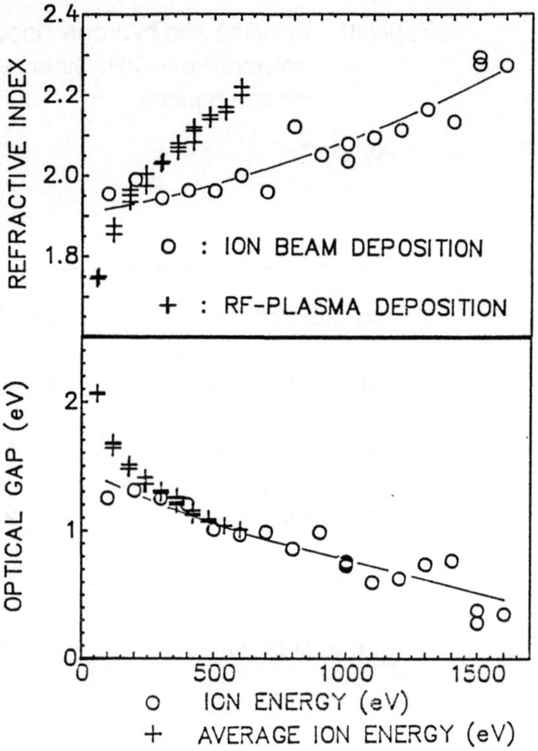

Fig. 14. Refractive index and optical gap of a-C:H films vs. energy of film forming hydrocarbon ions; the circles represent data from direct ion-beam deposited films (vs. energy of ion beam); the crosses represent data from rf-plasma deposited a-C:H films (vs. average ion energy) [29].

IV. CHARACTERIZATION OF FILM PROPERTIES

IV.1 Bonding and structure

The macroscopic properties of diamondlike a-C:H as shown in Fig. 5 are determined microscopically by

i. the hydrogen content,
ii. the details of local carbon and hydrogen bonding
iii. and the medium range structure of the a-C:H network.

Among other techniques, nuclear reaction analysis (NRA) and infrared spectroscopy have been used for the study of hydrogen incorporation [17]. The total hydrogen concentrations as shown in Fig. 5 have been determined by NRA [18]. Using IR absorption spectroscopy [30, 31] it has been shown that hydrogen in hard a-C:H is mainly incorporated as monohydride while the larger hydrogen content in

Table II. Bonding and hydrogen incorporation in hard and polymerlike a-C:H. Other relevant properties are included for comparison.

property	a-C:H (hard)	a-C:H (polymerlike)
$sp^3:sp^2:sp^1$	68 : 30 : 2	53 : 45 : 2
$CH^3 : CH^2 : CH$	0 : 40 : 60	25 : 60 : 15
optical gap (eV)	1.1	3.0
refractive index	2.1	1.65
density (g/cm^3)	1.9	1.3
[H] total (at.%)	27	50

polymerlike hydrocarbon films results in a larger relative concentration of dihydride. Typical data are given in Table II.

Quantitative data on carbon bonding in a-C:H have first been obtained by IR spectroscopy [31] and electron energy loss spectroscopy [32]. It has been shown that a-C:H contains both sp^2 and sp^3-hybridized carbon atoms, the latter being the dominant fraction (see table 2). These data have been confirmed later by nuclear magnetic resonance (NMR) [33, 34]. NMR also revealed [34] that tetrahedral sp^3-carbon atoms are bonded to at least one hydrogen atom, i.e. hydrogen stabilizes sp^3-hybridization.

It appears that the self adjustment of the a-C:H properties at a given impact energy, as shown in table 1, is related to the formation of a network with preferred average nearest neighbor coordination. It has been suggested [10, 35] that the network tends to form in such a way that stabilization by bonding and destabilization by strain are balanced. This balance is achieved by incorporation of hydrogen as a bond terminating atom and trigonally sp^2-bonded carbon into the carbon skeletal network.

If, with increasing deposition energy, the hydrogen content decreases (due to preferential sputtering), the development of medium range order appears to be another mechanism to relieve excess strain. By modelling the optical properties of diamondlike a-C:H, Bredas [36] and Robertson [5] have shown that clusters of sp^2-hybridized carbon atoms have to be assumed. Especially the explanation of the relatively small optical gap in a-C:H requires clustering of the sp^2-carbons into fused aromatic rings. It is the size of these π-bonded carbon clusters which determines the energy and strength of optical π-π^* transitions, i.e. the energy of the optical gap [5]. Recently, resonant Raman spectroscopy gave experimental evidence for the existence of these π-bonded carbon clusters [37, 38].

IV.2 Thermal decomposition of a-C:H

It is known that the beneficial properties of a-C:H get lost at elevated temperatures. Under annealing the hydrogen content decreases and the films undergo a transition towards a distorted graphitic material. This is shown in Fig. 15 where the type of carbon bonding, the C-H concentration and the optical gap of a typical diamondlike a-C:H film are plotted as a function of the annealing temperature. This particular film is thermally stable up to about 350°C. At higher temperatures chemically bonded hydrogen is lost and the resulting dangling bonds rearrange by forming graphitic sp^2-bonds. The optical gap decreases reflecting the growth of the graphitic sp^2-carbon clusters. Raman spectroscopy gave evidence for the formation of graphitic microcrystallites by the observation of disorder-activated scattering by optical zone edge phonons in graphite [39]. The graphitisation is reflected also in the drastic increase of the electrical conductivity (Fig. 16) which in untreated a-C:H is very low [40, 41]. The sudden increase of the conductivity at 350°C points towards a percolation of the π-electron clusters embedded in the non-conducting sp^3-bonded matrix.

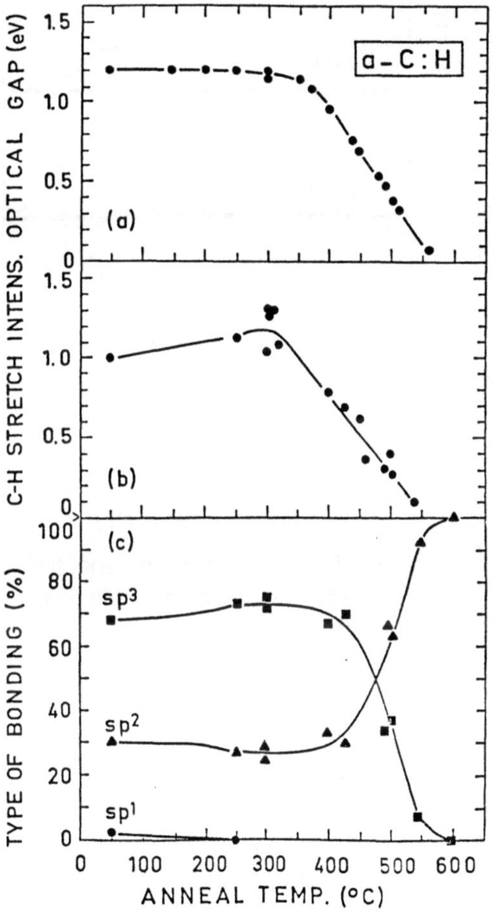

Fig. 15. Temperature induced changes in a-C:H. Optical gap (a), integrated CH stretch intensity (b) as a measure for bonded hydrogen and percentage of sp^3, sp^2, and sp^1 carbon (c) as a function of annealing temperature [19].

It should be noted, that the decomposition temperature of diamondlike a-C:H is not a universal constant. Using thermal desorption spectroscopy it has been shown that the thermal stability of a-C:H increases for films prepared with higher energies [20]. This is shown in Fig. 17, which also demonstrates that the decomposition of hard a-C:H proceeds by the effusion of hydrogen, while softer films,

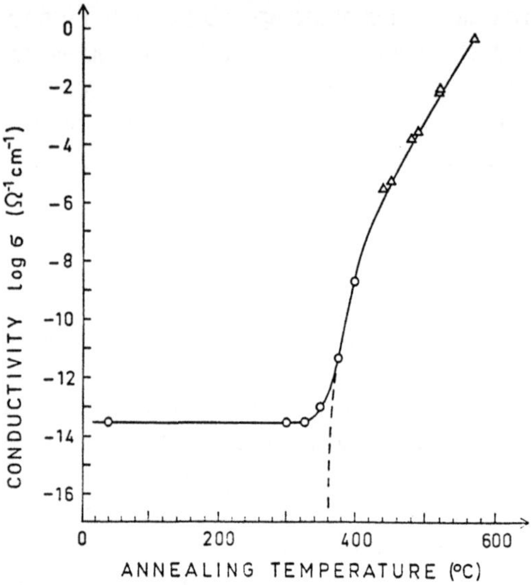

Fig. 16. Electric conductivity of a-C:H vs. annealing temperature [41].

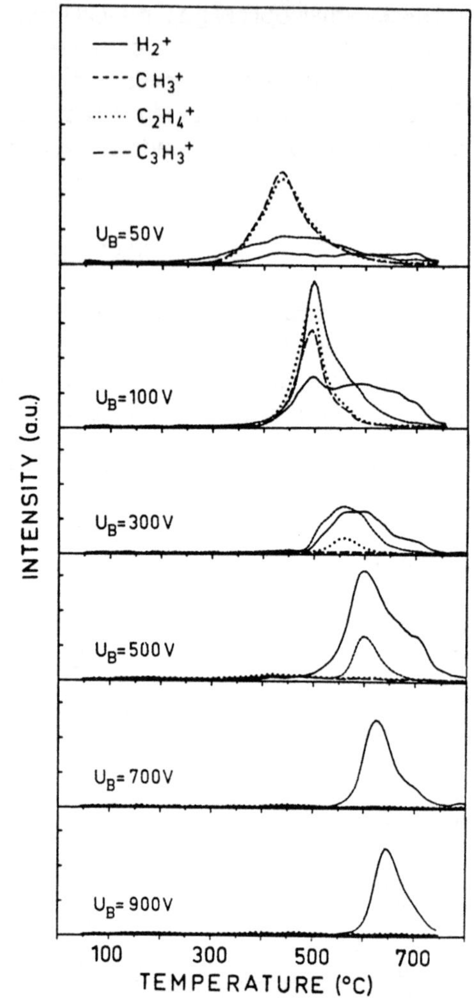

Fig. 17. Thermal desorption spectra of a-C:H films deposited at different bias voltages [20].

deposited at lower bias voltages, desorb not only hydrogen but a variety of hydrocarbons.

The data of Fig. 17 can be explained by an a-C:H network with bias dependent cross-linking and stability. While all films are permeable for hydrogen which migrates in molecular form through the a-C:H network [20], hydrocarbon diffusion and effusion is inhibited by the limited pore size in the denser and more strongly cross-linked films deposited at high bias voltages [20].

IV.3 Related materials

Under appropriate deposition conditions, very hard and dense, chemically inert and infrared-transparent a-C:H films can be prepared. There are many applications of these "diamondlike" films. Most advanced and commercially exploited are optical applications, e.g. as protective or interference coatings for IR optics, solar cells or solar heating panels [12, 42, 43].

Mostly for those optical applications it is advantageous that the optical constants of hard a-C:H can be fine-tuned within about ±10 percent [10, 12]. It would however be desireable to further extend the range of optical properties while retaining the beneficial mechanical and chemical properties of hard a-C:H. This has been tried by chemical substitutions in the hydrogen and carbon sub-networks of a-C:H.

IV.3.1 a-C:D

Deuterated amorphous carbon films (a-C:D) have been deposited from deutero-carbons such as C_6D_6 [19]. The chemical and physical properties of a-C:H and a-C:D are identical with the exception that all H-induced vibrational bands are substantially shifted to longer wavelengths in a-C:D. This has been used to identify H-induced vibrations and to extract structural information from functional group spectroscopy [19]. From an applicational point of view deuteration is a way to shift unwanted IR absorption bands. The transparency of a-C:D in the near IR extends out to the onset of C-D stretch vibrations around 4.2 μm. a-C:D has been reported to exhibit a lower absorption in the 10 μm region as compared to a-C:H [44].

IV.3.2 a-C:H,F

Partially or totally fluorinated carbon films (a-C:H,F) have been prepared by rf-plasma deposition from fluorinated hydrocarbons [45] (e.g. fluorinated benzenes $C_6H_{6-m}F_m$ with $0 \leq m \leq 6$) or from mixtures of hydrocarbons and fluorocarbons [46]. The deposition rate under constant bias voltage has been found to rise significantly with increasing fluorination. Hydrogen in a-C:H can be replaced partially or totally by fluorine in this way. As a result the C-H vibrations vanish at the expence of strong C-F absorptions around 8 μm (Fig. 18). In contrast to the IR absorption, the optical gap did not vary significantly with fluorination.

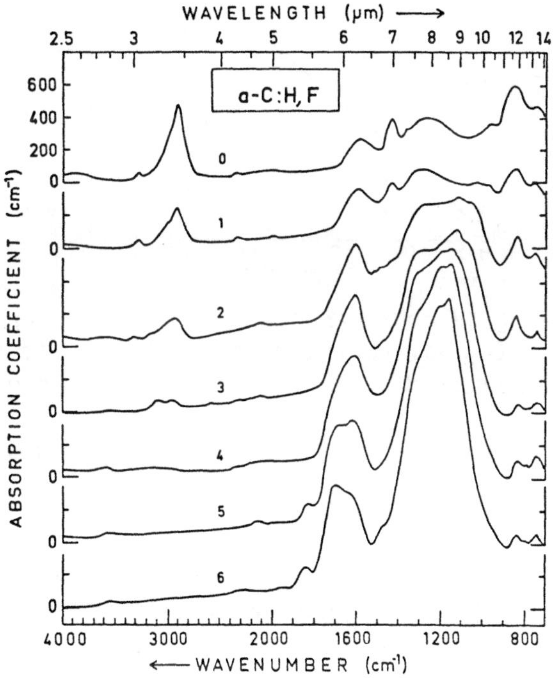

Fig. 18. Infrared absorption of a-C:H,F films deposited from fluorinated benzenes $C_6H_{6-m}F_m$ with fluorine content $0 \leq m \leq 6$ as indicated [45].

Composition and bonding in a-C:H,F has been studied by IR spectroscopy, nuclear reaction analysis and EELS [47]. The details of local bonding were found to be similar to a-C:H. Fink et al. [47] reported that in plasma deposited a-C:H,F films about 2/3 of the carbon atoms were sp^3 hybridized while 1/3 were sp^2 carbons, independent of the degree of fluorination. Relatively more F atoms were bonded to sp^3 carbon than to sp^2 carbon which is consistent with a model of isolated sp^2-carbon clusters in a matrix of (fluorinated) sp^3 carbons [47].

IV.3.3 a-$C_{1-X}Ge_X$:H

While hydrogen substitution in a-C:H modifies mainly the vibrational infrared absorption, more drastic changes of the material properties can be expected by substitutions in the carbon skeletal network.

Amorphous hydrogenated carbon/germanium alloys (a-$C_{1-X}Ge_X$:H) have been used to extend the range of refractive indices [48, 49]. By rf-plasma deposition from mixtures of germane and benzene [49] films with refractive indices between 1.8 and 4.1 have been realized (Fig. 19). Again, the IR absorption spectrum changes with increasing Ge substitution. With the exception of a Ge-H absorption band around 2000 cm^{-1}, a-$C_{1-X}Ge_X$:H films with large X are more transparent than corresponding a-C:H films [49]. Amorphous carbon germanium films have been used to realize multilayer interference coatings [48, 49, 50]. An example of a three layer design with a hard a-C:H top layer acting as a broad band antireflection IR-coating on germanium [49] is shown in Fig. 20.

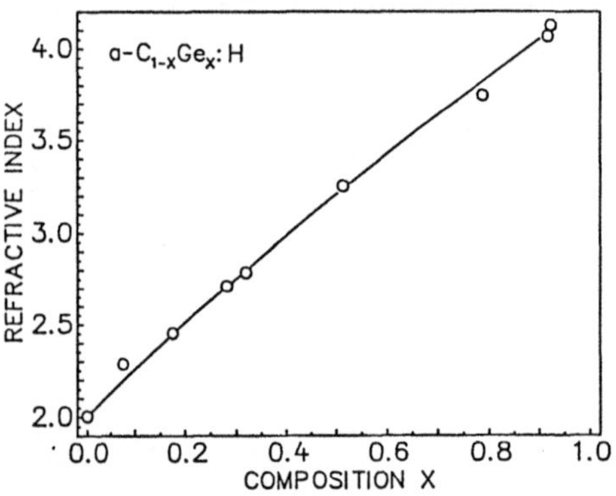

Fig. 19. Refractive index of a-$C_{1-X}Ge_X$:H films as a function of Ge content X [49].

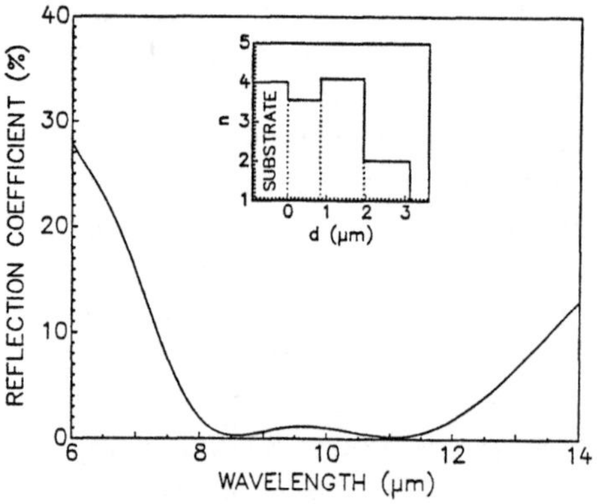

Fig. 20. Theoretical reflection spectrum of a three layer coating on germanium consisting of a-$C_{1-X}Ge_X$:H layers with a hard a-C:H top layer. Thicknesses and refractive indices of the individual layers are given in the inset [49].

IV.3.4 Metal containing a-C:H

A further class of related materials are metal containing a-C:H films [51]. Those films have been prepared by co-deposition of (hydro-) carbon and various metals, e.g. by sputtering a metal target in a hydrocarbon atmosphere. These materials have found great interest due to their favourable mechanical and tribological properties, as reported by Dimigen et al. [52]. Low friction at small metal concentrations and a maximum wear resistance at metal concentrations around 10-20 % have been reported [51].

V. CONCLUSIONS

It has been shown that amorphous hydrogenated carbon (a-C:H) constitutes a unique class of solids with widely varying properties. Under appropriate deposition conditions, very hard and dense, electrically insulating and infrared transparent films can be achieved by the condensation of hydrocarbon ions with energies of several tens to a few hundred eV. This is realized most easily by plasma deposition onto biased substrates.

The rf-plasma deposition process of a-C:H has been analyzed in detail. It is shown that the energy distribution of the film forming ions shows a detailed structure and covers a wide range from zero energy up to the bias voltage. Despite the complicated (but fully understood) ion dynamics, the self bias voltage of the rf discharge remains an important parameter for process control.

The properties of "diamondlike" a-C:H films can be fine-tuned by adjusting the average impact energy via the bias potential. The chemical nature of the hydrocarbon process gas plays a minor role in the deposition of hard a-C:H because of the impact induced fragmentation of the energetic hydrocarbon ions at the growth surface.

While the properties of a-C:H may be diamondlike, the structure of this material is far from diamond. a-C:H is a non-crystalline hard network solid with typical hydrogen concentrations around $H/C \approx 0.5$ and a mixture of tetrahedral (sp^3) and trigonal (sp^2) carbon bonding.

The substitution of hydrogen in a-C:H by fluorine and the alloying of carbon with germanium have been discussed as a way to further extend the range of properties of diamondlike carbon.

REFERENCES

1. for a comparison of CVD diamond and amorphous diamondlike carbon see: J.C. Angus and C.C. Hayman; Science 241, 913 (1988).
2. S. Aisenberg and R. Chabot; J. Appl. Phys. 42, 2953 (1971).
3. E.G. Spencer, P.H. Schmidt, D.C. Joy, and F.J. Sansalone; Appl. Phys. Lett. 29, 118 (1976).
4. "Amorphous Hydrogenated Carbon Films", P. Koidl and P. Oelhafen, Eds., Proc. European Mat. Res. Soc., Vol 17 (Les Editions de Physique, Paris, 1987).
5. J. Robertson; Adv. Phys. 35, 317 (1986).
6. J. Robertson, and E.P. O'Reilly; Phys. Rev. B 35, 2946 (1987).

7. H. Tsai and D.B. Bogy; J. Vac. Sci. Technol. A5, 3287 (1987).

8. "Properties and Characterization of Amorphous Carbon Films", J.J. Pouch, S.A. Alterovitz, Eds.; Materials Science Forum Vol. 52&53 (Trans Tech Publications, Switzerland, 1989).

9. "Diamond and Diamondlike Films and Coatings"; J.C. Angus, R. Clausing, L. Horton, P. Koidl, Eds.; NATO-ASI Series, (Plenum, New York, 1990).

10. J.C. Angus, P. Koidl and S. Domitz; chapter IV in "Plasma Deposited Thin Films", J. Mort, F. Jansen, Eds., (CRC Press, Boca Raton, 1986).

11. L. Holland and S.M. Ojha; Thin Solid Films 38, L17 (1976).

12. A. Bubenzer, B. Dischler, G. Brandt and P. Koidl; J. Appl. Phys. 54, 4590 (1983).

13. B. Chapman; "Glow Discharge Processes", (Wiley, New York, 1980).

14. J.M.E. Harper, J.J. Cuomo, R.J. Gambino and H.R. Kaufmann, in "Ion Bombardment Modification of Surfaces: Fundamentals and Applications", edited by D. Auciello and R. Kelly (Elsevier, Amsterdam, 1984), p. 127.

15. J. Winter, H.G. Esser, F. Waelbroeck, and P. Wienhold;in Ref. 4, p.405.

16. C. Wild, J. Wagner, and P. Koidl; J. Vac. Sci. Technol. A5, 2227 (1987).

17. P. Koidl, C. Wild, D. Dischler, J. Wagner and M. Ramsteiner; Materials Science Forum Vols. 52&53, pp. 41-70, (Trans Tech Public., Switzerland, 1989).

18. H. Baumann, T. Rupp, K. Bethge, P. Koidl, and C. Wild; Ref. 4, p. 343.

19. B. Dischler; Ref. 4, p. 189.

20. C. Wild and P. Koidl; Appl. Phys. Lett. 51, 1506 (1987).

21. C. Wild and P. Koidl; in Ref. 4, p. 207.

22. J. Wagner, C. Wild, F. Pohl, and P. Koidl; Appl. Phys. Lett. 48, 106 (1986).

23. C. Wild, P. Koidl, and J. Wagner; in Ref. 4, p. 137.

24. C. Wild and P. Koidl; Appl. Phys. Lett. 54, 505 (1989).

25. J.W. Coburn and E. Kay; J. Appl. Phys. 43, 4965 (1972).

26. W.D. Davis and T.A. Vanderslice; Phys. Rev. 131, 219 (1963).

27. C. Wild and P. Koidl; J. Appl. Phys., to appear.

28. K.S. Fancey and A. Matthews; Surface and Coatings Technology 33, 17 (1987).

29. R. Locher, C. Wild, and P. Koidl; Proc. 1st European Conf. on Diamond and Diamond-like Carbon Coatings, Crans-Montana 1990, to appear in Surface and Coatings Technology.

30. B. Dischler; in Proceedings of the European Materials Research Society Vol. XVII, "Amorphous Hydrogenated Carbon Films", edited by P. Koidl and P. Oelhafen (Les Edition de Physique, Paris, 1987), p. 189.

31. B. Dischler, A. Bubenzer, and P. Koidl; Solid State Commun. 48, 105 (1983).

32. J. Fink, Th. Müller-Heinzerling, J. Pflüger, B. Scheerer, B. Dischler, P. Koidl, A. Bubenzer, and R.E. Sah; Phys. Rev. B 30, 4713 (1984).

33. S. Kaplan, F. Jansen, and J. Machonkin; Appl. Phys. Lett. 47, 750 (1985).

34. A. Grill, B.S. Meyerson, V.V. Patel, J.A. Reimer, and M.A. Petrich; J. Appl. Phys. 61, 2874 (1987).

35. J.C. Angus; in Ref. 4, p. 179; J.C. Angus and F. Jansen; J. Vac. Sci. Technol. A6, 1778 (1988).

36. J.L. Bredas and G.B. Street; J. Phys. C, Solid State Phys. 18, L651 (1985).

37. M. Ramsteiner and J. Wagner; Appl. Phys. Lett. 51, 1355 (1987).

38. J. Wagner, M. Ramsteiner, C. Wild, and P. Koidl; Phys. Rev. B 40, 1817 (1989).

39. R.O. Dillon, J.A. Wollam, and V. Katkanant; Phys. Rev. B. 29, 3482 (1984).

40. P. Koidl; Electrochem. Soc. Proc. 89-12, 237 (1989).

41. E. Klausmann, R.E. Sah, P. Koidl; unpublished.

42. A.H. Lettington; in Ref. 9.

43. J.C. Angus, C.C. Hayman, R.W. Hoffman; in "Diamond Optics I", Proc. SPIE Vol. 969, 2 (1988).

44. B. Dischler, R.E. Sah, P. Koidl, W. Fluhr and A. Wokaun; Proc. 7th Int. Symp. on Plasma Chemistry, C.J. Timmermans, Ed., Eindhoven (1985).

45. R.E. Sah, B. Dischler, A. Bubenzer, and P. Koidl; Appl. Phys. Lett. 46, 739 (1985).

46. R.E. Sah, B. Dischler, and P. Koidl; Le Vide, les Couches Minces, Suppl. 235, 447 (1987).

47. J. Fink, N. Nücker, R.E. Sah, P. Koidl, H. Baumann, and K. Bethge; in Ref. 4, p. 475.

48. A.H. Lettington, J.C. Lewis, C.J.H. Wort, B.C. Monachan, and A.J.N. Hope in Ref. 4, p. 469.

49. R.E. Sah, C. Wild, P. Koidl, and H. Baumann; Proc. SPIE Vol. 1275, 59 (1990).

50. C.J. Kelly, J.S. Orr, H. Gordon, L.T. Traub, and A.H. Lettington; Proc. SPIE Vol. 1275, 122 (1990).

51. For a review see: C.P. Klages and R. Memming; in Ref. 8, p. 609.

52. H. Dimigen, H. Hübsch, and R. Memming; Appl. Phys. Lett. 50, 1056 (1987).

MASS-SPECTROSCOPY OF SPUTTERED NEUTRAL PARTICLES DURING THE GROWTH OF a-C:H FILMS

V. von Bonin, K.G. Tschersich

Institut für Grenzflächenforschung und Vakuumphysik
Association Euratom-KFA, KFA Jülich, D-5170 Jülich
Fed. Rep. Germany

INTRODUCTION

Despite of the scientific interest in amorphous hydrogen containing diamondlike films (a-C:H), their growth mechanism is not fully understood. One of the unclarified aspects is the mass balance during film growth. It has been observed that only a fraction of the carbon atoms hitting the sample surface are incorporated into the growing film. Angus et al. /1/ found carbon sticking-coefficients between 0.13 and 0.41 using hydrocarbon/argon-beams at 100 eV while Miyazawa et al. /2/ reported sticking-coefficients of 0.53 and 0.26 using a mass-separated C^+-beam of 300 eV and 600 eV, respectively. In both investigations the deposition coefficient was determined from film thickness measurements. Also our previous measurements /3/ with 600 eV ethane ions incident on various substrates raised questions concerning the nature of the channels which may balance the incoming carbon flux.

A theoretical treatment of film deposition by energetic particles was proposed by Berg et al. /4,5,6,7/ which took into account both deposition and sputtering. Because of several unknown parameters, the model is not able to make quantitative predictions. Eckstein /8/ has simulated numerically the growth process using a program based on the TRIM code. All basic processes known from the theory of ion-solid-interaction, such as reflection and implantation of primary ions and sputtering of film and substrate material were included. However, the theoretical predictions were not quantitatively compared with experimental results.

A complete experimental characterization of the various particle fluxes during the a-C:H film deposition is still missing. The aim of this investigation is to measure directly the particle fluxes from the substrate back into the gas phase during the initial film deposition.

Diamond and Diamond-Like Films and Coatings
Edited by R.E. Clausing *et al.*, Plenum Press, New York, 1991

For this purpose, a-C:H films are deposited onto silicon using a low energy (0.9-2.5 keV) benzene ion beam. The neutral particles sputtered into the gas phase are detected and analyzed during deposition. The measurements are performed by detecting the sputtered neutral particles in order to exclude matrix effects (see, e.g. /9/) known to be associated with secondary ion mass spectroscopy. The main advantages of this method, especially in the particular case of the present study, are :

- the measured signal is produced by the deposition process itself: no supplementary excitation is necessary, nor must the deposition process be interrupted for measurements.
- the small escape depth of sputtered particles (typically one or two monolayers) ensures a high surface sensitivity.

EXPERIMENTAL PROCEDURE

The experiments are performed in a combined SIMS/SNMS-instrument schematically shown in fig. 1 and previously described in ref. 10. In contrast to common SIMS/SNMS work, a hydrocarbon beam is used here to deposit an a-C:H film onto the sample. This method of film deposition is known as direct ion beam deposition /11/. Contrary to other (e.g. plasma assisted) methods the direct ion beam deposition allows one to deposit the film using selected ions of well defined energy and flux density. In this experiment benzene molecules are ionized by electron bombardment in a plasma ion source. The mass resolution of the subsequent Wien-filter of dm/m=0.1 is sufficient for removing all ions with less than six carbon atoms. By deflecting the ions neutral particles are removed from the beam. The ion beam with diameter 0.6 mm is scanned over a sample area of typically 4 mm^2 by means of deflection plates. Crater rim effects are suppressed by an electronic window which only selects signals originating from the center of the growing film. The benzene ion beam current density is typically 1 $\mu A/cm^2$ corresponding to $3.75 \cdot 10^{13}$ C atoms/cm^2 s. The benzene ion energy is varied between 0.9 and 2.5 keV corresponding to 138 eV and 385 eV per carbon atom, respectively. The angle of incidence is 45° with respect to the surface normal.

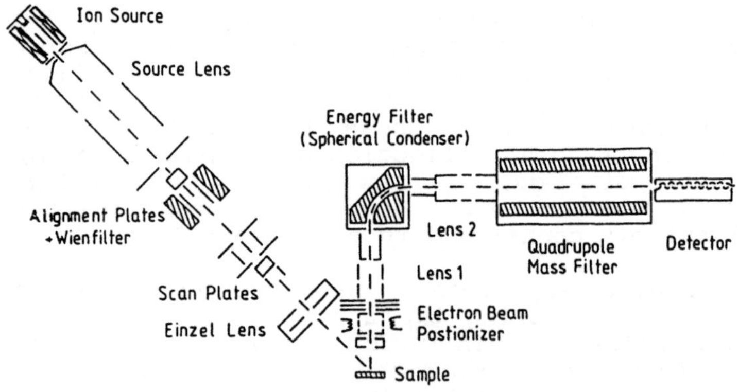

Fig. 1 Schematic layout of the experimental setup

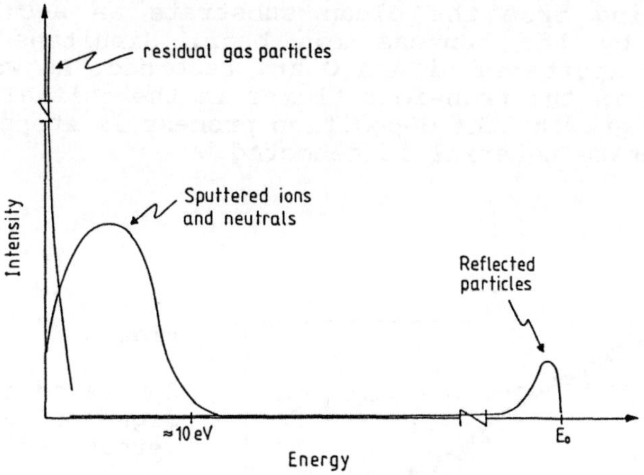

Fig. 2 Schematic energy distribution of particles created by particle bombardment. E_0 denotes the energy of the primary carbon ions.

The sputtered neutral particles are first ionized in an electron beam postionizer, then energy filtered and mass analyzed and finally counted. A crucial function of the detection system is to exclude all but the sputtered neutral particles, such as sputtered ions, residual gas atoms and reflected atoms. A schematic energy distribution of the latter particles is given in fig. 2. Sputtered ions are removed by deflection plates in front of the ionizer. An energy window is adjusted to cut off the low energy residual gas ions as well as the high energy reflected ions. Moving the energy window to higher energies ($E \approx 20eV$) yields a decrease of the measured signal to zero, which supports the assumption, that the contribution from fragmented and reflected primary particles of low energy are of no importance. The particle separation obtained in this way is sufficient for the suppression of reflected particles, but not for that of the residual gas particles. Therefore, residual gas particles are additionally suppressed by placing a positive bias on the first diaphragm behind the ionizer with respect to the ionizer. The stringent energy-filtering results in a reduction of the useful yield to about $2 \cdot 10^{-10}$. As a consequence the measurements had to be restricted to the most abundant sputtered neutrals C (amu 12) and Si (amu 28).

The substrate material is silicon. It is cleaned by heating to about 900°C until the fraction of impurity atoms (mainly carbon) is only a few percent as measured with Auger Electron Spectroscopy (The sample can be transferred to an AES system attached to the SIMS/SNMS system without breaking the vacuum). The working pressure in the deposition chamber is about $2 \cdot 10^{-9}$ mbar. Between cleaning the substrate and starting the measurement the sample is exposed to this pressure, which may cause a certain initial concentration of carbon at the surface.

Starting from the clean substrate an a-C:H film is deposited by the benzene ion beam. Simultaneously, the signals of sputtered Si and C are recorded. As we are only interested in the transient fluxes in the initial stages of a-C:H film growth, the deposition process is stopped when no more substrate material is detected.

a

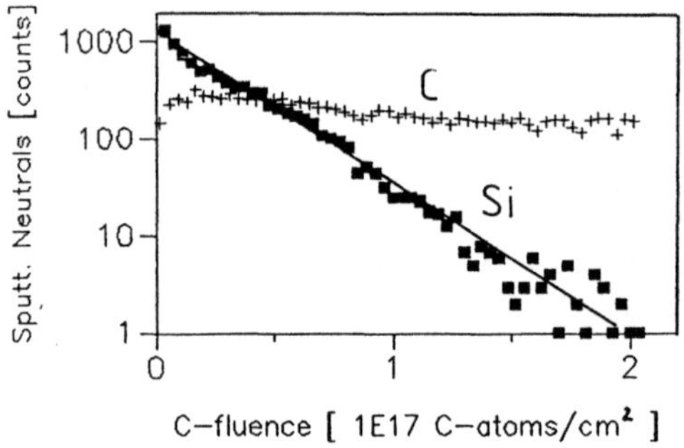

Fig. 3

Sputtered neutrals signal of Si and C versus C-fluence

a) logarithmic scale indicating the exponential decrease of the Si signal

b

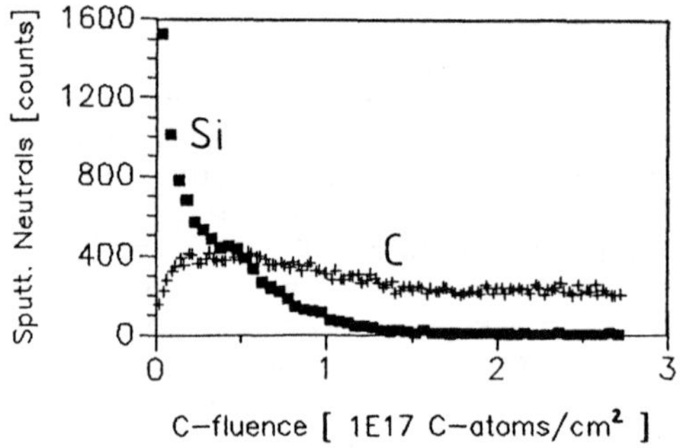

b) different measurement with the carbon signal recorded with double frequency (linear scale)

RESULTS AND DISCUSSION

The signals of the sputtered neutral particles, plotted versus the carbon fluence, are subsequently referred to as 'growth profiles'. Fig. 3 a) shows the growth profiles for masses 12 (C) and 28 (Si) measured for a deposition energy of E=250 eV per C-atom on a logarithmic scale. The number of sputtered silicon atoms decreases exponentially with the carbon fluence. The carbon signal reveals a strange behaviour. Fig 3 b) shows the growth profiles of another deposition at the same energy with the carbon signal recorded with higher frequency. In the beginning, the carbon signal shows a steep increase due to the enrichment of carbon at the surface and reaches a maximum. It then falls off to a constant value while the silicon signal continues to diminish to nearly zero. This final stationary phase corresponds to the continuing growth of the a-C:H film.

Before interpreting these results we regard Auger measurements recorded on the same system. Fig. 4 shows the surface composition of silicon and carbon versus the carbon fluence as measured with Auger electron spectroscopy. The Auger signals are normalized to those of the pure elements. The Auger measurements also yield an exponential decrease of the silicon signal while the carbon signal behaves just complementarily. We have generally observed this complementarity of substrate and film signal in our Auger measurements /3,12/, if we compare Auger signals of about the same information depth, i.e. the same Auger electron escape depth. It is reasonable to assume that these Auger signals directly reflect the relative concentrations of substrate and film material within the information depth.

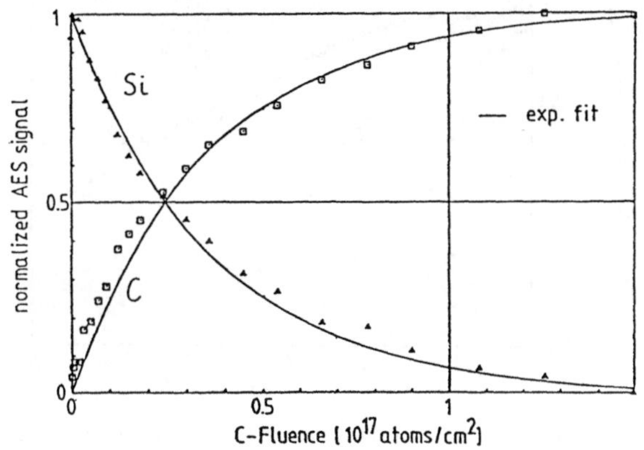

Fig. 4 Normalized AES signals of Si and C versus C-fluence, the solid lines represent exponential fits to the data

Comparing now figs. 3 and 4 we conclude that the decrease of the sputtered neutral signal of Si and the initial increase of the C signal is attributed to the dilution of the Si-concentration in the uppermost atomic layers as a result of C-concentration increase. This increase shows that the detected particles originate from the surface of the growing film and not from fragmentation of primary molecules. The fact that the carbon signal does not start at zero is attributed to a certain initial concentration of carbon at the surface. The reason for the subsequent decrease of the carbon signal will be discussed later. From the equality of the Si signals in both figures we also infer that the sputtered neutrals Si signal I_{Si} reflects the Si concentration c_{Si} within the information depth of sputtered neutrals mass spectroscopy:

$$(1) \qquad c_{Si} = I_{Si}/I_{Si}^0$$

where I_{Si}^0 is the signal of the pure Si substrate. The complementary C concentration c_C is then obtained from

$$(2) \qquad c_C = 1 - c_{Si} = 1 - I_{Si}/I_{Si}^0 \ .$$

In fig. 5 'growth profile' curves for Si are shown for different benzene ion incident energies (expressed in energy per C-atom). In view of eq. (2), we infer that the carbon net deposition rate decreases with increasing benzene ion energy. This is attributed to an increase in the sputtering yield of the deposited C-atoms with increasing deposition energy. Simulation calculations show that the shoulders observed on the curves in fig. 5 measured at the higher benzene ion energy can be attributed to the higher sputtering rate, which at a certain stage almost compensates the deposition rate /13/.

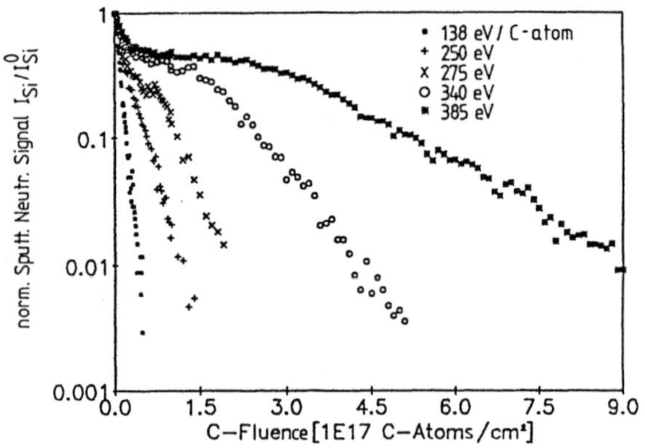

Fig. 5 The normalized sputtered neutrals signals of Si (i.e. Si-concentration c_{Si}) at different deposition energies between 138 and 385 eV / C-atom.

The characteristic shape of the carbon 'growth profile' curve obtained for 250 eV/C-atom in fig. 3 suggests that the probability of a C-atom to be sputtered by an incoming C-ion decreases during the film growth. This probability, subsequently denoted Y_c^*, is in fact a sputtering yield per C-atom. It is obviously related to the measured sputtered neutrals C signal I_c by

$$I_c = a \cdot Y_c^* \cdot c_c \ . \qquad (3)$$

In fig. 6, the dependence of Y_c^* (normalized to $Y_c^*(c_c=1)$) on the C-concentration c_c is plotted for different benzene ion incident energies. The values of c_c are obtained via eq. (2) from the I_{Si} signal (fig. 5) measured simultaneously with the I_c signal. The result is interesting for the following reasons:

1) The probability of a carbon atom being sputtered decreases with increasing carbon concentration. At low carbon concentrations it is at least five times higher than at high concentrations.
2) The dependence of this probability appears to be qualitatively the same at all energies, suggesting that this dependence is a fundamental one.

According to the theory of sputtering /see, e.g. 14,15/ the sputtering probability mainly depends on the binding energy of the atom in question. Changes of shape and extent of the collision cascade during deposition cannot explain the extent of the decrease of the carbon sputtering probability.

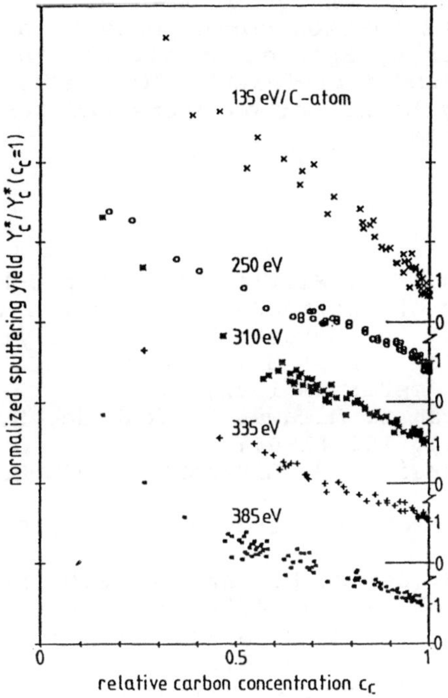

Fig. 6 Normalized probability of C-atoms sputtered by incident benzene ions of different energies (expressed in eV / incident C-atom) versus relative C-concentration in the film.

Thus, the physical content of the first statement above is that the average binding energy of the carbon atoms increases with carbon concentration. At the beginning of the deposition, the carbon atoms are loosely bound, and thus easily sputtered away, leaving only the strongly bound C-atoms at the surface. During further carbon deposition, the carbon atoms become more and more tightly bound, leading to a decrease of their sputtering probability.

This result supports the idea of Spencer et al. /16/, that a hard carbon film develops by the preferred erosion of weak bonds: carbon atoms are bound in various binding states. Preferential erosion of weak bonds causes the average carbon binding to become stronger in the course of film deposition.

SUMMARY

The initial stages of a-C:H film growth on a silicon substrate were examined by mass-sensitive detection of neutral particles sputtered by the deposition ion beam. From the experiments, it is inferred that the sputtering probability of a silicon atom by a primary carbon atom remains constant during the deposition process. In contrast, the sputtering probability of a carbon atom decreases with carbon concentration. This is attributed to an increase of the average binding energy of the carbon atoms during film growth. This observation corresponds to a growth mechanism proposed earlier by Spencer et al. stating that weak bonds are generally eroded preferentially leading to an accumulation of tightly bound carbon atoms and resulting in a hard carbonaceous film.

REFERENCES

/ 1/ J.C. Angus, M.J. Mirtich, E.G. Wintucky, 'Ion beam deposition of amorphous carbon films with diamondlike properties' in 'Metastable Material Formation by Ion Implantation', Eds: S.T. Picraux, W.J. Choyke, Elsevier, Amsterdam 1982, S.433

/ 2/ T. Miyazawa, S. Misawa, S. Yoshida, S. Gonda, J. Appl. Phys. 55(1), 188 (1984)

/ 3/ K.G. Tschersich, U. Littmark, E-MRS-Meeting, Vol.XVII, June 1987

/ 4/ L.P. Andersson, S. Berg, Vacuum 28(10,11), 449-451 (1978)

/ 5/ S. Berg, B. Gelin, M. Östling, S.M. Babulanam, J. Vac. Sci. Technol. A2(2), 470-473 (1984)

/ 6/ S. Berg, B. Gelin, A. Svärdström, S.M. Babulanam, Vacuum 34(10,11), 969-973 (1984)

/ 7/ S. Berg, C. Nender, B. Gelin, J. Vac. Sci. Technol. A4(3), 448-452 (1986)

/ 8/ W. Eckstein, Surface and Interface Analysis, Vol. 14, 799-808 (1989)

/ 9/ A. Benninghoven, E.G. Rüdenauer, H.W. Werner in Second Ion Mass Spectrometry, Chemical Analysis Vol. 86, John Wiley & Sons (1987)

/10/ H. Gnaser, J. Fleischhauer, W.O. Hofer, Appl. Phys. A 37, 211-220 (1985)

/11/ J.C. Angus, P. Koidl, S. Domitz in 'Plasma Deposited Thin Films' (Ed. J. Mort, F. Jansen), Chapter 4 : 'Carbon Thin Films', CRC Press, Boca Raton, Florida, USA (1986)

/12/ K.G. Tschersich, J. Nucl. Mater. 162-164, 887 (1989)

/13/ to be published

/14/ P. Sigmund in Topics in Applied Physics, Vol. 47 : Sputtering by Particle Bombardment I, R. Behrisch (Ed.), Springer Berlin 1981, S.9

/15/ P.C. Zalm, J. Vac. Sci. Technol. B2(2), 151 (1984)

/16/ E.G. Spencer, P.H. Schmidt, D.C. Joy, F.J. Sansalone, Appl. Phys. Lett. 29, 118 (1976)

PREPARATION AND CHARACTERIZATION OF DOPED *a*-C:H FILMS

K. Rohwer, P. Hammer, A. Helmbold,
D. Bahnemann, and D. Meissner

*Institut für Solarenergieforschung (ISFH),
Sokelantstraße 5, D-3000 Hannover 1, FRG*

W. Hedderich, R. Hartmann, and W. Ronge

*Battelle-Institut e.V.,
Am Römerhof 35, D-6000 Frankfurt am Main 90, FRG*

INTRODUCTION

Amorphous hydrogenated carbon films are known to be extremely hard, wear-resistant and chemically inert[1]. In general, they are electrically insulating. Meyerson and Smith[2,3] and Jones and Stewart[4] showed that *a*-C:H can be made conductive by addition of the "classical" doping agents, *i.e.* PH_3, B_2H_6, and N_2. It has been discussed[4,5,6], whether the term "doping" can be used with this materials like with crystalline semiconductors.

Another way to synthesize *a*-C:H films which are electrically conductive is to incorporate carbide or metal clusters. This has been shown by Köberle[7] for carbide forming tantalum (*a*–C:H(Ta)) and gold (*a*–C:H(Au)), respectively.

In the following we describe a newly built plasma CVD reactor at the ISFH and its advantages over previous facilities of this kind. With this apparatus we were up to now able to prepare only undoped films. In a subsequent section we report on temperature dependent electrical conductivity measurements performed with nitrogen doped films prepared at the Battelle Institute using a rather similar apparatus.

EXPERIMENTAL

A radio frequency plasma enhanced chemical vapour deposition apparatus (rf-PECVD) was built at the ISFH which is shown schematically in figure 1. It consists of a 24.5 liter stainless steel vacuum chamber with two circular, water-cooled electrodes with vertical surfaces of 10 cm diameter. The electrode distance can be varied between 5 and 12 cm.

Diamond and Diamond-Like Films and Coatings
Edited by R.E. Clausing *et al.*, Plenum Press, New York, 1991

There is an electro-pneumatically controlled gas supply system for argon, two process gases (*e.g.* ethylene and acetylene), and two dopant gases (*e.g.* phosphane and diborane). A vapour source for liquid organic precursors and for metal-organic compounds is still under construction. Furthermore, there is the possibility to sputter the rf driven electrode during plasma deposition in order to incorporate metals in the film growing on the counter-electrode.

RF power of 13.56 MHz is supplied by a 1.5 kW generator via a matching network. The dc-bias and rf voltages can be measured, and the generator can be operated in a constant dc-bias mode. The pumping system consists of an electrically controlled throttle valve, a wide range turbomolecular pump and a membrane roughing pump. The turbo pump used has the advantage that there is no need for a rotary roughing pump, so there is no oil in the system, which could

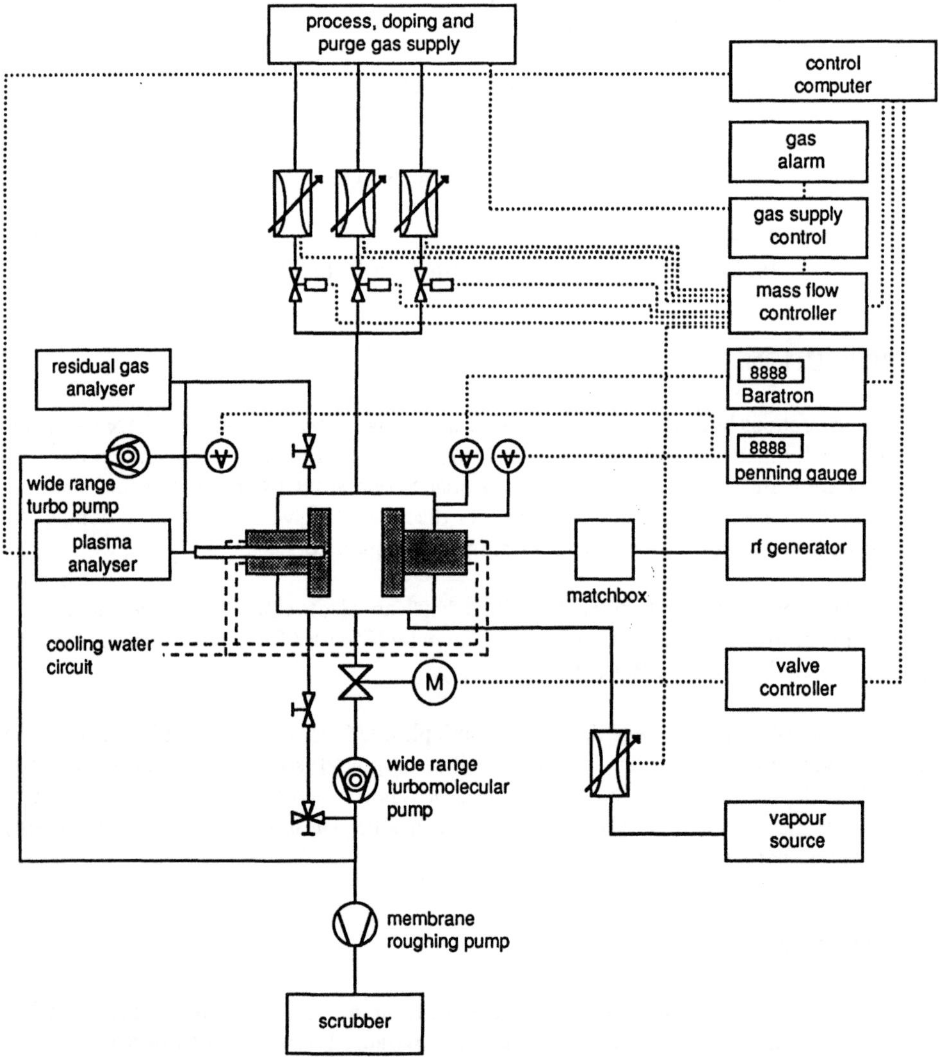

Fig. 1. Schematic drawing of the plasma enhanced CVD facility at the ISFH which in part is still under construction. The vacuum chamber is in the centre. The shaded areas represent the electrodes.

cause problems due to contamination by plasma products or residues of dopants. The waste gas will be cleaned by a scrubber which consists of a chemical oxidation and a combustion stage.

One of the electrically insulated electrodes can be replaced by a grounded one with a small aperture in the centre. (This configuration is shown in the figure.) Behind the aperture there is a differentially pumped vacuum space containing a quadrupole mass analyser specially designed for plasma diagnostics. This electrode as well as other plasma diagnostic facilities (optical emission spectroscopy and electrical probes) are still under construction. The whole system will be controlled by a personal computer.

So far the samples produced in this reactor were made from pure acetylene and turned out to be electrically insulating with conductivities below the detection limit of our measurement configuration ($10^{-16}\ \Omega^{-1}cm^{-1}$).

Furthermore, we investigated samples prepared on a similar but simpler apparatus at the Battelle Institute. These were also made from acetylene, but "doped" with nitrogen. The rf frequency was 13.56 MHz as well, the bias voltage amounted to about -700 V on the driven electrode which carried the substrates. One sample ("Battelle 116") was separated from the plasma by a copper mesh (electrically connected to the electrode). The flow rates of acetylene and nitrogen were 23 sccm and 60 sccm, respectively, so there was a large excess of the dopant gas over the monomer. The total gas pressure during deposition was 0.02 mbar. The films had a thickness of 200 nm.

One sample ,"Battele 116", was deposited on a polycarbonate substrate, another one, "Battelle 480", on a silica substrate covered with indium tin oxide (ITO) as a transparent bottom electrode for electrical measurements. Top electrodes were made by evaporating gold patterns onto the films, which consisted of a 1 cm diameter circle for through-the-film measurements, and of four stripes of 7 mm length with distances of 0.3 mm for measurements along the film surface. With this configuration we were able to perform 2-probe as well as 4-probe measurements as shown in figure 2.

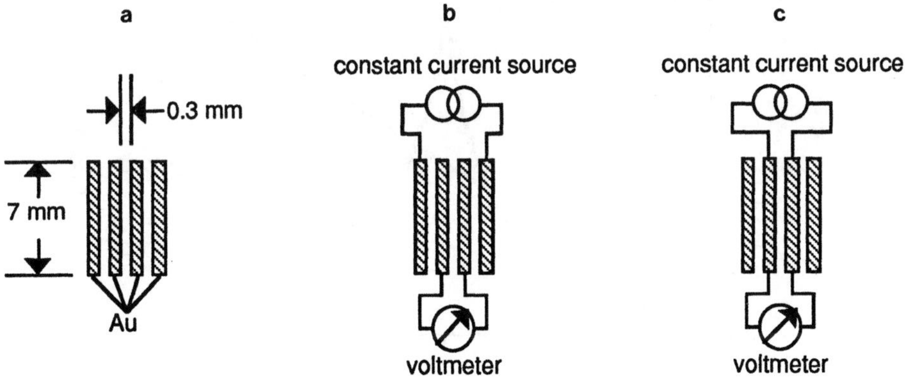

Fig. 2. a) Geometry of the gold stripe pattern evaporated onto the sample surface for electrical conductivity measurements; electrical connections for b) 4-probe and c) 2-probe measurements.

The electrical measurements were carried out using a constant current source and an electrometer, both controlled by a computer. Up to now 4-probe measurements were done at room temperature only, under air at atmospheric pressure.

On the Battelle 480 sample temperature dependent measurements were carried out using a continuous flow liquid helium cryostat, but liquid nitrogen as coolant. As ambient medium of the sample helium gas was used. The resistance was measured through the film at temperatures from 100 K to 280 K in steps of 20 K. Here the electrometer alone was used in the current mode, applying the integrated voltage source in order to be able to measure at constant fieldstrength.

RESULTS AND DISCUSSION

The first feature we noticed was that the resistivity of the samples depended strongly on the electric fieldstrength applied during measurement, as displayed in figure 3. This occured independent of the direction of measurement (along or through the films), so surface effects can be excluded. A comparison of 2-probe and 4-probe measurements showed an almost voltage-independent contact resistance which evinces that this is also not a contact phenomenon.

In figure 4 the temperature dependence of the conductivity is displayed in the form of an Arrhenius plot. It shows that either an activated conduction with a temperature dependent activation energy (E_{act} = 0.08 eV ... 0.16 eV at low and room temperature, respectively) or a different conduction mechanism can be assumed. With regard to the work of Köberle[7] we evaluated the data in terms of hopping models which lead to

$$\sigma = \sigma_0 \cdot \exp(-(T/T_0)^{-p}),$$

where σ is the electrical conductivity, T the temperature and p a dimensionless exponent. Mott's variable range hopping model[8] predicts p = 1/4, while its modification, the model of Efros and Shklovskii[9], predicts p = 1/2.

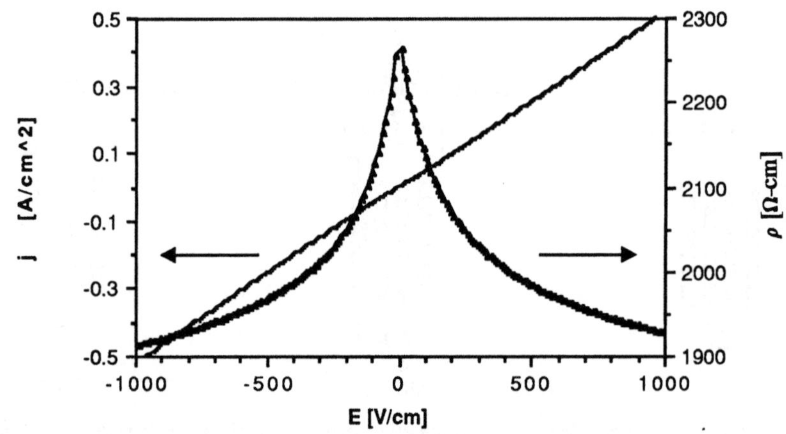

Fig. 3. Current density and resistivity vs. fieldstrength applied during measurement parallel to the surface of a nitrogen doped a-C:H sample (Battelle 116).

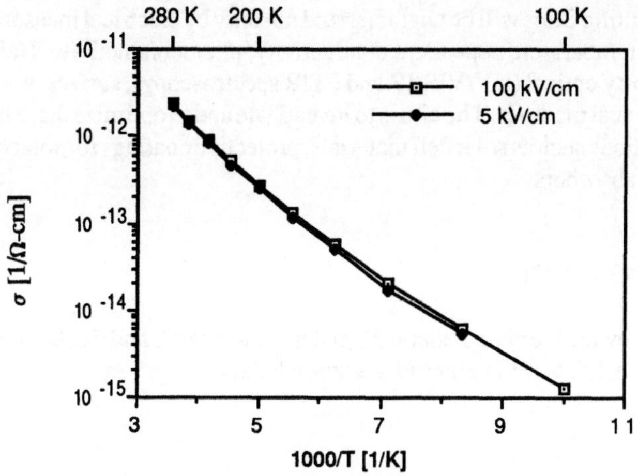

Fig. 4. Arrhenius plot of the electrical conductivity (through-the-film) of a nitrogen doped a–C:H sample (Battelle 480) taken at two different field strengths.

The slope in a plot of ln(-dlnσ/d(1/T)) *vs.* ln(T) yields a value for 1-p (fig. 5). From this we derived $p = 0.34 \pm 0.17$. Thus an unambiguous decision between the two models cannot be made. This is in accordance with the results of Köberle[7], though the dopants, N_2 and Ta (forming TaC), respectively, were quite different in both cases.

OUTLOOK

When the PECVD facility at the ISFH will be complete, we will be able to prepare films doped with classical dopants as well as with metals. During deposition we will be able to investigate the processes occuring in the plasma and, hopefully, in the growing film.

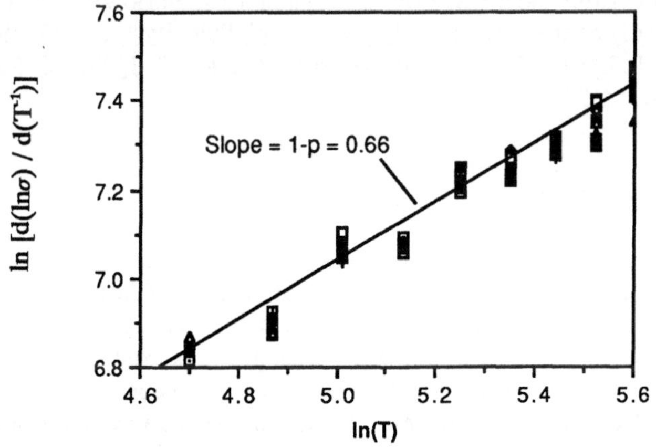

Fig. 5. Determination of the exponent p *(as defined in the text) from the data displayed in figure 4.*

The resulting films will be characterized not only by electrical measurements (current-voltage curves, temperature dependent conductivity, photoconductivity, Hall effect, thermopower), but also by optical (UV/VIS/IR and FTIR spectroscopy), surface analytical, chemical and electrochemical methods. The aim is to investigate and to optimize these materials with respect to applications such as solar cell materials, protective coatings for solar cells, or selective layers for solar absorbers.

ACKNOWLEDGEMENT

Support by the German Federal Minister for Research and Technology (BMFT) under contract No. 13 N 5575 1/2 is gratefully acknowledged.

REFERENCES

1. C.P. Klages and R. Memming, "Microstructure and Physical Properties of Metal-containing Hydrogenated Carbon Films" in: "Amorphous Hydrogenated Carbon", J.J. Pouch, S.A. Alterovitz, eds., TransTech Publications (1989)
2. B. Meyerson and F.W. Smith, "Chemical modification of the electrical properties of hydrogenated amorphous carbon films", Solid State Commun. 34: 531 (1980)
3. B. Meyerson and F.W. Smith, "Thermopower of doped semiconducting hydrogenated amorphous carbon films", Solid State Commun. 41: 23 (1982)
4. D.I. Jones and Stewart, "Properties of hydrogenated amorphous carbon films and the effects of doping", Philos. Mag. B 46: 423 (1982)
5. J. Robertson, "Clustering and gap states in amorphous carbon", Philos. Mag. Lett. 57: 143 (1988)
6. J. Robertson and E.P. O'Reilly, "Electronic and atomic structure of amorphous carbon", Phys. Rev. B 35: 2946 (1987)
7. H. Köberle, *"Struktur und elektrische Leitfähigkeit von HF-Plasma-erzeugten metallhaltigen Kohlenwasserstoffschichten"*, PhD Thesis, University of Hamburg (1989)
8. N.F. Mott, "Conduction in glasses containing transition metal ions", J. Non-Cryst. Sol. 1: 1 (1968)
9. A.L. Efros and B.I. Shklovskii, "Coulomb gap and low temperature conductivity of disordered systems", J. Phys. C: Sol. State Phys. 8: L49 (1975);
A.L. Efros, "Coulomb gap in disorderd systems", J. Phys. C: Sol. State Phys. 9: 2021 (1976)

NOTE ADDED IN PROOF

In the meantime we became aware that our 4-probe measurement configuration is not suited to avoid contact effects completely, because at least part of the measurement current will flow through the inner two contacts and produce an according voltage drop unless the sum of the interface resistance and the resistance between the edges in the contact material is sufficiently high. However, the fact that we found a difference between 2-probe and 4-probe measurements at all shows that at least an (unknown) portion of the contact resistance can be observed this way.

IN SITU PLASMA AND SURFACE DIAGNOSTICS OF C:H DEPOSITION FROM ECR PLASMAS

A. Koch, M. Engelhard, W. Jacob,
W. Möller and R. Wilhelm

Max–Planck–Institut für Plasmaphysik,
EURATOM–Association, D–8046 Garching/München, F. R. G.

INTRODUCTION

In order to afford a better understanding of the deposition of C:H films from plasmas heated by the electron–cyclotron–resonance (ECR) both plasma and surface diagnostics are employed. Important parameters as film thickness and refractive index, plasma density and plasma composition are measured by means of *in situ* and *non–intrusive* techniques using microwaves and visible light. A two–beam laser interferometer independently monitors film thicknesses and refractive indices during the deposition process. In this way deposition rates can be evaluated. The line integral of the electron density is evaluated from phase shift measurements using a 35 GHz microwave interferometer. Furthermore, optical spectroscopy in the visible spectral region is used as a plasma probe.

EXPERIMENTAL ARRANGEMENT AND RESULTS

Polymer–like hydrocarbon films are deposited from ECR heated methane plasmas. The plasma composition varies from pure methane to CH_4/H_2 mixtures. The particular advantages of ECR plasmas compared to RF or DC plasmas are
(i) the high degree of ionisation yielding high deposition rates,
(ii) a high power absorption for even meter–sized substrate dimensions[1] and
(iii) a low sheath potential.

Two different ECR reactors used here are shown schematically in figure 1. One reactor is a H_{10} mode waveguide with magnetic field coils coaxially arranged around the waveguide. The high–frequency power at 2.45 GHz is fed into the discharge waveguide through a matched vacuum–tight microwave window. A local magnetic field of 87.5 mT is generated in the interior of the waveguide to satisfy the ECR condition. The inner walls of the waveguide are coated by moving the magnetic field coils along the waveguide axis. In principle, this moving coil arrangement can be used to coat the interior of arbitrary metallic tubes[2].

The other plasma chamber is of cylindrical shape with an inner diameter of approximately 26 cm. Here, the magnetic field configuration may be varied to adjust the position of the ECR zone and to form magnetic mirrors in order to change particle fluxes[3]. The coupling of the 2.45 GHz microwave power is accomplished in this case by a cylindrical slot antenna of 13 cm inner diameter (Lisitano coil[4]) mounted inside the vessel and fed via a coaxial feedthrough. The samples are coated in the downstream plasma.

Diamond and Diamond-Like Films and Coatings
Edited by R.E. Clausing *et al.*, Plenum Press, New York, 1991

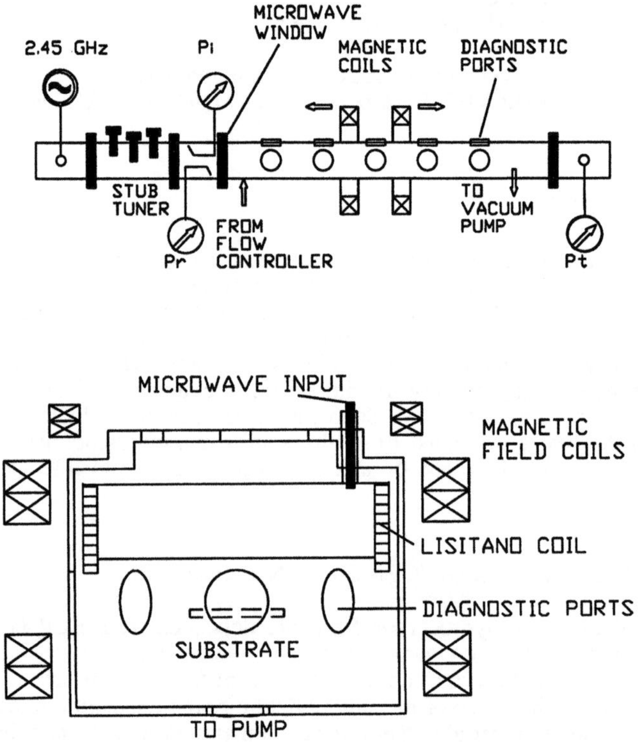

Fig. 1. Experimental ECR reactors used in this work. (a) Waveguide reactor; (b) cylindrical reactor.

In order to provide useful data for process modelling, in situ characterisation of plasma and film properties is of particular interest. Here, special attention is paid to non–perturbing measurements. Due to the comparatively high ambient pressure of 0.04 to several Pa many well known in situ diagnostics like SEM, AES, SIMS etc. have to be excluded. A retarding field analyser is used to measure the ion energy at the location of the substrate[5]. Furthermore, a Langmuir probe and a quadrupole mass spectrometer yield qualitative plasma data. In the following three sections the installed optical and microwave diagnostics are briefly described.

Laser Interferometry

A simple interferometric device using a HeNe laser as light source is well suited to study thin and optically transparent films. As shown by figure 2 two laser beams are directed to the probe at two angles. In order to perform an accurate measurement the difference between the two angles and hence between the periodicities of the detected signals must be large. Due to experimental restrictions, α_1 and α_2 are fixed at 8.2 and 73 degrees, respectively. The interferences of the beams reflected at the film surface and at the substrate result in a periodic change of the detected intensity depending on the film thickness d_f, the film refractive index n_f, the laser wavelength λ_0 and the angles of incidence α_i (i=1,2; see figure 2). Thus two known angles unambiguously yield both refractive index and film thickness at a local point with the dimensions of the

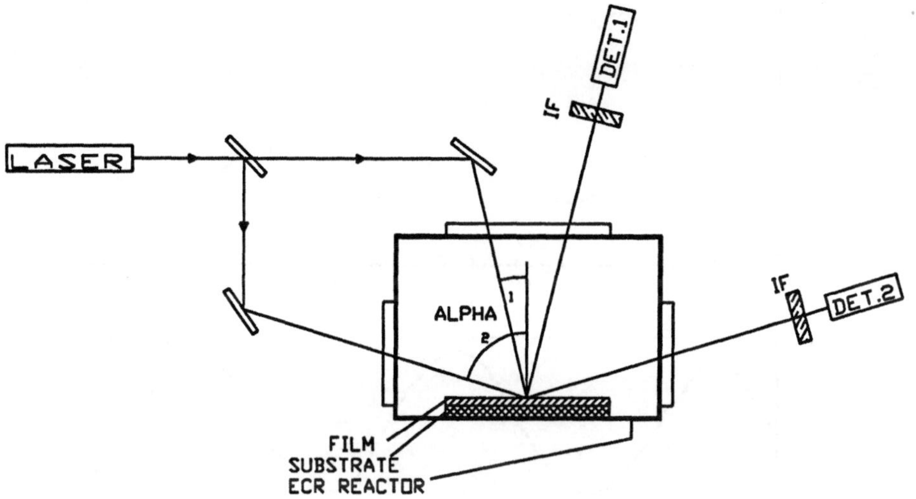

Fig. 2. Two beam laser interferometer for film measurements.

laser focus spot size, approximately 0.5 mm in diameter. The evaluation of the film parameters is straightforward when introducing the ratio of the measured number of periods p_1 and p_2 for beam 1 and 2, respectively: $P = p_1/p_2$. Hence the refractive index becomes

$$n_f = \left[(P^2 \sin^2\alpha_2 - \sin^2\alpha_1)/(P^2 - 1)\right]^{1/2} .$$

Either detector signal may be used to obtain the film thickness:

$$d_f = \frac{1}{2} p_i \lambda_0 \left[n_f^2 - \sin^2\alpha_i\right]^{-1/2} ; \quad i=1,2 .$$

In some cases interference filters (IF) at 632.8 nm have been proven useful to block the plasma background emission. The deposition rates found range from 0.2 to 2 nm/s

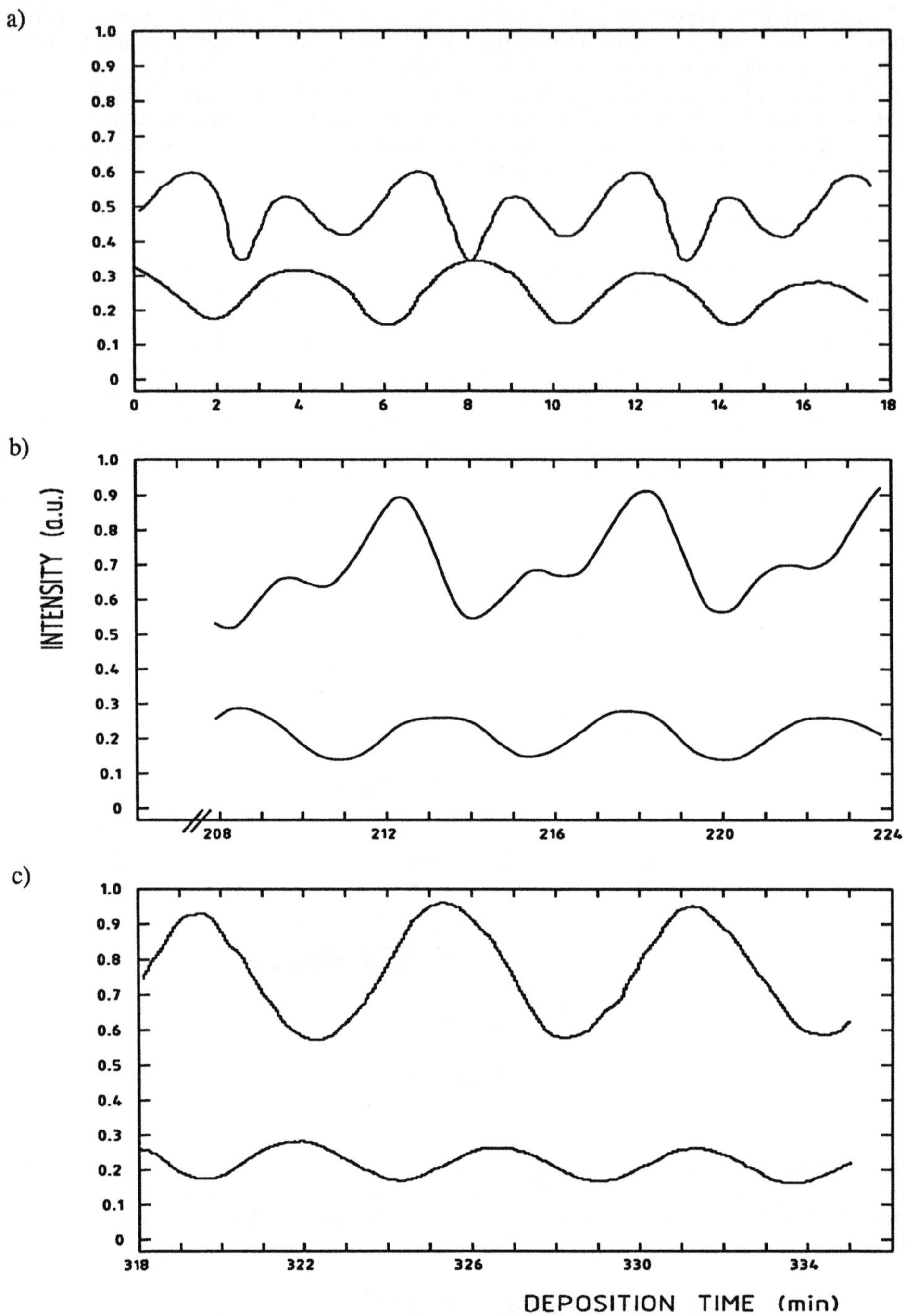

Fig. 3. In situ optical film monitoring during the C:H deposition process in the waveguide reactor. Experimental parameters are: Absorbed microwave power ≈ 110 W, pure methane at 2.50 Pa, gas flow = 10.0 sccm, sample = V2A–stainless steel. Three time intervalls are shown exemplary; n_f=1.54±0.01. The corresponding film thicknesses d_f and deposition rates R are (a) d_f=0–936 nm, R=52.0 nm/min; (b) d_f=8.84–9.57 µm, R=42.7 nm/min; (c) d_f=13.28–14.01 µm, R=41.8 nm/min.

depending on the substrate material, the absorbed microwave power and the methane partial pressure.

The interferometric curves exhibit characteristic shapes with intermediate peaks, especially for large angles of incidence (beam 2). The different shapes are attributed to the influences of the substrate and/or intermediate reflecting layers. It has been found that these special features of beam 2 change during the deposition and even vanish for thick films ($\approx$ 10 µm). Three interferograms at different stages of the deposition process are depicted in figure 3. Further investigations have to show whether additional information can be obtained from these measurements.

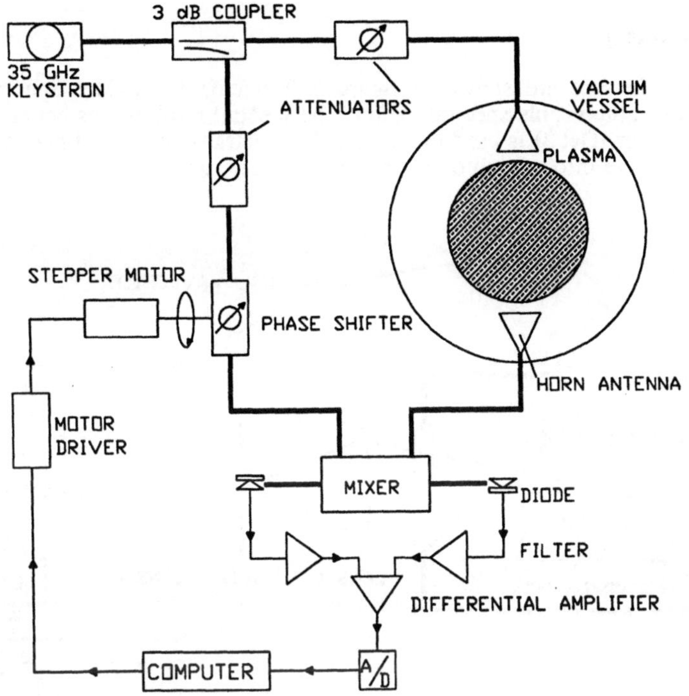

Fig. 4. Setup of microwave interferometry for the measurement of the plasma density, shown schematically.

Microwave Interferometry

In the cylindrical reactor, a microwave interferometer at $f_0 = 35$ GHz has been installed as shown in figure 4. The measurement provides the line integrated density $\int n_e dl$ along the propagation path L through the plasma. With ε_0 being the vacuum permittivity and m_e and e being electron mass and charge, respectively, the cut–off density in electrons per m^3 becomes

$$n_{eco} = \varepsilon_0 \, m_e \, (2\pi f_0/e)^2 \ .$$

For low electron densities ($n_e \ll n_{eco}$) the line integral reads

$$\int n_e \, dl \approx n_{eco} \frac{c}{\pi f_0} \Delta\varphi ,$$

with $\Delta\varphi$ being the measured phase shift caused by the plasma electrons. This phase shift caused by the plasma electrons results in a difference signal. This signal controls the computer–aided adjustment of the phase shifter in the reference branch. Data storage, averaging and the calculation of the electron density is performed by the computer, too. For typical operating conditions, i.e. 40 W of absorbed microwave power and pure methane at a pressure of 0.04 Pa, the electron density of the downstream plasma has been evaluated to be approximately

$$\langle n_e \rangle \approx n_{eco} \frac{c}{\pi f_0 L} \Delta\varphi \approx 2\text{-}3 \cdot 10^{10} \text{ cm}^{-3} .$$

Optical Spectroscopy

Two optical detectors are shown in figure 5. Detector 1 provides spectrally resolved measurements in the visible spectral region. The spectral resolution is better than 0.15 nm. The other detector (Det.2) is used to monitor the spectrally integrated plasma emission. In this way qualitative information on the plasma stability can be derived.

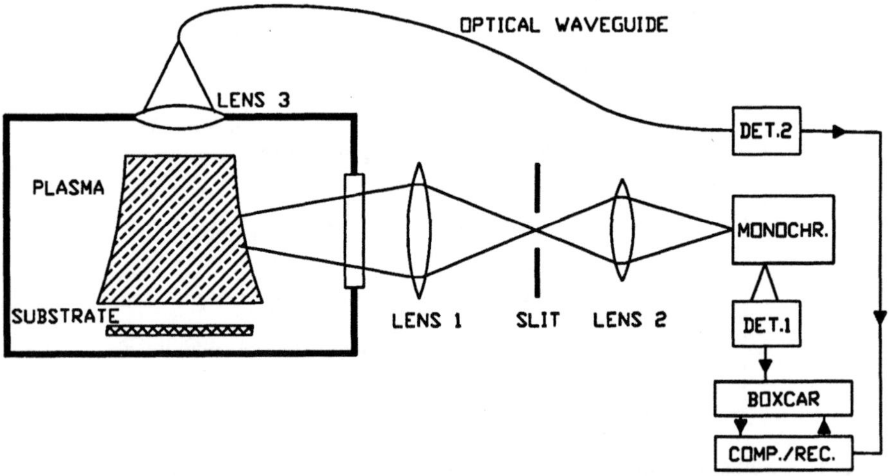

Fig. 5. Experimental arrangement for optical diagnostics.

A survey for the spectral region between 385 nm and 490 nm is depicted in figure 6. Characteristic changes of spectral details are elucidated by figure 7. Here, the plasma composition is a CH_4/H_2 mixture. The rotational spectrum of the CH $A^2\Delta$–$X^2\Pi$ (0–0) band can be identified up to R = 11. However, the spectral resolution has to be improved in order to obtain unambiguously CH rotational temperatures. Further information can be inferred from intensity measurements. Here, changes of the emission of atomic and molecular hydrogen can be correlated with the hydrogen flow in respect to the methane flow. Increasing methane partial pressure results in a decrease of the CH and an increase in the H and H_2 emission intensities.

These first measurements are used as a qualitative plasma probe and as an aid to set and control reproducible plasma conditions during the deposition process.

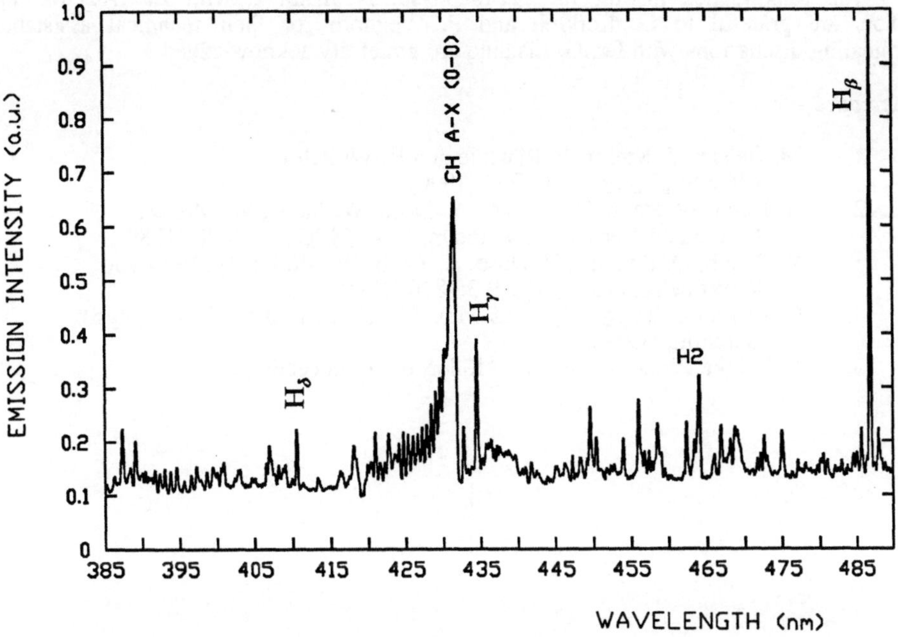

Fig. 6. Methane spectrum, power ≈ 40 W, CH_4 flow = 5.5 sccm, p ≈ 0.10 Pa.

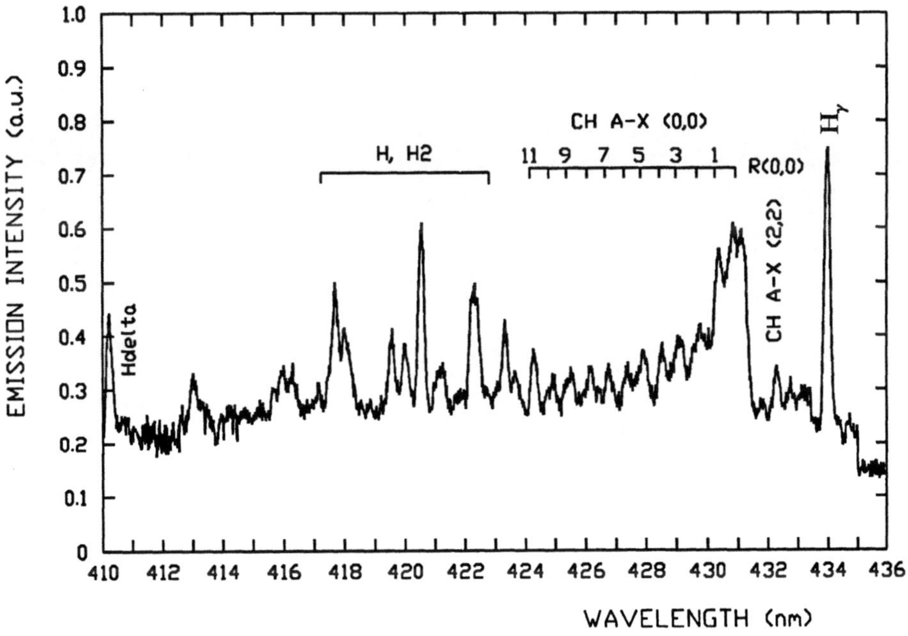

Fig. 7. Details for power ≈ 40 W and CH_4/H_2 flow = 2.5/10.0 sccm, p ≈ 0.19 Pa.

Acknowledgements

The Max–Planck–Institut für Plasmaphysik is affiliated with EURATOM. The authors are grateful to G. Kerkloh and P. Weghorn for their technical assistance. Stimulating discussions with Dr. G. Lisitano are gratefully acknowledged.

References

1. M. Geisler, J. Kieser, E. Räuchle and R. Wilhelm, J.Vac.Sci.Technol. A8:908 (1990).
2. A. Koch, W. Renz, D. Boutard, V. Dose, W. Jacob, W. Möller, J. Perchermeier and R. Wilhelm, Proc. ISPC–9 3:1820 (1989).
3. W. Jacob, D. Boutard, V. Dose, A. Koch, W. Möller, W. Renz and R. Wilhelm, Proc. ISPC–9 3:1826 (1989).
4. G. Lisitano, Report IPP 3/145 Max–Planck–Institut für Plasmaphysik, Garching (1989).
5. P. Reinke, W. Jacob and W. Möller, these proceedings.

PRODUCTION OF CARBON CLUSTER BEAMS AND THEIR CHARACTERISATION BY TIME–OF–FLIGHT MASS SPECTROSCOPY

H.–G.Busmann, H.Gaber, T.Müller and I.V.Hertel

Freiburger–Material–Forschungszentrum (F–M–F) and
Fakultät für Physik, Albert–Ludwigs–Universität
Hermann–Herder–Str.3, D 7800 Freiburg, F.R.G.

Introduction

Carbon clusters are of considerable interest because of their chemical and physical properties/1/, their occurrence e.g. in combustion flames/2/ and in laser ablation of solid carbon materials/e.g. 3–7/. Also, carbon clusters are discussed as being nuclei in CVD–diamond film growth and for their potential in a–C:H film growth by ion cluster beam deposition (ICBD). ICBD /8,9/ is an interesting alternative to established methods for thin film technology and has found applications e.g./9/ in silicon and compound semiconductor technology, opto–eléctronics and optical films. It has been suggested/10/ that ICBD is special in that cluster fragmentation at the surface occurs and the fragments move on the surface without penetration into or destruction of the surface, but they do have enough energy for chemical reactions.

For studying an effect of carbon clusters in CVD–diamond film growth and for carbon–ICBD, it is desirable to have carbon clusters available which are different in average cluster size, energy (neutrals and ions) and hydrogen content. In the present paper we describe various methods for carbon cluster production from materials originally in the liquid or solid phase and characterise their cluster beams by time–of–flight mass spectroscopy.

Experimental

Carbon cluster beams are formed in an expansion of partly ionised compressed gases. From material originally in the gaseous phase, an adiabatic expansion is produced by a pulsed valve and ionisation is achieved by either electron beam injection into the expansion (PVEI) or by a hollow cathode discharge (HCD). Cluster formation from solid material is obtained by excimer laser ablation (ELAP). A hybrid source, in which the expansion of a pulsed valve picks up an ionised carbon plasma, produced by an electrical arc discharge, is also applied (PACIS). In the following, a short description of the various cluster sources will be given.

Pulsed valve with Electron Injection (PVEI)

Ions which are the nuclei of ion clusters are produced by electron injection into the expansion of a pulsed valve (General Valve, left part in fig 2a)/11/. To this end a dc electron beam (2mm dia., 300μA, 2kV) crosses the cluster beam line perpendicularly. Positioning of the electron beam to the desired position in front of

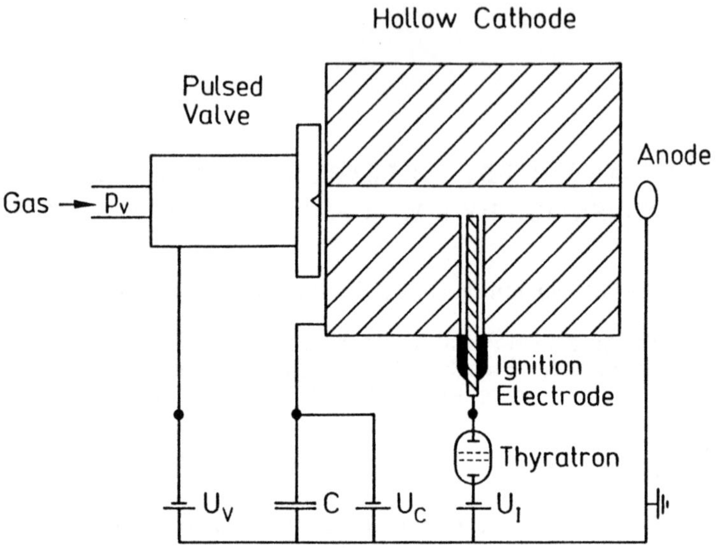

Fig.1a Scheme of the hollow cathode discharge (HCD) cluster source.

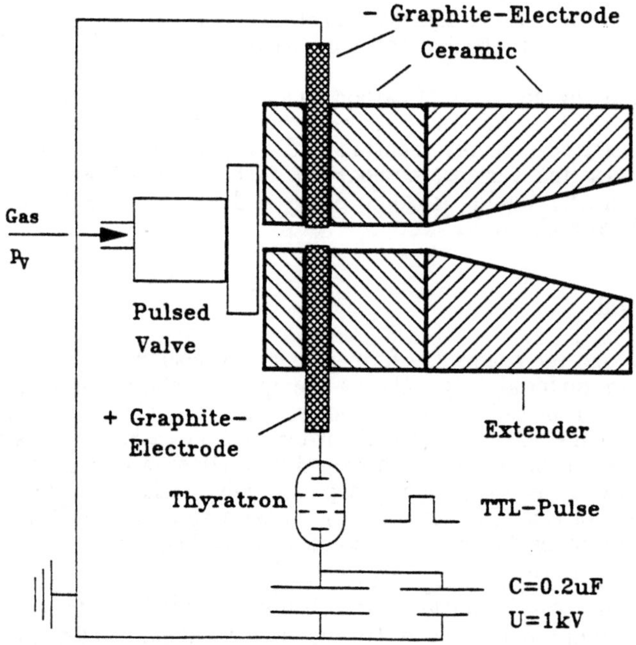

Fig.1b Scheme of the pulsed arc cluster ion source (PACIS) /11/.

the pulsed valve is achieved by manipulation with the electron optic system of a low cost xy–oscilloscope. However, the oxide cathode of this oscilloscope does not operate at the ambient pressure conditions and thus has been replaced by a thorium doped tungsten wire.

Hollow Cathode Discharge (HCD)

Fig.1a shows a hollow cathode discharge device suitable for continuous and pulsed operation. In continuous operation, the gas flow is regulated by an adjustable valve and the cathode anode current is limited to 40mA. For pulsed operation, the gas expansion of a pulsed valve is initiated by a high voltage pulse on a separate electrode where a capacitance is discharged across the cathode to the anode.

Excimer Laser Ablation of Polymers (ELAP)

Under particular conditions an intense carbon cluster beam could be observed using excimer laser ablation of polymers, especially from polyimide/4–7/. The clusters are produced by focusing the light from an excimer laser (Lambda LPX200 or EMG101MS) onto a piece of solid polymer.

Pulsed Arc Cluster Ion Source (PACIS)

Ganteför et al./12/ have developed a pulsed arc cluster ion source for producing metal clusters. Briefly, a pulsed high current pulse produces a plasma between two face–to–face metal electrodes separated by a distance of 2mm. This plasma is ejected by the gas expansion of a pulsed valve. With the intention of making hydrocarbon clusters in our device (fig.1b), we use graphite instead of metal electrodes and a hydrocarbon gas in the expansion. In some cases argon has been used in order to produce carbon clusters without hydrogen. The power supply of an old flash lamp pulsed dye laser (Lambda FL3000) has been used to generate the plasma.

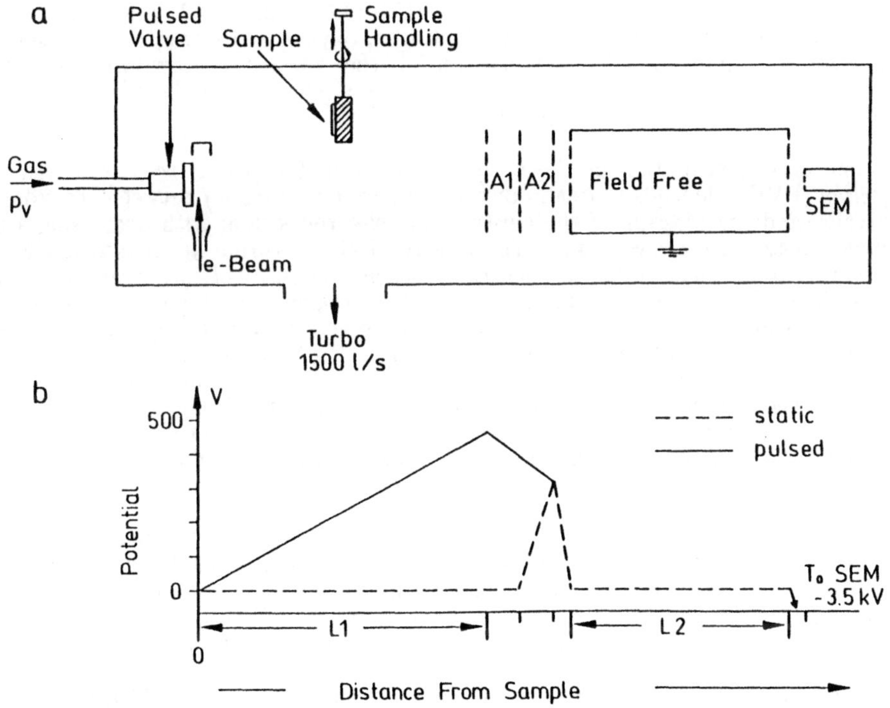

Fig.2 (a) Scheme of the experimental setup. As one of the cluster sources the pulsed valve with electron injection (PVEI) is shown.
(b) Electric potentials for mass analysis along the cluster beam line.

Fig.2a shows the experimental setup for investigating the characteristics of the cluster beams: a cluster source (the PVEI source is shown), a time-of-flight (TOF) mass spectrometer for analysing mass and velocity distributions of the cluster beam and a sample holder are combined in a high vacuum chamber. The base pressure is less than 10^{-6} mbar (turbo pump, 1500 l/s) which during operation is maintained by a liquid nitrogen cryo trap if necessary.

Fig.2b shows the potentials applied for mass analysis of the cluster beam. In regions A1 and A2 the ions are accelerated by pulsed potentials (dashed lines) and are detected with a secondary electron multiplier (SEM). Appropriate values of the pulsed potentials result in a mass separation at the SEM/13,4/. The signal of the SEM is recorded by a transient recorder (100MHz), Camac–Crate and AT personal computer. CO_2 clusters, which are well known and easy to produce, serve for calibration of the TOF. The velocities ($v=L1/\Delta t$), and hence energies ($E={}^1/_2mv^2$), of the different masses are determined by tuning the time delay Δt between the start of the cluster beam and pulsing the potentials at A1,A2.

For studying the clusters produced by excimer ablation, a similar setup has been used but mass resolution was improved by using a reflecting time of flight spectrometer (RETOF) /14,5/ with mass resolution $m/\Delta m \approx 2000$. Neutral spectra are investigated by deflecting the ions out of the beam with an electric field and using a second excimer laser to ionise the neutral species.

Results and Discussions

The composition of the cluster beams from the various sources will now be described. Fig.3 shows typical mass spectra. Roughly, the mass distribution of the clusters can be divided into two groups: smaller clusters with less than 50 carbon atoms are either hydrocarbons or pure carbons depending on the original material used for cluster formation. The heavier clusters with up to 400 carbon atoms/6,7/ obtained from ELAP applied to polyimide (C:H:O:N = 22:10:5:2) are pure carbons without any foreign addition, although hydrogen, oxygen and nitrogen are present in the beam. A more detailed comparison of the various sources will be given elsewhere/15/. Here we will focus on some aspects which might be of relevance for application of carbon clusters in thin film deposition.

The mass distribution of the cluster beam from the *pulsed valve with electron injection* (PVEI) depends critically on the distance s (see fig.3c) between the front of the valve and the crossing of the cluster– and electron–beam; with decreasing s the average cluster size grows (fig.3). The growth of clusters from an ionised nucleus is more favorable than from a neutral nucleus because of stronger attractive interactions. Operating the pulsed valve with methane at a stagnation pressure of p_v=8bar clusters up to 50 carbon atoms are observed with s=1mm; Fig.3a shows a typical distribution peaking at 9 carbon atoms. Fig.3c shows the mass spectrum obtained with distance s=10mm at which clusters are grown from a neutral nucleus and the electron–beam ionises the already formed neutral clusters. Within the extrema shown in fig.3a and 3c, a beam of desired average cluster size can be formed by tuning the distance s(e.g. fig.3b). The hydrogen content is highest compared to all other cluster beams studied: the H:C ratio is greater than 2.

We have found the *hollow cathode discharge* (HCD) difficult to operate in such a manner that clear mass spectra are obtainable. In the pulsed mode this is mainly a problem of the time delay between pulsing the valve and the ignition breakdown. However, in single shot mass spectra small clusters up to $C_{10}H_n$ have been observed. In dc–mode the pressure in our chamber is too high for TOF–spectroscopy.

Both positively charged and neutral carbon clusters formed by UV *excimer laser ablation of polymers* (ELAP) have been characterised. Campbell et al./6,7/ have investigated in detail the cluster spectra produced by 308nm ablation of polyimide. An example of a neutral cluster spectrum is shown in fig.3a. The mass distribution

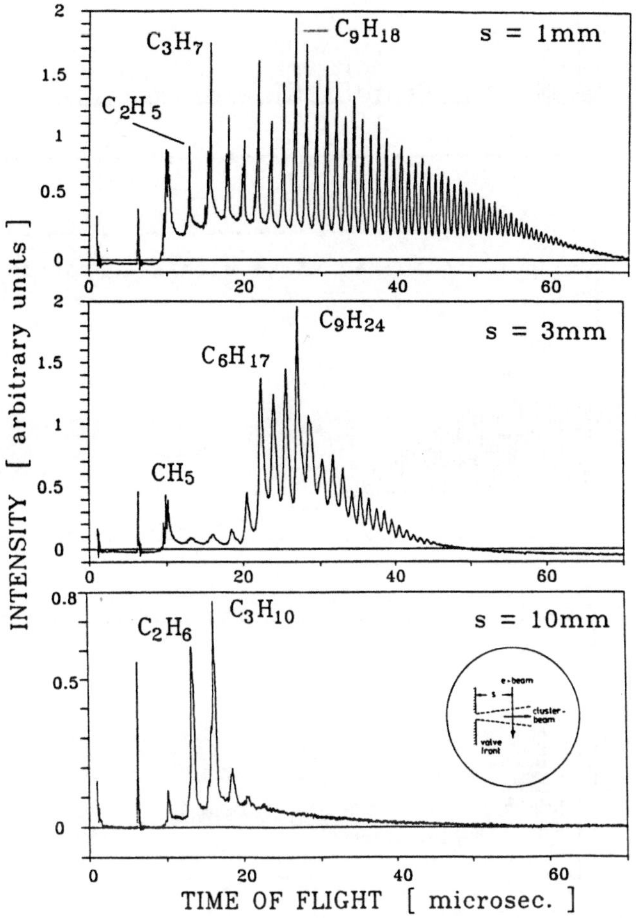

Fig.3 Mass–spectra of positive charged carbon clusters from a pulsed valve with electron beam ionisation(PVEI). The valve was operated with methane at 9 bar stagnation pressure. Shown are the intensities of the various masses as a function of the time of flight in our mass spectrometer. Parameter is the distance s between the front of the valve and the crossing of the cluster– and electron–beam (see insert).

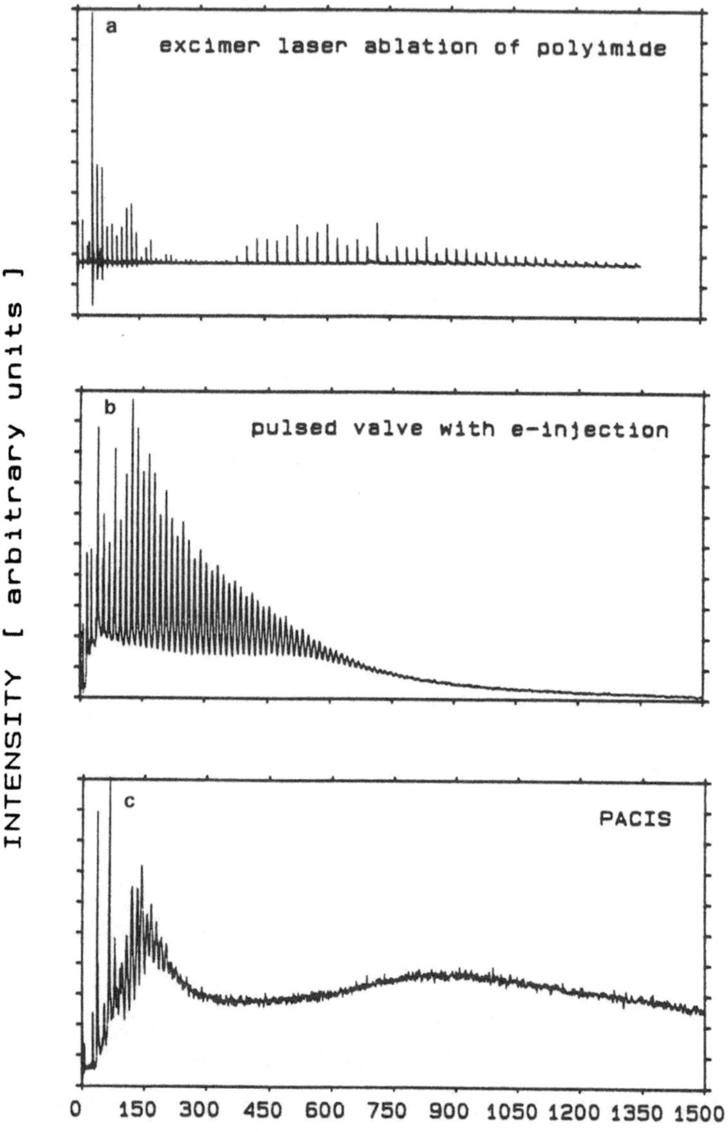

Fig.4 Mass–spectra of carbon clusters from (a) excimer laser ablation of polyimide, ELAP (neutrals, 308nm, 250 mJ/cm^2), from (b) pulsed valve with electron injection, PVEI (positive ions, methane) and from (c) pulsed arc cluster ion source, PACIS (positive ions, ethylene, graphite electrodes). Shown are the intensities of the various masses as a function of atomic mass units (amu). The spectra shown in fig.3a and fig.4b correspond to the same mass distribution.

can be tuned to the range of interest by varying the fluence of the ablating laser and the lack of any hydrogen in the large clusters (40...400 atoms/cluster) has been shown.

Carbon clusters from the *pulsed arc cluster ion source* (PACIS) operated with ethylene at 9bar stagnation pressure are shown in fig.3c. In comparison to the cluster distribution from the PVEI source (fig.3b), a broad distribution at higher masses and with its maximum at about C_{80} is observed. By comparing the spectrum to those obtained from our high resolution spectrometer using the laser ablation of polyimide (fig. 3a), it is tentatively suggested that this broad distribution consists of unresolved peaks of carbon clusters. The H:C ratio for the clusters in the range up to 50 carbon atoms/cluster is about one. Pure carbon clusters with a maximum cluster intensity at 14 carbon atoms have been formed by PACIS operated with argon in which argon acts as an inert seeding gas.

Velocities of the positive clusters obtained by ELAP applied to polyimide and PACIS operated with ethylene have been found to correspond to supersonic expansions with velocity distributions much narrower than corresponding to thermal equilibrium. However, the velocities of the heavier clusters obtained from ELAP are higher than from PACIS; typical values are 4000 m/s and 800 m/s, respectively. For C_{60} this amounts to energies of 60 eV and 25 eV. Velocities of the neutral clusters obtained by ELAP applied to polyimide are also supersonic; A typical value is 2000 m/s. Again for C_{60}, this amounts to energies of 15 eV/cluster or 250 meV/(atom and cluster) which is about 10 kT at room temperature (=26meV). From this one might expect that these neutral species are interesting candidates for low temperature deposition experiments.

In conclusion, the ability to produce carbon cluster beams different in hydrogen content, average cluster size and energy has been shown. In the mass range up to 50 carbon atoms/cluster the hydrogen is highest in the clusters formed by PVEI operated with methane (H:C $\geq$ 2), medium by PACIS operated with ethylene ($\approx$ 1) and pure carbons are formed by PACIS operated with argon. The heavier clusters obtained by ELAP with up to 400 C–atoms are also pure carbons. Primarily cluster deposition experiments performed in our laboratory show a dependence of the qualities of films grown (hardness, opacity) on the characteristics of the cluster beam used. A systematic investigation of film properties as a function of cluster composition and energy is underway.

Acknowledgements

We thank E.E.B.Campbell for very helpful discussions and, together with G.Ulmer and B.Hasselberger, for performing the laser ablation experiments. This work has been supported by the Bundes Ministerium für Forschung und Technologie (BMFT) under contract number 13N5026 "Erzeugung, Charakterisierung und Abscheidung von Ionen Clustern".

References

/1/ R.F.Curl and R.E.Smalley: Science 242,1017(1988)
H.Kroto, Science 242,1139(1988).

/2/ Ph.Gerhardt, S.Loffler and K.H.Hormann: Chem.Phys.Lett. 137,3061(1987).

/3/ H.W.Kroto, J.R.Heath, S.C.O'Brien, R.F.Curl and R.E.Smalley: Nature **368**(1985)6062.

/4/ E.E.B.Campbell, G.Ulmer, K.Bues, and I.V.Hertel: Appl.Phys. **A48**(1989)543.

/5/ E.E.B.Campbell, G.Ulmer, B.Hasselberger and I.V.Hertel: Appl.Sur.Sci. **43**(1989)34.

/5/ E.E.B.Campbell, G.Ulmer, B.Hasselberger, H.–G.Busmann and I.V.Hertel: J.Chem.Phys., in press.

/7/ G.Ulmer, B.Hasselberger, H.–G.Busmann and E.E.B.Campbell: Appl.Sur.Sci., in press.

/8/ T.Takagi: Thin Solid Films, **39**(1976)207.
T.Yamada, T.Takagi: Thin Solid Films **80**(1981)105.

/9/ Proceedings of the Thirteenth Symposium on Ion Sources and Ion–Assisted Technology: Special Seminar (ISIAT'90); edt. by T.Takagi, Advanced Material–Processing and Machining Technology Research Association, June 1990.

/10/ Y.Yamamura: Nucl.Inst.Meth. **B45**(1990)707.

/11/ M.A.Johnson, M.L.Alexander and W.C.Lineberger: Chem.Phys.Lett. **112**(1984)285.

/12/ G.Ganteför, H.R.Siekmann, H.O.Lutz, and K.H.Meiwes–Broer: Chem.Phys.Lett., **165**(1990)293.

/13/ W.C.Wiley, I.H.McLaren: Rev.Sci.Instr. **26**(1955)50.

/14/ B.A.Mamyrin, V.I.Karataev, D.V.Shmikk and V.A.Zagulin: Soviet Phys. JETP **37**(1973)45.

/15/ H.–G.Busmann, E.E.B.Campbell, H.Gaber, B.Hasselberger, T,Müller, G.Ulmer and I.V.Hertel: to be submitted.

CHARACTERIZATION OF DIAMOND-LIKE FILMS PREPARED BY LASER ABLATION OF GRAPHITE

E.B.D. Bourdon, W.W. Duley, A.P. Jones and R.H. Prince

Department of Physics, York University
Toronto, Canada, M3J 1P3

INTRODUCTION

Deposition of diamond-like amorphous carbon films has been carried out primarily by high frequency plasma discharge and/or ion beam methods using hydrocarbon gases. Laser ablation has been used recently for film deposition of elemental materials[1,2] and the very rapid energy input can potentially produce novel materials due to the non-equilibrium conditions created at the evaporation source and the possibility for photo-chemical processes.

These amorphous films are interesting for their bonding characteristics, inherently complex due to the number of ways carbon atoms may be coordinated. Generally, the structure is defined as that of a random covalent network of aromatic regions linked by polymeric networks[3,4]. Laser ablation introduces another dimension to the growth mechanism in that the nature of the carbon vapour itself may be separately modified. Clusters containing a large number of carbon atoms in caged arrangements are formed in the laser ablation plume[5] which may lead to the formation of new materials.

The laser ablation deposition technique is inherently clean since thermal changes occur at the target rather than in its supporting structure. Introduction of buffer gases is quite unrestricted in that partial and total pressures are independently controlled. In particular, the partial pressure of carbon vapour is independent of the buffer gas mixture. For amorphous carbon films (a-C), the complete elimination of hydrogen is advantageous for interpretation of its role in the growth of hydrogenated amorphous carbon films (a-C:H).

In this work, films prepared by laser ablation of graphite were characterized by their optical band gaps, infrared absorption spectra, and morphology. Measurements of reactivity with atomic oxygen are also presented and compared with existing data for amorphous carbon films and graphite.

EXPERIMENTAL

Output from a Lumonics HyperEx 400 laser at 308 nm was focussed on a spectroscopic grade graphite rod using a 5 cm diameter, 15 cm focal length quartz lens. Laser fluence was estimated to be 500 MW/cm^2. Substrates were positioned 0.64 cm away from, and parallel to the graphite surface. The laser beam entered the vacuum chamber through a 0.32 cm thick quartz window at an angle of about 60 degrees with respect to the graphite surface.

Base pressures below 1 milliTorr were obtained by evacuation with a single stage mechanical pump having a liquid nitrogen foreline trap. Research purity hydrogen and helium gases were metered through Matheson 604 flowmeters such that the chamber pressure could be set to the desired pressure. The gas flow lines were evacuated and the whole apparatus was purged with the desired gas for about 10 minutes prior to introducing the laser beam.

Choice of substrate depended on the information being sought: polished quartz discs for VIS-UV, polished sodium chloride discs for FTIR and quartz crystals for reactivity measurements. Differing requirements in the amount of material necessary for each technique meant that analysis of a single sample by all techniques was not feasible. However, films were made in batches under the various gas environments to reduce variabilty due to differing geometries and laser parameters.

A Minuteman vacuum uv monochromator with a Hammamatsu deuterium lamp was scanned from 1500 to 5000 A. Radiation passed through the sample and was sensed by a Hammamatsu R189 phototube. The total uv-induced luminescence signal was observed using the uv setup. Measurements between 4000 and 8000 Å were performed with a Spex monochromator, a tungsten-halogen lamp, and an EMI S-20 phototube. Films were typically 150 nm thick for these VIS-UV transmission measurements. Transmission FTIR data was collected using a Nicolet 20SX system on films of about 1 micron thickness. Spectra were the average of 100 scans from 400 to 4000 cm^{-1} at a resolution of 1 cm^{-1}.

Reactivity of the films was determined by exposure to a constant flux of atomic oxygen, typically 10^{15} cm^{-2} sec^{-1}. Films were deposited on gold coated quartz crystals and the resonant frequency shift due to the areal mass could be followed as a function of time or atomic oxygen fluence. Translational energies as high as 2.5 eV for atomic oxygen were achieved by the supersonic expansion from a He/O_2 plasma[6]. In situ microweighing via the quartz crystal technique[7] used 6MHz, AT-cut, plano-convex crystals having a mass sensitivity for carbon of 12ng Hz^{-1} cm^{-2} or 0.1 monolayers Hz^{-1}. Pre- and post- deposition weighing of the quartz crystals allowed calibration of frequency shift vs. mass change, along with the deposit area.

RESULTS

Optical Band Gap Measurements

Table 1 summarizes the optical band gap data. Values from this work were obtained from a Tauc plot of transmission data. Comparing with band gaps from other studies, the data are seen to be consistent in range with previous results for films produced by plasma discharge or ion beam techniques[8].

The size of the band gap may be correlated with the size of graphitic regions in the film[3,4]. Films made in helium having a 0.69 eV gap have regions consisting of 34 to 60 fused 6-membered rings in a compact arrangement 1.5 nm in size. Vacuum deposited films then have graphitic islands consisting of over 60 rings with bonding between islands involving C-C single bonds with the planar graphitic sheets at random angles to each other. This accounts for the metallic appearance of these films.

As the size of graphitic clusters decreases, the band gap increases[3]; this is accomplished by inhibiting the growth of clusters with a reactive reagent or breaking up clusters with ion bombardment. Films produced in 5 Torr H_2, having a 1.5 eV band gap, would have clustering into 2 to 8 rings of 0.3 to 0.6 nm size. Very large band gaps of 3 eV for films made in 100 Torr H_2 may still contain sp^2 clusters, however these would be widely separated by an sp^3 matrix. Uv-induced luminescence observed for these wide band gap films is also a consequence of very localized graphitic regions[3].

Table 1. Summary of Optical Band Gap Data

Deposition Method	Band Gap,eV
Laser Ablation in Vacuum (this work)	0.14±0.06
Laser Ablation in 5 Torr He (this work)	0.69±0.13
Laser Ablation in 5 Torr H_2 (this work)	1.56±0.04
Laser Ablation in 100 Torr H_2 (this work)	3.01±0.04
Rf Plasma Discharge[9]	1.1-3.0
DC Glow Discharge[10]	2.6
Rf Plasma Discharge[11]	3.02
RF Sputtering, H_2 reactive[12]	1.4-2.8
DC Sputtering[13]	0.4-0.74
Ion Beam Deposition[14]	0.96-1.44
Vacuum UV Laser Ablation[15]	0.24
Low Temperature Vapour Deposition[1]	0.82-1.32

Films made by laser ablation have the ingredients required to decrease cluster size via the ionic component of the plasma at higher laser power. The presence of uv radiation,however, tends to enhance the formation of graphitic clusters since radiation above 3.5 eV is sufficient to photolyze C-C or C-H single bonds. This is demonstrated qualitatively by observing laser treated films during film ablation for thickness measurements: treatment of hydrogenated films caused the films to change to a black deposit indicating the loss of hydrogen and formation of a more graphitic material subsequently more easily removed by further laser irradiation.

FTIR Absorption Assignment

Figure 1 displays the IR transmission spectra for the various films produced. Absorptions observed in 0% hydrogen are taken to be due to the skeletal vibrations of carbon. This conclusion is further strengthened by the spectra of films formed in D_2 gas where the frequencies below 1700 are not altered, except for the expected shift of the 1450 and 1375 cm^{-1} C-H bending modes. H_2 impurity in the D_2 gas shows up by the persistence of C-H stretch frequencies in figure 1a. The assignment of absorptions in the 1300 to 900 cm^{-1} range to oxygen containing species is ruled out since such species would enter as a small impurity; the assignment of features in the fingerprint region must be furthermore consistent with features which occur in the functional group region, above 1600 cm^{-1}. Absorptions due to oxygen or nitrogen containing groups above 1600 cm^{-1} are absent.

In common with previously given assignments[16], absorptions between 1300 and 850 cm^{-1} are due to mixed sp^2/sp^3 bonded regions and sp^3 C-C bonds. Branched chain structures and non-aromatic ring systems would absorb in this range. The sharp feature at 800 cm^{-1} is not previously assigned as a C-C absorption. Noting that in hydrogenated films, the sharp features prominent in the CH stretch region near 2900 are narrower, the 800 cm^{-1} feature may also be related to sp^3 or olefinic sp^2 carbon. The narrow bandwidths of the 1260 and 800 cm^{-1} absorptions possibly indicates high film homogeniety in the structures characterized by these absorptions.

C-H stretching vibrations of saturated structures are evident at 2850, 2870, 2920, 2960 cm^{-1}. Weak absorptions above 3000 cm^{-1} are attributed to stretching vibrations of aromatic sp^2 C-H, cycloalkane structures, and sp^1 C-H at 3300 cm^{-1}. On the basis of these assignments, substitution of deuterium for hydrogen gas at 5 Torr pressure produces features for C-D stretch frequencies at 2583 and 2117 cm^{-1} consistent with reduced mass changes.

Increasing the hydrogen gas pressure to 100 Torr gives IR absorption spectra which differ greatly over the region below 1700 cm^{-1}. Very few features appear except for a broad absorption centered at 1080 cm^{-1} with fwhm of 200 cm^{-1}. This is indicative of increased reactive encounters between the buffer gas and carbon species within the laser ablation plume, thereby prohibiting the formation of structures present without the reactive component.

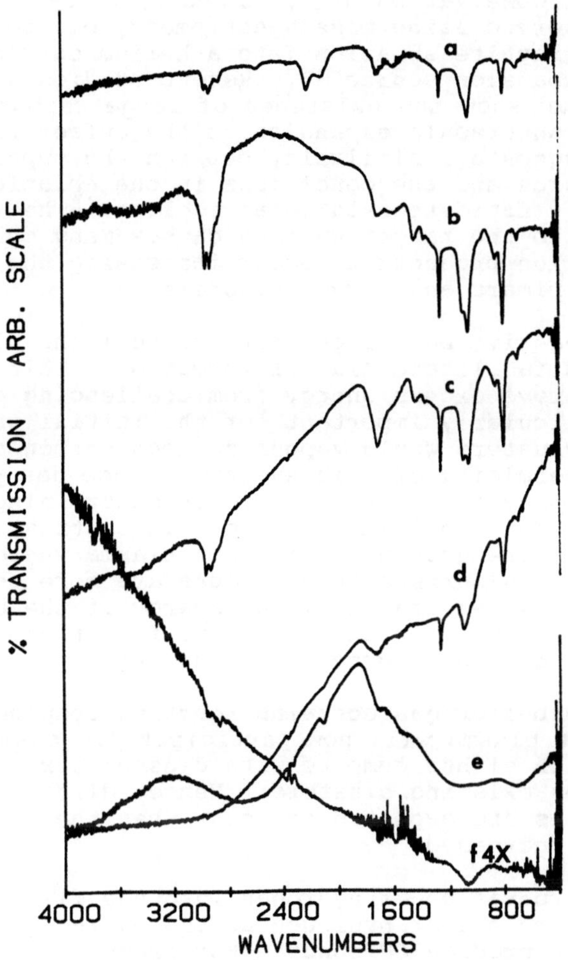

Figure 1. FTIR transmission spectra for films made at : a) 5 Torr D_2, b) 5 Torr H_2, c) 5 Torr 1:1, He:H_2, d) 5 Torr He, e) vacuum deposit, f) 100 Torr H_2.

Role of the Buffer Gas

Optical emission studies of the luminescent plume over laser ablated graphite surface have been reported[17]. Prominent emission features were the Swan bands from C_2, and from excited C, CI, CII, and CIII via electron-ion recombination. The Swan bands were seen to increase in intensity in a higher pressure background of helium probably due to recombination of C monomers. The existence of monomers and small carbon units is also verified by the mass spectrometric observation of C, C_2 and C_3[15]. Carbon clusters have been observed using mass spectrometry of the vapour stream from graphite ablation into a helium carrier gas in a supersonic expansion source[18,19]. Recent studies of the laser ablation plume[5] show the existence of large carbon clusters even without supersonic expansion of the buffer (carrier) gas. This suggests a similarity between the supersonic expansion source and the conditions in the ablation plume itself, namely rapid translational cooling. The very high density next to the target surface causes many collisions between ablation products to occur increasing the likelihood for dimers, trimers and larger clusters.

A non-reactive buffer gas acts to cool the plume particles and facilitate cluster formation. Third body collisions remove excess energy from coallescing clusters which is particularly important for the initial growth, since small clusters would vapourize upon carbon addition via the exothermicity of this event[20]. Hence gas phase interactions play an important role in determining the growth medium for the film produced. In contrast, films made in vacuum are growing from a hot plasma where smaller particles and gas-surface interactions are more common. In addition, the excess translational energy of the carbon species is available for annealing the film thereby enhancing the growth of graphitic clusters.

When the buffer gas contains reactive components such as H_2, the hot plasma will now participate in chemical reactions which either compete with cluster formation or react with the existing clusters. Hence, differences in film properties are expected to occur when the reactive component is increased.

The incorporation of hydrogen into the films is possible with molecular hydrogen as the buffer gas via reactions that produce CH bonds. For example it is known that molecular hydrogen reacts with ground state carbon atoms to produce CH + H at high incident kinetic energies, and to CH_2 at low kinetic energies[21]. Reaction with ^1D carbon (1.263 eV above ground state) produces CH + H with no activation energy required)[22]. Carbon cluster cations may be hydrogenated by reactions with H_2 which proceed at or near the collision rate[23]. Cation cluster-H_2 collisions may also be a mechanism for charge neutralization.

It is suggested that the films made at 5 Torr in helium or hydrogen will contain a larger proportion of larger carbon clusters as compared to those produced at 100 Torr hydrogen. The sharp C-C skeletal IR features in the fingerprint region may be attributed to vibrational frequencies of specific carbon cluster structures which have condensed onto the substrate. These features weaken at higher hydrogen pressure as $H_2 + C$, $H_2 + C_n$, and $H_2 + C_n^+$ reactions will be more frequent, inhibiting formation of larger clusters.

Reactivity with Atomic Oxygen

The observed linear frequency shift with exposure to atomic oxygen indicates a chemically homogeneous film which is consumed by zero order kinetics, hence the rate of change in mass is directly proportional to the incident atomic oxygen flux. Data showing the measured atomic oxygen reaction probability are shown in table 2 where calculations from the present work presume atomic oxygen reacts to form volatile CO.

Since the reactivity at 5 eV is similar to that at 2.5 eV and much larger than that at 1 eV the activation energy for the reaction should be between 1 and 2.5 eV. The decreased reactivity for vacuum deposited films may be due to their higher density or surface morphology compared to the high pressure amorphous films. SEM photos reveal the vacuum films to be very smooth compared to the globular appearance of films made in He or H_2. A large graphite network would be quite unreactive since this would require activation by the energy necessary for breaking aromatic sp2 C=C bond of 6.3 eV/bond, not including the energy necessary to disrupt the long range lattice field.

CONCLUSIONS

The present study has found that films produced by laser ablation of graphite are similar to films made by other methods. These films are taken to be diamond-like in that there is a substantial amount of non-aromatic carbon

Table 2. Summary of reaction probability of atomic oxygen with carbon and carbon films.

Material	Atomic Oxygen Source	Reaction Probability
pyrolytic graphite	rf O_2 plasma, low translational energy	0.01[24]
a-C film	supersonic beam, 1 eV translational energy	0.01[25]
a-C film	low earth orbit exposure on STS-8, 5 eV translational energy	0.2[26]
a-C, vacuum deposit	supersonic beam,2.5 eV translationl energy	0.071±10% this work
a-C, 5 Torr He and a-C:H, 5 Torr H_2	supersonic beam,2.5 eV translational energy	0.156±13% this work

accounting for increased band gaps. Infrared features in the fingerprint region have been attributed to skeletal carbon in structures which are abundant in the laser produced carbon vapour. Raman and NMR spectroscopy of the films would be useful tools in further determining the bonding nature of the films. Since the plume contains large carbon clusters, examination of the plume by IR techniques would be enlightening with regard to the the assignment of IR absorption frequencies. The buffer gas is demonstrated to be vital to the control of film properties by modifying the growth medium above the substrate via translational cooling and chemistry within the ablation plume. The effect of uv radiation on film properties is indicated as an important factor both during and after growth. A quantitative study of the laser ablation plume and its interaction with the buffer gas is essential to understanding and controlling film attributes.

Data for the reactivity of these amorphous films with atomic oxygen indicate results similar to other measurements obtained at large atomic oxygen kinetic energies. Realizing that the reaction is activated, it would be interesting to map out the reactivity as a function of incident kinetic energy. In addition, it would be necessary to conclusively determine the reaction products in order to understand the reaction mechanism. Vacuum deposited films appear to be less reactive due to their larger graphitic regions.

REFERENCES

1. M. Ogmen, and W. W. Duley, Astrophys. J. 334:117 (1988).
2. T. Sato, S. Furuno, S. Iguchi, Jpn. J. Appl. Phys. 26:L1487 (1987).
3. J. Robertson, Phil. Mag. Lett. 57:143 (1988).
4. J. L. Bredas and G. B. Street, in: "Hydrogenated Amorphous Carbon Films," E-MRS Symposia Proceedings, Vol XVII Strasbourg, France (1987).
5. J. T. Brenna and W. R. Creasy, J. Chem. Phys. 92(4):2269 (1990).
6. W. D. Morison, R. C. Tennyson, J. B. French, T. Braithwaite, M. Moisan, and J. Hubert, in: "15th Space Simulation Conference," NASA Conference Publication 3015, Williamsburg, VA (1988).
7. C. Lu, J. Vac. Sci. Technol. 12:578 (1975).
8. J. C. Angus, P. Koidl, and S. Domitz, in: "Plasma Deposited Thin Films," CRC Press, Boca Raton, Fla (1986).
9. R. H. Jarman, G. J. Ray, R. W. Standley, and G. W. Zajac, Appl. Phys. Lett. 49:1065 (1986).
10. D. A. Anderson, Phil. Mag. 35:1 (1977).
11. J. J. Zelez, Vac. Sci. Technol. A1:305 (1983).
12. T. Miyasoto, Y. Kawakami, T. Kawano, A. Hiraki, Jpn. J. Appl. Phys. 23:L234 (1984).
13. N. J. Savvides, J. Appl. Phys. 59:4133 (1986).
14. J. Ishikawa, Y. Takeirei, K. Ogawa, and T. J. Tagagi, J. Appl. Phys. 61:2509 (1987).
15. W. W. Duley, Astophys. J. 287:694 (1984).
16. B. Dischler, in: " Amorphous Hydrogenated Carbon Films," E-MRS Symposia Proceedings, Vol XVII Strasbourg, France (1987).

17. G. Koren and J. T. C. Yeh, J. Appl. Phys. 56:2120 (1984).
18. D. M. Cox, K. C. Reichmann, and A. Kaldor, J. Chem. Phys. 88:1588 (1988).
19. E. A. Rohlfing, D. M. Cox, and A. Kaldor, J. Chem. Phys. 81:3322 (1984).
20. J. Bernholc and J. C. Phillips, J. Chem. Phys. 85:3258 (1986).
21. H. O'Kabe, "Photochemistry of Small Molecules," J. Wiley & Sons, New York (1978).
22. W. H. Fisher, T. Carrington, C. M. Sadowski, C. H. Dugan, Chem. Phys. 97:433 (1985).
23. D. K. Bohme and S. Wlodek, in press : Int. J. Mass Spectrom. Ion Processes (1990).
24. G. S. Arnold, and D. R. Peplinski, AIAA J. 24:673 (1984).
25. C. Park, AIAA J. 14:1640 (1976).
26. L. E. Bareiss, R. M. Payton, and H. A. Papazian, in: "NASA Contract Report 4053," pages 16-21, sect.3 (1987).

[illegible] *Appl.* [illegible] (1981).

18. [illegible] Cox, [illegible] *[illegible]* 88: 508 (1962).

19. [illegible] *J. Chem. Phys.* 31: 1158 (1959).

20. [illegible] Phillips, *J. Chem. Phys.* [illegible] (1961).

21. [illegible]

22. [illegible]

23. [illegible]

24. [illegible]

25. [illegible]

26. [illegible]

RF PLASMA DEPOSITION OF a-C:H FILMS: DIAGNOSTICS AND MODELING

V. Barbarossa, O. Martini, S. Mercuri,
R. Tomaciello and F. Galluzzi

Eniricerche S.p.A., via E. Ramarini, 32
00015 Monterotondo (Rome), Italy

INTRODUCTION

The low temperature deposition of amorphous carbon films from hydrocarbons plasmas has been the subject of several recent works[1-3] but the relationship between plasma processes and film growth is still not completely understood. As a contribution to this topic we report an experimental and theoretical study of plasma processes in the RF deposition of a-C:H films from methane/hydrogen gas mixtures. The plasma is characterized by both electrical probes and optical emission spectroscopy, obtaining electron density, electron mean energy, atomic hydrogen and CH radical concentrations. Deposition rates of carbon layers are measured as a function of feedstock gases composition. A simplified kinetic model of plasma bulk is also developed and calculated equilibrium distributions of chemical species are compared with experimental data.

EXPERIMENTAL

The deposition apparatus is shown in Fig. 1a. The glow discharge reactor is constituted by stainless steel vacuum chamber. In order to obtain a large self-bias voltage a small area RF powered plate is surrounded by a large area grounded electrode. The electrode asymmetry ratio is about 0.09 and the maximum self-bias voltage is about -500 V. A background pressure of 10^{-6} torr before each deposition run and a dynamic vacuum of 0.1-0.2 torr during the process are obtained by turbomolecular and rotary pumps. Methane and hydrogen are regulated by MKS mass flow controllers. The total flow is maintained at 50 sccm, while the methane concentration is changed from 1 to 100 %. A π matching network connects the RF power supply (13.56 MHz) to the electrode.
A schematic view of the diagnostic set-up is shown in Fig. 1b. Tungsten electrical double probes (l=2 mm; d=0.3 mm) are connected to voltage/frequency and frequency/voltage double converter in order to keep them floating. The plasma emission is analyzed by a high-luminosity grating monochromator (1800 g/mm) and a multiarray detector (OMA III, EG&G).
All the samples were deposited at the experimental conditions listed in Table 1.

Diamond and Diamond-Like Films and Coatings
Edited by R.E. Clausing *et al.*, Plenum Press, New York, 1991

Table 1. Experimental conditions for deposition

Pressure	0.1-0.2 torr
RF power	25-100 W
% CH_4 in H_2	3-100

RESULTS AND DISCUSSION

Deposition rate

Carbon films were deposited on Si substrates and films thicknesses were measured by using a profilometer (Alpha-Step 200 Tencor). Deposition rate is affected by both RF power and plasma pressure at constant gas feed composition. The effect of RF power at total pressure of 0.15 torr and 5 % of methane concentration is reported in Table 2. The behaviour of deposition rate versus methane percentage at plasma power of 50 W and pressure of 0.2 torr is plotted in Fig. 2. The deposition rate increases up to about 1.5 Å/s when the film is produced at 20% of methane; then it remains approximately constant up to methane content of 70 %, and finally rapidly increases at higher CH_4 percentages. The etching rate of a a-C:H film, produced at 100% of methane, has also been measured. The value obtained in a pure hydrogen plasma at 50 W and 0.2 torr is shown as an open circle in Fig. 2.

Plasma diagnostics

Two I/V characteristics of tungsten electrical probe at different total pressure are reported in Fig. 3. Mean electron energy, obtained under the assumption of Maxwellian distribution[4], are listed in Table 3.

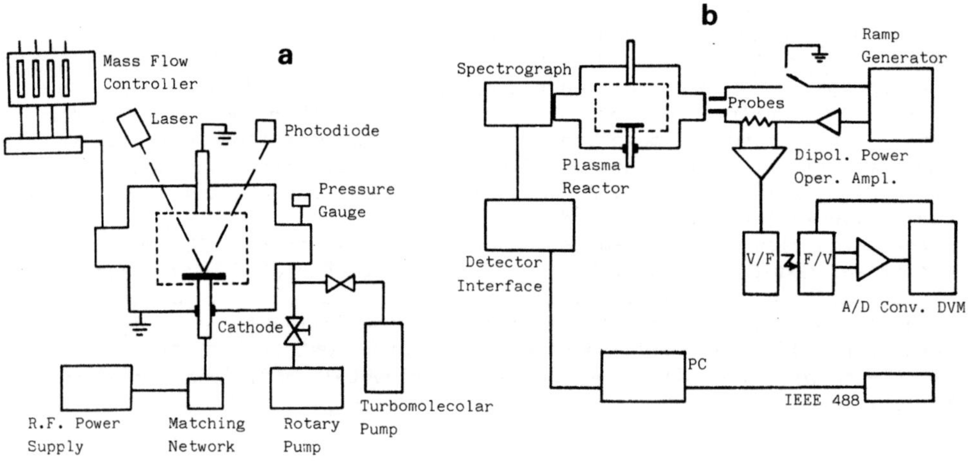

Fig. 1. Schematic diagram of the experimental apparatus: a) deposition system; b) diagnostic system.

Table 2. Deposition rate vs RF power

P=0.15 torr; CH_4= 5%			
W(W)	25	50	100
Rd(Å/s)	0.2	0.3	0.007

Table 3. Mean electron energy and electron density values

W= 50 W; CH_4= 5%		
P(torr)	0.15	0.5
Te (eV)	5	3

Emission spectra obtained from a plasma of pure CH_4, 10% of CH_4, and 3% of CH_4 are reported in Fig. 4a,b,c respectively. H_2 and H_α lines are evident also in 4a spectrum, while the CH 4300 Å system corresponding at $A^2\Delta \rightarrow X^2\Pi$ transition is the strongest signal from carbon-containing species. For CH 3900 Å (B->X) system, the (0,0) lines are also clearly detected.
The emission intensity of CH (A->X) and H lines as a function of methane fraction are shown in Fig. 5. CH and H_α intensities exhibit opposite trends: the former increasing with the CH_4 content, the latter decreasing.

Modeling of the plasma

In our model we focus the attention on the chemical processes in the plasma bulk where electrons, radicals, ions and neutral species are homogeneously distributed. We consider a closed system where the diffusion and drift effects are completely neglected. Under these assumptions, a rate equation of the form:

$$dN_i/dt=\Sigma_j\, n_e K_{ij} N_j + \Sigma_{j,l}\, K_{ijl} N_j N_l - N_i(\Sigma_j\, n_e K_{ji} + \Sigma_{j,l}\, K_{jil} N_l) \qquad (1)$$

can be written for each neutral and positive ion species included in the model. In the equation (1) n_e is the electron density, N_i is the number density of i-th species, K_{ij} the electron impact reactions rate coefficient for the production of i-th species by collision with j-th species, K_{ijl} the rate coefficient to produce the ith species by collision between j-th and lth species. The first and second terms describe the formation channels while the third and the last the loss channels. To keep charge neutrality condition, our model includes an electron rate equation of the form:

$$dn_e/dt=\Sigma_{i,j}\, n_e K^*_{ij} N_i - K_r n_e^2 \qquad (2)$$

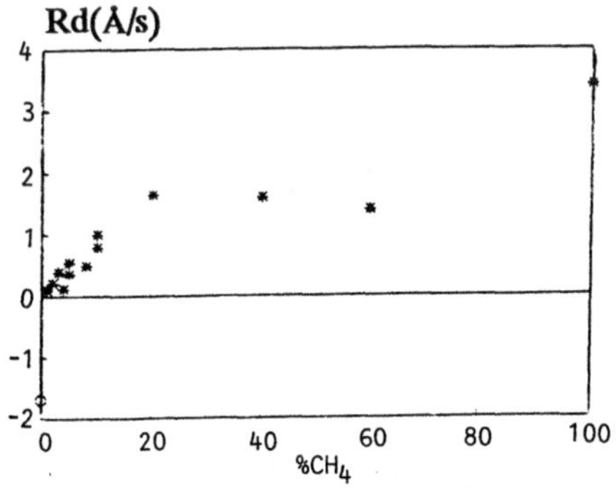

Fig. 2. Deposition rate vs. gas mixture composition measured at RF power of 50 W and pressure of 0.2 torr.

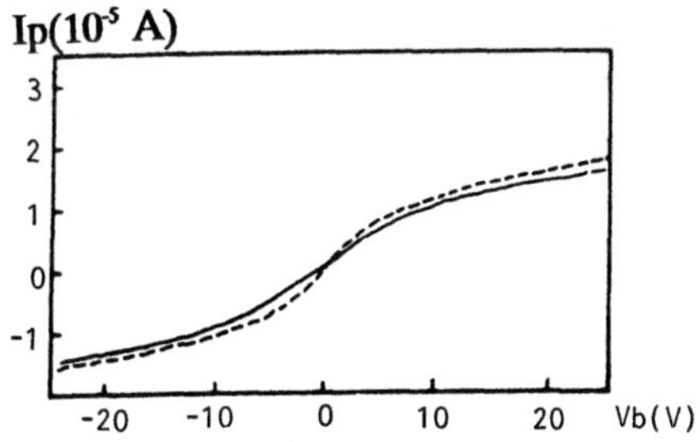

Fig. 3. I/V electrical double probe characteristics of 5% CH_4 plasma at 50 W in different total pressure: (——) 0.15 torr; (– –) 0.5 torr.

where K^*_{ij} is the ionization electron impact reactions rate coefficient and K_r the rate coefficient for ions electron attachment reactions. This value is assumed the same for all the ionic species and has been calculated from experimental electron density. The gas phase reactions of eighteen neutral and ionic species which have been considered in the present work are listed in Table 4 together with their rate coefficients[5]. The electron impact reaction rate coefficients are kept constant. A gas temperature of 300 K, a gas pressure of 0.2 torr and a RF power of 50 W are assumed throughout the modeling. Solution of the stiff ordinary differential equations set is obtained by Gear method. Calculations have been performed over different gas mixture compositions. It can be seen from the results of the calculations that the most abundant radical species in the plasma are H and CH_3 over the whole composition range. Similar results are reported by Kojima et al.[6] and Tachibana et al.[7], although they used different methods. At high CH_4 percentages the most abundant neutral species is C_2H_4. The most abundant ionic species predicted by the model are CH_3^+ and CH_4^+ over the whole composition range. An example of the results for the number densities of dominant species calculated as a function of the gas mixture composition is shown in Fig. 6. We also calculated the ratio CH/H as a function of methane concentration, finding a good agreement with experimental data (Fig. 7).

CONCLUSIONS

The low temperature deposition of amorphous carbon films by RF decomposition of methane-hydrogen mixtures has been studied as a function of CH_4 content. Experimental data and calculated results from a theoretical model of the plasma bulk have been compared.

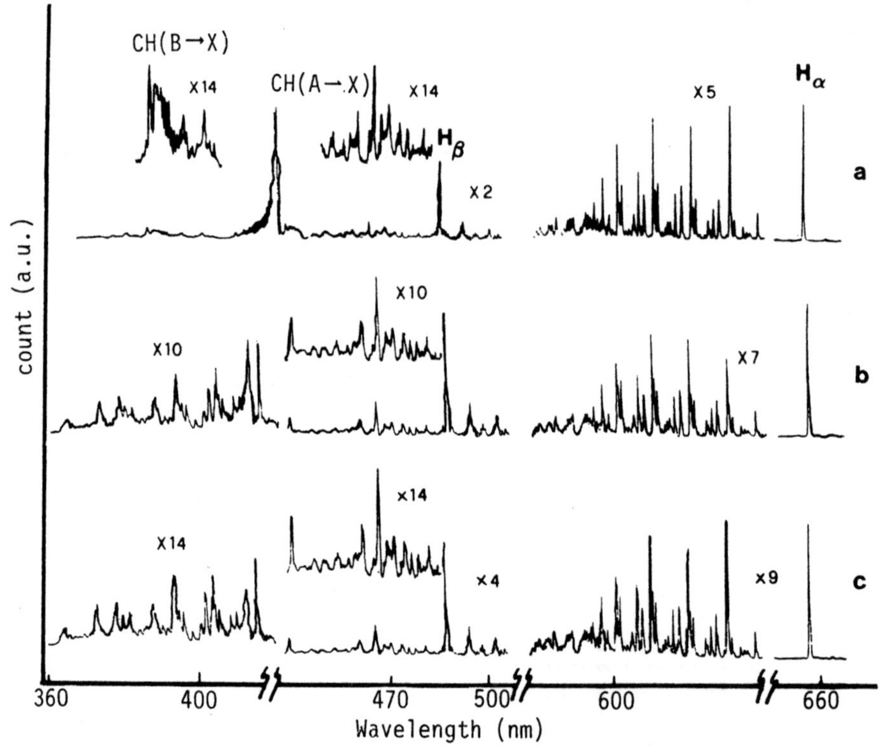

Fig. 4. Plasma optical emission spectra of: a) pure CH_4; b) 10% of CH_4; c) 3% of CH_4.

Table 4. Chemical reactions and relative rate coefficients (cm^3/s)

$e + CH_4 \rightarrow CH_3 + H + e$	$4.5\ 10^{-8}$	$C_2H_5 + H \rightarrow CH_3 + CH_3$	$6.0\ 10^{-11}$
$e + CH_4 \rightarrow CH_2 + H_2 + e$	$7.3\ 10^{-9}$	$C_2H_5 + CH_3 \rightarrow C_3H_x$	$4.2\ 10^{-11}$
$e + CH_4 \rightarrow CH + H_2 + H + e$	$3.7\ 10^{-9}$	$C_2H_4 + H \rightarrow C_2H_5$	$2.8\ 10^{-13}$
$e + CH_4 \rightarrow CH^* + H_2 + H + e$	$4.0\ 10^{-10}$	$CH_2 + H \rightarrow CH + H_2$	$2.7\ 10^{-10}$
$e + CH_4 \rightarrow CH_3 + H^* + e$	$8.0\ 10^{-11}$	$CH + CH_4 \rightarrow C_2H_5$	$1.0\ 10^{-10}$
$e + C_2H_6 \rightarrow C_2H_5 + H + e$	$1.7\ 10^{-7}$	$CH + C_2H_4 \rightarrow C_2H_x$	$2.0\ 10^{-10}$
$e + C_3H_x \rightarrow C_2H_4 + CH_4 + e$	$2.3\ 10^{-7}$	$CH + C_2H_6 \rightarrow C_3H_x$	$4.0\ 10^{-10}$
$e + C_4H_x \rightarrow C_3H_4 + CH_4 + e$	$2.8\ 10^{-7}$	$CH + C_3H_x \rightarrow C_4H_x$	$3.0\ 10^{-10}$
$e + H_2 \rightarrow H + H + e$	$2.0\ 10^{-8}$	$H + H \rightarrow H_2$	$1.0\ 10^{-15}$
$e + CH_4 \rightarrow CH_4^+ + 2e$	$3.4\ 10^{-8}$	$CH_4^+ + CH_4 \rightarrow CH_5^+ + CH_3$	$1.5\ 10^{-9}$
$e + CH_4 \rightarrow CH_3^+ + H + 2e$	$3.2\ 10^{-8}$	$CH_3^+ + CH_4 \rightarrow C_2H_5^+ + H_2$	$1.2\ 10^{-9}$
$e + C_2H_4 \rightarrow C_2H_4^+ + 2e$	$3.0\ 10^{-7}$	$CH_5^+ + C_2H_6 \rightarrow C_2H_5^+ + H_2 + CH_4$	$5.0\ 10^{-10}$
$e + C_2H_6 \rightarrow C_2H_4^+ + H_2 + 2e$	$3.0\ 10^{-7}$	$C_2H_4^+ + C_2H_4 \rightarrow C_4H_x^+$	$4.3\ 10^{-10}$
$e + C_3H_x \rightarrow C_2H_5^+ + CH_3 + 2e$	$4.0\ 10^{-7}$	$C_2H_4^+ + C_2H_6 \rightarrow C_3H_x^+$	$1.3\ 10^{-11}$
$e + C_4H_x \rightarrow C_3H_x^+ + CH_3 + 2e$	$5.0\ 10^{-7}$	$C_2H_5^+ + C_2H_4 \rightarrow C_3H_x^+ + CH_4$	$3.1\ 10^{-10}$
$CH_3 + CH_3 \rightarrow C_2H_6$	$3.7\ 10^{-11}$	$C_2H_5^+ + C_2H_4 \rightarrow C_4H_x^+$	$3.0\ 10^{-10}$
$CH_3 + H \rightarrow CH_4$	$7.0\ 10^{-12}$	$C_2H_5^+ + C_2H_6 \rightarrow C_4H_x^+ + H_2$	$1.0\ 10^{-11}$

The deposition process can be seen as the competition between sticking processes of carbon radicals and ablation processes by both carbon ions (sputtering) and atomic hydrogen (etching). Since a two-order magnitude change in the methane content (from 1 to 100%) affects only slightly the atomic hydrogen concentration (its change being about a factor 1.5-2.5), the contribution of the etching process can be considered constant and the variations of deposition rate with methane fraction can be mainly correlated to the variations of carbon radical and ion concentrations. Following Kobayashi et al.[8], we checked for the correlation between deposition rates of cathodic film and CH concentration. Indeed, as shown in Fig. 7, a quite

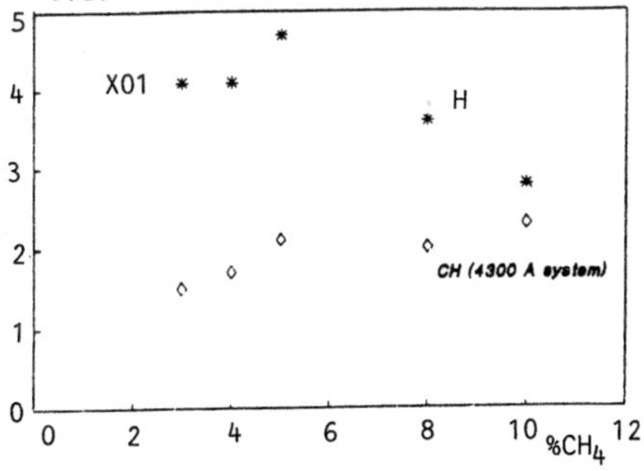

Fig. 5. Emission intensity of CH and H as a function of CH_4 percentage.

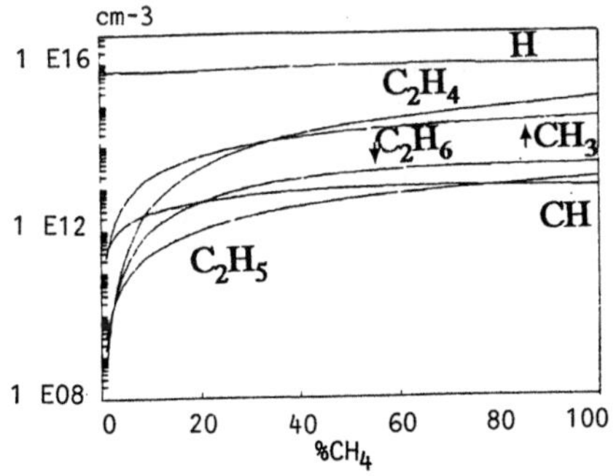

Fig. 6. Predicted species densities vs gas mixture composition.

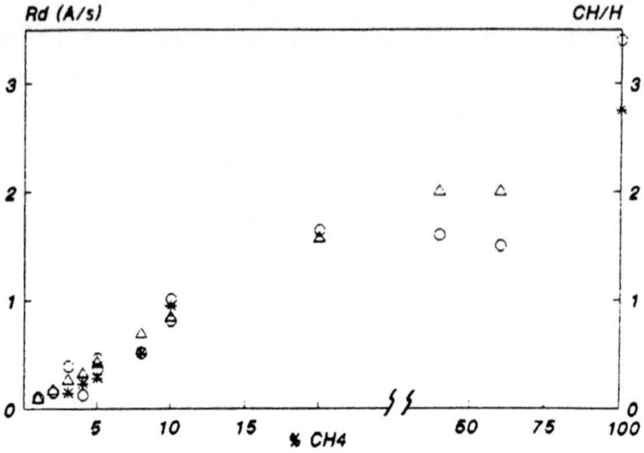

Fig. 7. Comparison between deposition rate and CH/H ratio trends: (o) Rd; (*) optical data: (Δ) theoretical data.

good agreement is obtained between experimental deposition rates and experimental CH/H values as a function of feedstock composition. The agreement is good also with theoretical CH/H trend, although the model is approximate. This result could suggest a direct role of CH radicals in the film growth, as reported by Koidl et al.[9], and a strong linkage between the processes in plasma bulk and those on electrodes.

REFERENCES

1. K. E. Spear, Diamond-Ceramic coating of the future, J. Am. Cer. Soc. 72:171 (1989).
2. J. C. Angus, and C.C. Hayman, Low-pressure,metastable growth of Diamond and "Diamondlike" phases, Science 241:913 (1988).
3. A. Bubenzer, B. Dischler, G. Brand, and P. Koidl, RF-plasma deposited amorphous hydrogenated hard carbon thin films: Preparation, properties, and applications, J. Appl. Phys. 54:4590 (1983).
4. N. Matsukura, M. Ohuchi, S. Satoh, and Y. Machi, The analyses of an SiF_4 plasma in an R.F. glow discharge for preparing fluorinated amorphous silicon thin films, Thin Solid Films 109:47 (1983).
5. L. E. Kline, W.D. Partlow, and E. Bies, Electron and chemical kinetics in methane RF glow-discharge deposition rates, J. Appl. Phys. 65:70 (1989).
6. H. Kojima, H. Toyoda, and H. Sugai, Observation of CH_2 radical and comparison with CH_3 radical in a RF methane discharge, Appl. Phys. Lett. 55:1292 (1989).
7. K. Tachibana, M. Nishida, H. Harima, and Y. Urano, Diagnostics and modelling of a methane plasma used in the chemical vapour deposition of amorphous carbon films, J. Phys. D: Appl. Phys. 17:1727 (1983).
8. K. Kobayashi, N. Mutsukura, and Y. Machi, Deposition of hard carbon films by RF glow discharge method, J. Appl. Phys. 59:910 (1986).
9. J. Wagner, Ch. Wild, F.Pohl, and P. Koidl, Optical studies of hydrogenated amorphous carbon plasma deposition, Appl. Phys.Lett. 48:106 (1986).

FROM AMORPHOUS CARBON TO AMPHOROUS DIAMOND-LIKE CARBON

B. André, J-Ph. Nabot, L. Lombard and P. Martin*

CEREM-SGS
*LETI-CMO
CEA-Centre d'Etudes Nucléaires de Grenoble
BP 85 X, 38041 GRENOBLE cedex, France

1 INTRODUCTION

Among the various carbon and hydrocarbon PVD coatings, only a few may be called diamond-like carbon (DLC) films. Angus proposed a classification based on the density and hydrogen content of the coatings determining their diamond-like character[1]. The high density is related to the large fraction of tetrahedral chemical bonds (diamond sp3 hybridization) observed in DLC and the correlatively reduced amount of trigonal bonds (graphite sp2 hybridization). The hydrogen content directly parallels the increase of the average coordination number.

The purpose of this study was to analyze more carefully the deposition parameters acting strongly on the film properties and to determine, within these parameters, critical values for which the diamond-like character was enhanced. The hydrogen content, measured with Elastic Recoil Detection Analysis and infrared spectroscopy techniques, was maintained as low as possible in order to avoid its influence on film properties. The working pressure, the deposition rate and the characteristics of the ion beam assisting the growth of the films were selected.

The effect of the pressure and deposition rate was studied, either with an High Pressure DC Magnetron (HPM) operating in the $5x10^{-1}$ to 3 Pa range, or with low pressure devices: ion beam sputtering and low pressure DC magnetron working at $4x10^{-2}$ and $6x10^{-2}$ Pa respectively.
The assisting bombardment parameters were varied using ion beam devices. The ion beam source, thanks to a remote plasma confi-

Diamond and Diamond-Like Films and Coatings
Edited by R.E. Clausing *et al.*, Plenum Press, New York, 1991

guration, offered a better and more independant control of bombardment parameters than a DC biased substrate. First results were obtained with a Dual Ion Beam System (DIBS). In order to investigate a wider range of the energy per deposited carbon atom, an Ion Beam Assisted low pressure DC Magnetron (IBAM) was choosen for its higher deposition rate.

2 APPARATUS AND PROCEDURE

The ion beam assisted devices, DIBS and IBAM were described elsewhere[2]. The common feature of both systems is a broad beam ion source of the Kaufman type bombarding the film during growth. Sputtering of a graphite target was performed either with an ion gun of the same type (DIBS), or with a low pressure planar DC magnetron (IBAM). Sputtering occurred with argon ions, whereas ion bombardment was carried out with various inert gases. The deposition rate was lowered by the bombardment which induced sputtering of the film, and thus was related to the flux, energy, and mass of the impinging ions. All these experiments were performed at a working pressure of a few 10^{-2} Pa.

Low pressure deposited films (LP films) were obtained with the DIBS and IBAM facilities without assistance. The deposition rate was 2.5 $nm.min^{-1}$ for the former and 6.5 $nm.min^{-1}$ for the latter. High pressure deposited coatings (HP films) were produced by sputtering graphite and pyrocarbon targets with a planar DC magnetron under an argon pressure ranging from $5x10^{-1}$ to 3 Pa. Under these conditions, high deposition rates were achieved: 30 to 45 nm min^{-1} for a 1,000 to 1,500 W discharge power. The main operating conditions are summarized in table 1.

Table 1. process parameters for carbon film deposition.

		HPM	LOW PRESSURE DEVICES	
			IBAM	DIBS
SPUTTERING PARAMETERS	Working pressure Pa	$6.5.10^{-1}$ to 3	6.10^{-2}	4.10−2
	Discharge power W	1000−1500	500 to 1000	(focused beam) 1000 eV 5 $mA.cm^{-2}$
	Deposition rate $nm.min^{-1}$	33−45	0 to 6.5	0 to 2.5
ASSISTING PARAMETERS	Ion energy eV	0	0 to 300	0 to 200
	Ion flux $mA.cm^{-2}$	0	0 to 0.25	0 to 0.11

3 RESULTS AND DISCUSSION

3.1 Effects of the Working Pressure and Deposition Rate

Figure 1 presents the morphology of two coatings obtained under high (2.6 Pa) and low pressure (4×10^{-2} Pa). The latter shows a very compact structure without any columnar tendency, while the HP film appears in the form of a densely packed columnar structure. The maximum adherent thickness was also dependant on the pressure: 1.5 µm for HP films instead of 0.3 µm for LP films on steel. Beyond these values, high compressive stresses induced film cracking and peeling.
These results are consistent with Thornton's diagram illustrating the dependance of the film morphology on the deposition pressure[3]. The better adherence for high pressure coatings can therefore be explained by recovery processes which occurred for the columnar layers and limited their intrinsic stress level.

Density was calculated from Rutherford Backscattering Spectrometry (RBS) and thickness determinations. The density of high pressure coatings (0.5 Pa) was equal to 1.5 $g.cm^{-3}$ as compared to 2 $g.cm^{-3}$ for low pressure conditions (6×10^{-2} Pa). This last value corresponds roughly to the lower limit of the amorphous diamond-like carbon domain as stated by Angus. The hardness, obtained with a Knoop device under a load of 5 g, was 450 MPa for the HP films and 17,000 MPa for the LP ones. Thus, beyond critical values of the deposition pressure, the amorphous carbon exhibits diamond-like properties.

Microscopic investigations were performed using Transmission Electron Microscopy (TEM) and High Resolution TEM (HRTEM)

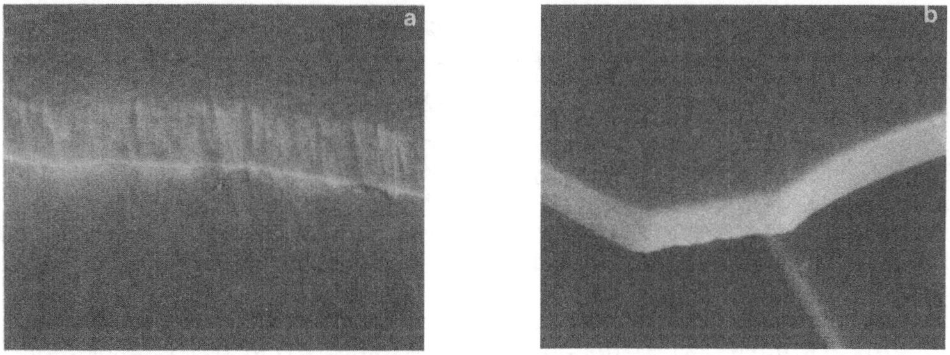

1 µm　　　　1 µm

Figure 1. Fractographs of carbon films
a. HP film (Pr.= 2.6 Pa)　　b. LP film (Pr.= 4×10^{-2} Pa)

in order to combine these dramatic changes of the macroscopic properties with structural modifications of the material.
The diffraction pattern of the LP material was representative of amorphous carbon. An additional diffraction ring was observed on the HP film diffractogram and was attributed to the presence of crystallites. HRTEM actually confirmed the existence of nano-crystallites in the HP film with a size ranging from a few tenths of nm to 3 nm. Two interatomic distances of about 0.2 and 0.14 nm were clearly deduced from optical diffractometry performed on the micrographs. The determination of the nature of this crystalline phase is planned. At the present time, chemical analysis showed that it was not due to film contamination.

The consequences of the deposition rate were determined using the IBAM device without assistance. Increasing the discharge power from 500 to 1,000 W raised the deposition rate from 2.8 to 6.5 $nm.min^{-1}$. The SEM morphologies of the layers were not affected, but the density decreased from 2 to 1.8 $g.cm^{-3}$. This last effect might be induced by a lesser atom mobility resulting from the reduction of the plasma to the growing surface interaction period. The condensation of further carbon atoms might therefore freeze voids and vacancies in previously deposited layers.

3.2 Effects of the Bombardment Parameters

Pressure and, to a lesser extent deposition rate, allowed to state roughly the deposition parameters favouring amorphous diamond-like with respect to amorphous carbon. Furthermore, it is well known that an ion bombardment during the film growth gives the opportunity to study more precisely the variation of the diamond-like properties[4,5]. Its effects were observed through hardness, density and optical indexes of the coatings. This analyse contributes to a larger investigation of the growth mechanisms.

Hardness, measured with a Knoop device under a load of 5 g, rose regularly with increasing bombardment intensity from 17,000±4,000 Mpa without assistance to 40,000±9,000 Mpa for beam parameters of 200 eV and 0.1 $mA.cm^{-2}$.
Different effects induced by the ion assistance might be responsible for this trend: density enhancement and related chemical bonding changes (increase in the sp3/sp2 ratio), compressive stress augmentation. To decide between or estimate the amount of each mechanism, further results are discussed hereafter.

The increase of the density up to 2.8 $g.cm^{-3}$ with the argon ion flux (figure 2a) made obvious the densification phenomenon due to the mobility enhancement induced by the ion bombardment of the deposited atoms. Surprisingly high values of the density were measured for high fluxes, but these results took into account an amount of about 2.5 at% argon incorporated in the

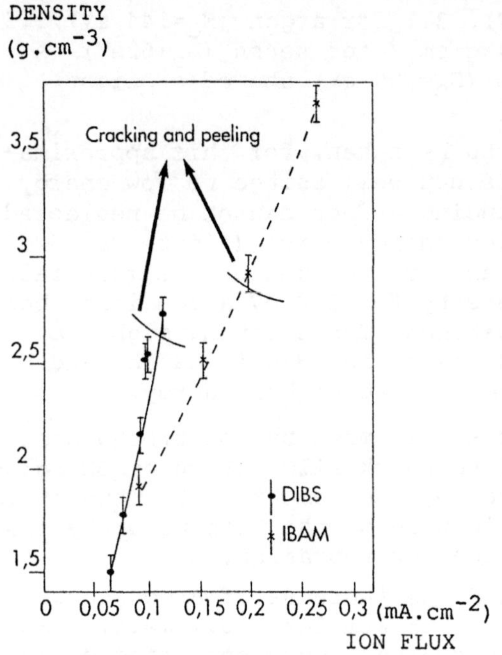

Figure 2a.
density vs. ion flux
(200 eV argon ion beams)

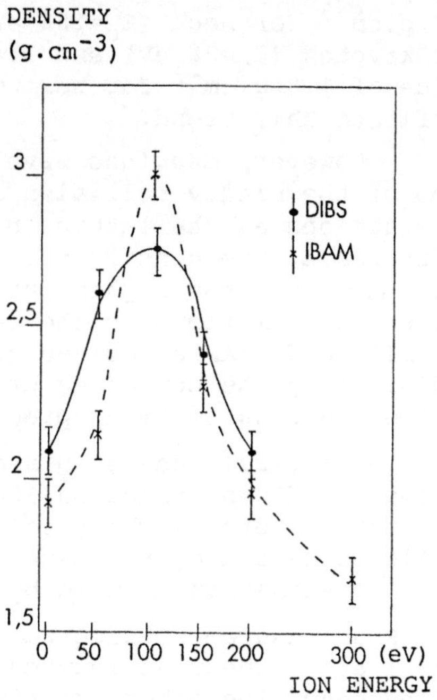

Figure 2b:
density vs. ion energy
(0.1 mA.cm^{-2} argon ion beams)

carbon, which contribution to the film density was approximatively 10%. For the higher deposition rates obtained with the IBAM device, a stronger flux is needed to observe a similar densification. Film cracking was observed for the higher fluxes: it illustrated the influence of the ion peening on intrinsic stresses already stated in previous work[6].

The film density depended also strongly on the mass of the bombarding ions :the maximum was observed with neon (2.3 g.cm^{-3} for 200 eV and 0.1 mA.cm^{-2}) which atomic mass is the closest to the carbon. These experimental results suggest a relationship with the binary collision efficiency: according to the principle of conservation of energy and mass of the colliding incident ion and condensed carbon atom, the highest energy transferred to this atom, E_c, is related to the energy of the impinging ion before collision as follows:

$$E_c = 4\,\alpha\,(1+\alpha)^{-2}\,E_i$$

where $\alpha = M_c/M_i$, and M_c= mass of the carbon atom, M_i= mass of the impinging ion. Thus, E_c is the highest when α approaches 1, which is the case for carbon and neon, and decreases rapidly for lighter and heavier ions. For the beam parameters set to 0.1 mA^{-2} and 200 eV, the density decreased with E_c:

2.3 g.cm^{-3} for neon (E_C=188 eV), 2.1 for argon (E_C=144 eV), 1.5 for krypton (E_C=88 eV) and 1.4 g.cm^{-3} for xenon (E_C=62eV). A value of 1.8 g.cm^{-3} for helium (E_C=150 eV) showed a slight misfit to this trend.

However, cautions have to be taken, for this approximation of the binary collision is not well suited to low energy interactions as the lattice binding effect cannot be neglected. Furthermore, the density vs. ion energy curve (figure 2b) did not show that the higher the energy, the denser the material, but showed a maximum in the density for 100 eV argon ions, both for DIBS and IBAM deposited coatings. These optima might be explained by the competing effects of ion densification and damage creation becoming predominant at higher energy.

The optical indexes n and k, measured by spectrophotometry in the 400-700 nm range on 100 nm thick films grown on BK7 glass substrates, were slightly raised by the bombardment. This could be either caused by the densification of the coating or by the film contamination induced by ion incorporation.

RBS measurement ascertained the increase of argon content from 1 at% without assistance to 2.5 at% for a 0.1 mA.cm^{-2} and 200 eV beam. The other impurity contents were not affected by the film bombardment.

3.3 Tribological Behaviour

As reported in previous studies, DLC may be considered as an attractive lubricating material[7,8].

The friction coefficient (μ) of ion beam assisted low pressure deposited films are reported figure 3. It was measured with

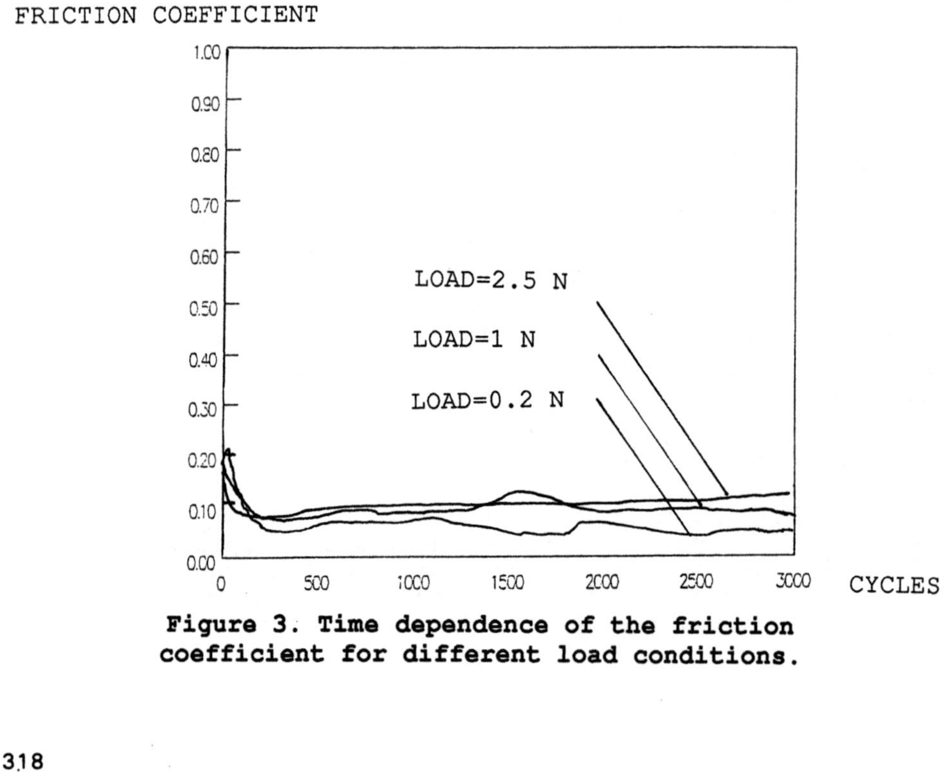

Figure 3. Time dependence of the friction coefficient for different load conditions.

a ball on disk apparatus under the following conditions: room temperature, 45 % relative humidity, disk of AISI 420 stainless steel, Al_2O_3 ball, linear speed of 4 $m.min^{-1}$, loads of 0.2, 1 and 2.5 N. These tests resulted in low friction coefficient ranging from 0.08 to 0.1, which remained stable during the run. The wear of all the samples was very slight, and could not be evaluated with Scanning Mechanical Microscope. μ increased with substrate roughness.

High temperature tests are in progress and satisfactory results have not yet been obtained as the films cracked when heated. Substrate and film expansion coefficient mismatch has to be reduced.

4 CONCLUSION

The working pressure and deposition rate were shown to be the chief parameters to observe a transition from amorphous carbon to amorphous diamond-like carbon. High pressure films presented a columnar morphology, with a nanocrystallized structure, poor mechanical properties and low internal stresses. Low pressure deposited films were denser, amorphous, with mechanical properties of diamond-like carbon. Reducing the deposition rate enhanced this diamond-like character which was explained in terms of plasma-surface interactions.

The effects of an ion beam bombardment during growth on the diamond-like properties were examined and related to the densification of the films. An optimum was ascertained both with the increase of impinging ion energy and mass. These phenomena seemed to be related to conflicting effects between densification induced by atom mobility enhancement and vacancy creation occurring at high energy for the former parameter, and to ion collision efficiency for the latter.

Acknowledgments

The authors are indebted to J-M. Penisson (DRF-SPH), J-C Deutsch (LETI-CMO) and Ph Renaux (CEREM-SECM) for their assistance in HRTEM and spectrophotometry analyses as well as in tribology.

REFERENCES

1. J.C. Angus and C.C. Hayman, Low Pressure, Metastable Growth of Diamond and "Diamondlike" Phases, Science 241:913 (1988).
2. J.Ph. Nabot et al., Diamondlike Carbon Films Prepared by Various Ion Beam Assisted Techniques, 17th ICMC and 8th ICTF, San Diego, USA, (1990), to be published in Surf. Coatings Technol.

3. J.A. Thornton, High Rate Thick Film Growth, Ann. Rev. Mater. Sci. 7:239 (1977).
4. V.I. Orlinov et al., Ion-Assisted Deposition of Thin Films by Magnetron Sputtering in Neon and Argon, 7th IPAT, Geneva, Switzerland 312 (1989)
5. N. Savvides, Optical Constants and Associated Functions of Metastable Diamondlike Amorphous Carbon Films in the Energy Range 0.5-7.3 eV, J. Appl. Phys. 59(12):4133 (1986)
6. D. Nir, Intrinsic Stress in Diamond-Like Carbon Films and its Dependance on Deposition Parameters, Thin Sol. Films 146:27 (1987)
7. K. Enke, Some New Results on the Fabrication and the Mechanical, Electrical and Optical Properties of i-Carbon Layers, Thin Sol. Films 80:227 (1981)
8. Y. Kokaku and M. Kitoh, Influence of Exposure to an Atmosphere of High Relative Humidity on Tribological Properties of Diamondlike Carbon Films, J. Vac. Sci. Technol. A7(3):2311 (1989)

THE STRUCTURE OF a-C:H THIN FILMS

P.J.R. Honeybone and R.J. Newport

Physics Laboratory, The University of Canterbury
Caonterbury, Kent

P.J.R. Honeybone and R.J. Newport

Physics Laboratory, The University of Canterbury
Kent, CT2 7NR UK

W.S. Howells

Neutron Scince Division
Rutherford Appelton Laboratory
Chilton, Didcot, Oxon OX11 0QX UK

J. Franks

Ion Tech Ltd.
2 Park Street
Teddington, Middx. TW11 0Lt UK

INTRODUCTION

The structure of amorphous hydrogenated carbon (a-C:H), prepared under certain conditions, is such that it is harder, denser and more resistant to chemical attack than any other solid hydrocarbon. This, coupled with its high infra-red transparency and the ability to control the as-deposited properties to suit specific requirements, has led to many applications [1].

The material is thought to consist of clusters of sp^2 hybridised carbon interconnected by regions of sp^3 carbon in a continuous random network, with the sp^2:sp^3 ratio varying between 1:2 to 2:1, depending on exact deposition conditions. The high degree of sp^2 clustering suggested by optical gap measurements indicates that intermediate, as well as short range order.are important. The reviews by Angus et al [2] and Robertson [3] provide an excellent outline of this and other model structures.

There have been few direct diffraction experiments performed on a-C:H and these have been limited mainly to electron diffraction (e.g. McKenzie et al. [4]) in which the K-space range of available data is too restricted to extract high resolution real-space information Neutron diffraction, especially from a pulsed source, overcomes this resolution problem by taking a significantly wider data range than is possible by electron or X-ray diffraction. Neutron diffraction also probes the hydrogen environment in a direct way.

Complementary information concerning the local bonding configurations can be obtained from the vibrational properties of the material. Optical probes (IR and Raman spectroscopy) have been widely used in this way, and comprehensive tables have been published [7]. The incoherent neutron scattering cross-section of hydrogen is several orders of magnitude greater than that for carbon. Incoherent inelastic neutron scattering takes advantage of this with the effect that the observed spectra are, to a very good approximation, purely from the hydrogen vibrations.

EXPERIMENTAL DETAILS

The amorphous hydrogenated carbon used in these experiments was produced using a saddle field ion beam source [6]. Samples 1 and 2 (prepared from propane and acetylene gases respectively) were deposited in the conventional way onto copper substrates (- as a-C:H does not adhere to copper, this proved an ideal method for providing the large, ~ 2-5 gm, powder samples required for neutron scattering experiments). Sample 3 was collected from within the source chamber and had been deposited over a period of time, from a mixture of propane, butane and acetylene gases.

The neutron diffraction and inelastic neutron scattering data were collected at the ISIS pulsed neutron source (Rutherford Appleton Laboratory) using the LAD diffractometer and the TFXA spectrometer respectively (see Figure 2; detailed information can be found elsewhere [7]). The analysis of the neutron diffraction data is only fully complete for samples 1 and 3.

A spallation neutron source, such as the one at the Rutherford Appleton Laboratory produces a pulse of neutrons with a spread of velocities, and thus a spread of arrival times at any one detector. This gives a smooth variation of neutron wavelength as a function of time-of-flight ($\lambda \propto t/l$ where t is the time-of-flight over path length l)

The LAD detectors are arranged so that, on each side of the instrument there were detectors at fixed scattering angles. Spectra were recorded separately, as a function of time-of-flight, for each of these detector pairs and also for monitors in the incident and in the transmitted beam. Spectra were recorded for each of the samples together with their respective empty containers, the no-sample background and for a vanadium rod. The vanadium rod is used as a "standard" and provides the information necessary to put the sample scattering on absolute scale [8]. TFXA uses a time-focussed pyrolitic graphite crystal analyser to resolve the scattered beam, with an extended detector array covering 0.06 steradians.

A Carlo Erba CHN combustion analyser was used to determine the compositions of the samples. Infra-red measurements were performed using a Mattson Instruments Alpha Centauri FTIR and densities were determined using a residual volume technique.

Table 1. Compositions and densities of samples

		C (at.%)	H (at.%)	ρ (g cm^{-3})	ρ_N (gram atom cm^{-3})
Sample 1	Propane	0.68	0.32	1.66	0.20
Sample 2	Acetylene	0.65	0.35	1.51	0.19
Sample 3	Mixture	0.71	0.29	1.65	0.19

THEORETICAL BACKGROUND

The object of performing a diffraction experiment is to determine the structure factor, S(Q) where, for an amorphous material (i.e. an isotropic scatterer) [8]:

$$S(Q) = 1 + 4\pi\rho/Q \int_0^\infty r dr (g(r) - 1) \sin(Qr) \qquad (1)$$

where ρ is the average number density of atoms in the material, Q is the wave vector transfer associated with the diffraction experiment - which for a liquid or amorphous solid, is defined as $Q = 4\pi \sin \theta/\lambda$ where 2θ is the scattering angle - and g(r) is the pair correlation function, which is a measure of the atomic density at a distance r from a given atom at the origin. The pair correlation function may therefore be obtained by Fourier transformation of the structure factor.

In a multi-component system, there are contributions to the total structure factor from each atom-type pair. For a binary alloy a-C:H this yields three terms, which are weighted thus to give a total structure factor, F(Q):

$$F(Q) = \sum_\alpha c_\alpha b_\alpha^2 + \sum_{\alpha,\beta} c_\alpha b_\alpha c_\beta b_\beta S_{\alpha\beta}(Q)-1) \qquad (2)$$

where c_α is the atomic fraction, and b_α the coherent scattering length respectively of element α. The first summation represents the "self" or "single atom" scattering while the second is the "interference" or "distinct" scattering term and contains the basic information on atomic positions. A Fourier trans-formation of $S_{\alpha\beta}(Q)$ leads to the total radial distribution function, $G(r) = \sum_{\alpha\beta} [c_\alpha c_\beta b_\alpha b_\beta g_{\alpha\beta}(r)]$where the $g_{\alpha\beta}(r)$ represent the partial terms in G(r). Weightings of individual terms in the structure factor and radial distribution are displayed in Table 2.(It should be noted that the C-H partial term is equivalent to the H-C term). It can be seen that the C-C and C-H terms will dominate the F(Q) and G(r).

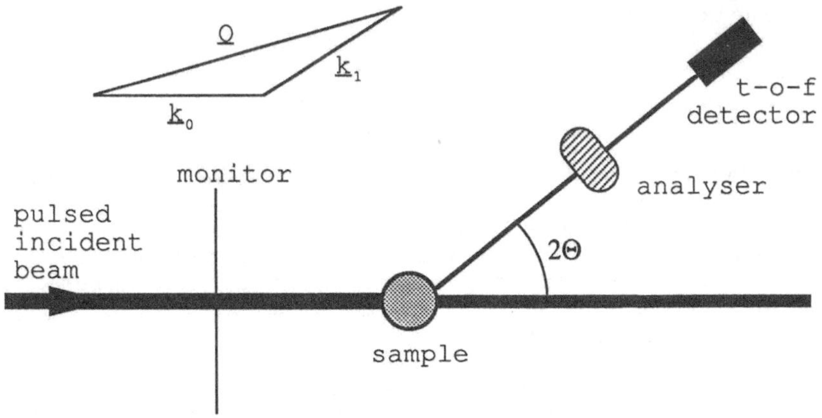

Figure 1. Schematic diagram of neutron scattering instruments. (The analyser crystal is absent in the diffraction case) With inset scattering triangle.

Table 2. Partial Contributions to F(Q) and G(r)

Term	C-C	C-H	H-H
Sample 1	20.4	-5.4	1.4
Sample 3	22.3	-5.1	1.2

A neutron interacting with matter transfers energy, as well as momentum to the system. This results in the need for a detailed inelasticity correction for neutron diffraction experiments, but provides the basis for inelastic neutron scattering experiments in which the energy transfer, $\hbar\omega$, as well as Q is measured (- in the case of TFXA this is achieved by determining the energy of the scattered neutrons using a crystal analyser). Given hydrogen's large cross-section for incoherent scattering, which dominates the total scattering cross-section for these samples (see Table 3), the measured scattering from TFXA is, to a very good approximation, given by the incoherent term. In this situation, the scattering function can be related directly to the vibrational density of states, $g(\omega)$. A fuller description of inelastic scattering theory can be found elsewhere [9,10].

DATA ANALYSIS

Each of the diffraction data sets is in the form of a number of spectra, one for each detector, consisting of counts as a function of time-of-flight. Before obtaining the F(Q) several corrections need to be applied, full details of which can be found elsewhere [11]. The major corrections are for background, container and multiple scattering, attenuation and the effects of inelasticity. . The vanadium is used to put the sample scattering on an absolute scale.

In hydrogenous materials, the inelasticity correction is in many ways the most problematic step in the correction procedures. The simplistic approaches to the problem [12] break down for low atomic mass systems, and alternative routes need to be adopted. Jung et al. [13] suggested a semi-empirical technique for their work on methylene chloride which is based on estimating an effective average molecular mass. The principal elements of their approach are used here to provide a model for the effects of inelasticity.

RESULTS

Angus et al [2] suggest that, for a hard carbon film, a gram atom density greater than 0.2 is required, with lower densities corresponding to a softer, more polymeric structure. This indicates that these samples are intermediate between the soft and hard a-C:H. Confirmation of this can be found from analysis of the C-H stretch, which indicates the presence of small quantities of CH_3 and olefinic CH_2 groups, which only appear in the polymeric films.

Examination of the radial distribution function G(r) from sample 1 (see figure 2b) reveals a split C-C peak at 1.39 and 1.52 ± 0.02Å. Comparison of the peak positions with known C-C bond lengths (see table 3) shows that the 1.52Å peak is unambiguously the single bond length and that the 1.39Å bond length

Table 3. Bond Lengths of Carbon and Hydrogen

C-C	Diamond	1.54Å
C-C	Graphite	1.42Å
C-C	Benzene	1.3954Å
C-C	Ethene	1.34Å
C-H	Methane	1.094Å
C-H	Ethene	1.074Å
H-H	Hydrogen	0.754Å

corresponds closest to that found in aromatic compounds, by contributions from the 1.52Å peak. The ratio 1.39Å : 1.52Å bonds has been found to be 1:4 and the average number of carbon atoms bonded to each carbon atom is estimated to be about 2.5 (the exact uncertainty is unknown, but it is expected to be quite large). The ratio of sp^2:sp^3 carbon is unknown, although the predominance of σ-bonding and the relatively high average C-C coordination both suggest a high sp^3 concentration.

The other features in the G(r) can be ascribed to a C-H bond (1.07Å), with a coordination number of 1.0 (due to the close proximity of all C-H bond distances it is not possible to identify the dominant hydrogen location), an H-H peak at 0.8Å which is either molecular hydrogen or H-H distances between neighbouring CH_n groups (it is hoped that molecular modelling will cast more light on this in the near future).

Moving beyond the first coordination shell, the broad feature between 2.3 and 2.7Å is a combination of sp^2 and sp^3 C-C-C distances, with an average carbon coordination number of 4.0. Resolving the individual contributions is not possible because of the proximity of the two distances, and the likely spread of these distances as bond angles vary between local configurations.

Preliminary analysis of the diffraction data from sample 2 shows that the structure factor is very similar to that of sample 1. The only difference seems to be at low Q, which would correspond to differences in the intermediate range order in the continuous random network. A full analysis of this system should be completed in the near future.

Comparison of the structure factors from samples 1 and 3 shows that it is only at low Q that the structure differs. This would suggest that the structure is very similar for the samples prepared in different parts of the vacuum chamber. Studying the radial distribution functions, again, many similarities are evident; both samples have the same C-H and first and second shell C-C coordination, numbers and have a feature at low $r(< 1Å)$ which corresponds to a H-H distance.

The inelastic data confirms that there is very little difference in the vibrational structures of the three samples (see figure 3), with the three spectra being equivalent (to within experimental error) for frequencies in the 200 to 4000 cm^{-1} range. Within this range, there are clear peaks at ~870cm^{-1}, 1250cm^{-1} and 3000cm^{-1}. Comparison with published IR work [5] shows that all three vibrations are consistent with olefinic C-H

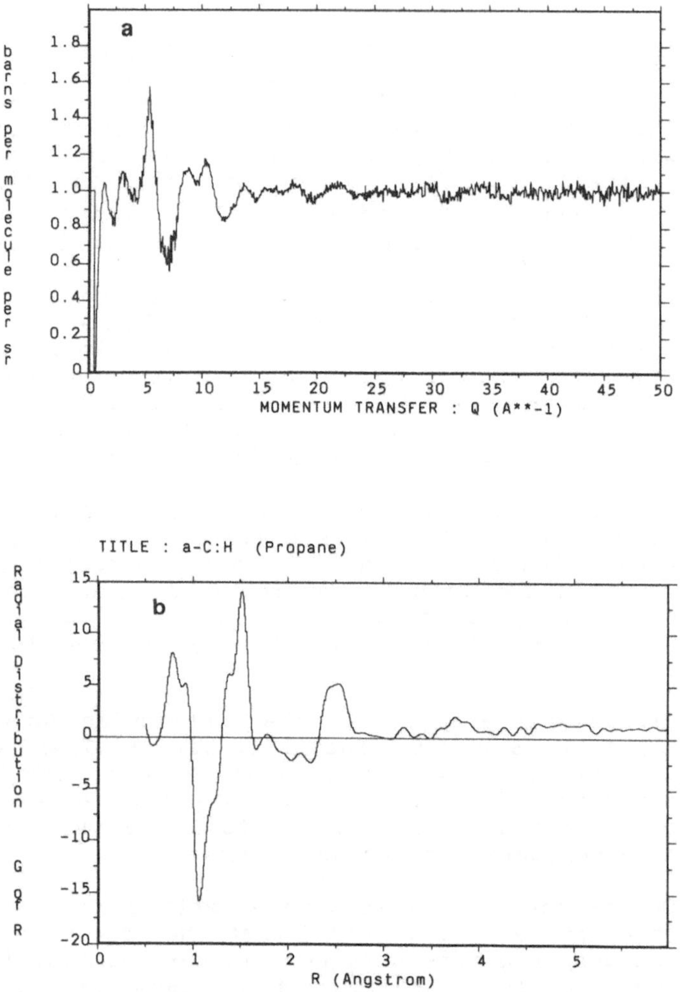

Figure 2. (a) S(Q) and (b) G(r) for sample 1

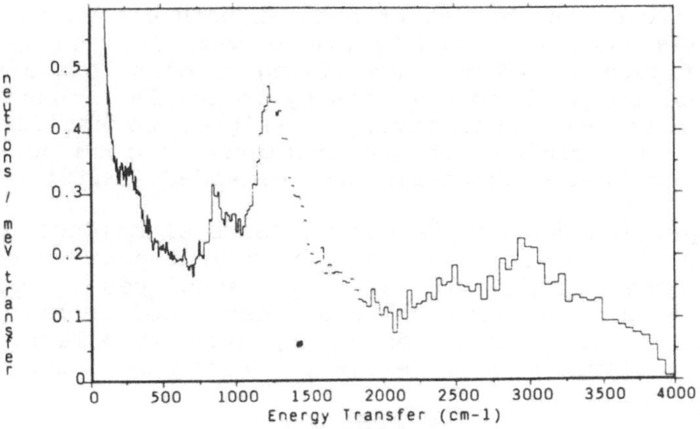

Figure 3. IINS spectrum for Sample 1

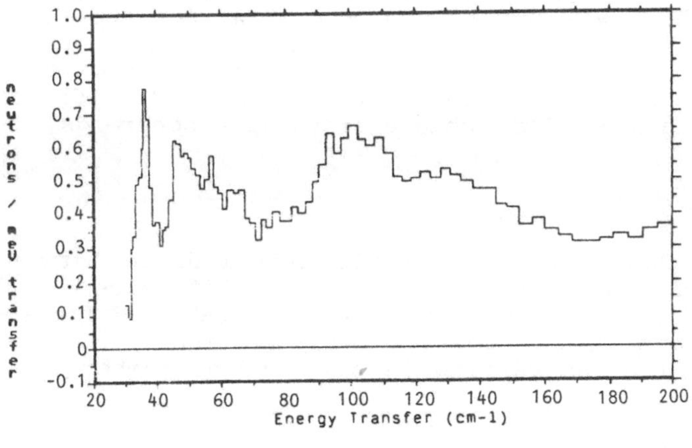

Figure 4. IINS spectrum for sample 3 in the 30-200 cm^{-1} range

group vibrations, although the poor resolution at ~3000cm^{-1} precludes any definitive assignment of this vibration. Although the olefinic C-H groups dominate the inelastic spectrum, there are signs that small contributions are made by other CH_3 groups, giving a general background to the 600-1800cm^{-1} region. The inelastic data seems to contrast with NMR work [14] which shows preferential bonding of hydrogen to sp^3 sites. Further work is required to clarify the situation.

The differences that occur between both conventionally prepared samples (samples 1 and 2) are at very low energy ($\omega < 120$ cm^{-1}) (see figure 4). The modes giving rise to this are seen in both samples, but with more intensity in sample 1 than in sample 2. These have been tentatively assigned to oscillations of relatively large regions of the continuous random network, and would be consistent with torsions of extended graphitic regions.

A surprising feature is the presence of molecular hydrogen in sample 3. Its presence in amorphous hydrogenated solids has previously been indicated by many techniques (e.g., IR on a-Si:H by Chabal and Patel [15]) but never before seen directly with neutrons. In a linear molecule, such as molecular hydrogen, the rotational energy levels are given by [16]:

$$E_J = hcBJ(J + 1) \qquad (3)$$

where E_J is the energy of the Jth rotational level, and $B = \hbar/4\pi cI$, with I being the moment of inertia. Thus, the transition from the J = 0 to J = 1 energy levels requires 2hcB cm^{-1}. The rotation for molecular hydrogen can be seen at ~110cm^{-1}. This can be used to determine the moment of inertia, and hence the bond length. This method was used to determine the hydrogen bond length, which was found to be ~ 0.72 Å, c.f. the accepted value of 0.75 Å for free hydrogen and 0.65 Å as determined by diffraction. The splitting of the INS band is indicative of asymmetry in the hydrogen environment, resulting in the hydrogen molecules being contained within oblate or prolate spheroids.

CONCLUSIONS

The three samples are all intermediate between soft and hard a-C:H and show little if any difference in their short range order, although there is evidence of differences in the intermediate range order. This is likely to be due to the graphitic regions varying in size and/or number from sample to sample.

From the diffraction data it is it can be deduced that the single bonds form the bulk of the random network, with a smaller, but nonetheless significant proportion of double or aromatic bonds. This data also shows that bonded hydrogen dominates over free hydrogen, although small quantities of molecular hydrogen have been detected in one sample by IINS.

The inelastic experiment, in particular, probed the hydrogen environment. This data suggests that hydrogen is bonded mainly at olefinic carbon sites, with smaller amounts at other sites. Also, molecular hydrogen has been discovered in one sample, although the quantity is small. Hopefully, molecular modelling of the diffraction data, which is now in progress using a Monte-Carlo based technique [19], will yield more information.

There remain many unanswered questions concerning the structure of a-C:H. Clarification of the hydrogen bonding environment is required and questions remain concerning the nature and extent sp^2 and sp^3 of regions. Further work using neutron and synchrotron X-ray techniques together with computer modelling is in progress which will address these outstanding issues.

ACKNOWLEDGEMENTS

We would like to thank Mr. A. Evans and Dr.P.J.Revell of Ion Tech Ltd., for their help in depositing the a-C:H, Mr. A. Fassam (Chemistry Department, UKC) for his help in determining the compositions and members of the Kent group for helpful discussions. One of us (PJRH) acknowledges the receipt of a SERC studentship.

REFERENCES

1. A.H.Lettington, "Amorphous Hydrogenated Carbon Films", E-MRS Proceedings, **17** P.Koidl and P.Oelhafen, eds. (1987).
2. J.C.Angus, P.Koidl, and S.Domitz, Chapter 4 in: "Plasma deposited thin films", J.Mort and F. Jansen, eds., CRC . Press, Boca Raton (1986).
3. J.Robertson, Advances in Physics, **35** 317 (1986).
4. D.R.McKenzie, L.C.Botten,and R.C.McPhedran, Physics Review Letters, **51** 280 (1983).
5. B.Dischler, "Amorphous Hydrogenated Carbon Films", E-MRS Proceedings, **17** P.Koidl and P.Oelhafen, eds. (1987).
6. J.Franks, Vacuum, **34** 259 (1984).
7. ISIS Annual Report 1989, Rutherford Appleton Laboratory Report RAL-89-050.
8. R.J.Newport, Chapter 13 in "Neutron Scattering at a Pulsed Source", R.J.Newport, B.D.Rainford and R.Cywinski, eds., Adam Hilger, Bristol, (1988).
9. R.N.Sinclair, J.of Non-Cryst.,So., **76** 61, (1985).
10. J.Tomkinson, Chapter 18 in "Neutron Scattering at a Pulsed Source", R.J.Newport, B.D.Rainford and R.Cywinski, eds., Adam Hilger, Bristol, (1988).
11. A.K.Soper, W.S.Howells and A.C.Hannon, ATLAS - Analysis of time-of-flight diffraction data from liquid and amorphous samples, Rutherford Appleton Laboratory, Report RAL-89-046, (1989).
12. G.Placzek, Physical Review, **86** 377, 1952.
13. W.G.Jung, M.D.Zeidler and P.Chieux, Mol.Phys., **68** 473 (1989).
14. M.A.Petrich, Materials Science forum, **52** 377 (1989).
16. J.Howard, C.J.Ludman and J.Tomkinson, Fuel, **62** 1097 (1983).
17. P.W.Atkins, Physical Chemistry, OUP, Oxford (1978).
18. Y.J.Chabal and C.K.N.Patel, Phys.Rev.Letts., **53** 210 (1984).
19. R.L.McGreevy and L.Putsztai, Mol.Simulation, **1** 359 (1988).

STRUCTURE AND ELECTRONIC PROPERTIES OF DIAMOND-LIKE CARBON

J. Robertson

National Power Labs
Leatherhead, Surrey, KT22 7SE, UK

INTRODUCTION

This paper reviews the structure and electronic properties of the various forms of amorphous carbon (a-C) and hydogenated amorphous carbon (a-C:H). It is shown that a-C:H consists of clusters of 3-coordinated sp^2 carbon embedded in a sp^3 bonded matrix. The sp^2 regions are found to control the electronic properties such as band gap while the sp^3 regions control mechanical properties such as rigidity and hardness. A model for the mechanical properties is introduced. The various experimental methods of determining atomic structure are critically reviewed and deposition processes favouring sp^3 bonding are discussed. Recent data indicate that a-C deposited from mass-selected ion beams is highly sp^3 bonded and has a hardness close to diamond.

Carbon forms a number of crystalline and non-crystalline solids with very diverse properties, as summarised in Table 1. The deposition of carbon under the appropriate conditions can produce a hard, chemically inert, optically transparent and electrically dopable thin film material of great interest and often known as diamond-like carbon. Diamond-like carbon was originally prepared by ion-beam deposition from a graphite arc source[1] and could contain in general both amorphous carbon and diamond crystallites. Addition of hydrocarbon gases to the ion source increases the deposition rate and produces hydogenated a-C. A-C:H can also been prepared by magnetron sputtering and plasma-deposition. Its properties have been reviewed by Angus et al[2], Robertson[3,4], and Koidl et al[5]. We concentrate here on the amorphous phases and describe their properties in terms of their underlying chemical bonding.

The two crystalline forms of carbon are diamond and graphite. Diamond is a hard, wide band gap semiconductor in which each atom is 4-fold coordinated and sp^3 bonded. In this configuration, each of carbon's four valence electrons is assigned to a tetrahedrally directed sp^3 hybrid orbital, which makes a strong σ bond with an adjacent atom. Graphite is an anisotropic metal, in which each atom is sp^2 bonded and 3-fold coordinated in a hexagonal layer structure. Here, three of each carbon's valence electrons are assigned to trigonally directed sp^2 hybrid orbitals, which form strong intra-layer σ bonds, and the fourth electron is in a $p\pi$ orbital which lies normal to the σ bonding plane. This orbital forms weak, delocalised π bonds with its neighbours. π bonding is often called unsaturated bonding. Finally, in the sp^1 configuration as found in acetylene, two valence electrons are assigned to linearly-directed sp^1 hybrids which form σ bonds, and the other two electrons are placed in each of the $p_y\pi$ and $p_z\pi$ orbitals.

Diamond and Diamond-Like Films and Coatings
Edited by R.E. Clausing *et al.*, Plenum Press, New York, 1991

Table 1. Properties of various forms of carbon

	Density gm cm^{-3}	Hardness kg mm^{-2}	% sp^3	at.% H	Gap eV	Ref.
Diamond	3.515	10^4	100		5.5	
Graphite	2.267		0		-0.04	
Glassy C	1.3-1.55	200-300	$\simeq 0$		0.01	(6)
a-C, evap.	1.9-2.0	200-500	1		0.4-0.7	(7,8)
a-C, MSIB	3.0	10^4	90 ± 5	<9	0.5-1.5	(8-13)
a-C:H, hard	1.6-2.2	1000-2000	40-70	10-35	0.8-1.7	(5)
a-C:H, soft	0.9-1.6	<500	50-80	35-50	1.6-4	(5,14)
Polyethylene	0.95	1-4	100	67	6	

The various forms of non-crystalline carbon can be considered to be intermediate between the extremes of diamond, graphite and hydrocarbon polymers in that they contain variable amounts of sp^3 carbon, sp^2 carbon and hydrogen. Glassy carbon consists of sp^2 sites organised in convoluted graphitic layers, with a correlation length within layers of $L_a \simeq 50$Å[6]. This high degree of medium-range order implies that glassy carbon cannot be classed as truly amorphous. Amorphous carbon can be prepared by evaporation or sputtering of graphite. This material is essentially pure sp^2 bonded, but with a much more disordered, graphitic-based structure.

The desireable properties of diamond-like carbon require a high content of sp^3 sites. However, a reference form of a-C which is both purely sp^3 bonded and unhydrogenated did not appear to exist. It is now recognised that this role has been fulfilled by a-C prepared by mass-selected ion beam deposition (MSIB) of C^+ ions from a carbon arc source[8-13]. This material will be described in detail later, together with a discussion of the deposition mechanisms that promote sp^3 bonding.

A-C:H itself can be prepared by a variety of techniques such as plasma deposition (PD), ion beam deposition or magnetron sputtering. The PD material has been studied in greatest detail, as reviewed by Koidl[5]. Three general types of material can be formed, depending strongly on the deposition conditions. The key deposition parameters are the depositing ion energy and the substrate temperature. The ion energy is a fraction of the bias voltage in plasma deposition. As discussed in detail later, moderate bias voltages of $\simeq$400-1500 V produce the desired hard network solid, denoted hard a-C :H. It has mixed sp^3, sp^2 bonding, a band gap under 1.7 eV and considerable chemically bonded hydrogen mainly as $\equiv$CH. Bias voltages below about 400 V favour a soft, wide band gap polymeric material, denoted soft a-C:H, containing both sp^3 and sp^2 carbon and more hydrogen, largely in the form of $=CH_2$ groups[14-16]. Substrate temperatures above 200°C or bias voltages above 1.2 kV cause a gradual increase in graphitic character, with more sp^2 sites, less hydrogen and a lower hardness than hard a-C:H.

THE Π ELECTRONIC STRUCTURE

The key to understanding the electronic structure of all amorphous carbons is to realise that this is controlled by the π states of the sp^2

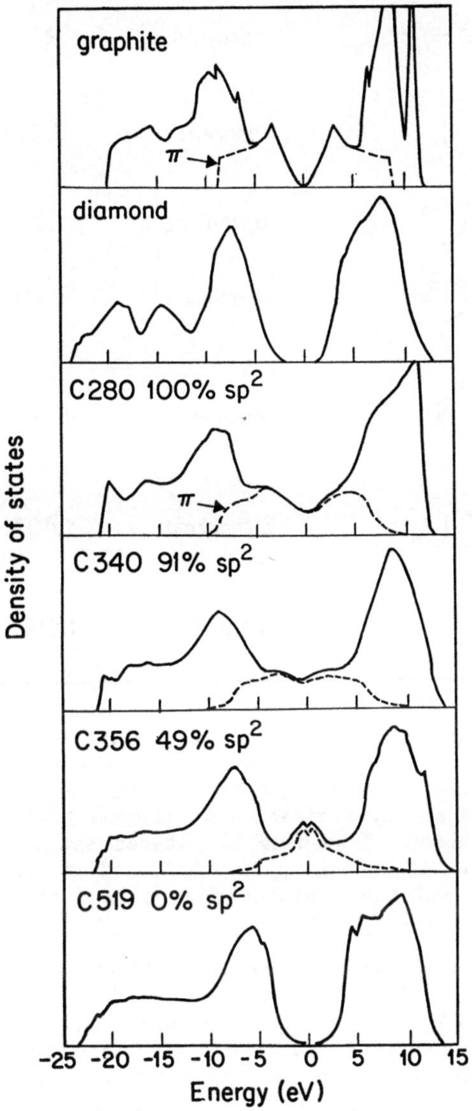

Fig. 1. Calculated DOS of graphite, diamond and the Beeman models of a-C

Table 2. π electron binding energies for various sp^2 configuations

group	name	E_{tot}/β
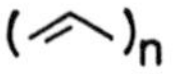	polyacetylene	1
$(\equiv-)_n$	carbynes	2
	benzene	1.333
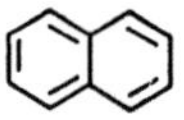	napthalene	1.368
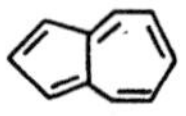	azulene	1.336
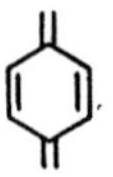	quinoid	1.240
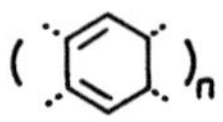	polybenzoid	1.403
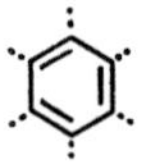	graphite	1.616

sites, as these states lie closest to the Fermi level, E_F[3,17,18]. This is seen in Fig. 1, which shows the density of states (DOS) calculated for diamond and a single layer of graphite, with the π states of graphite shown arrowed. The sp^3 hybridisation of diamond leads to a wide band gap between a valence band of bonding (σ) states and a conduction band of antibonding (σ^*) states. In graphite the σ and σ^* states lie at similar positions to this and the π states form a band which lies across the σ-σ^* gap, with a minimum DOS at E_F.

A standard approximation is to decouple the π and the σ states and to treat them separately. This is possible because their orbitals lie perpendicular to each other at any atom, which minimises their coupling. The π and σ states behave very differently. The σ states are the 2-centre bonds which form the back-bone of the network. Their bonding is very localised and their bonding energy can be written as the sum of the energies of each bond. This causes the σ electrons to control the Short-Range Order (SRO) of the network, ie bond lengths and bond angles, but leaves the longer range order undefined. The π states on the other hand can form multi-centred bonds, also known as resonant bonds. Their bonding energy is longer-ranged and cannot be subdivided into the contributions from individual bonds. They therefore introduce Medium-Range Order (MRO) into the structure.

The π states possess one electron per orbital and form a half-filled band. Their bonding energy is maximised if a gap is created in their spectrum at E_F, as this lowers the total energy of the occupied states and stabilises the resulting structure. It is possible to analyse the π electron bonding by using the Huckel model. This is a simplified tight-binding model, which retains only the π orbitals and their nearest neighbour interaction, $\beta = V(p_z\pi)$. The absence of π states on sp^3 sites tends to split the original network into a series of separate π-bonded clusters. The stable configuration of clusters are then those with the highest total π binding energy per site, E_{tot} , and are of course similar to the analogous organic molecules.

E_{tot} of various cluster configurations is given in Table 2. The simplest configuration is the ethylenic double bond, C=C, which has $E_{tot}/\beta=1$. E_{tot} of a row of double bonds as in alternant polyacetylene is also 1 in this model, so polyacetylene is not a particularly favoured group. Such chain species are called olefinic. E_{tot} however is dramatically increased by combining three C=C bonds into a planar 6-fold or benzene ring. Such rings are called aromatic. E_{tot} is further increased by fusing benzene rings into aromatic sheets, and in this way aromatic clusters are built up. E_{tot} is also larger for compact (graphitic) clusters than linear (acenic) clusters. Ring orders except six are generally unfavoured as they give states near E_F. However, 'dipole' pairs of 5- and 7-fold rings are allowed, as in azulene, as they keep states away from E_F and have only a slightly higher E_{tot} than naphthalene (Table 2). Finally, π-bonded clusters all tend to be planar, as this keeps the π orbitals parallel and maximises their interaction.

Summarising, π bonding strongly favours aromatic rings over olefinic chains and also favours the clustering of separate rings into graphitic sheets. Hence we expect aromatic ring clusters to be the dominant species of sp^2 sites. Also, for any given concentration of sp^2 sites, the sp^2 sites have a strong tendency to segregate into sizeable clusters rather than spread homogeneously throughout the sample. The σ bonds remain neutral in this process as their bonding energy is localised and independent of clustering effects.

We are therefore led to a 2-phase model of amorphous carbons. The first phase is the π-bonded clusters. It turns out that these determine the electronic properties. This phase is embedded in a second phase. In a-C the second phase consists of a rim of debf*a or possibly sp^3 sites which form a boundary between adjacent clusters. In a-C:H the second phase is the major sp^3-bonded matrix, which can be either a highly cross-linked network of sp^3 sites, as in hard a-C:H, or a more hydrogenated and polymeric phase, as in soft a-C:H. This second phase largely determines the mechanical properties, as shown later.

OPTICAL GAP

The optical gap of all amorphous carbons is determined by their π states, because these states form both the valence and conduction band edges, as seen in Fig. 1. The factors which control the optical gap are very different in π-bonded systems than in the more familiar σ-bonded systems like a-Si. In a σ system, the gap is determined by the short-range order, that is the coordination number, bond length and bond angles. In π system like a-C the gap depends on the medium-range order, that is the degree of clustering. Fig. 2 shows the calculated band gap of compact (upper line) and linear (lower line) aromatic clusters as a function of M the number of rings in the cluster. The gap E_g of compact clusters, the most stable sort, declines slowly and irregularly with M, according to the approximate relation[3,17]

$$E_g = 2\beta/M^{1/2} \simeq 6/M^{1/2}\ \text{eV} \tag{1}$$

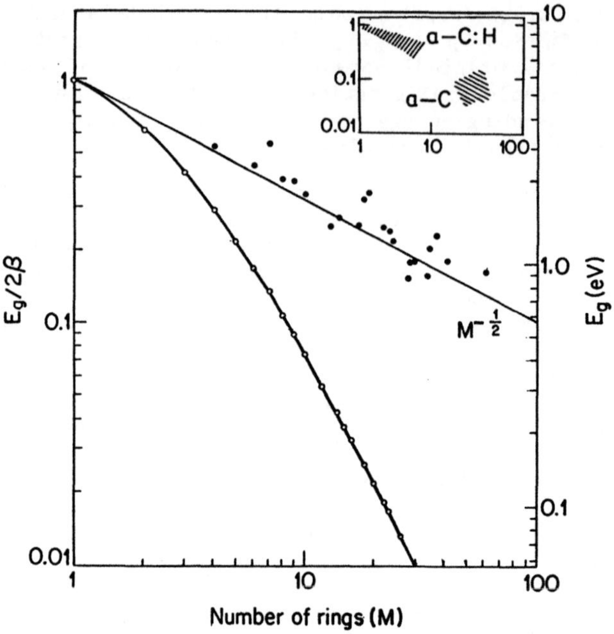

Fig. 2. Band gap of compact (upper line) and linear (lower line) aromatic clusters.

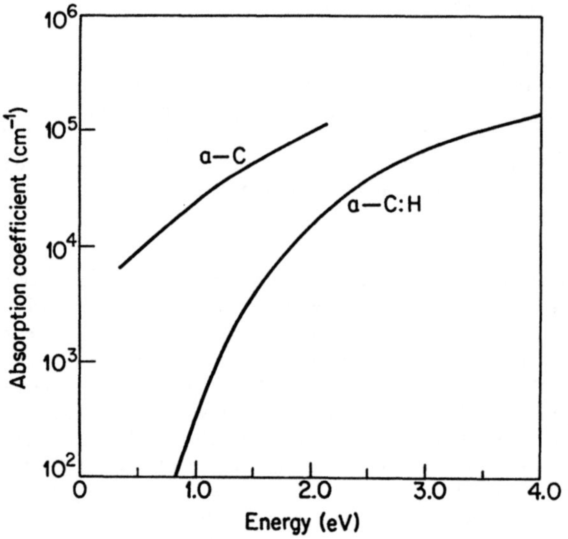

Fig. 3. Optical gap of sputtered a-C after Hauser[7] and a soft a-C:H after Smith[19].

where β is set to -2.9 eV from fitting the π bands of graphite. The gap of linear clusters falls off more rapidly, as roughly

$$E_g \simeq 60/(M + 2.16)^2 \text{ eV} \tag{2}$$

In the absence of sharp band edges the optical gap of an amorphous semiconductor is defined in an empirical manner as either E_{04}, the energy at which the optical absorption coefficient $\alpha = 10^4 \text{ cm}^{-1}$, or as the Tauc gap E_T found by fitting α to

$$\alpha E = B(E - E_T)^2 \tag{3}$$

The absorption edges of a-C and soft a-C:H shown in Fig. 3 are much broader than for example that of a-Si:H, suggesting that all amorphous carbons contain a range of cluster sizes with a range of local band gaps. The overall gap will be controlled not by the average clusters but by the largest significant clusters with the smallest gaps. As the maximum value of $\alpha \simeq 3.10^5 \text{ cm}^{-1}$, we expect the overall band gap to be given by the gap of the $\simeq$3% of largest clusters, From equation (1), E_{04}=0.5 eV[7] for a-C corresponds to M=140 for the typical larger clusters, and perhaps M=35 for a typical average cluster, corresponding to $L_a \simeq 9$Å.

The gap of a-C:H is strongly dependent on deposition conditions, but it does not exceed about 4 eV, as seen from Table 1. Fig. 1 shows that the gap of σ states in carbon always exceeds 5 eV. Therefore even in the absence of other information we conclude that a-C:H contains sp^2 sites and that the gap is also determined by the π-bonded clusters. For a hard a-C:H, a typical gap of 1.2 eV corresponds to M=25 for the larger clusters and perhaps M=6 for a typical cluster, or $L_a = 3.8$Å. For a soft a-C:H, a gap of 2 eV gives M=9 and perhaps M=2 for a typical cluster. This model suggests that the optical gap will depend on the size of π-bonded clusters, and not explicitly on hydrogen content or even the sp^2/sp^3 ratio. A dependence on hydrogen content is still expected but it is indirect, to increase the proportion of sp^2 sites and thereby reduce the mean cluster size.

STRUCTURE SIMULATIONS

The first structural simulations were the random network models of a-C of Beeman et al[20]. These contained various proportions of sp^2 and sp^3 sites. The radial distribution function (rdf), the probability of finding an atom at a given distance from a given atom, of each model was calculated and compared to experimental rdfs. It was concluded that evaporated a-C contained 1-5% sp^3 sites. Unfortunately, these networks were built before the bonding rules for π networks were known, and they incorrectly allowed the sp^2 sites to form many odd-membered rings.

The most sophisicated simulation is that of Galli et al[21] using the first-principles Car-Parrinello molecular dynamics method which treats both the nuclear and electronic degrees of freedom. A small sample of 54 atoms in a periodic lattice with a density set at 2 gm.cm^{-3} and given a particular quenching sequence was found to consist of 85% sp^2 sites and 15% sp^3 sites. The sp^2 sites were found to be clustered into buckled, largely graphitic layers, which also contained some adjoining 5- and 7-fold ring configurations, as in azulene. The sp^3 sites were also found to be slightly clustered. These results generally support our model of disordered graphitic clusters for a-C, and it was noted that the sample size was too small to study MRO effects.

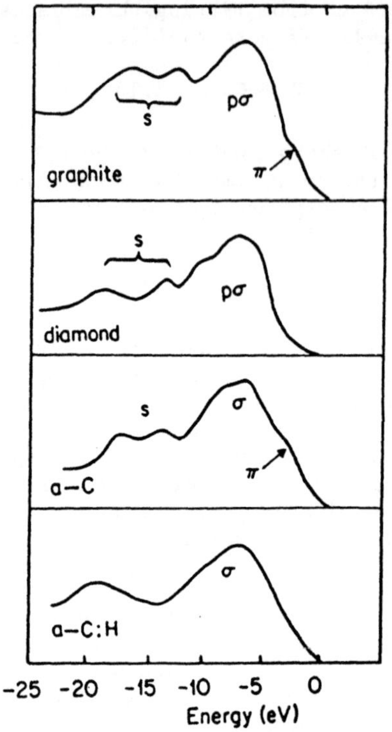

Fig. 4. Photoemeission spectra of graphite, diamond, a-C and a – C:H[25]

Tersoff[22] used an empirical interatomic potential allowing variable coordinations. With a larger cell of 216 atoms, he found a-C to consist of ≃91% sp^2 sites in a much more recognisably graphitic structure.

BAND STATES

The density of states of a-C:H has been calculated by the tight-binding method with a sp^3s^* basis and mainly nearest neighbour interactions[3,17]. The interactions for C-C bonds in Table 3 were found by fitting to the band structures of diamond[23] and graphite[24]. It was found possible to use a single set of parameters for both sp^3 and sp^2 sites despite their slightly different bond lengths. $V(p_{z\pi})$ is the interaction between π states and $V_2(p\pi)$ is a small second neighbour interaction between π states. The C-H interactions were fitted to methane.

Table 3. Tight-binding parameters, eV

	E(s)	E(p)	E(s*)				
C	-5.35	0.45	14.0				
H	-2.30						
	V(ss)	V(sp)	V(pσ)	V(pπ)	$V(p_{z\pi})$	$V_2(p\pi)$	V(s*p)
C-C	-4.55	5.2	5.45	-1.6	-2.9	0.15	4.50
C-H	-7.5	8.9					

Fig. 1 shows the calculated DOS of diamond, graphite and the Beeman models of a-C containing various proportions of sp^2 sites. The valence band consists of s-like states at -10 to -22 eV and a peak of $p\sigma$ states around -7 eV. The conduction band consists of π^* states from 0 to 8 eV and σ^* states above 2 eV. The DOS of the C519 network with only sp^3 sites resembles that of diamond, as expected. Its gap is slightly less than that of diamond due to bond angle distortions. The unrealistic peaks in the π DOS around E_F of the C356 model arise from odd-membered rings of sp^2 sites included in the Beeman models before the bonding rules for sp^2 networks were known. C-H bonds give states well away from the gap[17].

Fig. 4 shows the valence band DOS graphite, diamond, evaporated a-C and a-C:H, as seen by photoemission at 120 eV[25,26] The cross section of hydrogen is sufficiently low that H-derived states are not seen. The DOS of diamond and graphite are seen to be similar to the calculated DOS in Fig. 1. The π states of graphite and a-C are seen to give rise to a knee in the leading edge of the valence band at -2 eV. This confirms that the valence band maximum of a-C is π-like. The π states of a-C:H are seen in the photoemission spectra of Oelhafen[27].

The shape of the valence s DOS at -15 to -20 eV gives information on the topology of the network. The s DOS of a network containing only 6-fold rings consists of two peaks, as seen for diamond or graphite. Introducing 5- or 7-fold rings, as in the C519 network, washes out the dip between the two peaks. Thus it is interesting that the s band of a-C shows two peaks in photoemission, while the s band of a-C:H shows one peak. The DOS of a-C is consistent with it being predominantly graphitic, whose π bonding strongly favours 6-fold rings. On the other hand, the single peak in a-C:H is consistent with the presence of 5- and 7-fold rings which are now allowed in the sp^3 bonded portion of its network.

STRUCTURE OF a-C

In this section we compare the structures of a-C produced by the familiar electron beam evaporation or sputtering of graphite and by mass-selected ion beam deposition (MSIB). The MSIB method minimises the deposition of graphitic fragments and neutral species given off by an evaporation or sputtering source, which are thought to increase the amount of graphitic bonding in the final film. A number of workers have studied MSIB films deposited from an ion source[12,13,28,29] but in general their deposition rates were low. Much higher rates of $0.1\mu m.hr^{-1}$ are possible for a graphite arc source used by Green et al[8-10].

Fig. 5 compares the interference function, F(k), of glassy carbon measured by neutron diffraction by Mildner and Carpenter[6], of evaporated carbon from electron diffraction by Boiko et al[30] and of MSIB a-C from neutron diffraction by Gaskell et al[11]. A similar rdf. was found for MSIB a-C by x-ray diffraction by McKenzie[9]. Table 4 compares the peak positions in F(k) with those of graphite and diamond. Fig. 6 shows the respective radial distribution functions (rdf), G(r), and Table 5 compares the structural characteristics.

The F(k) of glassy carbon has very sharp peaks which can be indexed as the (hk0) and (002) peaks of graphite. This emphasises its close structural relationship to graphite. The F(k) of evaporated a-C has much broader peaks, emphasising its more disorder nature. The (002) peak in F(k) at 1.8 $Å^{-1}$ is missing, showing that there is little inter-layer correlation. Beeman et al[20] estimated that a-C contained about 5% sp^3 sites by comparing the rdfs of their random network models with the experimental ones. Berger et al[8] estimated a 1% sp^3 content from XANES.

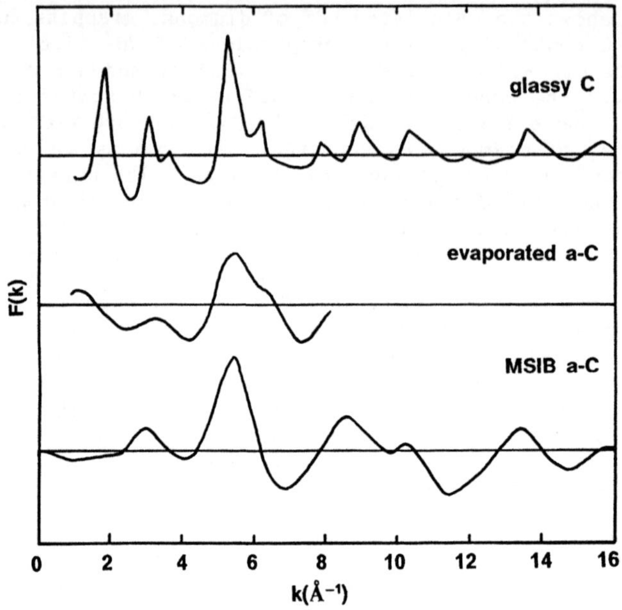

Fig. 5. Structure factors of glassy C[6], evaporated a-C[30] and mass selected ion beam deposited a-C[11]

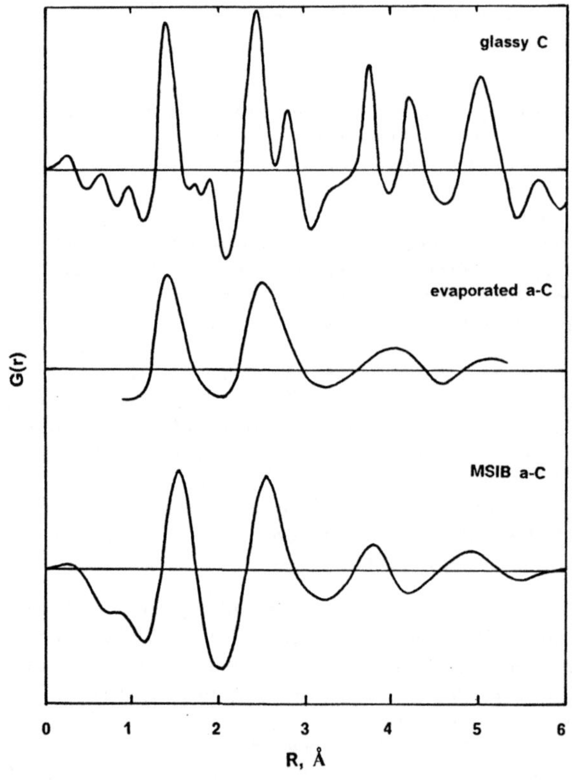

Fig. 6. Radial distribution functions of glassy C[6], evaporated a-C[30] and mass-selected ion beam deposited a-C[11]

Table 4. Peak positions in F(k) of various forms of a-C, compared to the indexed reflections of graphite and diamond

	Peak position, $Å^{-1}$							
glassy C	1.8	2.98		5.11	5.96	7.8	8.8	10.2
a-C (evap)	1.0	2.9		5.1				
a-C (MSIB)	1.2	2.9		5.1		7.8		
Graphite (hkl)	(002) 1.88	(100) 2.95		(110) 5.11	(200) 5.90	(210) 7.82	(300) 8.86	(220) 10.2
Diamond (hkl)		(100) 3.06	(200) 3.53	(220) 5.00	(311) 5.86	(331) 7.71		

Table 5. Distances(r), coordination numbers(n) and densities of various forms of C

	r_1, Å	n_1	r_2, Å	n_2	density, $gm.cm^{-3}$
Graphite	1.42	3	2.45	6	2.25
Diamond	1.544	4	2.512	12	3.51
glassy C	1.425	2.99	2.45	6.1	1.49
a-C (evap)	1.43	3.3	2.53	8.8	2.1 (imposed)
a-C (MSIB)	1.534	4.0	2.517	11.47	3.0

Gaskell et al[11] measured F(k) of MSIB a-C to $16Å^{-1}$. The rdf. shown in Fig. 6 is terminated by a sinx/x function. Its peak positions and areas are those expected for a sp^3 bonded network. The first coordination number is 4 and the second neighbour distance gives a 110.3° bond angle, close to the tetrahedral value. The first neighbour distance is a little short for sp^3 bonds and indicates the presence of $10\pm5\%$ of sp^2 sites. A similar value of 15% was deduced from XANES by Berger et al[8]. A density of 3.0 $gm.cm^{-3}$ was deduced from the rdf., similar to that found by float-sink methods by Green[9]. This is less than that of diamond but well above that of graphitic carbons. The second peak is found to be proportionally narrower than in a-Si, indicating a narrower spread of bond angles, and consistent with the stronger bond bending forces of C. A 9% hydrogen content was found from its q-dependent effect on the raw F(k), not show. However, this hydrogen must be largely H_2 in voids or interstitials, as it has not reduced the first neighbour coordination.

The values in Table 5 demonstrate that MSIB is indeed a tetrahedrally bonded form of a-C. Martin et al[10] find that its hardness is in the range of that of diamond (Table 1), but they note that there are graphitic inclusions which lower hardness locally. These inclusions cause MSIB a-C to have a lower band gap than diamond. McKenzie[9] found E_T=1.0 eV and E_{04}=1.75 eV for this material, while Ishikawa et al[13] found E_T=0.5-1.5 eV depending on ion energy. Clearly, these low gaps are determined by residual π states. It is interesting that the peak positions of F(k) are quite similar for sp^2 and sp^3 bonded networks, showing the difficulty of using F(k) to differentiate these structures. The key difference is in the density and the peak areas in the rdf.

STRUCTURE OF a-C :H

The measurement of the structure of a-C:H is more difficult than that of a-C alone because of the effects of hydrogen and of medium-range order (clustering). The minimum requirement is a value for the hydrogen content and carbon coordination number. It is also of interest to know the medium-range order, in so far as it affects physical properties such as optical gap. Clearly, there is no unique structure of a-C:H, it depends strongly on the deposition conditions.

The H content can be found by nuclear reaction methods[5] or by nuclear magnetic resonance (NMR)[31,32]. The thermal evolution method used in a-Si:H is less reliable as much H is evolved as hydrocarbons in soft a – C:H[5]. Many techniques have been employed to characterise the carbon bonding; diffraction, extended x-ray absorption fine structure (EXAFS), XANES, optical spectroscopy, Raman, infra-red and NMR. Each technique tends to have its strong and weak points, such as being sensitive mainly to sp^2 or sp^3 sites, needing hydrogen or not detecting hydrogen. On balance in this author's opinion, C^{13} NMR comes out ahead for coordination number, while IR has been the most popular to date.

Diffraction

Electron and x-ray diffraction were used on a-C:H by McKenzie et al[33]. There are two difficulties in applying these techniques to a-C:H. They do not detect hydrogen so they only give the partial C-C cordination not the total C coordination, a nontrivial issue given its sizeable hydrogen content. The structure must therefore be deduced by a more sophisticated means such as the positions of the higher order peaks. However Table 4 shows that F(k) peak positions differ little between sp^3 and sp^2 networks. Neutron diffraction can detect H and the C-C and C-H coordinations should be separable from the different bond lengths[78].

X-ray Absorption

Excitation of the carbon 1s core electrons into unoccupied conduction band states gives an absorption above 285 eV. The first 10 eV of this is called XANES and the subsequent oscillations beyond the edge are called EXAFS. EXAFS is unsuitable for a-C:H because it also does not detect hydrogen. The interpretation of EXAFS for elements of low atomic number is difficult. The coordination number is roughly proportional to the size of the oscillations, extrapolated back to the edge. Thus, using the graphite spectrum as a reference, the spectrum of Fink et al[34] suggests a coordination number of $\simeq 3$ for a-C, while the spectrum of Comelli et al[35] of their a-C was less damped, giving a coordination number nearer 2. This value seems unphysical in the light of other data, and may indicate that the sample deteriorated during the sputter cleaning process.

XANES spectra are a very sensitive to the presence of sp^2 bonding. The XANES spectra of graphite, diamond, a-C and a-C:H of Fink et al[34] and of MSIB a-C of Berger et al[8] are shown in Fig. 7. The π^* states lie lowest in the conduction band so the first peak at 285 eV arises from $1s \rightarrow 2p\pi^*$ transitions and the edge at 289 eV is due to the higher $1s \rightarrow 2p\sigma^*$ transitions. The presence of a 285 eV peak in the a-C:H spectrum is unequivocal evidence of the presence of sp^2 sites. With care, this peak can be used to extract the sp^2 content. This is particularly valuable as the answer is independent of H content. The XANES spectrum is proportional to the conduction band density of states, but distorted by the interaction of the excited electron and the core hole. The proportion of sp^2 sites can be estimated from the area of the 285 eV peak by normalising the spectrum to that of graphite and assuming that the core hole interaction

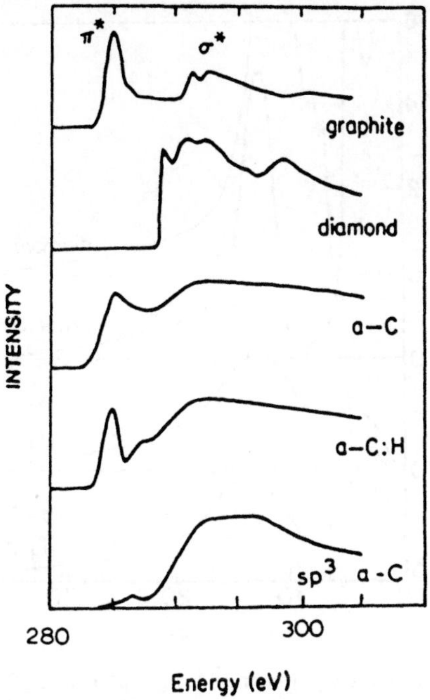

Fig. 7. XANES of graphite, diamond, a-C, a – C:H[34] and MSIB a-C[8]

effect is constant. In some cases the 285 eV peak tails into the 289 eV step, and its area can then be estimated by folding over its leading edge. In this manner Berger et al[8] estimated the sp^2 content of evaporated a-C as 99% and that of MSIB a-C as 15%. Jarman et al[36] used an alternative means of normalisation to deduce values of sp^2 content which agreed remarkably well with those deduced from NMR.

Optical Spectra

The excitations of π and σ electrons show up as two largely separate contributions to the optical spectra of graphite and they have therefore been used as a means of estimating the sp^2 content of a – C:H[34,37,38]. Fig. 8 shows the imaginary part of the dielectric function, ε_2, for diamond, graphite, a-C:H deposited at 25°C and a-C:H annealed at 540°C, the latter curve representing sp^2 a-C. Each curve consists of two peaks, at $\simeq$4 eV and 13 eV, with the lower peak absent in diamond. The absorption due to σ electrons is very weak below 8 eV in both diamond and graphite. A similar behaviour in amorphous carbons would allow absorption in the low energy peak below 8 eV to be assigned solely to $\pi \rightarrow \pi^*$ transitions and excitations above 8 eV to $\sigma \rightarrow \sigma^*$ transitions. Quantitatively, the number of electrons contributing to ε_2 between cutoff energies E_1 and E_2 is given by the sum rule

$$N_{eff} = \left(\frac{m}{2\pi^2 N_A e^2 \bar{h}^2}\right)\int_{E_1}^{E_2} E'\varepsilon_2(E')dE' \quad (4)$$

where N_A is the number of carbon atoms per unit volume and m and e are the electron mass and charge. The prefactor takes the value 0.766 for N_A in m^{-3} and E in eV. This approach was used by Fink et al[34,37] to estimate $N_{eff}(\sigma)$ and $N_{eff}(\pi)$ and thus the proportion of sp^3 and sp^2 sites for various amorphous carbons. They found $\simeq$100% sp^2 for evaporated a-C and $\simeq$32% for 25°C deposited hard a-C:H, with an accuracy of order 30%.

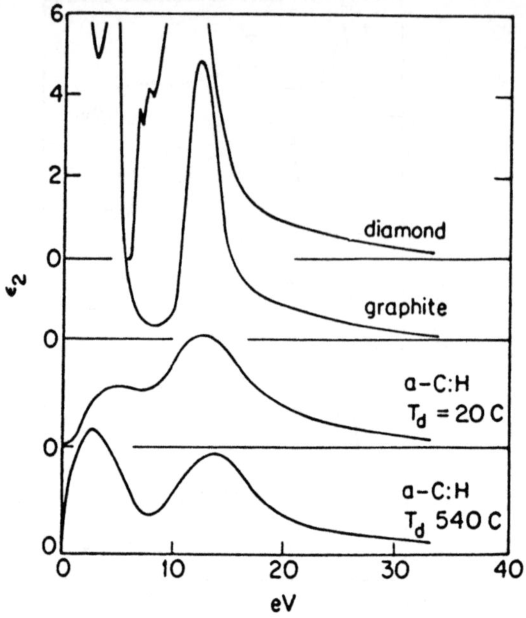

Fig. 8. ε_2 spectra of diamond[39], graphite $E \perp C$[40] and a – C:H[37] .

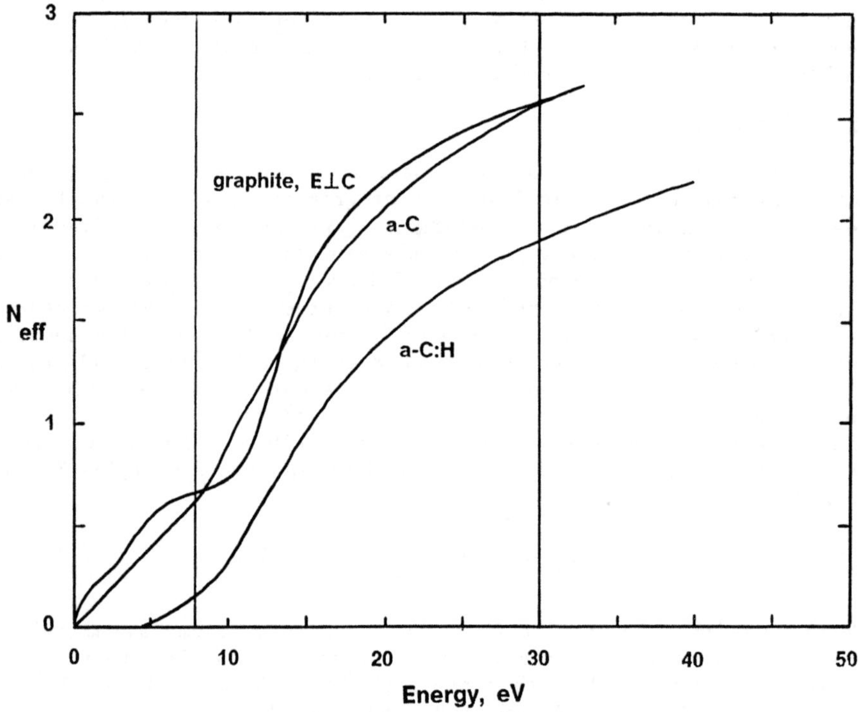

Fig. 9. N_{eff} spectra of diamond, graphite ($E \perp C$), a-C and a – C:H[42]

This approach was criticised by Gao et al[41] on two counts. First, an error was detected[42] in the derivation of N_{eff} of graphite by Taft and Phillip[40], resulting in N_{eff} being reduced by a factor of $2/\pi$. Fig. 9 shows N_{eff} for graphite for $E \perp c$ (corrected), a-C and a sample of a-C:H[43]. N_{eff} now no longer reaches 1 at 10 eV or 4 at 40 eV, so the π and $\sigma+\pi$ oscillator strengths are no longer exhausted at the expected energies. This is even more apparent in the $E//c$ spectrum of graphite (note that a-C can be considered to $2/3\ E \perp c + 1/3\ E//c$). Thus, the fundamental basis of the 8 eV cutoff is removed.

A second problem is that the π and σ excitations are poorly separated at 8 eV. A separation requires that N_{eff} reach a plateau around 8 eV, which in turn requires that $\varepsilon_2 \to 0$ in the dip at 8 eV. It is clear from Fig. 8 that the dip in ε_2 is much weaker in a-C and a-C:H than in graphite itself, so the separation is poorer and indeed N_{eff} is seen to only possess an inflexion rather than a plateau around 8 eV. Neverthelesss, we note that $N_{eff}(0-8\text{eV}) / N_{eff}(0-30\text{eV}) \simeq 1/4$ for both graphite and evaporated a-C, as expected for sp^2 bonding. Thus it is possible that the sp^2 contents deduced from the N_{eff} method are still reasonable, although they have a less secure theoretical basis. This would be useful, because as remarked earlier this method is not susceptible to the problems of hydrogen distribution or MRO.

NMR

C^{13} NMR is the most direct method of measuring the sp^3/sp^2 content of a-C:H[31,32,36,44,45]. The sp^3 and sp^2 sites give rise to two separate chemically-shifted peaks whose area is directly proportionally to their concentrations. NMR experiments have shown directly the increase in sp^2 content with increasing bias voltage in plasma-deposited films[44,45].

The presence of hydrogen is preferable for this technique as the peaks can then be sharpened by applying a proton decoupling signal. This feature can be used to advantage; the C-H dipole interaction is very short-ranged, decaying as r^{-3}, so turning off the decoupler eliminates from the spectrum those sites bonded to a hydrogen (or perhaps first or second neighbours of a hydrogen). This allows the distribution of hydrogen to be obtained. sp^3 sites are found to be much more heavily hydrogenated than sp^2 sites in soft a-C:H[32]. One may wonder whether NMR may miss those sp^2 sites within large sp^2 clusters due to a lack of nearby hydrogen[46]. In practice, this seems not to occur, as NMR has been successfully used to determine fine details of the site distribution in coals[47], in which sp^2 clustering is even stronger than in a-C :H. It appears that the extreme of all sp^3 sites being hydrogenated and all sp^2 sites being unhydrogenated is not reached.

IR

The C-H bond stretching modes produce a series of distinct infrared absorption peaks near 3000 cm^{-1}. Each peak can be assigned to a different C-H configuration, both in terms of the carbon configuration - sp^3, sp^2 aromatic and sp^2 olefinic - and the number of hydrogen neighbours. Dischler[14] has assigned the peaks and used their intensities to deduce both the sp^3/sp^2 ratio and H content. A thorough analysis of the sp^3 and H content was carried out for plasma-deposited a-C :H as a function of bias voltage V_b, source gas and pressure. He found the sp^3/sp^2 ratio and H content to fall with increasing V_b. Values for sp^3 content were similar to those found by the N_{eff} method[34]. Above V_b=400 V the films were found to be independent of source gas, while below 400 V the films depended on source gas due to incomplete molecular dissociation in the plasma.

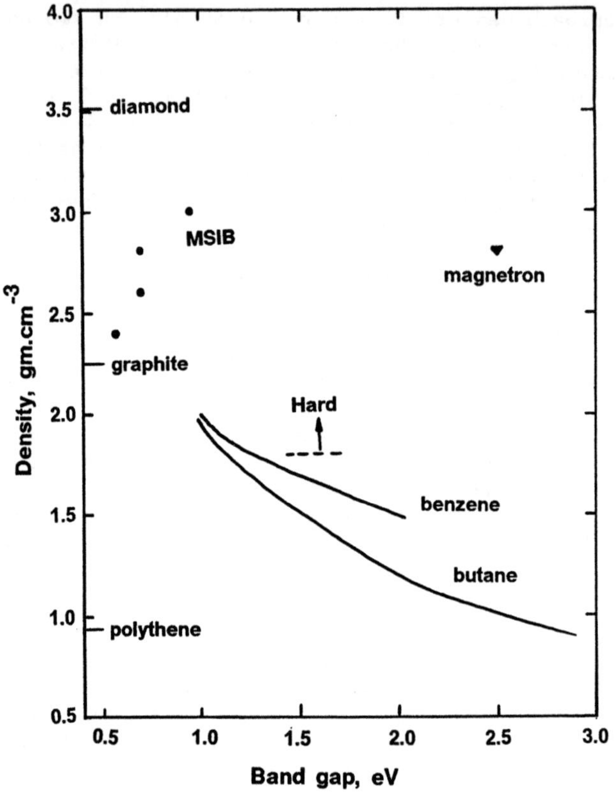

Fig. 10. Optical gap vs. density for PD a-C:H prepared from benzene[5] and butane[48] and a-C[13,60]

The IR method has been a very popular method because of its wide availability and ease of use. However, it clearly needs a uniform distribution of hydrogen over the sp^3 and sp^2 sites to give a reliable concentration values. NMR suggests that this is untrue[32]. The method is likely to omit those sp^2 sites within clusters which are unhydrogenated.

Tamor et al[48] in particular have noted these problems. They noted that a-C:H derived from benzene (as used by Dischler[14]) and butane had a different band gap - density relationship, particularly at large E_g. The benzene films were noted to contain an increasing proportion of IR modes attributed to 'pendant benzene', side-chain benzene molecules undissociated by the plasma which do not contribute to the band gap, connectivity or hardness of the main network, and serving only to change the density. It could even be suggested that the sp^2 sites of the irrelevant pendent benzene would dominate the IR spectra, and would overshadow the more important sp^2 sites in the main network clusters, which may even be unhydrogenated. In this way the IR method might be of little use for finding the sp^2 content of the extented network.

The situation is less serious in practice. First, considerable amounts of hydrogenated sp^2 sites have been observed by the IR method in films prepared from methane and acetylene[49,50], even at high V_b. Thus, IR will detect many sp^2 sites even if it does favour sp^3 sites. Second, the divergence between films prepared from benzene and butane occurs mainly for soft films with $V_b<400$ V and $E_g>1.7$ eV, consistent with the original classification of Koidl et al[5]. This is seen when the more

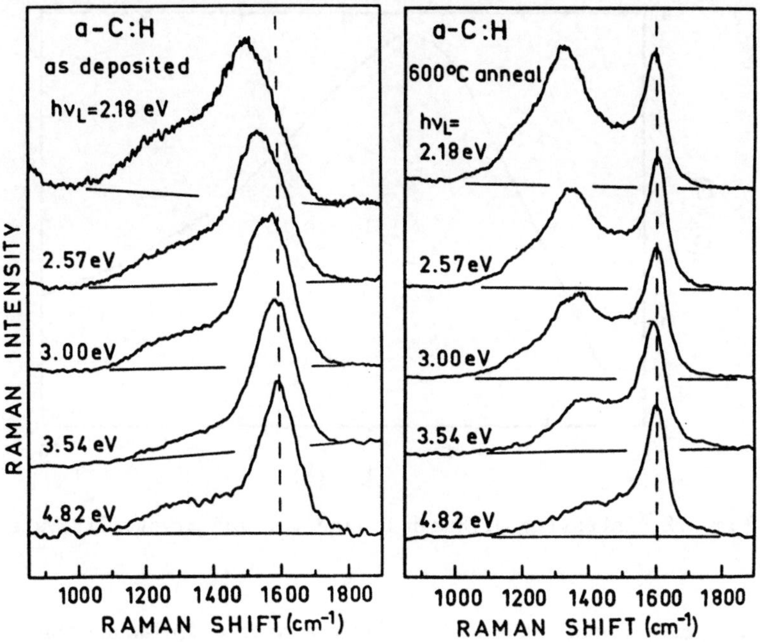

Fig. 11. Raman spectra of a-C:H and annealed a-C:H vs. laser frequency[56]

complete data for E_g and density are plotted in Fig. 10. Thus, the IR method is useful for the more important hard a-C:H films. Finally, the dominant mode in the IR spectra of wide-gap films prepared from benzene by Tamor[48] is at 3030 cm^{-1} and is assigned[14] to an olefinic not aromatic $C(sp^2)-H_2$ group, ie not pendant benzene if the assignment is correct. The other mode at 3060 cm^{-1} is an aromatic $C(sp^2)-H$ group and is related to benzene.

<u>Raman</u>

Raman spectroscopy is a popular tool for characterising the crystallinity of diamond thin films[51] Crystalline diamond gives a single sharp line at 1332 cm^{-1}. Graphitic carbon gives two peaks; the G peak centred on 1550 cm^{-1} is the zone-centre E_{2g} Raman line of a perfect graphite crystal while the D peak centred on 1350 cm^{-1} is a zone edge A_{1g} mode activated by disorder[52]. The D mode is a common feature of all disordered graphitic carbons (coals, glassy carbon, etc) whose intensity relative to the G peak, I(D)/I(G), has been shown to vary inversely with the size of the graphitic crystallites, L_a[53]

$$I(D)/I(G) = k/L_a \tag{5}$$

Raman spectroscopy of a-C:H is of interest in its own right as it gives evidence of the clustering of carbon sp^2 sites.

A key point in discussing the Raman spectra of carbons is that the Raman scattering cross-section Ω of the E_{2g} mode in single crystal graphite is 30-60 times that of single crystal diamond[54,51]. Raman can be modeled atomistically as the change in bond polarization due to atomic displacement. However, the polarizabilities of bonds in diamond and graphite are often treated as equal, as in for example Beeman et al[20].

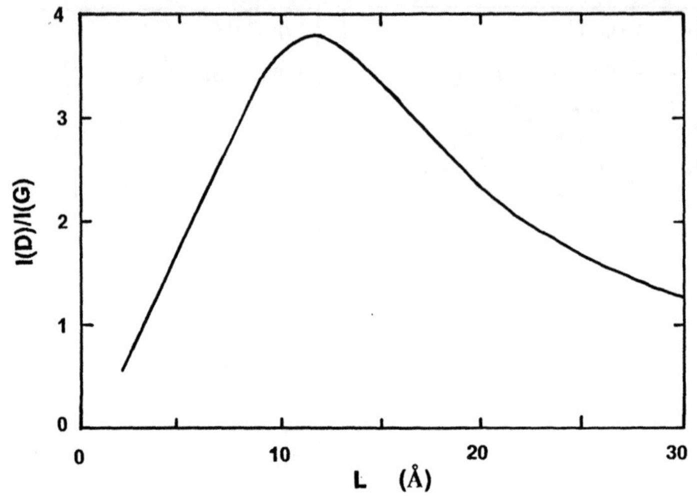

Fig. 12. Intensity of Raman D line vs. sp^2 crystallite size

This would give roughly the same Ω for graphite and diamond and is wrong. Clearly, an extra contribution to Ω must be included from π bonds which would be large due to their smaller band gap. A satisfactory microscopic model of π polarizability and Raman scattering has yet to be derived.

Fig. 11(a) shows the Raman spectrum[55,56] of as-deposited hard a-C:H as a function of laser frequency, ν. Fig. 11(b) shows the spectrum of a film annealed to 600°C, which is essentially sp^2 a-C. The spectrum of as-deposited a-C:H at $h\nu$=2.18 eV is seen to consist of a broad peak centred on 1520 cm^{-1} with a shoulder around 1300 cm^{-1}, perhaps related to the G and D modes of graphite. The upper peak sharpens into the G peak at 1600 cm^{-1} and the D peak develops around 1350 cm^{-1}. This behaviour is consistent with a transformation of a-C:H into disordered graphitic crystallites by annealling.

Tamor et al[57] studied changes in the intensity of the D peak in PD a-C :H as a function of bias voltage V_b. They found that I(D)/(I(G) increases with increasing V_b. They noted that the D peak arises from sp^2 clusters which are also believed to control the optical gap E_g. Now, E_g decreases with increasing V_b and as $E_g \propto L_a^{-1}$ from equation (1), we expect L_a to rise with V_b rather than fall as implied by equation (5). They therefore suggested that equation (5) fails at low L_a and that the dependence of I(D) on L_a follows that shown in Fig. 12, reaching a peak around $L_a = 12$Å . This is entirely reasonable; I(D)/I(G) cannot increase without limit at low L_a. The D peak is variously interpreted as a disorder activated zone-edge mode or as a density of states feature activated by a loss of periodicity. The Raman spectrum of an amorphous solid tends to equal the phonon density of states, weighted by a factor $\Omega(E)$. Clearly, I(D) will increase in microcrystalline graphite up to a maximum set by the phonon DOS in the region of K point. I(D) may then be forced to decrease again in very disordered cases for $L_a < 12$Å, as DOS is forced into modes with lower Ω values.

Wagner et al[56] noted the intensity changes of the G and D peaks arose from the Resonant Raman effect. The Raman cross-section Ω arises from changes in bond polarizability, $\varepsilon_2 - 1$, with atomic displacement. ε_2 arises from electronic transitions across the optical gap, so Ω will vary as

$$\Omega = (\partial \varepsilon / \partial v)^2 \qquad (6)$$

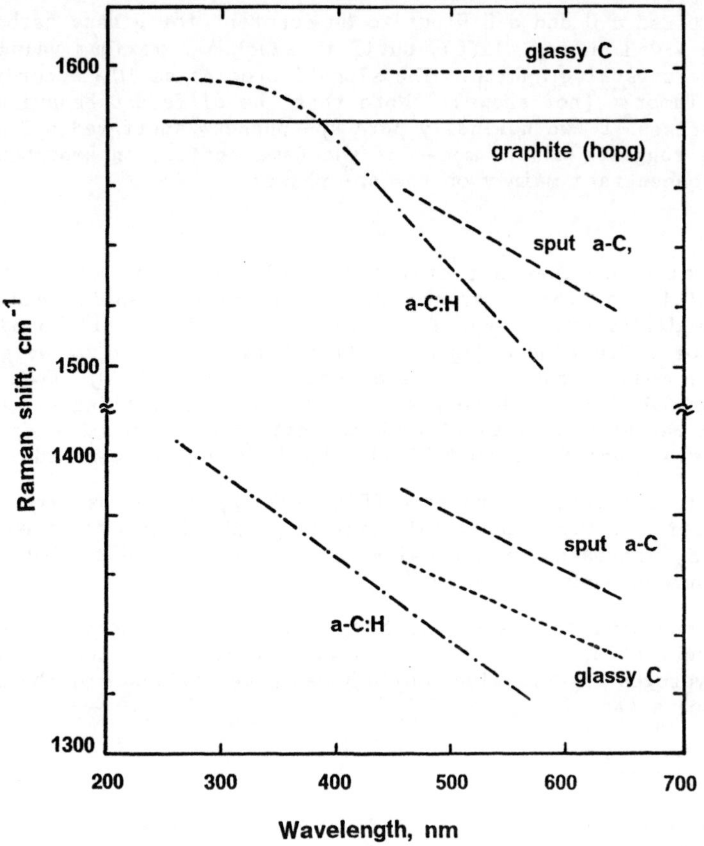

Fig. 13. Raman peak shift vs. laser wavelength for graphite, glassy C and sputtered a-C[58] and a-C:H[56]

If the excitation frequency is below the gap in the transparent region, Ω is independent of ν. Moving ν towards a band edge causes a resonant increase in Ω, giving the resonant Raman effect.

Closer inspection of Fig. 11 shows that the G peak moves upwards with ν in as-deposited a-C :H and that the D peak moves upwards in the annealed film while the G peak does not[55-58]. Wagner et al[55,56] noted that the peak shifts in the Raman spectra were consistent with the π bonded cluster model of a-C:H, with Ω varying as ν resonates in turn with the gaps of the various cluster sizes. They interpreted the overall peak shifts in terms of a 2-phase model, noting the various shifts and lack of shifts could not be described by a single phase. Phase 1 was assumed to give peaks at 1300 and 1500 cm^{-1} and was attributed to the sp^3 network. Phase 2 was assumed to give a peak at 1600 cm^{-1} and was attributed to sp^2 clusters. The peak shifts then arose from relative changes in Ω of each phase associated with resonances.

This interpretation may be partly wrong. The shifting of G and D peaks is a common feature of disordered sp^2 carbons[59]. In view of the 30-60 fold greater Raman cross-section of π bonds it seems more likely that essentially all features of the Raman spectrum of a-C:H arise from sp^2 sites. Fig. 13 shows the variation of G and D peaks vs. $1/\nu$. The D peak is seen to vary linearly with the same slope for glassy carbon, sputtered a-C and (some[58]) a-C :H. The G peak is seen to vary linearly for sputtered a-C and a-C:H but to be constant for glassy carbon and the annealed a-C:H in Fig. 11(b), until it reaches a maximum value set by G in single crystal graphite. The slope increases as $1/L_a$ according to the data of Tamor[57] (not shown). Note that the different behaviours of the G peak arises it two nominally pure sp^2 phases, sputtered a-C and glassy C. This suggests that a model of the Raman effect in amorphous carbons should concentrate mainly on the sp^2 phase.

Density

Density provides a useful and simple check on coordination data. The densities of the reference phases, diamond, graphite and polythene are very different as seen from Table 1. Koidl et al[5] noted that the density of a-C:H plasma-deposited from benzene, found by weight gain - thickness measurements, increased continuously with V_b, from polymeric a-C:H through hard a-C:H to graphitic a-C:H. Interestingly, the density of a-C:H measured by Rutherford backscattering of PD a-C:H from methane and acetylene was found to fall with V_b[49,50], which seems odd.

Overall trends can be seen (Fig. 10). E_g decreases with increasing density for PD a-C:H because polymeric or graphitic phases often dominate, whereas E_g increases with density for unhydrogenated a-C[13,60], as the density now exceeds 2.2 $gm.cm^{-2}$.

Summarising this section, NMR, XANES and IR spectra can give the sp^3/sp^2 content of a-C:H films. NMR gives evidence of the 2-phase nature of the hydrogen distribution, while Raman gives evidence of the clustering of the sp^2 sites.

DEPOSITION MODELS

The desired porperties of diamond-like carbon arise from a high concentration of sp^3 sites. In PD films, it is generally assumed that hydrogen promotes sp^3 bonding as a byproduct of saturating the bonding. High hardness in particular becomes a trade-off between too many sp^2 sites at low H content, and too many $=CH_2$ sites at high H content, as discussed more rigorously shortly.

A second mechanism must promote sp^3 bonding in unhydrogenated diamond-like carbon. Various mechanisms have been proposed; the thermal spike model where a pressure front promotes the formation of the higher density sp^3 phase[60,61], or the prefential sputtering of sp^2 sites[2]. The preferential displacement model of Lifshitz et al[29] seems the most likely. Concentrating on aspects relevant to thicker films, this model notes that the displacement threshold energy, the angle-averaged energy needed to create a permanent vacancy-interstitial pair, is much lower for graphite (25-30 eV) than for diamond (80 eV). The difference arises from the ease of displacement into the empty inter-layer region of graphite, so it should be retained for sp^2 and sp^3 sites in general. The difference causes the number of displacements per incident ion to be 20 times higher for sp^2 sites than sp^3 sites for 100 eV ions and roughly 3 times higher for ions over 200 eV (Fig. 14).

A dynamic balance model can be proposed for this process, a solid state analogue of the Deraygin model of gas phase diamond deposition. Displacements create a reservoir of weakly bonded atoms which may then trap into either sp^2 or sp^3 sites with a similar rate. However, the displacement rate back up to the reservoir is much larger for sp^2 atoms than for sp^3 atoms, so the sp^3 sites become favoured. The optimum incident ion energy is one higher enough for the displacement rate to exceed the deposition rate to create a sufficient reservoir population, but low

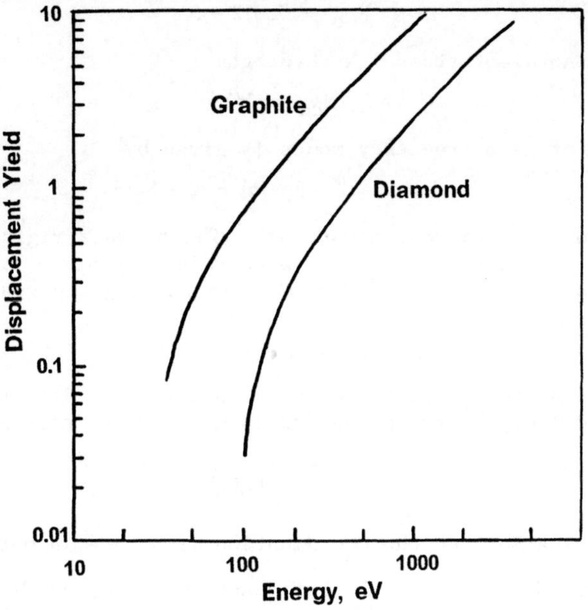

Fig. 14. Displacement yield for sp^2 and sp^3 sites vs. incident ion energy

enough to give the larger preferential displacement rate of sp^2 sites, ie 100-200 V. Clearly the preferential displacement mechanism also operates during hydrogenated conditions. It accounts for the optimum V_b for hard a-C:H being 400-1000 V, as the mean ion energy is about $1/3V_b$. It seems unnecessary to invoke a 'mold' effect as in Lifshitz et al[29] to define the different trap sites. Atoms in random networks have a very strong tendency to satisfy their valence. Their configurations appear to be determined by a combination of Maxwell-Boltzman statistics and thermal quenching. There appears to be no role for a 'template' effect of the rest of the network helping to fix the coordination of an arriving atom, analogous to epitaxy on crystal surfaces. For example, the proportion of substitutional donors or acceptors in a-Si:H is largely controlled by MB statistics.

MECHANICAL PROPERTIES

The key property of diamond-like films is their hardness, H. H is measured by indentation and is the pressure applied over the indented area. It is related empirically to the yield stress Y of the material which in turn is proportional to the Youngs modulus E. The elasticity of diamond-like carbon is related to the rigidity of an individual bond and the degree of connectivity of its bonding network.

Phillips[63] and Thorpe[64,65] introduced a mean field theory which related the rigidity of a network to its mean coordination. Consider a covalent network of N atoms whose energetics obey a valence force field of bond stretching and bond bending forces. When the mean coordination n is low, there are many ways in which the network can be deformed at zero energy cost, ie. leaving bond lengths and bond angles unchanged. The number of these deformations or zero-frequency vibration modes is given by the number of degrees of freedom (3N) minus the number of constraints. The number of constraints varies with coordination. There is one constraint associated with a bond (shared between two atoms) and $2n_i-3$ constraints with the angles of a n_i coordinated atom, giving a total of

$$N_{co} = 5n_i/2 - 3 \tag{7}$$

except for monovalent atoms like hydrogen

$$N_{co} = 1/2 \tag{8}$$

The fraction of zero-frequency modes is given by

$$f = \Sigma c_i(2 - 5n_i/6) = 2 - 5n/6 \tag{9}$$

if $n_i>1$, where c_i are concentrations. The network rigidity is proportional to -f. Thus a network of low coordination has a finite number of zero-frequency modes and is 'floppy'. Increasing the mean coordination above

$$n = 2.4 \tag{10}$$

gives a rigid network, and rigidity is said to percolate across the network. He and Thorpe[65] found the Youngs modulus to increase for n>2.4 as

$$E = E_0(f/f_0)^{1.5} = E_0\left(\frac{n-2.4}{n_0-2.4}\right)^{1.5} \tag{11}$$

to reach a value E_0 when the coordination is n_0, =4 in our case.

Angus and Jansen[66] found that condition (10) could be used on a network containing hydrogen if n represented the mean C-C coordination, by substituting c_4 by $c_4 - 2c_1$ in (9), ie by treating each C-H bond as a broken bond which contributes no rigidity. Indeed, all pendant groups like $-CH_3$ or contribute nothing to network rigidity, but it is difficult to allow for their effects.

The constraint model must be further modified to take into account π bonding. A π bond between two sp^2 sites as in ethylene adds torsional rigidity to the bond, that is it adds a 1/2 constraint per atom, increasing N_{co} from 4.5 to 5 for a sp^2 site. Angus and Jansen[66] noted that a π bonded reduced N_{co} to 3.5 by treating the atom pair as a quasi 4-coordinated site, but they omitted to include the reduction of 3 in the number of degrees of freedom due to the loss of 1 site. Thus, a sp^2 C atom is more constrained than a trivalent P atom.

The Thorpe[65] model is a mean field theory which assumes a network with no MRO. For instance, the bonds of a 3-coordinated network are assumed to be randomly connected in all three directions. This is not true if the network contains benzene rings or graphite layers. A graphite layer uses all its constraints in only two of the three directions and is floppy along Oz, despite being trivalent. Many of the 5 constraints per site of a graphite layer are <u>redundant</u> for the rigidity of the overall network. The number of non-redundant constraints of a 6-fold π-bonded ring is 15, 6 for the bonds, 6 for angles and 3 for bond torsions, giving

$$N_{co} = 5/2 \tag{12}$$

These constraints can be assigned differently because of the redundancy. This can be checked by adding further rings. A second ring of 4 sites can be added and made rigid with 5 bond constraints and 5 bond torsion constraints and no bond angle constraints.

This value (12) can also be checked against the vibrational frequencies of graphite[67]. A single graphite layer has 1 doubly-degenerate E_{2g} (Raman) mode, 1 out-of-plane B_{2g} rigid mode, 1 doubly degenerate E_{1u} in-plane acoustic mode with non-zero frequency away from Γ and 1 out-of-plane A_{2u} acoustic mode in which the atoms move in-phase along Oz. It has zero frequency because there are no inter - planar forces. Thus,

$$f = 1/6 \tag{13}$$

for graphite, consistent with (12). Thus, graphite is floppy because its bonds are ordered into only two dimensions, despite having n=3 which normally gives a rigid network.

The 6-fold rings can be joined to the network. N_{co} is then

$$N_{co} = 5/2(N' + N'') \tag{14}$$

where N' is the number of atoms in the cluster and N'' is the number of externally connected atoms. N'' is much less than N' for a compact cluster. For a circular cluster $N'' \simeq \sqrt{N'}$, giving a contribution to f of

$$f_3 = (3N' - 5/2(N' + \sqrt{N'}))/3N' = 1/6 - 1/(3\sqrt{N'}) \rightarrow 1/6 \tag{15}$$

as $N' \rightarrow \infty$. If the sp^2 clusters are quite large, we may approximate that these sites make no contribution to the network rigidity, giving finally

$$-f = 4(c_4 - 2c_1)/(1 - c_1) - c_3/6(1 - c_1) \tag{16}$$

Thus graphitic and polymeric bonding both tend to reduce rigidity.

These results can be confirmed by considering E of graphitic glassy carbon (Table 6). E of glassy carbon is seen to be only about 1/20 of that of the in-plane E of graphite. This is consistent with an E derived only from the inter-plane cross links, every $L_a \simeq 20$ rings apart.

The Youngs moduli of PD a-C:H has been measured as a function of V_b (Table 6). The C-C coordination may be estimated from equation (11), using E_0 for diamond, to give n=2.83 for the 100V film and n=2.63 for the 1 kV film, which is consistent with their measured[49] H and sp^3 contents. These values show that the networks have coordinations well below the ideal limit of 4 but well above the floppy limit of 2.4. The void fraction also has an effect on the mechanical properties of a-C:H[69,75], but much of this effect is included in terms of C-C coordination number.

Table 6. Mechanical properties

	E, GPa	ν	H, GPa	Y, GPa	ref.
Diamond	1050	0.104	103	59	68
PD a-C:H, 100 V	145	0.4	16	9.7	62,69
PD a-C:H 1 kV	55	0.2	6.3	3.1	62,69
MSIB a-C			20-110	12-65	10
graphite, //a	686				70
glassy C, GC10	29	0.15	3.0	1.0	71,72
glassy C, GC20	32	0.17	2.2	0.73	71,72
Si	130	0.278	10.4	5.0	73
a-Si:H	100	0.32	10.0	4.9	74

Hardness is a measure of the yield stress, Y. It is found to be related to Y by the empirical relation[76]

$$H/Y = 0.07 + 0.6 \ln(E/Y) \simeq -0.04 + 0.77 \ln(E/H) \tag{17}$$

where the constants are found from fitting data for a wide range of materials[76,77], metal, polymers, glass, etc. This gives $H/Y \simeq 1.8$ for diamond-like materials (much lower than the typical factor 3 for metals).

The yield stress of a metal is controlled by dislocation flow, but it is close to the cleavage stress for covalent materials like diamond and diamond-like carbons in which dislocation flow is very limited. The cleavage stress is the stress needed to separate two adjacent planes of atoms of the solid. This in turn can be related to the Youngs modulus by approximating the interplanar force by a sine function and equating E to the initial slope, to give[70]

$$Y = E/\sqrt{10} \tag{18}$$

Table 6 shows that a factor of 6 gives a better fit for diamond, giving

$$H/E \simeq 0.1 \tag{19}$$

Hardness values have been measured experimentally for both PD a-C:H[62,69] and MSIB a-C[10,12]. In the measurement, the indenter depth should be less than the film thickness and allowance should be made for elastic rebound. The hardness values for PD a-C:H scale with E according to equation (19), showing that both H and E follow the mean coordination model. The hardness value of MSIB a-C is seen to reach values similar to that of diamond, emphasising its high sp^3 bonding.

This indicates that it is truly possible to prepare diamond-like thin films with a hardness similar to diamond. The simple plasma-deposited a-C:H does not reach this limit being less than 1/6 as hard. MSIB a-C can reach this limit but the technique is probably not suited to commercial scales. Dual-beam magnetron sputtering offers probably the best commercial route towards very hard a-C. The material is also relatively free of bonded hydrogen and so is likely to have a better high temperature stability than a-C:H. Finally, a-C can be readily doped both n- and p-type[3] unlike diamond itself, although carrier mobilities are much lower. Thus, a-C is truly becoming usefully diamond-like, being similarly hard, reasonably temperature stable, transparent and bipolar dopable.

Acknowledgements

The author is very grateful to P. H. Gaskell, J. Fink, D. R. McKenzie and A. Ritter for access to unpublished data and to P. Koidl and M. Tamor for discussions.

References

1. S. Aisenberg and R. Chabot, J App Phys 42 2953 (1971)
2. J. C. Angus, P. Koidl and S. Domitz, in 'Plasma Deposited Thin Films', Ed J. Mort (CRC Press, 1986)
3. J. Robertson, Adv Phys 35 317 (1986)
4. J. Robertson, Mat Sci Forum 52 125 (1990)
5. P. Koidl, C. Wild, B. Dischler, J. Wagner and M. Ramsteiner, Mat Sci Forum 52 41 (1990); P. Koidl, this volume; Ch. Wild, this volume
6. D. F. R. Mildner and J. M. Carpenter, J Non-Cryst Solids 47 391 (1982)
7. J. Hauser, J Non-Cryst Solids 23 21 (1977)
8. S. D. Berger, D. R. McKenzie and P. J. Martin, Phil Mag Let 57 285 (1988)
9. D. C. Green, D. R. McKenzie and P. B. Lukins, Mat Sci Forum 52 103 (1990); D. R. McKenzie, D. C. Green, P. D. Swift, D. J. H. Cockayne, P. J. Martin, R. P. Netterfield and W. G. Sainty, Thin Solid Films 193 xxx (1990)

10. P. J. Martin, S. W. Filipczuk, R. P. Netterfield, J. S. Field, D. F. Whitall and D. R. McKenzie, J Mat Sci Lett 7 410 (1988)
11. P. H. Gaskell, K. W. R. Gilkes and A. Saeed, to be published
12. J. P. Hirvonen, J. Koshinen, R. Lappalainen and A. Anttila, Mat Sci Forum 52 197 (1990)
13. J. Ishikawa, Y. Takeiri, K. Ogawa and T. Takagi, J App Phys 61 2509 (1987)
14. B. Dischler, 'Amorphous Hydrogenated Carbon' ed P Koidl, Proc Euro Mat Res Soc (1987)
15. K. Enke, Thin Solid Films 80 227 (1981)
16. K. Enke, Mat Sci Forum 52 559 (1989)
17. J. Robertson, E. P. O'Reilly, Phys Rev B 35 2946 (1987)
18. J. L. Bredas and G. B. Street, J Phys C 18 L651 (1985)
19. F. W. Smith, J App Phys 55 764 (1984)
20. D. Beeman, J. Silverman, R. Lynds and M. R. Anderson, Phys Rev B 30 870 (1984)
21. G. Galli, R. M. Martin, R. Car, M. Parrinello, Phys Rev Lett 62 555 (1989)
22. J. Tersoff, Phys Rev Lett 61 2879 (1988)
23. G. S. Painter and D. E. Ellis, Phys Rev B 1 4747 (1970)
24. G. S. Painter, D. E. Ellis and A. R. Lubinsky, Phys Rev B 4 3610 (1971)
25. D. Wesner, et al, Phys Rev B 28 2152 (1983)
26. B. B. Pate, Surface Science 165 83 (1986)
27. P. Oelhafen and D. Ugolini, in ref 14.
28. T. Miyazawa, S. Misawa, S. Yoshida and S. Gonda, J App Phys 55 188 (1984)
29. Y. Lifshitz, S. R. Kasi and J. W. Rabalais, Phys Rev Lett 62 1290 (1989)
30. B. T. Boiko, L. S. PaLantik and A. S. Deveryanchenki, Sov Phys Dokl 13 237 (1968)
31. A. Grill, B. S. Meyerson, V. V. Patel, J. A. Reimer and M. A. Petrich, J App Phys 61 2874 (1987)
32. M. A. Petrich, Mat Sci Forum 52 387 (2989)
33. D. R. McKenzie, L. C. Botten and R. C. McPhedran, Phys Rev Lett 51 280 (1983)
34. J. Fink, T. Muller-Heinzerling, J. Pfluger, A. Bubenzer, P. Koidl and G. Grecelius, Solid State Commun 47 887 (1983)
35. G. Comelli, J. Stohr, C. J. Robinson and W. Jark, Phys Rev B 38 7511 (1988)
36. R. H. Jarman, G. J. Ray, R. W. Standley and G. W. Zajac, App Phys Lett 49 1065 (1986)
37. J. Fink, T. Muller-Heinzerling, J. Pfluger, B. Scheerer, B. Dischler, P. Koidl, A. Bubenzer and R. E. Sah, Phys Rev B 30 4713 (1984)
38. N. Savvides, J App Phys 59 4133 (1986)
39. R. A. Roberts and W. C. Walker, Phys Rev 161 730 (1967)
40. E. A. Taft and H. R. Phillip, Phys Rev 138 A197 (1965)
41. C. Gao, Y. Y, Wang, A. L. Ritter and J. R. Dennison, Phys Rev Lett 62 945 (1989)
42. R. Sonnenschein, M. Hanfland and K. Syassen, Phys Rev B 38 3152 (1988)
43. J. Fink, private communication
44. S. Kaplan, F. Jansen and J. Machonkin, App Phys Lett 47 750 (1985)
45. K. Yamamoto, Y. Ichikawa, T. Nakayama and Y. Tawada, Jpn J App Phys 27 1415 (1988)
46. M. A. Tamor and C. H. Wu, J App Phys 67 1007 (1990)
47. D. E. Axelson, 'Solid State Nuclear Magnetic Resonance of Fossil Fuels' (Multiscience Publications, Canada, 1985)
48. M. A. Tamor, C. H. Wu, R. O. Carter and N. E. Lindsay, App Phys Lett 55 1388 (1989)

49. J. W. Zou, K. Reichelt, K. Schmidt and B. Dischler, J App Phys 65 3914 (1989)
50. J. W. Zou, K. Schmidt, K. Reichelt and B. Dischler, J App Phys 67 487 (1990)
51. R. E. Shroder, R. J. Nemanich and J. T. Glass, Phys Rev B 41 3738 (1990)
52. R. J. Nemanich and S. A. Solin, Phys Rev B 20 392 (1979)
53. F. Tuinstra and J. L. Koenig, J Chem Phys 53 1126 (1970)
54. N. Wada and S. A. Solin, Physica B 105 353 (1981)
55. M. Ramsteiner and J. Wagner, App Phys Lett 51 1355 (1987)
56. J. Wagner, M. Ramsteiner, C. Wild and P. Koidl, Phys Rev B 40 1817 (1989)
57. M. A. Tamor, J. A. Haire, C. H. Wu and K. C. Hass, App Phys Let 54 123 (1989)
58. M. Yoshikawa, G. Katagiri, H. Ishitani and T. Akamatsu, App Phys Let 52 1639 (1988); J App Phys 64 6464 (1988)
59. R. P. Vidano, D. B. Fischbach, L. J. Willis and T. M. Loehr, Solid State Commun 39 341 (1981)
60. S. M. Rossnagel, M. A. Russak and J. J. Cuomo, J Vac Sci Technol A 5 2150 (1987)
61. C. Weissmantel, Thin Solid Films 92 55 (1982)
62. X. Jiang, J. W. Zou, K. Reichelt and P. Grunberg, J App Phys 66 4729 (1989)
63. J. C. Phillips, J Non-Cryst Solids 34 153 (1979)
64. M. F. Thorpe, J Non-Cryst Solids 57 355 (1983)
65. H. He and M. F. Thorpe, Phys Rev Lett 54 2107 (1985)
66. J. C. Angus and F. Jansen, J Vac Sci Technol A 6 1778 (1988)
67. R. J. Nemanich, G. Lucovsky and S. A. Solin, Solid State Commun 23 117 (1977)
68. C. A. Brookes, Phil Mag A 43 529 (1981)
69. X. Jiang, K. Reichelt and B. Stritzker, J App Phys 66 5805 (1989)
70. A. Kelly and N. H. MacMillan, 'Strong Solids' (Oxford University Press, 1986)
71. H. M. Hawthorne, Carbon 13 215 (1975)
72. J. X. Zhao, R. C. Bradt and P. L. Walker, Carbon 23 15 (1985)
73. V. I. Trefilov and Y. V. Milman, Sov Phys Dokl 8 1240 (1964)
74. J. W. Zou, K. Schmidt, K. Reichelt and B. Dischler, J App Phys 67 487 (1990)
75. X. Jiang, K. Reichelt and B. Stritzker, J App Phys 68 1018 (1990)
76. D. Tabor, Rev Phys Technol 1 145 (1970)
77. S. S. Chiang, D. B. Marshall and A. G. Evans, J App Phys 53 298 (1982)
78. P. J. R. Honeybone, R. J. Newport, W. S. Howells and J. Franks, this volume

CONDUCTIVITY IN AMORPHOUS CARBON AND AMORPHOUS HYDROGENATED CARBON FILMS

D. Das Gupta^, F. Demichelis and A. Tagliaferro

Dip. Fisica - Politecnico Torino - Torino - ITALY
^ on leave from North Bengal Univ.-Siliguri-INDIA

INTRODUCTION

There has been a considerable recent interest in amorphous carbon (a-C) and amorphous hydrogenated carbon (a-C:H) films. These films present a combination of a number of useful properties such as a high value of hardness, transparency to visible radiation over a wide wavelength range and high corrosion resistance. Carbon atoms in these materials have both threefold (sp^2) and fourfold (sp^3) coordination, and the physical properties are determined by the proportion and distribution of sp^2 and sp^3 sites[1]. It is known, for example, that in evaporated a-C sp^2 sites are essentially uniformly distributed within the material, whereas in a-C:H these sp^2 sites form "islands" which are embedded in a matrix consisting essentially of sp^3 sites.

A study of the electrical conductivity, σ , of these materials as a function of temperature would be useful in extracting information regarding the distribution of energy states and also in justifying the above structural models. The log σ vs T^{-1} curve in this case approaches a straight line at high temperatures (>500 K) whose slope slowly decreases with decreasing temperature. The conductivity behaviour in the high-temperature region is explained by activated conduction in extended π -states, whereas the usual explanation for conduction near room temperature is variable-range hopping at the bandtails.

In this paper we report the conductivity measurements on our a-C and a-C:H samples. From the data for our samples and for a-C:H samples reported in the literature, we would like to conclude that conduction in these materials near room temperature is due to hopping between graphitic (sp^2) islands within the sp^3 matrix.

EXPERIMENTAL DETAILS

a-C and a-C:H films were deposited in an RF magnetron sputtering system (MRC Sputtersphere) by the sputtering of a graphite target in pure Argon and Argon + H_2 atmospheres respectively. Pressure was kept constant at 3.3 Pa, while RF power was varied between 300 and 500 W. The flow rate of Argon was maintained at a constant value of 80 sccm, and that of H_2 was varied from 5 to 35 sccm. All of our a-C:H samples were deposited at a constant RF power of 300 W. Films were deposited at room temperature on Corning 7054 glass substrates with pre-evaporated Al electrodes (8 mm wide by 1 mm spaced).

In the following, our samples have been identified by C or CH denoting a-C and a-C:H, respectively, followed by a number representing RF power for a-C and hydrogen flow rate for a-C:H.

RESULTS AND DISCUSSION

It is a general feature for all the films that heating of the as-deposited films produces a permanent increase in their room-temperature conductivity, whose value, however, remains unaffected by subsequent heating. We report here only on the equilibrium conductivity curves, as the slope of the curve for as-deposited samples depend, in addition to the activation energy, on the change of structure. The results are shown in Fig. 1, along with those for some of the samples of Meyerson and Smith[3].

It can also be seen from our results that the conductivities for a-C and a-C:H samples have comparable values, somewhat in disagreement with most of the reported work about a-C. We will argue that the crucial parameter in determining the electrical conductivity is the same for both types of material, namely, the proportion of sp^2 sites and not the hydrogen content. This is in agreement with the current understanding of the role of hydrogen in a-C:H[1] according to which the optical properties and bandgap are determined more by the amount and size distribution of the sp^2 islands than by the H content.

Fig. 1 shows that the log σ vs T^{-1} curves approach straight lines at low and high temperatures, with a transition region in between. The data can be fitted reasonably well with either of the following two equations

$$\sigma = \sigma_1 \exp(-\Delta E_1/kT) + \sigma_2 \exp(-B/T^{1/4}) \quad (1)$$

$$\sigma = \sigma_1 \exp(-\Delta E_1/kT) + \sigma_3 \exp(-\Delta E_2/kT) \quad (2)$$

One can reasonably fit the experimental data to either equation (1) or (2), by selecting appropriate pre-exponential factors and constants. The choice of which equation better describes the conduction process depends upon whether the parameters are physically acceptable or not. In the following, we give arguments to show that the second term of Eq. (1), representing variable-range hopping near the Fermi level and usually thought to correspond[1] to the conduction process near room temperature, is unsatisfactory from this point of view.

For the expression (1), the relevant parameters are[2]

$$\sigma_2=[3N(E_F)/(2\pi\alpha kT)]^{1/2}$$

$$B=1.66[\alpha^3/(kN(E_F))]^{1/4}$$

where α^{-1} is the decay length of the relevant wavefunction . From a plot of the experimentally observed conductivity values against $T^{-1/4}$, one can obtain values for α and $N(E_F)$. Most of these values are found to be unphysical, as observed by Meyerson and Smith[3], and also for our samples. In addition, these values are also inconsistent with $N(E_F)$ values obtained from ESR measurements[4]. We, therefore, discard the above mechanism, described by eq. (1), for a-C and a-C:H for the region near room temperature. It may be noted, however, that Hauser[5] has found evidence for variable-range hopping near Fermi level being effective for temperatures below room temperature for evaporated amorphous carbon films.

Fitting the data from our samples and from Meyerson and Smith's[3] by means of eq. (2), we obtained the values of the parameters σ_1, σ_3, ΔE_1, ΔE_2 . The values of σ_3 and ΔE_2 obtained for some representative samples are reported in Table I. It is observed that ΔE_2 , the activation energy for conduction near room temperature varies only slightly from sample to sample, while the pre-exponential σ_3 changes by several orders of magnitude.

In the lower temperature range, one must explain both the near-constancy of ΔE_2 and the large variation of σ_3 from sample to sample. A conductivity having a temperature variation corresponding to the second term of Eq. (2) is commonly attributed[2] to either

a) conduction by carriers excited into the band tail below the mobility edge;

b) hopping between states lying in a narrow peak in the DOS at the Fermi level;

However, neither of these mechanisms can easily account for the large change in the value of σ_3. For example, the sp^2/sp^3 bond ratio in a-C:H lies between 40% to 80%[1] for the usual deposition conditions. Even if we assume that only the sp^2 bonds give rise to the band tailstates or to states near the Fermi level, σ_3 can not change by such a large amount.

We, therefore, propose a model for conductivity which is in agreement with the well-known microstructure of a-C:H, which consists of graphitic sp^2 islands embedded in an sp^3 matrix. This same microstructure would appear to be applicable to our samples of a-C also. The sp^2 islands give rise to a number of states in the gap, whose exact position is dependent on the structure of the material. However, in general these states are much lower in energy compared to those in the sp^3 regions. For example, a 7-fold carbon rings give rise to a state about 0.3 eV above the Fermi level[1]. Near room temperature, according to our proposed model, electrons are excited into these low-energy states and then hop over the sp^3 gaps.

TABLE I

Conductivity parameters for a-C and a-C:H

Sample	T_d (°C)	σ_3 ($\Omega^{-1}cm^{-1}$)	ΔE_2 (eV)	R (Å)	Ref.
C300	room	$9.4\ 10^{-5}$	.30	5.3	p.w.
CH20	room	$4.0\ 10^{-3}$	.27	2.7	p.w.
a-C:H	200	$2.6\ 10^{-8}$	.29	10.0	A
a-C:H	75	$1.7\ 10^{-11}$	.31	14.0	A

p.w. present work
A Meyerson and Smith[3]

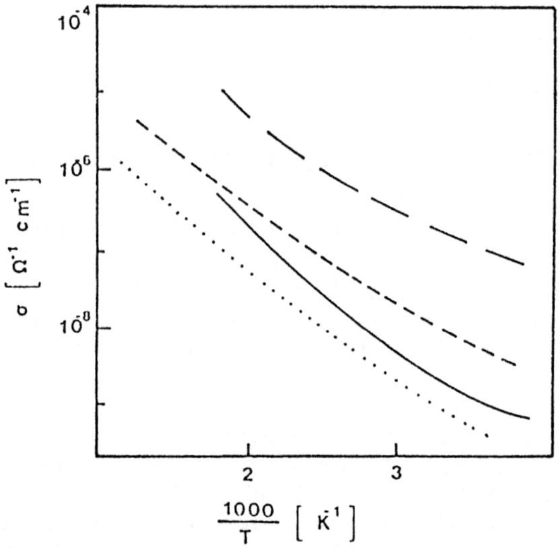

Fig.1 Arrhenius plot of conductivity data for some of our a-C samples and of Meyerson and Smith[3] a-C:H (H followed by deposition temperature):
C300 ———— C440— — — —
H200············ H250-----------

To see how this can explain the observed features of conduction, we note that[2] nearest-neighbour hopping has the same type of temperature dependence and the corresponding pre-exponential factor has the form

$$(2e^2R^2/kT)\nu_{ph}N_{\pi(tot)}\exp(-2\alpha R)$$

where R is the average hopping distance, $N_{\pi(tot)}$ the total number of such states, ν_{ph} is the phonon attempt-to-escape frequency and the other symbols have their usual meanings. The corresponding wavefunctions are strongly localized which means that α^{-1} is very small compared to R.

The large change in the pre-exponential factor is due to the presence of the term exp(-2αR) which is very strongly dependent on R. We can get an idea of the magnitudes involved by assuming the following reasonable values for the parameters

$$\nu_{ph}=10^{12}s^{-1} \quad R=10Å \quad \alpha^{-1}=1Å \quad N_{\pi}=10^{20}cm^{-3} \quad T=300K$$

This yields a value of σ=2.54 10^{-8} Ω^{-1} cm^{-1}, which is in the experimentally observed range. The large variation in the pre-exponential factor is also easily explained if one keeps in mind the strong dependence of the term exp(-2αR) on R; in fact, if R is changed from 6 to 14 Å, σ_3 varies from 7.6 10^{-6} to 8.5 10^{-12} Ω^{-1} cm^{-1}, a change of six orders of magnitude. We also report in Table I the values of R evaluated on the basis of the above assumed values for the relevant parameters. Although R values are only rough estimates, it is clear that similar large changes in σ_3 with R would be maintained in any case.

To explain the near-constancy of ΔE_2 at a value of about 0.3 eV, we note that π and π* states arising out of the usual 6-fold carbon rings in the sp^2 islands are more than 2 eV apart, and thus are unlikely to be responsible. However, rings having a different number of carbon atoms are also expected to exist, and it is theoretically found[1] that, for instance, 7-fold rings give rise to a state which lies 0.25-0.35 eV above the Fermi level. The exact position will, of course, depend on the configuration of such a ring.

It should be pointed out that the above argument will be valid only when the microstructure of the material has the assumed form, i.e., sp^2 islands embedded within an sp^3 matrix. When such a segregation between sp^2 and sp^3 regions is absent, and/or when the sp^2/sp^3 ratio exceeds the percolation threshold value, conduction will be through variable-range hopping at the band tails. Even for the case considered here, the latter would be the predominant mechanism effective at high temperatures[1].

SUMMARY AND CONCLUSION

In amorphous carbon with or without hydrogen, when the sp^2 bonded carbon atoms are mostly segregated into graphitic islands, two different conduction mechanisms are active at room temperature and above. At low temperatures, the excitation of electrons to states lying about 0.3 eV above the Fermi level and their hopping through the sp^3 matrix would be responsible. Such states are sitting near the border of the graphitic-like islands and might, for instance, be produced by 7-fold carbon rings. At higher temperatures, on the other hand, the predominant process is variable-range hopping by carriers excited into band tail states.

REFERENCES

1. J. Robertson, Adv. Phys. 35, 317 (1986)

2. N.F. Mott and E.A. Davis, Electronic Processes in Non-crystalline Materials, 2nd edition, Oxford University Press (1979)

3. B. Meyerson and F.W. Smith, J. of Non-Cryst. Solids, 35, 435 (1980)

4. D. Dasgupta, F. Demichelis, R. Spagnolo and A. Tagliaferro, MRS Proceedings 188, W. Oliver, M. Doerner, G. Pharr and F.R. Brotzen (eds.), In press (1990)

5. J. Hauser, J. of Non-Cryst. Solids, 23, 21 (1977)

INDENTATION TESTING OF THIN FILMS AND HARD MATERIALS

Carl J. McHargue

Metals and Ceramics Division, Oak Ridge National Laboratory
P.O. Box 2008, Oak Ridge, TN 37831-6118 USA

INTRODUCTION

Indentation testing techniques can provide important information about the near-surface mechanical and deformation properties of solids. Properly executed and interpreted, such techniques can provide information regarding hardness, elastic modulus, compressive yield strength (for metals), fracture toughness (in ceramics) and residual stresses (in ceramics). However, the apparent simplicity of the test procedures often belie the difficulties of obtaining quantitative information on specific properties of interest.

Although most people have an intuitive appreciation for the concept of hardness, it is a poorly defined term. It is generally understood to denote the resistance to local deformation as measured by forcing an indenter of a specific geometry into a free surface under a defined load. Hardness is not a fundamental property of matter but is some complex combination of properties such as elastic deformation, yield point (stress), ductility, work-hardening characteristics, etc. The hardness number or index is very much a function of the way the test is conducted and the character of the test material.

Static indentation hardness is conventionally determined by making a permanent indentation in the surface of the test material and determining a mean contact pressure by dividing the maximum applied load by the area of the residual indent. As will be discussed, this procedure is often adequate to provide a quantitative or semi-quantitative measure of the mechanical properties of metals that are not subject to severe work-hardening. Tabor[1] has shown that the mean contact pressure (load divided by the projected contact area) is approximately three times the compressive yield stress for work-hardened metals. Hardness is sometimes defined as the resistance to penetration, but it is now recognized that a measure of this resistance involves both hardness and elastic modulus.

Much of the information regarding mechanisms active during indentation arises from studies of relatively large scale indentations, i.e., with dimensions of approximately 100 μm. As the interest in the properties of thin films and surface layers has increased, the use of microindentation techniques has been employed on a finer and finer scale. There have been few studies on the mechanisms involved in very small

Diamond and Diamond-Like Films and Coatings
Edited by R.E. Clausing *et al.*, Plenum Press, New York, 1991

indentations and the results from macroindentation tests have been extrapolated downwards to interpret the measurements. As Samuels has noted, the principle of geometric similarity is essential for this purpose but it has not been tested for the scale of indentation currently being studied.[2]

The development of load-depth sensing instruments for indentation testing allows us to examine the manner in which the indenter load varies with indenter penetration. Figure 1 shows one loading-unloading cycle for a single crystal Al_2O_3 specimen where the load is applied parallel to the

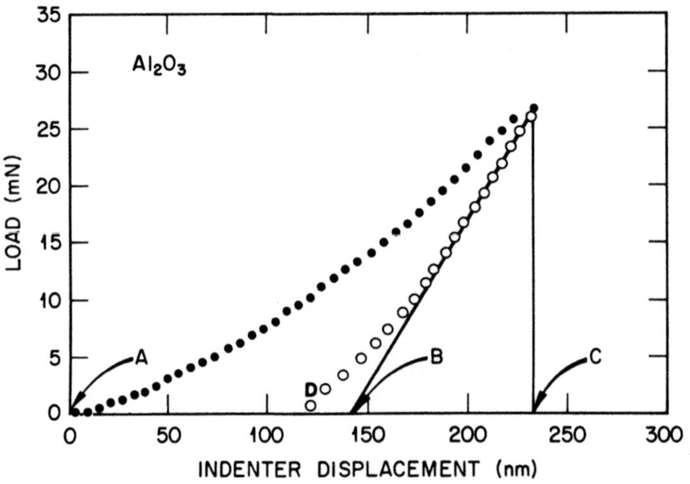

Fig. 1. Load/unload-displacement curve for single crystal Al_2O_3 with load applied parallel to the c-axis. AC = total indenter displacement; BC = elastic displacement; AB = plastic displacement; DB = elastic after-effects.

c-axis. Lawn and Howes[3] have analyzed the case for a sharp indenter being forced into a brittle substrate. The loading portion is controlled by the hardness and the deformation has both elastic and plastic components whereas the unloading is totally elastic. As shown in Fig. 1, anelastic effects may be apparent toward the end of the unloading curve.

The type of deformation that occurs during the formation of the indent depends upon the nature of the solid. This fact needs to be considered both in the design of the indentation experiment and in defining the hardness index derived from the experiment. Since Tabor[4] has discussed these differences for several classes of solids, only a few examples will be given here.

The response of <u>ideally elastic solids</u> (rubber, diamond ?) is reversible and totally elastic. The indentation vanishes upon removal of the load and "hardness" must be defined in terms of the loaded condition, e.g., the load to give a specific penetration depth. In this case, the mean contact pressure is a direct measure of the elastic properties of the material. Figure 2 illustrates the response of a Type IIa diamond crystal indented on a (100) face.[5] Within the resolution of this instrument (0.16 nm in depth, 0.3 μN in load) the response is totally reversible to the maximum penetration depth of 275 nm.

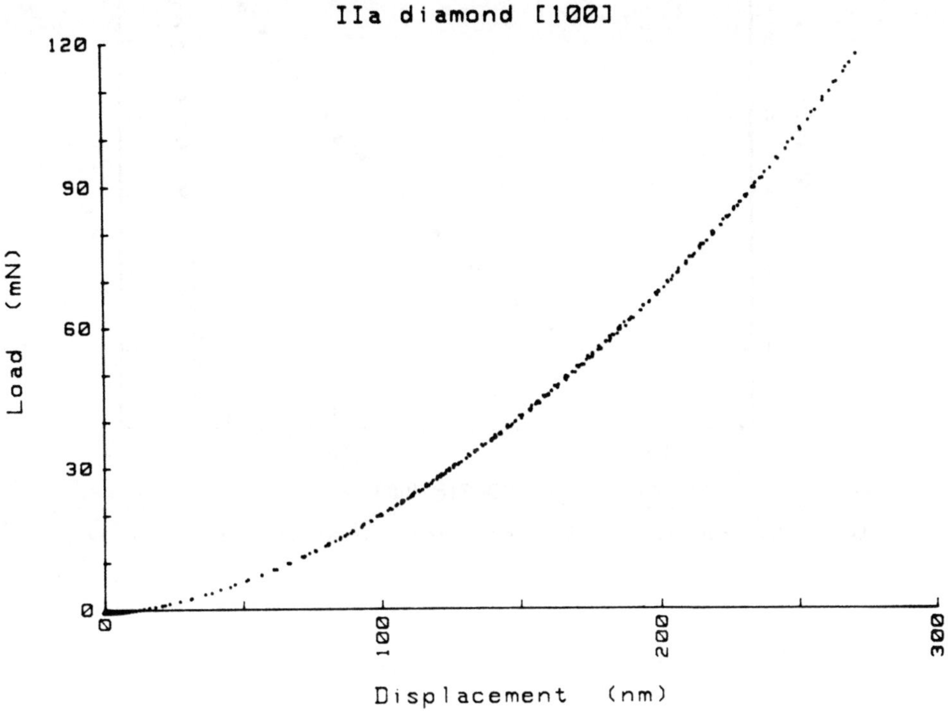

Fig. 2. Load/unload-displacement curve for a Type IIa diamond indented on a (001) face. The behavior is totally elastic.

Metals generally exhibit an <u>elastic-plastic</u> response under the loaded indenter. In the ideal case, the metal first deforms elastically and at some critical stress it deforms plastically at a constant yield or flow stress. A fully work-hardened polycrystalline cubic metallic specimen responds in such a manner. Figure 3 shows the response of an annealed copper single crystal loaded with a triangular pyramid-shape diamond indenter. The amount of elastic recovery is small relative to the plastic portion, and the conventional hardness number derived by dividing the maximum load by the residual projected contact area differs little from the true value.

<u>Elastic-brittle</u> solids such as ceramics and glasses may undergo brittle fracture in the elastic range during loading or due to tensile stresses under the indenter at the moment of unloading. Ideally, these materials first deform elastically until cracking occurs at some critical value of the tensile stress component. In most cases, some plastic deformation may occur in regions of high hydrostatic constraint where crack initiation is inhibited. Figure 4 gives one loading-unloading cycle for single crystal Al_2O_3 indented on the (0001) plane. Significant plastic displacement occurred in addition to the elastic displacement. Cracking

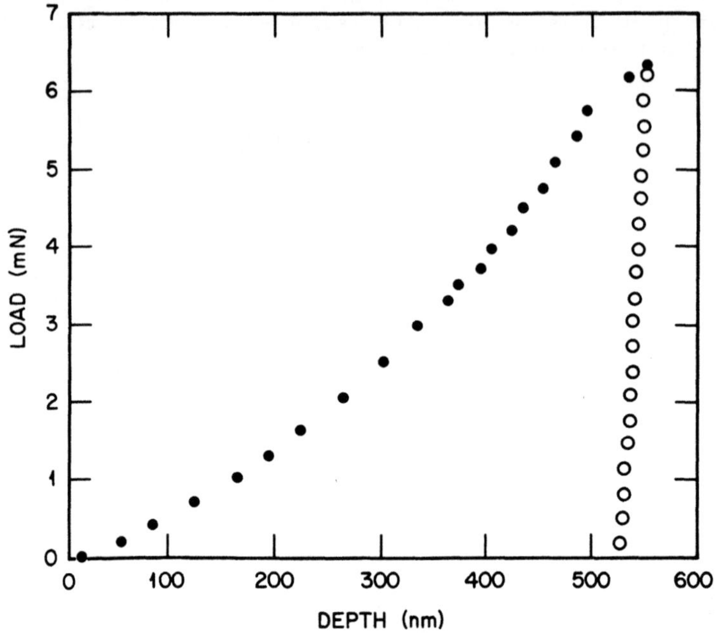

Fig. 3. Load/unload-displacement curve for a copper single crystal.

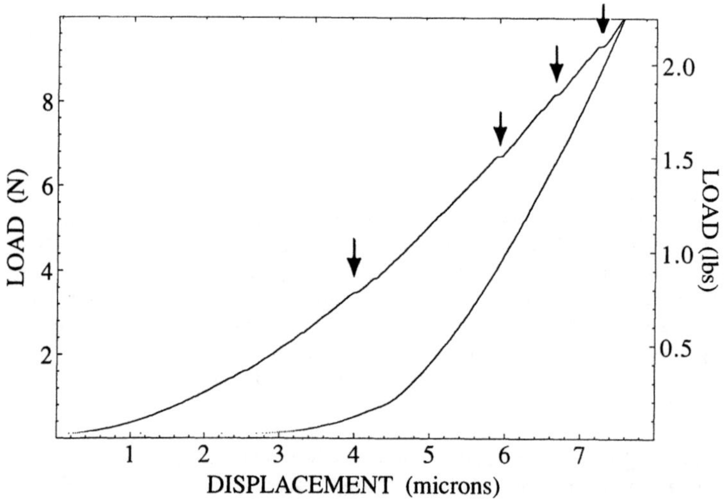

Fig. 4. Load/unload-displacement curve for single crystal Al_2O_3 indented normal to the (0001) plane. The horizontal displacements in the loading segment correspond to crack formation. Some of the discontinuities caused by cracking are noted by the arrows.

was evidenced by the several horizontal displacements in the loading segment and caused the large "swing-back" at the end of the unloading segment.

In addition to the elastic and plastic displacements, compaction or densification often occurs in glasses and other amorphous materials. Viscous or viscoelastic effects may also appear in the indentation testing of polymers and amorphous solids.

Some of the issues of particular interest in determining the mechanical properties of diamond and diamond-like films are: separation of the elastic and plastic contributions; depth of the indent relative to the thickness of the film; small sample volume effects; and Indentation Size Effects.

SEPARATION OF ELASTIC AND PLASTIC EFFECTS

In conventional hardness testing an indentation is made with the application of a constant load, the indenter is removed, and the lateral dimension of the residual indent is measured and used to calculate the mean contact pressure under load. An important assumption is that any elastic recovery in or around the indent can be ignored and the size of the "recovered" indent represents the plastic deformation under load. This assumption appears to be valid for large indents in metals; note that Fig. 3 shows only a small elastic recovery for copper. However, materials with high values of H/E (hardness/elastic modulus) do exhibit large amounts of elastic recovery (Fig. 1) which will have a large effects on the size of the residual indentations at the scale of nanoindentation now being used.

A schematic representation of an indentation by a sharp indenter is given in Fig. 5 and indicates the contribution of the elastic and plastic deformations. The separation of the loading-unloading curve into the various components is indicated in Fig. 1. The total displacement of the indenter, h_t, is given by AC; the elastic displacement, h_e, is given as BC; the plastic contribution, h_i, is given as AB. The displacement BD is due to elastic after-effects. Optical examination of the <u>residual</u> indents would have indicated that the plastic deformation was AD for the application of the maximum load.

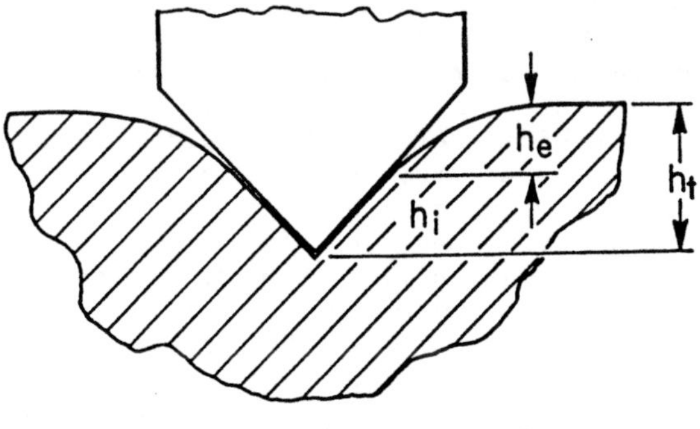

Fig. 5. Schematic diagram of a loaded sharp indenter showing the elastic and plastic contributions to the total indenter displacement. h_t = total displacement; h_e = elastic displacement; h_i = plastic displacement.

One of the major problems involved in the characterization of the properties of thin films is the elimination of substrate effects in the surements. Various models for the deformation zone under an indenter consider a hydrostatic core surrounded by a plastic zone and in turn by an elastically deformed region. The plastic/elastic boundary expands as a hemisphere as the indenter is forced into the surface and extends to several (7-10) times the depth of the indent. The "rule of thumb" offered in many text books and recommended by the National Physical Laboratory (UK) requires the film thickness to be ten times the indentation depth.[6]

Sargent[7] used a simple rule of mixtures to give the measured (effective) hardness, H_{eff}, in terms of the hardness of the film, H_f, and substrate, H_s, and the portion of the deformed volume contained in each the film (V_f) and the substrate (V_s). Thus,

$$H_{eff} = \frac{H_s V_s + H_f V_f}{V_s + V_f}$$

Burnett and Page[8] extended this model to analyze the properties of a soft amorphous layer on a hard substrate, produced by ion implantation, by using empirical fitting parameters. Bückle's empirical correlation suggests that the substrate contributes <2% to the measurement for a thickness ten times the depth but 30% at five times the depth.[9]

Lebouvier et al.[10] used a plain strain model to suggest that the factor of ten is required for cases where the strength of the film is much greater than that of the substrate. In the case of softer films on harder substrate the critical ratio appeared to be closer to four. Stone et al.[11] employed closed-form elasticity solutions to calculate the average displacement of the surface beneath a region of uniformly distributed pressure. The model was used to predict the effective modulus for a high modulus film on a low modulus substrate. This model predicts observable effects for ratios of depth/thickness of 0.1 to as high as 10.

Depth-sensing microindentation hardness testers[12,13] have provided data on hardness and elastic modulus on a fine enough scale to critically examine these models. Figure 6 shows the variation in the "effective" or measured hardness as a function of the plastic displacement of the indenter for a soft film on a hard substrate.[14] An amorphous surface layer was produced on a sapphire single crystal by implanting 4×10^{16} Al/cm^2 (90 keV) and 6×10^{16} O/cm^2 (55 keV) at 77K. The surface was amorphous to a depth of 155 nm. The hardness is shown as a relative value, i.e., hardness of the implanted (amorphous) region normalized by the hardness of the unimplanted (crystalline) region. A substrate effect on the measurements is apparent for indentation depths as small as 20 nm (≈12% of the thickness). An effect of the "composite" is still apparent to indenter depths as large as 150% of the layer thickness. The value of relative hardness as a function of depth should have a constant value of 0.45 from the surface to a depth of 155 nm and then show a step-wise increase to a constant value of 1.0 at depth greater than 155 nm. The variation in the measured elastic modulus follows a similar pattern, but approaches the substrate value at a faster rate although it has not reached a ratio of unity at a depth corresponding to 150% the layer thickness.

Figure 7 shows a second case of softer films on harder substrates. Diamond-like carbon films 100 nm thick were deposited on Al_2O_3 and AlON substrates under conditions that produced adherent (DLC 10 and 11) and non-adherent (DLC 2, 4, and 5) films. (AlON is an aluminum oxynitride compound with a spinel crystal structure.) Extrapolation of the data for the adherent films (DLC 10 on Al_2O_3 and DLC 11 on AlON) indicates significant substrate effects on the measurements of both hardness and elastic modulus at indenter depths less than 20% of the film thickness. The substrate hardness values were 25 GPa for Al_2O_3 (corrected for the indentation size effect) and 20 GPa for AlON. The elastic moduli were 400 and 150 GPa, respectively.

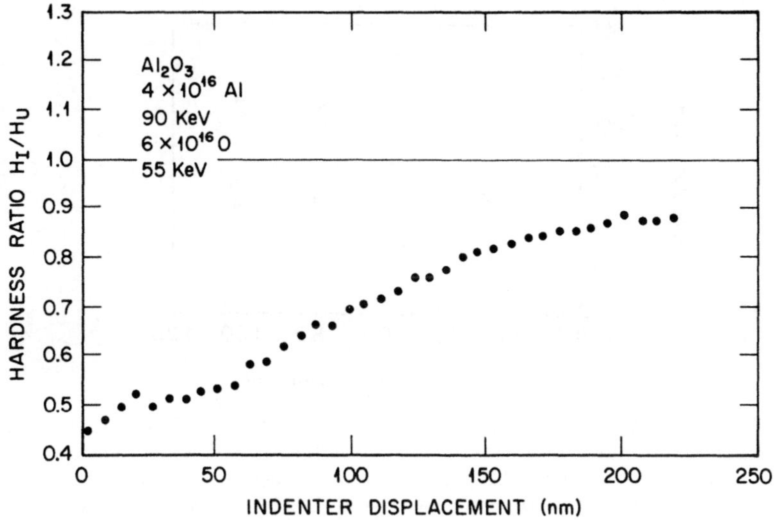

Fig. 6. Relative hardness vs indenter displacement for an amorphous surface layer produced by ion implantation of Al_2O_3. The amorphous layer extends to a depth of 155 nm.

Results for a harder film on a softer substrate are given in Fig. 8 for adherent 100 nm thick diamond-like carbon (DLC) films on glass.[15] The apparent hardness of the DLC films is approximately independent of the substrate for indentation depths less than 25 nm (25% of the film thickness) but show substrate effects at greater depths. The elastic modulus measurements, however, show substrate effects for penetrations as small as 10% and the apparent value reaches that of the substrate at a depth of 40 to 50 nm (40 to 50% of the film thickness).

For these examples, the critical film thickness/depth ratio for substrate-free hardness values is larger for soft films on hard substrates than vice versa, contrary to the argument of Lebouvier et al.[10] The critical ratio appears to be greater (i.e., depth of indent smaller) for modulus measurements than for hardness measurements for both hard and soft films.

Adherence of the film to the substrate affects the measurements of both hardness and elastic modulus. Shear tractions cannot be transmitted

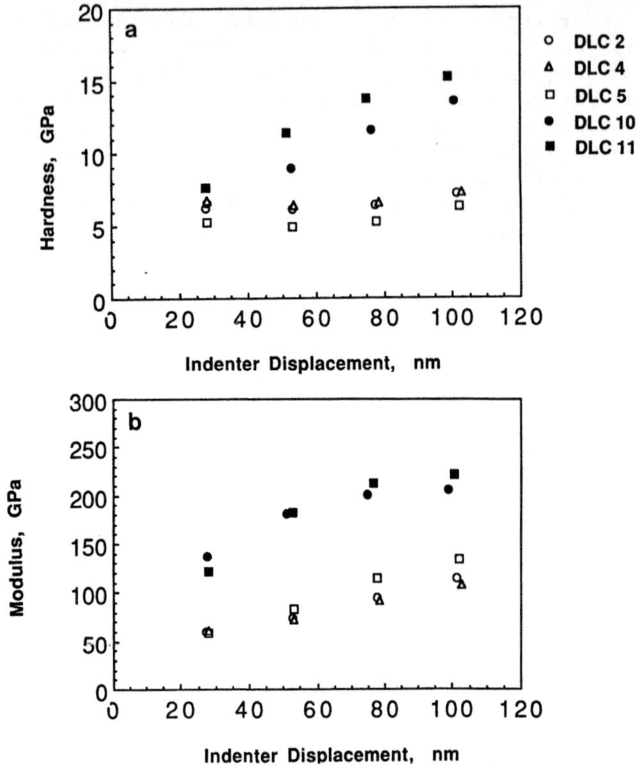

Fig. 7. (a) Hardness and (b) Elastic Modulus of diamond-like carbon films deposited on Al_2O_3 and AlON substrates plotted as functions of indenter depth. Films DLC 2, 4, and 5 were non-adherent with Al_2O_3 substrates. Films DLC 10 (on Al_2O_3) and 11 (on AlON) were adherent.

across non-adherent interfaces, but normal tractions (both tensile and compressive) can be effectively transmitted. The hardness measurements should then be more sensitive to adherence than the modulus measurements. Figure 7 shows the apparent hardness and modulus as a function of depth of indentation for non-adherent (DLC 2, 4, and 5) and the adherent (DLC 10 and 11) 100 nm thick diamond-like films discussed above.[15] The indicated hardness values for the non-adherent films are independent of the depth of indenter penetration, consistent with the above argument. As previously noted, substrate effects were evident for depths less than 25% of the film thickness for the adherent films (DLC 10 and 11). The modulus measurements showed substrate effects for both adherent and non-adherent films, but were more pronounced for the adherent films.

SMALL SAMPLE VOLUME EFFECTS

As the volume of material being sampled decreases with the use of nanoindentation techniques, the difficulty of assuring that the measurements represent the "bulk" properties increases. Surface preparation now becomes of paramount importance in order to eliminate the effects of cold-worked layers, thin oxide films, or segregated impurities. Pethica and Tabor[16] have demonstrated the difficulty of preparing metal surfaces that are free of chemical effects. It has long been known that

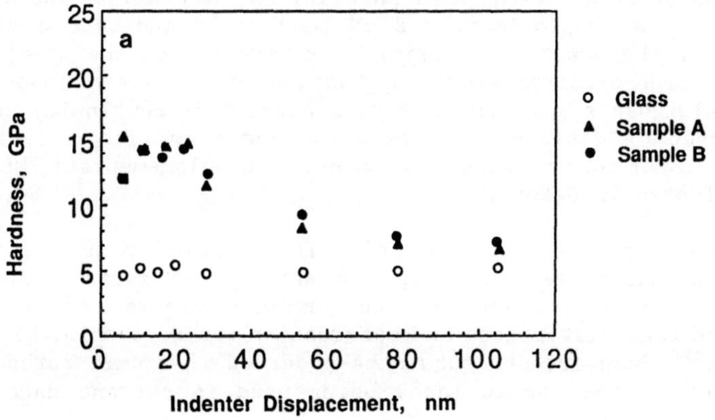

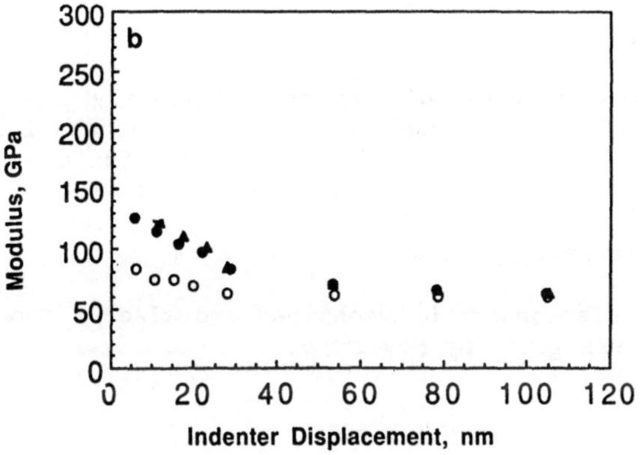

Fig. 8. (a) Hardness and (b) Elastic Modulus of DLC films deposited on glass substrates.

the presence of adsorbed contaminants affect the hardness measurements of sapphire (single crystal Al_2O_3).[17] Additionally, there is the areal sampling problem in multiphase specimens.

As the stressed volume beneath the indenter becomes small, the probability of finding a defect becomes small, particularly in high-quality single crystal specimens. Since some plastic flow contributes to the hardness measurement of most materials, the stress at which plastic flow begins is important. This stress depends on the density of defects in the stressed volume. Page et al.[18] have shown that in highly perfect single crystals of Al_2O_3, the deformation may be totally elastic until a "yield point" is reached at which stress dislocation generation is initiated. Figure 9 shows one loading-unloading cycle for an Al_2O_3 single crystal tested with the indenter parallel to the c-axis. The displacement discontinuity at approximately 8 mN load could indicate a yield point, twinning, or the onset of cracking. Examination of the specimen surface by transmission electron microscopy showed that at lower loads there were no residual indents and that this displacement discontinuity corresponded to the onset of dislocation generation and flow, i.e., a yield point. Thus, the deformation caused by indenter displacements up to 80 nm is totally elastic in nature.

Recent application of the molecular dynamics technique to simulate the interaction between a sharp tip and a flat surface suggests that material transfer between the two surfaces occurs before the actual contact and that distortions and deformation occurs within the body of the sharp tip.[19] Such effects should be observed in atomic force microscopy and perhaps at the lowest loads being used in current nanoindentation instruments.

INDENTATION SIZE EFFECT

According to the principle of geometric similarity cited above, the hardness index or number should not vary with size of the indent, i.e., with the applied load. However, it has been known for many years that materials often exhibit an indentation size effect (ISE) wherein the hardness appears to increase as the indent size (or load) decreases.[20] This effect is particularly pronounced in ceramics.

An empirical relationship between load and size of the indent for spherical indenters was given by Meyer[21] as

$$W = kd^n$$

where W is the load, d is the chordal diameter of the residual indent, and k and n are constants for a given material. For conical or pyramidal indenters, Sargent[7] used the empirical relationship

$$n = \frac{\delta F/F}{\delta d/d}$$

where n is a dimensionless index to describe the ISE, F is the force, and d is a measure of the size of the indent. In each model a value of n = 2 indicates the absence of such an effect.

Many explanations have been proposed for this effect including surface artifacts, "quantized" slip or deformation band formation, anelastic recovery of small indents, and the probability of finding a flaw in a small volume. As discussed above, the low probability of finding a deformation-initiating flaw in a small volume can lead to a yield point effect in Al_2O_3. However, it has been shown that even in the absence of the yield point effect, there is an ISE in single crystalline Al_2O_3.[22] Figure 10 shows the hardness as a function of plastic depth of indenter penetration for a single crystal which did not exhibit the yield point effect and an implanted single crystal that contained a large concentration of point defects and dislocations. The data are normalized to the unimplanted hardness value taken at 80 nm. The data indicate an ISE for both specimens although it was larger for the unimplanted material. It was concluded that the difference in the hardness values of the implanted and unimplanted specimens at 20 nm plastic depth was due to the difference in the concentration of defects in the stressed volume. Plastic flow was easier to initiate in the sample with the larger concentration of defects (the implanted specimen). However, even the sample with the very large number of defects exhibited a significant increase in hardness at very low indent depths, i.e., an ISE.

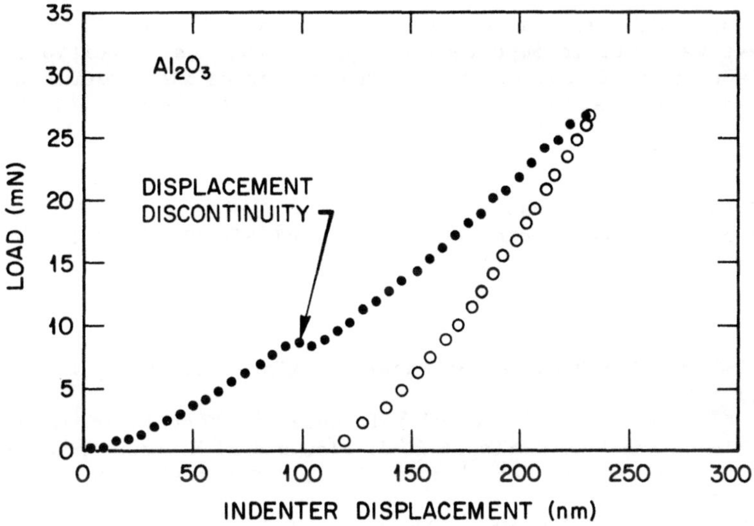

Fig. 9. A loading-unloading vs indenter displacement curve for a single crystal Al_2O_3 specimen showing a displacement discontinuity at ≈ 80 nm.

SUMMARY

Indentation test techniques can give important information about the near-surface mechanical properties and deformation behavior of all classes of materials. It is important, however, to take into account the many difficulties involved in obtaining artifact-free data. While the introduction of the depth-sensing techniques has increased the amount of

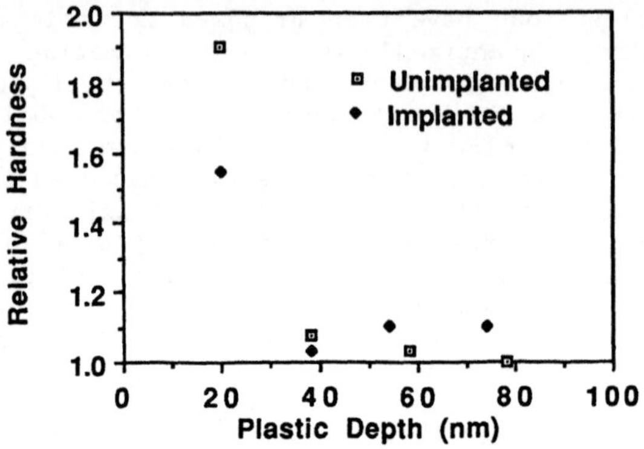

Fig. 10. The Indentation Size Effect in Al_2O_3 single crystals. The data are normalized to the hardness of the unimplanted specimen at 80 nm.

quantitative data obtained in such tests, their use requires great care and a recognition of the nature of the specific material's response to the applied stress.

Since different materials respond to the test conditions in different ways, it is important to specify the test conditions and the exact definition of "hardness" used when comparing one material to another or the same material tested under different conditions.

ACKNOWLEDGEMENTS

This work was sponsored by the Division of Materials Sciences, U.S. Department of Energy, under contract DE-AC05-84OR21400 with the Martin Marietta Energy Systems, Inc.

REFERENCES

1. D. Tabor, "The Hardness of Metals," Oxford University Press, London (1951).
2. L. E. Samuels, p 5 in "Microindentation Techniques in Materials Science and Engineering," P. J. Blau and B. R. Lawn, eds., ASTM STP 889, American Society for Testing and Materials, Philadelphia, PA (1985).
3. B. R. Lawn and T. R. Howes, J. Mater. Sci. 16:2745 (1981).
4. D. Tabor, pp 129-159 in "Microindentation Techniques in Materials Science and Engineering," P. J. Blau and B. R. Lawn, eds., ASTM STP 889, American Society for Testing and Materials, Philadelphia, PA (1985).
5. M. E. O'Hern and C. J. McHargue, this volume.
6. G. N. Peggs and I. C. Leigh, "Recommended Procedure for Micro-indentation Vickers Hardness Test', Report MOM 62, UK National Physical Laboratory, England (1983).
7. P. M. Sargent, "Factors Affecting the Microhardness of Solids," PhD Thesis, University of Cambridge, England (1979).
8. P. J. Burnett and T. F. Page, J. Mater. Sci. 19:845 (1984).
9. H. Bückle, pp 1453-494 in "The Science of Hardness Testing and Its Research Application," J. H. Westbrook and H. Conrad, eds., American Society for Metals, Metals Park, OH (1973).

10. D. Lebouvier, P. Gilormini, and E. Ferber, J. Physics D 18:199 (1984).
11. D. S. Stone, T. W. Wu, P-S Alexopoulos, and W. R. LaFontaine, pp 105-110 in "Thin Films:Stresses and Mechanical Properties," J. C. Bravman, W. D. Nix, D. M. Barnett and D. A. Smith, eds., Materials Research Society, Pittsburgh, PA (1989).
12. J. B. Pethica, pp 147-156 in "Ion Implantation into Metals," V. Ashworth, W. A. Grant, and R. P. M. Proctor, eds., Pergamon Press, Oxford (1982).
13. D. Newey, H. M. Pollock, and M. A. Wilkins, pp 157-166 in "Ion Implantation into Metals," V. Ashworth, W. A. Grant, and R. P. M. Proctor, eds., Pergamon Press, Oxford (1982).
14. W. C. Oliver and C. J. McHargue, Thin Solid Films 161:117 (1988).
15. D. L. Joslin, M. E. O'Hern, C. J. McHargue, and R. E. Clausing, pp 218-225 in "Infrared Systems and Components III, R. L. Caswell, ed., SPIE Conference Proc. Vol. 1015 (1989).
16. J. B. Pethica and D. Tabor, Surface Science 89:182 (1979).
17. A. R. C. Westwood, N. H. MacMillan, and R. S. Kalyoncu, J. Am. Cer. Soc. 56:258 (1973).
18. T. F. Page, W. C. Oliver, and C. J. McHargue, submitted to J. Mater. Res.
19. J. B. Pethica, to be published in "Thin Films:Stresses and Mechanical Properties II," W. C. Oliver, M. F. Doerner, G. M. Pharr and F. Brotzen, eds., Materials Research Society, Pittsburgh, PA
20. E. M. Onitsch, Mikroskopie 2:345 (1947).
21. E. Meyer, Z. Ver. Deutsch. Ing. 52:645 (1908).
22. C. J. McHargue, M. E. O'Hern, and D. L. Joslin, to be published in "Thin Films:Stresses and Mechanical Properties II," W. C. Oliver, M. F. Doerner, G. M. Pharr and F. Brotzen, eds., Materials Research Society, Pittsburgh, PA.

Electron Spectroscopy on Carbon Based Films: Bulk and Interface Properties

Peter Oelhafen, Dario Ugolini, Sabine Schelz and Jürgen Eitle

Institut für Physik der Universität Basel
Klingelbergstr. 82, CH-4056 Basel, Switzerland

1 Introduction

Photoelectron spectroscopy is one of the most powerful tools in order to analyse the electronic structure of matter. It has been extensively applied to the gas, liquid and solid state phases [1, 2, 3, 4]. Depending on the energy of the incoming photons and the detection mode of the emitted photoelectrons detailed information on the core electrons, dispersion relation E(k) of valence electrons, spin polarisation, local (site decomposed) and partial (angular momentum decomposed) density of electronic states (DOS) can be obtained. From these data many other properties can be determined: elemental composition, chemical state, depth and laterally resolved composition, structural information, Fermi level position, workfunction, magnetic moment and state density of the Fermi energy (E_F). The latest developments include the observation of the energy gap in high Tc superconductors [5] and the application of photoelectron diffraction [6].

It is the aim of the present paper to demonstrate the potential of photoelectron spectroscopy for the study of highly insulating carbon based films both in terms of bulk and interface properties. From the experimental point of view and for reasons discussed below the emphasis is on the combination of film preparation (by various methods) and film analysis on the same equipment in order to perform in situ electron spectroscopy measurements. Using this technique the electronic structure of amorphous hydrogenated carbon (a-C:H) films has been studied and by changing the deposition parameters the transition to soft polymeric films has been analysed. The interface properties of a-C:H on various metallic and semiconducting substrates have been analysed mainly in terms of band formation in the interface region. These properties have been correlated with the adhesion behaviour of a-C:H on the various substrates. The in situ film preparation applied in the present work is crucial for three reasons:

1. The sample surface of not in situ prepared films would have to be cleaned before measurements because of the very high surface sensitivity of 1 to 2nm of electron spectroscopy. However, no suitable cleaning procedure exists for a-C:H or polymeric films since sputtering, cleaving or extensive heating can not be applied without severely destroying the surface structure.

Diamond and Diamond-Like Films and Coatings
Edited by R.E. Clausing *et al.*, Plenum Press, New York, 1991

2. Photoelectron spectroscopy measurements of highly insulating samples ($> 10^{12}\Omega$cm) can severely be influenced by charging effects. Due to the photoelectron emission from the sample surface and the impossibility to pass a charge compensating current from the substrate to the sample surface, a positive charge is established within the region from where the photoelectrons originate. Energy shifts of several eV in the photoelectron spectra can then result. The charge compensation by impinging electrons from an external source (flood gun) is not feasible since we are interested in the whole energy range of photoelectrons down to some eV. The only method to overcome this problem is to perform the measurements on films with appropriate thicknesses, d_0. This thickness is essentially determined by the escape depth of the photoelectrons (which itself is determined by the mean free path for electron-electron scattering [7]). In the present study, d_0 has been determined by the disappearance of the photoelectron signal from the substrate monitored by depositing gradually a-C:H and intermittently recording photoemission data. A value of $d_0 = 10$nm has been found for a-C:H on Au. At these low film thicknesses the formation of a positive charge in the surface region of the insulating film is hindered by the injection of photoelectrons from the conducting substrate into the insulating film since the penetration depth of the incident light is much larger than the escape depth of the photoelectrons.

3. The direct observation of the interface formation by photoelectron spectroscopy is best performed through a step by step deposition (i.e. roughly atomic layer by layer) and measuring sequence. Such deposition-measurement cycles can only be performed in an in situ experiment, so as to preserve the inherent nature of the chemical reaction that takes place at the interface.

The electron spectroscopy measurements have been complemented in some cases by experiments performed not in situ such as Fourier transform infrared spectroscopy (FTIR), optical absorption, Rutherford backscattering (RBS), forward scattering in grazing incidence and transmission electron microscopy (TEM).

The paper is organised as follows: in section 2 a description of the experimental set up is given, section 3 contains the photoemission data which refer to the bulk properties of a-C:H and related films prepared with different deposition energies and using different process gases. In section 4 the data obtained from various a-C:H/substrate (Au, W, Si, Ge and GaAs) interfaces are presented. A general conclusion is given in section 5.

2 Experimental

2.1 Electron Spectroscopy

The electron spectroscopy measurements have been performed on Leybold spectrometers (EA10/100 and EA11/100) equipped for ultraviolet photoelectron spectroscopy (UPS), X-ray photoelectron spectroscopy (XPS) and Auger electron spectroscopy (AES). The photon energies used in the present study were $h\nu = 21.2$ and 40.8eV for UPS (HeI and HeII radiation from a discharge resonance lamp) and $h\nu = 1253.6$eV for XPS using Mg K_α radiation and $h\nu = 1486.6$eV for XPS using monochromatized Al K_α radiation (denoted as MXPS in the following). Depending on the analyser operation mode (constant pass energy or fixed retardation ratio) and the line width of the excitation radiation, energy resolutions of 0.1 and 0.2eV in UPS (for HeI and HeII radiation) and 0.5 and 0.9eV for MXPS and XPS have been obtained, respectively. A reduced energy resolution has been used in some cases in order to increase the sensitivity of the XPS or MXPS measurements so as to detect possible contamination . Core electron binding energies are referred to the Fermi level of a Au or Pd test sample and are calibrated by using the Au $4f_{7/2}$ core line for which a binding energy of 84.0eV is assumed. Core line shifts as determined by XPS or MXPS are measured with an accuracy of 0.05eV. The measuring time for UPS spectra is typically 10 to 30 minutes whereas XPS or MXPS spectra have been measured within 1 to 5 hours. The base pressure (residual gas pressure) of the spectrometers is below $1 \cdot 10^{-10}$mbar. The C KLL Auger spectra and electron energy loss spectra have been recorded using electron sources with primary energies ranging from 500eV to 2keV.

2.2 Film Preparation

Hydrogenated amorphous carbon (a-C:H) and polymeric films have been deposited in preparation chambers of the spectrometers using different deposition methods. Direct ion beam deposition has been performed for both the a-C:H and the polymeric films [8, 9]. A Penning ion source and a broad beam ion source of the Kaufman type [9] have been used and deposition rates of about $6 \cdot 10^{-4}$ and $5 \cdot 10^{-2} nm \cdot s^{-1}$, respectively, were obtained typically in the case of a-C:H films (using methane as a source gas). The penning source with its inherent low deposition rate was particularly well suited for depositing even fractions of monolayers, which was important for the interface studies. The gas pressure in the preparation chamber was in the 10^{-5} (10^{-4})mbar range for the Penning (Kaufman) source. Deposition energies (denoted beam energy E_I) between 70 and 1000eV have been used. A parallel plate reactor, also attached to the electron spectrometer for rf plasma CVD deposition (PACVD), has been used only recently in the authors laboratory and a-C:H and polymeric films have been deposited in order to compare the electronic structure of the IBD and PACVD films [10].

The substrates were cleaned by Ar^+ ion sputtering at ion energies of 3.5 and 5 keV until no contaminants like oxygen and carbon could be detected by XPS and UPS. Heat treatments have been applied in some cases in order to restore the atomic structure after ion etching. The as deposited a-C:H and polymeric films have been analysed for contaminants as well: no oxygen or any metallic elements could be detected in the coatings. In some cases a high reactivity of the interface phases (mostly carbides) with oxygen could be observed during formation and the presence of oxygen could not be completely avoided as e.g. in the case of a-C:H on Si(100). In other cases this increased reactivity was not present as e.g. in the case of a-C:H on Ge(100) or GaAs(100) [11].

3 Electronic Structure of a-C:H and Related Films

3.1 Valence Band Structure of a-C:H

In the following we are going to present UPS and XPS valence band spectra of a-C:H. As it has already been pointed out in earlier studies (see e.g. Bianconi et al. [12]) the photoionization cross sections for C 2s and 2p electron states depend strongly on the excitation energy: at low photon energies (< 50eV) the spectra are dominated by the 2p electron states whereas at high energies (> 1keV) the 2s electrons dominate the spectra. It is the aim of the UPS and XPS measurements to deduce partial (l = 0 and l = 1 projected) density of states of the valence band (VB). Since we are dealing with an amorphous material the VB spectra reveal essentially a sum of weighted partial DOS, weighted with the corresponding photoionization cross sections.

Figure 1 shows UPS VB spectra of a-C:H films measured with two different excitation energies of 21.2 and 40.8eV. For this comparison the spectra have been normalized to the same spectrometer transmission. The similarity of the two curves is a clear indication that the spectra can be interpreted in the 'density of occupied states'- picture and dispersion effects due to direct optical transitions are absent which is typical for amorphous samples. The valence band edge (E_V) is located at a binding energy of 0.5eV.

The XPS VB spectrum of a-C:H shown in Fig. 2 is dominated by an intense peak at a binding energy of 17.2eV. A weak kink is visible at 10.9eV. The bottom of the band can be located at about 24.3 ± 0.5eV. The comparison with published data of diamond and graphite shows a general similarity between the XPS VB spectra of a-C:H and those of diamond and graphite [13]. We note, however, one distinct difference; a peak at a binding energy of about 13eV clearly present in diamond and less pronounced in graphite, is completely missing in the spectrum of a-C:H.

As mentioned above the UPS and XPS spectra reveal essentially the density of p- and s-electron states, respectively. The combination of the UPS and XPS data is depicted in Fig. 3. The spectra of Fig. 1 and 2 have been used for Fig. 3 after the subtraction of a non-linear background.

The discussion of Fig. 3 is based on the comparison of the observed features in the a-C:H VB spectra with the corresponding spectra of graphite and diamond. The motivation for this comparison is given by the fact that our a-C:H films consist of an amorphous network of carbon atoms with sp^2 (graphite-like) and sp^3 (diamond-like) sites. In analogy to rf plasma deposited a-C:H films we assume that our films contain about 70% tetrahedrally coordinated (sp^3) carbon atoms and 30% threefold coordinated (sp^2) sites.

The width W of the valence band as deduced from the onset of the XPS valence band E_Γ is $W = E_\Gamma - E_V = 23.8 \pm 0.5$eV. This value can be compared with the corresponding VB widths of graphite and diamond which are 24eV and 24.2 ± 1eV, respectively [13]. The dominating peak at about 17.2eV also observed in graphite and diamond is due to emission from σ_1 bands near the symmetry point Q in graphite and correspondingly from band 1 in diamond with predominantly s character in both cases [12, 14]. As mentioned above the graphite and diamond XPS VB exhibit a peak not present in a-C:H at about 13eV. This peak has been attributed to emission from the σ_2 band close to the symmetry point Q and is formed by 2s, $2p_x$, $2p_y$ orbitals in the case of graphite and emission from band 2 with mixed s and p character in the case of diamond.

The p electron density of states (UPS spectrum) shows a pronounced peak at 7.6eV and a shoulder near 4eV. Similar features have been observed in the X-ray emission spectrum (XES) of graphite [15, 16] Due to the dipole selection rule $\Delta l = \pm 1$ the K emission spectrum shows the partial p electron density of states. This spectrum exhibits two bands which can be assigned to a p-σ contribution with a peak at about 8eV and a p-π band with a maximum at about 4eV. The shoulder in the UPS spectrum

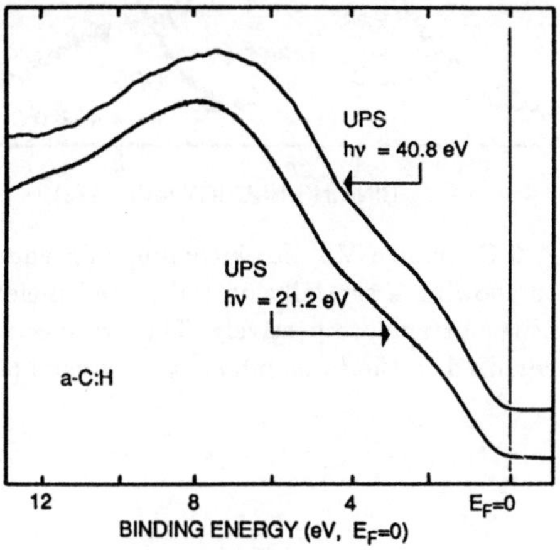

Fig. 1 UPS VB spectra of an a-C:H film measured with 21.2 and 40.8eV exitation energy (film thickness $d_0 \approx$ 11nm; E_I= 400eV; process gas: methane) [8].

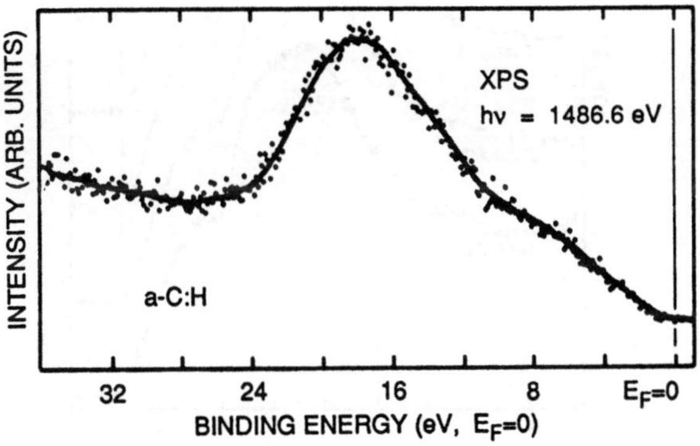

Fig. 2 MXPS VB spectrum of an a-C:H film measured with monochromatized radiation Al K_α (film thickness d_0 = 10nm, E_I = 140eV) [8].

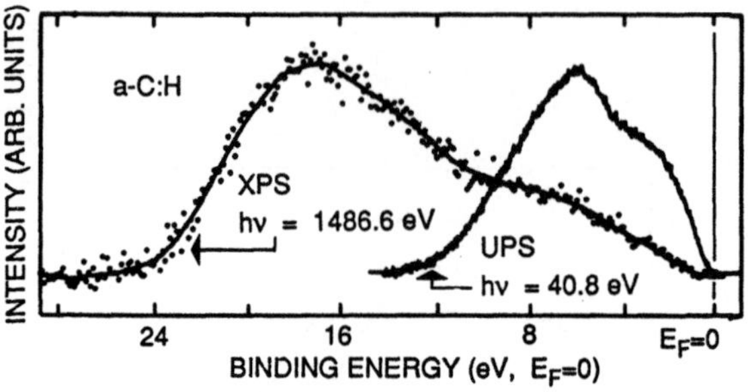

Fig. 3 Combined VB plot including XPS and UPS data showing essentially partial s- and p-electron density of states, respectively. The two spectra are normalized to the same intensity maximum [8].

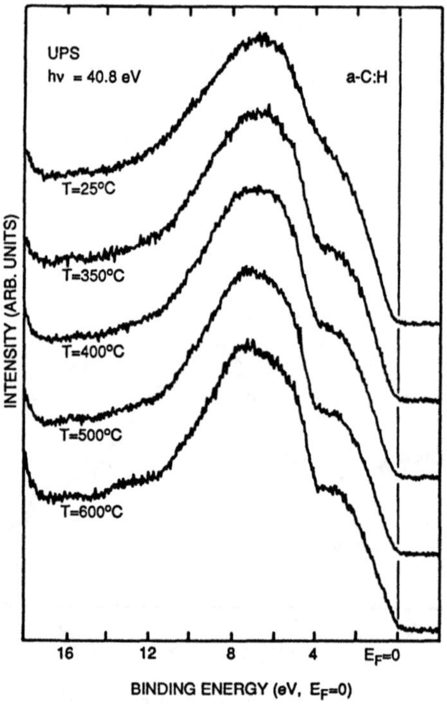

Fig. 4 UPS ($h\nu = 40.8$eV) VB spectra of a-C:H deposited at room temperature and spectra obtained after several annealing steps (E_I= 400eV) [8].

at 4eV shows a much weaker intensity as compared to the corresponding structure in the XES spectrum which can be explained by the fact that the major part of the carbon atoms in a-C:H are tetrahedrally (sp^3) coordinated, which do not give rise to π bonds. The XES spectrum of diamond [17] shows one prominent peak at a binding energy of about 6eV which is obviously related to our peak at 7.0eV in a-C:H. A second weak peak in the XES spectrum is located at 13eV showing the p character of the peak which is missing in the XPS spectrum of a-C:H.

More information about the influence of the ratio of sp^3 and sp^2 coordinated atoms on the VB structure can be obtained either by preparing films with different deposition parameters (see below) or by annealing experiments. It has been demonstrated that a heat treatment of a-C:H from room temperature up to 600 °C increases the amount of sp^2 sites from about 30% to 60% [18]. Annealing effects have also been observed by electron energy loss and work function measurements [19]. The structural changes upon annealing are accompanied by a hydrogen loss of a-C:H. Figure 4 shows VB spectra of a-C:H as deposited at room temperature and after several annealing steps up to 600 °C. The annealing time for each heat treatment was 15 minutes. Three distinct changes can be observed in the latter spectrum:
(i) The energy range in which a contribution of π band emission has to be expected exhibits the development of a distinct peak with a sharp kink at a binding energy of about 4eV.
(ii) A peak shows up at a binding energy of 13eV, also present in diamond and graphite.
(iii) The main peak at 7eV is slightly shifted to higher binding energy and exhibits a shoulder at about 5eV.
(iv) The position of the band edge E_V is shifted gradually to lower binding energy with increasing annealing temperature before finally reaching a similar position as in the semimetallic graphite.

3.2 Influence of Process Gas and Deposition Energy: Departure from a-C:H Structure

In this section photoemission data obtained from films deposited with different ion deposition energies E_I ranging from 145 to 1000eV and using different process gases such as methane (CH_4), benzene (C_6H_6) and styrene ($(C_6H_5)CH{=}CH_2$) are presented. In Figure 5 the deposition was performed with benzene as a source gas. Different ion energies E_I assumed to be determined by the anode voltages $U_A$of the Kaufman type ion source, have been used. For higher ion energies E_I the shape of the VB is quite structureless, forming essentially one prominent peak at 7eV and a shoulder at about 4eV. This VB spectrum is in close agreement with those discussed in the previous section. The shoulder at 4eV is due to emission from π states. The variation of E_I reveals a well defined threshold ion energy $E_{I,t}$ of 305 ± 20eV. For $E_I < E_{I,t}$ the VB spectra are quite different from those of a-C:H. Three obvious changes are visible: (i) The VB spectra at $E_I < E_{I,t}$ become more detailed in structure. (ii) The 4eV-peak is much more prominent for $E_I < E_{I,t}$. (iii) The position of the VB edge E_V is drastically shifted from 0.5eV for $E_I > E_{I,t}$ (a-C:H) to 1.6eV for $E_I < E_{I,t}$.

Films deposited at low ion energy using benzene are assumed to represent some kind of polymer, like polystyrene [20]. In fact, the similarity of the UPS VB spectrum of polymeric films prepared from styrene (Fig. 7) and benzene (Fig. 5) deposited at low ion energy supports this assumption.

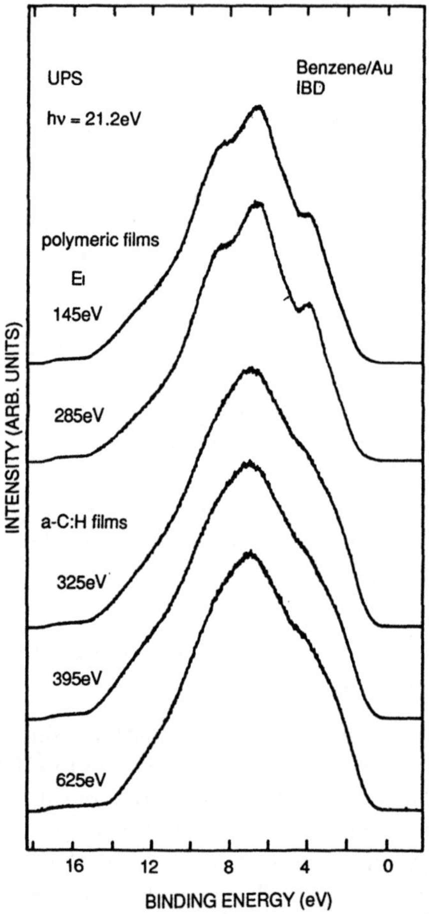

Fig. 5 UPS ($h\nu = 21.2$eV) VB spectra of IBD films using benzene as a process gas deposited at different ion energies E_{I} [34].

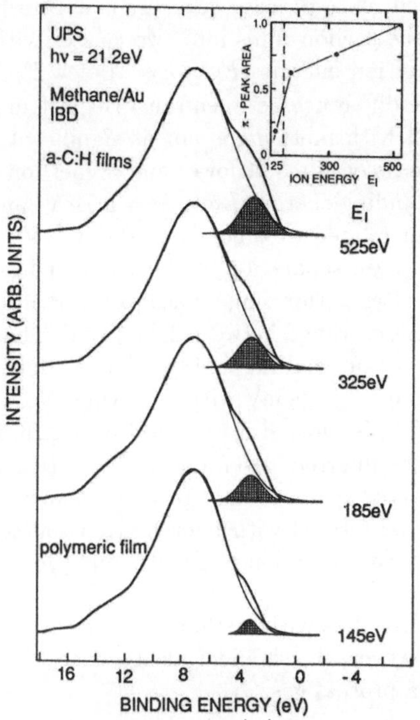

Fig. 6 UPS VB spectra as a function of E_I for IBD films using methane as a process gas. A deconvolution of the VB band spectra into the main peak and the shoulder related to π states is performed. The integrated intensity of the shoulder as a function of E_I is depicted in the inset [33].

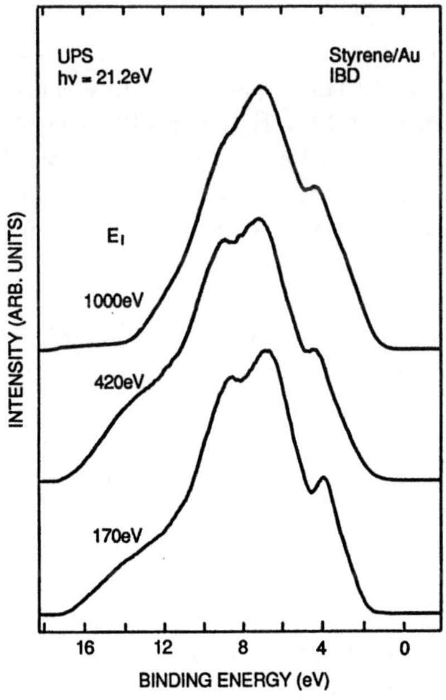

Fig. 7 UPS VB spectra of IBD films prepared with styrene at different ion energies E_I [33].

If methane is used as a process gas, again a threshold ion energy of $E_{I,t}$ = 165 ± 20eV separates the carbon films into two classes with different properties (Fig. 6). a-C:H is obtained for ion energies $E_I > E_{I,t}$. Below $E_{I,t}$, again a polymer-like film is formed. However, the difference between the VB spectrum of the polymeric film and that of a-C:H produced with methane is not as significant as in the case of benzene. The main differences between films at lower and higher ion energy are the shift of the VB edge E_V to higher binding energy (E_V - E_F = 1.54eV) and a marked decrease in the intensity of the shoulder related to π bonds at about 4eV for films at $E_I < E_{I,t}$. This is clearly demonstrated by a separation of the shoulder from the main peak at about 7eV which can be seen after performing a deconvolution into two gaussian peaks fitted to VB spectra in the energy range between 1 and 9eV. The resulting π peak (hatched area) is located at a constant binding energy of 3.2eV independent of E_I and reveals peak widths ranging from 1.3 to 2.3eV with increasing E_I. The relative intensity of the π peak as a function of E_I is plotted in the inset of Fig. 6, from which the position of $E_{I,t}$ at 165 ± 20eV can be inferred. Even for $E_I > E_{I,t}$ (i.e. a-C:H) the intensity of the shoulder related to π states at about 4eV grows continuously with increasing E_I. The same investigation was performed with benzene as a source gas but no change in the intensity of the shoulder as a function of E_I was found for $E_I > E_{I,t}$.

The deposited a-C:H films with methane and benzene yield quite similar VB spectra (Figs. 5 and 6). However, the shoulder at about 4eV is slightly more prominent in the case of benzene as a process gas.

Films prepared with styrene reveal a distinctly different dependence on E_I (Fig.7). Comparing these VB spectra with those of benzene or methane, the lack of an a-C:H like VB spectrum is obvious. In fact, even a relatively high ion energy of 1000eV is not sufficient for preparing an a-C:H like film. The VB structure of these polymeric styrene films is discussed in detail elsewhere [21].

XPS measurements were also carried out. No significant differences were visible by measuring the binding energies of the C 1s core level peak even when different process gases were used. In fact, for all three gases a binding energy of the C 1s core level of 284.80 ± 0.10eV was obtained for $E_I > E_{I,t}$. For all three gases, a constant C 1s binding energy of 284.7 ± 0.1eV was found for E_I below and above $E_{I,t}$. The same FWHM of the C 1s peak for methane and benzene was found and amounts to 1.40 ± 0.1eV. The FWHM of the styrene C 1s peak (1.55eV) could not be compared directly because the data were recorded with lower instrumental resolution.

The measurements presented above (Fig. 5, 6 and 7) clearly reveal the high sensitivity of photoelectron VB spectroscopy in monitoring structural changes in carbon based films. In short a-C:H films could be prepared by exceeding a gas dependent ion energy threshold $E_{I,t}$ by using benzene or methane as a process gas. Values of 165eV and 305eV were found for $E_{I,t}$ in the case of methane and benzene, respectively. a-C:H films prepared using methane ($E_I > E_{I,t}$) exhibit a weak ion energy dependence with respect to the amount of sp^2 coordinated bonds in the deposited films. With increasing deposition energy the ratio of the sp^2 to sp^3 bonds in the films increases. No correlation between E_I and the amount of sp^2 bonding was found for a-C:H films prepared with benzene as a source gas. In addition, the amount of sp^2 bonding is somewhat larger for a-C:H films prepared with benzene than for films deposited using methane as a process gas. Thus, the precursor gas clearly determines the chemical structure of the a-C:H films deposited by IBD even above $E_{I,t}$. However, recent IR measurements performed on a-C:H films prepared by means of a rf plasma from various process gases did not reveal any differences for films deposited above a certain bias voltage [22]. This may be explained by the higher sensitivity of UPS measurements.

We failed to prepare a-C:H films with styrene as a source gas since even $E_I = 1000eV$ was not sufficient for obtaining the typical UPS VB spectrum of a-C:H. Polymer-like films were prepared with $E_I < E_{I,t}$. Quite large differences were found between films using methane and benzene as a source gas, indicating the strong influence of the monomer on the structure of the polymer-like films. This result is in agreement with that obtained by Wild et al. [22]. Very similar VB spectra were obtained for polymer-like films using benzene and styrene as a process gas. In the case of the polymeric styrene films prepared by IBD a strong similarity with polystyrene was found.

In conclusion, it has been demonstrated that the precursor gas and the deposition energy may have a considerable influence on the properties of a-C:H films.

4 a-C:H Films on Various Substrates: Interface Properties

The application of thin films in technology is determined in many cases by the interface properties between the film and the substrate. Both the electronic and atomic structure of the interface region play a crucial role. The former one can be described in terms of the local electron density of states across the interface, including information on band bending, interface states, valence band and conduction band discontinuities, and barrier heights. The atomic structure includes information on the nature and spatial distribution of the different phases found in the interface region. These strongly correlated main properties determine basically all other derived physical characteristics such as electrical transport, mechanical properties and interface chemistry. a-C:H films reveal very different adhesion properties on different substrates. In a number of publications we tried to correlate the macroscopic adhesion behaviour with the microscopic interface properties as observed by photoelectron spectroscopy, TEM, RBS and channeling [23, 24, 25, 26, 27].

4.1 a-C:H on Metallic Substrates

a-C:H/Au Interface

A sequence of UPS VB spectra measured after a number of deposition cycles of a-C:H on polycrystalline Au is shown in Fig. 8. The initial spectrum (deposition time t = 0s) shows the typical VB spectrum which is dominated by intense peaks due to the 5 d bands at binding energies between about 2 and 7eV and a shoulder near the Fermi level (E_F) formed by the s-p band. The following spectra up to t = 25 min. do not exhibit marked changes. The main features of the Au VB are preserved and particularly no binding energy shifts of the peak positions are observed. This is in clear contrast to the strong interactions observed in other cases (see below) and is a first indication that a-C:H on Au forms a **passive interface**. In fact, the only changes observed in the VB spectra upon a-C:H deposition is a decrease in intensity of the peaks related to the Au electron states due to the attenuation of the signal from the substrate by the growing overlayer and, complementary, an increasing admixture of a-C:H related features. Finally, after a deposition time of t = 300 min. a VB spectrum of a-C:H is obtained and the substrate is no longer visible. The XPS core level binding energy shifts of the C 1s and Au $4f_{5/2}$ and $4f_{7/2}$ as a function of deposition time is shown in Fig. 9. The Au 4f binding energy shifts remain rather small ($<$ 0.1eV) during the whole deposition procedure until the intensity of these core lines become vanishingly small and the photoelectrons originate essentially from the interface region. This is a further clear indication that the Au substrate does not react chemically with the a-C:H overlayer.

The C1s core level binding energies are shown as well in Fig. 9. For very thin overlayers (d$<$ 0.6nm, i.e. t$<$ 10^3s) the C 1s binding energy is reduced by -0.4eV as compared to the C 1s binding energy in 'bulk' a-C:H. Since the Au 4f core lines do not reveal a binding energy shift we do not attribute the C 1s binding energy shift to the usual chemical shift. We rather explain the relatively small shift (the corresponding shifts in the case of Si or Ge substrates are of the order of 1.2 eV) by the very small amount of a-C:H deposited at this stage leading to a very thin film with a thickness of about 2 to 3 monolayers, or by relaxation effects.

The above photoelectron spectroscopy measurements clearly reveal a passive interface between a-C:H and the Au substrate. This is in contrast to the distinct interface phase formation in the case of many other substrates which form reactive interfaces. The absence of a chemical interaction at the present interface is obviously responsible for the **very poor adherence** of a-C:H on Au substrates. Even very thin films with thicknesses of about 20nm could be wiped off from the substrate (see Table II).

a-C:H/W Interface

The following photoemission data exhibit a typical example of a metallic substrate which forms a reactive interface: a-C:H on polycrystalline W. Figure 11 shows UPS VB spectra of a clean substrate along with a sequence of spectra measured after increasing deposition time. The initial spectrum (t = 0s) reveals the W VB which is dominated by distinct peaks originating from 5 d electron states. Dramatic changes in the valence electron structure can already be seen after very short deposition times (t = 10s): the peak close to E_F is strongly attenuated and a new peak at about 7eV appears. This behavior is in strong contrast to that observed in the case of the Au substrate and as a consequence a strong chemical reaction can be inferred from the UPS data. With increasing deposition time the spectra resemble more and more the final spectrum representative of the pure a-C:H film in Fig. 8. However, a Fermi edge is visible even after a prolonged deposition time, t = 350 min. This observation will be discussed below.

The C 1s core line measured with XPS (excitation energy 1253.6eV) shows a similar behaviour as in the case of the Si(100) substrate (see below). A decomposition into interface and bulk contributions can be made [23]. Figure 12 shows the corresponding C 1s intensities as a function of deposition times for two different substrate temperatures. We note the strong temperature dependence in the ratios of the bulk and interface intensities. The formation of the interface phase is obviously strongly favoured by the higher temperature (T = 90 °C). The W 4f core levels exhibit a relatively small binding energy shift of the order of 0.5eV upon a-C:H deposition consistent with published binding energy shifts [28]. The binding energy shifts are summarized in Table I.

The attenuation of the W 4f core line intensities as a function of deposition time is displayed in Fig. 13. In contrast to the deposition of a-C:H on Au (Fig.10) the

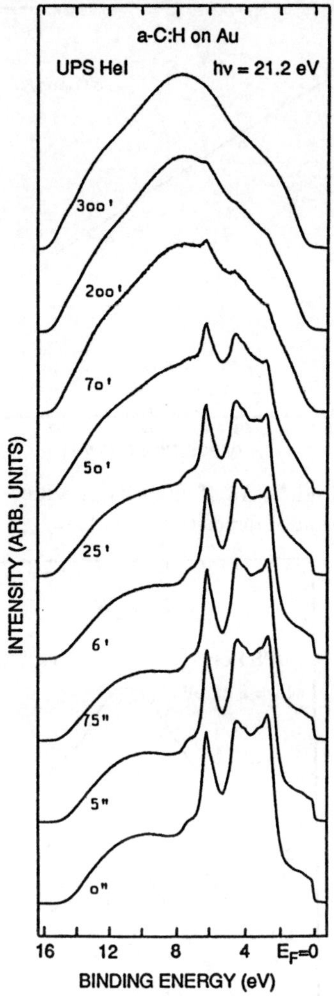

Fig. 8 UPS VB spectra of clean polycrystalline Au and a sequence of spectra measured after increasing deposition times t. The final spectrum (t = 300min.) shows the UPS VB spectrum of a-C:H without any contribution from the Au substrate (E_I = 400eV, process gas: methane) [24].

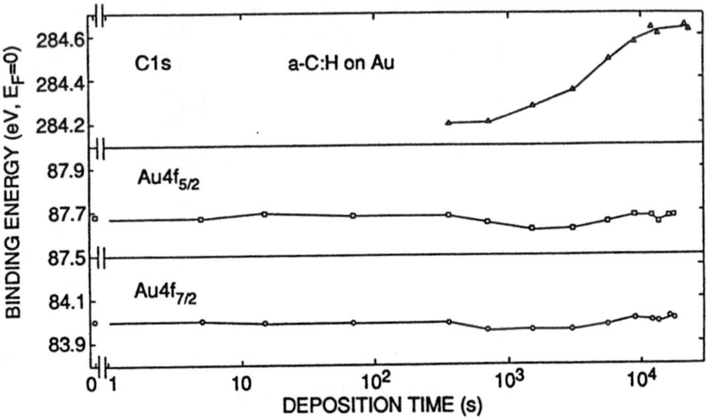

Fig. 9 Core level Au 4f and C 1s binding energies measured by XPS with an excitation energy of 1253.6eV as a function of deposition time [24].

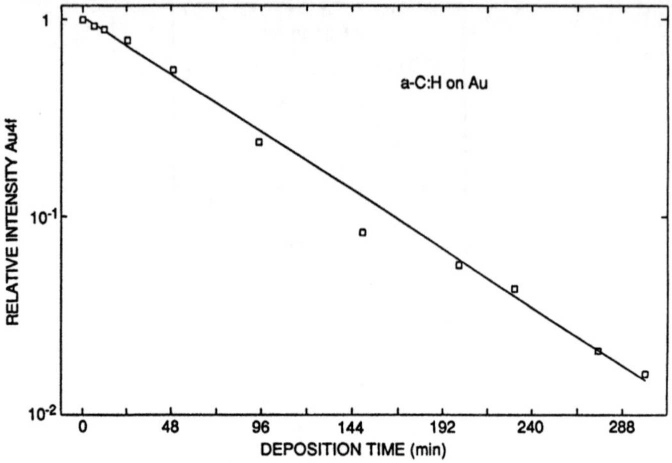

Fig. 10 Au 4f intensity as a function of deposition time [24].

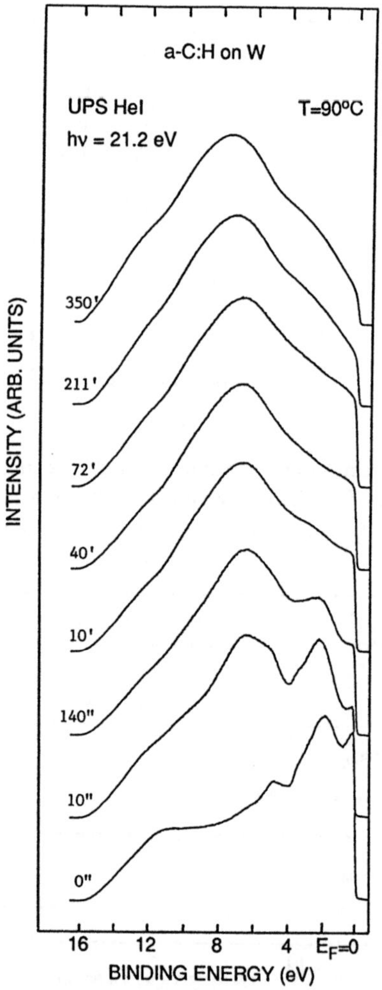

Fig. 11 UPS VB spectra of clean polycrystalline W and a sequence of spectra measured after increasing deposition times (t). The final spectrum (t = 350min.) still shows a Fermi edge from the carbide interface phase (E_{I} = 400eV, substrate temperature T = 90 °C) [24].

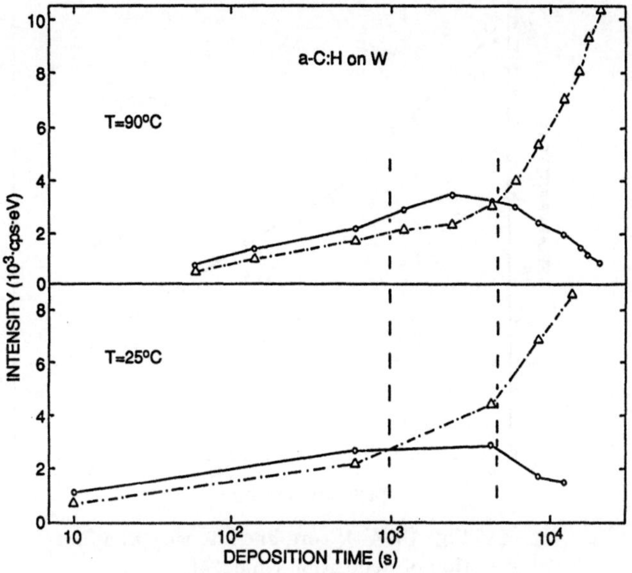

Fig. 12 C 1s core line intensities of the interface (cycles) and a-C:H bulk (triangles) contribution for two deposition temperatures T = 25 °C and 90 °C. Note the delayed cross over the interface and bulk intensity curves if deposition is performed at high temperatures [24].

Table I. XPS core level binding energies and binding energy shifts for a-C:H on Au and W (±0.1 eV).

Substrate	core level	Binding energy E_B (eV)		Shift ΔE
		bulk	interface phase	(eV)
Au	C 1s	284.6	284.2	-0.4
	Au $4f_{7/2}$	84.0	84.0	0.0
W	C 1s	284.6	283.3	-1.3
	W $4f_{7/2}$	31.4	31.9	0.5

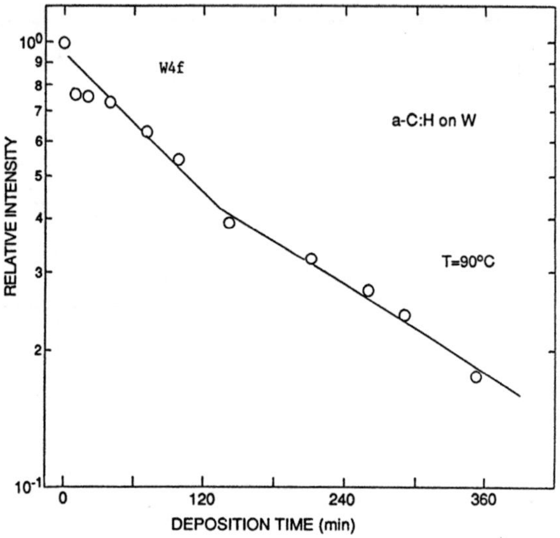

Fig. 13 W 4f core line intensity as a function of deposition time [24] .

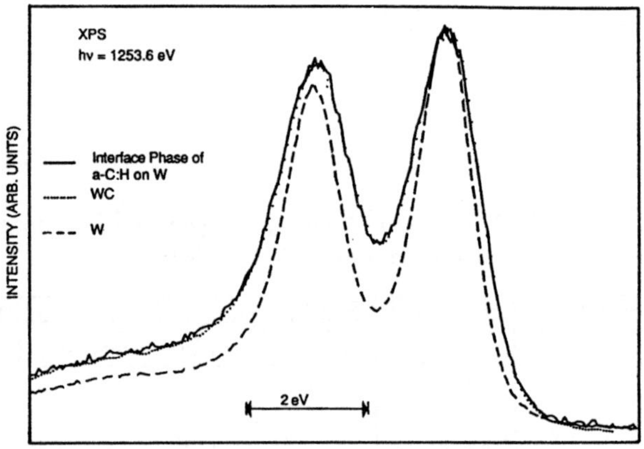

Fig. 14 W 4f core lines measured by XPS for the a-C:H/W interface phase, a WC reference sample, and polycrystalline W. The three spectra have been aligned to the same binding energy in order to compare the line shapes [24].

Table II. Characterization of a-C:H/substrate interface and related adhesion properties

a : amorphous
pc : polycrystalline
x : interface phase layer thickness †
d_{cc} : critical film thickness for spontaneous peeling off
PES : characterization by UPS/XPS
* : measurements not yet performed

Substrate	**type of interface**	**adhesion**	**d_c (nm)**	**PES**
Si(100)	reactive carbide formation $x = 4$ nm	good	>500	yes
Ge(100)	reactive carbide formation $x<4$ nm	good	>500	yes
GaAs(100)	—	poor	= 100	yes
pc-Au	passive	poor	<25	yes
pc-W	reactive carbide formation $x\approx8$ nm	good	≈50	yes
pc-Fe	—	poor	< 40	*
a-$Fe_{86}B_{14}$	reactive reaction at Fe and B sites observed $x\approx1$ nm	poor	≈20	yes

†x has been estimated from the deposition time t_x for which the C 1s interface peak intensity amounts to 10% of the total C 1s intensity and assuming a growth rate of $6 \cdot 10^{-4}$ nms^{-1}.

present curve shows two interesting features: a pronounced attenuation after very short deposition times and a kink close to a deposition time of t = 150min. The former is clearly related to the high reactivity of the clean W surface correlated with a high sticking coefficient for the incoming $C_mH_n^+$ ions. The kink indicates that the growth of the overlayer is not uniform. The formation of a-C:H islands cannot be ruled out since the attenuation after the kink at longer deposition times is reduced by about a factor of 4 as compared to the a-C:H growth on Au. In this case the final UPS spectrum of Fig. 11 has to be interpreted as being a superposition of two spectra originating from a-C:H (island contribution) and the interface phase (metallic tungsten carbide). This would also explain that a Fermi edge is still visible in this spectrum.

Though the growth mechanism is not evident from looking at Fig. 13, we can conclude from the data above that a pronounced interface phase is formed during the a-C:H deposition. In order to compare directly with a tungsten carbide we have measured the W 4f and C 1s core levels of a WC reference sample. The sample was prepared from WC powder pressed into an Al foil. The sample surface was cleaned by Ar^+ sputtering. Figure 14 shows the strong similarity in the W 4f core line shape for the a-C:H/W interface and the WC reference sample. Note the marked deviation of these two core line spectra from that of polycrystalline W. The three spectra have been adjusted to the same W $4f_{7/2}$ peak energy position. In addition, the W 4f/C 1s intensity ratios have been determined both for the a-C:H interface and the reference sample. The two ratios turned out to be identical to within 2%.

Finally we would like to point out that the adherence of a-C:H on W is rather poor. Thicker films than 50nm (deposited at room temperature, E_I = 200eV) peeled off spontaneously. This observation was not expected at all considering the strong chemical interaction at the interface leading to a fairly thick interface phase of about 6 to 10nm. In order to clarify this apparent contradiction we analyzed the W substrate surface after the removal of the a-C:H film. It turned out that the rupture takes place **between the interface phase and the substrate.** As a consequence the adherence of the a-C:H film on the substrate is limited by the interface phase properties and not by the a-C:H film itself. The interface and adhesion properties examined so far are summarized in Table II.

4.2 a-C:H on Semiconducting Substrates: a-C:H/Si(100) Interface

Figure 15 shows UPS VB spectrum ($h\nu = 21.2$eV) of clean Si(100) and a sequence of spectra measured after a number of deposition cycles making use of the Penning ion source. The spectrum of the clean Si(100) surface is dominated by two peaks at about 2 and 9 eV binding energy which are essentially due to the Si 3p and 3s electron states, respectively. Since the surface states close to the VB edge are not visible in the spectrum of our disordered Si surface, we are able to locate the position of the Fermi level E_F with respect to the VB edge (E_V). A low value (E_F - E_V) of 0.2 eV was found which indicates a Fermi level position close to the valence band edge. The other VB spectra reveal the gradual interface formation and finally that of an a-C:H layer thick enough for all the photoelectrons to originate from the film itself without being influenced by the interface region underneath. After a deposition time of 175 min a film thickness of about 7 nm is obtained. The spectrum measured after a very short deposition time of 0.1 min, which corresponds to a film thickness of a fraction of a monolayer, clearly shows the strong interaction with the incoming ion fragments. The Si 3s and 3p states are shifted to lower and higher binding energies, respectively, as compared to clean Si, revealing the strong chemical interaction. These VB spectra clearly tell us that a reactive interface is formed between Si(100) and a-C:H. This conclusion will be confirmed by the XPS core level data presented below. After longer deposition times ($\geq$ 5 min) the Si 3p band is gradually transformed into a shoulder at about 2eV binding energy and the peak close to 8eV becomes more and more dominant. The different features of the a-C:H VB spectrum are discussed in section 3.1.

The XPS data of the Si 2p core levels are depicted in the Figures 16 through 18. Here we would like to recall that we obtain information from the substrate at very low deposition times, whereas a-C:H/Si interface information is obtained at prolonged deposition times due to the limited escape depth of the photoelectrons. As we can see from Fig. 16 a marked chemical shift of the Si 2p core level is observed during the film deposition. The initial binding energy of the Si 2p electrons of the substrate amounts to 99.4eV and is shifted by 1.3eV when the photoelectrons originate from the interface region. In close agreement with the UPS data the marked core level binding energy shift is characteristic for the reactive interface.

The Si 2p binding energies as a function of deposition time are depicted in Fig.17 for two different substrate temperatures during a-C:H deposition. The evolution of the distinct chemical shift is clearly visible. The comparison of the two curves clearly shows the presence of a thicker reacted interface layer for the deposition at the higher temperature i.e. a higher binding energy shift is observed for equal deposition times. The initial binding energy shift of about 0.2 eV for deposition times below 100 s is most probably due to a change in band bending at the Si substrate surface leading to an increased binding energy. This change in binding energy is obviously induced by the deposition of a small amount of $C_mH_n^+$ ions.

Additional information can be obtained from the measurement of the full widths at half maximum (FWHM) of the Si 2p core lines as a function of deposition time as shown in Fig. 18. Relatively narrow lines are found at the beginning and at the end of the deposition process, and can be explained by the existence of a dominating single phase. At intermediate deposition times (around 10^3 s) an enhanced width of the core lines is observed due to the superposition of core lines originating from the substrate and the interface phase.

The C 1s core line is depicted in Fig. 19 as a function of a-C:H deposition time. In this case interface information is obtained at low deposition times, whereas the film properties themselves can be studied after prolonged deposition when the film thickness has reached is such that essentially all photoelectrons originate from the a-C:H film. From the topmost curve a C 1s binding energy of 284.6 eV is obtained which is typical for C in a-C:H. With respect to this value a core level binding energy shift of -1.2 eV is observed at low deposition times. The C 1s and Si 2p binding energies obtained from the interface phase are in close agreement with those reported for SiC. In order to make a more accurate identification of the interface phase XPS measurements on a sample of polycrystalline SiC have been performed using the same spectrometer settings as in the case of the interface measurements. The corresponding C 1s and Si 2p core line spectra of SiC and the a-C:H/Si interface are shown in Figs. 20 and 21. The data clearly reveal the close similarity of the interface phase and SiC as far as the binding energies, line widths and line shapes are concerned. The C 1s peak as shown in Fig. 19 can be decomposed in two contributions, the interface phase and an a-C:H related peak [23].

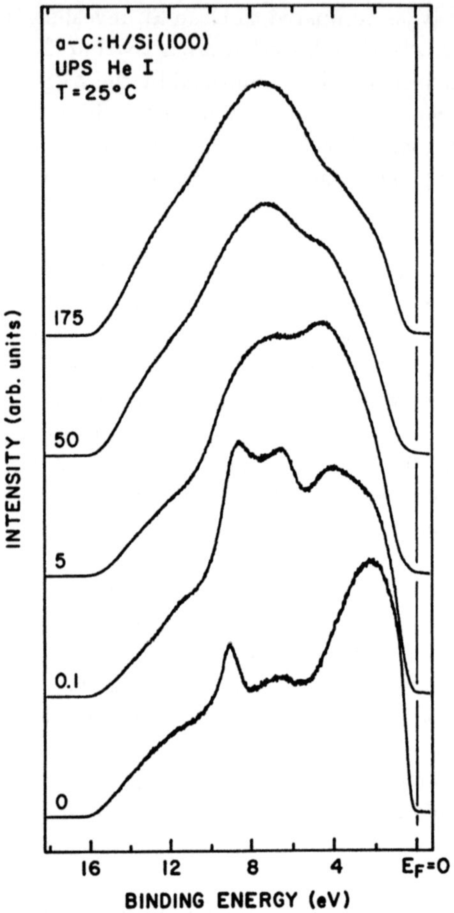

Fig. 15 UPS ($h\nu = 21.2$eV) VB spectrum of clean Si(100) and a sequence of spectra measured after the indicated a-C:H deposition time in minutes. The topmost curve represents the VB spectrum of a-C:H. The a-C:H deposition energy is 400eV (room temperature deposition, process gas: methane) [25].

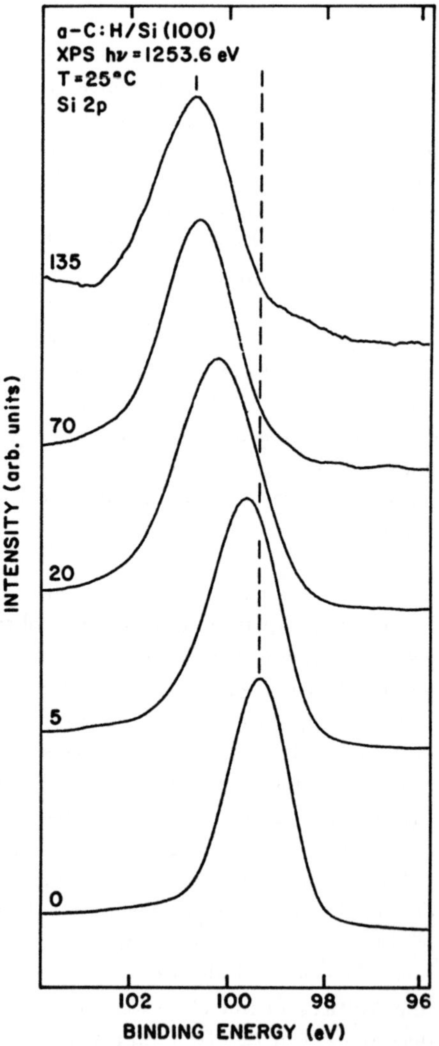

Fig. 16 XPS Si 2p core level spectrum of the clean Si substrate and a sequence of corresponding spectra after a-C:H deposition [25].

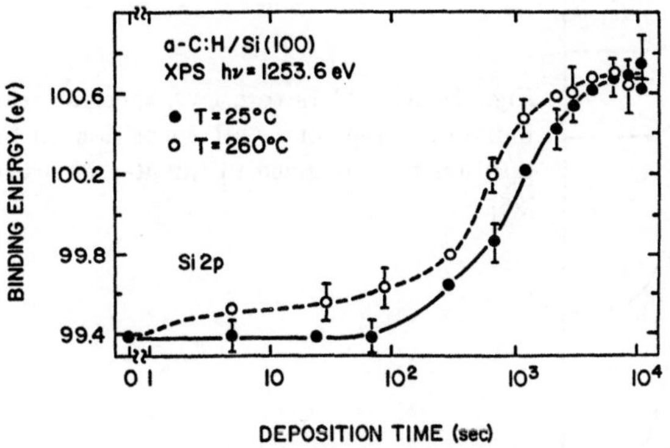

Fig. 17 Si 2p binding energy as a function of deposition time for two different substrate temperatures [25].

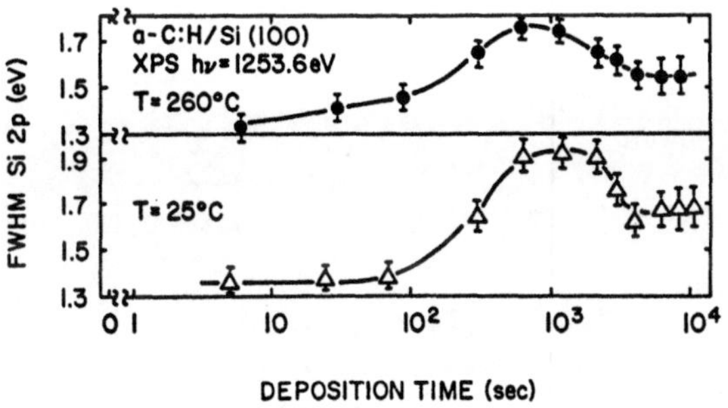

Fig. 18 Width of the Si 2p core line as a function of deposition time for two different substrate temperatures [25].

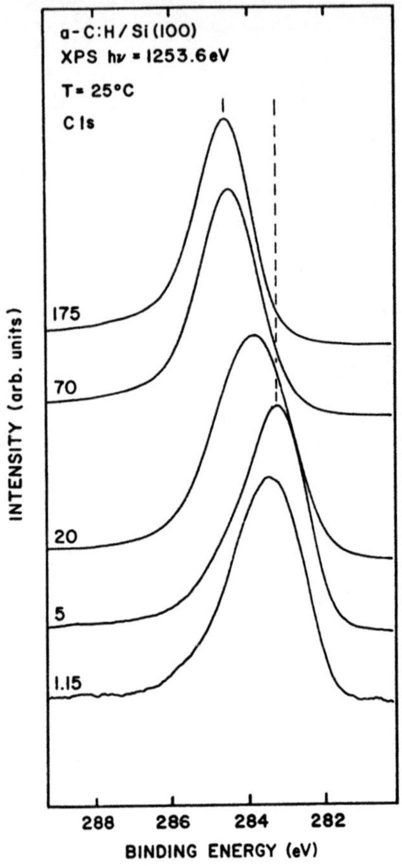

Fig. 19 XPS C 1s core level spectra measured after individual steps of a-C:H deposition on Si. The deposition time is given in minutes for each spectrum [25].

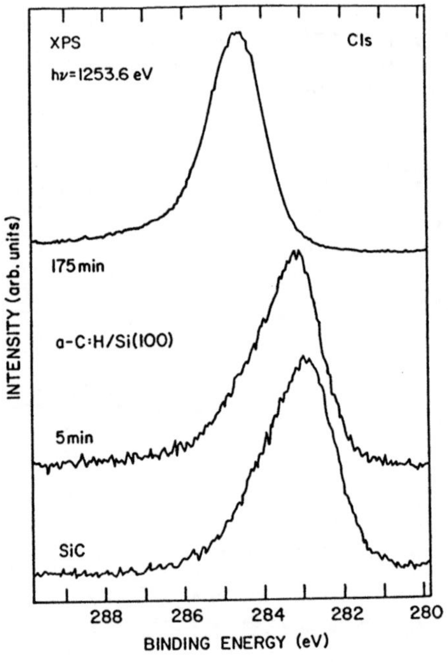

Fig. 20 XPS C1s core level spectra of polycrystalline SiC, a-C:H/Si interface phase (after a deposition time of 5min.) and a-C:H (175min.). Note the close similarity of C 1s core line in SiC and the spectrum related to the interface phase (5min.) [25].

The intensities of these two contributions are depicted in Fig. 22 for two deposition temperatures. It has to be pointed out that even at very low deposition times (500 s) a distinct contribution from the a-C:H related peak can be observed. From this we can conclude that the a-C:H overlayer grows on top of the interface phase. The interface phase is fully developed within a deposition time of 10^3 s or more and amounts to a thickness of the order of a few nanometers.

Figure 23 shows the integrated Si 2p intensity on a logarithmic scale as a function of deposition time. The intensity curve measured after room temperature deposition exhibits a kink at 3×10^3s due to a marked decrease in the slope for higher deposition times. This change in the slope does not seem to be related to island formation and subsequent homogenous film growth. The comparison with Fig.22 shows a distinct decrease of the C 1s intensity related to the interface phase at the time of about 3×10^3s and, correspondingly, an increase in the intensity of the a-C:H peak. As a consequence, we attribute the kink in the curve as being due to the presence of two different growth regimes. Below a deposition time of 3×10^3s the decrease of the Si 2p core line intensity is influenced by the formation of the SiC phase at the interface, whereas at higher deposition times a layer by layer growth of a-C:H takes place without any further development of the interface region.

Rutherford Backscattering and Transmission Electron Microscopy Rutherford backscattering (RBS), forward scattering in grazing incidence and high resolution ion channeling have been used in addition to characterize the films and the interfaces. Regarding the film properties the H content is of particular interest since it is (beside He) the only element which can not be detected with XPS or AES.

The RBS analysis of a film deposited at an ion energy of 100eV (process gas methane) is shown in figure 24a. The energy positions of Si and C surface atoms are indicated near 1.3MeV and 0.6MeV, respectively. The leading edge of the Si substrate signal, however, appears at 1.14MeV due to the energy loss of the $^4He^+$ particles in the a-C:H overlayer. The C signal is superimposed on the Si substrate signal because of the lower mass of C compared to Si. The amount of C in the a-C:H film can be determined quantitatively by subtracting the Si background signal with a smooth spline function in the energy interval of 0.4 to 0.6MeV. The value obtained from Fig. 24a amounts to 6.55×10^{17}C/cm^2. This area concentration corresponds to a film thickness in the order of 700Å.

The determination of the hydrogen content of the a-C:H film requires a forward scattering analysis because hydrogen is lighter than the $^4He^+$ particles. The scattering geometry is illustrated in the inset of Fig. 24b. A grazing incidence was chosen for the impinging $^4He^+$ particles and the recoiling hydrogen nuclei were measured in forward scattering direction. A mylar foil was placed in front of the detector to block forward scattered $^4He^+$ ions. The measured hydrogen profile is shown in Fig. 24b. Quantitative determination of the hydrogen content was carried out by comparing the spectrum of Fig. 24b with that of a Si calibration sample with a hydrogen implant of known dose. The hydrogen profile in Fig. 24b thus corresponds to a concentration of 3.2×10^{17}H/cm^2. From the above analysis a relative hydrogen concentration of $x_H = n_H/(n_H + n_C) = 0.33$ is found for the a-C:H film. Other a-C:H films were deposited at room temperature with ion energies of 200 and 400eV. However, we have not found a correlation between ion energy and hydrogen content in this energy range. The hydrogen content of our a-C:H films was generally in the range of 0.30 to 0.35. All a-C:H

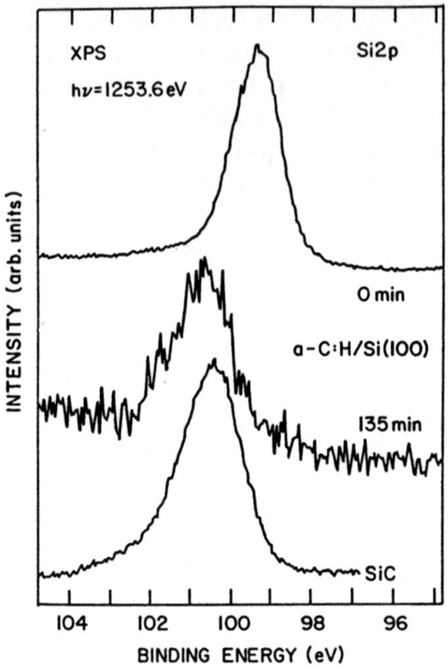

Fig. 21 XPS Si 2p core level spectra of SiC, a-C:H/Si interface phase (5min.) and Si (0min.). Again, the SiC and interface phase related spectra exhibit a strong similarity [25].

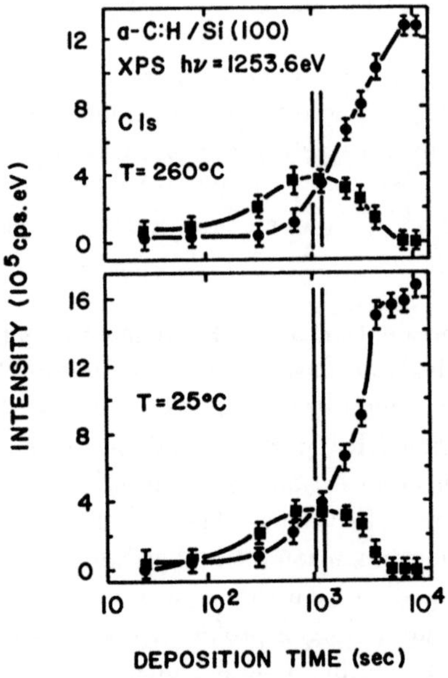

Fig. 22 Intensities of the two decomposed C 1s core line peaks which are related to the interface phase (squares) and related to the a-C:H phase (circles), respectively [25].

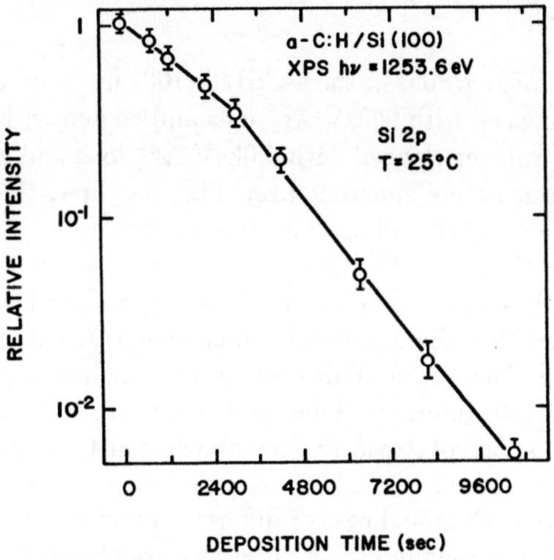

Fig. 23 Intensity of the Si 2p core line of the substrate as a function of a-C:H deposition [25].

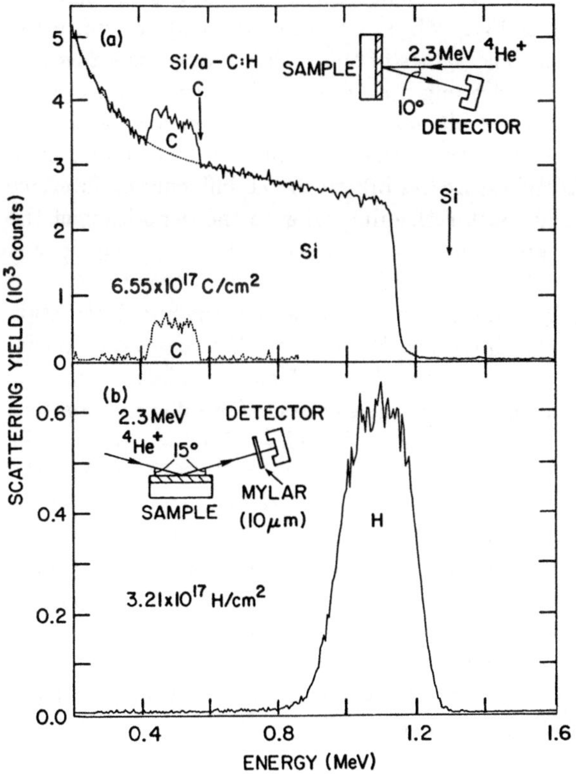

Fig. 24a RBS spectrum of a 60nm a-C:H film deposited at room temperature on Si(100). The vertical arrows indicate surface positions of the corresponding elements. **b** Hydrogen signal as determined by forward scattering at grazing incidence. Same sample as in **a** [26].

films investigated were amorphous. This has been confirmed with flat-on as well as cross-sectional TEM analysis.

Lattice imaging micrographs of the a-C:H/Si(100) interface are shown in Fig. 25a for a sample sputter cleaned with 900eV Ar^+ ions and coated with a-C:H at RT, and in Fig. 25b for a sample sputter cleaned with 5.0keV Ar^+ ions and coated at 300°C. Two characteristics are immediately apparent from Fig. 25: first, the a-C:H film is amorphous and second, the roughness of the interface is very different for the two samples. The microstructure of the a-C:H film is featureless, even when the micrograph of Fig. 25a is further enlarged as shown in Fig. 26. Actually, a close inspection of the micrograph of Fig. 26 shows that the amorphous microstructure extends right up to the Si single crystal interface. There is no indication of a crystalline or polycrystalline second phase at the a-C:H/Si(100) interface. Thus we have to conclude that the SiC-like phase identified by photoemission spectroscopy (see above) must be amorphous.

In order to compare the roughness of different interfaces we will use the following qualitative measure of the roughness: the interface roughness d_I is measured perpendicular to the interface from the first continuous row of lattice atoms in the crystal to the last discernible discontinuous set of lattice atoms in the amorphous a-C:H overlayer. With this measure at hand we can compare the various interfaces.

The sample in Fig. 25a exhibits an exceptionally smooth interface. By looking from the side of Fig. 25a along the [002] plane one finds that the roughness extends over mostly one, at some spots two, lattice spacings. The interface roughness is thus of the order of 3 to 6Å. Through the same analysis, the sample in Fig. 25b has an interface roughness of 9 to 15Å. The origin of the big indentations observed at some spots of the interface such as on the left side of Fig. 25b is not clear. It could be due to the sample preparation for TEM analysis. The values of the interface roughness together with the deposition parameters and the cleaning procedure of the substrates are listed in Table III. Included in Table III are the results for a sample sputter cleaned at 3.5keV and coated at RT. It has an interface roughness that is similar to that of Fig. 25b. This indicates that the deposition temperature has practically no influence on interface roughness. However, a higher energy sputter cleaning prior to the deposition of the a-C:H overlayer produces a rougher interface.

Judging from the interface roughness one would assume that the Si-C interaction is stronger for films that show a rougher interface. But even in these cases the interfacial compound cannot be detected because, though the interfacial SiC-like layer may be thicker, it is still amorphous. This is supported by the electron microdiffraction pattern taken of the interfacial layer that is shown in Fig. 25b. The diffraction spots in the lower part of Fig. 27 are the matrix spots of the single crystal Si(100) substrate. The labeling of some of the spots clearly reveals the [110] zone axis alignment. Apart from the Si matrix spot, no other diffraction spots are visible that could identify a second phase. However, Fig. 27 shows clearly a halo and two diffuse diffraction rings that point towards an amorphous microstructure of the interfacial layer. Finally, it is important to notice that neither the micrographs nor the diffraction pattern reveal the presence of any extended defects in the single crystal substrate near the interfacial region. This fact excludes high interfacial stresses due to the a-C:H overlayer. In fact, a low interfacial stress is a prerequisite for good adhesion properties.

The extent of the atomic intermixing in the interfacial layer can be investigated

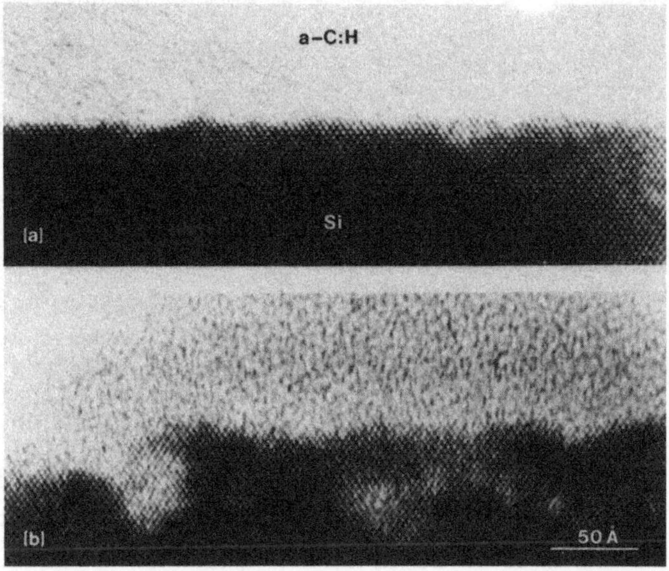

Fig. 25a,b Cross-sectional electron micrograph of the interface between an a-C:H film and bulk Si(100) viewed in the [110] direction: a sputter cleaned with 900eV Ar^+ ions and coated at room temperature, b sputter cleaned with 5.0keV Ar^+ ions and coated at 300 °C [26].

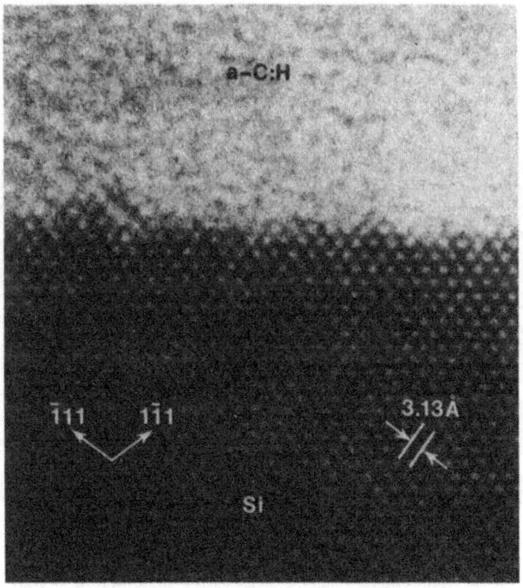

Fig. 26 Magnification of part of the interface of Fig. 25a. Note the smoothness of the interface and the absence of any second phase particles [26].

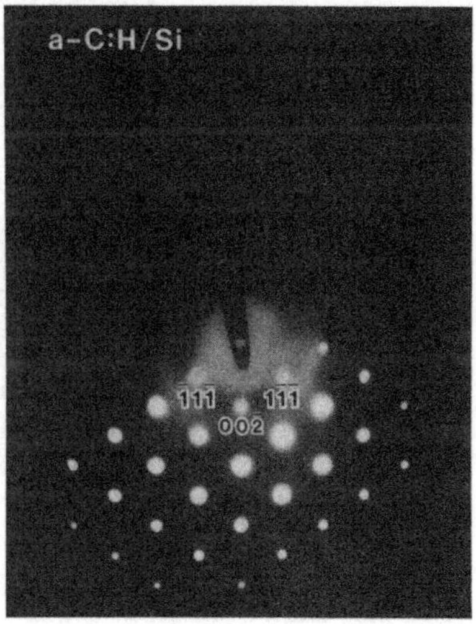

Fig. 27 Electron diffraction pattern taken at the a-C:H/Si(100) interface of the sample of Fig. 25b showing the matrix spots of the Si lattice and diffuse diffraction features of the amorphous interface layer [26].

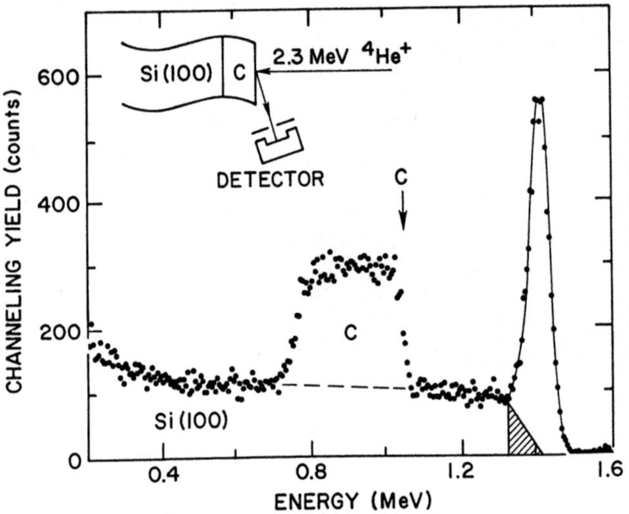

Fig. 28 High resolution 2.3MeV ion channeling analysis of the a-C:H/Si(100) interface of the sample sputter cleaned with 5.0keV Ar^+ ions and coated at 300 °C. Clearly visible are the carbon signal of the a-C:H overlayer and the Si peak associated with the interfacial layer. The method of background subtraction is indicated [26].

Table III. Properties of the a-C:H/semiconductor interface

Substrate	Cleaning		Deposition		Interface	
	E_{Ar}	T_A	E_I	T_A	d_I[Å]	$N_I[cm^{-2}]$
Si(100)	900eV Ar^+	RT	200 eV	RT	3–6	$3.2 \cdot 10^{16}$
Si(100)	3.5keV Ar^+	RT	200 eV	RT	9–15	$6.8 \cdot 10^{16}$
Si(100)	5.0keV Ar^+	RT	500 eV	RT	9–15	$6.7 \cdot 10^{16}$
GaAs(100)	150eV Ar^+	540 °C	200 eV	RT	7–10	$2.0 \cdot 10^{16}$
GaAs(100)	150eV Ar^+	540 °C	200 eV	300 °C	7–13	$2.5 \cdot 10^{16}$
GaAs(100)	150eV Ar^+	540 °C	500 eV	300 °C	16–23	$5.0 \cdot 10^{16}$

with high resolution ion channeling analysis. In Fig. 28 we present such an analysis of the a-C:H/Si(100) interface of the sample sputter cleaned with 5.0keV Ar^+ ions and coated at 300 °C. The major crystallographic axis of the Si substrate (in this case the [100] axis) has been aligned with the $^4He^+$ beam. The vertical arrow labeled C indicates the surface position of the a-C:H layer. The silicon surface position is beyond the right ordinate of Fig. 28. Again, the energy difference between the leading edge of the Si substrate signal at 1.5MeV and the Si surface position is due to the energy loss of the $^4He^+$ particles in the a-C:H overlayer. The area under the carbon peak corresponds to $5.8 \times 10^{17}C/cm^2$. This is equivalent to a thickness of about 600Å as measured with a Tali Step instrument. The accuracy of the analysis was checked by comparing the carbon peak in Fig. 28 with that of a regular RBS spectrum taken with random alignment of the sample. Since the a-C:H overlayer is amorphous one expects to obtain the same carbon atomic density for both spectra. The agreement was found to be within 5%.

The peak at the leading edge of the Si substrate signal is due to Si atoms in the interfacial layer. Since they are not in registry with the single crystal substrate they produce an increased scattering yield in the channeling spectrum. The contribution of the single crystal substrate to the Si interface peak can be subtracted by means of the dashed area shown in Fig. 28. This background subtraction corrects also to a first approximation for the straggling of the $^4He^+$ particles due to the amorphous a-C:H overlayer. In addition, we have taken into account the energy loss of the $^4He^+$ particles in the overlayer. Though more elaborate techniques of background subtraction can be employed, the present one should have only a small systematic error. However, the comparison between different samples is very accurate because the same technique of background subtraction has been employed in all cases. The analysis of the Si interface peak in Fig. 28 then yields an areal density N_I of $6.7 \times 10^{16}Si/cm^2$. The density of Si atoms at the interface is also listed in Table III.

Table III includes the results for samples that were sputter cleaned with low and high energy Ar^+ ions and coated at RT. For a sample sputter cleaned with 900eV Ar^+ ions we find a value of $3.2 \times 10^{16}Si/cm^2$ for the atomic intermixing whereas a sample sputter cleaned with 3.5keV Ar^+ ions has a value of $6.8 \times 10^{16}Si/cm^2$. From this comparison we conclude that a higher energy Ar^+ ion beam causes more damage at the Si surface, and initiates an enhanced intermixing of Si and C atoms at the interface during deposition of the a-C:H overlayer. From Table III we find that the same trend applies to the interface roughness.

a-C:H/Ge(100) Interface

A similar investigation as in the case of the a-C:H/Si(100) interface has been performed for the a-C:H/Ge(100) interface. These results have been published in detail in [27]. The main results can be summarized as follows:

1. The a-C:H/Ge(100) interface is as reactive as the a-C:H/Si(100) interface. Similar core level binding energy shifts are found for the C 1s core level and marked binding energy shifts of the Ge 2p, 3d core levels are observed upon a-C:H deposition.

2. A similar interface roughness has been observed by TEM as in the case of Si substrates.

3. The reacted interface region could be observed directly in TEM in the form of a darker band between the single crystalline substrate and the a-C:H overlayer. The

thickness of this reacted zone amounts to 30 to 40Å. An electron microdiffraction pattern of this zone revealed again an amorphous structure of this interface phase.

Due to the nonexistence of a stable germanium carbide compound it was speculated that a metastable compound is formed at the interface [27]. The formation of this phase was attributed to the impingement of energetic $C_mH_n^+$ ions onto the substrate.The excellent adhesion of a-C:H on Ge has been attributed to the formation of this metastable Ge-C phase.

a-C:H/GaAs(100) Interface

The UPS/XPS study described above has been applied to a-C:H overlayers on GaAs(100) substrates, too. Again, the deposition has been performed with a Penning ion source operated at 400 eV and using methane as a process gas. The substrates were maintained at room temperature during deposition. Figure 29 shows a sequence of UPS valence band spectra measured after the individual deposition steps. The spectrum of clean GaAs(100) is essentially determined by the density of occupied electronic states and is in agreement with corresponding published data on GaAs [29]. The final spectrum (topmost in Fig. 29) was obtained after a deposition time of 160 min which corresponds to a film thickness of 6 nm. This spectrum is identical to that obtained after a deposition time of 175 min on Si (Fig. 15). At low deposition times up to 2 min only minor changes in the VB spectra are observed. The peaks at a binding energy of about 2eV and 7eV remain almost unchanged in their position and intensity. More distinct changes are observed for deposition times greater than 10 minutes, where one of the two main peaks disappears.

The following XPS spectra have been measured with monochromatized Al K_α radiation ($h\nu = 1486.6$ eV). Figure 30 shows the evolution of the C 1s, As 3d, Ga 3d and Ga $2p_{3/2}$ binding energies as a function of deposition time. The first striking feature of Fig. 30 is the marked binding energy shifts observed in all measured core lines. The C 1s binding energy for deposition times below 600 s is 283.1eV which is close to the value of 283.4eV measured in the early stages of a-C:H deposition on Si(100). This indicates clearly that a chemical reaction between carbon and the substrate takes place in the interface region. The final value after a deposition time of 1×10^4s is 284.7 ± 0.1eV and is typical for the a-C:H film.

The initial binding energies of the As 3d, Ga 3d and Ga $3p_{3/2}$ core electrons correspond to bulk values of GaAs with a Fermi level position, E_F close to the valence band edge E_V (E_F - E_V = 0.27 eV). With increasing deposition time both the As and Ga core levels exhibit distinct shifts. Most pronounced is the evolution of the As 3d binding energy shift, which amounts 1.6eV, to a final position of 42.6eV. Less pronounced are the binding energy shifts of the Ga core levels. The Ga 3d core line has been measured up to a deposition time of 6×10^3s and could not be measured beyond that time due to insufficient intensity. Within this time period a binding energy shift of 0.7eV has been measured. Due to the higher intensity of the Ga $2p_{3/2}$ core line measurements up to a deposition time of 1×10^4s could be performed. A very peculiar behaviour of the binding energy as a function of deposition time has been observed: a maximum in the binding energy shift of 0.9eV is found for a deposition time of about 7×10^3s, followed by a rapid decrease resulting in a final measurable binding energy of 1117.7 eV, which is that of metallic Ga. After a deposition time of 1×10^4s the intensity of the Ga $2p_{3/2}$ core line is too low to yield binding energies with meaningful accuracy.

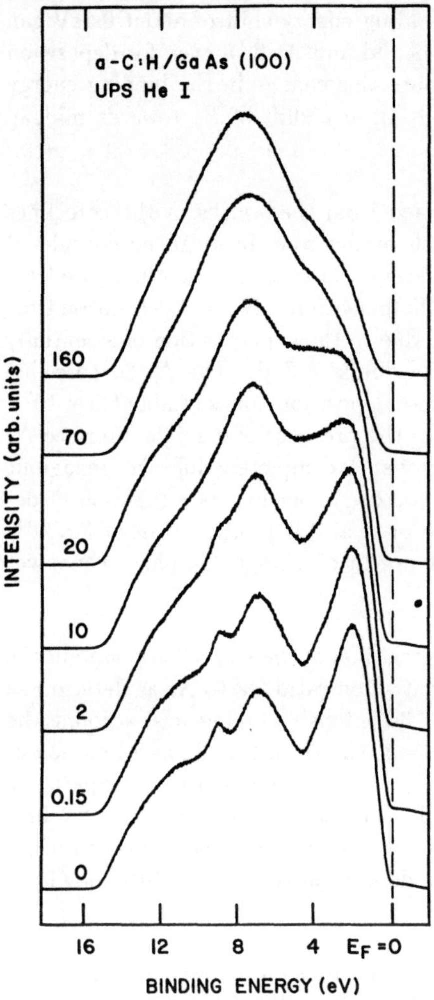

Fig. 29 UPS ($h\nu$ = 21.2eV) VB spectrum of clean GaAs(100) and a sequence of spectra measured after the indicated a-C:H deposition time in minutes. The topmost curve represents the VB spectrum of a-C:H. The a-C:H deposition energy is 400eV (room temperature deposition) [25].

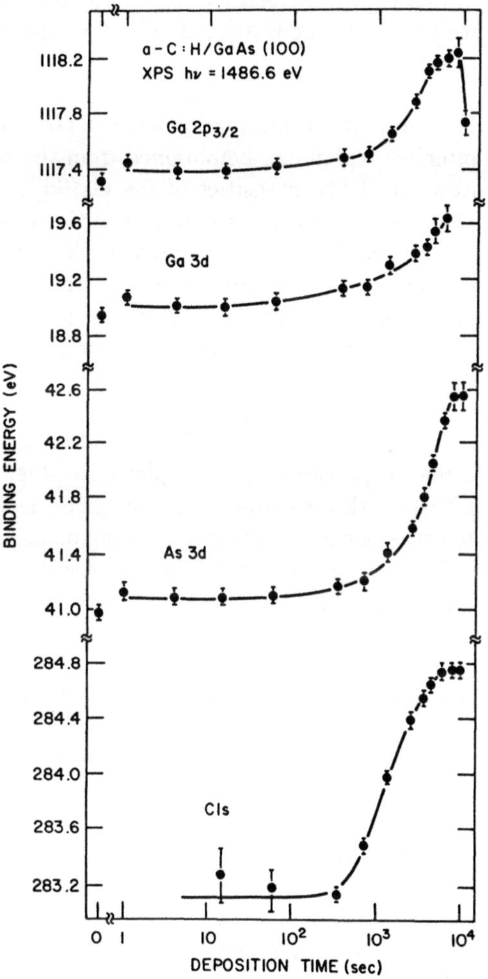

Fig. 30 XPS C 1s, As 3d, Ga 3d, and Ga $2p_{3/2}$ core level binding energies as a function of a-C:H deposition time [25].

As in the case of a-C:H/Si(100) a small binding energy shift of about 0.1eV can be observed in the substrate core levels (Ga $2p_{3/2}$, 3d and As 3d), even for deposition times as low as a few seconds. Again we attribute this change in the binding energy as being due to a change in band bending resulting in a shift of E_F towards midgap position.

Complementary information can be obtained from the widths of the core lines presented in Fig. 31. Again an increased width of the core lines can be considered as being due to the presence of a mixture of different phases with individual core level binding energies within the escape depth of the photoelectrons. Below a deposition time of 10^3s the C 1s core line is markedly broadened due to the superposition of essentially two peaks originating from the interface and the 'bulk' a-C:H. The As 3d core line clearly shows a similar broadening at intermediate deposition times of about 5×10^3s, whereas the final narrowing at 1×10^4s indicates the presence of a single interface As phase within the escape depth of the photoelectrons. A completely different behaviour is found for the FWHM of the Ga 3d and the Ga $2p_{3/2}$ core line as a function of deposition time. A distinct broadening is observed even at a deposition time of 1×10^4s which clearly shows that at this stage the development of a single Ga phase is not yet completed.

Important information about a possible segregation of the GaAs compound in the interface region can be obtained from the concentration ratio Ga to As as determined from the XPS intensities of the Ga 3d and As 3d intensity taking into account the photoionization cross sections. The Ga to As ratio is shown in Fig. 32 as a function of deposition time. It is evident that the Ar^+ ion sputter cleaning at room temperature produces a Ga rich surface with a ratio Ga/As of 1.7 in agreement with earlier findings [30]. With progressive a-C:H deposition, and as a consequence of the formation of interface phases, the As content increases. After a deposition time of 1×10^4s the Ga to As ratio is close to unity.

The dependence of the individual core line intensities on deposition time (using a semi-logarithmic scale) is plotted in Fig. 33. For a laterally uniform growth of the a-C:H overlayer without reaction with the substrate an exponential decrease of the intensities is expected with increasing deposition time. But as in the case of a-C:H on Si (Fig. 23) a change in the slope is observed taking place at about 3×10^3s, pointing to a reactive interface.

The core level binding energy shifts observed upon deposition of a-C:H on GaAs represent an additional indication that a chemical reaction takes place at the interface. The C 1s binding energy related to the interface phase (measured at low deposition times) is shifted by -1.6eV with respect to the corresponding value in a-C:H. This shift is even larger than those observed in the case of a-C:H on Si and on Ge, which amounts to −1.2eV and −1.3eV, respectively. The core level binding energy shift of the As 3d level amounts to 1.6eV (with respect to the GaAs substrate) and the final position of the core line, which is related to the interface phase, is at a binding energy of 42.6eV. This value is about 1eV above the value of elemental As. This distinct chemical shift indicates that a marked selective reaction between the impinging $C_mH_n^+$ ions and the As takes place. This reaction is responsible for the As enrichment in the interface region (Fig. 32).

The Ga 3d and $2p_{3/2}$ core lines show a smaller binding energy shift during a-C:H

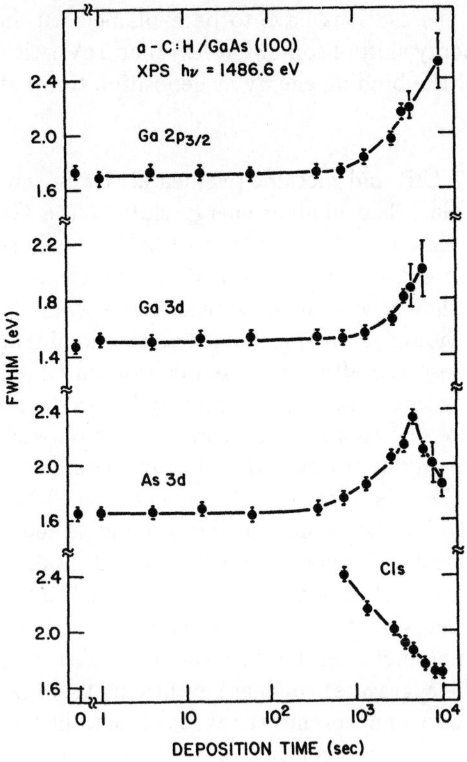

Fig. 31 Width of the XPS core lines as a function of a-C:H deposition time [25].

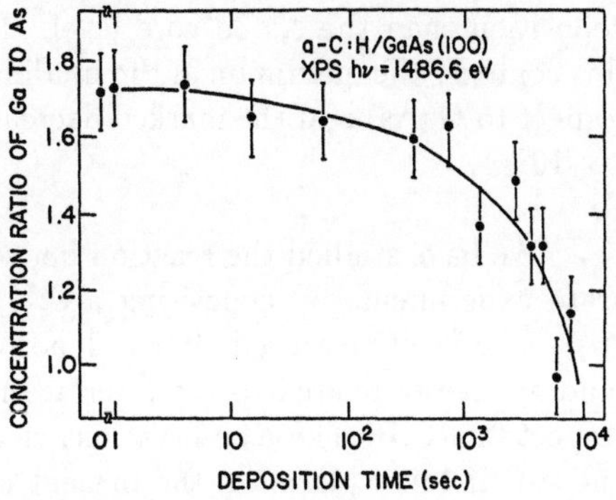

Fig. 32 Ga to As concentration ratio as obtained from integrated XPS core line intensities as a function of a-C:H deposition time [25].

Fig. 33 XPS C1s, As 3d, and Ga 3d core line intensities as a function of deposition time [25].

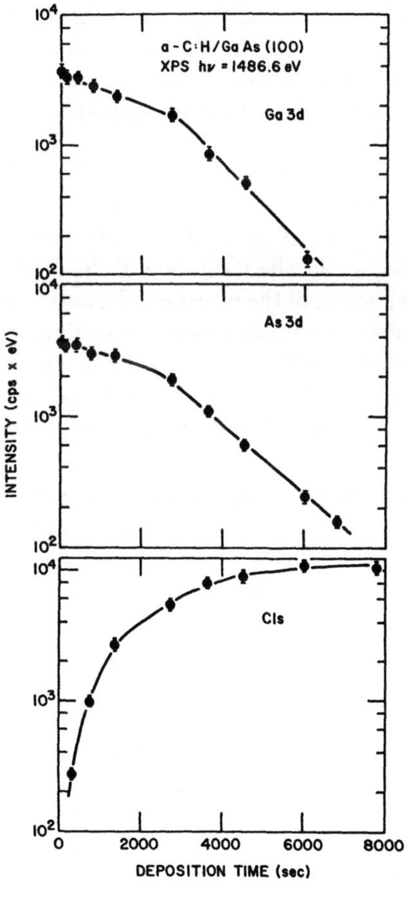

deposition than the As 3d core level. Two observations have to be explained within this context: the maximum in the binding energy shift of somewhat less than 1eV (with respect to GaAs) and the marked decrease of the binding energy at deposition times of 1×10^4s.

We have studied the reaction between a-C:H and metallic (elemental) Ga in separate experiments by depositing a-C:H on Ga. The binding energy shift of the Ga $2p_{3/2}$ core level measured after a deposition time of 10^4s amounts to 1.0eV. The C 1s binding energy related to the interface region exhibits a marked shift of -2.0 eV with respect to a-C:H. These observations clearly show that a chemical reaction between Ga and a-C:H takes place. In the present case, however, we conclude that the dominant reaction in the fully developed interface, as observed after prolonged deposition times, takes place between As and a-C:H. This conclusion is supported basically by two facts: An As enrichment takes place at the interface due to a strong chemical reaction with a-C:H (Fig. 32), and the Ga $2p_{3/2}$ binding energy approaches the value of metallic Ga after a deposition time of 10^4s. In addition, the existence of an As carbide (AsC_3) has been reported [31], whereas to our knowledge no Ga carbide has been found to date. Nevertheless, we cannot conclude unambiguously whether the Ga 2p and 3d binding energy shifts can be attributed solely to a chemical shift or to a size effect related to the formation of small Ga aggregates. Binding energy shifts of the order of 1eV have been observed in photoemission experiments on small metal clusters deposited on substrates [32, 33]. After a prolonged deposition time the aggregates condense eventually to larger clusters and therefore the Ga $2p_{3/2}$ binding energy approaches the value of metallic Ga, 1117.0eV. Finally, we would like to emphasize that the continuing increase of the line widths of the Ga 2p and 3d core levels with progressive a-C:H deposition (Fig. 31) clearly indicates a behaviour that is different from the As core levels and consequently is inconsistent with the formation of a single phase.

The reduced adhesion of a-C:H on GaAs as compared to Si substrates is related to the reactions observed at the interface as discussed above. It has to be emphasized that the a-C:H/GaAs interface does not represent a passive boundary zone and, consequently, the formation of chemical bonds at the interface could provide a reasonable adhesion of the overlayer. However, the chemical reaction of the overlayer with the As is obviously much stronger than that with Ga, which is responsible for the As enrichment at the a-C:H/GaAs interface. This process is strongly coupled with the breaking up of GaAs bonds and a subinterface enrichment of Ga. It is this decomposition process which we believe to be responsible for the observed poor adhesion of a-C:H on GaAs.

Rutherford Backscattering and Transmission Electron Microscopy We have carried out a microstructural analysis for the a-C:H/GaAs(100) interface. The GaAs samples were sputter cleaned under identical conditions but the deposition of the a-C:H film was performed with different ion beam energies and at different substrate temperatures. Figure 34a presents a lattice imaging micrograph of the interface for a deposit with 200eV ion beam energy at RT and Fig. 34b for a deposit with 500eV ion beam energy at 300 °C. Again, a difference in interface roughness is immediately apparent. Using the above definition we find an interface roughness of 7 to 10Å for the sample of Fig.34a and 16 to 23Å for the sample of Fig. 34b. A GaAs sample that was coated with 200eV ion energy at 300 °C had an interface roughness of 7 to 13Å, which is comparable to the one coated at RT. The higher interface roughness for the sample of Fig. 34b is not caused by the cleaning procedure in this case but rather by the enhanced C-GaAs interaction due to the use of a higher ion beam energy during deposition of the a-C:H film. The smoothness of the interface for the sample of Fig. 34a is well visible in the enlarged micrograph shown in Fig. 35. As in the case of Si, Fig. 35 demonstrates that the a-C:H overlayer is amorphous up to the GaAs single crystal interface. The presence of a second phase in the interfacial layer cannot be detected in Figures 34 and 35.

Again a darker band of amorphous material about 30Å wide can be clearly seen at the a-C:H/GaAs interface in Fig. 34a and 35. It runs parallel to the interface. The same band can also be seen in Fig. 34b, though it is more difficult to observe because of the larger interface roughness. But it follows the undulations of the interface too.

The electron diffraction pattern of the a-C:H/GaAs(100) interface closely resembles that of Fig. 27. Apart from matrix spots of the GaAs lattice and the diffuse diffraction features of the amorphous overlayer no other diffraction spots could be found. This supports the amorphous nature of the interfacial compound. As in the cases of Si and Ge we have not found any extended defects that would indicate a high stress level in the crystalline material.

The high resolution ion channeling analysis of the a-C:H/GaAs(100) interface of the sample coated with a 200eV ion beam at 300 °C is presented in Fig. 36. Clearly visible are the carbon peak from the overlayer and the interface peak of the GaAs interface layer. The carbon peak corresponds to $5.6 \times 10^{17} C/cm^2$ which is comparable to the coverage of the overlayer on Si. In case of GaAs the identification of the interface peak is difficult. Due to the very similar mass of Ga and As Rutherford backscattering technique is unable to distinguish between these two elements. Thus, the interface peak can be due to Ga and/or As atoms. Nevertheless, the analysis of the interface peak provides some information on the C-GaAs interaction during deposition of the a-C:H films. An analysis identical to that of the Si peak, employing the background subtraction discussed above, yields an areal density N_I for the peak in Fig. 36 of $2.5 \times 10^{16} at/cm^2$. For the RT deposit a value of $2.0 \times 10^{16} at/cm^2$ is found. The GaAs sample that was coated with 500eV ion energy at 300 °C had a density of $5.0 \times 10^{16} at/cm^2$. The results are listed in Table III. They indicate that deposition of a-C:H at elevated temperatures does not increase the intermixing with the GaAs substrate significantly. On the other hand, a higher ion energy for the deposition of the a-C:H film promotes extended atomic intermixing at the a-C:H/GaAs interface.

The microstructural investigation of the a-C:H/Si(100), a-C:H/Ge(100) and a-C:H/GaAs(100) interfaces showed that the properties of the interfaces depend on the energy of the ion beams used to either clean the substrates or to deposit the a-C:H overlayer. Generally, a rougher interface and extended atomic intermixing is caused by the use of a higher energy ion beam. In contrast, a higher deposition temperature has practically no influence on the interface roughness. This is true for both Si and GaAs substrates. We found that, in general, the interface roughness is less than 25Å for the samples investigated in this study. The compounds (carbides of Si and As) that have been detected with photoemission spectroscopy in the interfacial layer are found to be amorphous. In case of GaAs we have found additional evidence of the precipitation of metallic gallium in the interfacial layer. In neither case stress seems to be important because of the absence of any extended defects in both Si and GaAs substrates.

Relation to Adhesion Properties The quality of adhesion depends to a large degree on the properties of the interfacial layer that is formed between overlayer and substrate. In general, one can distinguish between the following types of adhesion

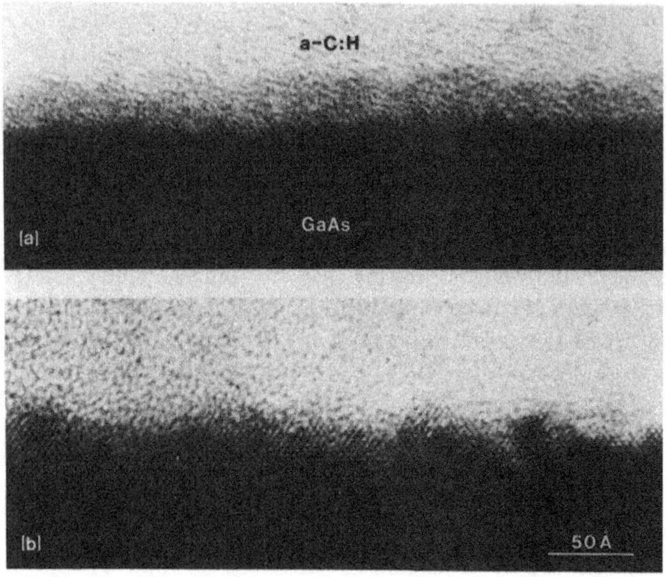

Fig. 34 Cross-sectional electron micrograph of the interface between an a-C:H film and bulk GaAs(100) viewed in the [110] direction: **a** coated with 200eV at room temperature, **b** coated with 500eV at 300 °C [26].

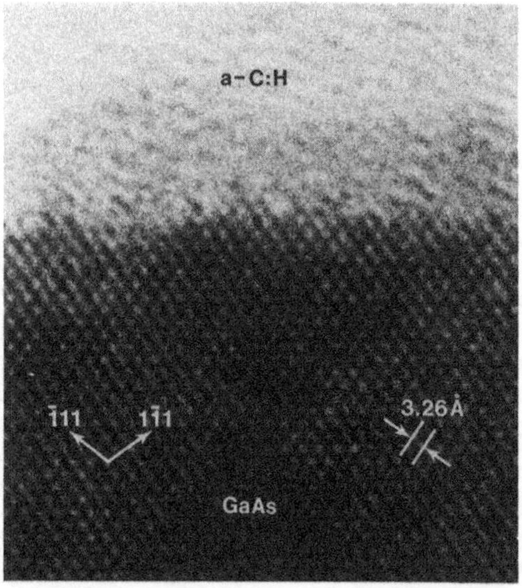

Fig. 35 Magnification of part of the interface of Fig. 34a. The relative smoothness of the interface and the absence of any second phase particles are apparent [26].

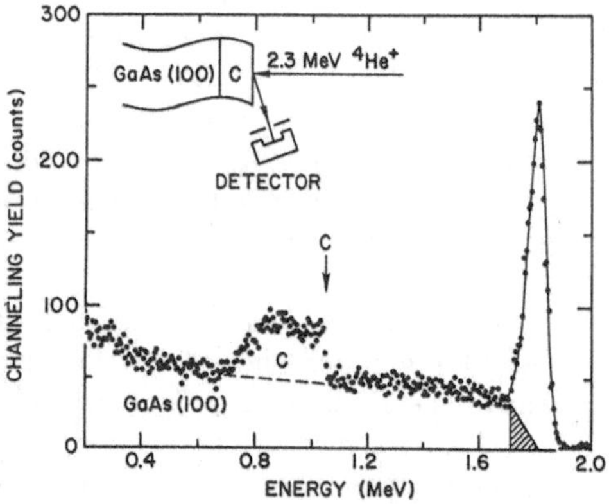

Fig. 36 High resolution 2.3 MeV ion channeling analysis of the a-C:H/GaAs(100) interface of the sample coated with 200eV ion energy at 300 °C. The carbon signal of the a-C:H overlayer and the peak associated with the interfacial layer are well discernible. The method of background subtraction is indicated [26].

mechanisms: mechanical interlocking, physisorption, chemisorption, chemical bonding, and atomic intermixing (diffusion). Mechanical interlocking plays an important role in the adhesion of films on ceramics and polymers. In our case, where the substrates are highly polished and the interfaces are comparatively smooth, mechanical interlocking can be ruled out. Physisorption contributes only small adhesion forces due to the small Van der Waals forces involved. Chemisorption provides stronger bonds and accounts for adhesion energies up to a few eV/atom. The dominant mechanism in the adhesion of a-C:H films on Si, Ge and GaAs substrates is chemical bonding. Due to the formation of a SiC like compound on Si substrates we expect an adhesion energy of about 1eV/atom. There is no doubt that the formation of the chemical bonds is achieved by the impingement of energetic $C_mH_n^+$ ions onto the substrates. It is precisely this mechanism which is responsible for the formation of metastable compounds in cases where no carbide phases exist in the equilibrium phase diagram, such as for a-C:H on Ge(100) substrates.

Our photoemission results have clearly indicated the formation of carbide compounds on Si and GaAs substrates. Unfortunately, we are not able to determine the stoichiometry of these compounds due to their amorphous microstructure. Therefore, it is not clear whether only binary phases like SiC and AsC_3 are formed or whether ternary compounds such as Si_xC_{1-x}:H, Ge_xC_{1-x}:H and As_xC_{3-x}:H are involved. The amorphous microstructure of the interface phases, however, is more likely indicative of the ternary than the binary compounds.

Atomic interdiffusion in the interfacial layer is certainly another factor that contributes to the adhesion of a-C:H layers on Si, Ge and GaAs substrates. In fact, adhesion is often the sum of different mechanisms such as interdiffusion and chemical bonding, which are interrelated because the formation of compounds in the interfacial layer necessitates the diffusion of atoms across the interface. Atomic interdiffusion in the interfacial layer can be greatly influenced by the use of higher ion energies either for surface cleaning or deposition of the a-C:H films.

There are factors that can deteriorate adhesion. Such is the case for the GaAs substrates. The precipitates of metallic Ga and the disruption of Ga-As bonds at the interface are deleterious to the adhesion of the overlayer. In comparison with the Si substrate we therefore conclude that chemical bonding and atomic interdiffusion are necessary but not sufficient conditions for strong adhesion properties of a-C:H overlayers on semiconductor substrates.

5 Conclusion

It was the aim of the present parer to demonstrate the potential of in situ electron spectroscopy measurements for the investigation of carbon based films in the form of a-C:H or related materials such as polymeric coatings. The main advantages of photoelectron spectroscopy in this field can be summarized as follows:

VB spectroscopy performed with UPS (or XPS) is primarily an analytical technique in order to characterize the film material itself (bulk properties). VB spectra can be recorded in relatively short measuring time. Small changes in the atomic and electronic structure of a-C:H such as a transformation towards a more graphitic modification in an annealing experiment or the growth of polymeric films can be detected.

The analysis of core level binding energies is especially useful in order to study chemical reactions at an interface. A careful analysis of the core line shape may yield information on the homogeneity of the phase under investigation. The cleanliness of substrates and coatings can be checked by wide scan XPS core level spectroscopy.

Acknowledgments

Financial support of the Swiss Bundesamt für Bildung und Wissenschaft, Bundesamt für Energiewirtschaft, as well as the Swiss National Science Foundation is gratefully acknowledged.

References

[1] In M. Cardona and L. Ley, editors, *Photoemission in Solids I and II*, Springer, (1985).

[2] In D. Briggs and M. P. Seah, editors, *Practical Surface Analysis*, Wiley and Sons, (1983).

[3] P. Oelhafen. In E. Lüscher, G. Fritsch, and G. Jacucci, editors, *Amorphous and Liquid Materials*, page 333, Martinus Nijhoff Publishers, Dordrecht, Boston, Lancaster, (1987).

[4] G. Indlekofer and P. Oelhafen. In M. Borrisov, N. Kirov, and A. Vavrek, editors, *Disordered Systems and New Materials*, page 707, World Scientific Publishing Co., Singapore, (1989).

[5] J. M. Imer, F. Patthey, B. Dardel, W. D. Schneider, Y. Baer, Y. Petroff, and A. Zettl. *Phys. Rev. Letters* **62**, 336 (1989).

[6] J. Osterwalder, T. Greber, S. Hüfner, and L. Schlapbach. *Phys. Rev. Letters* **64**, 2683 (1990).

[7] M. P. Seah and W. A. Dench. *Surface and Interface Analysis* **1**, 2 (1979).

[8] P. Oelhafen and D. Ugolini. In P. Koidl and P. Oelhafen, editors, *Amorphous Hydrogenated Carbon Films*, page 267, Les Editions de Physique, Les Ulis (France), (1987).

[9] H. R. Kaufman. *J. Vac. Sci. and Technol.* **15**, 272 (1978).

[10] S. Schelz, J. Eitle, R. Steiner, and P. Oelhafen. (1990). Contribution to ICSFS-5 conference.

[11] R. Zehringer, H. Künzli, P. Oelhafen, and C. Hollenstein. *J. Nucl. Mat.* (1990). in press.

[12] A. Bianconi, S. B. M. Hagström, and Bachrach R. Z. *Phys. Rev. B* **16**, 5543 (1977).

[13] F. R. McFeely, S. P. Kowlczyk, L. Ley, R. G. Cavall, R. A. Pollak, and D. A. Shirley. *Phys. Rev. B* **9**, 5268 (1974).

[14] F. J. Himpsel, J. F. van der Veen, and D. E. Eastman. *Phys. Rev. B* **22**, 1967 (1980).

[15] F. C. Chalkin. *Proc. Royal Soc. A* **194**, 42 (1948).

[16] Chr. Beyreuther, R. Hierl, and G. Wiech. *Ber. Bunsen ges. Phys. Chem.* **79**, 1081 (1977).

[17] G. Wiech and E. Zöpf. *Electronic Density of States*, chapter , page 335. Natl. Bur. Stand. Publ. 323, US GPO, Washington D. C., (1969).

[18] J. Fink, T. Müller-Heinzerling, J. Pfluger, A. Bubenzer, P. Koidl, and G. Greceling. *Solid State Commun.* **47**, 887 (1983).

[19] P. Oelhafen, J.L. Freeouf, J.M.E. Harper, and J.J. Cuomo. *Thin Solid Films* **120**, 231 (1984).

[20] C. Weissmantel, K. Bewilogua, D. Dietrich, H. J. Erler, H. J. Hinneberg, S. Klose, W. Nowick, and G. Reisse. *Thin Solid Films* **72**, 19 (1980).

[21] D. Ugolini, M. H. Tuilier, J. Eitle, J.Q. Wang, and P. Oelhafen. *Appl. Phys. A* (1990). in press.

[22] P. Oelhafen and D. Ugolini. In P. Koidl and P. Oelhafen, editors, *Amorphous Hydrogenated Carbon Films*, page 137, Les Editions de Physique, Les Ulis (France), (1987).

[23] D. Ugolini, P. Oelhafen, and M. Wittmer. In P. Koidl and P. Oelhafen, editors, *Amorphous Hydrogenated Carbon Films*, page 287, Les Editions de Physique, Les Ulis (France), (1987).

[24] D. Ugolini, P. Oelhafen, and M. Wittmer. In P. Koidl and P. Oelhafen, editors, *Amorphous Hydrogenated Carbon Films*, page 297, Les Editions de Physique, Les Ulis (France), (1987).

[25] D. Ugolini, J. Eitle, P. Oelhafen, and M. Wittmer. *Appl. Phys. A* **48**, 549 (1989).

[26] M. Wittmer, D. Ugolini, J. Eitle, and P. Oelhafen. *Appl. Phys. A.* **48**, 559 (1989).

[27] M. Wittmer, D. Ugolini, and P. Oelhafen. *J. Electrochem. Soc.* (1989). accepted for publication.

[28] C. D. Wagner, W. M. Riggs, L. E. Davis, J. F. Moulder, and G. E. Mullenberg. Eden Prairie, Perkin Elmer Corporation, (1979).

[29] L. Ley, R. A. Pollak, F. R. McFeely, S. D. Kowalczyk, and D. A. Shirley. *Phys. Rev. B* **9**, 600 (1974).

[30] W. Ranke and K. Jacobi. *Prog. Surf. Sci.* **10**, 1 (1981). and references therein.

[31] F. A. Shunk. *Constitution of Binary Alloys.* Mc Graw-Hill, New York, (1985).

[32] M. G. Crescenzi, P. Picozzi, S. Santucci, C. Battistoni, and G. Mattogo. *Solid State Commun.* **51**, 811 (1983).

[33] M. G. Mason. *Phys. Rev. B* **27**, 748 (1983).

[34] D. Ugolini, J. Eitle, and P. Oelhafen. *Vacuum* (1990). in press.

THE EFFECT OF DEPOSITION CONDITIONS ON THE OPTICAL AND TRIBOLOGICAL PROPERTIES OF ANNEALED DIAMOND-LIKE CARBON FILMS

A. Grill, V. Patel and B.S. Meyerson

IBM Research Division
T.J.Watson Research Center, Yorktown Heights, N.Y. 10598

INTRODUCTION

Diamond-like carbon, or DLC films, are metastable amorphous materials, containing a mixture of sp^2, sp^3 and even sp^1 coordinated carbon atoms in a disordered network. Since first deposited by Aisenberg and Chabot in 1971 [1], DLC films have been prepared by a variety of methods including DC or RF plasma assisted CVD, sputtering, or ion beam deposition, from a variety of carbon bearing, solid or gaseous source materials [2-5].

The metastable structure of DLC films most likely originates from the thermal and pressure spikes produced by impinging energetic species at the surface [6], characteristic of all DLC deposition processes. These films are characterized by extreme hardness, measured to be in the range 3000-9000 kg/mm^2 [2], a generally low friction coefficient, between 0.01[7] and 0.28[8], as well as very high internal stresses [7, 9-11]. The films typically have high optical transparency over a wide spectral range[4], high electrical resistivity, and chemical inertness to both acids and alkalis.

Depending upon the precursor materials and preparation technique used for deposition, DLC films contain a significant amount of hydrogen, whose concentration has been shown to vary between less than 10%[12] up to 50%[13]. The hydrogen content critically determines film structure, e.g. the ratio between the carbon atoms in the different coordinations, and thus the films properties. Hydrogen content is also key to obtaining a wide optical gap (E_{opt}) and high electrical resistivity, as it passivates the dangling bonds in the amorphous structure [3]. The properties of DLC films are known to depend strongly upon both the deposition method and the specific deposition conditions[4]. Although the properties of DLC films tend to be independent of the exact hydrocarbon precursor[5] employed, this is not strictly true (e.g.[14]).

Deposition is generally done at relatively low substrate temperatures, $T_d < 300$ °C, because it was found[15] that for substrate temperatures above 325 °C the films became graphitic in nature ($E_{opt} \to 0$ eV, $\sigma \to 10^2\ \Omega^{-1}cm^{-1}$) and softened significantly. Thermal stability has been previously examined for some DLC films. Dischler et al[16] reported changes in the 2900 cm^{-1} IR absorption peak, with the change occurring subsequent to a 300 °C anneal. Kanazawa and Ebihara[17] found a large decrease in the optical gap of a-C:H (DLC) even after annealing at 200 °C. Couderc and Catherine[14] reported complete removal of hydrogen from DLC films deposited at 0.5 torr after annealing at 590 °C, while hydrogen was still present in films deposited at 0.1 torr after annealing them at 800 °C. Gonzalez-Hernandez et al[18] reported that no changes occur in DLC films as a result of annealing up to 390 °C, but the IR carbon-hydrogen peaks disappeared

Diamond and Diamond-Like Films and Coatings
Edited by R.E. Clausing *et al.*, Plenum Press, New York, 1991

completely after annealing at 590 °C. Suffice it to say that no consensus exists, nor must there necessarily be one, regarding the thermal stability of DLC films.

The high hardness and chemical inertness of the DLC films make them good candidates for wear-resistant protective coatings on metals, optical or electronic components, and in particular for applications where layers of order of 100 nm thickness or less are required. Many applications are found where DLC films are used at about room temperature, but one may readily envision applications requiring higher usage temperatures, where the thermal stability of DLC films is therefore important. Due to the variability of the reported behavior of DLC films as a function of the specifics of the deposition system and chemistry, it is important to characterize a specific set films, and extract general trends from their behavior. In addition, we are not aware of a previous report on the effect of annealing on the wear behavior of DLC films, an important consideration if they are to be employed for wear protection above their deposition temperature.

This paper presents a study of the thermal stability of DLC films deposited by RF PECVD from acetylene and the effect of annealing on wear properties. Room temperature properties of these films have been reported previously[4, 11, 19].

II. EXPERIMENTAL

DLC films were prepared for the present study by the rf plasma decomposition of acetylene (C_2H_2). The plasma was sustained in a parallel plate geometry by a capacitive discharge at 13.56 MHz as described in detail elsewhere[19]. Si(100) substrates were mounted on the cathode, whose bias was controlled during deposition at a fixed value between -80 and -200 V using an rf choke isolated DC power supply. In the present system, the negative substrate bias could not be increased above 100 V at the typical deposition power of 15 W, corresponding to a power density of 0.1 $W.cm^{-2}$. Therefore, depositions were performed at 23 W (0.15 $W.cm^{-2}$) and biases of 80 and 150 V, and at 30 W (0.2 $W.cm^{-2}$) and biases 80 to 200 V. Deposition was performed at substrate temperatures in the range of 100 to 250 °C. Following deposition, the DLC layers were annealed under vacuum for four hours at 390 °C or 440 °C, or for three hours at 490 °C or 590 °C.

The index of refraction of the DLC films was measured by ellipsometry at λ = 6238 Å. The stress in the DLC films was determined employing a profilometer to measure the curvature produced by 500 nm thick films on 0.25 mm thick Si substrate. Infrared absorption was measured between 1000 and 4000 cm^{-1} using a Fourier transform infrared spectrometer, or FTIR, at a resolution of 4 cm^{-1} and signal averaging over 500 scans. The absorption through an uncoated Si wafer was taken as background. A base-correction was then performed on the obtained spectra, using the software of the FTIR apparatus. The hydrogen content of the DLC films was determined by the forward recoil elastic scattering method, or FRES, using a 2.3 MeV ^4He beam and fitting the calculated energy distribution of the recoiled atoms to the experimental data[20], by adjusting the hydrogen concentration in the calculations.

The wear resistance and friction coefficients of thin DLC films were investigated using a specially designed pin-on-disk type tribotester described in detail elsewhere[21, 22]. The Si wafers coated with DLC films were mounted on a turntable, rotating at a speed of 45 rpm, using stainless steel riders mounted on a high quality stereo tone arms immobilized so as to travel only in vertical plane. Stainless steel (440C) ball bearings with a diameter of 0.794 cm were cemented at the end of small steel rods to serve as riders, which were then dragged across the surface to be tested. A load of 0.088N (9 g), corresponding to an initial Hertzian stress[23] of 215 MPa was used to measure the sliding wear of the films. The wear test was performed by running the apparatus, for a number of rotations, under a constant load, and measuring the resulting wear tracks with the profilometer. Strain gauges mounted on a rider arm enabled the measurement of both the static and dynamic coefficient of friction between the surface of the rider and that of the DLC film. The wear tests were performed at zero humidity in a nitrogen purged dry box.

III. RESULTS and DISCUSSION

A. Index of Refraction and Hydrogen Content

The index of refraction of the films deposited at a substrate bias of -80 V and 0.1 W. cm^{-2} was found to be insensitive to deposition temperature, and was measured as n = 1.9 ± 0.2[24]. However, as shown in Figure 1, the index of refraction of DLC films increases with deposition power, at constant bias, and decreases with increasing bias, at constant power. Figure 2 presents the index of refraction as a function of the annealing temperature. As can be seen, the index of refraction remains nearly constant with annealing temperatures up to 390 °C, independent of deposition temperature and substrate bias. Annealing at temperatures above 390 °C results in an increase of the index of refraction of the films deposited at temperatures below 100 °C, while the index of refraction starts changing only at temperatures above 440 °C, for the films deposited at 250 °C. However, after annealing above 440 °C, the indexes of refraction of the film deposited at 100 °C and -80 V and of the film deposited at the larger negative bias of -150 V increase much faster than the index of refraction of the film deposited at 250 °C and -80 V.

Figure 3 shows the hydrogen concentration in DLC films deposited at a substrate bias of -80 V and three different substrate temperatures. The values are accurate to ± 2 %. The data shown for the film deposited at 250 °C is identical to the data obtained for the film deposited at same temperature and -150 V. It appears therefore that the change of the substrate bias from -80 to -150 V does not affect the amount of hydrogen incorporated in the film, although it affects the value of the index of refraction. As can be seen in Figure 3, the hydrogen concentration in the films is insensitive to annealing at 390 °C and starts to decrease only after annealing above 440 °C, reaching essentially identical values of hydrogen content in all films, independent on their deposition conditions. Even after annealing at 590 °C for 3 hours, DLC films retain approximately 22% hydrogen.

Figure 4 presents the correlation between the index of refraction and hydrogen content, for films deposited at differing temperatures and substrate bias. For a given set of deposition conditions, the index of refraction increases with decreasing hydrogen content, as was reported

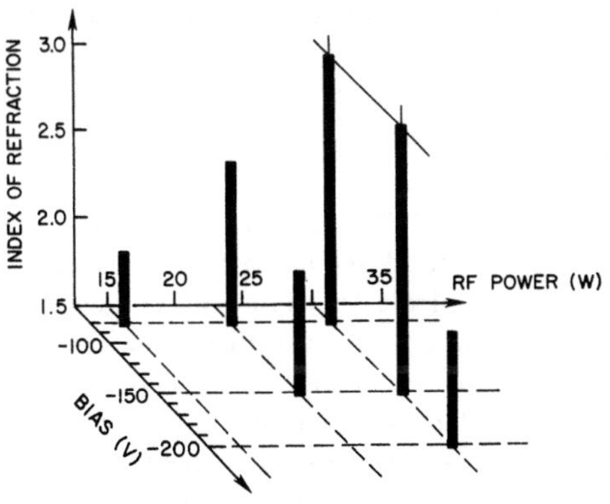

Figure. 1. Index of refraction vs RF power and bias.

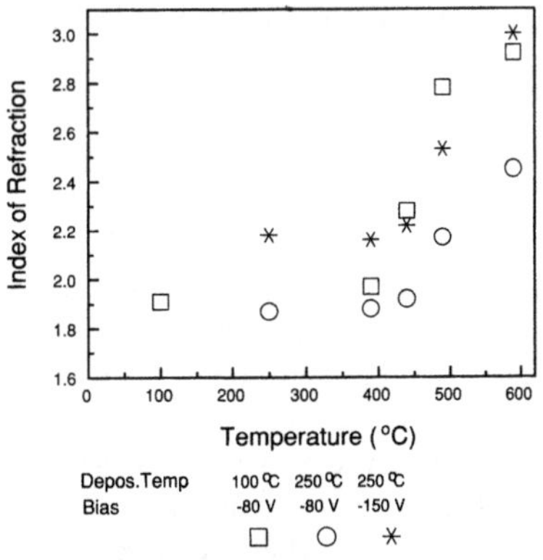

Figure 2. Index of refraction vs annealing temperature.

previously by other authors[25]. However, for the same hydrogen content, the initial value of index of refraction and the values obtained after annealing are a function of the deposition conditions. The behavior of the index of refraction of the film deposited at 250 °C and -150 V is almost identical to that of the film deposited at 100 °C and -80 V. These results indicate that DLC films deposited under different conditions of the present study most probably have different distributions of sp^2 and sp^3 hybridized carbon.

B. FTIR

Figure 5 presents the IR absorbance spectra of DLC films deposited under three different sets of conditions (100 °C, -80 V; 250 °C, -80 V and 250 °C, -150 V), as a function of the annealing temperature. Only those portions of the spectra relevant to the DLC structure are shown. These include the broad absorption band centered at about 2900 cm^{-1} and the absorption peaks between 1700 and 1300 cm^{-1}. The identification of the absorption peaks is detailed in Table I according to [26, 27]. The spectra of the as-deposited films are similar for all deposition conditions. The shape of the peak at 2800-3000 cm^{-1}, corresponds to a superposition of stretching vibrations of C-H with hydrogen bonded to sp^3 carbon[26, 27]. Absorption peaks at 1450 cm^{-1} and 1370 cm^{-1} correspond, respectively, to the asymmetrical and symmetrical deformation frequencies of $C\text{-}CH_3$[26]. A small absorption peak is observed in some cases at 1700 cm^{-1}. This peak is identified as corresponding to C=O vibrations in the -COOH group[26] and may be caused by oxygen adsorbed on the surface of the films. AES measurement showed that the bulk of the DLC films does not contain oxygen, both before and after annealing.

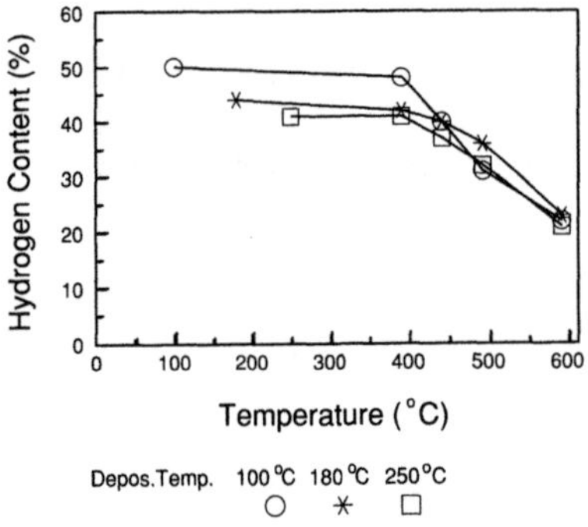

Figure 3. Hydrogen content vs annealing temperature

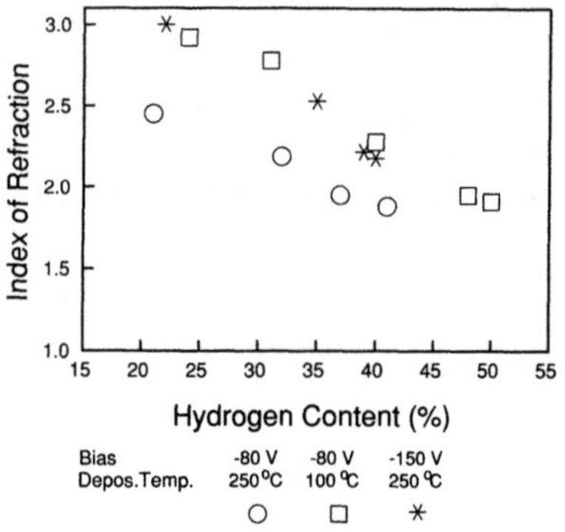

Figure 4. Index of refraction vs hydrogen content.

As can be seen in Figure 5, annealing the films at 390 °C does not cause a significant change in the intensity of the peak at 2800-3100 cm^{-1}. Instead, the appearance of the stretch peak of conjugated aromatic C=C sp^2 carbon at 1600 cm^{-1}, indicates that the major change resulting from the annealing at 390 °C is the reorganization of those carbon atoms in local graphitic coordination, without a significant loss of hydrogen (see Figure 3). This is consistent with the observed variations in the index of refraction after annealing at 390 °C (Figure 2). Only after annealing above 450 °C does the hydrogen peak corresponding to hydrogen bound to sp^3 C decrease, while the peak of the sp^2 CH stretch mode appears at 3045 cm^{-1}. This is accompanied by an increase of the C=C peak at 1600 cm^{-1} and a decrease in the peaks of $C\text{-}CH_3$ at 1450 cm^{-1} and 1370 cm^{-1}.

Table I. IR absorption frequencies in Diamond-Like Carbon.

Line	Wavenumber (cm^{-1})	Assignment	Type
a	3045	sp^2 CH	stretching
b	2960	sp^3 CH_3 asymm.	stretching
c	2920	sp^3 CH_2	stretching
d	2875	sp^3 CH_3 symm.	stretching
e	1600	sp^2 C=C conjugated	stretching
f	1450	C-CH_3 asymm.	deformation
g	1370	C-CH_3 symm.	deformation

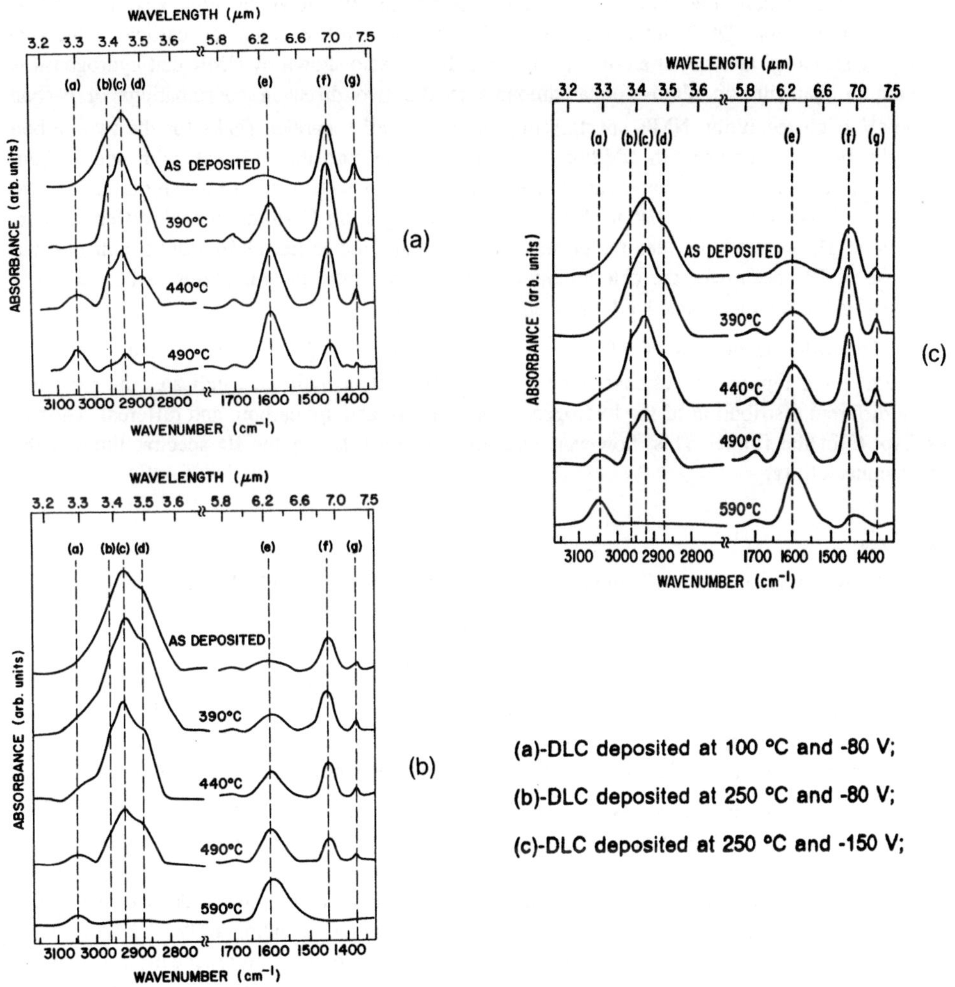

Figure 5. IR absorbance vs. annealing temperature.

The comparison of the IR spectra also show that:

1. the spectra of DLC films deposited at 250 °C, at biases of -80 and -150 V (Figure 5 (b) and (c)) are essentially identical, reflecting the fact that both films contain the same amount of hydrogen.
2. the spectrum of the film deposited at 100 °C, after annealing at 490 °C (Figure 5 (a)) is very similar to the spectra of the film deposited at 250 °C after annealing at 590 °C (Figure 5 (b)).
3. The IR spectra of the films annealed at 590 °C, or at lower temperature for the film deposited at 100 °C, show only a very small peak at 3045 cm^{-1}, corresponding to sp^2 CH, and the peak at 1600 cm^{-1}, corresponding to sp^2 C=C.

These results indicate that annealing above 590 °C causes the near but not yet complete dehydrogenation of the DLC films, accompanied by an increase in the concentration of sp^2 carbon. The changes taking place in the DLC films during annealing occur at lower annealing temperatures for films deposited at lower temperatures. The FTIR results, together with those for the refractive index presented before, demonstrate that DLC films deposited at high substrate temperatures are stable to higher annealing temperatures.

No absorption peak corresponding to the sp^2 CH stretching mode is observed at 3045 cm^{-1} for the as-deposited films, although it was previously shown by high resolution NMR spectroscopy[11] that DLC films deposited at 250 °C and -80 V contained carbon in the two bonding states sp^2 and sp^3, at a ratio sp^2:sp^3 ≈3:2. It was also shown by NMR that hydrogen was bound in that film in almost equal amounts to the two differing coordinations of carbon atoms[11] High resolution NMR spectroscopy provides well separated peaks for the two carbon coordinations, sp^2 and sp^3. In addition, the ability to measure the ^{13}C peak with and without proton decoupling allows the accurate determination of the ratio of carbon atoms in the two configurations, and of the relative distribution of hydrogen on C atoms in each different coordination. The contrast between the present IR spectra and the earlier NMR results indicates that the use of IR spectra for the determination of the sp^2/sp^3 ratio in amorphous carbon layers is suspect, given the clean, resolved NMR signals obtained.

The differing indexes of refraction seen in films deposited under various bias conditions (Figure 1 and Figure 2), yet containing the same amount of hydrogen, indicates the existence of a different distribution of the hydrogen among the sp^2 and sp^3 carbon, and different ratios of sp^2/sp^3 C in the films. This, however, was also not reflected in the IR spectra, limiting the techniques utility.

C. Stress

Stress values in the DLC films were calculated using the formula[28]:

$$\sigma_f = \frac{E t_s^2}{6(1-\nu) t_f R}$$

where E is Young's modulus of the substrate, ν is Poisson's ratio, t_s , t_f are the thickness of the substrate and film respectively and R is the radius of the curvature caused by the stresses in the film. The value E/(1-ν)=229 GPa was used for Si(111) wafers according to Brantley[29].

Figure 6 presents the stresses in DLC films deposited at a substrate temperature of 180 °C, in the as-deposited state and after annealing. The stresses are compressive and the values are accurate to ± 0.1 GPa. As can be seen, films deposited at higher bias have a slightly higher stress in the as deposited condition, but there is no evidence of the strong dependence of the stresses on bias as reported previously[3]. Stress decreases after annealing at 390 °C. Annealing at 440 °C reduces the stress in the film deposited at 15 W and -80 V to practically zero, within the measurement error, indicating near complete stress relaxation of the films. However, stresses

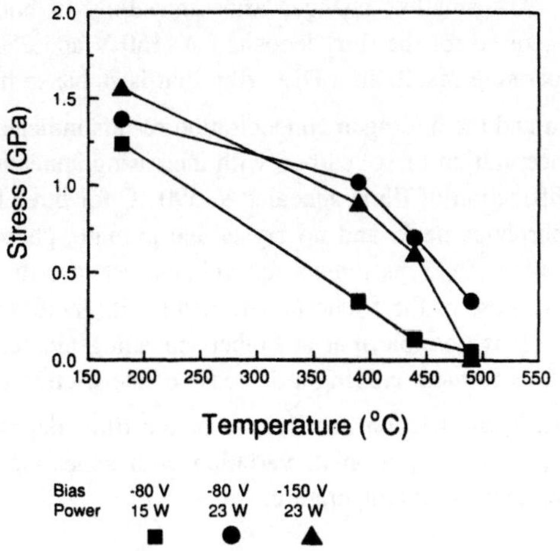

Figure 6. Stresses in DLC vs annealing temperature.

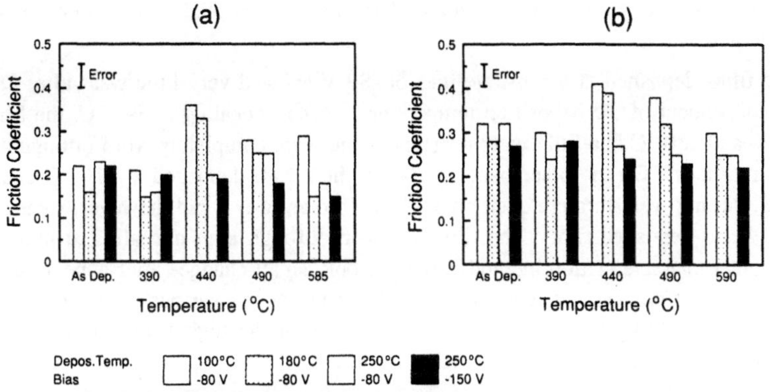

Figure 7. Friction coefficients vs. annealing temperature: (a) static friction coefficients; (b) dynamic friction coefficients.

in films deposited at the higher power of 23 W (even at the lower bias of -80 V) are relieved only after annealing at temperatures above 490 °C.

The results thus indicate that the power density during deposition has a stronger effect on the stresses in the films and their relaxation than the substrate bias. As the hydrogen concentration in the films after annealing at 400 ° C is practically independent on the deposition conditions (Figure 3), it appears that the hydrogen content is not a controlling factor of the stresses in the annealed DLC films.

D. Friction Coefficient

Figure 7 presents the static and dynamic friction coefficients of DLC films deposited under different conditions as a function of annealing temperature. The values of the static friction coefficients of the as-deposited DLC are 0.18±0.03, while the dynamic friction coefficient is 0.30±0.03. The friction coefficients of the films deposited at the lower bias of -80 V remain almost constant, or decrease somewhat, after annealing at 390 °C, reaching a maximum after annealing at 440 °C. The smallest changes after annealing, in both the static and dynamic friction coefficients, occurred for the film deposited at -150 V and 250 °C, indicating again that deposition at this temperature results in a DLC film that is stable to higher temperatures.

The FTIR spectra and the hydrogen concentration results indicate the loss of hydrogen and the increase in the concentration of sp^2 carbon with increasing annealing temperature. However, grazing angle X-ray diffraction of films annealed at 590 °C for three hours showed only a very shallow and broad amorphous peak, and no crystalline graphite phase was observed. This indicates that after annealing the films remain amorphous, yet lose their "diamond-like" quality. This may explain the increase of the friction coefficients with increasing annealing temperature, at least up to 440 °C. It is possible that at higher annealing temperatures a small amount of nanocrystalline graphite is formed, causing a decrease of the friction coefficient.

As seen in Figure 7, the friction coefficients of the films deposited at the higher bias of -150 V and at 250 °C show only a small variation with annealing temperature, slightly decreasing with increasing annealing temperature.

E. Wear

The wear characteristics of annealed films is presented in Figure 8, where the wear is expressed, on a logarithmic scale, in track depth per 10^4 rotations on the pin-on-disk apparatus. The bars extending over the whole height of the box indicate that the respective DLC films were completely worn-through after a few thousand rotations. The other values were obtained after 260,000 rotations.

The films deposited at a substrate bias of -80 V showed very little wear after 260,000 rotations, independent of the deposition temperature. After annealing at 390 °C, the films deposited at 100 and 180 °C had little wear resistance, and were completely worn through after a few thousand rotations. On the other hand, a 1000Å thick film deposited at 250 °C and annealed to 440 °C, did not wear-through even after 260,000 rotations. It did, however, wear at a higher rate than the as-deposited film. The film was worn-through after annealing at 490 °C. Thus, while the other characterization methods revealed only small changes after annealing at 390 °C with almost no differences between the films deposited at the various temperatures and same bias, wear testing showed that only the film deposited at the higher temperature is still wear resistant after annealing for four hours at 390 or 440 °C.

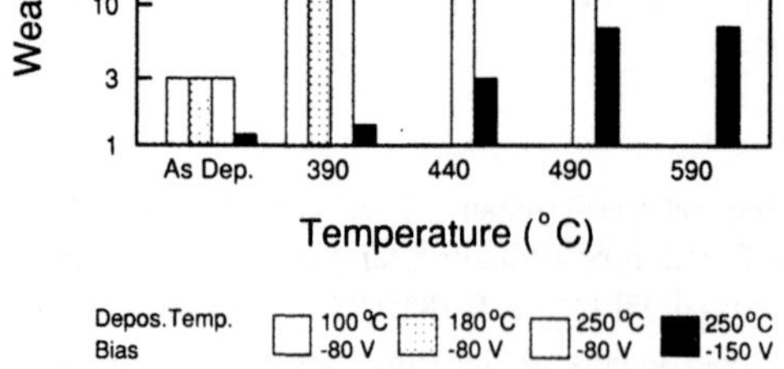

Figure. 8. Wear of DLC films vs. annealing temperature.

Few papers report on the effect of the deposition temperature on the properties of DLC films. It was for example reported[2, 15] that the optical gap decreases with increasing substrate temperature, with a rapid decrease above 250 °C. As asserted by Weissmantel et al[10] in 1982, the thermal and pressure spikes produced on the surface by the impact of energetic ions are necessary to obtain diamond-like properties in amorphous carbon films. In the present case, using a substrate bias of -80 V, inelastic gas phase collisions are expected to reduce the average ion energy to about 10 eV[2]. These energies appear to be sufficient to produce DLC qualities in the deposited films, but it is expected that they are insufficient to control the growth of the films. The deposition is most probably controlled by the interaction between the radicals produced in the plasma and the surface. This in turn may be affected by the coverage of the surface with atomic hydrogen, as in the case of a-Si:H deposition[30]. Higher deposition temperatures apparently reduce the hydrogen coverage of the surface which then affects the growth rate. However it appears that the surface radical interactions also affects the structure of the deposited film.

The wear of DLC films deposited at 250 °C and -150 V was negligible after 260,000 rotations, both in the as-deposited state and after annealing at 390 °C. The wear rate increased after annealing at higher temperatures, but it remained low even after annealing at 590 °C for three hours (see Figure 8). The results show that the increase of the negative surface bias from 80 to 150 V causes significant improvement in the wear resistance of the films, making films less sensitive to annealing temperature. Though the index of refraction (Figure 2), the hydrogen content (Figure 3) and the IR spectra (Figure 5 (c)) of the film deposited at the higher bias all change with increasing annealing temperature, indicating modifications in the structure of the film, these modifications have little effect on the friction coefficients and the wear behavior of the films. On the other hand, the comparison of the mentioned physical properties of the films deposited at the higher bias with those of the films deposited at the lower bias does not indicate that one has to expect a different wear behavior of the DLC films. The observed behavior of the friction coefficients and of the wear rates indicates that the films deposited at the higher bias have significantly different initial and final local structure. As the films deposited at same temperature but different biases contain the same amount of hydrogen, it seems again that the hydrogen is bound differently in the films, and that the films may have significantly different ratios of sp^2 and sp^3 carbon.

The effect of the bias on DLC properties has been investigated before by different authors[31-33], and the bias has been found to affect the structure or hardness of the films. However, there appears to be no systematic study of the effect of the bias on the structure and bonding (e.g. sp^2/sp^3) in the films. To what extent this happens has to be determined by a method more sensitive to structure than FTIR, or even Raman spectroscopy. High resolution NMR, with the proton decoupling option[11], may eventually be able to elucidate the difference in the structure and its relation to wear behavior between DLC films deposited under different conditions.

IV. CONCLUSIONS

Annealing in vacuum up to 390 °C of DLC films deposited by RF decomposition of acetylene causes a reduction in their internal stresses, but has only slight effects on their optical properties or hydrogen content. Annealing at higher temperatures results in increases in the index of refraction, with larger changes found for films deposited at lower temperatures or higher biases. The index of refraction is dependent on the hydrogen content of the films but also on the deposition conditions themselves at a fixed hydrogen content.

After annealing at 390 °C, films deposited at 15 W and -80 V, and at temperatures of 100 °C and 180 °C provide no wear resistance. Films deposited at 250 °C are still wear resistant after annealing at 440 °C. Films deposited at 23 W, -150 V and 250 °C remain wear resistant even after annealing at 590 °C.

Although IR absorption indicates similar structures for the films deposited under different conditions, the wear behavior of the as deposited and annealed films is strongly dependent on the deposition conditions, reflecting differences in the structures of the films.

REFERENCES

1 S.Aisenberg and R.Chabot, J.Appl.Phys. **42**, 2953 (1971).
2 J.C.Angus, P.Koidl, and S.Domitz, in *Plasma Deposited Thin Films*, edited by J.Mort and F.Jansen, (CRC Press Inc., Boca Raton, FL. U.S.A., 1986), p. 89.
3 Hsia-chu Tsai and D.B.Bogy, J.Vac.Sci.Technol. **A5**, 3287 (1987).
4 A.Grill, B.S.Meyerson, and V.Patel, in *Diamond Optics, Proc.SPIE*, edited by Albert Feldman and Sandor Holly, (SPIE, Bellingham, WA, 1989), Vol. 969, p. 52.
5 John C.Angus and Cliff C.Hayman, Science **241**, 913 (1988).
6 C.Weissmantel, in *Thin Films From Free Atoms and Molecules*, edited by K.J.Klabunde, (Acad.Press, New York, 1985), Vol. Part 2C, p. 153.
7 K.Enke, H.Dimigen, and H.Hubsch, Appl.Phys.Lett. **36**, 291 (1980).
8 L.Holland and S.M.Osja, Thin Solid Films **58**, 107 (1979).
9 J.Zelez, J.Vac.Sci.Technol. **A1**, 305 (1983).
10 C.Weissmantel, K.Bewilogua, K.Breuer, D.Dietrich, U.Ebersbach, H.J.Erler, B.Rau, and G.Reisse, Thin Solid Films **96**, 31 (1982).
11 A.Grill, B.Meyerson, V.Patel, J.A.Reimer, and M.A.Petrich, J.Appl.Phys. **61(8)**, 2874 (1987).
12 P.V.Koeppe, V.J.Kapoor, M.J.Mirtich, B.A.Banks, and D.A.Gulino, J.Vac.Sci.Technol. **A3**, 2327 (1985).
13 M.J.Mirtich, D.M.Swec, and J.C.Angus, Thin Solid Films **131**, 245 (1985).
14 P.Couderc and Y.Catherine, Thin Solid Films **146**, 93 (1987).
15 B.Meyerson and F.Smith, J.Non Cryst.Solids **35/36**, 435 (1980).
16 B.Dischler, A.Bubenzer, and P.Koidl, Solid State Commun. **48**, 105 (1983).
17 Seiji Kanazawa and Kenji Ebihara, in *Proc. 9th Int.Symp. on Plasma Chemistry*, edited by R.d'Agostino, (IUPAC, Sept.,1989), p. 1473.
18 J.Gonzalez-Hernandez, B.S.Chao, and D.A.Pawlik, J.Vac.Sci.Technol. **A7**, 2332 (1989).
19 A.Grill, B.Meyerson, and V.Patel, J.Mater.Res. **3**, 214 (1988).
20 A.Turos and O.Meyer, Nucl.Instr.and Metods in Phys.Res. **B4**, 92 (1984).
21 A.Grill, B.S.Meyerson, and V.V.Patel, in *IBM Report RC 13117*, (IBM Research Division, Yorktown, NY, 1987).
22 A.Grill, B.S.Meyerson, and V.Patel, IBM J.Res.Dev. **(in press)**, (1990).
23 J.Hamrock and D.Brewe, J.Lubrication Tech. **105**, 171 (1983).
24 A.Grill, V.Patel, and B.S.Meyerson, J.Mat.Res. **(in press)**, (1990).
25 T.J.Moravec and J.C.Lee, J.Vac.Sci.Technol. **20**, 338 (1982).
26 L.J.Bellamy, *The Infra-red Spectra of Complex Molecules*, (Chapman and Hall, London, 1975), p. 14, 38, 74, 184.
27 B.Dischler, R.E.Sah, P.Koidl, W.Fluhr, and A.Wokaun, in *Proc.7th Int.Symp. on Plasma Chemistry*, edited by C.J.Timmermans, (IUPAC, Eindhoven, July,1985), p. 45.
28 E.Kobeda and E.A.Irene, J.Vac.Sci.Technol. **B6**, 574 (1988).
29 W.A.Brantley, J.Appl.Phys. **44**, 534 (1973).
30 C.C.Tsai, J.C.Knights, G.Chang, and B.Wacker, J.Appl.Phys. **59**, 2998 (1986).
31 S.Kaplan, F.Jansen, and M.Machonkin, Appl.Phys.Lett. **47**, 750 (1985).
32 R.Kato, K.Hiraoka, and K.Masuhara, in *Proc. 9th Int.Symp. on Plasma Chemistry*, edited by R.d'Agostino, (IUPAC, Sept.,1989), p. 1456.
33 P.Wood, T.Wydeven, and O.Tsuji, in *Proc. 9th Int.Symp. on Plasma Chemistry*, edited by R.d'Agostino, (IUPAC, Sept.,1989), p. 1485.

BAND EDGES AND GAP STATES BY OPTICAL ABSORPTION AND ELECTRON SPIN RESONANCE IN AMORPHOUS CARBON (a-C) AND HYDROGENATED AMORPHOUS CARBON (a-C:H)

D. Das Gupta, F. Demichelis, C.F. Pirri, R. Spagnolo,* A. Tagliaferro

Dip. Fisica - Politecnico Torino - Torino - ITALY
* IEN Galileo Ferraris - Torino - Italy

INTRODUCTION

There have recently been many experimental and theoretical studies of amorphous carbon (a-C) and hydrogenated amorphous carbon (a-C:H)[1,2]. In a-C, the carbon atoms form both fourfold (sp^3) sites (diamond-like) and threefold (sp^2) sites (graphite-like).

In the sp^3 configuration each electron of the carbon atoms forms a σ bond with an adjacent atom; in the sp^2 configuration only three of four electrons form σ bonds, the fourth electron lies in a $p2(p\pi)$ orbital normal to the σ bonding plane: The π states, being more weakly bonded than σ states, are situated in the valence and conduction bands near the Fermi level E_F and, when due to defects, in the mid gap.

The electron structure of a-C and a-C:H has been widely investigated through calculations on a number of model structures containing different proportions of sp^2 and sp^3 sites. In contrast, there have been few experimental studies of the density of π states especially around E_F.

The density of states $N(E)$ in the region of the pseudogap can be measured by both optical absorption and the Electron Spin Resonance (ESR). Previous measurements (1) of the optical absorption coefficient α of amorphous carbon have been extended to low energy and they don't appear to show the characteristic Urbach tail found in other amorphous semiconductors, in which

Diamond and Diamond-Like Films and Coatings
Edited by R.E. Clausing *et al.*, Plenum Press, New York, 1991

α decreases exponentially into the gap. Hence, the literal extension of many of the usual concepts of amorphous semiconductors must be made with some care.

In tetrahedral amorphous semiconductors, the spin density is found to be related to the optical absorption deep in the gap. An excess absorption, above the exponential tail, is found to be proportional to the spin density[3]. This absorption arise from deep state - band state transitions.

Since the details of the density of states distribution in carbon films is still considered an open question, we present and discuss experimental data obtained by measurement of the optical absorption coefficient and the spin density.

In the following the paper deals with the deposition technique of a-C and a-C:H, the absorption coefficient measurements over a wide range of energy and ESR measurements. The results from our laboratory are also compared with those reported in literature in order to contribute to a better understanding of the electron structure of the films. Finally, the above quoted procedure for extracting the gap state density from the absorption curves is examined and seen not to apply to a-C and a-C:H.

EXPERIMENTAL

The methods of deposition for carbon films have been thoroughly discussed in literature[1]. Generally a-C is deposited by Chemical Vapor Deposition (CVD) or by Sputtering, while a-C:H is obtained by RF glow discharge of gases such as methane, acetylene, ethane and ethylene.

Our films were prepared in an RF magnetron sputtering system by sputtering of a graphite cathode, 20 cm in diameter, in an Argon atmosphere (Ar flow kept at 80 sccm), with a pressure of 3.3 Pa. Substrates were placed on a plate at 2.5 cm from the target. For a-C films the RF power was varied from 300 to 500 W. For a-C:H films the RF power was kept at 300 W and a mixture of Ar and H_2 was used, varying the H_2 flow from 5 to 35 sccm.

The film properties were analyzed by optical, PDS and ESR spectroscopy. Transmittance and reflectance measurements in the 0,2 - 3 μm wavelength range were recorded on a Perkin Elmer UV-visible-NIR Lambda 9 spectrophotometer. Transmittance measurements were taken at normal incidence in the double beam mode. Reflectance measurements from both the air and the substrate sides of the film were made with an angle of incidence of 7 degrees and compared with the reflectance of a standard mirror calibrated at near normal incidence at the National Bureau of Standards (NBS).

The extraction of the optical constants, that is the index of refraction n, the extinction coefficient k and then the absorption coefficient $\alpha = 4\pi k/\lambda$ followed the procedure reported in reference[4].

The absorption coefficient α is a fundamental quantity for the characterization of materials, but its measurement in this film by conventional optical measurements (transmittance T and reflectance R) is limited to high absorbance A = 1 - R - T. To measure extremely low optical absorbance we used the Photothermal Deflection Spectroscopy (PDS). PDS consists in measuring heat associated with light absorption and more specifically with the deflection of light beam caused by refractive index gradient in a heated medium. Specially sensitive techniques to detect the small quantities of heat deposited by light in weakly absorbing materials are needed[5]. Values of α as low as 0.1 cm^{-1} can be reliably measured using this technique.

Unpaired spins associated with half-filled states around E_F are detected by ESR. We used a Varian EPR 109 spectrometer. The g factor and the density of spins were determined by calibration against a reference material (DPPH) and a Varian pitch.

RESULTS

Typical values α for a-C and a-C:H measured by many workers (see ref.1 for a review) vary between 10^2 and 10^5 cm^{-1}. They show a very broad absorption edge.

We show in Fig.1 the absorption coefficient α vs energy for some of our samples of a-C and a-C:H, measured by the optical and PDS complementary techniques. There is a remarkable difference in shape from the Urbach edge observed in other amorphous semiconductors, such as a-SiC:H, as no sharp absorption edge is observed.

To obtain a possible value of the pseudogap in such films we have applied the Tauc's relation[6], written as

$$[\alpha(E)]^{1/2} = B^{1/2}(E - E_g) \tag{1}$$

where B is a constant and E_g is the optical gap. The resulting values of E_g are consistent with values reported in literature and are reported in Table I.

Fig. 2 shows examples of the spin resonance peaks of a-C and a-C:H (derivative signals). In Table I the spin density (Ns), the g-values and the peak-to-peak amplitude (δH_{pp}) of all our samples are reported. For a-C:H a spin density of $\sim 10^{16}$ cm^{-3} was reported by Jansen et al[7] and of $10^{18} cm^{-3}$ by Gambino et al[8] and by Miller et al[9]. Our values are at least one order of magnitude higher and, contrary to what reported by Gambino et al.[8], no saturation effects were observed, except for the sample CH05.

If one looks at Fig. 1, it is not possible to see any excess absorption, so that the procedure quoted above for amorphous tetrahedral semiconductors can not be applied. In the following paragraphs, we will try to explain the peculiar behaviour of a-C and a-C:H, in order to identify the origin of the states contributing to the different properties.

OPTICAL ABSORPTION

The parameters which characterize a material from an optical point of view are refractive index (n), extinction coefficient (k), absorption coefficient (α) and the real and imaginary parts (ϵ_1 and ϵ_2) of the dielectric constant. However, the fundamental quantity correlated to the structure of the material is ϵ_2, from which ϵ_1 can be derived through the Kramer-Koenig relationship

$$\epsilon_1(E) = 1 + \frac{1}{\pi}\int_0^{\infty} \frac{\epsilon_2(E')}{E'^2 - E^2} dE' \tag{2}$$

n and k by means of the couple of equations

$$\epsilon_1 = n^2 - k^2 \tag{3}$$

$$\epsilon_2 = 2 \cdot n \cdot k$$

and α using

$$\alpha = \frac{2\pi}{hcn} E\,\epsilon_2(E) \tag{4}$$

As far as amorphous materials are considered, the relevant expression that gives ϵ_2 is[10]

$$\epsilon_2(E) = C \cdot Q^2 \int_{E_f}^{E_f+E} N_f(Z-E) \cdot N_e(Z)\,dZ \tag{5}$$

where E is the photon energy, Q the (constant dipole) matrix element of the transition, N_f (N_e) the density of filled (empty) states, E_f the Fermi level energy and C is a constant.

A different matrix element can be used in the description of the optical transitions, that is the constant momentum matrix element P, related to Q through the following relationship[11]

$$P^2 = \left(\frac{m \cdot E}{\hbar^2}\right)^2 Q^2 \tag{6}$$

where m is the electron mass. While the independence of P on photon energy is adequate to describe optical transitions in crystalline materials[11], an energy independent value of Q is found for a-Si:H[11].

Tauc[6] defined a value for the optical gap by assuming valence and conduction bands to be parabolic near their edges, and by removing the rule that electronic momentum has to be conserved, but mantaining the assumption that P is energy independent. One would then expect Cody's[10] definition of the gap

$$\sqrt{\alpha / E} = B' \cdot (E - E_g') \tag{7}$$

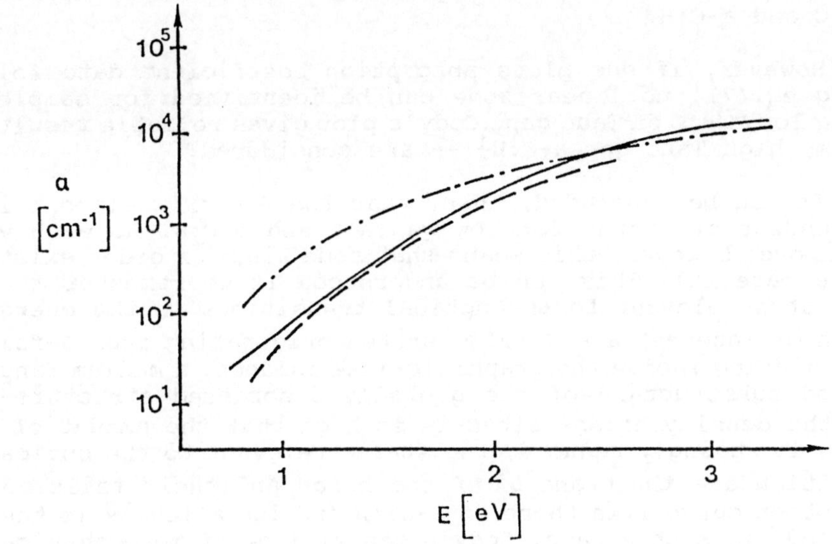

FIGURE 1 - Absorption coefficient vs.energy for samples:
CH05 (-·-·-·), C300 (——), C500 (- - - -).

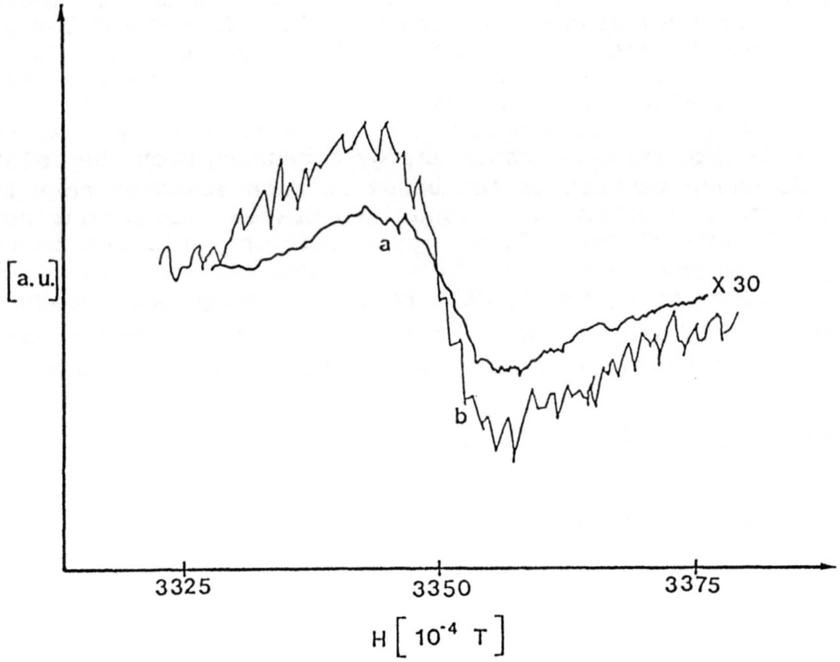

FIGURE 2 - Tipycal ESR derivative signal of a-C:H (line a, CH05) and a-C:H (line b, C500).

that differs from Tauc's only in considering Q as the energy independent matrix element, to give better results when applied to a-C and a-C:H.

However, if one plots absorption coefficient data following eq.(7), no linear zone can be identified for samples having low values of Tauc gap. Cody's plot gives reliable results if some high Tauc gap a-C:H[1,12] are considered.

It can be concluded, then, that the P matrix element is independent of energy for low gap a-C and a-C:H. As we have said above, however, this means that some kinds of order exists in the material. This can be understood if we remember that the states relevant to the optical transitions in the energy region of interest are π and π^* states originating from 6-fold rings sitting inside the graphitic-like islands, a medium range ordered substructure of the globally disordered structure[2]. When the density of sp^3 sites is so high that the amount of π states is strongly reduced, the states relevant to the optical transition are the σ and σ^* of the bands and their tails. An absorption curve like those of a-Si:H and its alloys[10] is then expected. In such a case, Cody's gap will be of more physical significance and an excess absorption over an exponential tail can be seen (see Fig.1a of Ref. 11).

As noted above, another assumption made by Tauc is that the bands are parabolic near their edges. This assumption applies well to crystalline semiconductors and with some limitations to tetrahedral amorphous semiconductors[10]. The sharp edge observed in the log α vs E curve in a-Si:H is due to the presence of an exponential tail of the bands, originated by the structural disorder of the material. In a-C and low gap a-C:H, however, no sharp edge is observed and this can be connected with the fact that the states sitting over and below the "optical edge" of the bands have the same origin. In fact, all of them are originated by the 6-fold rings of the graphitic-like islands, their energy depending upon the island size. No sharp feature of the bands is then expected near the transition from extended to localized states, but a continous curve. We assume, as a first approximation, this continuous curves to be gaussians, simmetric with respect to Fermi level. In the following E_π and σ_π will be the gaussian parameters of the "valence" band, $E_{\pi*}$ and σ_π those of the "conduction" band. Substituting in eq.(4) and by using eq.(3), we can write:

$$\alpha(E) = K' \cdot \sigma_\pi \cdot \exp\left\{-\left(\frac{2 \cdot E_{\pi*} - E}{2 \cdot \sigma_\pi}\right)^2\right\} \cdot \frac{erf(E/2\sigma_\pi)}{E} \qquad (8)$$

Tauc's function then becomes:

$$\sqrt{\alpha(E) \cdot E} = K'' \cdot \exp\left\{-\frac{1}{2} \cdot \left(\frac{2 \cdot E_{\pi*} - E}{2 \cdot \sigma_\pi}\right)^2\right\} \cdot \sqrt{erf(E/2\sigma_\pi)} \qquad (9)$$

If we look at eq.(9), we can see that the exponential is a gaussian centered at $2 \cdot E_{\pi*}$ and $2\sigma_\pi$ wide, modified by the error function in the low energy region. Around the point of inflexion of this curve, a linear zone is expected, the slope of the curve in that point being the Tauc slope $B^{1/2}$ of the fitting curve. A straight line having that slope and passing through the point of inflexion will intercept the energy axis in a value which represents the optical energy gap E_g'.

To check the validity of the approach, we fit the low absorption region by means of eq.(8), so obtaining $E_{\pi *}$ and σ_π values, which determine the shape of α. We, then, use these values to apply eq.(9), relative to the high absorption region and find E_g'. E_g' values turn out to be in good agreement with experimentally determined E_g values. This indicates that, as supposed above, there is no change in the shape of the bands near their "optical edges".

ESR STUDY

As noted above, we observed spin density values larger than some literature ones, and too high to derive from the gaussian tails of π bands. The presence of unpaired electron can arise from odd aromatic ring with states laying energetically near the Fermi level[1]. It is possible to introduce such states in our optical simulation (see figs. 3 and 4) without affecting the absorption coefficient vs energy curves as in amorphous tetrahedral semiconductors with parabolic and exponential bands.

In fact in amorphous tetrahedral semiconductors the absorption coefficient in the very low energy range (under the Urbach tail) arises from transition dangling bonds - parabolic conduction band. The integral of the excess absorption is related to the spin density measured by ESR[12]. So ESR and PDS can give the same information on neutral defects. In a-C and a-C:H we do not expect such a relation. In table I, we can also observe a change of two orders of magnitude in Ns for hydrogenated and unhydrogenated samples. Probably hydrogen saturates part of the spin states or, decreasing the coordination number, releases internal stresses responsible for unpaired spins.

Information about electron delocalization and relaxation behaviour can be obtained studing saturation effects. In a previous work[7] saturation was found in a-C:H films with 10^{18} spin cm^{-3} at about 10 mW. The authors attributed it to the delocalization of electrons expected in a-C. In fact localized electrons are characterized by a strong spin-lattice coupling, and thermal relaxation at room temperature should be sufficient to prevent saturation. On our samples saturation was found, at about 5 mW, only in the film with the lowest spin density; all the other measurements gave no saturation up to 30 mW. This result can be explained considering the high spin density ($> 10^{19}$ spin cm^{-3}): at this concentrations, in fact, spin-spin coupling could prevent saturation.

DENSITY OF STATES (DOS) NEAR AND BELOW BAND EDGES

As a conclusion of the above discussions, we will try to sketch a picture of the DOS of sp^2 rich a-C and a-C:H near and below the band edges. We expect the bands to be nearly gaussian in shape (broad solid lines peaks in Fig.3) on both sides of

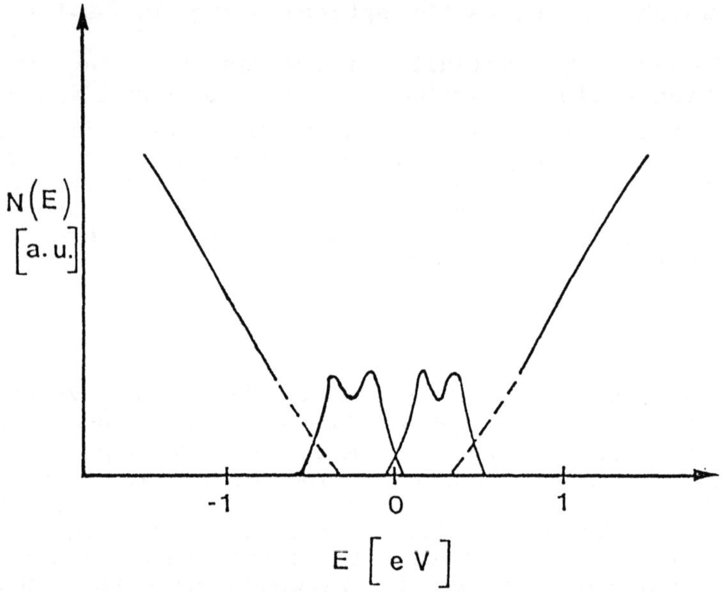

FIGURE 3 - Qualitative behaviour of density of states in sp^2 rich amorphous carbon. The tails are due to gaussian π states bands, the fine structure around the Fermi level is due to 5-fold and 7-fold aromatic rings.
(The figure is not to scale.)

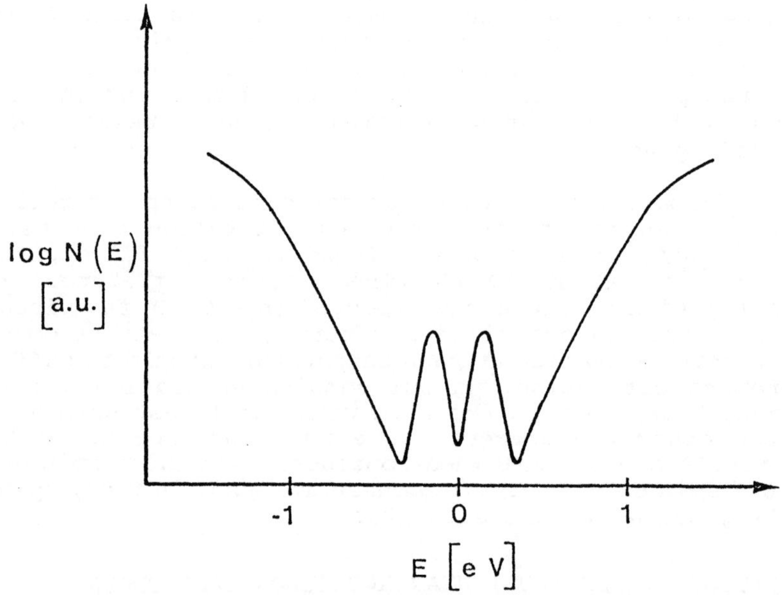

FIGURE 4 - Global density of states of amorphous carbon in logarithmic scale.

TABLE I

Optical and ESR properties of a-C and a-C:H films.

Sample	E_g (eV)	N_s (cm^{-3})	g	ΔH_{pp} (10^{-4} T)
C300	1.46	$6\ 10^{20}$	-	6
C350	0.97	$1\ 10^{21}$	2.0035	7
C400	0.98	$1\ 10^{21}$	2.0029	10
C440	1.24	$8\ 10^{20}$	2.0030	7
C500	0.90	$9\ 10^{20}$	2.0023	8.5
CH05	1.45	$1\ 10^{19}$	-	8.5
CH12.5	1.37	$1\ 10^{20}$	2.0026	7.5
CH20	1.24	$8\ 10^{19}$	2.0026	5.5
CH27.5	1.21	$1\ 10^{20}$	2.0023	7
CH35	1.18	$2\ 10^{20}$	2.0023	4.5

Cxyz = carbon film prepared at xyz watts RF power.

CHxyz = a-C:H films prepared at xyz sccm H_2 flow rate.

the "optical edges". However, we will not expect peaks to reach Fermi level, since only 6-fold rings pertaining to infinite size islands can produce π and π^* states having no difference in energy[2]. However, the fall-off of the bands will be strictly dependent on the island size distribution, so we limited ourselves to draw a qualitative behaviour on Fig. 3 (the dashed lines falling to zero without reaching Fermi level).

As deduced from ESR measurements analysis, two features related to 5 and 7 fold rings are placed near Fermi level in Fig. 3 (not in scale). It has to be noted the double peak of each feature, expected because in 5 and 7 fold rings two levels very near in energy are present (one with one electron and one empty)[2].

The global DOS, on a logarithmic scale, is reported in Fig. 4.

We would expect, as said above, this sketch to lose its validity when the number of sp^3 sites is large. In such a case, the gaussian bands are covered by the exponential tails of the σ and σ^* bands, wich will conserve their parabolic character near their edges. In such a case most of the concepts used in a-Si:H can be applied, while this is not the case when π and π^* states control the electronic properties of a-C and a-C:H.

CONCLUSION

From the discussion of optical absorption and ESR data, of a-C and a-C:H samples, we deduced a DOS shape. The main features of this DOS are:

a) no sharp features are present near the band edges
b) two peaks near Fermi level are originated by odd-fold rings

While the first feature is responsible for the broad absorption edge of a-C and sp^2 rich a-C:H, the second takes care of the ESR signal.

The applicability of Tauc assumptions to a-C and a-C:H has been verified. The origin of the validity of the constant momentum matrix element approach has been considered. It has been argued that it is due to the quasi-ordered nature of the graphitic like islands.

REFERENCE

1. J.C. Angus, P. Koidl, S. Domitz - Carbon thin films, in "Plasma Deposited Thin Films", J. Mort and F. Jansen Eds, CRC Press Inc, Boca Raton (FL, USA), Chap. 4 (1986)
2. J. Robertson - Adv. Phys. 35, 317 (1986)
3. W.B. Jackson and N.M. Amer - Phys. Rev. B 25, 5559 (1982)
4. F. Demichelis, G. Kaniadakis, A. Tagliaferro and E. Tresso - Appl. Optics 26, 1737 (1987)
5. N.M. Amer and W.B. Jackson - Optical properties of defect states in a-Si:H, in Semiconductors and Semimetals Vol. 21B, J.I. Pankove Ed., Academic Press, Chap. 3 (1984)
6. J. Tauc - in Amorphous Semiconductors, M.H. Brodsky Ed, Plenum, New York (USA), (1973)

7. F. Jansen, M. Machonkin, S. Kaplan and S. Hark - J. Vac. Sci. Techn. A3, 605 (1985)
8. R.J. Gambino and J.A. Thompson - Solid State Comm. 34, 15 (1970)
9. D.J. Miller and D.R. McKenzie - Thin Solid Films 108, 257 (1983)
10. G. Cody - The optical absorption edge in a-Si:H, *in* Semiconductors and Semimetals Vol. 21B, J.I. Pankove Ed., Academic Press, Chap. 2 (1984)
11. W.B. Jackson, S.M. Kelso, C.C. Tsai, J.W. Allen, S.J. Oh - Phys. Rev. B 31, 5187 (1985)
12. D.A. Anderson - Phil. Mag. 35, 17 (1977)

CHARACTERIZATION OF THE COMPOSITION, ELECTRONIC STRUCTURE AND HARDNESS OF CARBON FILMS OBTAINED BY ION IMPLANTATION, LASER ABLATION, AND DUAL ION BEAM SPUTTERING

M. Allouard, J. C. Pivin

C.S.N.S.M., BP 1, 91405 Campus d'Orsay, FRANCE

Diamond-like films were grown by C^+ ion beam implantation at low energies (0.1 to 2.0 keV) on various substrates (Si, SiC, metals). The film properties were compared to those of films prepared either by dual ion beam sputtering (DIBS) of an a-C target and the substrate, magnetron sputtering or laser ablation.

The composition and depth profiles of carbon, oxygen, hydrogen, and substrate elements in the film and at the interface were quantitatively determined by Rutherford and non-Rutherford Backscattering (RBS), Nuclear Reactions Analysis (NRA) and Secondary Ion Mass Spectrometry (SIMS). The deposition efficiency, characterized by the self-sputtering yield of the incident ions, increased only slightly with the incident energy for ion energies of 500 eV and above (Table I), contrary to what is generally postulated. The extent of mixing with the substrate was limited to the ion range (2 to 10 nm) except when the substrate was SiC, where the mixing extended to 30 nm. Similarly, profiles of implanted ^{12}C and ^{13}C evidenced little diffusion of carbon for all ion energies (1 to 2 nm), except for a value of 5 nm for a film implanted at 100 eV that was found to contain hydrogen (Fig. 1 and 2). Special attention was paid to the hydrogen contamination: for all ion implantation and laser ablation specimens, it was much less than in films deposited by DIBS or magnetron sputtering (Table I). Note that the ion implanted and laser ablated films also contained $<5 \times 10^{15}$ O/cm^2, which was segregated at the interface with the substrate.

X-ray Photoelectron Spectroscopy (XPS) and RBS did not permit the establishment of a scale of merit for the diamond-like character of the films. Soft X-ray emission spectroscopy (SXS) was more selective. The most diamond-like film was obtained by laser ablation (C_{laser}) (See Fig. 3 and Table I). The basis of comparison is i) the significant (2.0 eV) difference in the $C(K_\alpha)$ peak position for diamond (CD) compared to graphite (CG) or amorphous carbon and ii) a 20% narrower peak for CD than for CG. The binding energy of valence electrons in the implanted films was more dependent on the energy and the fluence. (Table I)

Ellipsometry also proved to be very useful. From measurements of the polarization of reflected light at more than 20 wavelengths over the range 0.2 to 5 μm and for three different angles of incidence, independent values of the thickness, t, and refractive index (n + ik) were derived. Knowing the areal density of carbon atoms (measured by RBS), we could first establish that the atomic density was $0.95 \pm 0.05 \times 10^{23}$ C/cm^2 for all deposition processes and conditions. The comparison of raw data (n and k variations) showed indisputably that DIBS

Diamond and Diamond-Like Films and Coatings
Edited by R.E. Clausing *et al.*, Plenum Press, New York, 1991

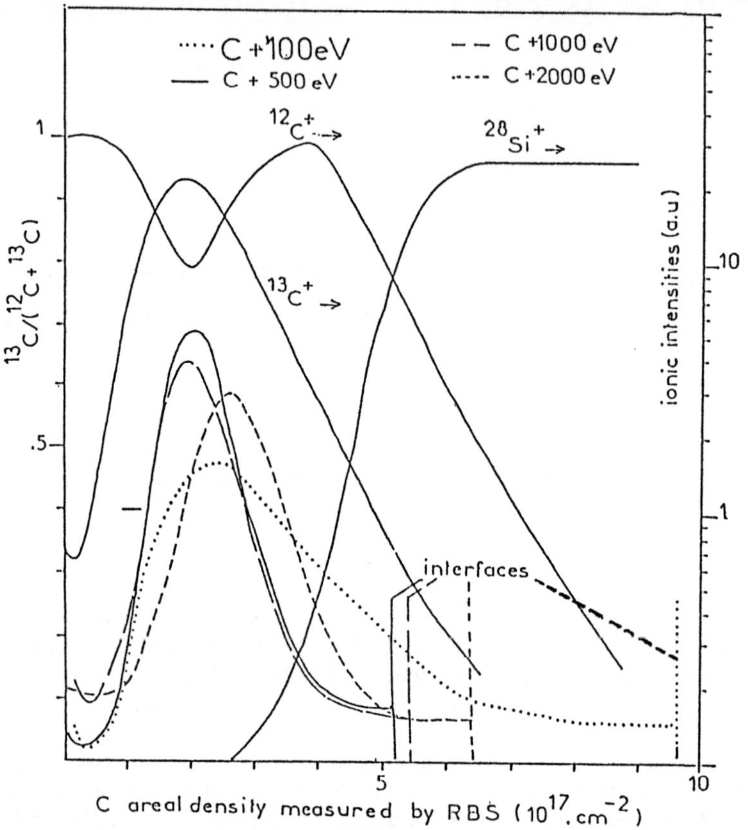

Figure 1. SIMS profiles for C and Si in silicon specimens successively implanted with 5 X 10^{17} $^{12}C/cm^2$, 2 X 10^{17} $^{13}C/cm^2$, and 5 X 10^{17} $^{12}C/cm^2$. Profiles at these fluences were chosen because interdiffusion effects and the contamination with hydrogen are more dependent on ion energy during the preliminary stages of implantation.

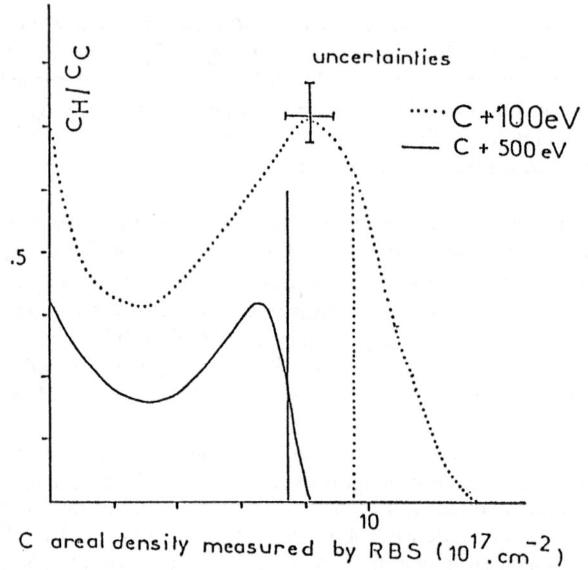

Figure 2. Profiles of hydrogen determined by means of the $^{1}H(^{15}N,\alpha\gamma)^{12}C$ nuclear resonance for 6.38 MeV ^{15}N ions, with a resolution of 2 keV, corresponding to a range of 1 X 10^{16} C atoms/cm^2.

Table I. Film Properties

FILM	COMPOSITION	DEPOSITION EFFICENCY	OPTICAL INDEX				SXS	
	amount of contaminant per C atom	self sputerring	Variation between 1.5 - 5eV n	k	Emax absorption (eV)	Eoptics (eV)	Relative of shift of CK peak E-ED/EG-ED	FWHM CD 8.8eV 0.1 CG 10.3eV 0.2
DIBS 1	0.04H 0.02Ar	---	2.6↘1.8	1.0↘0.8	1.5	0	---	---
DIBS 2	0.07H	---	2.6↘1.9	0.5↗0.8	3.5	0.1	1.0	10.0
MG	0.02 H 0.05F	---	2.2↘1,7	0.8↘0.7	2.0	0	---	---
LASER	0.02H	---	2.6↘2.2	0.5↗0.9	4.2	0.4 0.1	0.12	9.0
C^+ 100eV								
5E17	0.05H	0.05	2.3↘1.5	0.6↗0.9	3.0	0	0.42	10.7
1e19	N. D	0.01	2.1↘1.4	0.9↘0.5	1.5	0	0.75	10.5
C^+ 500eV								
5e17	0.03H	0,5	2.1↘1.6	0.5↗0.7	4.0	0.5 0.1	0.42 0.02	10.2
1e19	0.02H	0.15	2.4↘1.5	0.9↘0.6	1.5	0	0.90 0.05	10.5 0.2
C^+ 1000eV								
5e17	0.02H	0.5	2.4↘1.9	0.5↘0.7	3.5	0.3 0.1	0.42	10.0
1e19	n.d	0.25	2.5↘2.0	0.6↗0.7	3.0	0	---	---
C^+ 2000eV								
5e17	0.02H	0.5	2.5↘2.1	0.5↗0.9	3.2	0.5 0.1	0.42	10.5
1e19	0.02H	0.25	2.6↘2.1	0.3↗1.0	4.8	0.5 0.1	0.52 0.1	10.5 0.2

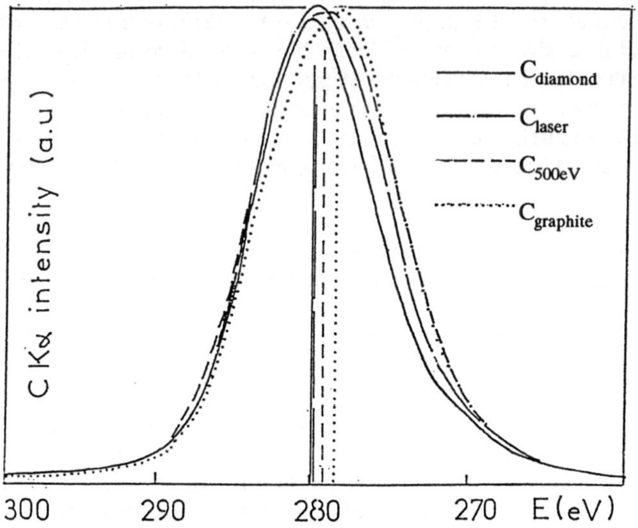

Figure 3. $C(K_\alpha)$ peaks recorded on C_{laser}, films implanted with 500 eV C^+ (C_{500eV}), graphite ($C_{graphite}$) and diamond ($C_{diamond}$) under irradiation with 10 keV electrons in an electron microprobe. No peaks were recorded on the virgin substrates.

and magnetron-sputtered films, as well as films implanted at energies <500 eV, often exhibited dielectric properties which are characteristic of a metallic material (Fig. 4). The most semiconducting films were those deposited by laser ablation or implantation at 2000 eV. Values of the optical gap (E_{optics} in Table I) appear to be very low with respect to published data. However, these data were deduced from the imaginary part of the dielectric constant, ϵ_i, instead of the absorption coefficient, α, i.e. from plots of $\epsilon_i^{0.5} = f(E)$ [Fig. 5] instead of $(\alpha E)^{0.5} = f(E)$ as in most published investigations. The later do not take into account variations of n(1,2). For instance, an optical gap of 1.0 eV was established by means of the second type of Tauc plot on the film formed by laser ablation in this study. This compares to the value of 0.1-0.4 eV determined by the first type of plot (Table I).

One could conclude from these experiments that the most suitable process for preparing diamond-like carbon (DLC) films is laser ablation. These films may also be more adherent to the substrate since a SiC layer with a thickness of 22 nm at the Si-C interface had to be assumed in our fits of the RBS spectra. However, this result could be an artifact. Davanloo (1), who provided this film, determined by Scanning Tunneling Microscopy that this type of film contained inclusions with a diameter of the same order of magnitude as the thickness of the interfacial layer we assumed; this value is much lower than the lateral resolution of RBS, NRA, or even secondary ion microscopy. Particles of 100 nm were also detected at the external surface by classical scanning electron microscopy.

Such a heterogeneity is inherent to the minimal power density and to the associated growth rate for obtaining DLC films, which has been established by several authors [e.g. ref. (1)]. Below a threshold of 5×10^{10} W/cm^2, the films are soft, graphitic, and highly contaminated with hydrogen. On the other hand, the poor properties of DIBS or magnetron-sputtered films are certainly related to their high contamination level. Even the published values of optical gaps, which range between 0.5 and 1.0 eV, could be due to the stabilizing effect of hydrogen on dangling bonds. Such an effect is not specific to DLC films but is general for any amorphous semiconductor. The change in the dielectric properties of the films implanted at 100 eV is attributed in part to a hydrogen gradient and to a higher overall hydrogen concentration than at other energies (Fig. 2). Probably the sputtering yield of hydrogen and of clusters of CH and SiH increases with hydrogen content. Another factor accounting for the gradual loss of semiconducting properties of the films implanted at 500 or 1000 eV is the catalytic effect of a SiC interfacial layer of DLC films during the preliminary stages. However, the implantation energy must be over 1000 eV to maintain a beneficial effect of ballistic cascades at the later stages; simulation by means of a molecular dynamics model would be helpful to analyze the mechanisms. Moreover, since the deposition rate does not decrease drastically with the implantation energy, ion implantation at energies between 1000 and 2000 eV appears to be perfectly suitable for preparing pure films with a high degree of homogeneity on a nanometric scale. Such films are good candidates as subjects for fundamental studies and for selected applications (for instance in the biomedical field where any roughness of porosity of the films is unacceptable).

Preliminary indentation tests at the submicroscopic level, with continuous recording of the indentation resistance as a function of the penetration depth, were also performed. Despite the high resolution of the technique (50 nm) (3), quantitative values of hardness and elastic modulus could not be directly deduced from the curves (Fig. 6). A large number of loading-unloading curves that were taken with incremental values of the maximal depth, are required on each film because of the large amount of elastic recovery that occurs upon unloading. It accounts, in part, for the fact that the sharp diamond behaves much like a flat punch: the load varies proportionately to the depth, d, instead of the expected variation as the square of the depth for a pyramidal indentor. Another factor which must be considered for an indentation of a hard, thin film is that the indentor itself can deform. This deformation is not considered in most analyses. Values of hardness derived by other authors from the on-load pressure have little meaning.

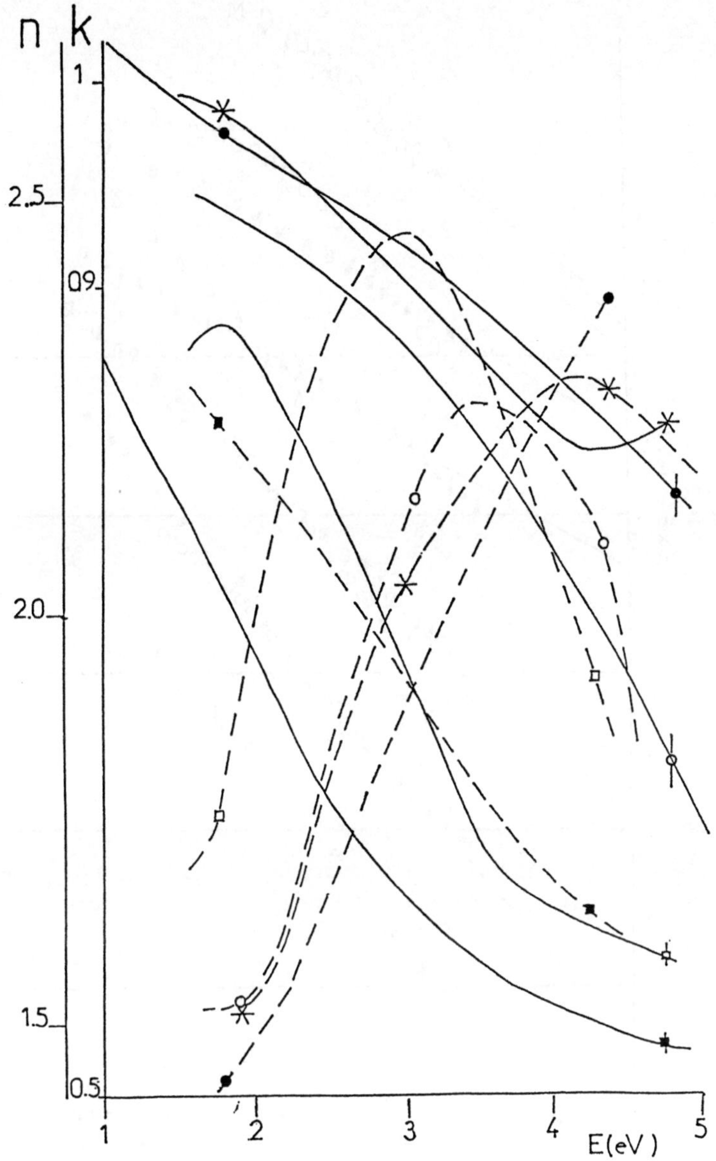

Figure 4. Typical variations of n(—) and k (---) deduced from ellipsometry measurements. For symbol definitions, see the caption of Fig. 5.

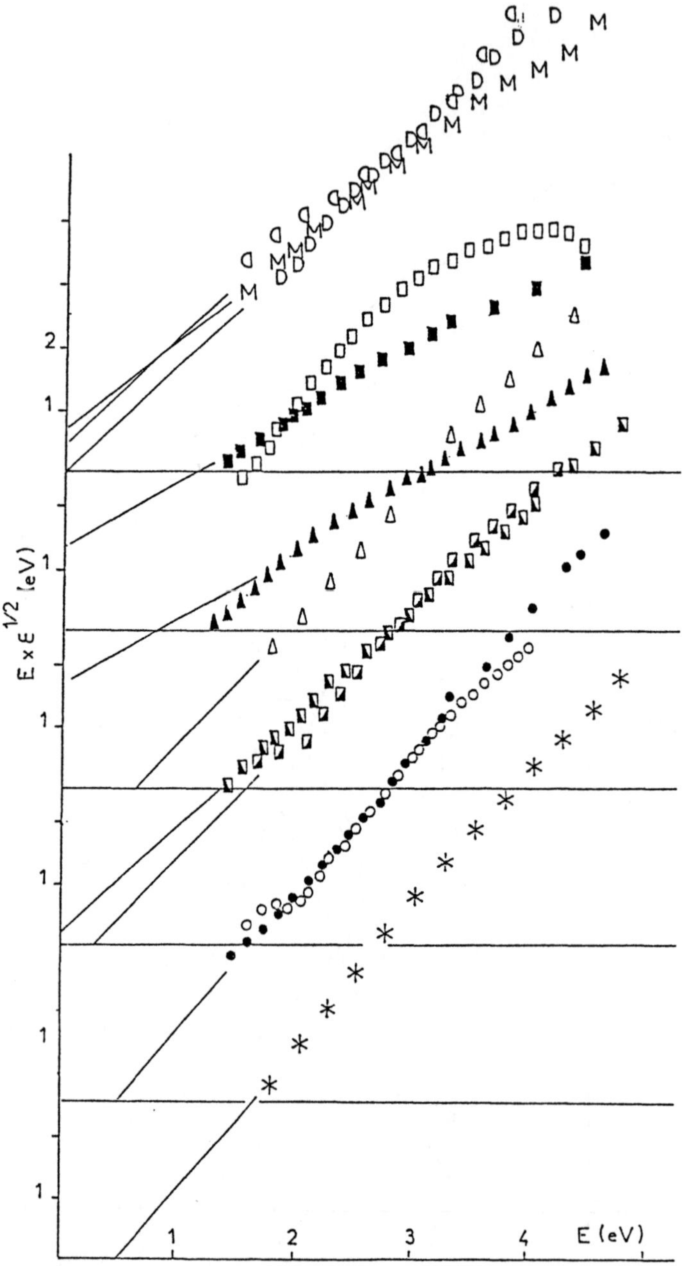

Figure 5. Tauc plots characteristic of optical absorption in an amorphous semiconductor. Symbols are relative to:

DIBS 1	D DIBS 2
□ C^+, 100 eV, 5 X $10^{17}/cm^2$	■ C^+, 100 eV, 1 X $10^{19}/cm^2$
△ C^+, 500 eV, 5 X $10^{17}/cm^2$	△ C^+, 500 eV, 1 X $10^{19}/cm^2$
C^+, 1000 eV, 5 X $10^{17}/cm^2$	C^+, 1000 eV, 1 X $10^{19}/cm^2$
○ C^+, 2000 eV, 5 X $10^{17}/cm^2$	● C^+, 2000 eV, 1 X $10^{19}/cm^2$
* C_{laser}	M Magnetron Sputtering

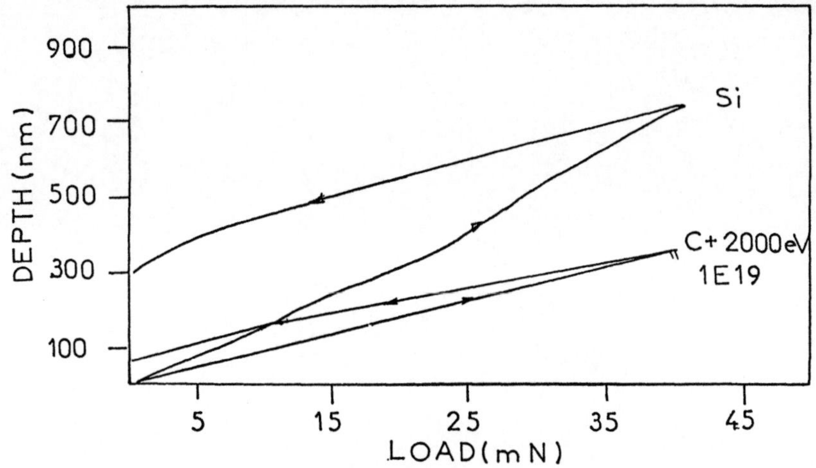

Figure 6. Indentation curves recorded with a trigonal diamond while loading (→) and unloading (←). The plastic work of indentation (area between loading and unloading) is highly reduced when the silicon substrate is covered by a DLC film.

Our preliminary data can be summarized by stating that the on-load pressure was identical to within 5% for all films and are thus characteristic of the films. Indeed, the hardness values did not depend on the film thickness and a steep change of slope was recorded at an indentation depth corresponding exactly to the interface for a film with a thickness of 90 nm. The data in Fig. 6 are for a film with a thickness of 900 nm.

References

1. F. Davanloo, E. M. Juengerman, D. R. Jander, T. J. Lee, C. B. Collins, *J. Appl. Phys.* 67(1990)2081.
2. A. P. Mashel, *J. Mater. Res.*, 4(No. 5)(1989)1358.
3. J. C. Pivin, D. Lebouvier, J. M. Pollack, E. Felder, *J. Phys. D. Appl. Phys.* 22(1989)1443; J. C. Pivin, F. Pons, J. Bernas in "Materials Science by High Fluence Ion Beams," ed. R. Kelly and M. F. da Silva, NATO ASI Series 155, 547; and other references in these texts.
4. SIMS profiles were recorded with the CAMECA IMS 3F microscope of ETCA, 16 bis Av. Prieur de la Cote d'Or, ARCUEIL, FRANCE. Ellipsometry data were obtained with one of the MOSS spectrometers of SOPRA, 68 Rue P. Joigneaux, BOIS COLOMBES, FRANCE.

POST DEPOSITION TREATMENTS AND STABILITY OF DIAMOND-LIKE FILMS

R. Kalish

Solid State Institute and Physics Department
Technion-Israel Institute of Technology
Haifa 32000, Israel

The effects that energy deposition by ions, electrons and photons have on the structure and properties of Diamond-Like Carbon (amorphous hydrogenated carbon) films are reviewed. In all cases the treated films turn electrically conductive and optically opaque; nevertheless it is shown that different physical processes are responsible for the observed changes for the different modes of energy deposition. Results obtained by a variety of experimental techniques which probe the structural and compositional changes are described. Models which fit the experimental data on hydrogen loss and on changes in electrical and optical properties as a result of ion irradiation are proposed.

1. INTRODUCTION

Diamond-like carbon films are, in a sense, the analogy to amorphous hydrogenated silicon, both being an amorphous material based on a group IV element and both containing substantial amounts of hydrogen. However, the basic difference between ${}^{6}C$ and ${}^{14}Si$, namely the possibility of carbon to form both sp^2 and sp^3 bonds, in contrast to Si which binds only in its tetrahedral form, is reflected in many of the properties of a-C:H films. Depending on preparation method, the ratio of sp^3 to sp^2 bonds in the material can vary, and with it, also the properties of the film. Furthermore, like amorphous Si, which contains certain amounts of hydrogen, diamond-like amorphous carbon films (DLF) have H atoms incorporated in the material in varying amounts and their presence much influences the properties of the material. The amount of hydrogen incorporated in the films may vary from a few atomic percent up to $\sim 50\%$. Typical films contain 30-40% of hydrogen. In this respect a-C:H differs from a-Si:H which contains only a few ($\sim$ 5) atomic percent of H, an amount which is sufficient to saturate all dangling bonds in Si. As we shall show below, hydrogen loss from DLF induced by ion bombardment and other energy deposition mechanisms is responsible for the changes in many of the properties of Diamond-like, a-C:H, films subjected to ion implantation.

Energy may be delivered to the near surface region of solids in several ways, including direct heating (in a furnace), irradiation by photons, electrons or ions. As we shall show below, all these processes eventually lead to hydrogen loss from the affected volume and through it to dramatic changes in most physical and chemical properties of the materials.

Diamond and Diamond-Like Films and Coatings
Edited by R.E. Clausing *et al.*, Plenum Press, New York, 1991

Below we first briefly describe the physical mechanisms related to the interaction of ions, electrons and photons with matter, emphasizing the peculiarities of these in relation to carbon-based materials; we then review the data available on the modifications induced by the exposure of DLF to ion bombardment or electron and laser irradiation; we describe a model for ion-induced hydrogen loss in DLF, offer some possible explanation for the structural transformations which the films undergo as a result of energy deposition, and finally mention some potential uses. As the bulk of the available data are for ion-implantation induced damage in DLF we will mainly concentrate on these.

2. BASIC INTERACTIONS

2.1 Interactions with energetic ions

An energetic ion penetrating into a solid will lose its energy through scattering events involving the Coulomb interaction of the ion with the atoms and electrons in the target. This energy loss will determine the final penetration of the projectile into the solid and the amount of disorder that is created in the lattice.

Each collision of the ion with constituents of the target is a complicated many-body event which is usually approximated by assuming that the interaction between the ion and the scatterer can be separated into two components, namely an ion (projectile)-electron (target) interaction and an ion (projectile)-nucleus (target) interaction. The slowing down process in a solid can thus be roughly divided into two dominant mechanisms for energy loss by the incident ions, namely electronic energy loss, involving the interaction between the incident ion and electrons of the host material, and nuclear energy loss, involving the interaction between the incident ions and the nuclei of the host material.

The general way to treat the slowing down of an ion in matter is through the "stopping power" (dE/dx), defined as the energy dE lost by an ion traversing a distance dx, where

$$\frac{dE}{dx} = N \int T d\sigma \tag{1}$$

in which $d\sigma$ denotes an element of the cross sectional area, T is the energy lost by the projectile in a particular collision and N is the atom density of the host material.

In the ion-electron interaction, the scattering angle θ is close to zero and the maximum transfered energy is $T_{max} = 4mE_0/M_1$ in which m and M_1 are the masses of the electron and the ion, respectively. This interaction induces small losses in the energy of the incoming ion as the electrons in the host atoms are excited to higher bound states or to ionization states. These ion-electron interactions do not produce significant deviations in the projectile trajectory. A more detailed description of the electronic energy loss has been given by Lindhard et al. [1], in the commonly called LSS theory which predicts for ion velocities of relevance to ion implantation that the electronic stopping is proportional to the ion velocity, i.e. dE/dx is proportional to $\sqrt{E_1}$ where E_1 is the energy of the projectile.

In contrast, the ion-nucleus interaction is given by the simple Coulomb interaction between two positive (screened) charges which results in both energy loss and significant deviation in the projectile trajectory. The cross-section for this process is given by the simple Rutherford Scattering cross-section. In the ion-nucleus interaction, not only do the colliding ions change directions, but the atoms of the host material are also significantly dislodged from their original positions, giving rise to lattice defects. The deviations in the projectile trajectory result in both a lateral spread and a depth distribution of the implanted species, while the displacements of the host atoms will give rise to lattice damage.
The total stopping power is due to both electronic and nuclear processes

$$\frac{dE}{dx} = \left(\frac{dE}{dx}\right)_e + \left(\frac{dE}{dx}\right)_n \tag{2}$$

At low projectile energies ($E_1 < 15$ keV for carbon ions slowing down in diamond) nuclear stopping is dominant, while at higher energies electronic stopping dominates the energy loss of the moving ion (see Figure 1).

From the energy loss we can obtain the ion range by

$$R = \int \frac{dE}{dE/dx} \tag{3}$$

where the integration limits are from the initial ion energy to zero. Since the energy transferred to the lattice (including both phonon generation and permanent displacement of the host ions) is known one can calculate the energy $E(x)$ of the incoming ions as a function of distance along their trajectory. It is important to remember that only in the energy regime in which nuclear stopping dominates are the host atoms severely dislodged. Electronic stopping will usually not create extensive damage to the host crystal, except in those cases in which electronic excitation and bond breaking is possible, so that some structural rearrangement may take place.

The slowing down processes of an ion moving in an amorphous solid is a statistical process. Thus the locations at which the implants finally come to rest and the damage they create along their paths are also of a statistical nature, and are thus expressed using statistical variables. The ion range R is related to the mean track length of the ion before coming to rest, while the projected range R_p gives the mean penetration depth of the ion relative to the surface. The projected range distribution can therefore be approximated by a Gaussian distribution giving the impurity density $n(x)$ as

$$n(x) = \frac{\phi}{\sigma_p\sqrt{2\pi}} exp\left[-\frac{(x-R_p)^2}{2\sigma_p^2}\right] \tag{4}$$

where x is measured along the direction of the beam, ϕ is the ion dose/cm^2, and σ_p is the standard deviation in the projected range R_p, (also referred to as ion straggling ΔR_p).

As for the damage created in the host material by the traversing ion, the statistical nature of the process is even more pronounced. The elastic (nuclear) collisions which occur between ions and target nuclei may result in significant energy transfer. Provided the recoiling lattice atom receives energy in excess of the so-called "displacement energy" which is about 20 eV for a-C:H it will become permanently displaced in the solid. In many cases the transferred energy is far in excess of the displacement energy and a cascade of secondary and higher generation collisions is initiated in the vicinity of the primary collision event leading to a certain volume around the ion track in which many host atoms are in motion for a short ($\sim 10^{-13}$ sec) time. This is the "thermal spike" which, once quenched, may leave the affected volume in a different (damaged) structural state as compared to that of the host material. This process is illustrated in Figure 2 which shows the results of Monte-Carlo simulations of collision cascades for several 50 keV carbon ions (left) and of a single 1 MeV gold ion (right) slowing down in a-C:H as obtained from the TRIM program [2].

The number of vacancies created per incident ion may be on the order of hundreds or even thousands and is an important characterization parameter of the radiation damage. As the ion slows down, it interacts more strongly with the lattice of the host and dislodges more atoms, thereby creating more vacancies. The damage profile for low dose implantation gives rise to isolated regions of damage within the damage-cascade volume. As the fluence is increased these damaged regions overlap and may eventually coalesce to form a new phase, such as graphite in the case of heavy dose implantation of diamond or diamond-like materials. Some of the damage created by ion-implantation can be annealed, however, it should be clear that if stoichiometric changes have resulted from the ion irradiation the modifications will be irreversible. As will be described below, loss evolution is perhaps the most prominent ion-induced effect in a-C:H.

Electronic processes may also be responsible for radiation damage, in particular in insulators. In analogy to the above described nuclear collision cascade, energetic electrons produced by the passage of an ion through a solid may go on to excite other target electrons. The result

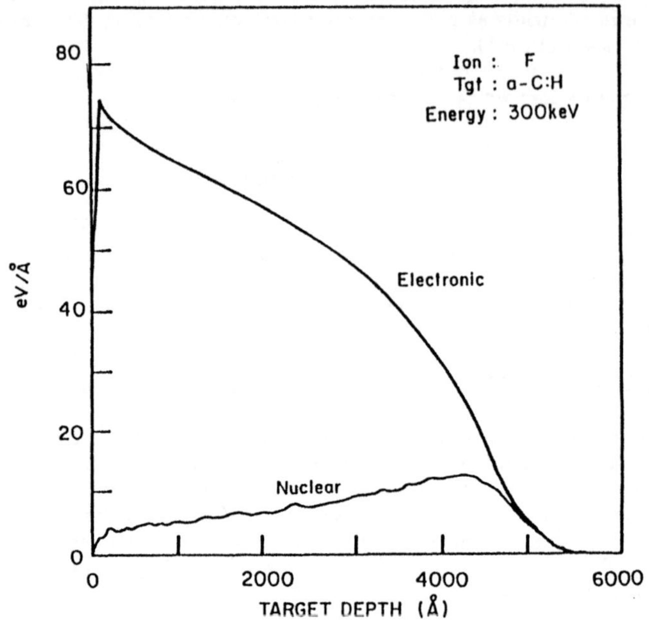

Fig. 1. Energy deposited by electronic and nuclear processes for 300 keV flourine ions in a-C:H, as simulated by computer program TRIM [2] with parameters: density 1.8 g/cm^3, composition 62 % C & 38 % H, displacement energy 20 eV & binding energy 1 eV.

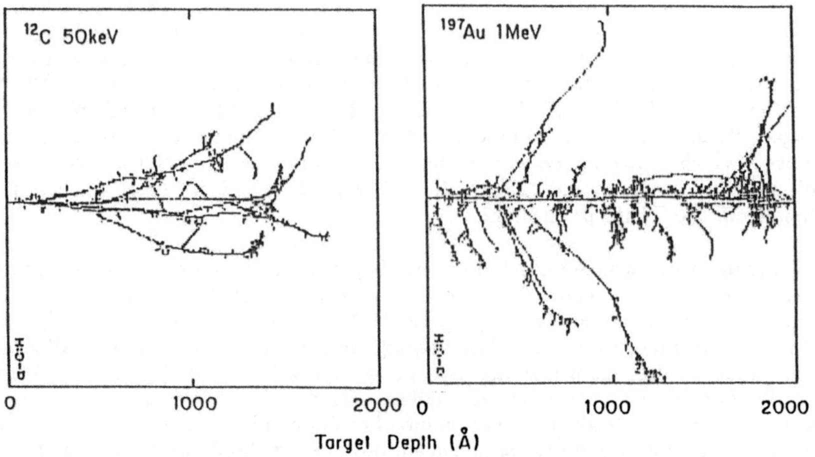

Fig. 2. Left: Collision cascades (dotted line) produced by several 50 keV carbon ions (solid line) impinging on a-C:H Right: Collision cascade for a single 1 MeV gold ion. Simulations by computer program TRIM [2]. Parameters as in Figure 1.

is a dense cascade of secondary electrons, which may also be responsible for the breaking of bonds. Considering carbon based materials, where different forms of bonding coexist, ion irradiation may result in bond restructuring. Simple kinematics, however, precludes the possibility of atomic displacement by electrons with energies in the range of tens or hundreds of electron volts.

2.2. Interactions with Energetic Electrons

When an energetic electron impinges on a solid the processes by which it transfers energy to the solid may again be roughly categorized as electronic or nuclear. Looking at the electronic interactions, the nature of the energy transfer process is very similar to that of electronic stopping of ions in solids; i.e. excitation and ionization of both core and valence target electrons. Secondary electron cascades may also result with typical energies in the range from ten to a hundred electron volts. For electron kinetic energies E_k around 5 keV the amount of energy lost in each scattering process relative to E_k is quite large such that the number of encounters between projectile and target electrons are fewer. Therefore, the statistical fluctuations (i.e. straggling) in range of the incident electrons are greater than in the case of ion implantation. Furthermore, the density and overlap of the secondary electron cascades are expected to be less for electron irradiation, since the energy loss per unit path length is much lower than in ion implantation.

In crystalline materials, a wide variety of defects may result from the above described processes including bond breaking, creation of colour centres and charge trapping.

The effects of electron irradiation in amorphous materials do not appear to have been extensively investigated. However, photoluminescence studies of electron irradiated amorphous hydrogenated silicon have indicated the creation of a thermally annealable metastable defect which quenches the photoluminescence intensity around 1.5 eV [3].

With regard to elastic collisions with target nuclei, kinematic calculations show that electron energies in the kilo-electron volt range are sufficient to produce atomic displacement of hydrogen due to its low mass and relatively small displacement energy. However, the cross-sections for these collisions are expected to be very small ($\sim 10^{-20}$ cm^2) and have been ruled out in explaining defect production in a-Si:H since the measured damage threshold energies for a-Si:H and a-Si:D are about equal, in contradiction to the expectation of a higher threshold for a-Si:D due to the more massive deuterium nucleus [3].

2.3 Interactions with Photons (Laser Irradiation)

The interactions of photons with matter differ basically from those of charged particles in that once a photon has undergone an interaction which may affect the material it has been annihilated (absorbed). Thus a photon beam of original intensity I_0 will gradually diminish in intensity upon penetration into the absorbing medium. The photon beam intensity at depth x ($I(x)$) is determined by the absorption coefficient α of the material to the particular photons, and is given by

$$I(x) = I_0 e^{-\alpha x} \tag{5}$$

The whole physics of the interaction is hidden in α, which may depend on temperature and on the absorbing material. Once the photon has been absorbed its energy will usually end up as heat in the absorbing material through electron-phonon interactions.

Focused laser radiation may interact with matter resulting in a broad range of transient effects. At sufficiently high intensities, photon irradiation may lead to irreversible changes in target properties. In semiconductors the most important factors in determining the dominant photon absorption mechanism are the photon energy, the optical band gap of the target and the laser intensity. If the photon energy exceeds the optical band gap of the target, absorption is primarily due to band to band excitation, i.e. electrons are promoted from the valence to the conduction band. In this case, the *damage threshold* is typically laser pulse energy dependent. If the photon energy is less than the optical band gap then at moderate laser intensities absorption is due mainly to the excitation of impurity or defect states within the gap. However, at sufficiently high intensities, non-linear mechanism such as two-photon absorption may occur and quickly dominate as the radiation intensity is increased. Furthermore, once a significant density of carriers has been excited into the conduction band, free carrier absorption may occur irrespective of photon energy as in the case of optical absorption in metals. In this case the peak power density is critical in determining the damage threshold [4].

Energy is therefore absorbed by the target via photon-electron interactions, however this energy is quickly transferred to the atomic lattice by electron-phonon interactions. These interactions occur on time scales much faster than typical laser pulse durations, which in turn are much shorter than characteristic thermal diffusion times.

Modification by laser irradiation may be due to a number of mechanisms. Firstly, defects may be introduced by direct interactions of photons with the electronic structure as is the case with photosensitive materials. Electromagnetic breakdown may occur if the local field strength is sufficiently high. Local heating at the beam spot may also induce modifications similar to those resulting from thermal annealing. This is of particular importance in metastable materials such as a-C:H since hydrogen effusion is known to commence at relatively low temperatures. Finally, intense thermal gradients in the vicinity of the irradiated region may cause catastrophic damage such as blistering and ablation. Such effects are critically dependent on surface conditions so that discontinuities such as cracks, groves or pores greatly reduce the damage threshold by causing local hot spots.

3. ION BEAM INDUCED CHANGES IN DLF

The changes that occur in a C:H films by ion bombardment have recently been studied by several groups and have been summarized in a review article by Kalish and Adel [5]. Hence there now exists a rather clear picture of the effects that the irradiation has on the material structure and composition. The most relevant parameters in these studies are (in close analogy to the case of ion-implantation in diamond), the ion mass and energy (both of which determine the dE/dx values), the ion fluence and the a-C:H film temperature during implantation.

The most pronounced and clear effects that ion-implantation induces in DLF are a dramatic drop in electrical resistivity and an increase in optical absorption in the visible (darkening of the irradiated region).

Resistivity changes as a result of ion irradiation are spectacular and are therefore the most intensively studied of ion induced effects. A wide variety of ions have been implanted into a-C:H and the resulting changes in conductivity were measured. Figure 3 [6,7] shows the dose dependence of the resistivity of a-C:H films irradiated with C^+ (50 keV), Ar^+ (110 keV) and Xe^+ (270 keV). The ion energies were chosen so that the projected range in each case was about $1000\AA$. The experiments were carried out with the samples held at 100°C so as to avoid blistering which otherwise occurred for high dose noble gas implantations. The room temperature resistivity of the unirradiated films, ρ_0, was typically about $10^7 \Omega cm$.

The dose dependence of the resistivity is qualitatively the same for each ion studied and is amazingly similar to that found for ion-implanted diamond crystal, i.e. above a certain critical dose the resistivity drops rapidly until it reaches an intermediate saturation level which appears as a knee in the curve. With still greater doses a further decrease in ρ is observed.

Ingram and McCormick [8] have measured the DC conductivity dependence on fluence in a similar geometry, but for 6.4 MeV flourine ions and 1 MeV gold ions. Their irradiation parameters were chosen so that both ions deposit roughly the same amount of energy via electronic processes, yet the gold ions deposited some 350 times more energy to nuclear processes than the fluorine ions. Their results shown in Fig. 4 display a similar threshold behavior in the onset of the conductivity increase. However the resistivity drop following this onset differs drastically for the two cases; the flourine irradiation results in a drop in resistivity by some nine orders of magnitude while the resistivity changes for the gold irradiation is much less pronounced, leveling off after a reduction of only two or three orders of magnitude.

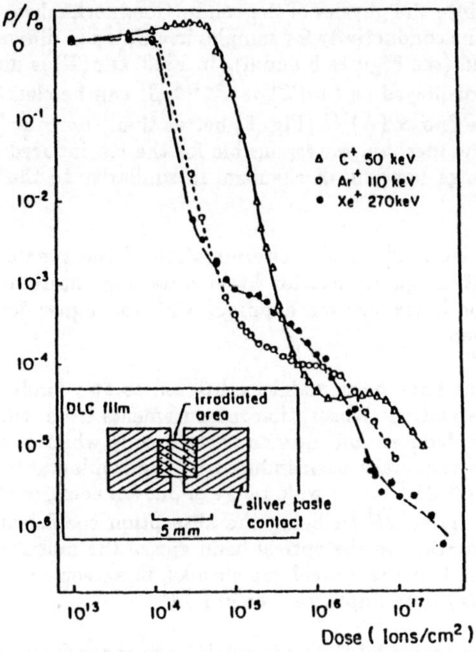

Fig. 3. Ion dose dependence of resistivity for C^+, Ar^+ and Xe^+ irradiation of an a-C:H film. Inset: the irradiation geometry. Taken from [7].

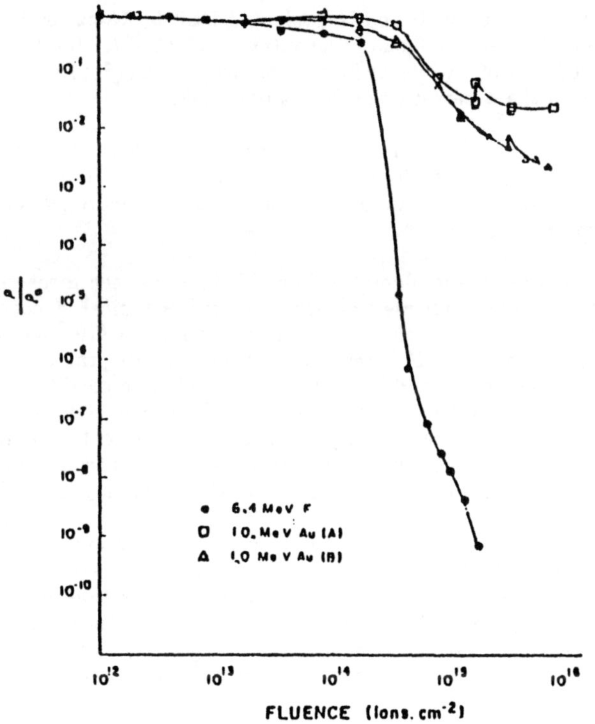

Fig. 4. Ion dose dependence of resistivity for 6.4 MeV flourine and 1 MeV gold, taken from [8].

In order to gain insight into the physics of the conduction mechanism, measurements of the temperature dependence of the conductivity for samples irradiated at different doses with 50 keV C^+ ions have been performed (see Figures 5 and 6). In Fig.5 $\ell n\sigma(T)$ is displayed vs $1/T$ while in Fig.6, the same data are displayed as $\ell n\sigma(T)$ vs $T^{-1/4}$. It can be clearly seen that the data fit the functional dependence $\ell n\sigma \propto (\frac{1}{T})^{1/4}$ (Fig. 6) better than $\ell n\sigma \propto \frac{1}{T}$ (Fig. 5), hence it has been concluded in [6] that the mechanism responsible for the ion induced conductivity in DLF is, most likely, a variable range hopping mechanism, in similarity to the case of ion-damaged diamond.

Thermopower measurements [9] for the determination of the type of carriers responsible for the electrical conductivity suggest that for lower dose implantations of C ions (70 keV, $5 \times 10^{15} cm^{-2}$) the conduction is via electron hopping, while for higher doses ($2 \times 10^{16} cm^{-2}$) it may be due to hole conduction.

Ion implanted DLC films have been mainly subjected to two kinds of measurements for quantitative optical characterization: Absorption measurements from which the optical band-gap of the material has been deduced and Raman spectroscopy which has yielded information on structural transformation that have been induced by the implantation. As for the optical band gap, it has been found [7] that over a wide range of photon energies the linear relationship between photon energy (E) and $\sqrt{\alpha E}$ (α being the absorption coefficient) holds, hence Tauc plots can yield reliable information on the optical band gap of the material as shown in Fig. 7. With increasing implantation dose the optical gap shrinks, in agreement with the darkening in color which accompanies heavy dose implantations in DLF.

The Raman spectra of as-grown DLF do not exhibit any sharp features, and are characterized by a broad peak at 1540 cm^{-1} (G peak). Upon successive irradiation a shoulder at lower wave numbers appears which can be associated with the D peak. Its growth is much enhanced for elevated temperature implantations as shown in Fig.8 [7].

Photoluminescence (PL) studies on H and N-implanted DLF by Gonzales-Hernandez and coworkers [10] show that for relatively low fluence implantations ($1 \times 10^{16} H^+/cm^2$ at 100 keV) the PL intensity increased by a factor of 2 while for higher dose irradiation (5×10^{16}, $1 \times 10^{17} cm^{-2}$) it sharply decreased, allowing the observation of the Raman peak.

The observed changes in electrical and optical properties of ion implanted DLF can be correlated with structural changes caused by the implantation damage. Two major structural modifications accompany ion-implantation of DLF. These are loss of hydrogen and increase in the density of dangling bonds.

Loss of hydrogen as a result of ion implantation in a-C:H film has been measured by several groups. The results of these studies indicate that for films with a typical H concentration of 30-40%, substantial H loss takes place at relatively low irradiation doses (10^{15} ions cm^{-2}), however a saturation level of about 5 at. % H is reached, at which H loss practically stops. This behaviour seems to be almost universal, and independent of irradiation conditions; MeV heavy ions, or sub-MeV light ions all exhibit a similar dependence for the H concentration on dose regardless of the dominant physical mode of energy transfer from the ion to the material. The question which one of the stopping mechanisms (nuclear or electronic) is responsible for the H loss is not entirely clear, however evidence will be given below that each may play a different role.

Baumann et al. [11] have fitted the measured hydrogen concentration versus dose curves with three different exponentials attributed to three different hydrogen release cross-sections. These may be interpreted as the result of different C-H bonding configurations in the amorphous carbon matrix. Sellschop and coworkers [12] proposed a similar interpretation for their data on ion-beam induced hydrogen release from natural diamond, suggesting that the changes in the slope in the logarithm of H concentration versus dose reflect different bonding states of H to C in diamond.

The mechanism by which H is lost from a-C:H films during thermal annealing has been

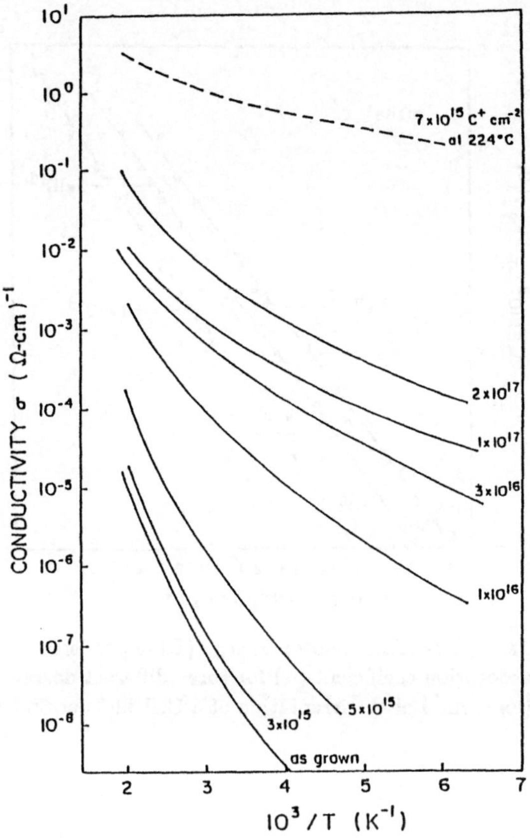

Fig. 5. Inverse temperature dependence of the conductivity (σ) for a-C:H film samples irradiated with different doses of C^+ ions. The broken line is from a sample irradiated with 7×10^{15} C^+ cm^{-2} at 220°C [7].

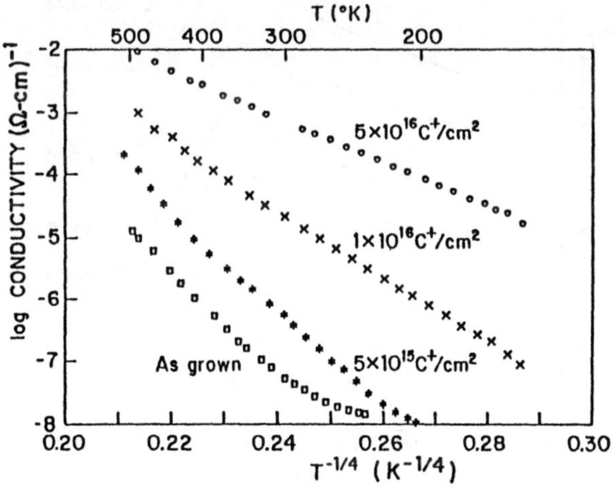

Fig. 6. Log of the conductivity versus $T^{-1/4}$ for the conductivity data shown in Fig. 5. for irradiated DLC films [7].

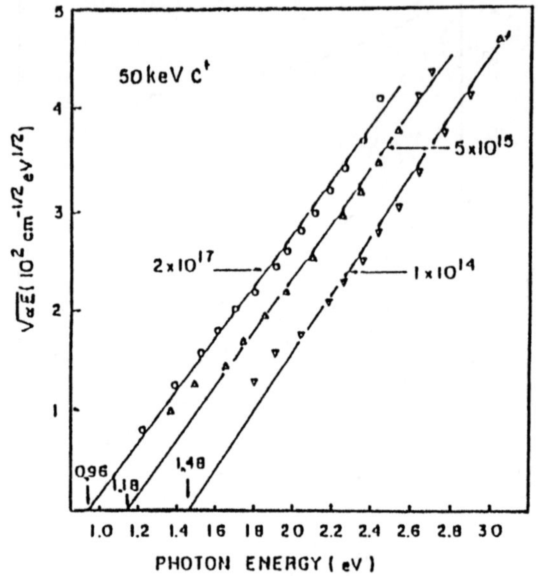

Fig. 7. Tauc plots of the photon energy (E) dependence of the optical absorption coefficient (α) for three different doses (ions/cm^2) of C^+ irradiation of a-C:H films at 50 keV [7].

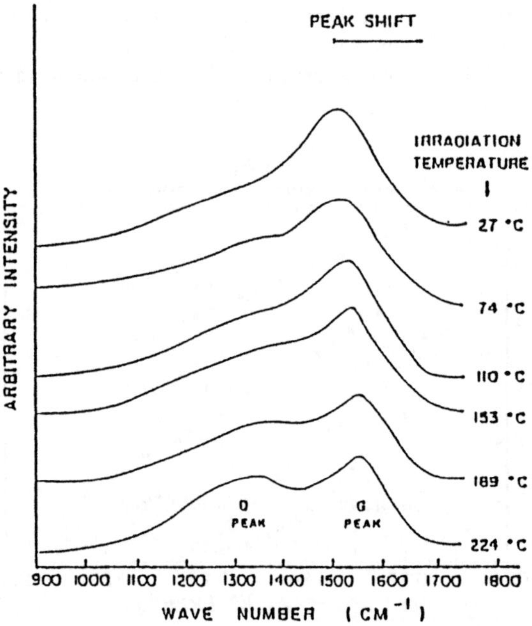

Fig. 8. Raman spectra for a-C:H irradiated with 50 keV C^+ ions to a dose of 7×10^{15} ions cm^{-2} at various substrate temperatures [7].

illuminated in a series of experiments by Wild and Koidl [13]. By performing mass resolved measurements of the species effused from sandwiches of deuterated and hydrogenated carbon films, they have shown that hydrogen leaves the a-C:H films always in molecular form, i.e. molecular recombination occurs in the bulk, followed by diffusion as molecules through the layer until they are lost into the vacuum. A similar conclusion was also reached by Moeller who has shown that during D^+ irradiation of H^+ implanted graphite, recombination of atomic to molecular hydrogen occurs in the bulk, and the effusing species are H_2 or HD molecules [14].

An alternative description of hydrogen effusion from a-C:H films induced by ion irradiation has recently been proposed [15]. It is based on the above described experimental evidence for molecular effusion and does not require the introduction of different C-H bonding configurations. The model is statistical in nature and is based on a second order kinetic process, namely the recombination of atomic to molecular hydrogen. The basic assumption in the model is that in order to create a hydrogen molecule in the bulk, two C-H bonds within some characteristic recombination length of each other must be broken due to the energy deposited by the passage of the same ion. As a consequence of this requirement the change in hydrogen concentration $d\rho$ due to a dose increase dD is proportional to the number of hydrogen *pairs* in a characteristic volume V, i.e.

$$\frac{d\rho}{dP} = -K\rho(\rho V - 1) \,, \tag{6}$$

where K is an effective molecular release cross-section.

This differential equation has the solution

$$\rho(D) = [1/\rho_f + (1/\rho_0 - 1/\rho_f)exp(-KD)]^{-1} \tag{7}$$

Here ρ_0 is the initial volume hydrogen density and $V = 1/\rho_f$, i.e. the inverse of the final hydrogen concentration. The details of this calculation have been published elsewhere [15].

The model has been applied to all available published data on hydrogen effusion from a-C:H using ρ_f and K as adjustable parameters. A typical example for the quality of the fits which can be obtained by this model is given in Figure 9 which shows the atomic hydrogen concentration dependence on N^+ ion fluence, adapted from Baumann and coworker's data. It is evident that the hydrogen concentration follows very closely the functional dependence given by the model; similar fits were obtained for all data analyzed, even for hydrogen release from diamond [12]. Table 1 summarizes the relevant experimental parameters for the various data including the release cross-section K and the final hydrogen concentration ρ_f used to generate the fits. The physics governing the release mechanism resides in the parameter K and perhaps the final H concentration ρ_f. The correlation between hydrogen loss during ion irradiation and the electronic stopping power of the incident ion has already been demonstrated by Baumann et al. [11]. The approximate correlation between K and the electronic stopping powers observed in Table 1 is illustrated in Figure 10 where the energy loss due to electronic processes is plotted versus the effective cross-section K. Although the points are somewhat scattered, a least squares fit to this data produced a line of slope 2.0 with standard error 0.3. This is in fact the value expected for a molecular release cross-section governed by the probability of independently freeing two hydrogen atoms within a characteristic recombination distance.

In almost all cases reported the nuclear energy loss is negligible as compared to the electronic, with the exception of Ingram and coworkers' data for 1 MeV gold ions [7]. In these data, a final H concentration of $\sim$ 20 at. %· was obtained, significantly larger than those observed to result from lighter ion induced effusion (typically 5-10 at. %, see Table 1). As pointed out in reference [7], the exceptionally high density of the collision cascade in this case results in a high density of H trapping centres. The characteristic volume (i.e. $1/\rho_f$) appears therefore to be governed by the density of retrapping centres which may compete with the molecular recombination process. This is reflected also in the abnormally small decrease in electrical resistivity induced by the Au implantation (see Fig. 4 above).

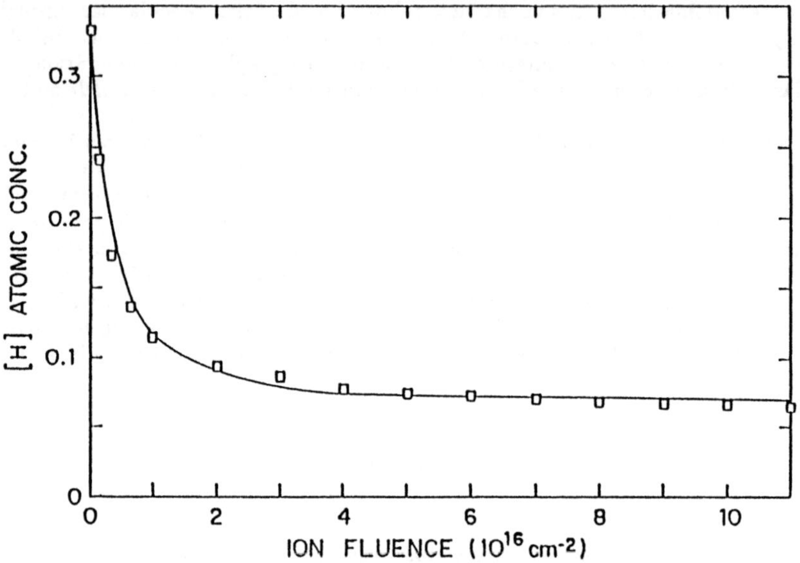

Fig. 9. Hydrogen atomic concentration in a-C:H as a function of irradiation dose with 6 MeV N^+ ions, according to Baumann et al. [11]. Solid line indicates fit using functional dependence from bulk molecular recombination model according to [15].

Table 1. Values of parameters used to fit bulk molecular recombination model for various experimental data on ion-induced hydrogen release from a-C:H. The electronic and nuclear energy loss for each case is also shown.

Author	reference	Ion	Energy (MeV)	release cross-section K (Å^2)	dE/dx electronic (eV/Å)	dE/dx nuclear (eV/Å)	final [H] ρ_f(at. %)
Baumann	[a]	^{15}N	7	0.5 ± 0.03	140	1.0	6.5
Baumann	[a]	^{15}N	3.5	4.5 ± 0.5	155	1.4	10
Baumann	[a]	^{20}Ne	1.6	4.5 ± 0.5	160	6	9
Baumann	[a]	^{40}Ar	2.15	4.5 ± 0.5	180	22	10
Zou	[b]	^{58}Ni	4	13 ± 3	192	17	7
Ingram	[c]	^{19}F	6.4	5	250	0.67	6
Ingram	[c]	^{197}Au	1	7	221	307	20
Fujimoto	[d]	^{12}C	12	5	100	0.14	9
Prawer	[e]	^{12}C	0.05	0.06 ± 0.02	15-27	18-14	5

(a) Ref. 11 (b) Zou J.W. et al., J. Vac. Sci. & Technol. **A6**, 3103 (1988). (c) Ref. 8 (d) Fujimoto, F. et al., Nucl. Instr. & Meth. **B33**, 792 (1988). (e) Ref. 7

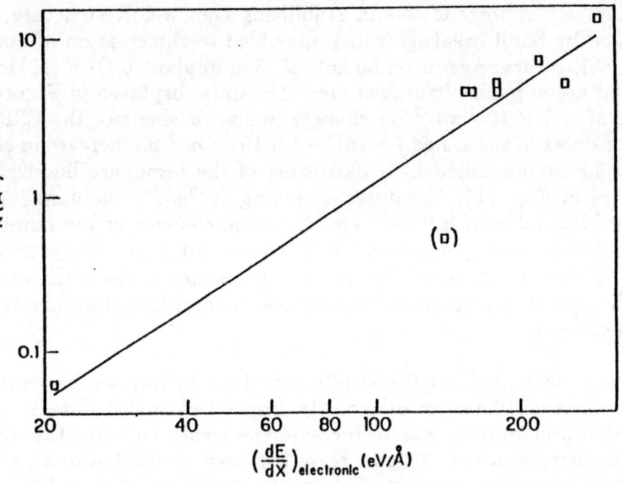

Fig. 10 **Log-log plot of molecular release cross-section K versus electronic energy loss $(dE/dX)_{electronic}$ as given in Table 1. The straight line has a slope of 2.**

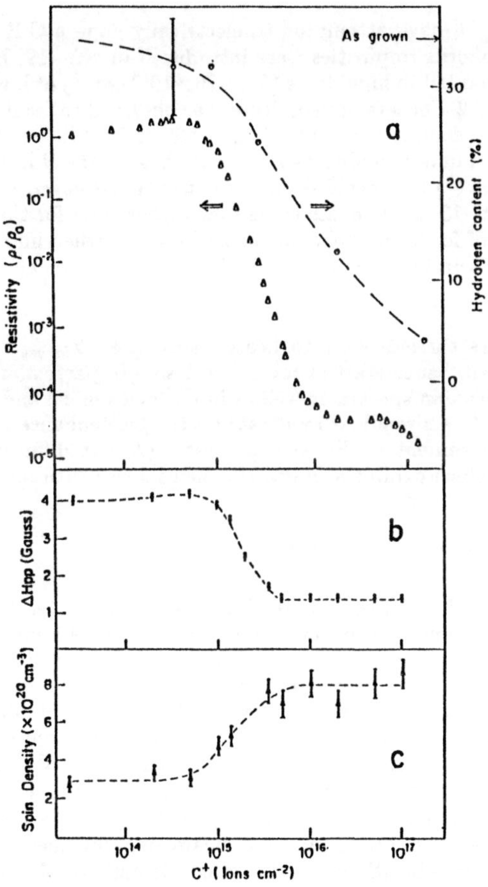

Fig. 11 **Variation with 50 keV C^{+} ion dose of (a) resistivity and hydrogen content, (b) EPR resonant linewidth and (c) spin density, from [16].**

As hydrogen plays an important role in stabilizing the a-C:H structure, hydrogen loss should be accompanied by bond breakage which may lead to the creation of unsaturated dangling bonds. Indeed ESR measurements on 50 keV C^+ ion implanted DLF [16] indicate that an increase in ESR signal accompanies hydrogen loss. The data displayed in Figure 11 show that below a critical dose of $\sim 7 \times 10^{14} cm^{-2}$ no changes in size or shape of the ESR signal can be detected, however, for doses in the range $7 \times 10^{14} - 1 \times 10^{16} cm^{-2}$ an increase in the spin density by about a factor of 3.5 accompanied by a narrowing of the resonance line-width is observed (see panels (b) and (c) in Fig. 11). For doses exceeding $10^{16} cm^{-2}$ the dangling bond density saturates at the very high value of 8×10^{20} cm^{-2}. These changes in the number of dangling bonds go hand-in-hand with hydrogen loss and with the sharp decrease in electrical conductivity as shown in panel (a) of Fig. 11. It should be noted that no shift in the EPR resonance position accompanies these changes at any stage, indicating that no graphitization has taken place even for the highest doses studied.

Some attempts have been made to electrically dope DLF by implanting acceptor and donor ions. Wong and coworkers [17] have, most recently, implanted a-C:H films by B^+, BF_2^+ and P^+. The effect of the implantation was to increase the conductivity of the film, once a certain threshold dose has been reached. This increase has been attributed to an electrical doping effect in Ref. 17. Similar experiments carried out by Amir and Kalish [18], in which DLF has been implanted with $^{11}B^+$ and as a control under identical conditions also with $^{12}C^+$ ions (both leading to similar damage due to the proximity in projectile mass), has not revealed any significant difference between the C and B implantation. It was therefore concluded in this work that the observed electrical effect was due to implantation induced damage of the kind discussed above, and not due to chemical doping effects caused by the particular implanted ions.

Cheng Shichang et al [19] have attempted to electrically dope a-C:H films p and n type. While aluminum and phosphorus impurities were introduced in ref. [19] into the films during growth, boron has been implanted to high doses ($0.5 - 20 \times 10^{16}$ cm^{-2}) at low energies (5 keV). A decrease in film resistivity by 2-7 orders of magnitude was observed to result from the B implantation which could be further enhanced by annealing at 500°C for 4 hours. Control experiments of C implantations performed under conditions similar to those of the B implantations have also lead at high doses ($> 10^{17} cm^{-2}$) to a rapid drop in resistivity. However, the onset of this drop was found, according to Ref. 18, for C doses somewhat higher than for the B implantations, a fact which was taken as proof for boron behaving as an impurity when implanted into DLF. No studies on the physical, structural or chemical state of the implantation affected layer have been reported in Ref. 18.

When implantations are carried out into heated DLF (i.e. $T_{target} \sim 200°C$) all effects described above are enhanced, hence start at lower ion doses. In particular, complete graphitization , as observed in the Raman spectra as well as in the mechanical and chemical properties of the irradiated films take place already at moderate doses. This enhancement should probably be traced back to the enhancement in diffusion processes (H out diffusion and vacancy and interstitial diffusion) all of which exhibit a sharp, exponential rise with temperature.

4. ELECTRON IRRADIATION EFFECTS IN DLF

As pointed out above, the basic interactions of electrons with matter do not differ in principle from those of ions - both being due to the coulomb interaction of a moving charged particle with the positive and negative charges of the substrate. The basic difference, however, has to do with the fact that electrons are substantially lighter than ions, therefore much less momentum transfer from projectile to target atom is possible. Hence, it may be expected that electrons will cause similar changes to DLF as ions do, but these will appear at higher electron bombardment doses than those needed with ion bombardment. Nevertheless, the possibility to modify DLF by electron bombardment is attractive because of the well developed technology of electron beam optics, which enables the irradiation by finely focused e-beams as those used in electron-microscopes. On the other hand, the small beam size, together with the higher doses required to induce noticable modifications in DLF makes it hard to obtain large enough electron irradiated areas to permit extensive measurements of their physical properties. These may be the reasons why electron irradiation effects in DLF have not been as extensively studied as those induced by ion bombardment. The available results, however, are interesting since some substantial differences between ion and electron irradiation of DLF have been found.

Adel and co-workers [20] have subjected a-C:H films $7500 \pm 500\AA$ thick deposited on glass and diamond substrates to electron bombardment in an ultra-high vacuum system, based on a scanning Auger microprobe. A beam energy of 5 keV was chosen corresponding to an electron range of $3000\AA$ in a-C:H. This range is substantially smaller than the film thickness, so that any irradiation effects in the substrate could be precluded. Areas 1×1 mm^2 were irradiated by raster scanning the beam at a current of 5 μA to fluences ranging from 2×10^{17} to 5×10^{19} electrons cm^{-2}. Under these conditions beam heating of the film is expected not to exceed 20°C. Due to the small irradiated area a number of analysis techniques which require significantly larger irradiated areas (i.e. ESR, thermopower) could not be applied.

The resistivity dependence on electron fluence is shown in Figure 12(a). No significant conductivity changes are observed below 10^{18} electrons cm^{-2}. Above this threshold, two distinct dose regimes in which the dose dependence of the resistivity is markedly different can be seen. In the first regime between 10^{18} and 10^{19} electrons cm^{-2} the resistivity follows quite closely an inverse exponential dependence [exp(-kD)] on electron dose. In the second regime, above 10^{19} electrons cm^{-2} the dose dependence can be described by a power law (D^{-3}).

Optical absorption measurements were performed on the same samples as used for the conductivity measurements. The optical band gaps, deduced from these measurements are shown in Figure 12(b). As can be seen from the Figure, bandgap shrinkage does occur, however it *preceeds* the onset of conductivity by electron doses by nearly an order of magnitude; the resistivty drop starting at a dose of $\sim 10^{18}$cm^{-2}, while the film starts to change its optical properties already at a dose of $\sim 10^{17}$cm^{-2}. In this respect electron irradiation of DLF differs from heavy ion implantation.

Other differences between electron and ion irradiated DLF found by Adel and coworkers [20] are in the Raman and in the RBS spectra (used to determine hydrogen content). No significant changes in the Raman spectra between as-grown and highly electron irradiated DLF ($2 \times 10^{19} e - cm^{-2}$) could be detected (Figure 13, lines a and b) indicating that no substantial graphitization has occurred. Also, RBS experiments, similar to those used to measure hydrogen loss for ion irradiated DLF, have shown that only little hydrogen (less than 25%) was lost from the irradiated films, even when exposed to very high electron fluences ($1.5 \times 10^{19} cm^{-2}$).

5. LASER IRRADIATION EFFECT ON DLF

The modifications that take place in a-C:H films as a result of intense photon irradiation have been studied relatively little, despite their potential practical applications. Prawer and coworkers [21] have subjected a-C:H films to short (70 ns) pulses from a frequency doubled Nd YAG laser ($\lambda = 0.53\mu m$, $E_{ph} = 2.35eV$) at ever increasing power densities. The response of the film to the laser depended on the laser power density and could be divided into three regimes. Below a threshold of about $0.2J/cm^2$, no changes could be detected in the films even if exposed to a large number of pulses. Hence, it seems as if at low energy densities there is no observable interaction between the laser radiation and the a-C:H material. However, above this threshold a transformation from a-C:H to some form of graphite takes place. With increasing energy density graphitization, accompanied by ablation, proceeds until at sufficiently high energy density ($1.5J/cm^2$ for a 1 μm thick layer) the material is totally ablated. Hence sharp trenches with some "spill over" shoulders can be created in the coating.

By subjecting the transformed material to some of the diagnostic methods used by the same group to study electron and ion induced effects in DLF, it was found that the irradiated material exhibits some physical properties which are characteristic of *graphite*: it has a low electrical resistivity of 0.2 Ωcm (as compared to 10^7 Ωcm for the untreated material) and its Raman spectrum, displayed in Figure 13, line c, exhibits graphitic features. The infrared absorption in the 400-4000 cm^{-1} range is increased by 20%, and the morphology, chemical resistance to acids,

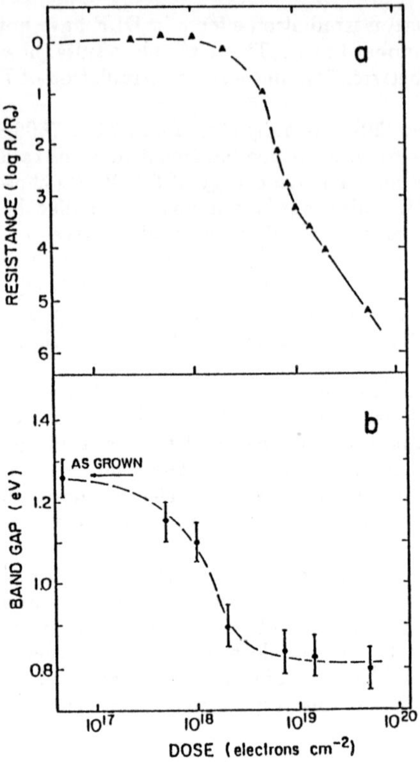

Fig. 12 a) Dependence of the resistance R on electron dose irradiation of a-C:H on glass with 5 keV electrons. R_0 was $1.2 \times 10^{11} \Omega$ for the unirradiated a-C:H film. b) Optical band gap as determined from Tauc plots, versus electron dose for irradiated a-c:H films, from [20].

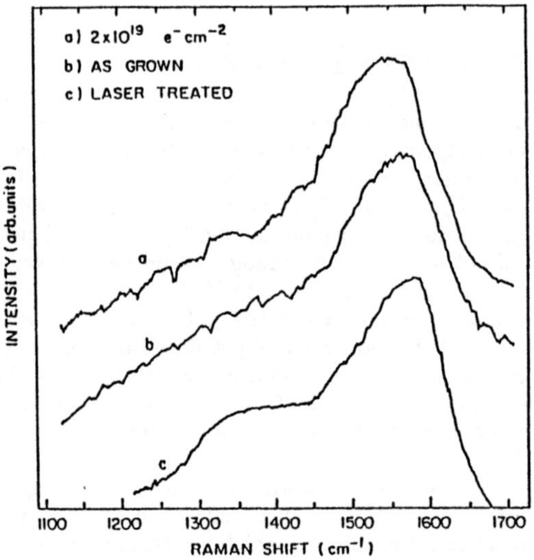

Fig. 13 Raman spectra of as-grown, high dose electron irradiated and laser-treated a-C:H films. Adapted from [20].

and hardness of the film deteriorate. These features resemble very closely those reported [22] for a-C:H heated to 400-500°C. Hence it was concluded that the pulsed laser irradiation has locally heated the a-C:H to ~ 500°C, a temperature at which partial graphitization takes place.

There are some technological implications to the laser irradiation induced modification of a-C:H. By judicious choice of the laser parameters it is possible to produce conducting pathways or well defined channels in the a-C:H layer. The former may have applications in device manufacture of micrometer scale thin-film resistors, while the latter may be used for high resolution lithography [23]. Rothschild and Ehrlich [24] have exposed a-C:H films to UV ($\lambda = .193\mu m$) radiation from an ArF excimer laser and have demonstrated that extremely fine lines can be dry etched in the film used as a "photo resist". As in the case for visible radiation, the results of ref. 23 show that a threshold in energy density exists above which thin a-C:H films are totally ablated. At energy densities of 0.13 J/cm^{-2}, a single 20 ns pulse was found to be sufficient to totally remove a ~ 1000Å thick a-C:H layer. This energy density is lower than that required to induce similar effects with visible radiation, probably due to the higher photon energy and hence the smaller absorption depth for the radiation. a-C:H films have also been exposed to laser radiation of energy well below the absorption edge. Dischler [25] has used 10.6 μm radiation and has found that the damage threshold depends on the laser power density (damage threshold ~ 300 MW/cm^{-2}). In this case it seems as if no simple direct absorption is responsible for the damage, but rather more complex power dependent processes must be considered.

6. DISCUSSION

As seen above, irradiation of any form causes rather extreme changes in a-C:H which are reflected in structural changes involving H loss, creation of dangling bonds and some transformation from sp^3 to sp^2 bonding. These, in turn, affect the electrical, optical, chemical and mechanical properties of the material.

The structure of the a-C:H can be described in terms of a two phase model along the lines suggested by McKenzie and coworkers, [26] as well as Robertson and O'Reilly [27]. In this model, hard a-C:H is considered to consist of sp^2 clusters, typically planar aromatic ring structures, which are interconnected by randomly oriented tetrahedral sp^3 bonds. The hydrogen may be bonded on either the tetrahedral sites, where they are required to reduce bond angle disorder, or on the edges of the ring structures where they cause termination.

Based on this model it is clear that significant loss of hydrogen must lead to major structural changes. Firstly, since a random tetrahedral phase is highly over constrained (analogous to a-Si:H), and since the hydrogen is responsible for reducing the co-ordination number, its removal is expected to destabilize the sp^3 phase. This would either increase the number of unsatisfied bonds or result in restructuring whereby two previously sp^3 C-H bonds "recombine" to form an sp^2 bond. The latter effect would result in a net conversion from sp^3 type bonding to sp^2 type bonding. The growth of the sp^2 clusters would be further enhanced by the removal of hydrogen from the ring edges, so that a net increase in cluster size is expected. The fact that ion and electron bombardment breaks bonds and thus leads to hydrogen release, as described above, suggests that growth of sp^2 (graphite-like) clusters will be found in damaged a-C:H. Thus it is expected that the major changes in electrical and optical properties in ion-implanted DLC should go hand-in-hand, and both should set-in at about the same damage levels. This indeed is the case for ion-induced damage, as can be seen in Figure 12. For 50 keV C^+ implantations hydrogen loss (Fig. 11) starts at the same dose at which changes in the EPR signal are noticable, and at the same dose regime other film properties also undergo dramatic changes - the electrical conductivity rises sharply and the optical band gap shrinks (Fig. 12). It is interesting to note the similarity in ion-beam induced electrical conductivity between implanted diamond crystal (Fig. 16) and implanted a-C:H films despite the basic differences that exist in their structure. One possible reason which requires further investigation, is that part of the onset of the observed conductivity has to do with the percolative formation of a conductive pathway due to the conductivity of the implanted ions (whether carbon or metals) and is not to be related with bond breakage, hydrogen loss, etc. This point was recently investigated by Prawer and Kalish [28] who have implanted diamond and quartz with Xe and C ions and have observed strong

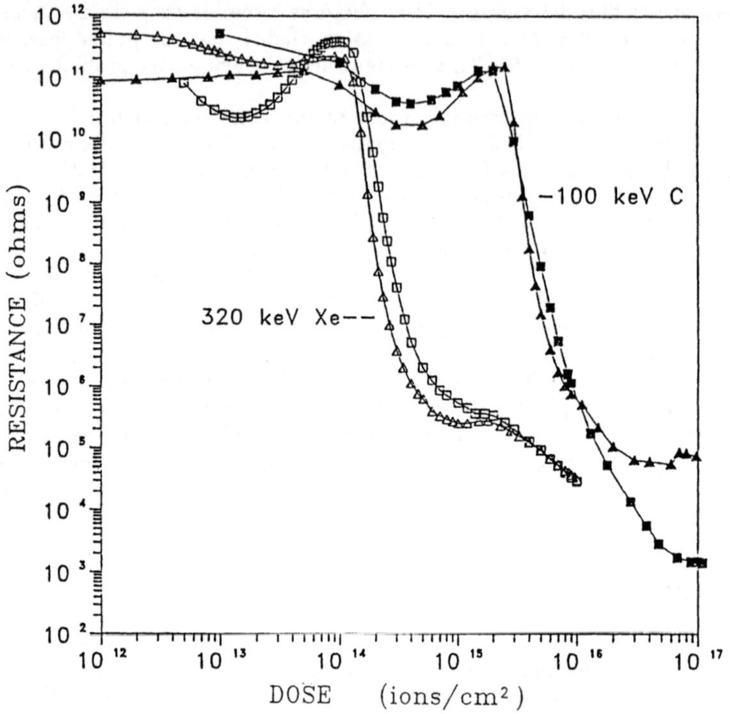

Fig. 14. **Resistance versus dose for irradiation carried out at 20 ± 2^oC, for CVD diamond films (squares) and natural type IIa single crystal diamond (triangles). The open symbols are for irradiation with 320 keV Xe, whilst the close symbols refer to irradiation with 100 keV C ions [31].**

increases in conductivity for the case of quartz implanted with C at about the same critical dose as found for diamond and DLC, while no change in conductivity could be detected for Xe implanted quartz.

The difference between the onset of electron and ion induced changes in the optical and electrical properties of DLF can, perhaps, be partially explained by the fact that e-irradiation does not introduce any foreign atoms into the lattice so that no conductive perculative pathway can be directly formed by the implants. Nevertheless, the fact that some light induced bleaching of the electron induced absorption could be observed indicates that at lower electron irradiation doses some defect other than a dangling bond (which is precluded due to its known effects on conductivity enhancement in both DLF and amorphous hydrogenated silicon) may be formed. Preliminary experiments indicate that this optically active defect state may be metastable since it could be "annealed" by visible laser radiation at moderate intensities (0.3W on $100 \times 100\mu m^2$). A thermally annealable metastable defect state of a similar nature has also been observed in electron irradiated amorphous hydrogenated silicon [29].

At higher electron irradiation doses, where the conductivity dependence follows a power law and where hydrogen release becomes significant, it is speculated that the main change in the nature of the film's defect structure is an increase in dangling bond density. This is analogous to the case of ion bombardment where the conductivity increase is attributed to a rise in the density of band tail states. These states result from increased local distortion since the dangling bond density rises as hydrogen is driven out from the film by the ion impact. The weaker dose dependence of the conductivity increase for electrons (D^3) as opposed to ions (D^4) [9] indicates a different, more gentle mechanism for hydrogen release.

As for photon induced changes in DLF, the fact that their onset depends on the power density, and the similarity of the transformed material to thermally treated DLF shows that mainly heating is responsible for the observed effects which can all be understood if laser irradiation transforms a-C:H to graphite.

The basic difference between graphitized a-C:H, caused by heat or laser irradiation and "impact-induced" changes in the material as caused by ion or electron implantation should be stressed. While the former material is graphitic in all of its properties, the latter (unless implanted to very high doses, or at elevated temperatures) is not.

Many of the attractive chemical and mechanical properties of diamond which diamond-like films exhibit are maintained also in the electron or ion treated material. It is therefore of particular interest, and possibly also of practical importance, to be able to change the electrical and optical properties of DLF using finely focussed beams or photolithographic methods to modify a coating of most attractive properties in selected areas on a micron scale.

REFERENCES

1. Lindhard, J. Scharff, M. & Schiott, H.E., Mat. Fys. Medd. Dan. Vid. Selsk., **33**, 14 (1963).
2. Ziegler, J., Biersack J.P. & Cuomo, J. (1980), Computer Code.
   Biersack, J.P. & Haggmark, L.G., Nucl. Inst. and Meth., **174**, 257 (1980).
3. Schade, H., in "Semiconductors & Semimetals", **21**, Part B, Chapter 11, edited by J. Pankove, Academic Press (1984).
4. Wood, R.M., " Laser Damage in Optical Materials", Adam Hilger, (Bristol & Boston, 1986), p. 427-474.
5. Kalish, R. & Adel, M.E., in "Properties and Preparation of Amorphous Carbon Films", Eds. J.J. Pouch & S.A. Alterowitz, Materials Science Forum, Vol. 52/53 (1990). Trans. Tech Pub. Switzerland.
6. Prawer, S., Kalish, R., Adel, M. and Richter, V., Appl. Phys. Letters **49**, 1157 (1986).
7. Prawer S., Kalish, R., Adel, M.E. and Richter, V., J. Appl. Phys. **61**, 4492 (1987).
8. Ingram, D.C. & McCormick, A.W., Nucl. Inst. & Meth. B, **34**, 68 (1988).
9. Adel, M.E., Ph.D. Thesis, Technion, unpublished (1989).
10. Gonzales-Hernandez, J., Asomoza, R., Reyes-Mena A., Richards J., Chao, S.S. & Pawlik, D., J. Vac. Sci. & Technol., A, **6**, 1798 (1988).
11. Baumann, H., Rupp, Th., Bethge, K., Koidl, P. Wild, Ch., Les Editions de Physique, EMRS Conference Proceedings, Vol. 17, 149 (Strasbourg 1987).
12. Sellschop, J.P.F., Madiba, C.C.P. & Annegarn, H.J., Nucl. Inst. & Meth. **168**, 529 (1980).
13. Wild, Ch. & Koidl, P., Appl. Phys. Lett. **51**, 19, (1987).
14. Moeller, W., Borgesen, P. & Scherzer, B.M.U., Nucl. Inst. & Meth., **B19/20**, 826, (1987).
15. Adel, M.E., Amir, O., Kalish, R. & Feldman, L.C., J. Appl. Phys. **66**, 3248 (1989).
16. Adel, M.E., Kalish, R. and Prawer, S., J. Appl. Phys. **62**, 4096 (1987).
17. Wong, S.P., Materials Research Society Symposium, Proceedings Series, Vol. 162, in press (1990).
18. Amir, O. and Kalish, R., to be published.
    Amir, O., M.Sc. Thesis, Technion (1990), unpublished.

19. Cheng Shichang et al., Nucl. Instr. and Methods **B39**, 692 (1989).

20. Adel, M.E., Brener, R. & Kalish, R., Les Editions de Physique, EMRS Proceedings, 334 (Strasbourg, 1987).

21. Prawer, S., Kalish, R. & Adel, M.E., Appl. Phys. Lett. **48**, 1585 (1986).

22. Nadler, M.P., Donovan, T.M. & Green, A.K., Appl. Surf. Sci. **18**, 10, (1984).

23. Rothschild, M., Arnone, C. & Ehrlich, D.J., J. Vac. Sci. & Technol. B, **4**, 310 (1986).

24. Rothschild, M. & Ehrlich, D.J., J. Vac. Sci. & Technol. B, **5**, 389 (1987).

25. Dischler, B., Bubenzer, A., Koidl, P., Brandt, G. & Schirmer, O.F., NBS Special Publication, **669**, 249 (1982).

26. McKenzie, D.R., McPhedran, R.C., Savvides, N. & Cockayne, D.J.H., Thin Solid Films **108**, 247 (1983).

27. Robertson, J. & O'Reilly, E.P., Phys. Rev. B, **35**, 2946 (1987).

28. Prawer, S. and Kalish, R., unpublished.

29. Thomas P.A. and Flachet, J.C., Philos. Mag. **B51**, 55 (1985).

30. Kalish, R., Bernstein, T., Shapiro, B. & Talmi, A., Rad. Effects **52**, 153 (1980).

31. Prawer, S., Hoffman, A. and Kalish, R., to be published 1990.

AGING PROCESSES IN DIAMOND-LIKE CARBON AND CARBON/METAL FILMS

Ludvik Martinu

"Groupe des Couches Minces" (GCM) and
Department of Engineering Physics, Ecole Polytechnique
Box 6079, Station "A", Montreal, Quebec H3C 3A7, Canada

1. INTRODUCTION

Thin films deposited by nonequilibrium processes, such as in low pressure plasmas, are sensitive to aging due to the presence of free radicals, dangling bonds, defects, voids, internal stress and other features. Generally, several aspects of film stability or aging can be considered:

A: spontaneous aging - internal material property characterized by the driving force to acquire minimum system energy without any intermediate agent;

B: environmental aging - reaction with the ambient atmosphere (not "deliberately severe");

C: accelerated aging - post deposition treatment (deliberately severe) using various thermal, chemical, radiative, mechanical and other agents.

This applies also to carbonaceous deposits. Data pertaining to aging A and B of diamond-like carbon (DLC) films are not very frequent in the published literature, and the "accelerated aging" has been mainly reported.

For the purpose of this work we will focus on carbonaceous materials deposited from hydrocarbon gases in a radio frequency (RF) glow discharge, among numerous other techniques[1]. These will range from plasma polymers (PP) to hydrogenated amorphous carbon (a-C:H, termed DLC throughout the text) and their modifications when ultrafine metal particles are imbedded in PP or DLC matrices[2]. We will relate the aging phenomena to the energy E_i of ions bombarding the growing films. In this context, the metal clusters can be considered as sensitive fingerprints of the film stability. The effect of ion flux ϕ_i related to the discharge excitation frequency will be particularly pointed out as an alternative to control the materials' features including aging.

Diamond and Diamond-Like Films and Coatings
Edited by R.E. Clausing *et al.*, Plenum Press, New York, 1991

2. PLASMA POLYMER AND DIAMOND-LIKE FILMS

2.1 Ion-induced surface reactions

In low pressure plasma in the presence of hydrocarbon gases or vapours, inelastic collisions of electrons with molecules lead to the formation ("initiation") of ions and radicals (chemically reactive fragments). The abundant radicals diffuse to the exposed surfaces ("propagation") where they are physisorbed and interact between each other and with the underlying surface forming a solid deposit ("termination"). The two latter processes proceed on the surface, and they are affected by simultaneous ion bombardment.

Ion induced phenomena have been described for ion beam assisted experiments[3]. In low pressure plasma, however, the range of impact energies is much broader, and the low energy ($\lesssim$ 1000 eV) ion bombardment influences in rather synergistic manner the surface effects, mainly:

a) preferential sputtering of loosely bound species,
b) mobility of adsorbed species,
c) initiation of chemical reactions,
d) sputtering,
e) direct ion incorporation.

Depending on the materials involved, on the average ion energy $\overline{E_i}$ and ion energy distribution function IEDF, on the flux of ions ϕ_i and neutral species ϕ_n, the ion bombardment affects the final film composition and microstructure and hence the features related to the aging process.

Carbonaceous thin films are usually deposited in an asymmetric RF parallel plate electrode system illustrated in Fig. 1. The maximum energy $E_{i,max}$ of ions hitting the surface is given by the difference between the plasma potential V_P and the substrate surface potential V_s:

$$E_{i,max} = e\,(V_P - V_s)\ . \qquad (1)$$

In such system V_P is typically few tens of volts positive with respect to ground, V_s reaches several volts positive to ground if an insulating substrate is placed on the grounded electrode (floating potential). On the RF powered electrode capacitively coupled to the generator, V_s adopts a high negative dc bias potential (V_B) due to the high mobility of electrons with respect to ions. In usual situations V_B is varied by changing the RF input power P_{RF}. The average ion energy on the target is smaller than $E_{i,max}$ given by eq. (1) due to the charge transfer within the sheath region. Recently, $\overline{E_i}$ yielding 0.4 eV_B has been reported for benzene at a pressure of 2.25 Pa and V_B = 500 V (ref. 4).

In most experiments the film characteristics have been related to V_B. Three categories of materials can be distinguished[1]: soft PP, hard PP and DLC. The transition between these groups is usually observed for V_B ~ 20-50 V and V_B ~ 150-200 V, respectively. However, these transitions are not abrupt but rather a continuous process during which the properties change upon the selective effects of ions on the surface events mentioned above. For example, when V_B is subsequently increased up to 1000 V, the film characteristics vary from PP to DLC as follows[1,4,5]: the hydrogen concentration from 50 to 20 at. %, density from 1.3 to 1.9 g/cm^3, optical gap from 3.0 to 1.0 eV, refractive index from 1.6 to 2.1, and the ratio of hybridizations sp^3:sp^2:sp^1 from 53:45:2 to 68:30:2. For

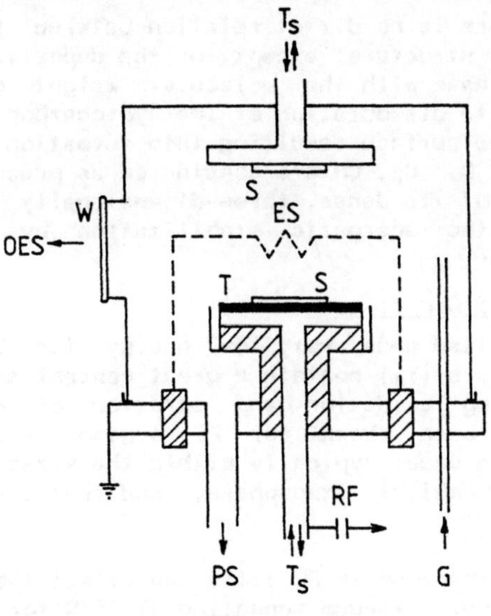

Fig. 1.

Schematic arrangement of a system for deposition of DLC, and PP, metal/DLC and metal/PP films: S - substrate, RF - to the power supply, T_s - substrate temperature control, T - target, G - gases, PS - to the pumping system, W - quartz window, OES - to the optical emission spectroscopy system.

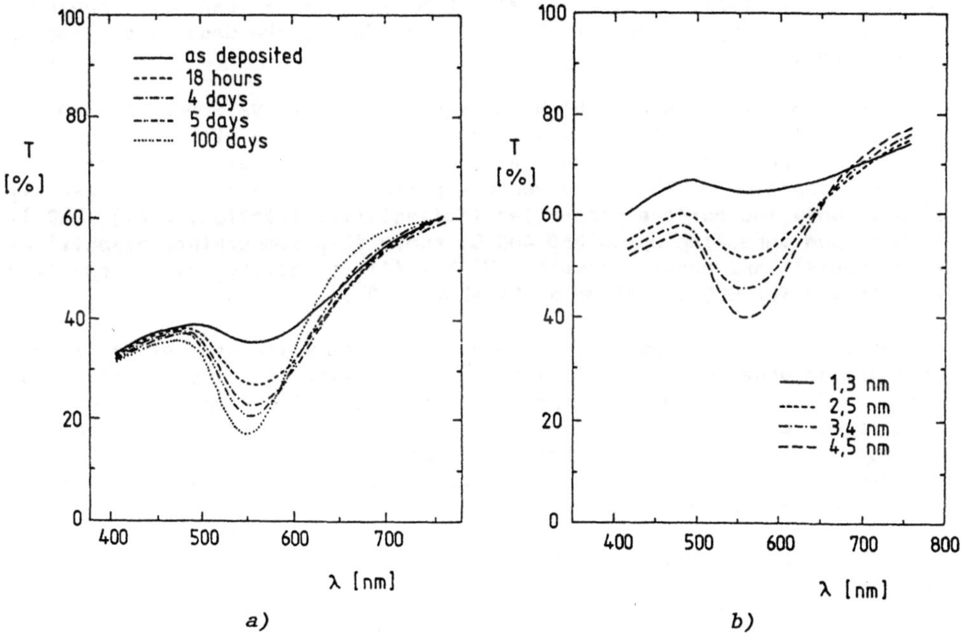

Fig. 2.

Aging effect in the transmission of gold-doped PP-C_2F_3 Cℓ films (f = 0.05). a) experimental spectra, b) calculated spectra for different grain diameters. (After ref. 16).

V_B above 200 V there is no direct relation between the hydrocarbon gas used and the DLC structure, except for the deposition rate r_D which has been found to increase with the molecular weight of the gas[6]. This documents a complete dissociation of the hydrocarbon species, mainly upon their impact on the surface resulting into formation of small fragments. Species such as C, C_2, CH are considered as precursors for DLC[4,6] and for the formation of its dense, three-dimensionally crosslinked network equilibrated by the energetic stabilization by bonding and destabilization by strain[7].

2.2 Aging of PP and DLC films

PP films deposited under very low energy ion bombardment and low substrate temperature (T_s) contain a great concentration of free radicals and voids[8] enhancing oxidation and sorption of water vapours. For example, it has been shown for PP organosilicones[9] that the most significant changes occur typically within the first 50 h of the sample exposure to the ambient atmosphere, and that they lead to hydrolysis reactions.

Water vapours absorbed in PP films can affect the properties such as dielectric behaviour. Vacuum annealing at 80°C for 30 minutes was found to desorb this water and to permit dielectric measurements of plasma polymerized fluorocarbons (PP-FC)[10]. The temperature of 70°C corresponds to a broad exothermic relaxation observed for PP - C_3F_8 accompanied by a thickness decrease and interpreted as a rearrangement and possibly a curing effect in the crosslinked structure[11].

In practical applications the vacuum post-annealing at 80-100°C for several hours or even days is seen as the most straightforward way for improving the stability of PP films. Similar effect can be, however, achieved also by increased T_s and/or V_B during the deposition process (see below).

DLC films grown under high energy impact (~ V_B > 200 V) and at relatively low r_D (typically 1-10 Å/s) are very stable with time compared to PP if proper substrates are selected and adequately pretreated to improve adhesion[1]. They are densely packed, and the atmosphere usually affects only the surface properties (wettability, friction, etc.). DLC is stable upon annealing up to 350-400°C. Above this temperature graphitization occurs[12] and gases evolve[13]. Higher film stability with annealing is observed for DLC deposited at V_B above 300 V.

Good stability and performance of DLC hermetic barriers on Si has been proved under severe conditions such as water at 100°C and steam jet[14]. Effective medium approach (EMA) using the Bruggeman model for spectroscopic-ellipsometric data has shown that moisture resides on the surface ($\lesssim$ 50Å) and fills only the upper-most voids. DLC coatings on magneto-optic data storage media showed a good stability against atmospheric oxygen and water vapour[15].

3. METAL-FILLED PLASMA POLYMERS

In an atttempt to further modify in a wide range the film properties (e.g. optical, electrical, mechanical), metal clusters have been imbedded in the matrix of PP using simultaneously a combination of techniques such as plasma polymerization, sputtering and/or evaporation, most often in a system similar to that in Fig. 1 (for a review see ref. 2). The process parameters (gas composition - monomer/inert gas mixture, V_B values, etc.)

must be chosen to favour the erosion of the target material (sputtering or sputter-etching) that can be incorporated in the films growing on the opposite electrode.

The resulting film properties are governed mainly by the microstructure of the metal component: metal volume fraction (f), cluster size, shape and their distribution. Depending on the metal and the deposition conditions, the grain diameters are typically 1-50 nm, the percolation threshold occurs for f between 0.3 and 0.4, and an anomalous optical absorption in the visible region is evoked resulting in optical selectivity and coloured appearance of the deposits[2].

Compared to PP and DLC films, the metal/PP two-component system is sensitive to aging in a more complex way, mainly due to (i) diffusion of ultrafine metal particles, their nucleation and the trend of metal clusters to adopt minimum surface energy, and (ii) ability of the PP matrix to impede these processes, for example by its "rigidity".

When metal-filled PP are grown under very mild conditions (low P_{RF}, V_s and T_s) pronounced aging effects can be observed, e.g. in optical properties as illustrated in Fig. 2a for a gold-doped PP-FC. The aging process has been studied by EMA using the generalized Maxwell-Garnett model[16,17]. The best fit was obtained when only an increase in the gold grain diameters was considered (see Fig. 2b). It has been found that the metal particles are dispersed in the PP matrix with a very broad size distribution. After deposition smaller clusters diffuse to their larger neighbours increasing their dimensions, as confirmed also experimentally by TEM. EMA has been suggested for contactless structural analysis of sensitive (e.g. thermally) composite materials if an appropriate microstructural model is chosen and the complex permittivity is known over a certain wavelength range[17].

Annealing substantially influences the film microstructure in terms of the cluster shape and size[2]. Temperature induced structural changes consist of the following two steps: (a) the motion of gold particles in the polymer as the rigidity of the matrix is relaxed by molecular segment mobility. This effect is most pronounced above the temperature of the glass-rubber transition in PP-FC, that was revealed around 160°C[2,10,11]; (b) sintering and coalescence of metal particles entering into contact.

Sintering, neck formation and spherodization are rapid processes requiring about 1 μs for forming a neck between individual grains at 180°C. Therefore, the motion of distant particles before touching each other is the limiting step dependent on the rigidity of the polymer matrix. The trend for irregularly shaped particles to reach minimum surface energy, i.e. spherodization, usually dominates leading to an increase in the intercluster distance. The time required for sintering increases rapidly with the particle size.

The aging process can be accelerated by annealing at 80-100°C[16]. However, elevated T_s and particularly V_s applied during the growth can also substantially improve the film stability. This is documented in Fig. 3 for Au-PP-C_2F_6 layers deposited in a triode RF system[18] where the third RF substrate-holder-electrode is placed at the edge of the plasma zone. Both, T_s and V_s affect individually r_D of each component and increase the surface mobility of species. V_s of 80 V appears to enhance crosslinking and rigidity of the PP matrix as well as coalescence of the metal particles[18,19]. However, the process conditions need to be calibrated with respect to the resulting film properties.

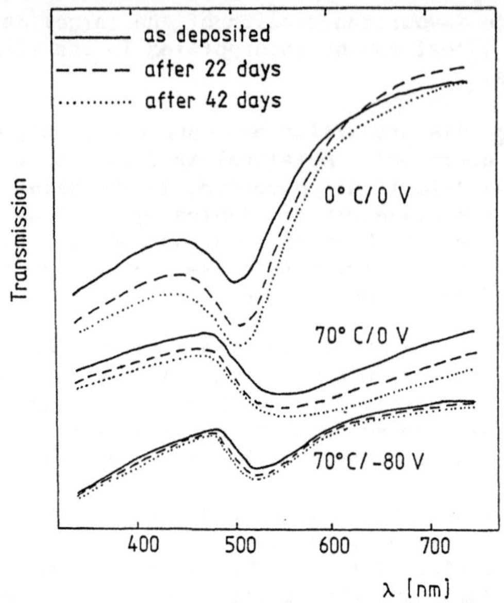

Fig. 3.
Aging effect in the optical transmission spectra for gold-doped PP-C_2F_6 films deposited at different T_s and V_B. (After ref. 18).

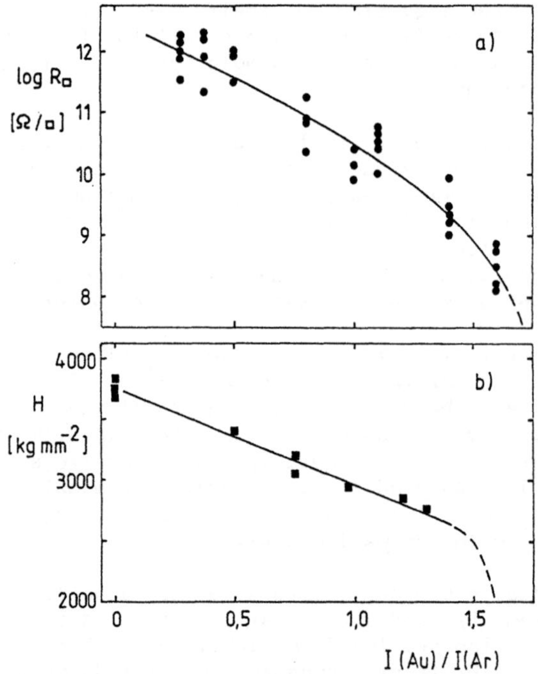

Fig. 4.
Sheet resistance R (a) and Vickers microhardness H (b) versus optical emission intensity ratio I(Au, 267.6 nm)/I(Ar, 418 nm) for gold-doped DLC films deposited from a C_4H_{10}(70%)/Ar(30%) mixture at V_B = 500 V. (After ref. 5).

Instead of affecting the whole bulk by T_s, it is useful to vary the film properties by V_s. This has been suggested e.g. for the metallization of plastics by metal-filled PP with a gradual increase of f (ref. 19) where good adhesion can be achieved by mechanical interlocking. Besides numerous other applications[2] metal-filled PP were tested for laser writing[20] when a conducting path is produced by local heating.

4. METAL-FILLED DLC MATRICES

Properties of DLC films have been modified by incorporation of additional substances. Doping by phosphine, diborane and nitrogen lead to an increase of conductivity by two orders of magnitude[21]. Application of fluorocarbon gases was aimed at an increase of optical transparency in the visible region but the DLC films became soft at higher fluorine concentrations[22,23]. Another way to alter the optical and electrical properties is to incorporate metal clusters or metal carbides[2,5,24] using simultaneously sputtering or evaporation in a system such as that in Fig. 1. Similarly to metal-filled PP, the metal concentration may be controlled in-situ by optical emission spectroscopy (OES), since the metal line intensities are proportional to the metal concentration in the gas phase and in the final deposits[2].

Different metals have been applied[2]. As an example, the effect of increasing gold concentration on the sheet resistance and microhardness of DLC films obtained from a butane/Ar mixture is shown in Fig. 4. The Ar emission line at 418 nm was used as a reference parameter in the OES data. Both the resistance and microhardness decrease when approaching the percolation threshold (around f ~ 0.4 for Au-containing matrices). It appears that after reaching a certain f the hardness drops abruptly, documenting that the DLC matrix can accommodate only a limited amount of metal. Above this f, depending on the metal, the DLC stability collapses, most likely due to an increase of the graphitic component. These transitions were observed e.g. for Sn[25] or Ta[25,26] around 14 at. % of the metal incorporated.

The presence of small metal clusters in the DLC matrix causes appearance of broad absorption maxima in the visible region[2,5] (see Fig. 5), the position of which depends on the metal and the grain characteristics. In contrast to metal-filled PP higher absorption of the DLC matrix is superimposed in the visible region.

Metal-containing DLC films show good stability with time and also during an accelerated aging process as illustrated in Fig. 6 for a Cu/DLC system. In this test, the absorbance is given for three different wavelengths (compare with Fig. 5) when the samples were kept at each temperature for 1 hour before T_A was further increased. No changes in the grain size and shape were observed by TEM after completing this test. Recent studies of electrical properties of metal/DLC films showed good thermal stability up to about 350°C when the graphitization commences.

Even if the film characteristics may be quite different, one can correlate the sensitivity to aging of metal/PP and metal/DLC with the incident energy of ions bombarding the growing composite structure. The values for E_i and IEDF are usually not readily available for various techniques and deposition conditions, but it can be concluded that the film stability subsequently improves with increasing $E_{i,max}$ derived from

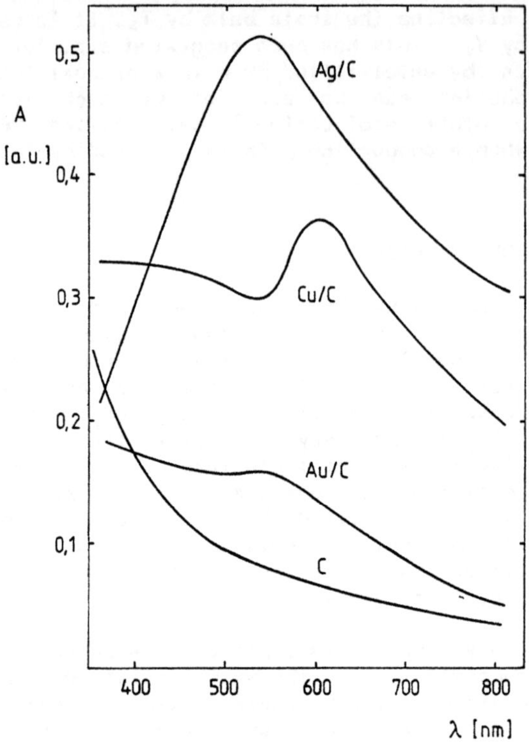

Fig. 5.

Optical absorbance spectra of DLC(C) films containing silver (Ag/C), copper (Cu/C) and gold (Au/C). (After ref. 5).

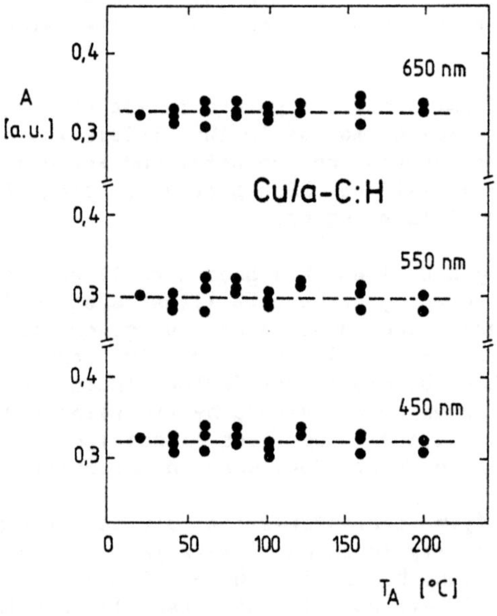

Fig. 6.

Optical absorbance at three selected wavelengths for a copper-doped DLC film versus annealing temperature T_A.

eq. (1). The characteristics of films deposited by other plasma techniques such as unbalanced magnetron[26] ($E_{i,max}$ ~ 20-50 eV[27]) also obey this dependence. Good film stability is reached usually for $E_{i,max} > 200$ eV.[2,24]

5. EFFECT OF FREQUENCY AND ION FLUX

The excitation frequency has an important effect on the plasma parameters, namely, the electron energy distribution function EEDF[28]. Particularly, in the microwave region higher population of electrons in the high energy tail of EEDF causes an increase in the rate of inelastic collisions. This leads generally to a higher concentration of active species in the plasma volume than at lower excitation frequencies and typically an order of magnitude higher deposition rates as documented for PP-isobutylene and PP-perfluorocyclobutane films[29] (see Fig. 7).

Recently, a dual-frequency (MW/RF) plasma deposition has been started in our laboratory[30,31]. This approach consists of the film growth in a large microwave (MW, 2.45 GHz) plasma (LMP®) apparatus using a periodic slow wave structure. Simultaneously, the RF (13.56 MHz) power is capacitively coupled to the substrate holder-electrode, inducing V_B. Adjustment of the incident MW (P_{MW}) and RF (P_{RF}) power values can be used for a relatively independent control of the energy and flux of ions. For a number of gases a two to five fold increase in the concentration of ions in the gas phase was observed when switching from the RF to the MW/RF mode at nominally equal conditions[32]. This was accompanied by a six to fifty fold increase in ϕ_i in the pressure range of 20-200 mTorr. High ϕ_i (10^{15}-10^{16} ions/cm^{2}s) and moderate E_i (for V_B between 100 and 300 V) were found to substantially improve the quality of silicon compound films (e.g. silicon oxide and nitride, a-Si:H, PP-HMDSO) permitting high r_D (20 - 100 Å/s) and low T_s[30,31]; this involves, for example, higher packing density, lower dielectric loss and conductivity or higher photoconductivity of a-Si:H.

The effect of ion flux on the characteristics of DLC has been recently studied using this MW/RF technique[33]. The deposition rate and ϕ_i for the RF and MW/RF modes in methane are plotted in Figs. 8 and 9 versus the total incident power ($P_{tot} = P_{MW}+P_{RF}$). In the MW/RF experiment P_{MW} = 100 W was kept constant and P_{RF} was subsequently increased leading to a rise of V_B up to 800 V.

The increase of r_D with P_{RF} and V_B in the RF mode is the consequence of enhanced surface processes activated by rising E_i. In this case, E_i increases with V_B more rapidly than does ϕ_i, since $\phi_i \sim \sqrt{V_B}$ is typically observed (Fig. 9). In the MW/RF mode, however, the values of ϕ_i are substantially higher. Therefore, the ions such as CH_n^+ (n = 0-4) and H_2^+ may more efficiently assist dissociation upon their impact on the surface promoting formation of atomic hydrogen that enhances etching of the carbonaceous deposit[27]. Also, higher degree of dissociation of CH_4 in the MW/RF mode was documented by a substantial increase in the concentration of CH radicals and atomic hydrogen in the gas phase as revealed by OES. Combination of these processes together with sputtering contribute to the decrease of r_D in Fig. 8. The occurence of a pronounced minimum in ϕ_i for the MW/RF mode (Fig. 9) is not yet well understood. This effect has been observed for numerous other gases[31,32] and it is believed to be associated with development of the sheath region during the transition from a "pure" MW to the MW/RF mode.

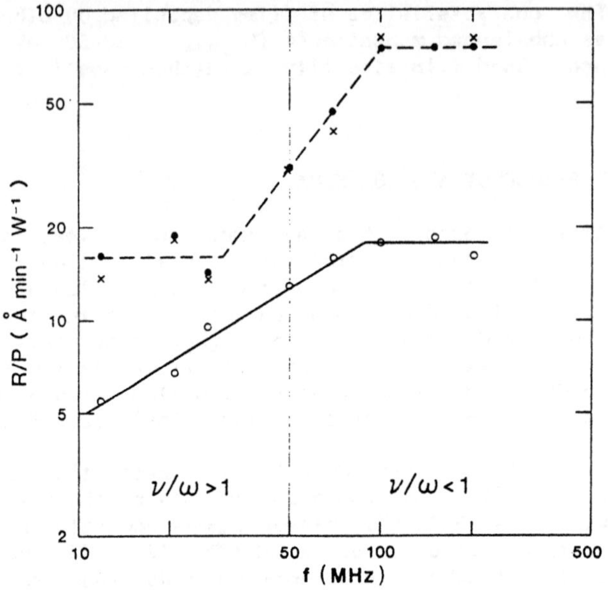

Fig. 7.
Normalized deposition rate (R/P) versus plasma excitation frequency. The upper curve: PP-isobutylene; the lower curve: PP-perfluorocyclobutane. (After ref. 29).

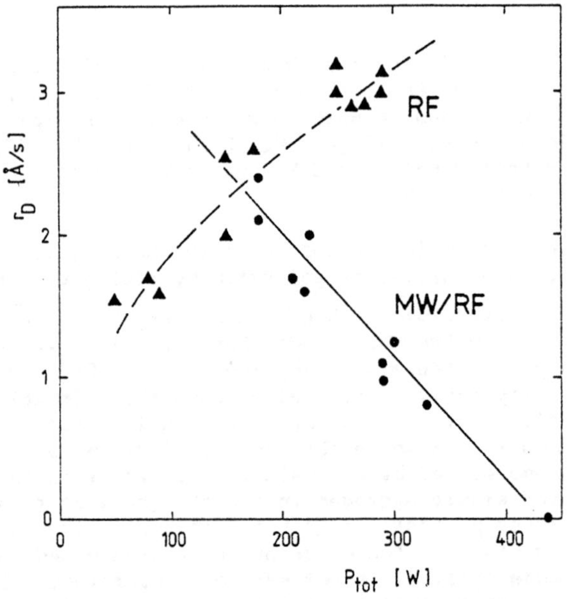

Fig. 8.
Deposition rate r_D versus the total input power P_{tot} for DLC films deposited in the RF discharge and in the dual-frequency MW/RF mode. Conditions were: gas-CH_4, flow rate-10 sccm, pressure-65 mTorr. (After ref. 33).

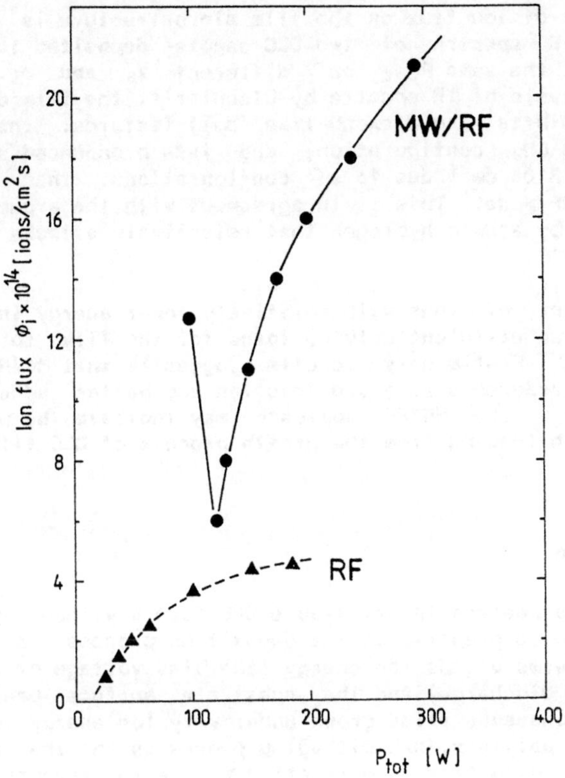

Fig. 9.

Ion flux ϕ_i versus the total input power during deposition of DLC films in the RF discharge and in the dual-frequency MW/RF mode as in Fig. 8. (After ref. 33).

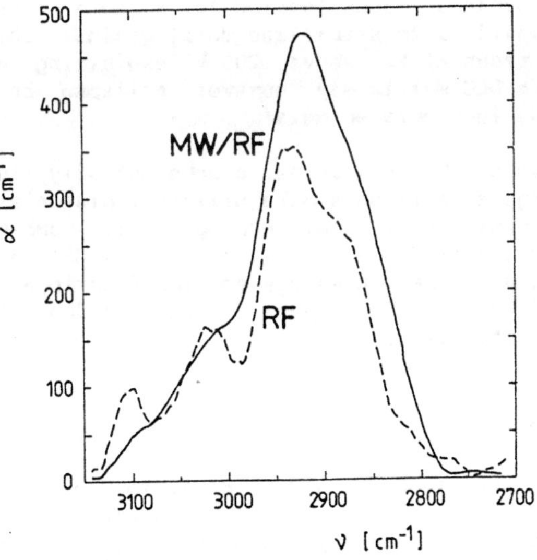

Fig. 10.

Absorption coefficient α in the infrared for DLC films deposited in the RF discharge (P_{RF} = 230 W, V_B = 750 V) and in the dual-frequency MW/RF mode (P_{MW} = 100 W, P_{RF} = 130 W, V_B = 350 V). P_{tot} was equal for both films, conditions were as in Fig. 8. (After ref. 33).

The effect of ion flux on the film microstructure is illustrated in Fig. 10 by IR spectra of two DLC samples deposited in the RF and the MW/RF modes at the same P_{tot} but different V_B and ϕ_i. Based on a detailed analysis of IR spectra by Dischler[12], the film deposited in the MW/RF mode exhibits more diamond-like (sp^3) features: the main peak at 2920 cm^{-1} (sp^3CH_2 configuration) and less pronounced shoulders around 3020 cm^{-1} and 3100 cm^{-1} due to sp^2 configurations, than does the film grown in the RF mode. This is in agreement with the argument of simultaneous etching by atomic hydrogen that selectively attacks the graphitic (sp^2) component[1].

Higher flux of ions with relatively lower energy in the MW/RF mode appears to be an efficient driving force for the films to adopt minimum surface area. Preliminary results suggest[33] that MW/RF-deposited DLC films possess reduced stress and friction and better adhesion, hardness and stability. The MW/RF approach may indicate the possibility of a continuous path leading from the growth process of DLC films to crystalline diamond.

6. CONCLUSIONS

Several micrometers thick, stable DLC layers without apparent spontaneous aging can be prepared if the deposition process is properly controlled in terms of the ion energy (the bias voltage of several hundred volts in an RF discharge) and the substrate surface pretreatment. In general, carbonaceous films grown under very low energy ion bombardment, such as plasma polymers, exhibit aging processes in the ambient atmosphere due to oxidation and reaction with water vapours. This type of aging can be largely suppressed by post-deposition vacuum annealing for several hours or even days. Improvement of the aging characteristics can also be achieved by increased substrate temperature ($\lesssim$ 100°C) and, substantially, by ion bombardment ($\lesssim$ 100 eV) during the film growth. This has been well documented for PP matrices containing metal clusters. The ion bombardment enhances the surface mobility of species and preferential breakage of loose bonds leading to a more crosslinked, densely packed polymer matrix with stabilized metal grains. This applies also to DLC/metal films grown at V_B above 200 V exhibiting negligible aging. Rigidity of the DLC matrix may, however, collapse above a critical concentration of the incorporated metal.

The microwave discharge provides a substantially higher ion flux than the RF discharge as well as a more efficient dissociation of the hydrocarbon gas and formation of atomic hydrogen. If combined with the RF deposition of DLC films in a "dual-frequency MW/RF" mode, it is possible to vary independently the ion energy and ion flux in a wide range. Lower ion energy but higher flux appear to enhance deposition of films with more diamond-like features.

ACKNOWLEDGEMENTS

The author wishes to express his thanks to Prof. M. R. Wertheimer for encouragement during the course of this work. Partial support by the Natural Sciences and Engineering Research Council of Canada and by the Fonds "Formation des Chercheurs et Aide à la Recherche" of Quebec is acknowledged.

REFERENCES

1 J.C. Angus, P. Koidl, S. Domitz, Carbon thin films, in: "Plasma Deposited Thin Films", J. Mort, F. Jansen, eds., CRC Press, Boca Raton (1986).
2. H. Biederman, L. Martinu, Plasma polymer-metal composite films, in: "Plasma Deposition, Treatment and Etching of Polymers", R. d'Agostino, ed., Academic Press, New York (1990).
3. J.M.E. Harper, J.J. Cuomo, R.J. Gambino and H.R. Kaufman, in: "Ion Bombardment Modification of Surfaces: Fundamentals and Applications", O. Auciello, R. Kelly, eds., Elsevier, Amsterdam (1984).
4. P. Koidl, Ch. Wild, B. Dischler, J. Wagner, M. Ramsteiner, in: "Amorphous Hydrogenated Carbon", J.J. Pouch, S.A. Alterovitz, eds., Mater Sci. Forum, Vol. 52/53: 41 (1990).
5. L. Martinu, Amorphous carbon films, in: "High Energy Density Technologies in Materials Science", F. Garbassi, E. Occhiello, eds., Kluwer Academic Publ., Dordrecht (1990).
6. C. Wild, P. Koidl, J. Wagner, Plasma deposition of a-C:H films: the role of process gas, plasma chemistry and plasma surface interactions, in: "Amorphous Hydrogenated Carbon Films", P. Koidl, P. Oelhafen, eds., Proc. E-MRS, Vol. 17, Les Editions de Physique, Paris (1987).
7. J.C. Angus, F. Jansen, Dense "diamondlike" hydrocarbons as random covalent networks, J. Vac. Sci. Technol. A 6: 1778 (1988).
8. H. Yasuda, "Plasma Polymerization", Academic Press, New York (1985).
9. A.M. Wrobel, Aging process in plasma-polymerized organosilicon thin films, J. Macromol. Sci.-Chem. A22:1089 (1985).
10. L. Martinu, H. Biederman, J. Nedbal, Dielectric properties of fluoro- and chlorofluorocarbon films plasma polymerized in an RF glow discharge, Thin Solid Films 136: 11 (1986).
11. J. Perrin, R.L. Siemens, E. Kay, Thermal transitions and dielectric relaxations in plasma-polymerized polytetrafluoroethylene, IBM Res. Rep. RJ 4792 (1985).
12. B. Dischler, Bonding and hydrogen incorporation in a-C:H studied by infrared spectroscopy, p. 189 in ref. 6.
13. C. Wild, P. Koidl, Network structure and thermal decomposition of plasma deposited a-C:H films studied by gas effusion spectroscopy, p. 207 in ref. 6.
14. S. Orzeszko, Bhola N. De, J.A. Woollam, J.J. Pouch, S.A. Alterovitz, D.C. Ingram, Thin film hermeticity: A quantitative analysis of diamondlike carbon using variable spectroscopic ellipsometry, J. Appl. Phys. 64: 4175 (1988).
15. C.J. Robinson, R.N. Payne, A.E. Bell, Hydrogenated amorphous carbon dielectric coatings for magneto-optic data storage media, J. Appl. Phys. 64: 4646 (1988).
16. H. Biederman, L. Martinu, S. Nespurek, Aging effects in composite plasma polymer/metal films, in: Proc. 8th Int. Symp. Plasma Chem. (ISPC-8), K. Akashi, A. Kinbara, eds., Tokyo (1987), p. 1364.
17. L. Martinu, Optical response of composite plasma polymer/metal films in effective medium approach, Solar Energy Mater. 15: 21 (1987).
18. R. d'Agostino, L. Martinu, V. Pische, Effect of bias and temperature on the bulk and surface properties of gold containing plasma polymerized fluorocarbons, Plasma Chem. Plasma Process., in press.

19. L. Martinu, V. Pische, R. d'Agostino, Surface Modification by plasma deposition of metal-filled polymers, in: "Metallisation in Polymers", E. Sacher, J.J. Pireaux, S. Kowalczyk, eds., ACS Symp. Ser. 440, (1990), in press.
20. E. Kay, Ion and laser interactions as encountered in synthesis and utilisation of metal-containing polymers, Proc. E-MRS, Vol. 15, p. 355, Les Editions de Physique, Paris (1987).
21. D.I. Jones, A.D. Stewart, Properties of hydrogenated amorphous carbon films and the effects of doping, Phil. Mag. B 46: 423 (1982).
22. R.E. Sah, B. Dischler, A. Bubenzer, P. Koidl, Amorphous carbon coatings prepared by high rate rf plasma deposition of fluorinated benzenes, Appl. Phys. Lett. 46:739 (1985).
23. H. Biederman, L. Martinu, J. Zemek, Carbon and composite carbon metal films deposited from an rf discharge operated in organic gases, Vacuum 35:447 (1985).
24. C.P. Klages, R. Memming, Microstructure in physical properties of metal-containing hydrogenated carbon films, p. 609 in ref. 4.
25. C. Weissmantel, E. Ackerman, K. Bewilogna, G. Hecht, K. Kupfer, B. Rau, Structure property relationships of carbonaceous films grown under ion enhancement, J. Vac. Sci. Technol. A 4:2892 (1986).
26. H. Biederman, K. Kahoutek, Z. Chmel, V. Stary, R.P. Howson, Hard carbon and composite metal/hard carbon films prepared by a dc unbalanced planar magnetron, Vacuum 40:251 (1990).
27. N. Savvides, Diamondlike films and properties, p. 408 in ref. 4.
28. M.R. Wertheimer, M. Moisan, Comparison of microwave and lower frequency plasmas for thin film deposition and etching, J. Vac. Sci. Technol. A 3:2643 (1985).
29. R. Claude, M. Moisan, M.R. Wertheimer, Z. Zakrzewski, Comparison of microwave and lower frequency discharges for plasma polymerization, Plasma Chem. Plasma Process. 7:451 (1987).
30. L. Martinu, J.E. Klemberg-Sapieha, M.R. Wertheimer, Dual-mode microwave/radio frequency plasma deposition of dielectric thin films, Appl. Phys. Lett. 54, 2645 (1989).
31. J.E. Klemberg-Sapieha, O.M. Küttel, L. Martinu, M.R. Wertheimer, Dual microwave/radio frequency plasma deposition of functional coatings, Thin Solid Films, (1990) in press.
32. O.M. Küttel, J.E. Klemberg-Sapieha, L. Martinu, M.R. Wertheimer, Energy fluxes in mixed microwave radio frequency plasma, Thin Solid Films 193, (1990), in press.
33. O.M. Küttel, L. Martinu, D. Poitras, J.E. Klemberg-Sapieha, M.R. Wertheimer, Diamond-like carbon films deposited in dual microwave/radio frequency plasma, Proc. E-MRS Meeting, Strasbourg, November 1990, submitted.

APPLICATIONS OF DIAMOND-LIKE (HARD CARBON) FILMS

Alan H. Lettington

Royal Signals and Radar Establishment
St Andrews Road, Malvern, Worcs WR14 3PS, UK

ABSTRACT

Pure diamond has excellent mechanical and infra red properties and would make an ideal window material were it readily available. Synthetic diamond films can be grown with good infra red transmission and used as protective coatings. Many coating requirements can already be met using diamond-like carbon films. This paper reviews the progress.

1. INTRODUCTION

Diamond is a metastable form of carbon which reverts to graphite only at elevated temperatures. It has quite remarkable physical properties. It is exceedingly hard, has a high Young's modulus and tensile strength combined with high thermal conductivity and thermal shock properties. See Table 1. It is chemically inert and in pure form has excellent infra red transmission in the 8-14 μm waveband. It would make an ideal infra red window material for high speed airborne applications, and for use with high power CO_2 lasers.

TABLE 1
PHYSICAL PROPERTIES OF DIAMOND

Hardness	10 Mg/mm^2
Young's Modulus	945 GNm^{-2}
Thermal Conductivity	20 W/cm.deg K
Refractive Index	2.42
Bandgap	5.45 eV

There is interest in growing this material at low temperatures and pressures, either as a bulk material, or else as a protective coating on existing materials. In those applications where thin layers are adequate then diamond-like carbon (DLC) may provide an immediate solution.

2. OPTICAL PROPERTIES OF DIAMOND AND DIAMOND THIN FILMS

Diamond has a cubic structure with the space group O_h^7. It has a very wide electronic band gap ~ 5.5eV and is transparent from about 2265Å in pure material out to the far infra red. In pure diamond first order phonon infra red absorption processes are disallowed from symmetry considerations and the material is transparent beyond 7 microns in the all important atmospheric window in the 8-14 μm waveband. Second order phonon processes are infra red active and result in the absorption processes present in the middle infra red 3-5μm waveband.

With increasing impurity concentrations symmetries are broken and forbidden transitions become allowed increasing the infra red absorption beyond 7 microns. This effect is illustrated in Figures 1 and 2. Type 2A diamond, which is a rare reasonably pure form of diamond, shows only the two phonon absorption processes while the type 1A diamond, which is more abundant, shows both two phonon and nitrogen impurity activated single phonon processes.

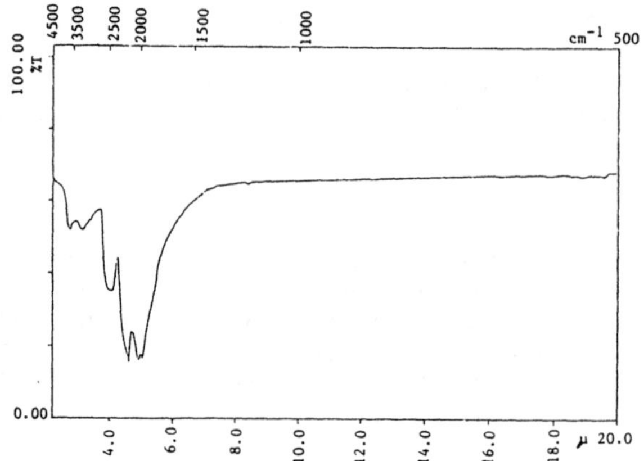

Figure 1. Transmission of Type 2A Diamond 1 mm thick

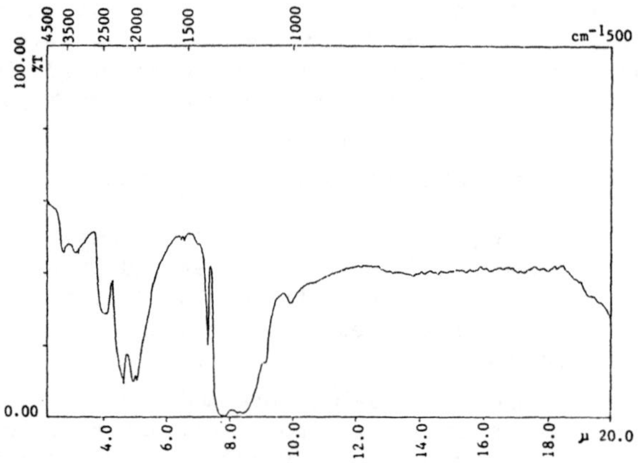

Figure 2. Transmission of Type 1A Diamond 1mm thick.

Many pieces of natural diamond can be totally opaque in the far infra red. The cost and availability of pure natural diamond has limited its infra red uses, and considerable attention is being paid to synthetic forms. Diamonds may be grown artificially[1] but still in limited sizes.

Diamond thin films have been deposited at low pressures by a variety of plasma assisted and other CVD processes under non equilibrium conditions using a hydrocarbon feedstock (see reviews[2,3]). One of the key features is to employ excess hydrogen which reacts preferentially with any graphite that might otherwise be formed. The thin diamond layers can contain a variety of defects which can increase their infra red absorption. These can include atomic impurities, fine grain size and other disorders such as the inclusion of trigonally bonded carbon atoms.

The Raman spectra of diamond are well known[4] and such measurements have been used to characterise the structure of diamond thin films. The zone centre optical phonon Γ^{25+} is Raman active in first order and produces a sharp line with a shift of 1332cm^{-1}. Second order Raman processes have a peak in their spectrum at 2458cm^{-1}. These processes can be seen in Figure 3 for both type 1A diamond and a sample of microwave PACVD diamond thin film. In comparison the 1332cm^{-1} line has disappeared in a sample of diamond-like carbon coating which contains a significant amount of trigonally bonded carbon.

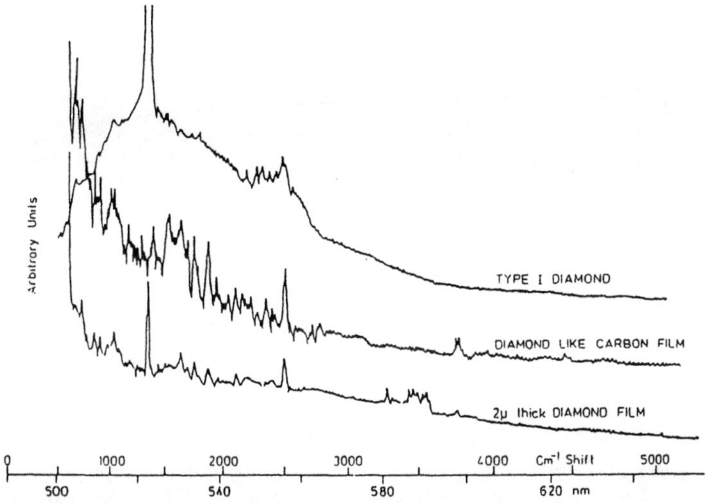

Figure 3. Raman Spectra of Type 1A Diamond, DLC and a 2μm Diamond Film

The possibility of being able to grow diamond films uniformly over large areas as thin protective coatings on conventional infra red optical materials is being investigated[5]. In spite of the absorption mechanisms present in diamond thin films it is unlikely that infra red absorption will be significant in layers a few tens of microns thick since the internal absorption is probably less than 20cm^{-1}. The major loss mechanism at present is scattering at the surface of the layer since this is quite rough (see Figure 4) and has a sooty black appearance. The reflectivity from such a layer on a silicon substrate is illustrated in Figure 5.

It can be seen that there is little difference in reflectivity between the coated and uncoated surfaces at 1300cm^{-1} where the optical thickness of the coating is $\lambda/2$. This implies that there is little absorption or scattering (ie ~ 25cm^{-1}) in this spectral region which is at the peak of the infra red absorption at 8 μm (see Figure 2). These scattering layers were deposited at a rate of 1μm/hr with a substrate temperature of 1000°C.

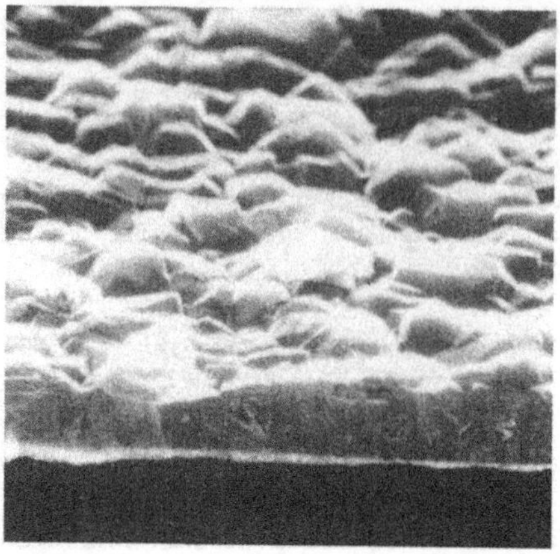

Figure 4. Electron micrograph of a section of a diamond thin film on a silicon substrate (x 5000)

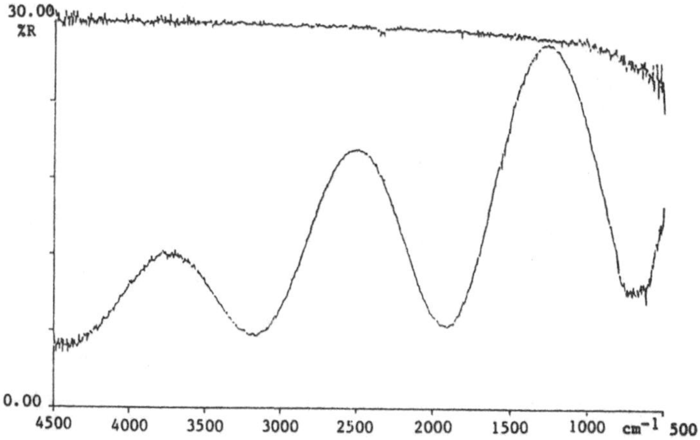

Figure 5. Reflectivity of uncoated silicon substrate and of a 2μm thick layer of a diamond thin film on silicon

In more recent experiments workers at Plessey Research Caswell have achieved visibly transparent thin self supporting layers having excellent infra-red transmission (see Figure 6) using a substrate temperature nearer 500°C. At this temperature it is possible to coat most infra-red transparent components without damaging the components.

The deposition rate has now dropped to around 0.1μm/hr and is the subject of further research.

There is, however, the prospect of being able to grow thicker self supporting diamond layers that can act as infra red transmitting windows. At present the thermal CVD deposition process for diamond

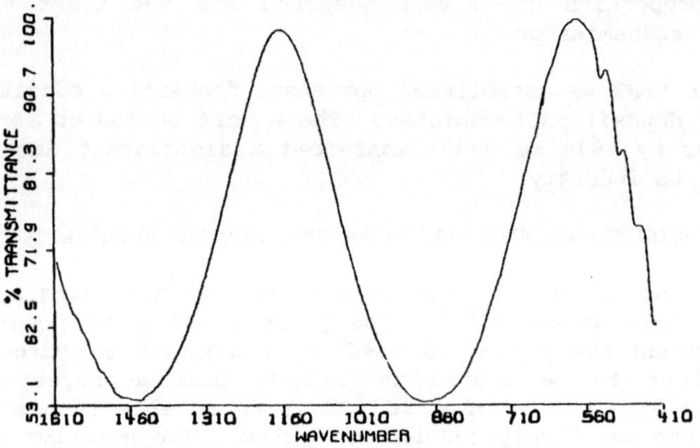

Figure 6. Transmission of thin self supporting diamond layer

thin films appears to have greater potential for scaling up to commercial sizes than most other processes.

3. DIAMOND-LIKE CARBON COATINGS

Diamond-like carbon coatings are far easier to grow than diamond films and have been applied more widely. They contain a mixture of sp^3 and sp^2 coordinated carbon as well as a significant amount of hydrogen so that any possible $1332cm^{-1}$ Raman line is completely swamped (see Figure 3). The absorption of these coatings decreases with increasing wavelength. This is illustrated in Figure 20 for a particularly absorbing layer. Typically the absorption at 10 μm wavelength is around $250cm^{-1}$. Although this is high it accounts for no more than 3% absorption in a coating 1μm thick. This coating is smooth, hard and durable and we have used it for a variety of infra red applications using a range of RF and DC plasma deposition techniques. Originally our requirement was for a durable anti-reflection coating for germanium infra red optics in the 8-13 μm waveband.

Germanium is the most commonly used window and lens material in this waveband. It is easily scratched by sand and readily attacked chemically by salt water. In addition it has a refractive index of 4 which means that 36% of all incident radiation is reflected at each germanium surface. Anti-reflection coatings based on ZnS and Ge multilayers were commercially available. They had good optical performance but their mechanical and chemical durability was poor.

We decided to investigate within RSRE the possibility of using "diamond-like carbon" (DLC) coatings to anti-reflect germanium optics, since pure diamond was known to have the correct refractive index for a single layer coating and to have good optical properties and excellent durability.

Professor Spear at Dundee University and Professor Holland at Sussex were known to be working on glow discharge methods. We supplied them with polished germanium samples to investigate their processes. The films had variable hardness and adhesion and the infrared properties which were measured for the first time showed reasonable transmission.

Within RSRE we established optimised deposition conditions for a variety of deposition techniques. These were scaled up to coat large samples and by 1978 we had transferred a significant amount of this technology to industry.

4. DEPOSITION TECHNIQUES FOR INFRA RED OPTICAL APPLICATIONS

In all of our DLC deposition processes we have used a version of the glow discharge method[6,7,8] using either DC or RF power supplies. In this method the plasma is used to dissociate a hydrocarbon gas. The component to be coated is usually biassed negatively either through the application of a DC potential or else through the diode action of a capacitively coupled RF system. The positive carbon ions are attracted to it producing the deposited coating.

The optical and mechanical properties of the deposited carbon layers are strongly dependent on the deposition conditions and the nature of the substrate. In the RF glow discharge process there is a balance between film growth and back sputtering from the substrate. With many substrate materials it is difficult to obtain an initial nucleation of the film and instead the substrate is eroded away or else becomes covered with a sooty deposit. Once nucleation has occurred, lateral growth can be very rapid but can be impeded by scratches on the surface of the substrate.

Where necessary we have used a variety of intermediate bonding layers on metallic and non-metallic substrates using either a DC or RF deposition process.

The properties of diamond-like layers using an RF glow discharge technique have been described elsewhere[9]. In early unpublished work we explored the range of hydrogenated carbon deposition from soft cross-linked plastics to hard black graphitic layers. The plastic nature occurred at low potentials (~ 100V) and temperatures while harder coatings were achieved with potentials nearer 1KV.

With this high negative potential on the substrate, positive carbon ions are attracted to it producing the DLC coating. During deposition the energy of the incident carbon ions tend to desorb the stable and most abundant graphitic form of carbon leaving the more strongly bonded diamond-like coating. This high impact energy has tended to build stresses in the coating limiting its maximum thickness to about 2-3 μm. This material contains very fine diamond crystallites, a few tens of nanometers large, which are embedded in a matrix of hydrogenated heavily cross-linked polymer containing graphitic impurities. Investigations at Loughborough University and more recently at Cambridge University have identified the presence of these small diamond crystallites in diamond-like carbon.

We have studied a modified RF glow discharge technique in which excess hydrogen is added to the hydrocarbon feedstock, in a similar way to the deposition of pure synthetic films. The effect of the excess hydrogen is to increase the efficiency of the process, and although the main flow of hydrocarbon has been reduced the deposition rate of the coating has remained unchanged at about 1μm per hour. In our improved deposition technique we have increased the temperature of the substrate in order to compensate for the otherwise increased hydrogen content in our films. The resulting coatings have shown a reduced graphitic scattering while still retaining their excellent IR transmission and durability. These coatings are still highly disordered which we believe is an advantage since it has suppressed or broadened the otherwise dominant impurity activated phonon absorption lines.

In the early stages of this work we investigated a variety of deposition techniques including ion beam processes but favoured the glow discharge technique because it is able to coat uniformly around curved shapes cover all sides of an object and can readily be scaled up to cover large areas.

The main requirement for anti-reflection coatings for germanium infrared optics called for good transmission in the 8-13 μm spectral region combined with good mechanical and chemical durability and strong adhesion to the germanium substrate. There were also the other considerations that the process had to be cheap, reliable and capable of coating large samples rapidly.

The variation of the hardness of the coating with deposition conditions tended to be in the opposite direction to the variation of the infrared transmission, and a compromise was adopted. The first decision was to relax the hardness requirement so that the coating was harder than sand but not exceptionally so. Then the requirement was for best possible transmission while maintaining the hardness. To this end about 30 at.% hydrogen remained in the films, and another compromise was achieved between removing spectral absorption lines and reducing the amount of scattering from small graphitic particles.

5. PROPERTIES OF DIAMOND-LIKE CARBON COATINGS ON GERMANIUM

Initially 25mm diameter test samples of germanium were coated and subjected to laboratory tests. The process was then scaled up to larger components which were subjected to various field trials. Finally the process was used to coat windows and lenses for project applications.

The testing was to RSRE Specification TS 1888 and is described in greater detail elsewhere[10]. For the laboratory testing of the abrasion and wear resistance of coatings we developed the abrasion tester illustrated in Figure 7.

The test sample is placed at the bottom of a cylindrical container and must survive being scoured under load in a slurry of sand and water at a 1000 rpm for 5 minutes. The equivalent field trial involved a simulated windscreen wiper mounted on the outside of a vehicle undergoing extensive driving trials, see Figure 8.

Sea trials involved a salt water immersion test. In the winter of 1978 two of our coated discs of germanium were mounted at sea level on a fort in the Solent along with an uncoated square of germanium. They remained at sea continuously for 4¼ months. The effect of the sea water at the end of the trial can be seen in Figure 9. The coated samples were virtually unaffected while the uncoated sample was badly corroded and etched away. Examples of the use of diamond-like carbon anti-reflection coatings on germanium FLIR windows are shown in Figures 10 and 11.

Figure 7. RSRE Abrasion Tester

Figure 8. Windscreen Screen Test

Figure 9. Results of Salt Water Imersion

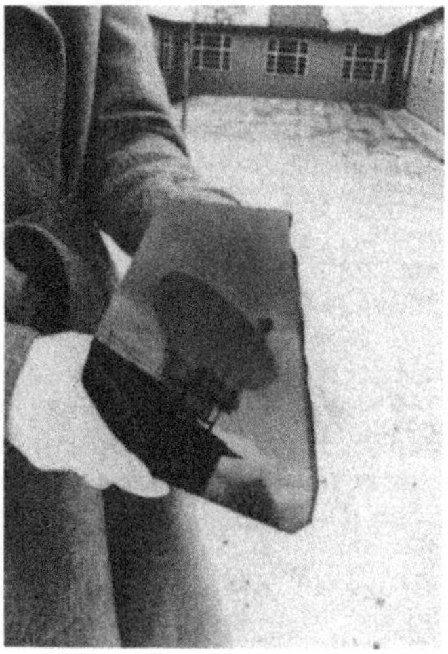

Figure 10. A coated window for a Helecopter Pilot Aid

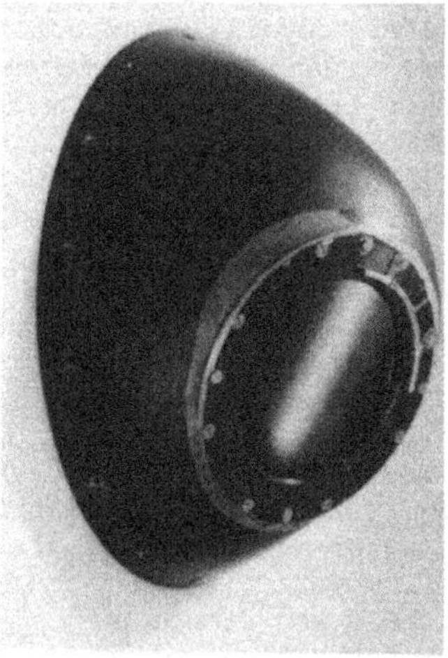

Figure 11. A coated FLIR window on an Aircraft nose cone

A typical optical performance of the now commercially available diamond like carbon coating on germanium is illustrated in Figure 12. This has a high efficiency coating on the reverse side. There is a five percent loss in transmission at the peak wavelength due partly to the refractive index mismatch and partly to absorption in the coating.

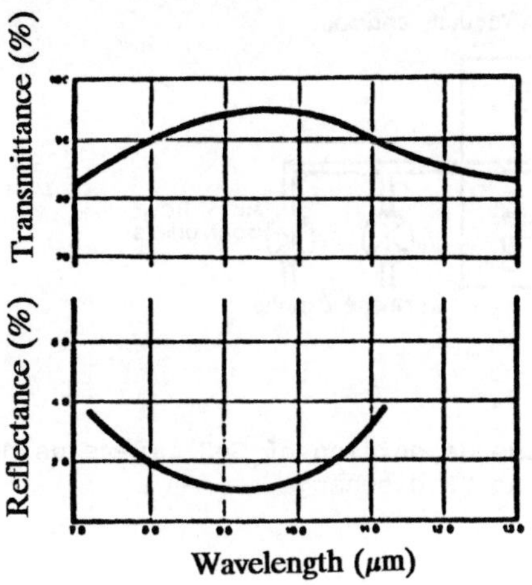

Figure 12. Typical Performance of a Commercially Available Hard Carbon Coating

This spectral graph shows transmittance of a 1mm thick germanium substrate coated on the front surface with Hard Carbon coating to OCLI specification 6040011 with a High Efficiency Antireflection coating to OCLI specification 6040008 on the rear surface with the following performance:

T ≥ 90% average 8-11 microns Peak T. of 95% (Typical)
R ≤ 3.5% average 8-11 microns Minimum R. of 1% (Typical)

6. RAIN EROSION STUDIES

Tests on our DLC coated germanium samples were carried out at RAE Farnborough. They were exposed to their standard rain conditions for a few minutes at normal incidence to the rain (1 inch/litre) and at 500 mph impact velocity. Pitting did occur in our samples but the extent was considerably reduced compared to uncoated germanium. The effect of rain damage can be reduced still further through inclining the window to the direction of flight. Our coated germanium samples have also been subjected to single liquid drop impact testing at Cambridge University[11].

Another of our coating requirements was to protect zinc sulphide infra red windows from rain impact damage. DLC coatings on their own were not adequate since the stress and absorption in the coating limited the thickness that could be used.

This problem was relieved to a certain extent through our development of GeC[12]. It is a tough durable coating material with a refractive index a function of composition, that can be designed into multilayer structures. It bonds well to DLC which is frequently used as the outermost layer. In this form it can act as a stress relieving coating for zinc sulphide windows.

In early work two plasma deposition methods were investigated. One used a mixed gaseous feed stock of Germane (GeH_4) and butane in a capacitively coupled RF discharge - the deposition occurring on the cathode. The other was a reactive sputtering technique in which a mixture of butane and argon was used to sputter from a germanium target - the deposition now occurring on the anode. These two processes are illustrated in Figures 13 and 14.

The gaseous technique could produce material of varying composition and refractive index throughout the entire composition range. The sputtering technique was limited to germanium concentrations greater than about 40 atomic %. Attempts to produce lower concentrations coated the germanium target with diamond-like carbon. These techniques were developed further and exploited in coating designs in the UK industry, mixed gases in Barr and Stroud and reactive sputtering in Plessey Research Caswell.

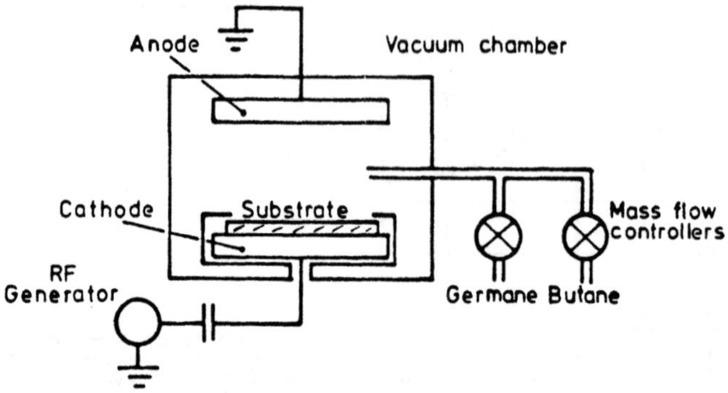

Figure 13. Mixed gas method for the deposition of GeC layers using germane and butane in an RF discharge.

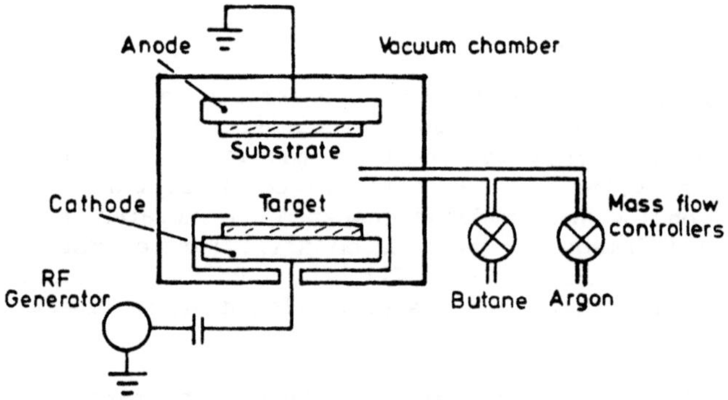

Figure 14. Reactive sputtering technique which uses a mixture of butane and argon gases and a germanium target.

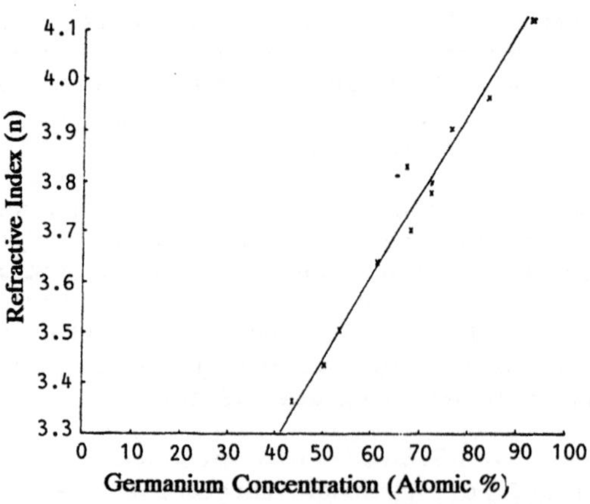

Figure 15. Refractive index as a function as germainium concentration for reactively sputtered layers

The refractive index is a complex function of layer stoichiometry, void content and included gas concentrations. Figure 15 shows the effect of composition on refractive index for reactively sputtered layers containing less than 1 at.% H and less than 0.4 at.% Ar. The equivalent refractive index variation for mixed gas deposition is shown in Table 2 as a function of gas flow ratio.

Table 2. Relationship between refractive index of Germanium Carbide and Butane: Germane gas flow Ratio.

Butane: germane gas flow ratio	Refractive index of germanium carbide
0	4.1
0.25	3.25
0.5	3.0
1.0	2.85
2.0	2.65
∞	2.0

High durability multilayer antireflection coatings for ZnS have been developed incorporating germanium carbide layers produced by both deposition techniques. In both cases the outside layer consists of diamond-like carbon.

These coatings are extremely durable and easily pass the RSRE sand/water wiper test of TS 1888. Thicker GeC layers may be used in coatings produced by reactive sputtering because of the lower absorption produced by this method. Both coatings have good transmission at elevated temperatures.

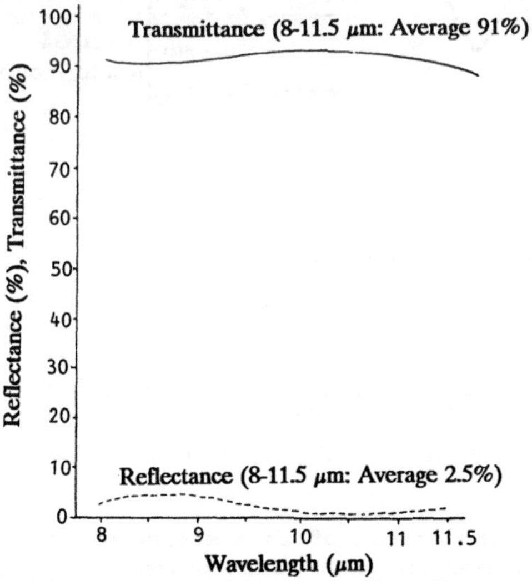

Figure 16. IR transmittance and reflectance of a ZnS substrate coated with Barr and Stroud ARZ5.

The optical characteristics of the Barr and Stroud coatings (ARZ5) on ZnS is shown in Figure 16. A full environmental test programme is in progress and initial results suggest that ARZ5 will provide a similar level of protection to zinc sulphide to that achieved with a diamond-like carbon coating on germanium. It has also enhanced the rain erosion resistance windows.

7. APPLICATION OF CARBON COATINGS TO FRONT SURFACE ALUMINIUM MIRRORS

In current thermal imaging systems, front surface mirrors produced by single point diamond machining of bulk aluminium are used for rotating polygons, flapping mirrors and for relay elements with optical power. The optical performance of these components tends to deteriorate with time and with exposure to the atmosphere. This process can be prevented through the use of a suitable optical coating. Unfortunately the reflectivity of these coated surfaces can be low when used at oblique incidence. This effect has been demonstrated[13,14] in aluminium mirrors protected with thin overcoatings of SiO_x and intended for use in the 8-12μm spectral band. This effect occurs for only one direction of polarisation, R_p, parallel to the plane of incidence. Similar effects are observed for many other protective coatings and other metallic reflectors[14,15] (see Figure 17) making these coatings unsuitable for use in 8-12 μm thermal imaging systems on 45 degree mirrors or scanning polygons.

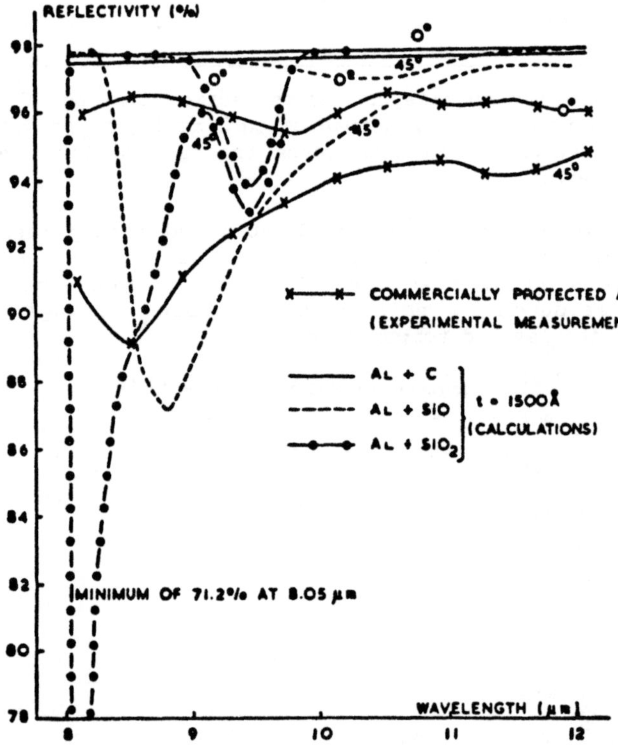

Figure 17. Reflectivity of Protected Aluminium Surfaces

We[16] identified the origin of this effect and demonstrated that it does not occur in diamond-like carbon protective coatings on these front surface mirrors (see Figure 17). The problem with coatings such as SiO_x is that they have strong optical absorption lines in the spectral region of interest. The coatings are sufficiently thin for this to produce negligible absorption. It does however, affect the values of n and k in the absorbing region such that the Brewster angle at the air-coating interface occurs at very low angles of incidence. There is destructive interference between this reflection and that at the coating-metal boundary resulting in a loss of reflectivity. This effect does not occur with diamond-like carbon coatings on these mirror surfaces. An example of an RSRE coated open catadioptric infrared telescope is illustrated in Figure 18.

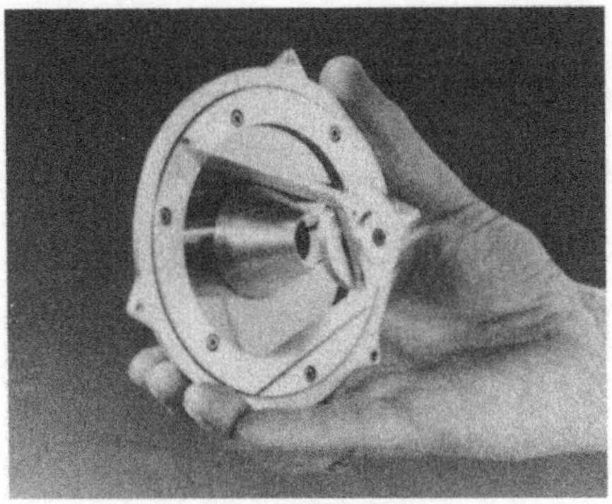

Figure 18. RSRE Coated Catadioptric Telescope
(Courtesy GEC Sensors Pullin Division)

8. THE USE OF DLC COATINGS FOR PHOTOTHERMAL CONVERSION OF SOLAR ENERGY

The main aim of photothermal solar energy conversion is to collect solar radiation and to convert it into useful heat.

There are two main types of converter: the flat plate collector, where an area of an absorbing material is placed so as to collect the solar radiation; and the focussing collector, where solar radiation is condensed on to a smaller absorbing area.

Heat losses from the flat plate collector are high and in the UK the operating temperature is rarely higher than 70°C. The focussing system has higher thermal efficiencies and operating temperatures of several hundred degrees centigrade are possible. It is usual to remove the heat from both systems with circulating water.

For either collection process to be effective there must be maximum absorption of the solar radiation and minimum heat losses.

The absorption of solar radiation of a coated surface is determined by measuring or calculating the solar absorptance α, of the coating, which is defined by

$$\alpha = \frac{\int_0^\infty a(\lambda)\ H(\lambda)\ d\lambda}{\int_0^\infty H(\lambda)\ d\lambda} \tag{1}$$

where $H(\lambda)$ is the solar flux and $a(\lambda)$ is the spectrally dependent absorptance of the coating.

The most widely employed solar flux data is that due to Moon[17] and for solar coatings calculations, the spectrum for an air mass of 2 is frequently used. This is equivalent to an elevation of the sun of 30° above the horizon. The data tabulated in Ref 18 is illustrated in Figure 19 and has been used when calculating α.

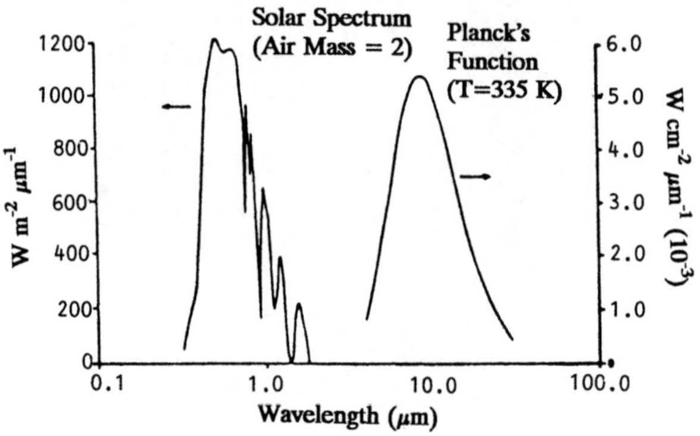

Figure 19. Solar and Re-radiated Spectra

With regard to heat losses there are three types: convection, conduction and radiation. The first two are under the control of the system designer and must be minimised for maximum efficiency. It is the radiation loss from the absorber which is of consideration here. This is defined by ϵ, the thermal emittance of the coating and given by

$$\epsilon = \frac{\int_0^\infty e(\lambda)\,\omega(\lambda,T)\,d\lambda}{\int_0^\infty \omega(\lambda,T)\,d\lambda} \qquad (2)$$

where $e(\lambda)$ is the spectral emissivity of the coating at wavelength and $\omega(\lambda,T)$ is Planck's function at wavelength λ and temperature T.

The variation of $\omega(\lambda,T)$ with wavelength is shown in Figure 19 for a blackbody temperature of 325 K - a typical operating temperature for a flat plate collector.

An important quantity which must be maximised for any photothermal solar energy converter is the ratio α/ϵ where the separate quantities are defined by equations (1) and (2). Both $a(\lambda)$ and $e(\lambda)$ in these equations are given by $1-R(\lambda)$ where $R(\lambda)$ is the hemispherical reflectivity of the coated surface. So for the highest α/ϵ ratio the surface must be spectrally selective.

From Figure 19 we see the surface must have high absorption (low reflectivity) from 0.3 to about 1.7μm and low emission (high reflectivity) above 2μm, with a sharp transition between these two regions.

Several spectrally selective coatings have already been proposed and made using silicon or germanium layers deposited on to polished metal substrates[18-21], and some time ago we[22] proposed the use of diamond-like carbon for this purpose. We measured the absorption coefficient of our diamond-like carbon coatings over the visible and infra red regions (see Figure 20) and from this and the measured refractive index of about 2.2 calculated the values of α, ϵ and α/ϵ for C, Si and Ge single layer coatings of varying thickness deposited on to aluminium (see Table 3). The optical properties for Al and Si were taken from Ref 23 and those for Ge from Ref 24. From Table 3 it can be seen that a single layer of carbon has the highest efficiency. This value however is not high enough for most applications and we have sought ways to improve the α/ϵ ratio in multilayer designs.

Table 3. Solar Absorptance and Thermal Emittance of Single Layer C, Ge and Si Coatings on Aluminium

Thickness (μm)	C			Si			Ge		
	α	ϵ	α/ϵ	α	ϵ	α/ϵ	α	ϵ	α/ϵ
0.5	0.70	0.05	14.0	0.53	0.07	7.6	0.50	0.10	5.0
1.0	0.77	0.17	4.5	0.56	0.07	8.0	0.52	0.08	6.5
1.5	0.79	0.23	3.4	0.54	0.06	9.0	0.52	0.07	7.4

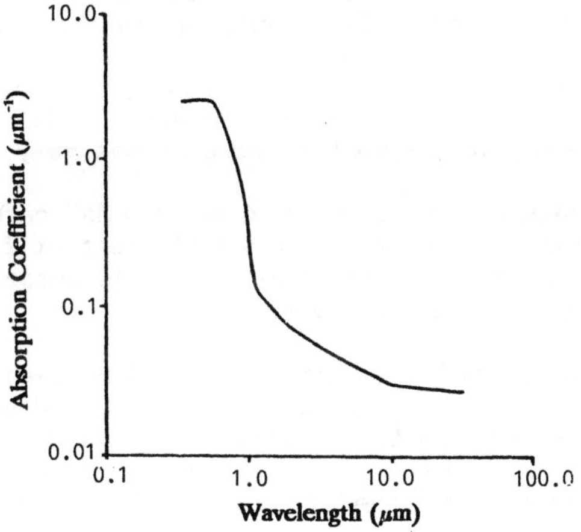

Figure 20. Absorption Coefficient of Carbon

The results for two layer coatings on aluminium are shown in Table 4, where we have compared the single layer carbon with anti-reflection coatings on silicon and germanium layers. The coated carbon layers were highest but may still not be high enough to be cost effective in temperate climates.

Table 4. Solar Absorptance and Thermal Emittance of Two Layer Coatings on Aluminium

Outside Layer	Outside Layer Thickness (μm)	Inside Layer	Inside Layer Thickness (μm)	α_n	ϵ_n (at 335K)	α_n/ϵ_n
-	-	C	0.5	0.70	0.05	14
		C	1.0	0.77	0.17	5
		C	1.5	0.79	0.23	3
n=1.45	0.09	C	0.4	0.75	0.04	19
n=1.45	0.09	C	0.5	0.79	0.06	13
n=1.45	0.09	C	0.6	0.80	0.08	10
C	0.06	Si	1.25	0.78	0.07	11
C	0.06	Si	1.40	0.79	0.07	11
C	0.06	Si	1.65	0.79	0.06	13
C	0.08	Ge	0.85	0.83	0.09	9
C	0.08	Ge	1.20	0.84	0.09	9
C	0.08	Ge	1.55	0.84	0.08	11

9. MECHANICAL APPLICATIONS OF DIAMOND-LIKE CARBON LAYERS

Apart from the infra red properties of diamond-like carbon, discussed in the previous section, the material is also hard and chemically durable and is useful for protecting metal objects from scratching and chemical attack. We have coated a variety of metal objects ranging from large sheets to nails, twist drills and machine tool inserts. Some of these have remained in the open exposed to the atmosphere for the last seven years without deterioration. The machine tool insert was used for the high speed cutting of aluminium and lasted longer than uncoated inserts. In these cases the commercial viability of the coating process was very finely balanced.

Other interesting features of diamond-like carbon are the fact that it is hydrophobic and has a low coefficient of friction (~ 0.14). We have coated a number of moving parts inside automobile engines and have successfully reduced wear rates.

Another area in which we had success has been in coating thin optical fibres. A carbon coating inhibited the attack of the silica fibre by moisture so preventing a site for subsequent brittle fracture from occurring. See Figure 21. These coatings were also able to restore the properties of aged fibres, probably as a result of the initial back sputtering stage removing some of the surface contamination.

10. MEDICAL APPLICATIONS OF DIAMOND-LIKE CARBON LAYERS

Early attempts to produce a carbon coated "non-stick" frying pan failed disastrously. The strong adhesion led one to think about its compatibility with skin and tissue.

One application that we investigated was coating orthopaedic pins, (see Figure 22). These pins are used to place a fractured bone under tension and pass through the surface of the skin. Unfortunately where they come into contact with the air is frequently a place where infection starts. In an attempt to overcome this problem we coated a set of pins which were used in sheep trials by Professor McGibben of Cardiff Royal Infirmary. The trials were successful and the tissue knitted to the coated pins.

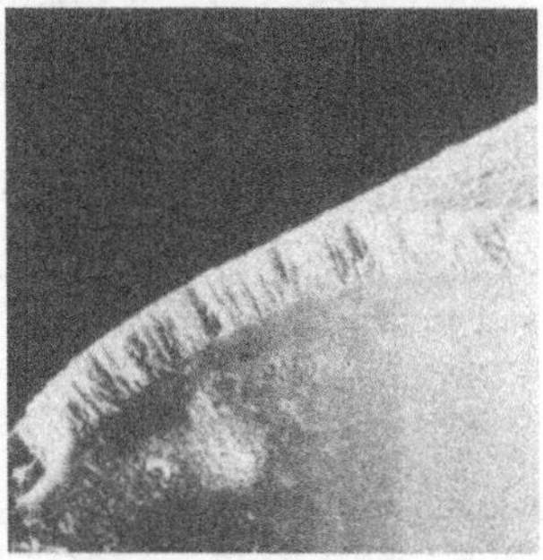

Figure 21. Cross section of a coated Optical Fibre

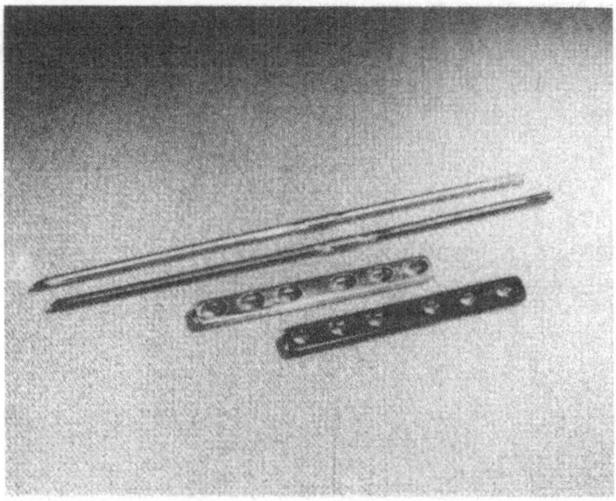

Figure 22. Carbon Coated Orthopaedic Pins

Far more labortory testing needs to be done before they could be accepted in human trials but if acceptable one could imagine coating many other implants such as the roots of false teeth.

REFERENCES

1. G. Davies, Diamond, Adam Hilger Ltd, Bristol (1984).
2. A. Badzian, B. Simonton, T. Badzian, R. Messier, K. Spear and R. Roy, Proc. SPIE 683, 127 (1986).
3. R. C. DeVries, General Electric, Corporate Research and Development, Technical Report 86CRD247 (1987).
4. S. A. Solin and A. K. Ramdas, "Raman Spectrum of Diamond", Phys. Rev. B. Vol 1, No 4, pp 1687-1698 (1970).

5. C. J. Wort, A. H. Lettington, C. Smith and A. V. Hetherington, "Optical and Other Properties of MPACVD Diamond", Proc. of SPIE Tech. symposia on Aerospace Sensing, Orlando (1989).
6. D. A. Anderson, Phil. Mag., Vol 35, pp 17-26 (1977).
7. S. M. Ojha and L. Holland, Thin Solid Films, Vol 38, L17 (1976).
8. D. S. Whitmell and K. Williamson, Thin Solid Films, Vol 35, pp 255-261 (1976).
9. B. Dischler, A. Bubenzer, P. Koidl and R. E. Sah, Spring Conference on Applied Optics, Paper TuA-D3, Monterey (1984).
10. A. H. Lettington, SPIE Conference Proceedings, Vol 590, pp 100-105, Cannes (1985).
11. S. Van Der Zwagg, J. E. Field, Philos. Mag., A(GB), Vol 48, pp 767-777 (1983).
12. A. H. Lettington, J. C. Lewis, C. J. H. Wort, B. C. Monachan and A. J. N. Hope, "Developments in GeC as a durable IR Coating Material", E-MRS Meeting, Vol 17, pp 469-474 (1987).
13. J. T. Cox, G. Hass and W. R. Hunter, Appl. Opt., Vol 14, p1247 (1975).
14. S. F. Pellicori, Appl. Opt., Vol 17, p3335 (1978).
15. J. T. Cox and G. Hass, Appl. Opt., Vol 17, p333 (178).
16. A. H. Lettington and G. J. Ball, RSRE Memorandum No 3295 (1981).
17. P. Moon, Jul Franklin Institute, Vol 230, p583 (1940).
18. R. E. Hahn and B. O. Seraphin, Physics of Thin Films, Vol 10, p1 (1978).
19. L. F. Drummeter and G. Hass, Physics of Thin Films, Vol 2, p305 (1964).
20. B. O. Seraphin, Thin Solid Films, Vol 57, p293 (1979).
21. M. A. Janai et al, Solar Energy Materials, Vol 1, Nos 1 & 2, p11 (1979).
22. G. J. Ball and A. H. Lettington, RSRE Memorandum No 3617 (1983).
23. D. E. Gray (Ed), American Institute of Physics Handbook, McGraw Hill, New York (1972).
24. W. H. Brattain and H. B. Briggs, "Optical Constants of Germanium", Bell Telephone Systems Technical Publications, Monograph B (1979).

THEORY AND MODELS FOR NUCLEATION AND GROWTH OF DIAMOND FILMS

Michael Frenklach

Department of Materials Science and Engineering
The Pennsylvania State University
University Park, PA 16802, USA

INTRODUCTION

Chemical vapor deposition (CVD) of diamond films at low pressures has been rapidly advanced in recent years.[1-3] It has been demonstrated that high-quality diamond films can be produced employing different experimental techniques and using different hydrocarbons as starting material. The deposition rates and film quality have been repeatedly shown to be sufficiently high for a number of commercial applications to be feasible. At the same time, it becomes evident that further technological advancement, particularly in such challenging areas as single-crystal growth for electronic applications and low-temperature deposition for coating of optic and plastic materials, requires detailed understanding of the fundamental phenomena responsible for diamond nucleation and growth.

Most research studies in the area of low-pressure deposition of crystalline diamond films have been focused on testing of various deposition techniques and characterization of the produced films, and less on the mechanism of nucleation and growth of crystalline diamond, although the efforts in the latter area have been steadily increasing. The objective of the present manuscript is: to review the principal theories and concepts advanced towards understanding the mechanism of diamond formation at metastable conditions; to provide a background information on high-temperature hydrocarbon chemistry required to follow the mechanistic arguments; and to summarize the results and conclusions of our most recent detailed modeling study of diamond CVD.

PROPOSED CONCEPTS AND IDEAS

The work of Derjagin, Fedoseev, Spitsyn and co-workers

A most extensive theoretical analysis was developed by the Russian school of Derjagin, Fedoseev, Spitsyn and co-workers.[4-10] They suggested the most insightful concept: despite the fact that diamond is a metastable form of carbon compared to graphite at low pressures, the growth of diamond is controlled by kinetics and not thermodynamics. Thus, for instance, Varnin *et al.*[11] wrote: "if the condensate as a whole under the regular conditions of crystallization mostly reveals properties close to those of graphite, then this is a consequence of not the energetic advantage of the graphite (which, by the way, is quite insignificant), but the kinetic preferences in the growth of the graphitic structure." The "kinetic advantage" was explained by Derjagin and co-workers, along with John Angus,[1] as the preferential etching of graphite over diamond by hydrogen atoms. This "etching theory" assumes that graphite and diamond are formed simultaneously but graphite (or, more generally, graphitic phases) is destroyed by reactions with H atoms whereas diamond is not, or at least not to the same extent. Deryagin and co-workers argued that hydrogen atoms should be in a superequilibrium state (i.e., higher than equilibrium concentration) for a successful growth of diamond.

As for the solid phase formation, the Russian school of thought is somewhat conflicting. On the one hand, they describe the growth of diamond as a set of chemical and physical processes taking place directly on the growing surface; for instance, Deryagin *et al.*[12] wrote that the surface of the growing crystal is covered with "hydrocarbon complexes having different numbers of C and H atoms. The processes occurring on the surface are: adsorption of hydrogen atoms and hydrocarbons C_nH_m; recombination of H atoms; hydrogenation and dehydrogenation of adsorbed complexes; formation and decomposition of diamond and non-diamond nuclei; desorption of H_2 and C_iH_j molecules." On the other hand, they advocate the importance of gas-phase nucleation, e.g.,[13] "in the vicinity of the growing crystal surface there is a hyper-concentrated layer, where homogeneous formation of nuclei and clusters of the new phase is possibly taking place, which then incorporate themselves into the lattice of the crystal."

The Russian scientists developed a global kinetic theory for diamond nucleation and growth based on macroscopic concepts of classical nucleation theory and the Langmuir adsorption-desorption kinetics and equilibrium. Thus, for example, Deryagin, Fedoseev and co-workers[6,8] developed the following equation for the rate of deposition of diamond in methane-hydrogen mixtures

$$v = \frac{(k_m + k_h P_h)\, P_m}{1 + P_m/b_m + P_h/b_h} ,$$

where v is the deposition rate, P_m and P_h are the partial pressures of methane and hydrogen, respectively; b_m and b_h are the respective adsorption coefficients; and k_m and k_h are the temperature-dependent rate coefficients of corresponding chemical reactions. In the limit of $b_m \ll P_m$ and $P_h/b_h \ll P_m/b_m$, the above equation becomes

$$v = b_m(k_m + k_h P_h) ,$$

meaning that when hydrogen is added to a methane mixture the rate of diamond deposition should increase.

Unfortunately, their theory identifies neither possible growth species nor the specific reactions taking place at the surface; it is formulated in rather general terms of surface coverage, formation of surface complexes, and etching of graphite by atomic hydrogen. Occasionally implied in their discussion is the role of CH_x radicals and unspecified C_nH_m surface complexes; such as, for instance,[8] that "graphite grows via methyl radicals, while diamond, at least partially, via metastable radicals CH_5."

Other classical nucleation and thermodynamic theories

Approaches based on classical nucleation theory and thermodynamic analysis, very similar to those of the Russian school, have been undertaken by several other researchers.[14-17] The results of these studies basically reiterate the conclusions of the earlier Russian work. The principal issue with these approaches is their validity. Thus, for example, applying classical nucleation theory (CNT) to diamond formation, Machlin[14] obtained that the critical cluster size is less than or equal to one carbon atom, in which case the concept of surface tension, one of the central postulates of CNT, cannot possibly be applied. Another shortcoming in many of these theoretical approaches is the assumption that diamond can be thought of as a cluster composed of only carbon atoms, whereas in practical situations we always deal with hydrogenated surfaces. Furthermore, the analysis of these idealized clusters is usually performed on energetic grounds only, whereas along with the potential energy change, it is the change in entropy (due to, e.g., steric hindrances and transition-state configurations) that determines the feasibility of specific transformations in many cases.

Equally inadequate are approaches based on the assumption that nucleation and growth of solid carbonaceous phases — diamond, graphite, amorphous carbon, etc. — are thermodynamically limited, that is, the film deposition proceeds so slow that all processes at the growing surface — physical adsorption and desorption and chemical reactions — are equilibrated, or in other words, the forming solid materials are in equilibrium with gas-phase species and the gas-phase species are in equilibrium with themselves. It has been demonstrated and emphasized by several researchers[18-21] that the state of equilibrium is not attained at the conditions typical for diamond CVD, and in fact, most gaseous species, including those suggested to be critical precursors, are in concentrations far removed from the corresponding equilibrium values. It is true, however, that some processes, like abstractions by free gaseous hydrogen atoms of hydrogen atoms from gaseous species, e.g.,

$$CH_4 + H\bullet \rightleftarrows CH_3\bullet + H_2 , \qquad \text{(g1)}$$

or from surface sites, e.g.,

$$C_{(surface)}\text{–}H + H\bullet \rightleftarrows C_{(surface)}\bullet + H_2 ,$$

are in the state of partial equilibrium, meaning that both forward and reverse directions of these reactions are much faster than other reactions in the system and therefore the concentrations of species involved in such reactions maintain an equilibrium among themselves. For example, concentrations of CH_4, H, CH_3, and H_2 obey relationship

$$K_{g1} = \frac{[CH_4]\,[H]}{[CH_3]\,[H_2]} ,$$

where K_{g1} is the equilibrium constant of reaction (g1), at any given location of the high-temperature part of a diamond deposition reactor. Such partial-equilibrium constraints can be effectively used in some modeling studies. However, in general, diamond deposition is not an equilibrium process and thermodynamic modeling alone is not capable, in principle, to predict kinetic behavior. This can be exemplified by a recent thermodynamic analysis of Piekarczyk *et al.*[17] which failed to reproduce one of the better established experimental facts[8,10,12,22-26] that the rate of diamond deposition has a pronounced maximum as a function of substrate temperature. It will be discussed later in the text that the kinetic approach is not only capable of predicting the experimental data correctly, but if performed properly incorporates the thermodynamic constraints as well.

Ionic mechanisms of Tsuda, Nakajima and Oikawa

These Japanese researchers[27,28] were probably the first to attempt a fundamental approach to search for the mechanism responsible for diamond formation in low-pressure plasma environments. They assumed a model compound — a hydrogenated carbon cluster — that simulates a part of the hydrogenated diamond (111) surface. In the course of the study, they provided an evidence that the assumed size of the model cluster does not significantly affect the results obtained. Assuming further that only C_1H_x radicals and ions can be the growth species in a CH_4–H_2 plasma, these authors performed a series of semi-empirical quantum chemical computations (MINDO/3) of potential energies, searching for the lowest energy path of diamond growth.

Based on these computations, Tsuda *et al.*[27] reported the following two-step reaction sequence. In the first step, the (111) plane of the diamond surface is covered by the methyl groups, either via methylene insertion into C–H surface bonds,

$$C_{(diamond)}\text{–}H + {}^1CH_2 \rightarrow C_{(diamond)}\text{–}CH_3\,,$$

or the abstraction of hydrogen from these surface bonds,

$$C_{(diamond)}\text{–}H + X \rightarrow C_{(diamond)}\bullet + XH\,,$$

where X = H, ^2CH, $^3CH_2$ and CH_3, followed by addition of free methyl radicals from the gas-phase to the created radical sites on the diamond surface,

$$C_{(diamond)}\bullet + CH_3\bullet \rightarrow C_{(diamond)}\text{–}CH_3\,.$$

Then, when the (111) diamond plane is covered with at least three neighboring methyl groups, these three neighboring CH_3 groups are bound together by a free gaseous methyl cation, CH_3^+, to form a part of the diamond lattice and release of three H_2 molecules.

In a subsequent publication, Tsuda *et al.*[28] extended their analysis and presented a similar reaction sequence, but now with the lattice growth accomplished entirely by reactions of free CH_3 radicals while the surface complexes maintain a positive electric charge. They computed a relatively large potential energy barrier for bringing the charge from the gas phase to the surface, about 30 kcal/mol, and thus concluded that "the most important requirement is the maintaining of positive charges on the growing surface so long as methyl radicals are supplied."

There are several difficulties with both of these proposals. First, if participation of ions is critical for diamond growth, then one would anticipate better results using an ion-producing technique, like microwave plasma, as compared to an essentially ion-free hot-filament reactor. However, the experiment shows that both environments produce very similar results. Second, bringing the charge to the surface encounters a higher potential energy barrier than activation of the surface by hydrogen atom abstraction. Third, it is not clear how the charge on the growing surface can be maintained, as recombination of surface ions with free electrons should be substantial in ion-producing environments. And fourth, there exists a significant repulsion between methyl groups placed on the diamond (111) surface.[29-31]

Acetylene addition mechanism

While single-carbon-atom species had been presumed to be the growth precursors for diamond growth by essentially all investigators, Frenklach and Spear[32] proposed that the main monomer growth species is acetylene, C_2H_2, and the reaction mechanism consists of two alternating steps: surface activation by H atom abstraction of a hydrogen atom from a surface carbon,

$$C_d\text{–}H + H\bullet \rightarrow C_d\bullet + H_2 , \qquad \text{(s1)}$$

where $C_d\bullet$ denotes a carbon atom of the hydrogenated diamond (111) surface, addition of a gas-phase acetylene molecule to the surface radical formed,

$$C_d\bullet + C_2H_2 \rightarrow C_d\text{–}CH{=}CH\bullet , \qquad \text{(s2a)}$$

and formation of a C–C bond between the created radical and a neighboring surface carbon site, C_d',

$$C_d\text{–}CH{=}CH\bullet + C_d'\text{–}H \rightarrow \begin{matrix}\text{hydrogenated}\\ \text{diamond surface layer}\end{matrix} + H\bullet . \qquad \text{(s2b)}$$

It was assumed that the starting position for this growth reaction sequence is a step fragment on the diamond (111) plane. In the addition reaction step (s2b), a number of solid C–C bonds are formed and hydrogen atoms migrate from a lower to an upper surface layer and are released into the gas phase.

The growth species according to this acetylene-addition mechanism does not need to be acetylene alone — the growth can be accomplished also by other "acetylenic" species, like C_2H, C_2H_3, C_4H_2, etc.[32] For instance, the addition of a vinyl radical to a diamond surface radical site,

$$C_d\bullet + C_2H_3\bullet \rightarrow C_d\text{–}CH{=}CH_2 , \qquad \text{(s3)}$$

followed by the H-abstraction,

$$C_d\text{–}CH{=}CH_2 + H\bullet \rightarrow C_d\text{–}CH{=}CH\bullet + H_2 , \qquad \text{(s4)}$$

results in the same surface species as the direct addition of acetylene in reaction (s2a). Thus, many acetylenic hydrocarbons, like C_4H_2, C_4H_4, C_6H_2, etc., can undergo surface growth reactions similar to those of acetylene, however their concentrations in the deposition zone of a typical CVD-type reactor, and hence their fluxes towards the growing surface, are much smaller than those of acetylene.

The proposed acetylene-addition reaction mechanism was motivated by the current knowledge of hydrocarbon chemistry, that will be discussed later in the text, and specifically, by a similar reaction mechanism identified for the formation and growth of polycyclic aromatic hydrocarbons (PAHs) in hydrocarbon pyrolysis and combustion, and the realization that acetylene should be one of the main gaseous species present in an activated hydrocarbon gas or plasma. The latter prediction was recently confirmed in experimental studies by several research groups.[16,19,33] A quantum mechanical study performed by Huang *et al.*[34] indicated that the acetylene addition mechanism of diamond growth energetically more favorable than the mechanism of Tsuda *et al.*[27,28] driven by methyl ions and radicals. It was found by Huang *et al.*,[34] however, that the acetylene additions to radical sites on diamond surface, like (s2a) and (s2b), should proceed as a single reaction step via a complex transition state rather than a sequence of two separate steps, as was proposed originally.[32]

The proposed mechanism offered explanations to many known facts:[32] that acetylene is the predominant species formed in harsh (energetic) environments explains the independence of the growth process from the nature of the hydrocarbon source or the method of gas activation; the critical role of atomic hydrogen follows from the abstraction reaction (s1), which plays the role of activating the diamond surface for subsequent growth; the coexistence of and competition between diamond and graphite growth follows from close kinetics of both, since graphite deposition also proceedsvia an H-abstraction–C_2H_2-addition reaction sequence that will be discussed later in the text; and finally, the proposed mechanism suggests that once a step is nucleated on the new surface layer, its growth propagates rapidly in the lateral direction.

Nucleation of a new surface layer

Starting with a smooth (111) surface layer, the initiation of the growth reaction sequence described above requires the formation of a step. The smallest step that is capable of initiating the growth is referred to as a *kernel.* The kernel can be made of two, three or four carbon atoms.[32] Figure 1 depicts an example of a three-membered kernel. Several reaction pathways leading to the formation of these kernels were discussed by Frenklach and Spear.[32] For instance, the kernel shown in Fig. 1 was suggested to be formed by the following reaction sequence:

$$C_d\text{–}H + H\bullet \rightarrow C_d\bullet + H_2 \qquad (s1)$$

$$C_d\bullet + CH_3\bullet \rightarrow C_d\text{–}CH_3 \qquad (s5)$$

$$C_d\text{–}CH_3 + H\bullet \rightarrow C_d\text{–}CH_2\bullet + H_2 \qquad (s6)$$

$$C_d\bullet + C_2H_2 \rightarrow C_d\text{–}CH_2\text{–}CH\bullet\text{–}CH_2\text{–}C_d' \qquad (s7)$$

$$C_d\text{–}CH_2\text{–}CH\bullet\text{–}CH_2\text{–}C_d + H\bullet \rightarrow \text{kernel in Fig. 1}. \qquad (s8)$$

In this reaction sequence, the hydrogenated diamond surface is activated by the H-abstraction, reaction (s1), forming a surface radical, $C_d\bullet$, to which a free methyl radical is added, reaction (s5). The surface complex formed in this manner, $C_d\text{–}CH_3$, is activated by the H-abstraction, reaction (s6), and an acetylene molecule from the gas phase is added to the radical formed, reaction (s7). The surface radical produced becomes the kernel shown in Fig. 1 upon addition of a hydrogen atom, reaction (s8). Recent quantum-chemical studies[35] support the feasibility of this reaction mechanism.

The key presumption in considering nucleation mechanisms such as one described above is that both nucleation and growth are governed by generally similar types of chemical reactions, although those responsible for the nucleation are slower than those for the growth. Following this presumption, starting with a distorted or non-diamond surface, the initial carbon deposition — i.e. nucleation — proceeds via the formation of a largely random network of *sp*, sp^2 and sp^3 hybridized carbon atoms. Others suggested that "the initial nucleation mechanism of a diamond crystallite is distinct from the mechanism for the extension of a preexisting diamond lattice."[1] Hydrocarbon cage compounds, like adamantane, tetracyclododecane and hexacyclopentadecane,[36] and multiply twinned saturated ring species[37] have been suggested as diamond precursors.

Search for gaseous precursors of diamond growth

Methyl radical has been the most frequently suggested growth species. In most cases, this is only a presumption based evidently on a "common-sense feeling" that diamond lattice should be built up one carbon atom at a time and that starting with methane, used by many as the hydrocarbon source, there should be plenty of methyl radicals formed after passage of the gas mixture through the plasma or hot-filament zone.

Kline *et al.*[18] carried out a mass-spectrometric analysis of the gas exiting an RF plasma reactor and performing a detailed chemical kinetic simulation. They concluded, assuming however that only hydrocarbon radicals and ions but not molecules can be effective growth species, that "CH_3 is the only radical species with a sufficient surface flux to explain the observed deposition rates." Similar approach and argumentation was presented by Harris *et al.*,[19] who concluded that diamond growth comes mainly from reactions of C_2H_2, C_2H_4, CH_4 or CH_3, and more recently by Goodwin and Gavillet.[38] CH_3 was proposed to be the main growth species in diamond CVD from methanol by Matsui and Sahara.[39] Mitsuda *et al.*,[40] based on an infra-red emission study on microwave-plasma deposition from methane-hydrogen mixtures, proposed CH as the critical species for diamond growth and C_2 for graphite. "C_2 radicals" were also suggested to be the source of graphite and diamondlike carbon by Cappelli and Paul.[41] The latter researchers reported a strong laser-induced fluorescence signal attributed to C_2H in diamond deposition zone of a premixed oxyacetylene flame. On the other hand,

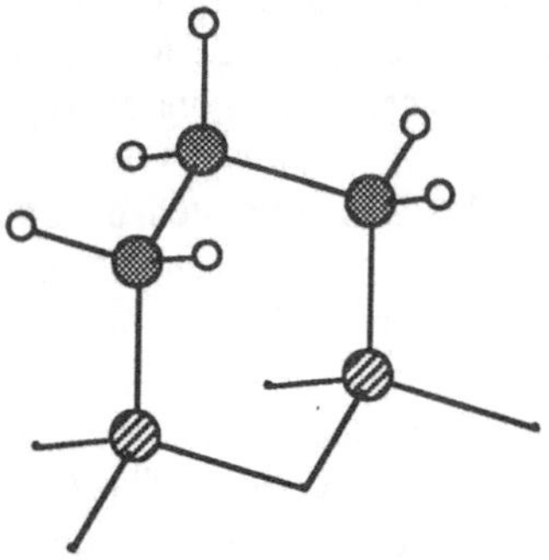

Figure 1. A schematic diagram of a three-membered initiation kernel.[32] Circles filled with diagonal lines represent the diamond (111) layer, circles with a fine mesh represent the carbon atoms of the newly formed kernel, and small open circles represent hydrogen atoms.

Matsui *et al.*[42] noted that "diamond growth rate is in good agreement with those of CH and C_2 concentrations in the feather" of an acetylene torch, whereas "the idea of the growth via C_2H_2 and/or C_2H is discounted." McNamara *et al.*,[43] applying nuclear magnetic resonance spectroscopy to study diamond films grown in a hot-filament reactor from ^{13}C-labeled acetone, concluded that "CO acts as a precursor for diamond growth in the absence of heterogeneous kinetics."

The question on methyl versus acetylene as a growth species has received a particularly considerable attention. Chu *et al.*[44,45] conducted ^{13}C-isotope labeling study in an attempt to resolve this issue. Using mixtures of $^{13}CH_4$ and $^{12}CH_4$ or $^{12}C_2H_2$ in H_2, they observed complete isotopic scrambling in a "typical" hot-filament reactor. However, when they injected the hydrocarbons just above the substrate, separately from the H_2 stream going through the filament, a non-scrambled distribution was observed. In the initial analysis,[44] based on simple stoichiometric relationships, the authors reported that the average of their four performed experiments amounts to about two methane carbons for every one acetylene carbon atom contributing to diamond film growth. However, applying this analysis to the individual experimental runs (Table II in Ref. 44), the following results are obtained: 2.7, 2.6, 1.4 and 1.1, with the 95% confidence interval ranging from 0.3 to 3.6. The obtained numbers indicate that with equal probability the results of these experiments can be interpreted as methane is more efficient than acetylene, that both are about equal, or even that acetylene is more efficient than methane in depositing diamond. In the following work, Chu *et al.*[45] reinterpreted their experimental data, assuming now that CH_3 is the growth species and that it is in equilibrium with C_2H_2 but not with CH_4. The latter assumption is highly questionable, taking into account the establishment of a partial equilibrium between CH_3 and CH_4, as was discussed above. On the other hand, assuming that CH_3 is equilibrated with both C_2H_2 and CH_4 implies that the system is in equilibrium and thus the isotopes should be completely scrambled. The main difficulty with the experiment of Chu *et al.* is that the species concentrations were not determined *in situ*, directly in the deposition zone, and the flow-field geometry is so complex that it cannot be deconvoluted computationally.

Regardless of the difficulties associated with interpretation of experiments such as those described above and others,[46] it is important to realize that under different experimental conditions different species may dominate the growth. It is conceivable, for instance, that at conditions far removed from those typically realized in diamond CVD reactors, as those employed by Chu *et al.*[44,45] and Martin and co-worker,[46,47] the concentrations of CH_3 and C_2H_2 are inverted to such a degree that the former may become the dominant growth species. It is more appropriate to ask, therefore, what is the growth mechanism and species responsible for diamond growth under specific conditions. Thus, although C_2H_2 was suggested by Frenklach and Spear[32] to be the main *growth* species under typical deposition conditions, the same authors suggested that CH_3 may play a major role in *nucleation*[32] and dominate, along with other C_1H_x species, the growth on diamond (100) surface.[48] Similar suggestions were made by others.[1,8] One possible reaction mechanism for the insertion of methyl on diamond (100) surface was recently presented by Harris:[49] a CH_3 radical is added to a surface radical followed by two sequential H-abstractions forming a diradical species which then forms a C–C bond. The main difficulty with the growth on (100) surface is that this surface is quite "crowded" for an "insertion" of such a relatively large fragment as CH_3. This manifests itself, for instance, in strong repulsion energies computed for adjacently placed CH_3 groups on the diamond (111) surface.[29-31]

Role of hydrogen and the aromatics theory

The role of hydrogen in vapor-activated deposition of diamond has been the focus of attention, because it is only after a large excess of hydrogen was used in the process that microns-a-hour growth rates were obtained.[1,7,8,50] Several factors have been suggested to explain the role of the hydrogen dilution. Most of the proposals implicate atomic hydrogen, most notably the preferential etching of graphite over diamond, advanced by John Angus[1] and the Russian scientists[6] and now adopted widely, and stabilization of the diamond surface by satisfying the dangling bonds of surface carbon atoms thereby keeping them in the sp^3 configuration and thus preventing the diamond surface from reconstruction into graphitic sp^2 (or carbynic *sp*) structures. Other suggested roles of atomic hydrogen are: to promote the gas-phase production of acetylene;[32] to promote the sp^3 configuration of gaseous radical precursors and surface clusters;[14] and to remove oxygen chemisorbed on the substrate.[14]

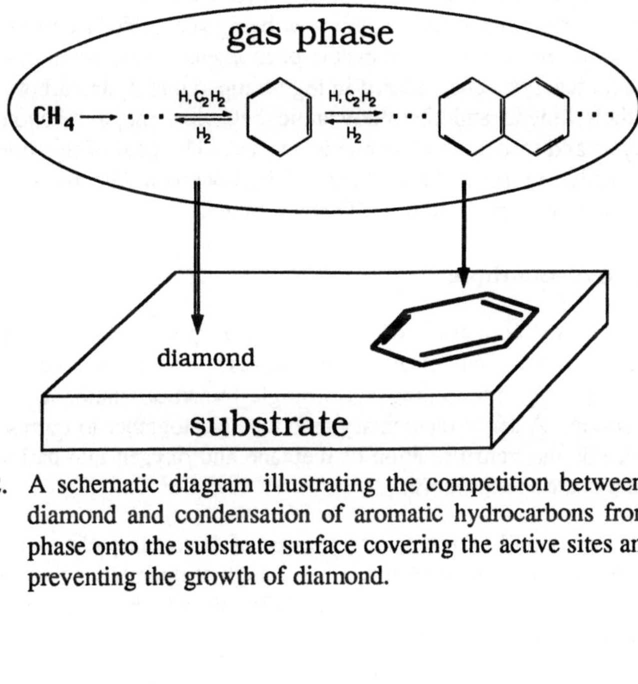

Figure 2. A schematic diagram illustrating the competition between CVD of diamond and condensation of aromatic hydrocarbons from the gas phase onto the substrate surface covering the active sites and thereby preventing the growth of diamond.

Performing a detailed computational investigation, it was concluded[20] that the key role that the hydrogen addition plays in diamond deposition is to suppress the formation of aromatic species by molecular hydrogen, H_2, in the gas phase and thereby to prevent the formation and growth of non-diamond, graphitic phases on the deposition surface. Figure 2 illustrates this theory: the formation of aromatic species in the gas phase, and most importantly that of benzene, is sufficiently fast to provide a substantial flux of these species to the growing surface where they condense covering the surface sites by a non–diamond component, thereby competing with the surface growth of diamond.

Role of oxygen

One of the most intriguing technological advancements made since the addition of hydrogen is the improved characteristics of diamond deposition with the addition of oxygen[51-55] and the growth of diamond films in hydrocarbon flames,[54,56,57] pioneered by Hirose and co-workers. Practically all researchers who studied the effect of oxygen addition found out the improved quality of diamond films at lower substrate temperatures and increased deposition rates, although some[55] begin to report that the addition of oxygen does not necessarily increase the deposition rate of diamond. The factors attributed to the role of oxygen are: reduction in the concentration of acetylene, which was assumed to be the precursor of graphitic and amorphous carbon;[51] reduction in the effective initial hydrocarbon mole fraction;[58] oxidation of non-diamond carbon by O and O_2[51,59] or by OH;[58,60] increase in the concentration of H atoms;[59] and removal of aromatic hydrocarbons in the gas-phase.[20,60]

HIGH-TEMPERATURE CHEMISTRY OF HYDROCARBON PYROLYSIS

Significant knowledge has been accumulated on elementary processes governing high-temperature pyrolysis and oxidation of hydrocarbon fuels.[61] Many complex systems, like laminar premixed flames, are now being successfully described in terms of elementary chemical reactions and transport phenomena. This section will review the common chemical reactions encountered in high-temperature hydrocarbon systems, basic principles of their kinetic and thermodynamic behavior, major reaction pathways of methane pyrolysis and formation of aromatic species. The goal of this brief review is to introduce the reader to some basic facts of hydrocarbon chemistry underlying the phenomena of diamond deposition in CVD environments.

Reactions of hydrocarbons

The act of transformation of one chemical species into another is called a *chemical reaction*. When a chemical reaction represents a very simple, elementary one-step transformation (e.g., proceeding via a single transition state), it is referred to as *elementary* reaction. A set of elementary reactions put together to represent a complex phenomena, such as the transformation of methane and oxygen into carbon dioxide and water, is called a *reaction mechanism*.

There are essentially three general classes of elementary chemical reactions that govern transformations of hydrocarbon species at high-temperatures; they are:
(1) thermal molecular decomposition or bond-fission, which can be exemplified by a C–H bond cleavage of a methane molecule,

$$CH_4 + M \rightarrow CH_3\bullet + H\bullet + M, \qquad (g2)$$

where "M" represents any chemical species present in the gas mixture;
(2) atom abstraction, e.g.,

$$CH_4 + H\bullet \rightarrow CH_3\bullet + H_2, \qquad (g1)$$

and (3) addition reactions, which can be of two types: radical–radical combination reactions, as e.g., the reverse of (g2),

$$CH_3\bullet + H\bullet + M \rightarrow CH_4 + M\,, \qquad \text{(g-2)}$$

and radical–molecule reactions, like acetylene addition to vinyl forming a $C_4H_5\bullet$ radical

$$C_2H_3\bullet + C_2H_2 \rightarrow CH_2{=}CH{-}CH{=}CH\bullet\ . \qquad \text{(g3)}$$

The dot "•" in the chemical notations identify an unpaired electron — a free vacancy; it can be usually inferred from the molecular notation itself and therefore omitted in most technical texts.

The rate of a chemical reaction is given by the product

$$\textit{reaction rate} = k \prod_i C_i^{\beta_i}$$

where k is the rate coefficient, C_i is the concentration of the ith species, β_i is the reaction order with respect to the ith species, and the index i span the reactant species. For an elementary reaction, the values of β_i are equal to the respective stoichiometric reaction coefficients. The rate coefficient is specific to each reaction, and its value depends on temperature and pressure. The temperature dependence takes the familiar Arrhenius form[62]

$$k = A\, e^{-E_a/RT}\,,$$

where A is the *frequency* or *preexponential* factor, determined by the frequency of collisions and steric hindrances or, equivalently, entropy changes; E_a is the *apparent activation energy*, which expresses the energetics of the reaction; T is the temperature; and R is the universal gas constant. This expression is typically valid only over a narrow temperature range. For wider temperature ranges, a modified Arrhenius form

$$k = A\, T^n e^{-E/RT}$$

is employed; in this expression A, n and E are simply fitted parameters.

Much less appreciated is the pressure-dependence of elementary reactions, which becomes a significant factor for reactions of small species at high temperatures.[62] It arises from the fact that most elementary reactions are themselves complex processes involving redistribution of energy among various modes of internal molecular motion (i.e., vibrations, rotations, etc.). For example, in the molecular decomposition reaction of methane, reaction (g2) above, a methane molecule acquires extra energy required to break the C–H bond by collisions with other molecules. Only when sufficient energy is collected and redistributed to the proper vibrational modes, in a large number of collisions with surrounding gaseous species "M", the C–H bonds breaks apart. Similarly, in the recombination reaction, e.g. reaction (g-2), the energy supplied by the collision of CH_3 and H is in excess of the CH_4 thermal energy and must be removed by a collision with "M" for a stable C–H bond to form (hence "M" is often referred to as the "third body"). The dependence of the energy redistribution on the collisions with the bath gas results in the dependence of the rate coefficient on the system pressure. Sometimes the adduct formed in the collision of reactant species may isomerize to a different transition state and form different reaction products; for example, parallel to reaction channel (g3), there also exists reaction channel

$$C_2H_3\bullet + C_2H_2 \rightarrow CH_2{=}CH{-}C{\equiv}CH + H\bullet . \qquad (g3a)$$

Such transformations are referred to as *chemical activation* processes. The competition between channels (g3) and (g3a) depends on both temperature and pressure.

We return now to the discussion of the three major classes of elementary reactions introduced earlier in this subsection. The activation energies of molecular decompositions, e.g. reaction (g2), are typically large, which results in low rate coefficient values. Therefore, this type of reactions have significant rates only for major molecular species: reactants and major products. For radicals, the activation energies of bond breaking, like for example

$$C_2H_5\bullet + M \rightarrow C_2H_4 + H\bullet + M ,$$

is relatively low, and these reactions are mostly responsible for the low stability of hydrocarbon radicals. The decomposition of larger hydrocarbon radicals follows the rule of β-scission, which states that the second bond (i.e., bond in the β position) from the unpaired electron is the one which predominantly breaks. Given a choice, a single C–C bond breaks before a C–H bond, because the former is weaker than the latter. For example, decomposition of a propyl radical

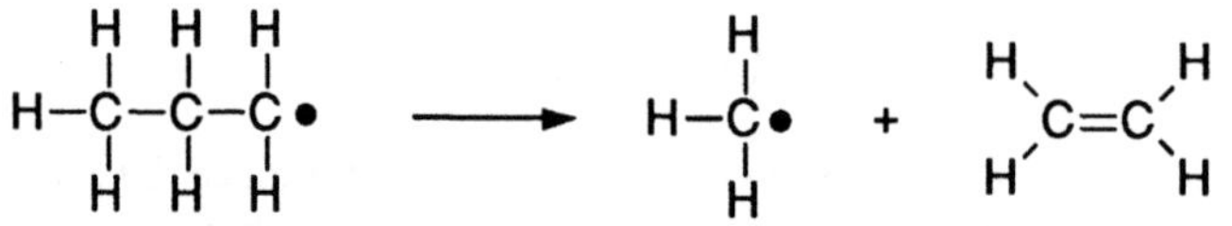

illustrates this simple but powerful rule.

Recombination of radicals has no or little activation energy. Counterbalancing this, however, is the need for the third-body collisions. As a result, recombination reactions of minor radicals are extremely slow. Yet, for most stable radicals, as e.g. CH_3, these reactions play an important role even at low pressures of diamond CVD reactors. With the increase of molecular size, the need for the third-body collisions declines because the increased number of vibrational and rotational modes begins to serve as the energy sink. Thus, combination reactions of larger stable radicals become pressure-independent and their rates may become significant.

Binary radical–molecule combination and abstraction reactions have low to medium activation energies (typically 10 to 30 kcal/mol), and therefore, their rates are mostly determined by the concentration product and steric/entropy (preexponential) factor. Abstraction and straight-forward addition reactions of major radicals and molecules present in the reactive mixture are usually the fastest reactions in the system. These reactions, however, may not be the rate limiting ones. When both forward and reverse directions are very fast, as in the case of H-abstraction (g1) discussed above, both rates are quickly equilibrated and a dynamic, or partial, equilibrium is established for such reactions. These dynamically equilibrated reactions serve then as thermodynamic constraints within a complex network of reactions.

Some isomerization reactions play an important role in high-temperature chemistry of hydrocarbons. A good example is molecular dimerization of acetylene:[63] an acetylene molecule upon collision isomerizes into a vinylidene,

$$HC{\equiv}CH + M \rightarrow :C{=}CH_2 + M, \qquad (g4)$$

which then inserts into a C–H bond of another acetylene molecule forming vinylacetylene,

$$HC{\equiv}C{-}H + :C{=}CH_2 \rightarrow HC{\equiv}C{-}CH{=}CH_2. \qquad (g5)$$

Isomerization reactions are often an essential element in cyclization of aliphatic hydrocarbons to form aromatic rings.

Pyrolysis and oxidation of methane

Methane is one of the most common hydrocarbon fuels used in diamond CVD; hence principal reactions governing its pyrolysis and oxidation are presented here. The pyrolysis in a homogeneous uniformly-heated environment is initiated by molecular decomposition

$$CH_4 + M \rightarrow CH_3\bullet + H\bullet + M \qquad (g2)$$

followed by H-abstraction

$$CH_4 + H\bullet \rightarrow CH_3\bullet + H_2. \qquad (g1)$$

In non-uniform environments, with the presence of large temperature and concentration gradients, the pyrolysis is initiated by reaction (g1), where the hydrogen atoms are supplied by diffusion from the hot zone.

The methyl radicals formed in reactions (g1) and (g2) undergo combination reactions

$$CH_3\bullet + CH_3\bullet + M \rightarrow C_2H_6 + M$$
$$CH_3\bullet + CH_3\bullet \rightarrow C_2H_5\bullet + H\bullet,$$

which initiate the formation of C_2H_x species. Ethane and ethyl formed in these reactions are then decomposed in a sequence of abstraction and decomposition reactions:

$$C_2H_6 + Y\bullet \rightarrow C_2H_5\bullet + YH$$
$$C_2H_5\bullet + M \rightarrow C_2H_4 + H\bullet + M$$
$$C_2H_4 + Y\bullet \rightarrow C_2H_3\bullet + YH$$
$$C_2H_3\bullet + M \rightarrow C_2H_2 + H\bullet + M,$$

where Y is primarily H and to some extent CH_3. Acetylene formed by these reactions is the most stable gaseous species at elevated temperatures.[64] At high temperatures, it produces polyacetylenes via

$$C_2H_2 + M \rightarrow C_2H\bullet + H\bullet + M$$
$$C_2H_2 + H\bullet \rightarrow C_2H\bullet + H_2$$
$$H{-}C{\equiv}C{-}H + \bullet C{\equiv}C{-}H \rightarrow H{-}C{\equiv}C{-}C{\equiv}C{-}H + H\bullet$$
$$H{-}C{\equiv}C{-}C{\equiv}C{-}H + \bullet C{\equiv}C{-}H \rightarrow H{-}C{\equiv}C{-}C{\equiv}C{-}C{\equiv}C{-}H + H\bullet$$

etc.

and aromatics, as will be discussed in the next subsection.

With the presence of oxygen, the main radical pool is maintained by chain-branching processes

$$H\bullet + O_2 \rightarrow OH\bullet + O\bullet$$
$$O\bullet + H_2 \rightarrow OH\bullet + H\bullet$$
$$OH\bullet + H_2 \rightarrow H_2O + H\bullet .$$

The radicals O and particularly OH formed in these reactions act as abstracting agents "Y" in reactions described above. The oxidation of gaseous carbon (i.e., the formation of C–O bond) is accomplished mainly via two reaction channels: via oxidation of CH_3 by O_2, OH and O forming formaldehide, $H_2C{=}O$, which after an H-abstraction followed by the HCO radical decomposition produces CO; and via oxidation of acetylene by O and OH forming CO and CH_2, and the oxidation of the latter by O_2. The details of these reactions are still debated and therefore left out of the present discussion. The produced CO is converted to CO_2 by reaction with OH,

$$CO + OH\bullet \rightarrow CO_2 + H\bullet .$$

In fuel-rich mixtures (i.e., when the amount of oxygen is insufficient to oxidize the hydrocarbon fuel into CO_2 and H_2O), CO cannot be entirely converted to CO_2 and is therefore present in large concentrations. In oxygen-enhanced deposition of diamond, the mixtures are not only fuel rich but also high in H atoms, which makes CO to be practically the only C-O-containing product of combustion.

In plasma environments, the decomposition of methane to form acetylene is further enhanced by reactions like[65,66]

$$CH_4 + e \rightarrow CH_3\bullet + H\bullet + e$$
$$CH_4 + e \rightarrow CH_4^+ + 2e$$
$$CH_4 + CH_3^+ \rightarrow C_2H_5^+ + H_2$$
$$C_2H_6 + e \rightarrow C_2H_4 + H_2 + e$$
$$C_2H_4 + e \rightarrow C_2H_2 + H_2 + e$$
$$C_2H_4 + e \rightarrow C_2H_2 + 2\ H\bullet + e ,$$

where C_2H_6 is formed by recombination of methyl radicals.

When larger hydrocarbons are subjected to pyrolysis, either thermal, oxidative or plasma-enhanced, they quickly decompose to smaller fragments by processes similar to those outlined above. The fragmentation eventually produces species CH_3, C_2H_5, C_2H_4, C_2H_3, etc., which follow the transformations discussed in this subsection.

Formation and growth of polycyclic aromatic hydrocarbons and soot

As was discussed earlier in the text, condensation of aromatics on a substrate surface was suggested to be one of the critical factors in deposition of diamond.[20] It is pertinent, therefore, to discuss here the formation of aromatic hydrocarbons under the conditions typical of diamond CVD. The following discussion is based on the results presented in Refs. 67–73.

The formation of aromatics begins by the addition of vinyl to acetylene, reactions (g3) and (g3a) discussed in the previous subsection. At high temperatures and low pressures, the chemical-activation channel, reaction (g3a), dominates. Vinylacetylene, $HC{\equiv}C{-}CH{=}CH_2$, formed in this reaction undergoes an H-abstraction, addition of acetylene to the radical formed, and cyclization to phenyl. The latter grows by sequential acetylene additions and H-abstractions. Figure 3 shows this reaction sequence, identified initially for high-temperature acetylene pyrolysis.[67] It was shown later[70] that at different conditions the formation of aromatics may be initiated by other reactions. However in all cases, as the reaction progresses, the acetylene-addition pathway emerges as the dominant one.

At low temperatures and high pressures, the formation of the first aromatic ring proceeds predominantly via the collisionally-stabilized adduct formed in reaction (g3) followed by the addition of acetylene and cyclization to benzene,

$CH_2{=}CH{-}CH{=}CH\bullet + C_2H_2 \rightarrow$  $+ H\bullet$,

the reaction suggested by Weissman and Benson[74] and Cole *et al.*[75] Benzene can be also formed by recombination of C_3H_3 radicals. Benzene and phenyl are interconverted to one another via H-abstraction and its reverse. At very low pressures, the reaction sequence shown in Fig. 3 is initiated[72] by the molecular dimerization of acetylene to form vinylacetylene,[63] reactions (g4) and (g5). Starting with an aromatic fuel, a "direct" condensation of intact aromatic rings becomes important.[70]

It may seem surprising at first that a single reaction pathway — sequential addition of acetylene as depicted in Fig. 3 — dominates the growth of aromatics. One may rather anticipate a multiplicity of possible reactions. However, this result was reiterated in all our numerical simulations[68-70,72,76] and confirmed by others.[77] The reason for this is a complex network of tightly-balanced chemical reactions, and when one reaction flux emerges as the dominant one, it effectively suppresses the fluxes of other competing reactions.

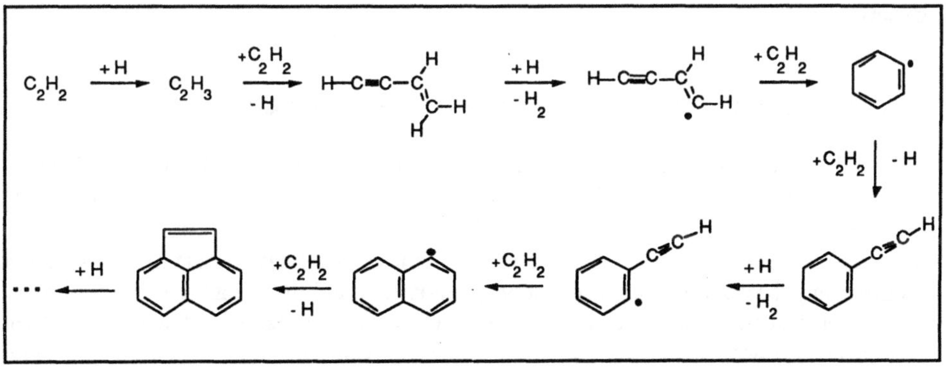

Figure 3. High-temperature acetylene-addition reaction pathway[67]

The dominant reaction sequence is a repetition of two principal steps:[67,71] hydrogen abstraction, which activates the aromatic molecule, and addition of an acetylene molecule to the radical formed. At the very high temperatures, the radicals formed decompose faster than they grow; and this defines the "ceiling" temperature for the formation of aromatics. At high temperatures, but below the ceiling temperature, the aromatics growth is controlled by the superequilibrium of hydrogen atoms and the formation of particularly stable aromatic molecules, like pyrene, coronene, etc. The change of the free energy in these latter reactions is so large that the reactions become practically irreversible, effectively "pulling" the reaction sequence forward, towards the formation of larger aromatic molecules. Other acetylene-addition steps are highly reversible, i.e., the rate of the forward reaction is nearly balanced by the rate of the reverse reaction. These steps with tightly balanced reaction fluxes create a thermodynamic barrier to PAH growth. The superequilibrium of hydrogen atoms acts as the force to overcome this thermodynamic barrier. At low temperatures, when the H superequilibrium is very large, the growth is determined by the rate of acetylene-addition steps.

The effect of oxygen addition is twofold.[68,69] On the one hand, because of the accelerated chain branching, it promotes fuel pyrolysis and hence the production of hydrocarbon radicals and hydrogen atoms, which has a promoting effect on aromatics formation and growth. On the other hand, it removes aromatic radicals and critical aliphatic hydrocarbon radicals, like C_2H_3 and C_4H_3, by reactions with molecular oxygen. The balance of these factors determines the net effect of oxygen addition. Thus, at low temperatures, when the process is kinetically limited, the addition of small amounts of oxygen has a strong promoting effect on aromatics formation, whereas at high temperatures, when the process is limited by the radical stability, the addition of oxygen suppresses it.

The PAH species formed in reactions described above can undergo *coagulation* which initiates the formation of soot particles.[78] That is, aromatic molecules collide with each other forming dimers; the dimers, in turn, collide with other aromatic molecules forming trimers or with other dimers forming tetramers; and so on. This coalescence does not result in a crystalline (or microcrystalline) graphite, but forms randomly oriented turbostratic layers, and therefore soot particles are electron amorphous. As the clusters grow and become solid (or liquid) phase, they begin to add and loose mass by surface reactions. It was recently suggested[73,76] that the surface reactions of soot particles are very similar to those of gas-phase aromatics: the mass is added by the H-abstraction/C_2H_2-addition reaction sequence discussed above. A detailed model developed following this assumption was found to be in good agreement with experiment.[76]

DETAILED MODELING OF DIAMOND DEPOSITION

Detailed modeling is a necessary tool in developing the fundamental understanding of diamond deposition and, ultimately, for design of commercial processes. Several research groups[18-21,36,39,41,46,49,58,79-81] have begun using detailed chemical kinetic modeling to describe gas-phase reactions occurring in diamond deposition processes, including transport phenomena, and the computational predictions for major gaseous species are usually found to be in good agreement with experiment. However surface processes in these studies were treated in a very general, phenomenological manner or not considered at all. Below is presented a summary of recent results obtained with a detailed kinetic model that includes gas-phase and surface processes and both are described in elementary terms using a rigorous formalism for gas-surface reactions developed for this purpose.[60,82]

The assumed model was an idealized constant-diameter hot-filament reactor and, computationally, consisted of two separate submodels, those simulating the gas-phase and the surface reactions. The gas-phase submodel was a constant-pressure laminar flow with an imposed temperature profile. The reactor conditions — initial mixture composition, pressure, initial gas velocity, and temperature profile — were founded on experimental data of Harris *et al.*[19] The "base" case was a 0.3 % $CH_4 - H_2$ mixture, pressure 20 torr, linear gas velocity 0.5 cm/s. We assumed that the temperature profile upstream the filament is symmetrical to the downstream profile and that the gas remains in contact with the filament for a finite period of time at a constant temperature of 2600 K. The inclusion of the upstream portion of the temperature profile was found to be necessary for correct prediction of downstream species concentrations.

A 158-reaction, 50-species chemical reaction mechanism was used to describe the gas-phase kinetics. This mechanism was based on two previous mechanisms developed in our laboratory for applications in combustion and flame-like environments: one which describes high-temperature pyrolysis and oxidation of methane and optimized to reproduce quantitatively a variety of ignition and flame experiments; and the other which quantitatively describes the formation and growth of polycyclic aromatic hydrocarbons in acetylene and ethylene flames. The principal reactions comprising these mechanisms are those described in the previous section. The reaction rate coefficients, usually known better than to an order of magnitude, included both temperature and pressure dependences. The reactive gas composition was computed using the Sandia burner code,[83] which solves the conservation equations for mass, momentum and species concentrations in a one-dimensional gas flow with an assumed temperature profile, accounting for chemical reactions, species diffusion (including thermal diffusion) and gas convection.

In the surface submodel, we assumed that the film is deposited on a substrate placed along the reactor wall and that either the deposition surface is in local thermal equilibrium with the gas or that the gas temperature is independent of the substrate temperature. The deposition process was described by a set of elementary chemical reactions of surface sites. The rates of these reactions were expressed in terms of two-dimensional concentrations of surface sites and diffusional fluxes of gaseous species produced in the gas-phase reactions. It was assumed that the gas-phase concentrations of chemical species are unaffected by their consumption or production in the surface reactions, and the computational results fully supported this assumption. The evolution of the surface sites was obtained by numerical integration of a set of simultaneous ordinary differential equations describing the kinetics of the developed reaction mechanism. The overall computational procedure was organized in two steps. First, for a given set of initial conditions, the concentration profiles of the gas-phase species were computed with the burner code. The obtained species concentrations were then used as an input to the surface kinetic code.

The surface processes were modeled in terms of elementary chemical reactions of surface sites. These reactions and associated rate parameters were founded on the premise that chemical reactivity of solid carbonaceous materials is localized on the carbon sites in a manner similar to that of the corresponding gaseous species. In other words, the kinetics of analogous elementary chemical reactions on a per site basis is the same for all forms of carbon.[20,32,73] Following this assumption, specific surface reactions were postulated. Most of them were founded on analogous gas-phase reactions, and their rate parameters were estimated based on closely related prototype gas-phase reactions. Specifically, the rates of the gas-surface reactions, the most common reaction type in the surface reaction mechanism, were formulated based on the assumption of equal reaction probabilities for similar gas-phase and gas-surface reactions. For example, the reaction probability for the abstraction of hydrogen atoms from the diamond surface, reaction (s1), was taken equal to the known reaction probability of gas-phase reaction

The reaction mechanism included a detailed description of the following surface processes: surface site activation by abstraction of the terminal H atoms leading to the formation of surface radicals,

$$C_d\text{–}H + H\bullet \rightarrow C_d\bullet + H_2 , \qquad \text{(s1)}$$

where $C_d\bullet$ denotes a carbon atom of the hydrogenated diamond (111) surface; deactivation of the surface radical sites by abstraction of H atoms from gaseous species and by recombination with free gaseous atoms and radicals, e.g.,

$$C_d\bullet + H\bullet \rightarrow C_d\text{–}H ; \qquad \text{(s9)}$$

growth of diamond by the addition of methyl, acetylene and other C_2H_X species — vinyl, ethylene, ethyl and ethane; condensation of aromatic species on the growing surface, thereby covering diamond *sp*^{3} sites with non-diamond *sp*^{2} sites;[20] growth of the *sp*^{2} sites by the acetylene-addition mechanism; gasification of *sp*^{2} carbon via reactions with hydrogen atoms, hydroxyl radicals, oxygen atoms, and oxygen molecules; and interconversion of *sp*^{2} and *sp*^{3} sites into one another.

The addition of acetylene to the surface radical sites was represented by reaction

$$C_d\bullet + C_2H_2 \rightarrow C_d\text{–}H + H , \qquad \text{(s2)}$$

which expresses the results of Huang *et al.*[34] The methyl addition reaction pathway leading to the growth of diamond (111) surface was represented as a sequence of the following reactions:

$$C_d\bullet + CH_3\bullet \rightarrow C_d\text{–}CH_3 \qquad \text{(s5)}$$

$$C_d\text{–}CH_3 + H\bullet \rightarrow C_d\text{–}CH_2\bullet + H_2 \qquad \text{(s6)}$$

$$C_d\text{–}CH_2\bullet + CH_3\bullet \rightarrow C_d\text{–}CH_2\text{–}CH_3 \qquad \text{(s10)}$$

$$C_d\text{–}CH_2\text{–}CH_3 + H\bullet \rightarrow C_d\text{–}CH_2\text{–}CH_2\bullet + H_2 \qquad \text{(s11)}$$

$$C_d\text{–}CH_2\text{–}CH_2\bullet + C_d'\text{–}H \rightarrow C_d\text{–}C_dH_2\text{–}C_dH_2\text{–}C_d' + H\bullet . \qquad \text{(s12)}$$

In this reaction sequence, a free methyl radical is added to the surface active site, reaction (s5). The surface complex formed in this manner, is $C_d{-}CH_3$, is activated by the H-abstraction, reaction (s6), and another free methyl radical is added, reaction (s10). The surface complex produced, $C_d{-}CH_2{-}CH_3$ is transformed into diamond lattice by the abstraction of a hydrogen atom from the terminal carbon atom, reaction (s11), and formation of a C–C bond between the created radical and a neighboring surface carbon site, C_d', by reaction (s12).

The surface reaction mechanism also included a combined $CH_3{-}C_2H_2$ addition reaction channel

$$C_d{-}CH_2\bullet + C_2H_2 \rightarrow C_d{-}C_dH_2{-}C_d\bullet{-}C_dH_2{-}C_d' + H\bullet , \qquad \text{(s7)}$$

in which an acetylene molecule is added to a surface radical created by methyl-addition reaction sequence (s1)–(s5)–(s6). The addition of C_2H_5 and C_2H_4 to $C_d\bullet$ forms species $C_d{-}CH_2{-}CH_3$ and $C_d{-}CH_2{-}CH_2\bullet$, respectively, followed by reactions (s11) and (s12). The surface species, $C_d{-}CH{=}CH_2$, formed by the addition of vinyl to $C_d\bullet$,

$$C_d\bullet + C_2H_3\bullet \rightarrow C_d{-}CH{=}CH_2 , \qquad \text{(s3)}$$

contributes to the growth of diamond by undergoing an H-abstraction, reaction (s4), followed by reaction (s2b) or H-addition followed by reaction (s12). The growth and gasification of "graphitic" carbon on the substrate surface was described by a reaction sequence based on the gas-phase chemistry of aromatics.

The model was tested by simulating diamond film deposition in a hot-filament reactor using methane–hydrogen, methane–argon and methane–oxygen–hydrogen gas mixtures. The gas-phase part of the model predicted correctly the concentrations of major gaseous species measured by Harris *et al.*[19] The surface part of the model reproduced correctly the dependence of the growth rate and film quality on the substrate temperature (Figs. 4, 5) observed in various experimental studies. Also reproduced by the model is the pressure effect reported by Derjagin and co-workers.[12] The model successfully predicted the effect of oxygen — the addition of small amounts of O_2 does not necessarily increase the deposition rate but rather improves the film quality, particularly at low temperatures, thus allowing low-temperature deposition of diamond films.

Analysis of the computational results, supplemented with various sensitivity tests, yielded the following conclusions:[60,82]

(1) Among several reaction pathways describing diamond (111) growth initiated by different gaseous species, including methyl, ethyl, ethylene and vinyl, the H-abstraction/C_2H_2-addition mechanism appears to dominate under the conditions typical of diamond CVD reactors. It is possible that at conditions significantly different from those, when the concentration of methyl radicals is much larger than that of acetylene, the contribution from other channels may become important;

(2) The reaction-path analysis indicated that diamond deposition by species other than acetylene proceeds via the formation of acetylenic or ethylenic intermediate complexes on the growing surface;

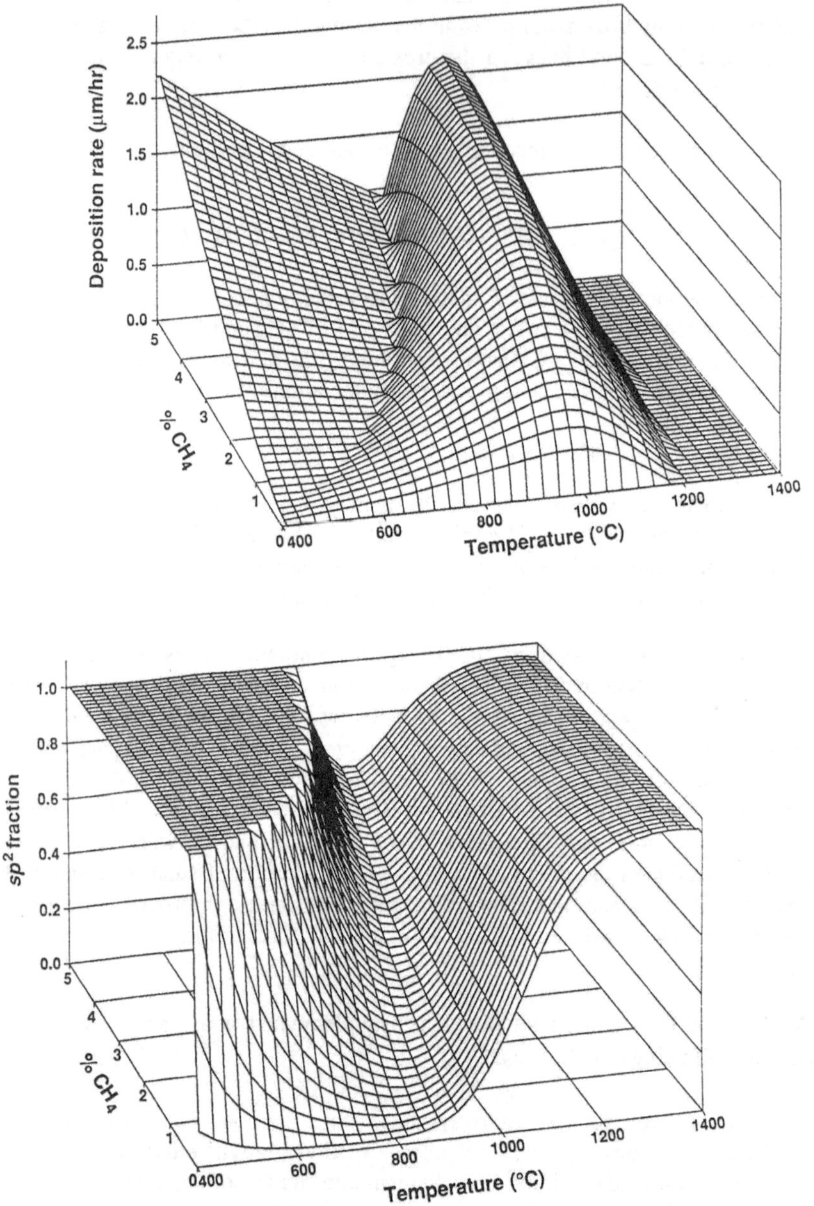

Figure 4. Deposition rate (top) and film quality (bottom) versus substrate temperature and initial methane concentration computed at a pressure of 20 torr. The substrate temperature is assumed to be equal to the temperature of the surrounding gas.

(3) The reaction "bottleneck" of diamond deposition at low substrate temperatures is the formation of $C_d\bullet$ radicals. Their formation rate is determined by the balance between the H-abstraction, $C_dH + H\bullet \rightarrow C_d\bullet + H_2$, and recombination, $C_d\bullet + H\bullet \rightarrow C_dH$, reactions. The number of available surface radicals $C_d\bullet$ increases with temperature because the former reaction has a substantial activation energy whereas the latter does not. At higher substrate temperatures, the thermodynamic stability of $C_d\bullet$ radicals becomes the limiting factor: the sp^3-hybridized $C_d\bullet$ sites decompose forming an sp^2 graphitic phase;

(4) The chemical growth of sp^2 graphite phase is much slower than the growth of diamond. The main source of sp^2 solid carbon is the surface condensation of aromatics at low temperatures and decomposition of $C_d\bullet$ sites at high temperatures;

(5) At low substrate temperatures, the condensation of benzene from the gas phase onto the growing surface competes with the diamond growth process by covering the available sp^3 diamond sites. The condensing aromatic molecules are converted into an amorphous sp^2/sp^3 carbonaceous network by the addition reactions of hydrogen atoms. The increase in the gas-phase temperature of the deposition zone suppresses the formation of aromatics and thus improves the quality of deposited film;

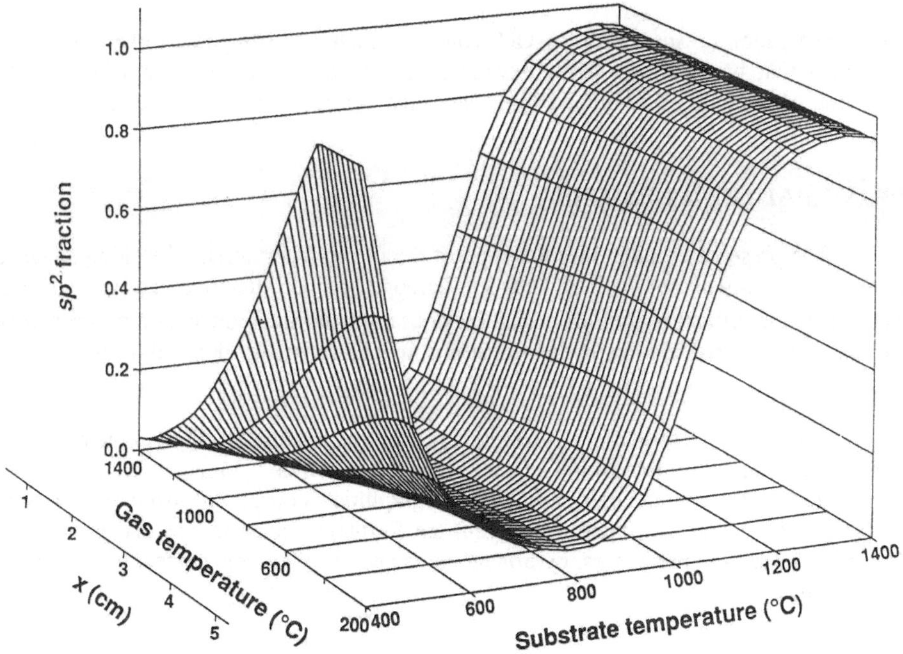

Figure 5. Film quality versus substrate and gas-phase temperatures computed for a 0.3% CH_4–H_2 mixture at a pressure of 20 torr.

(6) The main factors affecting the deposition characteristics with the increase in the initial concentration of methane are the increase in the aromatics formation and, to a lesser degree, the decrease in the H atom concentration. The former factor reduces the quality of the film, whereas the latter causes the shift of the peak deposition rate to higher substrate temperatures. Thus, at low CH_4 concentrations, when reaction (s2) dominates, the deposited film consists of diamond. At higher methane concentrations, when benzene condensation catches up with diamond deposition, the film begins to contain amorphous carbon, formed by the H addition to benzene condensed on the surface. At even higher methane concentrations, as the $sp^2 \rightarrow sp^3$ transformation by the H addition can no longer follow the benzene condensation, the depositing film should consist of predominantly graphite;

(7) The effect of pressure on diamond deposition is essentially the concentration effect: the absolute concentrations of the key hydrocarbon species are increased with the increase in pressure, thus following the phenomena described in the point above;

(8) The key role of hydrogen and oxygen in the diamond deposition process is to suppress the formation of aromatic species by H_2 and O_2, respectively, in the gas phase and thereby to prevent their condensation and subsequent growth into non-diamond, graphitic phases on the deposition surface;

(9) The major reactions of hydrogen atoms in the surface processes are: the H-atom abstraction, which activates the surface by creating radical sites; recombination with these surface radicals; and the addition to unsaturated carbon-carbon bonds, thereby converting aromatics into amorphous carbon at low temperatures and gasifying graphite at high temperatures;

(10) Gasification of sp^2 carbon by OH radicals, which was computed to be much faster than that by H atoms when oxygen is added to the mixture, contributes to the improved quality of diamond films deposited at lower substrate temperatures.

PRINCIPAL CONCLUSIONS

The presented model does not support the theory of preferential etching advanced to explain the kinetic competition between diamond and non-diamond phases. Instead, it establishes the critical role of aromatics condensation and interconversion of sp^2 and sp^3 carbon phases mediated by hydrogen atoms in gas-activated deposition of diamond films.

A larger concentration of OH and a higher temperature gradient in the deposition zone should improve deposition characteristics of diamond films at lower substrate temperatures: the OH radicals remove the "graphitic" component from the growing surface, and the high temperatures prevent the formation of aromatics and promote the formation of critical precursors, hydrogen atoms and acetylene molecules, for the growth of diamond.

ACKNOWLEDGEMENT

The work was supported in part by SDIO/IST via the Office of Naval Research, Contract No. N00014–86–K–0443.

REFERENCES

1. J. C. Angus and C. C. Hayman, *Science* **241**: 913 (1988).
2. K. E. Spear, *J. Am. Ceram. Sci.* **72**: 171 (1989).
3. *Diamond and Diamond-Like Films*, J. P. Dismukes, ed., The Electrochemical Society, Pennington, NJ (1989).
4. B. V. Deryagin and D. V. Fedoseev, *Russ. Chem. Rev.* **39**: 783 (1970).
5. B. V. Derjaguin and D. V. Fedoseev, *Sci. Am.* **233**: 102 (1975).
6. B. V. Deryagin and D. V. Fedoseev, *Growth of Diamond and Graphite from the Gas Phase*, Nauka, Moscow (1977); (English translation in *Diamond Growth and Films*, edited by UCFMG, Elsevier, New York, 1989).
7. B. V. Spitsyn, L. L. Bouilov, and B. V. Derjaguin, *J. Crystal Growth* **52**: 219 (1981).
8. D. V. Fedoseev, B. V. Deryagin, I. G. Varshavskaya, and A. S. Semenova-Tyan-Shanskaya, *Crystallization of Diamond*, Nauka, Moscow (1984); (Engl. translation in *Diamond Growth and Films*, edited by UCFMG, Elsevier, New York, 1989).
9. D. V. Fedoseev, V. P. Varnin, and B. V. Deryagin, *Russ. Chem. Rev.* **53**: 435 (1984).
10. B. V. Spitsyn, L. L. Bouilov, B. V. Derjaguin, *Prog. Crystal Growth and Charact.* **17**: 79 (1988).
11. V. P. Varnin, D. V. Fedoseev, and I. G. Teremetskaya, *Arch. Nauki Mater.* **7**: 121 (1986).
12. B. V. Deryagin, L. L. Bouilov, and B. V. Spitsyn, *Arch. Nauki Mater.* **7**: 111 (1986).
13. D. V. Fedoseev and B. V. Derjaguin, *Arch. Nauki Mater.* **7**: 213 (1986).
14. E. S. Machlin, *J. Mater. Res.* **3**: 958 (1988).
15. M. Sommer, K. Mui, and F. W. Smith, *Solid State Communications* **69**: 775 (1989).
16. J. E. Butler and F. C. Celii, in *Diamond and Diamond-Like Films*, J. P. Dismukes, ed., The Electrochemical Society, Pennington, NJ (1989), p. 317.
17. W. Piekarczyk, R. Roy, and R. Messier, *J. Crystal Growth* **98**: 765 (1989).
18. L. E. Kline, W. D. Partlow, and W. E. Bies, in Proceedings of the Sixth Symposium on Plasma Processing, G. S. Mathad, G. C. Schwartz, and R. A. Gottscho, eds., Electrochemical Society, Pennington, NJ (1987), Vol. 87–8; also *J. Appl. Phys.* **65**: 70 (1989).
19. S. J. Harris, A. M. Weiner, and T. A. Perry, *Appl. Phys. Lett.* **53**: 1605 (1988).
20. M. Frenklach, *J. Appl. Phys.* **65**: 5142 (1989).
21. G. Janssen, W. J. P. van Enckevort, and L. J. Giling, in *Diamond and Diamond-Like Films*, J. P. Dismukes, ed., The Electrochemical Society, Pennington, NJ (1989), p. 508.
22. B. V. Spitsyn and L. L. Bouilov, in *Extended Abstracts. Diamond and Diamond-Like Materials Synthesis*, G. H. Johnson, A. R. Badzian, and M. W. Geis, eds., Materials Research Society, Pittsburgh, PA (1988), p. 3.
23. D. V. Fedoseev, K. S. Uspenskaya, V. P. Varnin, and S. P. Vnukov, *Bull. Acad. Sci. USSR Div. Chem. Sci.* **27**: 1088 (1978).
24. W. Zhu, R. Messier, and A. R. Badzian, in *Diamond and Diamond-Like Films*, J. P. Dismukes, ed., The Electrochemical Society, Pennington, NJ (1989), p. 61.
25. T. McKenna, "Diamond Film Deposition and Microwave Plasma Processes in the C–H–O System," M.S. Thesis, The Pennsylvania State University, 1990.
26. K. M. Tomaswick and J. N. Bruggeman, in *Technology Update on Diamond Films*, R. P. H. Chang, D. Nelson, and A. Hiraki, eds., Materials Research Society, Pittsburgh, PA (1989), p. 83.

27. M. Tsuda, M. Nakajima, M., and S. Oikawa, S., *J. Am. Chem. Soc.* **108**: 5780 (1986).
28. M. Tsuda, M. Nakajima, M., and S. Oikawa, S., *Jpn. J. Appl. Phys.* **26**: L527 (1987).
29. D. W. Brenner, Presented at the Workshop on the Science and Technology of Diamond Thin Films, Concord, Ohio, May 1990.
30. M. R. Pederson, K. A. Jackson, and W. E. Pickett, Presented at the Workshop on the Science and Technology of Diamond Thin Films, Concord, Ohio, May 1990.
31. S. P. Mehangru and A. B. Anderson, Presented at the Workshop on the Science and Technology of Diamond Thin Films, Concord, Ohio, May 1990.
32. M. Frenklach and K. E. Spear, *J. Mater. Res.* **3**: 133 (1988).
33. F. G. Celii, P. E. Pehrsson, H.-t. Wang, and J. E. Butler, *Appl. Phys. Lett.* **52**: 2043 (1988).
34. D. Huang, M. Frenklach, and M. Maroncelli, *J. Phys. Chem.* **92**: 6379 (1988).
35. D. Huang and M. Frenklach, *J. Phys. Chem.*, submitted.
36. S. Matsumoto and Y. Matsui, *J. Mater. Sci.* **18**: 1785 (1983).
37. J. C. Angus, F. A. Buck, M. Sunkara, T. F. Groth, C. C. Hayman, and R. Gat, *Mater. Res. Soc. Bull.* **14**: 38 (1989).
38. D. G. Goodwin and G. G. Gavillet, *J. Appl. Phys.*, submitted.
39. Y. Matsui and M. Sahara, *Jpn. J. Appl. Phys.* **28**: 1003 (1989).
40. Y. Mitsuda, Y. Kojima, T. Yoshida, and K. Akashi, *J. Mater. Sci.* **22**: 1557 (1987).
41. M. A. Cappelli and P. H. Paul, *J. Appl. Phys.* **67**: 2596 (1990).
42. Y. Matsui, A. Yuuki, M. Sahara, and Y. Hirose, *Jpn. J. Appl. Phys.* **28**: 1718 (1989).
43. K. M. McNamara, K. K. Gleason, and M. W. Geis, in *Diamond, Boron Nitride, Silicon Carbide and Related Wide Bandgap Semiconductors*, J. T. Glass, R. F. Messier, and N. Fujimori, eds., Materials Research Society, Pittsburgh, PA (1990), in press.
44. C. J. Chu, B. J. Bai, M. P. D'Evelyn, R. H. Hauge, and J. L. Margrave, in *Diamond, Boron Nitride, Silicon Carbide and Related Wide Bandgap Semiconductors*, J. T. Glass, R. F. Messier ,and N. Fujimori, eds., Materials Research Society, Pittsburgh, PA (1990), in press.
45. C. J. Chu, M. P. D'Evelyn, R. H. Hauge, and J. L. Margrave, *J. Mater. Res.*, in press.
46. S. J. Harris and L. R. Martin, *J. Mater. Res.*, in press.
47. L. R. Martin and M. W. Hill, *Appl. Phys. Lett.* **55**: 2248 (1989).
48. K. E. Spear and M. Frenklach, in *Diamond and Diamond-Like Films*, J. P. Dismukes, ed., The Electrochemical Society, Pennington, NJ (1989), p. 122.
49. S. J. Harris, *Appl. Phys. Lett.* **56**: 2298 (1990).
50. J. C. Angus, in *Diamond and Diamond-Like Films*, J. P. Dismukes, ed., The Electrochemical Society, Pennington, NJ (1989), p. 1.
51. T. Kawato and K. Kondo, *Jpn. J. Appl. Phys.* **26**: 1429 (1987).
52. C.-P. Chang, D. L. Flamm, D. E. Ibbotson, and J. A. Mucha, *J. Appl. Phys.* **63**: 1744 (1989).
53. S. Matsumoto, in *Diamond and Diamond-Like Films*, J. P. Dismukes, ed., The Electrochemical Society, Pennington, NJ (1989), p. 50.
54. Y. Hirose, S. Amanuma, N. Okaba, and K. Komaki, in *Diamond and Diamond-Like Films*, J. P. Dismukes, ed., The Electrochemical Society, Pennington, NJ (1989), p. 80.
55. Y. Liou, R. Weimer, D. Knight, and R. Messier, *Appl. Phys. Lett.* **56**: 437 (1990).

56. W. A. Carrington, L. M. Hanssen, K. A. Snail, D. B. Oakes, and J. E. Butler, *Metal. Trans. A* **20**: 1282 (1989).
57. Y. Tzeng, C. Cutshaw, R. Phillips, T. Srivinyunon, A. Ibrahim, and B. H. Loo, *Appl. Phys. Lett.* **56**: 134 (1990).
58. S. J. Harris and A. M. Weiner, *Appl. Phys. Lett.* **55**: 2179 (1989).
59. J. A. Mucha, D. L. Flamm, and D. E. Ibbotson, *J. Appl. Phys.* **65**: 3448 (1989).
60. M. Frenklach and H. Wang, in *Diamond Technology Initiative Symposium*, Strategic Defence Initiative Organization Office of Innovative Science and Technology – Office of Naval Research, Crystal City, Virginia, July 1989, paper T10.
61. See, e.g., *Combustion Chemistry*, W. C. Gardiner, Jr., ed., Springer-Verlag, New York (1984); C. K. Westbrook and F. L. Dryer, *Prog. Energy Combust. Sci.* **10**: 1 (1984); J. A. Barnard and J. N. Bradley, *Flame and Combustion*, Chapman and Hall, London (1985); I. Glassman, *Combustion*, Academic Press, Orlando, FL (1987); and recent volumes of *Symposium (International) on Combustion*, The Combustion Institute, Pittsburgh, PA.
62. See, e.g., J. I. Steinfeld, J. S. Francisco, and W. L. Hase, *Chemical Kinetics and Dynamics*, Pentice Hall, Englewood Cliffs, NJ (1989).
63. J. H. Kiefer and W. A. von Drasek, *Int. J. Chem. Kinet.* **22**: 747 (1990).
64. R. E. Duff and S. H. Bauer, *J. Chem. Phys.* **36**: 1754 (1962).
65. H. V. Boenig, *Plasma Science and Technology*, Cornell U.P., Ithaca, NY (1982), Chap. 4.
66. N. G. Adams and D. Smith, in *Reactions of Small Transient Species*, A. Fontijn and M. A. A. Clyne, eds., Academic Press, London (1983), Chap. 6.
67. M. Frenklach, D. Clary, W. C. Gardiner, Jr. and S. E. Stein, in *Twentieth Symposium (International) on Combustion*, The Combustion Institute, Pittsburgh, PA (1985), p. 887.
68. M. Frenklach, D. W. Clary, T. Yuan, W. C. Gardiner, Jr., and S. E. Stein, *Combust. Sci. Technol.* **50**: 79 (1986).
69. M. Frenklach and J. Warnatz, *Combust. Sci. Technol.* **51**: 265 (1986).
70. M. Frenklach, D. Clary, W. C. Gardiner, Jr. and S. E. Stein, in *Twenty-First Symposium (International) on Combuistion*, The Combustion Institute, Pittsburgh, PA (1988), p. 1067.
71. M. Frenklach, in *Twenty-Second Symposium (International) on Combustion*, The Combustion Institute, Pittsburgh, PA (1989), p. 1075.
72. M. Frenklach and E. D. Feigelson, *Astrophys. J.* **341**: 372 (1989).
73. M. Frenklach, M., in *Carbon in the Galaxy: Studies from Earth and Space*, J. Tarter, ed., NASA CP, in press.
74. M. Weissman and S. W. Benson, *Int. J. Chem. Kinet.* **16**: 307 (1984).
75. J. A. Cole, J. D. Bittner, J. B. Howard, and J. P. Longwell, *Combust. Flame* **56**: 51 (1984).
76. M. Frenklach and H. Wang, in *Twenty-Third Symposium (International) on Combustion*, The Combustion Institute, Pittsburgh, PA, in press.
77. S. J. Harris, A. M. Weiner, and R. J. Blint, *Combust. Flame* **72**: 91 (1988).
78. B. S. Haynes and H. Gg. Wagner, *Prog. Energy combust Sci.* **7**: 229 (1981).
79. W. C. Roman and M. B. Colket, III, in *Diamond and Diamond-Like Films*, J. P. Dismukes, ed., The Electrochemical Society, Pennington, NJ (1989), p. 330.
80. S. J. Harris, *J. Appl. Phys.* **65**: 3044 (1989).
81. S. J. Harris and A. M. Weiner, *J. Appl. Phys.* **67**: 6520 (1990).
82. M. Frenklach and H. Wang, *Phys. Rev. B*, submitted.
83. R. J. Kee, J. F. Grgar, M. D. Smooke, and J. A. Miller, Report No. SAND85–8240, Sandia, Livermore, CA (December 1985).

56. W. A. Gambling, [illegible] A. Smith, D. A. [illegible], and [illegible] (1980).
57. [illegible] (1980).
58. [illegible] Appl. Phys. Lett. [illegible] (1982).
59. [illegible] J. Appl. Phys. [illegible]
60. M. [illegible] and H. [illegible], in Advanced [illegible] Infrared [illegible] Symposium, [illegible] Office of [illegible]
61. [illegible]
62. [illegible] and W. A. [illegible]
63. [illegible] J. Chem. Phys. [illegible]
64. [illegible] Science and Technology [illegible]
65. [illegible]
66. [illegible]
67. M. [illegible]
68. [illegible]
69. M. [illegible]
70. [illegible]
71. [illegible]
72. [illegible]
73. M. [illegible]
74. M. [illegible]
75. [illegible]
76. M. [illegible] Fourth [illegible] Symposium [illegible]
77. [illegible]
78. [illegible]
79. [illegible]
80. [illegible] J. Appl. Phys. 65 [illegible] (1989).
81. [illegible] Appl. Phys. [illegible]
82. M. [illegible] Rev. [illegible] (to be published).
83. [illegible] (December 1985).

A STEP BACK: HYDROGEN ABSTRACTION FROM METHANE USING A SEMIEMPIRACAL MOLECULAR ORBITAL METHOD

Steven M. Valone

Materials Science and Technology Division
Los Alamos National Laboratory
Los Alamos, New Mexico 87545

I. INTRODUCTION

Several studies have now been completed using semiempirical molecular orbital methods[1] to learn about various aspects of diamond growth.[2-4] Whole reaction mechanisms for diamond epitaxy have been proposed on the basis of these calculations with supporting evidence from experiment. Attempts have been made to understand the roles of hydrogen atoms, methyl groups, acetylene groups and charged species in diamond film growth from both hot filament and plasma sources. Based on previous testing of the semiempirical molecular orbital method used here[1] and in the previous studies, it seemed safe to assume that reasonable answers would be forthcoming.

For the semiempirical method to provide a reasonable description of a diamond film, it is essential that molecular species and configurations to be encountered in the system be included in the parameterization of the method, as described below. Otherwise, important physical effects may be missed. However, it is very expensive and time consuming to reparameterize the Hamiltonian for each new study. Nevertheless, some verification that the most important effects are correctly included is very necessary. For instance, in one of the previous studies,[4] the expected transition state barrier to hydrogen abstraction from diamond (111) surfaces was not found.

One way to verify appropriateness of the method is by testing it against simple systems for which detailed experimental measurements and *ab initio* calculations are available. Unfortunately, the semiempirical methods used in earlier studies have never been tested against simple known systems closely resembling the ones of interest to diamond films.

Here we propose to make just such a test, the simplest being the abstraction of hydrogen from methane by a hydrogen atom. Not only is the system simple,

Diamond and Diamond-Like Films and Coatings
Edited by R.E. Clausing *et al.*, Plenum Press, New York, 1991

but also both experimental and highly accurate theoretical results are known for the transition state energy and geometry, key parameters for comparison.[5] The simplicity allows the system to be examined in detail at a reasonable cost.

The results are reported after a brief review of the computational methods, for which more details are available in References 1-5. In essence, the transition states for hydrogen abstraction from methane by the semiempirical MOPAC method, to be described below, and by a sophisticated CI method, which is considered essentially exact, are vastly different. In addition, the heats of reaction by the two methods differ substantially.

II. COMPUTATIONAL METHOD

A. Semiempirical Molecular Orbital Methods

The basic strategy in developing semiempirical potential energy surfaces is to start with the functional forms suggested by a first-principles description of the system and, then, to fit the parameters with empirical information. The first-principles description is the Hartree-Fock expression for the energy of a molecule, cluster or solid, and more specifically, the linear-combination-of-atomic-orbitals (LCAO) representation of the Hartree-Fock equations as developed by Roothaan[6] and Hall.[7] The empirical data base is composed of molecular geometries, heats of formation, heats of atomization and reaction intermediates. The species appearing in the data base contains only the atoms H, C, N and O.[1]

Since the Hartree-Fock energy is taken as the key equation, the matrix elements of the various atomic orbitals act as the parameters of interest. Sometimes it is necessary to choose a functional form for the matrix elements which depends pairwise on nuclear separation. For more details, the original works should be consulted. The particular parameterization used here is referred to as the PM3 Hamiltonian[8] and the specific implementation of the method is the MOPAC package of subroutines written by Stewart.[9] The results of the calculations differ in detail depending on which of the Hamiltonians available in MOPAC is chosen, but the same conclusions apply.

B. POL-CI Calculation

Earlier work has been done on the CH_4 + H system at several different levels, including Hartree-Fock and CI. Perhaps the best, and certainly the most thorough is the work by Walch et al.[5] using what is referred to as the POL-CI method,[5] a total-energy electronic structure study with a basis of double zeta valence plus single zeta polarization quality. For this reason, we use their results for comparison.

III. RESULTS AND DISCUSSION

The physical system consists of a methane molecule and a free hydrogen atom. The CH bond being approached by the hydrogen atom, r_{CH} in Fig. 1, the HH bond, r_{HH}, and the HCH angle, α_{rxn}, are varied systematically to generate a grid of heats of formation values for the system. The range of each variable extends from reactant to product values. The nonreactive CH bond, R_{CH}, is optimized at each choice of the other variables and the nonreactive HCH dihedral angle, α_{nr}, is fixed at 120 °.

The POL-CI calculations give a heat of reaction for $CH_4 + H \rightarrow CH_3 + H_2$ of 2.2 kcal/mole, including zero-point corrections, compared to the 0.6 kcal/mole experimental value. The POL-CI transition state barrier height, 13.5 kcal/mole, agrees favorably with the experimental value, 11.9 kcal/mole. The rate constants derived from the calculation are in good agreement with experiment as well, suggesting that the description of the transition state geometry in this calculation is fairly realistic. In addition the isotope effect for deuterium is approximately correct.

On this basis, we take the POL-CI calculations as being correct. The MOPAC calculations, if reasonable, ought to give similar results. The semiempirical estimate of the heat of reaction is -22.7 kcal/mole, not including zero-point corrections on the order of -3 kcal/mole. In the transition state region, the values of r_{CH}, r_{HH} and α_{rxn} should be approximately 1.47 Å, 0.92 Å and 102.4 °, respectively, with a barrier of 10-15 kcal/mole.

Plots of the transition state region, as defined above, for the POL-CI calculation, Fig. 2a, and the MOPAC calculation, Fig. 2b, are vastly different. Figure 2a shows the familiar saddle shape of a transition state and there is a flat region in both coordinates. On the other hand, in Fig. 2b, the saddle region is

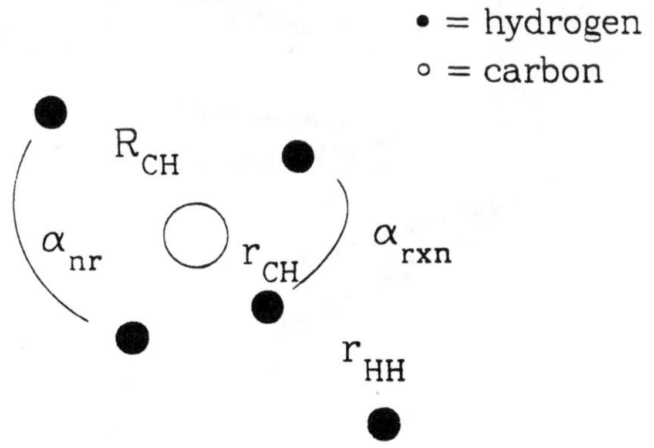

Figure 1. CH_4 - H system, reacting to yield CH_3 - H_2. The labeled variables are r_{CH} for the CH bond being attacked; R_{CH} for the bond between the C atom and the nonreactive hydrogens; α_{rxn} for the bond angle between the reactive H and the nonreactive ones, and α_{nr} for the HCH angle between nonreacting hydrogens. α_{rxn} changes from 109.47° to 90° during the reaction.

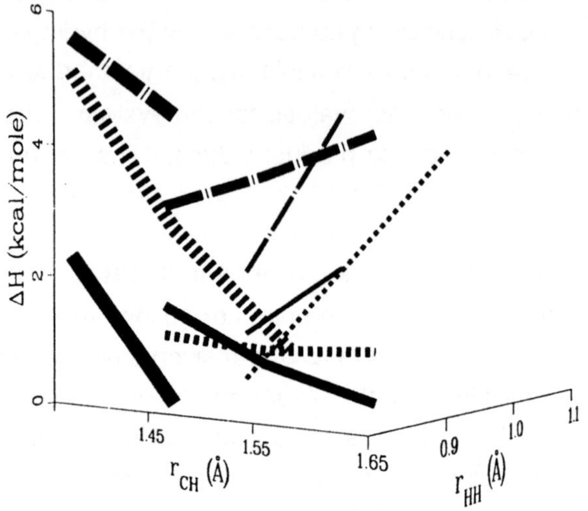

Figure 2a. Transition state region for the POL–CI calculation. The solid curve is for 100 °, the dot-dash curve for 105 ° and the dashed curve for 110 °. The transition state occurs when the surface is flat in both coordinates.

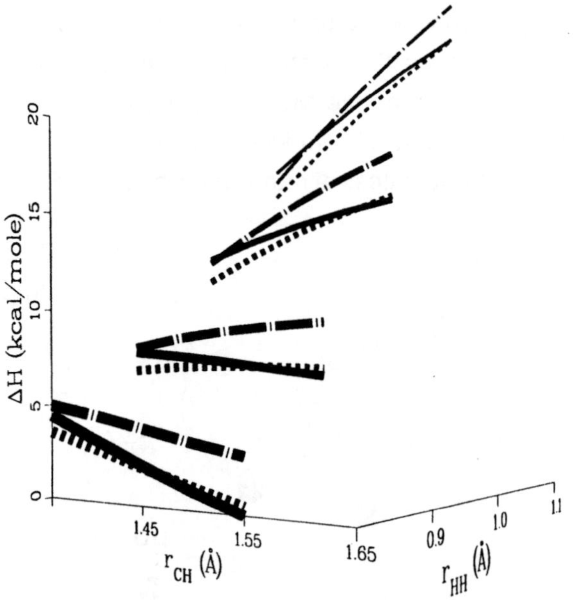

Figure 2b. Same region as in (2a) for the MOPAC calculation. The line types are also the same. The surface is never flat in the r_{HH} coordinate and no transition state occurs in this region.

flat only in the r_{CH} coordinate, not in both. Consequently, a barrier is absent and no estimate of the barrier can be made. Figure 3 shows larger portions of the potential energy surface where it is obvious that there is no transition state. In all the parameter ranges studied, the intervals are [1.0,1.6] (Å), [0.65,1.1] (Å) and [90,110] (°) for r_{CH}, r_{HH} and α_{rxn}, respectively. No evidence for a transition

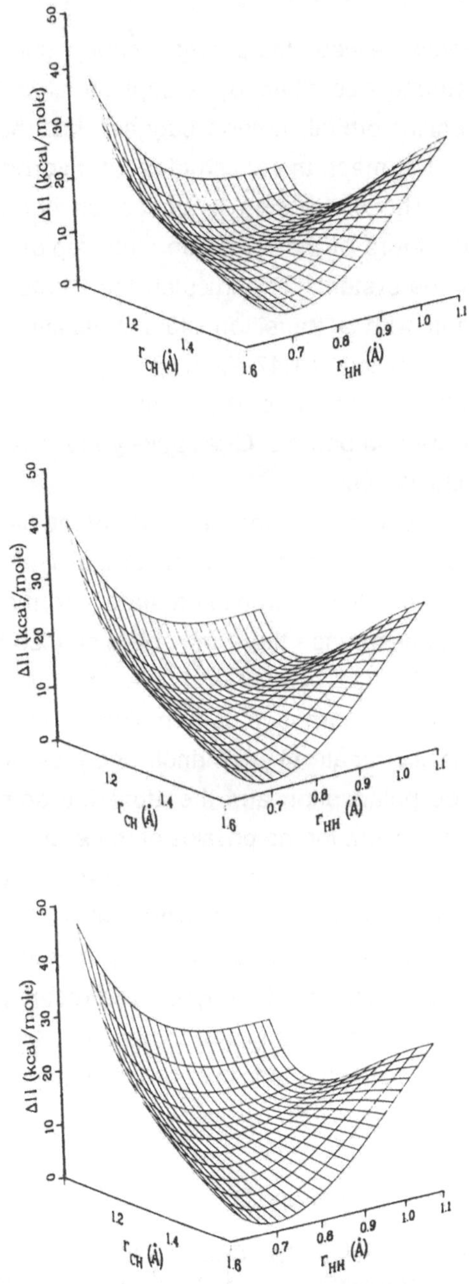

Figure 3. Full potential energy surface for both reactants and products. No transition state is apparent in these coordinates at any angle. The top frame is for α_{rxn} = 110°, the middle for α_{rxn} = 105°, and the bottom for α_{rxn} = 100°.

state is seen anywhere. On this basis, it is difficult to say that the MOPAC results can be relied on for the type of mechanistic insight that was originally hoped for.

IV. CONCLUSIONS

The vast differences between the potential energy surfaces for abstraction of hydrogen from methane described by a sophisticated CI calculation and a semiempirical molecular orbital method suggest that the latter, in its current form, is unreliable for mechanistic studies of hydrogen abstraction from diamond surfaces. The conclusion is based on comparisons of heats of reaction and on the nature of the transition state region, as defined by the CI calculation, for this one system. In particular, the CI method predicts a slightly endothermic reaction with a transition state consisting of a 13.5 kcal/mole barrier at a CH bond distance of 1.47 Å, an HH separation of 0.92 Å and a HCH bond angle of 102.4 °, whereas the semiempirical method predicts an exothermic reaction with no barrier. Qualitatively these results are independent of the specific hamiltonian used.

There are two possible reasons for the failure of the semiempirical method in the CH_4 + H system. One is that the database used to parameterize the hamiltonian contains insufficient transition-state information for hydrocarbon systems. If this is the case, then there may be a chance to reparameterize the hamiltonian for specific use in diamond, diamond-like carbon, amorphous carbon and hydrogenous amorphous carbon films. The other possible reason may be that the transition state phenomenon depends heavily on an accurate description of charge polarization and therefore the semiempirical method is intrinsically unable to account for the physics of the system. If this is the case, in order to gain the desired insight into carbonaceous film growth, it may be necessary to move toward other methods which fall into two categroies. One is more robust, but more expensive, electronic structure calculations. The other is the completely empirical methods of the type used recently by Brenner,[10] which are simplier and can sometimes be forced to have the correct physical behavior by choosing proper functional forms for the potential energy surface.[1,8,9]

REFERENCES

[1] R. C. Bingham, M. J. S. Dewar, and D. H. Lo, *J. Am. Chem. Soc.* **97**, 1285 (1975);
M. J. S. Dewar and W. Thiel, *J. Am. Chem. Soc.* **99**, 4907 (1977);
M. J. S. Dewar, E. G. Zoebisch, E. F. Healy, and J. J. P. Stewart, *J. Am. Chem. Soc.* **107**, 3902 (1985);
M. J. S. Dewar and D. M. Storch, *J. Am. Chem. Soc.* **107**, 3898 (1977).

[2] M. Tsuda, M. Nakajima, and S. Oikawa, *J. Am. Chem. Soc.* **108**, 5780 (1986); *Japan. J. Appl. Phys.* **26**, L527 (1987).
[3] D. Huang, M. Frenklach, and M. Maroncelli, *J. Phys. Chem.* **92**, 6379 (1988).
[4] S. M. Valone, "Possible Behavior of a Diamond (111) Surface in Methane/Hydrogen Systems," submitted.
[5] S. P. Walch, *J. Chem. Phys.* **72**, 4932 (1980);
G. C. Schatz, S. P. Walch and A. F. Wagner, *J. Chem. Phys.* **73**, 4536 (1980);
G. C. Schatz, A. F. Wagner and T. H. Dunning, Jr., *J. Phys. Chem.* **88**, 221 (1984);
P. J. Hay and T. H. Dunning, Jr., *J. Chem. Phys.* **64**, 5077 (1976).
[6] C. C. J. Roothaan, *Rev. Mod. Phys.* **23**, 69 (1951).
[7] G. G. Hall, *Proc. R. Soc. London, Ser. A* **205**, 541 (1951).
[8] J. J. P. Stewart, *J. Comp. Chem.* **10**, 209 (1989); 221 (1989).
[9] J. J. P. Stewart, MOPAC, A Semiempirical Molecular Orbital Program, *QCPE* **455** (1983).
[10] D. Brenner, "Empirical Potential for Hydrocarbons for Use in Simulating the Chemical Vapor Deposition of Diamond Films." In press, *Phys. Rev. B.*

INVESTIGATIONS ON GROWTH MECHANISMS OF DIAMOND THIN FILMS BY HOT FILAMENT ASSISTED CVD

J. Mercier, A.M. Bonnot*, E. Caignol and E. Gheeraert

LEPES - CNRS associated with University J. Fourier
B.P. 166 - 38042 Grenoble Cedex - France
*LEPES-CNRS and GPEC-URA CNRS 783, Case 901
13288 Marseille Cedex - France

INTRODUCTION

Since the pioneering work of Derjaguin [1], the development of various processes of diamond thin film deposition under subatmospheric pressure and relatively low temperature has been impressive. The state of the art has been reviewed [2, 3, 4]. Our experience is in the process of hot filament assisted CVD [5]. Here are discussed some kinetic aspects of the results of a parametric study carried out by our group [6, 7].

EXPERIMENTAL

Basically, the apparatus already described [5] is a hot filament CVD reactor (Fig. 1), under a reduced pressure (665 Pa - 13 kPa) of a CH_4 - H_2 reactive mixture in controlled amounts. Total pressure is controlled by a throttle valve. External heating of the substrate holder and substrate (a Si (100) wafer) is provided by a movable electric furnace. Central to the process is the tungsten filament, parallel to the substrate, a short distance (5 - 10 mm) apart. It is heated to T > 2000°C to activate the chemical reactions, by forming highly reactive H atoms. Its temperature is measured by pyrometry and checked by measurement of the wire resistance. The substrate temperature is calibrated from the value (reported in the text) given by a thermocouple inserted in the substrate holder. Continuous films can be deposited on finely abraded (0.1 μm diamond powder) surfaces.

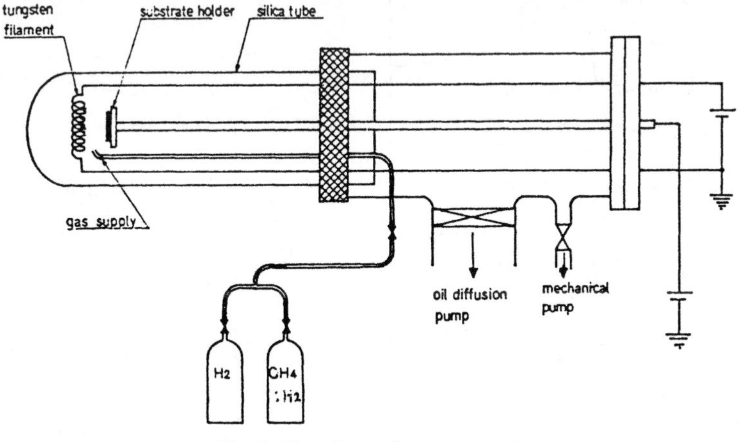

Fig. 1. Experimental arrangement.

Diamond and Diamond-Like Films and Coatings
Edited by R.E. Clausing *et al.*, Plenum Press, New York, 1991

PARAMETRIC STUDIES

The molar percentage $[CH_4] / [H_2] = x$ % of gaseous species concentrations and the total pressure P_t were changed successively. x changed from 0.1 to 2 % at $P_t = 4$ kPa and P_t varied from 665 Pa to 13.3 kPa at x = 0.5 %. Other conditions are given below :

- Filament temperature :	2200°C
- Filament substrate distance :	5 mm
- Substrate temperature :	750°C
- Total flow-rate :	45 sccm
- Run duration :	3 hours

Film properties were characterized by :
- SEM for the morphology, EDAX (tungsten : less than 0.5 at. %)
- Micro Raman spectroscopy for phase determination [8], crystalline quality and stress evaluation [9, 10].
- Auger electron spectroscopy [6] which proved unambiguously that the external surface is diamond only.

INFLUENCE OF CH_4 / H_2 RATIO

The Raman spectra (Fig. 2) indicate that the diamond formation occurs for all x's. But when x exceeds 0.6 %, crystalline quality related to the Full Width at Half Maximum (FWHM) of the diamond peak $\left(\overline{\nu_0}\right)$ as well as film purity degrade with increasing amounts of graphitic and a-C : H components [11, 12]. The intensity ratio $I\left(\overline{\nu_0}\right) / I\ (1550)$, related to purity and the FWHM values are plotted in Fig. 3.

SEM observations substantiate these results (Fig. 4). The well faceted cubo-octahedral crystals exhibit secondary nucleation at x = 1.2 % and become ball-like at x = 2 %. Fig. 5 shows the percentage of surface coverage of the diamond film and the [001] side length of the (001) square face of the cubo-octahedral crystals. Due to the diamond structure, it is safe to assume that the growth rate is well related to this last parameter.

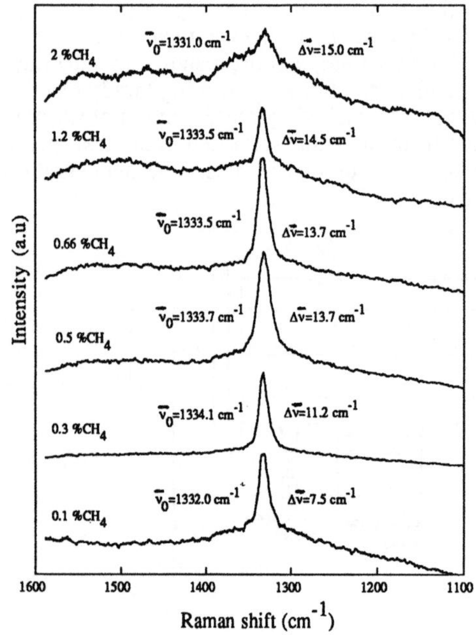

Fig. 2. Raman Spectra vs $x = 100 \times [CH_4 : H_2]$ (%)

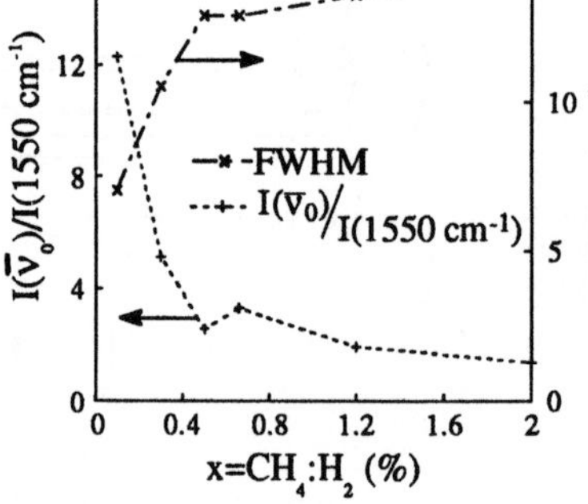

Fig. 3. FWHM $(\overline{\nu_0})$ and $I(\overline{\nu_0})/I(1550)$ versus $x = 100 \times [CH_4 : H_2]$ (%)

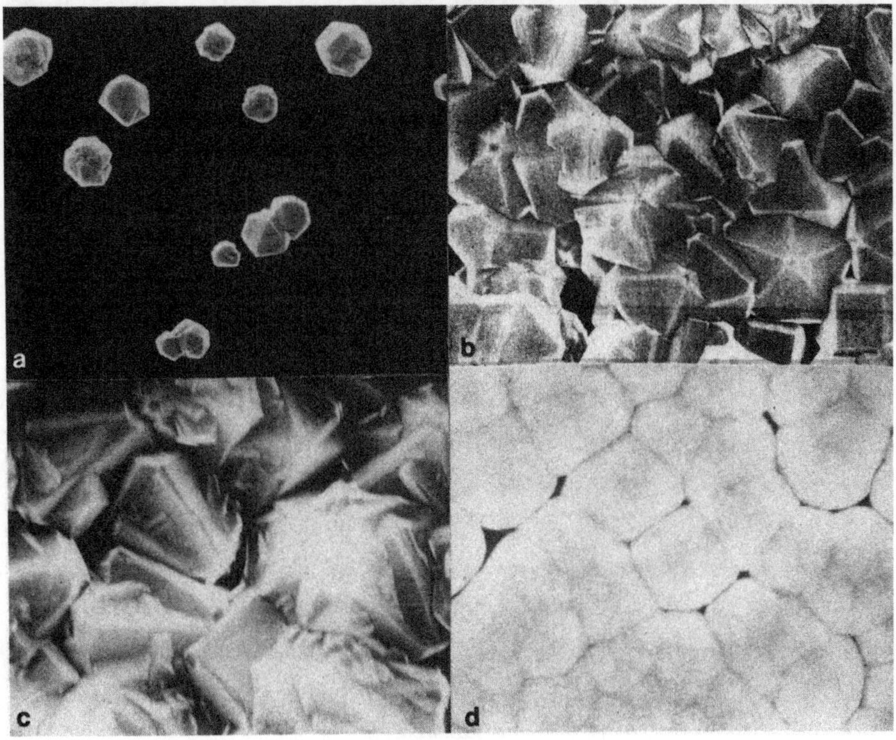

Fig. 4. SEM Micrographs for various x = $CH_4 : H_2$
a) x = 0.1 % b) x = 0.5 % c) x = 1.2 % d) x = 2 %
______ 1 μm except d) ______ 5 μm

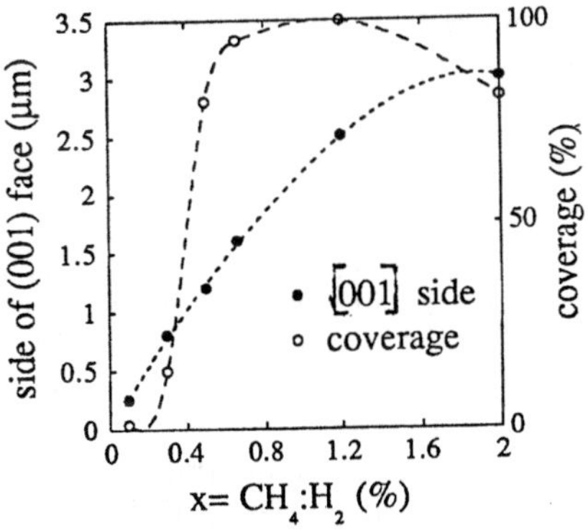

Fig. 5. Size of [001] side and film coverage versus x = 100 x [$CH_4 : H_2$] %

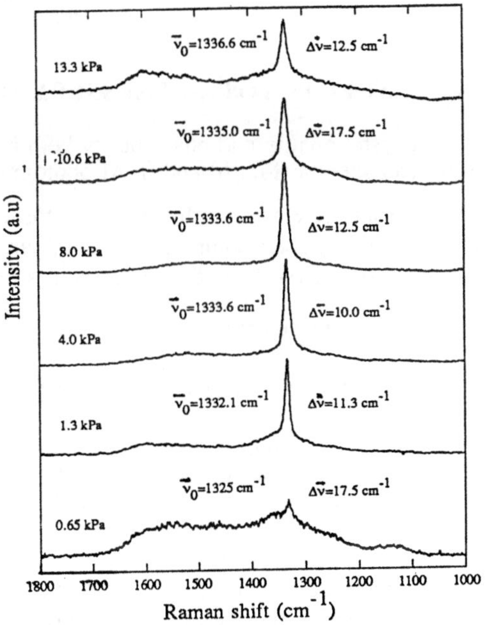

Fig. 6. Raman Spectra for various P_t

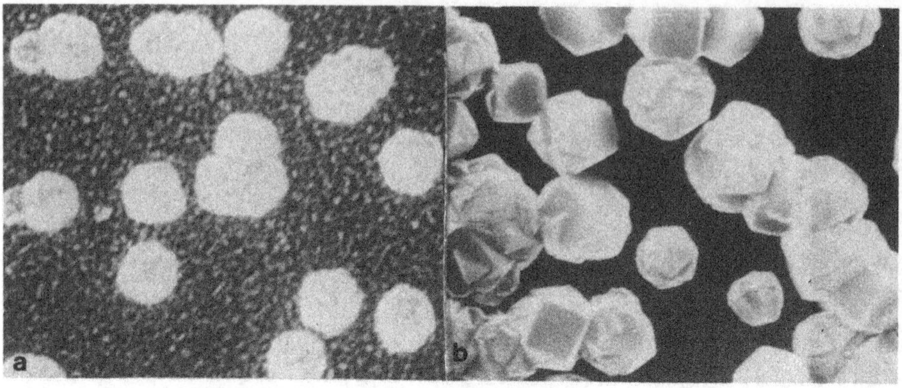

Fig. 7. SEM micrographs for P_t a) 0.65 kPa b) 10.6 kPa ______ 1 µm
For P_t = 4 kPa see Fig. 4 b)

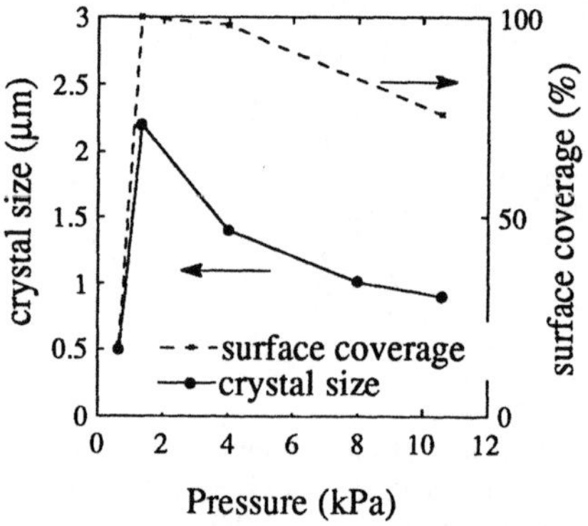

Fig. 8. Film coverage and crystal size versus P_t

INFLUENCE OF TOTAL PRESSURE (P_t)

Crystal quality and phase purity both are optimized around P_t = 4 kPa, as shown by the Raman spectra (Fig. 6) and SEM photographs (Fig. 7). At 665 Pa, an amorphous film covers the substrate and ball-like particles form. When P_t increases, $\overline{\nu_0}$ changes from lower to higher wavenumbers compared to that of pure diamond. These trends have been ascribed to microcrystallization [13] and compressive stresses induced by graphitic components [14], respectively. The size of the crystals (Fig. 8) shows a maximum at 1.3 kPa, in contrast to the monotonous increase with x.

DISCUSSION OF KINETICAL ASPECTS

A good introduction to kinetics of diamond thin film growth is found in [15]. It is now taken for granted that the overall process is surface reaction rate limited. Here we give figures to show that a diffusive mass transfer in the gas phase cannot be rate limiting. The injected CH_4 and H_2 species undergo many reactions[16,17,18] in the vicinity of the filament, the hottest part in the reactor. Above 2000°C, H_2 dissociates into H atoms. Their concentration varies as $P_{H2}^{1/2}$ [19]. At 4 kPa and 2500 K, the dissociation fraction is 6.5 %. Then, the H partial pressure is about 250 Pa. Due to the (P^{-1}) dependence of the diffusion coefficient, the diffusing flux of H towards the substrate varies as $P_{H2}^{-1/2}$ .To estimate the magnitude of the various diffusion fluxes, we consider only CH_4 and H as the diffusing species in H_2. The mutual diffusion coefficient of species 1 and 2 is expressed [20] as :

$$D_{12} = 0.002628 \frac{T^{3/2}}{P} [M_{12}^{1/2} \sigma_{12}^2 \Omega_{11}(T^*)]^{-1} \qquad D_{12} \text{ in cm}^2\text{/s, P(atm), T(°K)}$$

$\sigma_{12} = \frac{\sigma_1 + \sigma_2}{2}$ (Å) is the collision diameter

$M_{12} = 2 / [1/M_1 + 1/M_2]$ is the molar reduced mass

$\Omega_{11}(T^*) = kT/\varepsilon_{12}$ $\quad \varepsilon_{12} = (\varepsilon_1 \varepsilon_2)^{1/2}$ is characteristic of the potential of interaction between 1 and 2.

The σ's and $\frac{\varepsilon}{k}$'s for CH_4 and H_2 are found in [20] and for H in [21].

Hence :

At T =	1000	1500	2500 K
D_{CH_4/H_2} =	139.3	272.1	636.7cm²/s
D_{H/H_2} =	418.6	813.3	1900 cm²/s

The H diffusing flux is approximated by :

$$J_H = D_{H/H_2} \frac{P_H/RT}{d} \text{ mole/cm}^2\text{s}$$

d is the substrate - filament distance. D and T are to be taken at a mean value, 1500 K, if T_{sub} = 1000K and T_{fil} = 2500 K

Then : $J_H = 1.45 \times 10^{-4}$ H mole /cm²s.

Considering a site density of 10^{15} cm^{-2} for (100) Si, this flux would correspond to 10^5 monolayer/s !

Of course, H atoms (as well as CH_4 molecules) undergo many reactions in the vicinity of the substrate surface, such as[16,17]:

recombination : $H + H + (H_2) \rightarrow H_2 + (H_2)$

addition : $H + CH_n \rightarrow CH_{n+1}$ $\quad n \leq 3$

ablation : $H + CH_n \rightarrow CH_{n-1}$ $\quad n \leq 4$

and others with C_2H_2 [18].

This must lead to a reduction of J_H, accounted for by a reactive sticking coefficient α. (For silane on Si, a highly reactive species, α is of the order of 10^{-5}.) Even for $\alpha = 10^{-5}$, J_H would correspond to 1 monolayer/s.

Since D_{CH_4/H_2} is 1/3 D_{H/H_2} and the P_{CH_4} at the filament is $\frac{1}{100} \times 4000 = 40$ Pa, it turns out that J_{CH_4} must be about one or two orders of magnitude smaller than J_H.

Therefore it can be concluded that :

1) a copious supply of reacting species takes place by diffusion, lending support to surface reaction rate limitation.

2) H adsorption must be crucial in these surface reactions.

Considering now the chemistry of the system, it seems almost impossible to cope with its tremendous intricacy. Rather, we will discuss an (over) simplified model. Let CH_4 and H_2 eventually change into CH_y and H species at the vicinity of the growing interface. Let's assume the following steps :

- adsorption at vacant sites (*) :

$$* + CH_y \overset{K_c}{\rightleftarrows} CH_y^* \qquad (1)$$

$$* + H \overset{K_H}{\rightleftarrows} H^* \qquad (2)$$

- reactive desorption :

$$CH_y^* + H^* \overset{k}{\rightarrow} C^* + \frac{y+1}{2} H_2 \qquad (3)$$

Reaction (1) and (2) are fast with equilibrium constants K_c and and K_H respectively.
Reaction (3) is slow and rate limiting with a kinetic rate constant k.
Leaving aside the phase selection between graphite, a C : H and diamond, the diamond deposition rate G is related to [C*]. It reads :

$$G = k\,[CH_y^*]\,[H^*] \qquad (4)$$

We used the consistent units as follows : the bracketed concentrations $[CH_y^*]$, $[H^*]$ of adsorbed species are represented by the factional coverages θ_c and θ_H of CH_y and H species respectively.

Similarly the vacant site concentration is represented by its fractional coverage θ. Therefore :

$$K_c = \frac{\theta_c}{\theta} : [CH_y] \quad (5) \qquad K_H = \frac{\theta_H}{\theta} : [H] \quad (6) \qquad (\text{in } m^3/\text{mole})$$

$$G = k\,\theta_c\,\theta_H \quad (7) \qquad k \text{ and } G \text{ in } \mu m/h.$$

Taking into account that : $\theta + \theta_H + \theta_c = 1$ (8), G becomes :

$$G = \frac{k[CH_y][H]}{\left[\frac{1}{\sqrt{K_H K_C}} + \sqrt{\frac{K_H}{K_C}}[H] + \sqrt{\frac{K_C}{K_H}}[CH_y]\right]^2} \qquad (9)$$

This is a well known Langmuir-Hinshelwood model [22]. It expresses the competing influence of H^* and CH_y^* in the slowest process (reaction 3). It will be shown that $(K_H K_C)^{-1/2}$ can be neglected.

Now consider the first series of experiments. Assume that x % = 100 x $[CH_4]/[H_2]$ is directly related to $[CH_y]/[H]$. Then (9) can be written as :

$$\sqrt{\frac{x}{G}} = \frac{1}{k}\left\{\sqrt{\frac{K_H}{K_C}} + \sqrt{\frac{K_C}{K_H}}\,x\right\}$$

Indeed the experimental values (see Fig. 5) give :

$$\sqrt{\frac{x}{G}} = 1.05 + 0.105x \qquad (x\ \%\ \text{and}\ G\ \mu m/h)$$

Then k = 3 µm/h which is of the right order of magnitude and $\frac{K_H}{K_c} = 10$. It must be noted that there should be a maximum for G at $x = \frac{K_H}{K_c}$ (= 10). Due to the restricted range for x in our experiments, it cannot be observed.

From Giling [18] K_H can be calculated : $K_H = 10^{15}$ m³/mole.
Thence $K_c = 10^{14}$ m³/mole and $(K_c K_H)^{-1/2} << 1$.

On the contrary, considering the second series, and assuming that $P_t \sim P_{H_2}$ due to the H_2 molecules, is directly related to [H] at constant $[CH_y]$, G now has a maximum value for

$$\frac{[H]}{[CH_y]} = \frac{K_C}{K_H} \qquad \left(= \frac{1}{10}\right)$$

It is interesting to note that this maximum corresponds to an optimal quality determined from the Raman spectra :

- at high pressure, since the H flux decreases, the phase selection with respect to diamond is less efficient.
- at low pressure, the H flux increases, bringing about a stronger etching, but also a noticeable incorporation of hydrogen to form a - C : H, as pointed out by Buckley [23].

Russian searchers [24] have presented a different adsorption model. CH_4 is assumed to adsorb without competing against H. Then G is proportional to θ_c only and reads :

$$G = \frac{k\,[CH_4]}{1 + K_H[H] + K_c[CH_4]}$$

This model doesn't fit our experiments since it does not forecast maximum values for G versus $[CH_4]$ or [H].

CONCLUSION

Diamond thin films were prepared by hot filament assisted CVD, varying either the methane to hydrogen ratio or the total pressure. In both cases, optimized conditions with respect to deposition rate, crystalline quality and purity of the films obtained for intermediate values of the above deposition parameters. Diffusive fluxes of reacting species are estimated to show that mass transfer through the gas phase is not rate-limiting. Experimental results are well described by a Langmuir adsorption model. It stresses the competing influence of H-atoms and some carbon-bearing species on adsorption sites in limiting the deposition rate and controlling at the same time the film quality and purity. It emphasizes the crucial role of H-atoms in this whole process of diamond growth.

ACKNOWLEDGEMENTS

This work is supported by the Ministry of Research and Technology, the Ministry of Education and University J. Fourier.

REFERENCES

1. B.V. Derjaguin, D.V. Fedoseev et al, J. Cryst. growth 2 : 380 (1969).
2. J.C. Angus, F.A. Buck, M. Sunkara, T.F. Groth, C.C. Hayman and G. R. Gat, MRS Bull 38, oct. 89.
3. J.C. Spear, J. Am. Ceram. Soc. 72 : 171 (1989).
4. N. Sekata, J. Mater Res. 4 : 664 (1989).
5. A.M. Bonnot, Thin Solid Films 185 : 111 (1990).
6. A.M. Bonnot, to be published in Proceed E-MRS Symp. D Strasbourg (May-June 1990).
7. E. Caignol, E. Gheeraert, R. Cinti, J. Mercier, A.M. Bonnot, G. Lucazeau and L. Abello, to be published in Proceed E-MRS Strasbourg (May-June 1990).
8. D.S. Knight and W.B. White, J. Mat. Res 4 : 385 (1989).
9. A.M. Bonnot, in Proceed 1st International Symp. on Diamond and Diamond like films (L.A. 1989), J.P. Dismukes Ed. , The Electrochem Soc. Pub. 579.
10. A.M. Bonnot, Phys. Rev. B 41 : 6040 (1990).
11. M. Ramsteiner and J. Wagner, Appl. Phys. Letters 51 : 1355 (1989).
12. M. Yoshikawa, G. Katagiri, H. Ishida and A. Ishitani, Solid State Com. 66 : 1177 (1988).
13. A.M. Bonnot, Coll. Physique Suppl. to n° 5, 50 : 901 (1989).
14. W. Zhu, C.A. Randall, A.R. Badzian and R. Messier, J. Vac. Sci. Technol. A7 : 2315 (1989).
15. A.R. Badzian and R.C. De Vries, Mat. Res Bull 23 : 385 (1988).
16. S.J. Harris, J. App. Phys. 65 : 3044 (1989).
17. Y. Matsui and M. Sahara, Jpn J. Appl. Phys. 28 : 1023 (1989).
18. G. Janssen, W.J.P. Van Enckevort, L.J. Giling, in Proceed 1st International Symp. on Diamond and Diamond like films (L.A. 1989) Ed. J.P. Dismukes - The Electrochemical Soc. Pub. 508.
19. P. Pascal, Nouveau Traité de Chimie Minérale Vol. 1 Masson Pub. Paris (1956).
20. R.C. Reid, J.M. Prausnitz and B.E. Poling in "The Properties of Gases and Liquids" 4th Ed. Mc Graw Hill N. Y. (1986).
21. Y. Matsui, A. Yuuki, N. Morita and K. Tachibana, Jpn J. Appl. Phys. 26 : 1575 (1989).
22. D.W. Shaw in "Crystal Growth" Vol. 1 GHL Goodman Ed. , Plenum Press (1974).
23. R.G. Buckley, T.D. Moustakas, Ling Ye and J. Varon, J. Appl. 66 : 3595 (1989).
24. B.V. Derjaguin and D.V. Fedoseev, Growth of diamond and graphite from gas phase (in russian). Nauka Moscow (1977).

CARBON BONDING ENVIRONMENTS IN CVD DIAMOND FILMS INVESTIGATED VIA SOLID-STATE NMR

Karen Mary McNamara and Karen K. Gleason

Department of Chemical Engineering
Massachusetts Institute of Technology
Cambridge, MA 02139

INTRODUCTION

Diamond films are typically produced by both plasma and thermally assisted CVD using methane as the carbon source in an excess of hydrogen[1]. Although these films have been characterized using a wide variety of techniques, nuclear magnetic resonance (NMR) spectroscopy can yield valuable new information. Both hydrogen and carbon-13 NMR spectroscopy can be used to examine polycrystalline diamond films[2], even those with a hydrogen content of < 1%. NMR, however, has not been used extensively to date.

NMR is quantitative in that each nuclei gives rise to the same integrated NMR signal, regardless of its chemical environment, provided complete spin-lattice relaxation occurs between averages[3]. Gem-quality natural diamond can have a spin-lattice relaxation times > 1 day, making averaging painfully slow. The presence of paramagnetic defects decreases the relaxation time of neighboring ^{13}C nuclei to several seconds[4]. Increasing the ^{13}C concentration reduces the spin-lattice relaxation time of ^{13}C nuclei removed from the defect site through increased homonuclear dipolar communication as well as increasing the total signal. In addition, ^{13}C enrichment, if done selectively, can be used to gain insight into diamond chemistry.

Very high enrichments will, however, limit spectral resolution of sp^2 versus sp^3 carbon due to ^{13}C homonuclear dipolar line broadening. Resolution can be restored by magic-angle spinning, but this requires crushing the sample, destroying its physical integrity. Unlike magic-angle spinning, static NMR experiments do not require sample destruction. In fact, slivers of diamond films can be inserted directly into the NMR coil while intact on their substrate. Also, the low-energy radio-frequency irradiation used in NMR is unlikely to result in chemical alteration of the diamond samples.

In materials with high hydrogen concentrations, like polymers and diamond-like carbon films, cross-polarization

Diamond and Diamond-Like Films and Coatings
Edited by R.E. Clausing *et al.*, Plenum Press, New York, 1991

enhances the carbon-13 spectrum[5,6]. At the low hydrogen concentrations found in polycrystalline diamond films, carbon nuclei located far from hydrogen are less effectively cross-polarized, and thus, the quantitative character of the NMR experiment is lost. Direct-polarization ^{13}C NMR, while requiring larger samples, is a means of obtaining quantitative sp^2/sp^3 ratios.

To date, Raman spectroscopy is used to identify sp^2 and sp^3 bonded carbon in diamond films. However, the Raman results are difficult to quantify since the scattering efficiency of sp^2 bonded carbon differs from sp^3 bonded material. These scattering efficiencies may vary from film to film, as well. The sensitivity of Raman spectroscopy is also controlled, in part, by the spot size of the instrument. With a small spot size, it is possible to overlook inclusions of graphitic carbon and overestimate the quality of the film.

NATURAL ABUNDANCE CARBON-13 FILM

A free-standing polycrystalline diamond film was provided by Dr. James E. Butler of the Naval Research Laboratory. The film was produced from 0.5% methane in hydrogen in a hot-filament reactor. The total pressure was 34 torr, and the substrate temperature was ~1150K. After 93 hours, an opaque, almost black film was obtained. The micro-Raman spectrum, obtained by Dr. Butler, is shown in Figure 1a. Several locations were probed, and the micro-Raman indicated the presence of sp^3 bonded carbon, only. The static direct-polarization NMR spectrum of the ~0.1 g sample was obtained on a commercial 4.7 T IBM Instruments spectrometer. Two peaks are present, one centered at 34 ppm and the second centered at 120 ppm, indicative of sp^3 and sp^2 bonded carbon, respectively[4-6].

Thus, NMR has detected sp^2 carbon in a black, opaque CVD diamond film which was not apparent from the Raman spectrum.

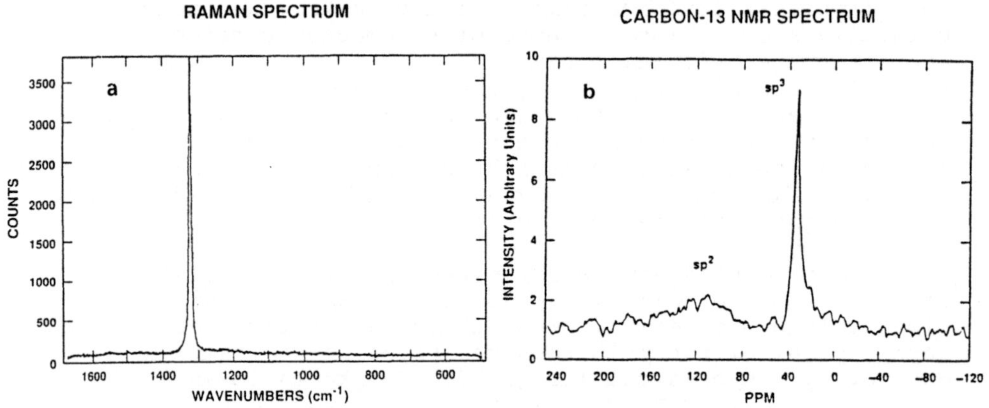

Figure 1. (a) The Raman spectrum of a natural abundance, hot-filament CVD film shows an intense adsorption for diamond at 1332 cm^{-1}. The presence of sp^2 bonded carbon is not apparent. (b) The NMR spectrum of the same sample clearly shows a peak at a chemical shift of 34 ppm with respect to TMS, indicative of diamond. The peak at 120 ppm indicates a significant amount of sp^2 bonded carbon in the sample.

Differences between the NMR and Raman spectrum could arise for several reasons, as discussed in the introduction. The NMR spin-lattice relaxation behavior for this sample has not yet been determined. When the period between averages is adjusted for the relaxation time, the relative intensities of the sp^2 and sp^3 peaks may vary. In order to reduce concerns about relaxation time and to decrease total deposition time, carbon-13 enriched films were examined.

CARBON-13 ENRICHED FILMS

Oxygen containing carbon sources such as acetone, ethanol, and carbon monoxide have been used to produce films at higher growth rates and/or with improved morphology[7,8]. Although there has been much speculation about the role of oxygen in such depositions, little has been proven at this time. Choosing acetone as a reactant and using selective isotopic labeling, resulted in higher growth rates, reducing reactant costs, and allowed insight into the role of oxygen in the deposition.

Two polycrystalline diamond films were deposited in a hot-filament reactor. The carbon source was 33% $^{13}CH_3CO^{13}CH_3$ or 66% $CH_3{}^{13}COCH_3$ in normal acetone, for an overall 22% gas-phase ^{13}C enrichment. The acetone and hydrogen flowrates were 2 sccm and 100 sccm, respectively. A 1" diameter silicon substrate was seeded with natural diamond grit prior to deposition. The filament was a 2 mm diameter tantalum wire maintained at a temperature of ~2600K, and the substrate temperature was ~1150K. The total pressure in the reactor was 150 torr.

Figure 2a shows the silicon substrate after 3.5 hours of growth in the carbonyl-enriched reactant. The area exposed during deposition is covered with a 12 μm thick polycrystalline film. Figure 2b is an optical micrograph of the film, showing a highly faceted surface with primarily <111> crystal planes and typical crystal sizes between 5 and 10 μm. Similar results were obtained with the methyl-enriched reactant. The 4 mm x 9 mm slivers, cut using a diamond saw, were inserted directly into the NMR probe, allowing non-destructive evaluation of the sample.

The static direct-polarization NMR spectra of the enriched films were obtained on a home-built, 6.3 Tesla spectrometer. Both spectra of the films produced from the methyl (Figure 3a) and carbonyl (Figure 3b) labeled reactants contain only one peak, centered at 34 ppm, indicative of sp^3 bonded carbon. Any signal upfield near 120 ppm, where sp^2 bonded carbon would be found, was below the noise level in our spectra.

The carbon-13 concentration in the films can be calculated from the second moment of the NMR lineshapes using the van Vleck equation for a spin-1/2 nuclei[9],

$$\langle \Delta\omega^2 \rangle = \frac{9}{20} C \gamma^4 \hbar^2 \Sigma_k r_{jk}^{-6}$$

Figure 2. (a) This photograph shows a 12 µm thick diamond deposit over a one-inch silicon substrate. The cuts show how slivers are prepared for intact insertion into the NMR probe where non-destructive evaluation takes place. The black edge regions appear where clips were used to hold hold the sample in place during deposition. Some of the film was chipped off the silicon substrate while under the diamond saw. (b) Optical micrographs show a highly faceted, continuous, polycrystalline morphology. The average particle size is 5-10 µm, and the film is ~12 µm thick.

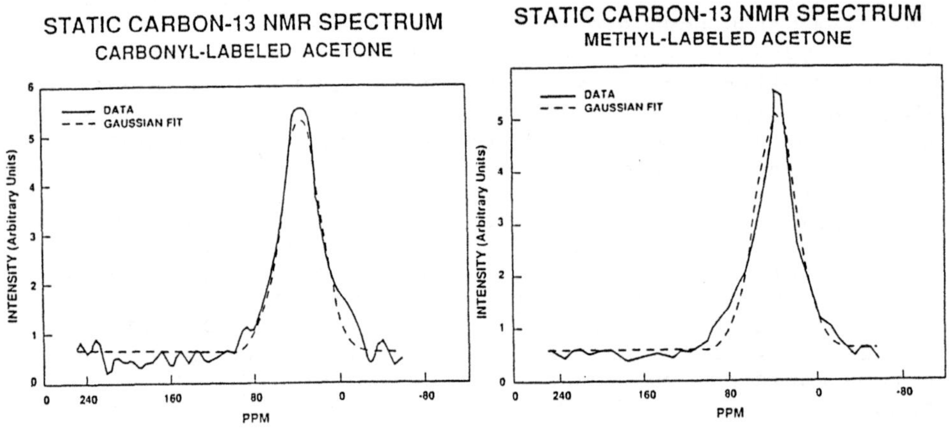

Figure 3. The carbon-13 NMR spectrum of both the film produced from the carbonyl and the methyl labeled acetone show only one peak, centered at 34 ppm, which is indicative of sp^3 bonded material. Relating the full-width at half-maximum of the best-fit guassian to the second moment gives a carbon-13 content of 22% in each film.

where

$<\Delta\omega^2>$ is the second moment;
C is the fraction of sites occupied by ^{13}C;
γ is the gyromagnetic ratio of the nuclei;
$\hbar$ is Plank's constant/2π; and,
r_{jk} is the internuclear distance between sites j and k, determined from diamond's crystallographic structure.

The dashed lines in Figure 3 represent the best Gaussian fit to these lineshapes. For a Gaussian, the full-width at half-maximum (FWHM) equals $2.35 <\Delta w^2>^{1/2}$. For both of the films, the FWHM is 45±2 ppm at 67.7 MHz corresponding to a carbon-13 concentration in the film is 22±1%, identical to the degree of gas-phase enrichment.

ANALYSIS OF HETEROGENEOUS CHEMISTRY

The incorporation of carbons from the carbonyl position of the acetone reactant raises questions about the role of oxygen and the chemistry within the reactor. There are three possible pathways for carbonyl-carbon incorporation into the diamond film - as CO incorporated directly from the gas phase, through homogeneous gas-phase conversion to a precursor species, or through heterogeneous conversion to a precursor species. The first, direct incorporation of CO from the gas phase, might be expected to follow a pathway analogous to the Boudouard reaction, ($2\ CO_{gas} \rightarrow CO_2 + C_{solid}$), for soot formation, although no evidence for this process in diamond growth has been found to date[10]. The kinetics of the Boudouard reaction are greatly enhanced by surfaces and iron contamination, suggesting the importance of heterogenous chemistry. There are many possible routes for the second method of incorporation, homogeneous carbon scrambling, but based on published homogeneous reaction rate constants for CO in the combustion literature, these are likely to be slow. In addition, through isotopic studies it has been shown that homogeneous scrambling of CO is minimal[11]. Scrambling of hydrocarbons, on the other hand, is essentially complete[12,13]. One final possibility is that the enol form of the acetone reactant provides a pathway for homogeneous conversion. Harris has measured a CO concentration of 0.018 mol% in the gas-phase when 1.0 mol% of O_2 enters the reaction zone[14]. This is close to the equilibrium CO concentration of 0.017 mol% at the filament temperature. Similar agreement was found at the other oxygen concentrations checked between 0% and 3%. However, the pathway by which this CO would form, directly from carbonyl carbons or through other homogeneous reaction, has not been determined.

We have used secondary ion mass spectrometry (SIMS) to evaluate the role of heterogeneous exchange between the carburized tantalum filament and gas-phase carbonyl-derived carbons. Figure 4a shows the SIMS spectrum obtained from a filament exposed for several hours to normal acetone in excess hydrogen. The peaks at mass 193 and mass 195 represent $Ta^{12}C$ and its hydride, $Ta^{12}CH_2$, respectively. The peaks at mass 197 and 198 are TaO and TaOH, respectively. Figure 4b shows the SIMS spectrum of a tantalum filament exposed to normal acetone and hydrogen in the same manner as the first and then exposed to carbon-13 enriched acetone labeled at the carbonyl carbon and an excess of hydrogen for several hours. The peaks at mass 193, 195, 197, and 198 appear as before. Two new peaks appear, however, at mass 194 and 196 indicating $Ta^{13}C$ and $Ta^{13}CH_2$. Depth profiling to > 2000 Å indicates that exchange has also occurred in the interior of the filament.

CONCLUSION

Nuclear magnetic resonance spectroscopy provided a non-destructive means of quantitatively evaluating diamond films. It was sensitive to the local environment throughout the sample and identified sp^2 bonded carbon in a film where it was not apparent from Raman analysis. Isotopic enrichment was used to decrease sample size requirements and spin-lattice relaxation times, increasing sensitivity. This will allow quantitative sp^2/sp^3 ratios to be obtained. Selective isotopic labeling was used to raise questions and gain insight into the chemistry of diamond formation as well. NMR has shown that the carbonyl carbons from acetone are incorporated into diamond films, and SIMS has been used to show that heterogeneous exchange reactions in the filament, involving the carbonyl carbon, occur.

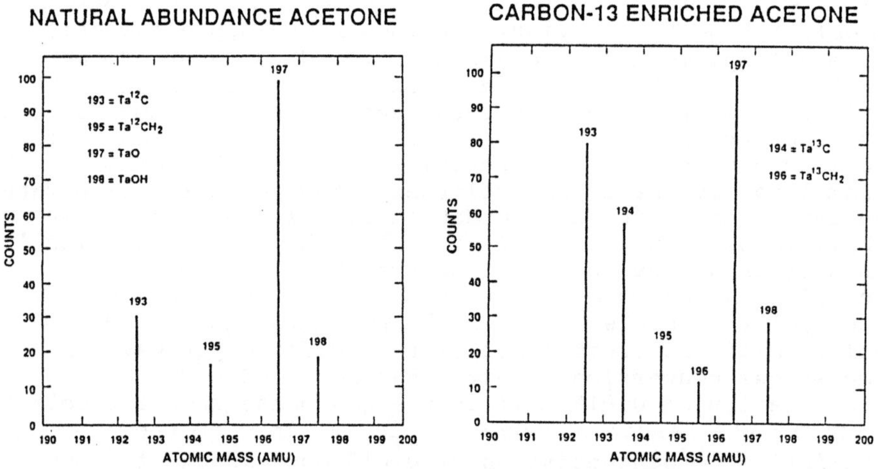

Figure 4. (a) SIMS spectrum taken from a tantalum filament exposed only to natural ^{13}C abundance acetone show peaks at 193, 195, 197, and 198 a.m.u. These are representative of $Ta^{12}C$, $Ta^{12}CH_2$, TaO, TaOH, respectively. (b) SIMS spectra taken from a filament initially carburized with normal acetone and then exposed to ^{13}C-labeled acetone, enriched at the carbonyl carbon, show the four peaks observed in spectrum (a) as well as peaks at 194 and 196 a.m.u. These represent $Ta^{13}C$ and $Ta^{13}CH_2$ and show that heterogeneous carbon exchange does take place at the filament.

ACKNOWLEDGEMENTS

We greatfully acknowledge Dr. Mike Geis and Mr. Keith Krohn of Lincoln Laboratories, MIT for their help in producing carbon-13 enriched samples. We also thank Dr. James E. Butler of the Naval Research Laboratories for providing the natural abundance sample and its micro-Raman spectra. Funding for this work was provided by ONR Grant #N00014-90-J-1901 and ACS-PRF #22595-G5.

REFERENCES

1. R.C. DeVries, Ann. Rev. Mater. Sci. **17**, 161 (1987) and references therein.
2. K.M. McNamara, K.K. Gleason and M.W. Geis, MRS Symposium Proceedings, **162** (1989). In press.
3. A. Abragam, "The Principles of Nuclear Magnetism" (Oxford, New York, 1983).
4. P.M. Hendricks, M.L. Cofield, R.H. Young, and J. M. Hewitt, J. Magn. Res. 58, 85 (1984).
5. S. Kaplan, F. Jansen, and M. Machonkin, Appl. Phys. Lett. **47**, 750 (1985).
6. M.A. Petrich, K.A. Gleason, J.A. Reimer, Phys. Rev. B, **6**(18), 9722 (1987).
7. Y. Hirose and Y. Terasawa, Jap. J. Appl. Phys. **25**, L519 (1982).
8. A. Hirayama, T. Ito, and K. Ito, Paper presented at Diamond Technology Initiative Symposium, SDIO/IST-ONR, Crystal City, VA, 12-14 July 1988
9. J.H. van Vleck, Phys. Rev. **74**, 1168 (1948).
10 A.G. Gaydon and H.G. Wolfhard, "Flames" (Chapman and Hall, London, 1979).
11. R. E. Ferguson, J. Chem. Phys. **23**, 2085 (1955).
12. R. E. Ferguson, Combustion and Flame, **1**, 431 (1957).
13. C.J. Chu, R.H. Hauge, M.P. D'Evelyn, and J.L. Margrave, MRS Symposium Proceedings, **162** (1989). In press.
14. S. J. Harris, J. Appl. Phys. **65**(8), 3044 (1989).

ACKNOWLEDGEMENTS

[illegible]

REFERENCES

[illegible]

DEPOSITION EXPERIMENTS WITH SEPARATED ATOMIC HYDROGEN AND CH_4 SOURCES

Pierre Joeris and Carsten Benndorf

Universität Hamburg
Institut für Physikalische Chemie
Bundesstr. 45
2000 Hamburg 13, FRG

INTRODUCTION

The mechanism of low pressure diamond deposition is a matter of high interest not only for researchers but also for those who want to apply these films. Only a deep insight into the deposition process will allow the adequate design (eg. transparency, graining etc.) of diamond films for the various fields of application.

In the present literature two main mechanisms for diamond film formation are discussed: First the deposition via methyl cations or methyl radicals as proposed by Tsuda and co-workers[1,2]. Second the growth mechanism suggested by Frenklach and Spear[3], which employs acetylene molecules as the main monomer growth species.

A lot of research has been done to identify the species present in deposition processes either in filament[4,5,6,7,8,9] or in microwave[10] assisted ones and to determine their respective concentrations .

One of the major problems encountered in investigating such plasmas is the great variety of hydrocarbons produced eg. acetylene, ethylene, ethane, propylene, propane, diacetylene, vinylacetylene, pentane, tri-acetylene, benzene, toluene etc. (from methane in a microwave plasma[11]). All of these species may be in different states of excitation. This complicates the modelling of such deposition processes which may be usefull in determining the required deposition parameters.

A possibility to avoid these difficulties is the utilization of a remote plasma system as proposed by Martin and Hill[12]. In such a system only the hydrogen is dissociated by plasma excitation or by thermal means (filament). The recombination of atomic hydrogen is very slow (rate constant[13] $k = 9.7e16\ T^{-0.6}\ cm^6\ mol^{-2}\ sec^{-1}$). This makes it possible to introduce the hydrocarbon eg. methane outside of the plasma downstream in order to produce methyl via the reaction $CH_4 + H \Rightarrow CH_3 + H_2$.

Under fast flow conditions there will be no thermodynamic equilibrium. Thus the number of reaction channels and reaction products is drastically limited. The deposition process of diamond films will become sensitive to the precursor hydrocarbon.

Reactions between atomic hydrogen and hydrocarbons have been described in literature eg. with methane[14] or acetylene[15]. Results from combustion chemistry (kinetics) will be helpful in understanding and modelling the system. The issue of growth species and which it may be should become clearer.

In the following we will describe our deposition system and report about first experiments.

EXPERIMENTAL

A microwave discharge is employed as source for the atomic hydrogen. The experimental setup (see Fig. 1) consists of a cylindrical cavity resonator with a directly coupled magnetron, similar to a system proposed by Meiners and Alford[16] for the generation of excited nitrogen. The magnetron has an output power of up to 750 W.

The cylindrical resonator is made of brass, it has a diameter of 203 mm and a length of 100 mm (Fig 2). The radial dimensions of the resonant cavity allow only a single halfwavelength between the plasma column and the side wall. This system allows a stable plasma to be maintained in a wide pressure range.

The tuning of the resonator is achieved by adjusting the length of the resonant cavity. For this reason one resonator plate is movable by turning the tuning knob. The plate is contacted to the side wall by a piston ring. Optimum tuning is controlled by observing the plasma column's extension and brightness.

A quartz tube with 20 mm o.d. is fed through the resonator and fitted to the gas inlet and the roughing pump with conventional high

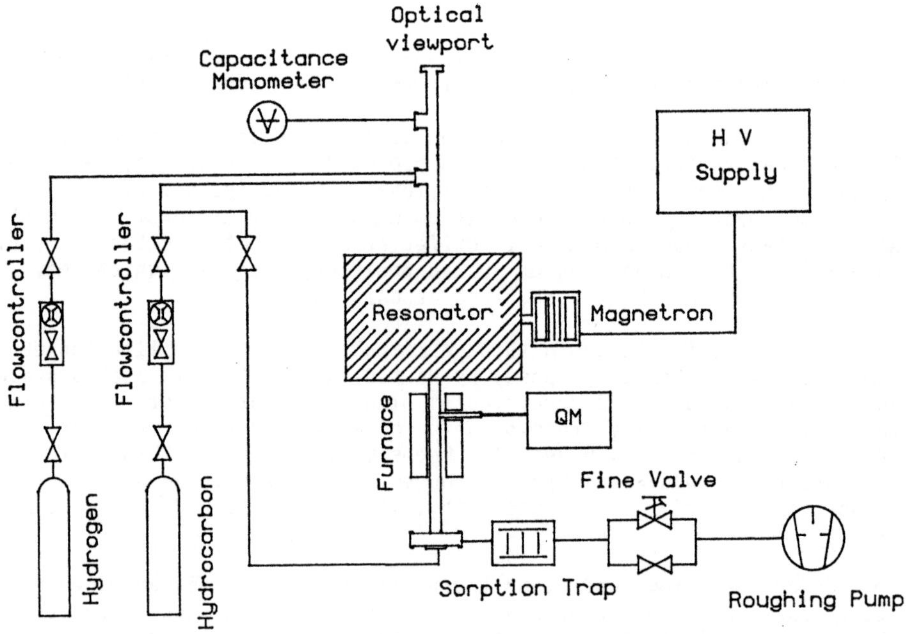

Fig. 1. Experimental Setup of the MPACVD System

vacuum parts. The walls of the tube have been passivated by soaking in NaOH and HNO_3 as described in literature[14].

Our deposition system was modified so that different modes of operation are possible without altering the setup. First it may be used in the standard MPACVD (Microwave Plasma Assisted Chemical Vapour Deposition) configuration. This means that the substrate is placed inside the plasma column and is heated by the plasma as shown in Fig. 2. The temperature is measured by an optical pyrometer. The reactive gases have to pass the resonator and are excited by the microwave discharge.

The second mode of operation is the remote plasma (Fig. 3). In this configuration the substrate is placed underneath the resonator outside the plasma. The distance between plasma and substrate may be changed easily. In these experiments it was chosen to be 20 to 30 mm depending on pressure and flow conditions.

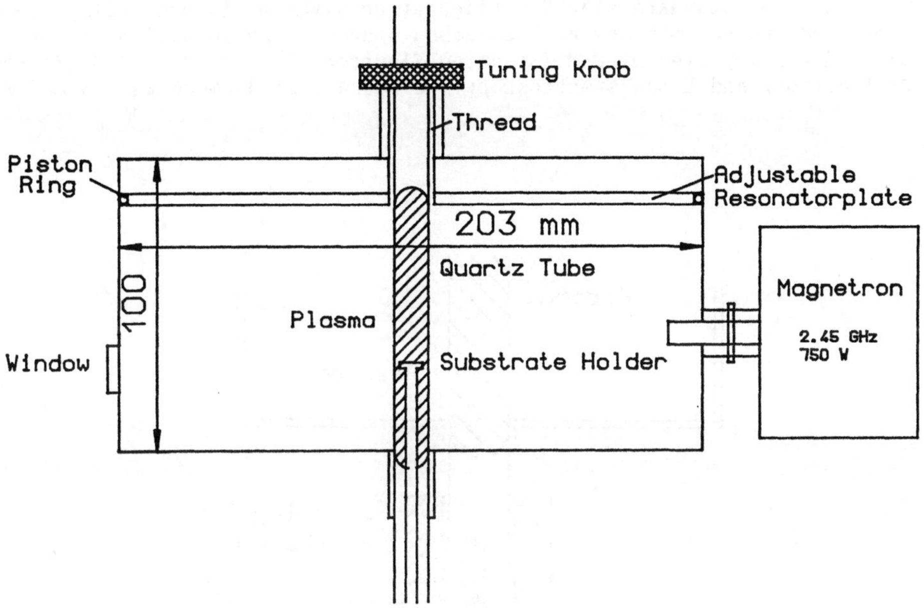

Fig. 2. Cylindrical Cavity Resonator

The substrate is heated by an external furnace. The substrate temperature is monitored by a thermocouple thus allowing a constant substrate temperature with changes less than two degrees centigrade.

The hydrocarbon gas is introduced into the system from below through the substrate holder which has a 0.5 mm nozzle at the upper end. The gas has to pass the furnace and reaches the desired temperature before it is injected thus avoiding possible quenching effects. The substrate is provided with a bore flush with the nozzle in the substrate holder to allow a steady flow of the hydrocarbon.

A quadrupole mass spectrometer (QM) is connected via a small quartz tube which is fitted to the main tube. This way it has direct access to the deposition zone. Optical spectrometers etc. may also be connected here.

The gas flows are controlled by calibrated MKS flow controllers and the pressure is measured by a MKS capacitance manometer. Depending on the flow rate, pressure and temperature in the heated part of the deposition system flow speeds are in the range of 1500 up to 3000 cm/s. Such high flow speeds were used to assure that hydrocarbons do not diffuse into the microwave plasma. In the standard MPACVD configuration these flow speeds have been proven not to hinder diamond deposition.

The preparation of the substrates (Si,Mo) was performed by polishing one side with 1/4 µm diamond spray followed by a cleaning procedure using an ultrasonic bath. After this procedure no residual diamond particles were detectable.

RESULTS AND DISCUSSION

In the standard MPACVD configuration diverse diamond films were deposited using methane as the carbon source. The formation of true diamond was approved by SEM (Scanning Electron Microscopy), XRD (X-Ray Diffraction) and Raman spectroscopy. We found that diamond was formed in

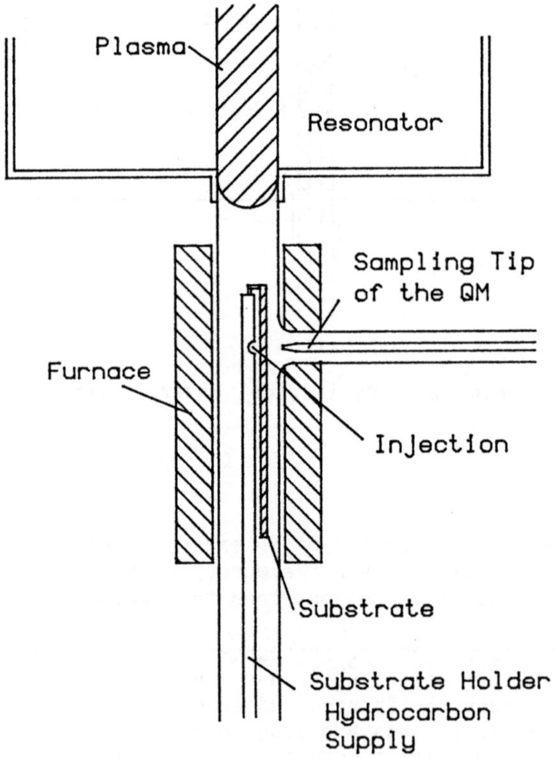

Fig. 3. Remote Plasma System

a temperature range reaching from below 700 up to 1050 degree centigrade. Films of best quality however were deposited at about 900 to 950 degree centigrade. Grain sizes were between 0.5 and 4 µm with run durations of one to eight hours using a mixture of 99.5% hydrogen and 0.5% methane.

The focus of present research is the remote plasma deposition system. Experiments were carried out in the temperature range from 800 to 1000 degrees centigrade. The runs usually lasted five to six hours. Pressures were varied from 1 up to 80 torr. The methane content was varied from 11% down to 0.1%.

Up to the present moment no diamond deposition could be detected by Raman spectroscopy and SEM. Raman spectra show only the G-band peak at 1590 cm^{-1} and a broad feature reaching from 1250 to 1500 cm^{-1}. XPS (X-Ray Photoelectron Spectroscopy) shows the presence of polymeric carbon and the formation of carbides.

Lowering of the power input into the resonator resulted in a lowering of the dissociation rate of hydrogen and to a decrease in deposition. Without the plasma no depositions could be observed. This demonstrates that atomic hydrogen is really reaching the deposition zone. This is confirmed by étching experiments using a graphite rod.

Reducing the methane content lead to a lowering in deposition rates also. If methane concentrations were below 0.1% no depositions could be detected. To exclude the influence of locally too high methane concentrations, the methane was mixed with argon which should not interfere with the chemical reactions going on in the deposition zone. The addition of argon to methane flow (90% Ar, 10% CH_4) resulted in significantly higher deposition rates but also in a higher diffusion of the hydrocarbon towards the plasma.

The quadrupole mass spectrometer measurements show the presence of methane and traces of acetylene. Theses measurements however have been made with the QM probing the exhaust gas. The configuration described above (see Fig. 3) will certainly lead to more detailed results.

In future experiments other hydrocarbons (acetylene !) will also be tested. In addition to the MPACVD system a filament reactor is presently under construction which will also allow a separate introduction of hydrocarbons.

REFERENCES

1. M. Tsuda, M. Nakajima, and S. Oikawa, Epitaxial Growth Mechanism of Diamond Crystal in CH_4-H_2 Plasma, J. Am. Chem. Soc. 108, 5780 (1986).
2. M. Tsuda, M. Nakajima, and S. Oikawa, The Importance of The Positively Charged Surface for the Epitaxial Growth of Diamonds at Low Pressure, Jap. J. Appl. Phys., Vol. 26, No. 5, May, L527 (1987).
3. M. Frenklach and K.E, Spear, Growth Mechanism of Vapour-Deposited Diamond, J. Mater. Res. 3(1), 133 (1988).
4. F.G. Celii, P.E. Pehrsson, H.-T. Wang, and J.E. Butler, Infrared Detection of Gaseous Species During the Filament-Assisted Growth of Diamond, Appl. Phys. Lett. 52(24), 2043, (1988).

5. F.G. Celii and J.E. Butler, Hydrogen Atom Detection in the Filament-Assisted Diamond Growth, Appl. Phys. Lett. 54(11), 1031 (1989).
6. S.J. Harris and A.M. Weiner, Measurement of Stable Species Present During Filament Assisted Diamond Growth, Appl. Phys. Lett. 53(17), 1605 (1988).
7. D.N. Belton, S.J. Harris, S.J. Schmieg, A.M. Weiner, and T.A. Perry, In Situ Characterization of Diamond Nucleation and Growth, Appl. Phys. Lett. 54(5), 416 (1989).
8. S.J. Harris and A.M. Weiner, Effects of Oxygen on Diamond Growth, Appl. Phys. Lett. 55(21), 2179 (1989).
9. S.J. Harris, D.N. Belton, A.M. Weiner, and S.J. Schmieg, Diamond Formation on Platinum, J. Appl. Phys. 66(11), 5353 (1989).
10. J.A. Mucha, D.L. Flamm, and D.E. Ibbotson, On the role of Oxygen and Hydrogen in Diamond Forming Discharges, J. Appl. Phys. 65(9), 3448 (1989).
11. Y. Kawahara, Decomposition of Hydrocarbons in a Microwave Discharge. I. Methane. Effect of Power, J. Phys. Chem. Vol. 73, No. 6, 1648 (1969).
12. L.R. Martin and M.W. Hill, Diamond Film Synthesis in a Chemically Simplified System, Appl. Phys. Lett. 55(21), 2248 (1989).
13. J. Warnatz, Rate Coefficients in the C/H/O System, in: "Combustion Chemistry", W.C. Gardiner, Jr., ed., Springer, New York (1984).
14. A.A. Sepehrad, R.M. Marshall, and H. Purnell, Reaction between Hydrogen Atoms and Methane, J.C.S. Faraday I, 75, 835 (1979).
15. E.L. Tollefson and D.L. Le Roy, The Reaction of Atomic Hydrogen with Acetylene, J. Chem. Phys., Vol. 16, No. 11, 1057 (1948).
16. L.G. Meiners and D.B. Alford, The Design of a Microwave Plasma Source, Rev. Sci. Instr. 57(2),164 (1986).

METHODS OF DIAMOND MAKING

Thomas R. Anthony

General Electric Corporate Research & Development Center
River Road, Schenectady, New York, 12301, USA

Diamond can be grown as an equilibrium phase at high pressures or grown metastably at subatmospheric pressures. Equilibrium growth at high pressures is carried out by both static and dynamic processes. The static process is an indirect process where metastable graphite is dissolved into a liquid metal solvent from which diamond precipitates out as the stable phase in a high pressure press. The dynamic high-pressure process directly converts graphite to diamond by going to very high pressures and temperatures for a short period of time by using explosives.

Metastable growth of diamonds occurs at subatmospheric pressures from hydrocarbon gases in the presence of atomic hydrogen. Many methods have been developed for metastable growth are in general based on different means of generating atomic hydrogen. The essential difference between most of these processes is how atomic hydrogen is generated and transported around in a particular system. Some processes that may make diamond by methods other than the high-pressure or atomic-hydrogen methods have been reported but so far remain unconfirmed by independent experiments.

I. HIGH PRESSURE SYNTHESIS

At room temperature and pressure, graphite, not diamond, is the stable phase of carbon because graphite has both a lower energy and higher entropy than diamond. However, the natural scarcity of diamond can not be accounted for by its small energy difference which is less than RT at room temperature or its small entropy difference.[1]

Diamond and Diamond-Like Films and Coatings
Edited by R.E. Clausing *et al.*, Plenum Press, New York, 1991

Graphite ⇔ Diamond : ΔG_{298K} = +693 cal/mol (1)

Diamond is probably rare in nature for two reasons. First, graphite itself is relatively scarce because of the reactivity of carbon with many other elements. Secondly, the kinetics of graphite formation under ordinary conditions are much faster than the kinetics of diamond formation. Thus competitive growth insures that graphite nuclei rapidly grow, overcome and bury any diamond nuclei that may form.

While synthesis of diamond has been attempted for hundreds of years[2], it was not until 1954, that a reliable process was invented for making diamonds. This process required the use of high static pressures[3] where diamond becomes the stable phase of carbon because the density of diamond is 1.56 times that of graphite. Most diamond produced commercially uses the indirect conversion process where metastable graphite at high pressures is dissolved in a liquid metal solvent and then precipitated as diamond. The majority of indirect high-pressure methods use the belt apparatus where two opposed anvils are pressed inwards by a large press against a diamond growth cell surrounded by a belt to contain the high pressure. A current is passed through the cell to heat it above 1200°C and a pressure greater than 60 kilobars is produced in the cell by the opposing anvils. At this temperature and pressure, the cell is in the region of the carbon phase diagram where diamond is the stable phase. The metastable graphite dissolves in the liquid metal catalyst and precipitates out as diamond. The use of a liquid metal catalyst in this process enables one to use relatively low pressures that are close to the diamond-graphite phase boundary. The pressures and temperatures (120 kbar and 2000°C) required for direct graphite-to-diamond conversion without a metal catalyst are approximately twice as high, making the direct static process economically unfeasible. The static indirect process has other advantages including a high diamond growth rate of up to 10,000 µm/hour and excellent crystal quality.

The invention of the static high-pressure, high-temperature process was followed by the development of a dynamic high-pressure process that uses high-energy explosives to generate a high-pressure high-temperature shock wave that directly converts graphite to diamond. Because of the short duration of the shock wave, this dynamic process produces very fine diamond powder (< 2 µm in diameter) that is useful mainly for polishing compounds. Consequently, 90% of commercial synthetic diamond is produced by the static high-pressure, high-temperature metal-catalyst process where a variety of crystals sizes up to gem-sized stones can be easily made.

Despite many advantages, the static high-pressure process has some limitations. First, the very high pressure and maximum yield strengths of materials limits the size of the growth cells and hence the maximum size of diamond objects that can be made to less than several inches across. Secondly, small amounts of the liquid metal catalyst can become trapped as inclusions in the diamond crystal unless growth is carried out at uneconomically slow rates. These metal inclusions have a different thermal expansion coefficient than diamond and cause internal stresses to develop and weaken the diamond. For some abrasive-tool applications, this weakening of the diamond by metal inclusions is undesirable. An alternative diamond-growing process [4] whose basis was first disclosed in 1960 [5] uses subatomspheric pressures and does not limit the size of diamond objects that can be made. Moreover, this low-pressure process produces diamond that is free of inclusions.

II. LOW-PRESSURE SYNTHESIS

In the original reports of the metastable growth of diamond at low pressures, the growth rate of diamond was very small and diamond substrates were required, making the process of scientific interest only. Most methods stemmed from the work of Eversole[5] who alternatively exposed diamond substrates to a hydrocarbon gas and then hydrogen at high temperatures and low pressures.[5-7] Under these conditions, the hydrocarbon was pyrolysized to form diamond and graphite and then hydrogen was used to etch the graphite away. This cycle (Reactions 2 and 3) was repeated over and over again until the diamond deposit grew to the desired thickness.

$$CH_4 \xrightarrow{\text{Heat}} \text{Deposit (Diamond \& Graphite)} + 2\,H_2 \qquad (2)$$

$$H_2 + \text{Deposit} \xrightarrow{\text{Heat}} \text{Diamond} + CH_4 \qquad (3)$$

In the late 1970's, an important invention changed this cyclic CVD (Chemical Vapor Deposition) process to a continuous one., increased the diamond growth rate by orders of magnitude, eliminated the need for diamond substrates and made the process technologically significant.[8] This new process[9] is shown in Reaction 4.

$$CH_4 \xrightarrow{\text{Heat \& Atomic Hydrogen}} \text{Diamond} + 2\,H_2 \qquad (4)$$

The key idea was the addition of atomic hydrogen to the reaction by Boris Spitsyn and his coworkers[9] following a suggestion by John Angus. Atomic hydrogen has many important roles in CVD diamond growth. These roles have been recently summarized and will not be covered here.[10] A large number of CVD methods based on the process of Reaction 4 have been developed over the last decade. All methods are basically just different ways of generating the atomic hydrogen required in Reaction 4.[11]

III. TRANSPORT OF ATOMIC HYDROGEN

IIIA. BASIC TRANSPORT EQUATIONS

Before going on to examine the different methods of generating atomic hydrogen, let us examine how atomic hydrogen is transported in a general gas reactor apparatus. We will then apply our findings to the hot filament system as an example. The transport of atomic hydrogen in a gas is governed by equation (5) for a mixture of atomic hydrogen and molecular hydrogen.

$$-\nabla J_H - k_3|H_2|\,|H|^2 - \nabla\{|H|(V_H+V_C)\} - \nabla\{|H|D(K_T/T)\,\nabla T\} = \delta|H|/\delta t \qquad (5)$$

where J_H is the diffusive flux of atomic hydrogen, k_3 is the kinetic rate constant for the trimolecular collision recombination of two atomic hydrogens with a hydrogen molecule, $|H|$ is the concentration of atomic hydrogen, $|H_2|$ is the concentration of molecular hydrogen, V_H is the flow velocity generated by the atomic hydrogen wind, V_C is the normal convective flow velocity imposed on the system, D is the diffusion constant of atomic hydrogen, K_T is the thermal diffusion ratio, T is the temperature and t is the time.

Equation (5) gives the change with time of the concentration of atomic hydrogen contained within a fixed elemental volume of gas. The first term on the left of Equation (5) is the difference in atomic hydrogen diffusion flux entering and leaving a volume element. The second term on the left-hand side of Equation (5) is rate of recombination of two atomic hydrogens aided by a molecular hydrogen that carries away the excess energy and momemtum that are released when two atomic hydrogens recombine (the collision and joining of two atomic hydrogens without involvement of a third moelcule results in their immediate breakup since the energy of recombination has no way to dissipate. Hence, no atomic hydrogen is lost in bimolecular collisions with itself). The third term on the left-hand side of Equation (5) is the convective flow term. If there is a gradient in the concentration of atomic hydrogen in the elemental volume and a convective flow of gas, the concentration gradient will move with time and change the concentration of atomic hydrogen in the elemental volume. The convective flow is comprised of two terms. The first term V_H is the atomic hydrogen wind that is produced when atomic hydrogen is generated from molecular hydrogen or recombined to form molecular hydrogen. At the generation source of atomic hydrogen, each mole of molecular hydrogen that is split produces two moles of atomic hydrogen. Conversely, at recombination points, two moles of atomic hydrogen recombine to form one mole of molecular hydrogen. This generation and disappearance of gas volume, respectively, at generation and recombination points of atomic hydrogen cause an atomic hydrogen wind that flows from generation points to recombination points in the gas. The second term V_C in the convective flow term is the ordinary gas flow term imposed on the system by external factors (eg, the experimental gas flow rate plus thermal convection). The last term on the left-hand side of Equation (5) is the concentration gradient produced by a thermal gradient in a gas.[12] If a homogeneous binary mixture of gas in a tube is heated at one end and cooled at the other end, the lighter and smaller molecules will drift towards the hotter end whereas the heavier larger molecules will drift towards the colder end. This drift is caused by a transfer of momentum from the small, light, fast-moving molecules to the larger heavier molecules which drives the larger heavier molecules down the temperature gradient. The small light molecules are reflected from these collisions up the temperature gradient and a concentration gradient results.

We will be principally interested in the steady state solution of Equation (5), so we set $\delta|H|/\delta t = 0$. The diffusive flux of atomic hydrogen is given by Ficks Law as $J_H = -D\nabla|H|$. Substitution of these conditions and division of Equation (5) by D yields:

$$\nabla^2|H| - (k_3/D)|H_2||H|^2 - \nabla\{|H|(V_H+V_C)/D\} - \nabla\{|H|(K_T/T)\nabla T\} = 0 \qquad (6)$$

Although Equation (6) looks formidable, it has analytic solutions in some simple cases.[13-15] Rather than pursue such solutions or fall back on numerical analysis, we can use inspectional and configurational analysis[16,17] to sift out most of the useful information contained in Equation (6).

IIIB. ATOMIC HYDROGEN RECOMBINATION LENGTH, LIFETIME AND DRIFT DISTANCE

Associated with Reaction (6) is a recombination length, L_r, which gives the average distance that an atomic hydrogen travels before it recombines in a trimolecular collision. L_r is given by:

$$L_r = \{D/(k_3\ |H_2|\ |H|)\}^{1/2} \tag{7}$$

where D is the gas diffusion coefficient of atomic H, $|H_2|$ is the concentration of molecular hydrogen and |H| is the concentration of atomic hydrogen and k_3 is the trimolecular recombination kinetic constant of atomic hydrogen with molecular hydrogen.[10] The kinetic constant varies with temperature approximately as $T^{-1/2}$ [18] and is independent of pressure. The gas diffusion constant varies as $T^{3/2}$ with temperature and as P^{-1} with pressure. Finally, the universal gas law shows that both $|H_2|$ and |H| vary as P/T. From these relations, the approximate variation of L_r with temperature and pressure is:

$$L_r = L_0\ (T/T_0)^2/(P/P_0)^{3/2} \tag{8}$$

where L_0 is the atomic hydrogen recombination length at a temperature T_0 and pressure P_0 and is a constant independent of pressure and temperature. For T_0 = 1000 K and P_0 = 10 torr, L_0 = 5.8 cm. In a CVD diamond apparatus, the pressure P is constant but the temperature decreases from the source of atomic hydrogen to the substrate on which the diamond is growing. Equation (8) shows that the recombination length decreases as one travels from the atomic hydrogen source towards the substrate.

The average lifetime τ of an atomic hydrogen atom in the gas of a CVD diamond deposition apparatus is also of interest since if large convective flows are present, the atomic hydrogen can be transported by convection as well as diffusion. Dimensional analysis shows us that τ is given by:

$$\tau = (L_r)^2/D = 1/(k_3\ |H|\ |H_2|) \tag{9}$$

The diffusion coefficient of atomic hydrogen varies inversely with the pressure and is proportional to the 3/2 power of the temperature. Equations (8) and (9) can be combined to obtain:

$$\tau = \tau_0\ (T/T_0)^{5/2} / (P/P_0) \tag{10}$$

where T_0 = 1000 K and P_0 = 10 torr and $\tau_0 = 3.9 \times 10^{-2}$ secs

The average drift distance, L_v, of atomic hydrogen in a convective flow before it recombines is given by:

$$L_v = (V_C+V_H)\ t = (V_C+V_H)/(k_3\ |H|\ |H_2|) \tag{11}$$

Equation (11) can be put in a more useful form for an experimentalist by combing Equations (10) and (11) to obtain:

$$L_v = L_{v_0} \; (V/V_0) \; (T/T_0)^{5/2} \; / \; (P/P_0) \tag{12}$$

where V_0 = 1 cm/sec, L_{v_0} = 3.9 X 10^{-2} cm, T_0 = 1000 K and P_0 = 10 torr and $V = V_C + V_H$.

The total average distance, L_T, that atomic hydrogen can travel from its generation source before recombining is given by the sum of the drift and diffusion distances:

$$L_T = L_v + L_r \tag{13}$$

IIC. THE ATOMIC HYDROGEN WIND

The term V_H in Equation (5) is the convective gas flow produced by the molar volume inbalance at sources of atomic hydrogen where one mole of molecular hydrogen splits into two moles of atomic hydrogen and at sinks of atomic hydrogen where two moles of atomic hydrogen recombine to form one mole of molecular hydrogen. This convective gas flow that is intrinsic to any system containing both atomic and molecular hydrogen is called the atomic hydrogen wind. The velocity of the atomic hydrogen wind, V_H is given by:

$$V_H = (J_H / 2) / |H_T| \tag{14}$$

where J_H is the flux of atomic hydrogen and $|H_T| = |H| + |H_2|$.

The flux of atomic hydrogen J_H is approximately equal to $D\nabla|H|$. $\nabla|H|$ is in turn equal to the change, $\Delta|H|$, in atomic hydrogen concentration between its source and sink locations divided by the separation distance, L_s, between the source and sink. Substitution into Equation (14) yields:

$$V_H = (D \; X_H) / (2 \; L_s) \tag{15}$$

where $X_H = |H|/(|H|+|H_2|)$ is the mole fraction of atomic hydrogen at the source. Equation (15) most accurately describes the atomic hydrogen wind velocity near the atomic hydrogen source. Atomic hydrogen recombination away from the source decreases the atomic hydrogen wind.

We can put Equation (15) in a parameterized form so its dependence on temperature, pressure and source-sink separation distance can be easily calculated by an experimentalist.

$$V_H = V_0 \; (L_{s_0}/L_s) \; (P_0/P) \; (T/T_0)^{3/2} \; \exp(-Q/RT) \tag{16}$$

where V_0 = 3.7 X 10^6 cm/sec , L_{s_0} = 1 cm, P_0 = 10 torr, T_0 = 1000 K, Q = 52 kcal/mol, the universal gas constant R = 1.987 cal/deg and T and P, respectively, are the gas temperature and pressure near the atomic hydrogen source.

An interesting ratio is the ratio of transport in the gas caused by diffusion versus transport by the atomic hydrogen wind. If $|X_i|$ is the ith species in a gas, then the ratio between atomic wind transport J_W and diffusion transport J_d is given by:

$$J_w/J_d = V_H\,|X_i| \,/\, (D\,\nabla|X_i|) \tag{17}$$

$\nabla|X_i|$ is equal to $\Delta|X_i|/L_S$. Substitution of this relation and Equation (15) into (17) yields:

$$J_w/J_d = \{|X_i|/(\Delta|X_i|)\}\,\{X_H/2\} \tag{18}$$

where $\Delta|X_i|$ is the concentration difference of the ith species between the atomic hydrogen source and sink. For the special case where the ith species is atomic hydrogen with 100% dissociation at the source, $J_W/J_d = 1/2$.

IIID. THERMAL DIFFUSION

The last term is Equation (6) is the thermal diffusion term whose direction and magnitude are important. Unfortunately, there is no available data on thermal diffusion in atomic and molecular hydrogen mixtures. However, experiments on deuterium and hydrogen mixtures have been done.[19] These experiments should give fairly accurate values for atomic-molecular hydrogen mixtures because: 1) K_T depends principally on the size and mass of the atoms of the mixture; 2) the mass ratios of the respective mixtures are the same (masses: $D_2/H_2 = H_2/H$); 3) all atoms are small. Typical values of K_T for these experiments run between 1 to 4 X 10^{-2}. Values for atomic-molecular hydrogen mixtures should be somewhat higher since there is a size difference between molecular and atomic hydrogen and virtually no size difference between molecular deuterium and molecular hydrogen. For purposes of this paper, we will assume K_T is about +6 X 10^{-2} (ie, atomic hydrogen is driven towards higher temperatures and molecular hydrogen is driven towards lower temperatueres by the thermal diffusion effect).

IV. CVD DIAMOND DEPOSITION PROCESSES

IVA. HOT FILAMENT PROCESS

Langmuir first reported the dissociation of molecular hydrogen on a hot tungsten filament in the early 1900's.[20-23] This was also the initial method that was used extensively to make atomic hydrogen in CVD diamond growth. Typically, a hot filament is heated to a temperature range of 1950-2300°C in the presence of molecular hydrogen and 1-2% hydrocarbon.[24,25] The molecular hydrogen adsorbs on the surface of the metal filament and dissociates into two atomic hydrogens that subsequently pass back into the surrounding gas.

$$H_2 \dashrightarrow 2\,H \quad : \quad \Delta H = +104 \text{ kcal/mole} \tag{19}$$

In addition, many other reactions[26,27] take place on the filament involving the hydrocarbon. The concentration of atomic hydrogen and other hydrocarbon species around the filament have been directly measured.[28,29] A general conclusion of these experiments is that thermodynamic equilibrium calculations give a reasonable guide to the type and number of species that are actually present.

There are some constraints on the electrically conductive refractory materials that can be used for atomic-hydrogen filaments. First, the melting point of the material must be higher than the operating temperature of the filament. Secondly, the binary eutectic temperature of carbon and the filament material must exceed the filament operating temperature to prevent the melting and breakage of the filament. (Table I compares the relevant parameters for various filament materials.) The second requirement makes molybdenum filaments marginal since a hot spot on a Molybdenum filament may take it above the Mo-C eutectic temperature of 2200°C and cause the filament to melt and fail before the filament is fully carburized.

While a carbon filament satisfies the melting-point and eutectic-point requirements, it does not produce CVD diamond because atomic hydrogen that forms on a carbon filament reacts with the carbon and releases a hydrocarbon species into the gas rather than atomic hydrogen. Without atomic hydrogen at the substrate, graphite is deposited instead of diamond. The failure of carbon filaments to produce diamond[30] suggests that the atomic hydrogen does not form in the hot gas surrounding the filament but actually forms on the filament surface[20-23] where it can react with the filament material.

Of the remaining electrically conductive refractory materials, tungsten, tantalum and rhenium have been used to produce CVD diamond. Both tantalum and tungsten react with the hydrocarbon in the gas to form carbides. Thus, the tungsten and tantalum filaments quoted in the literature are really tungsten carbide and tantalum carbide filaments. It is interesting to note that W_2C is a commonly used catalyst material in gaseous hydrocarbon chemistry. However, experiments that have compared the difference in diamond growth rates between tungsten or tantalum carbide filaments show no special catalytic effect for tungsten carbide.[31] Of course, the rate of diamond growth may not be controlled by the production of atomic hydrogen or carbon radicals on the filament surface but may instead be limited by some other process. Filament carburization can have a catalytic effect on the diamond nucleation rate by increasing the electron emission from the filament because of the lower work function of the carbide.[32] Electron bombardment of the substrate is known to increase the diamond nucleation rate.[33]

TABLE I.
Filaments for Diamond Synthesis

Material	Melting Point	Evaporation Rate	Carbon Eutectic Point
	°C	grams/cm^2 - sec	°C
Tungsten	3387	10^{-9}	2475
Carbon	3727	6×10^{-8}	3727
Rhenium	3180	7×10^{-5}	2486
Tantalum	2996	10^{-8}	2800
Molybdenum	2610	4×10^{-7}	2200
Niobium	2468	8×10^{-9}	2335

Carburization is a two stage process with M_2C forming first followed by MC where M is one of the refractory metals. Under typical hot-filament CVD diamond growth, M_2C forms in all cases. However, MC does not always form because the activity coefficient of carbon is too small under some conditions of CVD diamond growth.[34] Since the carbides of tantalum and tungsten have a molar volume that is 40-70% larger than the molar volume of the metal from which they were formed, the filaments swell, crack, bend, distort and embrittle as inward radial carburization proceeds. Because carburization occurs most rapidly along grain boundaries where carbon mobility is the highest, cracking usually follows the grain boundary patterns of the original metal filament. Carburization significantly decreases the grain size of the wire. This reduction in grain size can lead to superplastic deformation under some some conditions.[35] Carburization and cracking increase the electrical resistance of the filament[32] so the current and voltage must be monitored and changed to keep a constant filament temperature. Clausing[36] has noted that tungsten wire from different sources results in very different types of filament behavior and life in CVD diamond growth. Tungsten filaments are made by three different processes. Two processes involve sintering tungsten powder either with (218 tungsten lamp wire) or without the impurities that make a long stringy grain structure in the resulting tungsten wire. During wire drawing of 218 tungsten, the insoluble impurity precipitates are drawn out into strings which break down under surface tension forces into long linear arrays of small beads along the drawing direction. When the wire is recrystallized during heating, migrating grain boundaries become trapped by these long arrays of beads. Thus the final grain structure of 218 tungsten has long stringy grains along the major axis of the wire that resist the cross-axis grain boundary sliding that causes filament rupture. The grain structure of sintered tungsten wire prepared without added impurities is a more equiaxed struture. The third process to make tungsten wire uses zone-melted tungsten. The grain structure of each particular type of tungsten affects filament carburization and its cracking pattern. Although cracking embrittles the filament, it produces an increased filament surface area which may increase the production of chemically active species required for diamond growth. To avoid filament carburization, Aikyo and Kondo introduced the hydrocarbon downstream from the hot filament.[37] Diamond was successfully deposited and the filament life and the quality of the diamond film was improved. Unfortunately, the diamond deposition rate was decreased.

Rhenium is attractive because it does not form a carbide.[38,39] However, rhenium dissolves large amounts of carbon that causes rhenium to swell. In addition, the dissolved carbon may react with dissolved oxygen to form embrittling voids on grain boundaries. The dissolved carbon and the voids cause the filament to swell. In addition, the initially fine-grain structure that promotes wire ductility anneals into a large-grain non-ductile bamboo-structure. Thus, the annealed carbon-saturated rhenium wire is partially embrittled though much less so than either tungsten or tantalum. Rhenium, unfortunately, is two orders of magnitude more expensive than tantalum or tungsten.

A disadvantage of the filament process is that a filament is a *remote* source of atomic hydrogen. During the transport of atomic hydrogen from the filament to the substrate, a considerable fraction is lost through trimolecular collision between two atomic hydrogens and a third gas molecule[40]:

$$\mathbf{H + H + M \;\text{------>}\; H_2 + M} \qquad (20)$$

where H is atomic hydrogen and M represents a gas molecule or the walls of the reactor.

Associated with Reaction (20) and Equations (7) and (8) is a recombination length, L_r, which gives the average distance that an atomic hydrogen travels before it recombines in a trimolecular collision. For typical filament-method process conditions of T = 1500 K and P = 10 torr, the recombination length L_r is 13 cm. Matsumoto and coworkers[24] in the original filament paper found that the filament-substrate separation must be less than 3 cm to get diamond growth.

The loss of hydrogen by recombination can be decreased by reducing the filament-substrate separation distance. Hirose[41] used tiny filaments that were very close (<1 mm) to the substrate as well as oxygen additions to attain growth rates of 10 µm/hr which is 10 times the usual diamond growth rate for the filament process. Presumbably, the small separation between filament and substrate decreased the loss of atomic hydrogen and other active hydrocarbon species.

The temperature gradient in most of the gas between the hot filament and the substrate is a simple linear gradient. However, at the gas-solid interface at the filament or the substrate, there is a temperature jump δT given by Equation (21) [12]

$$\delta T = g\ \delta T/\delta x \tag{21}$$

where $\delta T/\delta x$ is the temperature gradient in the body of the gas extrapolated to the surface and g is some multiple of the mean free path λ of the gas molecule. g can range as high as 20 λ if the gas accomodation coefficient on the solid is small or the gas has a low mass or the temperature is high or the surface is clean and smooth.[12] Most of these high-g conditions are satisfied by a hot filament in hydrogen[20-23] so we should expect a large g at the filament in CVD diamond growth. In the case of a cylindrical hot filament, the thermal gradient near the filament surface is given by:

$$\delta T/\delta r = Q/(\ 2\pi\ a\ K\ ) \tag{22}$$

where T is the temperature, r is the radial vector, Q is the heat generated per unit length of the filament, a is the filament radius and K is the thermal conducitvity of the gas. If we use typical CVD diamond filament parameters (T = 2100°C, a= 0.05 cm, Q = 15 cal/cm-sec and K = 1.1 X 10^{-3} cal/cm-sec-°C), the temperature gradient in the gas near the filament is 4.3 X 10^4 °C/cm. Substitution of this value into Equation (22) and taking $g = 10\lambda = 5 \times 10^{-3}$ cm gives the temperature drop δT in the gas adjacent to the filament surface to be 215°C. This significant temperature drop coupled with the high activation energy of 52 kcal/mole to form atomic hydrogen means that the atomic hydrogen will not form in gas in the vicinity of the filament because the gas temperature is too low but must form instead on the filament surface. The failure of atomic hydrogen to form in the surrounding gas prevents a carbon filament from generating atomic hydrogen since any atomic hydrogen on the surface of a carbon filament will form a hydrocarbon instead of flying off as free atomic hydrogen. Thus, carbon filaments will not produce CVD diamond.[30]

The cylindrical geometry of the a CVD diamond filament apparatus means that there are large atomic hydrogen concentration gradients and diffusion fluxes in the vicinity of the filament. The radial flux, J_H, of atomic hydrogen near the filament is inversely proportional to the distance from the filament. Substitution of this flux term into Equations (14) and (15) yields the atomic hydrogen wind around a hot filament.

$$V_H = (D\, X_H) / (2\, \alpha\, a) \tag{23}$$

where α is a constant on the order of 10 and a is the filament radius. Substitution of typical CVD diamond filament parameters (T = 2300 K, X_H= 0.10, a = 0.05 cm , D = 3 X 10^3 cm^2/sec) gives an atomic hydrogen wind velocity of 300 cm^2/sec away from the filament. This atomic hydrogen wind will also drain heat away from the filament by convective flow. According to Equation (23), the heat drain will increase inversely with the filament radius. It is interesting to note that a recent study[42] on the thermal dissociation of hydrogen on tungsten filaments found an anomalous heat loss per unit area that increased as the filament size decreased.

Finally, the thermal diffusion effect will drive atomic hydrogen up the temperature gradient towards the the hot filament and against the chemical concentration gradient that is driving atomic hydrogen away from the filament. Because the temperature gradient is largest near the filament, this effect will be largest there. It is conceivable under certain conditions that the thermal diffusion effect could balance the chemical concentration difference near the filament. The chemical concentration gradient $\delta|H|/\delta r|_C$ near the filament is equal to:

$$\delta|H|/\delta r|_C = |H| / (2\, \pi\, a) \tag{24}$$

while the thermal diffusion gradient $\delta|H|/\delta r|_T$ near the filament is:

$$\delta|H|/\delta r|_T = (|H|\, Q\, K_T) / (2\, \pi\, a\, T\, K) \tag{25}$$

where Q is the power generated per unit length of the filament, K_T is the thermal diffusion constant, a is the radius, T is the temperature and K is the thermal conductivity of the gas.

The difference between these gradients (they oppose each other) is:

$$\delta|H|/\delta r|_C - \delta|H|/\delta r|_T = \{|H| / (2\, \pi\, a)\}\, \{1 - (Q\, K_T) / (T\, K)\} \tag{26}$$

If we use typical CVD diamond filament parameters near the filament (T=2100°C, a=0.05 cm, Q=15 cal/cm-sec and K = 1.1 X 10^{-3} cal/cm-sec-°C , K_T =6 X 10^{-2}), we find that the thermal diffusion gradient decreases the chemical concentration gradient by 30%.

IVB. HOT FILAMENT EACVD

The hot filament process has been modified by some investigators to include a positive or negative bias between the hot filament and the substrate.[43] The process has been named hot-filament EACVD (electron-assisted chemical vapor deposition). The applied bias has been reported to cause changes in the nucleation rate, the growth rate and the crystal quality of the deposited CVD diamond. A general remark that must be made here is that all filament processes have an EACVD character. Consider a filament with a current passing through it.

One end of the filament is at ground potential and the other end is at some applied voltage that is needed to drive the current through the filament. If the substrate is at ground potential, then the entire filament is biased with respect to the substrate except for the very end of the filament at ground potential. If EACVD had a large effect on CVD diamond growth, one would see a change in diamond growth along the substrate parallel to the filament. The lack of such reports in the literature imply that EACVD effects are not striking.

Generally, an increase in the nucleation and growth rate of the deposited diamond has been reported if the filament is held at a negative bias with respect to the substrate. Alternatively, if the filament is held at a positive bias with respect to the substrate, the diamond nucleation and growth rate decrease while the diamond crystal quality increases.[44, 45] In both cases, the bias potential must be kept below the breakdown point of the gases since the formation of a plasma seems to erase these favorable effects. There has been some confusion in the literature because some researchers heat their filaments with AC and others heat their filaments with DC. Identical filament-substrate biases with AC and DC heated filaments, respectively, can cause very different results on CVD diamond growth. Six separate cases should be documented: 1) AC heating, positive DC bias; 2) AC heating, negative DC bias; 3) DC heating, positive DC bias; 4) DC heating, negative DC bias; 5) AC heating, AC bias; 6) DC heating, AC bias.

The negative effects[44] of plasma formation in hot-filament EACVD is somewhat surprising since the next section will show that plasmas have been effectively used for CVD diamond growth. This contradictory evidence is not uncommon in CVD diamond deposition. Generally, it is not the result of erroneous experiments. Rather, there are dozens of independent parameters in an experiment and changing one parameter unavoidably changes other parameters and makes a clear interpretation of the data difficult. As an example, let us list some of the important parameters in a simple process such as hot-filament EACVD: 1) Filament composition; 2) Filament temperature; 3) Filament size and geometry, ie, wire, strip, mesh, etc; 4) Substrate composition; 5) Substrate temperature; 6) Substrate geometry; 7) Filament-to-substrate distance; 8) Gas pressure; 9) Gas temperature; 10) Gas flow rate; 11) Gas flow geometry; 12) Gas temperature gradient above substrate; 13) Gas temperature gradient near filament; 14) Hydrocarbon concentration in gas; 15) Impurity concentration in gas, ie, oxygen, water, nitrogen; 16) Filament-to-Substrate bias: positive, negative, AC; 17) Filament-to-Substrate current; 18) Filament Heating: AC or DC.

If an experimenter changes the filament-to-substrate bias, he will also change the gas temperature, the substrate temperature, the temperature gradient in the gas, the filament temperature and the filament-to substrate current. He may then report an increase of growth rate with increasing bias that may be just an increase in growth rate arising from an increase in substrate temperature rather than a new filament-to-substrate bias effect. Banholzer[46] has, in fact, recently carried out a very careful study investigating reports that EACVD increases the CVD diamond growth rate. He finds that there is no such effect when he keeps the filament and substrate temperature constant as he applies a filament-substrate bias. What previous investigators were seeing were effects caused by an increased filament temperature and/or an increased substrate temperature when the EACVD bias was turned on. Banholzer, however, did see an increased nucleation density of diamond when a negative bias applied to a DC heated filament.

The experimenter is not only confronted with interfering effects but also with an overwhelming number of parameters that should be tested for even a rough mapping of the depositon process. For example, if the 18 parameters above are each tested with 5 different values each, an experimenter would have to carry out $(19)^5$ or approximately 2 million experiments. Added to this problem is the discovery that when some parameters varied, qualitative as well as quantitative changes occur in the diamond deposition. Thus, the experimenter can not count on smoothly changing results as he changes even a single parameter.

This is not meant to be pessimistic portrayal of CVD diamond research. Rather, it is meant to encourage workers who may be now considering entering the field that there are still many simple experiments to be tried and many more discoveries to be made.

IVC. PLASMAS

Another popular method to generate atomic hydrogen is with gas plasmas. Plasmas can be formed by many methods. Typically, a gas is ionized by exposure to high temperatures or high electric fields to form a neutral mixture of electrons, negative ions and positive ions. In plasmas exposed to high AC electric fields, hydrogen dissociation occurs by electron impact. Although the dissociation energy of hydrogen is only 4.5 ev, the electron impact dissociation of hydrogen requires electrons with energies above 9.5 ev because of the large mass difference between the electron and a hydrogen molecule. Such dissociation reaches a peak at electron energies of 25 ev. Atomic hydrogen with a very high kinetic energy is consequently produced in plasmas since the difference between the dissociation energy and the electron impact energy must be taken up by the kinetic energy of the atomic hydrogen. This situation contrasts with thermal methods of generating atomic hydrogen where low kinetic energy atomic hydrogen is formed since a direct transfer of thermal energy breaks the hydrogen molecule apart without imparting additional excess energy that would appear as the kinetic energy of the atomic hydrogen product.

The typical charge density in a plasma is very low. Most molecules remain as molecules which are not reactive and do not take part in CVD diamond deposition. About 1% of the molecules in a plasma are converted to neutral radicals that are chemically reactive molecules with a dangling unsatisfied bond. These radicals are the main component of the gas that is active in CVD deposition. Finally, about 0.01% of the molecules in a plasma are ions. Like radicals, these ions are chemically reactive and can take part in CVD diamond deposition. However, the extremely small concentration of ions indicates that they are not the important depositing species in diamond deposition. Although charged ions have a very low concentration and are not the principal depositing species during CVD deposition, it is possible that they may have some catalytic effects.

Plasmas can be conveniently divided into low-pressure and high pressure plasmas. At low pressures in a gas excited by an electric field, the electrons and molecules are not in thermal equilibrium. Although electrons and molecular ions have the same charge, the mass of the molecular ion is typically 40,000 times that of an electron. Hence, the electron can accelerate rapidly in an electric field and gain energy while the ion is too massive to accelerate quickly and laggs behind the electron in energy gain from the electric field. If the mean-free path for collisions between electrons and molecules is large as in a low pressure gas, there will not be a redistribution of energy between the high energy electrons and the slowly

Microwave discharges are also very stable and provide up to 25% hydrogen atoms under typical CVD diamond growth conditions.[11] Moreover, microwave sources are very economical because of the mass production of microwave sources for microwave ovens. During CVD diamond deposition, the plasma must not contact the walls of the reactor since diamond and/or graphite can deposit on the walls. This carbon deposit will couple with the microwaves and heat up causing more deposition and the walls of the reactor (usually quartz) will gradually become opaque to microwaves. Under other operating conditions, the plasma can seriously erode the reactor walls by atomic-hydrogen reduction of the quartz. These problems can be avoided by magnetic confinement of the plasma.

A common magnetic confinement method uses an axial magnetic field in a cylindrical cavity with a microwave window and substrate, respectively, on opposite ends of the cylindrical cavity.[48,52,66] A gradient in the axial field is adjusted so electron-cyclotron-resonance (ECR) conditions exist near the substrate end of the cavity but not near the microwave window. Circularly polarized microwave electric fields drive free electrons in circular orbits around the axial field lines. Near the substrate, the ECR resonance conditions allow the electrons to receive resonance impulses during each orbit so these electrons acquire a higher energy than electrons elsewhere in the cavity. Thus, the plasma forms preferentially in the region of these high energy electrons adjacent to the substrate. No plasma forms at the microwave window at the opposite end of the cavity where ECR conditions do not exist. Hence, the microwave window is kept free of carbon deposits or erosion. An interesting observation in ECR experiments is that as the chamber pressure is lowered, the substrate temperature for diamond formation tends to decrease.[48]

Similar to the work of Aikyo and Kondo[37] for hot filaments, Martin and Hill[63] have introduced the hydrocarbon downstream from a microwave plasma hydrogen discharge and successfully grown diamond. These experiments again demonstrate that the hydrocarbon does not have to pass through the excitation source (filament or plasma) but only needs to be exposed to atomic hydrogen to grow CVD diamond . Much work has been done on the morphology of films grown in microwave discharges under varying conditions.[57-59,62,64] Low substrate temperatures and low hydrocarbon concentrations favor the development of (111) facets on the diamond crystals, indicating that the <111> direction is the slowest growth direction under these conditions. High substrate temperatures and high hydrocarbon concentrations favor the development of (100) faces showing the <100> direction is the slow growth direction. At intermediate hydrocarbon concentrations and temperatures, (110) facets will develop.

The lowest temperatures (365-500°C) for CVD diamond deposition have been reported for microwave plasma deposition techniques.[56,67] Additions of oxygen carrying species such as alcohols, water, etc can increase the diamond film growth rate or lower the pressure necessary for diamond growth. [49,56,60,61,65] Lower gas pressures are particulary important as they allow the plasma to spread and make a more uniform deposit over a wider area.[49,66] The electromagnetic waves used in microwave discharges are launched into the deposition chamber from outside sources. The penetration length of these waves into a low pressure chamber varies inversely with the intrinsic plasma frequency of the gas.[69] The plasma frequency, in turn, is directly proportional to the square root of the gas pressure. Consequently, lowering the gas pressure will result in a greater penetration length and a more uniform discharge. For a pressure of about 10 torr, the penetration length of microwaves in a plasma discharge is only a few centimeters.[70]

moving molecules and they will have different temperatures. As a result, a low-pressure plasma typically has a relatively cool gas temperature. Generation of atomic hydrogen and reactive molecular radicals will be accomplished principally by collisions with high energy electrons. As a consequence of the scarcity of electrons in the plasma and the large mean free path path for collision with gas molecules, the absolute concentration of atomic hydrogen and molecular radicals will be low in a low pressure plasma.

In high pressure plasmas on the other hand, the mean free path for collisions between electrons and molecules will be small and any excess energy absorbed by the electrons from the electric field will be quickly redistributed to the more massive molecules by electron-molecule collisions. Hence, the temperature of the gas molecules and electrons will equilibrate at a relatively high temperature. Generation of atomic hydrogen and molecular radicals can now occur as the result of either high-energy molecular or electron collisions in the plasma. Thus, the absolute concentration of atomic hydrogen and molecular radicals will be high in a high-pressure plasma.

Because of the higher concentration of atomic hydrogen and hydrocarbon radicals in a high-pressure plasma, the CVD diamond growth rate will be greater in a high-pressure plasma than in a low-pressure plasma. Such a correlation is observed experimentally where CVD diamond deposition rates in a high-pressure plasma are typically 500 μm/hour whereas the deposition rate in low-pressure plasmas are 0.2 μm/hour.

If the growth rate was the only parameter of interest in diamond deposition, low-pressure plasmas would be ignored. However, high-pressure plasmas have a very high energy content which necessitates elaborate cooling schemes for the substrate to prevent substrate melting. Furthermore, high pressure plasmas tend to be unstable and difficult to control. For many applications, these problems rule out the use of high-pressure plasmas.

IVC1. Low-Pressure Glow Discharge

A common low-pressure plasma method of atomic hydrogen generation is based on the Wood's Tube method[47] where a stream of low-pressure molecular hydrogen is passed through a glow discharge between two metal electrodes. Over a pressure range of 0.1 to 20 torr, up to 25% atomic hydrogen can be generated by this method. The substrate is placed in this glow discharge or one electrode serves as the substrate so that atomic hydrogen is generated close to the substrate surface. Although AC or DC can be used to form the discharge, DC is advantageous if an electrode is used as a susbtrate since diamond growth rates are higher on the positive electrode. This method has been used to produce very fine-grain CVD diamond films by Crystallume of Palo Alto, California. Disadvantages of this process include the limited pressure range of operation, the erosion of and contamination from the electrodes and the generally high stress and hydrogen content in the resulting diamond films.

IVC2. Low-Pressure Microwave Discharge

Low-pressure microwave discharges have been used more than any other method to generate atomic hydrogen for CVD diamond synthesis.[48-68] Since microwave discharges are electrodeless, electrode erosion problems are avoided.

IVC3. Low-Pressure RF Discharge

There has been much less work with low pressure RF discharges than with microwave discharges for CVD diamond growth. At first glance, this is somewhat surprising as RF is easier to scale to large volume and high power. Moreover, large RF plasma discharge units have been used in the semiconductor industry for some time. The scarcity of reported work in the RF area in CVD diamond growth may be the result of many unreported failures to achieve good quality diamond films rather than the result of not trying RF sources. The deposition rate of CVD diamond is closely correlated with the generation of atomic hydrogen and hydrocarbon radicals. The generation of such radicals seems to be more efficient in a microwave plasma than in an RF plasma. This is probably caused by both the higher energy and higher density of electrons in a microwave plasma. Average electron energies in a microwave plasma at 2450 MHz and 1 torr pressure average about 10 ev [71,72] which is enough energy (8.5ev) to dissociate hydrogen. On the other hand, average electron energies in an RF discharge at 13.56 MHz at 1 torr are about 4 ev [73] which is insufficient to dissociate hydrogen. In addition, electron densities are higher in microwave discharges.[74]

Nevertheless, low-pressure electrodeless radiofrequency discharges have also been used to grow CVD diamond.[75-81] Depending on the pressure, atomic hydrogen concentrations of 10 to 65% can be produced in a RF discharge.[11] These discharges are less stable than microwave discharges and Ar sometimes has to be added to the hydrogen-hydrocarbon mixture to stabilize the discharge. Similar to the microwave discharge, deposits of carbon on the walls of the reactor can be a problem with an RF discharge. In addition, the RF can electromagnetically couple with any electrically conductive body in the reactor chamber and heat it. Finally, RF power sources are more expensive than microwave sources per watt of delivered power.

Diamond made by RF discharges has been of poorer quality than diamond made by microwave discharges.[75] In addition, large well faceted diamond crystals are usually not obtained.

IVC4. High-Pressure DC Discharge

High-pressure DC discharges between fixed electrodes at atmospheric pressure have been used to grow good quality CVD diamond from hydrogen-hydrocarbon mixtures at very high rates of up to 200 µm/hour and thickness greater than 1 mm.[82-86] Because of technology developments over the years to make plasma torches, these DC discharges are now very stable. One disadvantage of this process is that the substrate must be directly cooled with running water to prevent it from melting because large amounts of power per unit of substrate area are used. Other disadvantages include the high power consumption, the erosion of the electrodes, the nonuniformity of the discharge and the small deposition area.

IVC5. High-Pressure Microwave Discharge Jets

A high-pressure microwave discharge jet has also been used to grow CVD diamond at moderately high rates of 30 µm/hour at atmospheric pressure over a square inch of area.[87] A gas mixture of Ar-H_2-CH_4 is used with a CH_4/H_2 ratio as high as 10%. Sustained operation is difficult because of plasma instabilities and the need to vigorously cool the substrate. The plasma is initially ignited by electric breakdown around the central electrode and then sustained by microwave electromagnetic waves generated between the central electrode and chamber walls. One advantage of low-pressure microwave discharges is that they are electrodeless. However, with high-pressure microwave jets, this advantage has been lost by using electrodes that are immersed in the plasma and that therefore can erode.

IVC6. High-Pressure RF Discharge

High-pressure radiofrequency discharges have also been used to grow CVD diamond.[88] These discharges are very unstable and large amounts of argon are routinely added to the hydrogen-hydrocarbon mixture to increase the stability of the discharge. Nevertheless, the discharge must be monitored continuously and is typically only run for short periods of time. High rates of diamond deposition are possible during these short runs. As with other high-pressure discharges, the high power input requires that the substrates be cooled directly with running water.

IVC7. High and Low-Pressure Flames

Hydrogen atoms are created during the burning of hydrogen or hydrocarbons in an oxygen flame.[89-91] With a suitable ratio of carbon, hydrogen and oxygen, Hirose[92] discovered that CVD diamonds can be grown with a simple plumbers torch at atmospheric pressures in the "feather" of the flame just downstream of the flame front.[92-101] High temperatures (2800-3400°C) in common flames cause a plasma to form and atomic species to be generated at the flame front.[102] The flames are operated under fuel-rich conditions so atomic hydrogen and hydrocarbon radicals are abundant in the feather of the flame where the substrate is placed. Although the flame process has the potential to be an inexpensive way to make diamond, very low conversion rates of carbon to diamond have thwarted this goal. The substrate must be vigorously cooled because of the high heat input rate of the flame.

Hydrogen atoms can also be generated in low-pressure hydrogen and hydrocarbon flames.[90,91] Advantages of low-pressure flames may be a higher carbon-to-diamond conversion rate, a wider spatial zone in the flame where diamond will grow and a much lower heat input to the substrate. Although work is underway at several laboratories, no successful CVD diamond growth has been reported yet in a low-pressure flame.

So far, work has been limited to hydrogen-hydrocarbon-oxygen flames. Flames with other chemical reactants may also generate CVD diamond growth. For example, hydrogen-hydrocarbon-fluorine, hydrogen-hydrocarbon-chlorine, hydrocarbon-fluorine, hydrocarbon-chlorine, halocarbon-fluorine and halocarbon-chlorine flames are promising candidates among many possibilities.

V. OTHER DIAMOND PROCESSES

VA. COMBINATION PROCESSES

Some investigators have used combinations of processes to see if any advantages would ensue. For example, a direct-current spiral hollow cathode combines features of the hot filament, electron beam and plasma process. Such a hollow cathode was used to grow CVD diamond in an Ar-H_2-CH_4 gas at a disappointing rate of one µm/hour.[31,103,104] The results of these experiments indicate that this particular combination of processes has no advantage over a simple hot filament process.

VB. THERMAL PROCESSES-SUBSTRATE ONLY

Rudder et al.[105] reported the growth of CVD diamond in a CF_4-F_2 mixture without the assistance of any external excitation such as a plasma or hot filament. Only heating of the diamond substrate was required. Although fluorine can stabilize the diamond surface in the same way that hydrogen does, the carbon-fluorine bond is much stronger than the flourine-fluorine bond. Thus, atomic fluorine can not produce vacant sites on a diamond surface covered with fluorine in the same manner that atomic hydrogen produces vacant sites on a diamond surface covered with hydrogen. Without vacant sites, carbon deposition should cease. It is not clear how vacant sites needed for carbon deposition are formed once a tenacious layer of fluorine covers the diamond surface. A few vacant sites should form by thermal activation but these are probably insufficient to maintain any reasonable diamond growth rate.

VC. LASER EXCITATION

There have been two reports[106-108] of the deposition of CVD diamond using laser excitation. One of these reports was later withdrawn.[107] There has been no additional successful work in this area so whether laser excitation is a viable method of CVD diamond growth is questionable.

VD. CVD PROCESSES IN STARS

Diamonds have been discovered in outer space by infrared absorbtion experiments and by an examination of meteorites which are an agglomeration of outer-space debris.[109-113] Two main sources of diamond generation have been identified. The first is a super nova which is an exploding star. Large quantities of hydrogen and carbon gases are thrown off during the explosion and as they expand and cool, diamonds are formed by a CVD diamond process. Because of the low gas density, the diamonds that are formed average only 10Å in diameter.

A second generator of diamonds in outer space are red giant stars. In the outer sheils of these stars, the conditions are very similar to those used for CVD deposition on earth. The carbon/hydrogen ratio is 1/2%, the pressure is 1 torr and the temperature is about 1000°C. Convection processes in the outer shell of the star cause temperature changes which cause carbon to precipitate from the gases as diamond because of the presence of atomic hydrogen. Photon pressure from the red giant imparts a push on the side of the diamond facing the star and propels the diamond into outer space if its area/mass ratio is sufficiently large. The largest diamonds that are recovered from meteorites from red giants are about 600Å in diameter. Larger diamonds, unfortunately, do not have a great enough area/mass ratio and fall back into the star as gravitational forces overwhelm photon pressure on them.

VE. Alternating Chemical Reactions

Recently, reports have circulated that diamond can be grown layer by layer by alternating exposures of diamond to different gases. No details are available about these experiments. However, the type of paired reactions that should work are shown in Equations (27a) and (27b) for CF_4 and CH_4 :

$$CF_4 + \text{Diamond–H} \text{ -----> } \text{Diamond–C–F} + HF \quad (27a)$$

$$CH_4 + \text{Diamond–F} \text{ -----> } \text{Diamond–C–H} + HF \quad (27b)$$

where Diamond-F and Diamond-H represent, respectively, diamond covered with a monolayer of fluorine and hydrogen. This idea is an extension of the alternating reaction work of Eversole[5] shown in Equations (2) and (3). This reaction works because the bond strength of H-F exceeds the bond strengths of C-F and C-H. Both reactions are exothermic with an enthalpy of reaction of about 8 kcal/mole.

One can generalize this type of reaction in a number of ways. One way is to use compounds of the type C_NX_M and C_LZ_P:

$$C_NX_M + \text{Diamond–Z} \text{ ------> } \text{Diamond–C–X} + ZX \quad (28a)$$

$$C_LZ_P + \text{Diamond–X} \text{ ------> } \text{Diamond–C–Z} + ZX \quad (28b)$$

where Diamond-Z and Diamond-X represent, respectively, diamond covered with a monolayer of Z and X. If the Z-X bond is stronger than the C-X and C-Z bonds, then alternate exposures of diamond to the respective gases will cause alternate exothermic chemical reactions where carbon is placed down during each exposure. These types of reactions inherently prevent the generation of double or sp2 carbon bonds and thus stop graphite formation. Examples of gases that can be used in such reactions pairs are all of the fluorochlorocarbons, fluorochlorohydrocarbons, alcohols, halogens and hydrogen.

If C_NX_M and C_LZ_P do not react in the gas phase, then a gas mixture of them can be used to grow diamond instead of the alternating exposure of one and then the other depicted in Reactions 28a,b. In this latter case, the reactions still alternate between (a) and (b) on a microscale but not on a macroscopic scale:

$$C_NX_M + C_LZ_P + \text{X–Diamond-Z} \text{ ------> } \text{Diamond} + ZX \quad (29)$$

where X-Diamond-Z is diamond covered with a monolayer of a mixture of X and Z.

VF. DEPOSITION FROM LIQUID SALT BATHS

Patel and Cherian[114] reported evidence of diamond growth on a diamond crystal that was placed in a liquid alkali salt bath which was subsequently cooled. Unfortunately, their paper did not disclose the most simple parameters of their experiment (e.g., the composition of the salt bath, temperature, etc.) so this experiment has never been successfully reproduced by other workers. If this experiment is valid, we can speculate as to why it may have worked by using a LiCl bath as an example. There is an analogy between conventional CVD diamond growth from an acetyline-hydrogen gas mixture in the presence of atomic hydrogen and diamond growth from a LiCl salt bath containing lithium acetylide.

Atomic hydrogen, lithium and chlorine all form single bonds with carbon and prevent the formation of carbon double bonds which might lead to graphite growth. The stabilization of the diamond surface by atomic hydrogen is analogous to the stabilization by either lithium or chlorine in a liquid salt solution. The main difference between the two cases is that the bonding with diamond is predominantly covalent in the case of hydrogen and predominantly ionic in the case of lithium and chlorine. Naturally, many other alkali-halide salts could be substituted for LiCl. The simplest method of diamond growth with this method would involve a temperature gradient in a molten alkali salt bath containing graphite and a diamond seed. Alternatively, the graphite could be replaced with an alkali acetylide.

VH. DEPOSITION FROM LIQUID METAL SOLUTIONS

Over the years, a number of patents have been issued claiming to have grown diamond from a liquid metal solution at low pressures. Many of these patents stemmed from the success of using metal catalysts in high-pressure, high-temperature diamond synthesis. Some of these patents are undoubtedly "paper" patents which were never really tried but which sounded logical and reasonable to a patent examiner. The success of using atomic hydrogen to stabilize diamond, however, reintroduces the question of whether a liquid metal bath can be used to grow diamond at low pressures. Diamond in a liquid metal bath may have its surface stabilized by metal atoms reacting with the diamond to form a carbide monolayer. If a metal with a metal-carbide bond energy equal to the hydrogen-carbon bond energy were used, one can argue that a low-pressure liquid-metal processes analagous to conventional CVD diamond deposition should be possible.

REFERENCES

1. Henry A. Bent, "Second Law of Thermodynamics", Oxford University Press, New York (1965)
2. Gordon Davies, "Diamond", Adam Hilger Ltd, Bristol (1984)
3. F.P. Bundy, H.T. Hall, H.M. Strong, and R.J. Wentorf Jr, Nature, 176, 51-54 (1955)
4. R.C. Devries, Ann Rev Mater Sci, 17, 161-187 (1987)
5. W.G. Eversole, U.S. Patent No. 3030188, Apr 17, 1962.
6. J.C. Angus, H.A. Will, and W.S. Stanko, J. Appl. Phys, 39, 2915 (1968)
7. D.J. Poferi, N.C. Gardner and J.C. Angus, J.Appl. Phys, 44, 1428 (1973)
8. K.E. Spear, J. Am. Ceramic Soc, 72, 171 (1989)
9. B.V.Spitsyn, L.L.Bouilov and B.V. Deryagin, J. Cryst. Growth, 52, 219 (1981)
10. T.R. Anthony,"Synthesis of Metastable Diamond" in "Diamond, Boron, Nitride, Silicon Carbide and Related Wide Band Gap Semiconductors", edited by J.F. Glass, R.F. Messier and N. Fujimori, Proc MRS 162, in press (1990)
11. W.E. Jones, S.D. Macknight and L. Teng, Chemical Revs, 73, 407 (1973)
12. M. Knudsen, "Kinetic Theory of Gases", Methuen & Co, London (1950)
13. H. Wise and C.A. Ablow, J Chem Phys 33, 10 (1960)
14. H. Wise and C.A. Ablow, J Chem Phys 29, 634 (1958)
15. J. Amdur, J. Chem Phys 4, 339 (1936)
16. H.A. Becker, "Dimensionless Parameters, Theory & Methodology", John Wiley & Sons, New York (1976)

17. J. Zierep, "Similarity Laws and Modeling", Marcel Dekker,Inc, New York (1971)
18. F.S. Larkin, Can J. Chem 46, 1005 (1968)
19. H. R. Heath, T.L. Ibbs and N.E. Wild, Proc Roy Soc (London) A178, 380 (1941)
20. I. Langmuir, J. Amer. Chem. Soc, 34,860 (1912)
21. I. Langmuir, J. Amer. Chem. Soc, 34,1310 (1912)
22. I. Langmuir and G.M.J. Mackay, J. Amer. Chem. Soc, 36,1708 (1914)
23. I. Langmuir, J. Amer. Chem. Soc, 37,417 (1915)
24. M. Kamo, Y. Sato, S. Matsumoto and N. Setaka, J. Crystal Growth, 62,642 (1983)
25. S. Matsumoto, Y. Sato, M. Kamo and N. Setaka, Japn. J. Appl. Phys, 21, L183 (1982)
26. H.J. Grabke,Berichte der Bunsengesellschaft, 69, 409 (1965)
27. H.J Grabke, Metallurgical Transactions 1, 2972 (1970)
28. F.G. Celii, P.E. Peterson, P.E. Wang and J.E. Butler, Appl Phys Letters 52, 2043 (1988)
29. F.G. Celii and J.E. Butler, Appl Phys Letters 54, 1031 (1989)
30. T.R. Anthony, MRS Fall Meeting, Nov30-Dec3 (1987) Boston
31. B.Singh, Y.Arie, A.W. Levine and R.O. Mesker, Appl Phys Letters 52, 451 (1988)
32. T.D. Moustakas, Solid State Ionics 32, 861 (1989)
33. A. Sawabe and T. Inuzuka, Thin Solid FIlms 137, 89 (1986)
34. T. A. Anthony, Unpublished work.
35. H.B. Vakil, Am Chem Soc Meeting, Mar (1989) Dallas
36. R.E. Clausing, personal communication (1990)
37. H. Aikyo and K. Kondo, Jap Jour Appl Phys 28, L1631 (1989)
38. J.E. Hughes, J. Less Common Metals,1,377 (1959)
39. G.B. Gaines, C.T. Sims and R.I. Jaffe, J.Electrochem Soc 106,881 (1959)
40. Benno Lux and R. Haubner, Proceedings 12th Int Plansee Seminar,C16, 461, (1989)
41. Y. Hirose, J Jpn Soc Precis Eng 53, 1507 (1987)
42. F. Jansen, I. Chen and M.A. Machonkin, J. Appl. Phys 66, 5749 (1989)
43. A. Sawabe and T. Inuzuka, Thin Solid Films 137, 89 (1986)
44. J.T. Glass, NATO Meeting on Boride, Carbides and Hard Materials, Manchester, UK (July, 1989)
45. K. Kobashi, NATO Meeting on Boride, Carbides and Hard Materials, Manchester, UK (July, 1989)
46. W.F. Banholzer, Diamonds 90 Conference, Crans-Montana, Switzerland (Sept, 1990)
47. R.W. Wood, Proc Roy Soc, Ser A, 102, 1 (1923)
48. J. Suzuki, H. Kawarada, K. Mar, J. Wei, Y. Yokota and A. Hiraki, Jap Journ Appl Phys, 28, L281 (1989)
49. I. Watanabe and K Sugata, Jap Journ. Appl. Phys., 27, 1397 (1988)
50. O. Matsumoto, H. Toshima and Y. Kanzaki, Thin Solid Films, 128, 341 (1985)
51. Y. Mitsuda, Y. Kojima,T. Yoshida and K. Akashi., Journ Mat Sci, 22, 1557 (1987)
52. H. Kawarada, K.S. Mar and A. Hiraki, Jap. Journ. App. Phys, 26, L1032 (1987)
53. M. Kamo, Y. Sato, S. Matsumoto and N. Setaka, J. Crystal Growth, 62, 642 (1983)
54. Y. Saito, S. Matsuda and S. Nogita, Journ. Mat. Sci. Letters, 5, 565 (1986)

55. A.R. Badzian and T. Badzian, Proceedings of the Fifteenth International Conference on Metallurgical Coatings Vol 1, 283 , Elsevier Applied Science (1988)
56. Y. Liou, A. Inspektor, R. Weimer and R. Messier, Appl Phys Letters 55, 631 (1989)
57. B.E. Williams, J.T. Glass, R.F. Davis, K.Kobashi and T. Horiuchi, J. Vac Sci Technol A, Vac Surf Films 6, 1819 (1988)
58. K. Kobashi, K. Nishimura, Y. Kawate and T. Horiuchi, J. Vac Sci Technol A, Vac Surf Films 6, 1816 (1988)
59. A.R. Badzian,T. Badzian, R. Roy, R. Messier, K.E. Spear, Mater Res Bull 23, 531 (1988)
60. K. Kobashi, K. Nishimura, Y. Kawate and T. Horiuchi, Phys Rev B38, 4067 (1988)
61. Y. Saito, K. Sato, H. Tanaka and H. Miyadera, J. Mater. Sci 24, 293 (1989)
62. L. Vandenbulcke, P. Bou, R. Herbin, V. Cholet and C. Beny, Colloq Phys C-5, 177 (1989)
63. R. Martin and M.W. Hill, Appl Phys Letters 55, 2248 (1989)
64. R. Haubner and B. Lux, Int J. Refractory & Hard Metals 6, 210 (1987)
65. C. Chen, Y.C. Huang, S. Hosomi and I. Yoshida, Mater Res Bull 24, 87 (1989)
66. H. Kawarada, K. Mar, and J. Suzuki, Jap J Appl Phys 27, L683 (1988)
67. W. Hsu, D.M. Tung, E.A. Fuchs, K.F. McCarty and A. Joshi, Appl Phys Letters 55, 2739 (1989)
68. W. Zhu, C.A. Randall, A.R. Badzian and R. Messier, J. Vac. Sci. Technol A Vac Surface Films 7, 2315 (1989)
69. J.D. Jackson, "Classical Electrodynamics", John Wiley & Sons, New York (1962)
70. J. Keiser and M. Neusch, Thin Solid Films 118, 203 (1984)
71. J.L. Franklin, S.A. Studniarz and P.K. Ghosh, J. Appl. Phys 39, 2052 (1968)
72. F.H. Dorman and F.K. Taggart, J. Microwave Power 5, 4 (1970)
73. K. Yanagihara and N. Niinomi, Proc Int Conf Ion Engineering Congr, pg 1475, Kyoto (1983)
74. M.M. Millard and E. Kay, J. Electrochem Soc 129, 160 (1982)
75. S. Matsumoto, J. Mater. Sci. Letters, 4, 600 (1985)
76. D.E. Meyer, R.O. Dillon and J.A. Woollam, J Vac Sci & Technol A 7, 2325 (1989)
77. K. Akashi, T. Yoshida, S. Komatsu and Y. Mitsuda, J. Korean Inst Metals 24, 712 (1986)
78. D.E. Meyer, N.J. Ianno, J.A. Woollam, A.B. Swartzlander and A.J. Nelson, J. Maters. Res 3, 1397 (1988)
79. H. Hayashi, Y. Ohnishi, K. Kobashi, K. Nishimura, K. Miyata and Y. Kawate, Kobelco Technol Rev 6, 14 (1989)
80. D.J. Vitkavage, R.A. Rudder, G.G. Fountain and R.J. Markunas, J. Vac. Sci Technol A6, 1812 (1988)
81. S. Matsumoto, M. Hino and T. Kobayashi, Appl Phys Letters 51, 737 (1987)
82. T. Chonan, M. Uenura, S. Futaki and S. Nishi, Jap. Journ. Appl. Phys 28, L1058 (1989)
83. K. Suziki, A. Sawabe, H. Yasuda and T. Inuzuka, Appl. Phys. Letters, 50,728 (1987)
84. F. Akatsuka, Y. Hirose and K. Komaki, Jap. Journ. Appl. Phys, 27, L1600 (1988)
85. K. Kurihara, K. Sasaki, M. Kawarada and N. Koshima, Appl Phys Letters, 52, 437 (1988)
86. K. Suzuki, A. Sawabe and T. Inuzuka, Jap. J. Appl. Phys, 29, 153 (1990)
87. Y. Mitsuda, T. Yoshida and K. Akashi, Rev Sci Instrum 60, 249 (1989)

88. S .Matsumoto, T. Kobayshi, M. Hino, T. Ishigaki and Y. Moriyoshi, 8th International Symposium on Plasma Chemistry, ISPC,8, Paper#S7-03, p2458,Tokyo, (1987)
89. R.R. Baldwin, D.E. Hopkins and R.W. Walker, Trans Faraday Soc., 66, 189 (1970)
90. C.P. Fenimore and G.W. Jones, J. Phys. Chem, 65, 2200 (1961)
91. C.P. Fenimore and G.W. Jones, 1962 Symp. Int. Combust. Proc.,9th, 597 (1963)
92. Y. Hirose and N. Kondo, Program and Book of Abstracts, Japan Applied Physics 1988 Spring Meeting, pg 434 (March 29, 1988)
93. S. Yazu, S. Sato and N. Fujimori, SPIE Proceedings (San Diego, August 16-20, 1988)
94. L.M. Hanssen, W.A. Carrington, D.B. Oakes, J.E. Butler and K.A. Snail, 1989 Diamond Technology Initiative Symposium, Paper T7, (July 11-13, 1989)
95. D.G. Goodwin, 1989 Diamond Technology Initiative Symposium, Paper T15, (July 11-13, 1989)
96. M.A. Cappelli and P.H. Paul, 1989 Diamond Technology Initiative Symposium, Paper W1, (July 11-13, 1989)
97. M. Murakawa, S. Takeuchi and Y. Hirose, Surface & Coatings Technol 39-40, 1 (1989)
98. R.G. Vardiman, C.L. Vold, K.A. Snail, J.E. Butler and C.S. Pande, Mater Letters 8, 468 (1989)
99. L.M. Hanssen, W.A. Carrington, J.E. Butler and K.A. Snail, Mater Letters 7, 289 (1989)
100. Y. Matsui, A. Yuuki, M. Sahara and Y. Hirose, J. Appl. Phys 28, 1718 (1989)
101. W.A. Carrington, L.M. Hanssen, K.A. Snail, D.B. Oakes and J.E. Butler, Metall Trans 20A, 1282 (1989)
102. Irvin Glassman, "Combustion", Academic Press, New York (1977)
103. Y. Tzeng, P.J. Kung, R. Zee, K. Legg, H. Solnick-Legg, Appl Phys Letters 53, 2326 (1988)
104. P.J. Kung and Y. Tzeng, J. Appl. Phys 66, 4676 (1988)
105. R.A. Rudder, J.B. Posthill and R.J. Markunas, Electronic Letters 25, 1220 (1989)
106. K. Kitahama, K. Hirata, H. Nakamatsu,S. Kawai, N. Fujumori, T. Imai, H. Yoshino and A. Doi, Appl Phys Letters 49, 634 (1986)
107. K. Kitahama, Appl Phys Letters 53, 1812 (1988)
108. I.G. Varshavskaja and A.V. Lavrientiev, Arch Nauki O Mater 7, 127 (1986)
109. R.S. Lewis, T. Ming, J.F. Wacker, E. Anders and E. Steel, Nature 326, 160 (1987)
110. T.J. Wdowiak, Nature 328, 385 (1987)
111. J.A. Nuth, Astrophysics and Space Science 139, 103 (1987)
112. R.S. Lewis, E. Anders and B.T. Draine, Nature 339, 117 (1989)
113. P. Badziag, W.S. Verwoerd, W.P. Ellis and N.R. Greiner, Nature 343, 244 (1990)
114. A.R. Patel and K.A. Cherian, Indian J. Pure & Appl Phys 19, 803 (1981)

NUCLEATION AND GROWTH OF LOW-PRESSURE DIAMOND

Benno Lux and Roland Haubner

Institute for Chemical Technology of Inorganic Materials
Technical University of Vienna, A-1060 Vienna, Austria

1. INTRODUCTION

The great breakthrough in low-pressure diamond synthesis research came with the discovery that diamond growth is also possible on non-diamond substrates /1,2/. Since then these methods have developed rapidly /3-6/. Despite intensive work by hundreds of scientists, in-depth knowledge about the diamond nucleation and growth mechanism is still lacking. Several authors /7-12/ have discussed atomic attachment mechanisms, but more systematic experimental work and quantitative data which could provide practical guidance is still needed.

Most industrial applications /13/ such as wear and optical coatings, including single-crystalline layer growth by homeoepitaxy for semiconductors, as well as coating of powder particles /14-16/ or their direct formation from the gas phase /17,18/ need controllable nucleation conditions and predictable early diamond growth stages leading to well-defined final diamond products. Perfect and reproducible adhesion of layers is required for many applications /13/.

It is known qualitatively that substrates of different chemical natures promote diamond nucleation to varying degrees /19/. This can be explained in part by their surface properties and simple chemical and physical gas/surface reactions, some of which can also interfere strongly with diamond layer adherence /20/.

2. DIFFERENCES BETWEEN PVD AND CVD

2.1 Carbon Products

The element carbon is unique in its ability to form an amazing variety of products /21/. Besides the well-known crystalline phases of diamond, graphite and the carbines /22/, carbon can form an extremely wide range of different materials, including such products as pyrolytic CVD carbons and graphites, turbostratic coals, cokes, soots, industrial carbon and graphite products, fibers of carbon or oriented graphite, etc. /23/. Some of these products are expensive specialities, others are commodities produced by the ton in precisely defined industrial qualities.

Diamond and Diamond-Like Films and Coatings
Edited by R.E. Clausing *et al.*, Plenum Press, New York, 1991

Here two very special carbon products, with quite different properties, are briefly compared:

(1) "low-pressure diamond" (or CVD-diamond) synthesized with a modified CVD process, and

(2) "diamond-like carbon" /24/ or "a-C:H" /25/ and the recently described "tetrahedral sp^3-carbon" or "a-C type carbon" /26/ (also sometimes misnamed "amorphous diamond"). They are typically produced by physical methods (PVD, sputtering, molecular beam techniques, etc.) /27/.

An excellent criterium for classifying the quality of these products is the ratio $R = sp^2/(sp^2+sp^3)$, as proposed by P.Koidl /25/ for "a-C:H" and recently by Y.Sato /28/ for low-pressure diamonds. The R values for these two carbon products are quite different and can be closely linked with many of their intrinsic properties.

Most properties of "low-pressure diamond", including X-ray diffraction pattern, Raman peaks, etc., are identical with natural and high-pressure diamonds. Its R values /28/ are situated between 10^{-1} and less than 10^{-8}. A useful indication of nuances in quality based on lattice perfection is possible. Diamonds with R values $>10^{-2}$ are not transparent, but can still be suitable for certain wear applications. An R between 10^{-3} and 10^{-4} is needed for optical/thermal applications and R $<10^{-7}$ is required for semiconductor qualities. However for advanced evaluation the specific kind and location of the defects (grain boundaries, linear cis- or trans-type sp^2 defects or benzene rings, etc.) must be identified and judged separately /29/.

For "a-C:H" products this ratio is much higher, since they have about 1/3 sp^2 bonds. Depending on the degree of hydrogen saturation of the remaining sp^3 bonds, the hydrogen content can vary over a wide range, but is always high (up to 60 at%) /26/. They are X-ray amorphous, extremely fine grained, and without clear diffraction lines. "a-C" shows predominantly sp^3 bonds in its Raman spectrum just like the low-pressure diamond /30/.

In contrast to this low-pressure diamond or CVD-diamond generally forms coarse facetted crystals, has hydrogen contents below 0.5 at%, and shows clear X-ray diffraction patterns. Fine-grained thin films, known as "nanocrystalline" CVD-diamond, can also be produced at high growth rates with nucleation enhancement /31/, or by specific methods such as DC-glow discharge /32/.

2.2 Methods

The essential difference between PVD and CVD lies in the gas activation.

The a:C-H formation by PVD needs a high-energy bombardment of the substrate with charged gas species to form the characteristic ultrafine-grained carbon layers. Physical methods such as PVD can create the necessary electrons and charged molecules and their acceleration in the vacuum. Though atomic hydrogen can form, its concentration is low.

The low-pressure diamond CVD principle is based on a thermally activated gas mixture of substantial amounts of atomic hydrogen, together with molecular hydrogen and smaller amounts of hydrocarbons. Remote growth experiments /33,34/ clearly proved that even thermal cracking of the hydrocarbon is not an essential step for the diamond formation, as the needed C radicals can form by reaction with the atomic hydrogen, without prior thermal cracking. Solid substrate surfaces are not needed either, as diamond particles can nucleate and grow on melts /35,36/ or directly from the gas phase /17,18/. Nor are substrate surface bombardments with electrons or ions necessary. They would in fact be detrimental for the growth of facetted crystals.

CVD diamond synthesis does need an additional gas activation step to form appreciable amounts of atomic hydrogen. Of the many theoretically possible ways to achieve the H_2 decomposition and to form sufficient H^o, only a few are used today:

- heated solid surfaces >2000°C and
- hot plasmas created by chemical reactions or electric arc or gas discharges /13/.

Other methods using dissolved hydrogen in liquids or solids, or catalytic low-temperature chemical decompositions to produce atomic hydrogen, have generally not yet been successfully applied for diamond synthesis, although they were described long ago in the patent literature /35,37,38/. Very recently however a low-pressure diamond synthesis based on the use of halogenides -- a possibility long under discussion /39/ -- was reported to operate at moderate growth temperatures and to lead to transparent diamond particles of substantial size /40/. The process, which seems to operate without any particular gas activation other than the temperatures needed in normal CVD processes, could bring about major changes in this technology.

Besides the basic differences between PVD and CVD described above, their experimental parameters differ quite substantially /41/ (Table 1). Similar gas compositions and the addition of neutral gases, such as argon, are possible for both methods, but are less frequently applied for CVD.

Table 1. Comparison between PVD and CVD for cutting tool coatings /41/

	PVD		CVD
A. DEPOSITION PARAMETERS			
pressure	lower	1 torr	higher
temperature of the substrate	lower	500°C	higher
deposition rate	high		low
layer crystal structure	amorphous to fine		fine to facetted
adhesion	difficult		easy
porosity/cracks	less		possible
B. PRODUCTION PARAMETERS			
size of charge	small		large
throwing power	bad		well
equipment costs per unit	high		low
costs per piece	high		low

Physical methods typically need pressure ranges below one torr, as they use ions, electrons and/or molecular beams. CVD works with reasonable growth rates best above this pressure. Since it is based on chemical reactions, its deposition rates increase with pressure, as do the concentrations of the gaseous reactants. Although industrial CVD is frequently operated at atmospheric pressure, for special purposes, CVD can also be applied at rather low pressures /42/. It is doubtful whether the low-pressure diamond synthesis can be applied in high and ultra-high vacuum, and wether this would lead to a more perfect diamond lattice than the usual methods /43/.

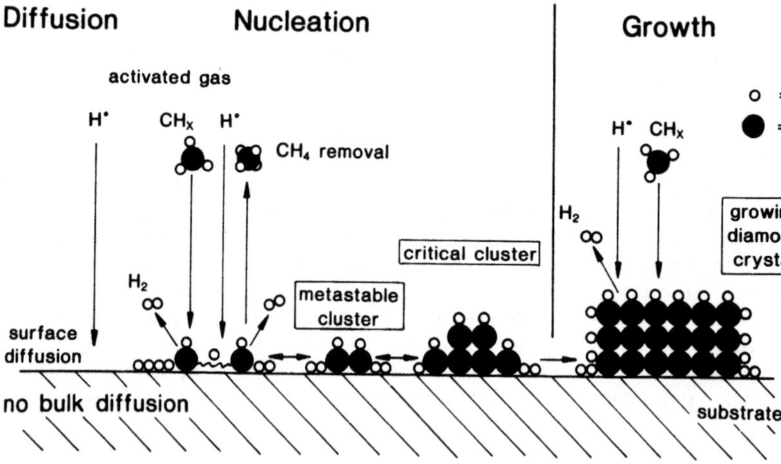

Fig.1. Nucleation and growth kinetics of low-pressure diamond deposition on a non-reacting surface

3. NUCLEATION

3.1 PVD and CVD Nucleation and Early Growth

The classical nucleation theories based on the gas/liquid transformation were extended to the gas/solid sublimation /44,45/. Specific theories on the nucleation occurring in PVD or CVD processes are rare /46-49/.

The principal pre-nucleation steps to be expected for diamond nucleation are illustrated in Fig.1. Surface diffusion and subcritical cluster formation should precede the stable growth stage. Yugo /50/ suggested the size of a homogeneous diamond nucleus to be about 30 Å. The expected theoretical size and structure of a homogeneous copper nucleus grown from an undercooled melt are shown in Fig.2 /51/. The comparison with the cubic Cu also suggest that the critical homogeneous diamond nucleus in a super-equilibrium /52/ surrounding can be expected to be rather small.

The early growth stages of diamond can be seen easily by SEM /53/. If it is assumed that each of the visible tiny crystallites originated from a surviving, super-critical diamond nucleus, the nuclei number can also be counted easily. A random nucleation can be deduced from such pictures for PVD /49/ and for CVD /48,54/. Fig.3 show schematic comparisons of such early growth stages for two experimental cases:

- PVD: precious metal (Au) deposited on ionic salt crystals /49/; (Nuclei number: $2.10^{11}/cm^2$)
- CVD: Al_2O_3 deposited on TiC /48/; (Nuclei number: $2.10^9/cm^2$)

Based on a model PVC experiment and on a typical industrial CVD coating application respectively, these examples show that the nuclei numbers of PVD and CVD differ by an order of magnitude. The number is even lower for diamond (around 10^8-$10^9/cm^2$) deposited with hot filament or microwaves on substrates without nucleation enhancement /55/.

Clusters of subcritical size should form continuously from the incoming flux on the uncovered substrate surface (Fig.3). In both cases stable nuclei develop, grow to crystals, and enlarge with time, resulting finally in the coating /48,54/.

Au and Al_2O_3 differ in their surface mobilty. Gold atoms and clusters on the clean ionic salt, repulsed by the electrically charged surface, can be expected to be highly mobile and, due to the shorter distance to their next neighbors, to undergo rapid coalescence. Also $AlCl_x$ or oxychloride on the substrate can be expected to diffuse.

Once Al_2O_3 and diamond, both having strong chemical bonds, are formed, their surface mobility will be low. Due also to the greater distance between clusters in CVD than in PVD (see Fig.3), layer formation proceeds differently. The "coalescence" will essentially consist of a lateral growth of the nuclei, until they join. This may be linked with void formations at the interface /48,56,57/, which can be detrimental for adhesion. As both diamond and Al_2O_3 can establish local chemical bonding with certain substrates, detrimental effects might be compensated to a certain extent.

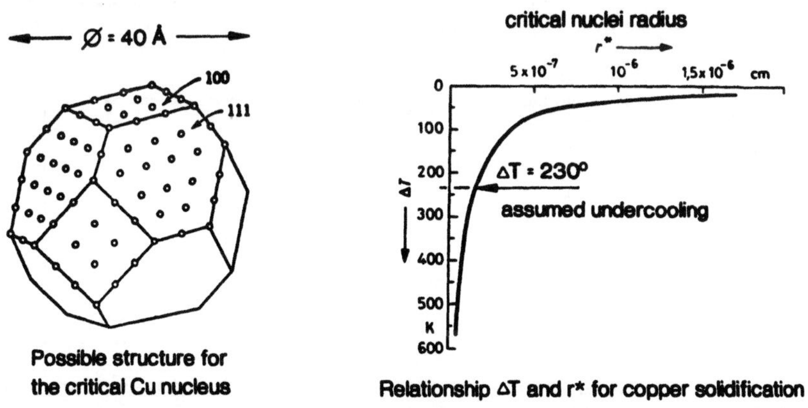

Fig.2. Critical size of a copper nucleus bonded with (100) and (111) planes during liquid/solid transformation /51/

3.2 Normal and Activated CVD Kinetics

The low-pressure diamond synthesis differs basically from a normal CVD process in its need for gas activation. The comparison of its chemistry with "normal CVD processes" however reveals various similarities and differences (see also the Sections 6 and 8).

The well-known industrial CVD coatings of cemented carbide tools with Al_2O_3, TiC, TiN, etc., are based on the following reactions type /41/:

$$AB + CD \longrightarrow AD + BC \qquad (1)$$

Applied to the Al_2O_3 deposition:

$$2\,AlCl_3 + 3\,H_2O \longrightarrow Al_2O_3 + 6\,HCl \qquad (2)$$

Here the thermodynamics are dominated by the reaction kinetics. H_2O is slowly released by a sluggish secondary reaction between gaseous CO_2 and H_2 /58/. Excess hydrogen acts as a carrier gas. For the Al_2O_3 deposition H_2O diffuses towards $AlCl_3$ molecules adsorbed at the substrate surface to form intermediate Al-oxychlorides and finally solid Al_2O_3 /59-61/.

For CVD diamond growth the reactants are hydrogenated carbon radicals and atomic and molecular hydrogen. The connection of carbon radicals to surface-bonded hydrogenated carbon atoms and the formation of H_2 molecules diffusing from the surface back into the reaction gas is illustrated in Fig.1 and shown in equations (3a-c). Various carbon species could be involved in the atomic attachment /7-12/.

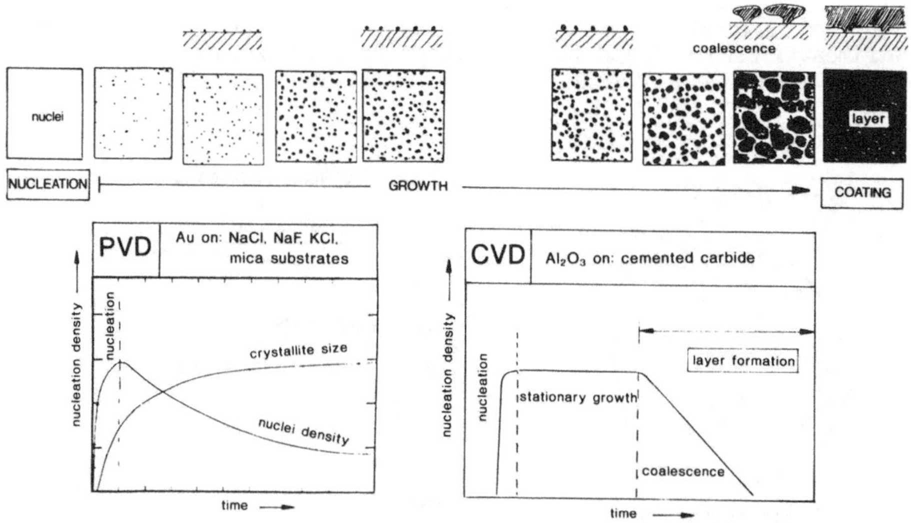

Fig.3. Schematic comparisons of the early growth stages for PVD and CVD for two experimental cases
PVD -- Au deposited on ionic salt crystals /49/
CVD -- Al_2O_3 deposited on TiC /48/

$$H_2 \longrightarrow 2\,H^o \qquad (3a)$$

$$CH_4 + X\,H^o \longrightarrow CH_{4-x} + X\,H_2 \qquad (3b)$$

$$-\overset{|}{\underset{|}{C}}-H + CH_{4-x} \longrightarrow -\overset{|}{\underset{|}{C}}-\overset{|}{\underset{|}{C}}-H + y\,H_2 \qquad (3c)$$

The global reaction of CVD diamond growth is a pyrolytic decomposition of hydrocarbons according to equation (4). As in other chemical reactions the solid carbon formation is always linked with a dissolution reaction by the hydrogen. At about 1000°C this equilibrium is very much on the side of the solid carbon and usually pyrolytic carbon or graphite forms.

Pyrolytic solid carbon formation:

$$CH_4 \Longleftrightarrow C_{(solid)} + 2\,H_2 \qquad (4)$$

Solid carbon dissolution with atomic hydrogen:

$$C_{(solid)} + 4\,H^o \longrightarrow CH_4 \qquad (5)$$

However, for CVD diamond, the thermodynamics are dominated by reaction kinetics. For diamond deposition the continuous formation of atomic hydrogen is important. These conditions can be realized by different gas activation methods and were called "super-equilibrium" conditions /52/. The presence of atomic hydrogen leads to a "selective etching". Under "super-equilibrium" conditions and certain temperature, pressure, etc., the sp^2-bonded carbon (graphite) is either removed as methane or transformed to C-H sp^3 bonds. The sp^3-bonded carbon (diamond) is stable. These kinetic reactions are essential for CVD-diamond formation /62/.

The overall reaction type is important for gas volume changes occurring during CVD deposition, which can be linked with the pressure dependence of a reaction. For the Al_2O_3 deposition in equation (2) only a small volume change occurs (ratio 6:5 = 1.2), indicating only a slight pressure dependence.

For the solid carbon formation according to equation (4) the gas volume ratio is 2, and for the dissolution the ratio is 0.5. This volume effect leads to "better" dissolution at high pressures and "better" deposition at low pressures. In addition to the higher reactivity of H^o the "C dissolution" would be even higher according to equation (5). In "super-equilibrium" envirements diamond deposition is possiby independent of the gas pressure.

Thermodynamically carbon deposition is possible up to high pressures. The amount of the carbon deposition at a given temperature, pressure and carbon concentration in the gas phase is regulated by the corresponding equilibrium conditions.

Due to the kinetic control of the reaction, conditions such as stability, lifetime of the atomic hydrogen, reaction rates during the solid carbon redissolution, etc., become dominant and override the global thermodynamic considerations.

The available experimental facts known sofar show that diamond synthesis at moderately low pressures (around 100 torr) is relatively easy, but that too low pressures may become detrimental /43/. One of the reasons for this is certainly the decreasing growth rate at lower pressures.

At "higher" pressure ranges the synthesis with flames and other hot gas sources operates surprisingly well, at least up to atmospheric pressures. As yet little is known about the upper pressure limits. In principle it seems to be possible to increase the pressure range above one atmosphere /63/. It could be an interesting challenge to explore this range further in future work.

Another phenomenon occurring during diamond deposition is the so-called atomic hydrogen wind /35/, based on equation (3a). For example, when using a hot filament for hydrogen activation the volume increases due to the formation of 2 H^o from one H_2 molecule, while it decreases on the substrate surface by a factor of two when it is retransformed. This leads to a "wind" between the source and the sink, which has no influence on the pressure dependence of the global reaction, but should influence the gas convection.

3.3. Nucleation on Virgin Substrates

3.3.1. Introduction

If nucleation were independent of the substrates, the weight increase during the nucleation period would be identical for all substrates. Experiments showed however (Fig.4) /64/ that the weight change during diamond deposition depends on the substrate used. Only after the surface is completely covered with diamond does the weight increase with time become constant.

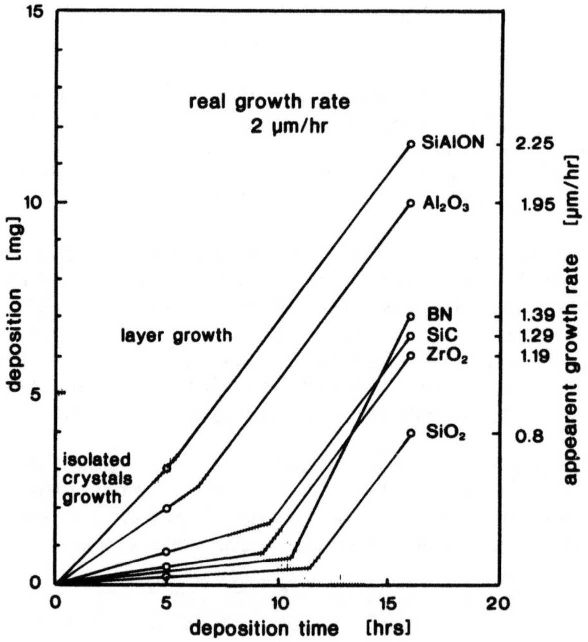

Fig.4. Nucleation and growth of low-pressure diamond on different substrates /64/

Not only the weight change but also the nuclei number, as seen by SEM, and the time to obtain full coverage of the surface (= layer formation) depend strongly on the chemical nature of the substrate and its surface condition (see Section 4.) /20,53,64,65/.

There are several reasons for this. The C oversaturation needed for heterogeneous nucleation to occure should depend on the chemical nature of the substrate. Or if, for example, the substrate surface cannot nucleate the diamond itself, but must first undergo a chemical reaction to produce a suitable basis for heterogeneous diamond nucleation, the longer incubation time is logical. For example, it takes hours for diamond crystals to become visible on the clean surface of a Si wafer, while nucleation on single-crystalline SiC takes place rapidly /64/. As known for the refractory metals /55,66/ (see Section 4.), apparently Si must also form its carbide before nucleation can occur /67,68/.

3.3.2 Substrate/reaction gas interactions

The main reason for the incubation times frequently observed for diamond nucleation is a chemical interaction on the substrate surface with the synthesis gases used, i.e.:

- atomic and molecular hydrogen,
- hydrocarbides or CH_x radicals and
- oxygen, dopants, etc., (if applied).

Depending on the nature of the substrate, different chemical reactions and/or diffusion with hydrogen and/or carbon are possible.

Chemical reactions can occur with compounds such as oxides, nitrides, sulfides, carbides, etc. Fig.5 shows thermodynamic calculations made under the extreme assumptions that, in addition to CH_4, the hydrogen in the gas mixture is present either as molecular H_2 or exclusively as atomic H^o /64/.

Obviously formation of new compounds on the substrate surface can be detrimental or beneficial for the diamond nuclei formation as such. Also gas formation can strongly interfere with the nuclei formation. Both can prolong the time to develop the early growth stages (see Fig.5) and also influence the adhesion of the final diamond coating (see Section 6.).

Regarding the carbon/substrate interactions only, metals, alloys and pure elements can be classified in three groups:

1. Little or no solubility or reaction
   diamond, graphite, carbons, Cu, Ag, Au, Sn, Pb, etc.

2. C-diffusion only, whereby carbon dissolves in Me(C) mixed crystals (Pt, Pd, Rh, etc.) /69/.

3. Carbide formation /70/
   metallic: Ti, Zr, Hf,
   V, Nb, Ta,
   Cr, Mo, W, etc.
   Fe, Co, Ni (metastable), etc.
   covalent: B, Si, etc.
   ionic : Al, Y, rare earth metals, etc.

Besides the surface modifications discussed above which lead to compounds promoting heterogeneous diamond nucleation, certain substrates delay diamond nucleation by acting as carbon sinks. At deposition temperatures large amounts of carbon are transported into the bulk where it forms either carbides or solid solutions.

This may lead temporarily to a substantial reduction of the carbon surface concentration, causing considerable delay in the onset of nucleation. For substrates with very high C diffusion rates their thickness becomes a relevant parameter as far as the onset of nucleation is concerned. Thin foils reach their carbon saturation more rapidly than bulky crystals /71,72/. If the carbon solubility decreases with temperature, formation of metastable carbides or precipitation of graphite (Me+C) usually occurs during cooling /69/.

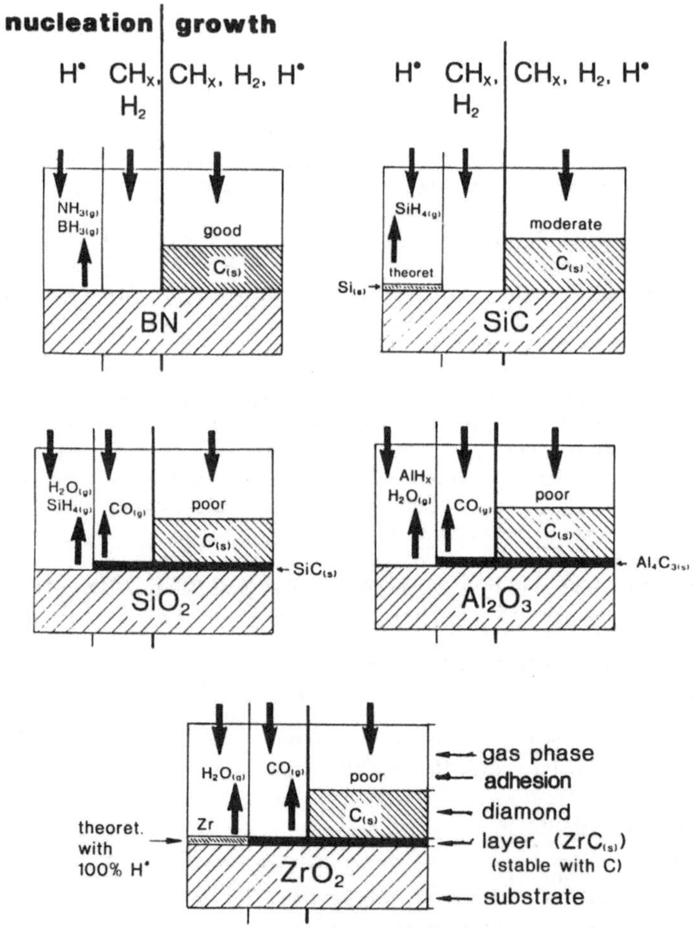

Fig.5. Chemical interactions between different substrates and the gas phase containing CH_x, H_2, H^o /64/

In the following the **case of stable carbide formation** is shown **for** the **refractory metals.** The carbon of the reaction gas leads to the formation of stable, high melting, metallic carbide surface layers during diamond deposition /73/. The C diffusion rate through the carbide layers determines, in this case, both the carbide growth rate and the diamond nucleation rate (Fig.6).

During the early stages the metal carburization uses all available carbon to form the carbide, and the minimum C surface concentration required for diamond nucleation cannot be reached on the pure metal. However, with increasing carbide layer thickness the carbon transport rate slows down and the C surface concentration increases, until the critical level for diamond nucleation is reached (Fig.7).

Hf and Ta carbides have very low C diffusion rates and form diamond nuclei rapidly. During the following growth step the number of growing diamond particles remains practically constant, i.e. no new nuclei are formed after the initial diamond nucleation period /55/.

On niobium carbide, which has an intermediate C diffusion coefficient, in certain CH_4 concentrations and pressure ranges a very striking observation was made. Tiny diamond crystals already formed dissolved again. The nuclei number decreased with time while the size of the remaining crystals increased (Fig.7). These observations can be explained as follows:
The amount of C transported per unit time into the metal depends not only on the carbide layer thickness, temperature and diffusion rate, but also on the C concentration at the carbide surface. Fig.8 shows the C concentration profiles at the substrate surfaces before and after diamond nucleation. The surface carbon concentration increases from about 0.5% at the gas/carbide interface to 100% at the diamond/carbide interface. This means that underneath the diamond crystals, the carbon transport rate will be strongly increased, leading to a permanent carbon dissolution at the diamond/carbide interface.

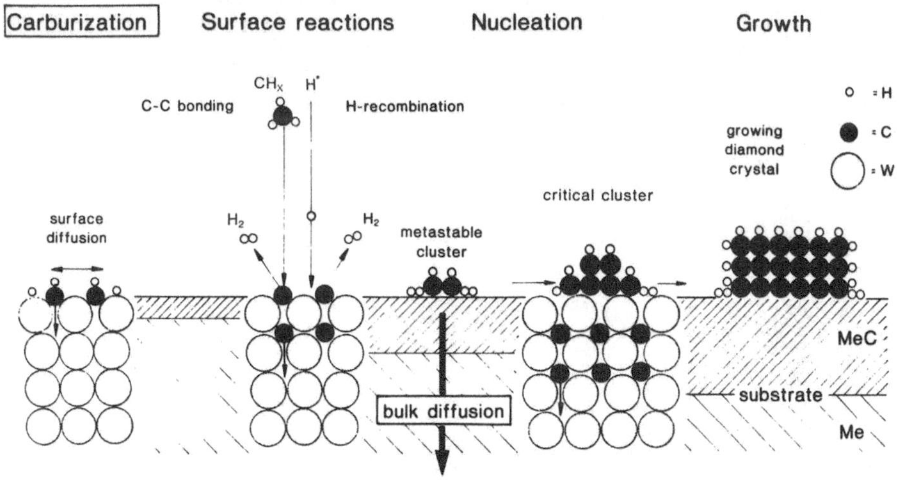

Fig.6. Nucleation and growth kinetics during low-pressure diamond deposition on a carbide-forming substrate surface

A tiny diamond crystal can survive on the substrate surface only as long as the C surface diffusion feed rate is higher than the dissolution losses at the interface. With increasing carbide layer thickness and a constant carbon flux to the surface, the surface carbon concentration will increase. At the same time the areas of surface diffusion will increase and start to overlap. If neighbouring particles exist very close to each other the feed rate of a particle can fall below its dissolution rate. This interplay between high and low surface diffusion and bulk diffusion rates during nucleation at the early crystal growth stage is illustrated in Fig.9.

TiC has the highest C diffusion rate of the refractory metal carbides and its dissolution tendency must be even greater than for the carbides of Hf, Ta, Nb, Mo and W. Contrary to NbC on TiC fewer diamond nuclei formed at the early growth period. Since only the evolution of a large population can be observed, but not the lifetime of an individual particle, the very high dissolution rate of the TiC/Ti substrate only becomes apparent through a steady increase of the surviving diamond crystals until a diamond layer covers the carbide surface. The relatively high diffusion rate of C in the TiC leads, -- contrary to the Hf, Ta, Nb, Mo and W cases --, to a continuous formation of nuclei with time /55/.

Although in all cases the dissolution will continue at the diamond/carbide layer interface as long as the substrate is heated during the deposition, it becomes slower and slower with increasing carbide thickness. The carbon diffusion at the diamond/substrate interface most likely also has sometime negative consequences for the interface bonding and adhesion. This has however not yet been investigated in detail.

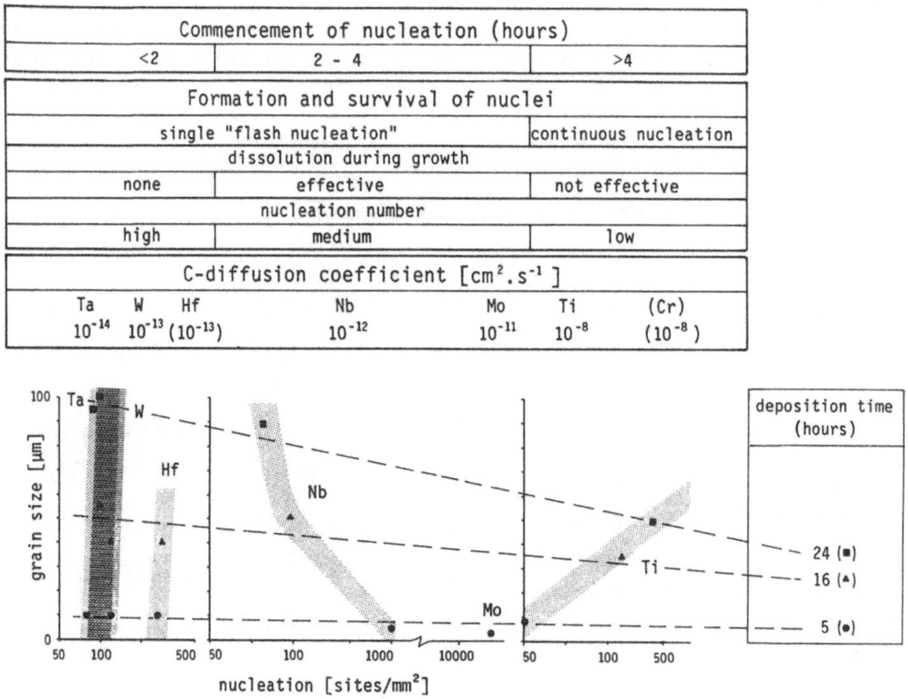

Fig.7. Deposition of isolated particles on Ti, Hf, Nb, Ta, Mo and W Grain size versus nucleation number and growth time /55/

4. NUCLEATION ENHANCMENT OF DIAMOND BY SURFACE MANIPULATIONS

Many research groups were using highly polished, defect-free, single-crystalline Si wafers as substrates. Silicon's perfect lattice seemed to be the ideal substrate for fundamental studies without unwanted side effects. Since its lattice is similar to that of diamond it was also of interest for heteroepitaxial single-crystalline diamond growth. However, on these surfaces diamond nucleation occurs readily only along scratches or sharp fracture edges. Long exposure times are required before the smooth Si surface starts to transform into SiC, and apparently only after this occurred did the diamond crystals start to grow /67/.

The scratching and polishing of Si wafers with diamond grits or other hard particles, such as SiC, B_4C, c-BN, etc., produces a spectacular nucleation promotion /50,65,74-77/. But also precoating with diamond-like carbons /77/ or even cracked pump oil /78/ were reported to promote diamond nucleation on Si wafers, as do fingerprints on refractory metals /79/. The importance of such miscellaneous "carbon sources" seems to confirm recent observations that, at the very onset of the diamond nucleation, carbon or graphite, rather than diamond is first deposited /71,72/.

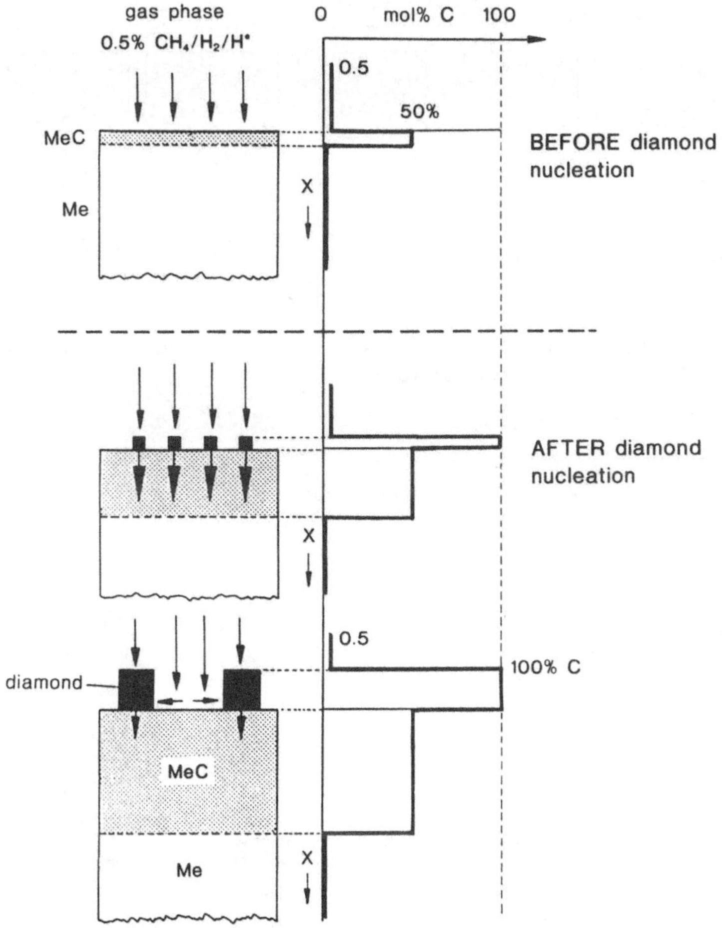

Fig.8. Carburization of refractory metals and C-concentration at the solid/gas interphase during diamond deposition /55/

Nucleation enhancements by surface treatments are however neither specific for diamond nucleation nor for Si substrates. Various sintered oxides and nitrides /53/ as well as refractory metals (Fig.10,11,12) /55/ showed analogous enhanced diamond nucleation as well as rapid film formation after polishing or scratching with diamond, SiC, Al_2O_3 and other particles /53/. As shown in Fig.12 /53/ the treatment of various refractory metals with different coarse and fine hard materials produces different nucleation enhancements. Scratching with coarse particles, leading to increased surface roughness, increased the nuclei number.

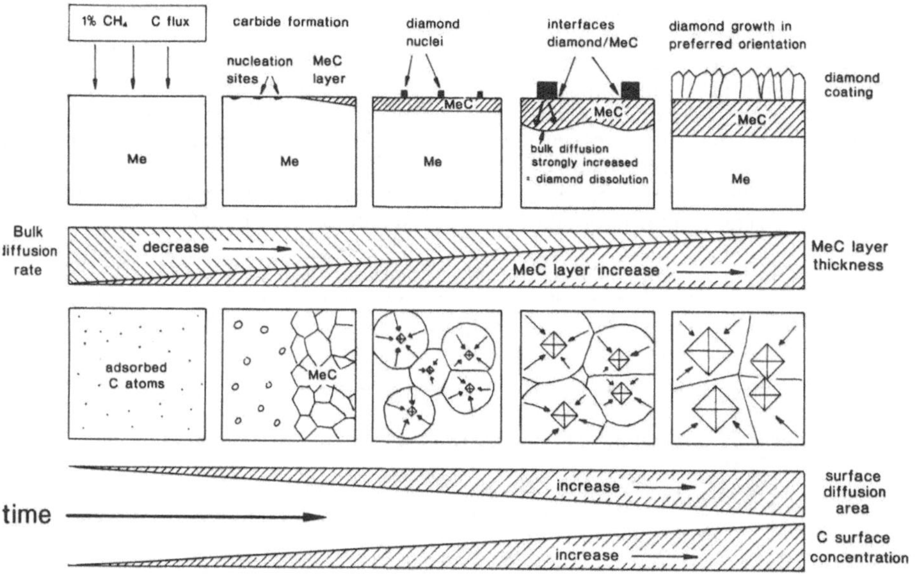

Fig.9. Important sequential steps in the formation of a diamond coating on refractory metals

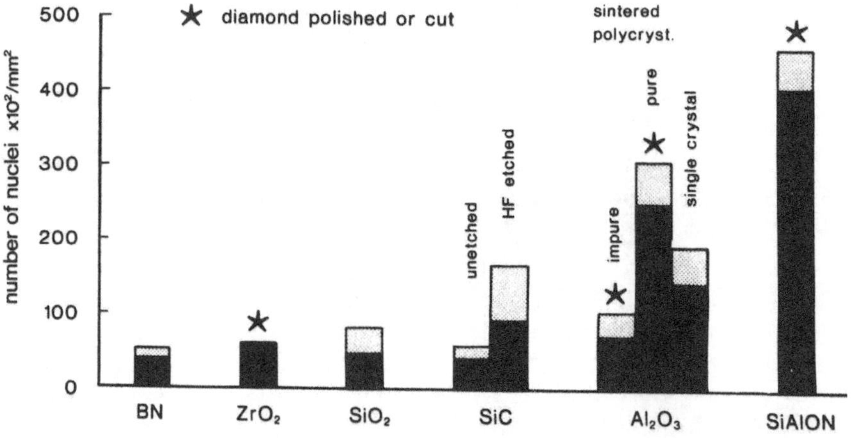

Fig.10. Diamond crystal density on various substrates after 3 hrs deposition (BN: 5 hrs) /64/
(0.5% CH_4/H_2, 20 Torr, $T_{Fil.}$=2000°C, $T_{sub.surf.}$=1050±20°C)

Polishing with 1 µm Al_2O_3 particles does not enhance nucleation, while polishing with 1 µm diamond did enhance diamond nucleation on Nb,Ta and W substrates. Mo was however an exeception: Only the 7 µm polish/scratch was effective. As Mo_2C has the highest C diffusion rate of the metals examined in this study it seems possible that very small remaining fractured diamond particles were dissolved in the Mo before diamond deposition could occur (Fig.12) /53/. In fact heat treatment applied during the deposition process could facilitate this dissolution. But whether the surface roughness, grooves or the defects created by scratching, or the particles themselves, or even tiny amounts of reaction products locally formed during the scratching are the actual nucleation promoters is still a controversial matter.

The comparison of different polishing and grinding materials and their interaction with different substrates should one day yield valuable information about the mechanism involved. Fig.13 shows different possibilities. Thin corners are prone to be carburized more rapidly than a flat surface, which could explain the early nucleation on exposed groove walls or fractured surfaces. Remaining fragments of the grinding materials are another possible nucleation source, as the critical homogeneous nuclei size for fractured diamond particles to act as nuclei can be expected to be rather small (see Section 3.3).

Although at present a complete picture is not yet available as to which materials would favor diamond nucleation most, the following preliminary conclusion can be drawn:
Diamond seems to provide the best nuclei. Heterogeneous growth can apparently take place on c-BN /65,80,81/. Heterogeneous nucleation was observed on the metallic (refractory metal) carbides, and some covalent carbides (SiC, B_4C) seem to have a positive influence. Little is known yet about ionic carbides (Al_4C_3, liquid salts, etc.).

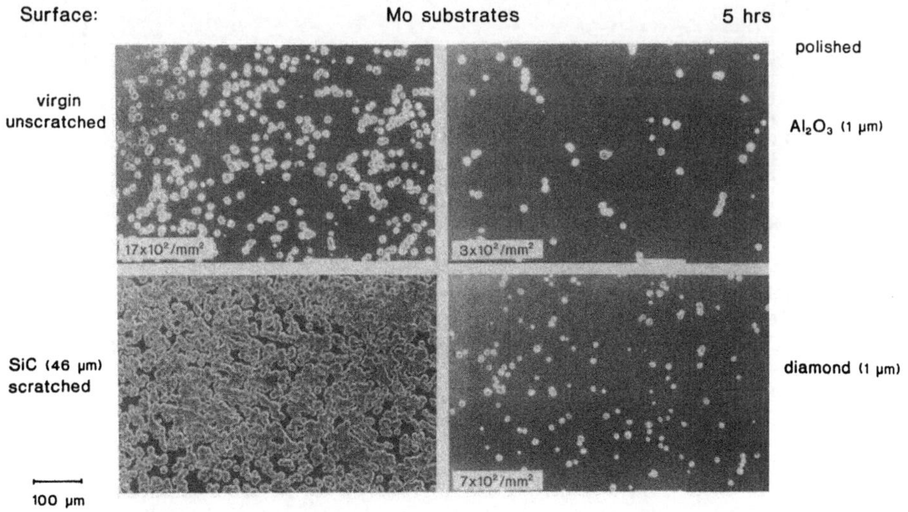

Fig.11. Influence of substrate surface treatment on low-pressure diamond deposition /53/
(2.1% CH_4/H_2, 80 mbar, 1250 W microwave, $T_{sub.surf.}=1000\pm20°C$)

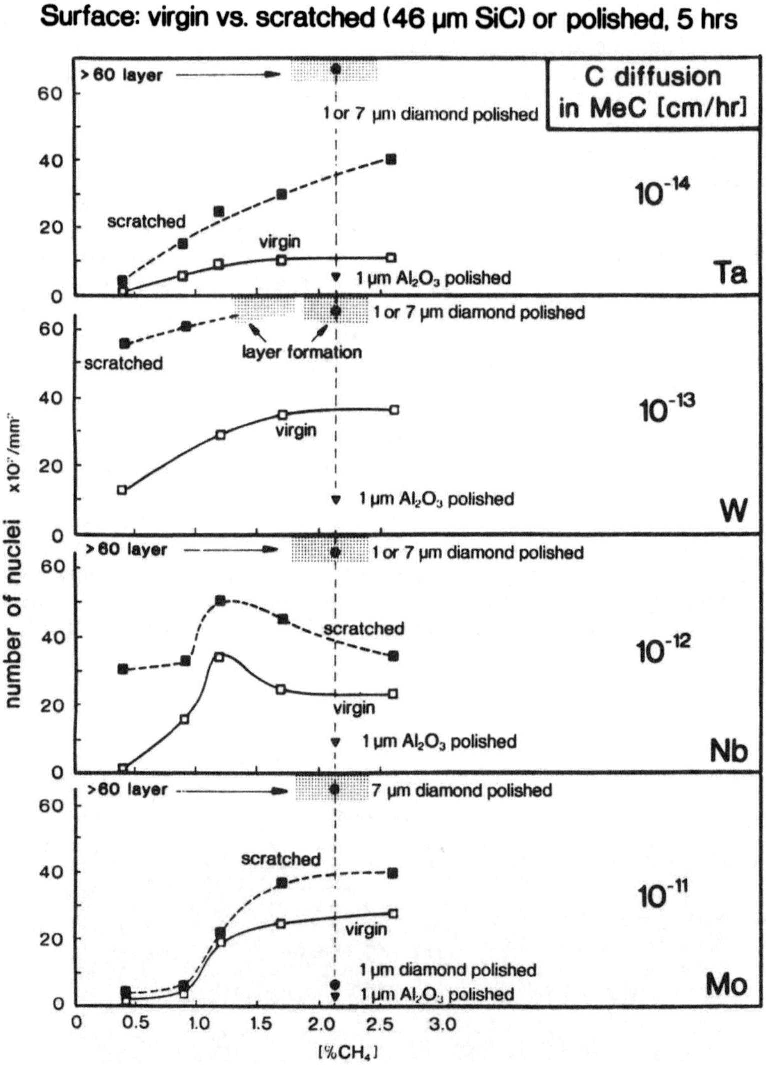

Fig.12. Comparison of different surface treatments on diamond nucleation and growth using refractory metals /53/ (2.1% CH_4/H_2, 80 mbar, 1250 W microwave, $T_{sub.surf.}=1000\pm20°C$)

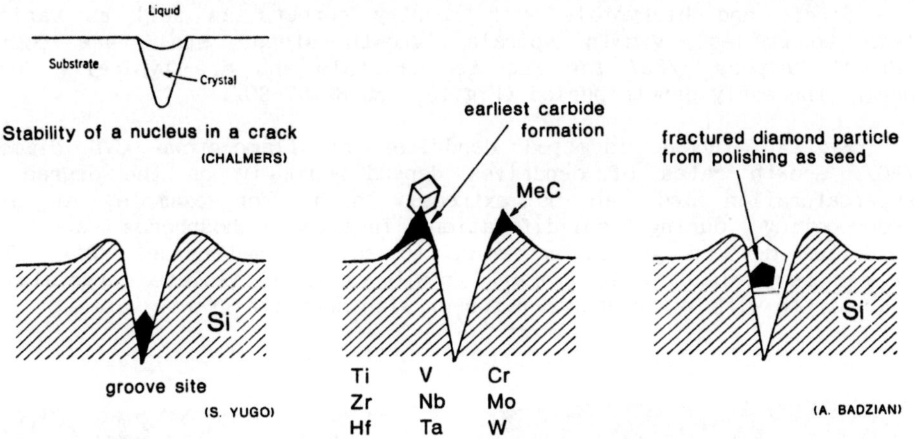

Fig.13. Different mechanisms for diamond nucleation on scratched substrates /19,50,51/

5. DIAMOND GROWTH

5.1. Nucleation Number and Layer Growth Rates

Fig.4 shown above indicates that for the coating formation two growth regimes must be distinguished. During the early growth period the nuclei grow to isolated crystals, which first join by growth to form larger islands and then finally form a continuous film completely covering the substrate surface. The initial increase in size of the isolated crystals and islands is due to abundant C surface diffusion from the relatively large diamond-free area surrounding them /82/. After layer formation, the increase in thickness is proportional to the incoming carbon flux however with a very low carbon yield. The time needed to form a continuous film depends not only on the growth parameter chosen but also on the number of nuclei formed. As outlined above, the substrate and its surface conditioning are very important additional parameters for this. Usually the growth rate of the film itself is independent of the substrate.

5.2 Growth Features of CVD-Grown Facetted Crystals and Coatings

As for many other materials, CVD diamond growth is characterized by a facetted crystal formation. SEM pictures illustrate the high degree of similarity with typical growth features occurring during the growth of other facetted crystals /83/.

Scratches are preferred nucleation sites for diamond growth, for example for CVD Al_2O_3 on sapphire (Fig.14) /84/. Five-fold symmetries are common not only in diamond (Fig.15) /20,59,66,85/. Since this symmetry apparently never develops on homeoepitaxially grown crystals /86/, it could perhaps be used to determine whether scratch-induced diamond growth originated from fractured diamond particles or not.

Single and bicrystals with reentry corners as well as various twin boundaries, growth spirals, growth steps, etc., are common "growth helpers" /83/ for facetted crystals and are typically found during the early growth period (Fig.16) /50,80,87-89/.

Fig.17 shows facetted dendrites of flame-grown CVD diamond /90/. Growth rates of dendrites depend strongly on the degree of supersaturation and can be extremely high. For example, at high undercooling during solidification facetted phosphorus and tin dendrites develop growth rates up to several meters/sec /91/. Thus it can be misleading to compare growth rates of isolated crystals on a substrate with layer growth rates or with dendrite growth rates.

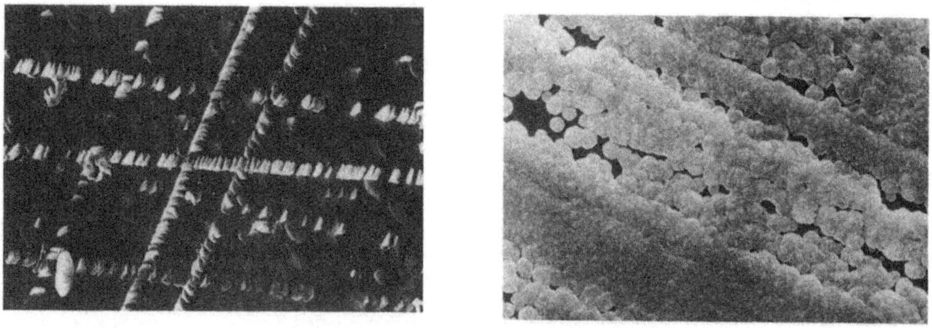

CVD Al_2O_3 10 μm
Al_2O_3 single crystal substrate

CVD diamond 10 μm
SiAlON substrate

Fig.14. Examples of preferred CVD nucleation on scratches and crystal defects /20,84/

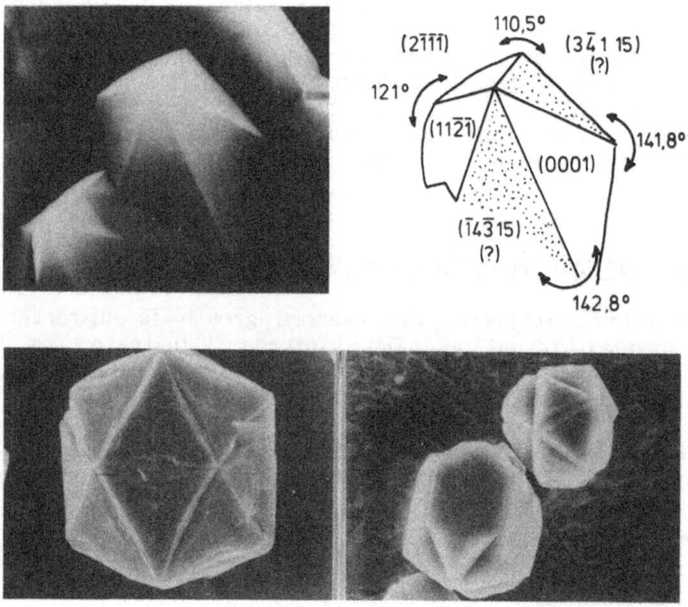

CVD Al_2O_3

1 μm

CVD diamond

10 μm

Fig.15. Five-fold symmetries formed during Al_2O_3 and diamond deposition /55,59/

Spherulitic "ballas-type" CVD-diamond crystals (sometimes called ball-shaped or even cauliflower-like) are frequently formed under non-optimal growth conditions during the low-pressure synthesis. "Ballas diamonds" are spherulites grown as a polycrystalline crystal and found in nature. They are inexpensive and often used for grinding. Also "carbonados" are found in nature but are sintered agglomerates with silicate and other inclusions /92/. Spherulites are polycrystalline, spherically grown crystals which can form from the gas phase, from liquids or in solids. A major industry, "Nodular Cast Iron", is based on the controlled spherulitic graphite formation in castings. In graphite spherulites the c-axes of all graphite crystals point to the center of the sphere /93,94/. Fig.18 shows the well-known corresponding diamond structures which frequently are hemispheres if grown on a substrate /95/ together with a hemispherical Si spherulite, grown from a hypereutectic Al/Si/Zn melt /96/. Some CVD-grown ballas diamonds have a star-like dendritic structure, with diamond branches and the spaces in between filled with "carbons" which can be readily removed by oxidation /97/.

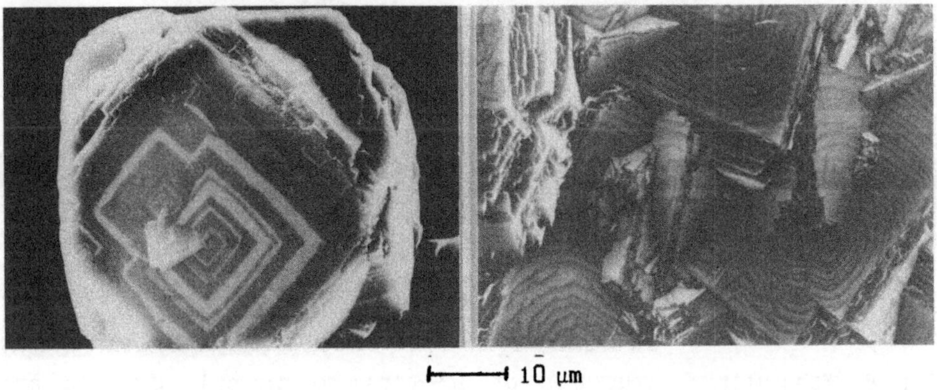

Fig.16. Growth spirals and growth steps formed during diamond growth /80,88/

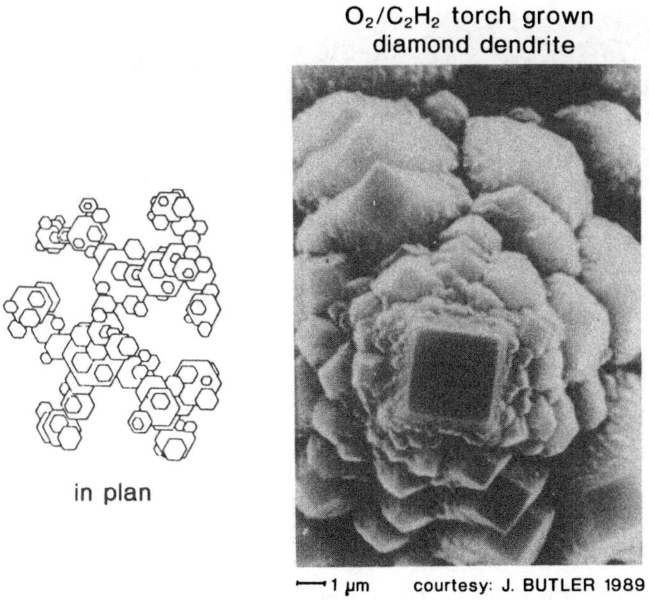

Fig.17. Faceted dendritic diamond growth

5.3 Formation of Layers and Coatings

The isolated crystals of the early growth stages (see Fig.3) form films and layers by the mechanism described above for facetted crystals. The preferred growth directions lead to "columnar" structured layers, well-known from such examples as pyrolytic CVD carbon, graphite and SiC /98/ and CVD coatings on tools (Al_2O_3, TiC, etc.) /41/ (Fig.19). CVD diamond can also develop columnar growth, as shown on a fracture surface in Fig.19. Recently such growth was simulated by a computer model /99/.

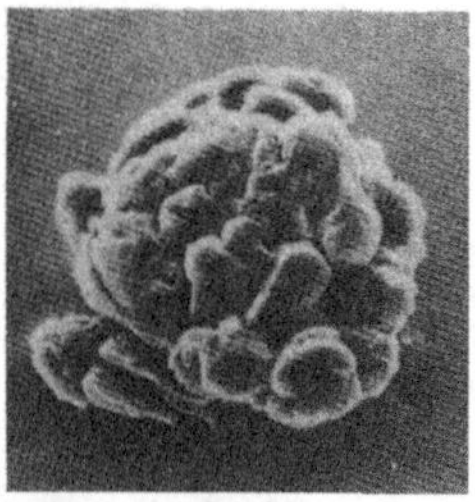

graphite in cast iron 10 µm

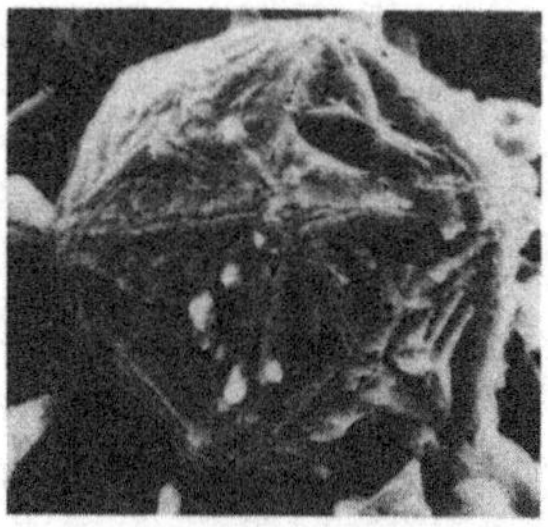

Si from a Al/Si/Zn melt 10 µm

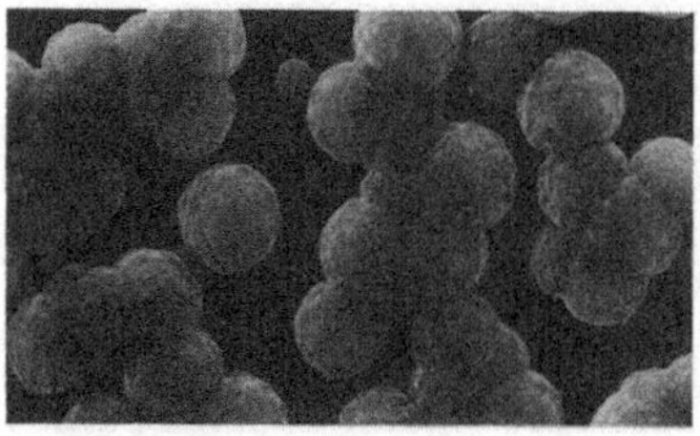

low-pressure diamond
10 µm

1 µm

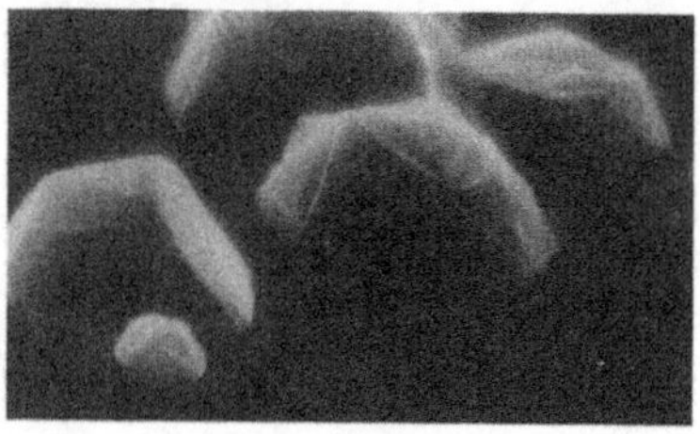

Fig.18. Spherulitic structures of low-pressure diamond, graphite and Si /41,94,96/

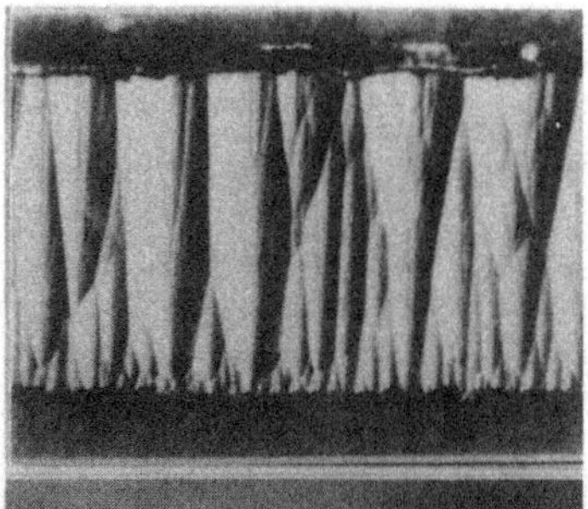

CVD-SiC
1 cm

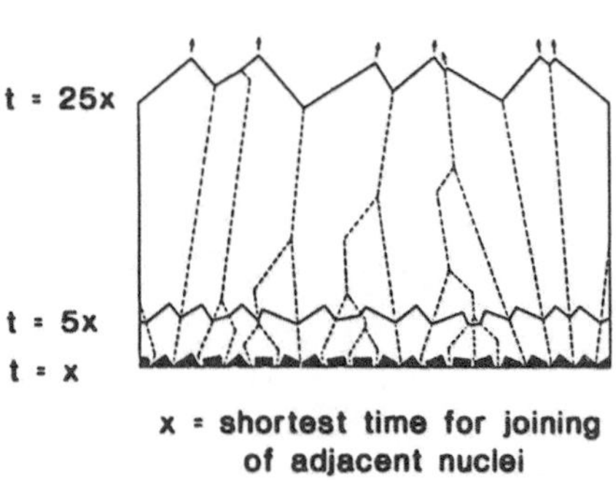

Low-pressure diamond
10 µm

Fig.19. CVD-SiC /98/ and low-pressure diamond /95/ columnar layer structures

Homeoepitaxial CVD Al_2O_3 and diamond single-crystal growth is shown in Fig.20 /84/. On the $(10\bar{1}0)$ facet of a sapphire crystal steps developed, as the cut was slightly disoriented. Growth in the (0001) direction is enhanced by screw dislocation defects in the sapphire, leading to growth spirals. Homeoepitaxial diamond single-crystal growth should look quite similar. It allows not only the study of the growth mechanism, but growth rates of specific crystallographic orientations can also be easily determined /84,100/. Healing of imperfect diamond powders is shown in Fig.21 /20,66/.

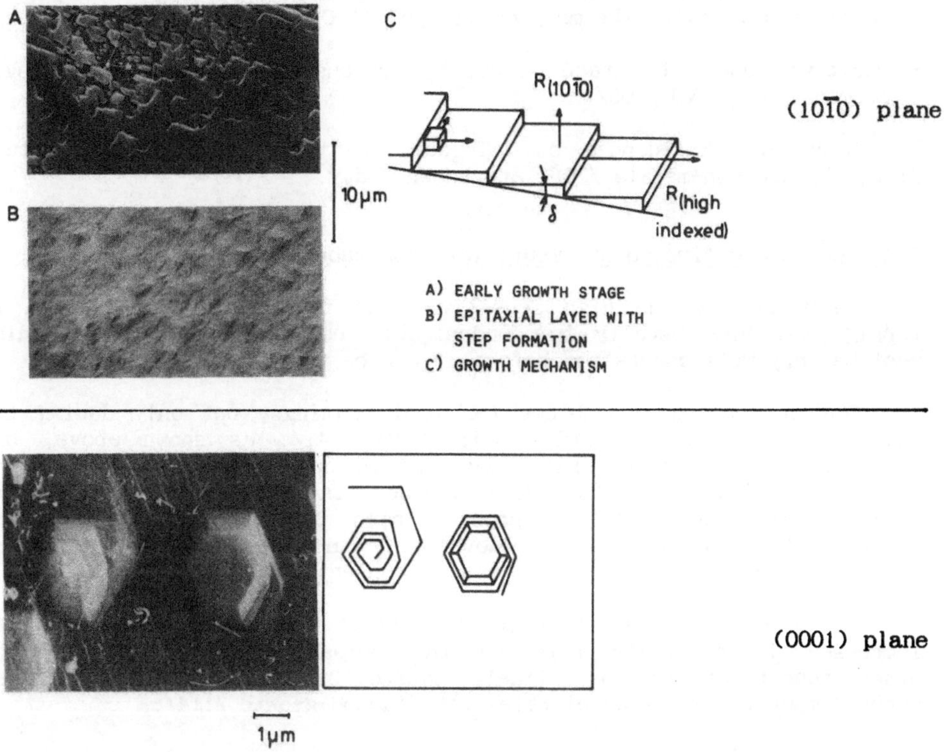

Fig.20. Epitaxial CVD growth of α-Al_2O_3 on different sapphire planes /84/

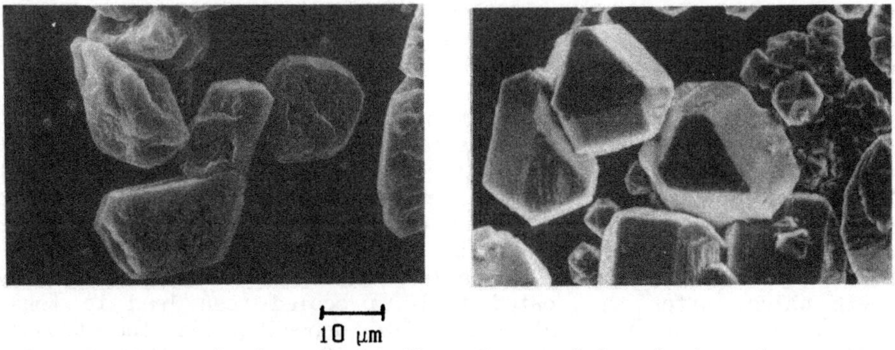

Fig.21. Healing of imperfect diamond crystals during low-pressure diamond deposition (substrate Ta) /66/

5.4. Impurity-Induced Changes in Facetted Crystal Growth

5.4.1 Changes in crystal habits and morphologies

It is well known that small amounts of impurities can strongly influence the CVD growth of facetted crystals. For example:

- During industrial Al_2O_3 coating small amounts of sulphur are routinely used today in large CVD reactors to control the uniformity of the Al_2O_3 layer thickness of hundreds of specimens /101,102/.

- Traces of Co diffusing out of cemented carbide substrates can detrimentally alter the morphologies of Al_2O_3 coatings /103,104/.

- Positive use of trace element actions is made for whisker production by CVD /105/.

Many such influences on CVD due to traces of various metals /106,107/ and non-metals /108/ are known today.

5.4.2 Outward diffusion of impurities from substrates

Unfortunately the high sensitivity of facetted crystal growth to foreign elements usually has undesirable consequences. Its mechanism must be very well understood before it can be applied beneficially.

One of the early problems for tool coatings, not only during the coating of cemented carbides with Al_2O_3 /41/, as shown above, but also during diamond coating, was the outward diffusion of Co, the binder normally used for the carbides in hard metals /109-111/. Various solutions to the problem have been published, including intermediate layers /112/, removal of the excess Co by etching /110/, and using substrates with low Co contents /109,113/.

An example of a "beneficial" effect is seen in Fig.4. The increased layer growth rate on the BN-substrate relative to the other substrates is most likely caused by an "outwards" diffused boron "impurity" which accelerates the diamond growth /114/.

6. ADHESION OF DIAMOND COATINGS

Adhesion is strongly associated with interfacial stresses, depending mainly on the different expansion coefficients of layer and substrate. Interface stresses and stresses within the layers were determined recently /115,116/.

Surprisingly X-ray measurements /117/ revealed tension in diamond layers on SiAlON substrates, where -- based on room temperature expansion coefficients -- compression would be expected /20,64/. The substrate surface temperatures during deposition and thus the interfacial temperatures during the diamond formation are however different from the bulk temperatures. This results in a thermal gradient at the diamond/substrate interface during layer growth which, after the coated tool is cooled, can lead to tension or compression at the interface /20/ depending on the expansion coefficient of the substrate. Fig.22 shows the expansion coefficients of various substrates and the range for which neither tension nor compression can be expected for a substrate surface growth temperature of 1050°C /20,64/. Fig.23 gives the theoretical substrate surface growth temperature needed to obtain a tension/compression-free interface after cooling (T_{bulk}=800°C) /20,64/. The experimentally observed adhesion is also given for several substrates. The experimental results support the assumption that if tension is acquired adhesion is good. As well known for the Al_2O_3 coatings on hard metals, cracking of the coating provides good stress relief /41/. In agreement with the theoretical predictions, it was shown recently that lowering the surface temperature during the diamond deposition of a cemented carbide tool, which has a slightly higher expansion coefficient than diamond, can significantly improve the diamond coating adhesion /118/.

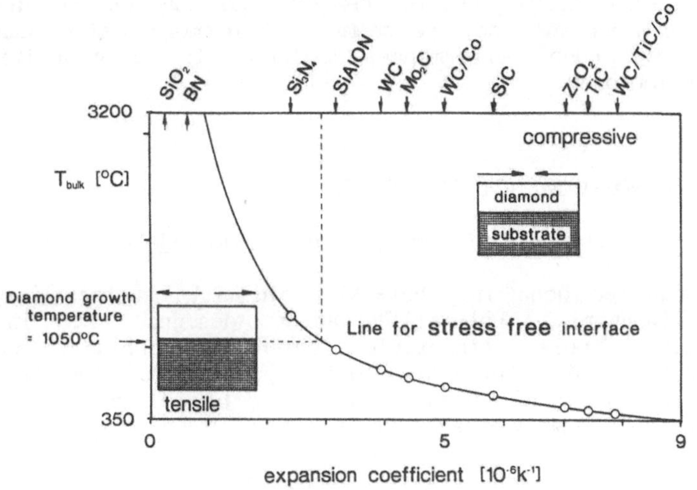

Fig.22. Influence of temperature and expansion coefficients on the interfacial stress an constant substrate surface temperature of 1050°C /20,64/

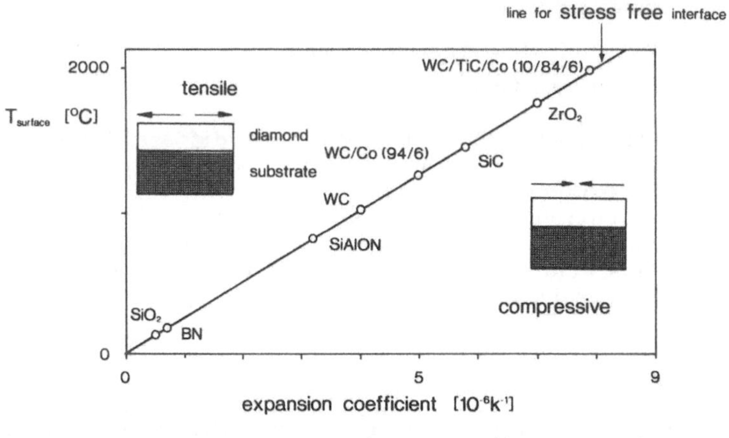

Fig.23. Theoretical substrate surface growth temperature needed to obtain a tension/compression-free interface after cooling a substrate with bulk temperature of 800°C /20,64/

In Fig.5 above a number of different substrate interactions deduced from thermodynamical calculations are given. Formation of gaseous species or solid intermediate layers not only influences the nucleation but also the adhesion of diamond layers. The exception, SiO_2, which forms an intermediate SiC layer (Fig.5), can be explained by a surface reaction, whereby the expansion mismatch occurs not at the SiC/diamond, but at the SiO_2/SiC interface.

A reasonable match of the expansion coefficients and the mastering of substrate reactions alone are however not sufficient to meet the high degree of adhesion absolutely necessary to satisfy for example a wear coating application for tools /57,115,118/. At least some local chemical bonding between the diamond coating and the substrate surface also seems to be necessary. Only through chemical bonding of the Al_2O_3, TiC or TiN CVD coatings on cutting tools can they perform as they do. It seems that diamond-coated substrates can also achieve such performances although the reproducibility still has to be improved.

7. TODAY'S INDUSTRIAL CVD TECHNOLOGY

7.1 Design and Products of "Normal" Industrial CVD Reactors

CVD can be done in hot- or cold-wall reactors (Fig.24) /41/. Typical industrial hot-wall reactor designs are in use for wear/cutting tools (indexible cemented carbide inserts) and semiconductors. The Van Arkel iodide process for the fabrication of pure "iodine Ti/Zr/Hf" uses a typical cold-wall reactor, as do most low-pressure diamond syntheses /119/.

7.2. Convection

Gas flows and convection streams are important for the uniformity of layer formation. Convection is easier to control in upstream reactors than in downstream reactors /120/. Most hot-filament diamond reactors are however downstream reactors. Small ones function well though -- despite the enormous temperature gradient between filament and substrate -- because atomic hydrogen diffuses over large distances /35/.

7.3. Boundary Layers and Coating Thickness

During CVD growth the reactants must diffuse through boundary layers. Different types of boundary layers and the reasons for their formation are summarized in Table 2 /121/. The growth kinetics of CVD layer formation are closely linked with this boundery layer transport mechanism in conjunction with the specific chemical reaction kinetics (Table 3) /121/. For example: To regulate the thickness of the coating independently of the specimen's position in the reactor or its shape, a surface-controlled reaction mechanism is ideal /122/. In Fig.25 two examples of the extent to which the growth can be influenced are given /123/.

Both the proper CVD parameters and the CVD system-specific reaction type (which depends on the nature of the chemical reactants involved) must be chosen.

Table 2. Various transport processes and boundary layers near the substrate surface during CVD deposition /121/

BOUNDARY LAYERS	MASS TRANSPORT PROCESSES	DRIVING FORCES GRADIENT ACROSS BOUNDARY LAYER
VELOCITY	STEFAN FLUX	velocity or gas volume changes during chemical reduction
NERNST or CONCENTRATIONE	FICK-DIFFUSION ADVECTIVE FLUX	concentration density
THERMAL	SORET-DIFFUSION THERMAL DIFF.	temperature gradient

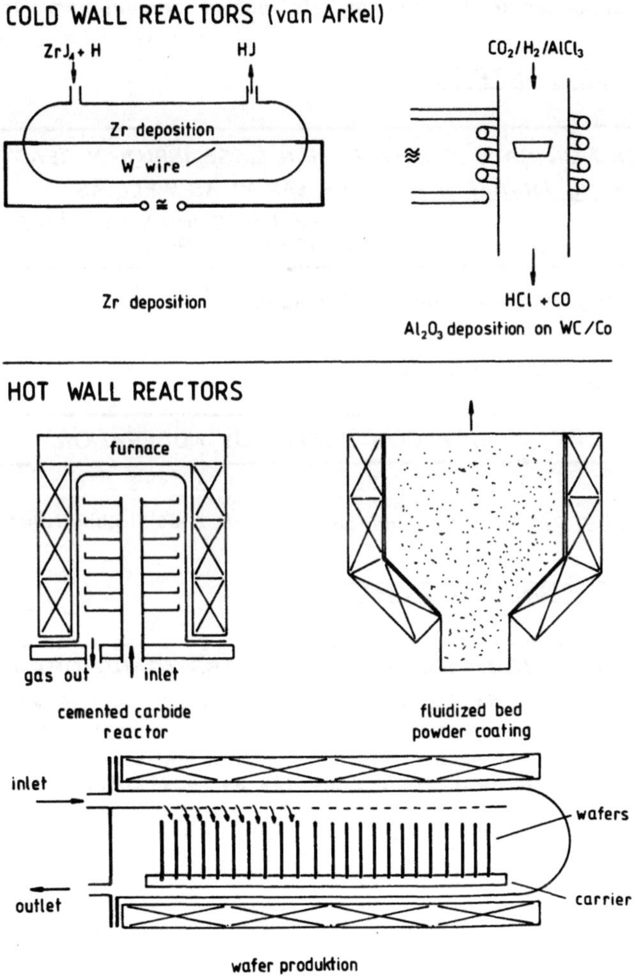

Fig.24. Principles of technologically used CVD reactors /41/

Table 3. Rate-controlling mechanisms for CVD depositions /121/

1) THERMODYNAMIC CONTROL:

DEPOSITION RATE EQUAL TO MASS INPUT INTO REACTOR
(corrected for the yield of the process)

Typical for extrems in: low gas flow, low concentration, low pressure, high temp., etc.

Temp. dependence of deposition rate follows thermodynamic laws.

2) SURFACE CONTROL:

DEPOSITION RATE IS LOWER THAN MASS INPUT IN REACTOR AND MASS TRANSPORT: IN THE VAPOR AS WELL AS TO/FROM THE BOUNDARY LAYER OF THE SUBSTRATE SURFACE

Typical for: complex geometrical shapes

3) NUCLEATION CONTROL:

RATE CONTROLLED BY NUCLEATION

Typical for: extremly low supersaturation or special substrates.

4) MASS TRANSPORT CONTROL:

RATE CONTROLLED BY THE MASS TRANSPORT RATES TO/FROM THE SUBSTRATE SURFACE IN THE VAPOR

Typical for: high pressure, high temperatures and with **very instable compounds as reactants!**

Surface controlled growth (Nu < 1):

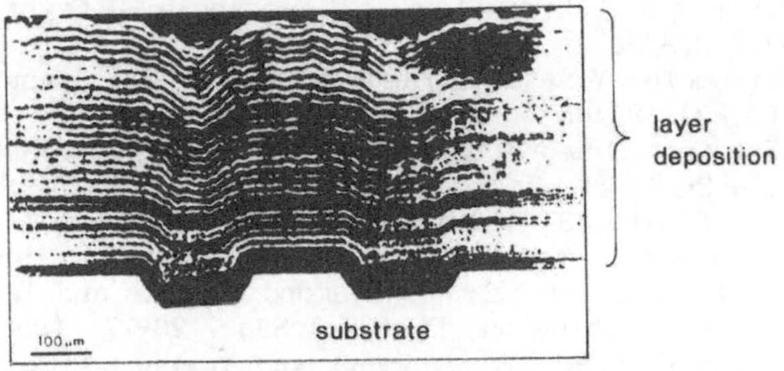

Gas diffusion controlled growth (Nu > 1):

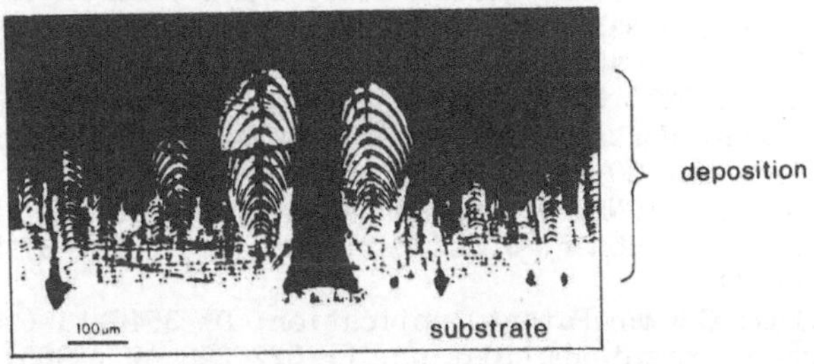

Fig.25. Surface- and diffusion- controlled CVD deposition of polycrystalline Si from $SiHCl_3/H_2$ mixtures /123/

Additional and specific problems exist for diamond CVD, such as:

- the convection is strongly influenced by the atomic hydrogen wind /35/;
- in hot-filament reactions the Sorret effect can become unusually important /121/;
- the metastable nature of the atomic hydrogen imposes design problems for reactor upscaling, etc.

8. ACKNOWLEDGEMENTS

The authors would like to thank the Austrian "Fonds zur Förderung der Wissenschaftlichen Forschung" which is sponsoring diamond work at the Technical University of Vienna (Project P6031, P7274).

Thanks are due also to the management of Sandvik Stockholm, particularly to the Hard Materials and Coromant divisions, which have sponsered research work for many years.

The authors would like to express their gratitude to the students and to all the other members of the research team at the Technical University who have been so enthusiastically studying the problems of diamond deposition all these years.

Special thanks go to DI.Angela Lindlbauer for her help in listing the literature for this work.

9. LITERATURE

1. B.V.Spitsyn, L.L.Bouilov, B.V.Derjaguin: J.Cryst. Growth 52 (1981) 219/26
2. S.Matsumoto, Y.Sato, M.Kamo, N.Setaka: Jpn.J.Appl.Phys. 21 (1982) 183/85
3. 1st Int.Conf.Diamond Science and Technology, (JNDF Tokyo) Okt.24-26, 1988, Tokyo, Japan
4. Proc. of Nato ASI (Advanced Study Institute) Il Ciocco, Castellveccio Pascoli, Italy, July 22-Aug.3, 1990, Plenum Press
5. 2nd Int. Conf. on the New Diamond Science and Technology., ICNDST-2, Washington, DC (USA) Sept. 23-27, 1990
6. 1st Europ. Conf. on Diamond and Diamond-like Carbon Coatings, DIAMOND FILMS 90, Sept.17-19, 1990, Crans-Montana, Sitzerland
7. S.Matsumoto, Y.Matsui: J.Mater.Sci. 18, (1983) 1785
8. M.Frenklach, K.Spear: Manu. Dept. Mater. Res. Laboratory, Pennsylvania State Univ. Park PA 16802
9. M.Tsuda, M.Nakajima, S.Oikawa: J.Am.Chem.Soc.108(9)(1986) 5708-3
10. M.Tsuda, M.Nakajima, S.Oikawa: Jap.J.App.Phys. Part 2, 26(5) (1987) L527-L529
11. J.Angus: Workshop on the Science and Techn. of Diamond Thin Films, May 20-24, 1990, Cleveland, Ohio
12. J.Harris: ICNDST-2, presentation 7.1, see Ref./5/
13. B.Lux, R.Haubner: J.Ref.Met.-Hard Mat., Vol 8, Nr.3, (1989) 158-174
14. B.Lux: German Patent Application: DE 3546113 (1985); Europ. Patent Application: EP 0226898 A2 (1986)
15. K.Komaki, M.Yanagisawa, I.Yamamoto, K.Krashida, T.Fujimaka: JNDF Tokyo, P2-11, see Ref./3/
16. J.M. Pinneo: Nato ASI, see Ref./4/
17. M.Frenklach et al: J.Appl. Phys., 66 [1] 395 (1989)
18. P.R.Buerki, S.Leutwyler: DIAMOND FILMS 90, presentation 2.3, see Ref./6/, (to be published: J.Appl.Phys. 1990)
19. A.R.Badzian, T.Badzian: Surface and Coatings Technology, 36, (1988) 283-293
20. R.Bichler, R.Haubner, B.Lux: High Temperature-High Pressure, Vol.22, 1990, pp. 99-118
21. Gmelin: Handbuch der anorg. Chemie, 8.Auflage, Kohlenstoff, Teil B, EG 14, Lief. 1(1967)
22. Winnacker Küchler: "Chemische Technologie", Band 3, pp.279 4. Auflage, Carl Hanser Verlag München, Wien 1983
23. Holleman Wiberg: Lehrbuch der Anorg. Chemie, 81-90 Auflage, Walter de Gruyter, Berlin-New York 1976
24. C.Weissmantel et al.: Thin Solid Films, 58, 101 (1979)
25. P.Koidl: 1st Int. Symp. on Diamond and Diamond-Like Films, May 7-12, 1989, Los Angeles, 237-249
26. J.Angus: Nato ASI, see Ref./4/
27. Y.Catherine: Nato ASI, see Ref./4/
28. Y.Sato: Workshop on the Science and Techn. of Diamond Thin Films, May 20-24, 1990, Cleveland, Ohio
29. Y.Sato, M.Kamo: ICNDST-2, presentation 4.1, see Ref. /5/
30. D.McKenzie, P.Swift, P.Martin, P.Gaskell: DIAMOND FILMS 90, presentation 11.1, see Ref./6/
31. W.A.Yarborough, R.Ray: Diamond and Diamond-Like Mat.Synth., Apr. 5-9, 1988, Reno
32. J.M.Pinneo: Nato ASI, see Ref./4/
33. P.Joeris, C.Benndorf: Nato ASI, see Ref./4/
34. L.R.Martin: ICNDST-2, presentation 4.3, see Ref. /5/
35. T.R.Anthony: Nato ASI, see Ref./4/
36. J.M.Pinneo: Private Communication
37. C.V.Burton: Nature, No 1869, Vol. 72, 1905, pp.397

38. K.A.Cherian: DIAMOND FILMS 90, presentation 5.4, see Ref./6/
39. B.V.Derjaguin, D.V.Fedoseev: "Diamond wrought by man", Mir Publishers Moscow, revised from the 1980 Russian edition, first published 1985
40. D.E.Patterson, B.J.Bai, C.J.Chu, R.H.Hauge, J.L.Margrave: ICNDST-2, presentation 3.5, see Ref./5/
41. B.Lux, R.Haubner, H.Altena, C.Colombier: Fischmeister/Jehn; Hartstoffschichten zur Verschleißminderung, pp.71-112, Verlag DGM, Oberursel BRD, 1987
42. B.Loisel, L.Haji, M.Guendouz: J.de Phyique, Colloque C5, suppl. ou n°5, Vol. 50, May 1989, 467-478
43. M.N.Yoder: ICNDST-2, PANEL Discusssion, 27.9.1990, see Ref./5/
44. Th.Wilke, J.Boehm: "Kristallzüchtung" VEB Deutscher Verlag der Wissenschaften Berlin, 1988
45. M.C.Flemmings: "Solidification Processing", McGraw-Hill, 1974
46. J.E.Sundgren: Nato ASI, see Ref./4/
47. H.Jehn: Fischmeister/Jehn; "Hartstoffschichten zur Verschleißminderung", pp.45-71, DGM, Oberursel Germany, 1987
48. T.R.Johansson, J.N.Lindström: J.Vac.Sci.Technol., 12 (1975) 854
49. K.Takayanagi, K.Yagi, K.Kobayashi, G.Honjo: J.Phys.E11(1978) 441
50. S.Yugo, T.Kimura, H.Kanai: JNDF Tokyo, P2-20, see Ref./3/
51. B.Chalmers: "Principles of Solidification", New York 1934
52. B.V.Derjaguin, D.V.Fedoseev, I.G.Varshavskaya: Journal de Physique, Colloque C5, suppl. to n°5, Vol.50, May 1989, 893-900
53. A.Lindlbauer: Doctoral Thesis, TU-Vienna, in preparation
54. M.Kornmann, H.Schachner, R.Funk, B.Lux: J.Cryst.Growth 28 (1975) 259-262
55. P.O.Joffreau, R.Haubner, B.Lux: J.Ref.Met.-Hard Mat., Vol.7, Nr.4, 1988, 186-194
56. C.Chatfield, J.N.Lindström, M.E.Sjöstrand: Journal de Physique, Colloque C5, suppl.au n°5, Vol.50, May 1989, 377-389
57. S.Söderberg, K.Westergren, M.Sjöstrand: DIAMOND FILMS 90, presentation 15.2, see Ref./6/
58. B.Lux, H.Schachner: High Temperature-High Pressure, Vol.10, 315-323, 1978
59. C.Colombier, B.Lux, J.Lindström: J.Ref.Met.-Hard Mat., Vol.5, Nr.4, 1986, 222-229
60. J.Lindström, K.Stjernberg: 5th EURO-CVD, Uppsala, 169-182 (1985)
61. S.S.Lin: J.Elektrochem.Soc., Vol.122, No.10, 1405-1408 (1985)
62. A.R.Badzian, R.C.DeVries: Mat.Res.Bull., Vol.23, (1988) 385-400
63. C.J.Brannon: ICNDST-2, poster presentation, see Ref./5/
64. A.Lindlbauer: Diploma Thesis, TU-Vienna, 1989
65. P.K.Bachmann, W.Drawe, D.Knight, R.Weimer, R.F.Messier: Diamond and Diamond-Like Mat. Synth., April 5-9, 1988, Reno, 99-103
66. P.O.Joffreau, R.Bichler, R.Haubner, B.Lux: J.Ref.Met.-Hard Mat., Vol.7, Nr.2, 1988, 92-97
67. B.E.Williams, J.T.Glass, R.F.Davis: SPIE, Vol.877, Micro-Optoelectronica Materials (1988) pp.56-63
68. B.E.Williams, D.A.Asbury, B.R.Stones, J.T.Glass: Nato ASI, see Ref./4/
69. W.G.Moffatt: "Binary Phase Diagramms Handbook", Vol.1-3 Genium Publishing Corporation, New York, 1984
70. H.Holleck: Fischmeister/Jehn; "Hartstoffschichten zur Verschleißminderung", pp.25-45, Verlag DGM, Oberursel BRD, 1987
71. D.N.Belton, S.S.Schmieg: J.Appl.Phys. 66 (9), (1989) 4223-4229
72. S.J.Harris, D.N.Belton, A.M.Weiner, S.J.Schmieg: J.Appl.Phys. 66, Dec.89, pp. 5353-5359
73. Kieffer-Beneschofsky: "Hartstoffe", Springer, Vienna, 1963
74. A.R.Badzian, T.Badzian, L.Pilione: 3rd Int.Conf.on Surf.Mod. Techn., pp.51-61, Aug.28-Sept.1, 1989, Neuchatel, Switzerland

75. K.Hirabayashi, K.Ikoma, Y.Taniguchi, N.I.Kurihara: JNDF Tokyo, P 2-03, see Ref./3/
76. L.Vandenbulk, P.Bou, R. Herbin: Journal de Physique, Colloque C5, supp. to n°5, Vol.50, May 1989, 177-188,
77. W.A.Yarbrough: MRS 1987, Nov.30 - Dec.5, N 3.8, Boston, MS
78. P.E.Pehrsson, A.A.Morrish: ICNDST-2, poster pres., see Ref./5/
79. B.DeVries, T.Anthony, R.Haubner: private communication
80. R.Haubner: J.Ref.Met.-Hard Mat., Vol.9, No.2 (1990) 70-77
81. H.Sei, M.Akaishi, H.Kanda, T.Osawa, S.Yamado: ICNDST-2, poster pres., see Ref./5/
82. T.Anthony: MRS 1987, Nov.30-Dec.5, N 3.8 Boston, Massachusetts
83. W.Kurz, D.J.Fisher: "Fundamentals of Solidification", Trans. Tech. Publications 1986, Switzerland-Germany-UK-USA
84. H.Altena, C.Colombier, B.Lux: Proc. of 4th EURO-CVD, Eindhoven 1983, pp. 435-443
85. R.Litos, R.Haubner, B.Lux: High Temperature-High Pressure, 1990, Vol.22, pp. 99-118
86. J.B.Schuster: private communication
87. D.R.Hamilton, R.G.Seidensticker: J.Appl.Phys., 31, 1165 (1960)
88. R.Haubner, B.Lux: Proc.3rd Int.Conf.on Surf.Mod.Technology Aug.28-Sept.1, 1989, Neuchatel, Switzerland pp.835-841
89. General Electric: Report No.72 CRD 174 June 1972
90. D.D.Saratovkin: "Dendritic Crystallization" Consultants Bureaus Transl., New York, 1959
91. M.E.Glicksman et al: J.Cryst.Growth **67**, 297 (1967)
92. R.C.DeVries, C.Robertson: J.of Mat.Sci.Letters 4 (1985) 805-807
93. B.Lux: Gießereiforschung 22(1970) Vol.4, 161-168; Vol.2, 65-68
94. B.Lux, A.Vendl, H.Hahn: Radex-Rundschau, Vol.1/2, 1980
95. R.Bichler, R.Haubner, B.Lux: 6.EURO-CVD Jerusalem, (1987) 413-22
96. I.Minkoff, B.Lux: Proc. of the 3rd Annual Scanning Electron Microscope Symposium, April 1970, Chicago, IL, USA
97. P.K.Bachmann: ICNDST-2, presentation 9.4, see Ref./5/
98. A.Auriol: privat communication
99. C.Wild, N.Herres, P.Koidl: J.Appl.Phys. (1990), to be published
100. M.W.Geis: Diamond Technology Initiative Symposium; July 11-13. 1989, Virginia, USA
101. U.Smith, J.Lindstrom: ER 0045291, 1982
102. P.Wilhartitz, M.Grasserbauer, H.Altena, B.Lux: Surface and Interface Analysis, Vol.8, 1986, 159-165
103. H.Altena, C.Colombier, A.Lebl, J.Lindström, B.Lux: Proc. of 4th EURO-CVD, Eindhoven, 1983, pp. 428-434
104. H.Altena, C.Colombier, B.Lux: Proc. of 4th EURO-CVD, Eindhoven 1983, pp. 451-458
105. H.Schachner, G.Horlaville, P.Fontaine: Journal de Physique, Colloque C5, suppl. to n°5, Vol.50, May 1989, 219-228
106. J.Peng: Doctor Thesis, TU-Vienna, 1984
107. J.Peng, M.Danzinger, B.Lux: in print High Temp.-High Pres. 1990
108. M.Danzinger: 3rd Int.Conf.on Surface Modification Technologies, pp.829-834, Aug.28-Sept.1, 1990, Neuchatel, Switzerland
109. R.Haubner, B.Lux: Journal de Physique, Colloque C5, suppl. to n°5, Vol.50, May 1989, 169-176
110. X.X.Pan: Doctoral Thesis, TU-Vienna, 1989
111. H.Matsubara, J.Kiahari: JNDF Tokyo, P 2-12, see Ref. /3/
112. H.Schachner, H.Tippmann, B.Lux, K.Stjernberg, A.Thelin: Swedish Patent 8403428 (1984)
113. K.Saijo, M.Yagi, K.Shibuki, S.Takatsu: DIAMOND FILMS 90, presentation 16.2, see Ref./6/
114. S.Matsumoto: ISPC Eindhoven (July 1985) Paper Nr.A-1-10
115. S.Söderberg, I.Reineck, K.Westergren, H.Shahani: ICNDST-2, presentation 13.4, see Ref./5/

116. M.Murakawa, S.Takeuchi: DIAMOND FILMS 90, presentation 16.3, see Ref./6/
117. O.Meier, U.Schlaak: Härtereitechnik-Bremen, unpublished measurements 1988
118. J.Oakes, X.X.Pan, R.Haubner, B.Lux: DIAMOND FILMS 90 presentation 14.14, see Ref./6/
119. R.Kieffer, G.Jangg, P.Ettmayer: "Sondermetalle", Springer, Wien/New York, 1971
120. G.Wahl: Thin Solid Films, **40** (1977) 13-26
121. J.O.Carlsson, M.Boman: Proc. 9th Int. Conf. CVD, 1984, 129-149
122. C.A.M.Mulder, W.H.M.M.Van de Spijker,J.Jensma, G.Vespui, G.H.C.Heijnen: Journal de Physique, Colloque C5, suppl. to n°5, Vol.50, May 1989, 241-248
123. Van der Breckel: Philips Res. Repts 32, 118-146, 1977

CONTROL OF TEXTURE AND DEFECT STRUCTURE FOR HOT-FILAMENT-ASSISTED CVD DIAMOND FILMS

R. E. Clausing, L. Heatherly, and E. D. Specht

Metals and Ceramics Division, Oak Ridge National Laboratory
P. O. Box 2008, Oak Ridge, TN. 37831-6093

INTRODUCTION

The morphology of diamond films grown by activated chemical vapor deposition, (CVD), varies greatly depending on the deposition conditions [1–6]. Small changes in the deposition conditions can cause major changes in the film morphology and crystal perfection. Areas a few millimeters apart in a film may be quite different unless considerable care is taken to ensure uniform deposition conditions. Because the properties of diamond films are strongly dependent upon the crystallite size, orientation, and perfection, it is important to understand the way in which growth conditions affect the film structure in order to design and optimize films for specific applications.

Nucleation and growth processes are both important factors in determining the structure and thus the properties of the films. When polycrystalline diamond films are grown thicker than a few µm, the fastest growing crystallites (those with favorable orientations) outgrow the others and grow into long parallel columns. The surviving grains have preferred orientations, and the film is said to have developed a "texture." The determination of the film texture and stresses by X-ray diffraction or other methods has been reported only infrequently [7–10].

In this paper, we describe the development of textures and discuss how the nucleation and growth processes may operate to produce the textures and internal crystallite defect structure of the films described above. We show that the textures develop as the result of growth competition and suggest that the internal defect structure is the direct result of the details of growth on {111} or {100} facets. We further show that growth conditions can be controlled to manipulate the film texture and internal crystallite defect structure. Using these concepts, a film with a <100> texture and only {111} facets was grown.

EXPERIMENTAL

The hot filament activated CVD apparatus used in this experiment has been described previously [6]. The filament consisted of a coiled tungsten wire, and in these experiments it was heated to approximately 2000°C. The filament and substrate were contained in a quartz envelope at a pressure of 40 torr. The gas composition was 1% methane in hydrogen. A flow rate of 55 std. cm^3/min was used. The substrate was a piece of a (100) silicon wafer positioned 0.5 cm below the filament. The substrate temperature was varied from 900° to 1100°C to produce the changes in morphology we sought. The temperature was measured by an optical pyrometer focused on the back surface of the silicon substrate. No corrections were made for the emissivity of the substrate or absorption in the quartz window. Textures were determined for films with thicknesses of 100 to 150 μm.

The films were characterized with scanning electron microscopy (SEM) and transmission electron microscopy (TEM) to determine the microstructure, growth rates, grain size, and morphology; X-ray diffraction to determine the texture and internal stresses; and Raman spectroscopy to determine the bond character. The techniques used for SEM, TEM, and Raman spectroscopy are described in a prior publication [6]. The X-ray measurement techniques used on these films and the resulting data will be reported elsewhere [9].

RESULTS AND DISCUSSION

Texture Development

The development of texture in vapor-deposited polycrystalline films may be considered to be the result of competitive growth between randomly oriented crystals. The crystals which are growing fastest in the direction perpendicular to the plane of the substrate grow ahead of and over the others. This leads to the development of a strongly oriented columnar polycrystalline film. This concept was modeled by Van der Drift [11], who assumed that growth occurs on all faces uniformly. The model must be modified somewhat if some crystal faces grow faster than others or if surface or bulk diffusion is important; but the final conclusion, that the texture reveals the fastest growing crystallographic direction, remains unchanged. Figure 1 shows the fractured edge of of a typical CVD diamond film. The columnar structure is consistent with a competitive growth mechanism.

In the following sections, we discuss different textures in detail and make reference to the growth processes on various crystallographic faces. We will specify textures according to the crystallographic faces exposed on the surface and the crystallographic direction normal to the substrate, which is also the growth direction. The notation is thus {111}{100}<110> for films composed of crystallites with exposed {111} and {100} facets and with <110> directions normal to the substrate. It is possible to estimate the ratio of growth rates in different crystallographic directions from the shape of individual crystallites. Figure 2 is a set of cubo-octahedral crystallite shapes and the ratio of <100>/<111> growth rates which will produce them. The calculated shapes can be compared to the shape of individual crystallites to determine the relative <100>/<111> growth rates.

Development of the{111}{100}<110> Texture. The morphology most often shown in the literature leads to this texture. Micrographs of the development and features of a <110> textured film are shown in Fig. 3. The initial nucleation on substrates, under conditions which led to the development of a <110> texture, was more or less random [10]. Figure 3(a) shows the shape of several of the individual crystallites. By comparison with Fig. 2, these crystallites appear to have had a <100>/<111> growth rate ratio between 0.9 and 1.0. There were single and multiply twinned crystallites as well as untwinned cubo-octahedra. Many of the square {100} faces of the cubo-octahedra had interpenetration twins [10]. These are apparently secondary nuclei. The twin plane is {111}. Even the exposed {100} faces of these twins often showed smaller twins on them. It is apparent that there was a strong tendency for twins to nucleate and grow. As shown in Fig. 3(b), the {111} faces, those which show a threefold surface patterning, are very rough compared to the {100} faces. They are microfaceted. The microfacets appear to exhibit the same faces as those of the initial crystallites. Since the twin plane is the same as that of the facet, twins might not be obvious. Figure 3(c) shows the surface of a film grown to a thickness of about 100 μm. The exposed facets are {111} and {100}, and nearly all of the grains have their <110> directions normal to the substrate. X-ray and electron diffraction confirmed the SEM observations [6, 9]. These X-ray diffraction and TEM observations also verify the prevalence of twinning.

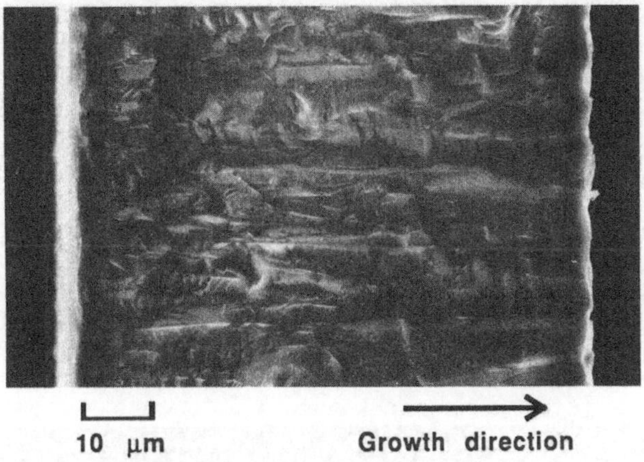

Fig. 1. Crystals with the fastest growth in the direction normal to the substrate outgrow those less favorably oriented, and a columnar structure develops. The fracture surface of a CVD diamond film grown at 930°C reveals its columnar grain structure.

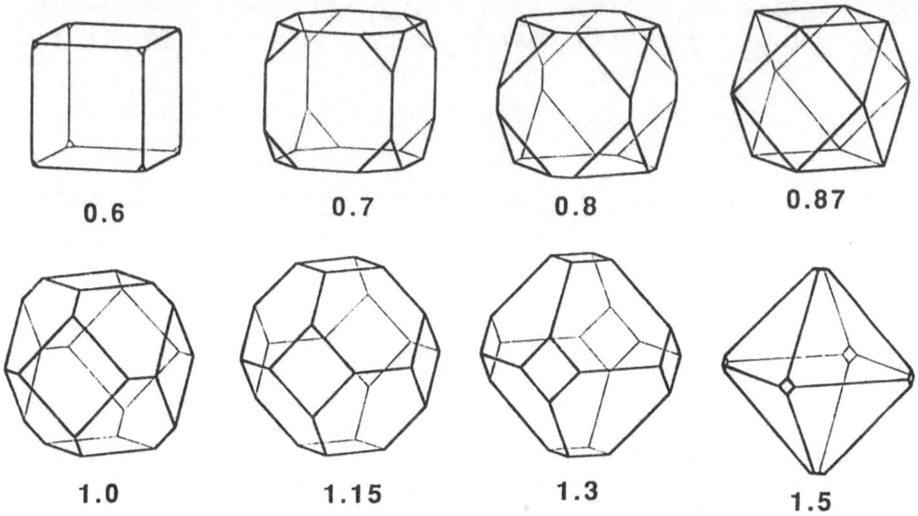

Fig. 2. Cubo-octahedral shapes reveal the ratio of growth rates in the <100> and <111> directions. These ratios are given for each shape.

Fig. 3(a) Initial growth at 1050°C shows randomly oriented singly and multiply twinned crystals as well as untwinned cubo-octahedra. (b) {111} faces are micro-faceted and very rough compared to the {100} faces. (c) A 100-µm-thick film shows grains with rough {111} and smooth {100} facets. Their <110> directions are normal to the substrate, i.e., the film has a {111}{100}<110> texture.

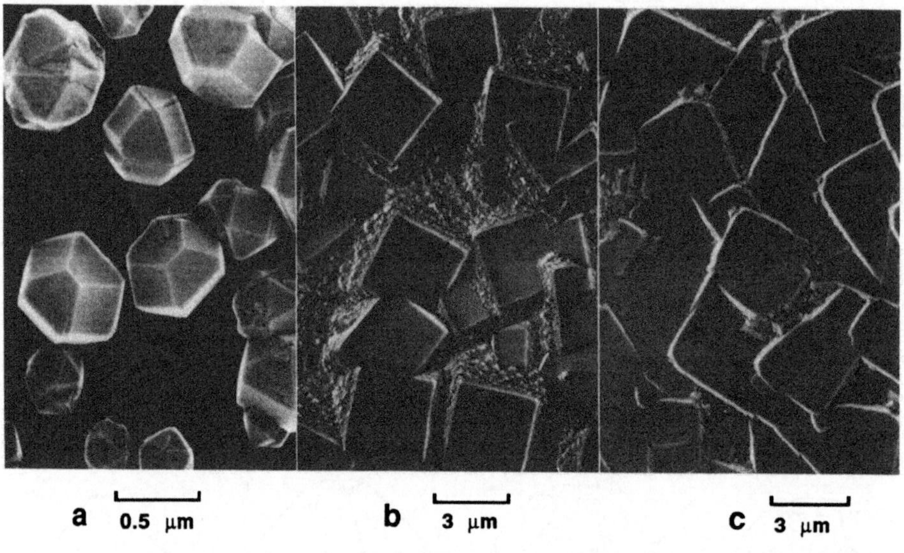

Fig. 4(a) Growth for 2 h at 950°C showed diamond crystallites with nearly random orientations. The <100>/<111> growth rate ratios were slightly greater than 1.0 . (b) Film topography after 20 h was dominated by crystallites with flat faces approximately parallel to the substrate. Crystallites initially oriented with <100> directions more nearly perpendicular to the substrate have crowded out those with less favorable orientations. (c) After 50 h of growth only <100> facets approximately parallel to the substrate remained. Fig. from ref.[10]

Development of the {100} <17 3 0> Texture. The development of this unusual and very interesting "near <100>" texture is shown in the series of scanning electron micrographs of Fig. 4. Samples were grown at 950°C. Initial growth consisted of crystallites with nearly random orientations as shown in Fig. 4a. By comparing the shapes of these crystallites with the reference crystallites in Fig. 2, we estimate the<100>/<111> growth rate ratio for these crystals to be slightly greater than 1.0 compared to 0.9 to 1.0 for the <110> texture.

Fig. 4b, a micrograph taken after 20 h of growth shows the development of a highly oriented film structure. The crystallites initially oriented with <100> directions more nearly perpendicular to the substrate have crowded out those with less favorable orientations. The surface is dominated by crystallites with flat {100} faces more or less parallel to the substrate. Finally, after 50 h of growth, the {100} facets have filled the entire surface, making it relatively flat. This series of SEM micrographs clearly shows the evolution of this texture and surface morphology. The fractured edge shown in Fig. 1, which is from a film of this type, reveals the resulting columnar structure. X-ray texture determination confirms that the resulting texture is "near <100>" (actually <17 3 0>), as was expected from the scanning microscopy. According to the Van der Drift model, this means that growth is occurring most rapidly in the <17 3 0> direction. The {100} facets do not, however, disappear in this particular case. This is apparently contrary to the simple axiom that the fastest growing directions grow the associated faces out of existence.

Development of {100}{111} <100> and {111} <100> textures. We have made use of the models above to grow other films with <100> textures, leading finally to a film with only {111} facets. By varying the growth conditions, we increased the ratio of the growth rate in the <100> to the <111> direction even more than was required for the development of the {100} <17 3 0> texture. The morphology of the {100} <17 3 0>, "near <100>" texture was first changed to a <100> texture with both {100} and {111} facets and finally,with a <100>/<111> growth ratio of 1.5, the {100} facets are completely eliminated, leaving the structure shown in Fig. 5.

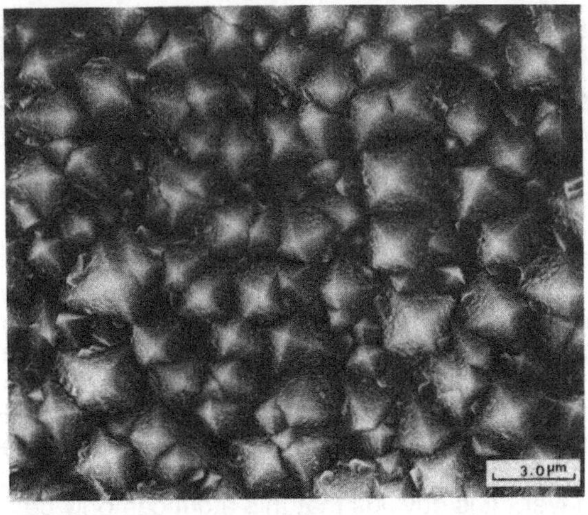

Fig. 5 Film grown with a {111} <100> texture by changing the growth conditions to increase the growth rate ratio of the <100> to the <111> direction to about 1.5.

Crystal Defect Structure Related to Growth Habit

TEM and SEM observations indicate major differences in the crystal defect structure of the above films, which seems to be directly related to the dominant growth direction. Faster growth in the <111> is accompanied by the creation of planar defects, i.e., twins and stacking faults, while growth in the <100> direction may have more point defects but no planar defects in the grains. We discuss here the differences in growth which might lead to this effect. In order to add carbon to a flat {111} face, one must add three carbon atoms to form a complete structural ring. Twins and growth steps make it possible to grow on {111} faces with the addition of only two carbon atoms, as is shown at the left edge of the twin in Fig. 6a . After the two-atom addition has been made, growth may continue in a chain by single-atom addition. At the inside edge of the twin (see Fig. 6(a) at the right), crystal growth may also proceed by single-atom addition. Thus, we see that growth on {111} faces will be facilitated by the presence of stacking faults , twins, and highly faceted surfaces, i.e., surfaces where planar faults are easily generated and propagated along the {111} twinning plane. By comparison, growth on atomically flat {111} faces would be difficult. SEM observations of films with a <100>/<111> growth rate ratio between 0.9 and 1.0 (faster growth on the {111} than on {100} facets) do indeed show that the growing {111} facets are rough and microfaceted (see Fig. 3b). TEM of the <110> textured film which results from this growth ratio shows many twins and stacking faults [10]. Meakin et al. [12] have developed a model for growth of highly twinned <110> CVD diamond cubic silicon films. This model should apply equally well to CVD diamond films and explain the characteristic highly faulted microstructure associated with the <110> texture.

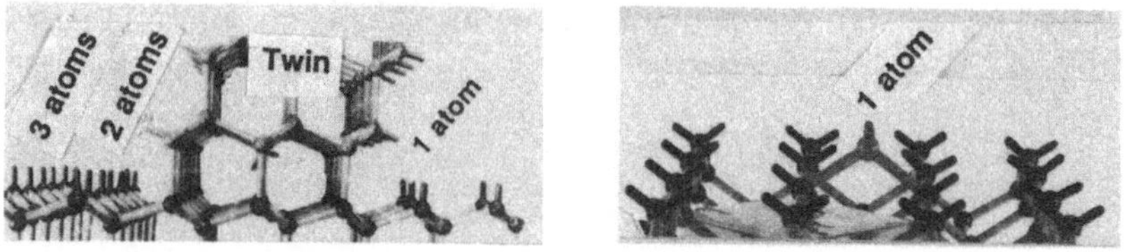

a {111} facet **b {100} facet**

Fig. 6. Models show the addition of carbon to the {111} and {100} facets of unreconstructed diamond crystals.

TEM of the <17 3 0> "near <100>" textured film shows no intragranular twins [6, 10]. Grain boundary regions, however, are wide and filled with crystal defects. Since only {100} facets remain on the surface, one must assume that growth is occurring on these facets. Examination of the facets on these films with an SEM having a resolution of about 10 nm showed no steps or ledges. However, material with {100} faces grown under other conditions exhibits pronounced growth steps. Figure 6(b) shows a model of the {100} surface with one atom of carbon added to begin a new {100} layer. It is obvious that this atom can only be added in one way and that no twins or stacking faults can arise. It can only go into the correct lattice position. Point defects may be incorporated into these growing facets, but planar defects cannot.

SUMMARY

We have been able to control the relative rates of crystal growth in the <100> and<111> directions by simple changes in growth parameters. This in turn dramatically changes the texture of the film from <110> to <100> or "near <100>." At the same time the internal structure of the columnar grains changes so that the stacking faults and twins which are so prevalent in the <110> textured material are not present at all in the grains of the "<100>" textured material. However, grain boundaries in the "<100>" textured material still contain highly defective regions. These regions may form relatively thick intergranular layers which can adversely influence the optical, electrical, and other properties of the material. Preliminary results indicate that by adding oxygen and carefully controlling the growth parameters, these intergranular layers can be reduced or eliminated. Further experimental verification, however, is required.

Evidence to support the hypothesis that twining plays an important role in the nucleation and growth of the (111)(100)<110> films is provided by X-ray diffraction texture determination, TEM microstructural analysis, and scanning electron micrographs of the topography of the growth surface. Crystallites in these films contain numerous stacking faults and twins which appear to be instrumental in the film growth and texture development in the same mannner as has been proposed to explain a similar texture seen in silicon film growth.

The "<100>" texture develops under conditions in which this is the direction of fastest growth. The films nucleate as more or less randomly oriented cubooctahedra with many single and multiply twinned nuclei, but growth competition soon leads to the crystals with the fastest growth directions normal to the substrate outgrowing and overgrowing the less favorably oriented crystals. The facets may be (100) or (111) depenuing on growth conditions.

Growth on (100) facets does not appear to require special nucleation sites such as steps or ledges. Single carbon additions to double-bonded sites are possible on the flat (100) face, but only to the correct position to propagate a proper continuation of the existing crystal and thus would not permit twins or stacking faults to nuceate. No twins are seen inside such crystallites, but the grain interfaces often contain a very high density of stacking faults and twins.

We have learned to control the development of crystal texture in polycrystalline films deposited by hot filament activated chemical vapor deposition. We can control the relative rates of growth in the <100>/<111> direction and thereby cause the desired texture and defect structure to grow.

Acknowledgement

This research was jointly sponsored by (1) the Exploratory Studies Program of Oak Ridge National Laboratory; (2) the U.S. Department of Energy, Division of Materials Sciences, Office of Basic Energy Sciences, and (3) the U.S. Department of Energy, Assistant Secretary for Conservation and Renewable Energy, Office of Transportation Systems, as part of the High Temperature Materials Laboratory Program. The research sponsored by the U.S. Department of Energy was under contract DE-ACO5-840R21400 with Martin Marietta Energy Systems, Inc.

References

1. B.V. Spitsyn et al. ., Prog. Crystal Growth and Charact.17, 79 (1988).

2. R.C. DeVries, Annu. Rev. Mater. Sci., 17, 161 (1987).
3. K. Kobashi et al , Phys. Rev. B, 38, 4067 (1988).
4. S. Matsumoto and Y. Matsui, J. Mater. Sci., 17, 3106 (1982).
5. S. Matsumoto and Y. Matsui, J. Mater. Sci., 18, 1785 (1983).
6. R.E. Clausing, L. Heatherly, Jr., K. L. More, and G. M. Begun, *Surface and Coatings Technology*, 39/40, 199-210 (1989).
7. Ch. Wild, N. Herres, J. Wagner, P. Koidl, and T. R. Anthony, Electrochem Soc. Proc. 89-12, 283 (1989).
8. Ch. Wild, N. Herres, and, P. Koidl, J. Appl. Phys. (Aug, xx (1990).
9. E. D. Specht, R. E. Clausing, and L. Heatherly, J. of Mater. Res. 5, 2351 (1990).
10. R.E. Clausing, L. Heatherly,and Jr., E. D. Specht, and K. L. More, to be published in the Proceedings of 2nd International Conference on the New Diamond Science and Technology, Washington, DC, September 24, 1990.
11. A. Vander Drift, Philips Res. Rep. 22, 267 (1967).
12. D. Meakin, J. Stoemenos, D. Miglierate, and N. A. Economan, J. Appl. Phys. 61, 5031 (1987).

PHYSICO-CHEMICAL ASPECTS OF SURFACE TREATMENTS FOR DIAMOND NUCLEATION

J.J. Dubray[1], W.A. Yarbrough[2], and C.G. Pantano[1,2]

The Pennsylvania State University
[1] Department of Materials Science and Engineering
[2] Materials Research Laboratory
University Park, PA 16802

1. INTRODUCTION

Diamond has been shown to nucleate and grow on a wide variety of materials.[1,2] It is known that nucleation density can be increased by several orders of magnitude when the substrate is roughened by abrasive powders (diamond, silicon carbide, cubic boron nitride, alumina).[2,3] Several alternative reasons for this effect have been suggested. One is the argument that diamond, diamondlike carbon or other carbonaceous residues from polishing or abrading with diamond powder are left adherent to, or embedded in, the polished surface and that these supply nucleation sites for subsequent diamond growth.[4] Alternatively, it can be argued that highly disordered surface sites or microscopic crater edges in the polished surface create *de facto* high energy sites which can be preferred nucleation sites for diamond.[5] Interestingly, Yarbrough[2] reported that when a diamond polished silicon surface was subsequently polished with cubic boron nitride, the observed nucleation density was reduced to well below that obtained using only a diamond polish. This in spite of the fact that it was also shown that diamond readily nucleates on c-BN, including the particles used to polish the samples. We report in this paper an Auger Electron Spectroscopy (AES) investigation of the residue left in the scratches after polishing silicon substrates with diamond, and cBN powders as well as graphite surfaces. Diamond films were subsequently grown using a conventional hot filament CVD process.

2. EXPERIMENTAL

<100> Silicon substrates were polished with diamond and cubic boron nitride (Borazon™, General Electric, Worthington OH) powders of 0.5 to 1 μm. Freshly cleaved {0001} surfaces of highly oriented pyrolytic graphite were used to polish silicon as well.

Diamond was grown, after cleaning, in a hot filament reactor. Process conditions were : 1% of methane, 99% hydrogen, at a pressure of 30 torr and flowing at 200 sccm. The tantalum filament was heated to 2100° C (measured with an optical pyrometer). The sample was mounted 4.4 mm above the filament which resulted in a back temperature of approximately 950° C. Every sample ran for one hour. A CAMECA NANOSCAN™ 50 was used to record spectra from locations inside scratches formed on silicon by various abrasive powders. The field emission gun enables us to reach spatial resolution as low as 50 nm. The gun was operated at 8 keV, and the pressure in the analysis chamber was $5x10^{-10}$ torr. The total acquisition time was kept to 5 minutes for C KLL windows and 8 minutes for a survey. Auger spectra were subsequently deconvoluted using Sherwood's model[6] in order to calculate the position of the valence band maxima[6,7]. The latter can be related to the structure of the carbon present on the surface.

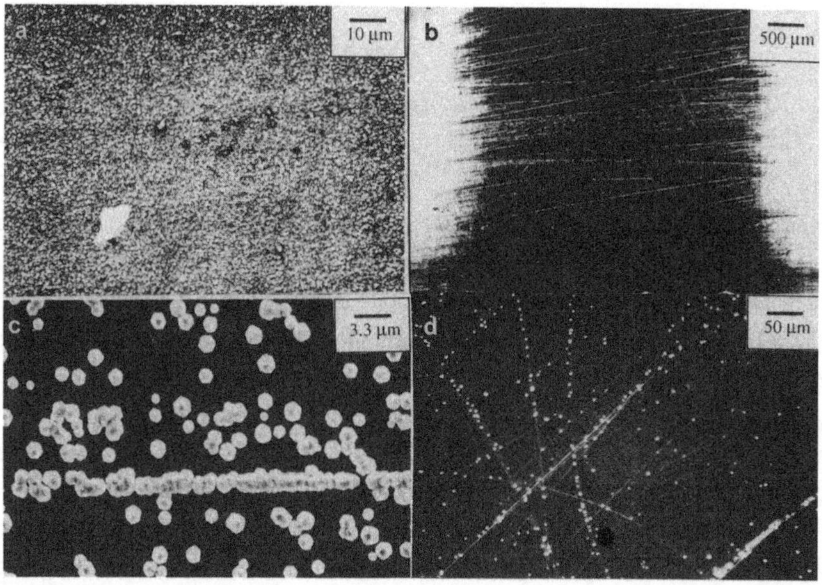

Figure 1. SEM photomicrographs of diamond crystals grown on <100> silicon wafers polished with (a) diamond, (b) diamond followed by c-BN (low magnification), (c) idem (high magnification), (d) {0001} graphite surfaces.

3. RESULTS AND DISCUSSION

SEM photomicrographs of diamond crystals grown on surfaces polished with diamond, diamond followed by cBN, and graphite are featured in figure 1. The second sample reproduces successfully Yarbrough's experiment and shows that nucleation density is highest inside certain diamond scratches. In the case of polishing with graphite, crystals are again

Table 1. C KLL component positions (eV) derived from Sherwood's model[6] for diamond, {0001} graphite, amorphous carbon and CVD monocrystalline silicon carbide.(See figure 3)

Surfaces\Components	A_0	A_1	A_2	A_3
Diamond	261.9	250.2	236.0	224.6
A_i-A_0		-11.7	-25.9	-37.3
Graphite {0001}	264.9	255.5	249.7	239.1
A_i-A_0		-10.3	-16.5	-25.8
Amorphous carbon	262.5	251.5	239.2	232.7
A_i-A_0		-11.0	-24.2	-34.9
CVD silicon carbide	265.0	257.3	248.5	239.5
A_i-A_0		- 7.7	-16.5	-25.5

Table 2. C KLL component positions (eV) for roughened silicon surfaces (inside the scratches, corresponding to spectra of figure 2.)

Treatments\Components	A_0	A_1	A_2	Structure	Nucleation
Diamond	264.4	253.5	240.3	Amorphous	Highest
A_i-A_0		-10.9	-24.1		
Graphite {0001}	263.7	251.7	237.6	Amorphous	Highest
A_i-A_0		-12.0	-26.1		
cubic boron nitride	265.7	258.0	248.4	Carbidic	Lowest
A_i-A_0		- 7.7	-17.3		
Graphite domain	263.2	255.5	246.8	Carbidic	Lowest
(unscratched) A_i-A_0		- 8.2	-16.4		

aligned preferentially along scratches. Figure 2 features typical C KLL spectra recorded inside the different scratches as well as the spectrum of carbon impurities present on the silicon wafer previous to any treatment. The reference spectra for different carbon materials (diamond, graphite, silicon carbide ...figure 3,), as well as table 2 that reports positions of the maxima of the valence band, indicate that amorphous carbon is present inside diamond or graphite scratches, where nucleation density is highest, while carbidic carbon appears to be present on an untreated wafer or inside c-BN scratches, where the nucleation density is lowest. Amorphous carbon, or boron carbide spectra could also detected inside c-BN scratches but only rarely, while diamond scratches consistently showed the presence of amorphous carbon. In all cases the native SiO_2 layer is removed by polishing. However, for extended polishing

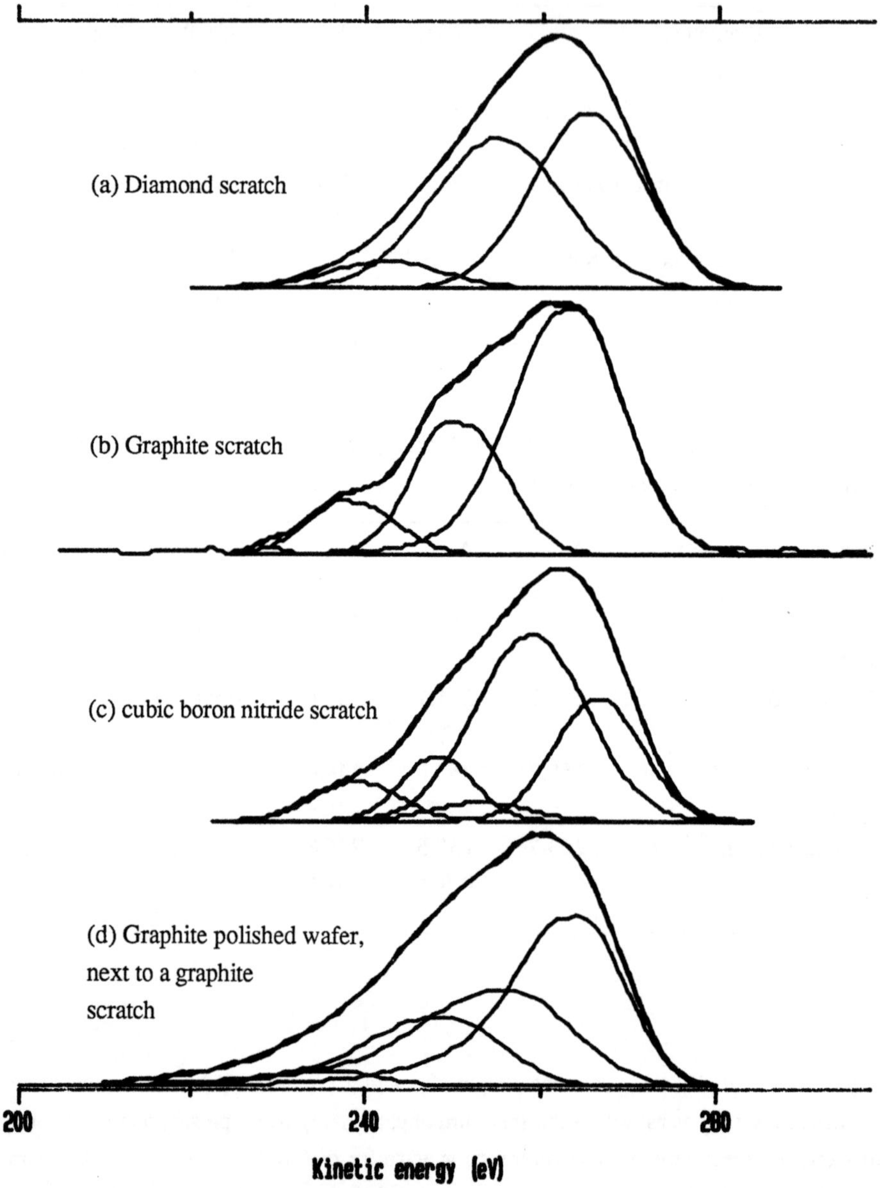

Figure 2. C KLL Auger spectra recorded in direct mode inside (a) a diamond scratch located on the area that was polished with both diamond and cBN powders, (b) a graphite scratch, (c) a cBN scratch perpendicular to (a), and (d) on the silicon wafer, next to (b).

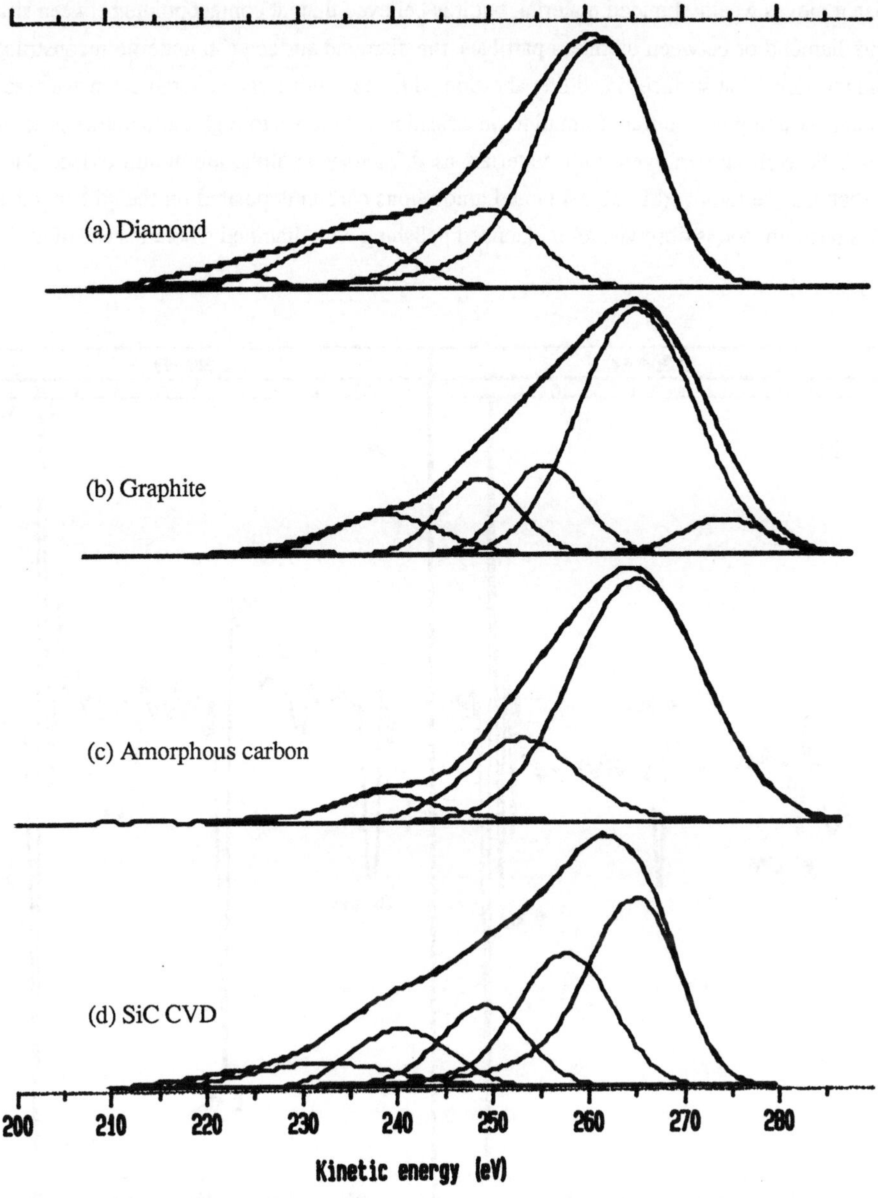

Figure 3. C KLL reference Auger spectra recorded in direct mode of (a) diamond, (b) {0001} graphite, (c) amorphous carbon and (d) CVD silicon carbide. (See table 1 for the energy positions of the components derived from Sherwood's model[6].)

time, the formation of oxidized silicon was observed (figure 4.a.). The effect of polishing with submicron c-BN appears to be to remove or destroy any weakly bonded material, such as silica or amorphous carbon, leaving surface carbon primarily as carbidic islands. The effects produced by diamond particles appear more complicated. Polishing with submicron diamond also removes weakly bonded material, but it is believed that, at contact points between silicon and diamond or between diamond particles, the diamond surface can undergo reconstruction and reaction. That surfaces in sliding abrasion with one another can undergo extensive reaction chemistry and physical transformation, in a fashion analogous to high temperature processes, has been well known even with materials as refractory as alpha aluminum oxide.[8] Hence carbon can be removed in abrasion and amorphous carbon deposited on the silicon surface. This phenomenon is observed after extended polishing with diamond where instead of a

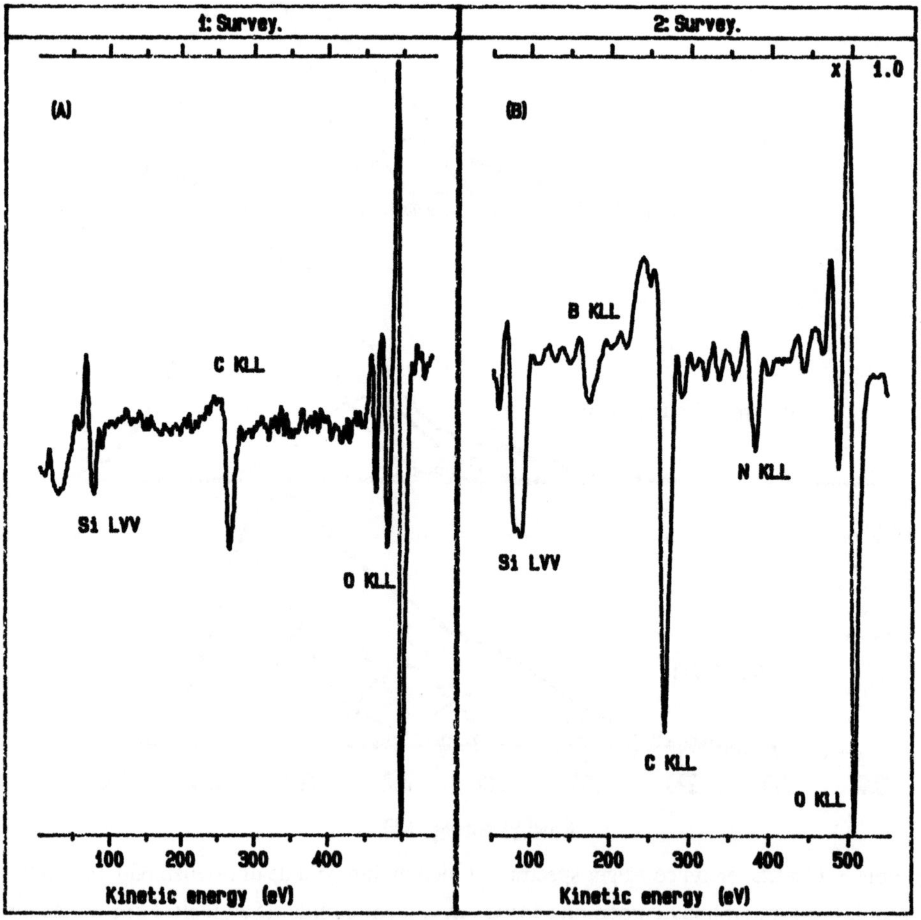

Figure 4. Differentiated Auger spectra recorded (a) on a silicon wafer after extended polishing with diamond powder, (b) on top of diamond particles left on the silicon substrate after diamond polishing and cleaning.

cleaner substrate, a brownish layer forms and the Auger signal from the substrate (Si LVV, 92 eV) disappears. This surface shows a high carbon and oxygen content as seen in figure 4 and also exhibits the highest nucleation density. Evidence of diamond particles, adherent to a diamond polished silicon surface after cleaning, was also obtained using the 50 nm resolution Auger microscope. Their spectra are featured in figure 5 and evidence a highly damaged surface relative to that of pure diamond.

4. CONCLUSION

These results suggest a model for the efficacy of diamond polishing treatments, consistent with much other published data, which supposes that diamond preferentially nucleates on either diamond detritus or highly disordered carbon left on the surface by polishing with diamond and not removed by simple cleaning. From the experimental results it can be seen that a highly disordered carbon is deposited on silicon by polishing with diamond, or other abrasives in the presence of carbon, and we suggest that it is this carbon which supplies most of the nucleation sites for the growth of diamond.

5. ACKNOWLEDGEMENTS

This work was supported by the Office of Naval Research (with funding from the Strategic Defence Initiative Organization's Office of Innovative Science and Technology), The Diamond and Related Materials Consortium of The Pennsylvania State University, the Ben Franklin Partnership Program and Thomson CSF Inc.

6. REFERENCES

1. K.E. Spear, *J. Am. Ceram. Soc.*, 72, 171 (1989).
2. W.A. Yarbrough and R. Messier, *Science*, 247, 688(1990).
3. K. Hirabayashi, Y. Taniguchi, O. Takamatsu, T. Ikeda, K. Ikoma, N. Iwasaki-Kurihara, *Appl.Phys.Lett.*, 53, 1815 (1988).
4. W.A. Yarbrough, A. Kumar, R. Roy, Mat. Res. Soc. Fall 1987 meeting, Boston, MA.
5. S. Yugo and T. Kimura in "First Int. Conf. on the New Diamond Sci. and Techn., Program & Abstracts," 24 to 26 September, 1988 (JNDF, Tokyo), p. 130.
6. D. Briggs and M.P. Seah, in *Practical Surface Analysis by Auger and X-Ray Electron Spectroscopy*, Wiley & Sons, New York, 1983, p 445.
7. P.G. Lurie, J.M. Wilson, *Surf. Sci.*, 65, 476 (1977).
8. A. G. King, "The Influence of Microstructure on the Mechanical Properties of Dense Polycrystalline Alumina, in "*Mechanical Properties of Engineering Ceramics*, W. W. Kriegel and H. Palmour III, eds., Interscience, New York, 1961, pp. 333-347.

LOCAL HETEROEPITAXIAL DIAMOND GROWTH ON (100) SILICON

*David G.K. Jeng, *H.S. Tuan, **James E. Butler,
+Robert F. Salat and +Glenn J. Fricano

*Dept. of Electrical Engineering, State University of New York, Stony Brook, N.Y. 11794
**Chemistry Division, Code 6174, Naval Research Lab, Washington, D.C. 20375-5000
+Vactronic Lab Equipment Inc., 160 Wilbur Place, Bohemia, N.Y. 11716

INTRODUCTION

Diamond film has been successfully grown on various substrates in its metastable state (low pressure and low temperature) via the chemical vapor deposition (CVD) process[1]. However, successful epitaxial diamond growth was limited only on diamond substrate (homoepitaxy), which are very costly. The advent of CVD diamond has stimulated great interests in various applications primarily due to the unique combination of electronic, mechanical and optical properties of diamond. Unfortunately, the grain boundaries existing in polycrystalline type diamond film will limit the performances in electronic devices as well as in optical components[2]. To overcome this disadvantage, several researchers have attempted for growing textured or oriented diamond on foreign substrates, and each effort has resulted in various degree of success[3,4,5]. The maturity in silicon technology and the near defect-free quality of silicon material available, together with its carbide (SiC) forming property has made silicon the choice as the substrate material for diamond film growth[6]. However, as high as 50% lattice mismatch between silicon and diamond has lowered the hope of diamond/silicon heteroepitaxy[7]. In this paper, we are presenting the phenomena of local heteroepitaxial diamond grown on silicon (100) substrates. The in situ pretreatment of the substrate surface prior to the diamond growth, has resulted in an off axis growth of high purity local epitaxial diamond pyramids over various parts of the substrate surface.

EXPERIMENT

The apparatus used in this experiment was a Microwave Plasma Chemical Vapor Deposition (MPCVD) system provided by Vactronic Lab Equipment Inc. As shown in Fig. 1 is a schematic diagram of the MPCVD system. The overhead 600 watt, 2.45 GHz microwave power was directly coupled through a cylindrical waveguide and a transition horn into the 6-1/2 inch ID process chamber.

A mirror surfaced silicon (100) wafer which had been polished with

Diamond and Diamond-Like Films and Coatings
Edited by R.E. Clausing *et al.*, Plenum Press, New York, 1991

Al_2O_3 was used as the starting material. The substrate had been cleaved into 2x2 cm^2 sections along its (010) and (001) crystalline orientations and cleaned with a typical chemical cleaning process prior to loading. The substrate surface was not polished by any mechanical means or seeded with diamond powder. The process was divided into two steps including an in situ pretreatment process, and a typical diamond growth process. Both the pretreatment and the subsequent diamond growth conditions are outlined in table I. The processing temperature, as shown in Table I was the substrate temperature monitored by an optical pyrometer. A thermocouple was used separately for a feedback control unit to monitor the substrate holder temperature, which always showed a temperature 200-250°C lower than the pyrometer output. During the pretreatment, the substrate was not uniformly immersed in the plasma for the plasma ball was smaller than usual due to the relatively high processing pressure such that the subsequent crystal growth was rather non-uniform. The sample was inspected under the optical microscope after the pretreatment step. Small etching pits were found randomly distributed across the substrate. No further characterization was performed in this step. When the growth process was completed, highly reflecting concurrent randomly-oriented diamond crystals were found in some parts of the substrate surface, however, in many locations, a large amount of pyramid-shaped cubic crystals were observed. All these pyramid-shaped crystals had almost the same crystal shape and size, each having a top surface and bottom surface area of approximately 50x50 and over 100x100 micron square, respectively. Each pyramid-shaped crystal was highly faceted and growing local epitaxially along the (110) silicon substrate orientation. Unlike the reflecting randomly oriented crystals as mentioned previously, the local epitaxial diamonds showed almost no reflection and clear under an optical microscope viewing through the top facets, one can even observe the detail morphology at the silicon interface even though the pyramids were 100 microns high.

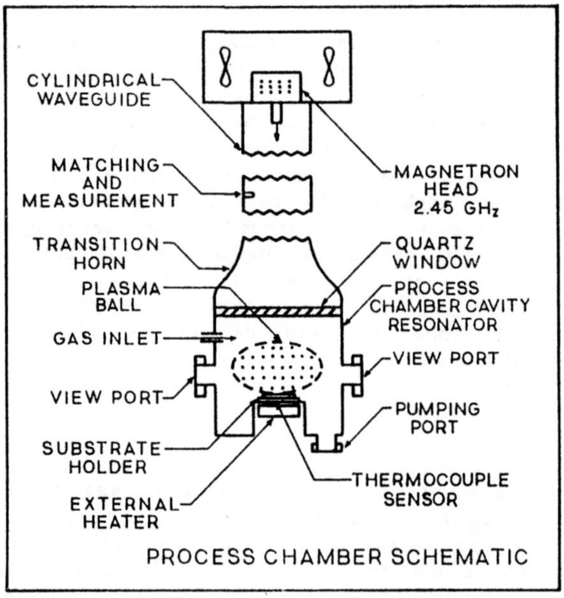

Fig. 1. The schematic diagram of the Vactronic Lab 6-1/2 inch ID resonant cavity type MPCVD system.

Table I. Process conditions of the pretreatment and the conventional diamond growth.

	Pretreament	Growth
Gas H_2/CH_4 (sccm)	200/0.2	200/2
Pressure (torr)	100-120	50-60
MW power (watt)	600	600
Temperature (°C)	1040-1050	850-880
Process time (Hrs)	10	40

CHARACTERIZATIONS AND RESULTS

X-ray diffraction

Characterizations made by x-ray diffraction were carried out in a traditional method. No intention of microfocusing the x-ray beam at any specific spot had been tried. The XRD spectrum showed a highly preferrential growth on the diamond (400) plane. Fig 2(a) is the reference XRD spectrum obtained from a continuous polycrystalline diamond film grown on the silicon (100) substrate that had been polished with diamond powder. Fig. 2(b) is the x-ray spectrum of the local epitaxial diamond sample being discussed.

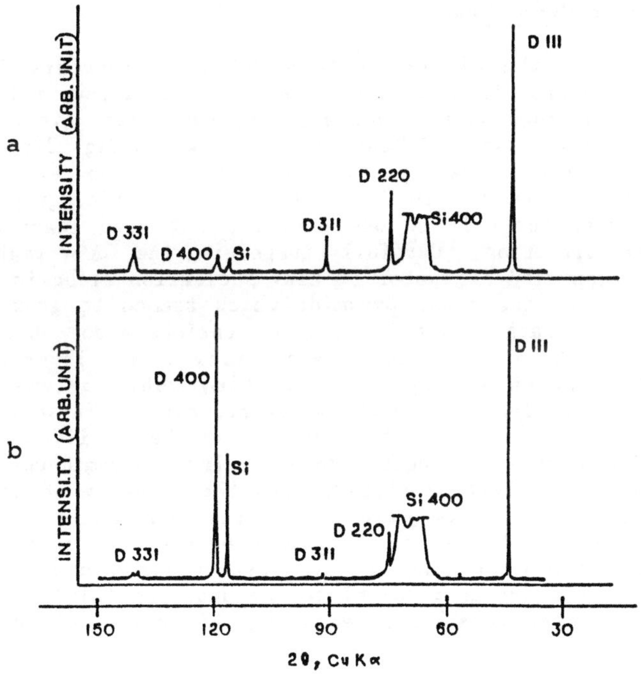

Fig. 2(a). The typical polycrystalline diamond film XRD spectrum, in which the (111) is the most intense peak and the diamond (400) is the weakest among the five peaks.

2(b). The XRD spectrum of the local epitaxial diamond grown on (100) silicon substrate, in spite of the random oriented growth, the diamond (400) peak emerges even stronger than the (111) main peak.

Table II. Lattice spacings (angstrom) comparison of the local epitaxial diamond to ASTM reference.

(hkl)	ASTM 6-0675	Local Epi Diamond
(111)	2.06	2.054
(220)	1.261	1.261
(311)	1.0754	1.0754
(400)	0.8916	0.8912
(331)	0.8182	0.8181

The polycrystalline diamond XRD spectrum as illustrated in Fig. 2(a) always showed diamond (111) dominant, whereas the diamond (400) was usually the least intense among the five diamond peaks. In contrast to the polycrystalline diamond XRD spectrum, the local epitaxial sample showed a very unusual (400) preferrential characteristics, the (400) peak, which was otherwise the least intense among the others, now emerges even stronger than the (111) main peak. It has to be pointed out that one of the XRD peaks shown near the diamond (400), i.e., the 117° peak in both Fig 2(a) and 2(b), was a signal diffracted by the silicon substrate. This peak was later found in the post-characterizations made on the same batch of the virgin silicon wafers. No evident sign of the SiC signal had been detected in the XRD spectrum. Corresponding lattice spacings to the heteroepitaxial diamond samples are listed in table II. In addition to the XRD, characterizations have also been made using SEM and micro-Raman spectroscopy, each has brought on a consistent result.

Scanning Electron Microscope

Fig. 3(a) is the diamond film morphology corresponding to the reference XRD spectrum in Fig. 2(a), i.e., the continuous polycrystalline type diamond film grown on the same starting substrate, i.e., the silicon (100), which showed no special feature. As shown in Fig. 3(b) are typical oriented growths of the local epitaxial diamond pyramids, which correspond to the unusual XRD spectrum in Fig. 2(b). In this micrograph, two giant pyramids, oriented not only to one another, but also exactly along the substrate (110) direction. Carefully inspecting the lower right corner in Fig. 3(b), one can find a cluster of tiny nucleation sites in exactly the same orientation as the giant pyramids which seemed to grow epitaxially extending from the large crystal. The nucleations also showed a uniform size between 1 to 2 microns. An enlarged view of the lower right corner in Fig. 2(b) is shown in Fig. 4. The tiny cubic oriented crystals nucleated in another location on the substrate surface is shown in Fig. 5. The outgrowth from the substrate surface can be easily determined by tilting the SEM observation angle. Both the large pyramid-shaped crystals and the 1-2 micron nucleation sites showed the tendency of growing along the substrate (110) orientation. Determination of the crystal orientations were made by careful alignment between the substrate cleaved edges and the SEM monitor frame before inspections started. In Fig. 5, the micrograph edges are in parallel to the substate (010) cleaved edges, therefore the nucleations oriented along silicon (110) direction is clearly shown.

Micro-Raman Spectroscopy

Strong evidence indicating the diamond properties of the local epitaxial growth was confirmed by the micro-Raman spectroscopy. Fig. 6(a) shows a typical Raman shift of these giant, highly faceted local epitaxial crystals centered at 1333 cm^{-1}, as well as the low fluorescence background in the Raman spectrum further fingerprint the local epitaxial diamond

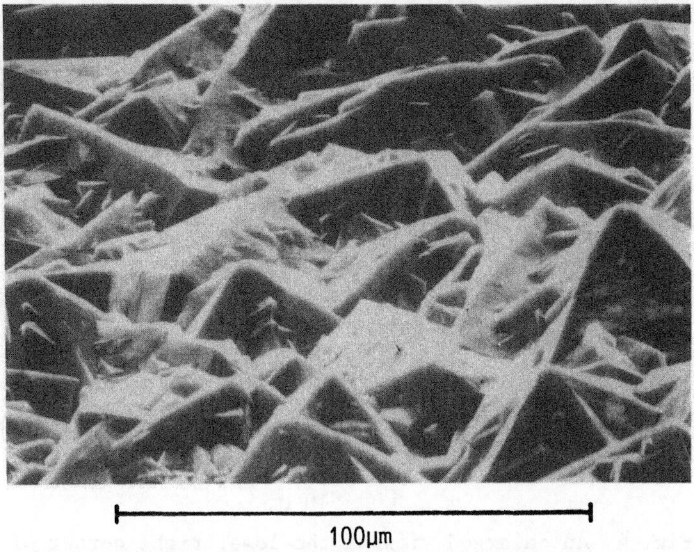

Fig. 3(a). The surface morphology of the continuous poly-crystalline type diamond film.

Fig. 3(b). Two neighboring giant pyramid-shaped diamonds of over 100x100 micron square grown epitaxially in the (110) direction of the (100) silicon substrate surface. In the lower right corner clusters of 1-2 micron nucleations were found to grow in exactly the same orientation as the giant pyramids.

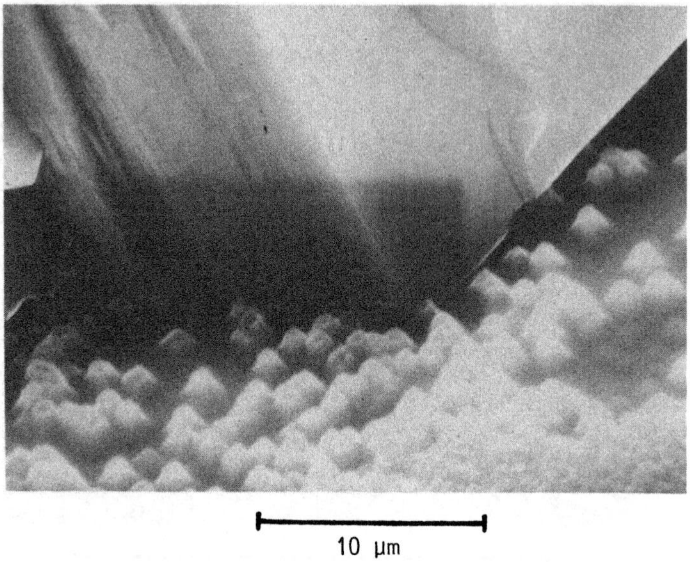

Fig. 4. An enlarged view at the lower right corner of the local epitaxial diamond in Fig. 3(b) showing outgrowth of the nearby oriented nucleation sites are also pyramid-shaped and co-oriented to the large local epitaxial diamond pyramids.

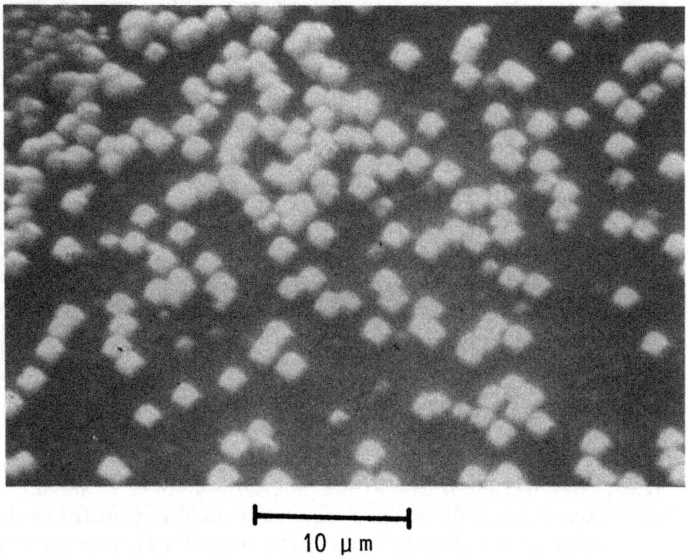

Fig. 5. Uniform distribution of the tiny nucleations found in various parts of the substrate surface, each is grown in the (110) direction of the (100) silicon surface.

quality. In an expanded micro-Raman spectrum as shown in Fig. 6(b), the peak is centered at 1332.8 cm^{-1} with a FWHM of 5.5 cm^{-1} which is close to the reported FWHM of natural diamond[8]. Low background and low FWHM are typical characteristics of the local epitaxial diamond crystals. We tried

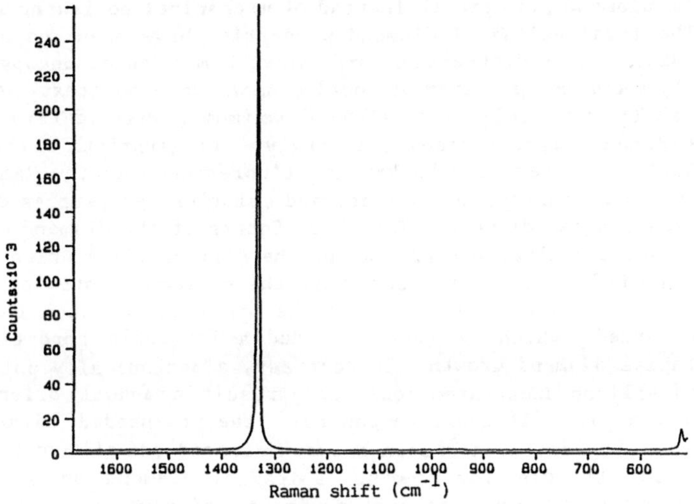

Fig. 6(a). A typical micro-Raman spectrum of the giant local epitaxial diamonds centering at 1333 cm^{-1} and the low background fluorescence strongly fingerprint the diamond quality.

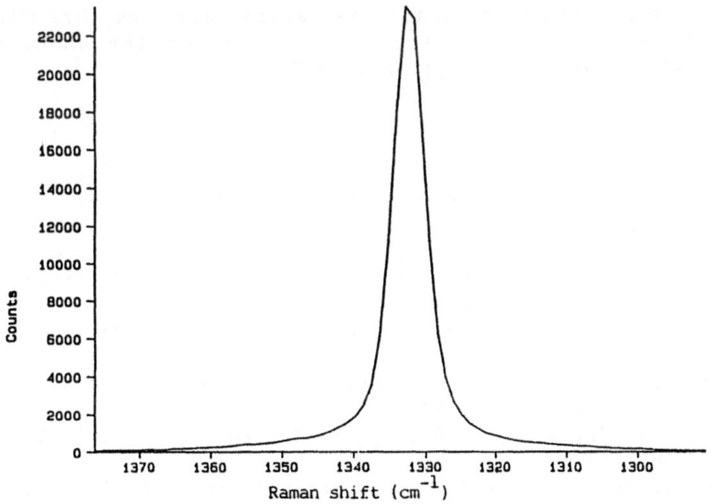

Fig. 6(b). An expanded micro-Raman spectrum showing a FWHM of 5.5 cm^{-1}.

very carefully to search for the trace of a SiC layer by focusing the micro-Raman laser at the interface between the silicon substrate and the giant diamond pyramids, but the results were inconclusive. There could be only a few atomic layers of SIC, if any exist, between the epi-diamond and the silicon substrate such that it was difficult for the Raman detector to pick up the signal.

DISCUSSION AND CONCLUSIONS

High quality local epitaxial diamond has been grown in a two-step process in which the substrate surface pretreatment was carried out in-situ prior to the diamond film growth instead of mechanical polishing with diamond powder. The local epitaxial diamond properties have been characterized by means of SEM, x-ray diffraction and micro-Raman spectroscopy; each has consistently shown the property of local diamond heteroepitaxy. A sharp 1333 cm^{-1} Raman shift and a full width at half maximum (FWHM) of 5.5 cm^{-1} obtained by micro-Raman spectroscopy strongly fingerprints the diamond characteristics. A very low background fluorescence in the Raman spectrum indicated the high purity, low defect, and colorless properties of the grown local heteroepitaxial diamond. The (010) facets of the diamond crystals and the tiny oriented nucleations grew along the silicon (110) orientation which greatly reduced the lattice mismatch with the silicon substrate. We believe that the surface had been modified by the pretreatment such that oriented nucleation formed, which in turn provided a favorable condition for the heteroepitaxial diamond growth. In contrast, a mechanically polished or ex-situ seeded silicon substrates could only result in randomly oriented diamond growth since it is difficult to control the pre-seeded diamond residue orientation. The lattice mismatch between diamond and silicon has long been considered an obstacle for growing epitaxial diamond on silicon. Our observations indicate growth of heteroepitaxial diamond crystals is possible. In this experiment, we have demonstrated a process of growing high quality heteroepitaxial diamond on silicon with a simple in-situ process of pretreatment, which conditions the silicon for heteroepitaxial growth.

ACKNOWLEDGEMENT

This work was supported by Vactronic Lab Equipment Inc. The authors are grateful to Mr. Ken Gilmore for his careful drawings, Mr. Chris Salat and Dan McKeown for technical support. One of the authors (D.J.) apprecited for the partially support by NATO.

REFERENCES

1. B.V. Spitsyn and B.V. Derjaguin, Pat. USSR 339,134 (1956)
2. A. Collins, Semicond. Sci. Technol. 4, 605 (1989)
3. M. Geis, Appl. Phys. Lett. 55, 550 (1989)
4. J. Mar, H. Kawarada, T. Yonehara, J. Suzuki, J. Wei, Y., Yokata and A. Hiraka, Appl. Phys. Lett. 55, 1071 (1989)
5. R. Rudder, J. Posthill, G. Hudson, M. Mantini and R. Markunas, SPIE 969, 72 (1988)
6. T. Hartnett, R. Miller, D. Montanari, C. Willingham, and R. Tustison, J. Vac. Sci. Techlol. A8(3) 2129 (1990)
7. J. Nanayan, A. Srivasta, M. Peters, S. Yokota and K. Ravi, Appl. Phys. Lett. 53, 1825 (1988)
8. A.M. Bonnot, Proc. of The 1st International Symp. on "Diamond and Diamond-like Films", 579 Ed. J.P. Dismukes, A.J. Purdes, B.S. Meyerson, T.D. Moustakas, K.E. Spear, K.V. Ravi and M. Yoder, The Electrochem. Soc. Inc., (1989)

STUDIES ON THE FORMATION OF DIAMOND NUCLEATION SITES ON <100> SILICON SUBSTRATES

Joseph A. Baglio, Barry C. Farnsworth, Sandra Hankin,
Changmo Sung, Jesse Hefter, and Marvin Tabasky

GTE Laboratories Incorporated, 40 Sylvan Road
Waltham, MA 02254, USA

INTRODUCTION

A number of challenges present themselves in successfully growing diamond on non-diamond substrates. Among them are the formation of a high density of nucleation sites and the creation of appropriate nucleation sites for films of desired orientation and perfection. Abrasion techniques, used to increase nucleation densities, are not amenable to normal semiconductor processing sequences. The focus of this investigation is to evaluate techniques to increase the nucleation density while maintaining the perfection of diamond deposited on <100> oriented Si wafers.

EXPERIMENTAL

Substrates were 5 cm diameter, <100> oriented n-type Si wafers ($r \approx 7$ ohm-cm). All samples were initially cleaned using standard semiconductor cleaning techniques. Diamond grains were deposited using a 1.5 kW Applied Science and Technology (ASTeX) HPMS microwave source incorporating a RF heated motor driven substrate stage.

Initial attempts were made to grow diamond on Si wafers processed using standard semiconductor processing techniques. These included: (1) forming 200 nm deep square trenches using either ion etching or (2) chemical etching, (3) using Si ion implantation, (4) using Si ion implantation within the 200 nm trenches formed by prior ion etching, and (5) by depositing polysilicon (250 nm)/Si_3N_4 (60 nm) on Si. A 600 W plasma containing methane (0.5 %) in H_2 with the substrate temperature of 850 °C (uncorrected optical pyrometer reading) was used to deposit diamond on this first set of samples. In addition, Si wafers abraded with 1 μm diamond paste (followed by cleaning in alcohol) were also subjected to these plasma conditions.

In situ CVD surface treatment was accomplished, on another set of samples, by a two-step procedure. Prior to plasma treatment, sections of Si substrates were abraded with 0.25 μm diamond paste for 10 min while other sections of the wafers were left untreated. The diamond was then removed from the surface by wiping with alcohol and/or ultrasonic cleaning in DI water. The substrates were then subjected to a 6 h *in situ* microwave surface pretreatment step under the following conditions: CH_4 (0.3%) in H_2, 300 cc/min flow, 800 W microwave power, 30 torr pressure, and a 1000 °C substrate surface temperature.

Following the pretreatment, samples were subjected to 15 h plasma processing steps to produce diamond nuclei. Both continuous and sequential processes were investigated. The continuous case involved the use of CH_4 (2%)/H_2 or CH_4(2%)/O_2(1%)/H_2 and 600 or 1000 watts microwave power. Total flow rate of reactant gases was maintained at 460 cc/min for 15 h. A pressure of 50 torr and a substrate temperature at 850 °C (uncorrected optical pyrometer reading) was used. In addition, three sequential flow cases were investigated (In all cases, 180 sequences (15 h) per run were used.):

1) CH_4(2%)/H_2 (1 min), H_2 (4 min),
2) CH_4(2%)/H_2 (1 min), O_2(1%)/H_2 (4 min), and
3) CH_4(2%)/O_2(1%)/H_2 (1 min), O_2(1%)/H_2 (4 min).

RESULTS AND DISCUSSION

Experiments aimed at growing diamond on wafers treated with the various semiconductor processing methods described above were unsuccessful; no diamond growth was observed by Raman spectroscopy. In contrast, Si wafers pretreated by abrasion (1 μm diamond paste, followed by an alcohol cleaning) and subjected to the same plasma growth conditions were found to support diamond growth.

SEM and RED examinations, of Si wafers polished with 0.25 μm diamond paste and then cleaned in alcohol, failed to detect the presence of diamond. An RED pattern (Figure 1a) reveals diffuse rings indexed as polycrystalline Si.

Thermodynamics calculations indicate that, in the absence of a plasma, SiC as opposed to diamond or graphite is the only stable phase expected on a heated Si substrate (CH_4 (0.3%) in H_2, 30 torr pressure, 1000 °C substrate surface temperature). This *in situ* plasma preconditioning of the abraded and cleaned surface results in the formation of microcrystalline diamond (Figure 1b). No indication of a SiC pattern is observed even though the spot patterns reveal the underlying Si substrate and the diffraction rings from diamond are distinguishable from SiC. Kawarada et al.[1] have reported (based upon SEM data) that an interfacial layer, presumably of SiC, exists underneath their grown diamond except at the nucleation site. However, the present results indicate that:

1) Crystalline SiC layer is not necessary for diamond nucleation.

2) Since the coverage of diamond is incomplete, and since samples that were not abraded with the diamond paste did reveal the presence of crystalline SiC (Figure 1c); abrasion of the surface seems to promote diamond nucleation, thus interfering with microcrystalline SiC formation .

It is expected that a higher concentration of methane during growth can lead to a higher degree of carbon supersaturation and to the formation of a higher density of microcrystallites. Therefore, for diamond growth experiments in which the above pretreatment step was used, a higher concentration of CH_4 (2%) was investigated to increase the nucleation density. In order to control the perfection of the crystallites, the use of O_2 and sequential flow techniques[2] were investigated. Oxygen was used to control the degree of carbon supersaturation, with possible concomitant changes in the nature and/or concentration of diamond precursors[3], and also as an etchant.

The quality of the diamond grown on Si (abraded with 0.25 μm diamond paste) was assessed by Raman spectroscopy (Ar ion laser operating at 514.5 nm with ≈100 μm spot size). It is assumed that the intensity ratio of the 1332 cm^{-1} peak to the 1530 cm^{-1} peak is a measure of the quality of the diamond coating. A baseline spectrum showing high-quality diamond growth (with, however, low coverage) is shown in Figure 2a(2).

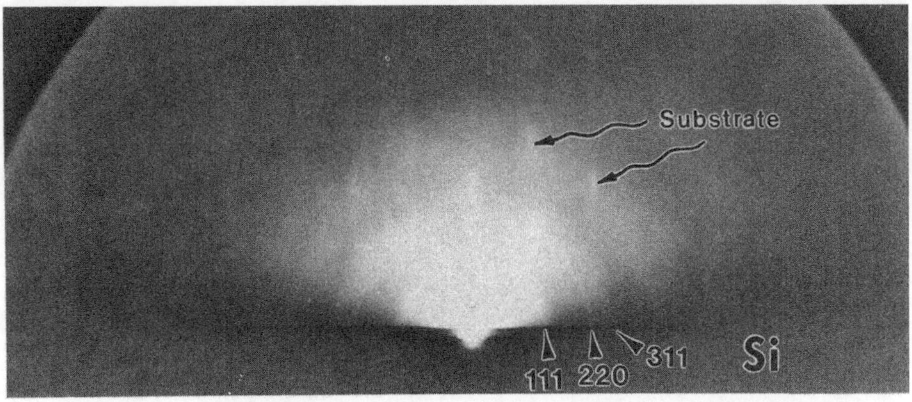

Figure 1a. Electron diffraction micrograph of <100> oriented Si 0.25 μm diamond abraded, followed by ultrasonic cleaning. Spots refer to Si substrate and rings index as polycrystalline Si.

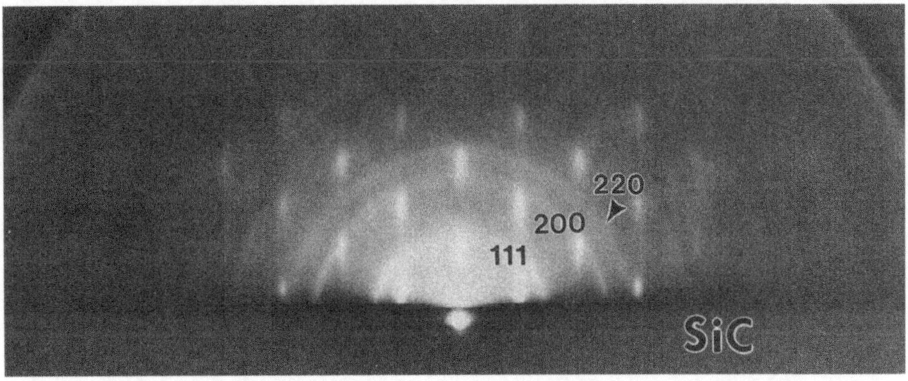

Figure 1b. Si wafer treated as in Figure 1a followed by a 6 hr. *in situ* microwave plasma pretreatment step. All rings index as diamond; no SiC rings observed.

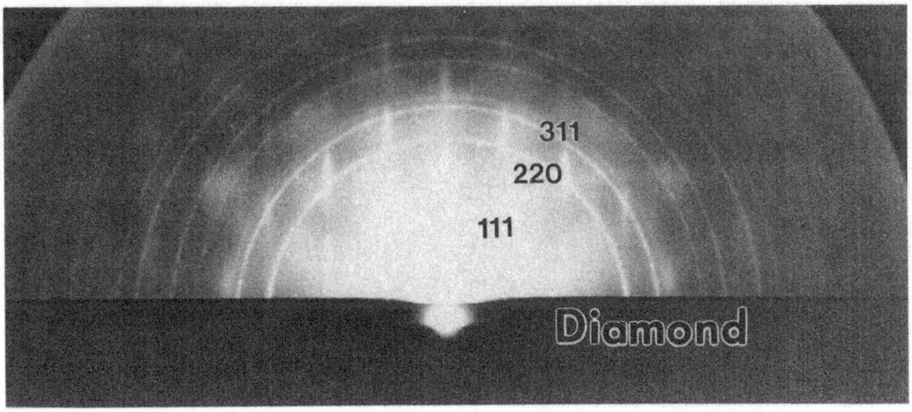

Figure 1c. Si Wafer (not abraded) subjected to pretreatment step as in Figure 1b.

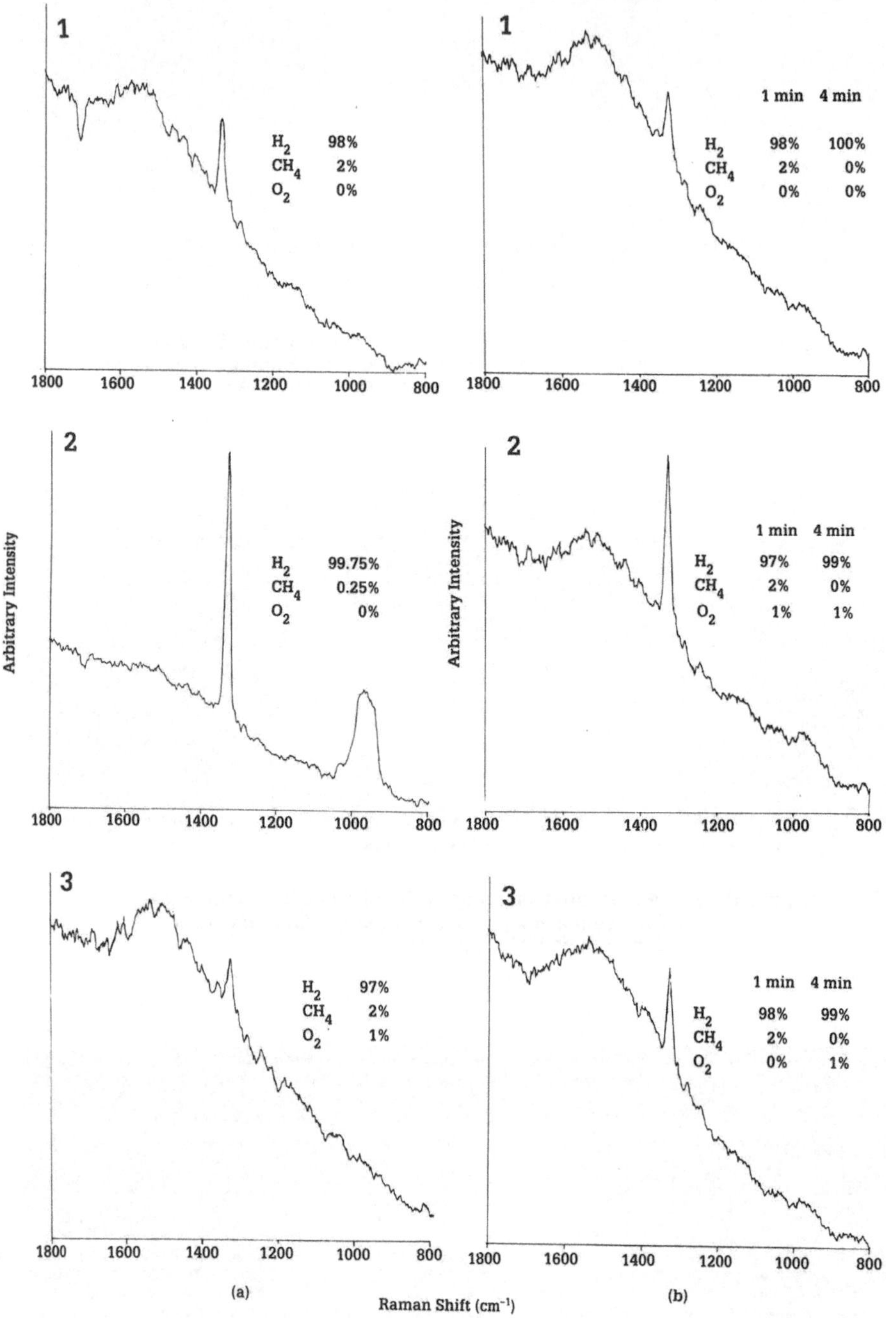

Figure 2. Raman Spectra of diamond synthesized on <100> oriented silicon (0.25 μm diamond abraded) substrates by Microwave Plasma CVD. (a) Continuous Flow and (b) Pulsed

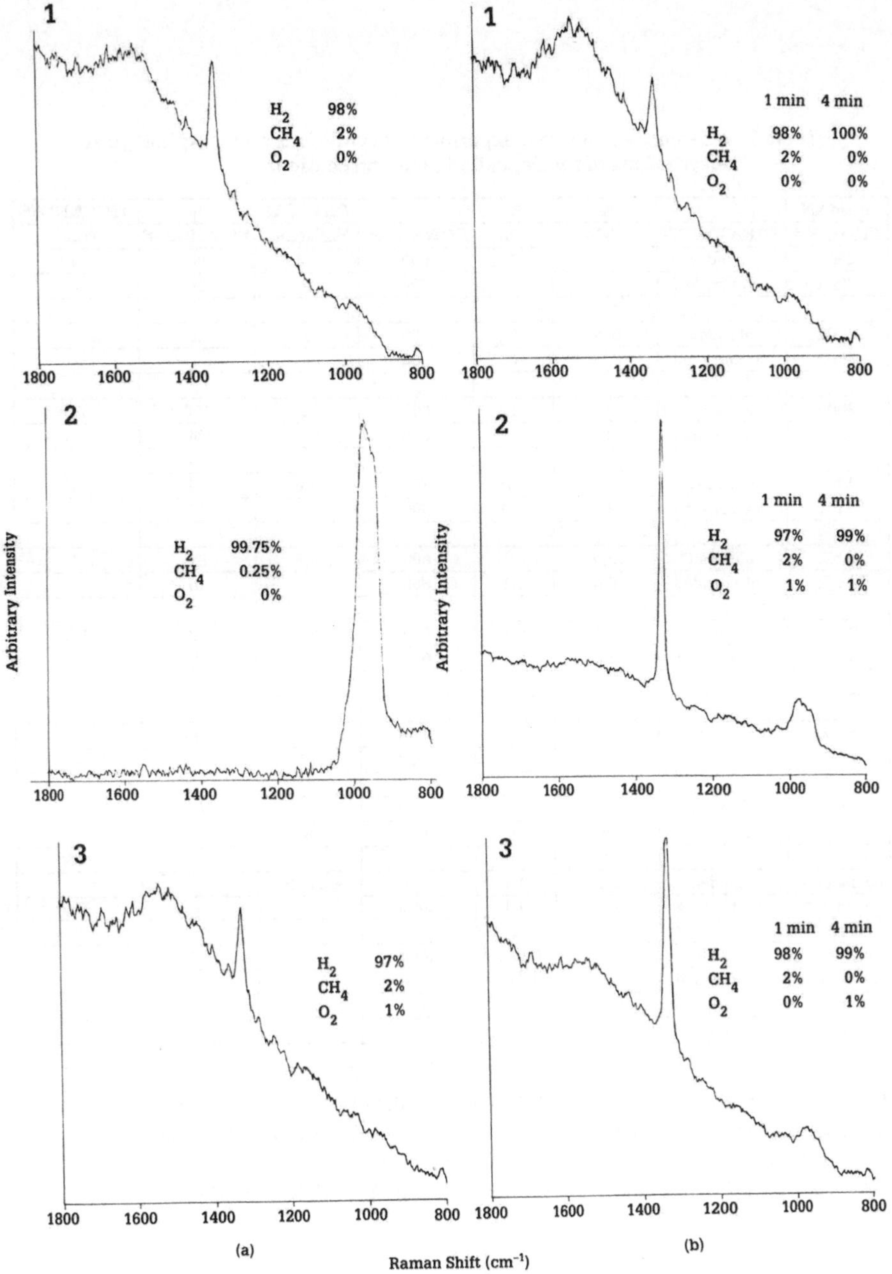

Figure 3. Raman Spectra of diamond synthesized on <100> oriented silicon (no diamond abrasion) substrates by Microwave Plasma CVD. (a) Continuous flow and (b) Pulsed

Table I. References to Raman, experimental conditions and morphological descriptions of the deposited diamond coatings.

RAMAN FIGURE #	TREATMENT	% CH4	% O2	POWER (W)	CRYSTAL SIZE(um)	% COVERAGE	THICKNESS (um)
2(a)1	abraded	2	0	600	2	98	2
3(a)1	not abraded	2	0	600	8	30	-
2(a)2	abraded	0.25	0	600	1	20	-
3(a)2	not abraded	0.25	0	600	-	-	-
2(a)3	abraded	2	1	600	2	100	7.5
3(a)3	not abraded	2	1	600	14.5	84	-
2(b)1	abraded	2,0	0,0	600	1.5	99	2
3(b)1	not abraded	2,0	0,0	600	3	14	-
2(b)2	abraded	2,0	1,1	600	3.5	81	2
3(b)2	not abraded	2,0	1,1	600	4.5	15	-
2(b)3	abraded	2,0	0,1	600	2.5	99	1.5
3(b)3	not abraded	2,0	0,1	600	5	12	-
2(b)2*	abraded	2	0	1000	2	100	4.5
3(a)3*	not abraded	2	0	1000	4.5	7	-
2(b)2*	abraded	2	1	1000	3	100	5.5
3(a)3*	not abraded	2	1	1000	6	100	5.5
no diamond	abraded	2,0	1,1	1000	etch pits	-	-
no diamond	not abraded	2,0	1,1	1000	-	-	-
no diamond	abraded	2,0	0.5,0.5	1000	-	-	-
no diamond	not abraded	2,0	0.5,0.5	1000	-	-	-
2(b)2*	abraded	2,0	0,0	1000	1	98	1
2(b)2*	not abraded	2,0	0,0	1000	1.5	9	

Raman spectra for depositions at 1000 W are not shown. The entries shown in Column 1 for 1000W (marked with*) are, in fact, for the 600 W since these spectra are indistinguishable. When there are 2 entries in Columns 3 and 4, they represent concentrations in the 1 min growth and 4 min etch cycles, respectively.

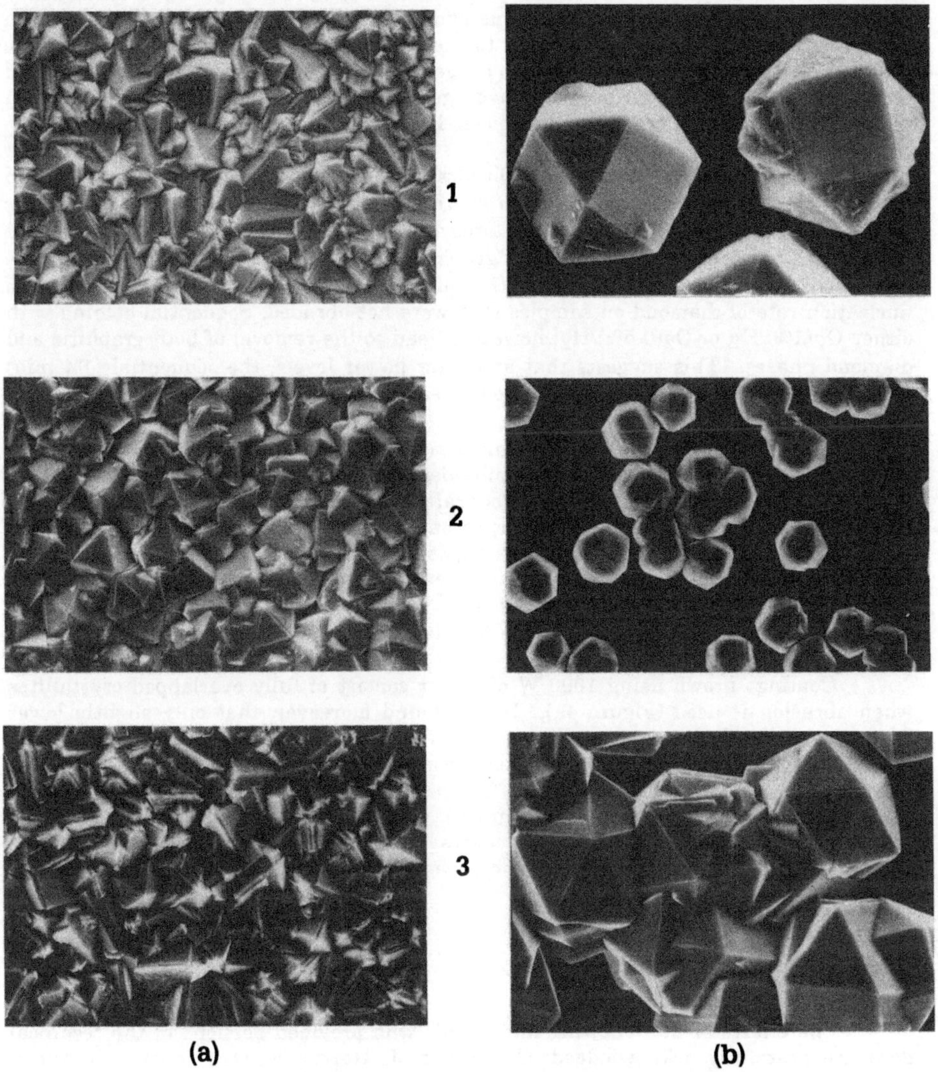

Figure 4. Electron Micrographs of Diamond Crystallites prepared at 1000 W (a) abraded, (b) not abraded. (1) continuous flow in H_2 (2) pulse–flow with H_2 etch (3) continuous flow with 1% O_2

Data obtained from wafers processed by both continuous and sequential growth schemes, at 600 watts, are shown in the balance of Figure 2a and 2b, respectively. As is seen, sequential mode growth using O_2, both during deposition and etching, led to the best-quality diamond coating on abraded substrates. Etching in hydrogen alone was found not effective in reducing the amount of graphitic phases (see Fig.2b(1)).

Raman spectra obtained from sections of wafers which were not abraded (but plasma-pretreated) prior to diamond deposition are displayed in Figure 3. Data obtained from wafers processed by both continuous and sequential growth schemes are shown in Figure 3a and 3b, respectively. Similar to the results obtained for abraded wafers, the best diamond growth was observed when oxygen was used both during the deposition and etching steps. In addition, etching in hydrogen alone was again found not effective in reducing the amount of graphitic phases (see Fig.3b(1)).

At higher microwave power (1000 watts) it was found that, without O_2 additions to the plasma, the graphite/diamond ratio at ≈100% coverage was as good as with lower microwave power with O_2 present. Moreover, sequential etching with H_2, even at 1000 W, did not decrease the amount of graphitic phase. The incorporation of O_2 in the continuous growth scheme improved the quality of all the films and improved the nucleation rate of diamond on samples that were not abraded. Sequential etching with either O_2(1%)/H_2 or O_2(0.5%)/H_2, however, lead to the removal of both graphitic and diamond phases. This suggests that at higher power levels, the sequential (1/4 min) etching with 0.5-1% O_2 is too high to allow selective removal of graphite.

The diamond grain sizes, film thicknesses, and coverage data of processed (600 W) wafers are listed in Table I. The results demonstrate that abrasion of the substrate leads to full coating coverage and generally smaller-sized crystallites than non-abraded areas of the substrates. Further, secondary nucleation and growth are observed on coatings prepared by the continuous deposition process. Coatings grown using the pulse technique are comprised of crystallites showing little evidence for secondary growth. Finally, growth of coatings at 600 W with continuous O_2 addition lead to the formation of the largest crystallites seen in this work.

Coatings grown using 1000 W of power consist of fully overlapped crystallites when abrasion is used (Figure 4a). It was found moreover, that only slightly lower density and slightly larger crystallites are grown in portions of the substrate which are not abraded prior to pretreatment and deposition (Figure 4b).

This investigation did reveal a growth process which minimizes the need of the abrasion step for the formation of high-coverage, high-quality diamond films. This was accomplished by using higher microwave power, higher CH_4 (≈2%) concentrations, and judicious (≈1%) use of O_2.

ACKNOWLEDGEMENTS

The efforts of the following individuals who provided portions of the technical data are gratefully acknowledged: G. Werber, J. Harris, K. Ostreicher. Technical discussions with C. Struck are greatly appreciated .

REFERENCES

1. H. Kawarada, K. Nishimura, K.S. Mar, Y. Yokota, J. Suzuki, T. Ito, A. Hiraki, "Diamond and Diamond-Like Materials Synthesis", Extended Abstracts, Materials Research Society, April 5-9, 89-92 (1988).

2. J.A. Mucha, D.L. Flamm, D.E. Ibbotson, *J. Apply. Phys.*, **65**(9), 3448-3452 (1989).

3. Y. Muranaka, H. Yamashita, K. Sato, H. Miyadera, *J. Apply. Phys.*, **67**(10), 6247-6254 (1990).

HYDROGEN DISSOCIATION AT HOT FILAMENTS: DETERMINATION OF ABSOLUTE ATOMIC HYDROGEN CONCENTRATIONS

L. Schäfer, U. Bringmann, and C.-P. Klages

Philips GmbH Forschungslaboratorium Hamburg
Vogt-Kölln-Straße 30, D-2000 Hamburg, FRG

U. Meier and K. Kohse-Höinghaus

DLR-Institut für Physikalische Chemie der Verbrennung, Pfaffenwaldring 38-40, D-7000 Stuttgart, FRG

INTRODUCTION

Several CVD methods, using thermal or plasma activation of the gas phase, are used for the formation of synthetic diamond films[1,2]. The deposition processes are performed at pressures of a few tens of mbar (microwave plasma-assisted CVD (MWPCVD)[3] or hot filament-assisted CVD (HFCVD)[4]) to atmospheric pressures (plasma jets[5] or flame synthesis[6]) with hydrocarbons and an abundance of molecular hydrogen in the gas phase. Although the detailed mechanisms are still in discussion, the importance of the presence of hydrogen for selective diamond formation is widely accepted. Atomic hydrogen, created by the activation of the gas atmosphere, can react with hydrocarbons to form hydrocarbon species like methyl or acetylene in the gas phase. These species are by some authors assumed to be key species in the process of diamond formation[7,8]. Furthermore the higher etching rate of H atoms for non-diamond carbon[9] is suspected to lead to the formation of thermodynamic metastable diamond as the only phase[10]. At last H atoms bonded to carbon on the surface may stabilize the sp^3-bonds necessary for diamond formation[11].

EXPERIMENTAL

The method of two-photon laser-induced fluorescence (LIF)[12] is based on the excitation of H atoms to the n = 3 level by absorption of two photons with 205 nm wavelength. Subsequent depopulation of the n = 3 level proceeds via radiationless quenching or transition to the n = 2 level with simultaneous fluorescence at 656 nm. A schematic view of the experimental set-up is shown in Fig. 1.

Diamond and Diamond-Like Films and Coatings
Edited by R.E. Clausing *et al.*, Plenum Press, New York, 1991

The LIF signals are calibrated by comparing the fluorescence intensities obtained in a filament reactor with those in a discharge flow reactor containing known H concentrations.[13] The different experimental conditions in both reactors are taken into account by temperature dependent calibration factors. These factors consider different quenching conditions[14] and the fact that at the high temperatures in the filament reactor the Doppler width of the H atom absorption line becomes larger than the bandwidth of the exciting laser system used. Radial temperature profiles, necessary for an evaluation of the calibration factors, are estimated using a stagnant film model introduced by Langmuir and coworkers for the description of heat losses from hot filaments to the surrounding gas phase.[15] With regard to statistical and systematical errors in the calibration the uncertainty in the determination of absolute values for atomic hydrogen concentrations is about a factor of 2, whereas the error for relative concentrations measured at equal pressures, that means equal quenching conditions, is approximately 10%, mainly given by the statistical error due to signal fluctuations.

The dependence of atomic hydrogen concentrations on methane addition and filament temperature in a filament-assisted diamond deposition system has also been investigated using resonance-enhanced multiphoton ionization (REMPI)[11]. Although this technique offers a good spatial resolution a quantitative determination of H concentrations was not possible because of the lack of a calibration for the REMPI signals.

RESULTS

Increasing the molecular hydrogen pressures from 1.5 mbar to 100 mbar shows a saturation of atomic hydrogen concentrations at pressures in excess of 10 mbar (Fig. 2; several data points measured at 100 mbar, yielding the same saturation values, are not shown). As the heating power was kept constant in these experiments, the filament temperatures, measured with a two-wavelength (0.87 μm and 0.90 μm) pyrometer, decrease with increasing pressure due to growing conductive heat losses to the gas phase. Temperature dependent profiles indicate values of c_H which decrease by about 20% for this temperature variation. This is not sufficient to explain the saturation of c_H because according to the trends of chemical equilibrium concentrations $c_{H,eq}$, calculated using thermodynamic data from literature[16], show a steady increase of $c_{H,eq}$ roughly proportional to the square root of pressure. At 1.5 mbar the measured concentrations near the filament are about 50% of the equilibrium values calculated for filament temperature. The deviations grow larger with increasing pressure until, at 100 mbar, experimental values are only about 12% of the equilibrium data. On the other hand the increase of c_H with temperature is nearly equal to what is suggested by equilibrium considerations, but absolute values at 30 mbar are only about 25% of $c_{H,eq}$, again calculated for filament temperatures. However the temperature dependence of c_H appears to be stronger than suggested by the relative H concentration measurements of Celii et al.[11].

Beside the saturation of $c_H(r)$ for pressures above 10 mbar and the deviation from $c_{H,eq}$ the kinetic limitations of hydrogen dissociation becomes evident from the dependence of atomic hydrogen profiles on filament diameter demonstrated by a com-

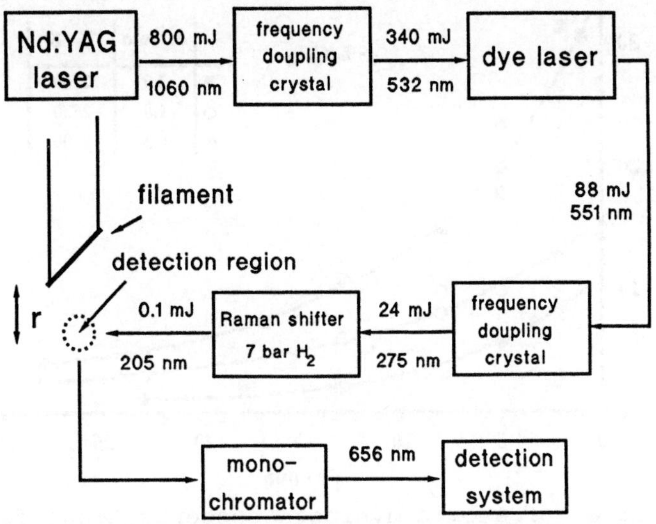

Fig. 1. Principle of the experimental set-up of two photon laser-induced fluorescence. The diameter of the detection region is about 0.25 mm. The distance r between filament and detection region can be varied between 0.4 mm and 28 mm by controlled movement of the filament.

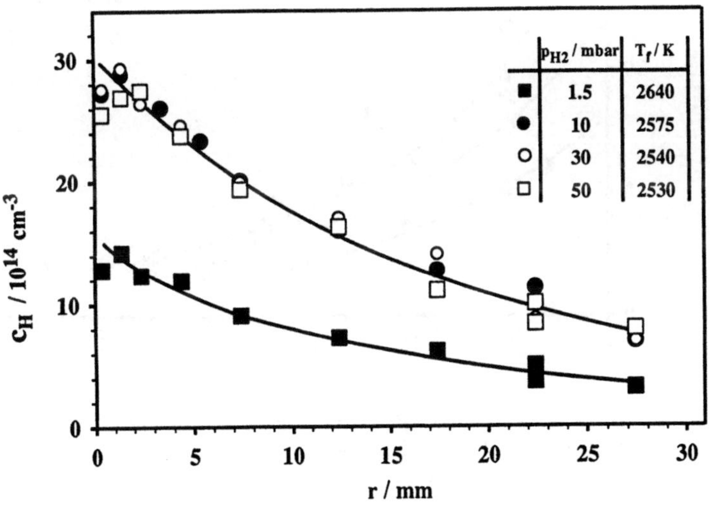

Fig. 2. Atomic hydrogen concentrations c_H versus distance r from the 2.0 mm diameter Ta filament surface for different molecular hydrogen pressures.

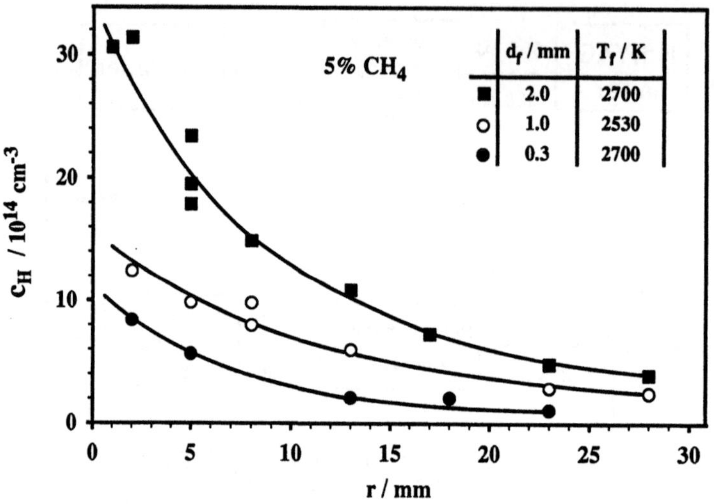

Fig. 3. Profiles of atomic hydrogen concentrations for Ta filaments with different diameter d_f. The gas phase contains 5% methane at a total pressure of 30 mbar.

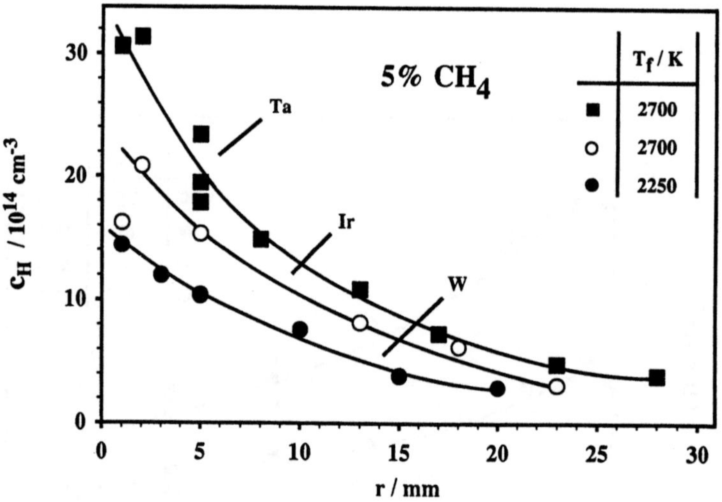

Fig. 4. Atomic hydrogen profiles for Ta, Ir and W filaments with 2.0 mm diameter. The filaments are placed in a gas phase containing 5% methane at a total pressure of 30 mbar.

parison of the concentration profiles for 0.3 mm and 2.0 mm diameter filaments, resp., (see Fig. 3). Because of their different filament temperatures the deviation between the profiles for 1.0 mm and 2.0 mm diameter, on the other hand, can be attributed to the temperature difference alone.

In chemical vapor deposition of diamond too large methane concentrations (typically 0.5%) have a detrimental effect on the phase purity of the growing film. This effect may be due to an acceleration of the hydrogen atom consumption in the gas phase by chemical reactions with hydrocarbons in addition to the relatively slow three-body recombination in pure hydrogen. To test the influence of methane on the atomic hydrogen profiles experiments with Ta filaments in a gas atmosphere containing 5% CH_4 were performed. Comparing the profiles for 2.0 mm filament diameter in Fig. 2 with that of Fig. 3 the maximum H concentration near the filament is more than 30% lower after the addition of methane. Simultaneously, the concentration profile becomes steeper, indicating the expected accelerated consumption of H in the gas phase.

The previous results show that kinetic factors determine the surface concentrations of atomic hydrogen on hot filaments. Therefore the question for filament materials with high efficiency for hydrogen dissociation is raised. Profiles presented in Fig. 4 indicate, that the Ta filament (carburized) generates somewhat higher H concentrations than the Ir filament, heated nearly to its melting point. In view of its relatively low temperature, however, the tungsten filament yields the highest H concentrations (smallest deviation from equilibrium value).

THEORETICAL CONSIDERATIONS

The diffusional collision integral quoted by Stefanov and Zarkova[18] can be used to calculate the atom/molecule diffusion coefficient D of hydrogen. The result is 3.95 cm^2/s at p = 1 bar and T = 273 K. D increases with 1/p and $T^{3/2}$. Its temperature dependence has to be taken into consideration if concentration profiles of hydrogen atoms around hot filaments are to be calculated. Within the Langmuir model[15] a relation between H atom concentrations, expressed as molar fractions x_H, and the temperature profile T(r) around the wire can be obtained as a solution of the differential equation

$$\mathrm{div}[D(T(r))*c(T(r))*\mathrm{grad}x_H] = 0. \quad (1)$$

The boundary conditions are $x_H=x_{H,0}$ at the filament surface and $x_H=0$ at the boundary of the Langmuir-type stagnant gas layer of radius r_L around the filament of radius r_f. c is the total concentration of atomic and molecular hydrogen. The temperature distribution within the stagnant layer is[19]

$$T(r) = \{(T_F^{\tau+1}-T_L^{\tau+1})*\ln(r/r_F)/\ln(r_F/r_L) + T_F^{\tau+1}\}^{1/(\tau+1)}, \quad (2)$$

τ being an exponent accounting for the temperature dependence T of heat conductivity (0.68 for molecular hydrogen[20]). The solution of eq. (1) is then

$$x_H(r) = x_{H0} * (T(r)^{\tau+0.5} - T_L^{\tau+0.5}) / (T_F^{\tau+0.5} - T_L^{\tau+0.5}), \quad (3)$$

demonstrating the temperature profile dependence of $c_H(r)$.

In order to include, in addition to ordinary diffusion, thermal diffusion as well as chemical reactions in the gas phase, eq. (1) has to be extended to

$$\mathrm{div}\{D*c*[\mathrm{grad}x_H + k_T*\mathrm{grad}(\ln T)]\} + 2*k_d*c^2 - 2*k_r*c^3*x_H{}^2 = 0 \quad (4)$$

For low concentrations of atomic hydrogen the thermal diffusion ratio k_T can be expressed as[21] $k_T = \alpha * x_H$, α being about -0.16 for H in H_2[20].

Eq. (4) holds for low degrees of dissociation where the total gas concentration c is nearly identical to the concentration of molecular hydrogen. For the constants k_d and k_r in the terms for homogeneous dissociation and recombination, resp., expressions recommended by Warnatz[22] were used:

$$k_d = 8.8*10^{14}*\exp(-48500/T) \ \mathrm{cm^3/mol/s}$$
$$k_r = 9.7*10^{16}*T^{-0.6} \ \mathrm{cm^6/mol^2/s} \quad (5)$$

Eq. (4) was solved numerically using a finite-difference method. The formula given by Covington[23] was used to estimate the Langmuir radius r_L necessary for the temperature profile calculation. Some results for a 2.0 mm diameter filament at 2620 K in hydrogen of 30 mbar pressure are given in Fig. 5 as $c_H(r)$ curves. Curve A was obtained assuming $\alpha=k_d=k_r=0$ except at the filament surface, where the ratio k_d/k_r was used to calculate the equilibrium hydrogen atom concentration, $2.2*10^{16}$ $\mathrm{cm^{-3}}$, in reasonable agreement with the value of $1.7*10^{16}$ $\mathrm{cm^{-3}}$ calculated from a thermodynamic data base.[16] Curve A is dominated by diffusion only and is identical to the result of eq. (3). Inclusion of thermal diffusion ($\alpha = -0.16$) yields curve B, curve C is obtained upon additional inclusion of homogeneous recombination. Taking into account the dissociation reaction (curve D) does not have any discernible effect on the appearance of the $c_H(r)$ distribution. Curve E corresponds to a hypothetical situation with $D=\alpha=0$. It represents the equilibrium distribution in a temperature field T(r). A comparison of these curves shows that the concentration distribution $c_H(r)$ is largely determined by the concentration diffusion of hydrogen atoms from the filament to the boundary of the stagnant layer while homogeneous recombination as well as thermal diffusion play only minor roles. The numerically calculated concentrations are, however, larger than the experimental ones (see Fig. 5), indicating a nonequlibrium situation at the wire surface under the conditions used in the experiments.

In order to discuss the influence of the filament radius on the hydrogen atom concentration profiles, it is useful to consider two limiting cases:

A) If chemical reactions at the filament surface proceed with infinite rate, chemical equilibrium with concentration $c_{H,eq}$ is always guaranteed at the hot wire, independent of its radius. Assuming a constant temperature and diffusion only the concentration distribution is given by

$$c_H(r) = c_{H,eq}*\ln(r/r_L)/\ln(r_f/r_L). \qquad (6)$$

If the wire is made thinner, the generation rate per unit area of the filament surface has to increase in order to keep step with the diffusion flux away from the filament. (The same is true for heat generation. The power input per unit area will therefore increase with decreasing filament diameter, if the temperature is held constant, see e. g. ref. 24.)

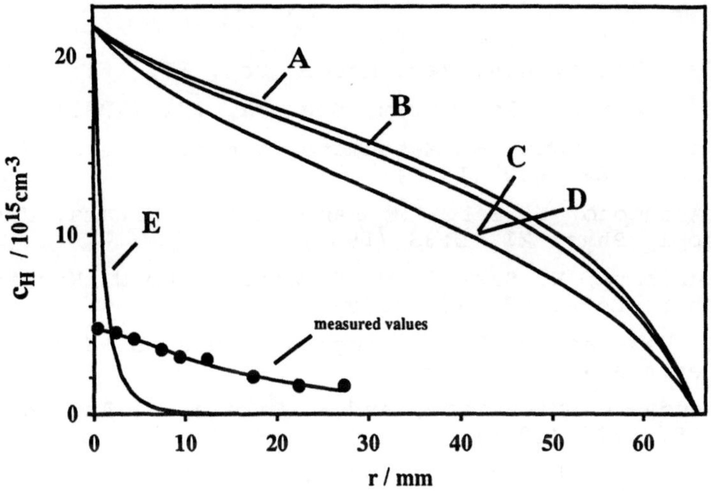

Fig. 5. Comparison of an atomic hydrogen profile measured for a Ta filament with 2.0 mm diameter at a temperature of 2620 K with different calculated profiles. The total hydrogen pressure is 30 mbar in each case.

B) If, on the other hand, the generation rate of hydrogen atoms q_H (atoms/cm^2/s) at the surface is kinetically limited, the concentration $c_{H,0}$ is determined by the competition between generation at the surface and diffusion to the film boundary at r_L. The profile is now

$$c_H(r) = (q_H*r_f/D)*\ln(r_L/r), \qquad (7)$$

satisfying the conservation equation $(D*dc_H/dr)_{rf} = -q_H$. The surface concentration $c_{H,0}$ is proportional to $r_f*\ln(r_L/r_f)$, i.e., it increases with r_f as demonstrated by Fig. 3.

CONCLUSIONS

Atomic hydrogen concentration profiles $c_H(r)$, determined by two-photon laser-induced fluorescence in a filament reactor show a saturation for pressures above 10 mbar, a deviation from calculated thermal equilibrium profiles and a pronounced dependence on filament diameter. The kinetic limitations of the hydrogen dissociation under the chosen conditions may be taken into consideration when selecting filament materials and geometries for hot-filament assisted CVD of diamond. Furthermore the accelerated decrease of atomic hydrogen concentrations in the presence of small amounts of hydrocarbons may be an explanation for the deterioration of the phase purity of CVD diamond at hydrocarbon concentrations of more than about 0.5%.

ACKNOWLEDGEMENTS

The work was supported in part by the Bundesministerium für Forschung und Technologie, grant No. 13N5607.

REFERENCES

1. R. C. DeVries, Ann. Rev. Mater. Sci. **17**, 161 (1987)
2. K. E. Spear, J. Am. Ceram. Soc. **72**, 171 (1989)
3. M. Kamo, Y. Sato, S. Matsumoto, and N. Setaka, J. Cryst. Growth **62**, 642 (1983)
4. S. Matsumoto, Y. Sato, M. Kamo, and N. Setaka, Jpn. J. Appl. Phys. **21**, L183 (1982)
5. K. Kurihara, K. Sasaki, M. Kawarada, and N. Koshino, Appl. Phys. Lett. **52**, 437 (1988)
6. L. M. Hanssen, W. A. Carrington, and J. E. Snail, Materials Letters **7**, 289 (1989)
7. M. Tsuda, M. Nakajiama, and S. Oikawa, J. Am. Chem. Soc. **108**, 5780 (1986)
8. D. Huang, M. Frenklach, and M. Maroncelli, J. Phys. Chem. 92, 6379 (1988)
9. N. Setaka, J. Mater. Res. **4**, 664 (1989)
10. B. V. Spitsyn, L. L. Bouilov, B. V. Derjaguin, J. Cryst. Growth **52**, 219 (1981)
11. F. G. Celii and J. E. Butler, Appl. Phys. Lett. **54**, 1031 (1989)
12. U. Meier, K. Kohse-Höinghaus and Th. Just, Chem. Phys. Letters **126**, 567 (1986)
13. U. Meier, K. Kohse-Höinghaus, L. Schäfer, and C.-P. Klages, submitted to Applied Optics
14. J. Bittner, K. Kohse-Höinghaus, U. Meier, S. Kelm and Th. Just, Combustion and Flame **71**, 41 (1988)
15. K. B. Blodgett and I. Langmuir, Phys. Rev. **40**, 78 (1932) and references therein
16. R. Duff and S. H. Bauer, J. Chem. Phys. **36**, 1754 (1962)
17. J. Crank, The Mathematics of Diffusion, Oxford University Press (1956)
18. B. Stevanov and L. Zarkova, J. Phys. D: Appl. Phys. **9**, 1217 1976
19. E. Schnedler, Philips J. Res. **38**, 224 (1983)

20. G. Dittmer, personal communication

21. R. B.Bird, W. E. Steward, and E. N. Lightfood, Transport Phenomena, John Wiley & Sons (1960)

22. J. Warnatz in Combustion Chemistry, W. C. Gardiner, Jr. (Ed), Springer-Verlag (1984)

23. E. J. Covington, Illuminating Engineering **63**, 134 (1968)

24. F. Jansen, I. Chen, and M. A. Machonkin, J. Appl. Phys. **66** 5749 (1989)

EMISSION SPECTROSCOPY OF THE MICROWAVE PLASMA USED FOR DIAMOND DEPOSTION: IMPORTANCE OF THE C_3 RADICAL

A. Campargue, M. Chenevier, and F. Stoeckel

Laboratoire de Spectrométrie Physique
Université Joseph-Fourier/Grenoble 1 (associé CNRS UA 08),
BP 87, 38402 Saint-Martin d'Hères Cédex, France

B. Marcus, M. Mermoux, and F. Vinet

S2MC (associé CNRS UA 143)
BP 75, 38402 Saint-Martin d'Hères Cédex, France

S. Ljungström

Department of Physics, Chalmers University of Technology
S-412 96 Göteborg, Sweden

I - INTRODUCTION

Knowledge of intermediate species present in the plasma during chemical vapor deposition (CVD) of diamond is a necessary first step to the understanding of this complex process. Several hydrocarbons have been proposed as such key intermediate species for diamond deposition. Frenklach and Spear [1] suggested a growth mechanism from acetylene, C_2H_2, while other authors assumed that the methyl radical, CH_3, plays the most important role [2-6]. A mechanism considering only the CH_n radicals has been recently proposed by Chen [7].

Remote gas sampling techniques cannot directly detect highly reactive chemical species and therefore in situ, real time experiments seem to be the only way to identify the consecutive reaction steps leading to diamond. Optical detection is probably one of the best suited methods to detect intermediate reaction products and a high sensitivity is normally needed because of the generally short lifetimes of the transient species. Furthermore, laser spectroscopy gives the opportunity to obtain the spatial distribution of the different species in the plasma near the substrate. Various optical emission measurements have been reported [5, 6, 8-10, 14] and emission from atomic and molecular hydrogen, CH and C_2 is generally observed. Infrared absorption spectroscopy is another powerful and quantitative technique as most species present in the plasma are advantageously detected by vibrational spectroscopy in the ground electronic state. For example, high concentrations of acetylene have been measured by this technique [11]. Acetylene has also been detected during filament assisted diamond deposition by on line mass spectrometry [8, 12]. Celii and Butler [13] have recently discussed the role of atomic hydrogen detected by resonance enhanced multiphoton ionization.

In order to improve the understanding of the diamond deposition mechanism we have performed a systematic study of the emission light from a microwave plasma in the CH_4/H_2 system and monitored the variations in the signals from H, H_2, CH, C_2, and C_3 when the methane concentration has been changed and at the same time we have also studied the diamond deposit on a silicon substrate with Scanning Electron Microscopy (SEM).

II - EXPERIMENTAL

An overview of the experimental set up is given in Fig. 1. The apparatus used in the present work was a standard plasma assisted CVD system using a microwave cavity. The reactor consists of a quartz tube (50 mm in diameter) crossing a rectangular wave guide (45 x 90 mm)connected to a 1.6 kW magnetron (2.45 GHz). The typical microwave power was varied from 200 to 400 W and the quartz substrate holder was placed in the centre of the discharge. The substrate was heated by the plasma and its temperature could be estimated to 800 - 950°C according to the deposition conditions. The flow rates of CH_4 and H_2 were controlled separately with mass flowmeters so that the total flow rate of the feed gases was 100 sccm with the CH_4 concentration varying from 0 to 10 vol. %. The total pressure was measured with a 0-1000 Torr absolute pressure gauge and adjusted from 1 to 30 Torr by a leak valve.

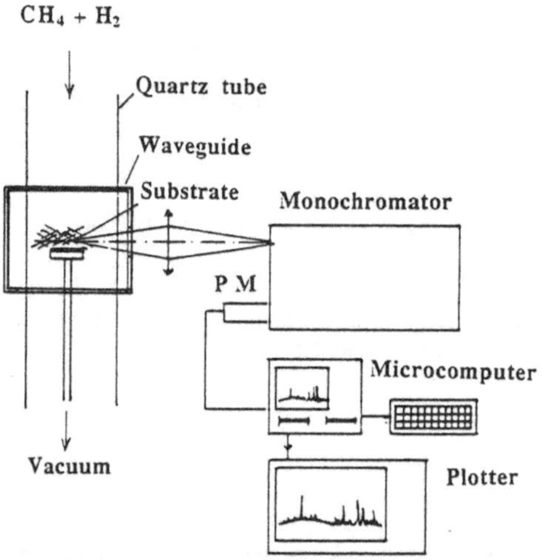

Fig. 1. Experimental apparatus used for diamond deposition and plasma emission spectroscopy.

Fig. 2. SEM picture of a diamond film grown at a CH_4 concentration of 0.5 % and a total pressure of 20 Torr (microwave power 300 W).

All the experiments presented here were carried out with 10 x 10 µm (100) silicon wafers (N type, 5.8 Ω cm). The silicon substrates were scratched with diamond paste (3 mm granulometry) , carefully ultrasonically cleaned until no diamond crystal was observable in an optical microscope and treated by a pure hydrogen plasma before deposition.

The morphological and structural quality of the different coatings were investigated by Scanning Electron Microscopy (SEM) and by Raman spectroscopy respectively. Fig. 2 shows an example of a polycrystalline well-faceted layer of diamond synthetized on a silicon substrate for a 0.5 % methane concentration in 20 Torr of hydrogen. The Raman spectrum (Fig. 3) presents the characteristic diamond sharp peak centered at about 1333 cm^{-1}. For methane concentrations higher than 1%, the deposits present a more spherical, also called ball-like morphology. The corresponding Raman spectra reveal an increasing content of amorphous carbon as shown by the appearance and growing of a large band around 1550 cm^{-1}.

The plasma emission spectra were recorded during the discharge through a 1 mm wide slit in the waveguide. The slit direction was parallel to the current lines in the waveguide and perpendicular to the substrate holder. The typical resolution of our monochromator (HRS type from the Jobin et Yvon company) was 1 Å. The spectra were obtained by scanning the wavelength of the monochromator from 300 to 700 nm and the signal delivered by the photomultiplier was stored and pre-analysed by a micro-computer.

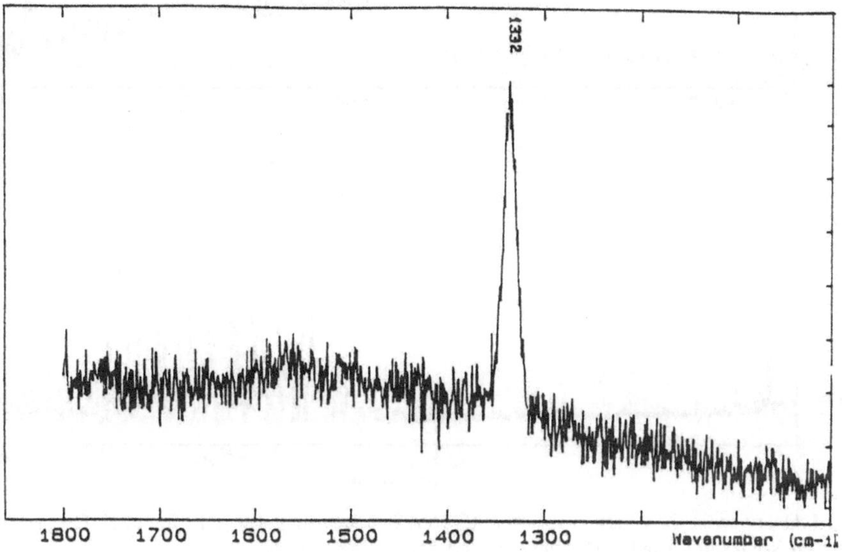

Fig. 3. Micro Raman spectrum of the diamond deposit shown in Fig. 2.

III - OPTICAL EMISSION SPECTROSCOPY OF THE MICROWAVE PLASMA

We are first going to discuss the results we obtained without substrate in the plasma.

Fig. 4 gives an overview of a typical emission spectrum of the plasma between 380 and 650 nm. In this experiment the pressure was 10 Torr of hydrogen with 1 % of CH_4 . The emission of five different species can be assigned in this spectrum :

Atomic hydrogen : six lines of the Balmer series are observed.

Molecular hydrogen : in the range of 600 nm, the Fulcher bands of H_2 are observed. They correspond to different rovibrational transitions of the 3p $^3\Pi_u$ - 2s $^3\Sigma_g^+$ electronic de-excitation. As shown by Baltayan and Nédélec [16], the relative intensities in the Q branches can be used to estimate the temperature of the discharge. However, in the present work, the resolution of the different rotational lines is not sufficient to use this method.

The C_2 radical : the Swan bands of the C_2 radical is frequently observed in flames and discharges involving hydrocarbons. These bands are assigned to vibrational sequences of the d $^3\Pi_g$ - a $^3\Pi_u$ transition. The (0,0) sequence at 516.52 nm has been previously reported in diamond deposition plasmas [8-10, 15] and appears in our spectra for high methane concentration (> 2 %).

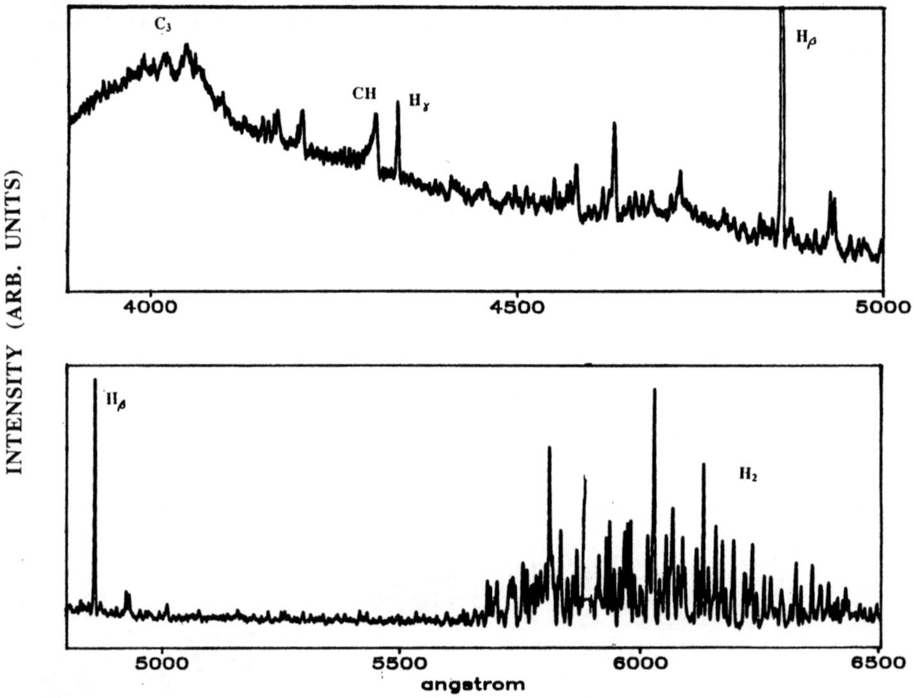

Fig. 4. Overview of the emission spectrum (5 Torr, 2 % methane). The emission of H, CH and C_3 are assigned (see text for details). The other lines arise from H_2 emission.

The CH radical : the two bands observed in the range of 430 nm and 390 nm are assigned to the A $^2\Delta$ - X $^2\Pi$ and B$^2\Sigma$ - X$^2\Pi$ transitions respectively. Fig. 4 shows the partly resolved rotational structure of these transitions. The signal results in fact of the superposition of the (0,0) and (1,1) vibrational sequences (see for example Refs. 17 and 18).

The C_3 radical : the three broad peaks observed at 399, 402 and 405.9 nm are assigned to the emission spectrum of C_3 [19]. These bands, also called the comet-heat group [20, 21], are usually observed in laboratory experiments in ac and microwave discharges, through flowing hydrocarbons. They are assigned to transitions from different vibrational states of the $\tilde{A}$ $^1\Pi_u$ excited state to the $\tilde{X}$ $^1\Sigma^+$ ground state. The analysis of this band is complicated by a large l-type doubling and a strong Renner-Teller effect due to the very low bending frequency [22].

To the best of our knowledge, the C_3 molecule has not been mentioned in the previous published studies of the emission spectrum of the plasma used in the synthesis of metastable diamond. However, the emission spectrum obtained by Saito et al [5, 6] shows a group of lines near 405 nm which they assigned to CH_2 but which is more likely due to the C_3 radical. The observation of the C_3 radical is not surprising since thermodynamical calculations [23] show that the mole fraction of C_3 at high temperature (more than 3000 K) is expected to be higher than the mole fractions of C_2 and CH.

IV - RESULTS AND DISCUSSION

Figure 5 shows the evolution of the emission spectrum in the range of 400 nm with an increasing percentage of methane up to 5 % in the gas mixture. The total pressure was 9 Torr

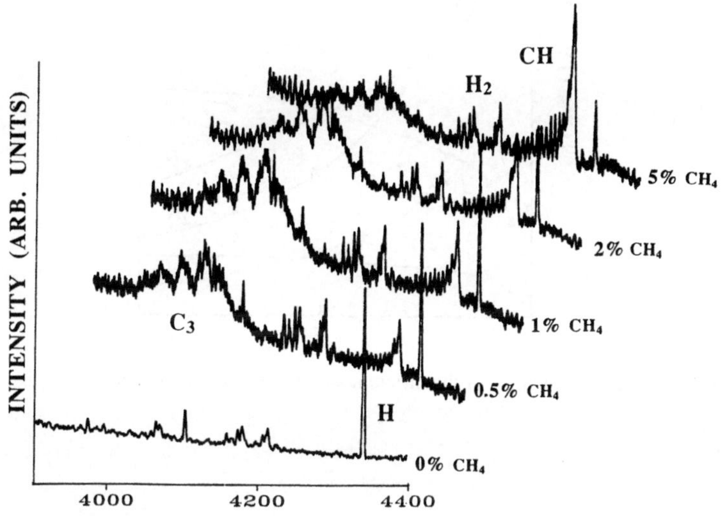

Fig. 5. Evolution of the emission spectrum in the range of 400 nm versus the methane concentration. The total pressure was 9 Torr.

and the different spectra were recorded in the central part of the discharge without substrate on the substrate holder. The influence of the presence of the substrate will be discussed below. These spectra clearly show the increase of the emission intensity of the CH and C_2 bands with the percentage of CH_4 while the emission intensity of the C_3 band is maximum for about 1 % of methane and decreases rapidly for higher percentages. At the same time the H_b intensity decreases monotonously with the CH_4 concentration. The emission intensity of the observed bands depends on the electronic quenching and is proportional to the population of the excited state and to the transition probability (i.e. the inverse of the lifetime of the excited state).

The population of the excited state is a function of the electron energy distribution, the excitation cross section and, of course, the population of the ground state. If one assumes that the conditions of the discharge (electron density and energy distribution) are not dramatically changed by the increase of the methane concentration, at a given total pressure, the emission intensity variation for the different species is supposed to reproduce, at least qualitatively, the variation of the total concentration of these species.

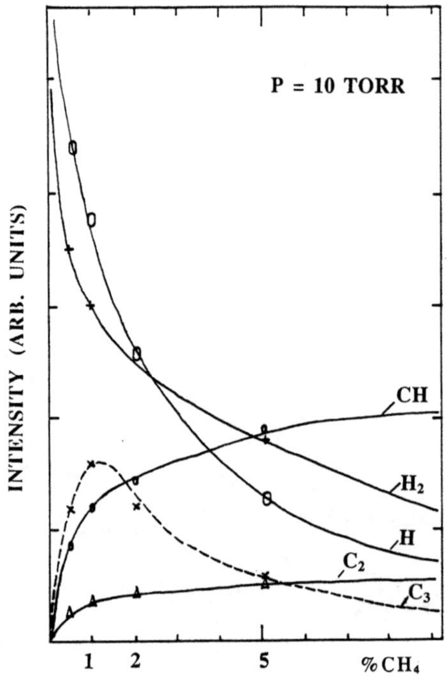

Fig. 6. Emission intensity of the different species versus the methane percentage.

The main striking feature in Fig. 6, which shows the evolution of the emission intensity of the different species versus the methane concentration, is the C_3 emission intensity variation. The other species have a monotonous variation while the C_3 emission shows a marked maximum at about 1 % methane concentration which is the standard concentration for a well faceted diamond deposition. Furthermore, the quick decrease of H_β intensity from 0 to 2 % methane concentration must be underlined. The etching rate by atomic hydrogen is supposed to be much faster for graphite than for diamond [25] and a high concentration of atomic hydrogen is expected to favour the diamond deposition

How is the emission spectra affected by the presence of the silicon substrate in the plasma? Fig. 7 shows two spectra recorded in the same conditions (5 Torr, 2 % methane) with and without substrate on the substrate holder. The C_3 emission appears much weaker when the substrate is introduced while the emission of the other species do not show noticeable changes. Spatial distribution studies of the C_3 emission intensity confirms this result. Our experimental apparatus does not allow a spatial resolution better than 1 cm. However, we have clearly observed an increase of the C_3 emission intensity when the distance between the observation and the substrate is increased. At a distance of 10 cm the spectrum is similar to the spectrum without substrate.

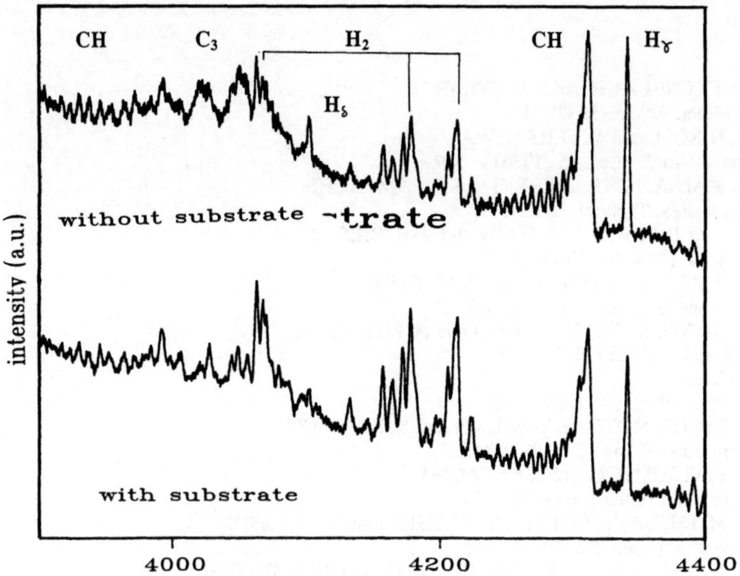

Fig. 7. Comparison of two spectra obtained in the same conditions (5 Torr, 1 % methane) with (a) and without (b) substrate on the substrate holder.

No time dependence of the emission spectrum has been observed within a time scale of an experiment of diamond deposition (several hours). Furthermore, the quality of the deposit is not only correlated to the emission spectrum of the plasma but depends on the substrate surface too.

V - CONCLUSION

Our observations clearly show that the C_3 radical is interacting with the substrate and that the mole fraction of methane producing the maximum C_3 emission corresponds to the best suited conditions for diamond deposition. However, the role of the C_3 radical must be clarified. Many attempts have been done and are reported in the literature to identify a key species in diamond deposition but we believe, in agreement with Harris and Weiner [24], that there probably is no single species acting as the growth species. Direct and in situ experimental data are really needed to understand this complex process.

Most of the modeling of the gas phase reactions during diamond film growth consider reactions exclusively involving C_nH_m species with $m \neq 0$ [1, 4, 7, 12, 24] i.e. neglecting C_n species. Our experimental results shows that C_n species (at least C_2 and C_3) are present in the plasma and cannot a priori be neglected.

ACKNOWLEDGMENTS

The authors would like to thank B. TRETOU for helpful discussions and L. ABELLO (Laboratoire d'Ionique et d'Electrochimie du Solide) for the micro Raman experiments.

[1] M. FRENKLACH and K.E. SPEAR
J. Mater. Res. 3 (1988) 133
[2] Y. HIROSE and Y. TERASAWA
Jpn. J. Appl. Phys. 25 (1986) L 519
[3] R. MANIA, L. STOBIERSKI and R. PAMPACHI
Cryst. Res. Technol. 16 (1981) 785
[4] L.E. KLINE, W.D. PARTLOW and W.E. BIES
J. Appl. Phys. 65 (1989) 70
[5] Y. SAITO, S. MATSUDA and S. NOGITA
J. Mater. Sci. Lett. 5 (1986) 565
[6] Y. SAITO, K. SATO, K. GOMI and H. MIYADERA
J. Mater. Sci. 25 (1990) 1246
[7] I. CHEN
J. Appl. Phys. 64 (1988) 3742
[8] O. MATSUMOTO, H. TOSHIMA and Y. KANZAKI
Thin Solid Films 128 (1985) 341
[9] O. MATSUMOTO and T. KATAGIRI
Thin Solid Films 146 (1987) 283
[10] Y. MITSUDA, Y. KOJIMA, T. YOSHIDA and K. AKASHI
J. Mat. Sci. Lett. 22 (1987) 1557
[11] F.G. CELII, P.E. PEHRSSON, H.-I. WANG and J.E. BUTLER
Appl. Phys. Lett. 52 (1988) 2045
[12] S.J. HARRIS, A.M. WEINER and T.A. PERRY
Appl. Phys. Lett. 53 (1988) 1605
[13] F.G. CELII and J.E. BUTLER
Appl. Phys. Lett. 54 (1989) 1031
[14] K. KOBAYASHI, N. MUTSUKURA and Y. MACHI
J. Appl. Phys. 59 (1986) 910
[15] J.A. MUCHA, D.L. FLAMM and D.E. IBBOTSON
J. Appl. Phys. 65 (1989) 3448
[16] P. BALTAYAN and O. NEDELEC
J. Physique 36 (1975) 125
[17] H. SEKIYA, T. HIRAYAMA and Y. NISHIMURA
Chem. Phys. Letters 138 (1987) 597
[18] J.J. HORVATH, K.M. PAMIDIMUKKALA, W.B. PERSON and A.E.S. GREEN
J. Quant. Spectrosc. Radiat. Transfer 31 (1984) 189
[19] K.H. BECKER, T. TATARCZYK, J. RADIC-PERIC
Chem. Phys. Letters 60 (1979) 502
[20] L. GAUSSET, G. HERZBERG, A. LAGERQVIST and B. ROSEN
Astrophys. J. 142 (1965) 45
[21] A.E. DOUGLAS
Astrophys. J. 114 (1951) 466
[22] A.J. MERER
Can. J. Phys. 45 (1967) 4103
[23] R.F. BADDOUR and J.L. BLANCHET
Ind. Eng. Chem. Process Design and Develop. 3 (1964) 258
[24] S.J. HARRIS and A.M. WEINER
J. Appl. Phys. 67 (1990) 6520
[25] T.R. ANTONY
To be published in Diamond, Boron Nitride, Silicon Carbide and Related Wide Bandgap Semiconductors, Edited bu J.T. Glass, R.F. Messier and N. Fujimori, Proceedings of the 1989 MRS Fall Meeting, Volume 162 (1990)

ION ENERGY DISTRIBUTION AT THE SUBSTRATE IN AN ECR-REACTOR

P. Reinke, W. Jacob, and W. Möller

Max-Planck-Institute für Plasmaphysik
Euratom-Association
D-8046 Garching, Federal Republic of Germany

1. INTRODUCTION

Electron cyclotron resonance (ECR) excited plasmas are especially useful for processes requiring low pressure and low gas temperature, and for the deposition of insulating layers [1,2]. In our case the ECR plasma is generated by coupling a circularly polarized microwave (2.45 GHz) into the reaction chamber. Resonance with the electrons gyrating in the overlaying magnetic field at 87.5 mT leads to an acceleration of the electrons, which in turn ionize the processing gas.
For the deposition of a-C:H and polymerlike hydrocarbon films it has been shown that the impact energy of ions on the substrate has a decisive effect on microstructure and macroscopic properties of the deposited films [3]. The possible control of ion energies by varying the magnetic field configuration makes the ECR plasmas attractive for a-C:H film deposition as it offers the possibility to intentionally alter the film properties. In order to investigate this relationship an ion energy analyzer (IEA) was introduced into the reaction chamber in place of the substrate. Ion energy distributions in argon and methane under various discharge conditions were measured and the results are discussed below.

2. EXPERIMENTAL SETUP

2.1. ECR-REACTOR AND MAGNETIC FIELD CONFIGURATION

The reaction chamber is a cylindrical stainless steel vessel with an inner diameter and height of 26 cm each and several diagnostic ports. The feeding of microwave power into the chamber is accomplished in two different ways. In a first set of experiments the microwave power is fed axially into the reaction chamber through a quartz window from a direction, in which the magnetic field strength exceeds the resonance value (fig. 1). Due to the large distance between resonance zone and IEA (about 30 cm) only a very low ion current density could be obtained (about 10^{-6}A/cm^2), in spite of a high microwave power input around 380 W. In a second set of experiments feeding of the microwave is done with a Lisitano coil [4]. This coil is a type of cylindrically shaped, slotted line radiator and directly inserted into the reaction chamber (fig. 2). It is connected to a magnetron microwave power generator via a coaxial feedthrough.

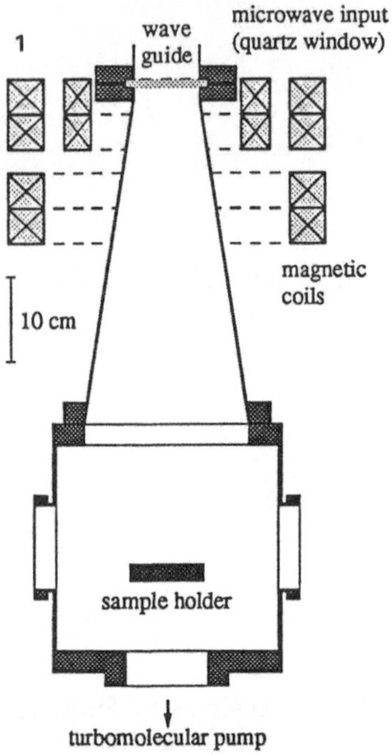

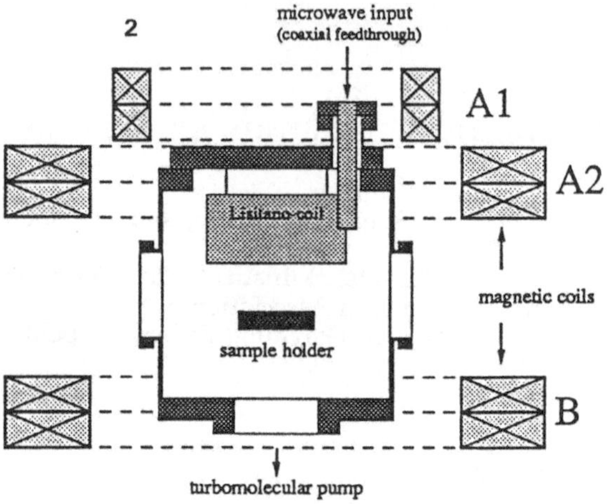

Fig. 1 left: conventional coupling (setup 1)
Fig. 2 right: Lisitano coil coupling (setup 2), the coils A1 and A2 are refered to as main coils, coil B is called subcoil

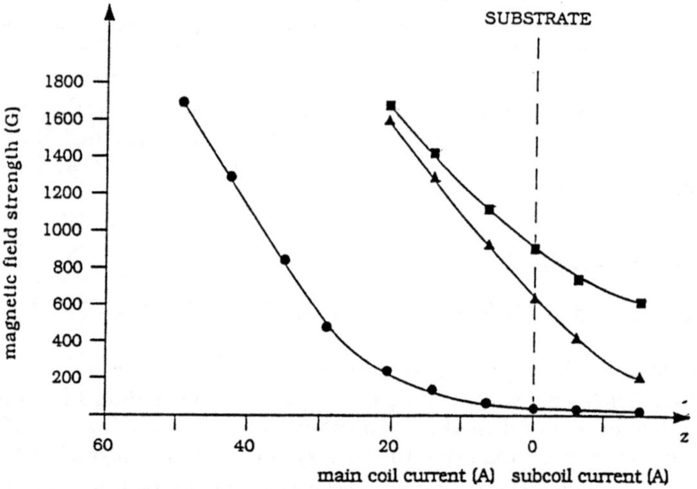

SETUP 2	▲	410	0
	■	410	200
SETUP 1	●	270	0

Fig. 3. Calculated magnetic field profiles on the center axis of the reactor for both setups with different coil currents

The distance between resonance zone and IEA is much smaller (about 15cm) so that the ion current density increases by a factor of 40 to 80. The chamber is pumped to a base pressure in the 10^{-6} mbar range. The gas flow is governed by flow controllers. The chamber is equipped with a differentially pumped quadrupole mass spectrometer connected to the chamber by a capillary of 0.3 mm diameter. Two sets of axially movable cylindrical coils are used to create the magnetic field necessary to establish the ECR condition as well as to vary the magnetic field configuration over a wide range. Three field strength profiles for the two different setups and variable coil currents are shown in fig. 3 (previous page).

2.2. ION ENERGY ANALYZER (IEA)

A retarding field analyzer (fig. 4) contained in a stainless steel cylinder of 22 mm in diameter and 35 mm in height is used to measure the ion energy distribution. It is located inside the reactor at the position of the substrate. Grid distances were chosen according to Prokopenko et al.[5,6]. The entrance aperture with a diameter of 0.1 mm is smaller than the estimated Debye shielding length, and differential pumping guarantees a collisionless movement of ions inside the analyzer. The analyzer consists of two electrodes and a collector. The first electrode behind the aperture is being used as an electron repeller, the second electrode and the collector define the ion retarding region.

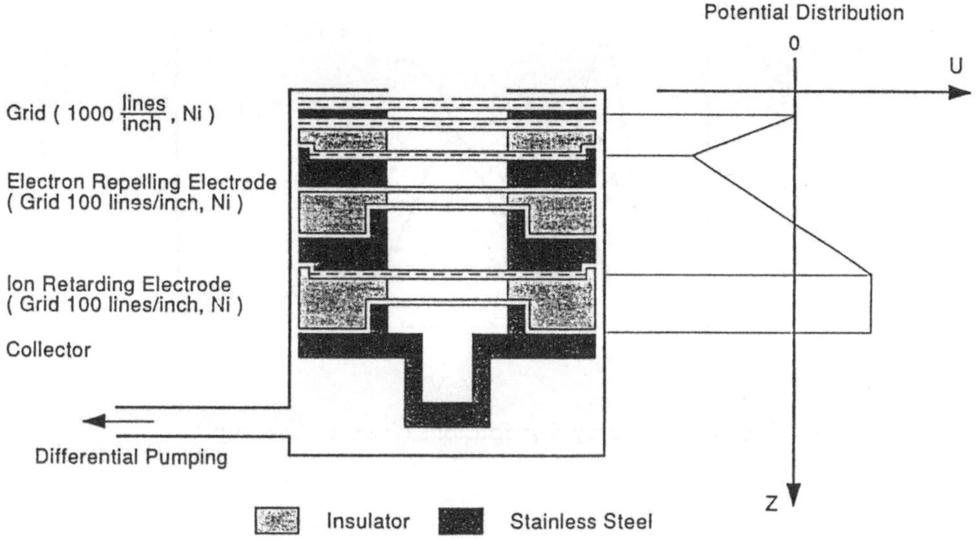

Fig. 4. The retarding field analyzer

3. RESULTS AND DISCUSSION

3.1 CONVENTIONAL COUPLING: SETUP 1

Ion energy measurements were performed as a function of pressure in the range from 0.15 to 0.35 μbar with argon as a sample gas keeping the other external parameters constant (microwave input of 380 W, coil current of 270 A).
The observed distributions are bimodal distributions (fig. 5) with a pressure dependent ratio of the peak maximum intensities, rather than sharp single peak distributions as described earlier [7,8]. Ion acceleration in front of a surface in any plasma is caused by the sheath potential originating from the differences in ion and electron velocities [9]. In magnetized plasmas confined in a divergent magnetic field another additional accelerating effect may be observed. Due to the adiabatic invariance of the magnetic moment the electrons experience an increase in their parallel velocity component with

respect to the magnetic field, while they move from a region of high field strength towards a region of lower magnetic field strength. The magnitude of this accelerating effect is strongly dependent on the magnetic field configuration. Through this electron drift an electrostatic drag is exerted on the ions [10].
The dominant scattering processes in argon are charge transfer collisions producing fast neutrals and slow ions. With this scattering cross section for argon [11] one can calculate the percentage of particles reaching the IEA without undergoing collisions on their way from the resonance zone (fig. 6). If we assume that the maximum acceleration experienced by the ions is located in a region close to the resonance zone (high field gradient), this will result in two groups of ions arriving at the analyzer. Fast non scattered ions with an energy resulting from the magnetic acceleration plus the sheath potential contribute to the high energy peak. Slow scattered ions mainly experiencing the sheath potential contribute to the low energy peak. Comparing maximum peak heights of the binodal distributions at different pressures with the ratio of scattered to non-scattered ions shows good agreement with the proposed mechanism (fig. 6).

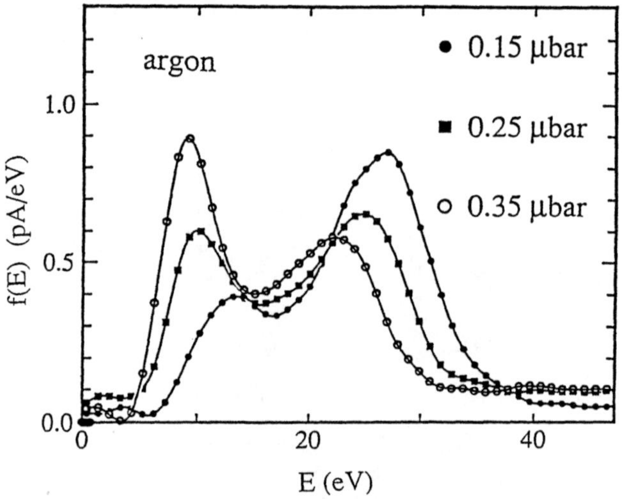

Fig. 5. Ion energy distributions observed in setup 1

Fig. 6. Comparison of: Experimental maximum peak intensities of the high energy peak E_m to max. peak intensity of the low energy peak E_m versus pressure and the ratio of non-scattered to scattered particles versus pressure. The scattering cross sections are taken from [11] and the distance from the resonance zone to the substrate is estimated from magnetic field configuration calculations .

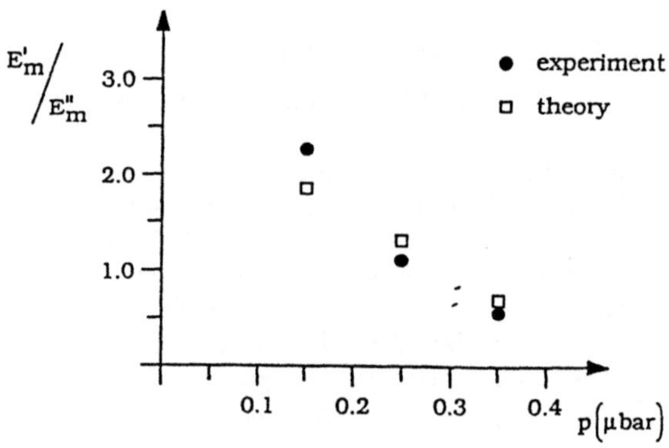

3.2. LISITANO COIL COUPLING: SETUP 2

In this setup argon as well as methane ion energy distributions were measured. The pressure was varied over a wide range without changing the chosen external discharge parameters: pressure 0.1 to 5.1 μbar main coil current 410 A, microwave input 24 to 28 W, subcoil current 0 A. Variations in the microwave power input were due to changes in the reflected power depending on the discharge. In addition for both, argon and methane discharges the subcoil current was varied between 0 and 200 A. All distributions show a tailing on the low energy side.

For the pressure dependence of distribution parameters such as full width at half maximum (FWHM) and the most probable energy (E_m), results are illustrated in fig. 7. E_m decreases with increasing pressure as expected, because a higher percentage of particles is scattered . Methane ion energies in the 0.1 to 1.0 μbar regime are higher than those of argon, which might be due to the different ionization and collision cross sections. Methane as a molecular gas dissociates into smaller particles with mostly unknown or uncertain scattering and reactional pathways. This makes it difficult to interpret these results without extensive calculations.

A decrease in FWHM with increasing pressure in the same regime is observed with argon and methane. At low pressures primary ionization is not confined in the region around the resonance zone. It rather takes place in an extended volume producing ions facing different potentials on their way to the substrate. As the ionization region becomes smaller through confinement via particle collisions, the majority of the ions experiences nearly the same accelerating potential, leading to smaller FWHM.

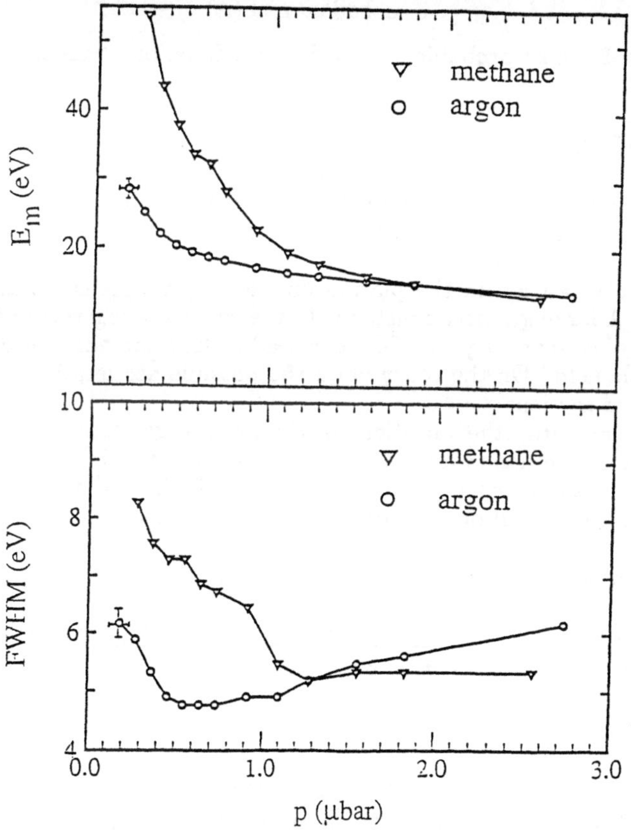

Fig. 7. Pressure dependence of the distribution parameters (FWHM) and E_m

The other set of ion energy distributions was obtained by varying the subcoil current maintaining constant maincoil current (fig. 8). The subcoils are used to influence the divergence of the magnetic field as illustrated in fig. 3. Discharge parameters were chosen as following: pressure 0.45 and 0.85 μbar for argon, 0.45 μbar for methane, maincoil current of 410 A, microwave input 28 to 24 W. Increasing the subcoil current from 0 to 200 A leads to a more homogeneous magnetic field distribution and lower E_m as expected from the acceleration mechanism explained above. The difference in E_m is about 8 eV.

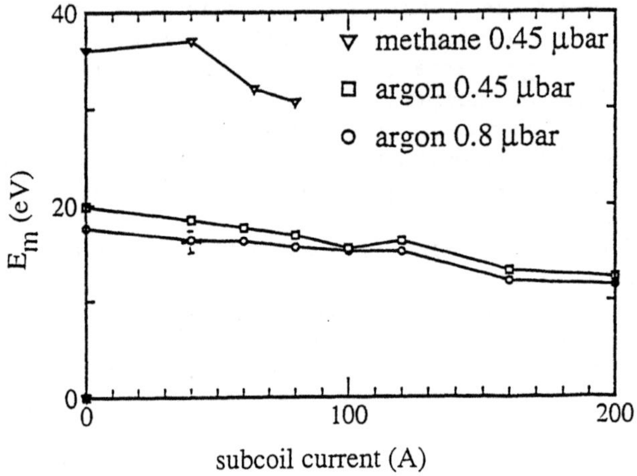

Fig. 8. Most probable energy E_m as a function of subcoil current

5. CONCLUSIONS

The experiments show that the different microwave coupling setups lead to very different results. The bimodal ion energy distributions, as observed in setup 1 were not apparent in the Lisitano coil setup.
The energy gain of an ion in the plasma on the way to the substrate is limited by several factors. A large geometric distance to the substrate together with magnetic acceleration by a divergent magnetic field can lead to high ion energies easily influenced via the magnetic field. On the other hand the possible acceleration is limited by the mean free path of the ions.
In the Lisitano coil setup the variations of the ion energy distributions with pressure for argon as well as methane were investigated. Ions extracted from a methane plasma at gas pressures below 1μbar have a higher E_m and FWHM than the extracted argon ions due to different ionization and dissociation mechanisms.

ACKNOWLEDGEMENTS

The Max-Planck-Institut fur Plasmaphysik is affiliated with EURAROM. We gratefully acknowledge the technical help of G. Kerkloh.

REFERENCES

1. J. Asmussen, J.Vac.Sci.Technol. A **7(3)**, 883 (1989)
2. W. Jacob, D. Boutard, V. Dose, A. Koch, W. Möller, W. Renz and R. Wilhelm, Proc. ISPC-9, 1826 (1989)
3. J.C. Angus, P. Koidl and S. Domitz in : "Plasma Deposited Thin Films", Eds. J. Mort and F. Jansen, CRC Press, Boca Raton 1986

4. G. Lisitano, IPP-Report III/145, May 1989 (Max-Planck- Institut für Plasmaphysik, Garching)
5. S.M.L. Prokopenko, J.G. Laframboise and J.M. Goodings, J.Phys. D: Appl.Phys **5**, 2152 (1972)
6. S.M.L. Prokopenko, J.G. Laframboise and J.M. Goodings, J.Phys. D: Appl. Phys **7**, 355 (1974)
7. M. Matsuoka and Ken'ichi Ono, Appl.Phys.Lett. **50**(26), 1864 (1987)
8. M. Matsuoka and Ken'ichi Ono, J.Vac,Sci.Technol.(1),**A6** 25 (1988)
9. P.C. Stangeby in: "Plasma Wall Interactions in Controlled Fusion", Eds. R. Behrisch and D.E. Post, NATO-ASI Series, Plenum Publishing Corporation,New York and London,1984, pg.41-99
10. H. Toyama,M. Okabayashi and H. Ishizuka, Plasma Phys. **10**, 319 (1968)
11. A. v.Engel in: "Electric Plasmas: Their Nature and Uses", Taylor and Francis Ltd,London and New York,1983, pg. 34-53

A NEW PROBE FOR IN-SITU CHARACTERIZATION OF DIAMOND SURFACES DURING LOW PRESSURE CHEMICAL VAPOR DEPOSITION*

Howard K. Schmidt, J. Albert Schultz**, and Zirao Zheng

Schmidt Instruments, Inc.
2476 Bolsover, Suite 234
Houston, Texas 77005 USA

Single crystal diamond has extremely high thermal conductivity, a large bandgap, high carrier mobilities, and low neutron and ionizing radiation dislocation cross sections. These physical properties make it an ideal material in which to fabricate electronic devices for high temperature, high frequency and/or high radiation service. Despite the significant body of work to date on low pressure chemical vapor deposition (LPCVD) of diamond, no methods now exist for manufacturing the large single crystal diamond substrates required to realize these potentials. It has become clear that the fundamental mechanisms of diamond's nucleation and growth must be determined before we can significantly improve its process technology. Prior research, focussing on the gas phase chemistry in diamond LPCVD systems, has provided some valuable clues to possible mechanisms. However, chemical transport through the boundary layer to the substrate is still rather mysterious, and any conclusions about the environment at the growing surface based upon gas phase data are speculative at best. Our contribution to the field is the first direct probe for diamond surface chemistry under LPCVD growth conditions: *in-situ* direct recoil spectroscopy (DRS). In conjunction with the earlier gas phase results, this new tool will give us the first comprehensive description of diamond LPCVD, enabling relatively straightforward determinations of both chemical mechanisms and improved process conditions for diamond growth.

I. Motivation and Scope

The research communities' current understanding of the chemical mechanisms involved in LPCVD of diamond is poor at best. In general, to characterize a chemical system, one needs information about both the reactants and the products. But prior to the development of high pressure direct recoil spectroscopy (DRS) described in this report, all of our information about *in-situ* process chemistry has come from the gas phase (reactants), mostly using infrared spectroscopy. Previously, no method for observing reaction chemistry at the surface (products) was available. This is understandable, since surface science is difficult even under the best ultra high vacuum (UHV) conditions, and the electron-based surface spectroscopies (e.g. XPS, AES, UPS, HREELS) would be subject to scatter and attenuation by process gas. If implemented, the first three of these would be of little value because they are completely insensitive to hydrogen and isotopic variations, while the notoriously difficult HREELS is very slow and would be severely compromised by inelastic electron scattering by the gas ambient.

DRS is a relatively new surface probe with sub-monolayer sensitivity that utilizes a pulsed energetic ion beam to simultaneously detect and resolve light elements, H through F, and their isotopes rapidly and quantitatively by time-of-flight (TOF) analysis.[1] Like the more common electron based surface spectroscopies (e.g. XPS, AES, UPS, HREELS), DRS has previously been used only under Ultra High Vacuum (UHV) conditions. Unlike electrons, however, energetic ions and atoms are not readily scattered or attenuated by gas molecules. Using existing literature on gas phase ion scattering cross sections, we estimated that the 3-10 keV Na^+ ion probes typically used in DRS would have a mean free path of at least 1 cm through 1 Torr of hydrogen, and subsequently proposed that *in-situ* analysis of growth surfaces under LPCVD conditions would be practical with DRS. The modest risk thus expected in extending DRS to operate in the LPCVD environment appeared small indeed in comparison to the potential value of the information it could produce. We have indeed successfully obtained DRS spectra from diamond surfaces under actual growth conditions, providing the the first demonstration that the operating environment of a proven UHV surface elemental analysis technique can be extended into the Torr regime - a pressure some billion-fold higher than typically used in surface science.

Typical conditions for diamond growth include a hydrogen:1% methane gas feed at 1 to 100 Torr, a substrate heated to about 950°C, and "activation" by an incandescent filament or electric discharge. Generally accepted features of low pressure diamond process chemistry are that atomic hydrogen must be present, along with a small carbon bearing growth species. Methyl radical and acetylene appear from gas phase diagnostics to be the only growth candidates sufficiently abundant to account for observed growth rates.[2] Speculations about the role of atomic hydrogen include: (1) formation of methyl radical by abstraction, (2) formation of active surface sites by abstraction, and (3) preferential etching of non-diamond deposits from the growth surface.

The critical role of the surface has largely been ignored theoretically, due to the lack of experimental data on it. The native surface of diamond is hydrogen terminated, and although UHV surface studies have shown diamond to desorb hydrogen and reconstruct at the usual growth temperatures, an implicit assumption in existing mechanistic theories is that diamond is fully hydrogen saturated under growth conditions. The degree of hydrogenation of the surface under process conditions has a large impact on growth mechanism theory, as the chemistry of saturated hydrocarbons and olefins are completely different. In keeping with this, atomic hydrogen has been assumed to activate the surface by H abstraction. Surface radical sites can reasonably react with either methyl radical (by recombination) or with acetylene (by polymerization).

Recognizing that the energetic, massive particles used in ion beam analysis techniques would be relatively insensitive to gas phase attenuation, we proposed that DRS could be used to observe the growing diamond surface *in-situ*, and resolve the above mechanistic issues. Therefore, the primary objectives of the Phase I feasibility study were to:

(1) demonstrate DRS under diamond LPCVD process conditions; and
(2) use this new tool to characterize LPCVD on the (100) surface.

II. Experimental

The UHV system, pulsed ion beam line, detectors and data acquisition electronics making up the basic direct recoil spectrometer were already in place at Ionwerks. The CVD cell and high vacuum interface were designed and constructed in-house and installed on the Ionwerks system. Details of this hardware will be published in a following paper.

Hydrogen and methane were introduced to the CVD cell separately. The particular arrangement allowed only hydrogen to pass directly over the filament; methane was injected downstream of this dissociator beyond a flow orifice. Flow and pumping rates were such that the gas residence time in the cell was about 0.5 second. This arrangement prevents carburization of the filament. Time-of-flight (TOF) spectra were recorded with a Schmidt Instruments TOF^+ data acquisition system running on an IBM-XT compatible computer. The beam was pulsed at a rate of 20 KHz, and 7.5 keV sodium ions were used as the primary probe. All data were acquired from (100) oriented 1.5 mm square type IIa diamond crystals.

CVD growth conditions at 0.3 Torr were verified by depositing polycrystalline diamond films onto resistively heated tungsten ribbon. The ribbon was pre-nucleated by scratching with diamond, and grey-white films about 1 micron thick were deposited in a four hour run.

III. Results and Discussion

Thermal desorption of native hydrogen on the (100) surface of diamond was first characterized by DRS in vacuo. These results are shown in Figure 1. Note that each trace in this and several following figures represent two closely spaced TOF/DRS spectra. The ion beam is pulsed by electrostatically sweeping it across a small aperture. In this work, we used a 200 volt step 10 microseconds in duration; this results in two pulses and therefore two spectra. Our pulse generator has a 25 ns leading edge and a 200 ns trailing edge, giving the first spectrum better temporal resolution, and the second improved signal intensity. The TOF peaks resulting from surface hydrogen and carbon are labeled "H" and "C" respectively, the intense scatter peak arising from reflected sodium ions is labeled "Na_{sp}". We have developed a procedure for calibrating the observed H and C DRS signal intensity to a surface stoichiometry. This procedure will be published elsewhere and is being refined. However, we are able to assign an approximate stoichiometry of CH_2 to the surface at 350° C shown in figure 1. The accuracy of this assignment will be improved with future works. At 725^oC a partially dehydrogenated surface with approximate CH_1 stoichiometry was obtained. At higher temperatures, the surface was denuded of hydrogen. The CH_1 surface could indicate a true stepwise reconstruction on this surface, or a equal mixture of bare and saturated surface sites. The intermediate state was generated from the saturated condition within two minutes at 725^oC, remained stable for over thirty minutes and progressed to the bare state within two minutes upon elevation to 815°; therefore, we suspect that a stable CH_1 surface

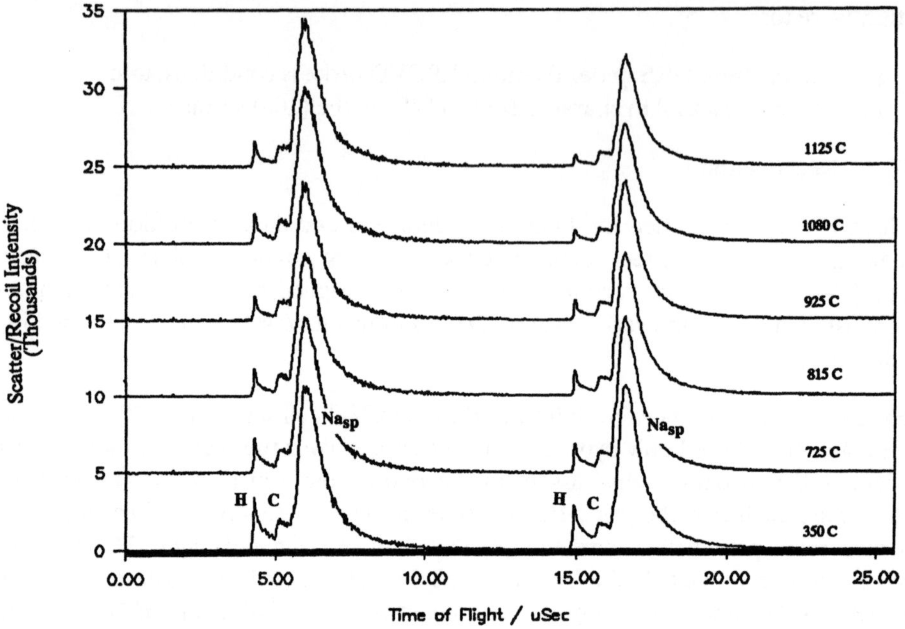

Figure 1. Thermal programmed desorption of surface Hydrogen on (100) diamond in high vacuum, determined by 7.5 keV Na^+ Direct Recoil Spectroscopy. The doubled TOF spectra arise from twin DRS probe pulses (see text).

does exist. Such a surface was recently proposed by Kubiak on the basis of intriguing but inconclusive LEED and ESDIAD results [3].

Previous LEED results concluded that the reconstructed (100) surface was chemically inert to molecular hydrogen; atomic hydrogen is required to transform the reconstructed 2x1 surface back to the original 1x1 LEED image. Note that LEED can only indicate the surface structure and not the amount of hydrogen present. The earlier conclusion is supported however, by the DRS results shown in Figure 2. This (100) sample was annealed at 1100°C to remove hydrogen, then cooled to 350°C for the experiment. At this temperature surface hydrides are stable (from above), and residual hydrocarbon contamination condenses very slowly (from previous experience). Indeed, a fully hydrogenated, CH_2, surface was obtained only after exposure to atomic hydrogen from the filament. A slight increase in the hydrogen coverage did occur over about twenty minutes, including exposure to 50 microns of pure hydrogen. There exists a small possibility that the reconstructed surface has a very slow reaction with H_2, however the residuals in the non-UHV CVD cell could easily account for the observed amount of surface H. The inert nature of the reconstructed surface was also supported by a series of thermal desorption runs in hydrogen ambient at pressures up to 1 Torr. The results were identical to those from the vacuum experiment above, supporting the notion that the reconstructed surface is inert to molecular hydrogen.

A similar series of experiments performed in a partially dissociated hydrogen ambient provide a view of the dynamics of desorption versus hydrogen addition under growth conditions. Results taken under 0.330 Torr of activated hydrogen are

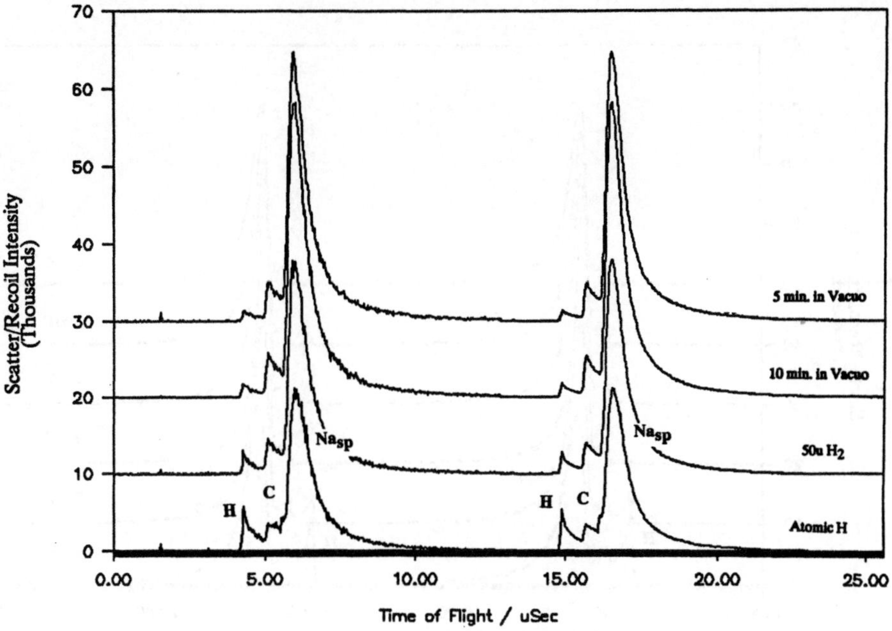

Figure 2. Rehydrogenation of reconstructed (100) diamond by residual molecular and atomic hydrogen at 50 mTorr, determined by *in-situ* DRS.

shown in Figure 3. There still exists a clear trend toward reduced hydrogen coverage on the (100) surface with increasing sample temperature. However, the surface never becomes completely denuded of hydrides. Plainly, at typical growth temperatures above 800°C, the (100) surface has at most one quarter of a monolayer of hydrogen. In analogy to established work on reconstructed silicon (100) surfaces, we assume that the dehydrogenated diamond (100) surface consists of carbon dimers which are chemically joined by a double bond. This leads us to the conclusion that the (100) surface at least has a predominantly alkenic character under process conditions. Noting that there is some hydrogen present, we can further conclude that at variable steady state portion of these double bonds have undergone addition of atomic hydrogen, and likely have an active free radical site as a result. Therefore, the surface has sites which can react readily with both methyl radicals and/or acetylene. At this point, both are still good candidates for the diamond growth species, and we infer that arrival and sticking of the carbon species is the rate limiting step in diamond CVD.

Reversible exchange of hydrogen and deuterium on the surface was studied to test the hydrogen abstraction hypothesis of surface activation. We expected that atomic hydrogen generated by the hot filament would abstract surface hydrides even at room temperature, leaving radical sites which would recombine with subsequent hydrogen atoms. This turned out not to be the case at all. We readily obtained DRS fingerprints for surface H and D by annealing the sample, then exposing it to either atomic H or D. Such spectra are shown in Figure 4. The smaller D peak is expected because the direct recoil cross section of deuterium is only about 60% as large as that of hydrogen. Poor counting statistics resulted in fairly noisy spectra in the isotopic

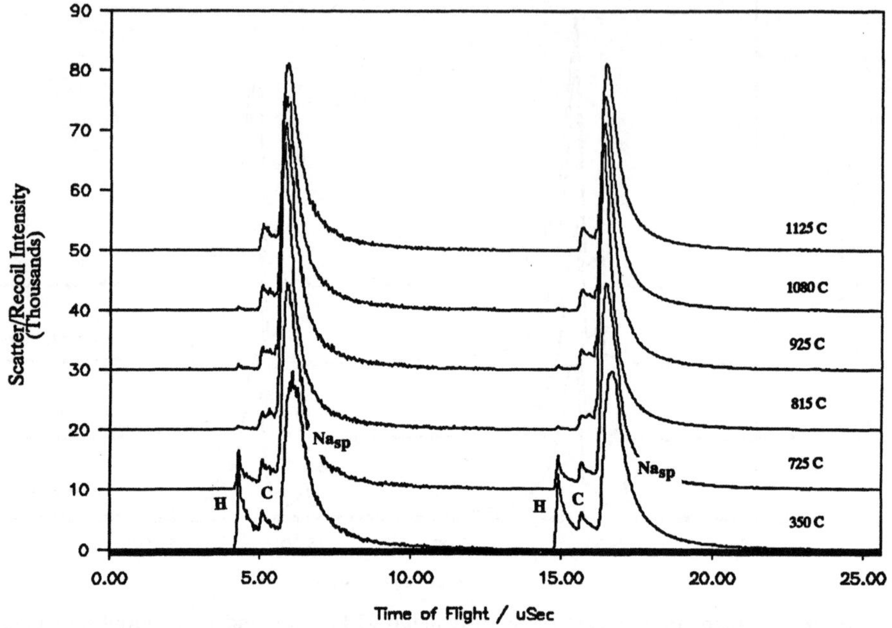

Figure 3. Steady state surface coverage of hydrogen on diamond(100) decreases with temperature under constant flux of activated hydrogen at 330 mTorr, determined by *in-situ* DRS.

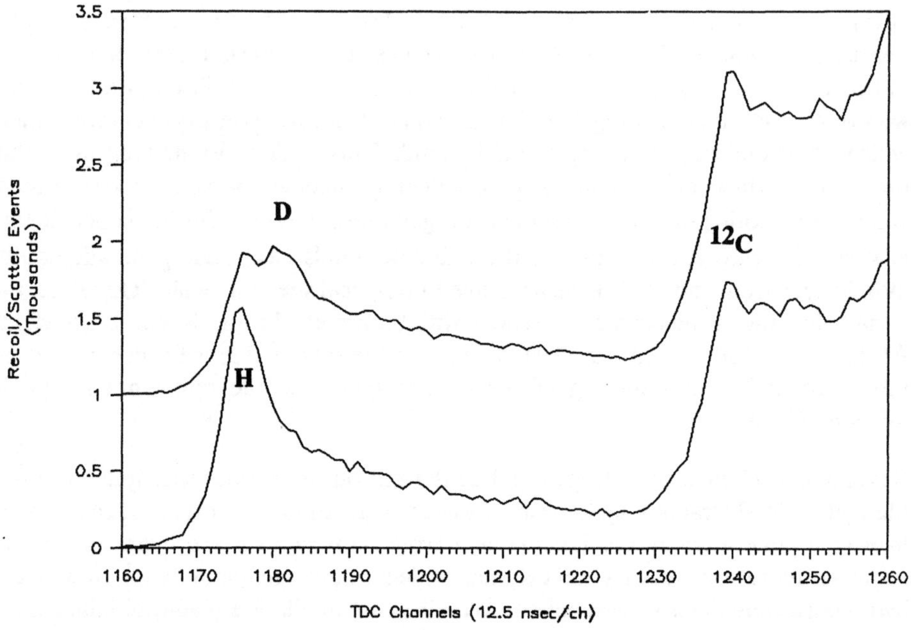

Figure 4. DRS spectra from natural and deuterated diamond (100) surfaces. H/D exchange on this surface under dissociated H_2 or D_2 flux at 1 Torr was only observed at or above substrate temperatures of 600C.

exchange work; the spectra displayed are from the second, more intense, DRS signal and have modest temporal resolution. We can estimate that the trace indicative of deuterium has no more than 25% hydrogen contaminating it. We performed exchange at a series of temperatures (300, 500, 600, 700, 800, 900°C) using both activated hydrogen and deuterium at a pressure of .3 Torr. After each 30 second exposure, the gas was evacuated prior to taking the DRS spectra. We found that complete exchange occurred only at or above 700°C; at 600°C there was partial exchange. At or below 500°C, there was no exchange between H and D at all. We repeated the entire sequence of tests three times to make sure of the results. The consequence of this data is that surface exchange and therefore activation of the surface for diamond growth must be mediated by thermal desorption of surface hydrides to generate unsaturated carbon sites. Atomic hydrogen is apparently therefore not the surface activator expected, unless the abstraction process is highly surface temperature dependant, and happens to have the same threshold as desorption.

IV. Summary and Outlook

The primary objectives of this demonstration project were met by successfully constructing a DRS spectrometer operating in the Torr regime and using it to study diamond surfaces during LPCVD. We have determined that with the current CVD cell, which can be improved substantially, we can collect high quality DRS spectra at pressures up to .3 Torr without significant data manipulation to subtract gas phase background. Ion scattering data have been obtained at pressures up to several Torr, and with appropriate data manipulation software, we may be able to collect useful DRS spectra in this pressure regime as well. We can likely extend the useful pressure regime of this technique by using smaller diameter CVD cells, or higher energy ions to probe the surface, or both.

Surface hydrogen coverage on diamond(100) was determined under high vacuum and low pressure deposition conditions; these experiments lead us to conclude that the diamond surface is primarily reconstructed during growth, and that there is a dynamic equilibrium between thermal desorption and free radical addition of atomic hydrogen on the surface. Temperature studies under activated isotopically labeled process gas further indicate that the rate limiting step in activating the surface for growth is thermal desorption of native hydrogen. The growth surface appears to possess both free radical and alkenic moieties, and may support growth by either methyl radical or acetylenic mechanisms or both. Arrival of the appropriate carbon bearing species appears to be the rate limiting step in diamond growth.

---

*Supported by SDIO-IST contract N00014-89-C-0152, managed by the Office of Naval Research.

** Ionwerks, 2215 Addison, Houston, TX 77030 USA

REFERENCES

1a. Y. S. Chen, G. C. Miller, D. A. H. Robinson, G. H. Wheatley, T. M. Buck, Surf. Sci. **62** (1977) 133.

1b. "Rapid and Sensitive Determination of Hydrogen Isotopes at a Surface by Li^+ Direct Recoil Spectroscopy," H.K. Schmidt, L.R. Anderson, and J.A. Schultz, Appl. Surf. Sci. **L35** (1988) 274.

2a. M.Tsuda, M. Nakajiama, and S. Oikawa, J. Am. Chem. Soc. **108** (1986) 5780.

2b. D. Huang, M. Frenklach, and M. Maroncelli, J. Am. Chem. Soc. **92** (1988) 6379.

3. G. Kubiak, Proc. SDIO-IST Diamond Workshop, Washington, D.C., 1989.

CHARACTERIZATION AND PROPERTIES OF ARTIFICIALLY GROWN DIAMOND

Peter K. Bachmann and Detlef U. Wiechert

Philips Research Laboratories
CVD-Technology Group
D-5100 Aachen, FRG

INTRODUCTION

The importance of thorough material characterization in CVD diamond research is excellently expressed by R.C. DeVries in his 1987 review of metastable diamond formation[1]: "Man-made diamond synthesis may yet be pushed back in time when a careful study is given to the soots from ancient cave walls". Nowadays such careful studies are possible and one of the major differences between the early days[2-5] of diamond CVD and the modern era[6-10] of low pressure diamond synthesis is probably the easy access to highly sophisticated, fast and extremely sensitive analytical tools to distinguish diamond from any other form of carbon as well as to determine the physical and chemical properties of the deposited material. In their 1968 paper, J. Angus and coworkers describe in detail the many tedious methods, including weight determination, visual inspection, traditional chemical analysis, chemical etching, density measurements, x-ray and electron diffraction, they applied. Their task was particularly difficult because they had to analyze diamond deposited onto diamond seeds. On the other hand, it is necessary to decisively characterize the deposited material, not only to distinguish it from other carbon modifications, but also to adjust the preparation conditions in order to tailor the properties of the diamond film to specific applications.

This paper reviews some of the most important characterization methods applied to vapor-grown diamond and summarizes the state-of-the-art of the properties of variously prepared diamond coatings. Both, literature data as well as previously unpublished results obtained by the authors and their collegues are included. Electrical and mechanical data may depend not only on the properties of CVD diamond but also on electrical or mechanical contacts and other extrinsic quantities and are therefore not included in this article.

Diamond and Diamond-Like Films and Coatings
Edited by R.E. Clausing *et al.*, Plenum Press, New York, 1991

A "WORKING DEFINITION" OF DIAMOND

"Graphite, diamond, amorphous carbon, pyrolytic graphite, glassy carbon, diamond-like carbon, a-C:H, a-C, i-C, soot". One could easily extend this list of names for different modifications of the element carbon. This variety, along with crystal size variations within crystalline phases ranging from nano-crystals to large single crystals, makes a precise definition of "diamond" very desirable but also extremely difficult. With their "working definition of diamond" R. Messier et al[9] tried to establish a guideline to separate diamond from any other form of carbon. Their definition demands :

- a crystalline morphology of the deposit discernible by electron microscopy,
- a single phase crystalline structure identified by diffraction techniques such as X-ray or electron diffraction and
- a Raman spectrum typical for diamond, i.e. a single narrow line at $\approx$ 1332 cm^{-1}.

Impurities that also occur in natural diamond, such as nitrogen, hydrogen, silicon, boron and others, are not excluded by this definition. However, "amorphous diamond" i.e. completely sp^3-hybridized carbon deposits without long-range structural order are excluded for good reason. Diamond is defined by its crystal lattice and an amorphous material is by definition non-crystalline. The border line between diamond nano-crystals and amorphous carbon material is, of course, not very well defined and depends somewhat on the characterization method used. The "working definition" also suggests the use of not only one but three different characterization techniques to unambiguously determine the presence of diamond. These and other important methods are, along with the measured properties of the deposited materials, discussed in the sections below.

CHARACTERIZATION METHODS AND DIAMOND PROPERTIES

Morphology by Optical Microscopy

Although not a very sensitive tool in case of lower levels of non-diamond carbon contaminations, it is very helpful to have ≥ 1000 x optical microscope close to the CVD unit for first inspection of the deposit. It is simple, easy to use, does not require sample preparation and is non-destructive. Where polycrystalline films with crystallite sizes >0.5 μm are desired, it allows the proper deposition conditions to be approached without devoting too much time to the characterization of obvious deposition failures. Fig.1 depicts optical micrographs of both diamond particles (1a and 1c) and continuous coatings (1b and 1d). The large faceted crystals (1a) and the polycrystalline surface with

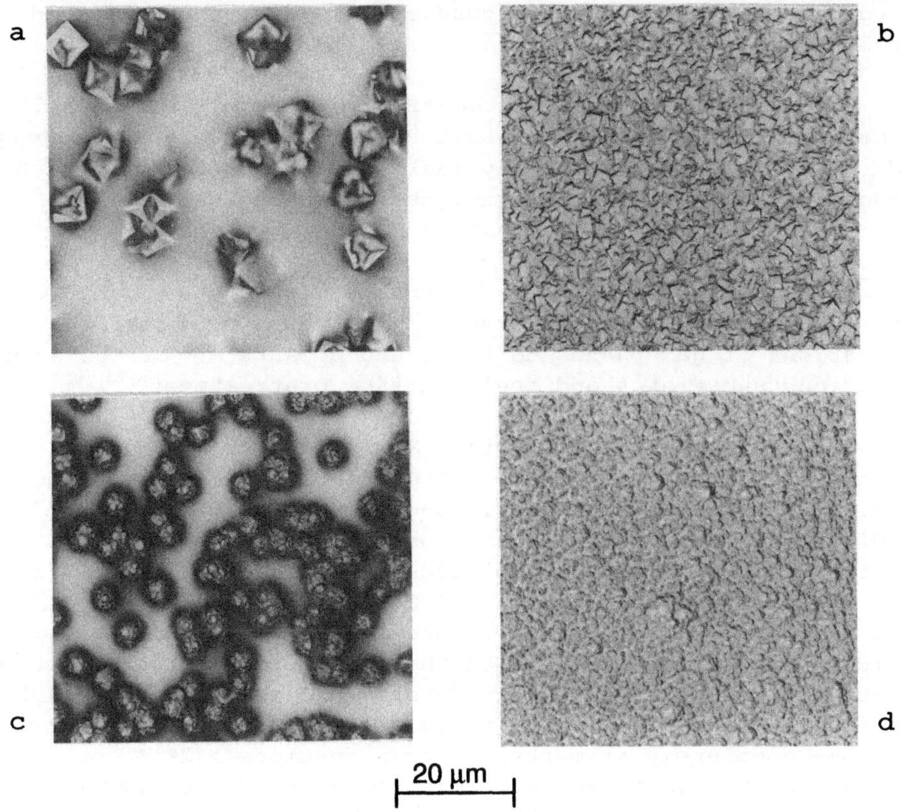

Fig.1. Optical micrographs of DC-plasma jet deposited carbon. (a) and (b) represent high-quality individual, faceted diamond crystals and a continuous coating with <100> facets. The materials in (c) and (d) consist of mostly non-diamond carbon resulting in ball-shaped particles (ballas) or films with a cauliflower morphology. Such morphological differences are caused by different pretreatments of the silicon substrate surface.

<100> features clearly visible (1b), are good hints for high quality material being deposited while ball-like structures (1c) and a cauliflower morphology of a film (1d) already indicate poor quality using this simple characterization technique.

Diamond Morphology by Scanning Electron Microscopy (SEM)

Optical microscopy is, of course, only useful for a first, quick inspection of the grown material. A more detailed study of the morphology of the deposit requires an electron microscope. Because of diamond's electrically insulating nature metallization of the diamond surface is required for graphite free material. However, in many cases attaching the sample to the microscope stage by means of a metal clip is sufficient to avoid severe charging because of residual conductivity due to non-diamond carbon impurities.

A detailed knowledge of the film morphology is of specific importance for applications where very smooth surfaces are required, e.g. substrates for X-ray masks or low scattering optical windows. SEM is helpful to study the morphological changes associated with certain deposition parameter variations. Examples include the influence of the concentration of the carbon carrier in the CVD gas phase[11-13], morphological changes due to the variation of the deposition temperature[14,15] or deposition time[11], concentrations of gas phase additions and dopants[16,17] or the effect of various substrates and their surface pretreatment[9,18,19].

The SEM micrographs in Figs. 2a-f and 3a-f, resulting from our own investigations[9,18,20,21], illustrate some of the morphological effects mentioned. Figs. 2a-d depict the well-known effect of different methane concentrations on the film morphology. The material was deposited onto silicon by means of a 2.45 GHz microwave plasma at ≈950°C from a 0.5% CH_4/H_2-mixture (2b and 2d) and from 1.6% CH_4/H_2 (2a and 2c), respectively. The resulting morphological changes agree well with the results of K. Kobashi et al[11]. The crystalline SEM morphology deteriorates rapidly with higher methane concentrations, and, at least for pure CH_4/H_2-mixtures, vanishes completely at methane concentra-tions > 2%. Interestingly, Y.Sato and M. Kamo[12] observed faceted crystals for methane concentrations as high as 4%. Well-faceted crystals for high methane concentrations were also observed by C. Chen[16] et al in a detailed SEM morphology study. However, they achieved this result only after adding substantial amounts of oxygen to the CVD gas phase. They found that a gas phase composition of 1% oxygen, 3% methane and a balance of hydrogen allows deposition of large, extremely well faceted idiomorphic crystals.

The influence of the substrate pretreatment on the film morphology is illustrated by comparing Figs. 2a and 2c for highly graphitic material and Figs. 2b and 2d for well-faceted crystalline diamond. The SEM morphology of a continuous film shown in Fig. 2b was obtained by means of scratching the silicon substrate surface with 1-2 µm size diamond powder prior to microwave deposition. This pretreatment was omitted for the sample shown in Fig. 2d, resulting in the formation of individual diamond crystals rather than a film. This nucleation density enhancement can be observed not only for microwave plasma deposited material. We found it in all of our diamond deposition experiments from microwave and DC jet plasmas as well as for our oxy-acetylene torch experiments. Fig. 2e depicts diamond particle formation from a DC cascaded arc plasma jet on untreated silicon sample[21]. Comparing the SEM micrographs 2d, 2e, and 2f reveals considerable morphological differences of particles formed by different deposition methods. Five-fold twinned crystals similar to the one shown in Fig. 2d are often found on low-pressure microwave or

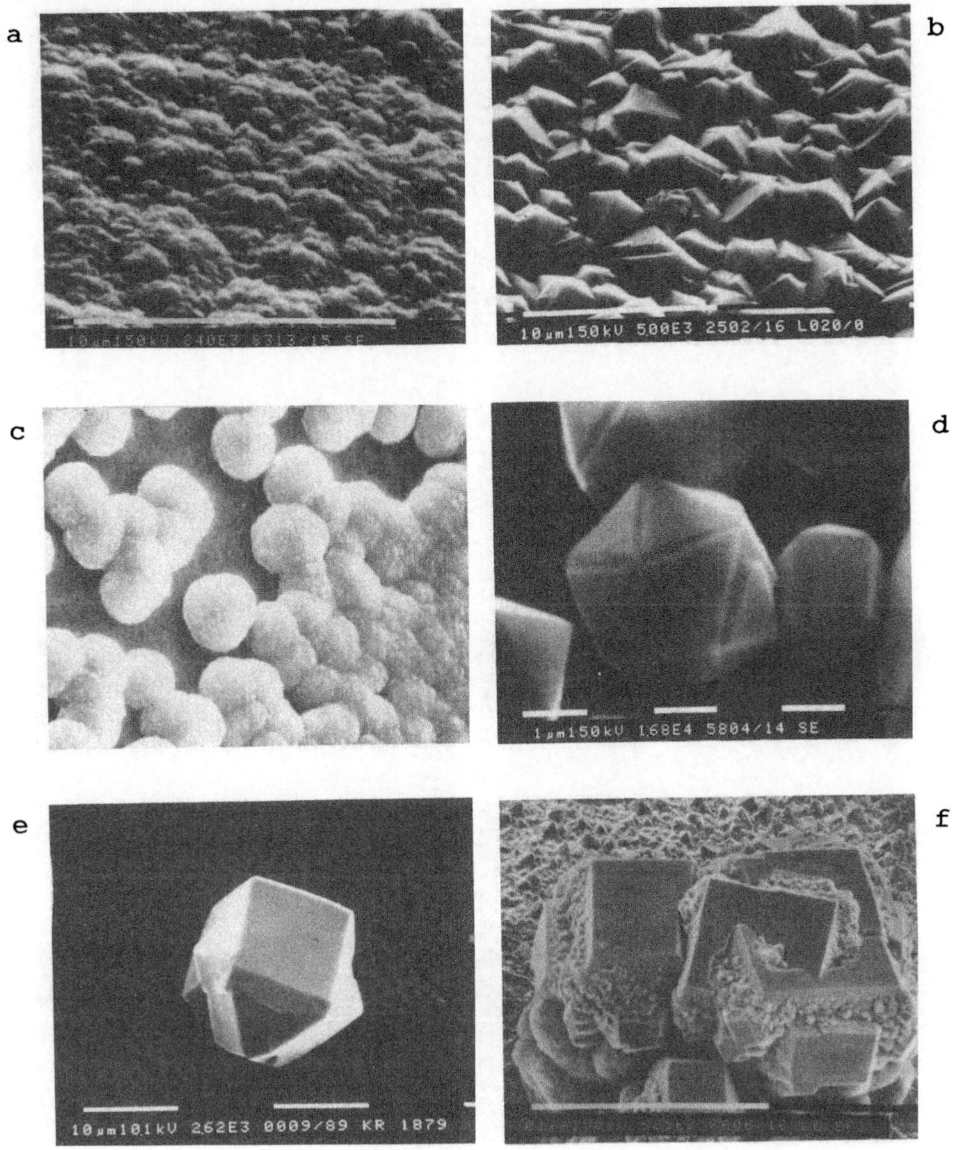

Fig. 2a-f. Examples of morphological studies by means of scanning electron microscopy (SEM): micrographs of deposits on diamond scratched (a,b) and untreated (c,d) single crystal silicon from a microwave plasma using 1.6% (a,c) and 0.5% (b,d) methane/hydrogen are shown. Five-fold twinned crystals (d) are frequently found in microwave and radio frequency plasma low-pressure deposits. The crystals in (e) and (f) are typically for DC plasma jets and chemical flames, respectively. The deposits in (a) and (c) contain large amounts of non-diamond carbon, while the materials of (b),(d),(e) and (f) are predominantly diamond.

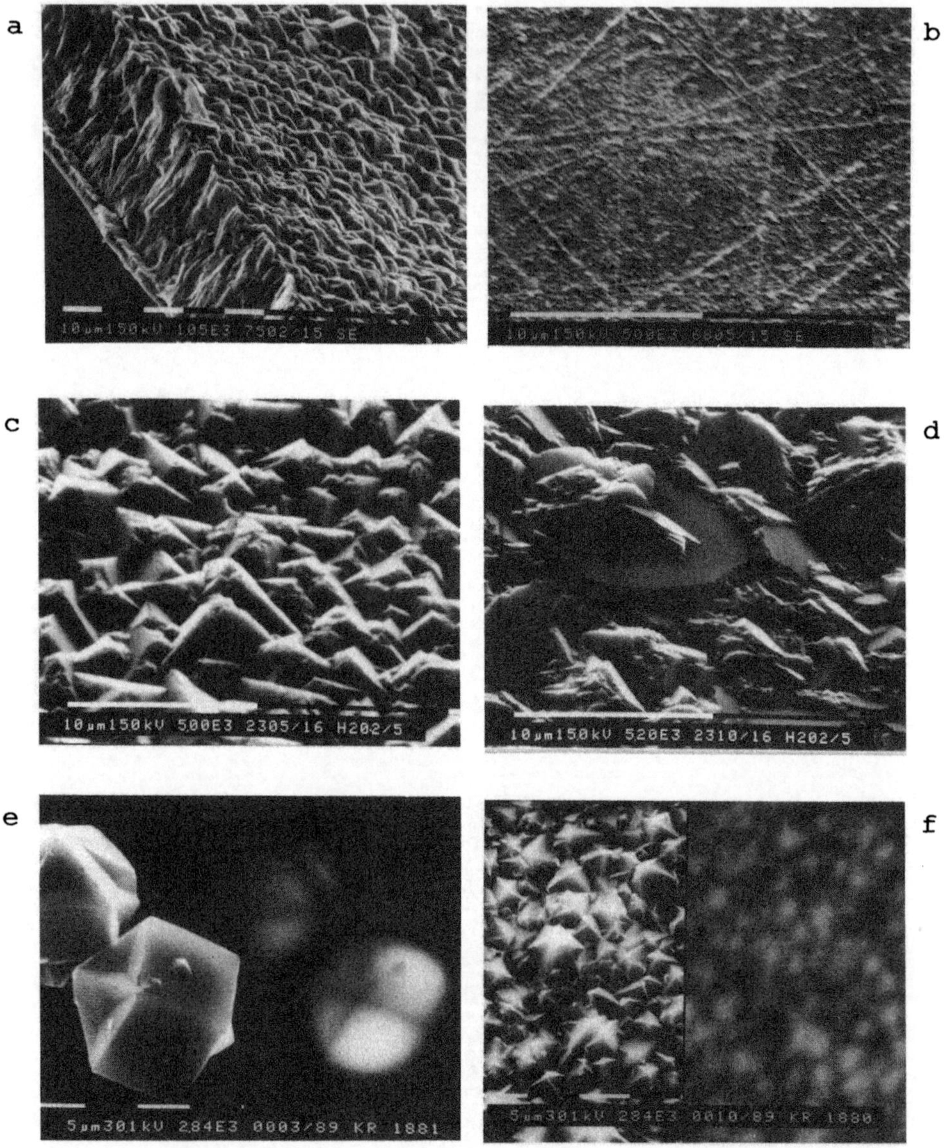

Fig. 3a-f. Examples of morphological studies by means of scanning electron microscopy (SEM). The evolution of the surface roughness of a film is illustrated by a top and cross-sectional view (a) and a view onto the substrate side (mirroring the substrate surface scratch marks) (b) of a free-standing diamond film. The effect of small variations of thenitrogen content of the CVD gas phase is represented by (c) (0.1% N_2) and (d) (0.2% N_2). Figs. (e) and (f) describe the use of SEM (left sides) to locate the centers of electron-beam induced cathodoluminescence (right sides)in CVD diamond particles and coatings, respectively.

radiofrequency plasma CVD samples[22,23] and heated filament deposited material. We have never seen such crystals on samples from our plasma jet[21] or oxy-acetylene torch experiments, i.e. from typical high rate deposition methods, where the CVD gas phase reaches high temperatures[22]. Typical SEM particle morphologies for these two methods are given in Figs. 2e and 2f, respectively.

Additional evidence for the importance of SEM as a tool in diamond thin film research is presented in Figs. 3a-f. The micrographs (3a) and (3b) illustrate that the size of the crystals forming a continuous film is time dependent. The film cross-section reveals the columnar growth characteristics of this microwave deposited, 25 µm thick, free-standing diamond membrane. The substrate side of the film (3b) mirrors the scratched silicon substrate surface (removed by etching with HF/HNO_3 mixtures). Y. Sato and M. Kamo[12] have shown the preferential etch of the deposit close to the substrate surface supporting the hypothesis that at the beginning of a deposition step, highly defective (diamond-like?), nano-crystalline material is first deposited, from which finally, crystalline diamond emerges.

SEM also manifests the impact of gas phase additives on the film morphology. The effect of oxygen additions was already mentioned above. Fig. 3c and 3d are taken from films, microwave deposited from a 0.5% CH_4/H_2-mixture (see Fig. 2b) with 0.1% (reference is H_2) (3c) and 0.2% (3d) nitrogen added to the CVD gas phase. With nitrogen added, the film morphologies not only get defective, but also large crystals develop for 0.2% N_2. K. Kobashi et al[17] investigated the morphological changes caused by ppma quantities of diborane added to the CVD gas phase and found needle-like structures replacing the commonly observed polycrystalline material.

Figs. 3e and 3f are examples for a different application of a scanning electron microscope, namely to map and measure cathodoluminescence radiation (CL) excited by the probing electron beam. Natural and high pressure synthesized diamonds have been extensively studied by CL and a complete chapter of J.E. Field's book[23] "The Properties of Diamond", the bible of diamond research, is solely devoted to this topic. CL of CVD diamonds was first investigated by H. Kawarada et al[24]. The luminescence studies are discussed separately below. Figs. (3e) and (3f) are merely to illustrate that the combination of SEM and CL imaging allows localization of the light emitting regions (3e and 3f; right sides) of particles and films. The dominating luminescence of the square <100> facets of the diamond particle (3e) is clearly visible.

Scanning electron microscopy is an extremely valuable tool to characterize the morphology of carbon deposits. The examples demonstrate that SEM studies are very helpful to evaluate preparation methods, to adjust for proper deposition conditions, and to allow tailoring of the film morphology according to specific applications. It is however, difficult to quantify the result or to get any depth information (except for cross-sectioned samples) of the deposit.

Diamond Crystal Lattice, Defect Structures, and the Diamond-Substrate Interface by Transmission Electron Microscopy (TEM)

One of the possibilities to get information complementing the SEM morphology data is a different electron microscopical technique, namely transmission electron microscopy (TEM)[25-34]. Other than SEM, this is an invasive probing technique because the samples have to be thin enough to allow for electron transmission. Sample preparation is an important and sometimes very time consuming part of TEM. Due to its hardness and chemical inertness this is especially true for diamond samples. Two different approaches, plan-view and cross section TEM (XTEM), are in use. In the plan-view mode, TEM observations through the film, i.e. parallel to the growth direction, allow investigation of the domain distribution and lattice defects such as twinning, stacking faults or dislocations. The structure of the diamond-substrate-interface regions can be studied by means of cross-section TEM. Details of suitable sample preparation techniques for both plan-view and XTEM are described in detail, e.g., in refs. 28,29, and 33. In the case of diamond coated silicon [28], a suitably sized piece of ≈3 mm diameter is cut out of the coated wafer. Most of the silicon is removed by means of 600 grit sandpaper and is subsequently thinned to 10-20 μm by means of a dimpling machine. Both, the rest of the substrate and the diamond film are further removed by Argon ion beam milling (6kV) thus forming a small hole surrounded by a very thin part of the film suitable for plan-view observation. Cross-section TEM samples include both substrate and deposit. They are cut from the sample and thinned similarly to plan-view samples. Great care is necessary to obtain a thin slice without peeling the film off the substrate.

Plan-view, cross-section, and high resolution TEM were utilized to image defects in vapor-grown diamond films and particles. Figs. 4 a-d illustrate some of the results obtained. The SEM micrograph in Fig. 4a depicts a typical multiply twinned particle (MPT) with five-fold symmetry deposited from a CH_4/H_2-microwave plasma. Figs. 4b and 4c (from ref.34) are excellent bright-field (utilizing the transmitted beam) and dark-field (using a <111> diffracted beam) TEM examples of such a particle, revealing its internal structure. Five tetrahedra with a common <110> axis are twinned on <111> relative to each other. In the dark field, <111> twins within the five sectors appear as bright streaks. The MTP internal structure is also depicted by means of a crystal model in (4d) (from ref. 1). This model of a five-fold twinned particle indicates a gap of ≈7.5° for an ideal particle consisting of five sectors with angles of 70.5° each. Up to a certain size of the nucleus, this misfit can be accommodated by internal strain[32], i.e. in the same way as is shown in the DeVries model.

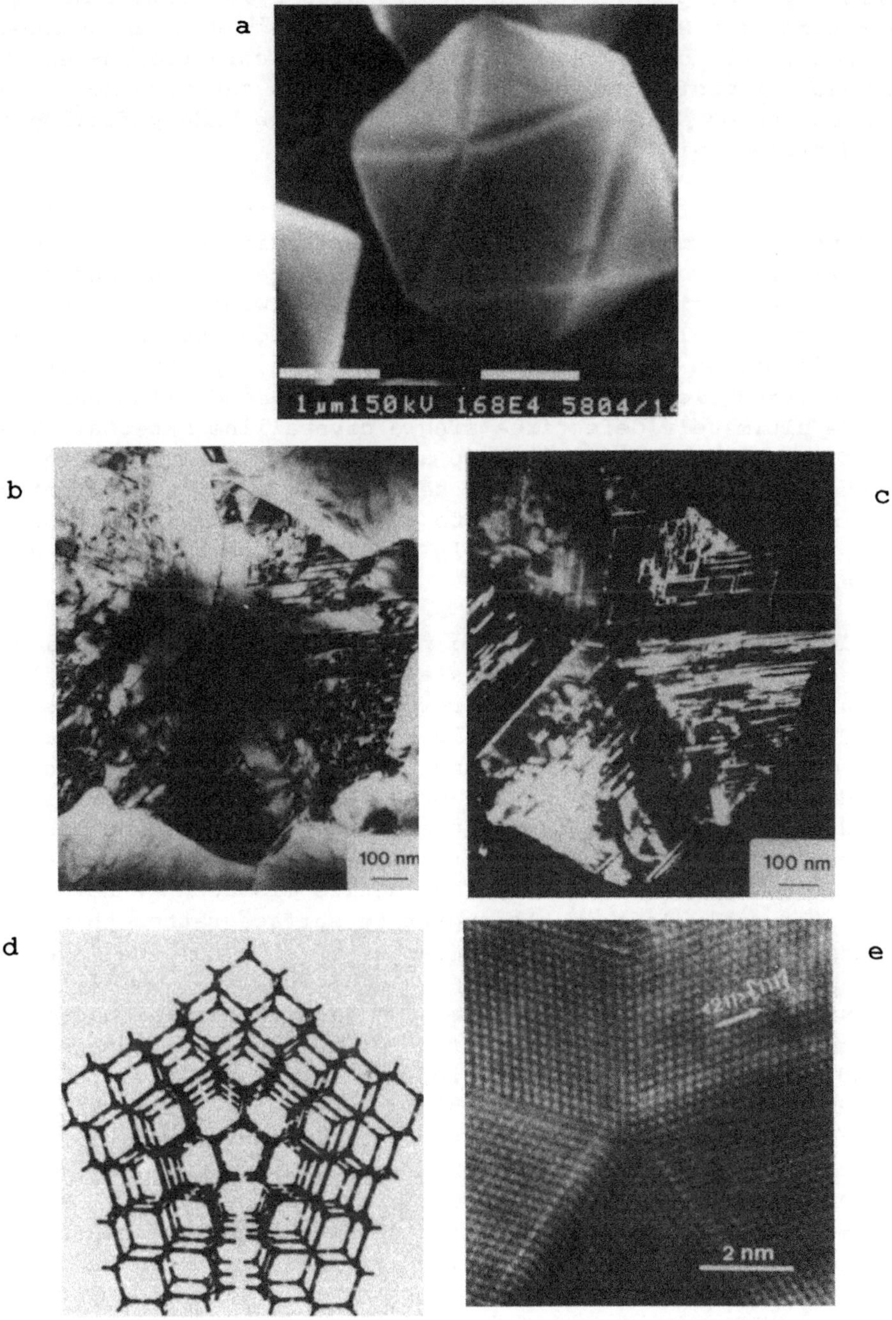

Fig. 4a-f. TEM investigations of five-fold multiply twinned diamond particles (MTPs): SEM micrograph (4a), bright-field (4b) and dark-field (4c) TEM images[34], and model[1] (4d) of an MTP. The HRTEM[32] micrograph (4e) images, with startling clarity, an undistorted central portion of an MTP.

The perfect core of such a particle is imaged with startling clarity by A.R. Srivatsa et al[32] using high resolution TEM. In their excellent HRTEM studies, B.E. Williams[34] et al have shown that not in all cases such perfect particle core regions exist and that the misfit between the twin sectors can also be accommodated by dislocations along the twin boundary forming so-called tilted-twin boundaries.

All of the TEM investigations performed on diamond have shown that CVD diamond -including homoepitaxially grown films-, high pressure synthesized diamond and also natural stones have a high defect density comprising twins, microtwins, dislocations, and stacking faults. Detailed characterization and understanding of the formation of such defects is of great importance for sophisticated electronic applications of diamond films which require ultimately defectfree single crystalline material. The five-fold twin shown in Figs. 4b and 4c, e.g. exhibits several defects within each of the five sectors. However, the defect density is surprisingly low close to the twin boundaries. If understood, this phenomenon could help develop methods to eventually grow electronic grade diamond.

Cross-section TEM data also revealed that on silicon and other carbide forming substrates a thin (50-100 Å) carbide layer may build up prior to diamond growth[25,35,36]. A.R. Badzian[37] suggested that, in the case of silicon carbide, this intermediate layer may act as lattice matching buffer layer that helps nucleating diamond on silicon. On the other hand, such a layer is not formed for all deposition conditions and the nucleation density was found to depend on the pretreatment of the substrate material. It correlates well with the lattice mismatch of hard materials presumably left in the substrate surface rather than the formation of scratches, lattice damage or of a carbide layer[18,38].

Similar to the morphological SEM data already discussed, XTEM confirms that the number of defects found in the deposit depend on the deposition conditions. In the case of microwave plasma CVD from CH_4/H_2 plasmas higher methane concentrations result in higher defect densities that are verified by XTEM[28,31].

Once the time consuming procedure of preparing a TEM sample is completed, electron diffraction studies to determine the lattice parameters of the deposited crystals can be made using the TEM. By means of the electron beam, selected area diffraction patterns (SAD) are obtained from small sample regions. Similar to X-ray diffraction, they allow determination of the lattice spacing, as well as detection of second phases such as graphite or silicon carbide.

D. Pickrell et al[36] found different compositions on top and bottom surfaces of some of their films by means of SAD. Fig.5 (from ref. 28) gives the diffraction pattern of a twinned diamond crystal along with its (hkl)-index values. The lattice constants

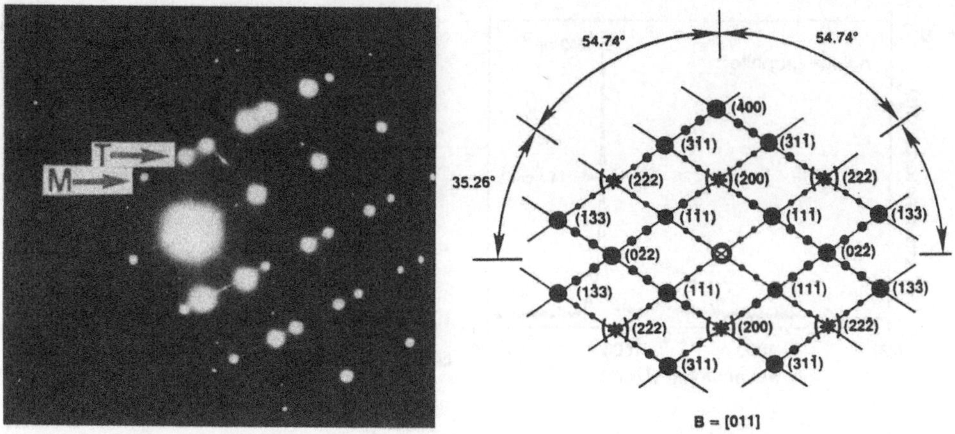

Fig. 5. Selected area electron diffraction (SAD) and indexing of the diffraction pattern: matrix (M) and twin (T) spots are indicated. The large dots (right side) correspond to matrix spots. Smaller dots are caused by twinning and double diffraction. Forbidden spots are marked by stars (from ref.28).

thus obtained usually agree well with diamond data from crystallographic tables[25,27]. However, as with other diffraction techniques, extreme care is necessary not to confuse diamond patterns with that of other oriented forms of carbon[39].

TEM and its varieties are important tools for the characterization of diamond deposits. However, the sample preparation is laborious and time consuming. Therefore, TEM characterization can only be used occasionally but not as a routine check similarly to SEM or Raman spectroscopy.

Diamond Phase Purity and Crystallinity by Raman Spectroscopy

Spontaneous Raman scattering, i.e. inelastic, incoherent scattering of light from lattice phonons of diamond was first observed by Ramaswamy in 1930[40] and investigated in more detail by Solin and Ramdas[41] in 1970. As a suitable characterization technique for diamonds in general, it has also been devoted a separate chapter in Field's book[23]. Like so many other important deposition and characterization techniques, Raman spectroscopy was introduced to CVD diamond research already in the early work of S. Matsumoto, Y. Sato, M. Kamo, and N. Setaka[7,8]. Raman data of CVD deposited carbon films are included in almost every paper published in this field. Excellent introductions and overviews of the specifics of this technique for CVD diamond research are given by D.S. Knight and W.B. White[42,43], or by R. Nemanich, J.T. Glass, and R.E. Shroder[44,45].

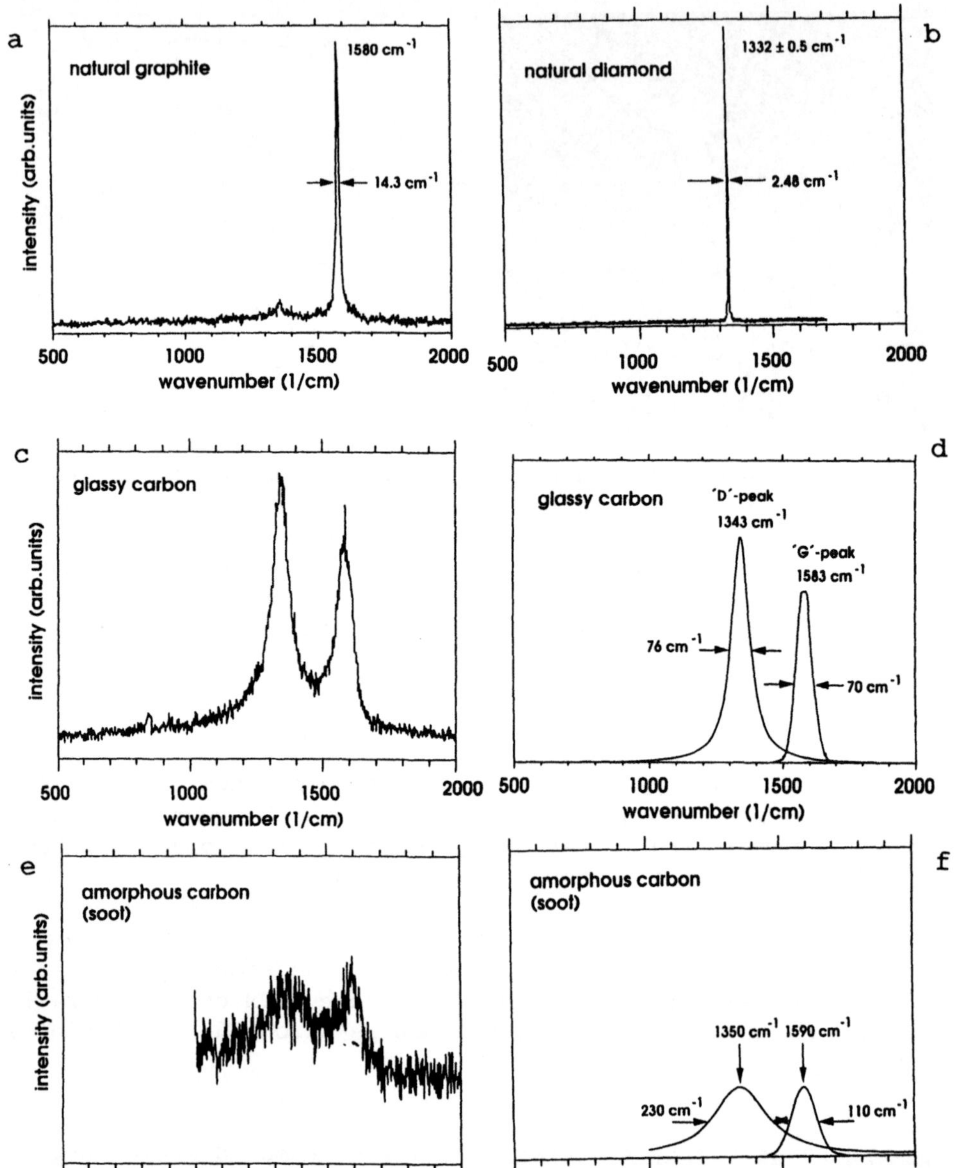

Fig. 6a-f. First-order Raman spectra for crystalline graphite (6a), natural single crystal diamond (6b), glassy carbon (6d,6e) and amorphous carbon soot (6e, 6f). Figs. 6d and 6f are not measured but synthesized data resulting from peak analysis to quantify the contribution of individually known peaks, to analyze their shape and, by subtracting a synthetic from a measured spectrum, to detect unknown peaks[21]. Such an analysis also allows for better correlation of Raman data and film preparation conditions.

The success of Raman spectroscopy in CVD diamond characterization is based on the negligible effort involved in sample preparation and the simple, fingerprint-like spectrum of genuine diamond that consists of only a single Raman line shifted with respect to the incident laser beam wavelength by 1332.5 ± 0.5cm^{-1} at room temperature[41]. For Lonsdaleite, the hexagonal crystalline form of sp3-hybridized carbon, a Raman shift of 1315-1320 cm^{-1} [42] is observed. The Raman-active phonon of cubic diamond is a triply degenerated zone-center vibrational mode [46]. It is, due to different binding and long range order in the solid, distinctly different from the spectra of other forms of carbon such as graphite, Lonsdaleite, diamond-like carbon (DLC), or glassy carbon. This allows non-invasive investigation of both, phase purity and crystal perfection of a carbon deposit.

In Fig.6a-f, the results of our own Raman measurements of various forms of carbon, namely crystalline graphite, natural single crystal diamond, glassy carbon and amorphous carbon soot are displayed within the range of the first-order Raman peaks. Natural single crystal graphite exhibits a single Raman peak, the so-called "G"-peak at 1580 cm^{-1} that is associated with the carbon-carbon stretching mode (Fig. 6a). If graphite becomes disordered within the carbon layers, the "G"-peak broadens[47,48]. For polycrystalline graphite, depending on the size of the crystallites, the Raman selection rule k=0 is violated and a second peak at 1357 cm^{-1} appears, confusingly termed "D"-peak. Again broadening is due to disorder phenomena. In Fig. 6a the "D"-peak is in fact only a little hump. The line width and the ratio of "D"- and "G"-peak highly depend on the preparation conditions of the carbon material. If the long range order of the crystalline material is lost and the carbon phase gets glassy, ""G"- and "D" broaden, the "D"-peak intensity increases and downshifts due to bond angle distortions are observable (Fig.6c). The lines are further broadened in the case of amorphous carbon soot (Fig. 6e). This spectrum also appears much noisier due to signal reduction caused by evaporation of the material during the laser measurement. The spectrum of Fig. 6b refers to natural single crystal diamond. The only first-order Raman peak visible appears at 1332 cm^{-1} and has a width (FWHM) of 2.4 cm^{-1}.

Figs. 6d and 6f are not measured, but synthesized spectra resulting from a detailed peak analysis[21] of the measured spectra. We routinely perform this analysis on-line on a computer by subtracting peaks known from the literature from the measured data. Peak positions, half widths, intensities and shapes (Gaussian or Lorentzian) can be adjusted on screen. In case of a complicated CVD diamond spectrum it is important to start this procedure with the widest of all peaks and continue with the next smaller one. If, finally, every contributing peak is located, its width adjusted, and then subtracted, the resulting difference of measured and synthetic spectrum should be zero. Any deviations are either due to false parameters, or due to unknown Raman or fluorescence peaks. Fig. 7 illustrates such an analysis.

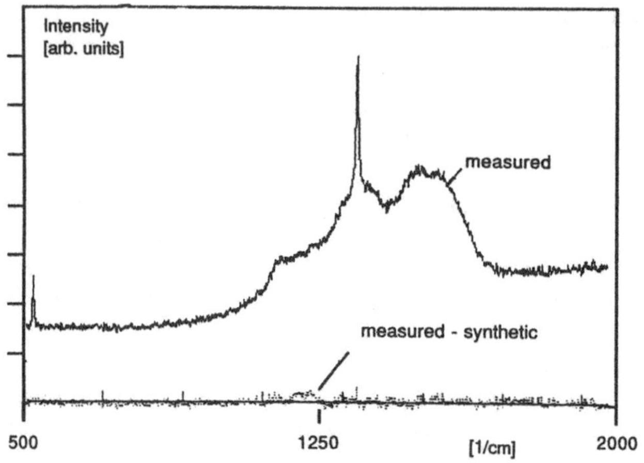

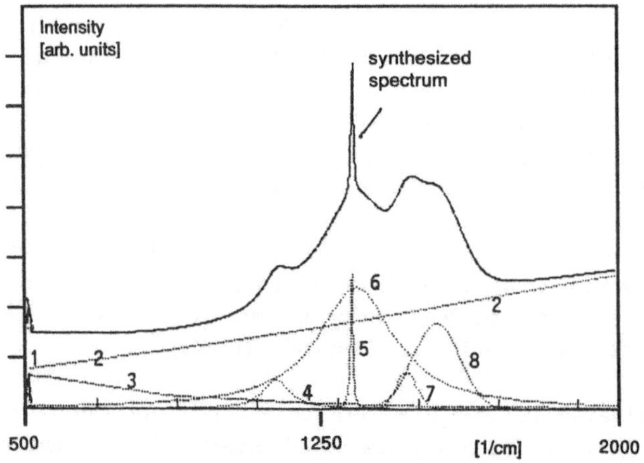

Fig.7. Raman spectrum of a carbon film containing diamond and non-diamond components (top). The spectrum was peak-analyzed, synthesized from identified, contributing peaks (bottom) and finally subtracted from the measured spectrum, yielding values close to zero over the whole range of interest, thus indicating complete and successful interpretation of the measurement. The assignments of peaks 1-8 (bottom) are listed in Table 1.

Table 1. Peak Assignment List for Raman Spectra

No. (Fig. 7)	position [cm^{-1}]	typical FWHM [cm^{-1}]	shape	assigned to
1	520	3.5	Lorentzian	silicon Raman peak
2	3143	3550	Gaussian	fluorescence background
3	≈500	550	Lorentzian	unknown
4	≈1150	80	Lorentzian	nanocryst. diamond
5	1332	7	Lorentzian	diamond Raman peak
6	1345	250	Lorentzian	"D"-peak
7	≈1470	80	Gaussian	diamond precursor (?)
8	1560	110	Gaussian	"G"-peak.

The measured Raman spectrum of a rather low quality diamond film was found to consist of 8 contributing peaks. They are listed in Table 1. The broadest peak is a Gaussian fluorescence peak centered around 3100 cm^{-1} that forms the background of the Raman spectrum within the range of interest. Next comes a so far un-identified, weak Lorentzian peak that sometimes occurs at ≈ 500 cm^{-1} and has a width of 500-600 cm^{-1}. The Lorenztian "D"-peak is, in this example, located at 1345 cm^{-1}. The "G"-peak, other than the "D"-peak, is best described by a Gaussian shape and appears, in this specific case, at 1560 cm^{-1}. Interestingly, even for quite good quality films, small contributions of the peaks at ≈ 1150 and ≈1470 cm^{-1} are often necessary to get minimum deviation from zero after subtracting the synthetic from the measured spectrum. These peaks were already mentioned in the publications of Nemanich et al[44] and Yarbrough et al[49]. Yarbrough attributes the peak at 1150 cm^{-1} to nanocrystalline diamond and a breakdown of the Raman selection rule for small crystallites, similar to the appearance of the "D"-peak in nano-crystalline graphite. The origin of the peak at ≈1470 cm^{-1} remains unclear. It cannot be associated with the diamond structure that does not allow any first-order Raman modes higher than 1332 cm^{-1}, however, it could be due to a tetrahedrally bonded diamond precursor[49].

In the range of 500-1800 cm^{-1} these 8 peaks are normally sufficient to synthesize any spectrum of materials deposited from C/H-, C/H/O-, and C/H/N- containing gas mixtures by means of either microwave plasma, DC jet plasma, or flame deposition.

Depth analysis by means of Raman spectroscopy is limited by the penetration depth of the laser light in the solid and thus depends on both its wavelength and on the extinction coefficient of the material under investigation. Raman spectroscopy is, therefore, a surface probing technique in case of graphite or soot, but a bulk characterization technique for diamond. If, e.g., an improper shut down procedure ends a deposition process by forming a thin glassy carbon layer on top of an otherwise perfect crystalline diamond film, it is not possible to detect the crystalline phase by means of Raman spectroscopy. Because of this problem it is also difficult to detect small amounts of diamond embedded in a non-diamond carbon matrix. On the other hand, small amounts of graphitic material are easily detected in a diamond matrix due to the higher Raman scattering cross section of graphite. N. Wada et al report the Raman cross section of crystalline graphite being 54 times that of a diamond crystal [50].

Raman measurements are commonly used to study the influence of various deposition parameters on the phase purity and crystallinity of the deposited material. Such studies include,e.g. the influence of the methane concentration in a methane/hydrogen CVD gas phase on the quality of the deposit[51]. Correlation with the thermal properties reveals that Raman data can be used to predict bulk properties of the deposits. Only films without second phases visible in the Raman spectrum approach natural diamond with respect to most of their other physical constants. Pressure dependence studies of the film quality deposited from DC plasma jet by means of Raman spectra reveal[21] that at 1 mbar only amorphous carbon is formed while, under otherwise similar conditions, diamond is found for $P > 30$ mbar in the reactor vessel.

K. Nishimura et al illustrate the effect of boron doping on the crystalline quality of diamond[52] by means of Raman data. They found that for small amounts of boron added to the CVD gas phase (< 400 ppma of carbon) the diamond Raman peak increases and gets narrower while for higher B-amounts the film graphitizes. We found similar effects when adding small amounts of nitrogen gas to the CVD gas phase while depositing from a methane/hydrogen mixture by means of microwave plasma CVD[20]. This effect is illustrated by the spectra in Fig. 8.

The peak width of the Raman line, indicating crystalline perfection of the probed sample region, may vary considerably depending on the preparation conditions. While high pressure synthesized single crystals may exhibit a line width of only 1.7 cm^{-1} [42] and natural diamond show 2-2.5 cm^{-1}, CVD diamond films normally exhibit a width of 5-10 cm^{-1} for the diamond signature peak. This is valid even for films with a very low level of contamination. However, it is possible to measure line widths as low as 2 cm^{-1}, if the deposited crystals do not form a film but remain separate[12]. For the DC jet deposited particle shown in Fig.2e, we measured, e.g., a line width of 3.3 cm^{-1}. Continuous films prepared under similar conditions are around 7 cm^{-1} [21]. This is probably due to strain and crystal defects in films consisting of intergrown particles. In this context it is important to mention that the probing beam needs, of course, to illuminate a

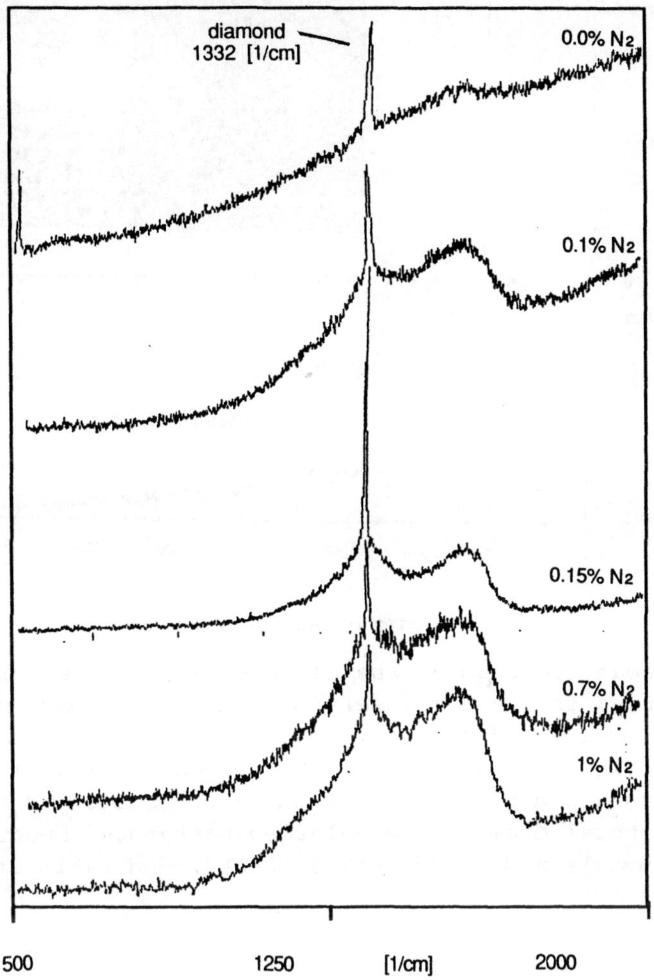

Fig. 8. Raman spectra of CVD diamond films deposited from a mixture of 0.75% methane in hydrogen by means of a microwave plasma. The spectra illustrate the effects 0-1% nitrogen (ref. H_2) added to the CVD gas phase. The ratio of diamond-to-"G"-peak is highest for ≈0.15% nitrogen added.
Nitrogen additions have also a prominent effect on the morphology, deposition rate and cathodoluminescence spectra of these films[20].

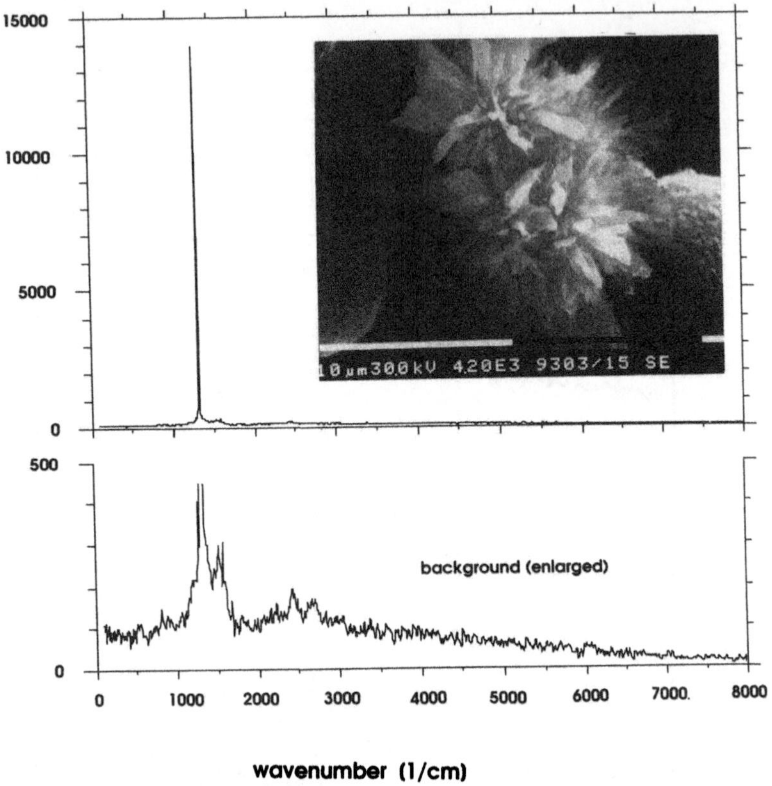

Fig. 9. Raman spectrum of the flower-like, filamentary material shown in the SEM insert[53]. These structures appear after laser-modification of the ball-shaped deposit also visible in the insert. The diamond peak at 1332 cm^{-1} has a width of only 2.4 cm^{-1} i.e. similar to that of natural diamond.The enlarged background spectrum reveals a diamond-peak-to-background ratio of ≈ 100:1.

representative sample area. The focus of the laser spot has to be considerably larger than the average size of the crystallites in the film in order to obtain film rather than particle characteristics. The use of a micro-Raman option needs, therefore, careful evaluation.

The "working definition" of diamond demands, simultaneously, a crystalline morphology as well as a Raman spectrum typical for diamond. Recently, we found that this definition can sometimes be rather misleading. Fig. 9 depicts ball-shaped plasma deposited carbon particles along with a flower-like webby structure. The fibrous substrate. We initially were not able to detect a diamond line in its Raman spectrum. However, after irradiation with the Ar-laser used for our Raman measurements the morphology of the particles changed from ball-shaped to flower-like and now exhibits a Raman spectrum (Fig. 9) similar to that of natural diamond with respect to both intensity and line width. (For a detailed explanation see ref.53). The odd morphology of this structure is hardly a "crystalline morphology discernible by SEM", nevertheless it is clearly diamond, thus suggesting the necessity to modify the "definition" e.g. by only using the Raman fingerprint or, at most, a combination of Raman and diffraction data as unambiguous proof of the presence of diamond in a deposit. This would also allow to include more easily the material obtained by microwave plasma CVD at substrate temperatures of less than 500°C[54] or the material often found on the substrate side of a free-standing film, were the Raman data indicate diamond, however, SEM data reveal only optically transparent material that consisting of submicron sized particles.

There are many more additional important studies that rely on Raman spectroscopical measurements, e.g. the application of a surface enhanced Raman technique to detect and investigate films of only 20-40 nm thickness, i.e. the initial stage of diamond growth by sputtering a 5 nm silver coating onto the carbon deposit[55]. However, for the sake of brevity we will now turn to a related characterization topic often associated with Raman spectroscopy, but also with SEM investigations, namely the investigation of luminescence light excited by either photons (e.g. during Raman measurements) or by the electron beam of an electron microscope.

Crystal Defects, Dopants and Impurities by Photo- and Cathodoluminescence Studies

Luminescence is the radiation emitted by relaxation and recombination processes of electrons following excitation by either photons, so-called photoluminescence, or by an electron beam, termed cathodoluminescence. Luminescence studies are a preferred tool to investigate natural and high pressure synthesized diamond and a vast literature on this subject exists [56a-e]. Similar investigations of CVD diamonds are still scarce[24,43,57-61].

Due to the high electron energies of the cathode ray applied, usually well above the diamond band gap of 5.4 eV, cathodo-luminescence (CL) is generally less selective than photoluminescence (PL), where the energy of the incident photons is normally lower than the band gap[60]. On the other hand, the penetration depth and thus the depth of the sample region probed can be easily varied in CL studies by varying the beam energy.

Photoexcitation followed by PL occur during Raman spectroscopy measurements and can easily obscure the Raman data because the PL light is picked up by the photodetector necessary to detect the Raman-scattered light. To separate the Raman from the PL contributions of a spectrum it is necessary to record the spectra at two different excitation wavelengths, e.g. the lines at 488 nm (2.54 eV) and at 514.5 nm (2.41 eV) of an Argon ion laser. While Raman signals exhibit a fixed frequency shift with respect to the excitation wavelength, the luminescence signal corresponds to a fixed absolute frequency. In the wavenumber scale usually used in Raman spectroscopy, all Raman signals still appear at the same wave number when varying the excitation wavelength while luminescence signals are displaced accordingly. Using the two wavelengths mentioned above results in a wavenumber shift of all luminescence lines by 1055.5 cm^{-1}. This is illustrated in Fig. 10a where both, Raman and luminescence spectra of a microwave plasma deposited diamond coating were recorded at excitation wavelengths of 514.5 nm and 488 nm, respectively. As required for this scale, the diamond Raman peak remains at 1332 cm^{-1} while the combination of broad luminescence features centered around 3000 cm^{-1} is shifted to ≈ 4000 cm^{-1} when the excitation wavelength is changed.

Our peak analysis routine (Fig. 10b) reveals that lines at 2000 cm^{-1}, 2800 cm^{-1}, 3140 cm^{-1}, and 3585 cm^{-1} contribute to this broad PL band, corresponding to absolute energies of approx. 2.16 eV, 2.06 eV, 2.02 eV and 1.96 eV, respectively. The 2.06 eV band may contain second-order Raman lines of diamond, but, according to our analysis, it is also due to a so far unidentified luminescence band.

In their extensive CL study, L.H. Robins et al[60] attributed the other peaks to a vibronic spectrum with a zero-phonon line at 2.156 eV and dominant electron-phonon coupling into an acoustic and an optic mode phonon. They correlated the broadening of these lines with data observed for non-CVD diamond and the residual stress in such material, concluding that in hot filament grown diamond the residual stress is ≈3 GPa. The nature of this luminescence center is attributed to a nitrogen atom in diamond. Similar to the studies of V.S. Vavilov et al[57] on homoepitaxially grown nitrogen doped CVD diamonds, we also found that the 2.16 eV PL intensity depends on the concentration of nitrogen added to the CVD gas phase[20]. Recently, G. Janssen et al[62] have further shown that these nitrogen associated PL bands dominate the PL spectrum of a thick homoepitaxial diamond layer deposited from an oxy-acetylene torch in air.

D.S. Knight and W.B. White[43] report that the luminescence of diamond films deposited from various carbon sources and different methane/ethanol or methane/CO_2 mixture does not vary much. However, these films were all prepared by means of microwave plasma CVD using a deposition system[18] with a base pressure of ≈10^{-2}mbar. Our own PL data for microwave plasma deposited material from a similar deposition unit look basically identical for undoped plasma deposition system with a base pressure of less than 10^{-6} mbar and equipped with a heavy pumping system did not show such PL spectra, indicating that residual nitrogen in the microwave deposition systems could be responsible for the broad band luminescence.

Another interesting and widely discussed feature visible in Fig. 10 is the narrow peak at 5900 cm^{-1} (514.5 nm excitation). This line is commonly termed GR1 (general radiation) center and is associated with a neutral atomic vacancy in the diamond lattice. GR1 is the zero-phonon line of a series of GR PL features and is located at 1.673 eV[60]. However, it is also possible[43] that silicon centers in diamond are responsible for this PL peak.

Clear evidence for the radiation induced vacancy interpretation of a peak around 5900 cm^{-1} or 1.68 eV (514.5 nm excitation) is given in Fig. 11, where the intensity of GR1 is varied by neutron irradiation[63]. Not only peak (e) of Table 2 increases, but also its shoulder (peak f) is enhanced by neutron irradiation. On the other hand, we also found that peak (e) increases when a silica rod is moved into the microwave plasma ball during diamond deposition. Perhaps both, vacancies and impurities are capable of producing luminescence centers around 1.68 eV.

Electron beam excitation followed by CL occurs e.g. during electron microscopy as already briefly mentioned (Figs. 3e-f). Other than in the case of Raman spectroscopy CL needs a special photosensitive detector and does not disturb electron microscopical measurements. PL is limited by the wavelength of the light source used. PL centers with excitations energies above the photon energy coressponding to the 488 nm line of an Ar^+-laser are e.g. not excited. With electron beam excitation, however, this is possible. Thus, CL and PL data partially overlap and are complementary, although CL and PL intensities may be different due to the different excitation mechanisms.

CL studies of CVD diamonds were first published by H. Kawarada et al[24]. The authors report that half of their microwave plasma deposited samples exhibited a green luminescence band with a peak energy of 2.45 eV, i.e. 505 nm. However, some of their samples had a a blue CL band with an intensity maximum at 2.8 eV (445 nm). Such differences also occur in high pressure synthesized diamond. In ref. 24, the authors associate the green CL with nitrogen present in the CVD system and incorporated into the coating. They also point out that <100> faces rather than <111> emit the CL light, a fact that has already been decribed in Fig.3e. Subsequently the same group[59] verified the association of the blue CL emission with undoped diamond. However, the band around 500 nm did not appear in slightly nitrogen-doped material but in boron-doped diamond. CL associated with intentionally added nitrogen seems to appear at around 600-650 nm instead.

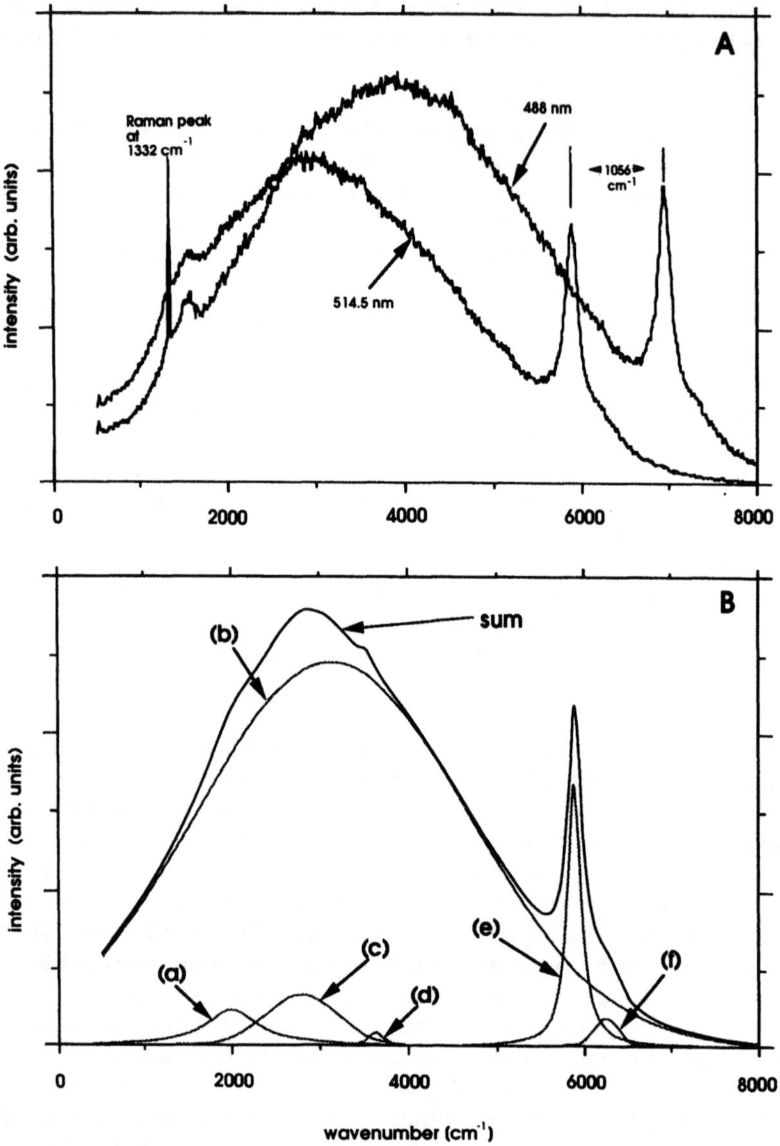

Fig. 10. (A) Raman and photoluminescence peaks obtained from a microwave plasma CVD diamond film. The spectra were recorded utilizing the 488 nm and 514.5 nm light of an Ar^+-laser. While Raman peaks appear at a fixed wave number, all PL peaks are shifted by $\approx$ 1056 cm^{-1} when changing the excitation wavelength. Analysis (B) reveals the contributing peaks that are listed in Table 2.

Tab. 2. Results of peak analysis of Fig. 10 B.

peak data in relative wavenumbers				absolute peak data			
No.	wavenumber (cm^{-1})	FWHM (cm^{-1})	intensity (cps)	peak position λ (nm)	peak position E (eV)	FWHM Δλ (nm)	FWHM ΔE (eV)
(a)	2000	700	350	573	2.16	23	0.087
(b)	3140	3650	3700	614	2.02	140	0.453
(c)	2800	950	500	601	2.06	35	0.118
(d)	3585	200	125	631	1.96	8	0.025
(e)	5900	190	2525	739	1.68	10	0.024
(f)	6258	330	275	759	1.63	19	0.041

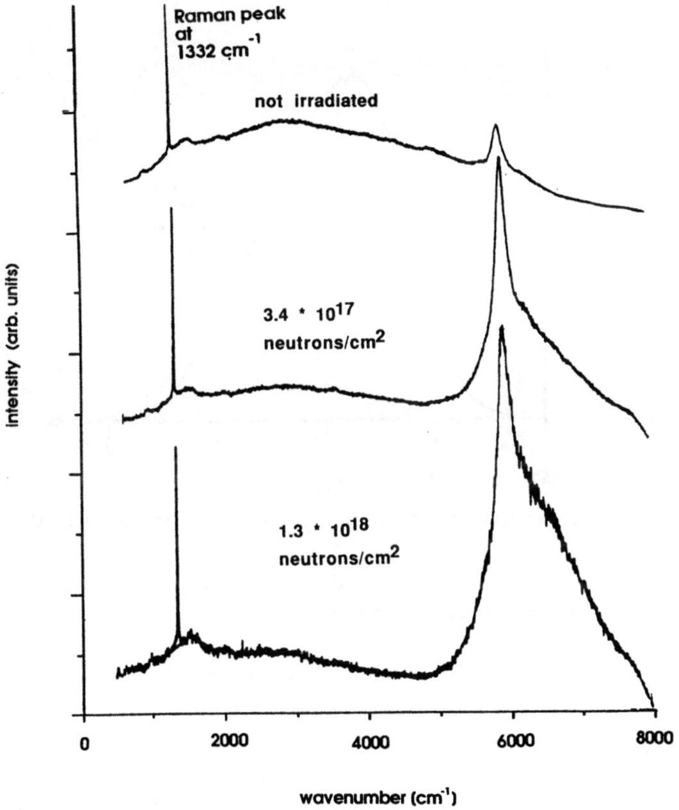

Fig. 11. The so-called GR1 PL center (peak (e) in Tab.2) and its shoulder (peak (f) in Tab. 2) are increased by irradiation a diamond film with neutrons[63].

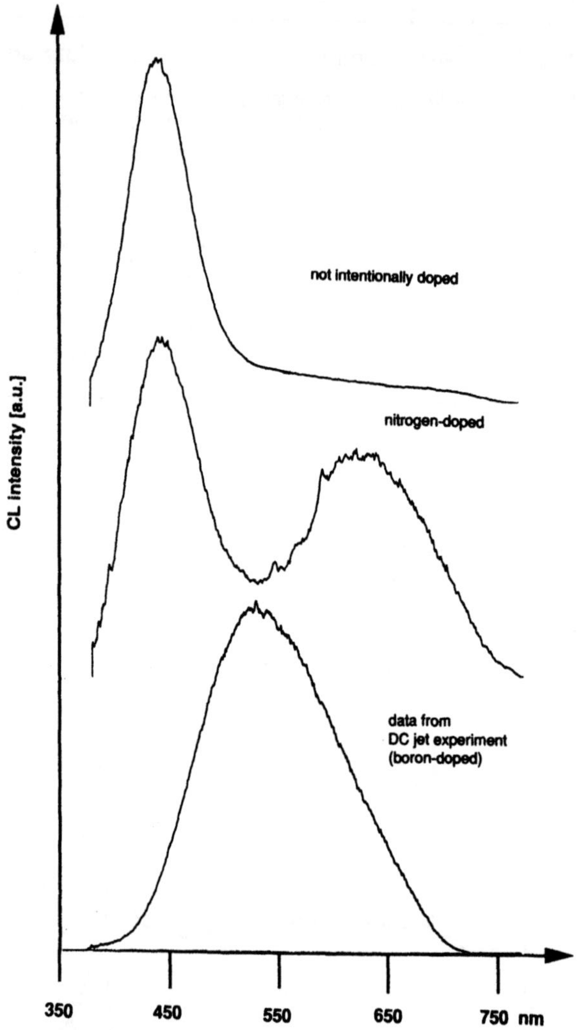

Fig. 12. Cathodoluminescence spectra of diamond films:

Top: spectrum of a film that was not intentionally doped.

Middle: data from a film deposited from a C/H/N gas phase.

Bottom: CL from DC cascaded arc deposited material where corrosion of boron-nitride parts of the plasma generator led to boron-doping.

In our own CL investigations[20], we found the blue diamond luminescence typical for undoped CVD material centered around 430 nm (2.9 eV). Fig. 12 (top) depicts such a CL spectrum. For natural diamond, this so-called "band A" emission is likely due to electron-hole recombination at donor-acceptor pairs trapped at dislocation lines[64]. The CL intensity depends highly on the preparation conditions. Films deposited from an oxygen-containing CVD gas phase show, for identical C/H-ratios, higher blue CL intensities, probably due to reduced non-diamond carbon contaminations as is also visible in the Raman spectrum. Intentional doping with nitrogen resulted in the red CL band with the plateau around 600-630 nm (2.07-1.97 eV) visible in Fig. 12 (middle). The intensity ratio of blue and red CL emission is highly dependent on the amount of nitrogen added to the CVD gas phase. We only rarely found green CL emission in material deposited from a cascaded arc DC plasma. There, the spacers isolating the copper plates of the cascaded arc are partly made from boron nitride[21] and are the most likely boron source.

CL and PL studies allow investigation of the optical properties of diamond associated with certain defects and impurities. However, it is sometimes very difficult to identify the true nature of the defect or impurity. Where spatial resolution is not mandatory, other characterization methods have proved more helpful. Actually, one of the results mentioned in the PL section of this paper originated from such an attempt. The neutron induced defects depicted by PL spectra in Fig. 11 were formed during irradiation in the course of neutron activation analysis to detect impurities in films.

Quantification of Impurities and Dopants and Lattice Matching by Neutron Activation Analysis, Secondary Ion Mass Spectroscopy, Rutherford Backscattering, Elastic Recoil Detection, and Nuclear Reaction Analysis

Neutron Activation Analysis (NAA)

Diamond deposition experiments are performed inside a reactor using a carbon source, an etchant gas, sometimes with additional diluting gas and dopants. The reactor, the gas feed stock as well as predeposition treatments of the substrate material may act as sources of contaminations of the deposited diamond. The filament in a hot filament reactor is, e.g., a very likely source of tungsten, tantalum or molybdenum in the film, depending on the filament used. M. Geis[65] reported considerable quantities of tungsten in his homoepitaxially "hot filament" grown diamond. Such contaminations may be of importance for the application of the deposited diamond. They may act as optical absorption centers, reduce the thermal conductivity or increase the electrical conductivity of a diamond film. Detection of such impurities is possible by neutron activation analysis (NAA).The material under test is irradiated by thermal neutrons in a nuclear reactor. Neutrons are captured by the atoms to increase the atomic mass by 1 thus often forming a radioactive nuclide. The concentration of such radioactive atoms is conveniently determined by measuring the γ-radiation emitted during the decay of the isotope. Carbon itself has a very low capture cross-section for thermal neutrons and is practically not activated. Therefore, it is an excellent matrix for NAA.

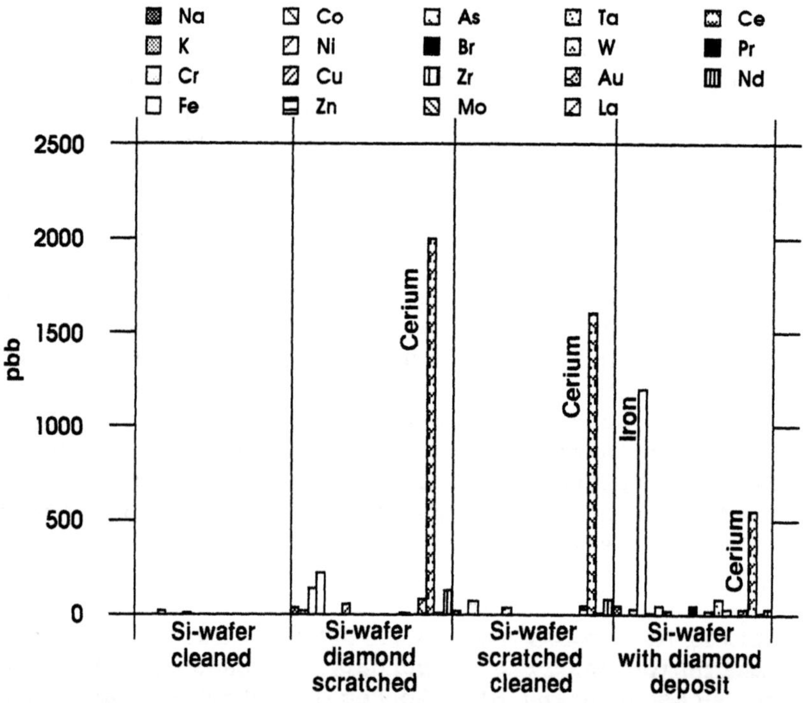

Fig. 13. Results of NAA measurements of impurities in a bare silicon wafer, a wafer after lapping with diamond powder, after lapping and subsequent cleaning, and after depositing a 5μm thick diamond film by means of microwave plasma CVD.
Cerium introduced during lapping stems from a contaminated lapping felt. The graphite heater cover in our plasma CVD reactor is the most likely iron source for the contamination during deposition.

In the course of investigating the effects of predeposition polishing with various hard materials[18], L. Pilione et al[63] observed by means of NAA that tantalum is incorporated in the deposit when TaC is applied for silicon substrate surface pretreatment. In some cases even a memory effect of the reactor was observed by NAA, i.e. contamination of several subsequent samples after once applying TaC.

NAA can also be used to trace sources of unknown contaminations as illustrated in Fig. 13. We activated cleaned, commercially available silicon wafers and found no impurities of any significance. After polishing the silicon surface with 1-2 μm natural diamond powder, 2 ppma of Cerium were detectable by means of NAA along with small amounts of other impurities. Most of the Cerium was still present after a subsequent cleaning step and was also found in the diamond-coated wafer. In addition, after deposition we also found iron in this sample. Obviously the Cerium stems from the polishing procedure, while the iron source is within the deposition reactor. We have identified the felt used for lapping as the source of Cerium (it was previously used for other polishing purposes).The iron source within the reactor is a little more difficult to trace, but the graphite heater cover in our set up is a likely candidate not only for iron, but also for other traces of metal impurities.

NAA is a helpful and sensitive tool for qualitative as well as quantitative determination of heavier elements. It is considered a non-destructive method, however, Fig. 11 provides evidence that some destruction of the film occurs. In addition it is very difficult or impossible to trace 1. and 2. row elements of the periodic system or to spatially resolve the collected information. On the other hand, 2. row dopants, e.g. Li, B, or N and their distribution are of great importance for electronic applications of diamond films. Secondary ion mass spectroscopy is a complementary tool in this case.

Secondary Ion Mass Spectroscopy (SIMS)

SIMS data are collected by destroying the deposit, utilizing an ion beam that sputters the material away. The ionized material from the deposit is then analyzed by means of mass spectroscopy. It is, of course, difficult to distinguish between chemical compounds and their constituent elements. However, it is possible to obtain depth information by successively sputtering deeper into the deposit. It is even possible to map out the elemental composition of a film by scanning the sputtering ion beam.

To quantify the result it is mandatory to use calibration samples. They are commonly prepared by ion implanting known amounts of a certain element to be detected into the diamond matrix. It needs to be mentioned that such an ion implantation process may result in a modification of the matrix, more specifically, in the graphitization of the diamond, thus changing the sputter rate of the calibration sample and obscuring the quantification.

J.T. Glass et al[26] investigated the depth profile of various impurities in plasma deposited diamond. The results are given in Fig. 14. The authors caution the readers not to take the quantitative results too seriously. They stress, however, that qualitatively the data are correct.

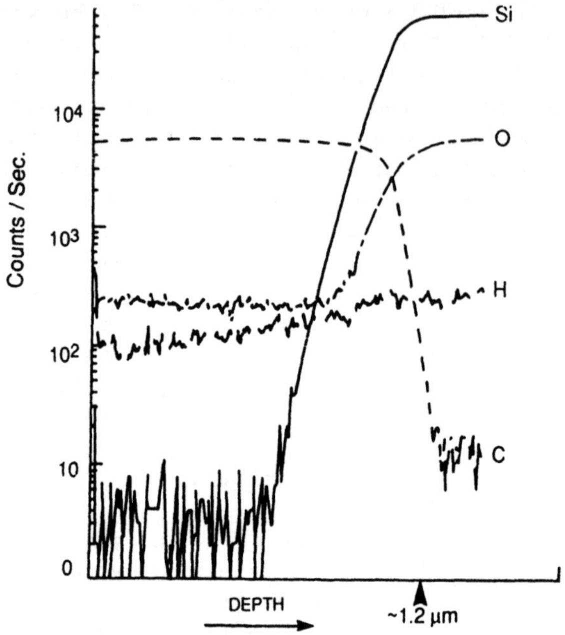

Fig. 14. SIMS depth profiles for the elements Si, O, H, and C in a plasma deposited 1.2 µm thick diamond film (from ref.26).

K. Nishimura et al[52] found by means of SIMS that the gas phase B/C ratio they used in their deposition experiments, is not reflected by the B/C ratio in the solid deposit. They found boron even in undoped films and concluded that a memory effect of the reactor caused B-contamination of "undoped" films deposited after using boron in the reactor. Detailed studies of diamond deposits near the substrate interface were performed by D. Pickrell et al[36]. They have shown that close to the interface impurities from the substrate material are detectable and by analyzing the sputter time-versus-depth during SIMS, they conclude that at the interface amorphous, weakly bound carbon is present in the film.

Rutherford Backscattering (RBS) and Elastic Recoil Detection (ERD)

To identify and quantify impurities in solids Rutherford Backscattering (RBS) and its variety, Elastic Recoil Detection (ERD)[66,67] are further techniques applicable. RBS relies on the detection of particles that are elastically backscattered by the atoms of the solid target. The energy of the backscattered particle is uniquely related to the mass of the target atom it has collided with. High energy ions, e.g. 2 MeV He^+, are used for RBS experiments and analyzing the energy of the backscattered ion allows to determine the nature and number of target atoms. RBS has the advantage that calibration is not mandatory (although sometimes used). ERD is a RBS variety specifically suited for the detection of hydrogen incorporated into the film. Here, not the scattered ion but the hydrogen knocked off the solid is detected.

We applied both methods to determine impurities in our films. All of our microwave plasma deposited films were, at least to the RBS probing depth of ≈1.5 μm, free of impurities other than hydrogen down to the ppma level. The ERD hydrogen profile of a 12 μm thick diamond film deposited on silicon is given in Fig. 15. Using the well-established scattering cross-sections of He with C/H/O it is possible to mathematically simulate the measurements and to quantify the result. The film has a hydrogen content of 2800 ppma. This level is comparable to data presented by W. Lanford in a hydrogen concentration study of diamond films from various sources[68]. He used the even more sensitive but also more complicated nuclear resonance reaction analysis method (NRA)[69], where 6-8 MeV ^{15}N of a primary ion beam react with H in the solid to form ^{12}C, ^{4}He. 4.43 MeV γ-radiation is emitted and quantitatively detected. In his study the hydrogen content of decent quality diamond films was determined ranging anywhere from 300-3000 ppma. Natural diamond typically contain 100-1500 ppma hydrogen.

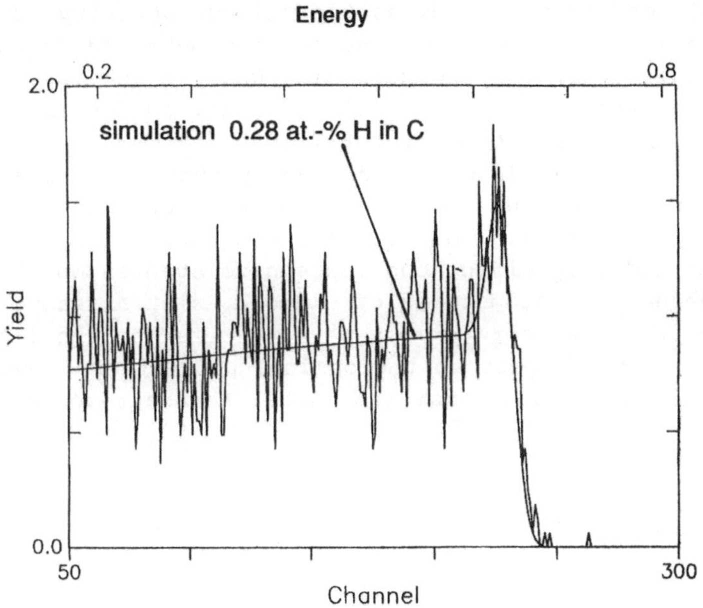

Fig. 15. ERD data obtained from a diamond film deposited by microwave plasma CVD from 0.7 % methane in hydrogen at 950°C. The hydrogen profile is best simulated by 2800 ppma hydrogen incorporated in the solid.
ERD at 700°C and 1000°C did not alter this profile.

The slight increase of the H-level close to the surface of our film (Fig. 15) points to either H-termination of the diamond or absorption of small quantities of hydrocarbons on its surface. In-situ heating to 700°C for 20 min and, in addition, to 1000°C for another 10 min did not change the hydrogen level of the film, thus indicating that the hydrogen is covalently bound to the carbon solid.

Recently, H.G. Maguire et al[70] have utilized RBS not to detect impurities, but to prove that a 1.5 μm thick diamond film deposited onto a single crystal diamond substrate is in perfect alignment with the substrate crystal. In this "channelling" RBS the substrate crystal directions are aligned with the beam. Perfect alignment allows the incident ions to travel along channels in the crystal and deviations from epitaxy are indicated by a higher scatter signal. The authors found perfect matching along the <111>, <100>, and <110> directions, indicating high quality homoepitaxy.

Many of the properties of CVD diamonds characterized by the methods discussed so far are of great importance for practical applications. It is, of course, possible to predict, e.g., poor electrical performance, reduced thermal conductivity or optical absorption due to impurities and defects detected in Raman, TEM or ERD studies. However, to determine what is tolerable for a specific application, it is still necessary to determine the quantities that are ultimately utilized.

On the other hand, data obtained by such measurements often consist not only of intrinsic diamond data but are obscured by the performance of, e.g. electrodes, contacts, or by adhesion properties under specific load. Extensive studies on the electrical and mechanical properties of diamond exist already. For the sake of brevity, we here simply have to refer to the literature[71,72,73]. With respect to the optical properties, however, some results are included as examples rather than attempting to be complete.

Optical Properties from UV- and IR- Spectroscopy

Diamond has a high potential for optical applications such as UV-VIS-IR windows, lens coatings, lasers etc[74,75]. Natural as well as high pressure synthetic material has been used for that purpose for many years. In order to substitute these materials and to find additional applications it is necessary to prepare phase pure material, minimize contamination, resolve adhesion problems on non-diamond substrates, as well as, in the case of polycrystalline material, to tailor the scattering properties of the films according to the wavelength region of use. The problems of phase purity and contamination have already been discussed in detail. The scattering issue and other optical properties are extensively covered in the proceedings of an SPIE conference

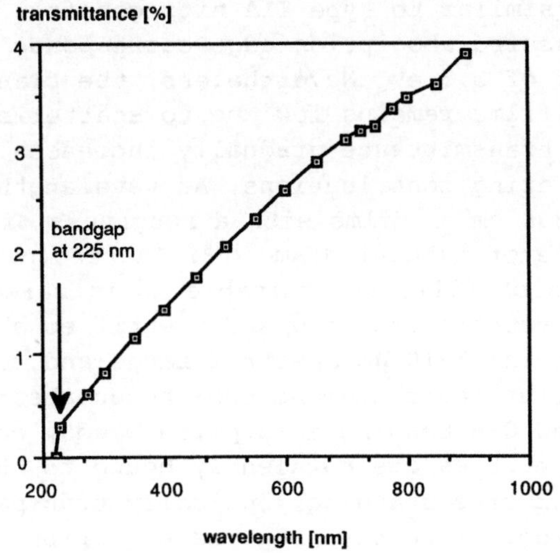

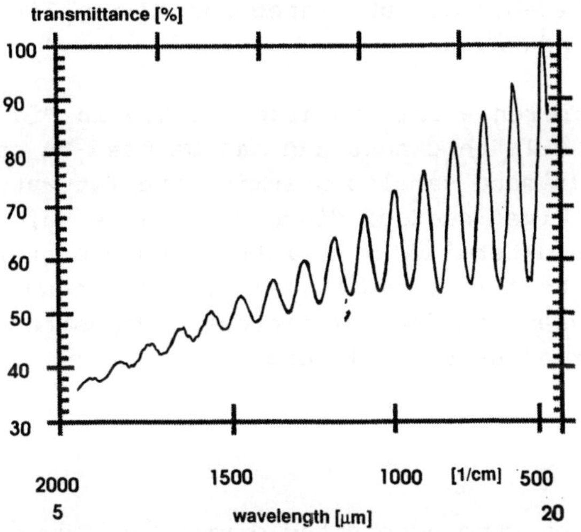

Fig. 16. UV-VIS (top) and IR (bottom) spectra of free-standing diamond membranes. The absorption at 225 nm agrees well with the diamond bandgap. Low transmission is due to scattering in polycrystalline material.

They are best illustrated by UV-VIS-IR spectra measured on films with the substrate being removed, e.g., by means of wet etching. Fig. 16 depicts a UV-VIS spectrum of an 8 μm thick diamond membrane (top) and a far-infrared spectrum of a 25 μm thick polycrystalline film (bottom). In the UV, a CVD diamond film may transmit at wavelengths as low as 225 nm, i.e. exhibit absorption characteristics similar to type IIA nitrogen-free natural diamond. The measured absorption edge corresponds well with the diamond band gap of 5.5 eV. Nevertheless, the transmittance of polycrystalline films remains low due to scattering. At higher wavelengths the transmittance gradually increases due to decreasing scattering contributions. At wavelengths > 8 µm, i.e. wavenumbers < 1300 cm^{-1}, films with a roughness of 500 Å approach the transmittance of natural diamond[78]. Therefore, as Ch. Wild et al point out[79], such films are suitable as infrared interference coatings. These authors isolated and identified absorption contributions in the 3-10 µm spectral range and found, in addition to the intrinsic diamond two-phonon absorption, small C-H streching and C-H bending absorption bands, confirming that the hydrogen impurities are covalently bound to the solid. Thin, only 2-3 µm thick free-standing, optically transparent films with low scattering where recently prepared by various groups[80] by controlling the particle size during the growth process. Such films are applicable as substrates for >X-ray lithography masks or as X-ray windows.

The interference fringes also visible in Fig. 16 (bottom) depend on the film thickness and can be used to calculate the thickness of diamond samples assuming the literature value of 2.4 for the refractive index of diamond. Vice versa, using thickness data otherwise determined, the interference pattern can be used to calculate the film refractive index. We found excellent agreement between the data of microwave deposited CVD diamond films and that of natural diamond.

SUMMARY

Research on vapor-deposited diamond has come a long way since the first reports in the 1950s and the revival of the field in the late 1970s. Deposition techniques have advanced dramatically over the past few years and with reported linear deposition rates of 500-1000 µm/h, CVD diamonds are getting more and more attractive not only for applications in sophisticated market niches, but also on a larger scale, competing with natural or high pressure synthesized products. Todays highly sensitive characterization techniques allowed to identify the weak spots of the vapor-deposited material and, feeding this information back into the preparation step, helped to drastically improve the quality of the material.

Although still plagued with defects and impurities, diamond films and coatings that are suitable for a number of mechanical, optical and electronical applications can be prepared today. That does not mean that this is the end of a development but, as R. Seitz puts it in his entertaining overview[81], we are just at "the end of the beginning".

ACKNOWLEDGEMENTS

The authors wish to express their appreciation to all the people involved within the Philips Research Organization and to the members of the "Diamond Thin Films" group at Penn State's Materials Research Laboratory.

Special thanks are to D. Leers for deposition experiments and SEM data, R. Raue and B. Busselt for CL measurements (all at Philips Research Labs, Aachen), to A. Jaspers for providing NAA data, to L. van Ijzendorn and A.T.E Kuiper for RBS and ERD results (all at Philips Research Labs, Eindhoven, NL), to D.S. Knight and L. Pilione for Raman and NAA data (both MRL), and to V. Zimmerman, R.F. Messier (MRL) and H. Lydtin (Philips Aachen) for constant support and many fruitful discussions over the past years. The loan of a natural diamond by D. Kersting-Wiechert is gratefully acknowledged.

REFERENCES

1. R.C. DeVries, Annu. Rev. Mater. Sci., 17, p.161 (1987)
2. B.V. Spitzyn and B.V. Derjaguin, author's certificate (patent application), July 10,1956; USSR Patent 339 134, May 5,1980.
3. W. Eversole, US Patents 3030187 and 3030188 (1962) (filed July 23, 1958).
4. J.C. Angus, H.A. Will, W.S. Stanko, J. Appl. Phys.39, p.2915 (1968).
5. H. Schmellenmeier, Z. Phys. Chem. 205, p.349 (1955).
6. B.V. Spitzyn, L.L. Bouilov and B.V. Derjaguin, J. Cryst. Growth, 52, p.219 (1981).
7. S. Matsumoto, Y. Sato, M. Kamo, N. Setaka, Jpn. J. Appl. Phys. Part 2, 21, p. L183 (1982).
8. M. Kamo, Y. Sato, S. Matsumoto, and N. Setaka, J. Cryst. Growth, 62, p. 642 (1983).
9. R. Messier, A.R. Badzian, T. Badzian, K.E. Spear, P. Bachmann, and R. Roy,Thin Solid Films, 153, p.1 (1987)
10. see, e.g., the following books, review articles and the references therein:
    (a) Proceedings of the "First International Symposium on Diamond and Diamond-Like Films",The Electrochemical Society,Pennington, NJ, Proceedings Vol. 89-12 (1989).
    (b) "Diamond and Diamond-Like Materials", (edited by G.H. Johnson, A.R. Badzian, and M. Geis) Materials Research Society, Pittsburgh, PA, Extended Abstracts,Vol.EA-15 (1988).
    (c) B.V. Spityn, L.L. Bouilov, and B.V. Derjaguin, Prog.Crystal Growth and Charact. 17, p. 79 (1988).
    (d) A.R. Badzian and R.C. DeVries, Mater. Res. Bull.23, p.385 (1988).
    (e) J.C.Angus and C.C. Hayman, Science, 241, p.913 (1988).
    (f) K.E. Spear, J. Am. Ceram. Soc. 72, p.171 (1989).
    (g) P.K. Bachmann and R.Messier, Chemical & Engineering News, May 15, p.24 (1989).

(h) "Diamond, Boron Nitride, Silicon Carbide and Related Wide Bandgap Semiconductors" (edited by J.T. Glass, R.F. Messier, and N. Fujimori), MRS Symposium Proc. Vol. 162, Materials Research Society, Pittsburgh, PA, (1989).
(i) B.V. Spitzyn, J. Cryst. Growth 99, 1162 (1990).
(j) "The Crystallization of Diamond" by D.V. Fedoseev, B.V. Derjaguin, and I.G. Varasavskaja (Part 1) and "The Growth of Diamond and Graphite from the Gas Phase" by B.V. Derjaguin and D.V. Fedoseev (Part 2) in "Diamond Growth and Films" edited by UCFMG, Elsevier Applied Science, London, New York (1989).

11. K. Kobashi, K. Nishimura, Y. Kawate, and T. Horiuchi, Phys. Rev. B, 38, no.6, p.4067 (1988).
12. Y. Sato, M. Kamo, Surf. Coat. Technol. 39-40,no. 1-3, p.183 (1989).
13. Y. Mitsuda, Y. Kojima, T. Yoshida, K. Akashi, J.of Mat. Science,22, p.1557 (1987).
14. Y. Liou, A. Inspektor, D. Knight, R. Weimer, D. Pickrell, A.R. Badzian, and R. Messier. Proc. SPIE 1146, p.12 (1989).
15. W. Zhu, R. Messier, and A.R. Badzian, Proceedings of the "First International Symposium on Diamond and Diamond-Like Films", The Electrochemical Society,Pennington, NJ, Proceedings Vol. 89-12, 61 (1989).
16. C. Chen, Y.C. Huang, S. Hosomi, and T. Yoshida, Mater. Res. Bull. 24, no.1, p.87 (1989).
17. K. Kobashi, K. Nishimura, K. Miyata, Y. Kawate, J.T. Glass, and B.E. Williams, Proc. SPIE 969, p. 159 (1988).
18. P.K. Bachmann, W. Drawl, D. Knight, R. Weimer, and R.F. Messier, in "Diamond and Diamond-Like Materials"(edited by A. Badzian, M. Geis, and G. Johnson), Mat. Res. Soc., Pittsburgh, PA, Extended Abstracts Vol. EA-15, p.99 (1988).
19. T. Hartnett, A.R. Badzian, and K.E. Spear, in "Diamond and Diamond-Like Materials"(edited by A. Badzian, M. Geis, and G. Johnson), Mat. Res. Soc., Pittsburgh, PA, Extended Abstracts Vol. EA-15, p.103 (1988).
20. P.K. Bachmann, D.Leers, H. Lydtin, R.Raue,C. Ronda, and D.U. Wiechert,accepted for presentation at the "2.Int.Conf. on the New Diamond Science & Technology" Sept.23-26, Washington, D.C. USA (1990).
21. P.K. Bachmann, H. Lydtin, D.U. Wiechert, J.J. Beulens, G. Kroesen, D.C. Schram, in "Proceedings of the Third International Conference on Surface Modification Technologies", Sept. 1989, Neuchatel,Switzerland, The Minerals, Metals & Materials Society (TMS), Warrendale, PA, (1990).
22. P.K. Bachmann and H. Lydtin, in "Characterization of Plasma-Enhanced CVD Processes" (edited by G. Lucovsky, D.E. Ibbotson, and D.W. Hess), MRS Symposium Proccedings, Vol. 165, Materials Research Soc. Pittsburgh, PA, USA, in press (1989).
23. J.E. Field (ed.) "The Properties of Diamond", Academic Press, London, New York, San Francisco, (1979).
24. H. Kawarada, K. Nishimura, K.S. Mar, Y. Yokota, J. Suzuki, T. Ito, and A. Hiraki, in "Diamond and Diamond-Like Materials"(edited by A. Badzian, M. Geis, and G. Johnson), Mat. Res. Soc., Pittsburgh, PA, Extended Abstracts, Vol. EA-15, p.89 (1988).

25. B.E. Williams, J.T. Glass, R.F. Davis, K. Kobashi, and Y. Kawate, in "Diamond and Diamond-Like Materials"(edited by A. Badzian, M. Geis, and G. Johnson), Mat. Res. Soc., Pittsburgh, PA, Extended Abstracts Vol. EA-15, p.59 (1988).
26. J.T. Glass, B.E. Williams, and R.F. Davis, Proc. SPIE, Vol.877, p.56 (1988).
27. J. Narayan, A.R. Srivatsa, M. Peters, S. Yokota, and K.V. Ravi, Appl. Phys. Lett. 53, no. 19, p.1823 (1988)
28. W. Zhu, A.R. Badzian, and R.F. Messier, J. Mater. Res. Vol.4, no.3, p. 659 (1989).
29. W. Zhu, C.A. Randall, A.R. Badzian, and R.F. Messier, J. Vac. Sci. Technol. A, Vol.7, no. 3, p.2315 (1989).
31. B.E. Williams and J.T. Glass, J. Mater. Res. Vol. 4, no.2, p. 373 (1989).
32. A.R. Srivatsa, J. Narayan, and K.V. Ravi, in Proceedings of the "First International Symposium on Diamond and Diamond-Like Films", The Electrochemical Society,Pennington, NJ, Proceedings Vol. 89-12, 215 (1989).
33. R.E. Clausing, L. Heatherly, K.L. More, and G.M. Begun, Surf. Coat. Technol. Vol. 39-40, no.1-3, p.199 (1989).
34. B.E. Williams, H.S. Kong, and J.T. Glass, J. Mater. Res. 5, no.4, p.806 (1990).
35. M.S. Wong, W.A. Chiou, F.R. Chen,D.X. Li, and R.P.H. Chang, MRS Symposium Proceedings, Vol. 139, p.363 (1989).
36. D. Pickrell, W. Zhu, A. R. Badzian, R. Newnham, and R.F. Messier, submitted to J. Mater. Res. (1990).
37. A.R. Badzian, Advances in X-Ray Analysis, 31, p.113 (1987).
38. P.K. Bachmann, W. Drawl, D. Knight, R. Weimer, Y. Liou, and R.F. Messier, Diamond Technology Initiative Symposium, July 12-14, 1988, Crystal City, Arlington, VA. Technical Digest, Paper T 20, 1988.
39. K. Kitahama, Appl. Phys. Lett. 53, no. 19, p. 1812 (1988).
40. C. Ramaswamy, Nature, 125, p. 704 (1930).
41. S.A. Solin and A.K. Ramdas, Phys. Rev. B1,p. 1687 (1970).
42. D.S. Knight and W.B. White, J. Mater. Res. 4, no. 2, p. 385 (1989).
43. D.S. Knight and W.B. White, Proc.SPIE,Vol.1055,p.144 (1989).
44. R.J. Nemanich, J.T. Glass, G. Lucovsky, and R.E Shroder, J. Vac. Sci. Technol. A6, 3, p. 1783 (1988).
45. R.E. Shroder, R.J. Nemanich, and J.T. Glass, Proc. SPIE, Vol. 969, p. 79 (1988).
46. M.J. Mehl and W.E. Pickett, Proc. SPIE, Vol. 1055, p. 181 (1989).
47. R.O. Dillon, J.A. Woollam, and V. Katkanant, Phys. Rev. B, 29, no. 6, p. 3482 (1984).
48. H.Tsai and D.B. Bogy, J. Vac. Sci. Technol. A5, no. 6, p. 3287 (1987).
49. W.A. Yarbrough and R. Roy, in "Diamond and Diamond-Like Materials"(edited by A. Badzian, M. Geis, and G. Johnson), Mat. Res. Soc., Pittsburgh, PA, Extended Abstracts Vol. EA-15, p.33 (1988).
50. N. Wada, P.J. Gaczi, and S.A. Solin, J. Non-Crystalline Solids, 35-36, p. 543 (1980)
51. A. Ono, T. Baba, H. Funamoto, and A Nishikawa, Japan. J. Appl. Res. 25, no. 10, p. L808 (1986).
52. K. Nishimura, K. Das, J.T. Glass, and R.J. Nemanich, "Proceedings of NATO Advanced Research Workshop on the Physics and Chemistry of Carbides Nitrides, and Borides", Manchester, U.K. Oct. (1989).

53. P.K. Bachmann and D.U. Wiechert,accepted for presentation at the "2.Int.Conf.on the New Diamond Science & Technology" Sept.23-26, Washington, D.C. USA (1990).
54. Y. Liou, R. Weimer, D. Knight, and R. Messier Appl. Phys. Lett. 56, no. 5, p. 437 (1990).
55. D.S. Knight, R. Weimer, L. Pilione, and W.B. White, Appl. Phys. Lett. 56, no. 14, p1320 (1990).
56. see e.g.
    (a) G. Davis, in "The Properties of Diamond" edited by J.E. Field, Academic Press, London, New York, San Francisco, p. 165 (1979).
    (b) J.P.F. Sellshop, in "The Properties of Diamond" edited by J.E. Field, Academic Press, London, New York, San Francisco, p. 155 (1979).
    (c) A.T. Collins, J. Phys.: Condensed Matter, 1, p. 439 (1989).
    (d) Y.K. Vohra, C. A. Vanderborgh, S. Desgreniers, and A.L. Ruoff, Phys. Rev. B, 39, no. 8, p 5464 (1989).
    (e) W.J.P. van Enckevort and H.G.M. Lochs, J. Appl. Phys. 64, no. 1 (1988).
    (f) P.J. Dean, Phys. Rev. A139, p. 588 (1965).
57. V.S. Vavilov, A.A. Gippius, A.M. Zaitsev, B.V. Derjaguin, B.V. Spitzyn, and A.E. Aleksenko, Sov. Phys. Semicond. 14, p. 1078 (1980)
58. A.T. Collins, M. Kamo, and Y. Sato, J. Phys.: Condensed Matter, 1, p. 4029 (1989).
59. Y. Yokota, H. Kawarada, and A. Hiraki, in "Diamond, Boron Nitride, Silicon Carbide and Related Semiconductors" (edited by J.T. Glass, R.F. Messier, and N. Fujimori), MRS Symposium Proc. Vol. 162, Materials Research Society, Pittsburgh, PA, (1989).
60. L.H. Robins, L.P. Cook, E.N. Farabaugh,and A. Feldman, Phys. Rev. B, 39, no. 18 p. 13367 (1989).
61. W.D. Partlow, J.Ruan, R.E. Witkowski, W.J. Choyke, and D.S. Knight, J. Appl. Phys. in press, June (1990).
62. G. Janssen, W.J.P. van Enckevort, J.J.D. Schaminee, W. Vollenberg, L.J. Gilling, and M. Seal, to appear in J. Cryst. Growth (1990).
63. L. Pilione, D.S. Knight and P.K. Bachmann, unpublished.
64. N. Yamamoto, J.C.H. Spencer, and D. Fathy, Phil. Mag. B, 49, no. 6, p. 609 (1984).
65. M. Geis, presented at the "Diamond, Boron Nitride, Silicon Carbide and Related Wide Bandgap Semiconductors" Symposium, MRS Fall Meeting, paper F4.1, Boston (1989).
66. for a recent review see, e.g., A.E.T. Kuiper, Surface and Interface Analysis, 16, in press (1990).
67. K. Schmidt, K. Reichelt, B. Strizker, and J. Zou, Fresenius Z. Anal. Chem. 333, p. 326 (1989).
68. W. Lanford, data presented at the 2nd Annual Diamond Technology Initiative Symposium, Durham, NC, 1988.
69. W. Lanford, Solar Cells, 2, p. 351 (1980).
70. H.G. Maguire, T.E. Derry, W.S. Brooks, J.P.F. Sellschop and M. Kamo, South African Journal of Science, 84,p. 696 (1988)
71. for thermal conductivity measurements of CVD diamond films, see e.g.
    (a) D.T. Morelli, C.P. Beet, and T.A. Perry J. Appl. Phys. 64, no. 6, p. 3063 (1988).
    (b) A. Ono et al, ref. 51.
    (c) R. Pryor, R.L. Thomas, P.K. Kuo, and L.D. Favro Proc. SPIE Vol. 1146, p. 68 (1989).
    and references therein.

72. for electrical properties of CVD diamond, see, e.g.
(a) G. Sh. Gildenblatt, S.A. Grot, C.R. Wronski, A.R. Badzian, T. Badzian, and R. Messier, Appl. Phys. Lett. 53, no. 7 p. 586 (1988).
(b) M.W. Geis, N.N. Efremow, and D.D. Rathman, J. Vac. Sci. Technol. A6, p.1953 (1988).
(c) G. Sh. Gildenblatt, S.A. Grot, C.W. Hatfield, C.R. Wronski, A.R. Badzian, T. Badzian, and R. F. Messier Mater. Res. Bull. 25, no.1 p. 129 (1990).
and references therein.
73. for mechanical properties of CVD diamonds, see e.g.
(a) B. Lux and R. Haubner, Proc. 12th Int. Plansee-Seminar, paper C2, p. 615 (1989)
(b) R.C. McCune, D.W. Hofman, T.J. Whalen, and C.O. McHugh, MRS Symp. Proc. Vol. 130, p.261, (1989).
(c) N. Kikuch and H. Yoshimura, in "New Diamond", Japanese New Diamond Forum, p.42 (1988).
and references therein.
74. M. Yoder, Proc. SPIE, Vol. 969 p. 106 (1988).
75. M. Seal and W.J.P. van Enckevort, Proc. SPIE, Vol. 969, p. 144 (1988).
76. K.A. Snail, L.M. Hanssen, and A.A. Morrish, Proc. SPIE, Vol. 1146, p. 144 (1989).
77. X. X. Bi, P.C. Eklund, J.G. Zhang, A.M. Rao, T.A. Perry, and C.P. Beetz, J. Mater. Res. 5, no.4, p. 811 (1990).
78. T. Feng, Proc. SPIE, Vol. 1146, p. 159 (1989).
79. Ch. Wild, N. Herres, J. Wagner, and P. Koidl, Proceedings of the "First International Symposium on Diamond and Diamond-Like Films",The Electrochemical Society,Pennington, NJ, Proceedings Vol. 89-12, p. 283 (1989).
80. J. Cuomo, IBM T. Watson Res.Lab. Yorktown Heights,N.Y., USA, D. Hoover, Air Products, Allentown, NJ, USA, personal communication.
81. R. Seitz, Proc. SPIE, Vol. 969, p. 124 (1988).

MECHANICAL PROPERTIES TESTING OF DIAMOND AND DIAMOND-LIKE FILMS BY ULTRA-LOW LOAD INDENTATION

Michael E. O'Hern* and Carl J. McHargue**

*University of Tennessee, Knoxville. Current address: Nano Instruments, Inc., P.O. Box 14211 Knoxville, TN 37914 USA
**Metals and Ceramics Division, Oak Ridge National Laboratory, P.O. Box 2008, Oak Ridge, TN 37831 6118 USA

INTRODUCTION

Potential applications of diamond-Like and diamond films are currently driving a concerted effort in the research and development community to further develop the deposition techniques by which the films are grown. Deposition parameters such as temperature, pressure, feed stock composition and flow rate can reasonably be expected to alter the properties of the resultant films.

Applications for these films often require the films to be resistant to wear and abrasion. Diamond-Like Carbon (DLC) films have found wide acceptance in aerospace and military applications where wear/abrasion resistance is of utmost importance [1]. Diamond has found its way into wear applications in cemented form, and the potential applications for smooth, adherent diamond films are virtually limitless.

It is important to correlate the mechanical properties (hardness and elastic modulus) to the wear/abrasion resistance of a material. The highly elastic response of diamond films during indentation calls for a reconsideration of the relative importance of hardness and other mechanical properties such as modulus or hardness to modulus ratio when one attempts to correlate wear resistance to such properties. As the relationship between the mechanical properties and wear characteristics is assimilated, the mechanical characterization of these films will prove to be of utmost importance. Therefore efforts are under way to establish a universal method by which these properties can be obtained and compared directly.

The indentation hardness of materials has traditionally been determined by introducing a normal load to the material by means of an indenter with a given geometry (e.g., Knoop microhardness). Typically, after removal of the load, the indentation left due to plastic flow about the indenter's contact is measured optically. This area is considered the "projected contact area." Hardness is then calculated as the load divided by this projected contact area.

In relatively soft, ductile materials, such as most metals, this calculation of the hardness is a good approximation of the mean pressure the material will support, since the area associated with the elastic contact is small compared to the plastic contact area.

Hard, brittle materials such as ionic- and covalent-bonded ceramics possess different characteristics which complicate the determination of the indentation hardness. Problems such as elastic recovery after removal of load and cracking during indentation may make the "projected contact area" approximation used for ductile materials inappropriate. The problem associated with making this approximation is dramatically illustrated by the example of rubber. Using the traditional hardness measurements, the hardness of rubber would be infinite.

In addition to the problems associated with brittle materials in general, measurement of the hardness of a material of similar hardness to that of the indenter penetrator presents some unique difficulties. Bound in the traditional definition of hardness as determined by indentation is the assumption that the penetrator is of "infinite" hardness and modulus, i.e., the penetrator is assumed to undergo no deformation during indentation. Clearly this assumption is not valid in the indentation of diamond and diamond films.

While the "absolute" value of the hardness (i.e., a quantitative value in units of pressure) cannot yet be determined with high precision due to problems discussed in this paper, useful results have been obtained by direct comparison.

MECHANICAL PROPERTIES MICROPROBE

The mechanical properties microprobe (MPM) negates the necessity of post-indentation imaging by recording the load applied by the indenter as well as its vertical displacement continuously during the indentation sequence, i.e., the entire time the indenter is in contact with the sample. Plotting the value of the load against the indenter displacement yields a curve representing the loading of the indenter, and a hysteresis which represents the load as a function of displacement during unloading (see figure 1). From these data, representing the elastic-plastic response of the material, and a functional relationship between the indenter's displacement and contact area, the hardness and elastic modulus can be calculated. The mathematical relationship between the data and the results has been described in detail elsewhere [2,3].

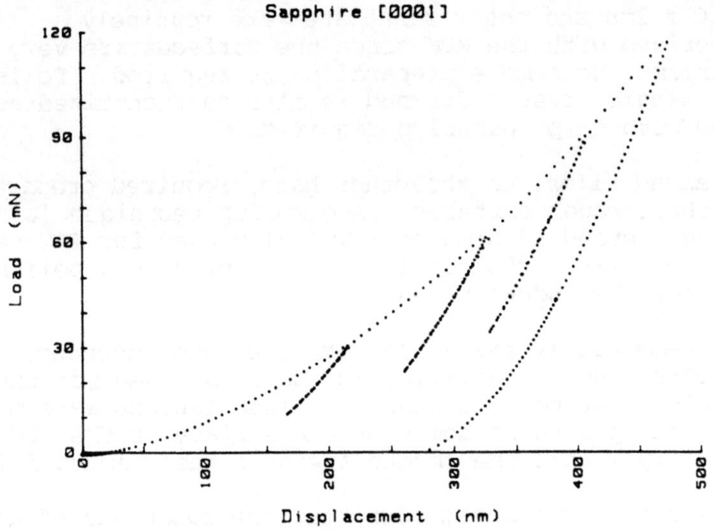

Figure 1. Load-displacement curve showing four distinct unloading segments (three partial, one complete). Reloading follows the previous unloading curve indicating elastic contact. This curve is for single crystal c-axis Al_2O_3. (from ref. 5).

For the purposes of this study, comparative values were used in order to relate the properties to materials which have been evaluated exhaustively. The DLC film properties were calculated with respect to single crystal Al_2O_3 due to their similar behavior in terms of elastic recovery. A single crystal IIa natural diamond was used as a reference for the diamond films due to their similar behavior in all respects.

EXPERIMENTAL PROCEDURE

The DLC film used for this study was deposited in a 13.5-MHz-RF plasma with a 1000-eV self bias at 6.6 Pa ($5x10^{-2}$ torr) total pressure consisting of a 1:1 ratio of Ar and CH_4. This produced a dark brown, slightly translucent film with a featureless surface typical of amorphous carbon films.

Two diamond films were examined. One (film 1) was deposited at Oak Ridge National Laboratory using a hot filament assisted chemical vapor deposition (CVD) technique with a feed gas composition of 1% CH_4 in H_2 and a substrate temperature of 1100°C. The second film (film 2) was deposited using a proprietary CVD process at the General Electric Research and Development Center in Schenectady, New York.

DLC films and their substrates are routinely characterized with the MPM since the surfaces are very smooth and uniform. No sample preparation is required. Polished natural single crystal diamond is also characterized as-received with no preparation required.

Diamond films, on the other hand, required preparation due to their rough surfaces. A dimpling technique [4,5] utilizing a standard dimpler (typically used for TEM sample preparation) was performed in order to provide a polished surface for the indentations.

A commercially available MPM (the Nano Indenter, Nano Instruments, Inc., Knoxville, TN, USA) was used for the indentation testing. At least ten indentations were made on each sample, placed randomly on the surface of the DLC films and randomly within the smooth areas of the diamond films.

The MPM is automated such that the locations of all of the indentations were chosen prior to testing. The indentations were then made sequentially in an enclosed chamber at constant temperature while unattended.

DISCUSSION

When made in comparison to a reference material, the characterization of DLC films is a relatively straightforward task. Since the elastic recovery in terms of load-displacement behavior is similar for the two materials, the indentations are made to specified depths, thereby making the indentations with similar contact areas. With this type of experiment the relative hardness is, to a first approximation, a ratio of the loads required to reach a certain depth into the material. In a similar fashion, the modulus is a ratio of the stiffnesses (the slopes of the unloading curves) [3].

DLC films grown by different methods and varying process parameters have been characterized. In general, the properties are similar regardless of the techniques used to deposit the films. Typically the DLC film has a modulus of about 0.2 and a hardness of about 0.3 with respect to Al_2O_3 (see figure 2).

Characterization of diamond films was originally carried out using single crystal Al_2O_3 as a standard of comparison in the same way that it is used in studying DLC films [5]. While the results were fortuitously consistent with data in the literature for conventional microhardness measurements [6], that data manipulation technique does not account for the problem of penetrator deformation. Indentation of diamond with diamond produces deformation of the indenter penetrator concurrent with the deformation of the sample under study. The deformation behavior of the penetrator can reasonably be expected to be the same during indentation of a diamond film as during indentation of natural diamond. This assumption is a good one in light of the similarity in the load-displacement curves (see figure 3). The displacement at

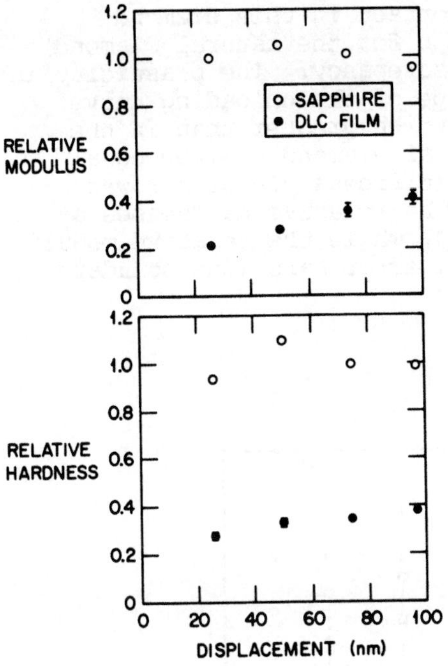

Figure 2. Typical results for a DLC film. (from ref 5).

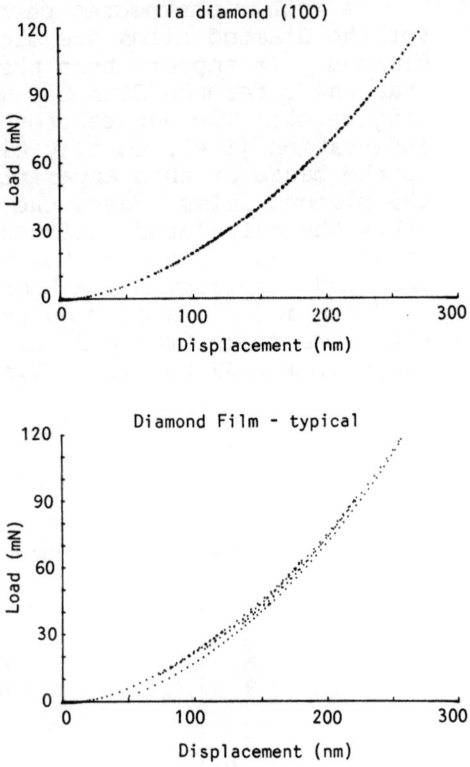

Figure 3. Load-displacement curves for single crystal IIa diamond (a) and diamond film (b). The loading/unloading sequence used was the same as that used for the Al_2O_3 in figure 1. (from ref 5).

full load and the nearly perfect elastic response of the films are almost identical to those of the single crystal diamond. A more reliable and consistent reference, then, is that of natural diamond.

Beyond these similarities, it is important to note that there does exist some plasticity in the films as seen in figure 3 when one compares 3a and 3b.. The unloading curves in 3b show a small hysteresis which indicates plastic deformation. In light of evidence found by other techniques (e.g., Raman spectroscopy), it is reasonable to expect that the grain boundaries are not strictly diamond, and therefore that the plastic deformation we see during indentation may be found in the grain boundaries. This plasticity has a bearing on the calculated results as described below.

A rather unexpected phenomenon is seen when the results for the diamond films are plotted with those of the natural diamond. It appears that the films have lower hardnesses than the reference diamond, while the moduli are higher (figure 4). The nature of the contact in this diamond indentation (i.e., totally elastic for the natural diamond) is the cause of this apparent discrepancy. The plasticity in the diamond films causes the slope of the unloading curve (thus the calculated stiffness) to be greater than in the totally elastic case of the natural diamond. Given equal load and displacement, a higher stiffness yields a lower hardness and a higher modulus. The relative hardnesses of the films range from 0.91 to 0.99, while the relative moduli range from 1.09 to 1.19. Typical error bars (one standard

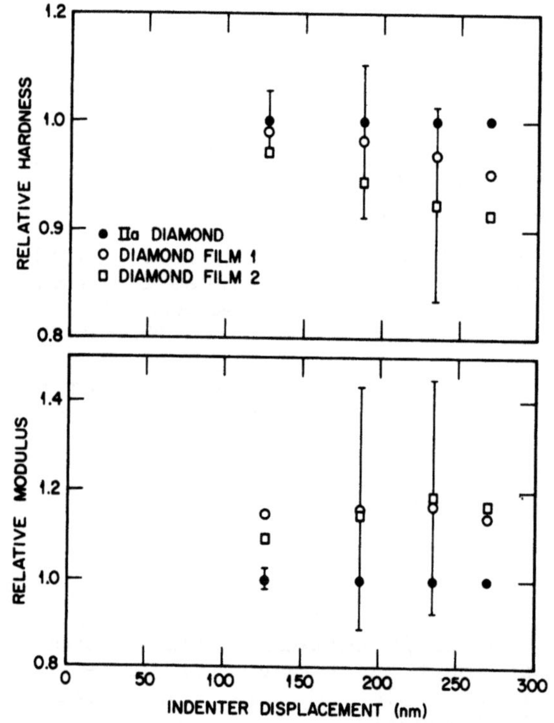

Figure 4. Hardness and modulus of elasticity for two diamond films with respect to single crystal IIa diamond. Error bars are typical standard deviations of the measurements for each sample at the following indenter displacements: 130 nm - IIa diamond; 185 nm - diamond film 1; 235 nm - diamond film 2. The error bars indicate that there is no statistical difference between the films and the reference diamond.

deviation) for each of the films show that statistically the films do not differ from the natural diamond. (The reference diamond used for the data shown is a IIa diamond in the [100] orientation.)

The maximum load of the MPM was used (120 mN), and was not great enough to induce plastic deformation in the natural diamond. It is expected that inducing plastic deformation in the natural diamond will give more directly comparable results. It has recently become possible to apply higher loads with the MPM, and future work includes the intent to compare these films to natural diamond when both the samples and the reference are in the plastic region.

SUMMARY

Hardness and elastic modulus of DLC and diamond films have been studied and characterized with a mechanical properties microprobe (MPM). The relative hardness and modulus of DLC films were determined using single crystal Al_2O_3 as a reference. DLC films have been found to be typically about 0.3 times as hard as Al_2O_3 with elastic moduli of about 0.2 times that of Al_2O_3.

The relative hardness and modulus of diamond films have been studied using single crystal, natural diamond as a reference. Diamond films studied thus far have been found to have hardnesses and moduli equal to those of natural diamond.

ACKNOWLEDGEMENTS

This research was performed at Oak Ridge National Laboratory and was sponsored by the Division of Materials Sciences, U.S. Department of Energy, under contract DE-AC05-84OR21400 with Martin Marietta Energy Systems, Inc.

REFERENCES

1. A. H. Lettington, SPIE Tech Symposium on "Window and Dome Technologies and Materials," Orlando, FL, March 27-31, 1989.
2. J. B. Pethica, R. Hutchings and W. C. Oliver, Phil. Mag. A48(1983)593.
3. M. F. Doerner and W. D. Nix, J. Mater. Res., 1(1986)601.
4. Michael E. O'Hern, Robert H. Parrish and Warren C. Oliver, Thin Solid Films, 181(1989)357-363.
5. M. E. O'Hern, C. J. McHargue,R. E. Clausing, W. C. Oliver and R. H. Parrish in *Technology Update on Diamond Films*, R. P. H. Chang, D. Nelson and A. Hiraki, eds., Materials Research Society, Pittsburgh, 1989, pp 131-137.
6. C. A. Brookes in *The Properties of Diamond*, J. E. Field, ed., Academic Press, New York, 1979, pp 383-402.

THERMAL WAVE AND RAMAN MEASUREMENT OF POLYCRYSTALLINE DIAMOND FILM QUALITY

R.W. Pryor, Lanhua Wei, P.K. Kuo, and R.L. Thomas

Institute for Manufacturing Research and Department of Physics
Wayne State University, Detroit, MI 48202, USA

Introduction

The measurement of the thermal properties of thin films and small samples are difficult, if not impossible, when using conventional thermal measurement techniques [1-3]. Such methods typically require extensive sample preparation and have many potential sources of error (e.g. contacts, absolute temperature measurement, emissivity, etc.). Measurement of the thermal conductivity of a thin film is especially difficult , when that film resides on a substrate that also has a reasonably good thermal conductivity, such as polycrystalline diamond on silicon. In this paper, we present a brief introduction to the Mirage Effect Thermal Wave Technique and demonstrate the application of that technique to the making of such thin polycrystalline diamond film measurements. We present experimental results for polycrystalline diamond films on silicon that show that a correlation exists between the "graphitic" content of a thin diamond film, as estimated by Raman spectroscopy, and the thermal diffusivity (conductivity) of those films.

The Mirage Effect Thermal Wave Technique is also applicable to the measurement of the thermal diffusivity in a large variety of bulk materials. In particular, the technique is well suited to the measurement of single crystal diamond. Some recent results are presented where measurements were made on the thermal diffusivity of natural, synthetic and isotopically enriched type IIA diamonds, using the thermal wave technique. The diffusivities obtained through those measurements have yielded thermal conductivities of 21.9, 22.3 & 33.2 W/cm-K, respectively. The first two values are in excellent agreement with measured values found in the literature for natural diamonds. The third value (33.2 W/cm-K), measured for the isotopically enriched type IIA diamonds (0.1% ^{13}C), reflects a 50% greater value for the thermal conductivity of single crystal diamond, at room temperature, than was previously found or predicted [4].

The thermal wave technique is a method for the contactless, direct, measurement of the thermal diffusivity of both bulk and thin film materials. This method makes use of the mirage effect in the measurement of the thermal diffusivity. It is applicable to all types of materials in both bulk and thin film form. This method is used here to determine the thermal diffusivity (conductivity) of

Diamond and Diamond-Like Films and Coatings
Edited by R.E. Clausing *et al.*, Plenum Press, New York, 1991

thin film, bulk, natural and isotopically enriched synthetic, diamond materials. A detailed description of the method appears elsewhere [5-6]. Briefly described, the mirage effect thermal wave technique uses a modulated laser beam, focused on the surface of the diamond material, to initiate a thermal wave [Fig. 1]. In the case of the thin diamond film, the thermal wave propagates along the thin film and into the underlying substrate and the overlying gas. For transparent, thin diamond films with an opaque substrate, the absorption of the energy of the heating laser occurs at the diamond-substrate interface. For opaque films, the absorption of the energy of the heating laser occurs at the thin film-gas interface. For the measurement of type IIA bulk diamond samples, an absorption layer is applied to one of the surfaces, since IIA diamond is transparent to the light of the currently used laser heating beam (5145Å). The thermal wave in the overlying gas creates thermal lensing (the mirage effect) by modulating the density of that gas. A second (probe) laser beam, skimming the sample surface, interrogates the changes in the overlying gas caused by the thermal wave by measuring the shift in the gradient of the refractive index of the heated air.

A quad-cell position sensor and two vector lock-in amplifiers are used to monitor the magnitude and phase variations of the periodic vector deflection (both transverse and normal) of the probe beam as a function of lateral offset between the heating and probe beams [Fig. 2]. The mirage signal versus offset position is then fitted to a theoretical solution of the three-dimensional thermal diffusion equation for the gas/film/substrate system [5].

Thermal Wave Theory

The time-dependent temperature field, T(**r**,t), produced by a time-dependent heat source, f(**r**,t), is governed by the diffusion equation,

$$\nabla\cdot[\kappa\nabla T(\mathbf{r},t)] - \rho c\frac{\partial T}{\partial t}(\mathbf{r},t) = - f(\mathbf{r},t) \tag{1}$$

where κ is the thermal conductivity, ρ is the density and c is the specific heat of the matter at position **r**. When the heat source is sinusoidal in time, as results when a periodically modulated laser beam is focused on the sample, once the dc component is removed, it can be described by

$$f(\mathbf{r},t) = f(\mathbf{r})e^{-i\omega t} \tag{2}$$

The heat diffusion equation then takes the form

$$\nabla\cdot[\kappa\nabla T(\mathbf{r})] + \kappa q^2 T(\mathbf{r}) = - f(\mathbf{r}) \tag{3}$$

Equation (3) admits (heavily damped) wave-like solutions, in plane-wave form,

$$\exp i(qx-\omega t), \tag{4}$$

with

$$q = (1+i)(\omega\rho c/2\kappa)^{1/2}. \tag{5}$$

These solutions are commonly called thermal waves. Analytic solutions can be obtained for a general situation in which the space is filled with a gas, a layered solid sample, and another gas, and the heat source is in the form of a focused gaussian laser beam on the layered sample. The mirage signal can be expressed in the form of a Fourier transform [6]. Theoretical expressions for the mirage signal permit a detailed comparison with experimental data which yields information on physical constants, such as the diffusivities,

$$\alpha\ (=\kappa/c\rho) \tag{6}$$

(where κ is the thermal conductivity, ρ is the density and c is the specific heat) and the thicknesses of each of the various layers. With the thermal wave tech-

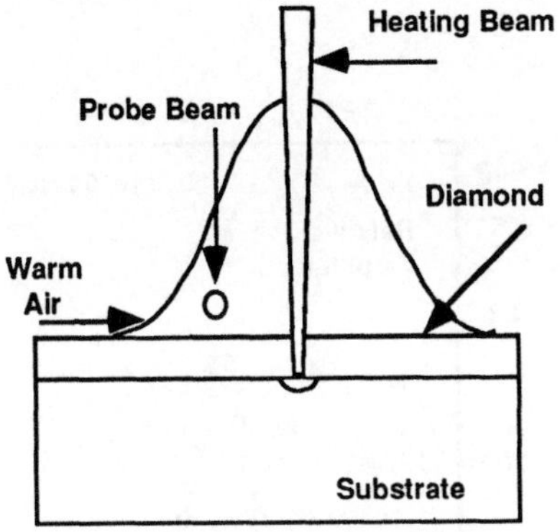

Figure 1. Mirage Effect Thermal Wave Technique

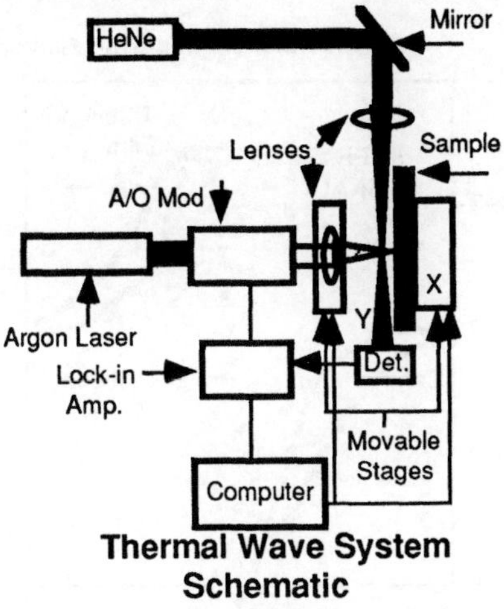

Figure 2. Thermal Wave System Schematic

nique, it is not required that the film be removed or thermally isolated from the substrate.

Experimental Results

We have extensively evaluated the above-mentioned theory for a medium of 4 layers, air/film/substrate/air. A thin interfacial layer, between the film and the substrate will not appreciably alter the measured diffusivity value and falls within the limits of error. In particular, we have utilized this theory extensively for application to the thermal wave study of diamond film-on-silicon wafers or thin-film-on-bulk diamond. Figure 3 shows a comparison between the mirage signals of a bare silicon wafer and a diamond-coated silicon wafer. As can be readily seen, there is a significant difference between the observed values for the transverse deflection for silicon and for diamond-coated silicon. Figures 4&5 show typical examples of the fit between the experimental mirage signal on a normal isotopic composition high pressure synthetic diamond crystal and the theoretical calculation for both the normal and transverse signals, respectively. There are a number of parameters used in the fitting calculation, whose values can be independently measured. Those parameters include the gaussian beam radii, film thickness, probe beam height above the surface, and the thermal diffusivities of the gas and the substrate. Various combinations of these parameters can be left as unknowns in the fitting routine and then the calculated values can compared later with the independently measured values to verify the validity of the fitting routine. The most physically important calculated parameter, for this paper, is the thermal diffusivity of the diamond material.

Estimates of the Quality of Polycrystalline Diamond Films

A precise analysis of this problem would, of course, take into consideration the size and distribution of the "graphitic" sites, their interaction with the laser beam, their scattering efficiency, etc. At this time, that information is exceedingly difficult to obtain and in some cases remain to be defined. Thus, as a means to evaluate the diamond films produced and measured here, a simple method has been developed. As a first approximation, let us assume that the

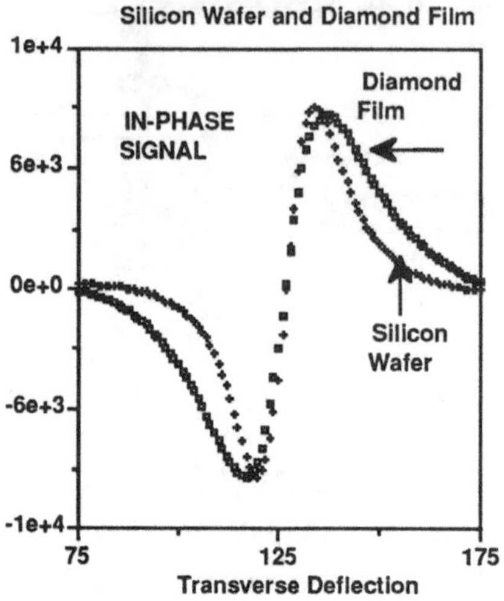

Figure 3. Thermal Wave In-Phase Signal for Silicon and Diamond on Silicon

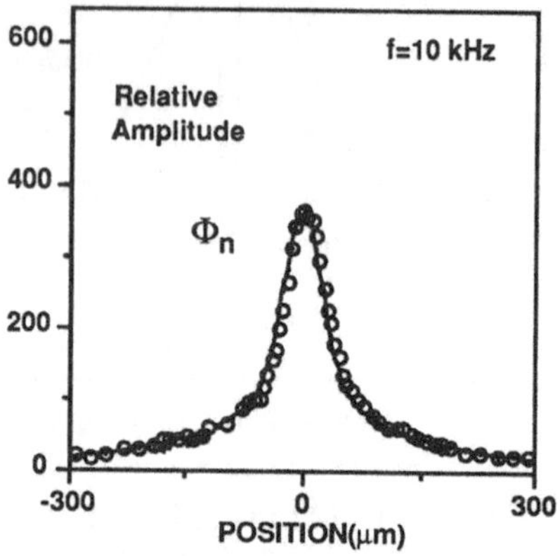

Figure 4. A Comparison of Thermal Wave Theory (lines) and Experiment (symbols) for Normal Isotopic Distribution High Pressure Diamond

thin, polycrystalline diamond film is of a uniform isotopic composition and comprised of two structurally different carbon based materials: diamond crystallites and a non-diamond carbon based "graphitic" material, which appear as defects in, and as binding material between, the diamond crystallites. The ratio of the "graphitic" material (G) to diamond material can be estimated through an analysis of the peaks in the Raman spectrum of the sample. We determine the relative amplitudes of the narrow diamond (1332 cm^{-1}) peak (A_D) and of the relatively broad "graphitic" material (~1550 cm^{-1}) peak (A_G). In estimating this ratio, the observed "graphitic" peak height is corrected by a factor (S) which takes into account the higher relative scattering efficiency (a factor of 50 to 80) of the "graphitic" material,

$$G = \frac{A_G}{A_D \cdot S} \qquad .(7)$$

Also, in this analysis, a signal-to-noise ratio improvement of the microfocus Raman spectrum is obtained by averaging multiple Raman scans for a given location and then averaging multiple scans at different locations for a given thermal diffusivity measurement. We use this procedure since the area (~1mm^2) of the diffusivity measurement is much larger than the area (~1μm^2) of the microfocus Raman measurement. The thermal diffusivity results presented in Fig. 6 represent measurements made on several polycrystalline diamond samples, fabricated over a period of about one year by three different diamond research laboratories. They show a reasonable inverse correlation between the thermal diffusivity and the percentage of "graphitic" material.

Figure 6 also shows that the value for the thermal diffusivity of the currently available thin diamond films differs significantly from that of the bulk crystals. The Raman spectra that we have measured also show differences among the various films, and significant differences between the film and bulk material. The bulk material shown here is a sample of synthetic, normal isotopic composition, type IIA, single crystal diamond. We have also made measurements, at room temperature, on the influence of changes in the isotopic composition of type IIA, single crystal diamonds. Those measurements are treated elsewhere [7]. The results are presented in Figure 7 and they show that a change from ~1.1% to ~0.1% ^{13}C in a ^{12}C single crystal diamond yields an improvement in the thermal diffusivity of ~50%.

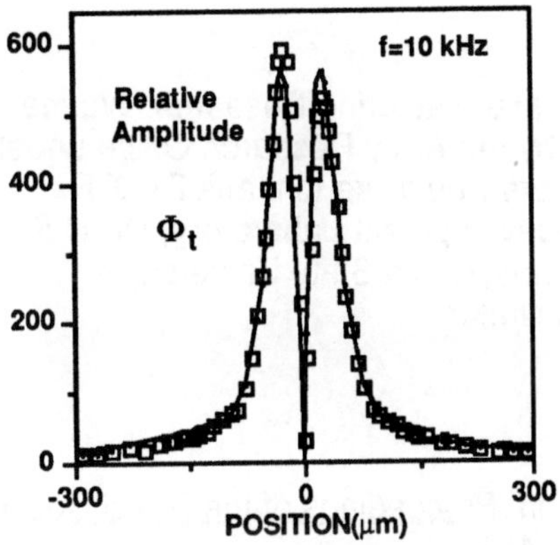

Figure 5. A Comparison of Thermal Wave Theory (lines) and Experiment (symbols) for Normal Isotopic Distribution High Pressure Diamond

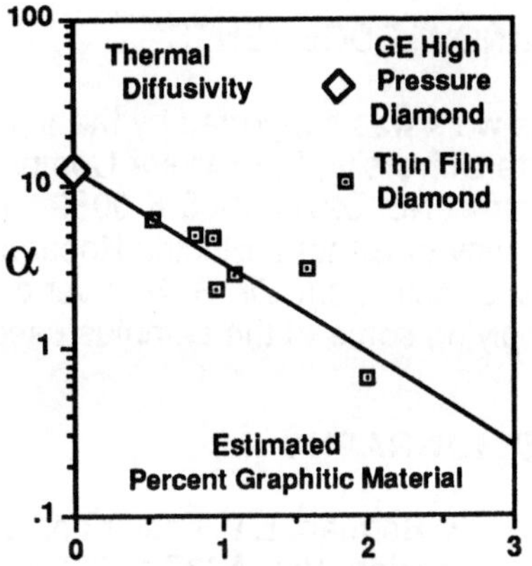

Figure 6. Thermal Diffusivity for Thin Polycrystalline Diamond Correlated with the Raman Estimated Percentage of "Graphitic" Material

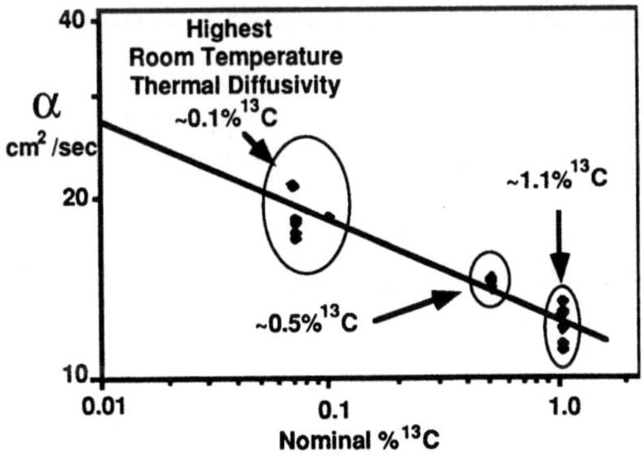

Figure 7. Room Temperature Thermal Diffusivity of Normal Isotopic Composition and Isotopically Enriched, Single Crystal Diamond

SUMMARY AND CONCLUSIONS

Thermal conductivity values herein reported for high-quality polycrystalline diamond films are of the order of 9 -12 W/cm-K. Measurements of normal isotopic composition single crystal diamond show a value of thermal conductivity on the order of 21W/cm-K. This agrees well with values found in the literature. Measurements of isotopically enriched single crystal diamond show a value of thermal conductivity on the order of 33 W/cm-K. This reflects an improvement of 50%, at room temperature, in the value of the thermal conductivity of single crystal diamond. This improvement was neither expected nor predicted by current theory. Analysis of the measured diffusivities of bulk diamond and diamond films with various "graphitic" contents show, as would be expected, that the thermal diffusivity decreases with increasing "graphitic" content.

ACKNOWLEDGEMENTS

This work was supported by the Institute for Manufacturing Research, Wayne State University, Ford Motor Company and by the Army Research Office under Contract No. DAAL03-88-K-0089. The authors would like to thank Dr. T.R. Anthony of General Electric Research and Development Laboratory, Dr. R.S. Post of Astex, and Dr. R. Messier of The Pennsylvania State University for supplying some of the samples used in this work.

BIBLIOGRAPHY

1. R. Berman, E.L. Foster and J.M. Ziman, Proceedings of the Royal Society, London, Vol. A237(1956), pp. 344-354

2. A.J. Schorr, Proceedings of the Industrial Diamond Conference, Chicago(1969), pp 185-190

3. A. Ono, T. Baba, H. Funamoto and A. Nishikawa, Japanese Journal of Applied Physics, Vol. 25, No. 10, October(1989), pp. L808-L810

4. G. Davies, Diamonds, Adam Hilger, Ltd., Bristol, UK, (1984), pp. 138

5. L.D. Favro, P.K. Kuo, and R.L. Thomas, Photoacoustic and Thermal Wave Phenomena in Semiconductors, ed. A. Mandelis, Elsevier, N.Y., (1978) pp 69-96

6. P.K. Kuo, L.D. Favro, and R.L. Thomas, Photothermal Investigations of Solids and Fluids, ed. J.A. Sell, Academic Press (1989), pp 191-212

7. T.R. Anthony, W.F. Banholzer, J.F. Fleischer, Lanhua Wei, P.K. Kuo, R.L. Thomas and R.W. Pryor, Physical Review B, Vol. 42, No. 2, 15 July (1990), pp 1104-1111

DIAMOND CRYSTAL GROWTH BY HOT FILAMENT CVD AND ITS CHARACTERIZATION

P. Ascarelli, S. Fontana*

C.N.R. I.M.A.I.
P.O.B. 10
00016 Monterotondo Scalo (Rome), Italy

E. Molinari, R. Polini, V. Sessa, M. L. Terranova

Dip. Scienze e Tecnologie Chimiche
II Universita' di Roma "Tor Vergata"
00173 Rome, Italy

E. Cappelli

ENIRicerche
via Ramarini
00015 Monterotondo (Rome), Italy

INTRODUCTION

An ever increasing research effort has been devoted to the low pressure and temperature deposition of thin diamond films, since its realization has been proved and widely confirmed in the last few years. The reason for this may be largely related to the expectation for a wide set of potential applications, ranging from diamond protective coatings, to heat sinks, to optical and semiconductor devices, all of them exploiting in some way the unique combination of mechanical, optical, thermal, chemical, and electrical properties of diamond.[1] Moreover, recent progress in the processing of diamond and diamond-like carbon coatings has already resulted in development and commercial success.[2]

However, several basic phenomena, occurring during the synthesis and growth of the thin films, need deeper understanding and better control before several important requirements associated with major application fields can be met.[3] For example, thin films of optical quality require optimum transparency; thus, films perfectly smooth and with very low concentration of light-scattering or adsorbing defects are needed. More or less the same requirements are necessary for good diamond thermal conductivity, where phonon scattering defects should be low. Other requirements include: (a) lower substrate deposition temperatures and (b) higher deposition rates and controllable and predictable substrate surface pretreatment to enhance nucleation and growth.

The phenomenology of diamond deposition may be subdivided into three major research areas:

*ENIRicerche Fellowship.

(1) Vapor phase: where the identity and role of the vapor precursor species are analyzed;

(2) Surface phenomena: where nucleation and growth are studied, together with the substrates surface characteristics which influence them;

(3) Diamond thin film characteristics: where the properties of the deposited film are studied in view of the intended use.

From this wide research panorama we focused our study on two main issues: (a) the role and optimization of the atomic hydrogen concentration, and (b) the analysis of the nucleation and growth mechanism together with the means for their enhancement on various substrates.

To meet these objectives, we devised an HFCVD (hot-filament-assisted chemical vapor deposition) apparatus that we intended to implement with an in situ REMPI (resonance enhanced multi-photon ionization) diagnostic spectroscopy tool to detect atomic hydrogen. We considered that the thermally activated gaseous reaction atmosphere, provided by the HFCVD, should be simpler to study and control, being largely composed of excited but not ionized species, in contrast to those operating with plasma-assisted techniques (i.e. microwave, radiofrequency, etc.).

We report in this paper some preliminary results obtained for thin diamond films deposited on Si (111) substrates making use of our HFCVD apparatus. The main features of the HFCVD apparatus are:

(a) a stainless steel cylindrical chamber of 12 cm radius and 20 cm height. The cylindrical wall is composed of two thin walls (about 1 cm apart) allowing cooling water circulation;

(b) a pumping apparatus which allows a pressure of 10^{-6} torr when gas fluxing is closed;

(c) a tantalum filament which can be set at a variable distance from the substrate surface;

(d) a sample substrate holder which consists of a thin molybdenum metal sheet that can be heated by a suitable electric current; and

(e) two gas inlets allowing the gases to flux into the chamber.

Thin films deposited for various times under several conditions of Si (111) substrate temperature and pretreatment have been characterized by XPS, SEM, and Raman spectroscopy, after exposure to air.

EXPERIMENTAL RESULTS

Several Si (111) substrates were scratched with 0.25 μ diamond paste. Deposition on these samples was stopped at various times (5, 20, 60, and 180 min.). The tantalum filament was kept at a temperature of T_f = 2200°C, and the substrate temperature was T_s = 750°C. The temperatures were measured by an optical pyrometer. The pressure was 76 torr and the gas composition was 2% methane in molecular hydrogen. Total gas flow was 100 sccm.

These samples, together with a standard undeposited scratched sample, have been observed by XPS, a technique that is sensitive to the first few nanometers from the sample surface. The results are summarized in Figs. 1(a),(b) and 2. In Fig. 1(a) an example of a general XPS survey spectrum obtained by a VG ESCALAB MK II apparatus using the Al X-ray

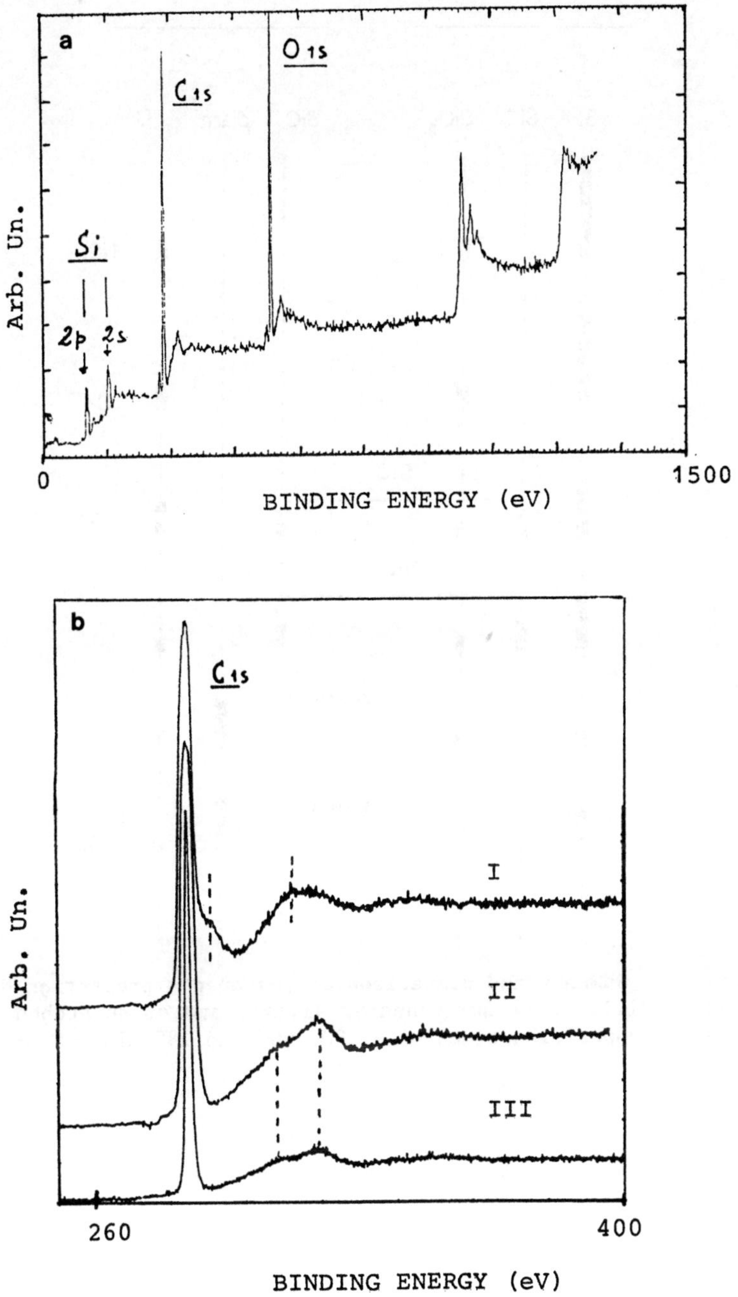

Fig. 1(a). XPS survey spectrum obtained from a diamond film after 180 min. deposition. (Note: O(1s) XPS sensitivity factor is three times larger than those of Si(2p) and C(1s.) (b). XPS-C(1s) electron energy loss spectrum comparing the graphite (I), the natural diamond (II), and the 180 min. sample (III). The dotted lines indicate the energy location of the major plasmon losses.

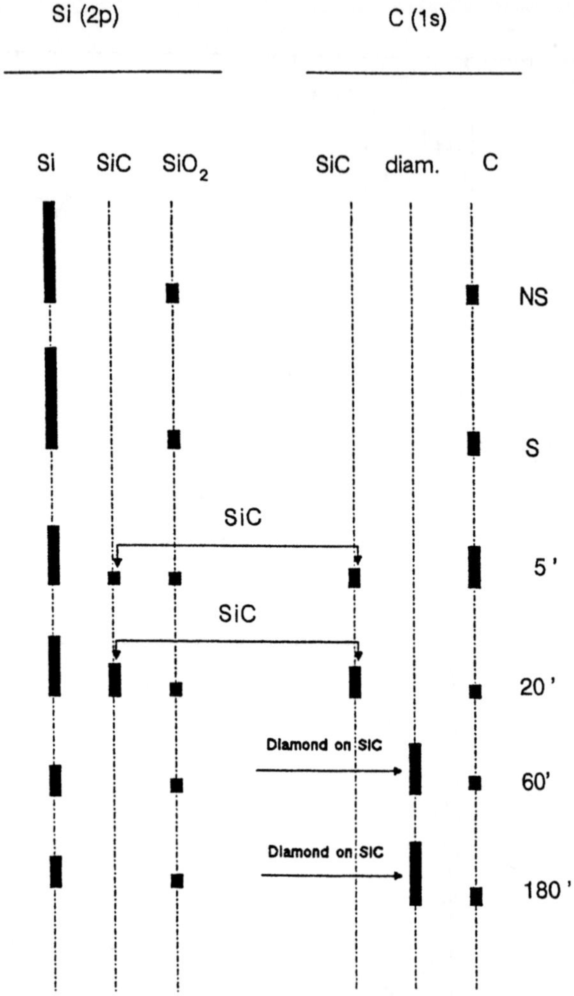

Fig. 2. Summary and comparison of components present on NS (= not scratched Si <111>, S (= scratched Si <111>), and on scratched Si <111> samples after deposition times of 5, 20, 60, and 180 min. T_s = 750°C.

K-alpha line is given and the atomic components present on the 180 min. deposited sample surface are shown.

For each sample we analyzed in detail the Si(2p) and C(1s) components of the XPS spectrum. In agreement with previous work,[4,5,6] the Si(2p)line was separated into three bonding components which were identified as Si-Si at 100.2 eV electron binding energy, Si-C at 102 eV, and Si-O at 104.2 eV. The binding energies are accurate to about ±0.2 eV. The weight percent of each component was derived assuming a Gaussian contribution to the total peak area.

An analogue procedure was used to extract information from the C(1s) line[4,5,6] that was considered to result from various components: a hydrocarbon at 285.5 eV, a Si-C at 283.5 eV, and a C diamond at 284 eV; and the presence and growth of the SiC was cross checked from the Si-C contributions from both Si(2p) and C(1s).

The diamond carbon contribution has been confirmed by the observation of the XPS C(1s) electron energy loss spectrum, as shown in Fig. 1(b), where the characteristics of the plasmon peaks are shown. These peaks are typical "fingerprints" distinguishing diamond from graphitic and from amorphous carbon species.[7,8]

Formation of SiC has been observed on the samples obtained by 5 and 20 min. deposition, together with a small percentage of silicon and of silicon oxide. No SiC was detected on the films deposited for 60 and 180 min., where diamond is instead largely present.

In Figs. 3 and 4 we show SEM views and Raman spectra of Si (111) scratched samples where the deposition has been conducted for different substrate temperatures (680 and 960°C) and for different times. We note rather different types of diamond crystals. At 960°C (Fig. 3) typical "urchin" structures are present, whereas a more definite crystalline morphology is apparent at 680°C (Fig. 4). A higher density of small crystals is moreover observed at low temperature while, at higher temperature, a reduced density of lager crystals is present.

CONCLUSIONS

An analysis of this preliminary set of results suggests the following observations that seem to be consistent with several of previously reported ones:[3]

(1) Diamond nucleation and growth is greatly enhanced by mechanical pretreatment (i.e., "scratching") of the Si substrate.

(2) The density of the nuclei is a decreasing function of the temperature. In fact, the nuclei density at 680°C has been found to be an order of magnitude larger than at 960°C.[9] Instead the radius of the crystal particles is an increasing function of temperature, that is larger particles are found at high temperature than at low temperatures.

(3) Crystal morphology is different at low temperature, where crystal boundaries are largely characterized by well defined crystalline surfaces. Instead, at high temperature, a disordered "urchin" structure is observed.

(4) The early stage of deposition on Si (111) indicates the formation of silicon-carbide layer some nanometers thick. Traces of unreacted carbon are also present in these spectra. We could not determine if this was due to a carbon environmental contamination or to an amorphous carbon layer deposition occurring in the early stage of deposition as some authors have previously suggested:[6]

(5) The fact that the SiC contribution disappears from the XPS spectrum when diamond is formed, while Si and Si-oxide are still present (Fig. 2), might suggest that diamond grows on a SiC layer only. A SiC buffer layer has been identified by cross-sectional transmission electron microscope examinations of diamond films on silicon substrates by Williams et al.[10]

Fig. 3. Sample deposited at T_f = 2200°C, T_s = 960°C for 60 min. (a) SEM view of nucleation density, (b) "urchin" morphology of a particle on the same sample, and (c) Raman spectrum of the sample.

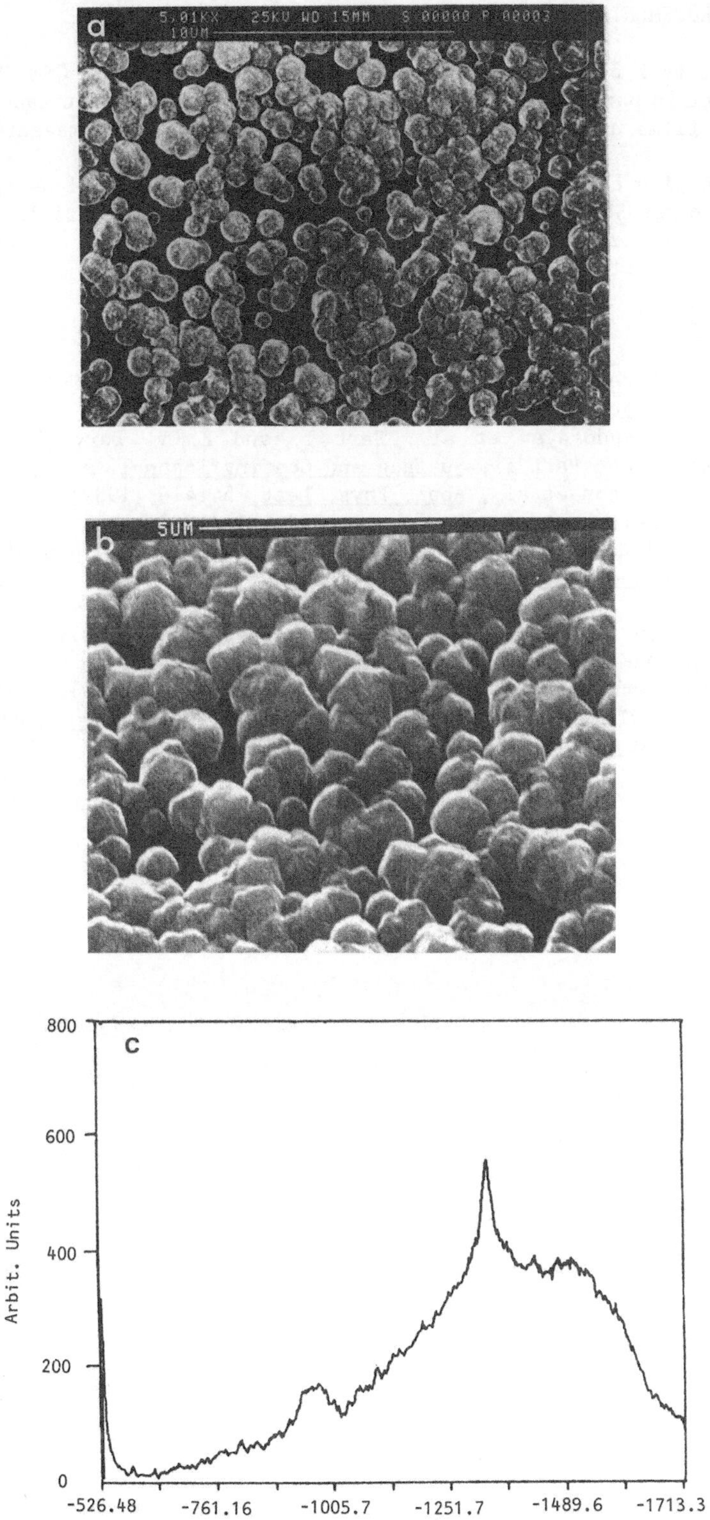

Fig. 4. Sample deposited at T_f = 2200°C, T_s = 680°C for 102 min. (a) SEM view of nucleation density, (b) "crystalline" morphology of the same sample, and (c) Raman spectrum of the sample.

ACKNOWLEDGEMENTS

It is a pleasure to acknowledge the Raman group of CNR IMAI and G. Mattei in particular for providing us with careful Raman spectra of the diamond films and N. Nistico of ENIRicerche for the XPS measurements.

We also thank Elettronica S.p.A. for allowing one of us (R. Polini) assess to the SEM. Finally, we want to acknowledge M. Lenzi for providing us with the Laser Area CNR Service Laboratory, a technical structure to set up our forthcoming REMPI experiments.

REFERENCES

1. W. A. Yarbrough and R. Messier, Science 247:688 (1990).
2. J. C. Angus et al., MRS Bull. 14:38 (1989).
3. D. V. Fedosayev et al., Part I. and B. V. Deryagin and D. V. Fedosayev, Part II, Surface and Coating Technol. 38: (1989).
4. D. N. Belton et al., Appl. Phys. Lett. 54:416 (1989).
5. D. N. Belton et al., Surf. Sci. 233:131 (1990).
6. B. Williams et al., in Proc. NATO-ASI on "Diamond and Diamond-Like Films and Coatings," Castelvecchio Pascoli, Italy, July 23-Aug. 3, 1990 (private communications).
7. P. G. Lurie and J. M. Wilson, Surf. Sci. 65:476 (1977).
8. F. R. McFeely et al., Phys. Rev. B9:5268(1974).
9. R. Meilunas et al., Appl. Phys. Lett. 54:2204 (1969).
10. B. E. Williams et al., pp. 202-214 in Proc. First Intern. Conf. Diamond and Diamond-Like Films, J. P. Dismukes, A. J. Purdes, B. S. Meyerson, T. D. Moustakas, K. E. Spear, K. V. Ravi, and M. Yoder (eds.), The Electrochemical Society, Inc., Pennington, NJ, 1989.

IN-VACUO SURFACE ANALYSIS OF DIAMOND NUCLEATION AND GROWTH ON SI(111) AND POLYCRYSTALLINE TANTALUM

B.E. Williams, B.R. Stoner, D.A. Asbury, and J.T. Glass

Dept. of Materials Science and Engineering
North Carolina State University
Raleigh, NC 27695-7907

INTRODUCTION

The growth of diamond thin films at low temperature and low pressure has made it an excellent candidate material for use in electronic and wear resistant coating applications. However, for diamond to reach its true potential in electronic applications, high quality monocrystalline diamond films must be grown on economically viable non-diamond substrates.

In order to achieve an understanding of the nucleation and growth processes of diamond films on different substrates, we have coupled a diamond growth chamber to an ultrahigh vacuum (UHV) system equipped with a variety of surface analytical techniques including Auger electron spectroscopy (AES), x-ray photoelectron spectroscopy (XPS), low energy electron diffraction (LEED), and electron stimulated desorption (ESD). In this research we stopped the growth at selected intervals and characterized the species present on the surface using XPS and AES without exposure to the air. Specifically, growth of diamond on silicon has been examined as a standard due to numerous reports of CVD of diamond on silicon. Much work has been accomplished on this system, but no study of this nature has been attempted with a microwave plasma CVD growth system. However, since the main interest of the present research is on heteroepitaxial growth of diamond, other substrates (silicon has been shown to be ineffective in this respect) and remote plasma deposition are of greater interest. Therefore, preliminary results for the growth of diamond on polycrystalline tantalum are also reported because single crystal tantalum is a candidate material for heteroepitaxy of diamond.[1] This was done in a remote plasma configuration to minimize ion and electron damage to the surface of the sample.

EXPERIMENTAL

The diamond films examined in this research were deposited using a microwave plasma CVD system. The plasma was formed using a 1 kW microwave source operated at 2.45 GHz with a rectangular waveguide coupled to a cylindrical cavity. The growth chamber is constructed of stainless steel to be suitable for UHV and has a base pressure of <1 x 10^{-7} Torr when evacuated with the attached turbopump. For growth of diamond films, a controlled mixture of CH_4 and H_2 delivered through mass flow controllers is fed into the chamber using a gas dispersal ring located at the end of the cavity where the waveguide joins the CVD chamber. The desired pressure is controlled by a throttle valve which is attached to a roots blower/mechanical pump assembly. Heating of the substrate is accomplished with an isolated, differentially pumped heating stage. The temperature of the substrate is measured using an infrared pyrometer which was calibrated by melting a bead of high purity aluminum bonded to a silicon wafer. After growth, the sample was cooled under vacuum to a temperature <200 °C before being transferred to the analysis chamber (base pressure 1 x 10^{-10} Torr).

The analysis chamber is equipped with a Riber Mac2 semi-dispersive electron energy analyzer which is used for the XPS and AES measurements. An argon ion sputtering gun is available for cleaning of samples as well as depth profiling. The x-ray source is a dual anode (Mg and Al) source from Riber. The electron source used for Auger is a VG LEG61 which was typically run at 3 kV and 200 μA emission. Acquisition of the XPS and AES data was achieved by controlling the electron analyzer voltages and detection electronics with an IBM AT computer, and the acquisition software was written by one of the authors (Asbury). For XPS, the spectra were obtained by pulse counting using a Riber pulse counter. For AES, standard phase-sensitive detection methods were used to obtain the spectra.

As shown in Figure 1, the analysis chamber and CVD system are connected by a central transfer tube (base pressure = 10^{-9} Torr) which is equipped with a metallization station. Introduction of samples is accomplished via the load lock attached to the transfer tube.

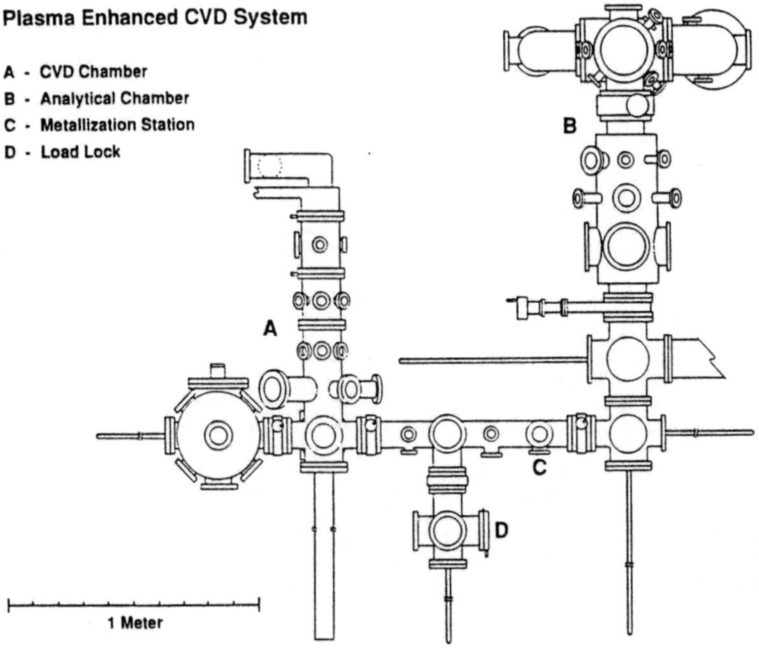

Figure 1. Schematic diagram of microwave plasma CVD chamber and surface analytical chamber.

The preparation of the silicon substrates consisted of an HF etch to strip the oxide from the surface followed by a polish with 0.25 μm diamond to enhance the nucleation density of diamond particles. The wafer was then cleaned in TCE followed by acetone and ethanol and finally rinsed in deionized water. The tantalum substrates were 10 x 10 x 0.5 mm sheets with a guaranteed purity of at least 99.95%. The sheets were polished using 600 grit SiC paper followed by 6, 3, and 1 μm Al_2O_3 suspended in water. The final polish was accomplished with 0.25 μm diamond paste as used on the silicon substrates. The Ta sheets were then cleaned using TCE, acetone, and ethanol followed by a deionized water rinse.

The growth conditions for the silicon and tantalum substrates are depicted in Table 1. The conditions used for the silicon substrate was used as a standard since the conditions listed are similar to those used in previous research.[2,3] The optimum growth conditions chosen for the tantalum substrate were determined from a parametric study of substrate temperature, pressure, methane concentration, and position relative to the plasma. For this work, the position relative to the plasma is defined as the distance from the edge of the glow discharge to the surface of the sample. The data presented in this research were obtained in the following manner: after exposure to the growth conditions for a given time, the sample remained in the growth chamber until the pressure was reduced to $<10^{-7}$ Torr. Then the sample was transferred to the analysis chamber for XPS and AES approximately 0.5 h after the growth was stopped. The XPS was performed first because previous work on diamond has shown that electron beam exposure can cause changes in the bonding on diamond surfaces. Cycles of growth and analysis on this sample continued until the surface was well covered with diamond.

Table 1. Growth parameters used for silicon and tantalum substrates.

Substrate	Temp. °C	Power (W) For/Rev	CH_4/H_2	Pressure (Torr)	Position relative to plasma
Silicon	800	850/100	1%	25	Immersed
Tantalum	600	850/150	0.5%	25	2.0 cm downstream

RESULTS

Surface Analysis of Diamond on Si (111)

Examination of the Si *2p* region at different growth times can reveal information about its bonding state as the nucleation and growth of diamond occurs. Initially, the Si *2p* indicates a mix of Si-O and Si-Si bonding as shown in Figure 2. The elemental Si *2p* peak at 99.0 e V was used for calibration of the energy scale. The SiO_2 peak is located at 102.7 eV.[4] Thus, prior to growth the silicon substrate is covered by a thin layer of SiO_2 which is expected since the substrate was not cleaned *in-vacuo*. After 0.25 h, some Si-C component is observed as shown by the presence of the new peak located at 100.3 eV. At 0.5 h, more of this carbide component has formed but the elemental Si peak is still dominant. After 1 h, almost all the Si is Si-C bonded. After 5 h, only a trace of Si is observed. The Si-O peak has a nearly constant intensity after the growth is begun, indicating that this Si-O component is confined to the surface. Separate studies have determined that the O component is not due to a leak in the vacuum chamber but rather may be present in the source gas. Efforts to eliminate the source of this oxygen contamination are ongoing.

The value of the C 1s binding energy for pure carbon species (C-C bonding) is not affected by the hybridization of the atoms. That is, the binding energies of graphite and diamond are indistinguishable from one another. However, the presence of the diamond phase on these diamond films was confirmed by other analytical techniques including AES, EELS, and Raman spectroscopy.

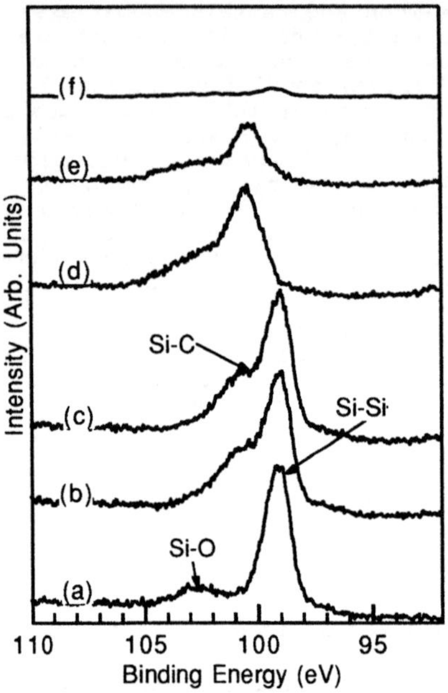

Figure 2. XPS spectra of Si *2p* (a) before growth, (b) at 0.25 h, (c) 0.5 h, (d) 1 h, (e) 2 h , (f) 4 h

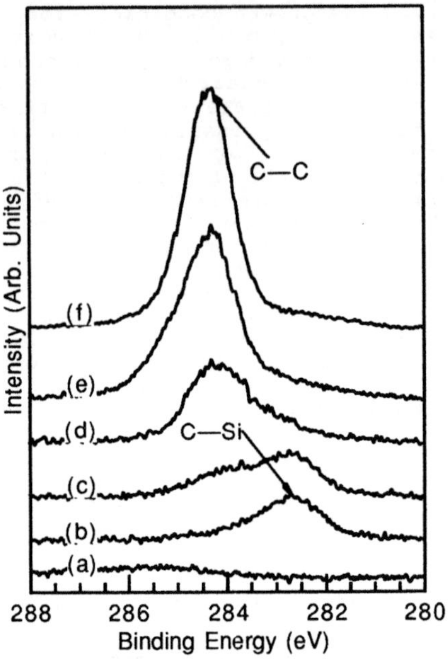

Figure 3. XPS spectra of C *1s* (a) before growth, (b) 0.25 h, (c) 0.5 h, (d) 1 h, (e) 2 h , (f) 4 h

XPS data obtained from the C1s core level at selected growth times is shown in Figure 3. The C observed prior to growth shown in Figure 3(a) is similar to that expected for hydrocarbon contamination on the surface (both C-H and C-O bonding are located at higher binding energy). After 0.25 h, this contamination is reduced and C-Si bonding is observed. After 0.5 h, approximately 40% diamond (C-C) and 60% carbide (C-Si) bonding is observed (determined by area calculation), indicating that nucleation of diamond particles has occurred. In the case of hot filament CVD, Belton found that a complete SiC layer was formed prior to nucleation of diamond particles. In the present work, at 0.5 h the intensity of the elemental silicon peak is higher then that of the SiC peak, indicating that an incomplete layer of SiC has formed, or at least the SiC is in the form of islands. But as observed in Fig.3(c), nucleation of diamond has already been achieved. This means that the diamond particles may be nucleating on silicon and/or SiC as opposed to only nucleating on the SiC as found by Belton. However, the XPS results cannot determine the actual site upon which the diamond is nucleating, but only those species present on the surface. A technique which has spatial capabilities, such as scanning tunneling microscopy, is necessary to determine the answer to this question. After 1 h growth, the C 1s shows an increasing amount of diamond component, and by 5 h only diamond bonding (C—C) is observed.

AES fine structure of the C *KLL* peak has been utilized previously to distinguish the diamond phase from graphite and amorphous carbon.[2,5] Differentiated C *KLL* Auger spectra obtained after different growth times on silicon are shown in Figure 4. After 0.25 h, the fine structure at lower energies of the major C *KLL* peak is indicative of the presence of silicon carbide. This is in agreement with the XPS results and with the results of previous TEM work on diamond growth on silicon. By 1 h, definite changes in the fine structure associated with the C *KLL* have occurred and the spectrum resembles a mix diamond and silicon carbide. The spectrum obtained at 4h has the fine structure expected for a nearly continuous diamond film.

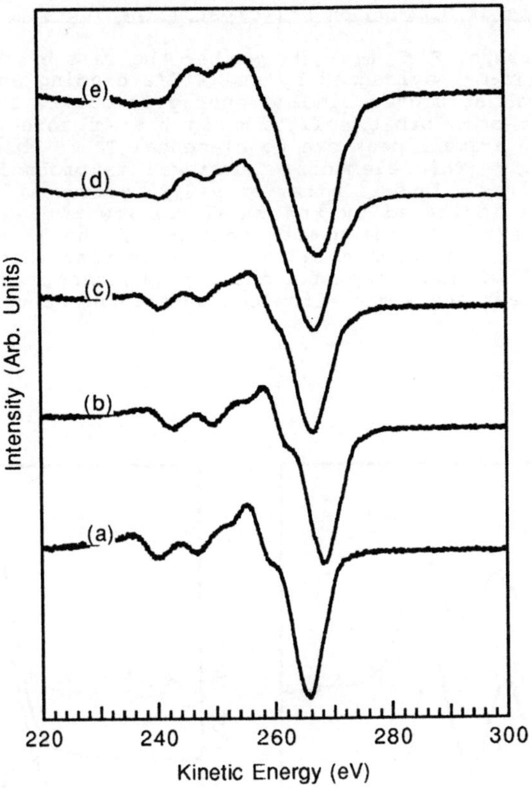

Figure 4. AES of C *KLL* for diamond grown on silicon after growth times of (a) 0.25 h, (b) 0.5 h, (c) 1 h, (d) 2 h, (e) 4 h.

Verification of the diamond phase is further evidenced by the crystalline morphology of the deposited diamond particles. Figure 5 is an SEM micrograph of the diamond film deposited on the silicon substrate. Note the multiple twinning of the particles. This twinning has been examined in detail in previous research and is a result of the structure of the nucleus for the diamond particles.[6]

Figure 5. SEM micrograph of diamond film deposited on silicon substrate.

Before growth, Fig. 6(a) shows that the Ta substrate is covered by a thin oxide (Ta_2O_5) evidenced by the 4f7/2 binding energy at 26.5 eV. The smaller peak at higher binding energy is due to the 4f5/2 electron as a result of spin orbit splitting in the 4f core level. At lower binding energy, a small peak due to elemental Ta is observed (denoted by Ta on Fig. 6(a). This elemental Ta signal is probably generated from underneath the Ta_2O_5 layer. After only 0.25 h growth, only Ta–C bonding is observed as indicated by the shift to lower binding energy (Ta-C 4f7/2 at 22.6 eV). As the growth continues, the Ta signal decreases. After 5 h, no Ta was detectable by XPS indicating that the coverage of the diamond film was complete and the thickness of the film was sufficient to mask any signal from the underlying substrate.

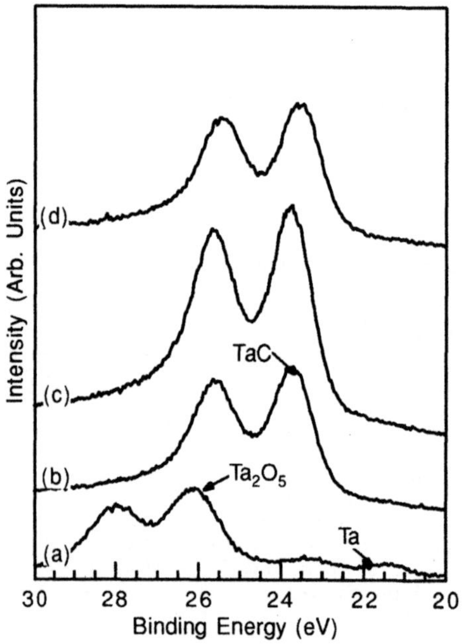

Figure 6. XPS of Ta 4f region at growth times of (a) 0 h, (b) 0.25 h, (c) 0.5 h, and (d) 1 h. Data for 5 h is not shown because no peak was detected.

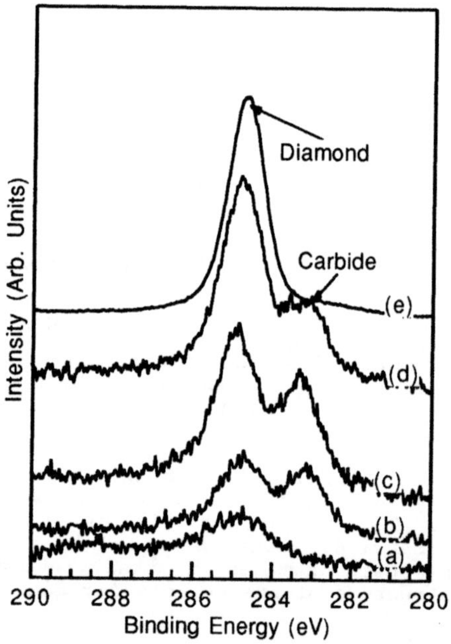

Figure 7. XPS of C 1s for diamond grown on tantalum at growth times of (a) 0 h, (b) 0.25 h, (c) 0.5 h, (d) 1 h, and (e),5 h.

Examination of the C 1s reveals that before growth (Fig. 7(a)) a small amount of carbon contamination is present on the surface, but it is not in the form of a carbide. After 0.25 h, a distinct carbide component is observed at 283.2 eV, while a roughly equal amount of carbon is C-C bonded. As the growth time is increased, the carbide component is reduced while the C-C component increases. At 5 h, a single peak is observed at 284.7 eV, associated with the C-C bonding of diamond.

AES of the C *KLL* after 0.25 h indicates the presence of a carbide (shown by the peaks at lower energy relative to the major peak), in agreement with the XPS observations. As the growth time increases, the peak closest to the low energy side of the *KLL* transition increases in intensity, and the carbide peaks at lower energies decrease in intensity. Finally, after 5 h, the C *KLL* has the shape associated with the diamond phase, and no carbide component is observed.

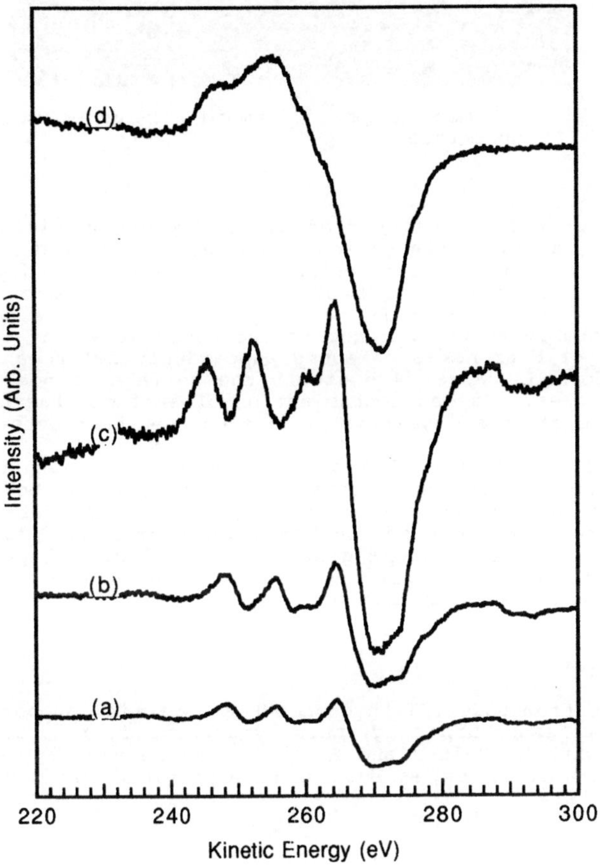

Figure 8. AES of C *KLL* for diamond grown on tantalum at growth times of (A) 0.25 h, (b) 0.5 h, (c) 1 h and (d) 5 h.

The morphology of the diamond film after 5 h growth can be seen in the SEM micrograph shown in Fig. 9. The (111) faceting is typical of high quality diamond films reported in the literature. The crystal size is small relative to silicon, indicative of a high nucleation density. The thickness of this diamond film is less than 1 μm and the film is transparent when viewed through an optical microscope.

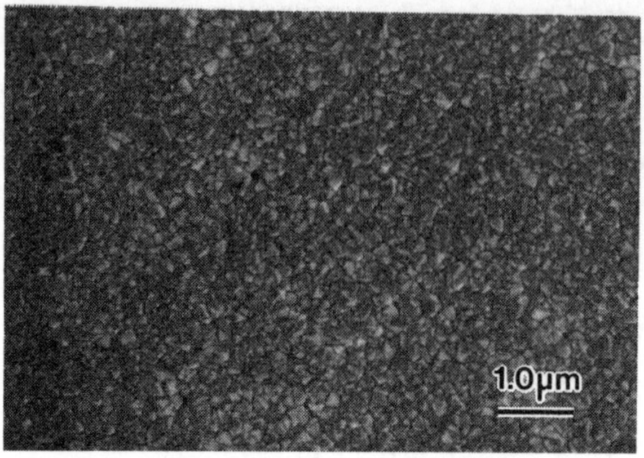

Figure 9. SEM micrograph of diamond film grown on tantalum substrate.

CONCLUSIONS

In-vacuo surface analysis of diamond grown on silicon substrates has shown that at first the silicon substrate is covered by a thin layer of oxide and some hydrocarbon contamination. After 0.25 h growth, SiC is formed at the expense of some of the native oxide observed prior to growth. However, oxygen is still present (as observed in survey spectra), probably at the surface of the newly grown SiC. After 0.5 h, the presence of diamond is detected, apparently before a complete film of SiC has formed. After 4 h growth the surface is nearly covered by diamond particles. Oxygen contamination of surfaces has been observed, but the source of this contamination has not yet been determined.

Diamond growth has also been achieved in a downstream mode on polycrystalline tantalum substrates. This method may be more suitable for heteroepitaxy since the crystallinity of the substrate may be affected by the ion and electron bombardment from the plasma. XPS and AES indicate that for diamond growth on tantalum, a layer of tantalum carbide is formed very rapidly (<0.25 h). The diamond then nucleates and grows on this tantalum carbide layer.

ACKNOWLEDGMENTS

Support from SDIO/IST through ONR and partial support of the Kobe Steel Ltd. Professorship is gratefully acknowledged. Useful technical discussions with D. Belton and R. F. Davis are also appreciated. T. Tachibana, S. Wolter, and A. Mundsen have provided invaluable assistance in the laboratory.

REFERENCES

[1] Y. H. Lee, P. D. Richard, K. J. Bachmann and J. T. Glass, Appl. Phys. Lett. 56, 620 (1990).

[2] B. E. Williams and J. T. Glass, J. Mater. Res. 4, 373 (1989).

[3] K. Kobashi, Phys. Rev. B 38, 4067 (1988).

[4] D. N. Belton, S. J. Harris, S. J. Schmieg, A. M. Wiener and T. A. Perry, Appl. Phys. Lett. 54, 416 (1989).

[5] P. G. Lurie and J. M. Wilson, Surf. Sci. 65, 476 (1977).

[6] B. E. Williams, J. T. Glass, R. F. Davis, K. Kobashi and Y. Kawate, MRS Proc. (1988).

TEM AND SEM INVESTIGATION OF THE GROWTH AND TEXTURE OF DIAMOND PARTICLES/FILMS SYNTHESIZED BY A FLAME CVD PROCESS

P.S. Nielsen*, K. Madsen**, P. Balslev*, P.L. Hansen**

*NKT A/S, Corporate R&D, Sognevej 13, DK-2605 Broendby, Denmark

**Laboratory of Applied Physics, Technical University of Denmark, DK-2800 Lyngby, Denmark

INTRODUCTION

The unique material properties of diamond[1], including hardness, low coefficient of friction, high thermal conductivity, electrical conductivity, and optical transmission, make diamond an attractive candidate for a wide range of industrial applications. As a consequence during the last decade, there have been extensive worldwide efforts toward the development of techniques for metastable synthesis. Diamond film synthesis has been demonstrated by using a wide range of low pressure chemical vapor deposition (CVD) techniques (see review by Spear[2]). Diamond synthesis by combustion of acetylene at atmospheric pressure was first reported by Hirose et al.[3] and has been confirmed by others[4,5,6,7,8]. This technique represents the technological advantage of a low initial capital investment and shows that diamond can be grown at 1 atmosphere with the ability to coat large substrate areas. The drawbacks of this method are that the reactive species in the flame are non-uniformly distributed, and also that the combustion in the flame is responsible for a considerable heat flux to the substrate. Today, only a few applications of diamond films are on the market. For most potential applications, it is very important to synthesize a homogeneous morphology as well as to obtain good adhesion of the film to the substrate. To control the morphology of the film and the adhesion to the substrate, the nucleation and growth of diamond have to be controlled.

In this paper, we report on a series of experiments on the synthesis of diamond particles/films with a pre-mixed oxygen-acetylene flame. A detailed investigation of the morphology and microstructure was carried out by transmission electron microscopy (TEM), scanning electron microscopy (SEM), and micro-Raman spectroscopy.

Diamond and Diamond-Like Films and Coatings
Edited by R.E. Clausing *et al.*, Plenum Press, New York, 1991

EXPERIMENTAL SET-UP

The flame CVD facility is schematically shown in Fig.1. A standard welding torch fitted with a 1 mm tip was mounted on a xyz translation stage for precise and repeatable positioning. The flow of acetylene and oxygen was separately measured by calibrated flow meters and controlled by needle valves. The gasses were mixed in a tube and transferred to the torch for formation of the combustion flame. Welder's grade acetylene of a purity above 99.5% and oxygen of a purity above 99.7% were used for these experiments. The substrate was placed on a water-cooled copper block. In order to obtain thermal contact, the cooling fixture was formed as a vacuum stage. The substrate temperature was monitored during deposition by a one band (width 2.0 to 2.7 μm) pyrometer calibrated by Si(111) at 750 °C. The heat emission from the flame did not significantly influence the pyrometer reading of the substrate temperature.

Silicon (111) wafers with a thickness of 0.4 mm were used as substrates. The SiO_2 film on the Si(111) surface was etched by a buffer solution of 3 wt% HF and 37 wt% NH_4F in water, and chemically cleaned by a procedure well known in the semiconductor industry[9]. To enhance the diamond nucleation[10], the substrates were ultrasonically scratched by a suspension of 1 μm diamond powder in ethanol and ultrasonically rinsed in ethanol. Prior to deposition, the substrates were dried by a nitrogen blow torch.

Diamond films were grown in the acetylene feather region of the flame in ambient laboratory atmosphere . To optimize the growth conditions, systematic experiments were carried out at deposition temperatures between 500 - 1000 °C, at C_2H_2/O_2 ratios between 0.80 - 1.10, and at distances of 0 - 9 mm between the inner cone of the flame and the substrate. The range of diamond deposition was in agreement with the results described by the NRL Group[4] and by Hirose et al.[11]. The experiments reported below were performed under optimum conditions for diamond growth: substrate temperature = 750 °C, C_2H_2/O_2 ratio = 0.92, and total gas flow = 0.123 standard m^3/h, while the substrates were placed in the acetylene feather 2.5 mm below the inner cone of the flame.

The deposition time was gradually increased from a few seconds to several minutes, and the films were characterized by SEM and TEM in order to investigate the nucleation and growth of diamonds and their morphology and microstructure. Micro Raman spectroscopy was used to confirm the SEM and TEM results.

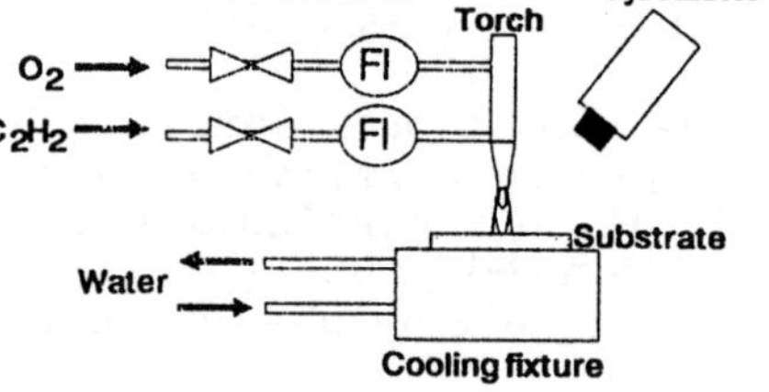

Fig. 1. Schematical illustration of the flame CVD deposition facility.

RESULTS

Scanning electron microscopy (SEM)

The SEM analysis was performed with a Jeol U3 microscope, operating at 25 kV. Prior to the investigation, a thin layer (50 Å) of platinium was deposited on the specimen in order to avoid charging.

The morphology of the film was found to vary across the growth zone of the substrate. {111} facetted, multiply twinned crystals were present in the center of the deposition zone(Fig. 2a). In the annular zone (Fig. 2b), located between the center and edge of the deposition zone, the crystals gradually changed from well facetted crystals to spherical structures (Fig. 2c). This variation of the crystal/particle structure across the deposition zone was characteristic for deposition times from 15 seconds to more than 16 minutes. The size of the particles seemed to be largest in the annular zone indicating a larger growth rate than in the center and at the edge. For deposition times of 1 and 7.5 seconds, no crystals were found on the substrate (x 60.000).

It was observed that the ratio of the nuclei density in the center to the nuclei density at the edge of the deposition zone was constant for different deposition times. But, the nuclei density was not the same during the deposition. Even for seemingly identical experiments, the nuclei density showed a wide variation. The nucleation as a function of the deposition time will be covered in more detail in a coming paper[12].

It was observed that the adhesion of the film to the Si(111) wafer in the center and at the edge was not as good as in the annular zone. In general, the crystals could be removed by scratching with a scalpel. The adhesion of the film to the substrate could vary for samples produced under seemingly identical conditions.

Transmission electron microscopy (TEM)

The TEM analysis was performed with a Philips EM430 microscope operated at 300 kV. The crystals/nuclei were scraped off

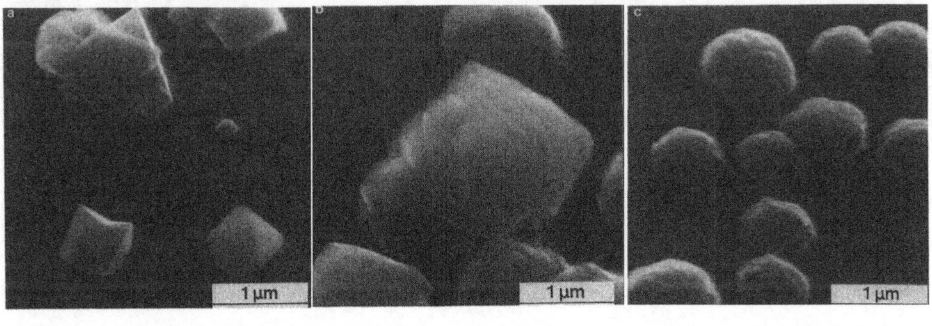

Fig. 2. Scanning electron micrographs showing diamond particles across the growth zone after 120 seconds of deposition. (a) In the center of deposition zone, (b) between the center and the edge of the deposition zone (annular zone) and (c) at the edge of the deposition zone.

the substrate by using a plastic scalpel and were then placed on an amorphous carbon film. Only the particles deposited up to 45 seconds were investigated because they were electron transparent without additional specimen preparation. The particles were examined by EDX, and it was verified that the particles only consisted of atoms lighter than sodium.

The microstructure of the particles showed two characteristics:

(1) The diamond particles in the center of the deposition zone were typically composed of only a few large crystals.

(2) The spherical structures at the edge of the deposition zone were composed of a large number of sub-micron sized diamonds crystals (10-100nm), Fig.3.

Diffuse scattering indicating amorphous carbon was observed in the selected area diffraction patterns (SADP) from spherical particles placed across holes in the amorphous carbon substrate film. It was not possible to quantify the amount of amorphous carbon in the these particles. The micrographs in Fig.3 are representative of a large number of spherical particles after deposition times of 15, 30, and 45 seconds.

Examination of the diamond crystals (both the sub-micron sized crystals in the spherical structures and the large crystals in the center zone) showed numerous inclusions,

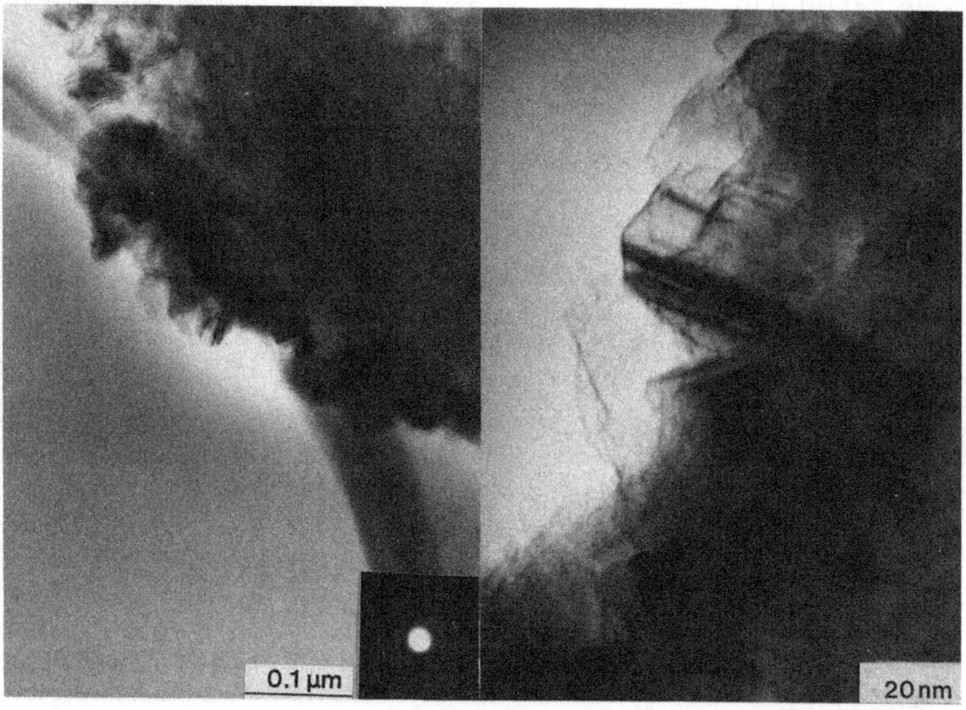

Fig. 3. Bright field transmission electron micrographs of a typical spherical structure after 30 seconds of deposition. It is evident that sub-micron sized crystals are present in the structure. The selected diffraction pattern (SADP) shows that all the crystal phases are diamond.

stacking faults, and twins. Fig.4a shows an aggregate of many crystals from the center zone. Fivefold twins are presented in Fig. 4b. Note that the multiply twinned particle is only 100 nm across. A magnification of another part of the aggregate (about 200 nm across) is shown in Fig. 4c. The dark markings in the figure is due to strains caused by inclusions. The inclusions were oriented along the {111} planes of the diamond lattice. In some places, the spacing between two defects was only as little as 10 nm. With respect to the number and to the characteristics of lattice defects, Fig. 4 is representative of particles investigated in the center and annular zones for deposition times of 15, 30, and 45 seconds although the macro-structure of the aggregate is atypical for the particles in the center zone. No significant diffuse scattering in the diffraction patterns was observed from the crystals in the center and annular zones.

Raman Spectroscopy

Micro Raman spectra were recorded by a Dilor XY spectrometer. Excitation was made by means of the 514.5 nm line of an argon ion laser at an output power of 100 mW. The spectra were obtained by micro-focusing to a circular sampling area with a diameter of 2 μm. Only one sample with a deposition time of 45 seconds was investigated. The Raman measurements were made on particles on the Si substrate.

Fig. 5a shows the Raman spectrum of a diamond crystal in the center of the deposition zone. The narrow peak at 1334 cm^{-1} corresponding to diamond[14] and the absence of peaks for non-diamond carbon confirmed that the purity of the diamond in the center zone was high.

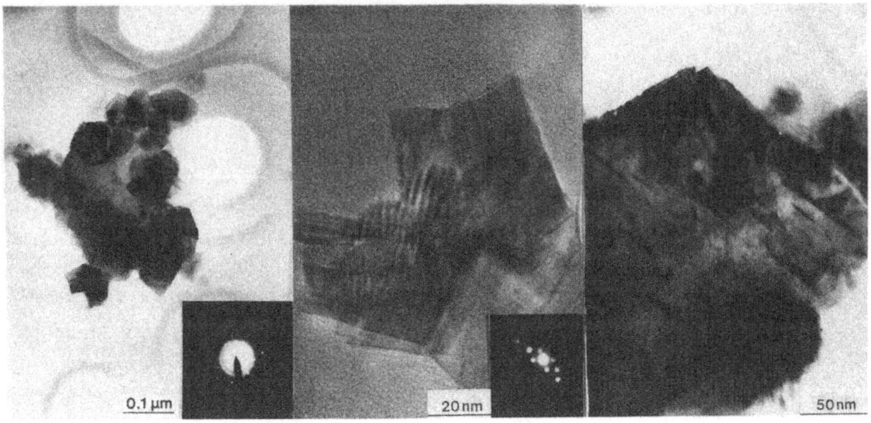

a b c

Fig. 4. Bright field transmission electron micrographs of an aggregate of many diamond crystals with different orientation. The aggregate was found in the center zone after 30 seconds of deposition. Same sample as in Fig. 3. a) The aggregate. b) Multiply twinned particle (MTP) which exhibits fivefold symmetry (magnification of part of the aggregate). c) A magnification of another part of the aggregate shows that the crystal is densely filled with stacking faults and inclusions (dark markings).

The Raman spectrum of the spherical structure (Fig. 5b) shows a broad peak centered at 1576 cm^{-1} corresponding to non-diamond carbon[14] and a broad peak at 1341cm^{-1} due to sub-micron sized diamond crystals.

DISCUSSION

The SEM investigation shows a wide variation of the nuclei density as a function of the deposition time. Furthermore, the nuclei density varied for experiments carried out under identical process parameters. This indicates that the flame process for synthesizing diamond is not reproducable. The reason for the variations in the nuclei density is not yet clear although we believe that these variations are due to the non-controllable dynamic conditions in the start-up phase of the process, such as the non-thermal equilibrium conditions in the ignition of the flame and the variations in thermal contact between the substrate and the cooling fixture.

The SEM investigation shows a variation of the particle structures across the deposition zone. This is due to the non-uniform temperature in the deposition zone and to the non-uniform distribution of reactive specimen in the flame as previously mentioned. This variation across the deposition zone was the same for different deposition times indicating approximately the same growth conditions for the experiments.

The TEM examination of the spherical particles (Fig. 3) showed that the structure consisted of many very small crystals, and the electron diffraction pattern showed these to be diamond.

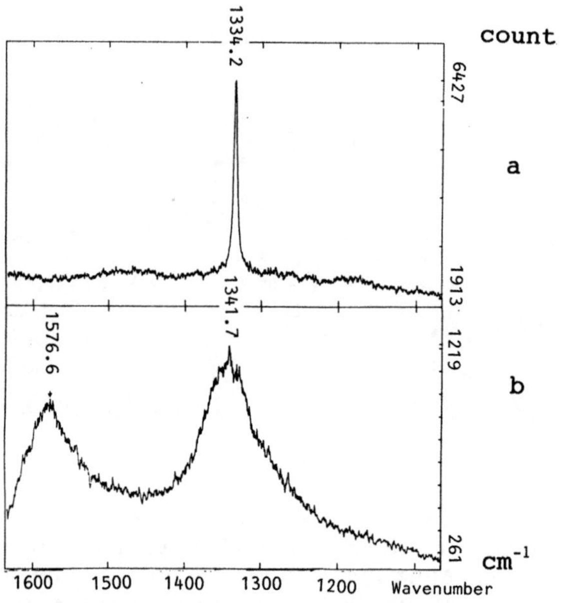

Fig. 5. Micro-Raman spectum at a) a diamond crystal in the center of the deposition zone and b) a spherical structure at the edge of the deposition zone.

Besides diamond, no other crystal phases were detected. Some diffuse scattering was observed in the diffraction patterns from parts of the particles placed over holes in the amorphous carbon film. This indicates that the diffuse scattering is due to amorphous carbon present in spherical particles and not due to the amorphous carbon film. The micro-Raman spectrum of the spherical particle (Fig. 5b) confirmed that an amount of non-diamond carbon was present in the structure. The amount of non-diamond carbon compared to the amount of diamond was rather small when taking into consideration that the Raman method for dectection of non-diamond carbon (for example graphite) is about 50 times more sensitive than the method applied for the detection of diamond[13]. Therefore, the spherical particle likely consists of mostly sub-micron sized diamond crystals and only a small amount of amorphous carbon. The amorphous carbon indications could result from inclusions and grain boundaries. A relative large grain boundary area (many sub-micron sized crystals) is maybe responsible for the amorphous carbon indications in these particle as no indications were observed in the crystals prsent in the center zone.

The investigation of the microstructure showed that the diamond lattice contains many defects. The main defects observed were inclusions, stacking faults, and twins. The exact nature of the inclusions was not identified even though it was observed that the inclusions were oriented along the {111} planes of the diamond lattice. The number of lattice defects was high both in the facetted crystals in the center and the annular zones as well as in the sub-micron sized crystals in the spherical structure. It means that high defect densities in the crystals were present over the entire deposition zone.

The high defect density in the diamond films synthesized by the flame CVD process is consistent with reports for diamond films synthesized by other processes[14].

CONCLUSION

The conditions for diamond growth by the flame CVD process were found to be in an agreement with those previously mentioned.

The nuclei density was not constant as a function of the deposition time. Even seemingly identical experiments showed different nuclei densities. We believe that these varations are due to the non-controllable, dynamic conditions in the start-up phase of the process, such as the non-thermal equilibrium conditions in the ignition of the flame and variations in thermal contact between the substrate and the cooling fixture.

In all experiments, the structure of the particles varied in the same way across the deposition zone. This variation was constant with the deposition time indicating approximately constant growth conditions across the deposition zone. Well {111} facetted, multiply twinned crystals were found in the center of the deposition zone and the structure gradually changed to spherical at the edge. The particles seemed to be largest in the annular zone, located between the center and the edge of the deposition zone, indicating a larger growth rate here.

The spherical structure at the edge of the films consisted of many sub-micron sized diamond crystals with indications of amorphous carbon. The amorphous carbon is maybe related to a large number of grain boundaries in the spherical structure.

The sub-micron sized crystals present in the spherical particles as well as the multiply twinned diamond crystals in the center of the deposition zone contained a high density of inclusions, stacking faults, and twins.

ACKNOWLEDGEMENTS

The authors acknowledge the kind assistance of Dr. S. Lindblom, the Geological Institute, University of Stockholm, Sweden, for the micro-Raman spectra and of Mr. F. Kragh, Department of Structural Properties of Materials, Technical University of Denmark, for taking the SEM micrographs.

The work was supported by the Danish Materials Technological Development Program, Center for Surface Technology/Dry Coating Processes.

REFERENCES

1. J. E. Field, "The Properties of Diamond", Academic Press, London (1979).

2. K. E. Spear, Diamond-Ceramic Coating of the Future, J. Am. Ceram. Soc. 72:2:171 (1989).

3. Y. Hirose and N. Kondohy, Synthesis of Diamond by Using Combustion Flame, Extended Abstracts, The 35th Spring Meeting, Japan, Appl. Phys. 434 (1988).

4. L. M. Hanssen, W. A. Carrington, J. E. Butler, and K.A. Snail, Diamond Synthesis Using an Oxygen-Acetylene Torch, Mater. Lett. 7:7/8:289 (1988).

5. P. G. Kosky and D. S. McAtee, An Experimental and Theoretical Investigation of Flame-Formed Diamonds, Mater. Lett. 8:9:369 (1989).

6. W. A. Yarbrough, M. A. Stewart, and J. A. Cooper, Combustion Synthesis of Diamond, Surf. & Coating Technol. 39/40:241 (1989).

7. Y. Tzeng, C. Cutshaw, R. Phillips, T. Srivinyunon, A. Ibrahim and B. H. Loo, Growth of Diamond Films on Silicon from an Oxygen-Acetylene Flame, Appl. Phys. Lett. 56:2:134 (1990).

8. M. A. Cappelli and P. H. Paul, An Investigation of Diamond Film Deposition in a Premixed Oxyacetylene Flame, J. Appl. Phys. 67:5:2596 (1990).

9. W. Kern and D. A. Puotinen, Cleaning Solution Based on Hydrogen Peroxide for Use in Silicon Semiconductor Technology, RCA Review 31:2:187 (1970).

10. B. V. Deryagin et al., Structure and Properties of Diamond Films Grown on Foreign Substrates, Dobl. Phys. Chem. 244:36 (1979).

11. Y. Hirose, S. Amanuma, N. Okada, and K. Komaki, The Synthesis of High Quality Diamond in Combustion Flame, in "Proc. 1st Int. Symp. on Diamond and Diamond-Like Films", J.P. Dimukes et al., eds., The Electrochemical Society Inc., Pennington (1989).

12. K. Madsen, P. L. Hansen, P. S. Nielsen, P. Balslev, "Nucleation and Growth of Flame CVD Deposited Diamond Films Investigated by TEM and SEM". To be presented at the Diamond Films '90, September 1990, Crans-Montana, Switzerland.

13. D. S. Knight and B. White, Characterization of Diamond Films by Raman Spectroscopy, J. Mater. Res. 4:2:385 (1989).

14. B. E. Williams, J. T. Glass, R. F. Davis, K. Kobashi and Y. Kawate, Electron Microscopy of Diamond Films Grown by Microwave PECVD, in "Extended Abstracts, Diamond-Like Ma-Materials Synthesis", Johnson, Badzian and Geis, eds., MRS, Pittsburgh (1988).

10. R. V. [illegible] "Structure and Properties of Diamond [illegible] Films Grown [illegible]" [illegible] Phys. Chem. [illegible] (19[illegible])

11. [illegible] "The [illegible] Diamond [illegible]" [illegible] Diamond and [illegible]

12. [illegible]

13. [illegible]

14. [illegible] Williams [illegible]

APPLICATIONS OF DIAMOND FILMS

J. Michael Pinneo

Crystallume
125 Constitution Drive
Menlo Park, CA 94025 USA

Rapid laboratory development of CVD diamond materials has already brought the material to military and commercial application in selected areas, and a wider variety of products based on diamond films is about to be introduced across a broad spectrum of technologies. Successful diamond film applications development requires resolution of significant technical and economic challenges not present in fundamental research on the material. We will examine the development of a diamond film product now in OEM sales, and of several products about to be introduced. Technology development details will occupy most of the discussion, but the impacts of financing and market development, and the importance of close collaboration with application end users will be examined as well.

Diamond Film X-Ray Windows

X-ray EDS systems, common accessories on scanning electron microscopes, often employ beryllium detector windows to provide vacuum isolation and x-ray transmission. A diamond window is now available to replace beryllium windows and provides significantly enhanced x-ray energy detection range. The structure and development of this window will be described.

Process technology developed to support this application includes control over nucleation density, crystallite size, large-area compositional and growth rate uniformity, and film gas permeability. Control over these material properties was achieved by appropriate choice of process parameters, including: plasma excitation means; plasma power density; substrate temperature; time variation of carbon concentration; reactor gas dynamics; reactor materials; and careful attention to the role of introduced and in situ process impurities.

Production technology was developed in parallel with process technology in a highly interactive manner. Methods of window fabrication, assembly, testing, and quality control were established and will be discussed. Production technology development improved yields by over three orders of magnitude from program beginning to current production levels.

X-Ray Lithography Mask Membranes

One critical missing link in bringing x-ray lithography processes into commercial use is the development of a suitable membrane for mask fabrication. Materials used to date include silicon, boron nitride, and others. Diamond's high modulus, stability under radiation, thermal conductivity, and x-ray transparency suggest its utility as an x-ray lithography membrane material, but significant technical uncertainties remain.

One of the most important of these, the requirement that a diamond membrane be under residual tensile stress, was predicted to be difficult to achieve. Simple modeling of residual stress states of diamond films deposited on non-diamond materials predicted residual compressive stress states due to diamond's low thermal expansion coefficient compared with most other materials. Experimental investigations initially supported these predictions. However, process parameters have been recently identified in which diamond films deposited at high temperatures on silicon substrates are found to be in residual tensile stress of ~$1X\ 10^8$ $dynes/cm^2$ after chemical removal of the silicon substrate. These films are flat and appear to be suitable for subsequent metallization form mask use.

While an understanding of the mechanisms underlying this effect is not yet in hand, the research questions which must be posed and answered are fairly clear. These will be discussed, along with data from experiments completed as of the conference date.

CVD Diamond Electronic Packaging Applications

CVD diamond's demonstrated high thermal conductivity (>15 W/cm/°C), coupled with its high electrical resistivity ($>10^{14}$ Ω-cm), indicate its advantageous use in electronic packaging applications, particularly those involving high power density devices. Process development issues for these applications include tradeoffs between deposition parameters which yield high growth rate and those which provide best thermal conductivity, surface roughness, expansion coefficient mismatch, dielectric constant and loss factor, and other material properties. Product development challenges include material deposition cost reduction, selection of fabrication and handling means, and development of adequate testing methods. Specific data from electronic packaging products now in development will be presented.

Diamond Film Friction and Wear Coatings

Many potential applications for CVD diamond films lie in applying the material to cutting tools, bearings, and other components whose performance is largely determined by resistance to wear. Enough is now known about the friction and wear behavior of CVD diamond to validate belief in the utility of this material in selected applications. Data will be presented which show dynamic friction coefficient variations in CVD diamond film as functions of surface morphology, sp^3/sp^2 Raman ratio, and exposing material composition (i.e., sapphire vs diamond film and single-crystal diamond vs diamond film). Two applications research efforts will be described, one aimed at applying CVD diamond film to carbide tools used for machining a specific aluminum alloy, and another undertaken to apply diamond coatings to selected turbine engine parts.

Turning of Aluminum Alloy Using an Insert with a Brazed Diamond Film

Masao Murakawa* and Sadao Takeuchi**

*Department of Mechanical Engineering
**Material Research and Testing Center
Nippon Institute of Technology
Saitama-Prefecture, Japan

1. Introduction

It has been found possible to coat polycrystalline diamond[1] onto different materials by use of vapor deposition methods. The coatings are expected to be put into practice in various industrial fields[2,3]. The technology has recently progressed to the point where diamond coated cutting inserts have begun to appear on the market[4]. However, since the coated diamond film inserts have relatively rough surface at the present time their performance is limited in certain cutting applications. Smooth mechanical finishing of the coating is difficult due to poor adhesion of the coating to its substrate. Because of this, the coated diamond inserts suffer not only from poor adhesion but also give a poorer surface finish of the workmaterial compared with that obtained with single crystal diamond or sintered diamond inserts. A so-called "brazed diamond film insert" has been developed[5] in an effort to solve this problem of conventional coated diamond inserts; it consists of a diamond film deposited onto a silicon wafer, which has been brazed to a tungsten carbide substrate with the diamond film side facing the substrate. This manufacturing process has a disadvantage in that the silicon wafer on which diamond is deposited must be removed by a separate processing step (e.g. grinding step). Further, in the previous study[5] no detailed results on the turning test of aluminum alloy containing Si have been disclosed. No detailed information, in particular, has been available comparing the performance of brazed, sintered and coated diamond inserts, which would be interesting from an industrial view point.

One objective of this study is to propose a method to eliminate the drawback of the conventional brazed insert which involves the removal, in a subsequent processing step, of the substrate on which the diamond was deposited.

Another objective of this research is to provide results of cutting performance tests using the above-mentioned inserts as the tools and a high silicon aluminum alloy as the workmaterial.

Table 1. Conditions for synthesizing diamond

Filament temperature (^0C)	2300~2400
Reaction pressure (torr)	100
Concentration of Ethanol in H_2 (vol%)	3~4
Flow rate of H_2 (standard cm^3/min)	150
Deposition speed (μm/hr)	4~6

2. The production of a polycrystalline diamond film-brazed cutting insert

2.1 Preparation of freestanding diamond film

In an effort to solve the problems with the conventional brazed diamond film insert, we propose an improved insert wherein a freestanding polycrystalline thick diamond film is made and is brazed to the tool substrate using the same technique as that used in making a single diamond bit. A thermal filament CVD apparatus[6], as shown in Fig.1, which utilizes ethanol as the precursor was used to make the diamond films. Table 1 shows the synthesis conditions. We used substrate made of tungsten carbide having a TiN coating, and a bare tungsten carbide substrate (conventionally used because of its good depositing capability with diamond).

In the case of tungsten carbide substrate which was used to obtain the freestanding diamond film, it was found that the degree of scratching the substrate surface has a big influence on the adhesion strength as well as the nucleation density of deposited diamond. The less the substrate is scratched the easier the film is removed from the substrate, but at the sacrifice of deposition rate. In view of this finding, the tungsten carbide was finally mirror lapped and lightly

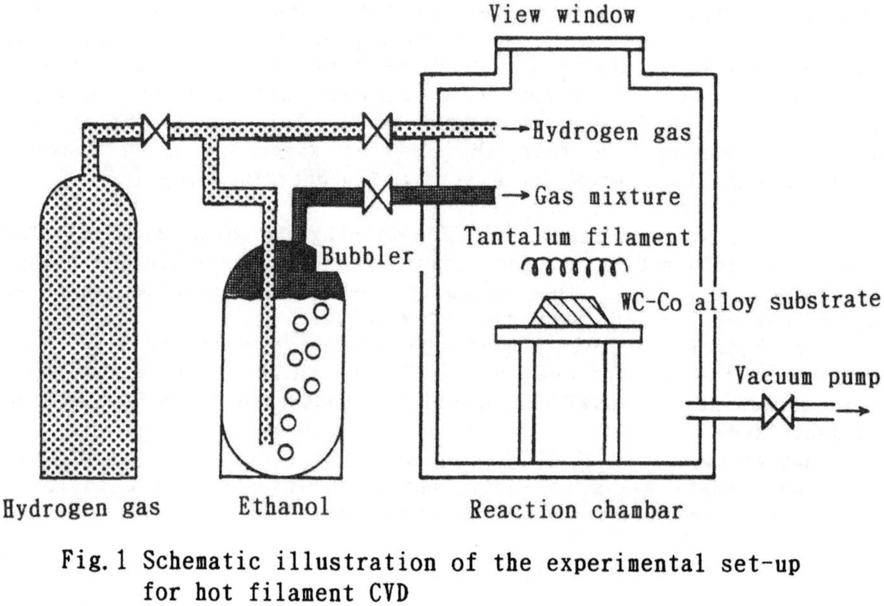

Fig. 1 Schematic illustration of the experimental set-up for hot filament CVD

scratched with diamond paste consisting of 10μm dia. diamond particles. This scratching condition gave a nucleation density of approximately $1\sim2\times10^6/mm^2$; the deposition rate was 6μm/hr. In the case of the TiN coated tungsten carbide substrate a similar scratching conditioning was employed.

We found that a TiN coated tungsten carbide substrate gave substantially no adhesion to diamond coating, resulting easily in a freestanding diamond film when the substrate is tilted after deposition. However, although a moderate thickness of diamond film up to 5μm can be easily obtained using this TiN coated tungsten carbide substrate, the far thicker film that would be required for the brazed insert in question would be impractical to make using this substrate because of the low deposition rate. In the case of the above-mentioned bare tungsten carbide substrate treated or scratched with diamond particles, a diamond film 20~50μm thick deposited onto a substrate up to 10×10mm in size can be easily separated after deposition by mechanical means, without any fracturing of the film. If one wants to get a thicker film (e.g. more than 80μm using this substrate) one can easily obtain it because the film will detach itself from the substrate during the cooling step. This thick freestanding diamond film has been found to have substantially no warping compared to the film prepared on a silicon

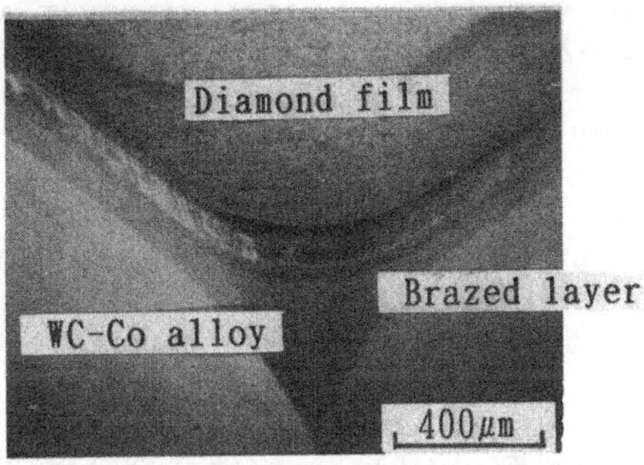

Fig. 2 SEM photo of the brazed diamond film insert

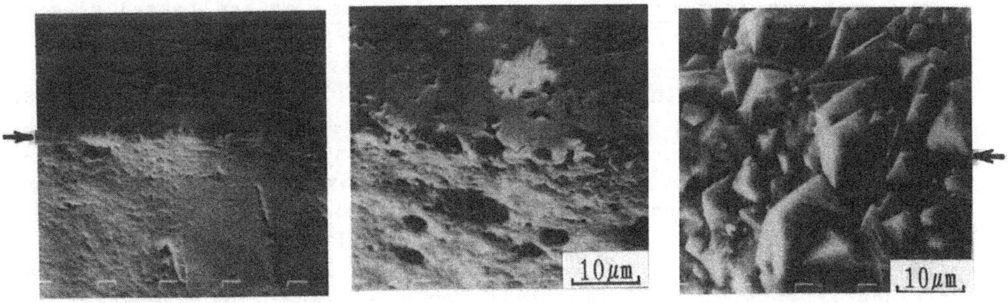

(A) Brazed insert (B) Sintered insert (C) Coated insert

Fig. 3 SEM photos of cutting edge of the diamond inserts
(→ ← Cutting edge line)

substrate of approximately the same size as the tungsten carbide substrate in question. This feature of less warping is very advantageous to the subsequent step of brazing. In the present experiment for making the brazed diamond film insert a polycrystalline film made with tungsten carbide substrate and having a thickness of 150～200μm was employed. No problem has been found even if the same tungsten carbide substrate is repeatedly used to obtain the freestanding film.

2.2 Brazing of the freestanding diamond film onto the tool substrate and finish grinding of the cutting edge

The brazed cutting tool or insert was made by first brazing a freestanding diamond film adjacent to the cutting edge portion of a commercial tungsten carbide insert(SPUN 120308). The brazing material, which consisted of a silver solder including a Ti component, was applied between the film and the substrate and the bonding operation was conducted within an atmosphere of Ar gas by use of a high frequency induction heater. In this experiment the diamond film surface of the side having a well crystallized face was bonded to the substrate in expectation of a larger bonding surface area which would lead to possible greater adhesion.

The cutting edge was made by first grinding and forming the film cutting edge using a #200 diamond wheel and then "finish grinding" both the cutting edge and rake face using a #1500 diamond wheel; an insert having a tip radius R=0.8mm and a relief angle of 11 degrees was formed, as shown in Fig.2. Fig.3(A) shows, in an enlarged scale, the tip portion or the ground cutting edge portion. In the same figure, (B) shows the tip portion of a sintered diamond insert and (C) shows the tip portion of a coated diamond film insert respectively for purposes of comparison. As shown, the brazed diamond film insert has no substantial chipping portion due to the falling out of crystal grains along the edge line, resulting in a sharp cutting edge. Furthermore, it will be obvious from Fig.3(C) that the coated insert consists of a well-crystallized surface peculiar to the diamond, resulting in a surface roughness value worse than that for (A) and (C) in the same figure.

3. A comparison test of the turning performance of the brazed diamond film insert

A conventional lathe was used for the test. The workmaterial was Al-Si alloy including 18%Si and having a size $\phi 80 \times \ell 400$mm. The inserts used for the comparison test had the same size and included:(a) tungsten carbide insert(SPUN 120308 H05 sold by Sandvik Co.,Ltd.); (b) brazed diamond film insert; (c) sinterd diamond insert(CD-10 sold by Sandvik Co.,Ltd.) and (d) coated diamond film insert with the film thikness of about 10μm. A tool holder CSTPR 2525M(Sandvik) was used, consequently resulting in an actual tool rake angle of 6 degrees and relief angle of 5 degrees. A cutting speed of 350～300m/min, a feed of 0.077mm/rev and a cutting depth of 0.2mm were employed with no lubrication. Fig.4 shows the relationship between the cutting time and the flank wear width for the above-mentioned inserts. As shown, a tungsten carbide insert resulted in a premature flank wear of up to

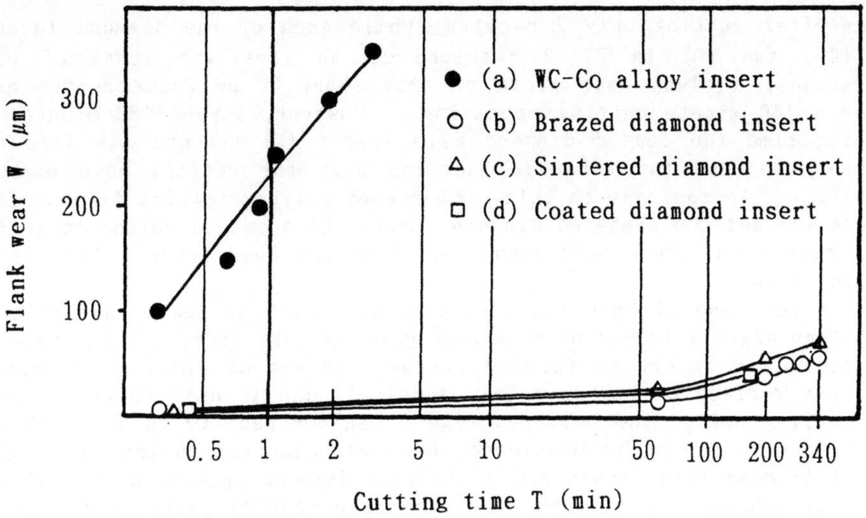

Fig. 4 Relation between cutting time and flank wear (Workmaterial: 18%Si-Al alloy)

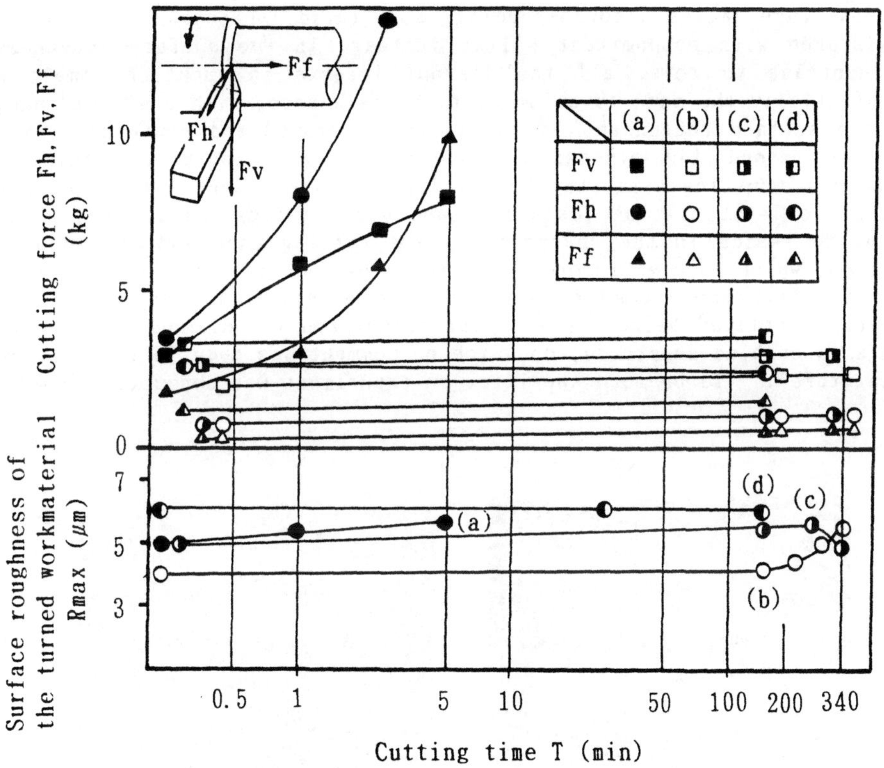

(a) WC-Co alloy insert (c) Sintered insert
(b) Brazed diamond insert (d) Coated diamond insert

(Fh: horizontal force, Fv: vertical force, Ff: feed force)

Fig. 5 Relationship between cutting forces and surface roughness (Workmaterial: 18%Si-Al alloy)

300μm after cutting only 2 minutes, while each of the diamond inserts (A),(B), and (C) in Fig.3 resulted in an excellent level of wear resistance; a flank wear width of only about 40 μm was observed even after a 180 minute cutting operation. However, beyond 180 minutes of cutting time the coated diamond film insert (C) reached its life end probably because the substrate portion near the cutting edge exposed itself. In contrast to this, the brazed polycrystalline diamond film insert (A) and the sintered diamond insert (B) showed a better tool life with respective flank wear amount of 70μm and 80μm after a 340 minute cutting time.

In the case of bare tungsten carbide insert we found instead of a "built-up edge" a fusion of aluminum onto the rake face. In the case of the coated insert we found a periodic change of gloss or roughness over the surface of workmaterial, seemingly due to an alternate growth and falling off of the built-up edge. In contrast to this, no change of surface roughness as mentioned above was observed in the cases of a brazed diamond film insert and a sintered diamond insert. Only a slight trace of adhesion of aluminum onto the rake face of each of the latter inserts was found by SEM observation after the cutting test.

Fig.5 shows the variation of values of cutting resistance and surface roughness of cut workmaterial versus cutting time. As shown, the bare tungsten carbide subjected to a relatively heavy wear of the cutting edge causes a corresponding large rapid increase in the cutting resistance with concomitant slight increase in the surface roughness. In contrast to this, all the diamond inserts in question remain at substantially the same level of cutting resistance, with a slight change of the surface roughness, in the cases of brazed and sintered diamond inserts beyond 180 minutes of cutting time. Further, within 180 minutes of cutting time, the brazed diamond film insert (b) is found to cause the least cutting resistance and surface roughness. The reason seems to reside in the sharpness of the cutting edge and its adjacent portion, which is high in order of (b), (c) and (d).

Finally, some comments will be made here on the strength of the brazed interlayer between the silver solder and the diamond film or the tungsten carbide substrate, on which no comment was made in the previous literature[5]. To be more specific, an experiment was conducted in which

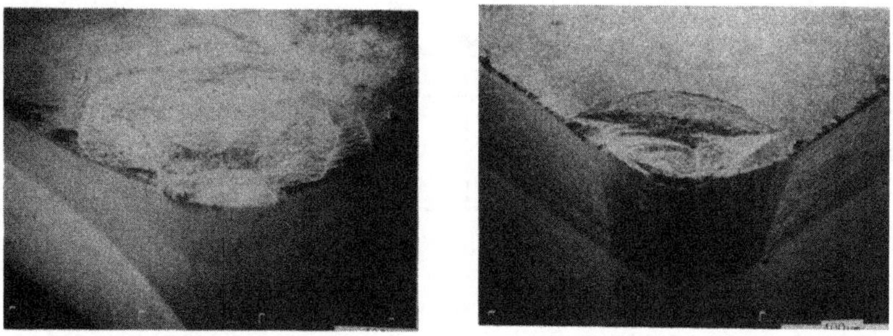

(a) Brazed diamond insert (b) Sintered diamond insert

Fig. 6 View of the inserts after cutting test (Workmaterial:WC-Co alloy, Hv=850) a workmaterial of WC-Co alloy(Co:21~25%,Hv=850) was turned with a cutting speed of 10m/min and a feed of 0.3mm/rev. When only the depth of the cut was gradually increased with the other conditions remaining the same, the eventual fracture of the diamond film was found at a certain constant cutting condition (a horizontal force of 34kg, a vertical force of 8kg, a feed force of 4kg and a depth of cut of 0.1mm).

A fracture of the sinterd diamond was also found at substantially the same cutting condition. Fig.6 shows the appearances of both the inserts which fractured. Fig.6 reveals that each fracture is of a cleavage nature with no separation through the interlayers themselves made with the silver solder. This should prove, from a practical viewpoint, that a diamond film brazed to a tungsten carbide substrate has an adequate bonding strength.

4. Conclusion

An improved manufacturing method for a cutting insert with a brazed diamond film on a tungsten carbide substrate has been proposed. The insert was compared in a performance test with three other inserts comprising (a) a bare tungsten carbide, (b) a sintered diamond and (c) a coated diamond film to turn a very-hard-to-turn workmaterial (high silicon aluminum alloy). The conclusions are as follows:

1) Polycrystalline diamond film can be ground more sharply than sintered diamond bulk using a diamond wheel, due to the strong adhesion of the film to the tool substrate.

2) The brazed insert gave a superior surface roughness of the machined Al-Si alloy workpiece than sintered and coated diamond film inserts.

3) Although the brazed diamond film insert gave substantially the same wear resistance as those of the sintered diamond insert and the coated diamond film insert, the brazed insert seems to be advantageous over the coated one in that the former can actually provide longer tool life because of the far greater thickness it can provide.

4) The adhesion or bonding strength of the brazed diamond film to the tungsten carbide substrate was found to be substantially the same as the fracture strength of the film itself, resulting in an adequate bonding strength of the brazed diamond insert for practical proposes.

Acknowledgment

The authors wish to thank Tokyo Diamond Tools MFG.Co.,Ltd., Sandvik Co.,Ltd. and Showa Denko K.K. for their cooperation in carrying out the experiment. Our thanks are extended to the then undergraduate students of our Institute, Messrs. Yuji Murakami and Tomoaki Shinoda who helped with the experiments.

References

1) S.Matsumoto, Y.Sato, M.Kamo and N.Setaka, Vapor deposition of diamond particles from methane, Jpn.J.Appl.Phys.,21:183 (1982).

2) N.Kikuchi and H.Yoshimura, Diamond coated inserts now, Japan New Diamond Forum (Japan Reviews in New Diamond),77 (1990).

3) K.Saijo, M.Yagi, K.Shibuki and S.Takatsu, The improvement of adhesion strength of diamond film, 17th Int.Conf.on Metallurgical Coatings Program and abstracts, A133 (1990).

4) T.Ito and N.Hayashi, Diamond coated cutting tools synthesized from

CO, Japan New Diamond Forum (Japan Reviews in New Diamond),83 (1990).

5) F.Okuzumi, Gaseous phase synthesis poly-crystalline diamond tool, Japan New Diamond Forum (Japan Reviews in New Diamond),80 (1990).

6) M.Murakawa, S.Takeuchi, H.Miyazawa and Y.Hirose, Chemical vapor deposition of a diamond coating onto a tungsten carbide tool using ethanol, Surface and Coating Tech.,36:303 (1988).

DEVICE APPLICATION FOR DIAMONDS*

M. W. Geis

Lincoln Laboratory
Massachusetts Institute of Technology
224 Wood Street
Lexington, MA 02173 USA

Diamond is a high band gap (5.5 eV) semiconductor that has superior properties to many of the commonly used semiconductors. The high breakdown voltage (ten times that of Si), exceptionally high thermal conductivity (five times that of copper), and hole and electron mobilities near 2000 $cm^2\ V^{-1}\ s^{-1}$ result in projections of higher operating frequencies and significantly higher output powers than those obtainable with devices made of Si or GaAs.

Fabrication technology has been developed to etch diamond and to form both ohmic and Schottky contacts. Permeable base transistors (PBTs) and diodes have been formed which exhibit some of the highest reported operating temperatures.

This presentation will discuss techniques used to obtain semiconducting diamond films, electrical properties of diamond, device results including diodes and PBTs, and future diamond applications.

*This work was sponsored by SDIO/OST through the Office of Naval Research.

Diamond and Diamond-Like Films and Coatings
Edited by R.E. Clausing *et al.*, Plenum Press, New York, 1991

SEMICONDUCTOR DEVICE DEVELOPMENT WITH HOMOEPITAXIAL DIAMOND FILMS

S.A. Grot,[1] G.SH. Gildenblat,[1] C.W. Hatfield,[1] A.R. Badzian,[2] and T. Badzian[2]

[1]Center for Electronic Materials and Processing and Department of Materials Engineering
[2]Materials Research Laboratory
The Pennsylvania State University
University Park, PA 16802

Introduction

Diamond has many physical properties that make it an attractive semiconductor material for high-temperature and high power electronic devices[1]. Natural diamonds and synthetic bulk diamond fabricated by the traditional ultra high-pressure high-temperature method have been used to produce rudimentary active devices[2,3]. Schottky diodes and point contact transistors have operated at high temperatures[3,4]. However, the high cost of fabrication combined with the difficulty in controlling dopants for either natural or synthetic (high-pressure, high-temperature) diamond severely limit the applicability of these devices. Recently, several groups succeeded in fabricating the first semiconductor devices (Schottky diodes) based on p-type thin-film diamond deposited using an activated CVD process[5-8]. However, until now, the quality of the diamond films did not allow investigation of high-temperature device characteristics.

In this work we describe the electrical characteristics of boron-doped homoepitaxial diamond films fabricated using a plasma assisted CVD process, formation of ohmic contacts, high temperature (580°C) Schottky diodes, and the operation of a rudimentary diamond MESFET. We also report reversible changes of the conductive state of the diamond surface by exposure to various surface treatments for both natural and thin-film diamond.

Film Deposition and Characterization

Homoepitaxial diamond films were grown in a microwave assisted chemical vapor deposition system. Details of the deposition system are described elsewhere[6]. Two types of natural diamond substrates, supplied by the Dubbeldee Harris Co., were used. The first were industrial type Ia crystals ("Congo octos") cut along the (001) plane. The second were type IIa crystals cut along the (001) plane with surfaces polished to optical window quality. Prior to growth of the diamond films, the diamond substrate surfaces were exposed to a hydrogen plasma similar to that present during growth. The purpose of this exposure is to remove any non-diamond carbon phase at the surface. The plasma for film growth was ignited inside a silica tube in a 100 sccm mixture of 1% CH_4 in H_2 at 80 Torr. An ellipsoidal ball of plasma is formed in the resonance cavity of the microwave applicator (2.45 GHz, 500 W). The position of the substrate was adjusted such that the plasma hovers above the substrate and barely touches its surface. The diamond substrate was heated by the plasma to a temperature of 850-950°C. Doping of the diamond films with boron was performed in two ways. In the first method, boron powder was placed onto the substrate holder. In the second method, a boron rod was periodically immersed for 15 seconds every 10 minutes into the plasma during the growth process. For both methods, boron is etched by the hydrogen plasma forming boron hydrides which react on the surface of the growing diamond film and boron is incorporated into the diamond lattice. Separate silica reactor tubes were used for the growth of boron doped and undoped films in order to eliminate boron contamination of undoped films. The diamond structure of the films was confirmed by Raman spectroscopy and x-ray diffraction. Doping with boron (either substitutional or clustered) causes displacement disorder of the diamond matrix. This results in a widening of the characteristic diamond Raman peak at 1332 cm^{-1}, as shown in Fig. 1. In the case of undoped films, the peak half-width was between 1.7 and 2.3 cm^{-1}, whereas for the doped films the half-width was between 3.4 and 5.5 cm^{-1}.

Surface Treatments

Boron doping did not effect the electrical measurements of untreated (*as deposited*) homoepitaxial diamond films used in this study. The measured sheet resistance, R_s, of a CVD diamond film with no surface treatment after growth (*as deposited*) does not depend on the boron doping in the film (see Table I). Furthermore, gold contacts to the *as deposited* diamond film exhibit linear I-V characteristics, while the I-V characteristics of aluminum contacts were rectifying, independent of the boron doping in the film (compare Fig. 2 and 3). This behavior is inconsistent with known properties of metal contacts to diamond which imply no difference between the electrical characteristics of Au and Al contacts[4-6]. After the films were cleaned in a solution of CrO_3 and H_2SO_4 at 160°C and rinsed in a boiling mixture of H_2O_2 and NH_4OH the effect of the doping become apparent. After the chemical cleaning, the undoped films became insulating ($R_s > 10^8$ kΩ/□) while for chemically cleaned boron-doped films R_s was found to be in a range of 5-60 kΩ/□. These results are summarized in Table I.

TABLE I

SHEET RESISTANCE (kΩ/□) OF HOMOEPITAXIAL DIAMOND FILMS AS A FUNCTION OF SURFACE TREATMENT

Sample	Film Thickness	Doping	Untreated Sample Surface	After Chemical Cleaning	After Chemical Cleaning and Hydrogen Plasma Treatment
S94	12 μm	none	9.6		
S96	5 μm	none	7.8	$>10^8$	10
S98	6 μm	none	7.6	$>10^8$	15
S112	14 μm	boron	5.0	45	
S126	15 μm	boron	3.1	5	
S135	4 μm	boron	6	60	
IIa Natural Diamond	—	none	$>10^8$	—	10

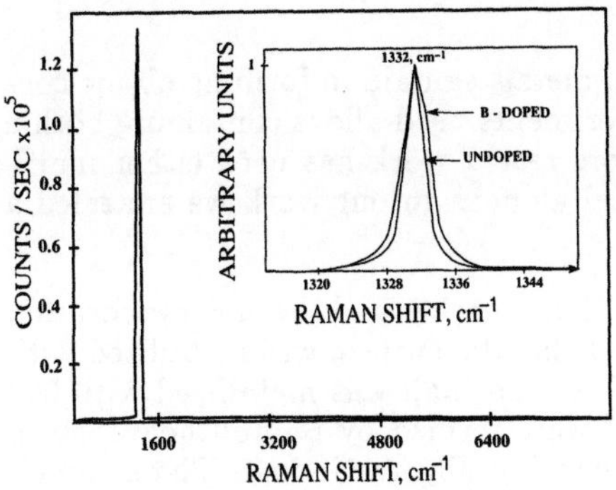

Fig. 1 Raman spectrum of a homoepitaxial diamond film. The insert shows the widening effect of the 1332 cm^{-1} peak from the boron-doped films.

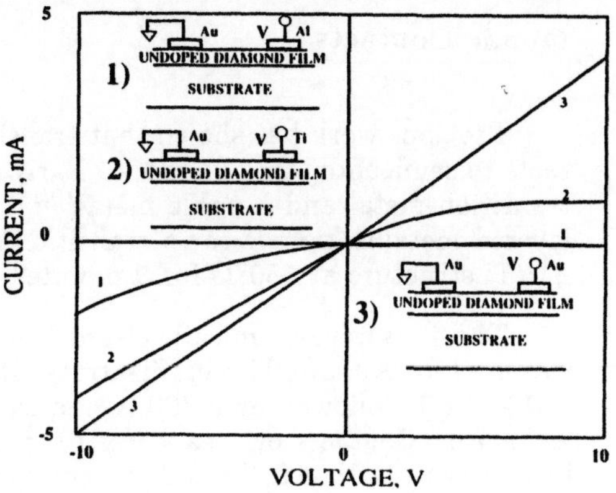

Fig. 2 I-V characteristics of metal contacts to the as-deposited undoped homoepitaxial diamond film. Contact area 0.07 mm^2.

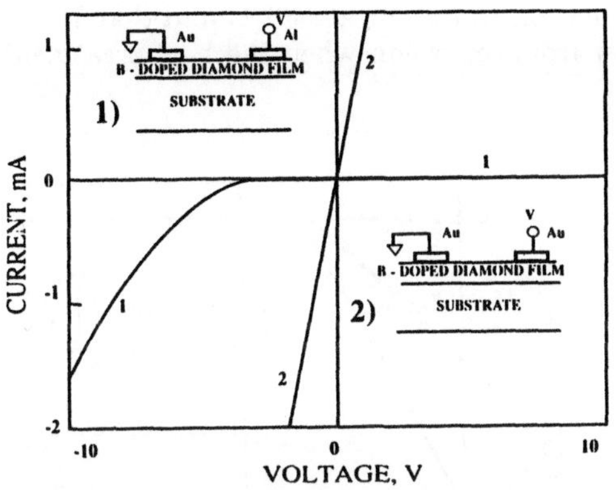

Fig. 3 I-V characteristics of gold and aluminum contacts to the as-deposited boron-doped homoepitaxial diamond film, contact area 0.07 mm^2.

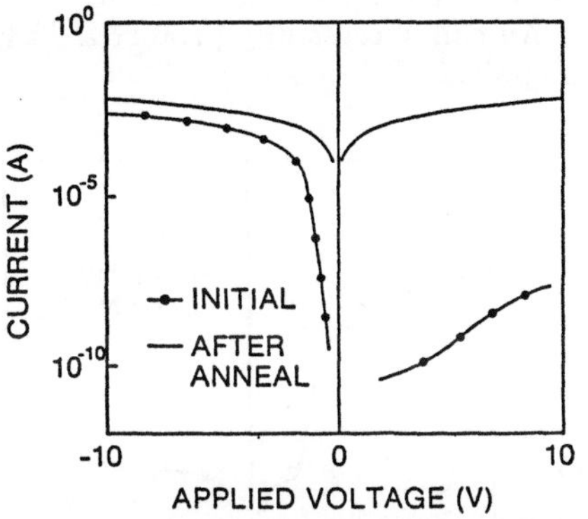

Fig. 4 Log I versus V characteristics of Au/Ti sandwich structure before and after annealing.

We found that for the chemically cleaned samples the highly conductive state of the diamond surface characteristic of the *as deposited* films can be reintroduced by exposing the films to a hydrogen plasma similar to that used at the end of the deposition process (2.45 GHz, 500 W, 900°C, 80 Torr, an 100 sccm flow of 100 % H_2). Furthermore, the same treatment of (001) oriented type IIa natural diamond also results in a conductive surface with $R_s = 10$ kΩ/□ (see Table I).

Thus, it seems that the formation of a conductive layer on the surface of the diamond film is related to the presence of the hydrogen plasma in the deposition system. Presumably, the role of the chemical cleaning is to remove this conductive layer. We also note that the switching between conducting and semiconducting (doped films) or insulating (undoped films, natural diamond) states of the diamond surface is reversible.

Ohmic Contacts

Previous work has shown that transition metals can aid in forming ohmic contacts to semiconducting diamond. Early experiments used alloys containing both a transition metal and a noble metal[10,11]. More recent work has used either multilayered metallizations[12], or a transition metal alone[8]. In our work we annealed a Au/Ti structure at 580°C for 5 minutes.

The films were chemically cleaned as mentioned earlier and thermal evaporation was used for Au and Ti metallizations. Half of the film surface was metallized with 100 Å of Ti followed by a 700 Å Au cap. The other half was metallized with 600 Å of Au. Contacts of area 1.9 x 10^{-2} mm^2 were formed by photolithography in both areas. Both Au/Ti and Au contacts were initially rectifying. The sample's temperature was raised to 580°C over a 20 minute period, maintained there for 5 minutes and then cooled to room temperature in 10 minutes. After this heat treatment, the Au/Ti contacts were virtually ohmic (Fig. 4). The most notable change in the Au/Ti contact I-V characterisitcs was the dramatic increase in the reverse current. However, the I-V's of the Au Schottky diodes were not significantly altered by the heat treatment (Fig. 5).

It has been suggested that the formation of ohmic contacts by this method can be attributed to low-resistivity titanium carbide formed by the heating[8,10-12] Our work demonstrates that this carbide formation can occur at 580°C and that 700 Å of Au can successfully protect a Ti layer from oxidation when the heat treatment is applied.

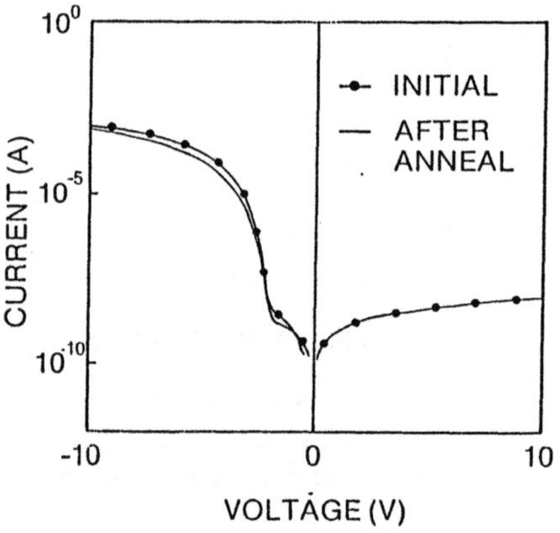

Fig. 5 I-V characteristics of a Au-gate Schottky diode before and after annealing.

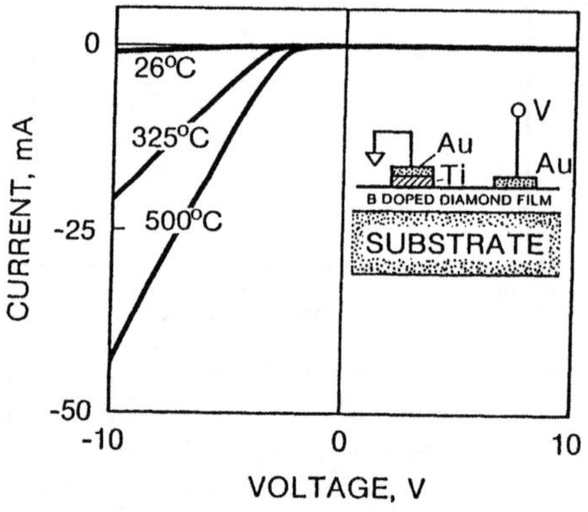

Fig. 6 I-V characteristics of a Au-gate Schottky diode (A = 1.2 x 10^{-3} mm^2) with a small area (A = 9.6 x 10^{-3} mm^2) Au/Ti ohmic back contact for different temperatures.

Schottky Diodes

Schottky diodes were formed by thermally evaporating small (<.02 mm^2) Au contacts on the surface of a boron-doped diamond film that was chemically cleaned as described above. The second, ohmic, contact of the diode was formed by heating a Au/Ti sandwich as described in the previous section. A Schottky diode with a Au gate of area 1.2 x 10^{-3} mm^2 and an annealed Au/Ti ohmic contact is shown schematically in Fig. 6. Highly rectifying I-V characteristics are observed in the temperature range 26-580°C. Analysis of the diode characteristics has shown that the reverse current *density* rapidly increases as a function of diode area, A, presumably due to the film defects encompassed by the gate[13]. For $A \geq 1\ mm^2$ the Au/diamond film contact characteristics are essentially linear so that Schottky diodes can also be fabricated using a combination of small area ($A = 2.2 \times 10^{-2}\ mm^2$) (rectifying) and large area ($A > 1\ mm^2$) (ohmic) Au contacts (Fig. 7). As shown in Figs. 6 and 7, both types of diodes, fabricated with either small Au/Ti ohmic contacts or large Au back contacts, exhibit significant rectification at temperatures in excess of 500°C.

The diode series resistance is thermally activated with activation energy E_A = 0.15 eV for samples 112 and 126. For sample 112 the activation energy of the film resistivity was measured separately using a four point probe method and found to be also 0.15 eV, indicating that the diode series resistance is dominated by the bulk resistance of the base. The activation energy of the reverse current density was found to be 0.25 eV (at V_R = 10 V) for sample 126 in the 26 - 580°C temperature range. Simple extrapolation of this result indicates that considerable rectification is expected at 800°C in an inert atmosphere.

To increase the area of the Schottky diode without adversely affecting the leakage current, films were grown on the optical quality type IIa diamond substrates. These films had fewer surface defects than films grown on the industrial type Ia substrates. Annular aluminum-gate Schottky diodes with gate area $3 \times 10^{-2}\ mm^2$ were fabricated with the structure shown in the insert of Fig. 8. The leakage current remained under 50 nA at reverse biases up to 16 V. This MESFET structure shows some degree of gate control over the drain current I_D. As can be seen in Fig. 8, a 6 volt reverse gate bias results in a reduction of the drain current.

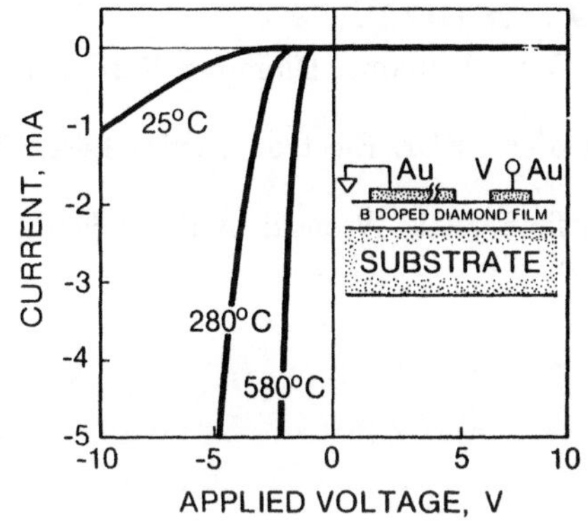

Fig.7 I-V characteristics of a Au-gate Schottky diode ($A = 2.2 \times 10^{-2}\ mm^2$) with a large area Au back contact ($A > 1\ mm^2$) for different temperatures.

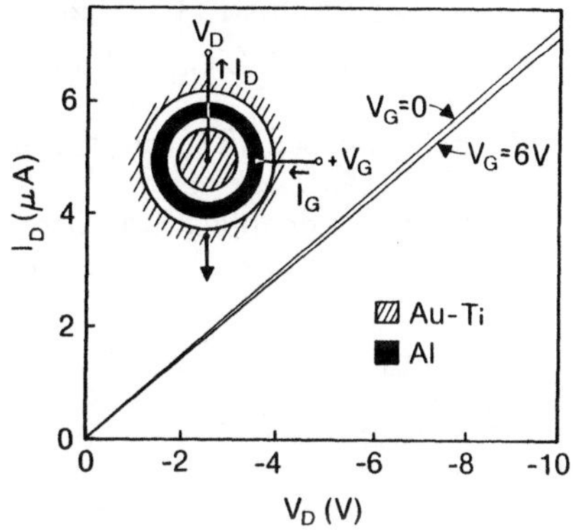

Fig. 8 Drain current vs. drain voltage characteristic for rudimentary MESFET.

Conclusions

Boron-doped homoepitaxial diamond films were grown using a PACVD process on natural diamond substrates. After growth, a conductive layer exists on the surface of the film. This layer can be removed by cleaning in a strong oxidizing chemical solution. Low resistance contacts can be formed on the surface of thin film diamond by heat treating a Au-Ti sandwich structure in air at 580°C for 5 minutes. The ohmic contacts and Schottky diodes were formed on the same film, allowing the fabrication of a rudimentary MESFET. Gold-gate Schottky diodes show rectifying characteristics at 580°C. The results of this study indicate the feasibility of thin-film diamond as a material for high temperature electronics.

Acknowledgement

This work was supported in part by the SDIO/IST through the Office of Naval Research under contract no. N00014-86-K-0443, and the Diamond and Related Materials Consortium at the Penn State University.

References

1. M.N. Yoder, Nav. Res. Rev. **39** , 27 (1987).
2. J.F. Prins, Appl. Phys. Lett. **41** , 950 (1982).
3. M.W. Geis, D.D. Rathman, D.J. Ehrlich, R.A. Murphy, and W.T. Lindley, IEEE Electron Device Lett. **EDL-8** , 341 (1987).
4. M. Bell, and W. Leivo, Phys Rev. **111** , 1227 (1958).
5. G.Sh. Gildenblat, S.A. Grot, C.R. Wronski, A.R. Badzian, T. Badzian, and R. Messier, Appl. Phys. Lett. **53** , 586 (1988).
6. M.C. Hicks, C.R. Wronski, S.A. Grot, G.Sh. Gildenblat, A.R. Badzian, T. Badzian, and R. Messier, J. Appl. Phys. **65** , 2139 (1989).
7. M.W. Geis, D.D. Rathman, J.J. Zayhowski, D. Smyth, D.K. Smith, and G.A. Ditmer, 3rd ONR Diamond Technology Initiative Symp. Abs., Crystal City, VA, 115 (1988).
8. H. Shiomi, H. Nakahata, T. Imai, Y. Nishibayashi, and N. Fujimori, Japan. J. Appl. Phys. **28** , 758 (1989).
9. G.H. Glover, Solid-St. Electron. **16** , 973 (1973).
10. M. Seal, Industrial Diamond Review **29** , 408 (1969).
11. A.T. Collins, E.C. Lightowlers, and A.W.S. Williams, Diamond Research, 19 (1970).
12. K.L. Moazed, R. Nguyen, and J.R. Zeidler, IEEE Electron Device Lett. **9** , 350 (1988).
13. G.Sh. Gildenblat, S.A. Grot, C.W. Hatfield, C.R. Wronski, A.R. Badzian, T. Badzian, and R. Messier, Mat. Res. Bull. **25** , 129 (1990).

Al METAL POINT CONTACT TO B-DOPED DIAMOND FILMS

Koichi Miyata, Yuichi Matsui, Kazuo Kumagai, Shigeaki Miyauchi, Koji Kobashi, and Akimitsu Nakaue

Electronics Research Laboratory, Kobe Steel, Ltd.
1-5-5 Takatsuka-dai, Nishi-ku, Kobe 673-02, Japan

INTRODUCTION

The mechanical, thermal, optical, and electrical properties of diamond have been shown to be superior to those of other materials[1,2,3,4]. These properties can be utilized in a wide range of optoelectronic device applications as well as for extreme environmental conditions[5,6]. Chemical vapor deposition(CVD) of diamond offers an attractive solution for obtaining semiconducting diamond thin films. In order to exploit the electronic properties of these polycrystalline films, it is critical to evaluate the properties of electrical contacts to the diamond surface. In particular, reliable ohmic and rectifying contact technologies have been difficult to develop and reproduce. Recently, the formation of Schottky contacts to polycrystalline diamond films have been reported[7,8], but the relationship between the rectifying characteristics and the quantity of B impurities incorporated into the films was not examined.

In this report, the relationship between the amount of B impurities and the current-voltage(I-V) characteristics is systematically investigated using Al point contacts to polycrystalline diamond thin films. As a preliminary measurement, the method of point contacts is simple and convenient. Although stable metal point contacts exhibit increased minority carrier injection and are often less reproducible than planar electrodes, carbides are not formed at the interface. Thus, the chemical interaction between the metal and diamond interface can be neglected.

EXPERIMENTAL

Undoped and B-doped diamond thin films were prepared by microwave plasma CVD[9]. CH_4, diluted in H_2, was used as the source gas, while B_2H_6 was used as the dopant gas. P-type (111) Si wafers polished with 0.25 μm diamond paste and cut to 20 mm x 10 mm were used as substrates. In order to minimize the influence of the interface between the diamond film and the substrate, Si substrates with resistivities of 1-10 ohm-cm were used. As listed in Table 1, five samples with varying amounts of B were prepared. The B/C ratios in the reaction gas were 0, 2, 4, 20, and 40 ppm. The CH_4 concentration with respect

Diamond and Diamond-Like Films and Coatings
Edited by R.E. Clausing *et al.*, Plenum Press, New York, 1991

to H_2 was 0.5%, and the total gas pressure was kept at 31.5 Torr during deposition. The substrate temperature was maintained at 800°C by adjusting the microwave power. The deposition time was 7 hours, which corresponds to approximately a 2 μm-thick diamond film. The films were characterized by Raman spectroscopy, Scanning Electron Microscopy (SEM) and X-ray diffraction. Quality of these films, as assessed by the above-mentioned techniques, were generally similar to those reported in reference [9].

Table 1. Preparation conditions of B-doped and undoped diamond films.

Substrate	p-Si(111) 1-10 ohm-cm
Reaction gases	CH_4(0.5%), H_2, B_2H_6
B/C concentration	0, 2, 4, 20, 40ppm
Gas pressure	31.5 Torr
Total flow rate	100 sccm
Substrate temperature	800°C
Deposition time	7 hrs

Figure 1 shows the schematic diagram of the I-V measurement system. Since the work function of Al is reported to be 4.2 eV, it is thought to be a good candidate for Schottky contacts to p-type diamond. The wires used as the metal point contacts consisted of Al(99.9%). At first, the Al wires were cut exposing a clean surface. After that, the Al wires were placed in physical contact with the diamond surface. Ohmic contacts to the back side of the Si substrates were made using Ag paste. Pt wire was also used as metal point contacts. In this case, I-V characteristics exhibited ohmic characteristics for all B/C ratios from 0 to 40 ppm. These results suggest that there does not appear to be any rectifying components between diamond films and p-type Si substrates. To isolate the experiment from random external noise, the measurement was performed in a shielded enclosure.

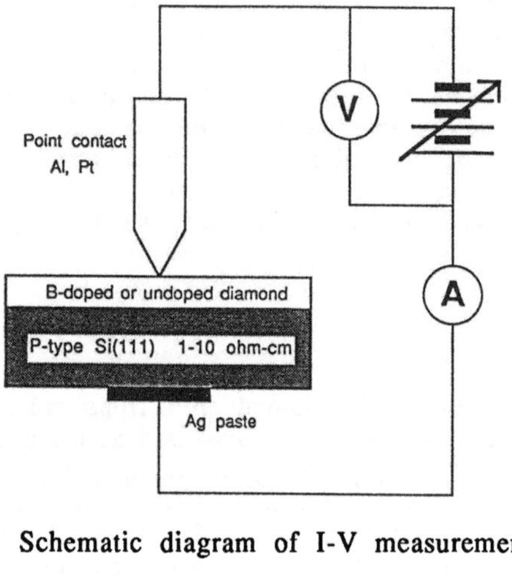

Fig. 1. Schematic diagram of I-V measurement.

RESULTS AND DISCUSSION

Figures 2(a)-(d) show typical I-V characteristics of Al point contacts to diamond films. The B/C ratios are 0 ppm (Fig. 2a), 2 ppm (2b), 4 ppm (2c), and 20 and 40 ppm (2d). For these I-V curves, a positive voltage on the horizontal axis corresponds to positive voltage applied to the p-type semiconducting films. At B/C ratios of 2, and 4 ppm (Fig. 2(b) and (c)), current begins to flow in the forward bias direction, almost no current flows in the reverse bias direction, and break-down doesn't occur for up to several tens of volts. Good rectifying characteristics could be obtained at a B/C ratio of 4 ppm and lower. Rectifying characteristics were also observed for the case of undoped diamond films. These films exhibited p-type semiconducting-behavior based on the direction of rectifying characteristics. This may be due in part to a small amount of residual B that may have been incorporated into diamond films from the inner wall of the gas inlet tube. Note, that the rectifying characteristics deteriorated for B/C ratios above 20 ppm, as shown in Fig. 2(d). Forward and reverse leakage currents increase drastically with increasing B/C.

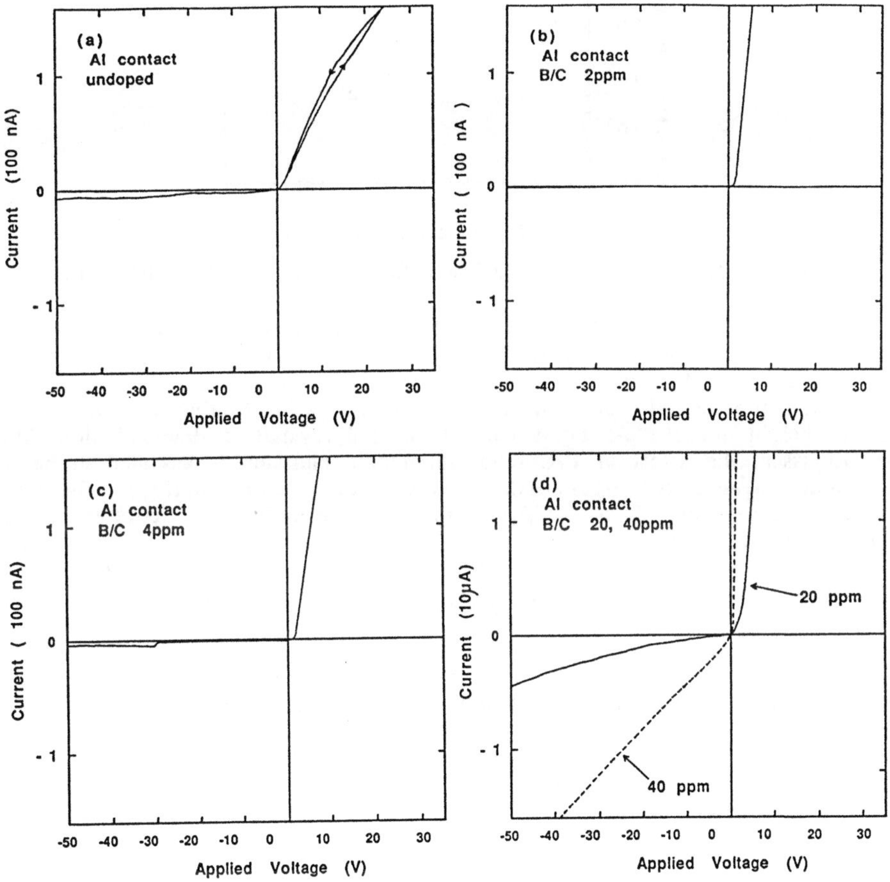

Fig. 2. I-V characteristics of Al point contact to B-doped and undoped diamond films (a) undoped, (b) B/C: 2 ppm, (c) B/C: 4 ppm, (d) B/C: 20 and 40 ppm.

The ratio, R= V/I, obtained at a forward bias of 5V is plotted as a function of B/C as shown in Fig. 3. It should be pointed out that, at this bias-level the current conducted by the point contact is not directly proportional to the applied forward bias. Thus, the ratio R is not a true resistance, but rather should be considered as a "characteristic inverse conductance" of the contact for a given value of B/C, regardless of the conduction mechanisms involved. A drastic change in R is observed as the B/C ratio is varied from 4 to 20 ppm. A similar large change in resistance is also seen in the case of GaAs as a function of carrier concentration. The current transport mechanism for the case of metal/GaAs contacts shifts from thermionic emission to field emission. Generally, the transport mechanism changes from thermionic emission to field emission with increasing carrier concentration and the contact resistance drastically decreases with increasing carrier concentration[10]. Thus, one explanation for the drastic change in the ratio R for diamond thin films may be due to such a shift in the dominant transport mechanism.

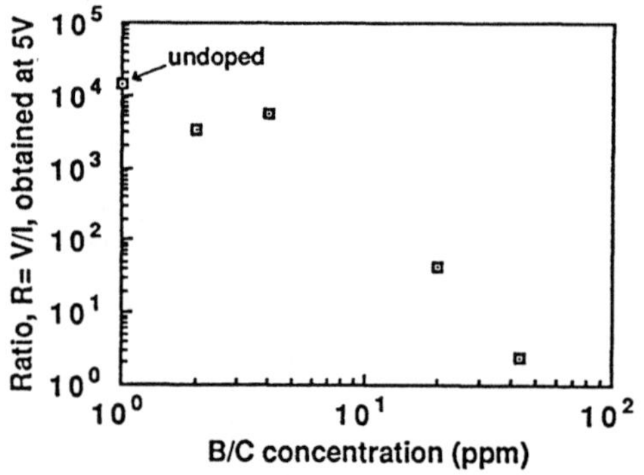

Fig. 3. Ratio, R= V/I, obtained at a forward bias of 5V vs. B/C ratio.

Figure 4 shows the log(I) vs. V plot for B/C ratio of 2 ppm. An ideality factor, n, of 4.0 was obtained for this contact indicating that mechanisms other than pure thermionic emission were also contributing to the current conduction process. However, in order to obtain a qualitative measure of the barrier height for the Al point contact on polycrystalline diamond thin films and to assess the effect of B/C ratio, thermionic emission is assumed to be the dominant transport mechanism. Thus, one can analyze the I-V characteristics (for V > 3kT/q) using the formula for Schottky metal-semiconductor contacts:

$$I = S\, J_s \exp\left(\frac{qV}{nkT}\right), \qquad (1)$$

where I is the forward current, S is the area of electrode, Js is the saturation current density, V is the applied voltage, n is the ideality factor, and k is Boltzmann's constant. In Fig. 4, the saturation current density, Js, can be obtained by extrapolating the current from the log-linear region to V=0, yielding Is, and using the following expression:

$$J_s = \frac{I_s}{S}. \qquad (2)$$

If the area of metal point contact S can be measured or estimated, the barrier height of the Schottky contact can be calculated. Unfortunately, it is difficult to know the exact area of the point contact. Due to surface roughness, multiple contacts are formed between the metal wire and the diamond film. In order to obtain a reasonable value for the contact area, the following experiments were performed. First, the I-V characteristic of Al point contact to GaAs was measured with the same Al wire used for the

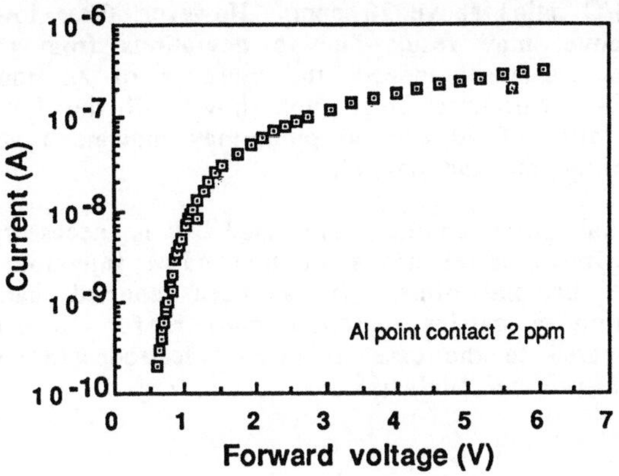

Fig. 4. Log(I) vs. V plot for B/C ratio of 2 ppm.

diamond films. The relation between barrier height of Al point contact to GaAs and contact area can be obtained from the following formula:

$$\phi_B = \frac{kT}{q} \ln\left(\frac{A^*T^2}{J_s}\right), \quad (3)$$

where A* is the Richardson constant. Figure 5 shows the barrier height as a function of the contact area obtained from the I-V characteristics of the Al point contact to GaAs. The experimental value reported by Sze[11] is also plotted in Fig. 5. From this result, it is reasonable to choose the contact area to be approximately 10^{-6} cm^2, this corresponds to a 10 μm diameter for a circular contact area.

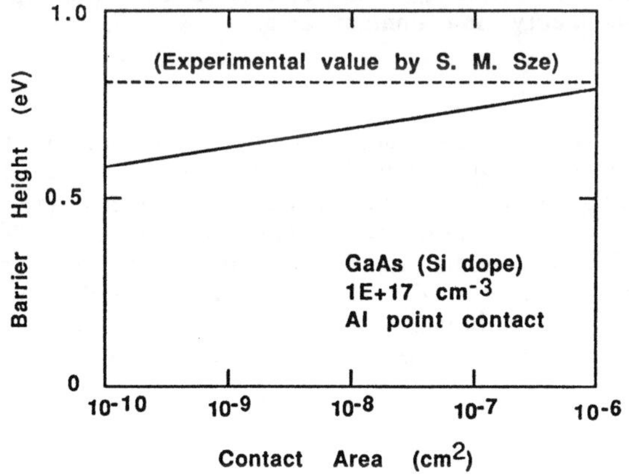

Fig. 5. Barrier height on GaAs vs. contact area from the I-V characteristics of the same Al point contact used for the diamond films. The experimental value by S. M. Sze is also plotted in this figure.

A characteristic barrier height of the Al/diamond point contact can be calculated using the above-mentioned contact area and expressions (2) and (3). The values of A* and Js used here are 120 $A/cm^2/K^2$ and 1.53×10^{-6} A/cm^2 respectively. Figure 6 shows the calculated barrier height as a function of B/C ratio in the reaction gas. The values of the barrier height are in the range between 0.7 and 0.8 eV for B/C ratios below 4 ppm. This value decreases to 0.5 eV for B/C ratios above 20 ppm. However, these low barrier height for 20 ppm and above, may result due to deviations from thermionic emission controlled current transport. Indeed, the presence of an ohmic regime at low forward biases in logarithmic plots (not shown) of the I-V characteristics for films with B/C ratios of 20 and 40 ppm, may indicate a space charge limited current conduction mechanism[12].

When metal point contacts are used, it is necessary to consider the effects of minority-carrier(electrons in this case) injection from metal point contacts into the diamond films. It has been reported that there is no large difference in minority carrier injection ratio r, for 10 μm diameter point contacts as compared to the case of planar electrodes[13], so this effect was neglected in the present analysis.

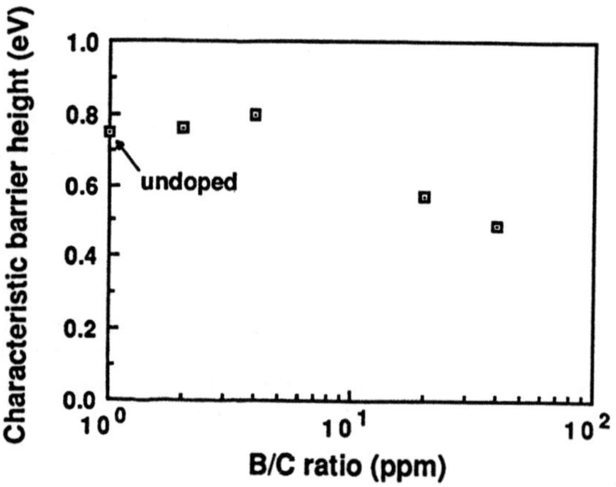

Fig. 6. Calculated barrier height vs. B/C ratio by assuming the thermionic emission theory and contact area.

Pale blue light emission near the Al point contacts to diamond films with B/C ratio below 4 ppm has been observed. When greater than 350V reverse bias is applied to the point contacts, and a few mA of current flows through the films, light emission was visibly detectable. When the voltage is increased slightly, the diamond film disintegrated probably due to the thermal expansion mismatch with the substrate. Similar light emission was also observed in the combination of bulk diamond and Ni contacts[14], and this light emission seems to be due to the generation of microplasma by impact ionization.

SUMMARY

The relation between the current-voltage characteristics of Al/diamond metal point contact and the quantity of B impurities were investigated using Al point contacts. Schottky contacts to diamond films could be obtained for B/C ratio below 4 ppm in the reaction gas, the barrier height of which were estimated to be in the range between 0.7 and 0.8 eV by use of thermionic emission theory. Rectifying characteristics drastically deteriorated for B/C ratio above 20 ppm. This may be because the current transport mechanism changes from thermionic emission to field emission. No rectifying characteristics were observed between diamond and p-type Si substrate for the case of Pt point contacts. Pale blue light emission could be observed near the Al point contacts for reverse voltages in excess of 350V applied to diamond films with B/C ratio's below 4 ppm. Although difficulties associated with the use of metal point contacts exist, as a non-invasive technique, metal point contacts to diamond thin films has, for the first time, provided qualitative electrical characterization of the diamond thin films.

ACKNOWLEDGEMENT

The authors wish to thank Prof. J. T. Glass, Dr. D. Dreifus, Dr. K. Das, and K. Nishimura for their useful discussions.

REFERENCES

1. J. E. Field, *The Properties of Diamond*, Academic Press(1979).

2. M. Seal, Interdisciplinary Science Reviews, Vol. 14, No. 1, 64 (1989)

3. V. K. Bazhenov, I. M. Vikulin, and A. G. Gontar, Sov. Phys. Semicond. 19(8), 829, August(1985).

4. V. S. Vavilov, and E. A. Konorova, Sov. Phys. Usp., Vol. 19, No. 4, 301, April(1976).

5. M. W. Geiss, D. D. Rathman, D. J. Ehrlich, R. A. Murphy, and W. T. Lindley, IEEE Electron Device Lett., EDL8, 341 (1987).

6. R. Wolfe, and J. Woods, Phys. Rev. Vol. 105, No. 3, 921, February(1956).

7. G. Sh. Gildenblat, S. Grot, C. R. Wronski, A. R. Badzian, T. Badzian, and R. Messier, Appl. Phys. Lett. **53**, 586 (1988).

8. Y. Mori, H. Kawarada, Y. Yokota, and A. Hiraki, Materials Research Society, Fall Meeting, Boston, December (1989) paper F5.7.

9. K. Kobashi, K. Nishimura, Y. Kawate, and T. Horiuchi, Phys. Rev. B, 38, 6, 4067.

10. E. H. Rhoderick, and R. H. Williams, *Metal-semiconductor contacts. Second edition*, Oxford University Press (1988).

11. S. M. Sze, *Physics of Semiconductor Devices 2nd edition*, John Wiley & Sons, Inc (1981).

12. M. A. Lampert, and P. Mark, *Current Injection In Solids*, Academic Press (1970).

13. R. A. Clarke, M. A. Green, and J. Shewchun, J. Appl. Phys., Vol. 45, No. 3, March (1974) pp. 1442-1443.

14. A. V. Bogdanov, I. M. Vikulin, and T. V. Bogdanova, Sov. Phys. Semicond. 16(6), June (1982) pp. 720-721.

TANTALUM OHMIC CONTACTS TO DIAMOND BY A SOLID STATE REACTION PROCESS

K.L. Moazed*, J.R. Zeidler**, M.J. Taylor**, and C.A. Hewett**

* North Carolina State University, Raleigh, NC 27695-7916
** Naval Ocean Systems Center, San Diego, CA 92152-5000

INTRODUCTION

The recognized need for devices that can perform under severe operating conditions and recent advances in the growth of diamond films at moderate temperatures and low pressures has generated great interest in the exploitation of the many unique properties of diamond for solid state devices. Among these properties are high strength, high thermal conductivity, high band gap energy, high carrier mobility, and high resistance to degradation by exposure to radiation, heat, and corrosives [1,2].

Although ohmic contacts to natural and synthetic diamond had been produced for many years, the techniques employed, while satisfactory for research purposes, were not deeemed suitable for large scale manufacturing. These techniques can be classified into four main categories, i.e. silver paint or tungsten probes on roughened or implanted surfaces [3-9],liquid phase reaction such as brazing [9-11], graphatization [11-13], and solid state reactions [1, 14-16]. Other techniques such as ion mixing have also been utilized [17].

In order to ensure that contributions of secondary factors, for example defects that might form during sputtering, etc. were minimized, the most benign technique for metallization to produce ohmic contacts, namely evaporation, was used in this study. Inasmuch as the techniques that were utilized and the results that were obtained in this study closely parallel those reported elsewhere [1] with molybdenum, molybdenum/gold, and molybdenum/nickel/gold; and in an earlier study with tantalum [14]; only a brief description of the experimental procedures will be provided below.

The diamond samples used in this study were naturally boron doped (p-type) semiconducting grade (Type IIb) with boron concentration of 10E15 to 10E16 per CC, carrier concentration of 10E13 to 10E14 per CC, and the dimensions of 5 mm by 5 mm by 1/4 mm. The samples were cleaned to remove organic surface contaminants and selectively metallized subsequent to masking the surface using slightly modified photolithographic techniques. The modification was required to accomodate exposure to high temperatures that were necessary to obtain

Diamond and Diamond-Like Films and Coatings
Edited by R.E. Clausing *et al.*, Plenum Press, New York, 1991

adherent ohmic contacts to diamond [1]. The metallized regions were a series of rectangles 0.12 mm by 0.3 mm. The spacings between adjacent rectangular pads were 5, 10, 20, 30, and 50 micrometers in extent. Ten nm thick tantalum films were deposited onto the partially masked diamond surface using an electron beam evaporation source in an ion-pumped ultra-high vacuum system. A 150 nm thick gold film was evaporated on the tantalum surface to improve its resistance to corrosion. The samples were annealed at 950°C in a tube furnace in an atmosphere of flowing purified hydrogen gas. Room temperature resistance of adjacent metallized pads were measured prior to annealing and as a function of annealing time. The diamond/metal interfaces were characterized using Auger electron spectroscopy (AES), metallography, scanning electron microscopy (SEM), and X-ray diffraction (XRD).

RESULTS AND DISCUSSION

The room temperature resistance of adjacent metallized pads for each of the five specified pad spacings as a function of annealing time at 950°C is shown in Figure 1. This figure shows that the room temperature resistance between adjacent metallized pads decreases sharply, reaches a minimum value at approximately 1/2 hour, and increases slightly with increasing annealing time at 950°C. This behavior closely parallels the results that were obtained for metallized molybdenum, molybdenum/gold, and molybdenum/nickel/gold contacts to diamond and reported earlier [1], except that the minimum in room temperature resistance occurred at approximately six to eight minutes for diamond/molybdenum. This type of behavior is indicative of a chemical reaction and in this instance clearly an interfacial chemical reaction between diamond and tantalum is the most likely event. Consequently, the interfacial region was examined and characterized to determine the reaction product and the reaction product morphology. The techniques that were used for characterization included Auger electron spectroscopy, metallography, scanning electron microscopy, and X-ray diffraction. Figure 2 is a scanning electron micrograph of the diamond/tantalum interface after the metallized and annealed surface was etched to remove the metal deposit. The micrograph shows a large number of bright spots and dark spots, however, the dark spots are less noticeable in the reproduction of the micrograph than the bright spots. These spots are images of protrusions and of intrusions, respectively, relative to the planar surface topography. Similar results were also observed for a diamond surface metallized with molybdenum after annealing and etching. The major difference between the diamond surface metallized with molybdenum compared to the diamond surface metallized with tantalum is the relative size of the individual spots which appear to be larger for diamond/molybdenum contacts [1]. X-ray diffraction of the sample, prior to etching, as a means of identifying the interfacial reaction products did not produce totally satisfactory results because of the extreme thinness of the film. However, Auger electron spectroscopy in conjunction with depth profiling from the vacuum/metal interface to the metal/diamond interface gave positive indication that the reaction product was a carbide of tantalum, as shown in Figure 3.

The Auger electron spectra has been used extensively and quite effectively to distinguish between graphitic carbon, diamond, and bound carbon in carbides [1,18,19]. As may be observed in Figure 3, the carbon spectra in the vicinity of the tantalum/diamond interface is indicative of carbon in the combined state, i.e. as tantalum carbide.

Based on the results of the characterization studies of the interface, it may be postulated that tantalum reacts with the diamond at

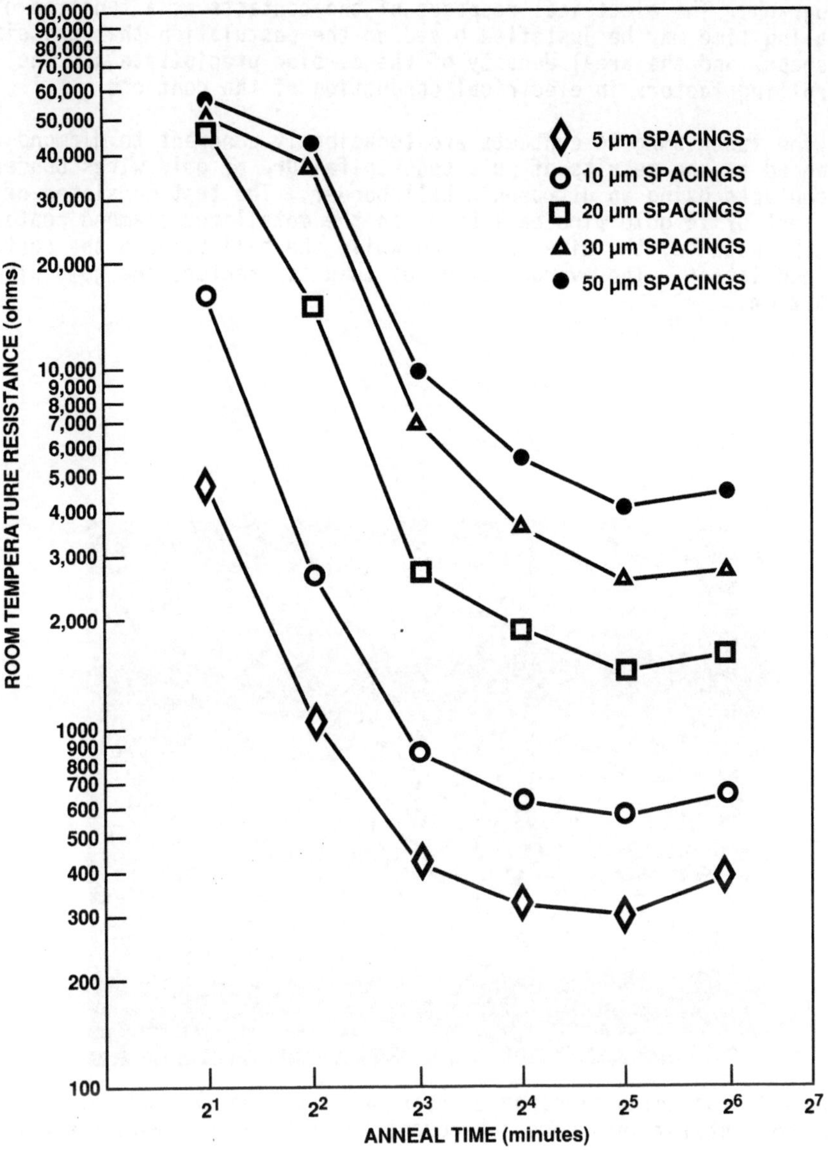

Figure 1. Room temperature resistance of metallized tantalum pads to diamond as a function of annealing time at 950°C.

the diamond/metal interface. This reaction produces tantalum carbide precipitates by a nucleation and growth kinetic process. Further, the carbide precipitates grow in both the lateral and normal directions to the interface. Based on the analysis of the SEM micrograph of the interface, it may be concluded that the carbide precipitates grow into the diamond lattice, hence the source of the intrusions observed on the micrograph. Further, the precipitates grow into the metal film and the projection of these precipitates are observed as protrusions on the micrograph. The electrical response of the contacts as a function of annealing time may be justified based on the postulation that the size, the shape, and the areal density of the carbide precipitate are the controlling factors in electrical conduction of the contacts.

The tantalum/gold contacts are tenaciously adherent to diamond as evidenced by the results of pull-test-to-failure of gold wires bonded to the contacts using an ultrasonic ball bonder. The test consisted of a pull test of 14 gold wire ball bonds to the metallized diamond contacts. In each instance, the wire fractured while the ball bond to the contact remained intact. The average value of load to fracture the gold wire was 4.2 gm.

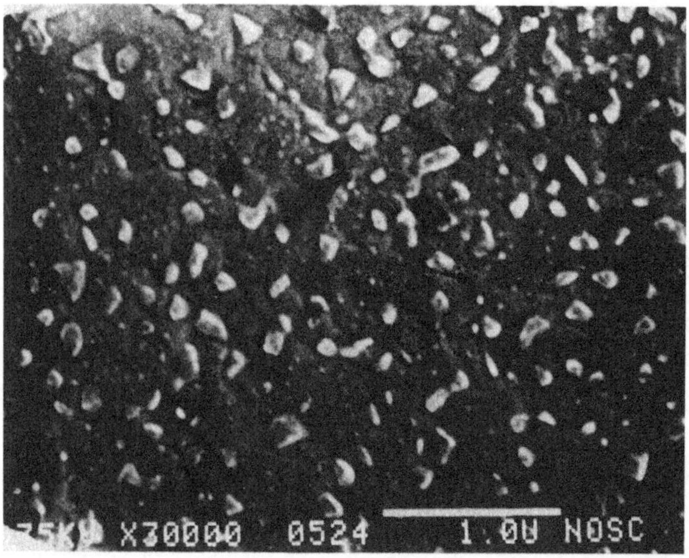

Figure 2. Scanning electron micrograph of the metallized and annealed diamond/tantalum interface subsequent to etching to remove the metal deposit.

CONCLUSIONS

Tantalum films (10 nm thick) evaporated onto diamond single crystal surfaces were shown to produce ohmic contacts subsequent to annealing at 950°C. An evaporated gold film (150 nm in thickness) on the surface of

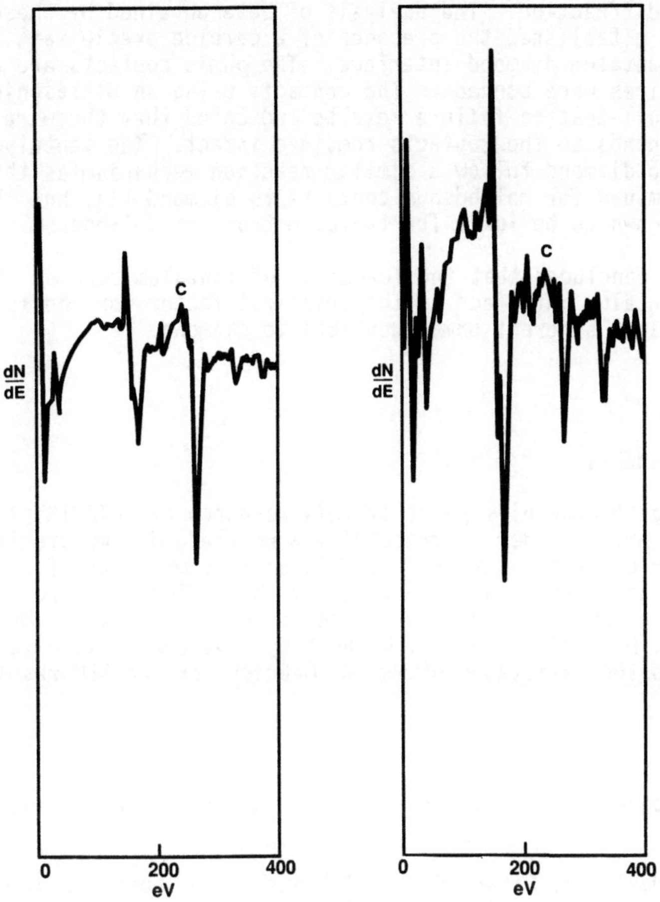

Figure 3. Auger electron spectra of a diamond/tantalum/gold sample after annealing and etching to remove the gold overlayer showing the vacuum/tantalum interface spectra (on the left) and the tantalum/diamond interface spectra (on the right). Note the marked difference in the carbon spectra in the 230-280 eV region.

the tantalum film decreases its susceptibility to corrosion. The annealed contacts have ohmic response and their room temperature electrical resistance decreases sharply with annealing time at 950°C. A minimum value for the room temperature electrical resistance was obtained for contacts annealed at 950°C for approximately 30 minutes. There was a slight increase in the electrical resistance as the time of annealing was increased further. The decrease in the resistance between pairs of contact pads after annealing was typically greater than ten orders of magnitude. The annealed contacts were characterized using Auger electron spectroscopy, metallography, scanning electron microscopy and x-ray diffraction. The analysis of data obtained by these techniques established the presence of a carbide precipitate at the original tantalum diamond interface. The ohmic contacts are adherent and gold wires were bonded to the contacts using an ultrasonic ball bonder. Pull-test-to-failure results indicated that the wires fractured while the bonds to the contacts remained intact. The tantalum ohmic contacts to diamond follow a similar reaction mechanism as those that were determined for molybdenum contacts to diamond [1], but the reaction rate was shown to be lower for tantalum than for molybdenum.

It is concluded that the formation of tantalum carbide at the diamond/tantalum interface is the principal factor for the formation of low resistance adherent ohmic contacts to diamond.

ACKNOWLEDGEMENTS

Partial financial support of this research by SDIO/IST through ONR managed by Mr. M. Yoder is gratefully acknowledged. We are indebted to many of our colleagues at the Naval Ocean Systems Center for their invaluable assistance in the pursuit of this effort. Among these colleagues a special note of gratitude is due Dr. H. Rast, Dr. C. Zeisse, Dr. D. Mullin, and Mr. K. Bentley. We are pleased to acknowledge the assistance of Ms. M. O'Brien for the SEM results.

REFERENCES

1. K.L. Moazed, J.R. Zeidler, and M.J. Taylor, J. Appl. Phys. 68, 2246 (1990)

2. R. F. Davis, Z. Sitar, B. E. Williams, H. S. Kong, H. J. Kim, J. W. Palmour, J. A. Edmond, J. Ryu, J. T. Glass and C. H. Carter, Jr., Mat. Sci. Eng., B1, 77 (1988)

3. H. Shiomi, H. Nakahata, T. Imai, Y. Nishbayashi and N. Fujimori, Jap. J. Appl. Phys., 28, 758 (1989)

4. P. T. Wedepohl, Proc. Phys. Soc., B70, 177 (1957)

5. I. G. Austin and R. Wolfe, Proc. Phys. Soc., B69, 329 (1956)

6. V. S. Vavilov, E. A. Konorova, E. G. Stepanova, and E. M. Trukhan, Sov. Phys. Semicond. 13(b), 635 (1979)

7. J. Prins, J. Phys. D, 22, 1562 (1989)

8. G. S. Sandhu, M. L. Swanson, and W. K. Chu, Appl. Phys. Lett 55, 2 October, 1397 (1989)

9. M. Seal, Industrial Diamond Review, 29, 408 (1969)

10. G. B. Rogers and F. A. Raal, Rev. Sci. Inst., 31, 663 (1960)

11. A. T. Collins, E. C. Lightowlers and A.W.S. Williams, Diamond Research, 19, 19 (1970)

12. E. A. Burgemeister, Phys. Med. Biol 26, 269 (1981)

13. A. T. Collins, E. C. Lightowlers and P. J. Dean, Phys. Rev. 183, 725 (1969)

14. K. L. Moazed, R. Nguyen and J. R. Zeidler, IEEE Elect. Dev. Lett., 9, 350 (1988)

15. G. Gildenblat, S. Grot, C. Wronski, A. Badzian, T. Badzian and R. Messier, Appl. Phys. Lett., 53, 586 (1988)

16. M. W. Geis, D. D. Rathsam, D. J. Erlich, R. A. Murphy, and W. T. Lindley, IEEE Electron Device Letters, EDL-8, 341 (1987)

17. F. Fang, C. A. Hewett, M. G. Fernandes and S. S. Lau, IEEE Trans, Elect. Dev., 36, 1783 (1989)

18. G. L. Gutseve, Y. M. Shulga, and Y. G. Borodko, Phys, Status Solidi B 121, 595 (1984)

19. Y. M. Shulga, G. L. Gustev and V. I. Rubstov, Phys. Status Solidi B 129, 683 (1985)

SOME PROPERTIES OF CVD-DIAMOND SEMICONDUCTING STRUCTURES

A.E. Alexenko and B.V. Spitsyn

Institute of Physical Chemistry of the Academy of Sciences of the USSR, Moscow, USSR

Semiconducting diamond structures - models of solid state electronic devices - were made by means of chemical crystallization of diamond films from the gaseous phase[1]. Doping by boron and phosphorus during growth made it possible to obtain homoepitaxial diamond films of p-and n-type conductivity[2]. The simplest device of this type is a thermistor. There are diamond thermistors, made of synthetic crystals doped by boron, obtained by the method of superhigh pressures. Diamond thermistors appear to be a good starting device owing to their ability to operate over a wide temperature range, high thermal stability and low inertia due to high thermal conductivity.

Our thermistor is designed for measuring the true temperature of a diamond crystal face grown by the method of the high-temperature high-gradient chemical transport reaction[3]. The measurement of surface

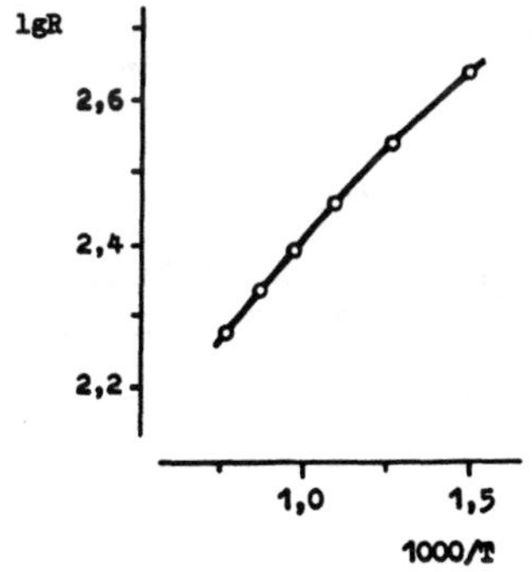

Fig. 1. The dependence of the resistance of the diamond thermistor on temperature.

temperature under growth conditions in a gaseous media with a temperature of 10^3-10^4 °C is of great difficulty, because of the large heat flux (a few of kW/cm^2), incident on the surface. The temperature of the latter may be hundreds of degrees higher than that of the substrate holder. The intensive light radiation produced by the source of carbon heated up to the temperatures above 2000°C, transparency and low radiation capacity of diamond hindered the use of optical pyrometry.

The thermistor is a 0.1 μm thick film of diamond moderately doped by boron in the form of a stripe, grown on the (111) face of a natural diamond crystal. To protect it from the gasifying effect of atomic hydrogen a 0.3 μm thick film of dielectric diamond was grown on it. This thermistor was used to study the temperature dependence of the growth rate of the diamond film in the method of chemical transport over the range, 600-1100°C [3].

The thermistor's resistance in this temperature range is shown in Fig. 1. The temperature coefficient of resistance in the middle of the above range was about -0.2 Ohm/°C or -0.1 %/°C

The next model had the MDS-structure (metal-dielectric-semiconductor). The use of diamond with differing specific resistance of 18 orders depending on the concentration of the dopant (boron) to make dielectric and semiconductor (and, possibly, "metal") layers in this structure would avoid the interface and, consequently, surface states, which usually degrade device characteristics.

To form the MDS-structure a film doped with boron was grown on a natural diamond crystal. Two or three films of dielectric diamond were then grown on it, which decreased the probability of the formation of reach-through microholes in the dielectric layer. Then the aquadag or gallium contact with an area of 1-2 mm^2 was deposited on the surface of the dielectric film (Fig. 2).

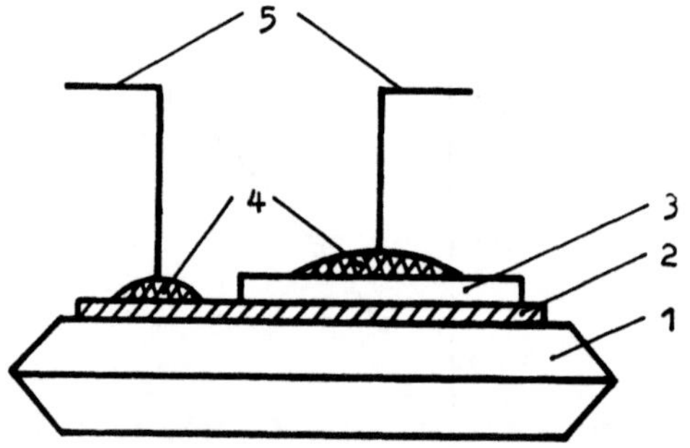

Fig. 2. General view of the diamond MDS-structure. (1) a natural diamond crystal; (2) a diamond film doped with boron; (3) a dielectric diamond film; (4) aquadag or gallium contacts; (5) wires.

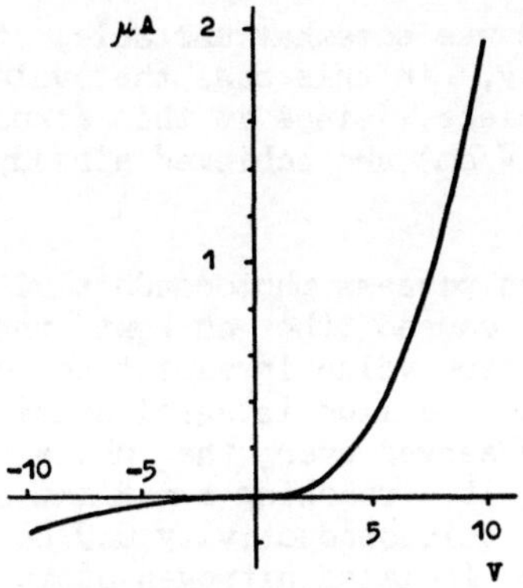

Fig. 3. I-V characteristics of the MDS-structure with a 1.4 μm thick dielectric film.

The structure with a 1.4 μm thick dielectric film had increased conductivity, when the current was passed from the p-film to the metal; the asymmetry coefficient was equal to 10 (Fig. 3). Both forward and reverse currents depend on temperature over the range 293-530 K following the Arrhenius law with the activation energy of 1.4 eV.

The MDS-structure with a thinner dielectric film (0.2 μm) exhibited increased conductivity with the reverse current (Fig. 4), moreover the current through the structure increased sharply at a certain specific

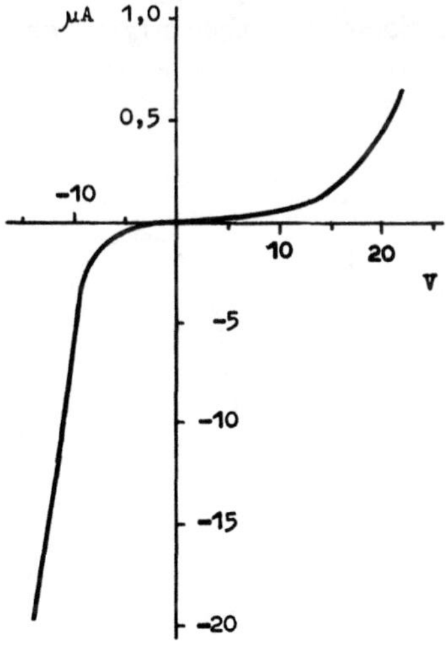

Fig. 4. I-V characteristics of the MDS-structure with a 0.2 μm thick dielectric film.

threshold voltage and was somewhat unstable; the asymmetry coefficient attained 150. Probably, in this case the avalanche breakdown of the dielectric film took place, since in this structure the critical values of the field (10^5-10^6 V/cm) are achieved already at a voltage of several volts.

The MDS-structures possess photoconductivity over a wide region of the spectrum. Light caused the current through the structure to increase by about the same value irrespective of the polarity of the external voltage. The maximum integral sensitivity was about 1 mA/lm. Photoconductivity is observed over the whole visible region of the spectrum and beyond it, reaching a maximum at 560 nm (Fig. 5). The levels responsible for this conductivity may be produced by dislocations or nitrogen impurities. Isolated nitrogen atoms in diamond are known to be donors with an ionization energy of 1.7-2 eV, and their aggregates are donors with an energy of about 4 eV (the absorption edge about 310 nm). If photoconductivity of the above MDS-structures is thought to be due to the nitrogen, the shape of the curve in Fig. 5 leads us to assume that the main amount of nitrogen in the synthesized layers is present in the form of isolated atoms - the photocurrent declines sharply in the ultraviolet region.

The possibility of synthesis from the gaseous phase of diamond with electron conductivity allowed the making of p-n diodes. Diodes had been made in diamond using ion implantation and doping in the process of synthesis at superhigh pressures. In the first case p-n diodes, with a small area of about $2 \cdot 10^{-4}$ mm^2, in type IIa natural diamond (the transition plane is perpendicular to the doped layers) were obtained by the implantation of boron and phosphorus[5] or boron and antimony[6]. Currents through the transition were equal to 10^{-9}-10^{-8} A, the asymmetry coefficient was 10-20.

The best results were obtained, when the p-n diode was formed by growing semiconducting diamond on a seed crystal of the opposite type of conductivity using the method of superhigh pressures[7,8]. Boron and arsenic were used as dopants. As a result diodes capable of conducting forward currents of 10^{-4}-10^{-3} A with the asymmetry coefficient 10^4-10^7 were obtained. However, the high resistance of diamond with

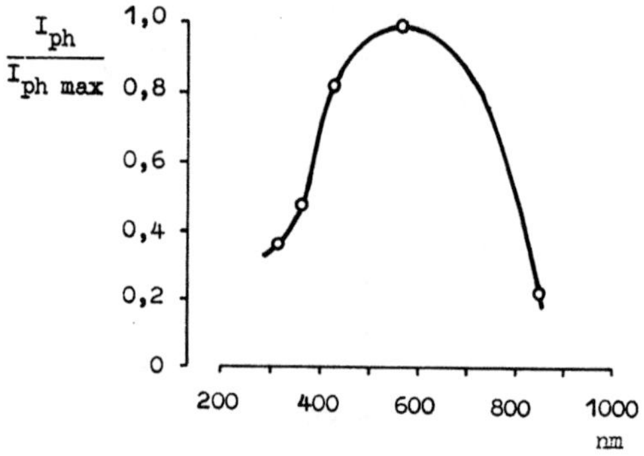

Fig. 5. The spectral dependence of the photocurrent (in relative units) of the MDS-structure with a 1.4 µm thick dielectric film.

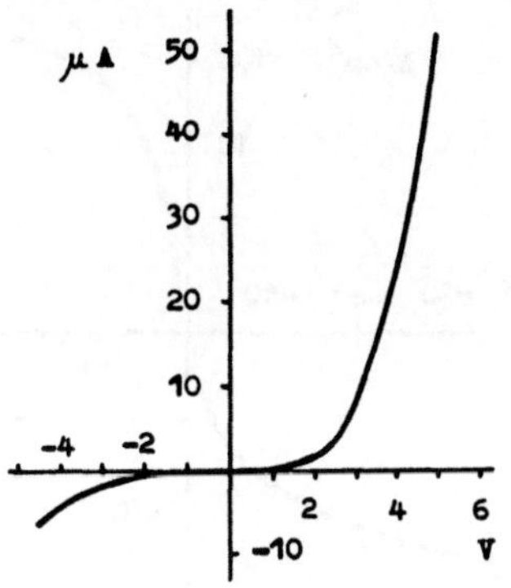

Fig. 6. I-V characteristics of the p-n diode between diamond films doped by boron and phosphorus.

electron conductivity (10^4-10^3 Ohm•cm) worsened the characteristics of the diode, increasing the degraded voltage in the forward direction up to tens of volts.

The p-n diodes made by us are two crossing stripes of different types of conductivity (doped by boron and phosphorus), grown on the (111) face of a natural diamond crystal. The specific resistance of the p-film was about 13, and that of the n-film 20 Ohm•cm. The diode area was equal to about 2 mm^2. The current through the diode attained $5 \cdot 10^{-5}$ A at a forward voltage of several volts (Fig. 6). However, the asymmetry coefficient was low, no more than 10. The I-V characteristics of this diode is similar to the I-V characteristics of the p-n diode obtained using ion doping[5]. Possible reasons for the increased reverse currents through the p-n diodes are the shunting of the diode by damaged regions formed during ionic bombardment[5], and the local breakdown of the diode due to the inhomogeneity of the doping level. In our case the shunting of the p-n diode is possible due to the parasitic diamond (or nondiamond carbon) growth or etching (gasification) of the aquadag mask used during the growth of films. A local breakdown of the diode (for example, near the boundaries of the layers, where the growth conditions may differ from those in other regions) can not be excluded.

Since the films of semiconducting p-type diamond of p-type are synthesized with less difficulty and have a more perfect structure than n-type films and since the combination of different semiconductors in one structure often allows interesting results to be obtained and widens the possibility of their application, we made an attempt to form a heterostructure on the basis of the p-type diamond and another wide-gap semiconductor - n-type alpha-silicon carbide.

To make a heterostructure, a flat crystal of alpha-silicon carbide of the 6H polytype doped with nitrogen to the concentration 10^{18} cm^{-3} and having the specific resistance about 0.1 Ohm•cm was used. The seed - a 30 µm diamond microcrystal with its (111) face turned to the substrate - was placed on the (111) "carbon" face of this crystal. This seed was grown from the gaseous phase during doping by boron to the size of about 200 µm. The crystal obtained was light-blue, which, with its size taken

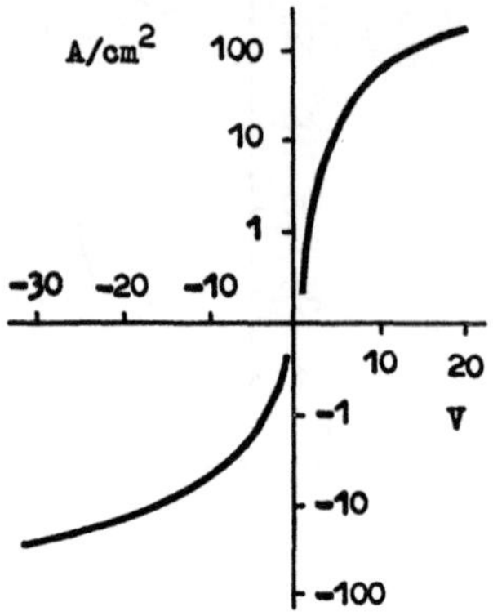

Fig. 7. I-V characteristics of the p-n diode between silicon carbide doped by nitrogen and diamond doped by boron.

into account, points to a relatively low doping level. The specific resistance of the grown diamond was about 300 Ohm•cm; it might have been that high due to the compensation of boron by nitrogen released during gasification of the silicon carbide substrate.

To study the electric properties of the structure metal contacts were used. The structure endures rather high current density; the asymmetry coefficient is equal to about 10 (Fig. 7). The fact, that much higher currents may be passed through the heterodiode than through diamond p-n homodiodes obtained by different methods at the same voltages, is, probably, due to the lower value of the potential barrier of the heterodiode, at least, for one type of carriers.

At a reverse voltage of 30 V, blue luminescence was observed. It should be noted, that there are effective diamond sources of light with the maximum in the blue-green region, made using boron and lithium ion implantation from two sides into a plate of natural type IIa diamond-dielectric (the p-i-n structures), in which double injection of charge carriers takes place[9]. However, the large thickness of the dielectric region (50-100 μm) requires relatively high voltages - about a hundred volts.

The authors want to thank A.D.Roenkov for the study of electrical properties of the heterotransition.

REFERENCES

1. B. V. Deriaguin, B. V. Spitsyn, L. L. Bouilov, A. E. Alexenko, and A. A. Botev, "Synthesis and some properties of films and crystals of gas-phase diamond", in: Sverkhtverdye materialy: sintez, svoistva, primenenie. Doklady mezhdunarodnogo seminara, N. V. Novikov, ed., Naukova Dumka, Kiev, 1983.
2. B. V. Spitsyn, and A. E. Alexenko, "Physico-chemical laws of diamond doping from the gas phase", Archivum Nauki o Materialakh, 7:201 (1986).

3. B. V. Spitsyn, L. L. Bouilov, and B. V. Deriaguin, "Diamond and diamond-like films: deposition from the vapour phase, structure and properties", Progr. in Cryst. Growth and Charact., 17:79 (1988).
4. L. L. Bouilov, A. E. Alexenko, A. A. Botev, and B. V. Spitsyn, "Some laws of the growth of diamond layers from the activated gas phase", Dokl. AN SSSR, 287:888 (1986).
5. V. S. Vavilov, M. A. Gukasyan M. A., M. I. Guseva et al., "Electron-hole transition in diamond, obtained by boron and phosphorus ion implantation", Dokl. AN SSSR, 200:821 (1971).
6. V. S. Vavilov, M. A. Gukasyan, E. A. Konorova, and Yu. V. Milyutin, "Ion implantation of stibium into diamond", Fizika i tekhnika poluprovodnikov, 6:2384 (1972).
7. Yu. M. Rotner, S. M. Rotner, V. A. Presnov et al. "The study of electric characteristics of p-n diodes in synthetic semiconducting diamonds", in: Sverkhtverdye materialy: sintez, svoistva, primenenie. Doklady mezhdunarodnogo seminara, N. V. Novikov, ed., Naukova Dumka, Kiev, (1983).
8. V. A. Laptev, V. P. Butuzov, Yu. M. Rotner et al., "The making and study of p-n diodes on the basis of synthetic semiconducting diamonds", Dokl. AN SSSR, 228:1080 (1976).
9. M. I. Guseva, E.A. Konorova, Yu. A. Kuznetsov, and V. F. Sergienko, "Double injection of a charge carrier in the p-i-n structure on the basis of ion-implanted diamond", Fizika i tekhnika poluprovodnikov, 12:505 (1978).

STATIC AND DYNAMIC ELECTRICAL CONDUCTIVITY OF POLYCRYSTALLINE DIAMOND FILMS

G. A. Sokolina, A. A. Botev, L. L. Bouilov,
S. V. Bantsekov, O. I. Lazareva, and A. F. Belyanin

Institute of Physical Chemistry of the Academy of Sciences of the USSR, Moscow, USSR

INTRODUCTION

At present, owing to the findings in the field of diamond synthesis from an activated gaseous phase, diamond polycrystalline films (DPF) can be obtained on foreign substrates on areas measured in square centimeters.[1,2] The high growth rates of DPF make it possible to obtain films and free diamond plates with thicknesses of tens and hundreds of microns. The total combination of valuable physical-chemical properties of DPF makes them applicable as new perspective material for creating various electronic devices.[3] This encourages the development and study of electrophysical properties of synthesized diamond films.

We studied the influence of temperature on the static, σ_{dc}, and dynamic, σ_{ac}, electrical conductivity of diamond films in the system: tungsten substrate-diamond film-aquadag electrode.

EXPERIMENTAL

DPF were obtained by the method of chemical crystallization from an electrically activated gaseous phase.[1] Tungsten substrates with an area of 0.5 cm^2 were used. The temperature of the substrates (T_c) was about 1000°C. The initial gas mixture consisted of hydrogen and methane, whose concentration was varied from 3 to 5 vol %, the pressure in the reactor was 0.1 to 0.2 atm. The thickness of the films was varied depending on the time of synthesis from a few to tens of microns.

The structure of the films was studied by the methods of fractography (study of fractures) and X-ray diffraction. Typical morphology of the growth surface of the film and the cleaved structure are shown in Fig. 1. The visible dimensions of the crystallites and the dimensions of the regions of coherent scattering (RCS) were determined. The size of the grains was determined by averaging the sizes of the grains measured using an electron microscope in different parts of the growth surface. The size of RCS in the direction perpendicular to the substrate surface was determined according to the broadening in the reflection diffractograms (111). The size of the grains was established to increase with increasing the thickness of the film, while the size of RCS practically did not depend on the thickness of the layer.

Diamond and Diamond-Like Films and Coatings
Edited by R.E. Clausing *et al.*, Plenum Press, New York, 1991

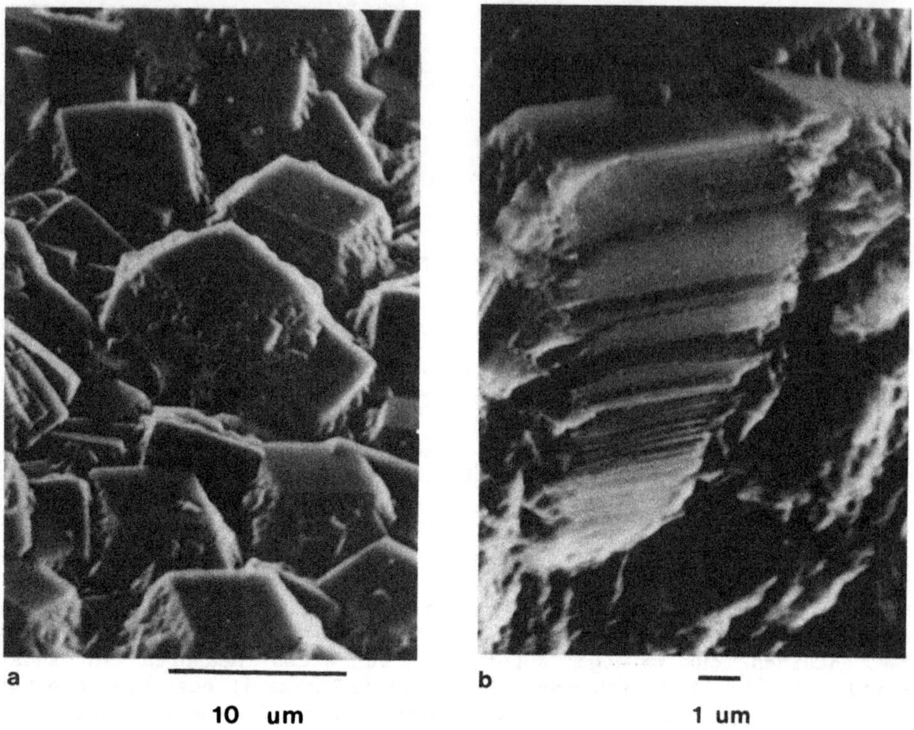

Fig. 1. (a) Morphology of the growth surface of the diamond film; (b) fracture of the film.

Static, σ_{dc}, and dynamic electrical conductivity, σ_{ac}, of DPF was measured in vacuum 10^{-4} Torr in the temperature range $150<T<800$ K. For all the samples under study the current-voltage characteristics (I-V) were recorded before measuring $\sigma(T)$. The $\sigma(T)$ dependencies were determined in the linear region of I-V.

It is noteworthy that the polarization processes are characteristic of natural diamond. They are also observed in high-ohmic DPF at low temperatures. In this case the σ_{dc} values were determined in terms of the stationary current values in the films.

In a series of the experiments on the measurements of $\sigma_{dc}(T)$ and $\sigma_{ac}(T)$ of the films, grown at a constant crystallization temperature T_c, but at a different content of methane in the gas mixture. It has been established that upon changing the concentration of methane from 3 to 5% the character of the dependencies of $\sigma_{dc}(T)$ and $\sigma_{ac}(T)$ change very little.

Figure 2 shows the typical dependencies of σ_{dc} vs 1/T, (curve 1) and σ_{ac} vs 1/T, (curves 2,3) for the films grown at the crystallization temperature, T_c, of 1050°C. As is seen in Fig. 2, in the studied temperature range of 150 to 800 K, there are three temperature regions: high (H), middle (M), and low (L), in each of these regions σ_{dc} is described by the exponent with its own activation energy value E:

$$\sigma_{dc}(T) = \sigma_{H}\exp(-E_H/kT) + \sigma_{M}\exp(-E_M/kT) + \sigma_{L}\exp(-E_L/kT),$$

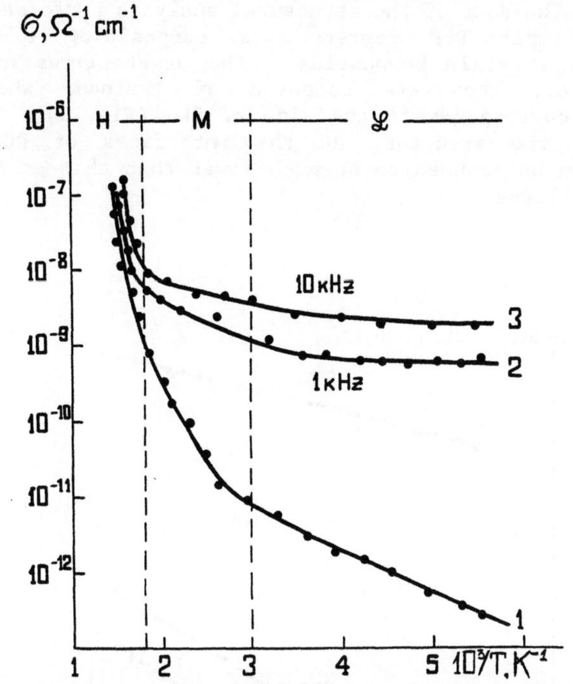

Fig. 2. Temperature dependence of the static (curve 1) and dynamic (curves 2 and 3) conductivity of the diamond film.

where k is Boltzmann's constant,

$$\sigma_H = 2.5\ 10^3\ \Omega^{-1}cm^{-1} \quad E_H = 1\pm0.1\ eV;$$

$$\sigma_M = 1\ 10^{-5}\ \Omega^{-1}cm^{-1} \quad E_M = 0.4\pm0.05\ eV;$$

$$\sigma_L = 4\ 10^{-10}\ \Omega^{-1}cm^{-1} \quad E_L = 0.1\pm0.05\ eV.$$

For the dependencies $\sigma_{ac}(T)$ measured at the frequencies 1 and 10 kHz in the low temperature region (T<250 K) the conductivity is weakly dependent on T: $\sigma_{ac}(T)$. At T>250 K the $\sigma_{ac}(T)$ dependence becomes nonlinear, and in the temperature range 300<T<600 K it can be described by the exponent with $E_m = 0.1$ eV for f = 10 kHz and $E_m = 0.15$ eV for f = 1 kHz. These E values differ markedly from E_m value 0.4 eV, measured in this temperature region for σ_{dc}; at high temperature they are about equal E = E_H.

The influence of f on σ_{ac} is illustrated by the plot σ_{ac} versus f, presented in Fig. 3 for the temperature of measurements: 150, 295, and 650 K. According to these data the law of the change of $\sigma_{ac}(f)$ can be written as $\sigma_{a\alpha}\ f^s$, where the power index s = 0.65 for T = 150 K and it decreases with increasing T.

Based on the data of the structural analysis performed by us it can be concluded that DPF represent a composition of crystallites interconnected at grain boundaries. The crystallites have a laminar structure, i.e., they are composed of laminas, whose thickness approximately corresponds to the size of RCS (Fig. 1). The degree of randomness of the structure at the interfaces of RCS within one crystallite can be assumed to be much lower than that at the boundaries between crystallites.

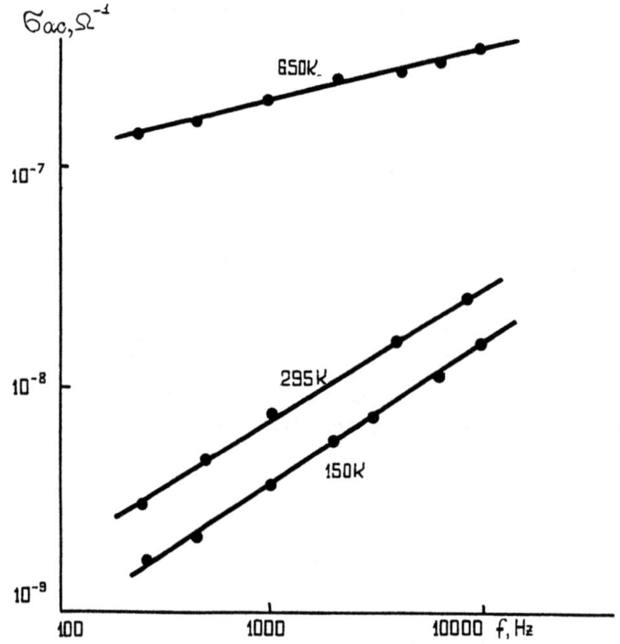

Fig. 3. Frequency dependence of the conductivity of the diamond film at different temperatures.

Though DPF consist of crystallites and RCS of various forms and sizes, in the first approximation one can suggest that the film is characterized by the averaged size of the crystallite, D, and by the size of RCS, d. At the boundaries between crystallites and between RCS, where the periodicity of the crystal structure is violated, and where the accumulation of impurities may take place, the strictly periodical potential of the lattice will be violated. This, in turn leads to the appearance of localized electron states in the forbidden band of diamond. The occupation of these states by current carriers causes the charge to appear at the interfaces of crystallites and RCS and promotes the appearance of energy barriers with the height W_D and W_d at the interfaces of crystallites and RCS, respectively.

Figure 4 shows the energy band structure of a diamond polycrystal where the bending of the bands at the interfaces of crystallites and RCS determines the height of the energy barriers for the transfer of current carriers through the film. This scheme refers to the particular case, when the localized states E_{td} and E_{tD} at the interfaces of crystallites and RCS, respectively, are negatively charged, and there appear barriers in the way of electrons. The positive charging of the interfaces will naturally result in the appearance of the barriers for holes.

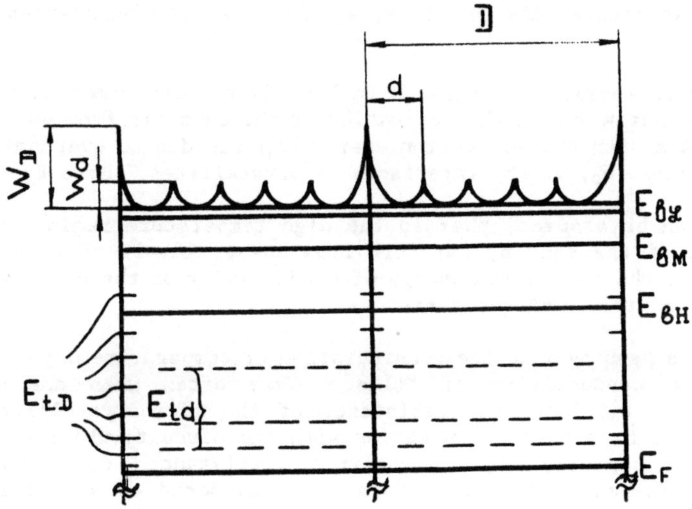

Fig. 4. The energy band structure of the polycrystalline diamond film.

D,d - the sizes of the crystallites and RCS, respectively;
E_F - the Fermi level;
E_{bL}, E_{bM}, E_{bH} - impurity levels;
E_{tD}, E_{tD} - trapping states at the interfaces of crystallites and RCS, respectively;
W_D, W_d - barriers at the interfaces of crystallites and RCS, respectively.

The transport of current carriers in such a polycrystalline film will be determined by the bulk conductivity of the synthesized diamond, σ_b, and will depend on the height of the barriers, W. The σ_b value in the diamond with a forbidden band (5.4 eV) depends on the presence of impurities and defects in it, and is determined by the concentration and energy position of the impurity electron states. The barrier heights at the interfaces of RCS, W_d, and crystallites, W_D, depend on the occupation by current carriers of traps located at the interfaces and on the concentration of current carriers in diamond.[4]

Based on this scheme, the typical for polycrystalline films dependencies, $\sigma_{dc}(T)$, which are presented in Fig. 2, can be explained in the following way. At low temperatures (region L in Fig. 2) σ_b is determined either by the band conductivity of the current carriers activated from lower levels or by hopping conductivity on the states in the forbidden band. The activation energy value of conductivity E_L in this temperature region will be determined by the activation energy value of bulk conductivity at low temperatures, E_{bL}, and by the height of the barriers, W, in the way of the carriers through the film. It can be suggested, that due to a higher degree of randomness and higher concentration of defects and impurities at the interfaces of crystallites compared to those of RCS the inequality $W_D > W_d$ is valid. In this case at low temperatures the transport of carriers will occur mainly through the barriers, W_d, connected in series. In particular, such a charge transfer may take place through a small number of large crystallites, which have grown all through the thickness of the films, for which $E_L = E_{bL} + W_d$. In the middle of Fig. 1b one can see one of such crystallites having a laminated structure; the barriers, W_d, arise at the boundaries of these laminas.

Upon increasing T (region M in Fig. 2) the experimentally measured σ value increases due to the activation of the carriers from deeper levels (σ_b increases with the activation energy E_{bM}) and due to overcoming of the higher barrier, W_D, at the interfaces of crystallites. Here $E_M = E_{bM} + W_D$.

It can be assumed, that in the high temperature region, where the filling up of the band by carriers from deeper levels (for which $E_{bH} > W$) takes place, the activation energy of conductivity of the polycrystal, E_H, will be close to E_{bH} of the grain.

Let us pass over to the consideration of temperature dependencies of the dynamic conductivity of films. Very often when analyzing the dependencies, $\sigma_{ac}(f,T)$, the resistance of the microcrystallite, R_b, is considered to be connected in series with the circuit, which consists of the barrier capacity, C_w, and barrier resistance, R_w, connected in parallel. Therefore the polycrystal is considered as a combination of such circuits. In this case the dynamic conductivity of the polycrystalline sample, σ_{ac}, will depend not only on T but also on f. As f increases, the contribution of the barrier resistance, R_w, to the total R value will decrease (the barriers are disconnected and at high frequencies the conductivity of the polycrystal becomes dependent only on R_b).

The $\sigma_{ac}(T)$ dependencies presented in Fig. 3 are consistent with these assumptions. In the region of low and middle temperatures σ_{ac} increases with increasing f. This means, that at these temperatures the intercrystalline barriers have a marked effect on the conductivity of the diamond polycrystal. The decrease in activation energy, E, for σ_{ac} as compared with σ_{dc} implies that the contribution of the barriers to the E value decreases with increasing f.

The weak effect of f on σ_{ac} in the high temperature region and the small difference in the activation energies, of E and E_H, in this temperature region means that the contribution of barriers to E_H is low. This may result from the activation energy of the carriers being high compared to the height of the barriers and/or from the decrease in W upon increasing the concentration of current carriers.[4]

Earlier[5] we have shown that the crystallization temperature, T_c, has a marked effect on the experimentally obtained E in the middle temperature region. Based on the above model we can conclude, that when changing T_c the energy position of deep impurity levels in diamond does not change, the only changing parameter being the height of the energy barriers W_D. This can be explained by the fact that changing T_c results not in the appearance of new impurity levels but in the change of the degree of randomness at the interfaces of crystallites.

Because of the great difference in the values of electrical conductivity of diamond and graphite (ratio of 10^{10}), conductivity is a highly sensitive indicator of the stability of diamond films. The absence of any changes in the $\sigma(T)$ dependencies after heating the films up to 800°C is good evidence of their high thermal stability.

REFERENCES

1. L. L. Bouilov, A. E. Alexenko, A. A. Botev, and B. V. Spitsyn, "Some laws of the growth of diamond layers from activated vapor phase," Dokl. AN SSSR287:888 (1986).
2. A. Sawabe, H. Yasuda, T. Inuzuka, and K. Suzuki, "Growth of diamond thin films in a DC discharge plasma," Appl. Surf. Sci. 33/34:539 (1988).
3. B. V. Spitsyn, "Crystal growth under thermodynamically metastable conditions," in Growth of Crystals, 13:58, E. I. Givargizov, ed., Plenum Publishing Corp., New York (1986).
4. John Y.W. Seto, "The electrical properties of polycrystalline silicon films," J. Appl. Phys. 46-5247 (1975).
5. G. A. Sokolina, S. V. Banstekov, A. A. Botev, L. L. Bouilov, and B. V. Spitsyn, "Electrical conductivity of polycrystalline diamond films," Izvestiva AN SSSR, Neorganicheskie materialy 24:1215 (1988).

ELECTRICAL AND OPTICAL PROPERTIES OF DIAMOND FILMS DEPOSITED FROM AN OXY-ACETYLENE FLAME

Yonhua Tzeng, Chin K. Teh, Richard Phillips, Alvin Joseph, Teradol Srivinyunon, and Calvin Cutshaw

Plasma Processing Laboratory
Department of Electrical Engineering
Auburn University, Auburn, Alabama 36849

Chin C. Tin

Department of Physics
Auburn University, Auburn, Alabama 36849

Richard Miller, Tom H. Hartnett, and Chuck Willingham

Advanced Materials Department
Research Division
Raytheon Company
131 Spring St.
Lexington, MA 02173

Amin Ibrahim and B.H. Loo

Department of Chemistry
University of Alabama, Huntsville, Alabama 35809

INTRODUCTION

Diamond exhibits remarkable physical and chemical properties that have attracted attention from scientists and engineers for a long time. The invention of chemical vapor deposition (CVD) techniques for the deposition of diamond on foreign substrates at low pressures, i.e., atmospheric or sub-atmospheric pressures, and low temperatures, i.e., below 1000°C, has further stimulated a great deal of interest for researchers to optimize the deposition processes for practical application of the remarkable diamond thin films.

Among a number of diamond CVD techniques, diamond synthesis from combustion flames in open air not only provides one of the most convenient ways of depositing diamond on various substrates but also leads to high quality diamond films. Diamond films have been deposited from a variety of hydrocarbon/oxidizer flames on various substrates including silicon wafers[1-3].

Among many outstanding properties of diamond films, the

Diamond and Diamond-Like Films and Coatings
Edited by R.E. Clausing *et al.*, Plenum Press, New York, 1991

high thermal conductivity, high electrical resistivity, and excellent infrared transmittance are the most significant ones for the very promising commercial electronic and optical applications of diamond films. The thermal conductivity of plasma-enhanced chemical vapor deposited (PECVD) diamond has been measured with values approaching that of natural diamond crystals. High thermal conductivity and high electrical resistivity are the two major requirements for the application of diamond films as heat sinks for electronic devices.

Diamond films deposited from a hydrogen diluted methane plasma display low electrical resistivity around 10^6 Ωcm. The low resistivity of the as-grown diamond films can be improved by more than six orders of magnitude by a heat treatment. High electrical resistivity on the order of 10^{13} Ωcm has been measured for dehydrogenated diamond films. The proposed cause for the observed change in resistivity is either the movement of hydrogen within or out of the film. This was confirmed by the hydrogenation and dehydrogenation of natural diamond crystals[4] as well as CVD diamond films[5] which show that current conduction in both kinds of diamond can be strongly influenced by subjecting the crystals to atomic hydrogen generated in a hydrogen plasma[6]. This discovery provides an approach to the improvement of CVD diamond films for achieving high electrical resistivity at the cost of an additional heat-treating process.

In addition to the heat sink application, diamond films made by chemical vapor deposition (CVD) are currently being considered for use as an optical window material for use in the long wavelength infrared (LWIR) band. All current optical materials (ZnS, ZnSe, Ge) used in this wavelength range are soft, and have relatively low strength and thermal conductivity which limit their use in severe environments such as rain erosion and thermal shock[7]. High purity type IIa diamond has no intrinsic absorption[8] in this wavelength range from 8 μm to 12 μm and because diamond has superior thermal and mechanical properties it would be unsurpassed in thermal shock and erosion applications for LWIR windows. The expected wavelength band of use for CVD diamond windows is in the 8-12 μm region. Although diamond has no intrinsic absorption in this region, impurity and defect induced absorption can render the material useless as an optical window. The majority (98%) of all clear diamonds found in nature have large absorption coefficients in the 7 μm to 12 μm range because of nitrogen impurity induced absorption. Absorptions in this range are used to distinguish Type I and Type II diamond[9]. Absorption in diamond films in the "C-H stretch" region has been identified by several researchers[10,11] as $-C-H_2$ stretching. The methylene group is a doublet with twin peaks at 2926 cm^{-1} and 2853 cm^{-1} $\pm$ 10 cm^{-1} due to symmetric and antisymmetric stretching vibrations[12]. These absorptions due to bonded hydrogen in diamond have been found to be associated with increased absorption in the one-phonon region of diamond. The amounts of hydrogen necessary to cause significant absorption are less than 1% atomic [H] determined by nuclear reaction analysis[13].

In this paper, the electrical and optical properties of diamond films deposited from an oxy-acetylene torch are presented. Excellent electrical resistivity as high as 10^{14} Ω-cm with little optical absorption in the long wavelength infrared range have been measured for as-grown diamond films.

DIAMOND DEPOSITION

Diamond films 20-50 μm in thickness are deposited on molybdenum, titanium, aluminum, and silicon using either a stationary[3,14] or a scanning[15] oxy-acetylene torch. The substrates are water cooled so that substrate temperature can be controlled between 400°C and 1200°C for diamond deposition. Diamond films deposited on metals have poor adhesion to the substrates. Free-standing diamond films can be readily obtained after the substrates cool down from the deposition temperature. The deposition of diamond films on substrates with low melting points such as aluminum demonstrates the capability of low-temperature diamond deposition by the oxy-acetylene flame. Diamond films deposited on silicon wafers adhere to the substrates very well. Diamond films deposited in open air from an oxy-acetylene flame with $O_2:C_2H_2$ ration between 0.9 and 1 are used for the electrical and optical measurements.

ELECTRICAL RESISTIVITY

Silver dots are coated on both the oxy-acetylene flame deposited (10^{14} Ωcm) and methane/hydrogen microwave plasma deposited (10^6 Ωcm) diamond films for electrical characterization. Figure 1 shows the current-voltage characteristics of an as-grown free-standing diamond film at room temperature. The film is optically transparent in the visible light range. Three voltage ramps are applied to the silver contacts on both sides of the diamond film. When the voltage is scanned from 0 to 100 volts for the first time, the diamond film shows electrical resistivity about 10^{14} Ωcm at bias voltages up to about 40 volts.

When the applied voltage increases further, sharp current jumps are measured at about 68 volts and 82 volts as shown in Figure 1. This current jump phenomenon has been reported for dehydrogenated diamond films by Landstrass and Ravi[4,5], who attributed it to the trap-filling effect. After the first

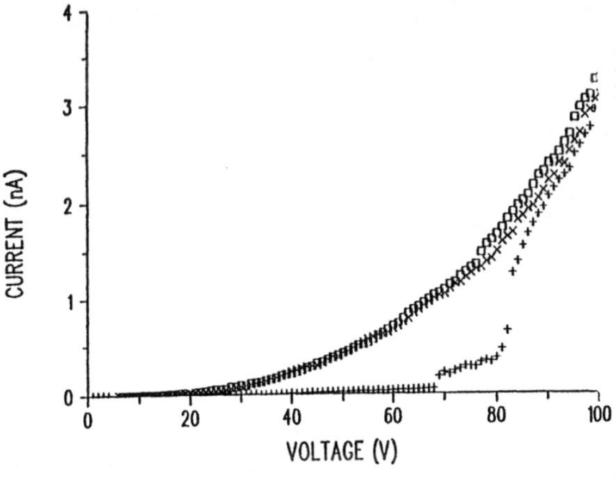

Figure 1. Current-voltage characteristics of a free-standing diamond film deposited by a stationary oxy-acetylene torch on a molybdenum substrate. The lower curve with clear current jumps is for the first voltage scan. The middle curve is for the second voltage scan. The upper curve is for the third voltage scan[14].

voltage ramp from 0 to 100 volts is completed, the second voltage ramp is applied starting from 0 volt again. The second scan does not cause current jumps but displays a higher conduction current for the same applied voltage than that of the first scan. The third voltage ramp causes only a slightly higher conduction current than the second one. The higher conduction currents for the second and the third voltage ramps are consistent with the interpretation based on trap-filling mechanism.

In order to assess the effect of hydrogen plasma on the electrical resistivity of oxy-acetylene deposited diamond films a diamond film is immersed in a hydrogen plasma powered by 50 watts 2.45 GHz microwave at 10 torr inside an Evenson cavity for one hour. The electrical resistivity of the diamond film decreases by three orders of magnitude[14]. According to calculation there exists a high partial pressure of atomic hydrogen in the oxy-acetylene flame. It indicates that the incorporation of atomic hydrogen into a diamond film deposited from an oxy-acetylene is different from that corresponding to hydrogen/methane plasma process.

RAMAN SPECTROSCOPY

Clear flame deposited diamond films display well-defined crystal characteristics as well as Raman spectra as shown in Figure 1a with a sharp peak near 1332 cm^{-1} due to diamond bonding. Figure 2b shows a Raman spectrum of typical black diamond films deposited either with a high C/O ratio or at too high or too low a substrate temperature. In addition to the smaller diamond peak at around 1332 cm^{-1}, the luminescence background and the intensity of the broad Raman feature at around 1536 cm^{-1} can be clearly detected.

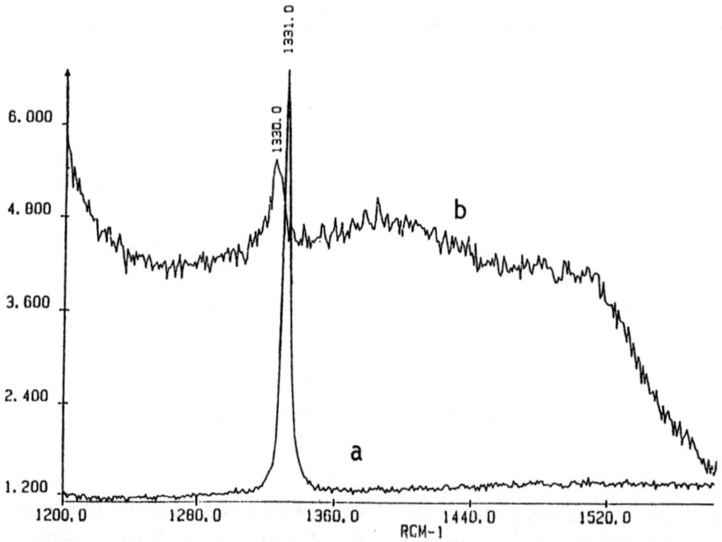

Figure 2. Raman spectra of diamond films deposited from an oxy-acetylene torch. (a) Clear and well-facetted diamond film deposited with O_2/C_2H_2 ratio close to 1. (b) A typical diamond film deposited with low O_2/C_2H_2 ratio.

FOURIER TRANSFORM INFRARED (FTIR) SPECTROSCOPY

Infrared spectra were collected on eight oxy-acetylene torch free-standing films deposited at varied substrate temperatures and C_2H_2/O_2 ratios. Two of these spectra were chosen as representative of a best case clear, highly facetted film and a worst case dark, more defective film. The results are plotted against each other as absorption coefficient versus wavenumber in Figure 3.

The spectra were collected in transmission on an FTIR spectrometer through a 4X microbeam sampling chamber (Bomem MB-120). The resolution was set at 4 cm^{-1} and in each case 20 scans were co-added. The absorption coefficient spectrum was obtained through conversion of the transmission data referencing each spectra with respect to the well known two phonon peak at 2000 wavenumbers with an absorption coefficient of 12 cm^{-1}.

In the best sample with the Raman spectrum shown in Figure 2a the "C-H stretch" band is less intense than the two-phonon intrinsic band and the associated single-phonon band is barely observable. In the dark sample with the Raman spectrum shown in Figure 2b the "C-H stretch" band is very strong with the 2850 cm^{-1} peak more intense than the 2915 cm^{-1} peak as has been observed in microwave plasma deposits with oxygen addition to the gas mixture. Also the single-phonon band is very strong which is only activated by a disturbance in the lattice symmetry. For diamond films deposited in the hydrogen/methane plasma without oxygen additive, the 2915 cm^{-1} peak is more intense than that of the 2850 cm^{-1} peak in the "C-H stretch"

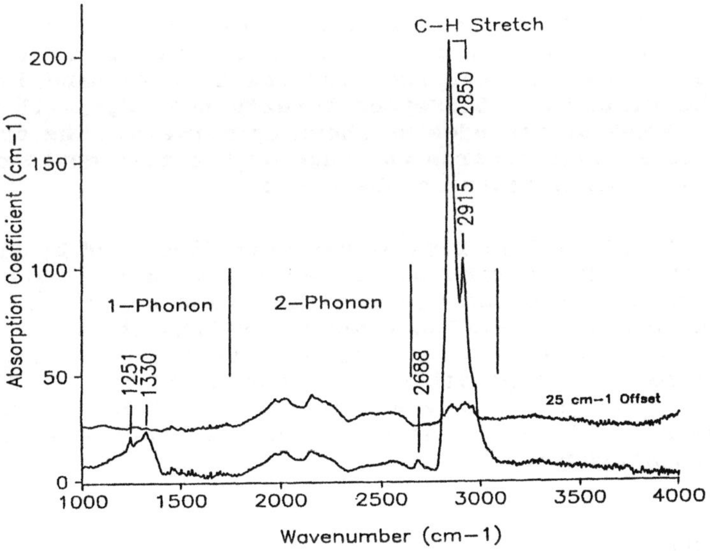

Figure 3. Comparison of the shape and intensity of the IR absorption of diamond films with their Raman spectra shown in Figure 2. The lower curve corresponds to that shown in Figure 2(b). The upper curve is offset by 25 cm^{-1} in absorption coefficient and corresponds to that shown in Figure 2(a).

band. From this data it can be concluded that good optical quality diamond films can be synthesized through the oxy-acetylene torch method, and the nature of the hydrogen incorporation is similar to that observed in films grown by Microwave Plasma with oxygen. The difference in the nature of hydrogen incorporation might be the reason for the much higher electrical resistivity of diamond films deposited from an oxy-acetylene flame than that from a hydrogen/methane plasma without oxygen.

PHOTOLUMINESCENCE SPECTROSCOPY

Photoluminescence (PL) measurements were carried out with the sample at 12°K using an argon ion laser tuned at 476.5 nm. Typical laser power used was 50 mW and the spot size of the focused beam was about 300 μm. The PL spectrum from the sample was dispersed by a 0.5 m Spex monochromator with a 1200 grooves/mm grating, and detected by a GaAs photomultiplier tube. The signal was processed using the lock-in amplifier and recorded by an IBM microcomputer. Measurements for each sample were made at 4 different spots, one of which was at the center and the other three at different radial distances from the center.

Figure 4 shows the spectra taken at different spots on the samples. It can be seen that the zero phonon line (ZPL) PL band at 1.68 eV band dominates at the center of the film (curves a) and decreases towards the edge. The intensities of the ZPL PL band at 2.16 eV and its phonon replicas increase towards the periphery as shown by curves a to c. Curve d, which was measured at the outer edges of the sample, shows the presence of the 1.95 eV band and a diminishing 2.16 eV band.

The full width at half maximum (FWHM) for the 1.68 eV band is about 3 meV and is independent of the radial distance from the center. However, the FWHM for the 2.16 eV band is about 7 meV at the center and increases towards the edge with value as large as 25 meV at the edge as shown by curve d. The broadening of the 2.16 eV band towards the edge of the film could be due to a strain-induced broadening phenomenon[16].

The ZPL PL band at 1.68 eV has been discussed by Freitas et al[17] and has been attributed to neutral vacancy. The 1.95 eV band has been attributed to a nitrogen-vacancy pair[18]. The band at 1.95 eV has also been observed by Freitas et al[17]. They also reported that it was more predominant at the outer edge of the film. There are two possible assignments for the ZPL PL band at 2.16 eV. One possible assignment is a divacancy[18]. Other researchers[20,23], however, have assigned it to a complex center related to nitrogen.

CONCLUSIONS

Diamond films with very high electrical resistivity and excellent infrared transmittance can be deposited from the oxy-acetylene flame. Oxygen not only lowers the minimum substrate temperature needed for obtaining high quality diamond films but also changes the nature of hydrogen incorporation, which

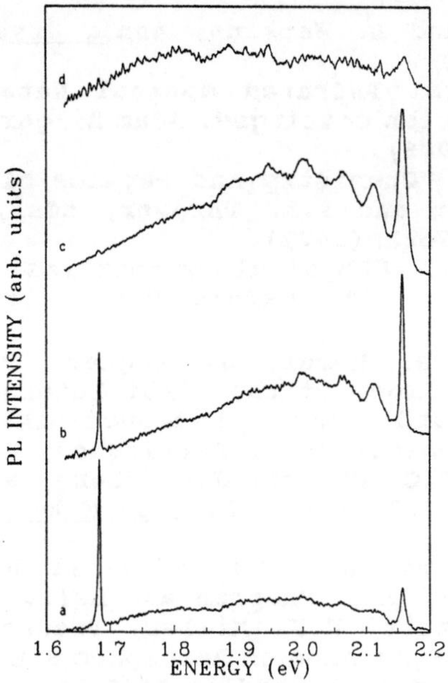

Figure 4. Photoluminescence spectra of a diamond film deposited from an oxy-acetylene flame. Curve a was taken at the center while curves b, c and d were taken at increasing radial distances, respectively, from the center of the film.

subsequently leads to a much higher electrical resistivity and less optical absorption in the long wavelength infrared range as compared to diamond films deposited from hydrogen/methane mixtures without oxygen.

ACKNOWLEDGMENT

The work done at Auburn University was supported in part by the Strategic Defense Initiative Organization's Office of Innovative Science and Technology (SDIO/TNI) through contract number N60921-86-C-A226 with the Naval Surface Warfare Center and in part by the Office of Naval Research through contract number N00014-90-J-1921. Richard Phillips is a Graduate Fellow supported by NASA-Lewis Research Center with Dr. Michael J. Mirtich, Jr. as the Center Advisor.

REFERENCES

1. Y. Hirose and N. Kondo, "Program and Book of Abstracts", 35th Japanese Applied Physics Spring Meeting, 434 (1988).
2. L.M. Hanssen, W.A. Carrington, J.E. Butler, and K.A. Snail, Matls. Lett. 7: 289 (1988).
3. Y. Tzeng, C. Cutshaw, R. Phillips, and T. Srivinyunon, Appl. Phys. Lett. 56(20): 134 (1990).
4. M.I. Landstrass and K.V. Ravi, Appl. Phys. Lett. 55(14): 1391 (1989).
5. M.I. Landstrass and K.V. Ravi, Appl. Phys. Lett. 55(10): 975 (1989).

6. A. Albin and L. Watkins, Appl. Phys. Lett. 56(15): 1454 (1990).
7. J.A. Savage "Infrared Optical Materials and their Antireflection Coatings", Adam Hilger Ltd, Bristol and Boston, (1985).
8. G. Davies, "Chemistry and Physics of Carbon Vol. 13" P.L. Walker and P.A. Thrower, eds., Marcel Dekker, Inc., New York (1977).
9. C.D. Clark, "Physical Properties of Diamond" R. Berman, ed., 295 Oxford University Press, London (1965).
10. C. Wild, N. Heres, J. Wagner, P. Koidl, and T. Anthony, "Proc. of the First International Symposium on Diamond and Diamonod-Like Films", The Electrochemical Soc., Pennington, NJ (1989).
11. X.X. Bi, P.C. Eklund, J.G. Zhang, and A.M. Rao, T.A. Perry and C.P. Beetz, Jr., J. Mater. Res. 5(40): 811 (1990).
12. L.J. Bellamy in, "The Infra-red Spectra of Complex Molecules", p14, Chapman and Hall, London (1975).
13. T.H. Hartnett, R.P. Miller, Presented at SPIE 1990 Technical Symposium on Aerospace Sensing, April 18-22, 1990. To be published in SPIE Proceedings Vol. 1307.
14. Y. Tzeng, C.C. Tin, R. Phillips, T. Srivinyunon, and Y. Chen, Appl. Phys. Lett. 57(8): 789 (1990).
15. Y. Tzeng, R. Phillips, C.C. Tin, T. Srivinyunon, and C. Cutshaw, "Proceedings of the Symposium on Diamond, Boron Nitride, Silicon Carbide and Related Wide Bandgap Semiconductors", MRS 1989 Fall Meeting, Boston, MA, Nov: 28, 1989.
16. T. Evans, S.T. Davey and S.H. Robertson, J. Mater. Sci. 19: 2405 (1984).
17. J.A. Freitas, Jr., J.E. Butler, S.G. Bishop, W.A. Carrington, and U. Strom in: "Diamond, Boron Nitride, Silicon Carbide and Related Wide Bandgap Semiconductors", J.T. Glass, R.F. Messier and N. Fujimori., eds., Mater. Res. Soc. 162, (1990) in press.
18. G. Davies and M.F. Hamer, Proc. R. Soc. Lond. A348: 285 (1976).
19. J. Walker, Rep. Prog. Phys. 42: 1605 (1979).
20. A.T. Collins and S.H. Robertson, J. Mater. Sci. Lett. 4: 681 (1985).
21. G. Davies, Rep. Prog. Phys. 44: 787 (1981).
22. A.T. Collins, M. Stanley, and G.S. Woods, J. Phys. D 20: 969 (1987).
23. L.H. Robins, L.P. Cook, E.N. Farabaugh, and A. Feldman, Phys. Rev. B 39: 13367 (1989).

BORONATED DIAMOND FILMS DEPOSITED BY RADIO FREQUENCY PLASMAS

T. B. KUSTKA*, R. O. DILLON*, AND T. FURTAK+

*ELECTRICAL ENGINEERING DEPARTMENT AND CENTER OF MICROELECTRONIC AND OPTICAL MATERIALS RESEARCH
UNIVERSITY OF NEBRASKA
LINCOLN, NE 68588

+PHYSICS DEPARTMENT
COLORADO SCHOOL OF MINES
GOLDEN, CO 80401

1. INTRODUCTION

P-type doping of diamond films is required to fabricate electronic devices. Several doping techniques have been used. Firstly, solid boron has been placed next to the substrate in microwave assisted chemical vapor deposition (CVD)[1-7]. Secondly, diborane gas has been used in reactant gases as a dopant. Thirdly, Okano *et al*[8] have dissolved boron oxide in methanol to produce a doping gas. Since the chemical reaction in this third method produces trimethyl borate (TMB), $(CH_3O)_3B$, we decided in this work to investigate the use of TMB directly as a dopant. This is important since TMB is much less toxic than diborane.

2. METHOD

A radio frequency (RF) was used to generate the plasma and heat the substrate. This system offers superior scale-up potential to other techniques, and is described in previous publications[9,10]. Hydrogen was bubbled through the liquid TMB and the known vapor pressure-temperature relation for the liquid allowed the volume fraction of TMB to be calculated. The substrates were 2.5 X 3.0 cm unpolished T-15 Homosil quartz plates from Heraeus Amersil that were seeded with sub-micron diamond powder.

Table 1 summarizes the deposition parameters for the preliminary runs and the three series of experiments. Series 2 spanned about the same range of TMB concentration as series 1, but had lower CO flow rates. In Series 3 the deposition area was moved downstream to reduce thermal decomposition of the TMB on quartz wool which was used to confine the plasma.

3. RESULTS

A. RAMAN SPECTROSCOPY

Exceeding a critical gas carbon concentration reduces the film quality as defined by appearance of Raman peaks at wavenumbers higher than the diamond line at about 1332 cm^{-1}. These peaks at higher wavenumbers are associated with graphitic forms of carbon. The preliminary runs 128 and 130 in Table 1 established that the critical carbon concentration for

Diamond and Diamond-Like Films and Coatings
Edited by R.E. Clausing *et al.*, Plenum Press, New York, 1991

TABLE 1
SUMMARY OF DEPOSITION PARAMETERS

RUN #	FLOW (SCCM) H2	CO	TMB	ATOMIC % CARBON	B/C (ppt)	POWER (Watts)	PRESSURE (Torr)
128	98.0	0	0.462	0.681	333	900	20.8
129	98.0	2.0	0.396	1.545	124	900	20.4
130	96.0	0	0.660	0.971	333	900	20.7
SERIES 1:							
131	97.5	2.5	0.000	1.250	0	900	20.6
134	98.0	2.0	0.132	1.185	55	900	20.5
135	98.0	1.5	0.264	1.128	115	900	20.2
136	98.0	1.0	0.396	1.071	181	900	20.7
SERIES 2:							
137	97.7	2.3	0.000	1.150	0	900	20.8
138	98.2	1.8	0.116	1.064	54	900	20.7
139	98.7	1.3	0.231	0.978	116	900	20.5
141	99.2	0.8	0.338	0.883	186	900	20.5
142	99.5	0.4	0.405	0.783	251	900	20.6
SERIES 3:							
148	97.7	2.3	0.000	1.150	0	1050	20.4
149	97.7	2.3	0.058	1.231	23	950	21.0
150	98.0	2.0	0.116	1.163	49	1000	20.6
151	98.5	1.5	0.173	0.996	86	1000	20.5

pure TMB was between 0.68 and 0.97%. We previously determined that the critical carbon concentration for pure CO was between 1.25 and 1.50%. The Raman results for run 129 indicated that the critical concentration increases when both TMB and CO are used.

Raman spectroscopy was done on the upstream (U), middle (M), and downstream (D) sections of most films. It indicated that the films were inhomogeneous with better film quality usually on the ends than in the middle. This is because the ends were 30-50 °C hotter than the middle which was usually below 900 °C. We did not observe a monotonic trend of increasing film quality with boronation as predicted by Bernholc *et al*[11] and observed by other workers[5,7]. This is probably because input power was the same for most runs and without independent substrate heating a 20-70 °C drop of substrate temperature occurred when the boron to carbon ratio increased. Following Shroder *et al*[12], curve fits and analysis of Raman intensities for film 142M yielded a diamond fraction of 98%.

B. MICROSCOPY

Optical and scanning electron microscopy showed that films with high quality Raman spectra contained many more (100) and (111) facets than others. In addition, typical grain sizes were 5 microns.

C. PROFILOMETRY

The films had typical dimensions of 1.5 X 3.0 cm. Although the films were rough, a Dektak mechanical profilometer was used to measure average thickness at the upstream, middle, and downstream parts of the films. The average extremal variation in thickness between the three sections was 27%. The middle was not always thinner than either end, so the variation was caused by factors other than substrate temperature. One of the factors may have been the geometry of the plasma with respect to the substrate.

Figure 1 shows, for series 1, the relationship between average deposition rate, boron to carbon ratio (B/C) in the reactant gas in parts per thousand (ppt), and atomic percent carbon in the gas. The maximum deposition rate occurred at a B/C ratio between 50 and 60 ppt for all three series and is attributed to boron reducing the activation energy for self diffusion in diamond films as in Okano *et al*[8]. The subsequent decrease in deposition rate with increasing B/C ratio

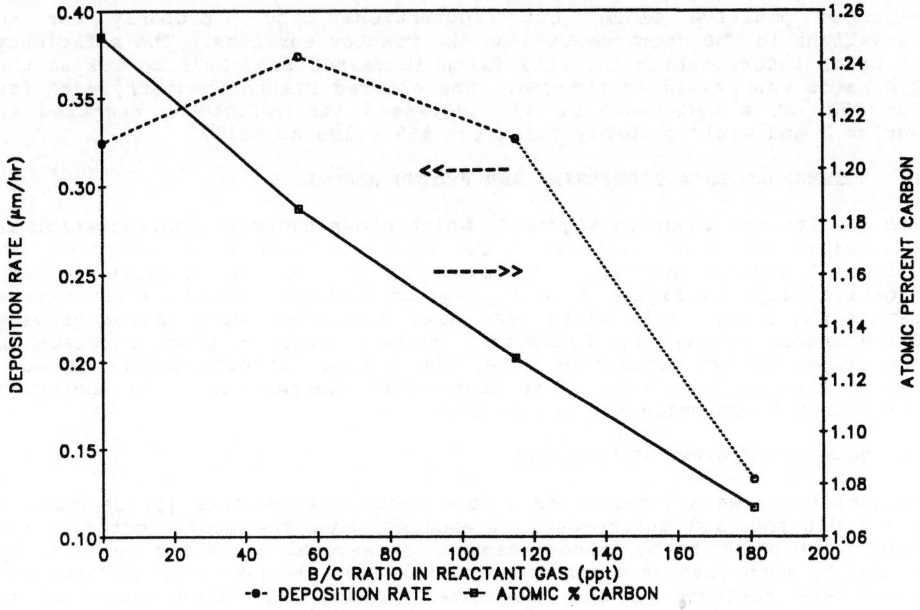

FIGURE 1, B/C RATIO VS. DEPOSTION RATE, AND PERCENT C

occurred in all three series and is due to three factors: 1) decreasing carbon concentration, 2) decreasing substrate temperature, and 3) higher percentages of TMB which partially decompose on hot reactor surfaces to form boron oxide and methane.

D. SECONDARY ION MASS SPECTROSCOPY

Boron concentrations were determined by Charles Evans and Associates for the series 2 films and run 150. Figure 2 shows the boron concentration versus B/C ratio in the gas for series 2 films. The trend is the

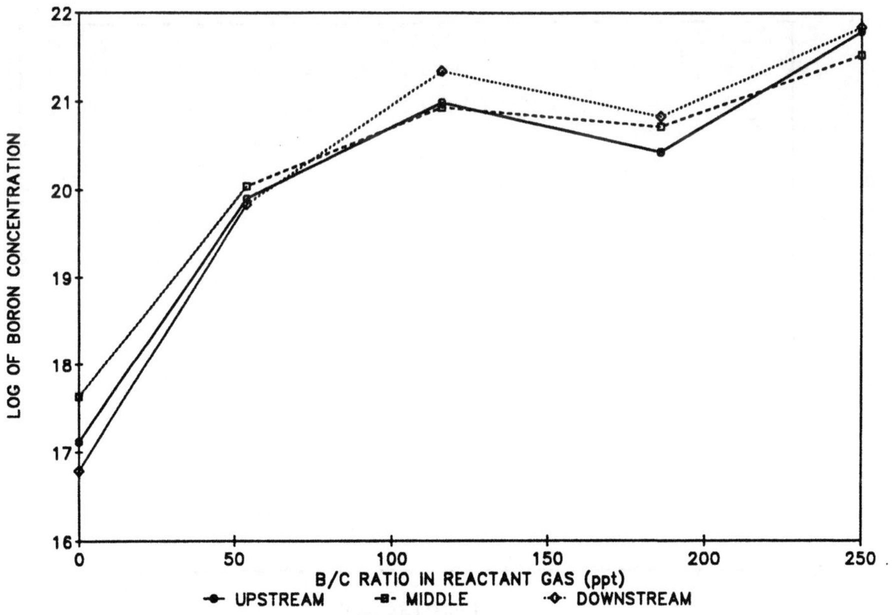

FIGURE 2, BORON IN SERIES 2 FILMS VS. BORON IN REACTANT GAS

expected positive slope, but fluctuations occur, probably due to variations in TMB decomposition on the reactor surfaces. The efficiency of boron incorporation into the films increased from 0.75 to 16% as the B/C ratio was raised in the gas. The altered reactor geometry used for run 150, at a B/C ratio of 49, increased its efficiency compared to series 2 and would probably raise the 16% value as well.

E. RUTHERFORD BACK SCATTERING AND PROTON RECOIL

The results are shown in Figure 3, which shows hydrogen concentration as a function of film thickness. The concentrations were determined at Universal Energy Systems. We expected to see the clearly defined negative slope of Figure 3 on a graph of hydrogen concentration versus deposition rate. This would have been consistent with faster growing films incorporating less hydrogen. While a negative slope persists on the graph against deposition rate, the scatter of data points is much larger than in Figure 3. This means that another factor is modifying the hydrogen concentration in our films.

F. ROOM TEMPERATURE RESISTIVITY

Resistivities were obtained from four probe measurements (probe spacing was 1.016 mm) and thicknesses determined with the profilometer. The films were shown to be inhomogeneous. In several cases the resistivity varied by more than an order of magnitude from the upstream, middle, and downstream portions of the substrate. Also, all films exhibited an increased conductivity as the electric field was raised. The average increase was about 5% per volt from 0.5 to 12 volts, but this was small compared with resistivity variations between films. This non-ohmic behavior is being investigated and we plan to discuss it in a future paper.

In spite of these problems, Figure 4 shows a clear relation between room temperature resistivity and boron concentration in a particular section of film. The slope of the logarithmic graph is -2.78 instead of -1.0 which would result if the carrier concentration were proportional to the boron in the film. The lowering of the medium-temperature activation energy with doping as discussed in the next section, is a mechanism that may explain this larger decrease in resistivity.

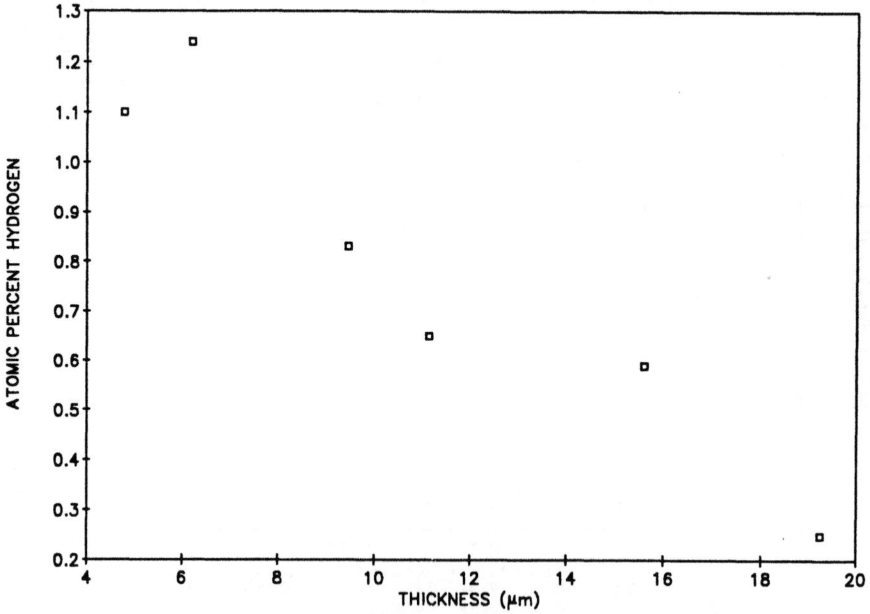

FIGURE 3, HYDROGEN CONCENTRATION IN FILM VS. THICKNESS

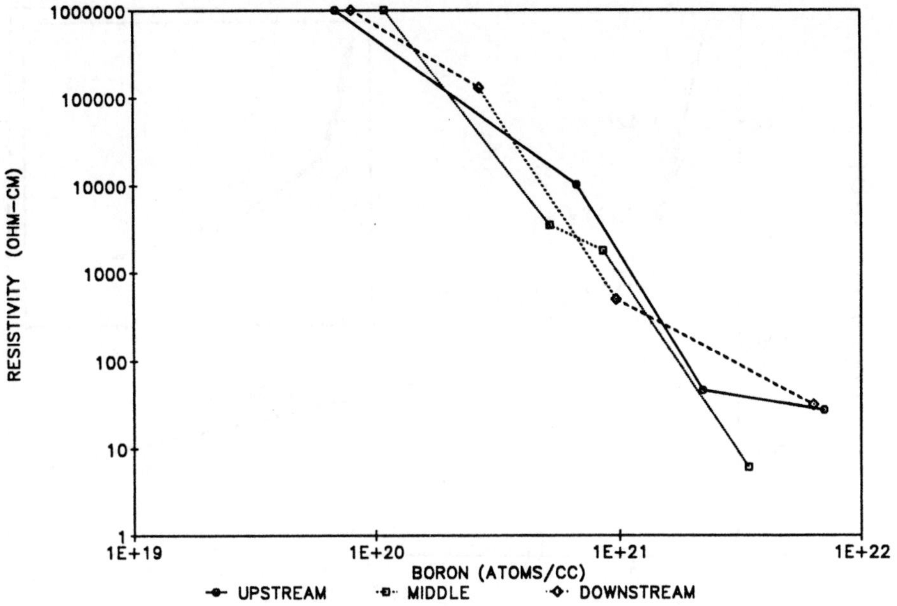

FIGURE 4, FILM RESISTIVITY VS. FILM BORON CONCENTRATION

G. TEMPERATURE DEPENDENCE OF RESISTIVITY

Resistivity from 80 K to approximately 1000 K, was measured on six sections of series 2 films. This more than doubled the temperature reached by other workers studying CVD films. The data could not be fit by a single Arrhenius activation energy. Four of the sections required four distinct activation energies to fit the data as shown in Figure 5 for film 139D, the other two sections needed three energies to fit the data. Figure 6 illustrates how these energies depend on temperature and boron concentration. Note that the activation energy decreases with increasing boron concentration at medium and high temperatures. This decrease has been observed in silicon and germanium as well as diamond films[3,6], and is discussed by Bourgoin *et al*[13]. Another feature that has been observed by other workers is the increase in activation energy that occurs as temperature is raised[14]. In addition, the medium-high-temperature energies agree well with the range of 0.12 to 0.20 eV reported by Okano *et al*[8].

The insert in Figure 5 shows the conductivity of film 139D, with a boron concentration of 9.7×10^{20} atoms/cc, fitted to

$$\sigma = \sigma_1 \exp(-\epsilon_1/kT) + \exp(A + BT^{-1/4}). \qquad (1)$$

The latter term in the equation, representing variable range hopping, fits the data well from about 80 to 500 K. The conductivities of three more highly doped films 139U, 142D and 142U were fit about as well as 139D with equation 1. However, more lightly doped films 141D and 141U with boron concentrations of 2.7×10^{20} and 6.7×10^{20} atoms/cc respectively, were not fit as well with equation 1 as shown in figure 7 for film 141U. For these films the Arrhenius term in equation 1 fits the data above 500 K, a second Arrhenius term is needed to fit the data for $250 \leq T \leq 500$ K, and the variable range hopping term can fit the data for 80 to 250 K.

We interpret the Arrhenius part of equation 1 in terms of transitions from the impurity levels to the conduction band. Activation energies higher than the accepted 0.37 eV may be caused by interaction between the boron acceptors and deep level traps caused by film defects.

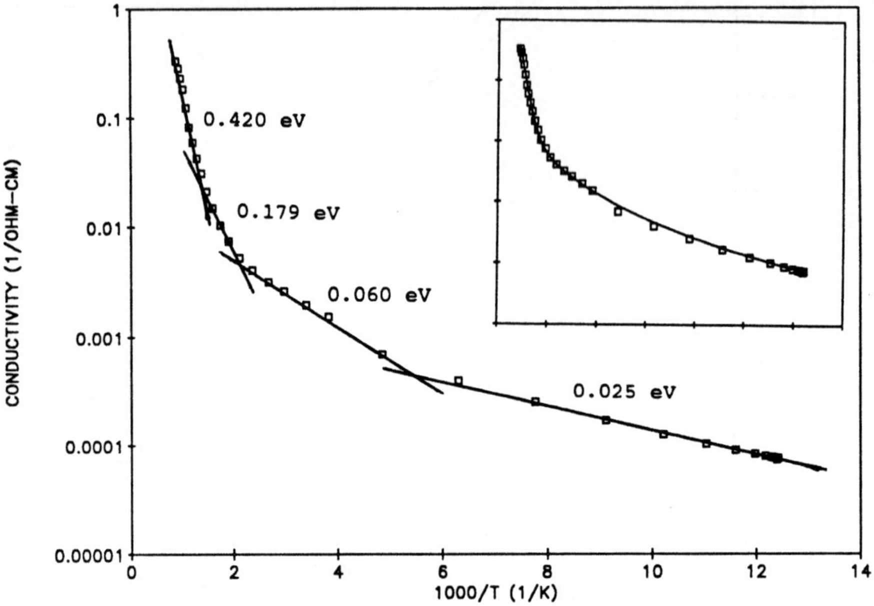

FIGURE 5, CONDUCTIVITY VS. INVERSE TEMP. OF FILM 139D WITH CURVE FIT IN INSERT

Similar activation energies were explained by Snadhu *et al*[14] for ion-implanted diamond in terms of interactions between the acceptor and radiation damage levels.

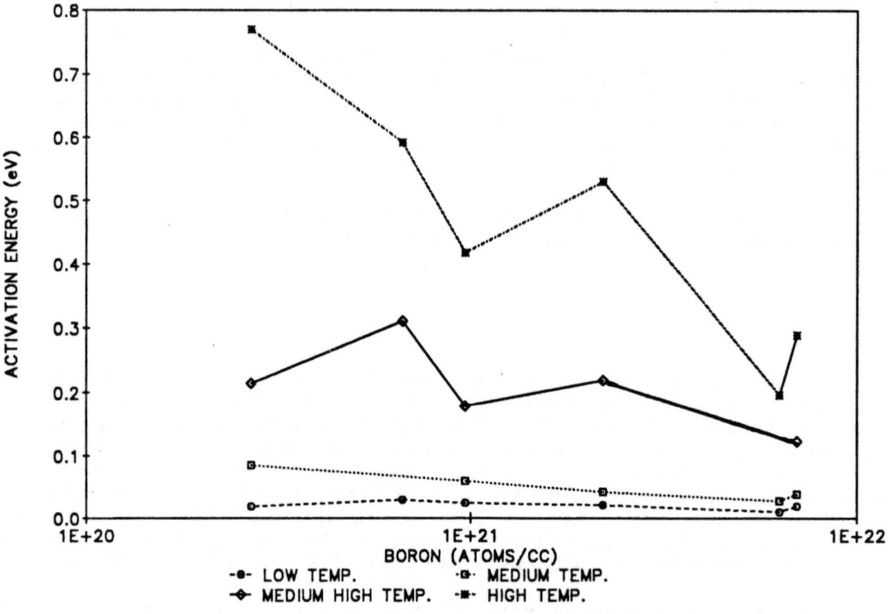

FIGURE 6, ACTIVATION ENERGIES AT DIFFERENT TEMP. VS. FILM BORON CONCENTRATION

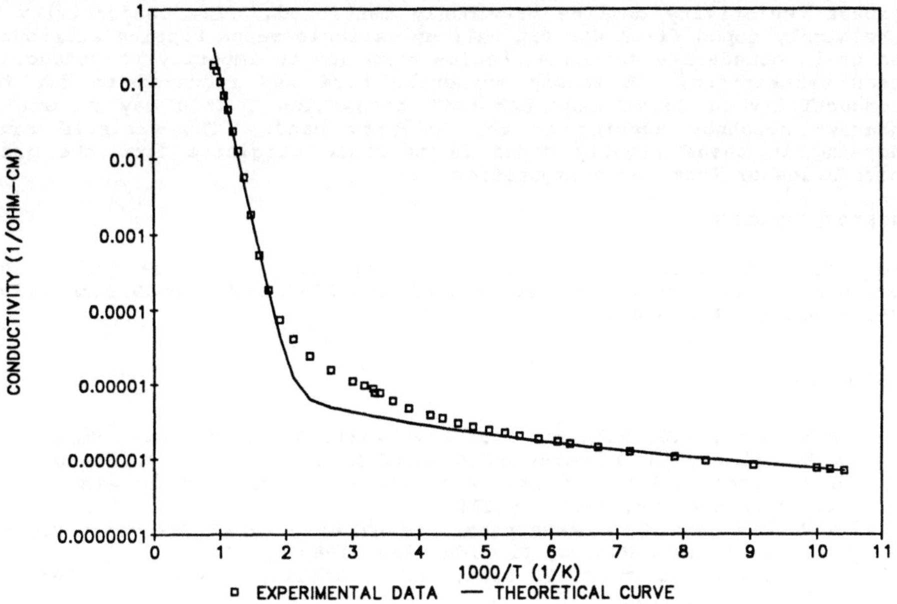

FIGURE 7, CURVE FIT TO FILM 141U USING EQUATION 1

The Mott insulator to metal transition for an impurity band in diamond with a hole effective mass of $0.93m_0$ is calculated to occur at a concentration of $2X10^{20}$ atoms/cc. This is near the lowest doping level in the six sections of the studied films. Even though the hole effective mass is not well determined, a Mott transition above 10^{21} atoms/cc is very unlikely. Therefore, impurity hopping conduction should not be observed in the more highly doped films. We then associate the variable range hopping in these films with grain boundaries.

The second Arrhenius term needed to fit the conductivities of films 141D and 141U, doped at $2.7X10^{20}$ and $6.7X10^{20}$ atoms/cc, has activation energies of about 0.09 eV and nearly zero, respectively Since this activation energy should approximate zero at the Mott transition, this term may derive from nearest neighbor hopping in the impurity band. A Mott transition at $6.7X10^{20}$ atoms/cc could be caused by boron inhomogeneities or a hole effective mass greater than $0.93m_0$. Repeated boron determinations for a given sample typically varied by only 25%. Thus, large boron fluctuations over scales of less than the 30 micron analysis region would be required to explain a $6.7X10^{20}$ atoms/cc Mott transition. In homogeneous diamond a Mott transition at this concentration would imply a hole effective mass of $1.32m_0$, which is not unreasonable given the range of reported effective masses[15,16]. Thus, we associate the second Arrhenius term, needed for the lightly doped films, with nearest neighbor hopping in the impurity band. The variable range hopping in the lightly doped films could originate from the boron impurities or grain boundaries.

4. CONCLUSION

We have used TMB to dope CVD diamond films at concentrations from $6.2X10^{16}$ to $7.0X10^{21}$ atoms/cc. Doping efficiencies up to 16% were achieved. The films were inhomogeneous due to substrate temperature gradients. They showed a room temperature resistivity decline in excess of that expected if the generated carriers were proportional to doping concentration.

The temperature dependence of the resistivity was examined over 80 K to about 1000 K, a wider range than previous workers. A decreasing high-temperature activation energy with doping would tend to produce the excess resistivity decline previously mentioned. The conductivity of the highly doped films was fit well by variable range hopping attributed to grain boundaries and an Arrhenius term due to impurity to conduction band transitions. A second Arrhenius term was required to fit the conductivity of films near the Mott transition. This may be due to nearest neighbor hopping in the impurity band. The variable range hopping in these lightly doped films could originate from the grain boundaries or from boron impurities.

ACKNOWLEDGMENTS

We thank Jeff Lewis and Mark Nielsen for untiring effort devoted to this project. This work was funded by contract F04704-89-C-0059 from Wright Patterson Air Force Base.

REFERENCES

1. M.W. Geis, D.D. Rathman, J.J. Zayhowski, D. Smythe, D.K. Smith, and G.A. Ditmer, in "Diamond and Diamond-Like Materials Synthesis", G.H. Johnson, A.R. Badzian, and M.W. Geis, eds., (Materials Research Society, 1988), p.115
2. K.V. Ravi and M.L. Landstrass, Presented at "High Frequency Power Conversion 89", Naples, Florida, May, 1989, p. 1
3. M. Kamo, H. Yurimoto, Y. Sato, and N. Setaka, J. Vac. Sci. Technol. A **6**(3), 1818 (1988)
4. G.S. Gildenblat, S.A. Grot,C.W. Hatfield, C.R. Wronski, A.R. Badzian, T. Badzian, and R. Messier, Mat. Res. Bull., **25**, 129 (1990)
5. A.W. Phelps and R. Koba, in "Proceedings of the First International Symposium on Diamond and Diamond-Like Films", J.P. Dismukes, A.J. Purdes, K.E. Spear, B.S. Meyerson, K.V. Ravi, T.D. Moustakas, M. Yoder, eds., **89**(12) (Electrochemical Society, 1989) p.38
6. K. Nishimura, K. Das, J. Glass, K. Kumagai, K. Miyata, K. Kobashi, and Y. Kawate, in SDIO/IST-ONR Diamond Technology Initiative Symposium Book of Abstracts, p.TH2, July, (1989)
7. J. Mort, D. Kuhman, M. Machonkin M. Morgan, F. Jansen, K. Okumura, Y.M. LeGrice, and R.J. Nemanich, Appl. Phys. Lett. **55** (11), 1121 (1989)
8. K. Okano, H. Naruki, Y. Akiba, T. Kurosu, M. Iida, and Y. Hirose, Jap. J. Appl. Phys. **27**(2), L173 (1988)
9. D.E. Meyer, R.O. Dillon, and J.A. Woollam, J. Vac. Sci. Technol. A **72**(3), 2325 (1989)
10. D.E. Meyer, R.O. Dillon, and J.A. Woollam, Proceedings of the First International Symposium on Diamond and Diamond-Like Films, Electrochemical Society, 494 (1989)
11. J. Bernholc, A. Antonelli, T.M. DelSole, Y. Bar-Yam, and S.T. Pantelides, Phys. Rev. Lett. **61**(23), 2689 (1988)
12. R.E. Shroder, R.J. Nemanich, and J.T. Glass, SPIE Diamond optics, Vol. 969 (1988) p.79
13. J. C. Bourgion, J. Krynicki, and B. Blanchard, Phys. Stat. Sol.(a), **52**, 293, (1979)
14. G.S. Snadhu, W.K. Chu, M.L. Swanson, and J.F. Prins, SPIE Diamond Optics, **969**, 37, (1988)
15. C. J. Rauch, in "Proc of Int. Conf. on the Physics of Semiconductors", A. C. Strichland ed, The Inst. of Phys. and the Physical Soc., London, 276-80 (1962)
16. A. T. Collins, and E. C. Lightowlers in "The Properties of Diamond", edited by J. E. Field, Academic Press, 82, (1979)

RADIATION RESPONSE OF DIAMOND FILMS

N.K. Annamalai, Reginald F. Blanchard, and Joseph Chapski

Rome Air Development Center
Solid State Sciences Directorate
Hanscom Air Force Base, MA 01731

INTRODUCTION

Diamond is expected to be a radiation hardened material due to its tight bonding. In this work, we are reporting the total dose response of CVD diamond films, 2 μm thick, grown on 3" p-type silicon wafers, in terms of electrical parameters such as conductivity and high frequency capacitance. It is proposed that diamond be used as a buried insulating layer in a silicon-on-insulator (SOI) technology.

SOI substrates are gaining importance for realizing very high speed and radiation hardened integrated circuits [1]. A silicon-on-insulator technology using CVD diamond film is proposed in this work. The structure of this technology is different from diamond film technologies proposed by others [2,3]. In the proposed technology (refer to Figure 1), a diamond film is grown on the whole silicon wafer, except for a rim about ¼" wide (Figure 1a). A polycrystalline silicon of 2 μm thickness or more is then deposited to cover the whole wafer including the rim area (Figure 1b). Using the rim area silicon as the seed, the polycrystalline silicon crystallizes into a single crystal by a zone melting recrystallization technique (Figure 1c). The rim area is scribed off from the wafer, leaving a recrystallized silicon layer on diamond insulator (Figure 1d). The structure is similar to ZMR SOI, but the insulator here is diamond instead of silicon dioxide. The devices are fabricated in the recrystallized silicon that sits on top of the diamond film.

It is also possible to grow a thick ($\geq$ 10 μm) diamond film on silicon, with a ¼" rim as discussed above. After top silicon recrystallization, the substrate silicon is etched off to yield a recrystallized silicon on a stand-alone diamond film. The rim is scribed off, and the structure is mounted on a heat sink. The final structure is recrystallized silicon on diamond mounted on a heat sink.

Hardness against radiation damage is important for electronics in space satellites, nuclear reactors, and military applications where nuclear weapons are used. MOSFETs experience a shift of threshold voltage V_T with radiation dose due to charge build-up in the gate oxide. They also experience sudden changes in channel current due to photo-generated carriers in the bulk silicon. SOI MOSFETs minimize these transient effects by greatly reducing the amount of silicon volume, because each device is fabricated on an insulating layer. But, radiation damage can cause charge build-up in the insulating layer, inducing a conductive back-channel in the MOSFET. It is hypothesized that using diamond film instead of SiO_2 as the insulating layer will minimize this radiation damage.

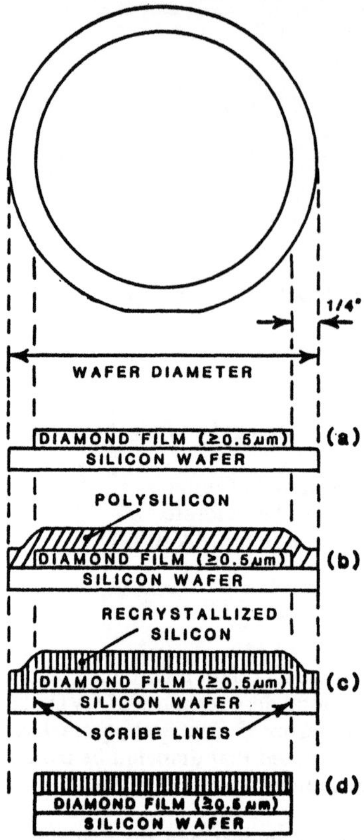

Figure 1. Proposed silicon-on-insulator technology using CVD diamond film as the buried insulator.

Another advantage of diamond film SOI integrated circuits is improved heat conduction. As device geometries become small, heat generation density increases tremendously, and necessity for heat conduction arises. Diamond, being a good thermal conductor, is an ideal candidate for meeting this heat conduction requirement. To allow maximum heat conduction, it is possible to etch away the bottom silicon substrate from the wafer, as described above. This processing would allow heat to be conducted directly through the diamond substrate to the heat sink on which it was mounted.

EXPERIMENTAL SET-UP

CVD diamond films grown on 3" p-type silicon wafers were chosen for this study. The diamond films were grown by two different vendors. One was prepared by DC plasma assisted CVD grown on 75 Ω-cm silicon (sample A), and the other was prepared by filament assisted CVD grown on 10 Ω-cm silicon (sample B). Capacitor devices were fabricated on the diamond films. The top contact was a 30 mil diameter dot of aluminum (sample A) or titanium (sample B) evaporated through a shadow mask onto the diamond surface. The bottom contact was aluminum evaporated onto the silicon substrate. Thus, the devices operated like a standard MIS capacitor, but with a very thick diamond dielectric.

A day-long total-dose radiation experiment was run on each device tested. Each device was irradiated in a Cobalt-60 chamber at a dose rate of 500 krad (SiO_2) per hour. During irradiation, the device was biased with +5 V on the top diamond layer and zero volts on the bottom silicon substrate. Irradiation was performed in several dosage steps within the day, and a complete set of measurements, taking 30-45 minutes, was made after each dosage step.

The measurements were performed by an automated system consisting of an HP 236 laboratory computer and HP digital meters. The computer controlled the test equipment, stored all measurement data, and analyzed and plotted the results. The complete set of measurements was as follows:

1. *Conductance vs. voltage.* Equipment: HP 4140B Picoammeter. Voltage was stepped through a range of -100 to +100 V, with a usual step size of 20 V. At each step, DC current through the dielectric was measured after applying the voltage and holding it for 30 seconds, allowing nonconductive transient currents to settle out. For the conductivity plot in Figure 3, conductivity values were calculated from the I-V relationship in the -20 to +20 V range.

2. *Leakage current vs. voltage.* Equipment HP 4140B Picoammeter. A triangular voltage sweep (TVS) [4] was applied, and the total current was measured as a function of voltage. The total current is the sum of the capacitive and conductive currents. The DC conductance measurements in No. 1 above showed that the total leakage current for the diamond films was almost entirely conductive current. For past SiO_2 SOI devices tested, the conductive component from No. 1 was of a similar magnitude to the capacitive component, and the capacitive current could be calculated by subtracting the conductive current from the total TVS current measured. Static capacitance was then calculated by dividing the capacitive current by the voltage ramp rate, dV/dt [5]. A voltage ramp rate of 1 V/s was used, which is the maximum rate available on the HP 4140B. The maximum rate was necessary to measure as much capacitive current as possible, since conductive current was high, and to minimize testing time.

3. *High frequency capacitance vs. voltage.* Equipment: HP 4280A 1 MHz Capacitance Meter. Bias voltage was stepped from -100 V to 100 V in 1 V steps and was held at each step for 0.5 seconds. At each step, the meter used a superimposed small-amplitude 1 MHz AC voltage to determine the high-frequency capacitance.

RESULTS

Total leakage current was measured as described in the Experimental Set-up section. A plot of total leakage current as a function of voltage, for various total doses, is shown for sample A in Figure 2a and for sample B in Figure 2b. The total leakage current (sum of the conductive and capacitive currents) increased exponentially. The increase was sharp at voltages from 20 to 100 V and increased even more sharply at voltages near zero. This increase is different from that of a SiO_2 buried layer, in which the current increase is nearly linear with voltage up to 100 V, indicating a constant conductance over the bias range [5]. The total leakage current was seen to be mostly conductive; hence, total leakage current variations are conductivity variations of the diamond film. The increasing conductance of the diamond film produces a very high current — about 1 μA — at the highest bias. The current is not symmetrical with bias polarity: current is higher at negative biases than at positive biases. These statements are true for both of the samples.

Comparing the two samples shown, sample A did not show any significant variation in the total leakage as a function of total dose even up to a dose of 10 Mrad (SiO_2). All curves lie one on top of another as seen in Figure 2a. Sample B showed a general increase in the total leakage current as the total dose increased. To avoid confusion, only two curves (pre-rad and 2 Mrad (SiO_2)) for each bias are shown in Figure 2b. The current minimum occurs at -4 V for sample A (Figure 2a) and at 0 V for sample B (Figure 2b). The only variations between the two samples which could be attributed to this difference are the metal contacts to the diamond film (aluminum in sample A and titanium in sample B) and the resistivities of the silicon substrates.

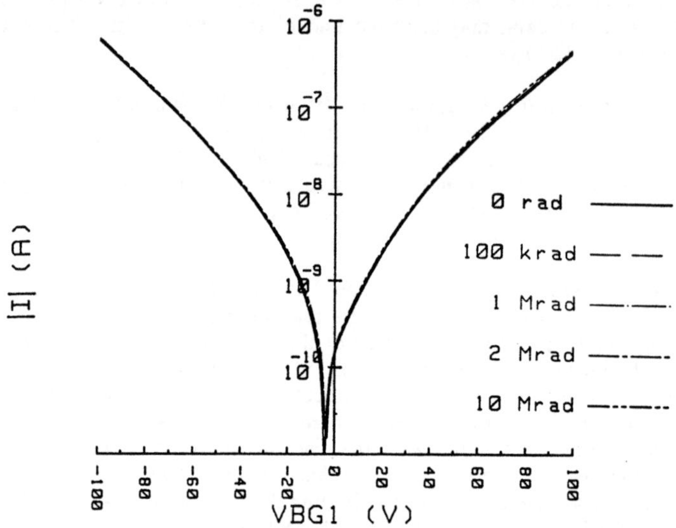

Figure 2a. Sample A: magnitude of total leakage current vs. voltage at various total doses.

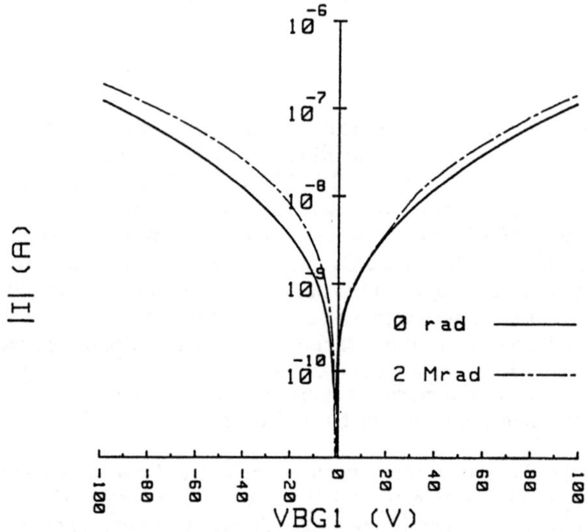

Figure 2b. Sample B: magnitude of total leakage current vs. voltage at various total doses.

A plot of conductivity as a function of total dose is shown in Figure 3. The conductivity variation is seen to be insignificant even up to 10 Mrad (SiO_2) for sample A and up to 500 krad for sample B. A number of other samples exhibited insignificant conductivity variation up to the final dose used (2 Mrad or 10 Mrad). There is a certain amount of uncertainty in the diamond film thickness (and conductivity) due to non-uniformity of its thickness over the wafer. A thickness value of 2 μm, quoted by the supplier, was used in the conductivity calculations.

High frequency capacitance versus voltage at various total doses is shown in Figure 4a for sample A and in Figure 4b for sample B. For both samples, the capacitance value changes rapidly beyond 30 V. For sample A, the capacitance value drops rapidly beyond -40 V and beyond 30 V. Diamond silicon heterostructure capacitance [6] is given by

$$\frac{C}{A} = \sqrt{\frac{qN_1N_2\epsilon_1\epsilon_2}{2(\epsilon_1N_1 + \epsilon_2N_2)} \frac{1}{(V_{Bi} - V)}},$$

depletion width into diamond is given by

$$X_1 = \sqrt{\frac{2}{q} \frac{N_2\epsilon_1\epsilon_2 \ (V_{Bi} - V)}{N_1 \ (N_1\epsilon_1 + N_2\epsilon_2)}},$$

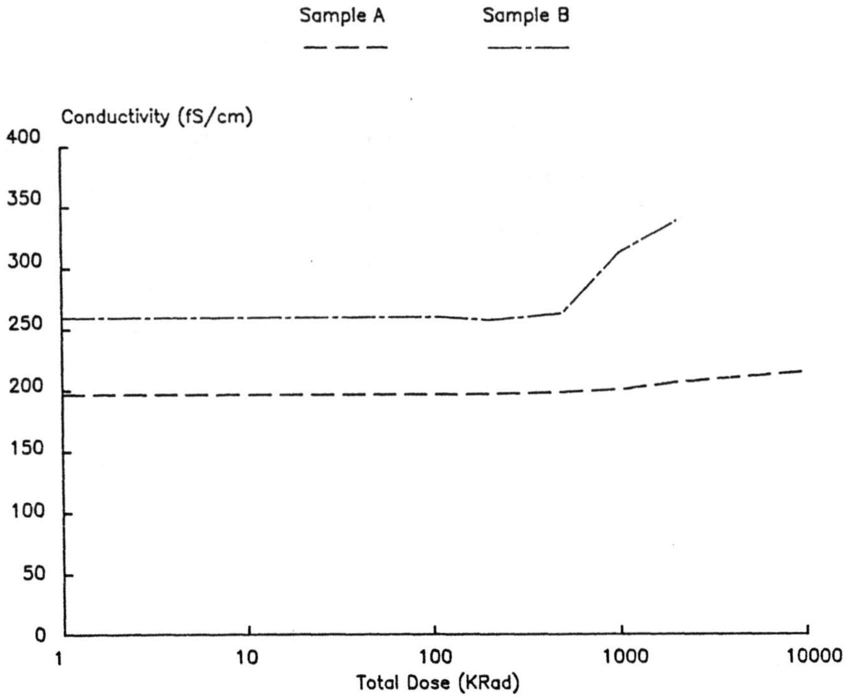

Figure 3. Conductivity vs. total dose for samples A and B.

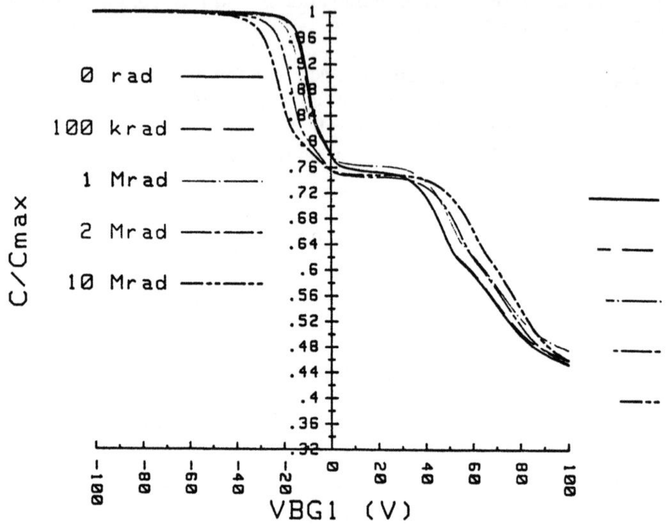

Figure 4a. Sample A: high frequency capacitance vs. voltage at various total doses.

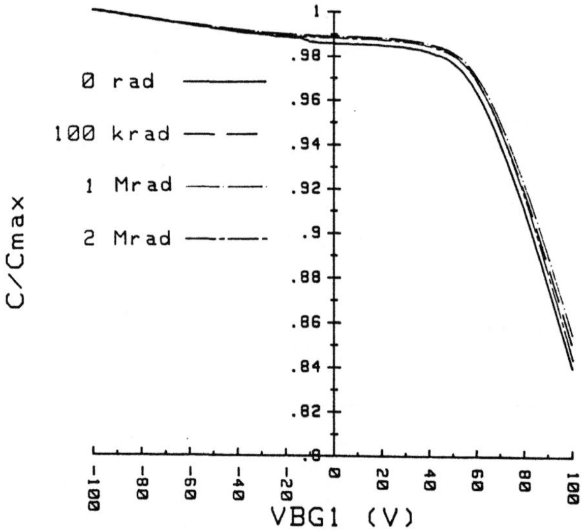

Figure 4b. Sample B: high frequency capacitance vs. voltage at various total doses.

and depletion width into silicon is given by

$$X_2 = \sqrt{\frac{2}{q} \frac{N_1 \epsilon_1 \epsilon_2 (V_{Bi} - V)}{N_2 (N_1 \epsilon_1 + N_2 \epsilon_2)}}$$

where:

q = electronic charge (C)
N_1 = doping concentration in diamond (cm^{-3})
N_2 = doping concentration in silicon (cm^{-3})
ϵ_1 = dielectric constant of diamond (F/cm)
ϵ_2 = dielectric constant of silicon (F/cm)
V_{Bi} = total built-in-potential (V)
V = total applied voltage (V).

Because diamond is highly resistive ($N_1 << N_2$), the capacitance is due to depletion into the diamond first and then into the silicon. Constant C/C_{max} (= 1) for sample A from -100 to -40 V may be due to a very thin oxide being present between diamond film and silicon. For sample A, the constant value at C_{max} may be due to oxide, the first drop may be due to depletion into silicon, and the constant value of 0.75 C_{max} from zero to 30 V may correspond to an inversion layer. For further increase in bias, the capacitance value drops. Is it possible that this shape is due to deep depletion [7]? For sample B, capacitance drop is very small, dropping from C_{max} at -100 V to 0.98 C_{max} at 40 V. Negative polarity implies a positive bias on the p-Si-intrinsic diamond heterostructure. When the bias is varied from -100 V to positive voltages, the device goes from forward bias to reverse bias. As the positive voltage is increased, the depletion region extends into the silicon (beyond 40 V for Si). The device Si never goes into inversion. C-V characteristics as a function of frequency will be of great help in improving the understanding of the diamond-silicon interface.

DISCUSSION

The conductive leakage current through diamond film is more or less the same as the total leakage current. Hence, static capacitance could not be obtained [5]. In order to satisfy the requirement that the capacitive component of leakage current be at least an order of magnitude higher than the conductive component, the resistivity of the diamond films should be increased by a couple orders of magnitude. The resistivity of the present diamond film is not sufficiently high for applying this technique nor for developing an SOI technology using diamond instead of silicon dioxide buried layer.

Conductivity does not show any significant variation as a function of total dose. There are a relatively few positive charges trapped in the diamond, as evidenced by the high frequency C-V plots. There may be few charge sinks resulting in no net variation in the conductivity of the diamond film. High frequency C-V plots show a very small threshold voltage shift compared to buried silicon dioxide SOI. The threshold voltage shift for diamond at a total dose of 10 Mrad (SiO_2) is approximately 15 V compared to a shift of 50 V at a dose of 1 Mrad (SiO_2) for SOI with and implanted buried oxide. The threshold voltage shift for diamond at 10 Mrad (SiO_2) total dose is about one-third the value for buried silicon dioxide SOI at one-tenth that dose.

CONCLUSION

The CVD diamond film is seen to be radiation hardened. The conductivity shows no significant variation when the total dose is increased. The high frequency capacitance shows a threshold shift which is considerably smaller than that of buried oxide SOI. Hence, diamond film will be an ideal candidate for developing radiation-hardened, high-speed, very large scale integrated circuits. But, to realize a viable SOI technology using diamond insulator, the resistivity of the diamond films are to be increased by at least two orders of magnitude.

ACKNOWLEDGEMENTS

Authors wish to thank Kenneth Vaccaro for depositing metal contacts on to the samples. We wish to thank Dr. Walter Shedd and Dr. Roth for their support of this research and Dr. Diamond for providing funds under lab director's fund.

REFERENCES

[1] Jean-Pierre Colinge, "Thin-Film SOI Technology: The Solution to Many Submicron CMOS Problems", IEDM Technical Digest, pp. 817-820 (1989).

[2] K.V. Ravi and M.I. Landstrass, "Silicon-on-Insulator Technology Using CVD Diamond Films", Proceedings of the First International Symposium on Diamond and Diamond-Like Films, The Electrochemical Society, (1989).

[3] M.I. Landstrass and D.M. Fleetwood, "Total Dose Radiation Hardness of Diamond-Based Silicon-on-Insulator Structures", Applied Physics Letters, Vol. 56, No. 23., pp. 2316-2318 (1990).

[4] E. H. Nicollian and J. R. Brews, *MOS Physics and Technology* (John Wiley & Sons, New York, 1982), pp. 435-440.

[5] N. K. Annamalai and J. Chapski, "SOI Buried Oxide Leakage Currents," Proceedings of the Fourth International Symposium on Silicon-on-Insulator Technology and Devices, The Electrochemical Society, Vol. 90-6, pp. 337-350 (1990).

[6] A. G. Milnes and D. L. Feucht, *Heterojunctions and Metal-Semiconductor Junctions* (Academic Press, New York, 1972), Chapter 2.

[7] Robert F. Pierret, *Modular Series on Solid State Devices, Volume IV, Field Effect Devices* (Addison-Wesley, Reading, Mass., 1983), p. 54.

HIGH RATE VERSUS LOW RATE DIAMOND CVD METHODS

Peter K. Bachmann and Hans Lydtin

Philips Research Laboratories
CVD-Technology Group
D-5100 Aachen, FRG

INTRODUCTION

A wide variety of methods to deposit diamond films and particles from the vapor phase emerged from different laboratories since the breakthrough publications of diamond deposition onto non-diamond substrates by B.V. Spitsyn et al [1], B.V. Derjaguin et al [2] of the USSR Academy of Science in Moscow as well as M. Kamo [3] and his colleagues at the Japanese National Institute for Research in Inorganic Materials (NIRIM). All of these methods have in common that diamond is deposited from a carbon carrier gas that is decomposed and the formation of non-diamond carbon is prevented by the presence of a selective etchant such as atomic hydrogen or oxygen.

In Tab. 1, the deposition methods are categorized according to the specific way of initiating the chemical reactions that lead to diamond formation. They are listed along with a number of references of original articles [4-36] as well as reviews [37-42].

Some of these diamond CVD methods are capable of producing the desired carbon phase at linear growth rates of 0.1 to 1 µm per hour while for other methods up to 900 µm per hour were reported.

In this paper, which is an updated and extended version of an earlier publication [43], we attempt to explain the rate differences by comparing such experiments, where only minor changes of specific deposition parameters lead to drastic changes in the deposition rates. In addition,the results of thermodynamical calculations for both C/H- and C/H/O-containing gas mixtures in the temperature range between 300 K and 6000 K are correlated with the approximate temperatures of the CVD gas phases (not the substrate surface temperature which has to stay below 1200 °C to allow for diamond formation!) in order to clarify the role of various gas phase species for high rates in CVD diamond synthesis.

Tab. 1. CVD methods for the preparation of diamond.

Thermal CVD

thermal decomposition [4,5,5a]
chemical transport reaction (CTR) [1]
hot filament technique [6-8,8a]
oxy-acetylene torch [9-11]

DC Plasma CVD

low pressure DC plasma [12-14]
medium pressure DC plasma [15,15a]
hollow cathode discharge [16]
DC arc plasma [17]
DC plasma jet [18-21]

RF Plasma CVD

low pressure rf glow discharge [22-24]
thermal rf plasma CVD [25]

Microwave Plasma CVD

915 MHz plasma [26]
low pressure 2.45 GHz plasma [27-32]
atm. pressure 2.45 GHz plasma torch [33]
2.45 GHz magnetized (ECR) plasma [34,35]
8.2 GHz plasma [36]

DIAMOND CVD METHODS AND RESULTS

Thermal Decomposition, Hot Filament and Flame Synthesis of CVD Diamond

All of the early CVD approaches to form diamond from the gas phase were characterized by the thermal decomposition of carbon-containing gases such as CBr_4, CI_4 [4], CH_4 [5], or CO [5,5a], They were carried out at gas temperatures of about 600°C to 1200°C. Gas temperature and surface temperature of the diamond seeds that were used as substrate material were the same and the linear diamond growth rates were in the order of approximately 0.01 µm per hour. Interestingly, the first breakthroughs [1,6] in diamond CVD occured, when the substrate surface temperature and the gas phase temperatures were decoupled, i.e. the decomposition of the carbon carrier together with the formation of the atomic hydrogen used as etching agent were performed at much higher temperatures than the substrate temperature that still had to remain below 1200°C to avoid graphitization of the growing diamond film. Both the "chemical transport reaction (CTR)" method [1] as well as the now widespread hot filament technique create a hot zone in the gas phase by either a hot graphite disk [1,1a] or by means of a hot tungsten, molybdenum or tanatalum wire [6-8a]. For CTR approx. 2000°C were reported [1a], the filament temperature in a hot filament diamond CVD reactor is usually between 2000°C and, at most, 2400°C [8a]. For ≈1% methane in hydrogen, the growth rates for both methods are in the order of 1 µm per

Tab. 2. Typical deposition conditions for hot filament growth of diamond.

substrate temperature [°C]	reactor pressure [mbar]	total gas flow [sccm]	filament temperature [°C]
850- 1050	20-40	100-200	2000 - 2400

linear deposition rate [µm/hour]	deposition area [cm 2]	types of substrates
0.3 - 2	2-100 (depending on filament structure)	Si, Mo, silica, alumina, diamond, etc.

quality/crystallinity [Raman data]	uniformity / homogeneity	special remarks
+++ (filament metal codeposited !)	- (mirrors filament shape !)	additional bias voltage affects quality and rate

advantages of the method	drawbacks of the method
simple set up large area possible	filament stability problematic, metal contamination, no long term growth process, higher oxygen conc. difficult

(- = poor quality , +++ = excellent quality)

hour, i.e. 1-2 orders of magnitude higher than for lower temperature thermal decomposition, while still maintaining the thermal initiation of the CVD reactions. Tab. 2 summarizes the reaction conditions and results for typical hot filament assisted diamond deposition.

Another big step forward in terms of increasing the linear growth rate of diamond was the introduction of the oxy-acetylene torch synthesis by Y. Hirose et al [9,10,10a,10b]. They used a ≈ 1:1 mixture of acetylene and oxygen to grow diamond from an ordinary welding torch. Hirose and coworkers point out that high linear growth rates of 50-100 µm per hour in combustion diamond synthesis are achievable using hot gas mixtures with the flame still showing a reducing acetylene feather. Gas phase temperatures of 3100°C to 3300°C are reported [10a]. Tab. 3 lists the typical conditions for combustion flame synthesis of diamond. Fig. 1 provides sketches of both a hot filament and an oxy-acetylene torch set-up for diamond synthesis.

Tab. 3. Typical deposition conditions for combustion flame synthesis of diamond.

substrate temperature [°C]	pressure [bar]	total gas flow [sccm]	flame temperature [°C]
400-1100 (cooling required !)	1	1000-200C	≈3100

linear deposition rate [µm/hour]	deposition area [cm 2]	types of substrates
50 (high quality) - 200 (poor quality	≈ 0.5 (fixed torch, larger for moving torch)	Si, Mo,alumina, TiN diamond, etc.

quality/crystallinity [Raman data]	uniformity / homogeneity	special remarks
+ to +++	-	rates and quality depend on nozzle used, distance nozzle-substrate critical

advantages of the method	drawbacks of the method
simple set up high rates	control of substrate temperature difficult, inhomogeneous dep. contaminations.

(- = poor quality , +++ = excellent quality)

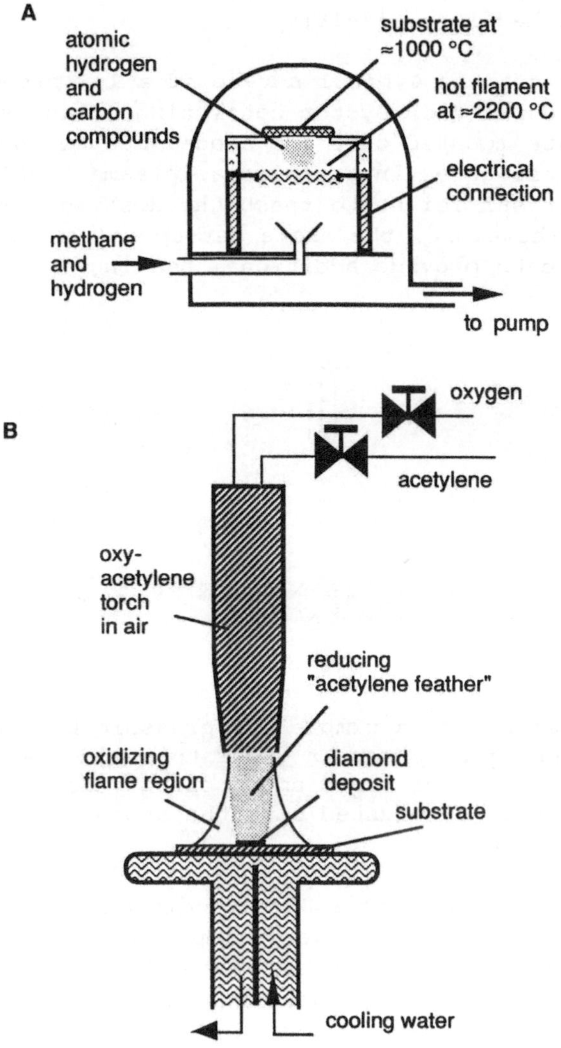

Fig. 1. Set-up of a hot filament reactor(A) and an oxy-acetylene torch deposition system (B) for diamond CVD.

Spitsyn, Derjaguin, Bouilov and Fedoseev [1,2] were the first to suggest the application of electrical discharges to create a superequilibrium concentration of atomic hydrogen in order to remove codeposited graphitic carbon from the growing film. A DC plasma discharge is probably the simplest way to form such an electrical discharge at low pressures and in the meantime, a number of papers covering the DC discharge method have been published [e.g. 12-16].

Fig. 2 depicts the typical set-up of a low pressure DC plasma diamond deposition system consisting of anode, cathode and the substrate mounted onto the anode. Due to the low temperature of the gases in the plasma, plasma heating of the substrate is not sufficient to reach the desired temperatures of 600-1000°C. Therefore, the whole set-up is usually mounted inside a furnace to provide additional heating.

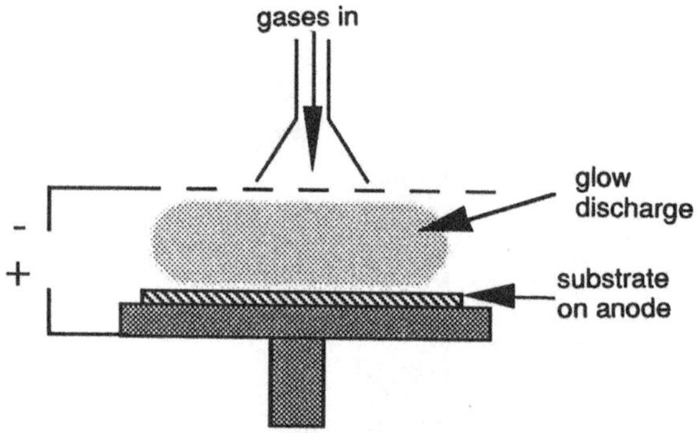

Fig. 2. Schematic of a simple low pressure DC plasma deposition system.In order to avoid the deposition of non-diamond carbon the substrate needs to be mounted onto the anode

One of the advantages of a DC glow discharge diamond deposition system is its relatively simple set-up. The discharge is formed between the cathode and the anode that carries the substrate. If the substrate is mounted onto the cathode, graphitic material is formed. Another advantage is the large discharge diameter that can be formed. Continuous, uniform coatings on more than 4 inch diameter wafers have been reported [13,14].

The quality of the "as deposited" diamond films, however, is only mediocre. They can take up substantial amounts of hydrogen thus detoriating the electrical properties of the material [14]. Another disadvantage of this method is its rather low deposition rate of less than 0.1 µm/hour. Tab. 4 gives the characteristics of the low pressure version of a DC discharge diamond deposition process.

Tab. 4. Typical deposition conditions for low pressure direct current plasma synthesis of diamond.

substrate temperature [°C]	reactor pressure [mbar]	total gas flow [sccm]	input power data
600-1000	20-40	100-200	40 mA/cm 2 250-500 V
linear deposition rate [µm/hour]	**deposition area** [cm 2]	**types of substrates**	
0.01-0.1	≈ 70	Si, Mo, silica, alumina, diamond, other	
quality/crystallinity [Raman data]	**uniformity / homogeneity**	**special remarks**	
0 / +	+++	substrate preferrably on anode to avoid graphitic material	
advantages of the method		**drawbacks of the method**	
simple set up large area		low linear dep. rate poor quality (H-content)	

(- = poor quality , +++ = excellent quality)

Both, the quality of the deposit as well as the linear deposition rates are substantially improved, if the reactor pressure is increased to ≈ 200 mbar [15,15a]. To maintain a plasma under such conditions, it is necessary to increase the power fed into the system by increasing both the current density and the discharge voltage. This also results in a much higher gas temperature with the plasma being closer to local thermal equilibrium (LTE-plasma, hot plasma). Therefore, in this variant of DC plasma diamond formation,the substrate has to be cooled rather than additionally heated. Key experiments in this field were published by K. Suzuki and coworkers in 1987 [15] and in 1990 [15a]. Their results are of specific interest because they used the same deposition set-up changing only one deposition parameter, the plasma power, (increasing the current density) and went from linear growth rates of 20 µm/h to more than 250 µm/h. Moreover, they achieved these rates with very low total gas throughputs of only 20-100 sccm and without particularly directing any of the reactants towards the substrate surface. By means of spectroscopical characterization of the discharge they concluded that the heavy particles , i.e.

the gas temperature in the plasma is approximately 4000 °C at 0.5 A/cm^2,5300 °C at 4 A/cm^2 and 6000 °C at 10 A/cm^2 thus considerably extending the gas temperature of ≈3200°C achievable for chemical flames. A major disadvantage of DC discharge plasma CVD of diamond at higher pressures is also obvious from refs.[15,15a]. The deposition area is severely reduced in this operating regime. In Tab. 5, typical conditions and results for medium pressure DC plasma diamond deposition are condensed.

Tab. 5. Typical deposition conditions for medium pressure DC plasma synthesis of diamond.

substrate temperature [°C]	reactor pressure [mbar]	total gas flow [sccm]	input power data
600-900 (cooling required for P> 100 mbar)	200	20-100	4 A/cm^2 at 1 kV to >10 A/cm^2 at 1 kV
linear deposition rate [µm/hour]	**deposition area** [cm^2]	**types of substrates**	
20-130 at 4A/cm^2 >250 at 10 A/cm^2	<1	Si, Mo, alumina	
quality/crystallinity [Raman-data]	**uniformity / homogeneity**	**special remarks**	
+++	? (small area)	substrate on anode to avoid graphitc material; cooling required !	
advantages of the method		**drawbacks of the method**	
high rates		small area	

(- = poor quality , +++ = excellent quality)

Diamond CVD from High Pressure DC Arcs, Plasma Flames and Jets

Further increase of the pressure in a DC discharge plasma system leads to a deposition approach introduced by S. Matsumoto as well as by researchers at Fujitsu Laboratories [18-20]. They fed methane, hydrogen and argon into a conventional plasma torch commonly used for plasma spraying. In our own deposition experiments, we used a cascaded arc plasma generator as plasma source [21]. In all cases, the plasma ignited inside the generator is usually very hot, i.e. gas temperatures of more than 5000 °C are easily obtained. The plasma is blown out of the generator thus forming, at least in a hydrogen-containing environment, a narrow hot lance that creates a relatively small, spot-like deposition zone. The basic principles of this approach are illustrated in Fig. 3.

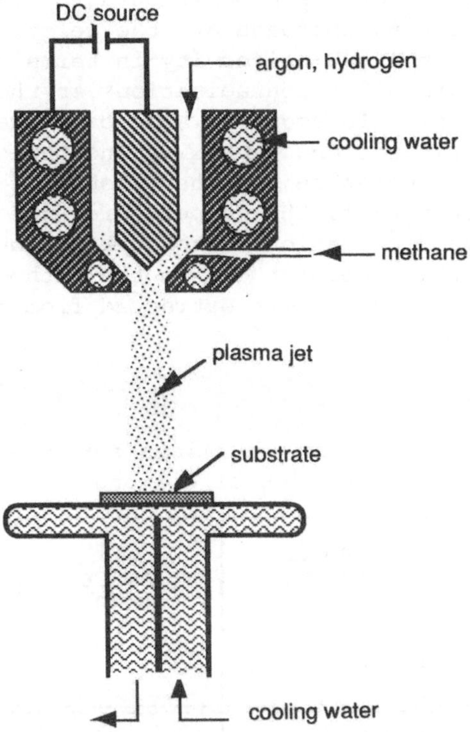

Fig. 3. Set-up for diamond deposition from a hot DC plasma jet.

The flame-like plasma expands from the plasma torch and intensively heats the spot-like deposition zone. Therefore, the substrate has to be cooled during deposition. The injection point of the gaseous feed stock is of vital importance to the deposition process. With DC plasma arc discharges at pressures between 0.2 and 1 bar, the highest diamond deposition rates for any method were achieved. The top value of 930 µm/hour has been reported by N. Ohtake et al [17]. Interestingly, they did not even directly activate the carbon carrier gas, methane, by injecting it into the plasma generator, but activated the hydrogen and argon to form the hot plasma that expanded into the reactor vessel containing the methane gas. The carbon carrier simply diffuses into the hot zone and is decomposed. This approach avoids plasma instabilities created by carbon deposits that form inside the generator when methane is injected directly into the discharge region [17,21] but still allows for such high deposition rates. At lower pressures, i.e. at correspondingly reduced gas temperatures, the deposition rates are much lower. We reported 25 µm/hour at pressures around 50-100 mbar and found only amorphous carbon being deposited at pressures of less than 1 mbar in the reactor vessel [21]. In terms of the Raman spectra, the quality of plasma jet deposited diamond can be very good, at least in the central portion of the deposition area. Major disadvantages of this approach are the relatively small deposition area, the non-uniformity in terms of the thickness as well as the graphitic contaminations at the periphery of the diamond deposit. In addition, the experimental set-up can be quite sophisticated and expensive and the stabilization of both the plasma generation and the substrate deposition conditions, specifically the temperature needs considerable effort. For some applications were bulk diamond is needed, e.g. as heat sinks, this approach can be the method of choice. Typical deposition parameters extracted from the literature are collected in Tab. 6.

Tab.6. Typical deposition conditions for DC arc discharge and plasma jet deposition of diamond

substrate temperature [°C]	reactor pressure [mbar]	total gas flow [sccm]	input power data
800-1100°C (cooling required!)	0.1-1000 (good quality only at P> 30 mbar)	3000 - 70 000 (!!)	3-18 kW (!)
linear deposition rate [µm/hour]	**deposition area [cm 2]**	**types of substrates**	
25 at P<100 mbar 500 -930 at higher P	<2	mostly Mo Si	
quality/crystallinity [Raman data]	**uniformity / homogeneity**	**special remarks**	
+++ (central part) - (periphery)	-	T-shock destroys Si substrates; severe deformation observed for Si-substrates larger than the plasma beam.	
advantages of the method		**drawbacks of the method**	
extremely high deposition rates of up to 930 µm/hour		small deposition area; contaminiations by graphitic and plasma gen. materials; stability and temperature control; thickness and quality variations.	

(- = poor quality , +++ = excellent quality)

Low Pressure Radio-Frequency (RF) Plasma CVD of Diamond

A driving force to utilize RF plasmas at low pressures for the preparation of diamond coatings is the availability of fully developed equipment proven capable of coating large areas of a variety of substrates with a variety of coating materials. From the equipment point of view, this method would have been the most likely candidate for scale-up and industrialization of diamond deposition. Both, the commonly used parallel plate reactor as well as inductively coupled low pressure RF plasmas have been tested [22-24]. Certainly, diamond can be formed from such plasmas, however, all attempts to synthesize high quality coatings at a reasonable rate failed so far. Frequently, amorphous carbon rather than diamond is deposited or, if diamond forms, only particles are detected. In parallel plate reactors, the electrodes need to be made from graphite in order to avoid contaminations by sputtered electrode material.

Tab. 7. Typical deposition conditions for low pressure RF plasma CVD of diamond.

substrate temperature [°C]	reactor pressure [mbar]	total gas flow [sccm]	input power data
700-1000 °C (heating required)	0.1-40	100-200	300 - 3000 W
linear deposition rate [µm/hour]	**deposition area** [cm 2]	**types of substrates**	
<0.1	- (particles)	Si, silica, BN, alumina, Ni, diamond, etc.	
quality/crystallinity [Raman data]	**uniformity / homogeneity**	**special remarks**	
-	-	inductively & capacitivly coupled plasmas used at frequencies between 0.4-13.56 MHz. **Higher pressures, higher power,& frequency work better**	
advantages of the method		**drawbacks of the methoc**	
easy scale-up option complex substrate shape option		low rates, poor crystalline quality, contaminations.	

(- = poor quality , +++ = excellent quality)

Similar to low pressure DC plasmas, the substrate has to be additionally heated in order to perform experiments at temperatures of ≈900°C. The low pressure plasma alone is not sufficiently hot enough for that purpose. Data used in attempts to deposit diamond from low pressure RF plasmas are given in Tab.7.

Atmospheric Pressure RF Plasma CVD of Diamond

Only little work has been done to deposit diamond films from yet another type of a hot LTE-plasma, namely an inductively coupled RF plasma operated at atmospheric pressure. It was again S. Matsumoto [25] and his colleagues at the NIRIM who utilized this method for diamond deposition for the first time. Such plasmas are typically created by inductively coupling RF power into a tubular silica reactor. At a frequency of 4 MHz Matsumoto used up to 60 kW of high frequency power. A suitable experimental set-up is depicted in Fig. 4. Similar gas mixtures as for the other diamond preparation techniques have been used, however, the total gas flow was as high as 80 slm.

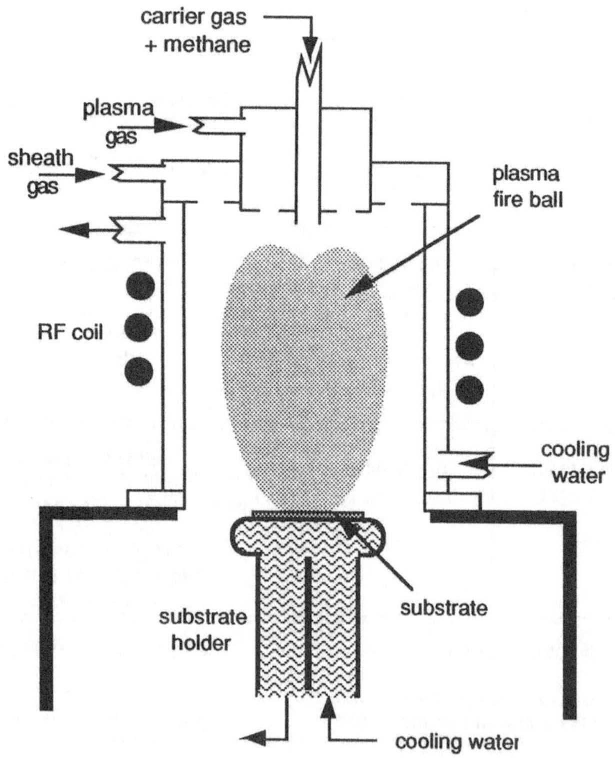

Fig. 4. Thermal RF plasma deposition set-up as described by S. Matsumoto [25].

At growth rates of 2-3 µm/min i.e. up to 180 µm/hour, high quality diamond films were obtained on molybdenum substrates. The deposition area is small but comparable to other high rate methods such as medium pressure DC discharges, DC plasma jets or chemical flames, i.e. less than 2 cm^2. Contamination of the deposit by reactor materials seems to be avoidable, however, the handling of such plasmas is not easy. They are very hot, 10 000°C and higher, and the key problems are to avoid any contact of the plasma fire ball with insufficiently cooled reactor walls, to position it correctly with respect to the substrate, to stabilize the substrate temperature at < 1100°C and to prevent the growing deposit from being burnt away. Other problems seem to be the adhesion of the film on the substrate, the choice of substrates applicable in such a harsh environment, and the poor uniformity of the deposit. At present only a very limited set of data is available for this method. The deposition conditions and results as well as the advantages and drawbacks of the method are outlined in Tab. 8.

Tab. 8. Deposition conditions for atmospheric pressure RF plasma CVD of diamond.

substrate temperature [°C]	reactor pressure [bar]	total gas flow [sccm]	input power data
≈ 700-1200 (cooling!! T-measurement difficult)	1	80 000 (!!!)	20-60 kW (!!!) at 4 MHz

linear deposition rate [µm/hour]	deposition area [cm 2]	types of substrates
120-180	<3	Mo

quality/crystallinity [Raman data]	uniformity / homogeneity	special remarks
+++	-	very hot plasma fire ball

advantages of the method	drawbacks of the method
high deposition rates	small deposition area; stability and temperature control problematic; thickness and quality variations; poor adhesion.

(- = poor quality , +++ = excellent quality)

Microwave Plasma CVD of Diamond

The diamond research group at the NIRIM were the first to publish the application of a 2.45 GHz microwave plasma for the preparation of diamond films [3]. There is no doubt that this specific deposition method (along with the heated filament technique [6]) has moved diamond thin films close to an industrially applicable technology. The majority of diamond related patent applications utilizes this approach. Many of the experimental set-ups used so far are similar to the apparatus originally described by Kamo et al [3]. Its cross-section is shown in Fig. 5, along with a sketch of a different design that was jointly developed by Applied Science and Technology of Woburn, MA and one of the authors (P.K.B.) then being on a sabbatical leave with the diamond thin film group at the Materials Research Laboratory, The Pennsylvania State University, PA.

In the original NIRIM set-up the plasma is formed inside a silica tube. The substrate size is limited by the tube diameter and plasma etching of the reactor wall can cause contamination of the growing film. The bell jar system [29,30] allows for 4″ diameter substrates, separate heating by a resistance or an induction heater rather than by the plasma itself and greatly reduces contamination problems by utilizing a ball-shaped plasma that only touches the substrate surface. 2-3 inch diameter silicon wafers can be coated in such a system with thickness variations of less than 5%. Gas temperatures of 2500-2900°C inside the plasma ball are sufficient to uniformly heat substrates of up to 2 inches in diameter by just the plasma if approximately 1.5 kW of microwave power are supplied.

2.45 GHz microwaves were also applied by Y. Mitsuda et al [33] in a system that operates at atmospheric pressure rather than at 50-100 mbar, the pressure range where the other units are normally used. At higher pressures, as already mentioned above, the plasma is closer to thermal equilibrium and allows for higher gas temperatures and, similarly to the trend already observed in comparing low and high pressure DC or RF plasmas, it is again capable of substantially increasing the growth rate of the diamond deposition process. While linear growth rates of 0.5-5 μm/h are common for both the tubular and the bell jar reactor, rates of up to 30 μm/hour were achieved using Mitsuda's microwave torch approach. A cross-section of this apparatus is also included in Fig. 5.

In Tab. 9, data for both the common low pressure 2.45 GHz microwave plasma CVD technique and the 2.45 GHz plasma torch experiments are given, indicating the advantage of the higher rates and the drawback of the smaller area for the latter set- up.

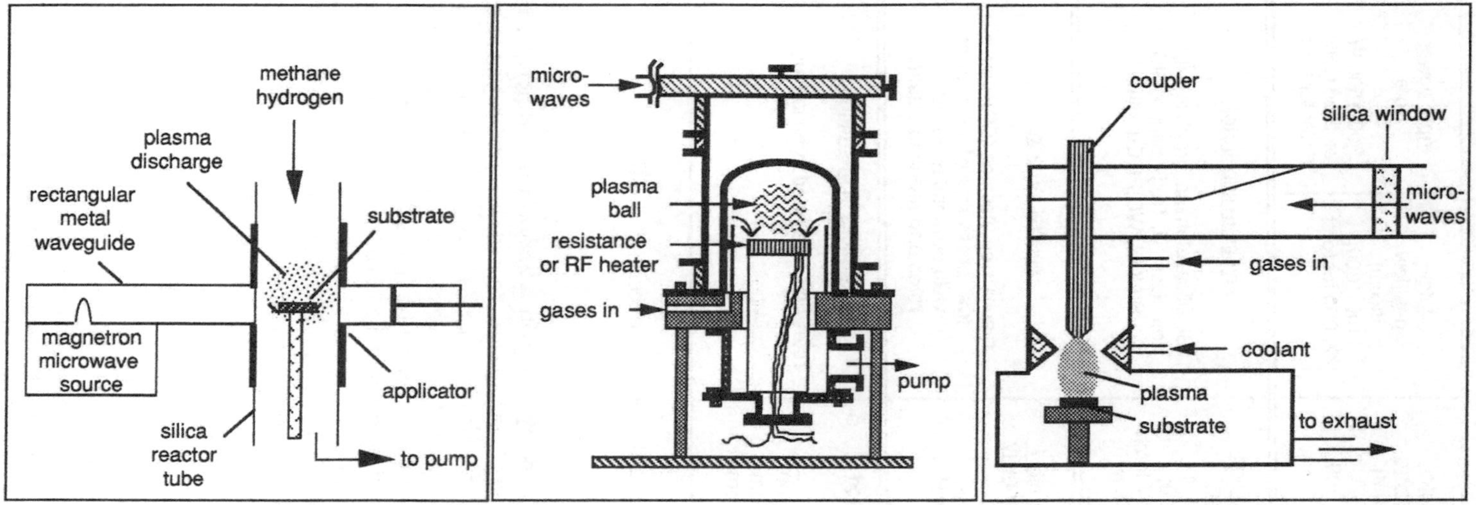

Fig. 5. Different set ups for microwave plasma CVD of diamond.

<u>left :</u> original tubular reactor developed by M. Kamo and collaborators at NIRIM [3].
<u>middle:</u> bell jar reactor with separate substrate heater suitable for substrates of up to 4 inches diameter;this unit has been jointly developed by Applied Science and Technology and P.K. Bachmann [29,30].
<u>right:</u> microwave plasma torch set up for atmospheric pressure microwave plasma CVD of diamond; this unit has been developed by Y.Mitsuda and collaborators [33].

Tab. 9. Low pressure microwave plasma and atmospheric pressure microwave plasma torch CVD of diamond

substrate temperature [°C]	reactor pressure [mbar]	total gas flow [sccm]	input power data
350 -1100 (optimum ≈ 1000; low rates at low T)	20-100	100-1000; 50 000 for torcl	300-1500 W at 2.45 GHz; 5 kW for torch

linear deposition rate [μm/hour]	deposition area [cm 2]	types of substrates
1-5; 10-30 for MW-torch at 1 bar	≈ 30; 5 for MW-torch at 1bar	Si, silica, alumina, graphite, Ni, soft glass, Ti, Ta, Si-nitride, SiC, SiAlON, WC,Pt, Cu, other

quality/crystallinity [Raman-data]	uniformity / homogeneity	special remarks
+++	+++ (low P)	magnetized plasmas (ECR) at low pressures of < 0.1 mbar yield diamond at lower rates and reduced quality

advantages of the method	drawbacks of the method
very stable (cont. operates for weeks) high quality material; uniform rates and area are not too small	rates and area need to be improvec

(- = poor quality , +++ = excellent quality)

Magnetized and electron cyclotron resonance (ECR) 2.45 GHz microwave plasmas have also been used for diamond deposition experiments [34,35]. In the normal pressure range of more than 10 mbar, the collision rate of the electrons in the plasma is far to high for full ECR action and the plasma is hardly affected by the presence of a magnetic field. At pressures of less than 0.1 mbar, where ECR is starting to be significant, both the quality of the deposit and the deposition rate are markedly reduced compared to conventional microwave plasma CVD of diamond, probably due the now non-isothermal plasma, were only the electrons are hot and the heavy particles remain low in temperature.

Microwave frequencies other than the common 2.45 GHz where also exploited, namely 0.915 GHz [26] and X-band frequencies around 8 GHz [36]. The X-band experiments were performed at rather low power levels of only 300 W and at pressures of approx. 0.1 mbar. On non-diamond substrates, only diamond-like (DLC) films were deposited. The 915 MHz experiments by S. Miyake and collaborators [26] were carried out at both 25 mbar and at atmospheric pressure using the same experimental set-up, however with 1 kW and 7 kW input power, respectively. At the low pressure of 25 mbar, rates of ≈ 1 μm/hour that are also common for the normal 2.45 GHz experiments, were achieved, while at 1 bar diamond grew at a rate of up to 50 μm/hour. This corresponds well with the results obtained when using the 2.45 GHz atmospheric pressure microwave torch rather than the conventional low pressure plasma and illustrates, that the excitation frequency of the plasma does not seem to be of specific importance for the deposition process. On the other hand, pressure increase accompanied by higher power input again results in higher deposition rates.

To conclude the comparison of methods,we have summarized the present status of diamond CVD in Tab.10. It illustrates that a variety of quite different methods are capable of producing high quality material on various non-diamond substrates. Which of the methods to choose depends on the specific application. For contamination-free, uniform coatings on medium size substrates, low pressure microwave plasma CVD is probably a good choice. For bulk material on small areas, DC arc discharges, DC jets and flames seem to be promising. In cases were contamination of the film by the filament material is not a drawback and large areas are required, the heated filament method has its advantages.

CORRELATIONS AND THERMODYNAMICAL CONSIDERATIONS

There are additional varieties of one or the other method published in the literature and in addition to the method of initiating the CVD reactions, additives to the CVD gas phase such as oxygen, boron or nitrogen do significantly affect the rate and quality of the deposited material. Nevertheless, there is undoubtedly a common trend for very different deposition techniques: higher temperatures of the CVD gas phase result in higher linear diamond deposition rates. The substrate temperature has, of course, to remain below 1200 °C to prevent graphitization of the growing film. This requires rapid quenching of the from very high to relatively moderate temperatures. Other deposition parameters such as e.g. a specific way of initiating the CVD processes or the excitation frequency to sustain a plasma seem to be less important for high rate diamond synthesis. Both less than 0.1 and ≈ 1000 μm/hour were reported for DC plasmas. Both, less than 0.1 and 180 μm/hour were reported for RF discharges. High rates of 100 μm/hour were reported for combustion flames.

In several cases, high deposition rates correlate with high total gas flows in the deposition zone. For atmospheric pressure RF plasma, DC jet, and microwave torch diamond CVD, flow rates of up to 80 slm were used. However, there are also very important exceptions. In both publications on medium

Tab. 10. Diamond Deposition Processes: Summary of Methods and Results

Method \ Results	rate [μm/hour]	area [cm^2]	quality [Raman]	substrates	advantages	drawbacks
flame	30-100	<1	+++	Si, Mo, TiN	simple	area, stability
heated filament	0.3-2	100	+++	Si, Mo, silica, etc.	simple, large area	contaminations, stability, no O_2
DC discharge (low pressure)	< 0.1	70	+	Si, Mo, silica, etc.	simple; large area	quality, rate
DC discharge (medium P)	20-250	<2	+++	Si, Mo, alumina	rate, quality	area
DC plasma jet	930	<2	+++	Mo, Si	highest rate, quality	area, stability, homogeneity
RF (low pressure)	< 0.1	?	-/+	Si, Mo, silica BN, Ni	scale-up	quality, rate, contaminations
RF (thermal,1 atm)	180	3	+++	Mo	rate	area, stability, homogeneity
Microwave (0.9-2.45 GHz)	1 (at low P) 30 (high P)	40	+++	Si, Mo, silica WC, etc.	quality, stability, (reasonable rate & area)	rate, area
Microwave (ECR 2.45 GHz)	0.1	<40	-/+	Si	area (?), low P	quality,rate, contaminations

pressure DC plasma diamond CVD by K. Suzuki et al [15, 15a], for instance, total gas flows of only 20-100 sccm resulted in rates of 20-250 μm/hour. Variations of the plasma power and hence the temperature of the CVD gas phase were sufficient to change the rates considerably. Taking all the experimental evidence into account we concluded [43] that a hot spot in the CVD gas phase with gas temperatures of more than 4000 °C is desirable for high rate diamond synthesis.

In Fig. 6, the correlation between the linear growth rates and the approximate temperatures of the CVD gas phases of basically all types of CVD methods, i.e. thermal decomposition, low pressure DC and RF plasma CVD, hot filament and microwave plasma CVD, flame synthesis and hot plasma CVD is depicted illustrating this link.

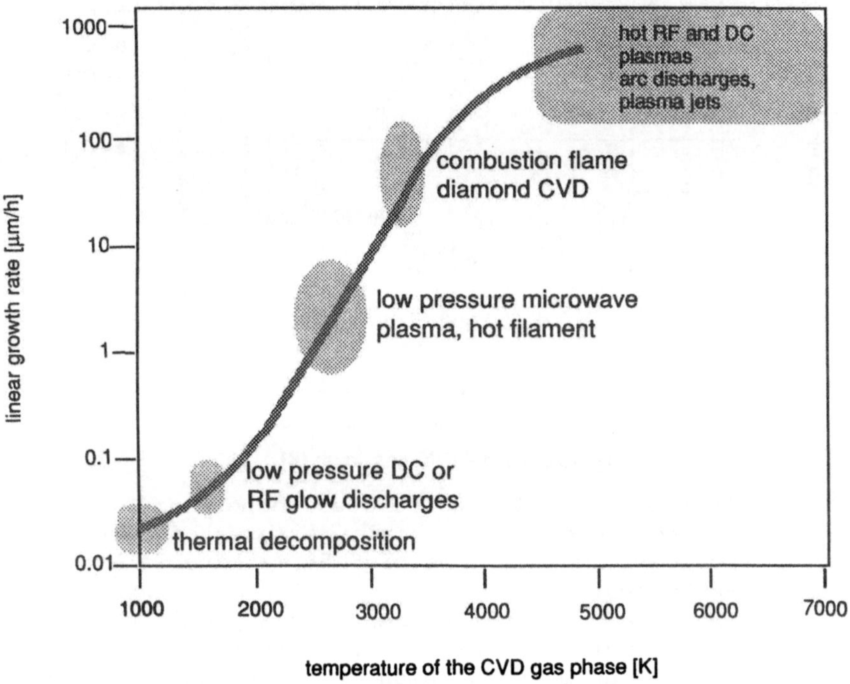

Fig.6. Correlation of the gas phase temperatures of various diamond deposition methods and the reported linear growth rates illustrating the importance of a hot spot in the gas phase for high rate diamond synthesis.

In order to clarify the role of various species under discussion for modelling diamond CVD [44,45], we performed thermodynamical calculations for typical C/H- and C/H/O-containing gas mixtures covering the temperature range of 300 K to 6000 K. The program we used for this purpose, its algorithm and its well maintained, frequently updated data base (including the latest JANAF thermochemical tables data) has been developed at the Philips Research Laboratories Aachen, FRG and is successfully applied to predict the performance of light sources as well as to model their high temperature chemistry at

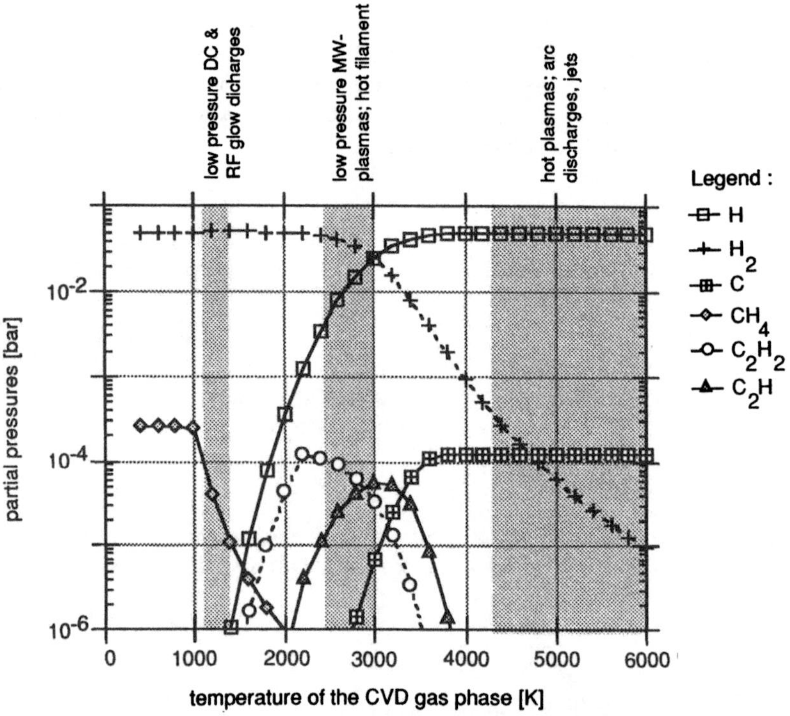

Fig. 7. Thermodynamically calculated temperature dependence of the partial pressures of species formed from a mixture of 0.5% methane in hydrogen at 50 mbar.
The shaded areas represent the approximate gas temperature regimes of the various diamond deposition methods. High rate methods obiously operate at temperatures where neutral C atoms dominate the CVD gas phase.

temperatures of up to 6000 K (see [46] for a comprehensive review). The results of these calculations are given in Fig. 7 (for a mixture of 0.5 % methane in hydrogen at 50 mbar) and Fig. 8 (for a mixture of 53 % acetylene and 47 % oxygen). Along with the partial pressures of the various gas phase components, the operating temperature ranges for the different deposition techniques are indicated by the shaded areas. Such calculations are meant as a guideline rather than quantitatively describing the reality for each and every component at each and every temperature. Evidently, high rate methods operate in a regime where neutral C atoms play an important role in the gas phase. In the case of the rather complex C/H/O-system there is still CO present at high temperatures. But the CO partial pressure hardly varies between 1000 and 6000 K and would not explain any rate changes.

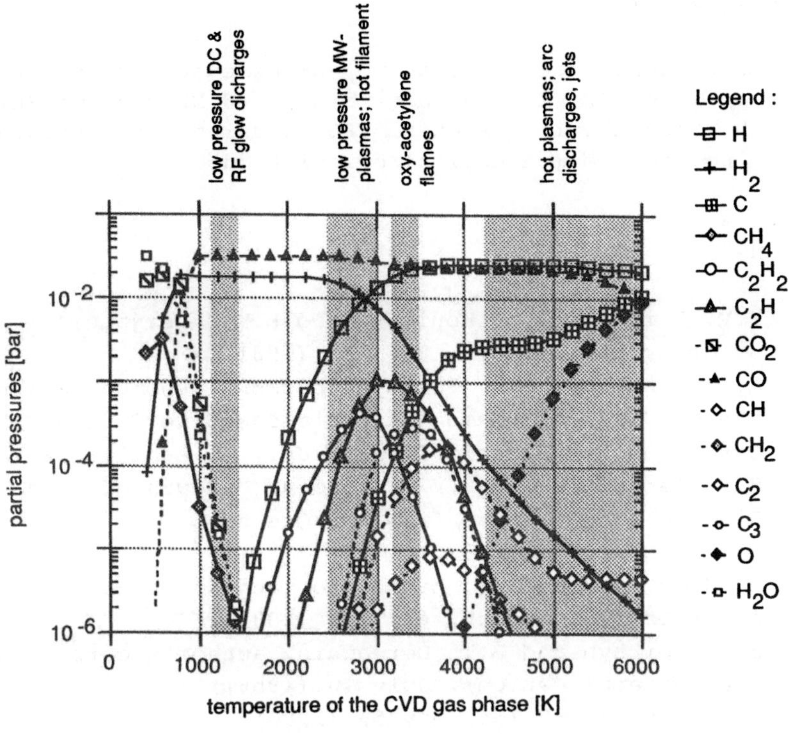

Fig. 8. Result of thermodynamical calculations for a mixture of 53 % acetylene and 47 % oxygen at 50 mbar.
Again, the shaded areas represent the approximate gas temperature regimes of the various diamond deposition methods.

Other frequently discussed species such as C_2H_2 or C_2H [45] exhibit a partial pressure maximum at temperatures that are normally reached in hot filament or low pressure microwave plasma CVD experiments. Their concentrations drop and finally vanishes at higher temperatures and would also not explain the rate increase observed for "hot" deposition techniques.
Our assumption [43] that neutral carbon atoms are important gas

phase species for high rate diamond deposition is obviously supported by the results of our calculations (no attempt to explain the details of the molecular steps on or very close to the very surface of the diamond film is made,though !) and is confirmed by the conclusions of K.Suzuki et al [15a]. In addition, the gas phase has to be rapidly quenched to temperatures below 1200°C and, no doubt, atomic hydrogen has to be present in order to either selectively remove graphitic phases to prevent the formation of sp^2-hybridized carbon atoms thus allowing for high rate synthesis of high quality diamond.

ACKNOWLEDGEMENT

It is with great pleasure that we acknowledge the assistance and support by Dr. E. Schnedler and Dr. U Niemann of Philips Research Laboratories, Aachen, FRG in using their program and performing the thermodynamical calculations.

REFERENCES

[1] B.V. Spitsyn, L.L. Bouilov, and B.V. Derjaguin, J. of Crystal Growth, 52, 219 (1981).

[1a] A.Alexenko and B.V. Spitsyn, personal communication at the NATO ASI on Diamond and Diamond-Like Carbon Films, Castelvecchio Pascoli, Italy, July 23-Aug.3, 1990.

[2] B.V. Derjaguin and D.V. Fedoseev, "Growth of Diamond and Graphite from the Gasphase", Nauka, Moscow (1977).

[3] M. Kamo, Y. Sato, S. Matsumoto, and N. Setaka, J. of Crystal Growth, 62, 642 (1983).

[4] B.V. Spitsyn and B.V. Derjaguin, author's certificate (patent application), July 10 (1956); USSR Patent 339 134, May 5, (1980).

[5] W. Eversole, US patents 3030187 and 3030188 (1962) (filed in 1959).

[5a] H.J. Hibshman, US patent 3371996 (1968) (filed in 1964).

[6] S. Matsumoto, Y. Sato, M. Kamo, and N. Setaka, Jpn. J. of Appl. Phys. 21, L183 (1982).

[7] Y. Hirose and Y. Terasaki, Jpn. J. of Appl. Phys. 25, L519 (1986).

[8] A. Sawabe and T. Inuzuka, Appl. Phys. Letters, 46, 2, 146 (1985).

[8a] M.Sommer and F.W. Smith, presented at the "Diamond, Boron Nitride, Silicon Carbide and Related Wide Bandgap Semiconductors" Symposium,MRS Fall Meeting, Boston (1989), MRS Symposium Proceedings Vol. 162 (1990).

[9] Y. Hirose and N. Kondo, Extended Abstracts, 35th Spring Meeting, Jpn. Appl. Phys. Soc. March 29, 434 (1988) .

[10] Y. Hirose, S. Ananuma, N. Okada, K. Komaki, Proceedings of the "First International Symposium on Diamond and Diamond-Like Films", The Electrochemical Society, Pennington, NJ, Proceedings Vol. 89-12, 80 (1989).

[10a] Y. Matsui, A. Yuuki, M. Sahara, and Y. Hirose, Japanese J. Appl. Phys., Vol. 28, no. 9, 1718 (1989).

[10b] Y. Hirose and S. Amanuma, Denshi Zairyo, Vol.28, no.8, 45 (1989).

[11] K. Snail, L. Hanssen, W. Carrington, D. Oakes, and J. Butler, Proceedings of the First International Conference New Diamond Science and Technology, Tokyo, Oct. 24-26, 38 (1988).

[12] B.V. Derjaguin, L.L. Bouliov, and B.V. Spitsyn, Arch. Nauki Mater., 7, 2, 111 (1986).

[13] J.M. Pinneo, 1st Diamond Technology Initiative Workshop, MIT Lincoln Laboratories, Boston, MA, Feb. 2, paper 4 (1987).

[14] K.V. Ravi and M.I. Landstrass,Proceedings of the "First International Symposium on Diamond and Diamond-Like Films", The Electrochemical Society, Pennington, NJ,Proceedings Vol. 89-12, 24 (1989).

[15] K. Suzuki, A. Sawabe, H. Yasuda, and T. Inuzuka, Appl. Phys. Letters, 50, 12, 728 (1987).

[15a] K. Suzuki, A. Sawabe, H. Yasuda, and T. Inuzuka, Japanese J. Appl. Phys, (1990).

[16] B. Singh, O.R. Mesker, A.W. Levine, and Y. Arie, Appl. Phys. Letters, 52, 1658 (1988).

[17] N. Ohtake, H. Tokura, Y. Kuriyama, Y. Mashimo, and M. Yoshikawa, Proceedings of the "First International Symposium on Diamond and Diamond-Like Films", The Electrochemical Society,Pennington, NJ, Proceedings Vol. 89-12, 93 (1989).

[18] S. Matsumoto, M. Hino, Y. Moriyoshi, T. Nagashima, and M. Tsutsmui, US Patent No.4 767 608, August 30, 1988 (filed Oct. 19, 1987).

[19] K. Kurihara, K. Sasaki, M. Kawarada, and N. Koshino, Appl. Phys. Lett., 52, 6, 437 (1988).

[20] S. Matsumoto, in Diamond and Diamond-like Materials Synthesis edited by G.H. Johnson, A.R. Badzian, and M.W. Geis, Mat. Res. Soc.Symp. Extended Abstracts, EA-15, Pittsburgh, PA,119, (1988).

[21] P. K. Bachmann, H. Lydtin, D.U. Wiechert, J.J. Beulens, G. Kroesen, D.C. Schram, Proceedings of the "Third International Conference on Surface Modification Technologies", Neuchatel,Switzerland, The Minerals,

Metals & Materials Society (TMS),Warrendale, PA, (1989) in press.
[22] S. Matsumoto, J. Mat. Sci. Lett., 4, 600 (1985).
[23] D.E. Meyer, R.O. Dillon, and J.A. Woolham, J.Vac. Sci.Tech. 7,3, 2325 (1989).
[24] P. Wood, T. Wydeven, and O. Tsuji, Abstracts of the 1st Int. Conference on the New Diamond Science and Technology, Oct. 24-26, Tokyo, P 100, Japanese New Diamond Forum (1988).
[25] S. Matsumoto, M. Hino, and T. Kobayashi, Appl. Phys. Lett., 51, 10, 737 (1987).
[26] S. Miyake, W. Chen, A. Hoshino, and Y. Arata, Trans. JWRI, 17, 2, 323 (1988).
[27] S. Matsumoto, Proceedings of the "First International Symposium on Diamond and Diamond-Like Films", The Electrochemical Society, Pennington, NJ, P Proceedings Vol. 89-12, 50 (1989); (see also ref. [3]).
[28] A.R. Badzian, B. Simonton, T. Badzian, R. Messier, K.E. Spear and R. Roy, Proc. SPIE, 683, 127 (1986).
[29] P.K. Bachmann, W. Drawl, D. Knight, R. Weimer, and R.F. Messier, in "Diamond and Diamond-Like Materials"(edited by A. Badzian, M. Geis, and G. Johnson), Materials Research Society, Pittsburgh, PA, Extended Abstracts Vol. EA-15, (1988) 99.
[30] P. K. Bachmann, R. Weimer, and R. Messier, Diamond Technology Initiative Symposium, July 12-14,1988, Crystal City, Arlington, VA. Technical Digest, Paper T 2.
[31] K. Ishibori and Y. Ohira, Abstracts of the 1st Int. Conference on the New Diamond Sci. and Tech. , Oct. 24-26, Tokyo, P104, Japanese New Diamond Forum (1988).
[32] D.M. Tung, W.L. Hsu, K.F. McCarty, Proceedings of the "First International Symposium on Diamond and Diamond-Like Films", The Electrochemical Society, Pennington, NJ, Proceedings Vol. 89-12, 500 (1989).
[33] Y. Mitsuda, T. Yoshida, and K. Akashi, Rev. Sci. Instrum. , 60, 2, 249 (1989).
[34] H. Kawarada, K.S. Mar, and A. Hiraki, Jpn. J. Appl.Phys. 26, 6, L1032 (1987).
[35] J. Suzuki, Jpn. J. Appl.Phys. 28, 2, L 281 (1989).
[36] M. Aklufi and D. Brock, Proceedings of the "First International Symposium on Diamond and Diamond-Like Films", The Electrochemical Society, Pennington, NJ, Proceedings Vol. 89-12, 114 (1989).
[37] Peter K. Bachmann and Russell F. Messier, Chemical & Engineering News, 67, 20, May 15, 24 (1989).

[38] K. Spear, J. Am. Ceram. Soc., 72 (2), (1989), 171.
[39] J.C. Angus and C.C. Hayman, Science, 241, (1988), 913.
[40] R.C. DeVries, Annu. Rev. Mater. Sci., 17, (1987) 161.
[41] A. Hara and N. Fujimori, Proceedings of the "First International Symposium on Diamond and Diamond-Like Films", The Electrochemical Society, Pennington, NJ, Proceedings Vol. 89-12, 14 (1989).
[42] A.R. Badzian, P.K. Bachmann, T. Hartnett, T. Badzian, and R.F. Messier, in "Amorphous Hydrogenated Carbon Films", edited by P. Oelhafen and P. Koidl, European Materials Research Society, Vol. XVIl, Les Editions de Physique, Paris, France, 63 (1987).
[43] P.K. Bachmann and H. Lydtin, presented at the 1989 MRS Fall Meeting, Nov. 27, 1989, paper I4.1; in "Characterization of Plasma-Enhanced CVD Processes" (edited by G. Lucovsky, D.E. Ibbotson, and D.W. Hess), MRS Symposium Proccedings, Vol. 165, Materials Research Soc. Pittsburgh, PA, USA, in press (1990).
[44] M. Tsuda, M. Nakajima, and S. Oikawa, Jpn. J. Appl. Phys., 26, L527 (1987).
[45] M. Frenklach and K.E. Spear, J.Mat.Res.,3, 1, 133 (1988).
[46] E. Schnedler, in Proc. of the Symp. on Sci. and Technol. of High Temperature Light Sources "High Temperature Lamp Chemistry" edited by E.G. Zubler, The Electrochemical Soc., Pennington, NJ, Proc. Vol. 85-2, p. 95 (1985).

ORIGIN AND EVOLUTION OF THE SCIENCE AND TECHNOLOGY OF DIAMOND SYNTHESIS IN THE USSR

Boris V. Spitsyn

Institute of Physical Chemistry USSR Academy of Sciences,
31 Leninsky Prospekt, 117915 Moscow, USSR

INTRODUCTION

Research and industrial development of methods of diamond synthesis at ultra-high pressures and high temperatures (the UHP/HT methods) have not reduced the interest toward developing other methods of diamond synthesis. There are several reasons for this, and the most important, whose value can only grow in the foreseeable future, are as follows.

1. Diamond is a crystal of pure carbon with a simple structure. Each carbon atom in the diamond structure has six electrons, four of which participate in the formation of chemical carbon-carbon bonds. This single sigma-bond determines the spatial skeleton of the diamond crystal. Thus, the diamond is one of the simplest crystals, and its fundamental importance to the physics and chemistry of solids lies primarily in the fact that theoretical calculations (in particular those starting from first principles) are becoming ever more realistic, which means that they produce data on the entire list of diamond properties and also predict the behaviour of the crystal in different physical and chemical conditions.

2. Diamond possesses, within a wide temperature range, a collection of properties many of which are extreme if compared with those of other known crystalline substances (Table 1). Some of these properties, say, thermal conductivity, can be exceeded in no other pure or composite substances and are only observed in pure diamond.

3. The UHP/HT methods employ conditions that require special equipment and materials. Diamonds synthesized by these methods require cleaning from impurities of graphite and metallic catalysts which usually constitute the synthetic raw material used to manufacture diamond and diamond-containing materials and products.

These three reasons, related both to diamonds and to traditional methods of synthesis, have generated interest in discovering more accessible and flexible methods of synthesis, methods that could ensure the manufacture of diamond at fairly medium temperatures,without using

Diamond and Diamond-Like Films and Coatings
Edited by R.E. Clausing *et al.*, Plenum Press, New York, 1991

enormous pressures or catalysts, and giving diamonds or diamond materials in a form most suited to the requirements of science and technology, for instance, diamond in the form of singlecrystalline and polycrystalline films.

Table 1. Some Physical Properties of Diamond

Property	Magnitude at 300 K	Source
Compressibility, mm^2/kgf	1.8×10^{-5}	1
Ultrasonic velocity, km/s	18.2	1
Thermal coefficient of linear expansion, K^{-1}	0.8×10^{-6}	2
Theoretical strength along <111>, kgf/mm^2	1.24×10^4	3
Experimental strength along <111>, kgf/mm^2	3.85×10^2	3
Thermal conductivity, W/m·K	2100	4,5
Band gap, eV	5.43	6
Specific resistance, Ω·cm	10^{12}-10^{16}	7
Electron mobility, $cm^2/V \cdot s$	2000	8,9
Hole mobility, $cm^2/V \cdot s$	1500	9

METASTABILITY OF DIAMOND

The metastability of diamond can be defined quantitatively, for instance, at a pressure of 1 atm as the difference in the standart Gibbs free energies[10] of diamond and graphite, the most stable phase of carbon At 300 K and a pressure of 1 atm this difference is about 0.5 kcal/mol. For the sake of comparison we note that the Gibbs free energy of white phosphorus is about 4 kcal/mol higher then that of red phosphorus at 300 K. Nevertheless, metastable white phosphorus can be obtained through simple sublimation of red phosphorus. Another example is the huge body of data gathered in the synthesis of organic compounds, whose metastability is a rule rather than an exception. Hence, one is inclined to think that the metastability of diamond cannot be an insuperable obstacle to its synthesis. On the other hand, the most natural path in searching for a new method of synthesizing diamonds is to employ carbon crystallization from the vapor or gas phase.

PHYSICAL AND CHEMICAL CRYSTALLIZATION

Processes that are promising for crystallization of diamond from the gas phase can be divided into two classes, physical and chemical. The first incorporates methods of synthesis in which the crystallization medium and the crystal have the same chemical composition, and the second deals with processes that involve a change in this composition during crystallization. In the latter case the phase transition can include both homogeneous and heterogeneous chemical reactions.

The simplest method for diamond crystallization could be the sublimation of graphite followed by condensation of carbon vapor on the surface of the seed diamond crystal. Although theoretically this process is possible, the following circumstances hinder its realization.

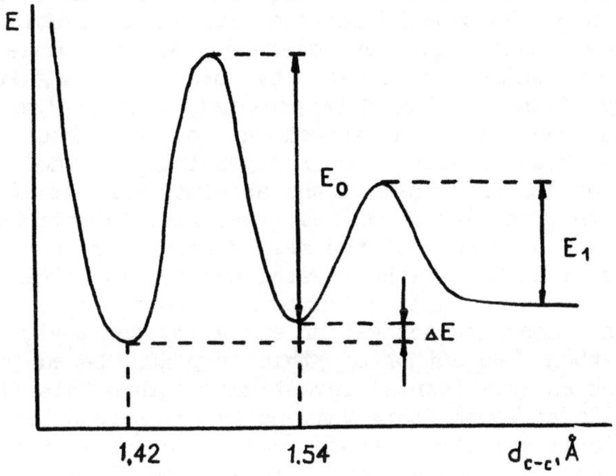

Fig. 1. Energy of the diamond crystal/carbon-containing compound molecule system as function of the distance between a carbon atom in the diamond crystal and it in the molecule.

1. According to LEED data[11], the diamond surface retains the bulk structure up to 1300-1400 °C in a ultra-high vacuum. At this temperature the equilibrium carbon vapor pressure above the diamond surface is approximately twice[12] the equilibrium vapor pressure above the graphite surface[13] and roughly amounts to 10^{-13} Torr. To achieve acceptable crystallization rates (of the order of 1 μm/h) the carbon vapor pressure above the diamond surface must exceed 10^{-6} Torr. Hence, to achieve sublimation growth with that rate a supersaturation of 10^{7} is necessary, with the result that the diamond films will, of course, defective. Besides, at such high supersaturations there is a noticeable nucleation and growth of the graphite phase[14], which is stable at low pressures.

2. The composition of the carbon vapor above the graphite surface corresponds to an excess of C_3 molecules with multiple bonds between the carbon atoms, which prevents these molecules from being incorporated in the diamond structure sufficiently smoothly.

3. Finally, even a single carbon atom in its ground state is bivalent, and transforming it into the tetravalent state, characteristic of the carbon atom in the diamond structure, requires 96 kcal/mol of energy[15].

The above reasons make physical crystallization of diamond by thermally vaporizing graphite highly improbable. At the beginning of our research we believed that the most promising processes for diamond synthesis would be those of chemical crystallization. These require as initial substances simple carbon-containing compounds in which carbon atoms are in the same sp^3-valent state as carbon atoms in the diamond structure. This process, obviously, may be assisted by a strong angular orientation of the bonds originating at the carbon atoms in the surface layer of the diamond crystal. Schematically the energy of the system consisting of a diamond crystal and a carbon-containing molecule is shown in Fig. 1.

Let us assume that the carbon-containing compound is decomposed to elemental carbon on the diamond surface with an activation energy E_1 considerably lower than E_1, the diamond-graphite phase transition activation energy, which is close, by order of magnitude, to the breaking energy of the C—C boud (approximately 85 kcal/mol). Then the chemical process may result in attachment of a carbon atom to the lattice at the bond distance corressponding to the local minimum (diamond structure) ruther then at the absolute minimum of energy, which corresponds to the graphite structure (see Fig. 1).Obviously, to ensure a fairly high rate of growth of the seed diamond crystal via chemical crystallization of carbon on the crystal surface it is resonable to use molecules of carbon-containing compounds that have low thermal stability, with a moderate activation energy E_1 for their decomposition to elemental carbon. The author of this report, based on the above ideas, undertook an experimental investigation into this field as early as in the mid-50s at Tomsk State University (Western Siberia, USSR). The least stable simple volatile carbon compounds, the tetrabromide and tetraiodide of carbon, were suggested in a filed Claim for an Invention during the same period. Subsequently, experimental studies were carried out in the laboratory of Prof. S.Z. Roginsky in 1956 and in the laboratory of Prof. B.V. Derjaguin in 1957 (both at the Institute of Physical Chemistry, Moscow, USSR).

CHEMICAL CRYSTALLIZATION OF DIAMOND

The first experiments in diamond regrowth were carried out at conditions where vapors carbon tetrabromide or tetraiodide interacted with the surface of a heated single crystal of diamond. To prevent homogeneous decomposition the substances were fed to the crystallization zone in molecular-flow conditions, with an effective pressure of about 10^{-3} Torr. When tetraiodide of carbon was used, special measures were taken to ensure that only single collisions of these unstable molecules with the diamond surface took place. Thus, the average temperature of the gas phase could be kept lower than the temperature of the seed crystal and decomposition was achieved only at the moment when a molecule collided with the surface. At these conditions it was discovered that after two hours of growth at 900 °C transparent protrusions up to 1-μm high and up to 10 μm in cross section on the polished surface (110) of natural diamond were formed. These protrusion were discovered to have the same blue cathodoluminescence (with an electron energy about 1 keV) as the natural diamond crystal. It was found, too, that protrusions produced scratches on samples of hexagonal silicon carbide, the hardest material after diamond. The studied properties of the grown formations, therefore, suggested that the protrusions were of diamond origin.

In 1956 the present author and in 1962 the present author and B.V. Derjaguin filed Claims for Inventions related to this process, which resulted in an Inventor's Certificate being issued on a method for growing diamond[16], with conserving of priority of the first file. A similar method is described in US patents obtained by William G. Eversole of the Union Carbide Corporation and published in 1962. Eversole suggested growing diamond films on seed diamonds by thermal decomposition on the surfaces of the seed diamonds of methane, other simple methyl-containing organic compounds, and carbon monoxide[17,18]. The linear growth rate amounts to several angstroms per hour. The growth of a new diamond was monitored by a weight method, with the seed crystals taken in powder form, with a particle size in the 0 to 1 μm range.

The use of a quartz microbalance made it possible for the first time to study the kinetics of isothermal chemical crystallization of diamond from methane at pressures ranging from 10^{-1} to 1 Torr and temperatures from 900 to 1100 °C.[19] Notwithstanding the simultaneous nucleation and growth of graphite, it was found possible to estimate the activation energy of diamond growth. Later, in the thorough experiments conducted by Prof. J.C. Angus with collaborators in the United States and by Prof. P.A. Tesner with collaborators in the Soviet Union, the activation energy of the heterogeneous reaction

$$CH_4 \longrightarrow C\ (\text{diamond}) + H_2$$

was estimated at a value ranging from 55 to 58 kcal/mol[25,26].

Initially the isothermal or near-isothermal chemical crystallization processes[16-18,21,26] had a number of restrictions and drawbacks, namely, they employed rather stable carbon compounds as source gases. The decomposition of these compounds occurred far from the state of thermodynamic equilibrium, but the rate of chemical crystallization was moderate owing to the rather high activation energy.[25,26] In addition, because of the process being highly nonequilibrium there was a considerable probability of spontaneous nucleation and growth of graphite, which seemed to be unavoidable.[27]

FILAMENTARY DIAMOND CRYSTALS

The works on filamentary diamond crystals, also performed at the Institute of Physical Chemistry of the USSR Academy of Sciences, occupy a special position. The discovery of these surprizing formations grown on the surface of natural diamond crystals,was first communicated by Prof.B.V.Deryaguin at the session of the Presidium of the USSR Academy of Sciences on 14 April, 1967. The report was followed by publications[20-22,14] including a description of the general conditions of the experiments. Like any other invention , the formation of filamentary diamond crystals was discovered unexpectedly. The initial objective of the experiments by D.V.Fedoseev, B.V.Deryaguin and co-authors was to carry out chemical crystallization of diamond under the conditions, when a co-deposition release of graphite is humpered or, possibly, even excluded completely. For this purpose the seed diamond crystal was heated not in the continuous but in the pulse mode; the pulse duration was of the order of 4-10 ms, the pulse recurrence frequency was 25-5 Hz and the on-off time ratio was of the order of 10. The main idea of the experiment was, that at the moment of heating the conditions for the release of carbon from the gas phase (contained CH_4, n-C_8H_{18} or other hydrocarbons, at pressure - of order of ten Torr) are provided. Diamond as a ready macrophase may grow practically without inertia, while the graphite nucleus requires restricted time for starting of growth. Specifing a rather low pulse duration for heating it is possible to eliminate graphite formation on the diamond surface in the process of its growth. A similar process was first described in, where the growth of the seed crystal was carried out by means of a pulsed increase in the temperature of the surface in the molecular flow of CCl_4, CBr_4, CI_4 or CSe_2. Consecutive heatings of small parts of the surface up to 1800-2500°C by a scanning electron beam leads to the formation of a continuous transparent diamond film. During the growth of filamentary crystals the optical radiation heating the seed crystal was obtained using an optical furnace[24]. The formation of elongated (the length-diameter ratio is more than 10) formations in a small number of the experiments was the most interesting but hardly reproducible result. The filamentary crystals grew from separate parts of the crystal at rate of up to 0.25 mm/h, and in a number of cases did not have outer facetting: their cross-section was usually constant and close to round and oval shape. In several cases the crystalline nature of the surface of the filamentary formations was revealed by the Laue microdiffraction of high-energy electrons (65 keV). However, the data available at present do not allow us to assert that all the filamentary crystals grown consist entirely of the diamond phase. Therefore, the nature of these curioues formations can not be established with a necessary degree of completeness.

THE CHEMICAL TRANSPORT REACTION

Carrying out chemical crystallization of diamond with carbon transport through the gas phase (the gas-transport chemical reaction) allows in principle to proceed with the reaction closer to equilibrium condition than isothermal chemical crystallization would permit. At the same time, one could expect a reduction in the hazards connected with graphite deposition. However, the well-known Boudard reaction, tested by the author as early as 1961, produced no positive results because of the highly retarded kinetics of carbon monoxide disproportionation (even at 1000 °C). For this reason, in 1966 the present author and the engineer A.V. Lavrentyev attempted to enhance the rate of diamond growth by use the chemical transport reaction

$$C(\text{graphite}) + H_2 \xrightarrow{2000\ ^\circ C} C_nH_m(+H) \xrightarrow{1000\ ^\circ C} C(\text{diamond}) + \frac{m}{2} H_2 \qquad (1)$$

for diamond growth.

The main idea of the experiment was to enhance up diamond growth by employing the active hydrocarbon particles that had formed on a graphite source instead of prefabricated materials. These particles were used for growing a diamond substrate directly after their formation. We were therefore able, using process (1), to raise the diamond film growth rate considerably (up to 1 μm/h). In addition for the first time epitaxial (single-crystal) films of synthetic diamond were manufactured on single crystals of natural diamond (Figs. 2 and 3). Finally, it was established that diamond growth via process (1) is selective, that is, the graphite phase was found not to crystallize on the surface.

A thermodynamic calculation of the equilibrium in the graphite-hydrogen system has shown that at high temperatures (approximately 2500 K) and at $p \sim 10^{-2}$ bar the equilibrium concentrations of acetylene (main equilibrium hydrocarbon) and atomic hydrogen may constitute 1 and 10% of the total number of particles in the gas phase, respectively[28].

The reason for the high selectivity of the diamond growth process (1) is that the atomic hydrogen has an etching effect on graphite and graphite-like clusters but has practically no such effect on diamond[32].

Thus, it was found possible to obtain a stationary gaseous medium that included active hydrocarbon molecules out of which a diamond film grew at a considerable rate and, at the same time, graphite was removed selectively at no less a rate at the expense of hydrogen atoms. The publication of these results in the USSR began from 1969[28,29-32].

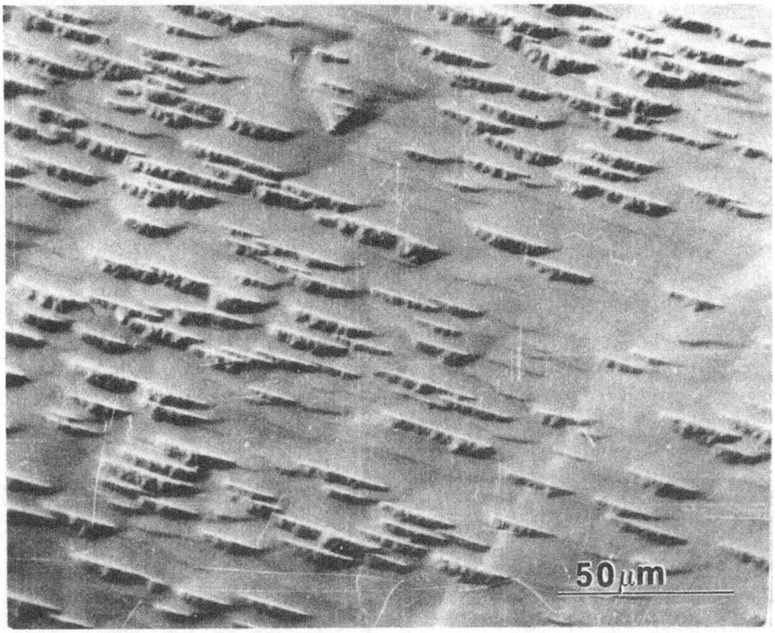

Fig. 2. Growth morphology of an epitaxial diamond film grown on the off 5° (100) face of natural diamond. The film is 2-μm thick.

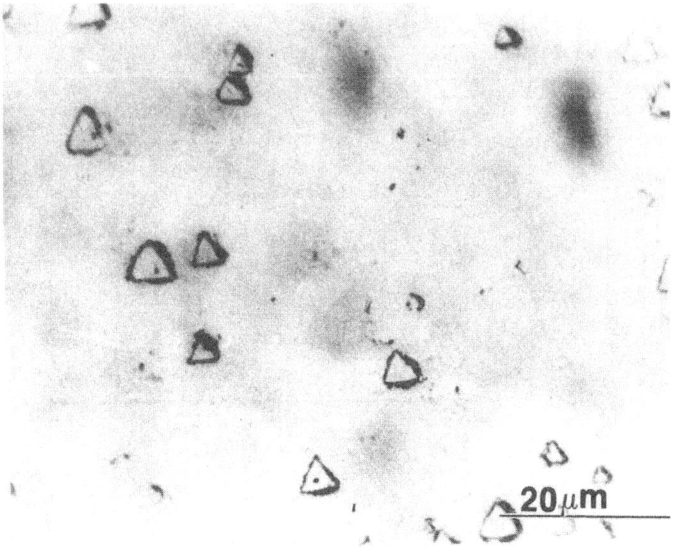

Fig. 3. An epitaxial diamond film 5.6-μm thick grown on a (111) face of a natural diamond octahedron.

Another method of creating a gaseous medium useful to grow fine-crystalline diamond films, with graphite practically not participating in the growth process, was discovered in 1970 and studied by V.P. Varnin and collaborators[33]. When activation of the gas phase was achieved in a glow discharge of mixtures of methane and other hydrocarbons with hydrogen, the hydrocarbons were converted to acetylene and other active hydrocarbon particles and simultaneously some part of the H molecules were dissociated into hydrogen atoms.

It must be noted at this point that the high activity of atomic hydrogen in its interaction with nondiamond forms of solid carbon (e.g. soot) has been known for a long time[34] and has even been used to clean grown diamond crystals from nondiamond carbon.[35] However, only in the experiments conducted by the author of the present review and then in Varnin's experiments was a gaseous medium used to synthesize diamonds, and this medium included both growth-hydrocarbon components and an etching component, atomic hydrogen[32].

The observed phenomenon can be formulated as follows: "A crystallization medium of a stationary chemical composition (C_nH_m, H, and H_2) was discovered in which the net rate of growth and etching of diamond was found to be positive (growth prevails) while the net rate of these processes for graphite was found to be negative (etching prevails)."

DIAMOND GROWTH KINETICS

The high selectivity of diamond growth made it possible to study the growth kinetics of epitaxial diamond films[28] For precise measurements of the temperature of the surface of the seed crystal at

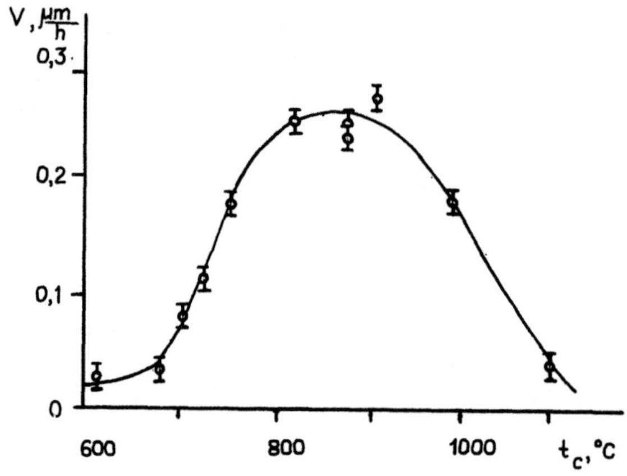

Fig. 4. Linear growth rate of diamond-film crystallized on (111) face of natural diamond vs. temperature of the seed crystal's surface. Chemical transport method; total pressure in gas phase were 12 Torr.

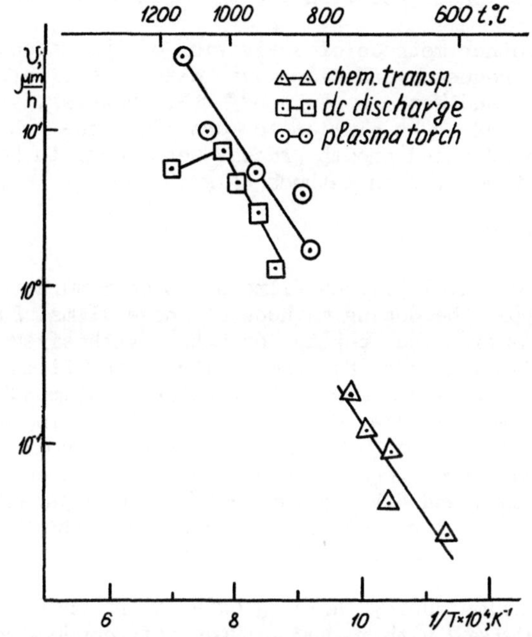

Fig. 5. The log of the linear diamond growth rate vs. inverse temperature (Arrhenius' coordinates) for various methods of chemical crystallization in activated hydrocarbon-hydrogen gas phase. Triangles, squares, and circles correspond to the chemical transport method[28], the DC discharge method[33], and the plasma torch method[46], respectively.

conditions of chemical transport we employed a specially designed diamond thermistor[51]. This enabled measuring the rate of diamond film growth at temperatures ranging from 620 to 1100 °C (Fig. 4)

The exponential dependence of the growth rate vs. crystalization temperature (Fig. 4) is typical for low-pressure/low-temperature conditions,for them is valied rate equition:

$$V(\mu m/h)=10^{5.83\pm0.39}\ P_{C_2H_2} \exp[-(25200\pm1800)/RT] \qquad (2)$$

Acetylene pressure(Torr)was established as equilibrium one at pressure-temperature condition for a carbon source in chemical transport reaction (1).Preexponentional factor in (2) became near gas-kinetical one, what say about first-order reaction for the diamond growth.

Starting in 1982, other methods of activation of the gaseous medium (heated filament, high-frequency and microwave discharges, etc.) were developed in Japan, USA, and Western Europe[36-41]. However, the main kinetic features were found to be similar to ours. For one thing, the activation energy of the diamond growth process were found to be roughly 25 kcal/mol in different activation methods (Fig. 5).

SEMICONDUCTOR DIAMOND FILMS

After dielectric epitaxial diamond films had been manufactured, it became possible to develop the doping methods of these films. But first it was necessary to determine the purity of the synthesized diamond films. According to SIMS data (Fig. 6), the purity of the films is not inferior to that of perfect single crystals of natural diamond[44]. The only exception is hydrogen (see Fig. 6). Its content in single-crystal diamond films ranges from 10^{-2} to 10^{-1} at.%. The forms in which hydrogen enters into the diamond have been studied by L.G. Karaseva, T.A. Karpukhina, L.D. Kislovsky, and B.V. Spitsyn[42,43], and apparently its main fraction is in the form of C—H bonds, by the observed characteristic frequencies in IR absorption spectra (Fig. 7).

By doping diamond films during their growth with phosphorus and boron a semiconductor diamond with *n*- and *p*-types of conductivity was manufactured[50]. In the latter case the measured Hall mobility was found to be moderate (Fig. 8) and increased with temperature. Possibly the data of Fig. 8 suggest a partial compensation of acceptors by intrinsic defects of the vacancy type.

NUCLEATION ON FOREIGN SURFACES

The high selectivity of the etching process[28] made it possible to ensure not only the growth of diamond films on a diamond surface but also spontaneous diamond nucleation on surfaces of nondiamond origin. As shown by L.L. Bouilov and his collaborators,[45] the origin, state, and treatment of the surface of the nondiamond substrate has a strong effect on the frequency of nucleation of diamond crystallization centers. On the surface of single-crystal copper the nucleation frequency was found to be several times lower than that on a polycrystalline surface, and annealing the polycrystal could considerably lower the number of centers of diamond nucleation and growth. While on copper and gold this frequency was found to be 10^2 $mm^{-2}h^{-1}$, on silicon, tungsten, and other carbide-forming substrates it was found to be of the order of 10^6 $mm^{-2}h^{-1}$. Bouilov's experiments established that during the initial period of saturation of the carbide-forming substrate with carbon the nucleation probability may be near to zero; as the surface becomes

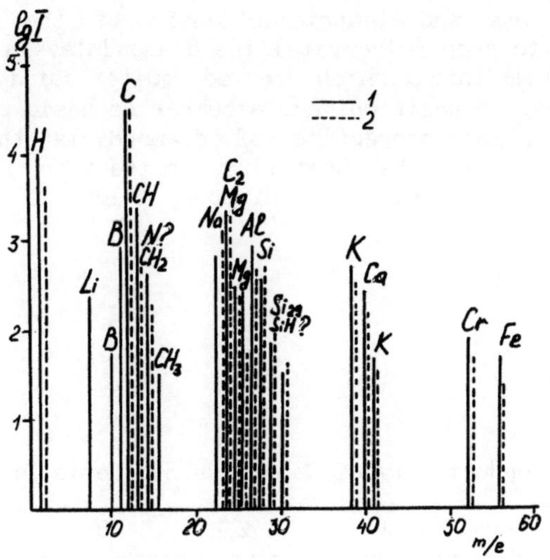

Fig. 6. SIMS data[44] for (1) diamond film, and (2) natural diamond.

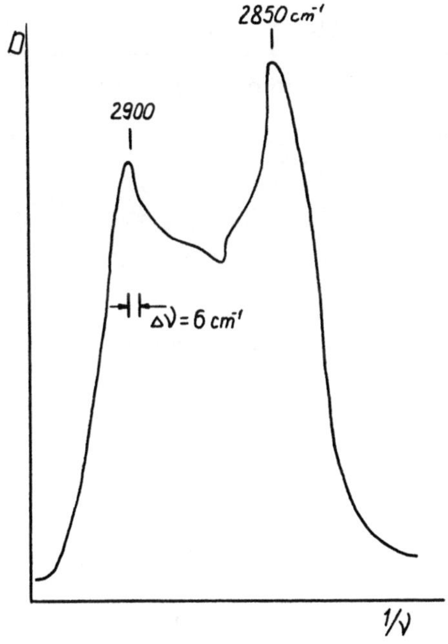

Fig. 7. IR spectrum of a synthetic diamond film[42].

saturated with carbon and after a layer of carbides have formed on it, nucleation continues with a stationary rate.

The use of both thermal and electric activation of the gas phase has made it possible[47] to grow polycrystalline diamond layers and free diamond films 10 to 300 μm thick, which proved quite sufficient for studying their structure, composition, and a number of basic properties. One of the most important properties of diamond is the thermal conductivity (see Table 1). As the heat-flux contraction method has revealed, the thermal conductivity of diamond films is also high (Table 2).

Table 2. Thermal Conductivity of Gas-Phase Synthetic Diamond

Number of Sample	1	2	3	4
Thickness of diamond film, μm	–	85	100	200
Temperature of sample, K	305.3	304.9	311.5	305
Thermal conductivity, W/m·K	128	418	500	630

Sample 1 in the Table 2 is nondoped single-crystal silicon; the value of its thermal conductivity is close to its tabulated value. Samples 2 and 3 are polycrystalline diamond films on silicon substrates. Sample 4 is a polycrystalline diamond plate.

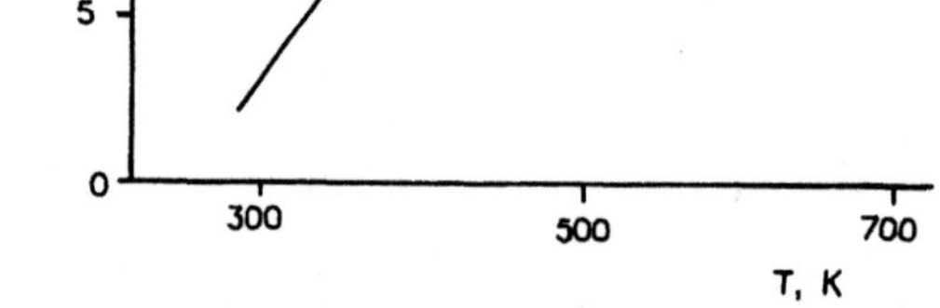

Fig. 8. Temperature dependence of the Hall mobility of acceptors in a diamond film doped by boron (concentration of about 10^{19} cm^{-3}), according to[44].

Table 3. Comparison of Properties of Synthetic and Natural Diamond

Property	Synthetic Diamond	Natural Single-Crystal Diamond
Lattice constant, Å	3.5664 ± 0.0001	3.5668 ± 0.0001
Carbon content, %	100	100
Microhardness on (111) face, kgf/mm^2	10 500 ± 400	9500-11 000
Specific resistance, Ω·cm	10^{12}-10^{13}	10^{12}-10^{16}
Dielectric constant	4-6	5.86
Dielectric loss tangent	10^{-3}	10^{-3}-10^{-4}
Breakdown voltage, V/cm	$> 1 \times 10^{6}$	10^{6}-10^{7}
Refractive index	2.38 ± 0.02	2.40
Transparency cutoff in UV region, nm	≥ 225	≥ 225
Thermal conductivity, W/m·K	≥ 630	400-2100

It should be emphasized that the heat-flux contraction method yields, for samples whose thickness is less than 0.5 mm. So highest thermal conductivity value in the Table 2 is, of coarce, underestimation.

According to indirect data obtained by testing a diamond-metal X-ray source (see the next section). the thermal conductivity of a diamond film about 30-μm thick is three to five times higher than that of copper.

The total list of data (Table 3) suggests that practically all the properties of single and polycristaline synthetic diamond films are identical to those of pure natural single-crystal diamond.

SIGNIFICANCE OF DIAMOND-FILM GROWTH PROCESSES FOR SCIENCE AND TECHNOLOGY

The appearance and development of growth processes for obtaining single-crystal and polycrystalline diamond films must undoubtedly be appraised as a significant breakthrough in scientific knowledge.

At the same time, diamond films are important both for improving the existing types of equipment and, which is especially important, for building entirely new equipment and even creating new technologies. International experts believe that diamond films are one of the five achievements that will be the most important factors in determining scientific and technological progress at the end of this century and the beginning of the next. We should particularly emphasize, however, the scientific importance of diamond films for solid-state physics, physical optics, the physics and chemistry of processes in which the solid state participates, and other fields of scientific research. Indeed, the synthesis of diamond layers and even free diamond films with controllable and given properties (thickness and shape, defect and impurity concentrations) provides means of obtaining data suitable for quantitative comparison with the results of theoretical calculations. These calculations, in turn, can be used for developing more sophisticated methods of experimental research necessary for obtaining new experimental and theoretical knowledge.

As for practical application of diamond films (in the form of layers and free films), this is determined, on the one hand, by the list of useful properties (see Table 3) and, on the other, by the possibilities of the processes of obtaining diamond films. Figure 9 shows some of such obvious applications[48].

Although practically all technological applications of diamond films are related to the necessity of providing for special preliminary studies and developments, there are first examples of progress in this field in the Soviet Union. In the next section we discuss using as epitaxial so polycrystalline diamond films.

EXAMPLES OF DIAMOND FILM APPLICATION.

There is report[28] of the growth of an epitaxial diamond film in the chemical gas-transport reaction mode on the surface of the drawhole of a single-crystal diamond die. An epitaxial diamond film approximately 2-μm thick was grown on the worn-out section of a cylindrical drawhole with initial 52 and final diameter 48 μm. Than diameter was turned to

50 μm by polishing the grown diamond film. The diamond die restored in this way was found to have, as testing has shown, practically the same operating characteristics as a new die. Twenty eight thousand kilometers of copper wire was drawn through the restored die, with a final diameter

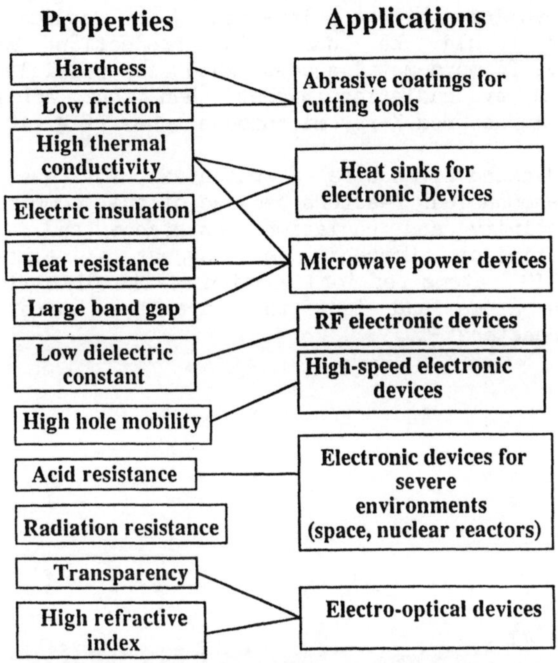

(K. Nishimura et al., Kobelco Technol. Rev., 1987, N2, p.49)

Fig. 9. Possible applications of diamond films[48].

of 50 μm. Such a long "service life" of the operating section of a diamond die consisting entirely of epitaxial diamond was provided, in addition to the high wear resistance, by the low friction of copper on epitaxial diamond (the friction factor was found to be close to that of copper sliding on natural diamond). Undoubtedly, this is also due to the low content of metallic impurities in the synthetic diamond film, comparable to the content of such impurities in natural diamond (see Fig. 6).

The very high thermal conductivity of diamond films and layers (see Table 2) makes it possible to employ them as components of anodes of X-ray sources. Anodes made of pure polycrystalline diamond films have a higher resistance to thermomechanical stresses than those made of single-crystal diamond. This makes it possible to obtain a higher intensity of K_{α} radiation of carbon at $\lambda = 44.6$ Å without risking destruction of the anode. This radiation has a definite potential for submicrometer X-ray lithography used in manufacturing IC and VLIC in solid-state electronics.

A double-layer anode made from a diamond film (about 30-μm thick) and covered by a thin (several micrometers) film of a heavy metal enabled manufacturing a bright point-like X-ray source of short-wave radiation that could be used in projection X-ray computer microtomography. Diamond-metal anodes make it possible to make the intensity of X-ray radiation higher than that achieved by using commercially manufactured X-ray microfocus tubes (Fig. 10).

Finally, diamond films have a potential for use as first-wall material in thermonuclear reactors because of the combination of high thermal conductivity and radiation resistance and a low thermal expansion coefficient. Current data suggest a relatively moderate (approximately 10^{-1} atoms per ion) erosion of the diamond layer caused by the hydrogen atoms and ions that are the operating medium of thermonuclear reactors[49].

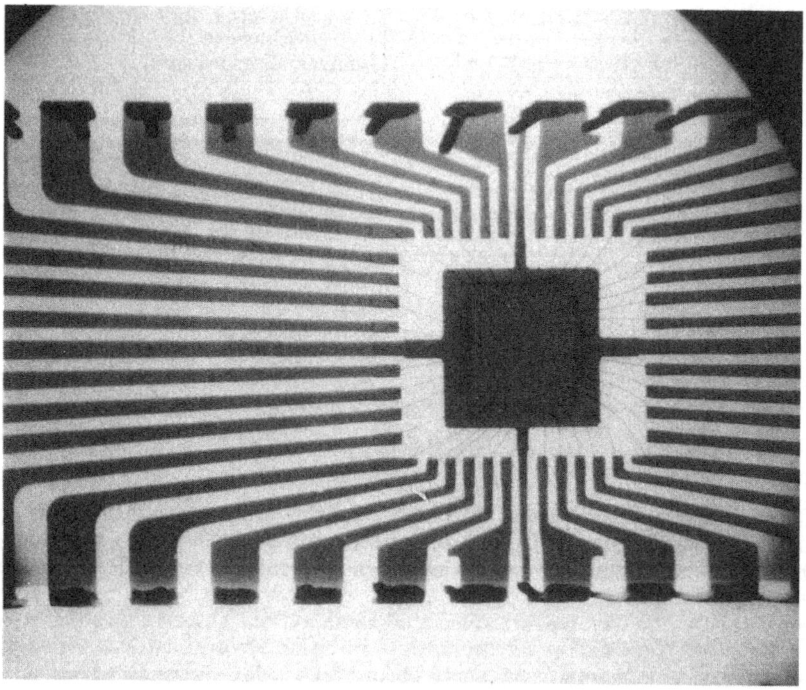

Fig. 10. A projection X-ray micrograph of an IC obtained by using a diamond-metal point-like X-ray source. Linear resolution is about 10 μm.

ACKNOWLEDGMENTS

I wish to thank my colleagues that participated in the synthesis of diamond films: L.L. Bouilov, A.A. Botev, and A.E. Alexenko, and in determining its properties and application potential: A.F. Belyanin, N.A. Bul'enkov, V.P. Martovitsky, A.E. Gorodetsky, T.A. Karpukhina,T.D. Ositinskaya, G.A. Sokolina, V.N. L'vov, V.P. Efanov, and V.M. Sharapov.

REFERENCES

1. H.J. McSkimin and W.L. Bond, Elastic moduli in diamond, Phys. Rev. 105:116 (1957).
2. S.I. Novikova, Thermal expansion of diamond from 25 to 750 K, Fiz. Tverd. Tela (Leningrad) 2:1617 (1960) (in Russian).
3. J.E. Field, ed., "The Properties of Diamond," Academic Press, London (1979).
4. R. Berman, F.E. Simon, and J.M. Ziman, The thermal conductivity of diamond at low temperatures, Proc R. Soc. London. Ser. A 220:171(1953).
5. R. Berman, E.L. Foster, and J.M. Ziman. The thermal conductivity of dielectric crystals: the effect of isotopes, Proc. R. Soc. London.Ser. A 237:344 (1956).
6. C.D. Clark, P.I. Dean, and P.V. Harris, Intrinsic edge absorption in diamond, Proc. R. Soc. London. Ser. A 277:312 (1964).
7. F.G. Champion, "Electronic Properties of Diamonds," Butterworths, London (1963).
8. A.G. Redfield, Electronic Hall effect in diamond, Phys. Rev. 94:526 (1954).
9. V.S. Vavilov, A.A. Gippius and E.A. Konorova. Electronic and process in diamond. Moscow, "Nauka" (1985) (Russ).
10. B.V. Spitsyn, Crystal growth under thermodynamically metastable conditions, in: "Growth of Crystals," 13:58, Plenum Press, New York(1986).
11. J. Lander and J.J. Morrison, Low energy electron diffraction study of the (111) diamond surface, Surf. Sci. 4:241 (1966).
12. J.A. Brinkman, C.J. Meechan, and H.M. Dieckamp, "Method for Artificial Synthesis of Diamond," US Patent N3, 175, 885, filed on July 1, 1960.
13. R. Hultgren, P.D. Desai, D.T. Hawkins, M. Gleiser, K.K. Kelley, and D.D. Wagman, "Thermodynamic Properties of Elements", US Atomic Energy Commission with Assistance of the American Society for Metals, MetalsPark, Ohio (1973).
14. B.V. Derjaguin and D.V. Fedoseyev, "Growth of Diamond and Graphite from the Gaseous Phase," Nauka, Moscow (1977) (in Russian).
15. A.G. Shenston,. ^5S state of carbon. Phys. Rev. 72:441 (1947)
16. B.V. Spitsyn and B.V. Derjaguin, "A Technique of Diamond Growth on a Diamond's Face," USSR Inventor's Certificate N339, 134 (1956) filed on July 10, 1956 (in Russian).
17. W.G. Eversole, "Synthesis of Diamond," US Patent N3, 030, 187 (1962), filed July 23, 1958.
18. W.G. Eversole, "Synthesis of Diamond," US Patent 3, 030, 188 (1962), filed July 23, 1958.
19. M.E. Belitsky, Ph.D. thesis, Chemical Department of Moscow State University (1970) (in Russian).
20. B.V. Derjagin, D.V. Fedoseev, V.M. Lukyanovich, B.V. Spitsyn, V.A.Ryabov, and A.V. Lavrentyev. Filamentary diamond crystals. J. Cryst. Growth, 2:380 (1968).

21. B.V. Derjagin, L.L. Bouilov, V.M. Zubkov, A.A. Kochergina and D.V. Fedoseev. On filamentary diamond crystals,Kristalografia 14:535 (1969).
22. B.V. Derjagin, B.M. Lukyanovich D.V. Fedoseev at al. About filamentary diamond crystals, Dokl. AN SSSR (Russ.) 181:1094 (1968).
23. Brit. Pat N 1001308, Method and apparatus for produsing crystalline structures (1963).
24. B.V. Spitsyn and G.G. Lopatina, An optical furnace, in: Proc. All-Union conf. on thermophys, prop. of substances at high temperatures. 1:468 ,Novosibirsk, July 25 to 30, 1966, Moscow, (1969).
25. J.C. Satya, P.S. Shauhan, J.C. Angus, and N.C. Gardner, Kinetics of carbon deposition on diamond powder, J. Appl. Phys. 47:4746 (1976).
26. P.A. Tesner, A.E. Gorodetsky, E.V. Denisevich, A.P. Zakharov, and T.V. Tekunova, Epitaxial growth of diamond from methane at atmospheric pressure, Dokl. Akad. Nauk SSSR 222:1384 (1975) (in Russian).
27. B.V. Derjaguin and D.V. Fedoseyev, Epitaxial synthesis of diamond in the metastable region, Usp. Khim. 39:1661 (1970) (in Russian).
28. B.V. Spitsyn, Chemical crystallization of diamond from activated vapor phase, J. Cryst. Growth 99:1162 (1989).
29. B.V. Spitsyn and A.V. Smolyaninov. USSR Author's Certificate N 987912, "Method of diamond layer growing", filed April 21, 1971.
30. B.V. Derjaguin, B.V. Spitsyn, et al., in: "2nd USSR Symposium on Growth of Crystals and Films of Semiconductor Compounds. Abstracts,"Novosibirsk (1969) (in Russian), pp. 36-37.
31. B.V. Spitsyn, "On the thermodynamics and kinetics of chemical crystallization of diamond, in: "Proc. 4th USSR Conf. on Crystal Growth," Part I, Yerevan Univ. Press, Yerevan (1972) (in Russian). p. 97.
32. B.V. Spitsyn, L.L. Bouilov, and B.V. Derjaguin, Vapor growth of diamond on diamond and other surfaces, J. Cryst. Growth 52:219 (1981).
33. B.V. Derjaguin, D.V. Fedoseyev, V.P. Varnin, A.E. Gorodetsky, A.P. Zakharov, and I.G. Teremetskaya, Growth of polycrystalline diamond films from the gas phase, Zh. Eksp. Teor. Fiz. 69:1250 (1975) (in Russian).
34. L.I. Avramenko, Reaction between H atoms and C, Zh. Fiz. Khim. 20:1299 (1946) (in Russian).
35. J.C. Angus, N.C. Gardner, et al., Growing of semiconductor diamond at subatmospheric pressure, Sintetich. Almazy No.6:38 (1971) (in Russian).
36. S. Matsumoto, Y. Sato, M. Kamo, and N. Setaka, Vapor deposition of diamond particles from methane, Jpn J. Appl. Phys. Part 2 21:183(1982).
37. M. Kamo, Y. Sato, S. Matsumoto, and N. Setaka, Diamond synthesis from gas phase in microwave plasma, J. Cryst. Growth 62:642 (1983).
38. K.E. Spear and M. Frenklach, Mechanisms for CVD at crystalline diamond, in: "Proc. 19th Biennial Conf. on Carbon," Extended Abstracts,Penn. State Univ (1980), p. 380.
39. A. Badzian, B. Simonton, T. Badzian, R. Messier, K.E. Spear, and R. Roy, Vapor deposition synthesis at diamond films, Proc. SPIE-Int.Soc. Opt. Eng. 683:127 (1986).
40. P.K. Bachmann, W. Drawl, D. Knight, R. Weimer, and R. Messier, Diamond nucleation and growth in bell jar microwave plasma CVD reactor, in: "Extended abstracts: Diamond and Diamond-like

Materials Synthesis," G.H. Johnson, A. Badzian, and M. Geis, eds., MRS, Pittsburgh (1988), p.99.

41. P.O. Joffreau, R. Haubner, and B. Lux, Low-pressure diamond growth on refractory metals, in: "Extended abstracts: Diamond and Diamond-like Materials Synthesis," G.H. Johnson, A. Badzian, and M. Geis,eds., MRS, Pittsburgh (1988), p. 15.
42. L.D. Kislovsky and B.V. Spitsyn, On the inclusion of hydrogen in synthetic diamond films, Kristallografiya 25:414 (1980) (in Russian).
43. L.G. Karaseva, T.A. Karpukhina, and B.V. Spitsyn, ESR spectrum of atomic hydrogen in synthetic diamond, Soviet Physical Chemistry, 1983: 302.
44. A.E. Alexenko, V.S. Vavilov, B.V. Derjaguin, M.A. Gukosyan, T.A. Karatygina, E.A. Konorova, V.F. Sergienko, B.V. Spitsyn, and S.D. Tkatchenko, Charge transfer and acceptor nature in semiconducting epitaxial diamond layers, Dokl. Akad. Nauk SSSR 233:334 (1977).
45 .B.V. Derjaguin, L.L. Bouilov et al., Structure and properties of diamond films grown on heterogeneous substrates, Dokl. Akad. Nauk SSSR 244:388 (1979) (in Russian).
47. L.L. Bouilov. A.E. Alexenko. A.A. Botev, and B.V. Spitsyn. Certain characteristics of the growth of diamond layers from an active gas phase. Soviet Physical Chemistry, 1986, p.302.
48. K. Nishimura, K. Cobashi, Y. Kawate, and T. Horiuchi, Growth of diamond using plasma CVD, Kobelco Technology Rev. No.2:49 (1987).
46. Y. Mitsuda, T. Yoshida, and K. Akashi, Diamond synthesis in MW plasma jet, in: "1st ICNDST. Program and Abstracts", Tokyo (1988), p. 44.
49. A.I. Fomitshev, V.M. Sharapov, A.P. Zakharov, B.V. Spitsyn, A.A. Botev, and A.A. Bouilov, Sputtering of diamond films by low-energy deuterium ions, in: "Proc 4th Int. Conf. on R&D in Constructing Materials for Thermonuclear Reactors," Dubna, 29-31 Jan. (1990) (in Russian), pp.6-7.
50. B.V. Spitsyn and A.E. Alexenko, Physico-chemical peculiarity of diamond doping from vapor phase, Archiwum Nauki o Materialakh (Warsaw) 7:201 (1986).
51. A.E. Alexenko and B.V. Spitsyn, Some properties of CVD-diamond semiconducting structures, This Proceedings (1991).

CRITICAL ASSESSMENT OF STATE–OF–THE–ART OF GROWING DIAMOND

Nobuo Setaka

National Institute for Research in Inorganic Materials
1–1 Namiki, Tsukuba, Ibaraki 305, Japan

INTRODUCTION

Currently, diamond synthesis has been performed by three methods: static high pressure, dynamic high pressure and vapor deposition at low pressure. Diamond films prepared by vapor deposition have received great interest in several industrial fields. The growth of diamond films has been made feasible by utilizing various techniques of vapor deposition. It is important to note that pure diamond films have been prepared by a modified chemical vapor deposition (CVD) process. The CVD technique is a versatile materials fabrication technique for protective coatings, epitaxial and whisker growth, coating of particles, preparation of powder, vapor forming, vapor consolidation, etc. Diamond has several unique properties that cannot be replaced by other materials. Therefore, diamond growth by CVD is a potentially valuable tool opening new areas of diamond technology.

The earliest substantial work on growing diamond from a gas medium at low pressure was presented by Eversole of Union Carbide Corporation in 1962 and independently by a group of researchers headed by Derjaguin. The work of Eversole was confirmed by Angus.[1] The Soviet group was successful in the growth of diamond films and particles with well–defined crystal habit on non–diamond substrates using a chemical transport process in a closed system.[2]

On the other hand, Aisenberg prepared a diamond–like carbon film by an ion beam deposition technique in 1973.[3] This work was confirmed by Spencer and co–workers in 1976.[4] A single crystal diamond film was formed on a diamond substrate by Freeman[5] using ion beam deposition with magnetic ion separation. Vora and Moravec[6] synthesized a diamond–like carbon film containing hydrogen by a biased high frequency discharge.

We have shown that diamond growth from a gas medium is feasible by a hot filament assisted CVD.[7] The diamond growth method has extended to a microwave plasma assisted CVD.[8] Though the diamond synthesis has been achieved using various techniques, there are some large barriers to commercialization and the development of opto–electronic and electronic devices based on diamond.

In principle, diamond synthesis by CVD is achieved by decomposing hydrocarbons thermally on a heated substrate and depositing diamond from the generated carbon. This method normally gives graphitic carbon. But modified CVD has made diamond growth possible.

There are early reports giving important suggestions about diamond growth from a gas medium. These are Veprek's and Wilson's reports.[9-10] Table I shows the early historical development of diamond synthesis from a gaseous medium. Veprek, at the University of Zürich, was investigating chemical transport of carbon with different gases, such as nitrogen, oxygen and hydrogen in a non–isothermal plasma. His purpose was to study kinetic and thermodynamic effects of the plasma, but not to deposit diamond. He pointed out in this experiment that production of atomic species in the plasma space was the rate controlling step for the chemical transport of carbon. Interestingly, the Soviet researchers headed by Derjaguin were successful in achieving diamond growth from a gas medium by a chemical transport process in a closed system. Wilson proposed a new idea for diamond growth on the basis of the experimental results of Eversole, Angus and also Derjaguin in 1976, when the diamond growth method had not yet been fully established. Wilson reported it in a popular magazine rather than a scientific journal. However, the idea contains significant matters for understanding diamond growth.

Table I. Historical Development of CVD Diamond in Early Stage

1949	W.G. Eversole	Initiated work
1953	W.G. Eversole (patent issued in 1962)	First successful in diamond growth
1956	B.V. Spitsyn B.V. Derjaguin	Initiated work on CVD Method
1966	J.C. Angus	Periodic CVD
1968	B.V. Derjaguin et al.	Pulse CVD
* 1972	S. Veprek	Chemical transport of carbon
1973	S. Aisenberg	Ion beam deposition
* 1973	W.D. Wilson	Proposed new idea
1976	B.V. Derjaguin et al.	Chemical transport in a closed system

ASSESSMENT OF DIAMOND GROWTH

Wilson's Idea

Wilson suggested that diamond growth would be possible if the following conditions were satisfied.

(a) A source of individual carbon atoms which can excite other carbon atoms, that is, having the reaction energy required for forming a chemical bond of diamond.

(b) A technique for extending the life time of an excited carbon or for increasing the reaction rate of individual carbon atoms.

According to Wilson's idea, high pressure is not a necessary and essential condition for diamond growth. High pressure may be required for a certain aspect of the problem, but it is not a basic requirement. Furthermore, he pointed out that the difficulty of diamond growth was due to the extremely short time of the excited carbon atoms, but it would be easy to produce the excited carbon atoms by several method as follows:

(a) Dissociation of a carbonaceous compound
(b) Electric discharge
(c) Electric bombardment
(d) Ultraviolet irradiation
(e) Shocks and shears
(f) X–ray

When considering the present techniques for chemical vapor deposition of diamond, they seem to have developed along the line of Wilson's idea. On the other hand, even in diamond growth at high pressure, we are required to provide excited carbon atoms.

Diamond grown under high pressure using metal catalysts is a commonplace item of commerce. The role of molten transition metals for diamond growth has not yet been sufficiently elucidated. However, a few opinions relating to that role have been proposed by several researchers.[11-12] One of them is the pure solvent hypothesis; the other is the solvent–catalyst hypothesis. The latter hypothesis is divided into two approaches. One of them is that the diamond deposits from positively charged carbon atoms. When carbon materials are dissolved in a molten metal under the conditions of high pressure and high temperature, if they are charged negatively or neutral, a graphitic carbon grows from these carbon atoms. The other hypothesis is that diamond growth rises from the excited carbon atoms produced by dissociation of intermediate carbides, such as nickel, cobalt or ferrous carbide. These unstable carbides are produced by a reaction of the metal with carbon under the conditions of high pressure and high temperature. When the carbide dissociates, the reaction energy required for diamond growth is supplied to the carbon atom. It seems to be a necessary condition to produce excited carbon atoms, even under the high pressure and high temperature conditions. The hypothesis of positively charged carbon may correspond to an ion beam deposition method, and the hypothesis of metal carbide may correspond to the CVD method. The molten metal plays the role of producing not only individual separated carbon atoms but also excited carbon atoms.

Mechanism of Diamond Growth on the Basis of Wilson's Idea

Diamond growth from a gas medium has been performed by various techniques. The problems encountered during vapor deposition of diamond are how to accelerate thermal decomposition of a carbonaceous compound at relatively low temperature and how to suppress the deposition of graphitic carbon. Also, an upper temperature limit of the substrate is fixed in order to protect the diamond nucleus from graphitization. We must consider the above conditions in order to realize diamond growth from a gas medium.

The deposition process considered from the standpoint of Wilson's idea and the actually encountered problems are shown in Fig. 1. In this case a methane–hydrogen system is used for illustration.

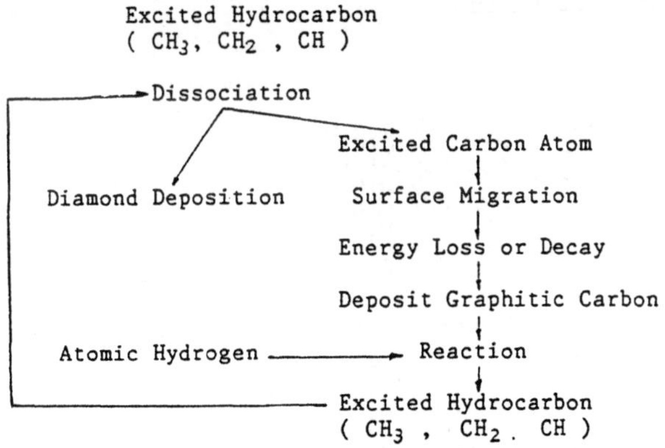

Fig. 1 Diamond growth process on the basis of Wilson's idea.

(a) First, we have to produce excited hydrocarbons, such as free radicals, in order to promote the dissociation of hydrocarbons at relatively low temperatures.

(b) The excited hydrocarbons are dissociated on a heated substrate, generating excited carbon atoms on the substrate.

(c) Some of the excited carbon atoms deposit as diamond, the others lose their stored energy during the migration on a substrate surface. They would deposit as graphitic carbon or non–diamond carbon.

(d) When a large quantity of atomic hydrogen exists on the substrate surface or close to it, the depositing non–diamond carbon would react with atomic hydrogen, and regasify as a hydrocarbon.

(e) If the products are free radicals, similar to the feed medium, the cyclic reaction process would have the effect as if the life time of the excited carbon atoms is extended, and also would result in increasing the supersaturation of excited carbon atoms.

As to nucleation and diamond growth, we have to consider the diffusion layer that forms an interface between the solid and gas phase, because the diamond syntheses have been performed at relatively low vacuum. Also, the concentration of carbonaceous compounds on the substrate surface is very low because the gas mixture is diluted by hydrogen.

Fig. 2 shows the diffusion layer. The curve Y_A indicates a concentration gradient of reactant gas and the curve Y_R shows that of by–product gas. A concentration of by–product on the substrate or close to it would suppress the diffusion of reactant gas. In order to increase the nucleation density and growth rate, we are required to provide methods for increasing the diffusion rate of activated hydrocarbons and free radicals.

This paper considers the current diamond synthesis from gas phase on the basis of Wilson's idea, the deposition process shown in Fig. 1, and also the relevant literature.

ASSESSMENT OF DIAMOND GROWTH FROM GAS MEDIUM

Diamond syntheses from gas media have been performed by various techniques using a gas mixture of methane diluted with hydrogen gas. There also have been reports of using organic molecules containing oxygen instead of methane. The Soviet research group has pointed out that atomic hydrogen plays a significant role in diamond growth from a gas medium.

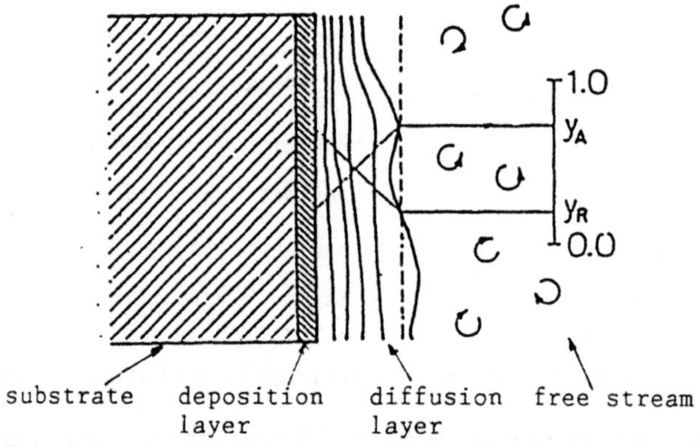

Fig. 2 Diffusion layer formed between the solid and vapor phases.

Hot Filament Assisted CVD (HF–CVD)

A schematic illustration of the apparatus is shown in Fig. 3. The diamond growth was achieved by using the apparatus with various substrates, for example, Si, Mo, W, various carbides, quartz glass and so on. The main difference from a normal CVD method is that a heated tungsten filament is set above the substrate and that a gas mixture of CH_4 diluted with H_2 is utilized as the reactant gas. Well-faceted diamond crystallites grow under the following growth conditions.[7]

Methane concentration (CH_4/H_2):	< 1%
Substrate temperature:	850–950° C
Pressure of reaction chamber:	20–40 Torr
Total gas flow:	100 sccm
Temperature of W filament:	2000° C

Under these experimental conditions, the distance between substrate and W filament is about 10 mm., and the average growth rate of diamond is about 1 μm/h or less. Unless the tungsten filament is heated, no diamond and no graphitic carbon deposit on the substrate. The growth rate of diamond depends remarkably on the filament temperature and the distance between the substrate and the tungsten filament. But the substrate temperature has to be maintained less than about 950° C by using a water–cooled pedestal. The above experimental results suggest that the heated tungsten filament plays a significant role for diamond growth.

Fig. 4 shows the diamond film deposited on a diamond substrate having a (111) face. A disadvantage of this method is that it has a strong tendency to deposit diamond particles isolated on non–diamond substrates. It is difficult to grow a diamond film. In order to raise the supersaturation value, increasing the methane concentration results in a great quantity of non–diamond carbon along with the diamond. However, growth of diamond films are easily achieved by using a substrate scratched with fine superhard materials, such as diamond or cubic boron nitride, as shown in Fig. 5.

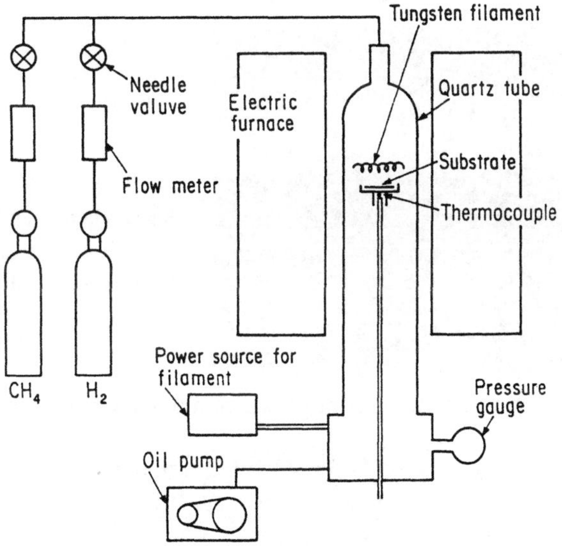

Fig. 3 Schematic drawing of hot filament assisted CVD apparatus.

Fig. 6 shows the calculated dissociation rate of molecular hydrogen as a function of temperature and pressure. The dissociation rate increases remarkably with raising the W–filament to a temperature greater than 2200K. The heated W–filament plays the role of dissociating molecular hydrogen to atomic hydrogen when the gas mixture passes the heated W–filament. That results in raising the concentration of atomic hydrogen on the substrate to a non–equilibrium condition. Furthermore, the atomic hydrogen can react with hydrocarbons, thereby promoting the formation of free radicals (for example, $CH_4 + H \rightarrow CH_3 + H_2$: 24.4 Kcal/mol).

Microwave Plasma Assisted CVD (MP–CVD)

We have also developed the microwave assisted CVD (MP–CVD). Because of its shorter wavelength, a microwave discharge has several advantages compared with a high frequency (rf) discharge.

(a) A very intensive localized discharge is obtained with a little tendency to spread.

(b) It is a more efficient process for producing atomic species: atomic oxygen, atomic nitrogen and atomic hydrogen.

(c) The maximum kinetic energy of ions produced in a plasma space is remarkably lower than those in a high frequency discharge.

(d) The plasma generated by microwave is very stable for a long time.

The maximum energy of ions is in inverse proposition to the squared frequency, as shown in Eq. (1)

$$\epsilon_{max} = \frac{(QE)^2}{8\pi^2 f^2 M} \qquad (1)$$

ϵ_{max} is the maximum energy of the ion; Q is the charge of the ion; M is the mass of the ion; f is the frequency and $E = E_0 \sin wt$.

Fig. 4. Diamond film grown on diamond substrate

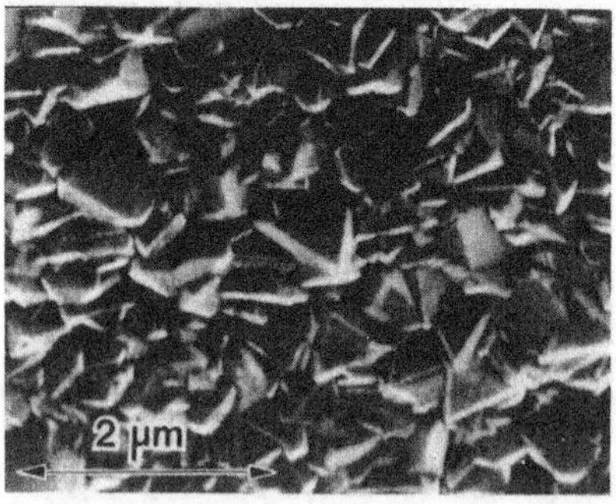

Fig. 5. Diamond film deposited on a surface–treated silicon wafer.

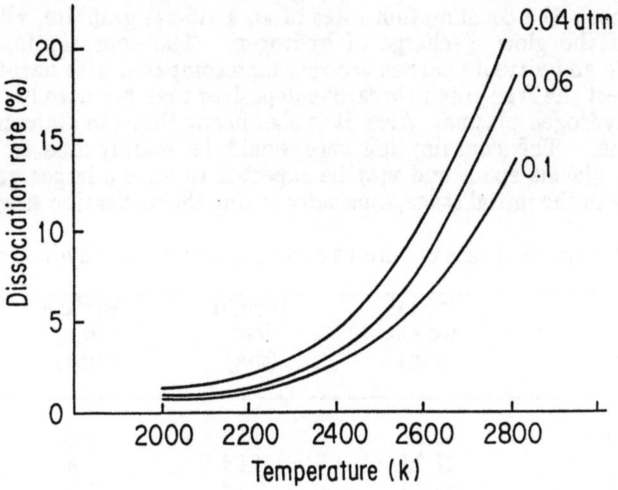

Fig. 6. Calculated thermal dissociation rate of a molecular hydrogen as a function of temperature and pressure.

It is significant that the maximum energy of the ions is small. Diamond CVD has been performed at relatively high temperature. If the ions generated in the plasma space have high kinetic energy, the depositing diamond would be etched by the ions, especially in the nucleation stage.

A schematic drawing of a microwave plasma assisted CVD apparatus is shown in Fig. 7. The growth conditions and morphology are nearly identical with those of the HF–CVD method. It is difficult to grow diamond films on non–diamond substrate as in HF–CVD.

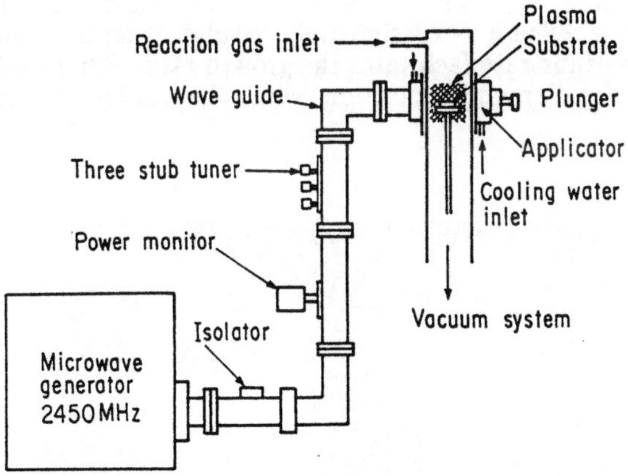

Fig. 7 Schematic drawing of microwave plasma assisted CVD apparatus.

In order to deposit well faceted diamond crystallites, the growth rate has to decrease to less than 1 μm/h. Fig. 8 shows particles deposited on a silicon wafer. The diamond films are able to grow on the diamond substrate or the surface treated substrate.

Table 2 shows the consumption rates of an artificial graphite, vitreous carbon, and diamond in the glow discharge of hydrogen. The consumption rates of the artificial graphite and vitreous carbon are very high compared with natural diamond.[14] The results suggest that the graphitic carbon deposited together with diamond is easily regasified in a hydrogen plasma. Also, it is significant that the diamond has a small consumption rate. The consumption rate would be mainly due to positive ions generated in the plasma space and may be expected to have a larger value than that shown in Table 2 in the initial stage, especially during the nucleation stage.

Table 2. Consumption rate of various carbon materials in a hydrogen plasma.

	Initial weight (mg)	Weight loss (mg)	Surface area (cm^2)	Consumpti rate (mg/cm^2h
Artificial graphite	40.97	4.17	1.8	0.13
Vitreous carbon	27.26	4.94	1.6	0.15
Natural diamond	25.28	0.13	1.1	0.006

Microwave power: 300 W, temperature 920° C, H_2 flow 10 ml/min; pressure 30 Torr; treatment time 20 h.

MODIFIED HF–CVD AND MW–CVD METHOD

The HF–CVD and MW–CVD methods have been further developed to electron assisted CVD or DC plasma assisted CVD. Interesting results are obtained from these methods. Sawabe and Inuzuka[15] have reported that the growth rate of diamond is accelerated when a positive voltage is applied between substrate holder and the W filament. When a negative voltage is applied at the substrate, only graphitic carbon deposits on the substrate. Similar experimental results are shown in the MW–CVD method.[16] When a negative voltage is applied at the substrate, diamond and silicon carbide co-deposit on a Si substrate. Conversely, if a positive voltage is applied at a substrate as shown in Fig. 9, the growth rate of diamond is accelerated. In the DC plasma CVD,[17] it is of great interest that the growth rate rises to about 20 μm/h and that a diamond film is able to grow on non-diamond substrates without surface-treatment. These experimental results suggest that the electron bombardment of the substrate facilitates the growth rate of diamond and plays a very important role for the diamond deposition, while the accelerated ions give a negative effect.

Fig. 8. Diamond particles deposited on a silicon wafer at CH_4/H_2 of 0.5%.

The onset energy to produce atomic hydrogen and its ions by the collision between electron and molecular hydrogen are as follows.[18]

$$H_2 \longrightarrow 2H \qquad 8.5 \text{ eV}$$
$$H \longrightarrow H^+ \qquad 13.5 \text{ eV}$$
$$H_2 \longrightarrow H_2^+ \qquad 15.4 \text{ eV}$$

In order to produce a great quantity of atomic hydrogen on the substrate, it is necessary to accelerate electrons to the range 8.5– 13.5 eV. The following effects may result from the electron bombardment.

(a) To promote the formation of atomic hydrogen and to hinder the recombination of atomic hydrogen on the substrate.

(b) To form a nucleation site on the substrate.

(c) To reduce the influence of the diffusion layer.

Raising the concentration of atomic hydrogen on the surface increases the concentration of free radicals, because it is possible to increase its concentration without diffusion of reactant. The effect would result in increasing the supersaturation value of the excited carbon. Ultraviolet irradiation of a substrate surface and a plasma discharge may give a similar effect.

The reaction of atomic hydrogen with graphitic carbon has been investigated by many researchers.[19-20] The major products of the reaction are methane and acetylene. Ashby[21] has reported evidence for major reactivity enhancement due to the effect of the bombarding electron.

ASSESSMENT OF THE CHARACTERIZATION OF DEPOSITED DIAMOND

The structural identification of materials deposited by diamond CVD has been normally by achieved using x–ray and electron diffraction or Raman spectroscopy. Diamond

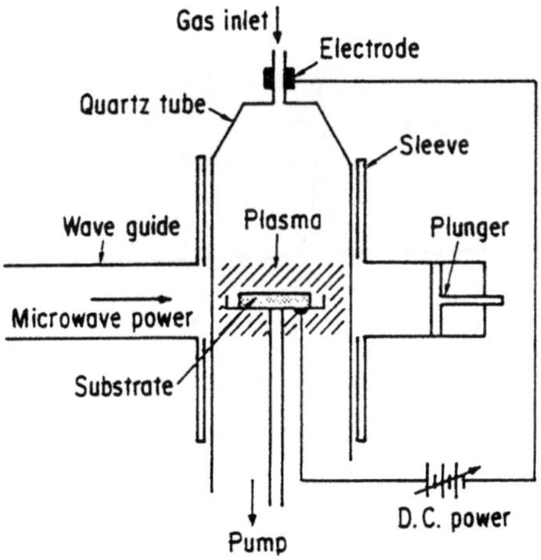

Fig. 9. Biased microwave plasma assisted CVD.

CVD is performed in the region where the graphite phase is thermodynamically stable. Therefore, there always exists the possibility of depositing graphitic carbon or other carbonaceous material together with diamond. It is a significant matter to find identification methods for these carbon materials. Raman spectroscopy is a suitable method for this purpose, but it is difficult to identify the structure of unknown materials. The Raman spectrum of mono–crystalline graphite appears at 1580 cm^{-1}.

A disordered graphite is composed by two broad bands at 1600 cm^{-1}. The broadness of both bands increases with increasing degree of disorder. The x–ray and electron diffraction methods are not suitable for identifying highly disordered graphite.

Some examples are shown in Fig. 10–13. Some inconsistencies are seen in the x–ray diffraction patterns and Raman spectra obtained from the same sample. Fig. 10 shows the x–ray diffraction patterns of the films deposited on a Si wafer at methane concentrations of 1% and 3%, using microwave assisted CVD method. The interlayer spacing obtained from diffraction patterns are in good agreement with the values reported for diamond. Fig. 11 shows the Raman spectrum of the sample deposited with a methane concentration of 0.5%. A single Raman line appears at about 1333 cm^{-1}. The position of the peak is in good agreement with the diamond (1332.5 cm^{-1}) reported by Solin and Ramdus.[22]

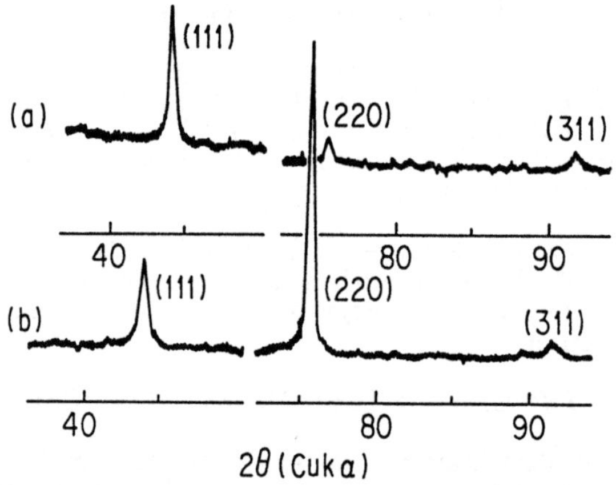

Fig. 10. X–ray diffraction pattern of the films deposited at CH_4 concentration of (a) 1%, (b) 3%.

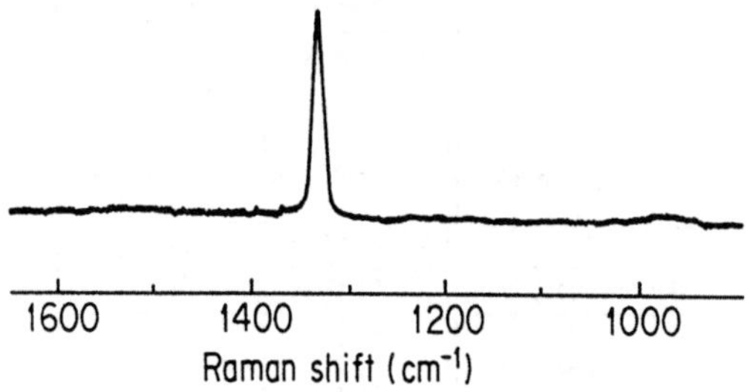

Fig. 11. Raman spectrum of the film deposited on Si wafer at CH_4 concentration of 0.5%.

At a somewhat higher methane concentration, a very broad band centered at around 1550 cm^{-1} appears along with the diamond line, as shown in Fig. 12 (a). The intensity of the broad bands have a tendency to increase with increasing methane concentration and the Raman band of graphitic carbon appears as shown in Fig. 12 (b), 3% methane. The Raman line at 1333 cm^{-1} is covered with broad bands of unknown carbon materials. The sample grown with 3% methane gives the well defined x–ray diffraction pattern of diamond (Fig. 10). This apparent phenomenon may be explained by a deposited layer that is composed of diamond and non–diamond carbon which strongly absorbs the incident laser beam and Raman light from diamond. The experimental results obtained from Raman spectra suggest the following:

(a) Raman spectroscopy is sensitive for detection of defective carbon materials compared with x–ray and electron diffraction methods.

(b) The regasification of the unknown carbon materials is difficult as compared with graphitic carbon materials. In Fig. 12 (a), the Raman band of graphitic carbon disappears and that of unknown carbon materials is observed at around 1550 cm^{-1}.

The complications of diamond synthesis from a gas medium seem to originate from the fact that there exist three allotropic forms of carbon, that is, diamond, graphite, and carbyne having sp–hybridization, i.e., linear chains of either conjugated triple bonds (poly–acetylene, polyyne) or cumulated double bonds (cumulene), as shown in Table 3[23-24].

Whittaker has proposed a new phase diagram of carbon including carbyne (carbin)[25]. The stable region of carbyne exists in the temperature range 2600–3800 K and it extends to the equilibrium line of diamond and graphite. When a solid carbon deposits from the excited carbon generated from a gas medium, the deposited carbon would contain fragments of structure having sp^3, sp^2 and sp^1 hybrid configuration. The proportion of sp^3, sp^2 and sp^1 in the depositing carbon would be due to differences in growth methods or growth conditions. That would be the reason for depositing the unknown carbon at 1550 cm^{-1}.

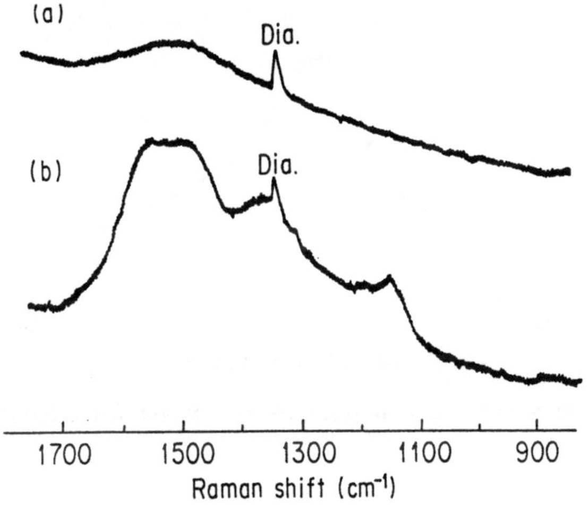

Fig. 12. Raman spectra obtained from the same specimens of Fig. 10.

The Raman spectrum of i–carbon films prepared by a biased rf–plasma method, is shown in Fig. 13. Only one broad band appears at around 1550 cm^{-1}. Wada has reported that the Raman spectra of a–carbon films prepared by electron beam vaporization or DC sputtering methods show an intense broad band at around 1550 cm^{-1} [26]. The a–carbon deposited at liquid nitrogen temperature is more amorphous than carbon films deposited at room temperature and does not contain 4–fold coordination carbon atoms. Shirakawa has reported that the Raman spectrum of carbyne prepared from poly–acetylene shows two intense peaks centered at around 1550 cm^{-1} and 1140 cm^{-1} as shown in Fig. 14. He assigned them to a C=C stretching vibration in trans conjugated polyyne[27]. These experimental facts suggest that the broad band centered at around 1550 cm^{-1} in CVD diamond is related to carbyne.

Table 3. Three Allotropes of carbyne

Allotrope	Polymorph	Bonding	Coordination number	Structure
Diamond	Cubic	Tetrahedral sp^3 hybridization	4	Zinc blend O
	Hexagonal		4	Wurtzite Hexagonal
Graphite	Hexagonal	Planar sp^2 hybridization	3	Hexagonal
	Rhombohedral		3	Rhombohedral D
Carbyne	α	Linear sp^1 hybridization	8	Hexagonal
	β		8	Hexagonal

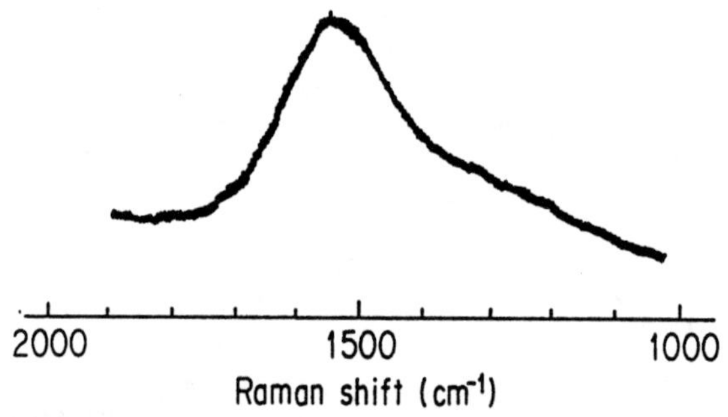

Fig. 13. Raman spectrum of a diamond–like carbon film deposited by a biased rf–plasma.

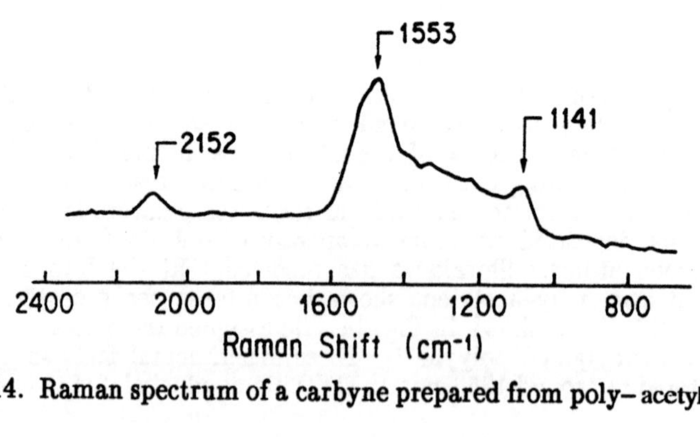

Fig. 14. Raman spectrum of a carbyne prepared from poly–acetylene.

Recently, deposition of diamond films has been achieved using oxygen containing gas mixtures, such as CO–H_2, CO–CO_2–H_2[28] CH_4–H_2–H_2O, or CH_3OH–H[29]. These gas mixtures seem to be effective in suppressing the appearance of the broad band at around 1550 cm^{-1} and promoting the growth rate of diamond. These experimental results suggest that the regasification of the unknown carbon materials with oxygen is more favorable than with atomic hydrogen alone.

ASSESSMENT OF WIDE BAND GAP SEMICONDUCTORS

Currently, considerable interest has been given to semiconducting diamond and also to cubic boron nitride. This is due to their unique properties and potential for devices operating at high temperature and high power or as a light emitter at short wavelength. However, there exists a large barrier to the realization of these applications.

The fabrication of semiconducting diamonds has been tried using several techniques. The p–type semiconducting diamond with low resistance is achieved by boron–doping. But it is not easy to produce n–type semi–conducting diamond with low resistance.

On the other hand, cBN has a diamond structure with alternating boron and nitrogen atoms, and it has the widest energy gap (> 6.6 eV) among group IV and III –V compound materials, including diamond. Wentorf[30] suggested that semiconducting cBN would be feasible when some impurities were added. Mishima[31] has reported growth of p– and n–type semiconducting cBN under the conditions of 55 Kbar and 1700 C, using LiCaBN solvent–catalyst, and doping with beryllium or silicon. Furthermore, he has shown that fabrication of a p–n junction diode is feasible. The diode indicates rectification characteristics up to 650 C. However, it is difficult to produce cBN films by vapor deposition and n–type cBN has a relatively high resistance.

A common problem for these materials is how to grow a large crystal which can be utilized as a substrate. Relatively large single crystal of diamond can be grown at high pressure and high temperature. But it is not easy to grow large single crystals of cBN at present. Single crystals of 2–3 mm in size have been grown. In order to grow large crystals, we have to develop new solvent–catalysts.

In order to develop wide band gap semiconducting materials we have to solve one of the following three problems.

(a) How to produce n–type diamond.
(b) How to grow cBN film by vapor deposition.
(c) How to grow large single crystal of cBN under conditions of high pressure and high temperature.

The fabrication of cBN films has been tried by using a gas mixture of diborane (B_2H_6) and ammonium or nitrogen gas. The method is meant to produce the individual excited boron and nitrogen atoms, and to deposit cBN by a reaction of these excited species. But it is a disadvantage that the collision probability is very small. Use of an excited BN compound as a source would seem to be an advantageous method. In order to realize that, we are having to use a volatile compound containing BN such as borazole or β–trichloborazole (BNHCl). The excited BN would be generated by thermal decomposition of the reactant.

On the basis of Wilson's idea, the other problems are how to extend the life time of the excited BN, and then to devise the regasification methods of graphitic BN (low pressure phase) which co-deposits with cBN.

REFERENCE

1. J.C. Angus, H.A. Will, W.S. Stanko, "Growth of diamond seed crystal by vapor deposition," J. Appl. Phys. 2 380 (1968).

2. B.V. Spitsyn, L.L. Bouilov, B.V. Derjaguin, "Vapor growth of diamond on diamond and other materials," J. Crystal Growth 52 219 (1981).
3. S. Aisenberg and R.W. Chabot, "Ion–beam deposition of thin film of diamond–like carbon," J. Appl. Phys. 42 2915 (1971).
4. E.G. Spencer, P.H. Schmidt, D.C. Joy, F.J. Sansalone, "Ion–Beam–deposition of polycrystalline diamondlike films," Appl. Phys. Letters, 29 118 (1976).
5. J.H. Freeman, W. Temple, G.H. Gard, "Epitaxial synthesis of diamond by carbon–ion deposition at low energy," Nature 275 634 (1978).
6. H. Vora and T.J. Moravec, "Structural investigation of thin films of diamond–like carbon," J. Appl. Phys. 52 : 6151 (1981).
7. S. Matsumoto, Y. Sato, M. Tsutumi, N. Setaka, "Growth of diamond particles from methane–hydrogen gas," J. Mat. Sci. 17 : 3112 (1982).
8. M. Kamo, Y. Sato, S. Matsumoto, N. Setaka, "Diamond synthesis from gas phase in microwave plasma," J. Crys. Growth, 62 : 642 (1983).
9. S. Veprek, "Chemical vapor and deposition of solids in a non–isothermal plasma:chemical transport of carbon," J. Crystal Growth, 17 : 101 (1972).
10. W.D. Wilson, "On the growth of diamond," Lapidary J. 27 : 982 (1973), ibid, 27 : 1096 (1973).
11. R.J. Wedlake, "Diamond synthesis," Industrial Diamond Review, 198 (1977).
12. R.H. Wentorf, Jr., "Solution of carbon at high pressure," Berichte der Bunsengesellschaft, 70 : 975 (1966).
13. S. Matsumoto, S. Morimoto, N. Setaka, Yogyo–Kyokai annual meeting. (1984).
14. N. Setaka, "Proceedings of the 10th International Conference of CVD," The Electrochemical Society, Pennington, NJ (1987).
15. A. Sawabe and T. Inuzuka, "Growth of diamond thin film by electron assisted chemical vapor deposition," Appl. Phys. Lett. 42 : 146 (1986).
16. K. Suzuki, A. Sawabe and T. Inuzuka, "Growth of diamond thin films by dc plasma chemical vapor deposition," Appl. Phys. Lett. 50 : 728 (1987).
17. S. Yugo and T. Kimura, "Nucleation process of diamond by plasma CVD," First Inter. Conf. New Diamond Sci, and Tech. Tokyo, Japan (1988).
18. C.C. Goodyear, A.V. Engel, "Dissociation and Ionization of Hydrogen in High Frequency Discharge," Proc. Phys. Soc., 79 : 732 (1962).
19. M. Balooch, D.R. Olander, "Reaction of modulated molecular beam with pyrolitic graphite," J. Chem. Phys. 63 : 4772 (1975).
20. W.L. Hsu, "Chemical erosion of graphite by hydrogen impact: A summary of the database relevant to diamond film growth," J. Vac. Sci. Technol. A 6 : 1803 (1988).
21. C.I.H. Ashby, R.R. Rye, "Synergistic of electrons on the reactivity of atomic hydrogen with graphite," J. Nucl. Mat. 92 : 141 (1980).
22. S.A. Solin, A.K. Ramadas, "Raman Spectrum of Diamond," Phys. Rev. B 1 : 1687 (1970).
23. R.J. Wedlake, "Three allotropic of carbon," Diamond Research, 2 (1974).
24. R.B. Heiman, J. Kleinman, N.M. Salansky, "Structural aspect and conformation of linear carbon polytypes," Carbon, 22 : 147 (1984).
25. A.G. Whittaker, "Carbon: a new view of its high–temperature behavior," Science, 200 : 763 (1978).
26. N. Wada, P.I. Gaczi, S.A. Solin, "Diamond–like 3–fold corrdinated amorphous carbon," 35–36 : 543 (1980).
27. Akagi, M. Nishiguch, H. Shirakawa, "One–dimensional carbyne–synthesis and properties," Synthetic Metals, 17 : 557 (1987).
28. C. Chen, Y.C. Huang, S. Hosomi, I. Yoshida, "Effect of additive gases on diamond
29. Y. Hirose, Y. Terasawa, K. Takahashi, "Synthesis of diamond thin film and diamond particles by thermal CVD using organic compound," Oyobutsuri, 56 : 247 (1987).
30. R.H. Wentorf, Jr., "Preparation of semiconducting cubic boron nitride," J. Chem. Phys., 36 : 1990 (1962).
31. O. Mishima, J. Tanaka, S.Y. Yamaoka, O. Fukunaga, "High temperature cubic boron nitride p–n junction diode made at high pressure," Science, 39 237 (1987).

SUMMARY OF PANEL DISCUSSIONS ON NEEDS AND OPPORTUNITIES FOR RESEARCH AND DEVELOPMENT OF DIAMOND AND DIAMOND-LIKE MATERIALS

R. E. Clausing

Panel discussions at the end of the Institute provided a forum for the lecturers and students to discuss the needs and opportunities for research and development in diamond and diamond-like materials technology. The key scientific needs are related to modeling of the nucleation and growth processes, to elucidation of the critical roles of atomic hydrogen, and to the mechanisms of carbon addition to the growing surfaces. The development and more extensive use of in situ diagnostics for both surface and gas phases are important to solving these issues. The more immediate practical questions concern the identification of the growth rate limiting steps, the relation of growth parameters to the resulting film structure and the dependence of properties on structure.

Three panels of experts were selected to discuss various areas. The first panel discussed the importance of considering structure when attempting to model or understand the growth of diamond films. It was pointed out that F.C. Frank in the 1950s explained many of the growth characteristics of natural diamond on the basis of surface geometrical effects. Surface structure and bonding are quite different on atomically flat {111} surfaces than on irregular {111} surfaces with steps or twins. It appears that growth on {111} surfaces often proceeds from surface irregularities, and as a result, stacking faults and twins are included in the crystal structure. Growth on {110} faces may be considered to be growth on fully stepped {111} faces and would be expected to be very fast. Growth on unreconstructed {100} faces should allow addition of single atoms anywhere on the surface, but only to crystallographically correct locations, resulting in a crystal which could contain dislocations or point defects but not stacking faults and twins. There was, however, a substantial difference of opinion as to experimental observation of growth details. It was pointed out repeatedly that the existing models for growth often ignore structural effects entirely. It was also pointed out that epitaxial growth of diamond should be compared to the more widely studied growth of silicon and other related semi-conductor materials. The several bonding states for carbon will, however, produce quite significant differences in the growth processes because many more structural arrangements can be accommodated by mixing the sp^2 and sp^3 bonding modes. The panel felt that the structural aspects of growth are an especially important, but currently neglected issue.

Diamond and Diamond-Like Films and Coatings
Edited by R.E. Clausing *et al.*, Plenum Press, New York, 1991

The second panel considered the future prospects for both diamond and diamond-like films. Diamond-like carbon magnetic storage disc coatings are commercially available while few diamond film products are at the same point of development. It was felt that non-hydrogenated "amorphous diamond" coatings may hold considerable promise. These coatings might be nearly as hard as crystalline diamond but would have inherently smooth surfaces and can be deposited on room temperature substrates, thus avoiding some of the manufacturing and processing problems which beset polycrystalline diamond films. Some of the discussion concerned the bright prospects for diamond products such as cutting tools, heat sinks, X-ray lithography masks and infrared optics, which most participants considered to be near term developments. A plea was made to look for innovative, and more economical growth processes for crystalline diamond in order to make high quality diamond films more attractive.

An attempt was made to summarize the level of support for research and development around the world. Much of this discussion concerned comparisons between research and development efforts in Japan and the United States. A team of U. S. experts visited Japan recently and concluded that (1) there is much more total industrial investment in Japan, (2) the quality of fundamental research in the U. S. and Japan is about equal, and (3) there is much more basic research in the U. S. than in Japan. There are about 180 members of the Japan New Diamond Forum compared with about 40 United States organizations addressing diamond research and development. The funding in the United States was estimated to be about $9,000,000 by government organizations. The General Electric Co. and Norton efforts in the United States were said to be larger than the U.S. Government supported efforts. The industrial effort in Japan is several times larger than the U. S. effort. Although individual Japanese companies are probably making investments comparable to those of General Electric and Norton, there are many more Japanese companies involved. It was reported that the effort in Europe appears to be much less than in the U. S. with relatively little industrially sponsored development. It must be kept in mind, however, that it is very difficult to estimate the level of proprietary industrial work or unpublicized government support especially when programs are in planning or early stages, as appears to be the case in the UK and the rest of Europe. There was a very active European participation in and contribution to this Institute. It was not possible to make a satisfactory comparison of the effort in the USSR. It was felt to be considerably less than that in the U S. The situation in the USSR is changing and somewhat confused due to the current economic situation and an increasing interest in commercialization.

The third panel concentrated on current and future needs. The materials requirements for diamond electronic devices were discussed along with the progress toward realization of prototype manufacturing methods. With regard to the development of diamond semi-conductors, it was felt that n-type semi-conductor diamond is not essential to the development of many useful devices. Two panelists stated that we need no fundamental breakthroughs, such as the fabrication of n-type semiconducting diamond, to make useful electronic devices. Most applications can be met without n-type material. Even so, cubic boron nitride grown heteroepitaxially on diamond (or diamond on cubic boron nitride) may satisfy this purpose. Both the homoepitaxial and heteroepitaxial growth of diamond are technologically very important needs. Several possibilities exist for growing large single crystal substrates although none of them have been realized at this time. An opinion was also expressed that some applications might be able to use good quality, oriented, polycrystalline diamond films for electronic devices.

Perhaps the most interesting discussions were related to the perceived needs for new information to advance the science of diamond growth. The modelers could point to no single key piece of data which is now needed. Several expressed the desire for a much larger data base of carefully controlled experimental observations on the effects of growth conditions on the rate and quality of film growth. In particular, data about interactions at the surface itself during the growth process would be very valuable. This includes local temperatures, species, flow profiles and fluxes at the surface; the presence of any higher molecular weight molecules on the surface; the effects of the structure of the deposit on growth rate; the roles of atomic hydrogen and oxygen; and the presence of aromatic hydrocarbons on the surface. It is quite obvious that available in situ diagnostics must be used to the fullest extent and that new techniques and techniques from other fields, such as the laser optical techniques, which have been so useful in combustion research, are needed to provide this data base.

All agreed that there are an abundance of opportunities and challenges in basic science, technology and applications and that the future for diamond and diamond-like materials will be full of many exciting developments.

PARTICIPANTS

Dr. Alexander E. Alexenko
Institute of Physical Chemistry
U.S.S.R. Academy of Science
Leninsky Prospect 31
Moscow 117915, USSR
Phone: 234-00-14, Ext. 427
FAX: 7-095-2302332

Mr. David B. Allen
Carnegie Mellon University
900 S. Negley Avenue
Apt. 19
Pittsburgh, PA 15232 USA
Phone: 412-441-9198

Ms. Marylu Allouard
C.S.N.S. M.
Bat 108 Campus D'Orsay
91405 Orsay, France
Phone: 33-69415212
FAX: 33-169415268

Dr. Marianna Ambrico
Instituto di Chimica
Universitat Della Basilicata
85100 Potenza, Italy
Phone: 39-971-474231 or 474225
FAX: 39-971-474223

Dr. Bernard Andre
Centre d'Etudes Nucleaires
de Grenoble
Department d'Etudes des Materiaux
Section de Genie des Surfaces
BP-85X
38041 Grenoble, Cedex, France
Phone: 33-76-884034
FAX: 33-76-885130

Prof. John C. Angus
Department of Chemical Engineering
Case Western Reserve University
Cleveland, OH 44106 USA
Phone: 216-368-4133
FAX: 216-368-3016
BITNET:ANGUS@CWRU

Dr. Nagappan K. Annamalai
U.S. Air Force - RADC/ESR
Hanscom Air Force Base, MA 01731 USA
Phone: 617-377-3047
FAX: 617-377-5041

Dr. Thomas R. Anthony
General Electric Company
GE Research Laboratories
P.O. Box 8, K1 1C30
Schenectady, NY 12301 USA
Phone: 518-387-6160
FAX: 518-387-5592
Email Anthony@GE.CRD.COM

Dr. Peter K. Bachmann
Philips Research Laboratories
P.O. Box 1980
Aachen D-5100, FRG
Phone: 49-241-600-3314
FAX: 49-241-600-3465

Dr. Joseph A. Baglio
GTE Laboratories, Inc.
40 Sylvan Road
Waltham, MA 02254 USA
Phone: 617-466-4224
FAX: 617-890-9320

Dr. William F. Banholzer
Manager, Advanced Inorganic
Materials Labortory
GE Corporate R & D
Building K-1; Room 5A59
P.O. Box 8
Schenectady, NY 12301 USA
Phone: 518-387-5951
FAX: 518-387-7563

Dr. Vincenzo Barbarossa
Eniricerche SPA
Via E. Ramarini 32
00015 Monterotondo (Rome), Italy
Phone: 39-690021279
FAX: 39-690021243

Dr. Alessandro Bellone
Olivetti Office
Via Jervis 20
10015 Ivres (Turin), Italy
Phone: 39-125528038
FAX: 39-125528116

Mr. Antonio Sao Miguel Bento
Physics Department
University of Coimbra
P-3000 Coimbra, Portugal
Phone:
FAX: 351-39-29158

Mrs. Ewa Bienk
Danish Technological Institute
Material Testing Department
Teknologiparken
DK-8000 Aarhus, Denmark
Phone: 45-86142400
FAX: 45-86147722

Dr. Bart Blanpain
Katholieke Universiteit Leuven
Department Metaalkunde en
Toegepaste Materiaalkunde
de Croylaan 2
B-3030 Heverlee, Belgie
Phone: 32-16220931
FAX: 32-16207995

Mr. Pierre Bou
C.N.R.S.
Centre de Recherches sur la Chimie
de la Combustion
et des Hautes Temperatures
1C Av. de la Recherche Scientifique
45071 Orleans Cedex 2, France
Phone: 33-38515489
FAX: 33-38642670

Dr. Elisa Bourdon
Department of Physics
York University
4700 Keele Street
North York, Ontario, Canada M3J IP3
Phone: 416-736-2100
FAX: 416-736-5516
@YORKU.SOL.FS300135

Dr. Hans-Gerd Busmann
Fakultät fur Physik und F-M-F
Albert-Ludwig-Universitat
Hermann-Herder Strasse 3
D-7800 Freiburg, FRG
Phone: 49-761-2033432
FAX: 49-761-2034527

Dr. A. Campargue
Laboratoire de
Spectrometrie Physique
BP-87
38402 St. Martin d'Heres
France
Phone: 33-76514319
FAX: 33-76514544

Dr. Yves Catherine
Laboratoire des Plasmas
et des Couches Minces
Institut de Physique
et Chimie des Materiaux
UMR C.N.R.S. 110
2 rue dela Houssiniere
44072 Nantes Cedex 03, France
Phone: 33-40373965
FAX: 33-40373991

Dr. P. R. Chalker
Materials Development Division
Harwell Labortory
Oxfordshire OX11 ORA
United Kingdom
Phone: 44-235-821111
FAX: 44-235-432348

Dr. Robert E. Clausing
Metals and Ceramics Division
Oak Ridge National Laboratory
P.O. Box 2008
Oak Ridge, TN 37831-6093
Phone: 615-574-4084
FAX: 615-574-7659
BITNET: CRL@ORNLSTC

Dr. John W. Coburn
Department K33/801
IBM Almaden Research Center
650 Harry Road
San Jose, CA 95120
Phone: 408-927-2414
FAX: 408-927-2100

Dr. G. Cocito
Head, Optical Technologies Dept.
CSELT
Via G. Reiss Romoli, 274
10148, Torino, Italy
Phone: 39-11-2169256
FAX: 39-11-2169520

Dr. Olinda Conde
Physics Department
University Lisbon
Campo Grande, Ed. C1
1700 Lisbon, Portugal
Phone: 351-1-7583141
FAX: 351-1-759771

Dr. Joseph B. Darby, Jr.
Division of Materials Sciences
Office of Basic Energy Sciences
Department of Energy
G-226, ER-13
Washington, D.C. 20585
Phone: 301-353-3884
FAX: 301-353-6594

Prof. F. Demichelis
Dipartimento di Fisica
Politecnico di Torino
Corso Duca degli Abruzzi, 24
10129 Torino, Italy
Phone: 39-11-5567326
FAX: 39-11-5567399

Dr. Paolo DeTassis
R&D Manager
Clevite Italy
Via B. Todesca 12
38100 Trento, Italy
Phone: 39-461-990321
FAX: 39-461-991860

Dr. Robert C. DeVries
17 Van Vorst Drive
Burnt Hills, NY 12027
Phone: 518-399-5225
FAX: 614-438-2235

Prof. Rodney Dillon
Department of Electrical Engr.
209N WSEC
University of Nebraska
Lincoln, NE 68588-0511
Phone: 402-472-5891
FAX: 402-472-4732

Ms. Concetta de Martino
Department of Nuclear Engr.
c/o Prof. F. Demichelis
Dipartimento di Fisica
Politecnico di Torino
Corso Duca degli Abruzzi, 24
10129 Torino, Italy
Phone: 39-11-3358240
FAX: 39-11-5567399

Dr. Jean-Jacques Dubray
Ceramic Science & Engr. Dept.
217, Steidle Building
University Park, PA 16802
Phone: 814-865-7131
FAX: 814-865-2917
Email: JJD4@PSUVM

Dr. Rainer Dŏhl-Oelze
VDI Technologiezentrum
Graf-Recke Strasse 84
4000 Dusseldorf 1, FRG
Phone: 49-211-6214567
FAX: 49-211-6214575

Prof. Helmut Ehrhardt
Fachbereich Physik
Universitat Kaiserslauten
Erwin-Schrodinger Strasse 46
6750 Kaiserslauten, FRG
Phone: 49-631-2052382
FAX: 49-631-2053300

Dr. Hermann Ferber
Daimler Benz
Department FVG/CH
Postfach 600202
7000 Stuttgart, FRG
Phone: 49-711-1722211
FAX: 49-711-1753267

Dr. John E. Field
Cavendish Laboratory
University of Cambridge
Madingley Road
Cambridge CB3 OHE
United Kingdom
Phone: 44-223-337318
FAX: 44-223-63263

Dr. Stefano Fontana
c/o Emilia Cappelli
Gruppo di Scienza
della Superfici
Eniricerche
00016 Via Ramarini
Monterotondo (Rome), Italy
Phone: 39-690021263
FAX: 39-690021243

Prof. Michael Frenklach
Materials Science and Engr.
Pennsylvania State University
University Park, PA 16802 USA
Phone: 814-865-4392
FAX: 814-865-3075
BITNET: IY2@PSUVM

Dr. Michael Geis
Microelectronics
Lincoln Laboratory
Massachusetts Inst. Tech.
224 Wood Street
Lexington, MA 02173 USA
Phone: 617-981-7810
FAX: 617-981-4983

Prof. Alfred Grill
IBM Research Division
T.J. Watson Research Center
23-227
P.O. Box 218
Yorktown Heights, NY 10598 USA
Phone: 614-945-1492
FAX: 914-945-2141
Email AGRILL@YKTVMT

Mr. Steve Grot
Department of Electrical Engr.
Center for Electronic
Materials Programs
210 EEW Building
Pennsylvania State University
University Park, PA 16802 USA
Phone: 814-865-9432
FAX:

Mr. Lee Heatherly, Jr.
Metals and Ceramics Division
Oak Ridge National Laboratory
P.O. Box 2008
Oak Ridge, TN 37831-6093 USA
Phone: 615-574-5088
FAX: 615-574-7659

Prof. Reinhard Helbig
Lehrstuhl fur Angewandte Physik
Universitat Erlangen-Nurnberg
Gluckstrasse 9
D-8520 Erlangen, FRG
Phone: 49-9131-858425
FAX: 49-9131-858423

Dr. Hans-Jurgen Hinneberg
ISI-2, KFA Julich
Postfach 1913
D-5170 Julich, FRG
Phone: 49-2461-613149
FAX: 49-2461-614673

Mr. Philip John R. Honeybone
Physics Laboratory
University of Kent
Canterbury, Kent CT27NR
United Kingdom
Phone: 44-227-764000, Ext. 3776
FAX: 44-227-762616
Email: PJRH@UK.AC.UKC

Dr. Linda L. Horton
Metals and Ceramics Division
Oak Ridge National Laboratory
P.O. Box 2008
Oak Ridge, TN 37831-6118 USA
Phone: 615-574-5081
FAX: 615-574-7659

Mr. David G.K. Jeng
Department of Electrical Engr.
State University of New York
Stony Brook, NY 11794 USA
Phone: 516-567-0520
FAX: 516-567-0684

Mr. Pierre Joeris
Universität Hamburg
Institut fur Physikalische Chemie
Bundesstrasse 45
2000 Hamburg 13, FRG
Phone: 49-40-41233455
FAX: 49-40-41232449

Prof. Rafi Kalish
Solid State Institute
& Physics Department
Technion Israel Institute
of Technology
Technion City
Haifa, 32 000, Israel
Phone: 972-4-293906
FAX: 972-4-221581

Dr. J. C. Knight
Department of Engr. Materials
University of Newcastle-upon-Tyne
Newcastle-upon-Tyne NE1 7RU
England
Phone: 44-91-222-6000, Ext. 7212
FAX: 44-91-2329259

Dr. Alexander W. Koch
Max-Planck-Institut
fur Plasmaphysik
Boltzmannstrasse 2
D-8046 Garching bei Munchen
FRG
Phone: 49-89-32992245
FAX: 49-89-32992200

Prof. Peter Koidl
Fraunhofer Institut fur
Angewandte Festkorperphysik
Tullastrasse 72
D-7800 Freiburg, FRG
Phone: 49-761-5159280
FAX: 49-761-5159200

Dr. Ritalba Lamendola
Centro di Studio Chimica de Plasmi
Dipartiment di Chimica
Universita degli studi Bari
Via Amendola 173
Bari, Italy
Phone: 39-80-242010
FAX: 39-80-242129

Prof. Mauro Lenzi
Institute Metodologie
Avanzate Inorganiche
C.N.R.
Area della Ricerca di Roma
Via Salaria Km 29 300
Casella Postale 10
00016 Monterotondo Scalo
(Roma), Italy
Phone: 39-6-90020218
FAX: 39-6-9005849/90020238

Dr. Alan H. Lettington
Royal Signals and Radar Estab.
St. Andrews Road
Great Malvern, Worcs.
WR14 3PS, United Kingdom
Phone: 44-684-894658
FAX: 44-684-894540

Mr. Roland Locher
Fraunhofer Institut fur
Angewandte Festkörperphysik
Tullastrasse 72
7800 Freiburg, FRG
Phone: 49-761-5159282
FAX: 49-761-5159200

Dr. Benno Lux
Institut für Chemische Technologie
Anorganischer Stoffe/Gust.161
Technische Universität Wien
Getreidemarkt 9
A-1060 Wien, Austria
Phone: 43-1-58801
FAX: 43-1-5874835

Dr. Kader Mansour
Physics Department
University of Basel
Klingelbergstrasse 82
CH-4056 Basel, Switzerland
Phone: 41-61-3212280
FAX: 41-61-3211440

Dr. Ludvik Martinu
Department of Engr. Physics
Ecole Polytechnique
Case Postale 6079, Succursale A
Montreal, Canada H3C 3A7
Phone: 514-340-4099
FAX: 514-340-4440

Prof. Allan Matthews
Department of Engineering
Design and Manufacture
The University of Hull
Cottingham Road, Hull
North Humberside, HU6 7RX
United Kingdom
Phone: 44-482-465073
FAX: 44-482-466205

Dr. Carl J. McHargue
Metals and Ceramics Division
Oak Ridge National Laboratory
P.O. Box 2008
Oak Ridge, TN 37831-6118 USA
Phone: 615-574-4344
FAX: 615-574-7659
BITNET: CRL@ORNLSTC

Ms. Karen McNamara
Department of Chemical Engr.
66-421
Massachusetts Inst. of Technology
Cambridge, MA 02139
Phone: 617-253-6480
FAX:

Dr. Ida Veronica Marotta
Via Adriatico No. 9
85100 Potenza
Italy
Phone: 39-971-45678
FAX: 39-971-474223

Mr. Jean Mercier
IEPES
BP 166
38092 Grenoble Cedex, France
Phone: 33-76-881183
FAX: 33-76-887988

Mr. Koichi Miyata
Kobe Steel Ltd.
79 TW Alexander Drive
Bldg. 4401, Suite 101
Box 13608
Research Triangle Park
NC 27709 USA
Phone: 919-549-9823
FAX: 919-549-9106

Prof. Wolfhard Möller
Max-Planck-Institut
für Plasmaphysik
Boltzmannstrasse 2
D-8046 Garching, FRG
Phone: 49-89-32991312
FAX: 49-89-32992591

Dr. Renato Mondini
APET Srl (ASTEX in Italy)
Via delle Volta, 94
I-2512 Brescia, Italy
Phone: 39-30-347491
FAX: 39-30-342906

Prof. Masao Murakawa
Department of Mechanical Engr.
Nippon Institute of Technology
4-1 Gakuendai
Miyashiro-town
Saitama 345, Japan
Phone: 81-480-34-4111
FAX: 81-480-34-2941

Mr. Peter S. Nielsen
NKT Corporate R & D
Building 61, Sognevj 13
DK-2605 Broendby, Denmark
Phone: 45-43432010
FAX: 45-43630099

Dr. Peter Oelhafen
Institut fur Physik
Universitat Basel
Klingelbergstrasse 82
CH-4056 Basel, Switzerland
Phone: 41-61-3212280
FAX: 41-61-3211440

Mr. Michael E. O'Hern
Department of Materials Science
and Engineering
Dougherty Engineering Hall
University of Tennessee
Knoxville, TN 37996 USA
Phone: 615-574-4344
FAX: 615-574-7659

Dr. Ahmet Oztarhan
Dokuz Eylul Universitesi
Mukendislik Mimarlik Fakultesi
Makina Muhendisligi Bolumu
35100 Bornova, Izmir, Turkey
Phone: 90-51-183138
FAX: 90-51-216148

Prof. Paolo Paroli
Dipartimento di Ingegneria
Meccanica
II Universita dgli Studi di Roma
Via Orazio Raimondo
I-00173 Roma, Italy
Phone: 39-6-79794664
FAX: 39-6-2491351

Deborah T. Peeler
University of Dayton
KL-444
300 College Park
Dayton, OH 45459 USA
Phone: 513-229-2577, 2627
FAX:

Dr. J. M. Pinneo
Crystallume, Inc.
125 Constitution Drive
Menlo Park, CA 94025 USA
Phone: 415-324-9681
FAX: 415-324-2958

Prof. Nicolaos S. Platakis
Chemical Engineering Department
Faculty of Technology
Aristotelian University
of Thessaloniki
Thessaloniki 54006, Greece
Phone: 31-991547
FAX: 31-206138

Dr. Roger W. Pryor
Inst. for Manufacturing Research
Wayne State University
666 West Hancock Street
Detroit, MI 48202 USA
Phone: 313-577-0846
FAX: 313-577-7743
BITNET: RPRYOR@WAYNEST1

Ms. Lou Pyatt
Institute Secretary
Metals and Ceramics Division
Oak Ridge National Laboratory
P.O. Box 2008
Oak Ridge, TN 37831-6118 USA
Phone: 615-574-4344
FAX: 615-574-7659

Ms. Petra Reinke
Max-Planck-Institut
fur Plasmaphysik
Boltzmannstrasse 2
D-8046 Garching bei Munchen
FRG
Phone: 49-89-32991757
FAX: 49-89-32992591

Dr. John Robertson
Central Electricity Res. Lab.
GB-Leatherhead KT22-7SE, UK
Phone: 44-372-374488, Ext. 2209
FAX: 44-372-374488, Ext.2515

Dr. Klaus C. Rohwer
Institut für Solarenergieforschung
Sokelantstrasse 5
D-3000 Hannover 1, FRG
Phone: 49-511-35850, Ext. 44
FAX: 49-511-35850-10

Mr. Louis St.Onge
Department de Physique
Univesite de Montreal
P.O. Box 6128
Montreal, Canada H3C 3J7
Phone: 514-343-6671
FAX: 514-343-2071

Ms. Filomena P. Santos
Physics Department
University of Coimbra
P-3000 Coimbra, Portugal
Phone: 351-39-28725
FAX: 351-39-29158

Dr. Nobuo Setaka
Director General
National Institute for Research
in Inorganic Materials
1-1 Namiki, Tsukuba
Ibaraki, 305 Japan
Phone: 81-298-513351
FAX: 81-298-527449

Dr. Lothar Schäfer
Fraunhofer Institut fuer Schicht-
und Oberflächentechnologie
Vogt-Kolln Strasse 30
D-2000 Hamburg 54, FRG
Phone: 49-40-5493-560
FAX: 49-40-5493-210

Dr. Howard K. Schmidt
Schmidt Instruments, Inc.
2476 Bolsover No. 234
Houston, TX 77005 USA
Phone: 713-529-9040
FAX: 713-529-8411

Dr. Jose F. Azevedo Silva
Departamento de Fisica
Faculdade de Ciencias
Campo Grande, Ed. Cl, Piso 4
1700 Lisboa, Portugal
Phone: 351-1-7583141
FAX: 351-1-7597716

Dr. Roger Smith
Department of Math. Sciences
University of Technology
Loughborough Leicestershire
LE11 3TU, United Kingdom
Phone: 44-509-263171
FAX:
BITNET: RSMITH@LUT.AC.UK

Dr. Galina A. Sokolina
Institute of Physical Chemistry
U.S.S.R. Academy of Sciences
Leninsky Prospect 31
Moscow 1117 915, USSR
Phone: 7-95-2340014, Ext. 427
FAX: 7-95-2302332

Dr. Boris V. Spitsyn
Institute of Physical Chemistry
U.S.S.R. Academy of Sciences
Leninsky Prospect 31
Moscow 117 915, USSR
Phone: 7-95-2340014, Ext. 427
FAX: 7-95-2302332

Prof. F. Stoeckel
Laboratoire de
Spectrometrie Physique
BP-87
38402 St. Martin d'Heres, France
Phone: 33-76514749
FAX: 33-76514544

Dr. J-E. Sundgren
Department of Physics
Linköping University
Linköping S-581 83, Sweden
Phone: 46-13281277
FAX: 46-13137568

Dr. Alberto Tagliaferro
Dipartimento di Fisica
Politecnico di Torino
Corso Duca degli Abruzzi, 24
10129 Torino, Italy
Phone: 39-11-5567329
FAX: 39-11-55677399

Dr. F. Teyssandier
IMP-CNRS
Universite
66025 Perpignan Cedex, France
Phone: 33-68-66-8649
FAX: 33-68668499

Dr. D. E. Tsatis
University of Patras
Physics Department
Patra 2610, Greece
Phone: 61-993133
FAX: 61-991980

Dr. Yonhua Tzeng
Department of Electrical Engr.
200 Broun Hall
Auburn University
Auburn, AL 36849-5201
Phone: 205-844-1869
FAX: 205-844-2432

Dr. Steven M. Valone
Materials Sci. & Technology Div.
Los Alamos National Laboratory
Los Alamos, NM 87545 USA
Phone: 505-667-2067
FAX: 505-665-2104

Mr. Victor von Bonin
IGV, KFA Jülich
Postfach 1913
D-5170 Jülich, FRG
Phone: 49-2461-615925
FAX: 49-2461-613907

Prof. M. R. Wertheimer
Department of Engr. Physics
Ecole Polytechnique
Case Postale 6079, Succursale A
Montreal, Canada H3C 3A7
Phone: 514-340-4749
FAX: 514-340-4440

Dr. Ch. Wild
Fraunhofer Institut fur
Angwandte Festkorperphysik
Tullastrasse 72
D-7800 Freiburg, FRG
Phone: 49-761-5159280
FAX: 49-761-5159200

Mr. Brad Williams
Dept. of Mater. Sci. & Engr.
Materials Research Center
Research Building 1
Box 7919
Raleigh, NC 27695-7919 USA
Phone: 919-737-2867
FAX: 919-737-3419

Mr. R. H. Wittstruck
SLCET-EA
Electronics Technology
& Devices Laboratory
U.S. Army Laboratory Command
Fort Monmouth, NJ 07703-5302 USA
Phone: 201-544-3319
FAX: 201-544-4306

Prof. Frank J. Worzala
Dept. of Mater. Sci. & Engr.
University of Wisconsin-Madison
1509 University Avenue
Madison, WI 53706 USA
Phone: 608-262-0389
FAX: 608-262-6707

Dr. Max Yoder
Office of Naval Research
800 N. Quincy Street
Arlington, VA 22217-5000
Phone: 202-696-4218
FAX: 202-696-3945, or 5383

Mr. Mark Zaleski
Department of Materials Science
Bard Hall
Cornell University
Ithaca,NY 14853
Phone: 607-255-4648
FAX: 607-255-2365
Email: ELMER@SNOOPY.TN.CORNELL.EDU

Dr. James R. Zeidler
Code 7601
Naval Ocean Systems Center
San Diego, CA 92152-5000
Phone: 619-553-1581
FAX: 619-553-6003

AUTHOR INDEX

INDEX

GPSR Compliance

The European Union's (EU) General Product Safety Regulation (GPSR) is a set of rules that requires consumer products to be safe and our obligations to ensure this.

If you have any concerns about our products, you can contact us on ProductSafety@springernature.com

In case Publisher is established outside the EU, the EU authorized representative is:

Springer Nature Customer Service Center GmbH
Europaplatz 3
69115 Heidelberg, Germany

Batch number: 10683448

Printed by Printforce, the Netherlands